AF829061

Innovative Renewable Energy

Series Editor

Ali Sayigh
World Renewable Energy Congress, Building, Brighton, UK

The primary objective of this book series is to highlight the best-implemented worldwide policies, projects and research dealing with renewable energy and the environment. The books are developed in partnership with the World Renewable Energy Network (WREN). WREN is one of the most effective organizations in supporting and enhancing the utilisation and implementation of renewable energy sources that are both environmentally safe and economically sustainable. Contributors to books in this series come from a worldwide network of agencies, laboratories, institutions, companies and individuals, all working together towards an international diffusion of renewable energy technologies and applications. With contributions from most countries in the world, books in this series promote the communication and technical education of scientists, engineers, technicians and managers in this field and address the energy needs of both developing and developed countries. Each book in the series contains contributions from WREN members and cover the most-up-to-date research developments, government policies, business models, best practices, and innovations from countries all over the globe. Additionally, the series publishes a collection of best papers presented during the annual and bi-annual World Renewable Energy Congress and Forum each year.

More information about this series at http://www.springer.com/series/15925

Ali Sayigh
Editor

Sustainable Energy Development and Innovation

Selected Papers from the World Renewable Energy Congress (WREC) 2020

Springer

Editor
Ali Sayigh
World Renewable Energy Congress
Brighton, UK

ISSN 2522-8927 ISSN 2522-8935 (electronic)
Innovative Renewable Energy
ISBN 978-3-030-76220-9 ISBN 978-3-030-76221-6 (eBook)
https://doi.org/10.1007/978-3-030-76221-6

© The Editor(s) (if applicable) and The Author(s), under exclusive license to Springer Nature Switzerland AG 2022
This work is subject to copyright. All rights are solely and exclusively licensed by the Publisher, whether the whole or part of the material is concerned, specifically the rights of translation, reprinting, reuse of illustrations, recitation, broadcasting, reproduction on microfilms or in any other physical way, and transmission or information storage and retrieval, electronic adaptation, computer software, or by similar or dissimilar methodology now known or hereafter developed.
The use of general descriptive names, registered names, trademarks, service marks, etc. in this publication does not imply, even in the absence of a specific statement, that such names are exempt from the relevant protective laws and regulations and therefore free for general use.
The publisher, the authors, and the editors are safe to assume that the advice and information in this book are believed to be true and accurate at the date of publication. Neither the publisher nor the authors or the editors give a warranty, expressed or implied, with respect to the material contained herein or for any errors or omissions that may have been made. The publisher remains neutral with regard to jurisdictional claims in published maps and institutional affiliations.

This Springer imprint is published by the registered company Springer Nature Switzerland AG
The registered company address is: Gewerbestrasse 11, 6330 Cham, Switzerland

Proceedings of World Renewable Energy Congress – 20

26–31 July, 2021
Instituto Superior Tecnico (IST), Lisbon, Portugal

Introduction

The World Renewable Energy Congress (WREC) and Network (WREN) is a non-profit company set up in 1990 to help foster transfer of renewable energy technology from developed countries to developing countries. For more than 10 years, their honorary president was the deputy director general of UNESCO, Prof Dr Adnan Badran. WREC is based in Britain and organizes conferences and seminars, as well as publishes books, journals and magazines around the world in the field of renewable energy. More than 40 countries have hosted the Congress, among them: UK, Germany, Italy, USA, UAE and Sweden. Congress participants have come from more than 80 countries. Regional conferences have been held in: Italy, Canada, Colombia, Algeria, Kuwait, Bahrain, Malaysia, Zambia, Malawi, India, West Indies, Tunisia, Indonesia, Libya, Taiwan, UAE, Oman, Czech Republic, Bahrain, Germany, Australia, Poland, Netherlands, Thailand, South Korea, Iran, Syria, Saudi Arabia, Singapore, Morocco, Sweden, France, USA, China, Sri Lanka and UK.

Among its eminent speakers have been H H Olafur Ragnar Grimsson, President of Iceland; Dr Romano Prodi, Prime Minister of Italy; Rt Hon Dr Stephen Timms MP; Rt Hon John Redwood MP; Rt Hon John Gummer MP; Sir Jonathon Porritt; Crown Prince Faisal of Jordan; H E Dr Abdulaziz Bin Othman Altwaijri, director general of ISESCO, now WREN honorary chairman; Rt Hon US Secretary of Energy Hazel R. OLeary and US Secretary of Energy and Nobel Prize laureate; Prof Steven Chu; Mrs Monika Ghandi, Minister of Environment, India; Admiral Truly; Dr Charly Gay and Dan Arvisu, directors general of NREL, USA; First Minister of Scotland Rt Hon Alex Salmond; Prime Minister of Sri Lanka H E Ranil Wickremasinghe; Hon Minister of Megapolis & Western Development Patali Champika Ranawaka; and H R H Princess Victoria of Sweden. WREN has worked with UN, UNESCO, WEC, UNIDO, UNEP, WMO, WHO, ESCWA, ISES, SER, EU, the Commonwealth Science Council and WB.

The chairman of WREC is Professor Ali Sayigh, B Sc, AWP, DIC, PhD, FIE, FIEE, C Eng, chairman of IEI, UK, member of WEC, and FRSA. The first Congress was held in Reading, UK, in 1990 and has been held every second year since then. In 2012, it was held in Denver where more than 3000 participants were present. WREN created the biannual "Ali Sayigh Trophy" to be given to the country that is most successfully promoting renewable energy. So far it has been awarded to Cyprus, Brazil, Italy, USA, Germany, France, Denmark, China and Morocco.

The 2020 Congress will be held in Lisbon, Portugal, in the Department of Civil Engineering and Architecture, Instituto Superior Técnico, from 18 to 23 July 2021.

Prof Ali Sayigh, chairman of the technical committee, is indebted to his colleagues at World Renewable Energy Network (WREN): Prof Robert Critoph – chairman, Prof Lawrence Kazmerski and Prof Kamaruzzaman Sopian – deputy chairmen, Prof Donald Swift-Hook and Mr Ken Stratton – trustees, and all members of WREN. Our thanks to H E Dr Abdulaziz Altwaijri, honorary chairman of WREN.

I am grateful to Prof Manuel Guedes, host and Congress chairman, and his entire team. I am grateful to all members of the technical committee who met regularly during the last 2 years. My thanks to all the sponsors and contributors, and to Lisbon Technical Superior Institute in having the Congress in their premises. The technical committee consisted of 15 members: Prof Ali Sayigh, UK chairman;

Mr Rainer Hinrichs-Rahlwes, Germany; Prof Antonella Trombadore, Italy; Dr Marta Szabo-Hungary; Prof Riadh Al-Dabbagh, UAE; Prof Mohsen Aboulnaga, Egypt; Dr David Goodfield, Australia; Prof Robert Critoph, UK; Prof Manuel C. Guedes, Portugal; Prof Kamaruzzaman Sopian, Malaysia; Mr Tony Book, UK; Dr Amina Balagarawa, Nigeria; Prof Don Swift-Hook, UK; Dr Salam Darwish, UK; Prof Anastasia Zabaniotou, Greece; and Dr Arthur Williams, UK.

The proceedings consist of more than hundred papers from more than ninety countries. They are divided into 12 parts: 1. Photovoltaic Technology, 2. Solar Thermal, Heat Pumps, and Geothermal Energy, 3.Sustanability and Low Energy Architecture, 4.Bio-energy, Biofuels, and Waste-to-Energy, 5.Policy, Finance, Social impact, Education, and Sustainability, 6.Energy Meteorology, 7.Wind Energy Technology, 8.Water, Hydropower, and Ocean Energy, 9. Hydrogen, Fuel Cells, and Transport, 10. Renewable System Integration, 11. Energy and Gender, and 12. Late Arrival Papers.

Ali Sayigh

5 January 2021

Contents

Part I
Photovoltaic Technology

Innovation in Solar Technology: Toward a 100% Renewable Electricity Future

Lawrence L. Kazmerski

1 Introduction and Historical Setting

A "100% renewable future" is now being envisioned [1–4]. Solar photovoltaics (PV) has undergone rather incredible growth over the past decade. Current annual solar PV markets approaching the 100-GW level have fostered multi-GW manufacturing facilities and approaching a TW of installations worldwide [5]. During this period, consumer prices dropped by more than an order of magnitude. When this author began his technology career more than 50 years ago, the group of contemporary PV researchers had the vision that this technology would eventually come to prominence, but at a more moderate scale, perhaps approaching 10% of electricity needs and much sooner time wise [6]. The Bell Telephone Laboratories discovery of the modern solar cell in 1954 [7] was foundational to the PV revolution, and the media announcement did provide foresight of what was to come:

> *New York Times, April 26, 1954: MURRAY HILL, N.J.—A solar battery, the first of its kind, which converts useful amounts of the sun's radiation directly and efficiently into electricity, has been constructed here by Bell Telephone Laboratories . . . It may mark the beginning of a new era, leading eventually to the realization of one of mankind's most cherished dreams—the harnessing of the almost limitless energy of the sun for uses of civilization.*

The initial market focus turned toward space, following the launch of the first solar-powered satellite, Vanguard, in 1958 [8]. Now PV is the power source of choice for almost every near-earth satellite and for major missions such are the Mars "rovers" [9, 10]. The tipping point for terrestrial PV came as the result of a world crisis—the Arab oil embargo in the early 1970s [11]. This led to the birth of the world solar programs and a surge in enthusiasm for these scientific device curiosities that produced electricity with no moving parts, silently, cleanly, and needing no fuel—except for the sunlight that fell on their surfaces. Major laboratories were established worldwide, including the Solar Energy Research Institute (later becoming NREL) in the United States and Fraunhofer ISE in Germany [12]. The terrestrial markets during this 1980s period were tens of MW size—limited by political influences and resurgence of the fossil fuels' availability—and PV of high costs. It took Japan to show the world how to manufacture in the 1990s, then Germany how to realistically drive consumer markets with incentives in the 2000s, and finally China to show how to scale up production and rapidly drive down prices in the period 2011 to present.

This chapter primarily examines the status and development of photovoltaics technology and markets over the period 2000 to present. Discussions of where this PV is manufactured show geographical shifts, ownerships, and technology investments. A look on where this PV product is being used worldwide provides insights on the markets, their trends, and some look at future world expansions—and finally, current materials and device technology and some projections of needs and potential trends. The possibility of a "100% PV electricity future" (or more) is postulated. Much of this is based upon the reality of the PV technology itself and the experiences of the author.

L. L. Kazmerski (✉)
Renewable and Sustainable Energy Institute (RASEI), University of Colorado Boulder, Boulder, CO, USA

© The Author(s), under exclusive license to Springer Nature Switzerland AG 2022

A. Sayigh (ed.), *Sustainable Energy Development and Innovation*, Innovative Renewable Energy,
https://doi.org/10.1007/978-3-030-76221-6_1

2 Evolution in Photovoltaics: Production, Shipments, and Installations

The commercial growth in PV and other renewable energy technologies can be represented by a number of metrics to highlight progress. Several organizations publish these on an annual basis to underscore how these clean energy resources continue to penetrate our energy systems [13–16]. One of these tracking parameters is the volume of the product shipped from the manufacturers into the worldwide markets [5]. Figure 1 presents recent data and information provided by Paula Mints in her SPV Market Research Report [5, 17], showing the shipments over the period 2001 through 2019 originating from manufacturing facilities from various geographical regions in the world. The reports for PV shipments in 2020 have been reported (P. Mints, Solar Flare, April 2021) and show a growth, despite the pandemic, of 8.2% over 2019; 131.7 GW of PVwere shipped. The cummulative installations reported by REN21 reached 737 GW worldwide though 2020 [18]. Since the 1970s, the growth in these shipments has shown annual positive growths—outside of the one outlier year, 2018. For the past 5 years, this growth has averaged more than a respectable 26%/year. With the indicated growth in shipments of 39% (i.e., 89.1 GW in 2018 to 123.5 GW in 2019), the calculated corresponding revenue in cell and module sales had a ~ $9B increase over this period. It should be pointed out that the average module prices decreased from $0.41/W in 2018 to about $0.36/W in 2019—a factor more important for the consumers. By comparison, the PV module price was between $3.00/W and $4.00/W just 10 years ago. PV has shown substantial progress in both getting product from the suppliers and the price for that product.

2.1 Where Is this PV Coming from?

In 2019, 63.3% of the PV shipped into the world markets came from China (Fig. 1). This continues the trend over 2010-present, during which China has dominated the world PV-manufacturing scene. In the decade before that, Europe and Japan were the major suppliers. China gained the advantage by investing in manufacturing ramp up, collaborating with intellectual resource support from other country centers, making use of cooperative support from government sources, standardization, and shifting toward automation—and complete indigenous control of the technology value chain from raw material through module design and production. In parallel, China has invested financial resources into compliance with international standards to provide the rest of the world a level of confidence in its PV product.

Certainly, Asia is the source for nearly 98% of the PV produced. Three other countries have shown significant emergence in their manufacturing status over the past 5 years—Malaysia, Vietnam, and India. The former two have benefitted from local subsidies to establish manufacturing plants and somewhat from the tariffs imposed directly by some countries on Chinese products. India, of course, is spurred by its government initiative—the National Solar Mission [20]—that has a new target to

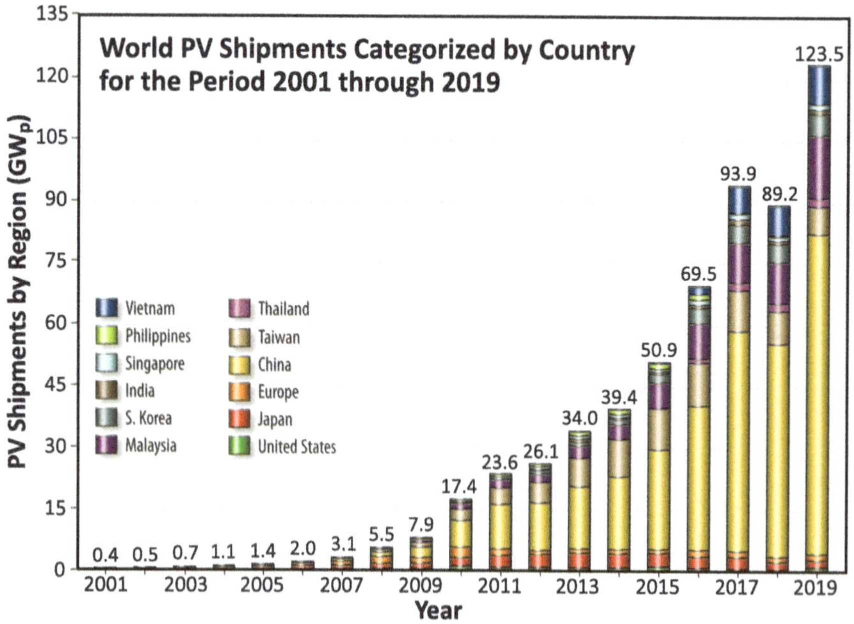

Fig. 1 Worldwide PV annual shipments over the period 2001 through 2019. These shipments are categorized by country of manufacturing origin. Growth in shipments from 2018 to 2019 was about 39% [5]

reach a 500 GW solar capacity by 2030. This ambitious program has some provisions to increase country-specific product on its Approved List of Modules and Manufacturers [21], meeting specification and reliability criteria. It should be noted that China organization "own" >85% of the world PV companies, though the manufacturing might be in Vietnam, Malaysia, or the United States.

2.2 Where Is this PV Product Being Used?

Figures 1, 2, and 3 provide some insight into who is supplying the PV and who is using it. The major markets remain "grid-connected" (~94%), but there is an increasing increment over the past 4 years of remote applications. Concerns with rural and remote populations in the world have provided some incentive to address these needs. When China began to lead in manufacturing in the period 2009–2012, very little of what they produced was used in China. This began to change with China's programs to increase clean energy shares within their electricity markets. Though their internal installations have declined over the past 2 years (mainly because of issues with incentives), they still lead the world in annual and cumulative installed PV, as represented in Fig. 2, showing the cumulative installation in 2018 and 2019. From various sources, the cumulative PV (producing electricity) now is between 620 GW and 660 GW [13]. In any case, the world electricity capacity for PV is approaching the TW level, about the current electricity generation capacity for the United States [13, 22].

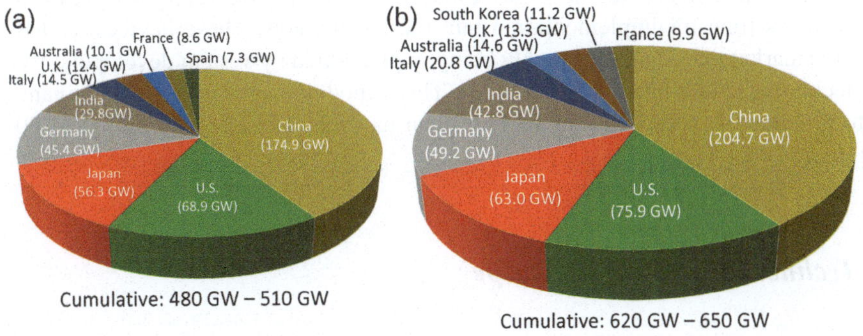

Fig. 2 Worldwide cumulative PV installations over the through 2018 and 2019. Note that there is some spread in reports, but the cumulative installed through 2019 is between 620 GW and 650 GW

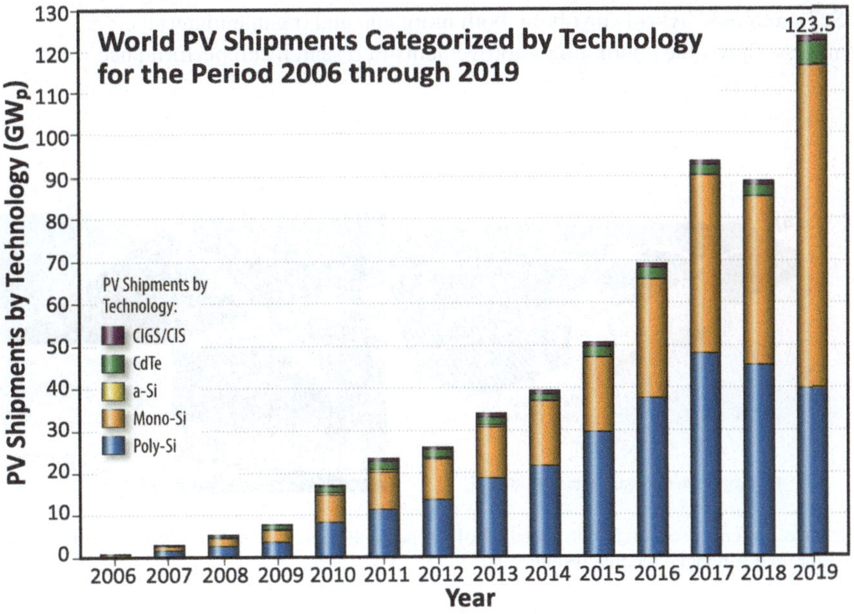

Fig. 3 World annual PV shipments categorized by technology (2006–2019)

3 PV Technology Trends

3.1 Current Commercial PV

For commercial PV technology, we are currently operating in a "silicon world." If we transform the shipment data in Fig. 1 to show the percentages of the major absorber technologies, it can be seen that Si accounts for more than 90% of what was shipped in 2019. Figure 3 shows that thin-film CdTe represented about 5%, thin-film CIGS about 1%, and a-Si:H based <1% of the total. This is still a respective number in "GW" and the percentage of the thin-film PV did increase in 2019 over 2018. The new, higher-efficiency (<17%) and higher-power (>400 W) thin-film CdTe modules [23] certainly have gained some respect, especially for better performances (kWh/kW) in higher temperature climate zones. The data in Fig. 3 show other trends as well in the Si materials of choice. Monocrystalline Si is trending to continue as the material of choice among the Si technologies, with 62% or about 77 GW of the 2019 shipments compared to 40 GW of multicrystalline (or polycrystalline) Si. Thus, the higher quality Si technologies have surpassed the multicrystalline ones for the first time in the last 6 years. For the device configurations, about 80% of the devices were higher efficiency "passivated-emitter and rear-contact" (PERC) solar cells [24]. The choice of materials and the PERC device (Fig. 4a) reflect the importance of performance (efficiency) in continuing to lower the cost of the module technologies. The monocrystalline Si PERC capacity was about 84 GW last year and is still growing. Of course, the major supplier is China, providing about two-thirds of this total. This movement toward more PERC and more monocrystalline Si will continue in the near future.

Another aspect is the interesting transformation of the module technologies. Discussions of solar cells that can collect light from both front and back go back to the 1980s. Pioneering R&D in Spain [25] brought this interesting concept—one that was far ahead of its time for implementation into the PV markets. These devices, called bifacial (Fig. 4b), have become reality in the market place and are quickly being implemented by system developers. Every PV manufacturer has moved into this technology with bifacial Si modules. These modules are becoming dominant, mainly because it adds almost no cost to produce the cells. And the modules can have enhanced energy output of 20–80% depending on the system implementation.

3.2 The PV Technology Device Landscape

A good and comprehensive overview of where we have been, where we are, where we are going, and what has been our performance progress with PV device technology is the NREL research solar cell efficiency chart (Fig. 5). The author first produced this chart in about 1984, early in his career at the Laboratory. NREL continues to update the progress of this chart available on its website [26]. Certainly, the chart depicts the growing complexity of the devices and the materials used. Figure 6 shows a group to show this—starting with elemental semiconductors (Si), through some binaries (GaAs and CdTe), ternaries (Cu(In,Ga)Se2), and now hybrids involving both inorganic and organic materials.

The efficiency chart also shows the continuing improvements in research cell performance.

Fig. 4 Technologies of choice: (**a**) cross-sectional representation of the passivated emitter and back contact (PERC) solar cell and (**b**) bifacial module configuration; background is bifacial module installation

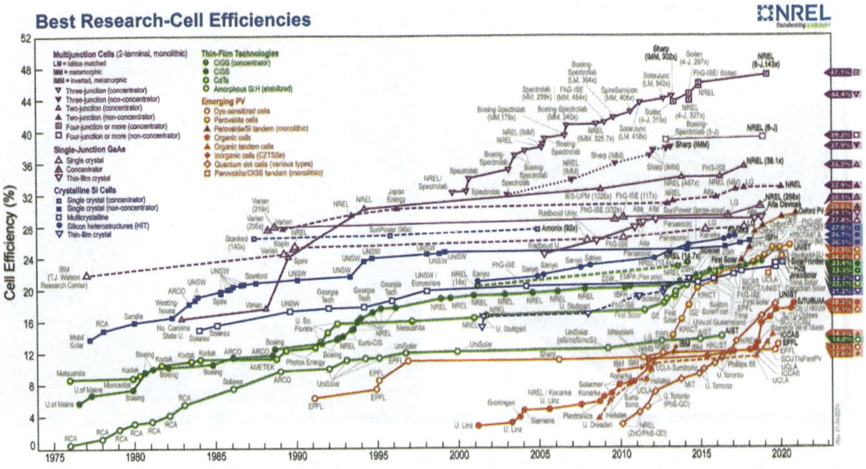

Fig. 5 Research cell efficiencies over the period 1975 through present showing evolution of "best" laboratory efficiencies over time. These cells have been validated in one of the world designated labs to perform this technical support. The chart is maintained by NREL [19, 26]

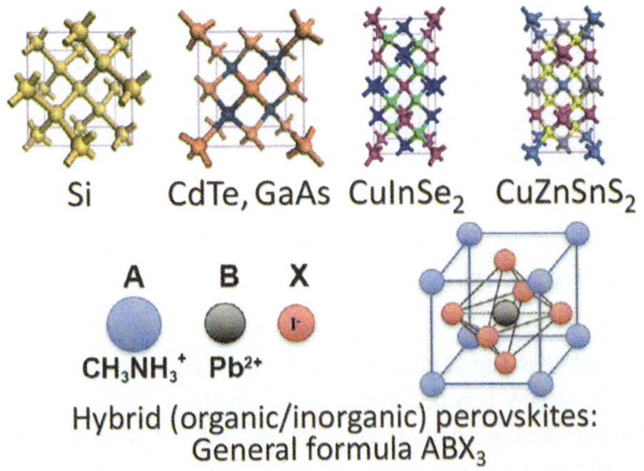

Fig. 6 Evolution of the complexity of PV materials from elemental semiconductors through hybrid (inorganic–organic) perovskites

With the experience gained with past technologies and the expanding availability of complex experimental characterization and processing tools, as well as computational science support through theory, the time from lab bench discovery to commercial readiness has contracted significantly. For example, the CdTe technology was first investigated for terrestrial use in the 1970s [6] and, finally, had commercial introduction in the early 2000s. The current craze with "perovskite" materials (Fig. 7) has compressed this time to less than a decade. PV research and commercial acceptance are both becoming smarter and more informed.

New materials and devices are not "dead" for PV technology. History shows that evolution, innovation, and creativity are required to bring about the next generation of solar conversion for the next generations of electricity consumers. Certainly, an examination of Fig. 5 highlights some recent advances. The first is that of the hybrid (organic–inorganic) perovskites (Fig. 7)—the subject of R&D (and funding) in almost every PV laboratory in the world [27]. Although there was some research on these materials in the 1990s and early 2000s, they really came on seriously with their device demonstrations—rapidly advancing from the 10% to the >20% regime for laboratory devices in 3–5 years. Now on the threshold of commercialization, the evolution is taking this material into two areas: (1) single-junction and multiple-junction "all-perovskites" (each cell tuned to a different portion of the solar spectrum); and (2) tandem cells [28], primarily with Si but also considered recently with other lower bandgap partners (e.g., CIGS). The Si tandem, in particular, is probably the closest to being introduced as a commercial product [27, 28].

The other cell to consider is the organic PV (OPV) device. This "plastic" solar cell offers flexibility in structure, ability to manufacture continuously by roll-to-roll processing, perhaps the least expensive materials, and offer benefits for other markets, such as architectural design. These cells have also recently jumped in performance from the 10% to beyond the 15% level. This is used as an example for another reason. It was a solar cell that was seriously researched, brought to commercialization, and well funded by various agencies.

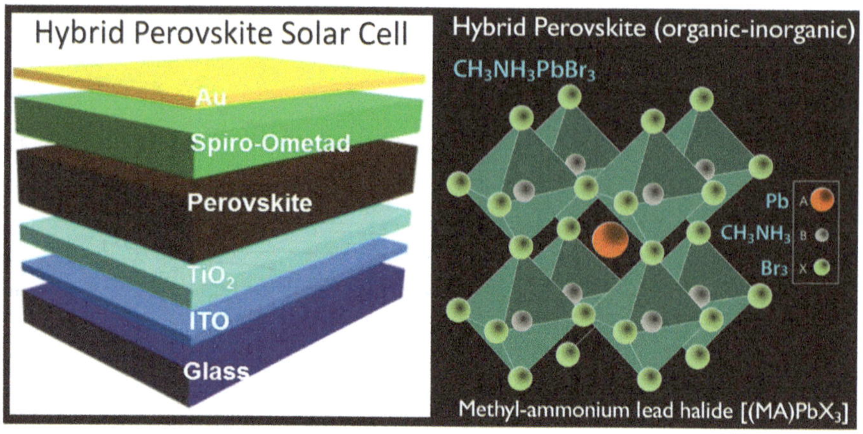

Fig. 7 Hybrid perovskite solar cells: (**a**) cross-sectional representation of the cell and (**b**) hybrid perovskite material showing atomic/molecular configuration

However, with the rise of the Si cell, the sources of support lost interest because of issues with being stuck at lower efficiencies at the time and issues with stability. The example is that we do not have sufficient patience with technology development in our PV sector and the worry is that we might prematurely throw out potential "killer technologies." The message is to keep tuned and have an open mind. What the future might hold is not what we have today. Innovation is moving up for a future passing rectangular-framed modules that are slapped on our rooftops or in our fields, toward a blend of energy-producing PV that is integrated into our living environment; architecturally beautiful and more than meeting our electricity requirements with exceptional performances.

4 The Future: Evolution or Revolution?

Photovoltaics has become a viable and growing segment of our electricity infrastructure worldwide. It is no longer a *future*, alternative energy source—it is *real now*. PV has become competitive to what have been termed conventional (coal, natural gas, and nuclear) electricity generation. And, in many parts of the world, PV is significantly less expensive than a utility connection for stand-alone, reliable, off-grid residential systems even with currently available storage on a levelized cost of electricity basis [29–31]. The shift toward renewable energy, especially the calls for a "100% renewable" future by 2050 (or sooner), includes increased components of solar PV in the electricity portfolios [1–4]. The rapid price reductions along with excellent progress in PV system reliability and functionality have brought PV to the forefront of the clean electricity options. Revolution based on innovation, not evolution, will bring about not only a 100% renewable future but also one dominated by solar PV electricity.

4.1 The Issues: Pandemic and the Economy

It has to be emphasized that the current world COVID-19 situation will take a toll on the immediate future of renewable energy growth. It has to be expected that the installations will not grow during the year 2020 and likely not in 2021 either. The industry (from manufacturers through installers) has been anticipating about a 20% decrease in the numbers (shipments and installations) in 2020. The recovery of the PV markets will depend upon several factors. Certainly, the diminishment of the virus and a possible effective vaccine will be an indicator of how rapidly the recovery can occur. One positive situation is that some of the major markets continue to publish tenders. The new published program in India now targets 500 GW installed by 2030. The good news is that this should provide increase in markets in the mid-future range, but it is also highly likely that the original "100 GW by 2022" will not now be met due to the slowdown caused by the pandemic situation. And the manufacturers seem to be pushing technology with more efficient modules and larger size modules exceeding 500 W. But, of course, PV and renewable energy are not alone in these unfortunate crisis situations. And if the economy in general is not stable and showing growth, many sectors, including solar PV, will slow down.

4.2 The Terawatt Challenge

This terawatt (TW) topic has been a concern for the solar electricity technologies for the last two decades. In the early 2000s, it was certainly a worry that the PV manufacturing could not grow fast enough to contribute to a sizeable fraction of the electrical power capacities of the world. As PV shipments did consistently exceed 30%/year levels, the concern shifted to political and policy incentives for renewables. How to get government commitments? With the period starting in about 2010, the solar PV climate changed in both manufacturing and policy areas. The first was driven, as already noted, by China who showed the world that multi-GW production facilities were feasible and cost effective. And parallel to this, consumers provided the impetus for government programs (India, China, United Kingdom, other European states, and even at the state level in the United States) because they wanted this clean energy source which has suddenly become affordable.

The world now has two-thirds of a TW installed. The TW (or even 10s of TWs) challenge is less of a barrier. The real issue is ***TWhr challenge***! This shifts the emphasis a bit toward (1) increasing the capacity factors for these energy sources that produce when the sun shines; therefore, efficient and cost-effective storage become the ally. And secondly, other markets than just the "power" segment have to be addressed.

4.3 Solar PV: Penetrating the Greater Energy Sectors

To this point, we have established where the solar PV technology and electricity markets stand with some assessment on how it has steadily progressed over time. How PV could have increased impact can be deduced from examining the major market segments. We have primarily considered *power capacities*, measured by MW, GW, or TW. Such is the representation in Fig. 8, comparing the relative estimated renewable energy power capacity percentages to the generation by nonrenewable sources. Using this metric, renewable energy power had about one-third of the world power generation capacities in 2019 (with some 2600 GW that includes 1150 GW from hydro, 651 GW of wind, and 627 of PV). Interestingly, the power from PV is about the same as that from wind, while the.

major renewable power source is hydropower. However, the capacity factors from the renewable source vary. PV only generates watts while the sun is shining, wind when the air movement is sufficient, etc. The important metric is the *power production* (or energy production) measured in MWh, GWh, and TWh. This is represented in Fig. 9 from the REN21 2020 Global Status Report [18], showing the electricity produced. Because of the differences in capacity factors, the contributions from the renewable sources are different, with wind contributing about twice that from PV. It can be seen that the "percentage" for hydro is about the same because of its high capacity factory. (Additionally, it is reported that both wind and PV increased their electricity contributions over 2018 with energy additions compared to fossil fuel capacities.)

To date, the major contributions of PV have been to the electricity sector. Figure 10 presents the REN21 evaluation of the buildings, transportation, and electricity markets this past year [18]. The renewable electricity contributions have been mainly in the power sector, which is the smallest of the three. The growth of PV in the power over the past decade is attributable to the price decreases already cited in the previous sections.

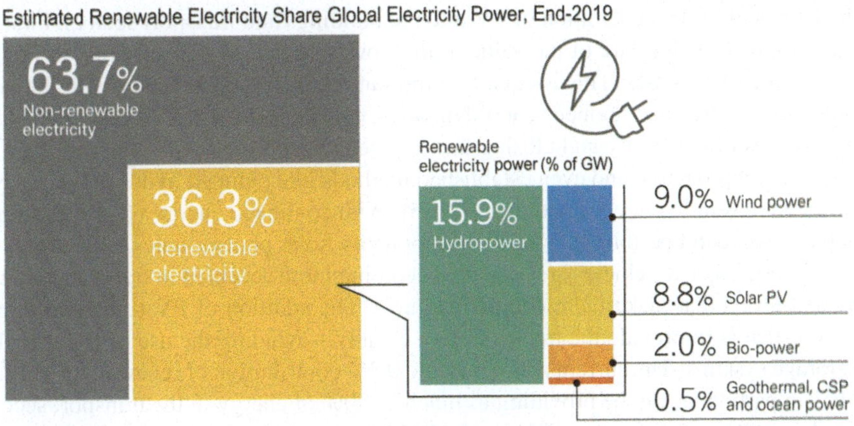

Fig. 8 Estimated renewable electricity share of worldwide installed electricity *power capacity* (based on GW) through 2019. This graphic is assembled from data in the REN 21 report [18]. *Note that this graphic is not in that report, but for comparisons, it is presented in the style of that publication*

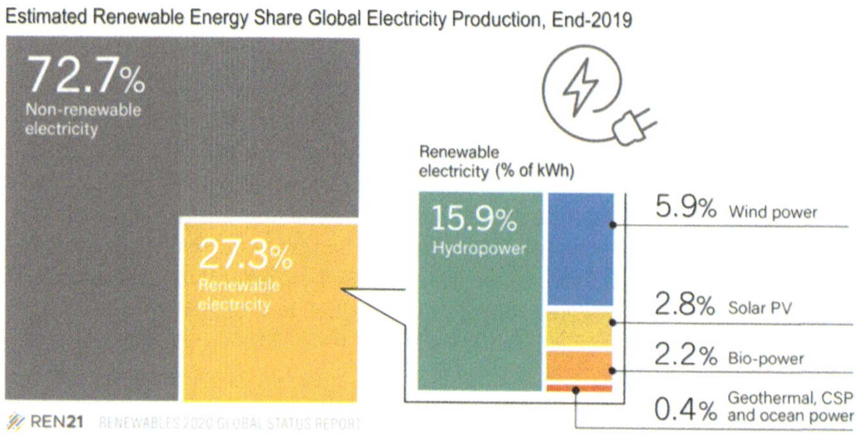

Fig. 9 Estimated renewable electricity share of worldwide installed electricity *power production* (based on GWh, for example) through 2019 from REN21 [18]

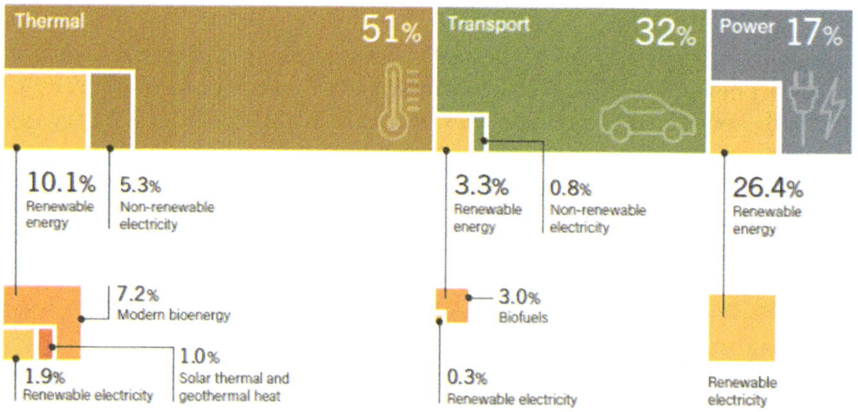

Fig. 10 Estimated renewable electricity share of worldwide installed electricity *power production* (e.g., GWh) in each of the three major sectors: (**a**) thermal, (**b**) transport, and (3) power, indicating the portions for renewable energy (2019) from REN21 [31]. PV is primarily assigned in the "Power" segment but has huge penetration potential for substantial increases in this segment and expansions in the thermal and transport segments as well

However, there are huge applications for photovoltaics in the other two markets as well. It should be cautioned that two factors will guide the extent of the penetration of PV into the buildings and transport sectors. First, *energy efficiency* is foremost. The buildings sector is well aware of the value of first lowering energy consumption through a myriad of approaches to lower energy use and energy losses. This is equally important in the transport sector with the improvement of the energy use of vehicles (engines, electrical components, aerodynamics, mechanical structures, etc.). Second, *energy storage* is essential to intermittent sources of electricity and closing that gap between "power capacity" and "power production." Batteries, hydrogen systems, novel approaches, and even established methods like pumped hydro will all play to make renewables and PV, in particular, more relevant in our energy living [32, 33]. With continued energy efficiency improvements and continued improvements in the costing and performance of PV components, solar photovoltaics will certainly become a bigger player in buildings and transport. Electric vehicles are not only becoming more accepted in our transportation systems but also have performance and cost levels that make them competitive now. The addition of PV fulfills an important objective of many countries' programs—that is to provide the transportation cleanly—avoiding the use of fossil fuel sources for replenishing the power in the storage systems. The current and miniscule 0.3% contribution of renewable electricity is expected to significantly rise. And PV, in particular, can be providing as much as 20% of energy in the transport sector by 2030 and more than 50% as we approach a 100% renewables portfolio by 2050. The aviation, heavy industry, and military sectors will still likely require liquid fuels, but the bioenergy sources will be major contributors.

Heating and cooling and buildings are the primary concerns in the thermal sector. This has been dominated in most parts of the world by nonrenewable energy sources. For example, with the low cost of natural gas, electricity for both heating and cooling has been dependent on fossil fuels. Recently, there have been some demonstrations of coupling into this segment by renewable electricity, and PV in particular. The price of PV modules has enabled the consideration of this source for solar cooking. Using batteries, the solar cooking can serve the remote household for near 24-hour periods [34]. Previously cited remote, off-grid energy-efficient homes are beginning to populate areas having high base electricity prices, such as Hawaii and California [29, 30]. Total electric heating was common in the United States in the 1950s and 1960s when the onset of nuclear power advertised "too cheap to meter." This turned out to be an overstatement, but many homes were built as "allelectric." If the innovation pathway for developing adequate storage (energy densities, low cost, lifetimes, safe materials, and minimizing environmental issues, as well as portability) continues to meet expectations, solar (and wind) can contribute to this thermal sector. Even now, the use of PV with "hot water on demand" systems [35, 36] is starting to replace the other solar thermal approaches and more than competes with utility electricity for this purpose. As the efficiency, performance, and prices of these electricity-enabled appliances continue to improve, a major ingression into the solar thermal sector can be realized by solar electricity. The thermal sector represents the biggest risk and challenge for PV, but it is also the one that has the biggest potential for increasing the solar PV share.

A final but increasingly important area that really bridges the "thermal" and "power" segments in Fig. 10 is that of the ***water-energy-food nexus*** [37]. Food and agriculture are already growing parts of the PV energy landscape. The "Agrivoltaics," the co-development of food and energy production on the same area, is now being incorporated worldwide to enhance crop production, lower water requirements, and provide electricity for local use [38].

(Note this is not a new idea, but one proposed ahead of its time by PV pioneer Adolf Goetzberger and Armin Zastrow in the early 1980s. [39]*)* Certainly, solar PV has already proved itself as viable for water pumping for irrigation and wells in rural and agricultural areas. In Egypt, for example, it has been employed for decades to transform the desert into fruit- and vegetable-producing regions. Solar has been increasing its presence in the desalination and water quality improvement and detoxification arenas. Reverse osmosis (RO) systems have become the products of choice for small, remote water purification or desalination. As populations continue to grow, the demand for water quality and quantity continues to increase, and this is an area that solar electricity has a strong position now and a bigger role in the coming decades. [*As a related point, many of the areas of the world that have the greatest solar resources and solar market growth are also those that have the greatest needs to clean module or mirror surfaces to ensure the power production. Water is currently the major cleaning agent. The combination of solar water pumping and solar water purification (because the water used to clean modules has some specifications) is a natural marriage for maintaining the reliability and cost effectiveness of power plants.*]

5 Summary

Will PV go further? The good news is that PV, whether on the system, module, cell, or material level, is driven by innovation, dedication, and competence. Cumulative TW levels will be reached by mid-2020s for sure. The pandemic is slowing down progress—but not enthusiasm and the reality of the clean energy. It is too big to stop. The ***TWhr challenge*** requires more investment, innovation, and technology development. Can a "100% PV electricity" future become reality? The answer is yes, BUT ... most of the renewable energy technologies have current limiting situation in that they are not "24-hour" producers. The key is the development of adequate storage technologies and efficient appliances that can couple with PV to supply a constant, reliable electricity system.

Renewable electricity will rule the future world's clean energy sector by midcentury. PV will be the major segment of any sizeable renewable electricity future. Innovation will take "evolution" to successful "revolution and our assured 100% renewable energy future.

Acknowledgments The author gratefully acknowledges the discussions and PV market information provided to him by Paula Mints, who has continued to provide the world PV community independent, traceable, and credible evaluations of the "world of solar PV." He also thank D. Gwinner and A. Hicks for helping with several of the graphic illustrations. The author also acknowledges discussions and inputs from Dr. Ali Sayigh of WREN on the renewable energy world scenarios.

References

1. Breyer C Towards 100% renewable energy, Proc. 2019 IEEE Photovoltaic Specialists Conf. (IEEE PVSC virtual conference). IEEE, New York.
2. Jacobson MZ, Delucci MA (2011) Providing all global energy with wind, water, and solar power: Part 1: Technologies, energy resources, quantities, and areas of infrastructure, and materials. Energy Policy 39:1154–1169
3. Bogdanov D, Farfan J, Sadovskaia K, Aghahosseini A et al (2019) Radical transformation pathway towards sustainable electricity via evolutionary steps. Nat Commun 10:1077
4. LUT/EWG (2019) Global Energy System based on 100% RE – Power, heat, transport and desalination sectors. http://energywatchgroup.org/wp-content/uploads/2019/04/EWG_LUT_Global100RE_AllSectors_global_results_overview.pdf
5. Paula Mints, The solar flame, Issue 2, SF-22020, SPV Market Research, April 2020.
6. Workshop Proceedings: Photovoltaic conversion of solar energy for terrestrial applications, Vol 1 and 2 (National Science Foundation, Research Applied to National Needs, AG-485), Cherry Hill, New Jersey, October 23–25, 1973.
7. See, for example, J. Perlin, L.L. Kazmerski, and S. Moon, Good as gold: silicon solar turns 50. Solar Today, January–February 2004, pp. 24–27.
8. See, for example, SciHi Blog on science, tech & art in history, Vanguard 1 – the first Solar Powered Satellite. (Referenced 2019): http://scihi.org/vanguard-solar-powered-satellite/
9. See, for example, TechAcute.com. How are satellites powered? (Accessed 2019). https://techacute.com/how-are-satellites-powered/
10. See, for example, NASA exploration rovers. Accessed 2019. https://mars.nasa.gov/mer/
11. U.S. Department of State, Office of the Historian, Oil Embargo, 1973–1974. https://history.state.gov/milestones/1969-1976/oil-embargo
12. See, www.nrel.gov and https://www.ise.fraunhofer.de/en.html. (Checked 7/2021).
13. International Energy Agency (IEA), World Energy Outlook 2020. https://www.iea.org/reports/world-energy-investment-2020
14. S. Philipps, W. Warmuth, Fraunhofer ISE Photovoltaics Report, June 2020. https://www.ise.fraunhofer.de/en/publications/studies/photovoltaics-report.html
15. Solar Power Europe, Strong global solar market outlook 2020. https://www.solarpowereurope.org/strong-global-solar-market-outlook-2/
16. International Renewable Energy Agency, Renewable Energy Statistics 2019. https://www.irena.org/publications/2019/Jul/Renewable-energy-statistics-2019. 13
17. NREL has used the analysis provided by SPV. The reason is partially historic because this source has been tracking for several decades, and mainly because of the documented care that is taken to avoid double- or triple-counting in the analysis. See, www.nrel.gov
18. REN21 Renewables 2020 Global Status Report. (Paris: REN21 Secretariat). ISBN 978–3-948393- 00-7 https://www.ren21.net/wp-content/uploads/2019/05/gsr_2020_full_report_en.pdf
19. L.L. Kazmerski, "Photovoltaics: Status and Assessment of the Technology," Proc. Third Arab Conference on Solar Energy, Vol.1, Baghdad, Iraq (Scientific Research Council, Baghdad; 1988) pp.10–44.
20. India National Solar Mission. See http://www.mnre.gov.in
21. Approved List of Modules and Manufacturers, India. https://mnre.gov.in/solar/manufacturersand-quality-control
22. U.S. Energy Information Administration, IEA. https://www.eia.gov
23. First Solar, Series 6 module. http://www.firstsolar.com/en/Modules/Series-6
24. ALEO, PERC cell technology explained. https://www.aleo-solar.com/perc-cell-technologyexplained/
25. Guerrero-Lemus R, Vega R, Kim T, Kimm A, Shephard LE (2016) Bifacial solar photovoltaics: A technology review. Renew Sustain Energy Rev 60:1533–1549
26. NREL efficiency chart. https://www.nrel.gov/pv/cell-efficiency.html
27. Wang R, Mujahid M, Duan Y, Wang Z-K, Xue J, Yang Y (2019) A review of Perovskites solar cell stability. Adv Funct Mat 29:1808843; Also, Melissa Davis, Zhibin Yu, A review of flexible halide perovskite solar cells towards scalable manufacturing and environmental sustainability, Journal of Semiconductors, https://doi.org/10.1088/1674-4926/41/4/041603, **41**, 4, (041603), (2020).
28. Oxford PV: The dance to deliver the world's most efficient solar panel. https://commercialsolarguy.com/2020/04/06/perovskite-silicon-tandem-solar-cells/
29. Denholm P, Eichman J, Margolis R Evaluating the technical and economic performance of PV plus storage power plants, **Technical Report.** NREL/TP-6A20–68737 August 2017. https://www.nrel.gov/docs/fy17osti/68737.pdf
30. Colthorpe A Proposed solar-plus-storage projects among Hawaii's lowest cost renewables ever. Energy Storage News 14(12), January 7, 2019. https://www.energy-storage.news/news/proposed-solar-plus-storage-projects-among-hawaiislowest-cost-renewables-e
31. Trahey L, Brushett FR, Balsara NP, Ceder G, Cheng L et al Energy storage emerging: A perspective from the Joint Center for Energy Storage Research. Proc Natl Acad U.S.A. 117(23):12550–12557; first published June 8, 2020. https://doi.org/10.1073/pnas.1821672117
32. Comello S, Reichelstein S (2019) The emergence of cost-effective battery storage. Nat Commun 10:2038. https://doi.org/10.1038/s41467-019-09988-z
33. Cano ZP, Banham D, Ye S et al (2018) Batteries and fuel cells for emerging electric vehicle markets. Nat Energy 3:279–289. https://doi.org/10.1038/s41560-018-0108-1
34. ISES-Elsevier Renewable Transformation Challenge (2019) SunSpot, Solar-powered home cooking system. https://www.ises.org/news/renewable-transformation-challenge-2019-winner-announced
35. See, for example, Rinnai. https://www.rinnai.us/tankless-water-heater
36. U.S. Department of Energy, Tankless or on-demand type water heaters. https://www.energy.gov/energysaver/heat-and-cool/water-heating/tankless-or-demand-typewater-heaters
37. IRENA, Renewable energy in the water, energy & food nexus, International Renewable. Energy Agency, Abu Dhabi, UAE, April 2015, ISBN 978-92-95111-55-4.
38. Dinish H, Pearce JM (2016) The potential of agrivoltaic systems. Renew Sustain Energy Rev 54:299–308. https://doi.org/10.1016/j.rser.2015.10.024
39. Goetzberger A, Zastrow A (1981) On the coexistence of solar-energy conversion and plant cultivation. International J Solar Energy 1:55–69. https://doi.org/10.1080/01425918208909875

Advancements in High-Performance Hybrid Photovoltaic/Thermal Solar Collector Technology

Kamaruzzaman Sopian, Ali H. A. Al-Waeli, and Hussein A. Kazem

1 Introduction

One of the crucial means of electricity production is the use of solar energy, which is mainly done through solar cells, or else known as photovoltaics (PV). These devices convert incident solar irradiance into electric power [1]. However, these devices have certain drawbacks that lead to not meeting their installed capacity and in some cases unreliable. Among those disadvantages is the increase in the cell temperature, which drives the voltage down and consequently causes reduction in PV power [2]. This problem can be solved using cooling mechanisms to maintain the operating temperature of the PV. Those mechanisms are classified as either passive or active and are used depending on the system requirements and their cost-effectiveness [3]. Passive cooling generally means the use of fins to dissipate heat or other forms of material such as phase change material [4]. Active cooling is commonly performed through hybrid photovoltaic thermal (PVT) collectors. In essence, the PV is attached to a solar thermal collector which will function as a heat exchanger; extracts waste heat from surface of PV into base fluid, thus producing heat and improving the production of electricity, simultaneously [5, 6]. Various types of PVT collectors have been established in literature and some have been adopted by the industry [7, 8]. Figure 1 shows the different classifications of photovoltaic/thermal collectors, and the different types of working fluids, absorbers, modes of cooling, etc. As depicted in Fig. 1, PVT active cooling can be done using working fluids operating in forced convection mode, such as air, water, nanofluids, and sometimes two different fluids are used [9, 10]. When the PVT operates in the passive mode, it either utilizes phase-change material (PCM) and fins or use a thermosiphon for the working fluid. The PVT can be classified depending on the type of channel or pipes used to carry the working fluid – for instance, tube-and-sheet type, evacuated tube type, and single-flow channel [11, 12], while the collectors used vary from flat plate (FPC) type to compound parabolic concentrator (CPC) type. Although we find a comprehensive body of literature in this field relative to the age of this technology, the need to establish reviews of conventional and advanced types of PVT is crucial to provide perspective and a forward direction toward standardization and industrialization. We present an updated review of the established and novel designs of PVT collectors and systems. Moreover, we discuss the current trends of research and development in this field.

2 Established Research and Designs of PVT Systems

Unlike water-based PVT systems, the air-based PVT systems do not require thermal collecting materials that have to be attached to the conventional photovoltaic unit. They are simple in design and the working fluid, that is, air, is easily available in nature, whereas water may not be available in some rural locations. Moreover, unlike the liquid types, air is not prone to freezing or boiling in cold or hot climates nor causes damage if leakage occurs [13]. However, the requirements of air-based PVT include the high volume of air flow to achieve a good thermal efficiency in addition to large diameter tubing and the common noise and fan losses. Nonetheless, air exhibits low heat conductivity and heat capacity which impedes its heat

K. Sopian (✉) · A. H. A. Al-Waeli
Solar Energy Research Institute, Universiti Kebangsaan Malaysia, Bangi, Selangor, Malaysia
e-mail: ksopian@ukm.edu.my

H. A. Kazem
Sohar University, Sohar, Oman

© The Author(s), under exclusive license to Springer Nature Switzerland AG 2022

A. Sayigh (ed.), *Sustainable Energy Development and Innovation*, Innovative Renewable Energy,
https://doi.org/10.1007/978-3-030-76221-6_2

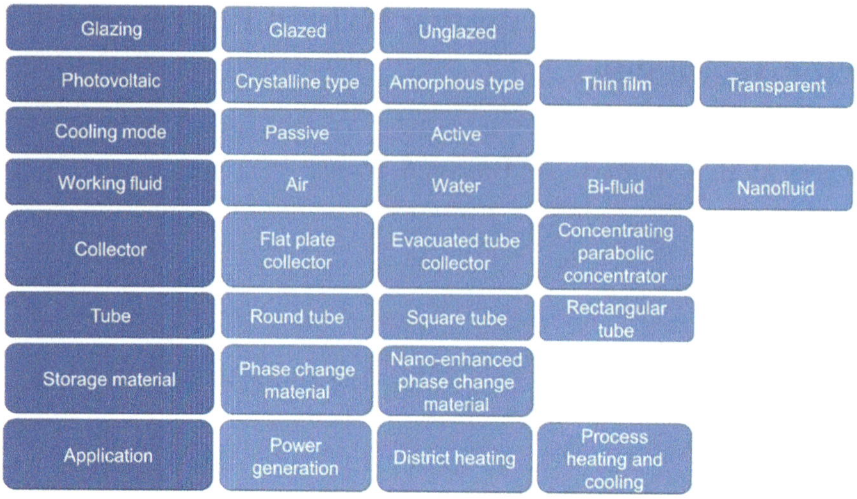

Fig. 1 Types of PVT systems

transfer and leads to needing higher volume transfer [14]. Barone et al. [15] proposed a low-cost innovative PVT and investigated its energy performance along with its economic feasibility. The authors used a dynamic simulation model and experiments for the investigation. An air-to-air heat pump was coupled with the air-based PVT for the application of space heating for a building. The PVT is composed of a PV and air ducts with fans and thermal insulation. The outlet hot air from the air ducts of PVT is injected into the evaporator part of the heat pump cycle. Moreover, the authors assessed the performance for this system by examining it under different climate conditions. The findings suggest that such system can achieve annual primary energy savings ranging between 11.0 MWh and 19.7 MWh. Other authors interested in air-based PVT also coupled the system with other components, namely Khanmohammadi and Shahsavar [16], who linked the collector to a thermal wheel (TW) for residential applications that provide preheating and precooling for both winter and summer, respectively. The TW system is generally positioned within and supply and exhaust air streams of an aid handling unit (AHU). Thus, during winter the ambient air passes through the TW, then enters through the PVT collector, and then to the AHU, and finally into and out of the building and back to the TW, and finally to the ambient. While in summer, the ambient air flows through the TW to the AHU, then into and out of the building, which then enters the TW, and finally into the PVT collector. A graphical schematic representation for PVT-HP and PVT-TW is provided in Fig. 2. Using PVT to assist heat pumps is also implemented for water-based PVT collectors. The reason water and other liquids such as motor oil and ethylene glycol are considered as conventional fluids is because of their high specific heat capacity. Obalanlege et al. [17] proposed a hybrid water-based PVT system coupled with a heat pump. Essentially the system consists of four loops: the first circulates water between the PVT and water tank, the second circulates water between water tank and evaporator, the third circulates a refrigerant (type R407c) through evaporator, compressor, and condenser, while the fourth circulates water through condenser, radiator, and condenser water tank. The fourth loop is also referred to as a "heat rejection loop." The PVT collector is a sheet and tube with a round cross section. The system is analyzed from thermodynamic and quasi–steady-state heat transfer standpoints. Moreover, climatic, design, and operational factors were taken into consideration such as the solar irradiance, size of water tank, and the water flow rate, respectively. The results indicated that increasing the solar irradiance leads to the highest overall PVT efficiency and that throughout time the efficiency will drop down. However, for low solar irradiance values, such as 250 W/m^2, the total efficiency will increase throughout time. Moreover, the highest combined coefficient of performance (COP) for the system is achieved at peak solar irradiance. The turbulent flow of water leads to higher drop in PV temperature than the laminar flow. Other applications of water-based PVT include the humidification-dehumidification (HDH) processes which are excellent solutions for small-scale desalination [18]. Meanwhile, plenty of case studies have examined the outdoor performance of water-based PVT systems for residential applications [19–21]. Water and air can be both utilized as working fluids in PVT systems and are commonly referred to as bifluid-based PVTs. These systems will have separate channels for water and air to flow through. For instance, air can be directed to flow above the PV module to achieve cooling on its front surface and avoid the reflections and other losses associated with using water. Meanwhile pipes could be attached to the back of the PV module for water or other liquid to flow through. Recent literature shows an interest in this type of collector for water and air heating purposes [22]. Jarimi et al. [23] investigated the performance of a bifluid PVT using air-only mode, water-only mode, and simultaneous cooling mode by testing it in indoor and outdoor conditions in Malaysia. The PVT was designed to have both fluid cooling and passive cooling through fins. In the indoor tests, where the control of climate conditions is easier, the PVT exhibits a thermal efficiency of 23–59% and 29–52% for the air mode and water mode, respectively.

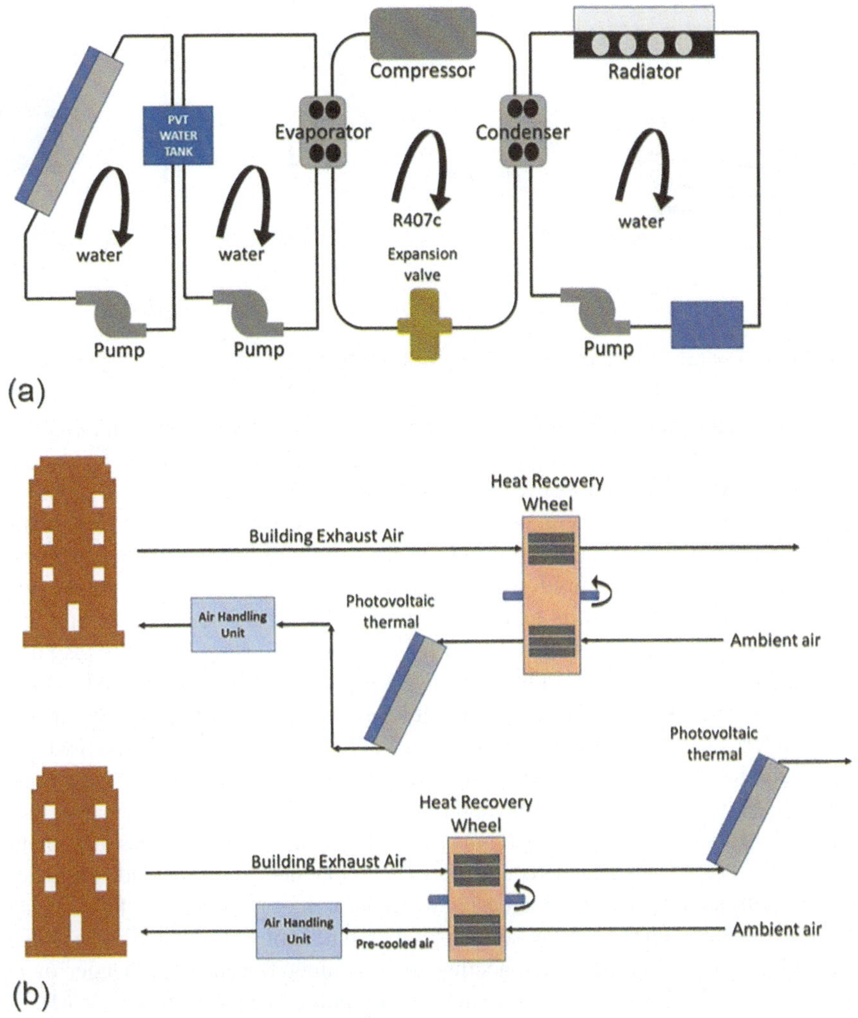

Fig. 2 Schematic drawing of (**a**) PVT-HP and (**b**) PVT-TW systems

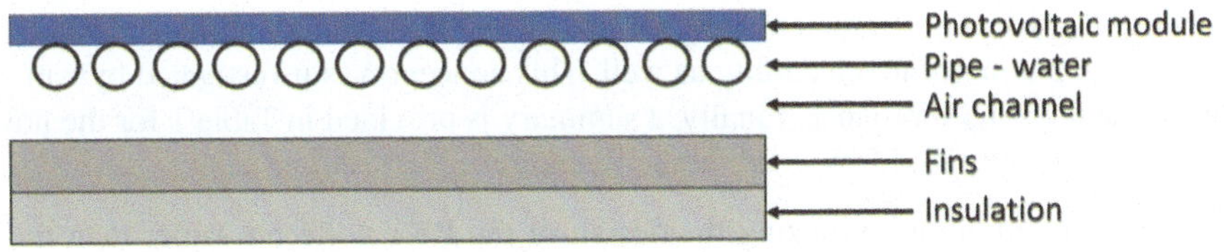

Fig. 3 PVT with fins and dual-fluid channel

Water as a working fluid can restrict the performance of heat exchangers due to its thermal conductivity. Higher thermal conductivity leads to better overall heat transfer, which is the motivation for using nanofluids as a working fluid in PVT systems. As established throughout the literature and extensively for the past 5 years, nanofluid-based PVT systems can outperform water-based PVT systems [24–26]. Nanofluids are produced through the dispersion of particles in the nanorange, with a base fluid to improve or change the thermophysical properties of these fluids [27]. The advantages of nanofluids make them excellent heat transfer agents for cooling and heating purposes. However, many issues are associated with the instability of nanofluids such as particle agglomeration and sedimentation [28]. In some research, nanofluids are used as working fluids, while in other they are used as beam splitters. The later technology allows for spectral separation which is useful to determining the input to the solar cell to ensure excluding the remaining spectrum which will convert into heat. Figure 4 showcases the difference in the schematic design of the two.

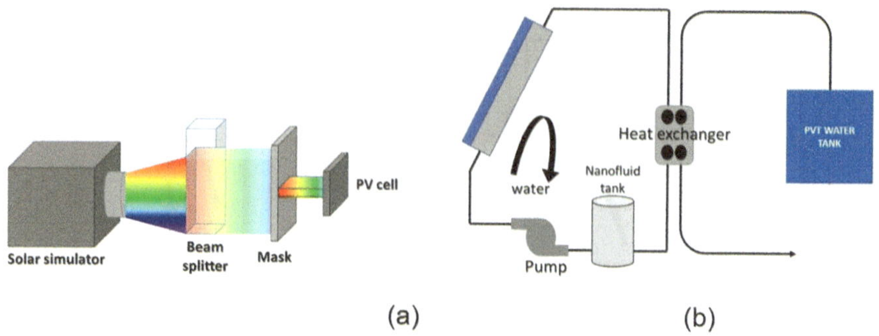

Fig. 4 Schematic cross-section view of (**a**) spectral beam splitter and (**b**) working fluid

He et al. [29] used Ag@TiO_2 nanoparticles for selective absorption of irradiance, while the base fluid used was ethylene glycol/water solution, which exhibits a low freezing point. The nanoparticle and base fluid were used for the beam splitter to maintain a low optical concentration. The authors used a beam splitter and a mask, in addition to the solar simulator and data acquisition device in the experiments. The function of the mask is to prevent unfiltered light from being absorbed by the cell, as the mask only provides a hole, allowing the flitered light to pass through, for the PV to absorb it. The results obtained showed that when using a nanoparticle concentration of 200 ppm, the overall efficiency of solar energy utilization peaks at around 83.7% at 1 kW/m^2. Moreover, the results indicated that increasing particle concentration led to steady increase in thermal energy conversion while the opposite was observed for the electrical. Nanofluids are also used in cooling channels to absorb the heat from the PV module. Research advances on this topic include the production of hybrid nanofluids and use of different surfactant material [30]. Also, use of novel nanofluids such as ferrofluids [31]. Heidari et al. [33] used rotating magnetic field (RMF) and a ferro-nanofluid which was composed of Fe_3O_4 nanoparticles mixed with water. The principle of the study is to use the magnetic field to actuate the nanoparticles and improve the overall heat transfer process. The suspended magnetic nanoparticles in ferrofluids have magnetic properties and the ability to enhance the heat transfer process. The enhancement in these fluids is possible when using magnetic fields such as static types and rotatory types. The authors [32] used a rotary magnetic field (RMF) by attached cubic permanent magnets to rotary blades actuated by a stepping motor. The results obtained by the authors point out that increasing the induction of the magnetic field leads to reduction in photovoltaic surface temperature. Further reduction in PV surface temperature is observed when increasing the ferrofluid concentration. The heat nanofluid could be utilized by connecting the container to a heat exchanger or using a thermoelectric module (TEM). Nanofluids have also been used in conjunction with phase change material (PCM). The use of PCM in PV application is aimed to absorb the waste heat and maintain an acceptable operating temperature. The use of PCM for PV heat energy storage is common; however, the use of nanofluid and PCM is considered a new development in the field [33, 34].

Another hybrid-type system is the PV with PCM and thermoelectric module (TEM). The use of TEM allows for conversion of waste heat into electricity. Hence, in principle, the PV will absorb sunlight and convert it into electricity while waste heat is absorbed by the PCM layer to maintain the voltage of the PV. The heated layer of PCM is then cooled by TEMs which, in turn, will convert the heat into electricity as well. This concept was investigated by Cui et al. [35], who also used a Fresnel lens to increase the solar irradiance. Finally, a summary is provided in Table 1 for the novel research discussed.

The following points were surmised from reading the literature:

1. The focus in literature is often on optimizing the design of the PVT collector rather than the system. Although some research presents new hybrid systems, it is necessary to emphasize optimizing the system as a whole.
2. Overengineering could be a problem and, hence, the control is ultimately done through techno-economic studies to ensure technical validity and economic viability for the proposed solution.
3. Current research has surpassed the industry for many reasons. Among those reasons are the lack of field studies at large scale and the lack of industrial application case studies.
4. Focusing only on improving the energy performance of the PVT collector may not be enough to fully understand and develop this technology. Hence, it is crucial to increase research directed toward applications and hybrid PVT systems such as the PVT-HP and PVT-TW systems.
5. Researchers focus on crystalline-based PV modules for research in PVT systems. This is, in part, due to the prevalence of first-generation PV in the market.

Table 1 Summary of novel research for hybrid PVT systems

Reference	PV type	Collector type	Cooling mode	Working fluid	System type/application
[15]	Polycrystalline PV	Corrugated steel sheets with air duct and fans	Active	Air	PVT coupled with a heat pump (HP)
[16]	Polycrystalline PV	Single channel collector	Active	Air	PVT coupled with a thermal wheel (TW) system
[17]	Monocrystalline PV	Sheet and tube collector	Active	Water	PVT coupled with heat pump (HP)
[18]	Polycrystalline PV	Roll-bond PVT	Active	Saline water	PVT coupled with a humidification-dehumidification (HDH) system
[19]	Monocrystalline PV	Sheet and tube – direct collector	Active	Water	PVT system
[23]	Monocrystalline PV	Serpentine-corrugated surface with fins	Active and passive	Air and water	PVT system
[29]	Silicon PV	Beam splitter	Passive	Ag@TiO_2 nanoparticles – ethylene glycol/water	PVT system using beam splitter
[32]	Monocrystalline PV	Fluid container	Active	Fe_3O_4-Water	PVT with a rotary magnetic field
[35]	Crystalline silicon, CIGS, single-junction GaAs, and GaInP/InGaAs/Ge (III–V)	PCM container, TEM module, and heat sink	Passive	Organic paraffin (solid-liquid)	PVT-PCM-TEM system

3 Conclusions

In conclusion, the main objective of this review is to provide a detailed study of novel photovoltaic/thermal solar collectors. Moreover, the review encompasses other critical aspects related to this technology such as conventional systems, hybrid systems, and the applications of PVT systems. From the reviewed research papers, we provide the following points to summarize the findings of this review:

1. (a) Air- and water-based PVT collectors can be coupled with thermal wheels, heat pumps, water pumping systems, etc.
   (b) The implementation of beam splitters allows selecting the appropriate spectrum for the solar cell to enhance its electrical performance and reduce waste heat.
2. The focus of literature is currently directed toward nanomaterial and novel nanofluids.
3. Rotary magnetic fields (RMFs) are used to apply external magnetic field to ferrofluids to enhance their heat transfer. This combination can be used to cool photovoltaics.

Finally, we suggest the following studies to be considered for future research:

1. To investigate advanced PVT collectors in industrial processes to achieve water desalination
2. To investigate the various types and qualities of ferrofluids and their use for heat transfer in PVT collectors
3. To investigate the cycles of heating and cooling that a PCM layer can sustain before degradation occurs in the process

Acknowledgments The authors of this manuscript would like to express their thanks and acknowledge the provision of DIP-2019-019 and Dana Modal Insan (MI-2019-011) and Solar Energy Research Institute – UKM for supporting this work.

References

1. Chaichan MT, Kazem HA (2018) Generating electricity using photovoltaic solar plants in Iraq. Springer, pp 47–82.
2. Pan R, Kuitche J, Tamizhmani G (2011) Degradation analysis of solar photovoltaic modules: influence of environmental factor. In 2011 proceedings-annual reliability and maintainability symposium. IEEE, pp 1–5
3. Sato D, Yamada N (2019) Review of photovoltaic module cooling methods and performance evaluation of the radiative cooling method. Renew Sust Energy Rev 104:151–166
4. Khanna S, Singh P, Newar S, Sharma V, Mallick TK, Radulovic J et al. (2019). Fins integrated phase change material for solar photovoltaic for South East United Kingdom. In AIP Conference Proceedings. AIP Publishing LLC, Vol. 2149, No. 1, p 060003

5. Al-Waeli AH, Sopian K, Kazem HA, Chaichan MT (2017) Photovoltaic/thermal (PV/T) systems: status and future prospects. Renew Sust Energy Rev 77:109–130
6. Chow TT (2010) A review on photovoltaic/thermal hybrid solar technology. Appl Energy 87(2):365–379
7. Al-Waeli AH, Kazem HA, Chaichan MT, Sopian K (2019) Photovoltaic/thermal (PV/T) systems: principles, design, and applications. Springer Nature
8. Ben Cheikh El Hocine H, Gama A, Touafek K (2018) The feasibility of new design of hybrid photovoltaic-thermal system–a theoretical approach. Int J Ambient Energy 39(5):496–507
9. Bakar MNA, Othman M, Din MH, Manaf NA, Jarimi H (2014) Design concept and mathematical model of a bi-fluid photovoltaic/thermal (PV/T) solar collector. Renew Energy 67:153–164
10. Jarimi H, Bakar MNA, Othman M, Din MH (2016) Bi-fluid photovoltaic/thermal (PV/T) solar collector: experimental validation of a 2-D theoretical model. Renew Energy 85:1052–1067
11. Al-Alili A, Hwang Y, Radermacher R, Kubo I (2012) A high efficiency solar air conditioner using concentrating photovoltaic/thermal collectors. Appl Energy 93:138–147
12. Sharaf OZ, Orhan MF (2015) Concentrated photovoltaic thermal (CPVT) solar collector systems: part II–implemented systems, performance assessment, and future directions. Renew Sust Energy Rev 50:1566–1633
13. Kim JH, Park SH, Kim JT (2014) Experimental performance of a photovoltaic-thermal air collector. Energy Procedia 48:888–894
14. Sathe TM, Dhoble AS (2017) A review on recent advancements in photovoltaic thermal techniques. Renew Sust Energy Rev 76:645–672
15. Barone G, Buonomano A, Forzano C, Palombo A, Panagopoulos O (2019) Experimentation, modelling and applications of a novel low-cost air-based photovoltaic thermal collector prototype. Energy Convers Manag 195:1079–1097
16. Khanmohammadi S, Shahsavar A (2018) Energy analysis and multi-objective optimization of a novel exhaust air heat recovery system consisting of an air-based building integrated photovoltaic/thermal system and a thermal wheel. Energy Convers Manag 172:595–610
17. Obalanlege MA, Mahmoudi Y, Douglas R, Ebrahimnia-Bajestan E, Davidson J, Bailie D (2020) Performance assessment of a hybrid photovoltaic-thermal and heat pump system for solar heating and electricity. Renew Energy 148:558–572
18. Gabrielli P, Gazzani M, Novati N, Sutter L, Simonetti R, Molinaroli L et al (2019) Combined water desalination and electricity generation through a humidification-dehumidification process integrated with photovoltaic-thermal modules: design, performance analysis and techno-economic assessment. Energ Conver Manage 1:100004
19. Kazem HA (2019) Evaluation and analysis of water-based photovoltaic/thermal (PV/T) system. Case Stud Therm Eng 13:100401
20. Sultan SM, Fadhel MI, Alkaff SA (2014) Performance analysis of the photovoltaic/thermal collector (PV/T) system for different Malaysian climatic conditions. In: Applied mechanics and materials, vol 467. Trans Tech Publications Ltd, Zurich, pp 522–527
21. Hossain MS, Pandey AK, Selvaraj J, Rahim NA, Rivai A, Tyagi VV (2019) Thermal performance analysis of parallel serpentine flow based photovoltaic/thermal (PV/T) system under composite climate of Malaysia. Appl Therm Eng 153:861–871
22. Jarimi H, Bakar MNA, Riffat S (2020) A bi-fluid PV/T solar collector and its potential application in solar drying. In: Renewable energy and sustainable buildings. Springer, Cham, pp 779–789
23. Jarimi H, Bakar MNA, Othman M, Din M (2017) Bi-fluid photovoltaic/thermal PV/T solar collector with three modes of operation: experimental validation of a theoretical model. In: Mediterranean green buildings & renewable energy. Springer, Cham, pp 445–464
24. Chaichan MT, Kadhum AAH, Al-Amiery AA (2017) Novel technique for enhancement of diesel fuel: impact of aqueous alumina nano-fluid on engine's performance and emissions. Case Stud Therm Eng 10:611–620
25. Sahin AZ, Uddin MA, Yilbas BS, Al-Sharafi A (2020) Performance enhancement of solar energy systems using nanofluids: an updated review. Renew Energy 145:1126–1148
26. Kumar R, Deshmukh V, Bharj RS (2020) Performance enhancement of photovoltaic modules by nanofluid cooling: a comprehensive review. Int J Energy Res
27. Dhahad HA, Chaichan MT (2020) The impact of adding nano-Al2O3 and nano-ZnO to Iraqi diesel fuel in terms of compression ignition engines' performance and emitted pollutants. Therm Sci Eng Progress. 100535
28. Ouikhalfan M, Labihi A, Belaqziz M, Chehouani H, Benhamou B, Sarı A, Belfkira A (2020) Stability and thermal conductivity enhancement of aqueous nanofluid based on surfactant-modified TiO2. J Dispers Sci Technol 41(3):374–382
29. He Y, Hu Y, Li H (2019) An Ag@ TiO2/ethylene glycol/water solution as a nanofluid-based beam splitter for photovoltaic/thermal applications in cold regions. Energy Convers Manag 198:111838
30. Alous S, Kayfeci M, Uysal A (2019) Experimental investigations of using MWCNTs and graphene nanoplatelets water-based nanofluids as coolants in PVT systems. Appl Therm Eng 162:114265
31. Ghadiri M, Sardarabadi M, Pasandideh-fard M, Moghadam AJ (2015) Experimental investigation of a PVT system performance using nano ferrofluids. Energy Convers Manag 103:468–476
32. Heidari N, Rahimi M, Azimi N (2018) Experimental investigation on using ferrofluid and rotating magnetic field (RMF) for cooling enhancement in a photovoltaic cell. Int Commun Heat Mass Transfer 94:32–38
33. Lari MO, Sahin AZ (2018) Effect of retrofitting a silver/water nanofluid-based photovoltaic/thermal (PV/T) system with a PCM-thermal battery for residential applications. Renew Energy 122:98–107
34. Al-Waeli AH, Sopian K, Chaichan MT, Kazem HA, Ibrahim A, Mat S, Ruslan MH (2017) Evaluation of the nanofluid and nano-PCM based photovoltaic thermal (PVT) system: an experimental study. Energy Convers Manag 151:693–708
35. Cui T, Xuan Y, Li Q (2016) Design of a novel concentrating photovoltaic–thermoelectric system incorporated with phase change materials. Energy Convers Manag 112:49–60

Photovoltaics Versus Fossil Fuel: Financial Feasibility Model for Expansion of Access to Electricity in Off-Grid Regions

Judith A. Cherni, Carlos J. Franco, Isaac Dyner, and Maritza Jimenez

1 Introduction

Global electricity access was 87% in 2016, an increase of 15% since 1993 [1]. While progress has undoubtedly been made, such improvement as there is should not obscure the fact that 1.1 billion people, mostly who live in rural areas of developing countries, still lack energy access [2]. Energy is fundamental to improve living standards [3–6] and enable basic services, such as clean water, energy for cooking and heat, personal safety at night and for health and education services, which all contribute to poverty alleviation [7]. The landmark Millennium Development Goals (by the United Nations, 2000), however, did not include energy needs as one of their eight stated objectives for eliminating world poverty. Nonetheless, in 2015, this oversight was remedied with the publication of Sustainable Development Goals [8], which lists universal energy access among its 17 objectives.

A large gap remains between urban and rural global electrification rates at 97% and 77% respectively [1]. For this reason, international development policy agenda often include rural energy access [9, 10]. The rural poor in developing countries still rely heavily on expensive and contaminating liquid fossil fuels, including kerosene, to meet their electricity needs, particularly in off-grid regions. Uncertainty over price and often sporadic availability, especially of oil-based products, threaten their viability in the medium and long term, often overlooked in the literature. Low income unquestionably deters poor users in rural towns and communities from increasing their energy usage. Moreover, the excessive prices which consumers are obliged to pay for liquid fossil fuels, coupled with limited opportunity to engage with alternative technologies, have restricted demand with the result that electricity consumption is often below the level required for subsistence.

Research into rural electrification is heavily focused on seeking feasible technical and financial energy solutions for the poor. Such significant research has looked into, for example, the costs of a wide range of energy solutions (i.e. grid connected and stand-alone options such as micro-grid, single or hybrid systems); and how to address limited resources, affordability and technical improvement of energy access and services (e.g [11–13]). However, such research often fails to appreciate the supply-demand dynamics of both local and regional markets, something which may well ultimately affect the limited electricity access of the poor. Low electricity utilisation, or cautious consumption (see [14]), is a subject that remains under-researched.

J. A. Cherni (✉)
CEP, Imperial College London, London, UK
e-mail: j.cherni@imperial.ac.uk

C. J. Franco
Universidad Nacional de Colombia, Medellín, Colombia
e-mail: cjfranco@unal.edu.co

I. Dyner
Universidad de Bogotá Jorge Tadeo Lozano, Bogotá, Colombia
e-mail: idyner@unal.edu.co

M. Jimenez
Universidad Nacional de Colombia, Medellín, Colombia

Universidad Pontificia Bolivariana, Medellín, Colombia
e-mail: maritza.jimenez@upb.edu.co

© The Author(s), under exclusive license to Springer Nature Switzerland AG 2022

A. Sayigh (ed.), *Sustainable Energy Development and Innovation*, Innovative Renewable Energy,
https://doi.org/10.1007/978-3-030-76221-6_3

This chapter discusses whether moving away from fossil fuels (diesel and gasoline) and employing instead renewable energy, specifically photovoltaics, could be financially viable and thus could be recommended to achieve electricity consumption at levels that permit subsistence. A design of a financial model that compares the cost viability of three stand-alone energy systems while also considering their respective capacities to increase energy demand is proposed in this chapter. This approach has been tested and modelled for poor rural towns in the Pacific region of Colombia.

2 Electricity Supply and Demand in Off-Grid Regions

Increasing access to energy for services such as cooking, heating, lighting and boiling water is crucial to reduce poverty. However, limited energy access means that restriction to such amenities persists. Factors such as remoteness [15] and limited investment in renewable energy technologies [16, 17] have severely constrained energy access in off-grid areas. Moreover, there is no guarantee that the poor would be able to afford adequate amounts of electricity should the grid be extended to remote rural and coastal regions [12, 18–20]. The high prices which the rural poor are obliged to pay for liquid fossil fuels such as diesel [4, 12, 21] and other energy sources [6, 22] seem to inhibit rather than promote development. It is postulated that lower energy prices and improved accessibility to energy sources would lead to higher electricity demand, resulting in such users saving on fuel cost which could free up cash for other spending.

As to renewable energy for off-grid regions, in Indonesia, studies have revealed that when compared to money spent on diesel and photovoltaics, micro-hydroelectricity is more cost-effective [23]. Even when allowing for grid extension, similar results were obtained for Kenya [24]. A case study in India demonstrated that, for communities located far from the grid and with low load demand, biomass gasification was a cheaper means of access to clean energy than PV or grid extension [25]. Stand-alone photovoltaics, as opposed to microgrids, was found to be most affordable for low-density rural areas [26]. Yet, renewable energy micro-grids had been adopted in remote communities in Colombia. Here, success has been linked to adequate public and private investment, appropriate regulation and more accessible terrain [27]. In remote communities in Choco, where inhabitants were worse off than their counterparts elsewhere, energy access was found to be contingent on the household income and affordability, available fuel and price satisfaction [12]. The widely employed terminology of energy affordability and availability, however, risks concealing the effect that regional markets have on local communities. If such effects go unheeded, even when energy services are expanded, levels of energy demand may well continue to be too low.

Solar photovoltaic systems are environmentally cleaner than liquid fossil fuels, easy to install and have the potential for increased power generation [21, 28, 29]. The costs of solar home systems have fallen considerably and, thus, they have become increasingly more competitive, including small-scale generation infrastructure for remote areas [6, 30]. Pilot cases as well as systems uptake have proliferated in developing countries [28, 31]. Careful innovative financing have led to greater implementation of PV in rural communities in South Asia [21], where the positive impacts of solar installations are evident, for example, on education and in women's empowerment. Nonetheless, supply has been found to be unreliable for some rural communities [17]. Further, in India, solar energy capacity often fell short of government subsistence consumption standards [18].

In Colombia, 52% of the territory is not connected to the national grid, and 1.1 million people lack access to electricity [2]. While urban electrification has reached 100%, only 96% of the total population have access to energy, placing Colombia among the lower ranks in terms of electrification for Latin American and Caribbean countries [1]. Within Colombia, electrification rates in the Department of Choco, on the Pacific coast, are among the lowest, 18% of households lack electricity and 28% only have partial access [32]. Choco is a highly marginalised region with 59% of the population living in poverty and 33% in extreme poverty [33]. For Colombia to achieve the targets of her National Development Plan on sustainable development and poverty reduction [34], electricity consumption in off-grid poor regions would have to expand considerably. For Colombia, studies indicate that technologies such as diesel, grid connection and use of renewables could successfully supply energy to poor rural communities of variable sizes and needs [11, 35], although any energy project in the unconnected areas must also demonstrate economic efficiency [36]. Abundant solar irradiance augurs strong potential for effective energy generation in many rural areas of Colombia [37].

3 Materials and Methods

3.1 *Model Development and Parameters*

The financial off-grid electricity access feasibility (OffG-El) model along with a sensitivity analysis which assesses the impact of potential changes on the outcomes have been developed to assess the suitability of diesel, gasoline and photovoltaic stand-alone systems to power local rural communities and towns and effectively stimulate medium- and long-term energy demand. The novelty of OffG-El lies in the use of levelised costs of energy (LCOE) to undertake an original comparative analysis of stand-alone technologies while also factoring in transportation costs originating in regional markets, and in regional conditions, such as solar irradiance.

The LCOE is employed here to compare the prices of the three technologies and their financial competitiveness in off-grid rural locations [38, 39]; and to calculate the costs of power (kWh) in net present value (NPV). Calculation of LCOE incorporated both the cost of generating electricity and the total electricity generated during the system's lifetime [39, 40]. Therefore, LCOE data for each of the three technologies studied here consider initial investment; total electricity generated within a defined timeframe; operation and maintenance costs; and, for photovoltaics, battery storage replacement costs. The components of electricity generation are defined as *installed capacity*, *degradation rate* and *use period*: (1)

$$\mathrm{LCOE} = \frac{\mathrm{In.Inv} + \sum_{t=1}^{n} \left(O\&M + \mathrm{St.Rep}\right) \Big/ (1+i)^t}{\sum_{t=1}^{n} \mathrm{InstCap} * \left(1-\deg_t\right)^{t-1} * UF * 8640 \Big/ (1+i)^t}$$

where LCOE is the cost of the selected technology; In.Inv represents initial investment in the technology; *O&M* are the operation and maintenance costs of each technology; St. Rep represents the storage replacement cost; i stands for the discount rate; n is the system lifetime; Inst Cap shows the installed capacity in kW; $\deg_t$ corresponds to the degradation ratio in the year t; *UF* indicates the usage factor (i.e. the ratio of the time that a piece of equipment is in actual use against the total time that it could be used); and 8640 is the number of hours in a year.[1]

The i discount rate parameter is 0.69%. This risk-free rate has been calculated by using Colombia's national economy annual rate of return (i.e. 5.31% according to 2019 Fixed Term Certificate Deposit [41]) without factoring in the average 2014–2018 inflation rate at 4.59% [42]. A 20-year lifetime, n, has been assumed for both PV (as per [19, 43, 44]) and fossil fuel systems [45]. The following four parameters vary according to the technology and regional circumstances:

(i) Usage factor (*UF*) compares required to installed capacity in the case of diesel and gasoline. For solar systems, *UF* is the available solar irradiance in a location on any single day and is calculated as the total irradiance per day/24 hours.
(ii) Initial investment cost (In. Inv.) is the amount necessary for the deployment of each technology; it is based on the specific population's electricity requirements and the characteristics of the studied location (i.e. solar irradiation availability and transport cost).
(iii) Installed and storage capacity (Inst Cap) represents information about locations where the energy systems may be installed. The model considers the electricity requirements, hourly availability of irradiance and costs of the systems. Each energy option is evaluated for its capacity to supply the full electricity requirement and avoid possible electricity shortages at the end of the 20 years system's life cycle, when generation capacity would have been considerably reduced. An optimisation sub-model calculates installed and storage energy capacity to minimise costs.

Operation and maintenance (*O&M*) annual expenses for diesel and gasoline systems represent 10% of investment [17]. O&M applies to these fuels, motor oil and their transportation costs from major cities to remote rural locations. For PV systems, O&M has nil costs.

In order to forecast prices in years 2026 and 2033, fuel [46] and transportation costs [47] are considered. The OffG-El model employs the 2019 prices which retailers in Bogotá charged their customers (as per the Unit for Mining and Energy Planning [46]) – using real prices is a feature that reinforces the reliability of the results given here. Calculation of the transfer costs of diesel and gasoline from Bogotá to the rural towns employs the *Origin and Destination Matrix* (Colombia Energy and Gas Regulation Commission (CREG) [47]). These costs are added to the fossil fuels' retail price.

[1] A yearly cycle of 12 months of 30 days each, totalling 360 days, has been used.

Table 1 Regional and energy characteristics of the case study, Choco, Colombia, 2019

	Solar energy technology*			Fossil fuels technology**	
Regional conditions	Selected locality	Solar irradiance (hours per day)	Solar usage factor UF	Selected locality	Transportation costs of diesel and gasoline (US$/gallon)
Favourable FL	*Quibdó rural*	*3.50*	*0.15*	*San José del Palmar*	*0.17*
Unfavourable UL	*Bahia Solano*	*2.19*	*0.09*	*Medio Atrato*	*0.27*
Standard SL	*Fictional town*	*2.76*	*0.12*	*Fictional location*	*0.22*

Authors' table data sources: * [37], ** [46, 47]
Using an exchange rate for the Colombian Peso ($) of $3400 to the U.S. dollar

4 Data Sources and Case Study

The model draws on both the Government of Colombia subsistence standards of electricity consumption, and knowledge that electricity utilisation in off-grid regions often falls below those standards. The official standard of minimum electricity consumption per household to ensure subsistence, established in 2004, is 173 kWh/month, or 130 kWh/month if altitude is taken into account (see Resolution No. 355[2] [48]). Analysis of the 2017 Colombia unified information system data [49] reveals that average consumption in households in the lowest socio-economic strata in the Department of Choco was, nonetheless, 68 kWh/month. That is, poorest people, who often live in off-grid areas, are likely to consume only just above one-third of the electricity deemed necessary for survival.

Six benchmark conditions have been considered: (i). *favourable* levels of solar irradiance *(FL)*; (ii). *favourable* prices paid for fossil fuels and transportation; (iii). *Unfavourable* (low) solar irradiance *(UL)*; (iv). *unfavourable* (expensive) fossil fuels and transportation costs *(UL)*; (v). *standard*, middle-ground solar radiance *(SL)* and (vi). *standard* middle-ground fossil fuels and transportation costs. The *standard* conditions are hypotheses since there is nowhere with precise middle-range characteristics (Table 1).

The outskirts of the city of *Quibdó* enjoy *favourable* high sunshine insolation at 3.50 hours/day ([37]); the selected *unfavourable* town is Bahia Solano with an average solar irradiance of 2.19 hours/day. The *standard* location enjoys 2.76 hours/day solar irradiance (Table 1). *Favourable conditions* for fossil fuels are found in San José del Palmar where transportation costs from Bogotá are US$ 0.17/gallon, compared to US$ 0.27/gallon in Medio Atrato (as transportation charges are not uniform even in these towns, the above fees represent the average market prices). *Standard conditions* represent the mid-range value of US$ 0.22/gallon. OffG-El forecasts the power capacity required to supply energy in the studied towns using these solar irradiance rates and fossil fuel prices.

5 Results and Discussion

This section discusses the results of the application of the OffG-El model to the case study in the Department of Choco, and elaborates the outputs through the sensitivity analysis.

5.1 Investment, Operation and Maintenance Costs

The model's basic scenario seeks to supply 173 kWh/month per household, which is the amount of electricity to enable subsistence as per the government's standards. The initial investment in *photovoltaic* systems has taken into account the anticipated fall in solar modules prices of 50% and 75% in 2026 and 2033 respectively; 2019 is used as the reference year and the price fall is 0% (price reduction rate results from technological improvement and the time taken for prices to fall

[2] The electricity subsistence level is defined as 'the minimum quantity of electricity that a user would typically consume in a month to satisfy basic needs which could only be satisfied through this type of final energy" [48, p.2; authors' translation].

Table 2 Installed PV capacity and initial investment for delivery of 173kWh/month (subsistence standard) in three locations, Department of Choco, Colombia

		Initial investment (US$)		
Regional conditions[a]	PV required installed capacity (kW)	2019	2026	2033
FL	2.80	4989.29	2961.64	1947.82
SL	3.36	5.671.64	3391.05	2250.76
UL	4.20	6606.94	3946.94	2616.94

Authors' table
Data Sources: [37]; commercial data [53–59]
[a]*FL* favourable, *SL* standard, *UL* unfavourable

Table 3 Operating costs of fossil fuel generators for delivery of 173kWh/month (subsistence standard) in three locations, Choco, Colombia

	Diesel initial investment (US$)	Diesel operation costs (US$/year)			Gasoline initial investment (US$)	Gasoline operation costs (US$/year)		
Regional conditions[a]		2019	2026	2033		2019	2026	2033
FL	342.94	1789.94	1595.33	1667.55	244.52	1664.56	1641.93	1718.16
SL	342.94	2210.96	2434.68	2594.93	244.52	2059.69	2527.63	2696.76
UL	342.94	3092.72	3677.13	4068.65	244.52	2890.16	4076.64	4583.93

Authors' table sources: [46, 47, 62]
[a]*FL* favourable, *SL* standard, *UL* unfavourable

[50]). Whereas initial investment is relatively high (see Table 2), operation and maintenance outgoings are very low, having a value of zero (see above [51, 52]). Also, as it rains very frequently in Choco, no additional expenses on solar panel cleaning are necessary. Nonetheless, batteries are required for storing power; their replacement every 5 years is an unavoidable expense and has thus been incorporated. The price of a storage battery was US$762.94 in 2019 and is forecasted to be US$341.47 and US$190.74 in 2026 and 2033 respectively [53].[3] The model calculates not just the required battery storage capacity in each of the three location but also allows for an anticipated fall in battery price which equals the predicted price cut of PV modules (Table 2).

In 2019, initial investment (including system and delivery charges) for 1.0 kW *diesel* and *gasoline* generators was US$342.94 and US$244.52 respectively [60, 61]. In contrast to photovoltaics, it is anticipated that the price for gasoline and gasoil equipment will remain unchanged because they are mature technologies [45]. In addition to the actual fuel, generators operate with *motor oil*, which also needs to be purchased through a third party invariably operating in large cities. Precise quantities of motor oil are needed to generate a specific amount of electricity and its cost is included in the O&M outlays, which is 10%/year of initial investment (see 3.1) (Table 3).

The transportation fees paid from Bogotá to the destinations in Choco (as per CREG *Matrix*) have been added to the fuels' retail prices. Transfer of these fuels to (unfavourable) Medio Atrato adds US$0.27 per gallon; if transferred to (favourable) San Jose del Palmar, the increment falls to US$0.17/gallon. Transportation to a *standard location* would amount to US$0.22/gallon. These transfer costs inexorably inflate prices which the poor would have to pay for electricity.

5.2 *Energy Consumption: Benchmarks and Costs*

The results obtained with OffG-El for rural towns in Choco indicate that, while fossil fuels and renewable energy systems could each produce enough power to supply basic electricity needs, it would be cheaper (and less polluting) to deploy solar PV than to continue relying on fossil fuels. The cost of diesel was 7–10 times higher than that of PV in 2019, and it is anticipated that the price will increase 12- to 20-fold and 20- to 35-fold by 2026 and 2033 respectively. The price of gasoline in 2019 was between 7 and 9 times higher than for PV, but rises to 12 and 22 times and 20 and 40 times higher in 2026 and 2033 respectively (Fig. 1).

[3]Authors' calculations are drawn on [53] commercial data.

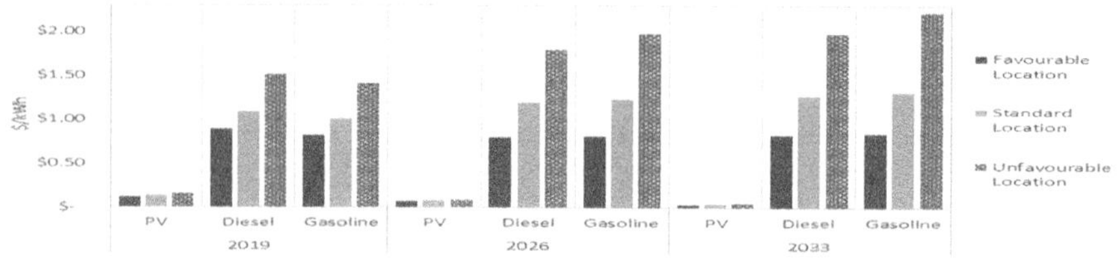

Fig. 1 Costs (LCOE) of delivering 173 kWh/month with PV, diesel and gasoline to satisfy subsistence consumption standard, Choco, Colombia. (*Source*: The authors)

Significantly, these forecasts hold true also for locations where solar irradiance is lower: electricity generation with *photovoltaics* in Bahia Solano cost US$0.12–US$0.16/kWh in 2019, and is forecasted to decrease to US$0.07/kWh–$0.09/kWh in 2026 and to US$0.04/kWh–US$0.06/kWh in 2033. The supply of electricity with *diesel* is notably more expensive for all three regional conditions and years, with prices projected to fluctuate considerably between the minimum and maximum for the same period; moreover, in contrast to PV, prices are projected to rise over time from US$0.89–US$1.52/kWh in 2019 to US$0.79–US$1.80/kWh in 2026, and to US$0.83–US$1.99/kWh by 2033. Finally, notwithstanding that *gasoline* prices in 2019 were lower than for diesel, they are significantly higher than for PV. Thus, gasoline emerges as the most expensive alternative, also with considerable variation between estimated costs by period, that is, US$0.82–US$1.41/kWh in 2019; US$0.81–US$1.98/kWh in 2026 and US$0.85–US$2.23/kWh in 2033 (Fig. 1). From the above, it would appear that solar energy represents the most viable technology of the three options. In terms of the ability to meet subsistence demand at least till 2033, PV offers the cheapest solution for all three scenarios in Choco. Moreover, in contrast to diesel and gasoline, PV technological advances and, therefore, price reductions are expected to continue. Improvements to the PV technology are likely to enhance the results of VE Off-Grid, particularly in the unfavourable and mid-range locations in the light of the projected 43% and 64% costs reduction for 2026 and 2033 respectively. Conversely, no pricing benefits are projected for diesel and gasoline since these are mature technologies with prices subject only to market fluctuations and/or political regulations.

5.3 Sensitivity Analysis

This following analysis explores uncertainties that may affect the results using the model basic setting, that is, changes to operation and maintenance costs.

(i) *Rise of photovoltaics O&M costs*
This scenario assumes that the O&M expenditures related to fossil fuels remain unchanged at 10%; however, O&M for photovoltaics increases by 10% of initial investment. Delivery of 173kWh/month with solar energy prices will, thus, be 2.3 to 2.5 times higher than those of the model's basic setting at 0% O&M cost. Specifically, generation prices for PV were US$0.27–0.37/kWh in 2019 (i.e. representing increases of 130.4–140.7% from the model's basic scenario) and are projected at US$0.16–0.22/kWh for 2026 (136.9–146.6%) and US$0.10–0.14/kWh for 2033 (146.1–154.7%) (Fig. 2). Yet, when compared to the fossil fuels, PV remains the cheapest option to supply viable energy in all the analysed scenarios; in addition, kWh prices decline over time (Fig. 2).

(ii) *Reduction of fossil fuels O&M costs*
A 50% reduction in fossil fuels' O&M expenses, from 10% to 5% of the initial investment, does not alter the outcome, so photovoltaics continues to represent a more affordable solution for the studied years and places. With lower O&M outlays, *diesel* prices would have been US$0.88/kWh–US$1.51/kWh in 2019, and are forecast to fall to US$0.79/kWh–US$1.79/kWh in 2026 and US$0.81/kWh–US$1.89/kWh in 2033 (representing reductions of 0.5–0.9%, 0.5–1.0% and 2.5–5.0% from prices which factor in 10% O&M outlays, i.e. the basic scenario). The predicted impact of reducing O&M outlays for *gasoline* mirrors that of diesel at US$0.81/kWh–US$1.40/kWh in 2019. However, gasoline prices are projected to be higher than for diesel and PV in 2026 at US$0.80/kWh–US$1.98/kWh, and in 2033 at US$0.81/kWh–US$2.10/kWh (gasoline prices represent reductions of 0.4–0.7%, 0.3–0.7% and 3.8–5.7% respectively) (Fig. 3).

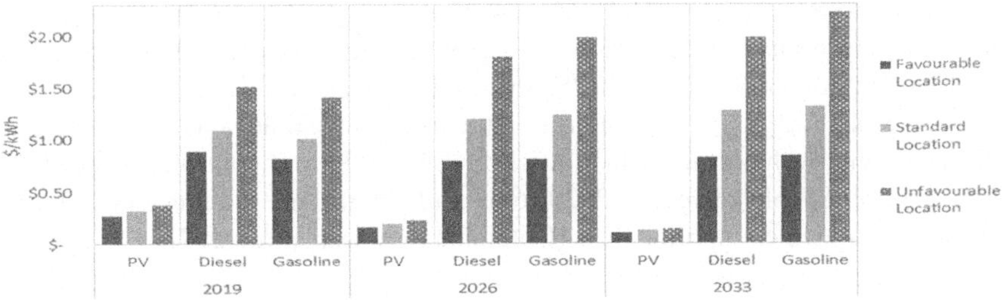

Fig. 2 Costs (LCOE) of supplying 173kWh/month when photovoltaics O&M have risen from 0% to 10% of initial investment, Choco, Colombia. (*Source*: The authors)

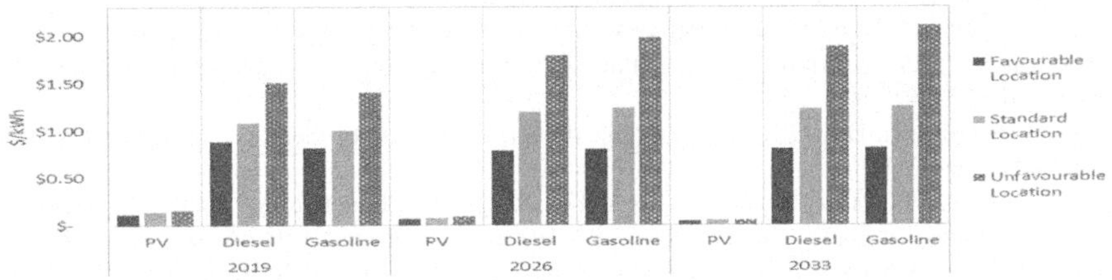

Fig. 3 Costs (LCOE) of delivering 173 kW/h per household with outgoings for diesel and gasoline systems reduced from 10% to 5% of initial investment, Choco, Colombia. (*Source*: The authors)

6 Discussion and Conclusion

Modelling of renewable versus fossil fuel stand-alone technologies indicates that solar photovoltaics consistently emerges as the most financially feasible solution to supplying electricity for households that consume below as well as above the subsistence standard. This finding holds true whether the town in question deploys favourable, unfavourable or standard irradiance, and for every studied period. In 2019, costs were higher where electricity was generated with diesel, but gasoline is predicted to be the costliest option for electricity delivery in 2026 and 2033. Furthermore, the findings also indicate that costs could be reduced if the energy systems were programmed to meet higher electricity demand. Consequently, it would be more effective (particularly if deploying photovoltaics) if energy systems were planned to generate and deliver a minimum amount of energy, that is, 173kWh/month. Such target would imply an additional 105kWh/month, or an increase of 154% over the current 68kWh/month. In turn, if energy generation were designed to deliver no less than the minimum amount for survival, consumers would be entitled to savings resulting from applying this economy of scale.

The sensitivity analysis indicates that even when reduced maintenance costs favourably impact diesel and gasoline prices, photovoltaic technology still offers the most affordable option to promote higher energy consumption among Choco's poorest households. These results are novel because financial viability has been examined in relation to finely elaborated composite parameters. Moreover, real and local information about naturally occurring radiance, regional market conditions, price inflation and measures of restricted electricity consumption have all been incorporated. This study fills a previous gap since it demonstrates the feasibility of lowering the price of electricity provision in off-grid Choco if PV were widely deployed and, where possible, if PV replaced existing expensive diesel and gasoline systems. A change to photovoltaic technology is vital to raise the very low levels of electricity usage.

While prospects for technical refinement of mature diesel and gasoline technologies are slim, as an evolving technology, advances to PV are expected, which will drive prices down. Of the parameters deployed in this study, the *usage factor* (UF) is particularly important; results may also be extrapolated to other aspects of the technologies. For example, the high modularity that characterises photovoltaic technology enables system components to be added incrementally. This means that should technological upgrades be required, it would not be necessary to have a complete system overhaul, which reinforces PV financial viability [29].

This work contributes to the understanding of the nascent notion of cautious energy consumption (see above [14]) through original analysis of regional market realities that threaten fuels viability for the poor in the medium and long term. Should PV be implemented in off-grid Choco, modularity will become is relevant because existing installed capacity might be increased easily and at relatively low cost. Fossil fuels, on the other hand, are not well-known for their modularity. Instead, selection of appropriate power capacity systems is always necessary upfront; and often, larger than required systems are often used to generate low amounts of energy. The poor modularity of fossil fuels points to their low UF (usage factor), which in turn increases the cost of electricity units. While the cost of fossil fuel kWh cost is lower if demand increases, price reduction is insufficient to bring the costs below those of PV because the installed capacity of diesel and gasoline systems remains unchanged. PV is a renewable energy solution offering attractive financial advantages. There are also environmental and social benefits in selecting photovoltaics rather than fossil fuel. PV solutions contribute towards the target of 30% off-grid electricity generation from renewable energy by 2020 [63], could contribute to reduce poverty, lower CO_2 emissions, and increase renewable energy deployment in rural areas (Colombia Renewable Energy Law 1715 [64]).

References

1. World Bank Group (2017) Access to electricity (% of population). http://data.worldbank.org/indicator/EG.ELC.ACCS.ZS
2. IEA (2017) Energy access database, World Energy Outlook. http://www.worldenergyoutlook.org/resources/
3. Adusah-Poku F, Takeuchi K (2019) Determinants and welfare impacts of rural electrification in Ghana. Energy Sustain Dev 52:52–62. https://doi.org/10.1016/j.esd.2019.07.004
4. Alstone P, Gershenson D, Kammen DM (2015) Decentralized energy systems for clean electricity access. Nat Clim Chang 5(4):305–314. https://doi.org/10.1038/nclimate2512
5. Ngowi JM, Bångens L, Ahlgren EO (Dec. 2019) Benefits and challenges to productive use of off-grid rural electrification: the case of mini-hydropower in Bulongwa-Tanzania. Energy Sustain Dev 53:97–103. https://doi.org/10.1016/j.esd.2019.10.001
6. World Bank Group (2017) Energy markets in Latin America: emerging disruptions and the next frontier. Washington, DC. [Online]. Available: https://openknowledge.worldbank.org/handle/10986/30211
7. Practical Action (2014) Poor people's energy outlook 2014: Key messages on energy for poverty alleviation, no. 1. Practical Action Publishing, Rugby
8. UNDP (2015) SDG 7: affordable and clean energy. Sustainable Development Goals. http://www.undp.org/content/undp/en/home/mdgoverview/post-2015-development-agenda/goal-7.html. Accessed 08 Oct 2015
9. IEA (2015) Modern energy for all: why it matters. World Energy Outlook. http://www.worldenergyoutlook.org/resources/energydevelopment/modernenergyforallwhyitmatters/#d.en.8612. Accessed 08 Oct 2015
10. World Bank Group (2015) World development indicators 2015. World Bank, Washington, DC
11. Henao F, Cherni JA, Jaramillo P, Dyner I (2012) A multicriteria approach to sustainable energy supply for the rural poor. Eur J Oper Res 218(3):801–809. https://doi.org/10.1016/j.ejor.2011.11.033
12. Tomei J et al (2019) Forgotten spaces: How reliability, affordability and engagement shape the outcomes of last-mile electrification in Chocó, Colombia. Energy Res Soc Sci 59:2020. https://doi.org/10.1016/j.erss.2019.101302
13. Madriz-Vargas R, Bruce A, Watt M (2018) The future of community renewable energy for electricity access in rural Central America. Energy Res Soc Sci 35:118–131. https://doi.org/10.1016/j.erss.2017.10.015
14. Tesfamichael M, Bastille C, Leach M (2020) Eager to connect, cautious to consume: an integrated view of the drivers and motivations for electricity consumption among rural households in Kenya. Energy Res Soc Sci 63:101394. https://doi.org/10.1016/j.erss.2019.101394
15. Jimenez R (2017) Barriers to electrification in Latin America: income, location, and economic development. Energ Strate Rev 15:9–18. https://doi.org/10.1016/j.esr.2016.11.001
16. World Bank Group (2010) Addressing the electricity access gap. [Online]. Available: http://siteresources.worldbank.org/EXTESC/Resources/Addressing_the_Electricity_Access_Gap.pdf
17. Harish SM, Morgan GM, Subrahmanian E (2014) When does unreliable grid supply become unacceptable policy? Costs of power supply and outages in rural India. Energy Policy 68:158–169. https://doi.org/10.1016/j.enpol.2014.01.037
18. Millinger M, Mårlind T, Ahlgren EO (Dec. 2012) Evaluation of Indian rural solar electrification: a case study in Chhattisgarh. Energy Sustain Dev 16(4):486–492. https://doi.org/10.1016/j.esd.2012.08.005
19. Ramadhan M, Naseeb A (2011) The cost benefit analysis of implementing photovoltaic solar system in the state of Kuwait. Renew Energy 36(4):1272–1276. https://doi.org/10.1016/j.renene.2010.10.004
20. Ranaboldo M, Domenech B, Reyes GA, Ferrer-Martí L, Pastor Moreno R, García-Villoria A (2015) Off-grid community electrification projects based on wind and solar energies: a case study in Nicaragua. Sol Energy 117:268–281. https://doi.org/10.1016/j.solener.2015.05.005
21. Palit D (2013) Solar energy programs for rural electrification: experiences and lessons from South Asia. Energy Sustain Dev 17(3):270–279. https://doi.org/10.1016/j.esd.2013.01.002
22. Olatomiwa L, Mekhilef S, Huda SN, Ohunakin OS (2015) Economic evaluation of hybrid energy systems for rural electrification in six geo-political zones of Nigeria'. Renew Energy 83:435–446. https://doi.org/10.1016/j.renene.2015.04.057
23. Blum NU, Sryantoro Wakeling R, Schmidt TS (2013) Rural electrification through village grids – assessing the cost competitiveness of isolated renewable energy technologies in Indonesia. Renew Sust Energ Rev 22:482–496. https://doi.org/10.1016/j.rser.2013.01.049
24. Moner-Girona M et al (Oct. 2019) Decentralized rural electrification in Kenya: speeding up universal energy access. Energy Sustain Dev 52:128–146. https://doi.org/10.1016/j.esd.2019.07.009
25. Mahapatra S, Dasappa S (2012) Rural electrification: optimising the choice between decentralised renewable energy sources and grid extension. Energy Sustain Dev 16(2):146–154. https://doi.org/10.1016/j.esd.2012.01.006
26. Chaurey A, Kandpal TC (Jun. 2010) A techno-economic comparison of rural electrification based on solar home systems and PV microgrids. Energy Policy 38(6):3118–3129. https://doi.org/10.1016/j.enpol.2010.01.052
27. Gaona EE, Trujillo CL, Guacaneme JA (Nov. 2015) Rural microgrids and its potential application in Colombia. Renew Sust Energ Rev 51:125–137. https://doi.org/10.1016/j.rser.2015.04.176

28. Solangi KH, Islam MR, Saidur R, Rahim NA, Fayaz H (May 2011) A review on global solar energy policy. Renew Sust Energ Rev 15(4):2149–2163. https://doi.org/10.1016/j.rser.2011.01.007
29. Toledo OM, Oliveira Filho D, Diniz ASAC (2010) Distributed photovoltaic generation and energy storage systems: a review. Renew Sust Energ Rev 14(1):506–511. https://doi.org/10.1016/j.rser.2009.08.007
30. REN21 (2015) Renewables 2015 Global status report – key findings. Paris
31. El Chaar L, Lamont LA, El Zein N (Jun. 2011) Review of photovoltaic technologies. Renew Sust Energ Rev 15(5):2165–2175. https://doi.org/10.1016/j.rser.2011.01.004
32. PERS, 'Plan de energización rural sostenible del departamento del Chocó', 2016. http://perschoco.com/Componentes_WEB/Demanda_WEB/PERSChocó_Demanda_DiagnosticoEnergetico_Oct2016.pdf
33. DANE (2017) Pobreza Monetaria Chocó año 2017. http://www.dane.gov.co/files/investigaciones/condiciones_vida/pobreza/2017/Choco_Pobreza_2017.pdf
34. DNP (2011) Conpes 140. Modificación a conpes social 91 del 14 de junio de 2005: Metas y estrategias de Colombia para el logro de los objetivos de desarrollo del milenio –2015. https://colaboracion.dnp.gov.co/cdt/conpes/social/140.pdf
35. Cherni JA, Dyner I, Henao F, Jaramillo P, Smith R, Font RO (Mar. 2007) Energy supply for sustainable rural livelihoods. A multi-criteria decision-support system. Energy Policy 35(3):1493–1504. https://doi.org/10.1016/j.enpol.2006.03.026
36. Smith RA, Mesa OJ (1996) A rural electrification expansion model. Int Trans Oper Res 3(3–4):319–325. https://doi.org/10.1016/S0969-6016(97)80001-R
37. IDEAM and UPME (2019) Atlas de radiación solar. http://atlas.ideam.gov.co/presentacion/. Accessed 21 Sep 2019
38. Ouedraogo BI, Kouame S, Azoumah Y, Yamegueu D (2015) Incentives for rural off grid electrification in Burkina Faso using LCOE. Renew Energy 78:573–582. https://doi.org/10.1016/j.renene.2015.01.044
39. Branker K, Pathak MJM, Pearce JM (Dec. 2011) A review of solar photovoltaic levelized cost of electricity. Renew Sust Energ Rev 15(9):4470–4482. https://doi.org/10.1016/j.rser.2011.07.104
40. Hurtado Munoz LA, Huijben JCCM, Verhees B, Verbong GPJ (2014) The power of grid parity: a discursive approach. Technol Forecast Soc Change 87:179–190. https://doi.org/10.1016/j.techfore.2013.12.012
41. Banco de la República de Colombia, Tasas de captación semanales y mensuales, 2019. https://www.banrep.gov.co/es. Accessed 18 Sep 2019
42. World Bank. International financial statistics and data files', Inflation, consumer prices (annual %) – Colombia, 2019. https://data.worldbank.org/indicator/FP.CPI.TOTL.ZG?fbclid=IwAR1T3oY42xcltyyVwUb3leENPEGlfaUuFyrIgauigFt7Kgra-fCGnGZbzG4&locations=CO. Accessed 18 Sep 2019
43. Cadavid L, Jimenez M, Franco CJ (2015) Financial analysis of photovoltaic configurations for Colombian households. IEEE Lat Am Trans 13(12):3832–3837. https://doi.org/10.1109/TLA.2015.7404916
44. Jiménez M, Cadavid L, Franco C (2014) Scenarios of photovoltaic grid parity in Colombia. Dyna 188(81):237–245. https://doi.org/10.15446/dyna.v81n188.42165
45. Lazard (2016) Lazard's levelised cost of energy analysis (version 10.0). [Online]. Available: https://www.lazard.com/media/438038/levelized-cost-of-energy-v100.pdf
46. UPME (2018) Unidad de Planeación Minero-Energética. Proyección de precios de los energéticos para generación eléctrica 2018–2040. http://www1.upme.gov.co/Paginas/Hidrocarburos.aspx. Accessed 20 Sep 2019
47. CREG (2019) Comisión de regulación de energía y gas. Matriz Origen Destino Transporte Combustible ZNI. http://www.creg.gov.co/sectores/energia-electrica/estadisticas/matriz-origen-destino-transporte-combustible-zni. Accessed 20 Sep 2019
48. UPME (2004) Resolución 355 de 2004. Por la cual se modifica el consumo de subsistencia del servicio de energía eléctrica. República de Colombia
49. SUI (2018) Sistema único de información de servicios públicos domiciliarios. Consolidado información comercial ZNI. http://reportes.sui.gov.co/fabricaReportes/frameSet.jsp?idreporte=ele_com_103. Accessed 21 Sep 2019
50. Partain LD, Fraas LM (2010) Solar cell electricity market history, public policy, projected future, and estimated cost. In: Solar cells and their applications, 2nd edn. Wiley, Hoboken, pp 17–43
51. Cadavid L, Jiménez M, Franco CJ (2015) Financial analysis of photovoltaic configurations for Colombian households. IEEE Lat Am Trans 13(12):3832–3837. https://doi.org/10.1109/TLA.2015.7404916
52. Bhandari R, Stadler I (Sep. 2009) Grid parity analysis of solar photovoltaic systems in Germany using experience curves. Sol Energy 83(9):1634–1644. https://doi.org/10.1016/j.solener.2009.06.001
53. MercadoLibre Colombia (2019) Batería. https://articulo.mercadolibre.com.co/MCO-526602332-bateria-de-gel-100amp-12volt-_JM?matt_tool=40815910&matt_word&gclid=EAIaIQobChMIoLmDwbnb5AIVU1mGCh3WWQjvEAQYASABEgKV4PD_BwE&quantity=1. Accessed 18 Sep 2019
54. Vivasolar Colombia (2019) Paneles solares. https://www.vivasolar-colombia.com/. Accessed 18 Sep 2019
55. ProViento (2019). http://www.proviento.com.co/inversores.html. Accessed 18 Sep 2019
56. MercadoLibre Colombia (2019) Cable solar fotovoltaico. https://articulo.mercadolibre.com.co/MCO-482037161-cable-solar-fotovoltaico-4mm2-x-40mts-_JM?quantity=1#position=16&type=item&tracking_id=ca93309a-919a-45b7-864d-366abb74e4d2%0D%0A/. Accessed 18 Sep 2019
57. MercadoLibre Colombia (2019) Controlador de carga solar https://articulo.mercadolibre.com.co/MCO-479099522-controlador-regulador-de-carga-solar-jarrett-30-amp-pwm-_JM?quantity=1#position=1&type=item&tracking_id=0a88c7ba-a06e-44d8-8908-cee947a24451. Accessed 18 Sep 2019
58. MercadoLibre Colombia (2019) Fusible energía solar. https://articulo.mercadolibre.com.co/MCO-481066567-10-unidades-1000vdc-fusible-energia-solar-fotovoltaica-a-_JM?quantity=1#position=22&type=item&tracking_id=df08a6fb-bd73-40f6-9384-3ee2e1d8e490. Accessed 18 Sep 2019
59. Homecenter (2019) Breakers https://www.homecenter.com.co/homecenter-co/. Accessed 18 Sep 2019
60. Alibaba Group (2019) Diesel generator price. https://spanish.alibaba.com/product-detail/cgf4000x-1kw-diesel-generator-price-60695467145.html?spm=a2700.8699010.normalList.4.549113c7AZbkT9 Accessed 18 Sep 2019

61. MercadoLibre Colombia (2019) Planta Eléctrica. Planta Electrica Jd1200 1 kW Sukra. https://articulo.mercadolibre.com.co/MCO-517056513-planta-electrica-jd1200-1-kw-sukra-_JM?matt_tool=59488733&matt_word&gclid=EAIaIQobChMImJGYwcPb5AIVgZ-zCh0yJQ83EAQYASABEgLcjfD_BwE&quantity=1&variation=38657723537. Accessed 18 Sep 2019
62. Energiteca (2019) Lubricantes. https://energiteca.com/. Accessed 20 Sep 2019
63. IEA (2013) Statistics and reports: Colombia. https://www.iea.org/statistics/statisticssearch/report/?country=COLOMBIA&product=electricityandheat&year=2013. Accessed 16 Jul 2017
64. Congreso de la República de Colombia (2014) Ley 1715. Por la cual se regula la integración de las energías renovables no convencionales al Sistema Energético Nacional. Colombia. pp 1–26

Comparison Analysis of Potential and Distribution of Photovoltaic Power in Five Open-Pit Mines in North Korea

Myeongchan Oh, Aran Yang, Yebin Kim, Han Jin Kim, Ji Yoon Ku, and Hyeong-Dong Park

1 Introduction

Mining industries account for approximately 14% of the total economy and 40% of the foreign export amount in North Korea. There are huge mineral resources and approximately 700 mines in North Korea. Magnesite ore reserves are the world's largest, and tungsten and graphite reserves are among the top 10 in the world [1]. However, operating these large-scale mines becomes difficult because of many social and economic reasons. Since the 1990s, North Korea has been experiencing problems such as reduction of national infrastructure, aging of mining production and mining facilities, and lack of mining technology development throughout the mining industry. In particular, a shortage of electricity in North Korea is a major risk that directly degrades the production capacity of mines. The electricity generation in North Korea is approximately 7% compared to that in South Korea according to the Electric Power Statistics Information System (EPSIS) in South Korea. Therefore, new solutions to supply power are required to operate mines in North Korea.

The photovoltaic (PV) power plants can be an alternative to previous power plants. Recently, the PV power plants have high economic feasibility and can be operated independently without facilities such as an existing power grid. Therefore, they can be suitable for North Korea where many facilities are old and damaged. They can be more efficient in terms of time and space than building a large power plant when a distributed power grid is properly designed. The evaluation of the potential for PV power in the area should be preceded to supply power to the mines. Therefore, the potential of PV power near large mines was studied to estimate the amount of possible power supply and feasibility.

2 Methodology

Five main mines, Musan, Daeheung, Ryongyang, Sangnong, and Eunryul, are selected based on capacity, mineral types, topography, and locations. The information of selected mines is listed in Table 1 and the locations of those mines are shown in Fig. 1. Each mine is analyzed with spatial data such as digital elevation map (DEM). Spatial data are acquired from various sources such as Information System for Resources of North Korea (I-RENK) and OpenStreetMap.

M. Oh
New and Renewable Energy Resource Map Laboratory, New and Renewable Energy Institute, Korea Institute of Energy Research, Daejeon, South Korea

A. Yang · Y. Kim · H. J. Kim
Department of Energy Systems Engineering, Seoul National University, Seoul, South Korea

J. Y. Ku
Department of Environmental Science and Engineering, Ewha Womans University, Seoul, South Korea

H.-D. Park (✉)
Department of Energy Systems Engineering, Seoul National University, Seoul, South Korea

Research Institute of Energy and Resources, Seoul National University, Seoul, South Korea
e-mail: hpark@snu.ac.kr

© The Author(s), under exclusive license to Springer Nature Switzerland AG 2022
A. Sayigh (ed.), *Sustainable Energy Development and Innovation*, Innovative Renewable Energy,
https://doi.org/10.1007/978-3-030-76221-6_4

Table 1 Information of the selected five main open-pit mines in North Korea [2–4]

Name	Location		Mineral type	Mining type	Reserves	Annual production
Eunryul	Hwanghaenam-do	Eunryul-gun	Iron	Open pit	100 million tons (Fe 48%)	1.6 million tons
Musan	Hamgyeongbuk-do	Musan-gun	Iron	Open pit	1.5 billion tons (Fe 25–35%) 100 million tons (Fe 48%)	8 million tons (30%) 3 million tons (60%)
Daeheung	Hamgyongnam-do	Dancheon-si	Magnesite	Open pit	880 million tons (MgO 46.7%)	1 million tons
Ryongyang			Magnesite	Underground Open pit	770 million tons (MgO 45.8%)	1 million tons
Sangnong		Heocheon-gun	Gold, copper	Underground	Gold 200 tons Copper 500,000 tons (Metal standards)	2.8 million tons (Capacity of mineral dressing Au lg/t, Cu 0.23%)

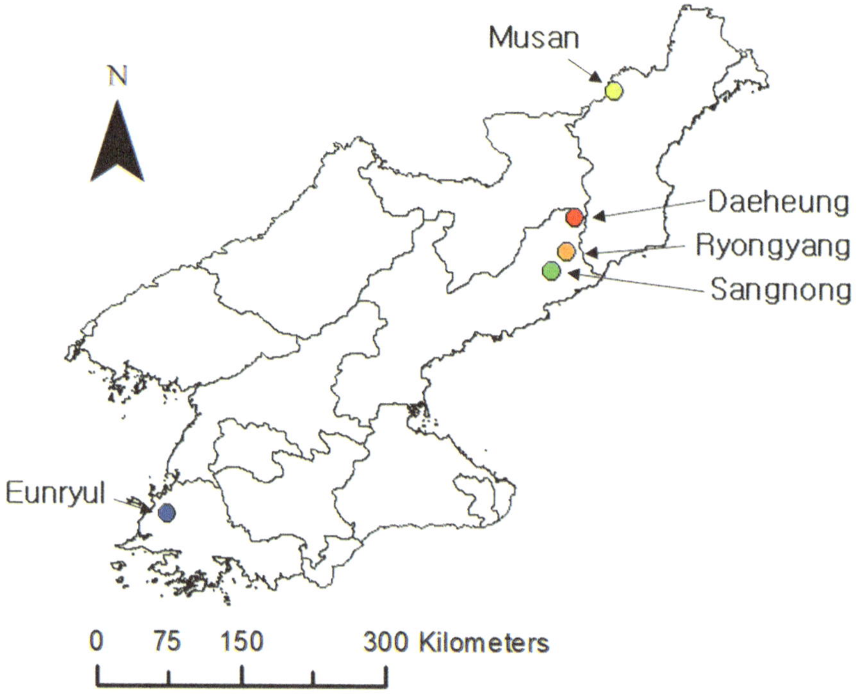

Fig. 1 The location of the selected five main open-pit mines in North Korea

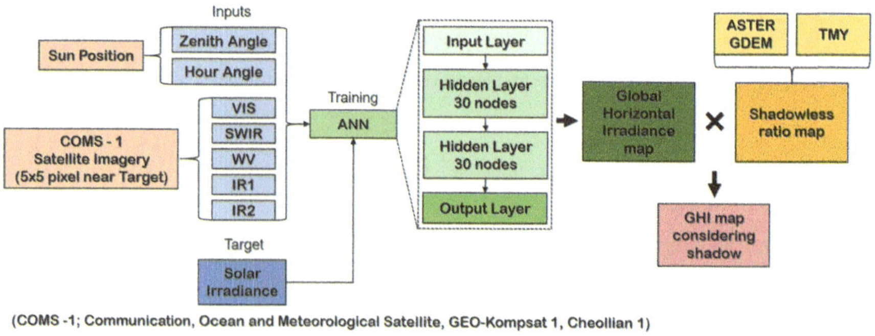

Fig. 2 Methodology for estimation of solar irradiance from satellite images

The potential of PV power is estimated considering weather and shadows from terrains. The spatial distribution of large areas is estimated using a machine learning technique. Artificial neural network (ANN) is used to estimate solar irradiance from satellite images. Figure 2 shows the process of large-scale potential estimation. Some variables for sun position and all-band images from weather satellites are used. The ANN model from a previous study [5], which is trained and validated in South Korea, is used. This model can estimate global horizontal irradiance (GHI) and have a relative root mean squared error (RMSE) of approximately 13%.

Shadows are calculated to modify the large-scale irradiance result. ASTER Global Digital Elevation Map (GDEM) is used to calculate shadows. Shadows are calculated in each hour based on the typical meteorological year (TMY) dataset in South Korea [6]. Not only shadows but the stability of ground conditions is also considered in the small-scale analysis. Unstable areas are extracted from the susceptibility analysis of landslides and floods. These areas are excluded from the site for PV power. Geographic information systems (GIS) is used in these extractions and exclusions. The landslide susceptibility map is calculated using ANN.

3 Results

As a result, the PV potential distribution of each mine is shown in Fig. 3. The results consider the shadows and all the areas with unstable ground conditions are excluded. Areas within three kilometer radius from each mine are visualized with three levels of irradiance. The statistics of PV potential is shown in Table 2. As Eunryul Mine is located at a high latitude and a flat area, it has a large area with high PV potential. However, huge mountainous areas have low PV potential. There is no area with more than 3.5 kWh/m^2 of PV potential near the Musan and Daeheung mines. The other two mines have moderate properties.

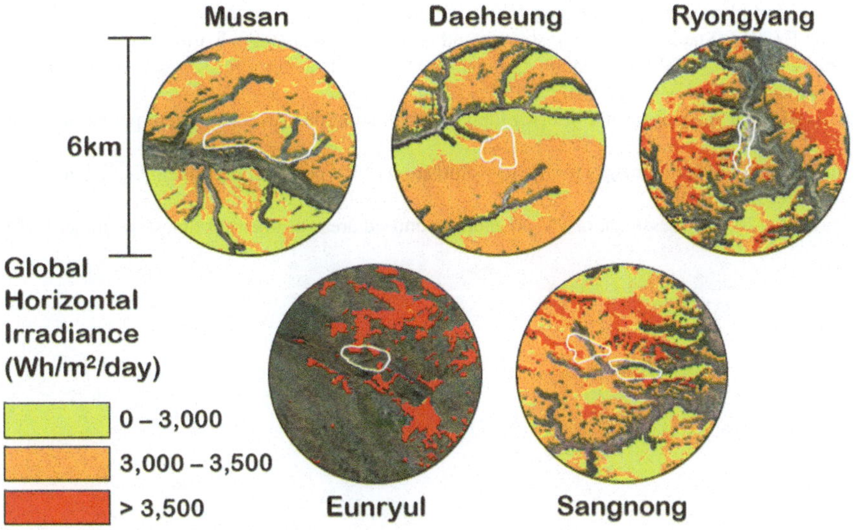

Fig. 3 The location of the selected five main open-pit mines in North Korea. (White line: mine)

Table 2 The estimated PV potential in selected five mines

	Area of global horizontal irradiance value (km^2)				
Name	>3500	3500–3000	3000-0	Summation	Mean solar irradiance (Wh/m^2/day)
Eunryul	5.19	0.02	0	5.22	3782
Musan	0	21.09	2.62	23.71	2850
Daeheung	0	16.66	6.45	23.11	3066
Ryongyang	3.00	9.89	4.37	17.27	3154
Sangnong	2.71	11.50	5.52	19.74	3116

An analysis of the TMY shows that azimuth of 200° for the best orientation of the PV panel in Yeongjong Sky City in South Korea [7]. In addition, open-pit limestone quarry can be a good potential site for temporary PV plant during quarrying work [8]. The results from both the recent studies can be used to simulate the potential PV generation in the mining areas in North Korea in much more detail.

4 Conclusion

The remote sensing, machine learning, GIS, and other techniques are used in this study to estimate PV potential near the mines in North Korea. Eunryul shows the highest PV potential in selected five main mines. However, in addition to GHI, other irradiance indices such as direct normal irradiance (DNI) are required to estimate actual PV power generation. Besides, as many trained models and data are validated in South Korea, further research on accuracy in North Korea is required. Nevertheless, this study shows the basic methodology for PV potential and presented the results. The results can be used in relevant economic evaluations and policy decisions.

Acknowledgments This work was supported by a grant from the Human Resources Development program (No. 20204010600250) of the Korea Institute of Energy Technology Evaluation and Planning (KETEP), funded by the Ministry of Trade, Industry, and Energy of the Korean Government.

References

1. USGS (2020) Mineral Commodity Summaries. 2020.
2. KIGAM (2016) Korean mineral resources database: https://www.kigam.re.kr/dmr/. Accessed 30 March 2020
3. Ministry of Unification, North Korea information portal: https://nkinfo.unikorea.go.kr/. Accessed 30 March 2020
4. Koh SM, Lee GJ, Yoon E (2013) Status of mineral resources and mining development in North Korea. Econ Environ Geol 46(4):291
5. Koo Y, Oh M, Kim S-M, Park HD (2020) Estimation and mapping of solar irradiance for Korea by using COMS MI satellite images and an artificial neural network model. Energies 13(2):301
6. Kim SM, Oh M, Park HD (2019) Analysis and prioritization of the floating photovoltaic system potential for reservoirs in Korea. Appl Sci 9(3):395
7. Ku JY, Park HD (2020) The use of aerial images for the fast simulation of PV potential in residential area in Yeongjong-do, Korea, PVSEC-30 & GPVC 2020, Jeju, Korea (submitted)
8. Yang A, Kim HJ, Park HD (2020) Assessment of PV potential in unused areas of open-pit limestone mine, PVSEC-30 & GPVC 2020, Jeju, Korea (submitted)

Optimal Ratio of Battery to PV Module Voltage for Efficient Operation of Direct-Coupled PV System

Sarah Paul Ayeng'o

Nomenclature

DC	Direct current
NCA	Lithium nickel cobalt aluminum oxide
Max	Maximum
Min	Minimum
MPP	Maximum power point
$P\text{-}PV_{dir.coup}$	Output PV power in a direct-coupled topology
$P\text{-}PV_{MPP}$	PV output power at maximum power point
$P\text{-}PV_{peak}$	Peak PV power of the module
PV	Photovoltaic
SHS	Solar home systems
SOC	State of charge
STC	Standard test condition at 25 °C and 1000 W/m^2

1 Introduction

In recent years, most of the off-grid communities in Tanzania have been educated and have realized the importance of using electricity for their daily activities. Electricity being an engine of development to any society can be used for different purposes like lightning, powering appliances, food storage, and cooking [3].

In general, electrification rate in Tanzania is still very low. Grimsby et al. [7] reported that 86% of Tanzanian population was not connected to the national grid, of which most live in rural areas. Bertheau et al. [3] also reported that electrification rate of the country is low of about 11%, whereby in rural areas it is falling below 2%. Another author [16] also reported that, in Tanzania, only about 14% of the population has access to electricity, with only 3% in the rural areas.

Recently, there has been an increase in electricity access to 32.8% of the whole Tanzanian population. Among this, 65% of the electrified population was from urban areas and 18% from rural off-grid area. Of the electrified population, 74.9% was electrified with national grid, 24.7% was electrified from solar PV systems, and 0.4% was electrified from individual small diesel generators [2, 4, 20]. This has shown a significant improvement and increase of off-grid electrification in Tanzania, especially with the use of solar PV systems.

Not only that the small percent of population using solar energy is expected to increase in the coming years due to the awareness which has been created on PV systems but also worldwide the cost of PV modules is decreasing [1, 11]. On the other hand, most of the off-grid population in Tanzania have no technical background on PV systems. Selection of correct PV module type, battery, and size is still a challenge. Not only that but also a large percent of the off-grid population lack knowledge on how to efficiently operate PV system. This leads to energy losses and early aging of some of the PV systems' components like batteries.

There are different types of PV panels which can be found in the market. They can be categorized depending on different technologies and they include first-, second-, and third-generation PV panels [6]. First-generation solar cells are also known as conventional cells. They are mostly made of silicon materials, including crystalline and amorphous cells, and are the ones which are mostly available in the market.

S. P. Ayeng'o (✉)
University of Dar es Salaam, Department of Mechanical and Industrial Engineering, Dar es Salaam, Tanzania

© The Author(s), under exclusive license to Springer Nature Switzerland AG 2022
A. Sayigh (ed.), *Sustainable Energy Development and Innovation*, Innovative Renewable Energy,
https://doi.org/10.1007/978-3-030-76221-6_5

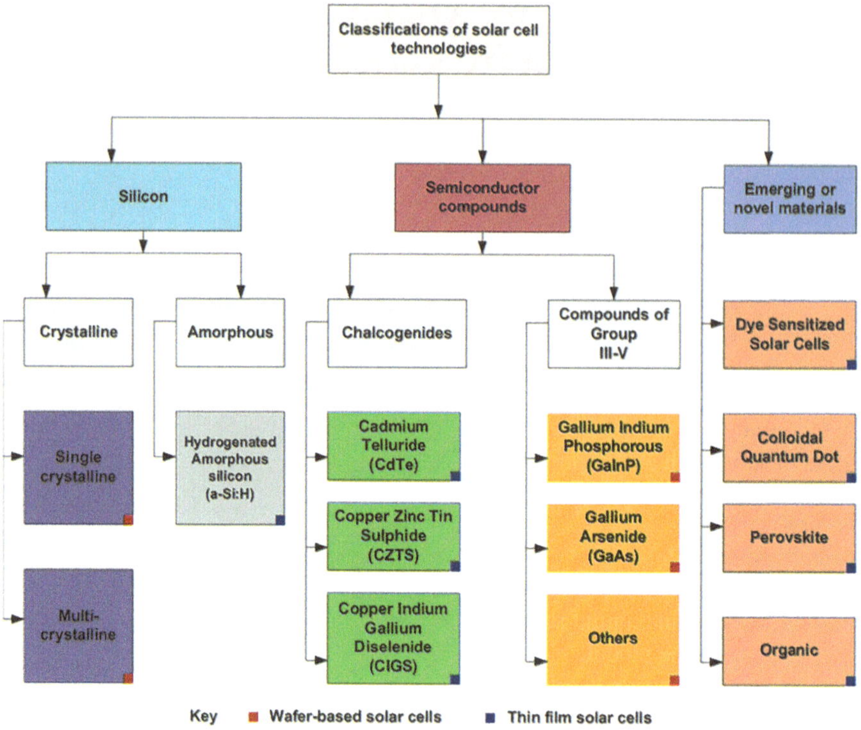

Fig. 1 Classification of solar cell technologies [9]

Second-generation solar cells are thin-film cells such as cadmium telluride (CdTe), copper indium gallium diselenide (CIGS), and gallium arsenide (GaAs) cells. Third-generation solar cells are mostly still in research phase. There is, furthermore, an emerging market of organic photovoltaic cells and perovskite [12, 15, 21]. Figure 1 shows in brief the classification of PV modules.

For solar PV systems, batteries are also of importance for electricity storage. Although in the market there are different types of batteries, most of the off-grid population uses lead-acid batteries. This is due to the fact that lead-acid battery is a mature technology and its investment cost is lower compared to other technologies like lithium-ion batteries [8, 13, 19].

Paul Ayeng'o et al. [18] did a study by comparing cost of lead -acid and lithium ion batteries for the lifetime of 20 years in the PV system. They found that solar home system (SHS) with lithium nickel cobalt aluminum oxide (NCA) battery generally had a lower LCOE compared to the SHS with lead-acid battery. This was mainly due to both the longer life time of NCA batteries and the reduction of the price of NCA batteries as a result of significantly increasing global production scales.

Due to the fact that market have different types of PV module and batteries, proper ratio between battery and PV module voltage during PV sizing has to be known, so as to easily select the system and have an assurance of operating the system efficiently. In this study, a solar PV model has been developed. A 24 V DC household using polycrystalline PV modules with lithium nickel cobalt aluminum oxide (NCA) has been studied.

By considering fluctuation of solar radiation, temperature, and load profile, the developed model can select an optimal range of maximum power point voltage (U_{MPP}) of different PV modules by selecting the optimal number of PV cells connected in series. The ratio between battery nominal voltage and maximum power point voltage (U_{MPP}) can help easy selection of PV module and battery voltage in the market and have confidence that PV system will always be operated at high efficiency. The results presented in this chapter are from the study which was done in Tanzania; however, the model can be adopted at any climatic area.

2 Methodology

2.1 Solar PV Model

The developed PV model calculates first PV output at an inclined plane. Then, it calculates current and voltage at maximum power point. The model uses solar radiation data (global, direct, diffuse, and reflected radiation), temperature, PV module electrical data, and inclination angle as input data [10, 14]. Input solar radiation was measured at horizontal plane. Hence, global solar radiation at an inclined plane was calculated as follows:

$$H_t = H_b R_b + H_d R_d + (H_b + H_d) R_r \tag{1}$$

Where R_b, R_d,and R_r are the ratios of beam (direct), diffuse, and reflected solar radiation on the inlined plane to that on the horizontal plane; H_b and H_d are direct and diffuse radiation measured at horizontal plane; and H_t is the total solar radiation at the inclined plane.

The calculation of current and voltage at maximum power point is as presented by Paul Ayeng'o et al. [17]. Optimization of PV model was also done using genetic algorithm. The optimization tool could select the optimal number of PV cells connected in series. By having the optimal number of PV cells connected in series, one can easily estimate the optimal current and voltage at maximum power point.

2.2 Electrical Data Sheet of Polycrystalline

In this study, polycrystalline PV module was used. The module has 54 number of cells connected in series. It has 0.489 V cell voltage. Table 1 below describes the electrical data at standard test condition (STC). The I-V-P curve at STC and the battery-operating window are presented in Fig. 2.

2.3 NCA Battery Model

Battery model was adopted from the model presented by [5]. Battery state of charge was calculated by using Coulomb counting equation. Refer Eq. 2 below.

$$\mathrm{SOC}(t) = \mathrm{SOC}_0 + \int_0^t (I_{\mathrm{eff}}(t) / C_{\mathrm{act}}(t) dt \tag{2}$$

Table 1 Electrical data sheet of polycrystalline module

Electrical data (at STC)	Value
Max. power (Pmax)	210 W
Max. power voltage (Umpp)	26.4 V
Max. power current (Impp)	7.95 A
Open circuit voltage (Uoc)	33.6 V
Short circuit current (Isc)	8.33 A

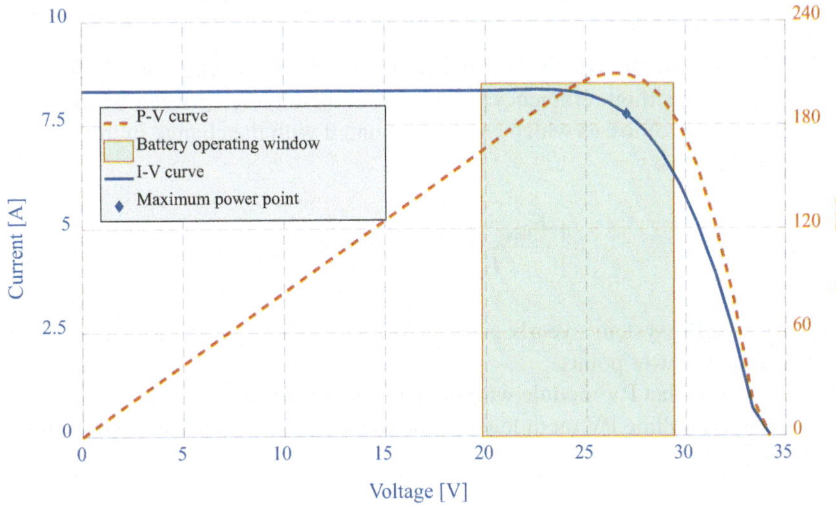

Fig. 2 I-V-P curve for polycrystalline module at 25°C and 1000 W/m²

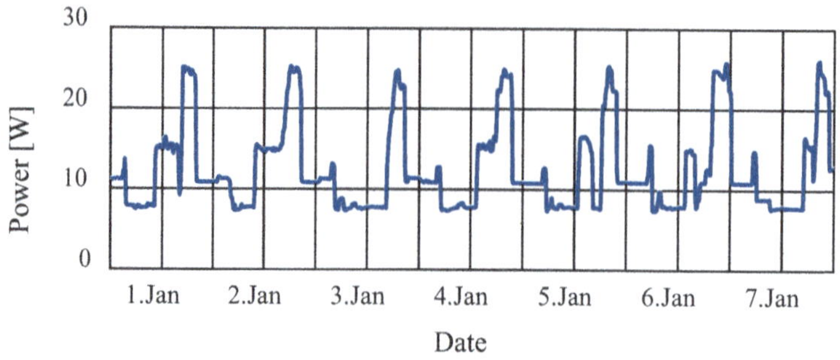

Fig. 3 Household load profile

SOC(t) is the actual state of charge in percentage, SOC_0 is the initial state of charge in percentage, $I_{eff}(t)$ is the effective battery current that is used to charge or discharge the battery, and $C_{act}(t)$ is the actual battery capacity in Ah.

2.4 Load Profile

Most of rural households have low electrical energy consumption. In this study, household which consumes 0.3kWh per day was used with its daily profile as presented in Fig. 3 below. It can be seen clearly from the graph that this household uses more electricity at night where there is no solar radiation. For this household, the use of battery is essential.

3 Results

3.1 Optimal Solar PV Module

The original simulated PV module has 54 cells connected in series. After optimization process, the optimal number of PV cells connected in series was 61. More optimal results are as presented in Table 2 below.

A small comparison was done between PV system which is directly coupled with the battery and PV system operated at maximum power point. It can be seen clearly from Fig. 4 below that, most of the time, direct-coupled system was operated near the maximum power point and, hence, efficiently operated.

4 Optimal Ratio of Battery to Maximum Power Point PV Voltages

Analysis was done to find the optimal number of PV cells connected in series where when direct-coupled system will be operated, it will operate at low losses (high efficiency). The optimal results lie between 53 and 65 number of PV cells in serial. Figure 5 shows the percentage of losses which can be accounted with the change in the serial number of PV cells.

Losses in this study is calculated as:

$$\left(\frac{E_{mpp} - E_{dircoup}}{E_{mpp}}\right) * 100\% \tag{3}$$

where $E_{dircoup}$ is the direct-coupled PV system's yearly generated energy and E_{mpp} is the yearly energy generated when the PV system is operated at its maximum power points.

From Fig. 5 above, it can be seen that PV module with number of PV cells connected in series from 53 to 67 can be operated at high efficiency. For polycrystalline PV module with cell voltage of around 0.489 V, this range of number of PV cells connected in series can be applied.

Table 2 Optimization results for polycrystalline PV panels

Parameter	Unit	Value
Inclination angle	°	7
Azimuth angle	°	182
SOC maximum	%	83
SOC minimum	%	0
Number of PV cells in series	–	61
Number of PV cells in parallel	–	5
Peak power	Wp	118.6

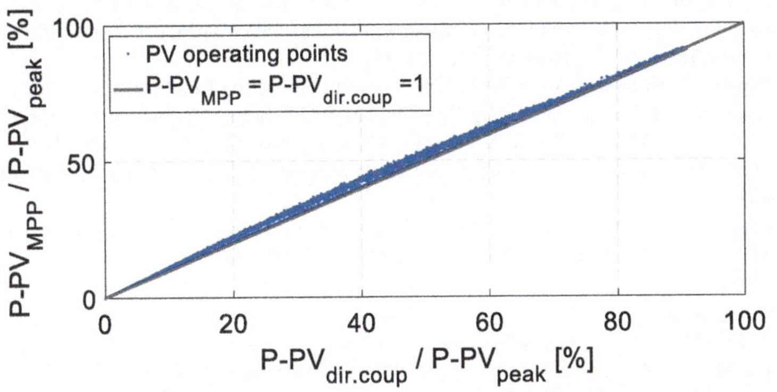

Fig. 4 PV modules operated at MPP and with a system operated at a direct-coupled topology

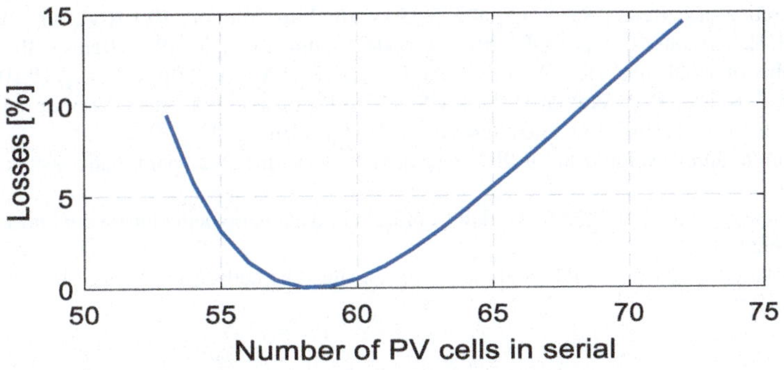

Fig. 5 Losses in PV modules with the change in number of PV cells connected in series

From this range, a ratio between battery nominal voltage and maximum power point at STC of 0.7 to 0.9 is proposed. Within this ratio, one can have an assurance when selecting the PV system that the system will be efficiently operated. This ratio can also be used by other types of PV modules and batteries during PV system design.

5 Conclusion

In this study, a simple method of selecting battery and PV module voltage has been presented. A ratio of battery to maximum power point voltage of PV modules has been proposed. This can help in the designing of PV system and have a confidence in efficiently operating PV system. The sizing of battery capacity and PV maximum power should also be done.

References

1. Barron AR (2015) Cost reduction in the solar industry. Mater Today 18(1):2–3. https://doi.org/10.1016/j.mattod.2014.10.022
2. Bastholm C, Fiedler F (2018) Techno-economic study of the impact of blackouts on the viability of connecting an off-grid PV-diesel hybrid system in Tanzania to the national power grid. Energ Convers Manage 171:647–658. https://doi.org/10.1016/j.enconman.2018.05.107
3. Bertheau P, Cader C, Müller H, Blechinger P, Seguin R, Breyer C (2014) Energy storage potential for solar based hybridization of off-grid diesel power plants in Tanzania. Energy Procedia 46:287–293. https://doi.org/10.1016/j.egypro.2014.01.184
4. Citizen Reporter @TheCitizenTz news@tz.nationmedia.com (4/23/2017): 33pc of Tanzanians have access to electricity: report. In The Citizen, 4/23/2017. Available online at https://www.thecitizen.co.tz/news/33pc-of-Tanzanians-have-access-to-electricity%2D%2Dreport/1840340-3900298-fis25i/index.html, checked on 5/27/2020
5. Dirk M (2009) 24th European photovoltaic solar energy conference and exhibition [electronic resource]. Conference 21–25 Septemb-er 2009 - Exhibition 21–24 September 2009, 2009 Hamburg proceedings. WIP-Renewable Energies, Munich
6. Grace T, Shearer C, Tune D, Yu L, Batmunkh M, Biggs MJ et al (2016) Use of carbon nanotubes in third-generation solar cells. In: Peng H, Li Q, Chen T (eds) Industrial applications of carbon nanotubes. Elsevier. (Micro and nano technologies series), Amsterdam, pp 201–249
7. Grimsby LK, Aune JB, Johnsen FH (2012) Human energy requirements in Jatropha oil production for rural electrification in Tanzania. Energy Sustain Dev 16(3):297–302. https://doi.org/10.1016/j.esd.2012.04.002
8. Hoppmann J, Volland J, Schmidt TS, Hoffmann VH (2014) The economic viability of battery storage for residential solar photovoltaic systems – a review and a simulation model. Renew Sust Energ Rev 39:1101–1118. https://doi.org/10.1016/j.rser.2014.07.068
9. Ibn-Mohammed T, Koh SCL, Reaney IM, Acquaye A, Schileo G, Mustapha KB, Greenough R (2017) Perovskite solar cells: an integrated hybrid lifecycle assessment and review in comparison with other photovoltaic technologies. Renew Sust Energ Rev 80:1321–1344. https://doi.org/10.1016/j.rser.2017.05.095
10. Ibrahim A, El-Sebaii AA, Ramadan MRI, El-Broullesy SM (2013) Estimation of solar irradiance on tilted surfaces facing south for Tanta, Egypt. Intern J Sustain Energy 32(2):111–120. https://doi.org/10.1080/14786451.2011.601814
11. Kavlak G, McNerney J, Trancik JE (2018) Evaluating the causes of cost reduction in photovoltaic modules. Energy Policy 123:700–710. https://doi.org/10.1016/j.enpol.2018.08.015
12. Krebs FC, Jørgensen M (2013) Polymer and organic solar cells viewed as thin film technologies: what it will take for them to become a success outside academia. Solar Energ Mat Solar Cells 119:73–76. https://doi.org/10.1016/j.solmat.2013.05.032
13. Kwiecien M, Schröer P, Kuipers M, Sauer DU (2017) Current research topics for lead–acid batteries. In: Lead-acid batteries for future automobiles. Elsevier, Amsterdam, pp 133–146
14. Merei G, Berger C, Sauer DU (2013) Optimization of an off-grid hybrid PV–wind–diesel system with different battery technologies using genetic algorithm. Solar Energy 97:460–473. https://doi.org/10.1016/j.solener.2013.08.016
15. Mohammad Bagher A (2015) Types of solar cells and application. AJOP **3**(5):94. https://doi.org/10.11648/j.ajop.20150305.17
16. Moner-Girona M, Ghanadan R, Solano-Peralta M, Kougias I, Bódis K, Huld T, Szabó S (2016) Adaptation of Feed-in Tariff for remote mini-grids: Tanzania as an illustrative case. Renew Sust Energ Rev 53:306–318. https://doi.org/10.1016/j.rser.2015.08.055
17. Paul Ayeng'o S, Axelsen H, Haberschusz D, Sauer DU (2019) A model for direct-coupled PV systems with batteries depending on solar radiation, temperature and number of serial connected PV cells. Solar Energy 183:120–131. https://doi.org/10.1016/j.solener.2019.03.010
18. Paul Ayeng'o S, Schirmer T, Kairies K-P, Axelsen H, Dirk US (2018) Comparison of off-grid power supply systems using lead-acid and lithium-ion batteries. Solar Energy 162:140–152. https://doi.org/10.1016/j.solener.2017.12.049
19. Pavlov DC (2017) Invention and development of the lead–acid battery. In: Lead-acid batteries: science and technology. Elsevier, Amsterdam, pp 3–32
20. Tanzania Energy Situation. energypedia.info (2020). Available online at https://energypedia.info/wiki/Tanzania_Energy_Situation, updated on 5/26/2020, checked on 5/26/2020
21. Wang D, Wright M, Elumalai NK, Uddin A (2016) Stability of perovskite solar cells. Solar Energ Mat Solar Cells 147:255–275. https://doi.org/10.1016/j.solmat.2015.12.025

Test of a Series of Solar Panels Set on a Parabolic Roof

Carlos V. M. Labriola, Jorge Lassig, Ubaldo Jara, and Juan Valle Sosa

1 Introduction

In recent years, numerous studies have been published about the wind effects on fixed solar panels due to the fast growth of photovoltaic installations. Some studies took an experimental approach using wind tunnels (Mohapatra and Pfahl [1, 2]) or CFD fluid dynamic simulations [3, 4].

Most of the studies were limited to the wind effects on a single solar panel, but in the last 5 years the growth of installations with solar panels with several arrays has been relevant. However, most of these installations of tens to hundreds of panels are placed on flat areas or roofs [5]. Other studies suggest wind shields or deviators before it reaches the solar panels [6].

In the case of Comahue, located in the Provinces of Neuquén and Río Negro, Argentina, there are a lot of warehouses, storehouses, and sheds of local industries (fruit packaging, oil services, etc.).

During the last 5 years, changes in electricity taxation and new regulations of the electrical market motivated users to make use of cogeneration with the grid or to be greater generator in the whole sale market (power generation more than 500KW).

This paper presents a model subjected to a wind tunnel test and CFD study for a shed with parabolic-shaped roof of Fruit Packaging Cooperative of General Roca, Río Negro, that want to generate around 500KW with solar PV panels on the roof.

In the bibliography viewed, the authors have not found references on wind loads in solar panels located on curved roofs, so a series of tests were carried out in the wind tunnel and simulated with CFD.

2 Studies in Wind Tunnel

The Laboratory of Environmental Fluid Dynamics (LaTiFA) of Faculty of Engineering of National University of Comahue has a wind tunnel that carries out fluid dynamic studies in buildings. It is 7.40 m long, with a test section of 0.90 × 0.90 m. It is open type (atmospheric boundary layer) with a 6 HP electric motor, with speed regulator/stabilizer.

The wind speed measurement inside the tunnel was made with a hot-film anemometer (CEM DT-8880). Static shots for measuring the loads on the solar panels were made to the model of panel installation. Piezoelectric sensors were connected through small tubes, allowing the reading of pressure on the model. All the data obtained were recorded in a data logger with a scanning frequency of 1000 Hz (Fig. 1).

C. V. M. Labriola (✉)
Department of Electrotecnia, Engineering Faculty, National University of Comahue, Neuquen, Argentina

J. Lassig · U. Jara · J. V. Sosa
Department of Mechanics, Engineering Faculty, National University of Comahue, Neuquen, Argentina

© The Author(s), under exclusive license to Springer Nature Switzerland AG 2022

A. Sayigh (ed.), *Sustainable Energy Development and Innovation*, Innovative Renewable Energy,
https://doi.org/10.1007/978-3-030-76221-6_6

Fig. 1 Wind tunnel of LaTiFA: view of left side (left), view of fiting models into the tunnel (right)

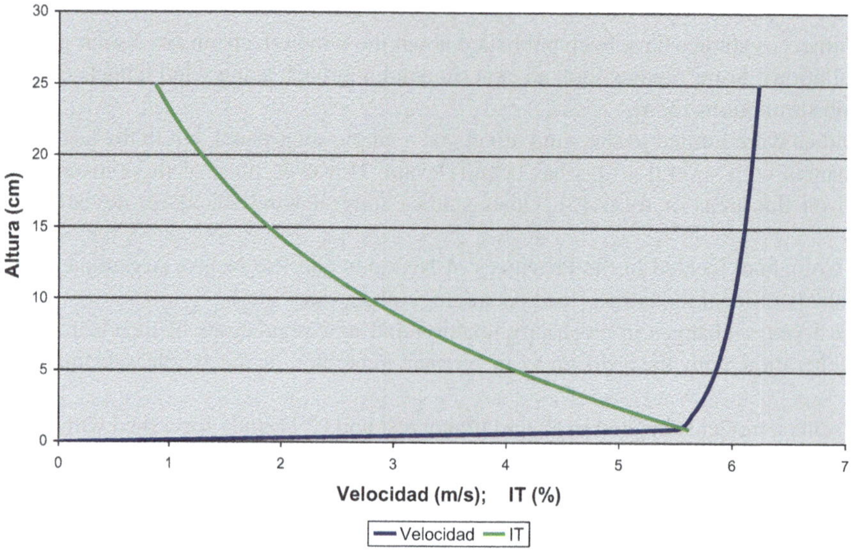

Fig. 2 Speed variation and turbulence intensity versus height into the wind tunnel

The simulated boundary layer profile for an exponent of Sutton's law of $P = 0.0389$ corresponds to a Class C exposure of the CIRSOC 102 Standard (2005); therefore, the resulting speed profile is shown in Fig. 2, as well as the IT turbulence intensity.

3 PV Installation Model

A parabolic shed model was built integrating five strips of solar panels with a scale of 1:40. This scale permitted to include plastic tubes within the strips of solar panels, which connect static intakes to piezoelectrical sensor. The sensors are also connected to the data logger that registers the measurements.

The dimensions of the model were 55 cm long and 50 cm wide, and the maximum height of the parabolic roof was 30 cm. Figure 3 shows the shed models and the location of the solar panels.

At the same time, simulations were made with commercial CFD software to compare results between the physical measurements in wind tunnel with the computational CFD model.

Fig. 3 Shed model tested in the wind tunnel

3.1 Results Obtained and Their Discussion

The roof has a curved shape, so the distance between it and the strip of solar panels varies, so the wind speed intensity is not constant and produces relevant variations in pressure on each panel strip.

The first panel strip facing the wind has a higher pressure load in its sides than its center, taking into account the pressure in the direction of wind (Fig. 4).

The other panel strips have the same variation of pressure along them but with less pressure as can be seen in Fig. 5, which is the top view of the roof. The effect of the higher wind speed on the center of panel strips is appreciated, and by the Bernoulli effect, there is less pressure.

At the end of the roof (right, Fig. 5), pressure in panel strips is slightly higher in the center and lower at the sides, which generates a wind load that pushes upwards in the center and downwards in its sides. Figure 6 shows the wind speed vectors, showing the vortexes generated in the three subsequent solar panels.

When the wind is from the opposite direction (Fig. 7), the panel strip have higher pressure at roof end (Fig. 8, left) and at the top there is depression (Fig. 8, right). It is clear that the highest loads are on the sides.

Also the effect in the center of the panels (upper part) has a higher pressure than its sides. This is due to the greater wake produced by the panels facing the wind as detailed in Fig. 9. The speed field vectors on those panels can be seen, where the opposite directions to the free current indicate wake vortexes produced in the leeward of each solar panel.

Finally, Fig. 10 shows pressure distribution when wind is 45° from the main direction of the Sun, where the loads are more asymmetric.

4 Conclusions

In the wind tunnel tests, because of the size of the scaled model, there were seven spots located on the central line of each panel. Results are labeled as pressure coefficients C_P.

The C_P values of the first panel are shown in Fig. 11, which is the case of the wind from the sun direction. It can be observed that in the center of the panel, loads are upwards (light blue areas), whereas on the sides are downwards (red areas).

The C_P values of the second panel are shown in Fig. 12, which is the case of the wind from the sun direction. You can see that, similar to the first panel, in the center of the panel, loads are upwards (light blue areas), whereas on the sides are downwards (red areas). The difference between the first and second panel is that in the first panel, there is a slightly bigger pressure on the center than on the second panel, making the camel hump.

When the wind goes in the direction of the sun, then all loads are upwards, like shown in Fig. 13, whose highest loads are on the sides.

The shape for the Cp values is similar between the first and second, but with lower values of Cp.

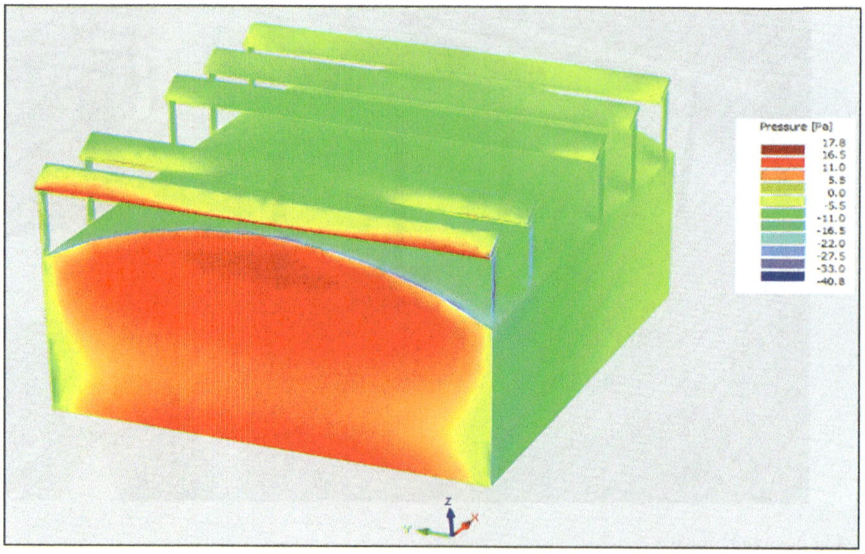

Fig. 4 Pressure distribution produced by wind fron North direction, on front face of the shed and on strips of solar panels

Fig. 5 Top view of shed roof and panel strip with described pressure variations

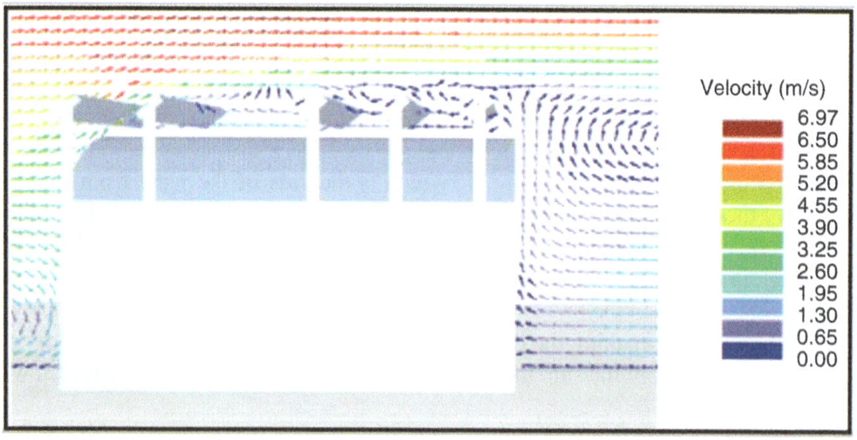

Fig. 6 View of wind vectors on strips of solar panels

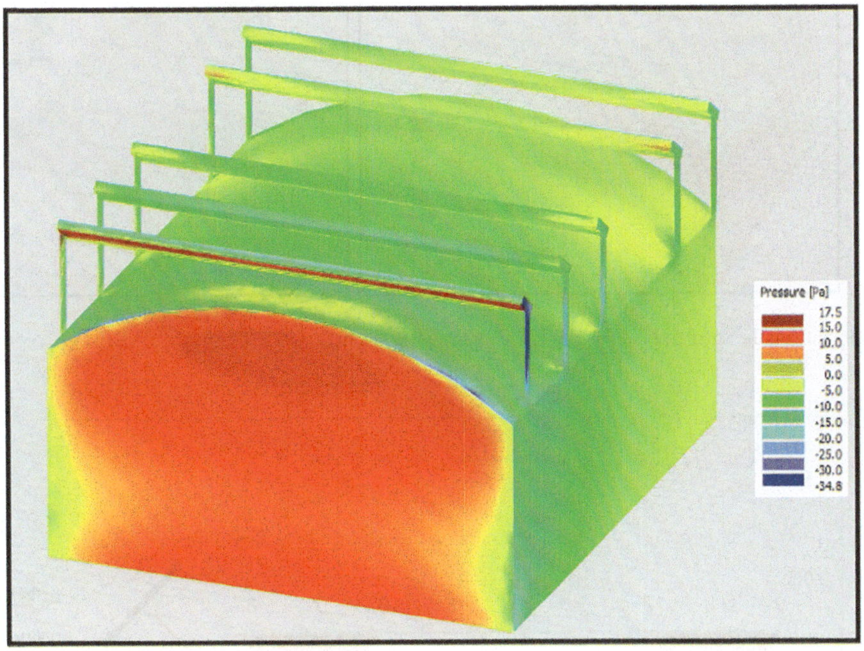

Fig. 7 Perspective view of pressure distribution on shed and solar panels

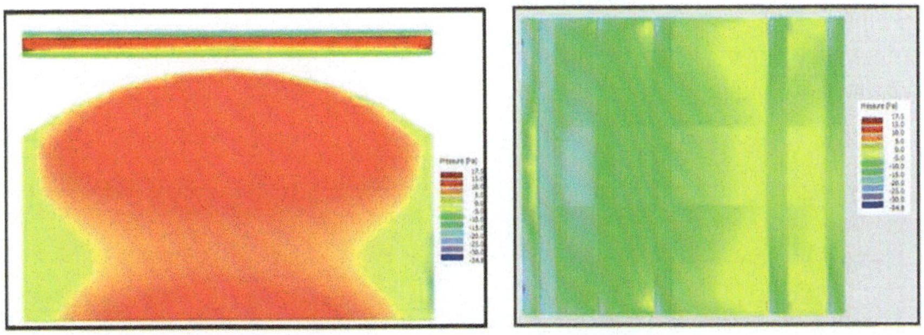

Fig. 8 (left) Pressure distribution on the first panel at the bottom; and (right) pressure distribution at the top of the panels and the roof

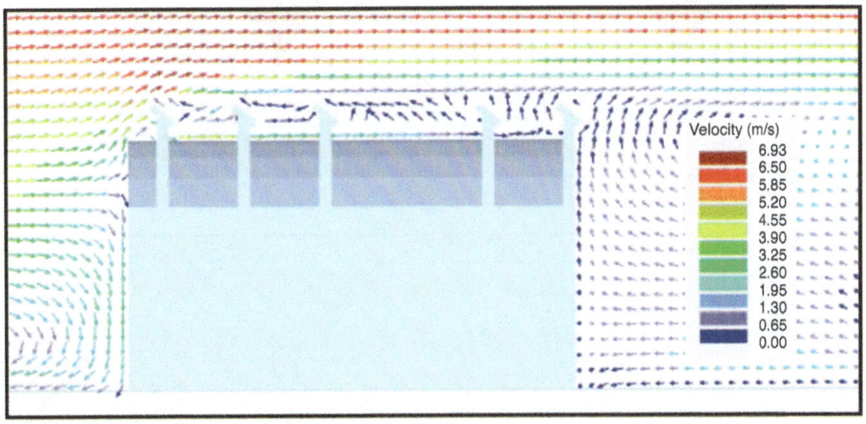

Fig. 9 Distribution of the wind vectors among the solar panels

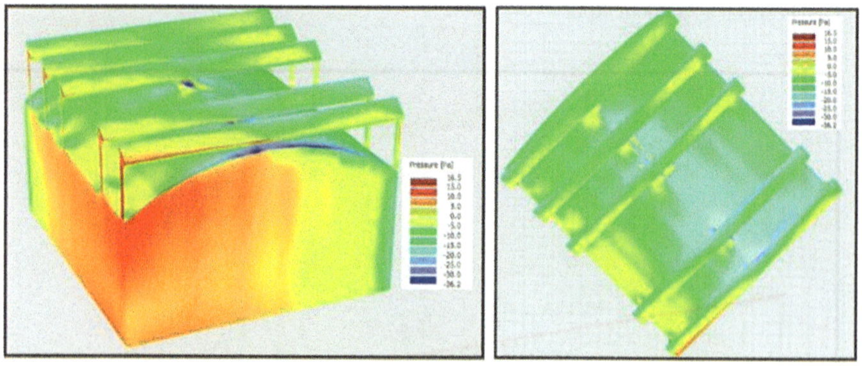

Fig. 10 Pressure distribution on the shed and the panels (left) and top view (right)

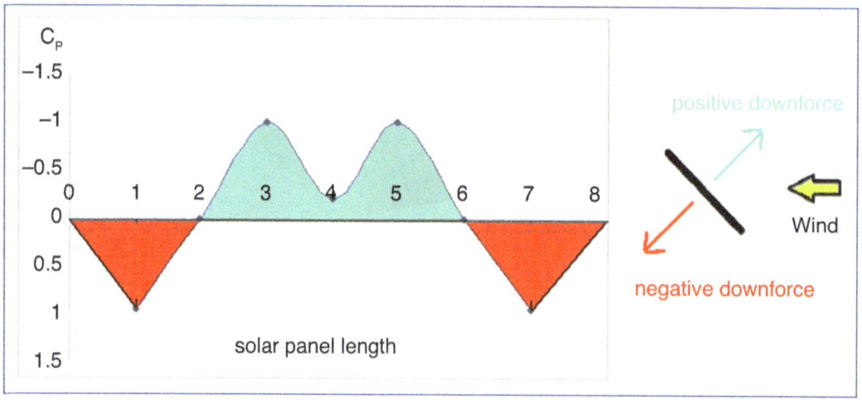

Fig. 11 Load forces on the first strip of solar panels

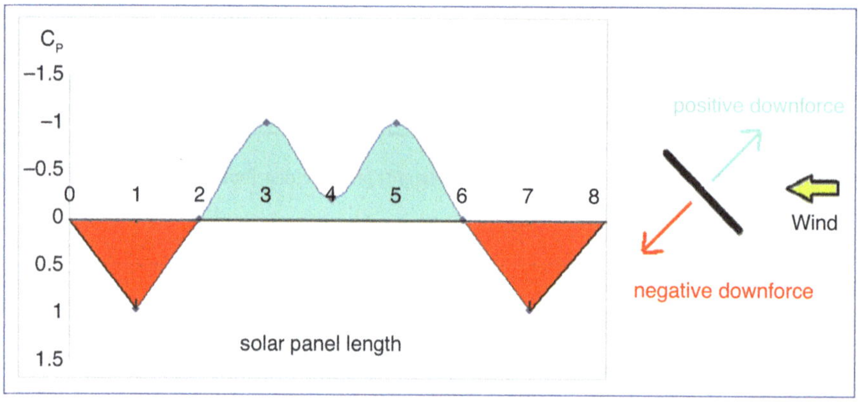

Fig. 12 Load forces on the second strip of solar panels

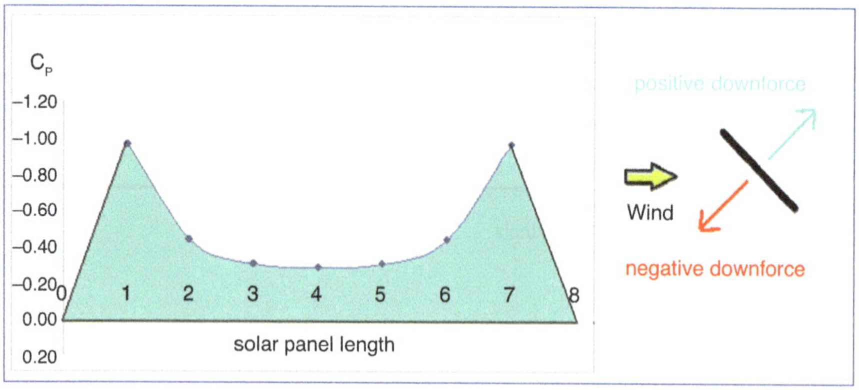

Fig. 13 Load forces with wind from the direction of the Sun

These results are very different from those reported by other authors on flat roofs (Jinxin Cao et al., 2013; Pratt and Kopp, 2013) [7, 8].

Acknowledgments To the Research Secretariat of the National University of Comahue for providing funds to Research Projects PI-04-I228 and PI-04-I242 and for contributing to the development of this work. To CYTED Organization and Iberoamerican "Microeolo" Net, that permit funds to this publication.

References

1. Mohapatra S (2011) Wind tunnel investigation of wind load on a ground mounted photovoltaic tracker, M.Sc. Thesis, Colorado State University, USA
2. Pfahl A, Buselmeier M, Zaschke M (2011) Wind loads on heliostats and photovoltaic trackers of various aspect ratios. Sol Energy 85:2185–2201
3. Hernández S, Méndez J, Nieto F, Jurado JA (2009) Aerodynamic analysis of a photovoltaic solar tracker. EACWE 5, Florence
4. Wang S, Sun Y, Wang Q, Wang Q, Wei M (2011) Limit requirements simulation of sundial solar tracking machine. Chin J Mech Eng 25:306–314
5. David Banks (2013) The role of corner vortices in dictating peak wind loads on tilted flat solar panels mounted on large, flat roofs. J Wind Eng Ind Aerodyn 123:192–201
6. Menacho J, Garcia-Granada AA (2013) A study about the protection of an array of solar panels from the wind load action. Proceedings of the world congress on engineering pp 1938
7. Cao J, Tamura Y, Yoshida A, Cao S 8th asia-pacific conference on wind engineering, December 10–14, 2013, Chennai, India
8. Nicolas Pratt R, Kopp GA (2013) Velocity measurements around low-profile, tilted, solar arrays mounted on large flat-roofs, for wall normal wind directions. J. Wind Eng. Ind. Aerodyn 123:226–238

Nanostructured P-N Junctions and Photoconductive Cells for Thermal Energy Conversion

Karen M. Gambaryan, Vladimir M. Aroutiounian, and Martin K. Gambaryan

1 Introduction

Thermophotovoltaic (TPV) power generation involves the conversion of solar or other thermal energy into electrical energy using PV converters that respond to middle-wavelength infrared radiation. Most of the photovoltaic devices developed for TPV applications have band gaps ranging from 0.5 eV to 0.75 eV [1, 2]. However, system modeling results have indicated the advantages of still lower band-gap (< 0.5 eV) TPV cells [3]. A maximum efficiency and maximum power density can be achieved with band gaps between 0.2 eV and 0.5 eV for black body sources temperature in the range 1200 K to 2500 K. Thus, there is a need for significant development in both new materials used for TPVs and in processing to produce high-performance TPV converters with lower band gaps. An alternative to InGaAs on InP and InGaAsSb on GaSb substrate are epitaxial InAsSbP lattice-matched structures on InAs or GaSb substrates. Lattice-matched InAs/InAsSbP TPV cells have variable band gaps ranging from 0.3 to 0.5 eV that displace the spectral response to the long-wavelength range.

TPV is a high-efficient, solid-state alternative to thermoelectric and thermionic converters [4]. In TPV, electricity is produced directly from radiant heat. Advantages of TPV with respect to competitors are the high efficiency (~25% in current state-of-the-art devices and potential to reach ~40%), the low weight, the independence of pressure and gravitational force, the absence of moving parts, and the fact that it directly produces DC power. The main concern of TPV is the low rejection temperature, which implies a large radiator area and limits the usefulness of TPV to relatively small power outputs, typically below 1 kW. This matches solar-powered systems, radioisotope (~100 W), and perhaps small (kilowatt-class) nuclear reactors. Other important concern of TPV is the unproven reliability for long-term missions in universe, especially concerning the possible contamination of the TPV optical elements by the evaporation of materials at very high temperatures [4].

On the other hand, in recent years, much research effort has been focused on quantum dots (QDs) [5, 6] and QDs-based devices due to carriers modified density of states, fascinating optoelectronic properties, and device applications in areas such as lasers, photodetectors, and other electronic devices. In general, quantum dots have better optical properties than other quantum nanostructures since electrons and holes are trapped in all three dimensions. This leads to a δ-like electronic density of states, which allows for better device performance, such as higher gain, lower threshold currents, and less temperature-sensitive threshold current for lasers, red shift in photoresponse spectra for photodetectors, as well as to increase operating temperature of infrared detectors and efficiency of TPV and PV solar cells.

As for photon detectors, the absorption of radiation depends on the semiconductor material used with the number of photons per second per watt being directly proportional to the wavelength. Since they utilize direct transition of electrons, they have high response speeds compared with thermal detectors. Thus, they find various applications where high response speed detector system is needed. Photon detectors can be fabricated in various material systems such as doped Si, II-VI compounds, III-V compounds, quantum well/dot, and superlattice structures. Such detectors commonly operate at cryogenic temperatures in order to reduce the thermally generated dark current. Optics and power supply system play an important part in the infrared (IR) sensor system and their cost and weight determine the overall cost of the system. Performance and cost are very important in the microscale dimension regime and are driving nanotechnology development, which is essential to continue the miniaturization process and efficiency improvement.

Another alternative to direct band-gap detectors is the subband transitions in quantum-confined heterostructures. Quantum well IR photodetectors have a narrow absorption spectrum that can be tuned by varying the quantum well width and barrier layer compositions. There is a large range of material combinations available allowing the tailoring of band gaps through

K. M. Gambaryan (✉) · V. M. Aroutiounian · M. K. Gambaryan
Department of Physics of Semiconductors and Microelectronics, Yerevan State University, Yerevan, Armenia
e-mail: kgambaryan@ysu.am

© The Author(s), under exclusive license to Springer Nature Switzerland AG 2022

A. Sayigh (ed.), *Sustainable Energy Development and Innovation*, Innovative Renewable Energy,
https://doi.org/10.1007/978-3-030-76221-6_7

stacking of thin layers. In particular, for advanced infrared focal plane arrays, a microscopic model of QWIP, multiband and broadband infrared detectors based on III–V materials for spectral imaging instruments, and the mid- and long-wavelength megapixel portable QWIP imaging cameras were developed and investigated.

Newest in the competition for infrared photodetector is the normal incidence quantum dot photodetectors (QDIPs) [7]. These and other QDs-based devices are expected to perform well at elevated temperatures due to their three-dimensional carrier confinement characteristics. QDIPs are inherently sensitive to normal incidence IR radiation. Dark current also is much lower than that of QWIPs. This technology is useful in remote sensing, chemical and biological detection application, as well as in PV and TPV cells. Some researchers use long-wavelength infrared QDIPs for spectroscopic applications. Bhattacharya et al [8] presented several heterostructure designs for obtaining improved responsibility in InGaAs/GaAlAs mid- and far-infrared QDIPs. It is noted that AlGaAs barriers when appropriately inserted in these heterostructures can perform wonders to the design and performance improvement of the QDIPs.

In this paper, we describe the growth and investigation of the p-InAsSbP/n-InAs diode heterostructures and the InAs-based mid-infrared photoconductive cells with QDs for TPV and other mid-infrared applications.

2 Experimental

Among QDs, wires, and other nanostructures fabricating techniques, self-organized Stranski–Krastanow growth mode [9] is an important one by which dislocation-free dots, elongated islands, and wires can be produced. For III–V compound semiconductors, such as InAs, InP, or InSb, one has to remark the very large difference in the respective diameters of the V element. One can, therefore, monitor the strain by adequate substitution between smaller/larger V element atoms. This could be used as mixed compounds such as InAsSbP, a promising candidate because it can cover the 3–5 μm mid-infrared regions adjusting its alloy composition with three V elements only. Actually, InAsSbP QDs have been successfully grown [5–7], while, in contrast, only a few studies on quaternary QDs/nanopits nucleation have been reported so far [5]. Monitoring the composition ratio between As, Sb, and P could lead to strain control in InAsSbP leading to QDs formation instead of/or together with nanopits.

The samples under consideration (QDs, epitaxial layers, and diode heterostructures) were grown by liquid-phase epitaxy (LPE) using a slide-boat crucible. In contrast to other growth techniques, such as molecular beam epitaxy or metal organic chemical vapor deposition, LPE operates comparatively closer to thermodynamically equilibrium. To ensure a high purity of the epitaxial layers and QDs, the entire growth process was performed under a pure hydrogen atmosphere. The InAs(100) substrates have 11 mm diameter, are undoped, with a background electron concentration of $n = 2\times10^{16}$ cm^{-3}. The quaternary alloy $InAs_{0,742}Sb_{0,08}P_{0,178}$ used here is conveniently lattice matched to InAs. To obtain strain-induced islands and QD formation, we used arsenic undersaturated and antimony supersaturated liquid phase. To provide a lattice mismatch up to 2% between the InAs substrate and the InAsSbP epilayer, we use mole fractions of $X_{InAs} = 0.0195$, $X_{Sb} = 0.1228$, and $X_{InP} = 1.7 \times 10^{-4}$ in the growth melt. The LPE growth solution components – undoped InAs, undoped InP, and Sb (6N) – were solved in a In (7N) solution which has first been homogenized for 1 h at $T = 580$°C and then 3 h at LPE initial growth temperature of $T = 550$°C to equilibrate the system thermodynamically. Then, the quaternary liquid phase was brought in contact with the InAs substrate. To initiate the growth of QD, we used an oversaturation by decreasing the initial growth temperature up to 1°C using the slower ramp rate.

3 Results and Discussion

First, we explore the resulting morphology looking at the QD size, shape, and density using SEM and AFM microscopes. At AFM characterization, images were acquired at a constant scanning force using a silicon nitride soft cantilever probe with a pyramidal tip. In order to avoid instrumental artifacts, each investigated area is scanned several times along different directions.

Figure 1 shows the SEM and AFM images of InAsSbP unencapsulated QDs grown by modified LPE on InAs(100) substrate.

The InAsSbP QDs are clearly visible and quite uniformly distributed over the substrate surface. The QDs average surface density is about (6–8) × 10^9 cm^{-2} with heights from 0.5 to 20 nm and diameters ranging from 15 to 70 nm. Because the difference in lattice parameter is large enough, the growth process is consistent with the Stranski–Krastanow growth mode [9]. The Gaussian distribution of the QDs number versus their average diameter calculated from the substrate surface $S = 10\ \mu m^2$ shows the optimum size of QDs to be at ~50 nm.

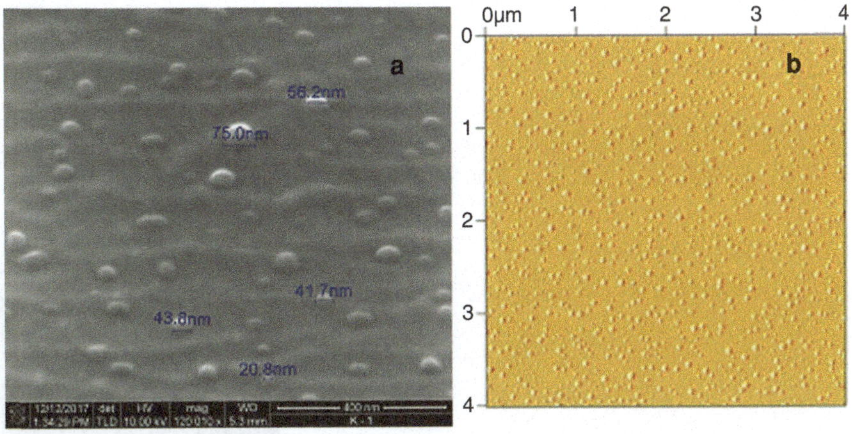

Fig. 1 SEM (**a**) and AFM (**b**) images of InAsSbP QDs grown by LPE on InAs(100)

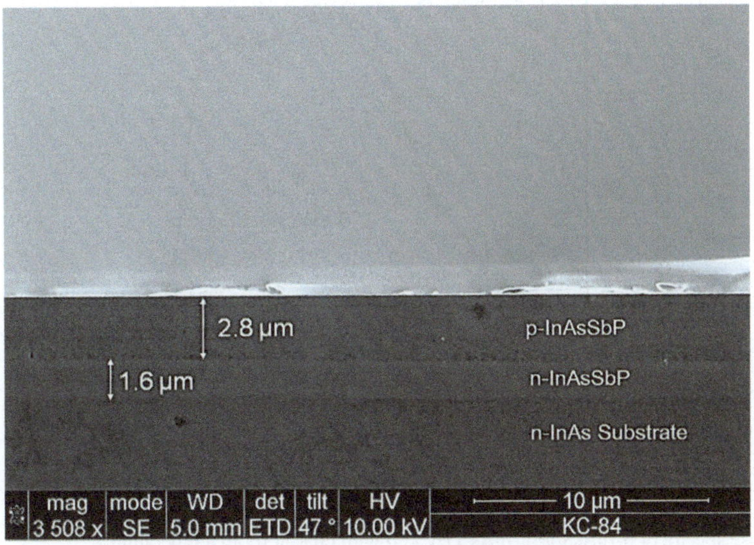

Fig. 2 SEM image of the cross section of p-InAsSbP/n-InAsSbP/n-InAs QDs TPV structure

In Fig. 2, the SEM image of the cross-sectional area of p-InAsSbP/n-InAsSbP/n-InAs QDs TPV structure is presented. Transmission electron microscopy (TEM) measurements reveal the presence of QDs in both substrate-layer and layer-layer interfaces. TEM image of the p-InAsSbP/n-InAs diode heterostructure's cross-sectional area with QDs is presented in Fig. 3.

The dark current-voltage (I-V) characteristic of a diode heterostructure n^+-InAs/n-InAs/p-InAsSbP with QDs on p-n junction interface was measured at room temperature. The experimental dependence of the dark current I on the voltage V applied to the diode is well described by the following equation.

$$J(V) = J_0\left[[\exp\left(\frac{e(V - JR_s)}{nkT}\right) - 1\right] + \frac{V - JR_s}{R_{sh}}$$

where J_0 is the saturation current density, R_S and R_{sh} are series and shunt resistances, respectively, k is the Boltzmann's constant, T is the absolute temperature, and n is an ideality factor. Fitting of the experimental I-V curves to the above-mentioned relationship yields the following values of diode parameters at $T = +20°C$: $R_S = 1.8\ \Omega$; $R_{sh} = 2700\ \Omega$; and $V_c = 0.14$ V, where V_c is the contact potential difference of the p-n junction. The ideality factor fitting the experimental data is close to unit, which indicates a diffusion mechanism of current transfer through the p-n junction. Mentioned above data allow to conclude about a good quality of grown heterostructures.

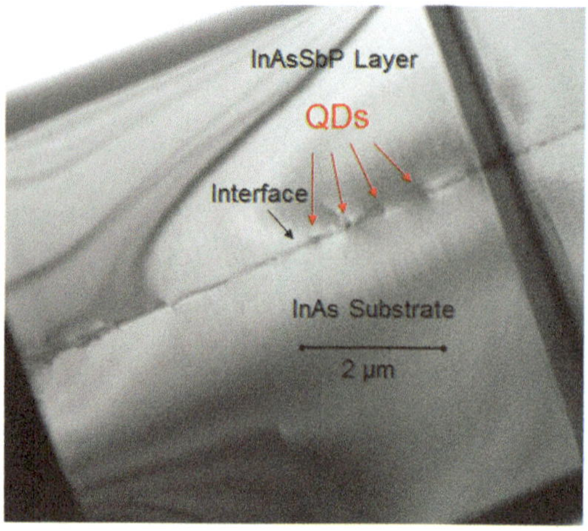

Fig. 3 TEM image of QDs p-InAsSbP/n-InAs diode heterostructure

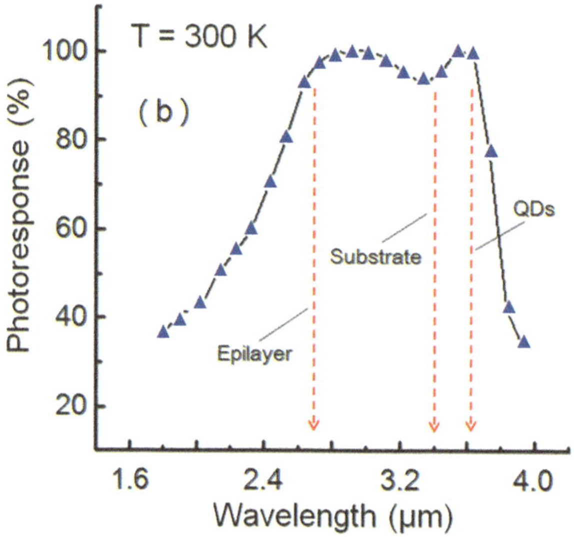

Fig. 4 Normalized photoresponse spectra of p-InAsSbP/n-InAs TPV cell with QDs at room temperature

Additionally, the photoresponse spectra of fabricated PCC and TPV structures were measured. The room-temperature, normalized photoresponse spectra of the p-InAsSbP/n-InAs diode heterostructure with QDs on the p-n junction interface are presented in Fig. 4.

The photoresponse spectra of fabricated PCC structures were measured at room temperature applying low biases up to 6 mV. It was found that the spectrum is extended up to 4 μm with main peak at 3.48 μm. This peak coincides with the energy band gap of InAs (E_g = 0.355 eV). Additional peaks were also revealed as a result of charge carrier's transitions through energy levels created by type-II InAsSbP QDs grown on InAs.

As for TPV diode heterostructures (see Fig. 4), the structure shows photosensitivity in a broad range, which is determined by the subband-gap widths of QDs and $InAs_{0.88}Sb_{0.04}P_{0.08}$ quaternary epilayer. The monochromatic radiation responsitivity measured at the maximum of the spectrum was 0.8–1.5 A/W, which corresponds to internal quantum efficiency of up to 40% (with allowance for ~30% losses for reflection). Actually, Fig. 4 shows the displacement of the absorption edge toward the long-wavelength region, as well as enlargement toward the short-wavelength region. Our quantitative calculation shows an increasd efficiency of QD-based TPV structures of up to 16% compared with the same structures without QDs.

4 Conclusion

Thus, the p-InAsSbP/n-InAs diode heterostructures and the InAs-based mid-infrared photoconductive cells with and without quantum dots for thermophotovoltaic and other mid-infrared applications were fabricated and investigated. For the nucleation of InAsSbP composition strain-induced QDs in Stranski–Krastanow growth mode, as well as for the growth of quaternary InAsSbP emitter cap layer lattice matched with the InAs(100) substrate, the modified version of liquid-phase epitaxy has been employed. The high-resolution SEM, AFM, and TEM techniques were used for the characterization of QDs morphology, dimensions, and distribution density, as well as for investigations of diode heterostructures' cross-sectional area. The QDs average density ranges from 3 to 8×10^9 cm^{-2}, with heights and widths dimensions from 4 nm to 15 nm and 10 nm to 35 nm, respectively. The electrophysical characteristics and optical response spectra of both TPV and photoconductive cells were also explored. The displacement of the absorption edge toward the long-wavelength region, as well as enlargement toward the short-wavelength region were detected for both QDs-based devices. The quantitative calculations show an increased efficiency of QD-based TPV structures of up to 16% compared with the same structures without QDs.

Acknowledgments This work was performed in the frame of the project № 18T-2J016 funded by the State Committee of Science of Armenia. The authors wish to thank Dr. T. Boeck and his colleagues from Leibniz Institute for Crystal Growth (IKZ, Berlin, Germany) for SEM and AFM characterization.

References

1. Wanlass MW, Ward JS, Emary KA, Al-Jassim MM, Jones KM, Couts TJ (1996) GaxIn1−xAs thermophotovoltaic converters. Sol Energy Mater Sol Cells 41/42:405
2. Coutts TJ, Ward JS (1999) Thermophotovoltaic and photovoltaic conversion at high-flux densities. Electronic Devices 46:2145
3. Gevorkyan VA, Aroutiounian VM, Gambaryan KM, Kazaryan MS, Touryan KJ, Wanlass MW (2004) Liquid-phase electroepitaxial growth of low band-gap p-InAsPSb/n-InAs and p-InAsP/n-InAs diode heterostructures for thermo-photovoltaic application. Thin Solid Films 451/452:124
4. Datas A, Martí A (2017) Thermophotovoltaic energy in space applications: Review and future potential. Sol Energy Mater Sol Cells 161:285
5. Gambaryan KM (2010) Interaction and cooperative nucleation of InAsSbP quantum dots and pits on InAs (100) substrate. Nanoscale Res Lett 5:587
6. Gambaryan KM, Aroutiounian VM, Boeck T, Schulze M, Soukiassian PG (2008) Strain-induced InAsSbP islands and quantum dots grown by liquid phase epitaxy on a InAs (1 0 0) substrate. J Phys D Appl Phys 41:162004
7. Gambaryan KM, Aroutiounian VM, Harutyunyan VG (2011) Photovoltaic and optoelectronic properties of InAs (1 0 0)-based photoconductive cells with quantum dots and nanopits. Infrared Phys Technol 54:114
8. Bhattacharya P, Su XH, Chakrabarti S, Ariyawansa G, Perera AGU (2005) Characteristics of a tunneling quantum-dot infrared photodetector operating at room temperature. Appl Phys Lett **86**:191106
9. Stranski I, Krastanow L (1938) Math-Naturwiss 146, 797

Evaluation of Omani Experience in Using Grid-Connected Photovoltaic Stations

Hussein A. Kazem, Miqdam T. Chaichan, Ali H. A. Al-Waeli, and K. Sopian

1 Introduction

In recent years, the need for electricity has multiplied as a result of the growth of economics and the global population. Electricity is generated, in general, using fossil fuels [1]. This causes a large and escalating consumption of a variety of fossil fuels (coal, oil, and natural gas) with the emergence of serious environmental problems such as acid rain, air pollution, and climate change [2]. To get rid of these problems, the world has become unanimous, to gradually shift to energy production from renewable sources such as solar energy, which can hamper the effects of greenhouse gases [3]. Solar energy can be used in many applications such as heating or electricity production applications.

Although solar energy currently represents 1% of global energy, it is expected that its contribution to electricity production will increase to 27% in 2050 [4]. Solar electricity can be produced by thermal applications such as concentrated power plants and a solar chimney [5, 6], or by using photovoltaic cells [7]. Several photovoltaic power stations have been established that contain monitoring systems and units of measurement and are equipped with sensors to preserve and evaluate the plant's daily production and the annual electricity.

Several studies associate the use of grid-connected PV power stations to provide clean and pollution-free electricity [8, 9]; for example, Ref. [10] examined the efficiency of converting sunlight into electricity and converting this electricity from direct current to alternating to feed the electrical grid. The inverters in such systems play an important role in converting direct current into alternating ones and its efficiency is up to 95%. The study listed some points that need more research and development such as for overscaling the PV generator for inverters.

Ref. [11] studied a grid-connected photoelectric system (GCPV) that can produce 2 kW. This system has managed to achieve from 4.1% to 8.0% of the monthly electricity demand. Ref. [12] examined the GCPV system operating under severe weather conditions of high temperatures and high relative humidity most days of the year with a large rate of volatilization of dust in the region. Researchers have found that dust accumulation causes a significant decrease in system performance and considered it a greater concern for the success of PV systems in desert areas. The study also showed that the high air temperature and relative humidity caused a decrease in the generated power by 27%.

In the Sultanate of Oman, some researchers have studied the usefulness of these systems when connected to the grid. Ref. [13] studied the economic feasibility of establishing such systems in 25 locations in Oman using RETscreen software to simulate the production of grid-connected PV. The results indicated that the cost of electricity for such systems can range from 0.210 to 0.304 USD/kWh. Ref. [14] relied on the HOMER software to access the best renewable energy systems that can meet the electricity needs of remote areas of Oman. The researchers found that the cost of electricity generated for the proposed systems ranged from 0.206 to 0.361 USD/kWh.

H. A. Kazem (✉)
Sohar University, Sohar, Oman

Solar Energy Research Institute, Universiti Kebangsaan Malaysia, Bangi, Selangor, Malaysia
e-mail: h.kazem@su.edu.om

M. T. Chaichan
Energy and Renewable Energies Technology Research Center, University of Technology, Baghdad, Iraq

A. H. A. Al-Waeli · K. Sopian
Solar Energy Research Institute, Universiti Kebangsaan Malaysia, Bangi, Selangor, Malaysia

© The Author(s), under exclusive license to Springer Nature Switzerland AG 2022

A. Sayigh (ed.), *Sustainable Energy Development and Innovation*, Innovative Renewable Energy,
https://doi.org/10.1007/978-3-030-76221-6_8

Oman needs for electricity has doubled from 2773 MW in 2007 to 5691 MW in 2014 and it is expected that the increase in electricity demand growth will continue at an annual rate of 17.8% due to the growing population and the acceleration of the industrial boom in the country. Unfortunately, electricity in Oman is produced entirely using fossil fuel (natural gas), which is produced locally. The government and privet sectors of Oman have carried out a pilot experiment in the Arab Gulf region to provide solar electricity in some schools with the aim of providing environmentally friendly electricity and developing and expanding work in the field of renewable electricity. These systems also contribute to developing the capabilities of the Omani cadres and providing them for future projects [15].

In 2016, the Omani government started with Shell Company to establish grid-connected photovoltaic power stations in several schools in various cities of Oman. The Omani state is still working on establishing 22 stations in multiple schools at the rate of two stations in each governorate. The most important idea here is to educate the community about the importance of using solar energy, which will save 30% of these schools' electricity consumption annually. This experience had the full support of all the official and popular authorities in the country. The roadmap included that Shell company will organize training courses to equip and prepare maintenance cadres and construct such stations.

The station whose performance will be evaluated in this study has equipped capacity of 151.2 kW and was established at Asmaa Bint Al-Harith School – Sohar, on August 9, 2018. The system was monitored and the measurements of all parameters lasted for a full year. This study focused on the effect of the Sohar weather conditions on the system's outputs to reach the extent of success of such systems in Oman.

The current work aims to assess the Omani experience described above by evaluating the performance of the 150 kW grid-connected PV system that was constructed in a school in the city of Sohar. This pioneering experience in Oman deserves to be monitored and evaluated annually and for many periods to ensure the success of building many of them in the future, which would reduce the consumption of natural gas and bring a significant economic class to the Omani treasury in addition to its positive effects on global health and the environment in this country.

2 Experimental Setup

2.1 System Description

The PV system that will be investigated and evaluated in this study is connected to the network and has been installed in the Asmaa Bint Al Harith School in Sohar. The system output is 150 kW and using an inverter to convert DC into AC. The area where the PV units spread is 428 m^2, of which 382 m^2 is the area of the PV system. The solar cells used are Si monocrystalline. Table 1 lists the specifications of the PV units used. The tilt angle of the PV modules is 5 degrees, and they are used to shade the parking garage in the school. The 216 PV modules were used to generate the necessary power. The station is equipped with accessories in addition to the PV panels for control purposes. Also, because the studied system is grid connected, it does not need batteries to store electricity. The system is equipped with an energy meter that records the produced electricity in addition to the meteorological station that records weather variables (solar radiation, wind speed, relative humidity, and air temperature).

Invertor converts DC to synchronous AC frequency, ensuring the safe transmission of electricity to the network. This device has the ability to detect network current fluctuation, and when necessary, it closes the transmission of electricity from the network to the system or vice versa immediately. The system is equipped with a heat-resistant wire system resulting from the resistance of the wires to the flow of electricity, solar radiation, and ambient air temperature.

2.2 PV Plant Performance Analysis

When analyzing the performance of a grid-connected PV station, it is important to calculate the performance parameters as it provides the information necessary for the actual operating losses of the station. Table 2 shows the performance parameters that were used in this study.

Table 1 The studied stations' PV module specifications

PV module	Monocrystalline 156.75 × 156.75 mm (6 inches)
Model	TSM-350DEG14C.07(II)
Manufacturer	Trina Solar
Cell orientation	72 cells (6 × 12)
Module dimensions	1985 × 998 × 28 mm
Weight	28.5 kg
PV module rated power (216 modules)	150 kWp
Maximum voltage	39 V
Maximum current	8.85 A
Open circuit voltage	47.4
Short circuit current	9.47
Efficiency	17.4
Temperature rating	
NOCT(nominal operating cell temperature)	44°C (±2°C)
Temperature coefficient of PMAX	−0.39%/°C
Temperature coefficient of VOC	−0.29%/°C
Temperature coefficient of ISC	0.05%/°C
Maximum ratings	
Operational	−40°C ~ +85°C
Temperature maximum system voltage	1500V DC (IEC) 1000V DC (UL)
Max series fuse rating	20A

Table 2 Performance parameters equations and description

No.	Parameter	Equation	Unit	Parameter description
1	Array yield (Y_A)	$Y_A = \frac{E_{PV}}{P_{nom}}$	h/d	Evaluate the system performance depending on the generated power
2	System yield (Y_S)	$Y_S = \frac{E_{AC}}{P_{nom}}$	h/d	The ratio of the energy output from the inverter (E_{AC}) to the nominal energy of the PV array used
3	Reference yield (Y_R)	$Y_R = \frac{H}{G_{STC}}$	h/d	The ratio of solar radiation intensity (H) to solar radiation intensity under STC conditions ($G = 1000\ W/m^2$)
4	Corrected reference yield (Y_{CR})	$Y_{CR} = Y_R(1 - \gamma(T_C - T_0))$	h/d	A correction to the reference yield due to the effect of temperature on the working PV module
5	Thermal energy losses (E_{therm})	$E_{therm} = E_{PV}\left(1 - \frac{1}{(1-\gamma(T_C - T_0))}\right)$	MWh	The losses in the PV performance due to temperature effects
6	DC performance ratio (PR)	$PR = \frac{Y_A}{Y_R} \times 100$	%	The proportion of useful electricity generated divided by the energy that is supposed to be generated in an ideal state without any waste of energy
7	Capacity factor (CF)	$CF = \frac{E_{AC}}{P_{nom}} \times 100$	%	The actual ratio of AC generated to the amount of electricity that PV arrays are supposed to generate if they operate under optimal conditions
8	Array capture losses (L_c)	$L_c = Y_R - Y_A$	h/d	All losses occurring during the operation of the solar panels array, and indicates the time required for the system to provide losses by operating at nominal power
9	Miscellaneous capture losses	$L_{CM} = Y_{CR} - Y_A$	h/d	All losses of PV modules for a variety of reasons except losses due to overheating of the unit, such as losses due to the accumulation of dust, contaminants, and dirt on the surface of the photovoltaic cells. Other losses can be added such as degradation losses, diodes losses, wire losses, and shading losses
10	Thermal capture losses (L_{tc})	$L_{tc} = Y_R - Y_{CR}$	h/d	The losses of PV modules resulting from the operation of these modules at higher degrees of nominal temperature under standard conditions (T0)

3 Results and Discussions

Figure 1 shows the net monthly demand for electricity in the school in addition to the power supplied from the studied station. The summer season in Oman begins early at the end of March and the hot days continue till October, so the demand for electricity increases during this period for air conditioning purposes, except for the summer vacation that reduces the need for electricity. Moderate temperatures in Sohar occur from November to January, so the demand for electricity decreases during this period as there is no need to use air conditioners. In December, due to the cold weather, the demand for electricity may increase in some cold days in which heating the halls and rooms is needed. As for the electrical energy produced by the GCPV station, it ranges from 16,240 kWh in December to 26,650 kWh in May. This fluctuation in the electricity produced is due to the change in the intensity of the solar radiation as the seasons change. In May, where the solar radiation intensity reaches its maximum value, accompanied by the wind speed, suitable for cooling the panels, the relative humidity is moderate, and the ratio of dust suspended in the atmosphere is small, resulting in an increase in the electricity produced from the station. As for the low productivity of the plant in the month of December (winter), knowing that the air temperature is suitable for the PV modules, but fog and clouds are causing misleading results that reduce the intensity of the solar radiation which reaches the PV modules [16].

Figure 1 shows that the tested station can meet the school's demand for electricity most of the year except for the months of September and October, but its production exceeds the need of the school during July and August, which represents the summer vacation.

Figure 2 shows the monthly amount of electricity supplied to the network and exported from the studied station to the grid. Electricity supplied from the grid increases during the months of September and October and decreases to its lowest levels in February and March and during the summer vacation days (June and July). The need for electricity to operate air conditioners increases during the hot months of the year, and in these months the productivity of the plant is insufficient for demand. As for the months of the moderate year of heat in which air conditioners are dispensed, the demand for electricity will decrease, resulting in a decrease in the electricity imported from the network. The figure shows the decrease in electricity exported to the grid on hot days, which increases the demand for electricity. Also, the electricity exported from the PV station increases during the moderate days, the atmosphere in which the demand for electricity decreases and during the summer vacation days.

Figure 3 depicts the self-study efficiency of the station (represents the electricity produced minus the electricity imported from the network, and the output divided by the amount of electricity produced). The curve shows high self-efficacy station most months of the year with fluctuation due to the change in the rate of electricity consumption in the school. It can be concluded that GCPV systems will reduce the load on the national network and at the same time supply them with redundant electricity. As a result, such systems are an ideal option that should be installed in all government buildings and schools in Oman.

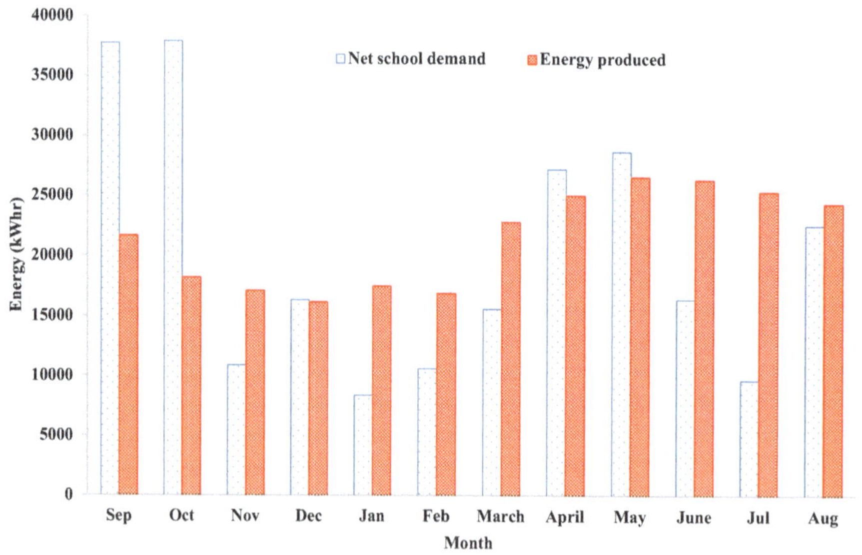

Fig. 1 The net energy demand for the studied school per month

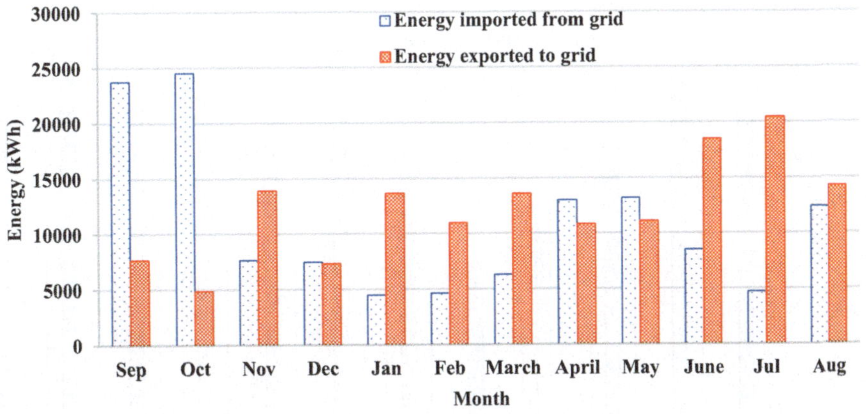

Fig. 2 The monthly electricity exported and imported

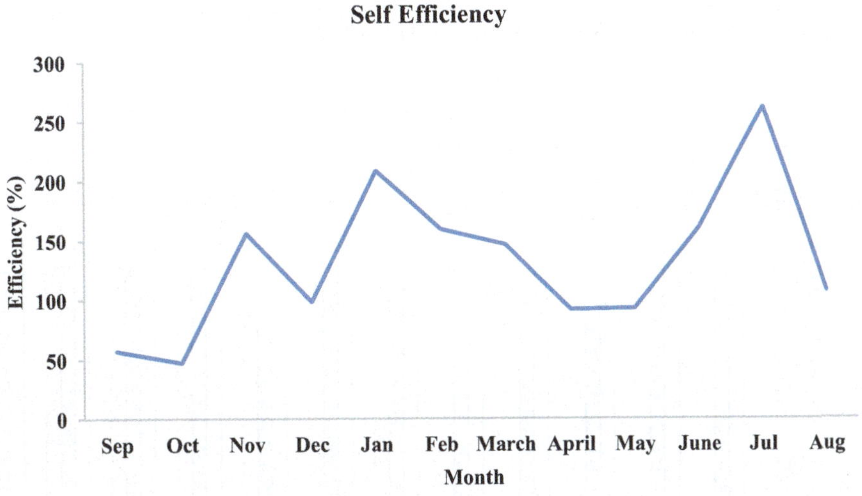

Fig. 3 The station self-efficiency variation through the studied year

Figure 4 clarifies the changes in the daily array yield, reference yield, and system yield monthly. The daily reference yield with the average daily brightness of the sun is directly related from 4.60 hours in January to 7.15 hours in October. The average daily yield curves of the array and system yield follow the same reference yield trend. The curves also show a great variation between the array and system productivity throughout the year. Readings show that the lowest daily return for the array is in January (3.05 h/d) and the lowest return to the system is 2.95 h/d for the same month. The maximum daily return for the array is 5.07 h/d that occurs in October when the system productivity is 4.94 h/d. The observed difference is normal and is due to losses in the inverter conversion from DC to AC.

Figure 5 shows the average capture loss for the array, various capture losses, and monthly thermal capture losses. Array capture losses represent the types of losses that result from converting sunlight into electricity in the semiconductor component of solar cells. The lowest of these losses is in January, at around 1.70 h/d, the highest in July, at around 2.40 h/d. It represents losses resulting from shading, wire loss, accumulation of dust and pollutants, loss of diodes, and the effect of aging with different picking losses. The figure shows that the lowest capture losses are in June (1.55 h/d) and the highest in July (2.19 h/d). Thermal capture losses are the losses caused by the unit temperature change. Therefore, we find thermal pickup losses that are high in the hot months of the year. These losses range from 0.04 h/d in January to 0.30 h/d in September.

Figure 6 shows the monthly changes in percentage of daily performance and average daily power factor. It is noted that in January, the PR is at the level of 63.59%, which represents the lowest level, while in June, the average daily direct performance is 73.56%, which represents the highest values reached during the observation year. As for the daily power factor, in January, it reaches its lowest value (11.74%), and in October it reaches its highest value (20.54%). The values of the capacity factor, as evidenced by the measured results, will be slight during the winter months of Oman (December, January, and February) and their values will rise during the rest of the year.

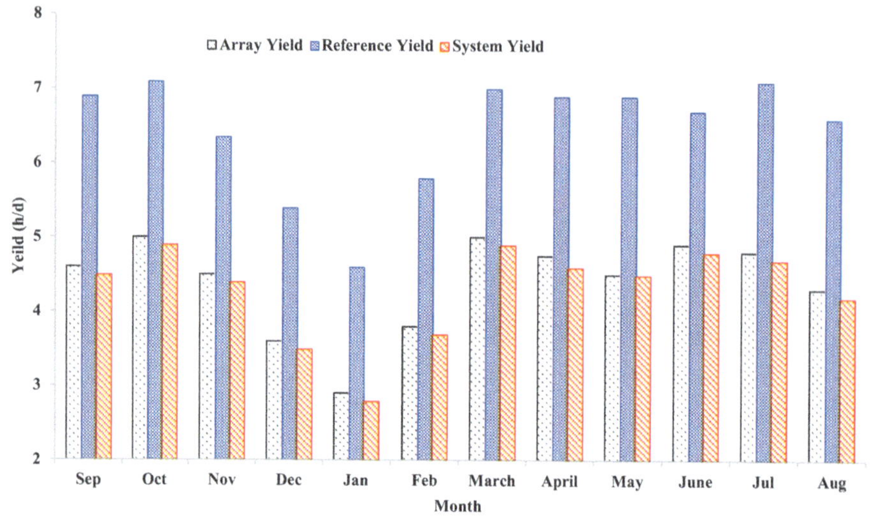

Fig. 4 Changes in the daily array yield, reference yield, and system yield per month

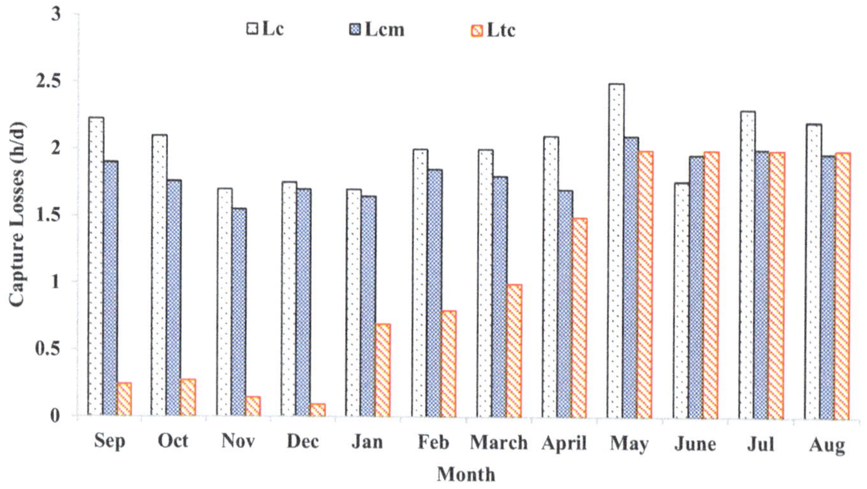

Fig. 5 Variable capture losses variation through the year

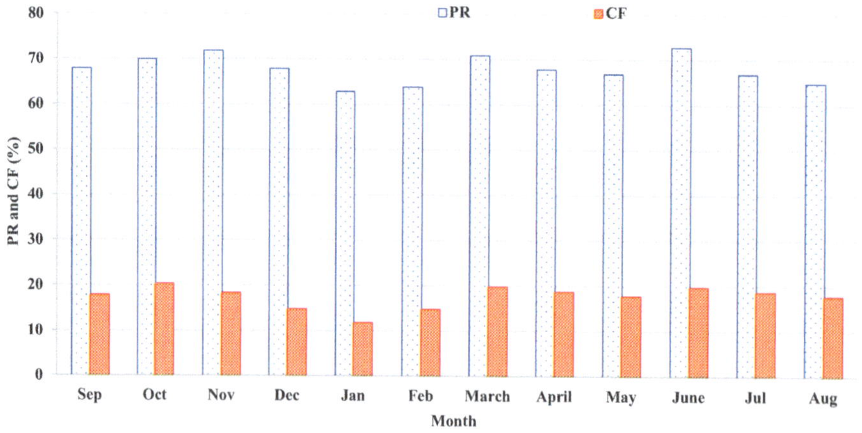

Fig. 6 The monthly changes in CR and PR

4 Conclusions

Omani government and Shell has conducted a pioneering experiment in the field of electric energy as it installed a number of grid-connected PV stations in some schools in Oman. Thus, these schools can produce clean electricity to cover their needs and send the surplus to the national grid. In this study, the performance of the PV power station connected to the grid established within this pilot project was evaluated in a school located in Sohar, Oman. The station's systems were used to record the performance of the system with all its variables and linked them to weather conditions. The results of the study showed that the PV power station changes its performance according to weather conditions, the period of sunlight rays. The highest average daily energy production is during the spring, summer, and then autumn months. It was noted that the productivity of the plant during the winter season is at its lowest level due to partial shading resulting from clouds and fog. The evaluation results showed that the average daily reference return ranges between 4.60 and 7.15 (h/d), the daily return from 3.05 to 5.07 (h/d), and the system return was 11.74–20.54 (h/d). The DC performance ratio ranged between 64% and 74%, while the power factor reached from 11.74% to 20.50%. The range of capture losses ranged from 1.55 to 2.19 (h/d), and system capture losses were 0.04 to 0.30 (h/d). The results of the study confirm that such systems are the best solution to produce clean and renewable electricity in Oman. Besides, the Omani experience will be successful if the government continues to deploy such systems in school buildings and government departments and encourage citizens to use such systems.

References

1. Chaichan MT, Kazem HA (2018) Generating electricity using photovoltaic solar plants in Iraq. Springer, ISBN: 978-3-319-75030-9. https://doi.org/10.1007/978-3-319-75031-6
2. Al-Waeli AHA, Kazem HA, Chaichan MT, Sopian K (2020) Photovoltaic/thermal system: principles, design and applications, 1st edn. Springer Nature, ISBN: 978-3-030-27824-3
3. US Energy Information Administration (2013) International energy outlook
4. Al-Waeli AHA, Sopian K, Kazem HA, Chaichan MT (2017) PV/T (photovoltaic/thermal): Status and Future Prospects. Renew Sust Energ Rev 77:109–130
5. Chaichan MT, Kazem HA (2015) Using aluminum powder with PCM (paraffin wax) to enhance single slope solar water distillator productivity in Baghdad-Iraq winter weathers. Int J Renew Energ Res 1(5):151–159
6. Chaichan MT, Abass KI (2012) Practical investigation for improving concentrating solar power stations efficiency in Iraqi weathers. Anbar J Eng Sci 5(1):76–87
7. Chaichan MT, Abass KI, Kazem HA, Hasoon F, Aljibori HSS, Alwaeli AAK, Raheem FS, Alwaeli AHA (2012) Effect of design variation on saved energy of concentrating solar power prototype. Lecture Notes Eng Comp Sci 3:1989–1993
8. Kazem HA, Albadi MH, Al-Waeli AHA, Al-Busaidi AH, Chaichan MT (2017) Techno-economic feasibility analysis of 1 MW photovoltaic grid connected system in Oman. Case Study Therm Eng 10:131–141
9. Al-Waeli AHA, Kazem HA, Sopian K, Chaichan MT (2018) Techno-economical assessment of grid connected PV/T using nanoparticles and water as base-fluid systems in Malaysia. Int J Sustain Energy 37(6):558–578
10. Lund H, Andersen AN, Østergaard PA, Mathiesen BV, Connolly D (2012) From electricity smart grids to smart energy systems market operation based approach and understanding. Energy 42(1):96–102
11. Li Z, Liu X, Tang R (2001) Optical performance of vertical single-axis tracked solar panels. Renew Energy 36(1):64–68
12. Narvarte L, Lorenzo E (2008) Tracking and ground cover ratio. Prog Photovolt Res Appl 16(8):703–714
13. Al-Ismaily HA, Douglas P (1998) Photovoltaic electricity prospects in Oman. Appl Energy 59((2–3):97–124
14. Al-Badi AH, Albadi MH, Al-Lawati AM, Malik AS (2011) Economic perspective of PV electricity in Oman. Energy 36(1):226–232
15. Al-Badi AH, Al-Toobi M, Al-Harthy S, Al-Hosni Z, Al-Harthy A (2012) Hybrid systems for decentralized power generation in Oman. Int J Sustain Energy 31(6):411–421
16. Kazem H A, Chaichan M T, Al-Waeli A H A, Mani K (2017) Effect of shadows on the performance of solar photovoltaic, Mediterranean Green Buildings and Renewable Energy: Selected Papers from the World Renewable Energy Network's Med Green Forum, pp 379–385. https://doi.org/10.1007/978-3-319-30746-6_27

Investigation of the Effect of Annealing Temperature of Perovskite Layer on Performance of Perovskite Solar Cells

Kamil M. Yousif, Issa M. Aziz, and Naseem S. Abdel

1 Introduction

The COVID-19 pandemic has brought massive changes to our lives and ways of thinking. Researchers and scientists of renewable energy (RE) can help or can contribute indirectly in combating the outbreak of the Corona virus (CRV). For example, hospitals are using medical equipment, such as respirators, ventilators, medical devices, robots and other smart applications technologies, with the aim of combating the outbreak of the CRV and to reduce the risks faced by doctors and health-care workers in hospitals, or to help treat CRV patients. However, all the above-mentioned devices need electricity to work. RE such as photovoltaic (PV) technology can do the job, thanks to the rapid expansion of PV. However, innovative activities or R&D implementing activities are important and particular efforts are needed in these areas to rise the efficiency of PV. Also, modernization of electricity can become the drivers of decarbonization in the energy sector to mitigate and avoid catastrophe of climate change that society is facing [1]. A perovskite solar cell (PSC) is a type of solar cell that includes a perovskite (PVK) structured compound – most commonly a hybrid organic-inorganic lead or tin halide-based material – as the light-harvesting active layer. PVK have good optical properties mainly due to their band gap that can be tuned by changing the alkyl group, the metal atom or the halide [2–4]. The materials also display a diffusion length for both holes (h) and electrons (e) [3]. Charges in the PVK material are predominantly present as free electron and hole rather than as bound excitons (EX) because the EX-binding energy is sufficiently low to enable charge separation at room temperature [5]. A PSC employing an organic-inorganic lead halide PVK light harvester, seeded in 2009 with power conversion efficiency (PCE) of 3.8% and grown in 2011 with PCE of 6.5% in dye-sensitized solar cell structure, has received great attention. In 2012, scientists succeeded in preparing PSCs that achieved efficiencies over 10%. Since then, efficiencies in new PSC designs have raised up. As of end of 2017, the highest PCE of 22.7% is listed in an efficiency chart provided by National Renewable Energy Laboratory (NREL) [6, 7]. In recent years, perovskite solar cells (PSCs) have attracted great attention in the PV research field, due to the PCE of PSC devices having been rapidly improved, from 3.8% in 2009, up to 22.6% in 2017. The progresses in PSCs have been discussed and reviewed comprehensively [8–12]. The general operating principle of PSCs involves the following steps: light absorption, EX generation and diffusion, EX dissociation to carriers at the hetrojunction (n-i-p) interfaces and carrier transportation and collection; these steps are already described elsewhere [11, 12]. Approaches or methods for fabricating (manufacturing) the lager-area PVK film layer, have been followed by many researchers, e.g. using many processes or methods, such as spin-coating, vapour deposition, etc. It is important to choose the method giving more uniform PVK layer or better surface coverage, which strongly affect consequent PV performance of devices [13], as well as improving the stability of the PSC devices. Despite positive traits, the PSCs have faced some issues, such as rapid degradation and stability in the presence of O_2, moisture, etc. The impact of these factors with various remedies adopted by researchers, scientists and engineers has been studied [10–15]. The degradation of PSCs device includes the degradation of the active layers [11, 13–16]. In this project, we tried to get rid of some structures or materials which are harmful to the environment. That is, using lead sheets from recycled car battery, the performance of PSC was investigated, together with the

K. M. Yousif (✉)
Department of Environment, Faculty of Science, University of Zakho, Kurdistan Region, Iraq
e-mail: kamil.yousif@uoz.edu.krd

I. M. Aziz · N. S. Abdel
Department of Physics, Faculty of Science, University of Zakho, Kurdistan Region, Iraq

© The Author(s), under exclusive license to Springer Nature Switzerland AG 2022

A. Sayigh (ed.), *Sustainable Energy Development and Innovation*, Innovative Renewable Energy,
https://doi.org/10.1007/978-3-030-76221-6_9

stability of PSC and the effect of annealing temperature of PVK layers on performance of PSC. Also, the effects of temperature, illumination and ambient (oxygen, moisture) upon performance and the stability of the PSCs were studied. Herein this paper, we will demonstrate the effect of relative humidity.

2 Methodology

2.1 *Sample Preparation*

Figure 1 shows the flowchart of the methodology for sample fabrication of perovskite solar cells.

The first step was harvesting material from the anodes and cathodes of car battery. The electrolyte acid was emptied out and prudently composed, and the electrodes in addition to the inner barrier of car battery were cleaned using clean water. Figure 2 shows some images of synthesizing PbI_2 from collecting anode and cathode materials of car battery and PbI_2 powder. The organo-halide PSC consists of five layers to form the first layer of PSC using an FTO glass substrate. The FTO glass should be prepared with IPA (isopropyl alcohol) by leaving it in IPA for 30 minutes. Ohmmeter was used to find the

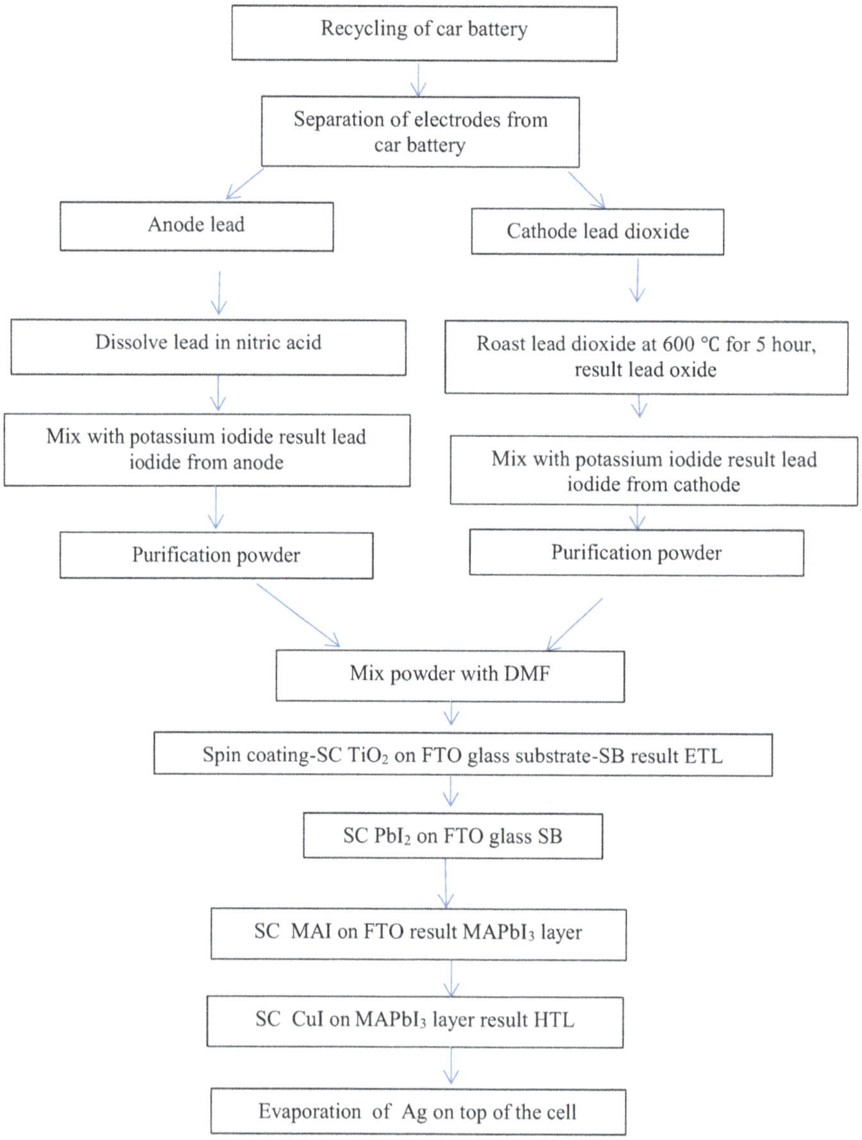

Fig. 1 Flow chart experimental techniques for sample fabrication of perovskite solar cells

Fig. 2 (**a**) Synthesis of PbI_2 by using the anode and cathode materials of car battery and (**b**) PbI_2 powder

conductive side. A 150-nm-thick electron transport layer (ETL) was used to produce cells from TiO_2 deposited by spin-coating process. More details about experimental techniques, preparation of chemicals materials and fabrication are given in Ref. [12].

2.2 Characterization of Perovskite Solar Cell

The efficiency of PSCs was measured using light source of AM1. 5 illumination (light intensity of 100 mW/cm^2). The morphology and thickness of PSCs were characterized using scanning electron microscopy (SEM, JSM-7600 F, Jeol). X-ray diffraction (XRD) was studied on a Rigaku ATX-XRD with Cu Ka radiation (k = 1.5405 Å). Spectrophotometer (Perkin-Elmer Lambda 25) was used to find the transmission.

3 Result and Discussion

3.1 Characterizations of the Fabricated Devices

3.1.1 X-Ray Diffraction and Optical Properties

Figure 3 shows the X-ray diffraction of $MAPbI_3$ thin film before and after annealing the film at a temperature of 25–120 °C for 12–15 minutes respectively. The results show that, a small amount of PbI_2, and new phase, in the fabricated film. This new phase is due to annealing at a higher temperature (i.e. 120 °C) which lead to PVK without PbI_2 traces (Fig. 3). We believe that the possible contamination from perovskite due to lead inside MAPbI3 could be reduced by encapsulation of PSC.

For optical properties, by using spectrophotometer, it can be seen that the transmission of the films of PSC was affected by temperature of annealing. More details about other results of transmission, band gap of the films, etc. will be published soon.

3.1.2 Morphological Analysis

Figure 4 shows a SEM image of the $MAPbI_3$ film obtained with a 3:1 PbI_2/MAI thickness ratio at room temperature and annealed in the range 25–120 °C. The SEM results suggest that the annealing treatment significantly changed the morphology of the $MAPbI_3$ film. Morphology of the film shows small grains (before annealing). After annealing at 100 °C, the grain size increased, whereas at 120 °C, the $MAPbI_3$ film showed dense pore-free morphology, but formed clumps in many areas.

3.2 Electrical Characteristics

Figure 5 shows the I-V curves of the solar cells (PSCs) that were fabricated with structure FTO/TiO_2/$MAPbI_3$/CuI/Ag before and after annealing at 100 °C and 120 °C.

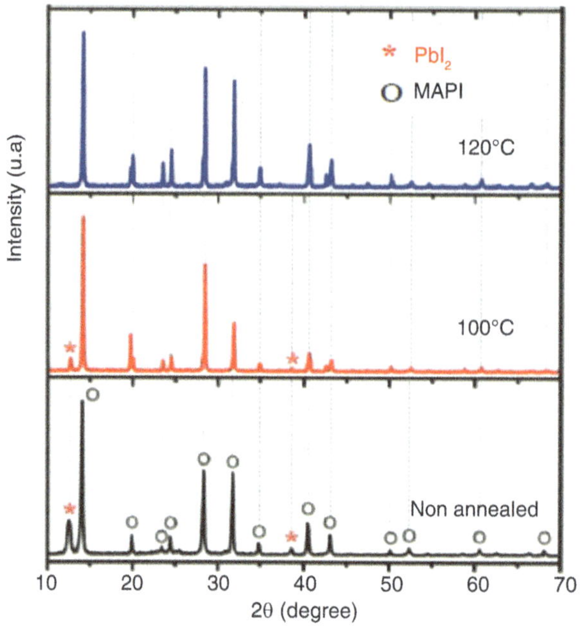

Fig. 3 The X-ray diffraction (XRD) spectrum corresponds to a $MAPbI_3$ film

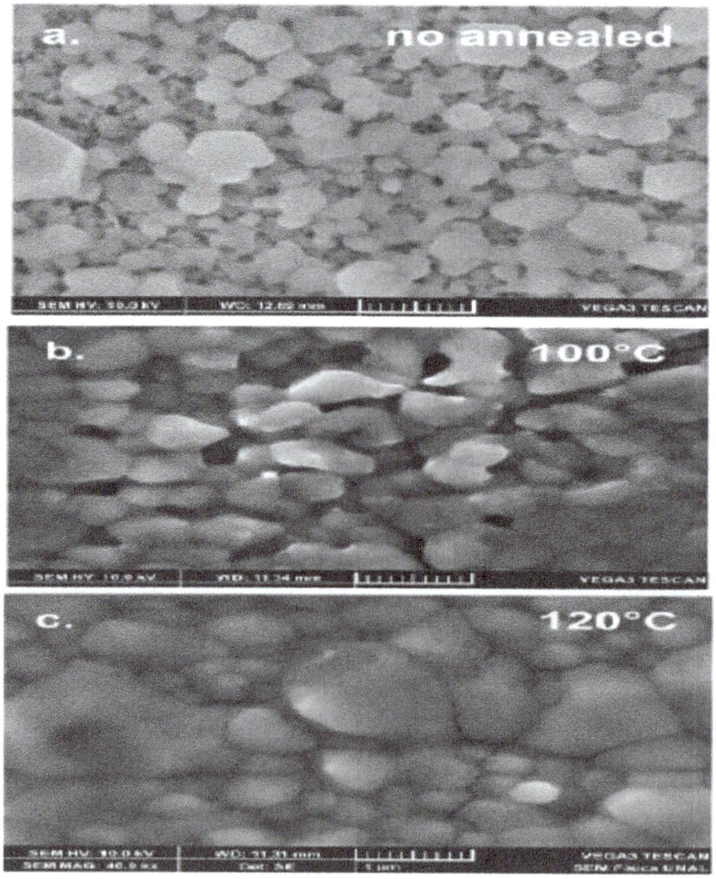

Fig. 4 Effect of annealing on the morphology of $MAPbI_3$ film

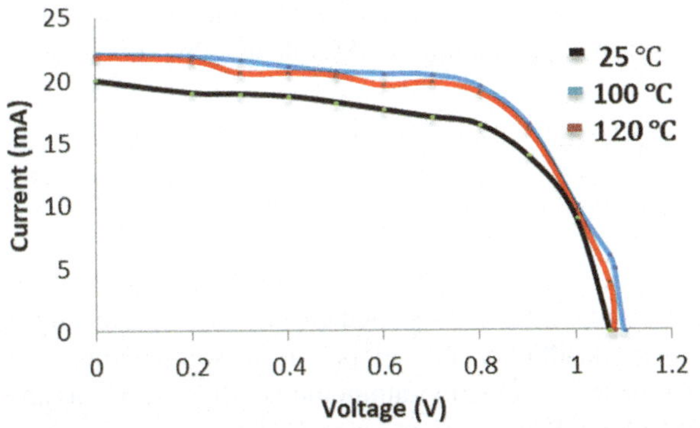

Fig. 5 I-V curves of the best solar cells prepared with structure FTO/TiO_2/$MAPbI_3$/CuI/Ag: (**a**) non-annealing, (**b**) annealing at 100 °C and (**c**) annealing at 120 °C

Table 1 Effect of annealing deposition on cell performance [short circuit current (I_{sc}), open circuit voltage (V_{oc}), fill factor (FF) and efficiency]

Annealing temperature	I_{sc} (mA)	V_{oc}(V)	FF	η (%)
(25 °C)	20	1.07	0.58	12.4
100 °C	22.1	1.1	0.61	14.8
120 °C	21.9	1.08	0.60	14.2

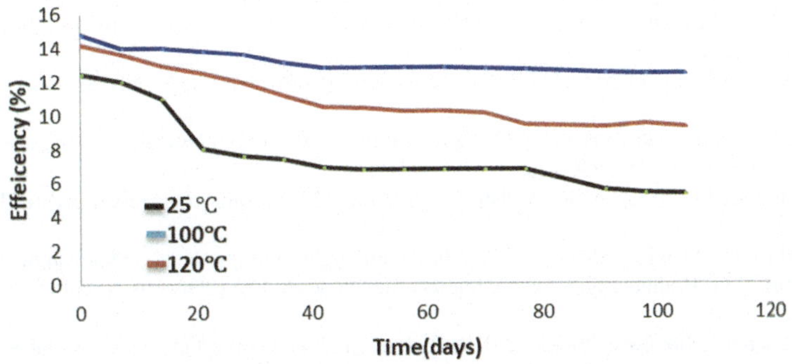

Fig. 6 The stability of corresponding perovskite solar cells during 105 days with 50–70% relative humidity at 25 °C

Table 1 shows the values of the corresponding performance parameters (short circuit current J_{sc}, open circuit voltage V_{oc}, fill factor FF and efficiency η). As can be seen from Fig. 5, we were able to produce a cell with an efficiency of 14.8%. The main reason for the low η is low FF and low short circuit current (J_{SC}), and part of the reason for low FF is the high series resistance of the device, Apparently, due to a poor charge transfer in the interfaces TiO_2 /$MAPbI_3$ (and CuI/$MAPbI_3$). High loss of photo-current by recombination in states of interfaces may be another factor that contributes to the losses of the short circuit current.

3.3 Stability of the Fabricated Cells

After plotting the I-V curve for PSCs before and after annealing for 100 °C, and 120 °C, under test conditions (1000 W/m2, 25 °C, humidity of 50–70%]. The efficiency-η was calculated. Figure 6 shows the efficiency-η for each PSC was plot as a function of various intervals of time. It clearly seen that the cell annealed at 100 °C is more stable than other cells. Also, Fig. 6 shows the influence of annealing on the films. The η of PSC annealed at 120 °C declined more than the PSC annealed at 100 °C and the film fabricated at room temperature have lower stability than other films, may be due to structure or morphology of the films.

The degradation of the PSCs device includes the deterioration of the active layer, the deterioration of charge transport layers and the deterioration of electrodes [17]. The $MAPI_3$ films are frequently used as absorber layer film. But the major problem with MAPbI is that it has thermal decomposition (exceeding 85) and water decomposition [18]. But we found thermal decomposition (i.e. degradation) occurred at temperature exceeding 100 °C.

We believe that the PVK degradation could be more severe with the incorporation of humidity and temperature. An effective encapsulation is crucial to prevent the degradation of PSCs device due to moister and oxygen to achieve the desired reliability and device lifetime [11].

4 Conclusion

This work demonstrates that the two-step spin-coating method can be a promising alternative to deposit high-quality perovskite solar cells [PSC]. In this work, PSC was developed using spin-coating solution method with precursors lead iodide synthesized from recycling car battery and methyl-ammonium iodide. TiO_2 was an electron transport layer (ETL), and $CH_3NH_3PbI_3$ a thin active layer and CuI a hole transparent layer (HTL).

The SEM results suggest that the annealing treatment significantly changed the morphology of the $MAPbI_3$ film. The XRD results suggest that the annealing treatment significantly changed the structure of the film. The experimental results showed significant improvements in the solar cell characteristics for annealed samples. The efficiency of the cell subjected to thermal annealing at temperature of 100 °C for 15 minutes was 14.8%. Then, the efficiency of the cell altered to 12.4% after 105 days of ageing at room temperature. The change in efficiency could be related to change in morphology and structure of the perovskite (PVK) films.

References

1. David Renné (2020) Is the global response to COVID-19 a prelude to the response to the climate crisis. International Solar Energy Society-ISES. Sunburst, Apr 2020
2. Tonuia P, Oseni SO, Sharma G, Yan Q, Mola GT (2018) Perovskites photovoltaic solar cells: an overview of current status. Renew Sust Energ Rev 91:1025–1044
3. Noh JH, Im SH, Heo JH, Mandal TN, Seok SI (2013) Chemical management for colorful, efficient, and stable inorganic–organic hybrid nanostructured solar cells. Nano Lett 13(4):1764–1769
4. Eames C, Frost JM, Barnes PR, O'regan BC, Walsh A, Islam MS (2015) Ionic transport in hybrid lead iodide perovskite solar cells. Nat Commun 6:7497
5. Stranks SD, Snaith HJ (2015) Metal-halide perovskites for PV and light-emitting devices. Nat Nanotechnol 10(5):391–402
6. Seok SI, Grätzel M, Park N-G Methodologies toward highly efficient perovskite solar cells. Small Jr 14(20):e1704177. https://doi.org/10.1002/smll.201704177
7. National Renewable Energy Laboratory (NREL) (2019) Best research-cell cell efficiencies. Available online: http://www.nrel.gov/pv/assets/images/effciency-chart.png
8. Kumari N, Patel SR, Gohel JV (2018) Current progress and future prospective of perovskite solar cells: a comprehensive review. Adv Mater Sci 53:161–186
9. Boix PP, Nonomura K, Mathews N, Mhaisalkar S (2014) Current progress and future perspectives for organic/inorganic perovskite solar cells. Mater Today 17(1):16–23
10. Roy P, Kumar N, Sanjay S, Khare TA (2020) A review on perovskite solar cells: evolution of architecture, fabrication techniques, commercialization issues and status. Sol Energy 198:665–688
11. Uddin A, Upama MB, Yi H, Duan L (2019) Encapsulation of organic and perovskite solar cells: a review. Coatings 9(2):51–68. https://doi.org/10.3390/coatings9020065. Pub. by MDPI
12. Aziz IM (2020) Synthesizing and characterization of Lead Halide Perovskite Nanocrystals solar cells from reused car batteries. MSc. thesis, Department of Physics Sciences, Zakho University, Zakho, Iraq
13. Chen Y, Zhanga L, Zhang Y, Gao H, Yan H (2018) Large-area perovskite solar cells – a review of recent progress. RSC Adv 8:10489
14. Pazoki M, Hagfeldt A, Edvinsson T (2019). books.google.com characterization techniques for PSC materials – books.google.com
15. Djurišić AB, Liu FZ, Tam HW, Wong MK, Ng A, Surya C, Chen W, He ZB (2017) Perovskite solar cells – an overview of critical issues. Quantum Electron 53:1–37
16. Lee K-T, Guo LJ, Park HJ (2016) Neutral- and multi-colored semitransparent perovskite solar cells. Molecules 21(4):475. https://doi.org/10.3390/molecules21040475
17. Philippe B, Park B-W, Lindblad R, Oscarsson J, Ahmadi S, Johansson EMJ, Rensmo H (2015) Chemical and electronic structure characterization of lead halide perovskites and stability behavior under different exposures – a photoelectron spectroscopy investigation. Chem Mater 27:1720
18. Frost JM, Butler KT, Brivio F, Hendon CH, van Schilfgaarde M, Walsh A (2014) Atomistic origins of high-performance in hybrid halide perovskite solar cells. Nano Lett 14:2584–2590

Characterization of Photovoltaic (PV) Modules Under Real Operating Conditions

Abdul Q. Malik

1 Introduction

Fiji is located in the Sun Belt region of the globe where sun shines for more than 6–7 bright sunshine hours daily that makes it an ideal location for photovoltaic (PV) energy generation. The PV technology is one of the clean sources of energy, and from the installed capacity over the globe, it is the third-most important source of energy [1]. Different types of photovoltaic cells/modules have been developed such as single-crystalline, multicrystalline, and thin-film amorphous silicon (a-Si, copper indium gallium selenium [CIGS], copper indium selenide [CIS], cadmium telluride selenium, and heterojunction with intrinsic thin layer [HIT]). Due to the variety of PV cells/modules, it becomes more difficult to choose the right type of solar cells/ modules for the right locality. This is because of the fact that, among other parameters, the optimal balance between module cost and energy yield is the deciding factor to determine the business.

The working of a photovoltaic solar module is described in terms of their electrical characteristics: open-circuit voltage (V_{oc}), maximum voltage (V_{max}), short-circuit current (I_{sc}), maximum current (I_{max}), maximum output power (W), and conversion efficiency (η). These are also known as rated performance information and manufacturers provide this information for each module. The rated performance data for each module is measured in a controlled environment known as standard temperature conditions (STC). These conditions correspond to the working temperature of a solar cell, the intensity of solar radiation, and the wind speed as 25 °C, 1000 Wm^{-2}, and 2 ms^{-1}, respectively. The Standard Temperature Conditions (STC) do not explain how can the energy yield of a modules effect due to the spectral variation, angle of incidence, and presence of atmospheric constituents.

When PV modules are installed in a natural environment, these conditions cannot be replicated and as a result PV system cannot deliver the power for which it was designed. It is, therefore, important that the characterization of PV devices can be carried out in the natural environment and the sizing of PV system would be based on the rating of these devices taken under such an environment rather than at STC [1–7]. The objective of this project is to study the influence of the localized climate on the performance of PV devices.

2 Effects of the Real Operating Conditions on the Performance of PV Modules

Natural environment offers entirely different working conditions to PV modules. These modules convert sun radiation into electrical current when placed under the sun and the angle of incidence of these radiation with PV modules is instantly changed from sunrise to sunset; the presence of clouds and other atmospheric constituents may or may not remain the same during the sunrise-to-sunset duration; the temperature of modules is no longer constant during the day; and seasonal change and unforeseen microclimatic conditions strongly affect the performance of modules.

A. Q. Malik (✉)
Department of Physics, School of Pure Sciences, College of Engineering, Science and Technology, Fiji National University, Lautoka, Fiji
e-mail: Abdul.Malik@fnu.ac.fj

© The Author(s), under exclusive license to Springer Nature Switzerland AG 2022

A. Sayigh (ed.), *Sustainable Energy Development and Innovation*, Innovative Renewable Energy,
https://doi.org/10.1007/978-3-030-76221-6_10

3 Effects of Solar Irradiance

The amount of solar radiation falling over the surface of PV modules has the greatest impact on PV power. Beyond these radiations, module temperature, angle of incidence (AOI), and air mass (AM) also strongly affect the power generation of PV module [8–10]. Photovoltaic modules perform differently under different environmental conditions. The performance of a building integrated photovoltaic system depends on local meteorological parameters, site conditions, and system characteristics as reported by Aaditya et al. [11]. They give an idea about how the solar cell performance changes with the change in mentioned parameters in reality and their results are shown by conducting a number of experiments. The outdoor performance of PV modules has been studied and compared with the parameters from the manufacturing companies for each module [12, 13]. It has been noted that the output power follows closely the irradiation level. Literature states a decrease in efficiency of PV modules by 55–90% from its STC value, and at irradiance levels below 200 W m^{-2} [14].

4 Effects of Temperature

The working temperature of a PV module is another important factor affecting its performance. The PV system operating in hot climates experience module working temperature significantly higher than 25 °C; as a result, these modules deviate from their performance to that of STC. An extensive research has been conducted to explore how an increase in the working temperature of module affect its performance [15, 16].

The current "*I*" passing through a PV device shown in Fig. 1 can be calculated using the following equation [17]:

$$I = I_L - I_D\left[\exp\frac{q(V + IR_S)}{mkT} - 1\right] - \frac{V + IR_S}{R_P} \tag{1}$$

where I_L is photocurrent generated by the device under sun, I_D the diode current, m is ideality factor that has value usually close to unity, R_S the series resistance, k the Boltzmann's constant, T the device temperature, and R_P the parallel or shunt resistance.

Solar cells only convert a small amount of absorbed solar radiation into electrical energy. The remainder is dissipated as heat in the bulk of the silicon solar cell which increases its temperature compared with the environment. An increase in the working temperature of a solar cell reduces the band gap allowing more energy to be absorbed which increases the short circuit current (I_{sc}) of a solar cell for a given irradiance. The band gap energy $E_g(T)$ of the material as a function of temperature can be written as [17, 19]:

$$E_g(T) = E_g(0) - \frac{aT^2}{T + b} \tag{2}$$

where $E_g(0)$ is band gap energy of the material at room temperature; a and b are constants. This effect alone rises the theoretical maximum output power of the solar cell.

At the same time, an increase in the temperature increases the population of electrons exponentially. This enhances the dark saturation current (I_D) that is a minority carrier current and its variation with temperature can be written as [17, 19]:

$$I_D = A_o T^3 \exp\left(-\frac{E_g}{k_B T}\right) \tag{3}$$

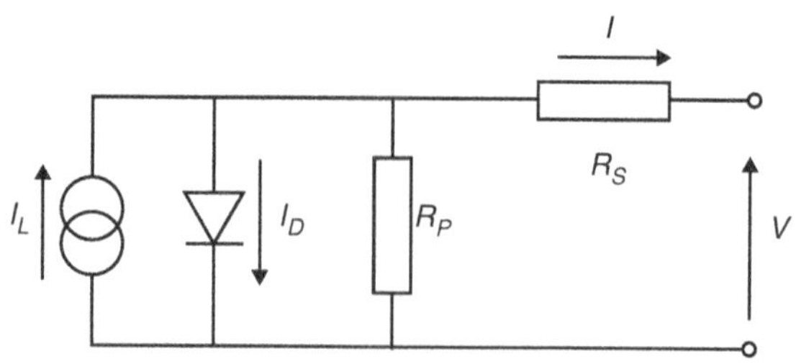

Fig. 1 The equivalent circuit of a solar cell. (Adapted from [18])

where A_o and k_B are the area of the device and Boltzmann's constant. The increase in the dark saturated current decreases the open circuit voltage, V_{oc}, of a device that may be expressed as:

$$V_{oc} = \frac{k_B T}{q} \ln\left(\frac{I_{sc}}{I_D} + 1\right) \approx \frac{1}{q}\left(E_g + k_B T \ln \frac{I_L}{I_{Do}}\right) \quad (4)$$

where T and q are the temperature and electronic charge, respectively. I_{Do} is called the prefactor of diode current that depends on the mobilities and lifetimes of the charge carriers and can be defined as [21, 22]:

$$I_{Do} = qN_c N_v \left(\frac{kT}{q}\right)^{1/2} \left(\frac{\mu_n^{1/2}}{\tau_n^{1/2} N_A} + \frac{\mu_p^{1/2}}{\tau_p^{1/2} N_D}\right) \quad (5)$$

where A, N_c, N_v, μ_n, μ_p, τ_n, τ_p, N_A, and N_D are the area of the pn junction, density of states of conduction band, density of states of valence band, mobility of negative charge carriers, mobility of positive charge carriers, lifetime of negative charge carriers, lifetime of positive charge carries, and concentration of acceptors and donors, respectively.

Theoretically, a decrease in the open circuit voltage would reduce the output power of the device.

The short circuit current and open circuit voltage are related to an important property of a PV device known as a fill factor, FF, which is defined as the ratio of the maximum power of the PV device to the product of $I_{sc}V_{oc}$:

$$FF = \frac{V_{max} I_{mx}}{V_{oc} I_{sc}} \quad (6)$$

The fill factor diminishes as the temperature of the device increases. The decrease in V_{oc} and FF with the working temperature of the device outweighs the slight increase in the short circuit current. The short circuit current of a device is sensitive to the parallel resistance as shown in Fig. 1 [20–22].

Louwen et al. [3] reported that the effect of temperature on PV modules performance decreases with the band gap of the material. High band-gap technologies, such as a-Si and CdTe, have lower-temperature coefficients than crystalline silicon and CI(G)S PV modules.

5 Spectral Effects

The rating of PV modules is related to the change and variation of the solar spectrum. The composition of the spectral contents of any locality is affected by more than factors such as air mass (AM), water vapors, clouds, aerosol particle size distribution, particulate matter, and ground reflectance [20]. The spectrum for a bright sky is dependent on air mass and relative humidity [21]. The spectral effects are more complex for cloudy weather and the light under these conditions is concentrated in the blue region of the solar spectrum than the standard AM 1.5 spectrum. Conversely, the blue region of the spectrum is attenuated as the sun moves lower in the sky [22]. A number of researchers studied the working of PV modules both indoors and outdoors to explore spectral response of PV modules [23, 24]. The red region of the solar spectrum generates low current because photons of this region have energy below the band gap and it provides low internal quantum efficiency and spectral response. It has been reported that the effect of spectral variation on the rating of PV modules is prominent for technologies with a narrow spectral response like CdTe, amorphous Si, and Si being very sensitive to the spectral response [24].

6 Angle of Incidence

The STC power ratings of PV modules are performed in a sun simulator in which simulated light is perpendicular to the PV module surface. PV modules in a real environment only receive sun radiation falling over perpendicularly on their surface at the solar noon due to the apparent motion of the sun over the sky. The inclination of earth gives strong seasonal variations in the course of zenith and azimuth throughout the day. Researchers reported a strong degeneration in the performance of PV modules at high angle of incidence [25].

7 The Experiment

One single-crystalline and one multicrystalline PV modules were tested in a natural environment located in the equatorial region (4.94° North and 115° East). The circuit diagram and methodology for this experiment have been discussed earlier [6]. The experiment was conducted twice a week and each time was performed at three different times of the day corresponding to the solar noon, 2 hours before, and 2 hours after solar noon. These three times provide the true variation in irradiance, the working temperature, and their effect on the performance of the module. Solar noon is the time when the sun is at a longitude of 115° east and this is the time that divides the day exactly into two equal parts.

8 Results and Discussion

The influence of the climate on the performance of this module is studied by comparing the electrical characteristics of the module with that reported at STC. The I-V characteristic of this modules at STC is given in Tables 1 and 2.

The monthly averaged amount of global radiation lies in the range of 17.46–20.88 MJm^{-2} throughout the calendar year. The bright sunshine hours are in the range of 6.97–8.47 hours per day and the diffused radiation ranges from 8.57 MJm^{-2} to 9.73 MJm^{-2} in a calendar year (Table 3).

The performance of single-crystalline silicon PV module of rating was investigated in the equatorial climate. The efficiency of the module in the field varied in the range 12–20% compared with that of 13% reported at STC by the manufacturer. The working temperature of the module, the irradiance, the open circuit voltage, the short circuit current, and the maximum output power of the module lie in the ranges 27–73 °C, 70–1263 Wm^{-2}, 18.3–19.6 V, 0.5–12.2 A, and 5.2–121 W, respectively. The temperature coefficients for various performance parameters have been calculated using measured data and tabulated in Table 3, which are qualitatively in agreement with the earlier studies [6, 26, 27].

The increase in the temperature decreases the bandgap. The electrons at standard temperature conditions could not manage to overcome the bandgap would be able to do as temperature increases. As a result, the increase in temperature would allow some electrons which could not manage to overcome the band gap in PV devices at standard temperature conditions would surmount the band gap at a higher working temperature. This increase in the lifetime of minority carriers results in a decrease in V_{oc}, an increase in I_{sc}, thereby, resulting in an overall decrease in P_{max} that can be seen in Table 3.

It has been observed that the polycrystalline module can work in a very low irradiance such as 60 Wm^{-2} with a decrease in the output power of 97% with reference to that at STC. The value of fill factor for this event was 0.69, indicating that the quality of the module is not deteriorated. The combinations of low working temperature with low irradiance and high working temperature with high irradiance increase the short circuit current of the module. An increase in the short circuit current in the polycrystalline module with reference to the STC has been observed at low as well as at high temperatures correspond

Table 1 I-V characteristics of single-crystalline silicon PV module at standard temperature conditions (STC) supplied by the manufacturer

V_{oc}	I_{sc}	V_{max}	I_{max}	P_{max}	Irradiance	η	Tc
21 V	54 A	16.5 V	4.6 A	76 W	1000 Wm^{-2}	12%	25–27 °C

Table 2 I-V characteristics of multicrystalline silicon PV module at standard temperature conditions (STC) supplied by the manufacturer

V_{oc}	I_{sc}	V_{max}	I_{max}	P_{max}	Irradiance	η	Tc
10.3 V	0.762 A	8.2 V	0.705 A	5.83 W	1000 Wm^{-2}	12.7%	25–27 °C

Table 3 Temperature coefficients for single-crystalline silicon and multi/polycrystalline PV modules

Temperature coefficient	Single-crystalline value	Multi/polycrystalline value
$\frac{dV_{oc}}{dT}$	−0.038 $V^{\circ}C^{-1}$	−0.019 $V^{\circ}C^{-1}$
$\frac{dI_{sc}}{dT}$	0.026 $A^{\circ}C^{-1}$	0.011 $A^{\circ}C^{-1}$
$\frac{dP_{max}}{dT}$	−0.016 $W^{\circ}C^{-1}$	0.059 $W^{\circ}C^{-1}$

to low and high irradiance, respectively. The maximum power, efficiency, and the fill factor of the module degraded as the working temperature increases. A linear relation could not be established to correlate these variables.

The dual behavior of the short circuit current with an increase in the working temperature of the module is due to the following facts:

- An increase in working temperature of the module increases the dark current that in turns increases the short circuit current.
- An increase in working temperature also increases the free carriers' losses.

The net effect of these two terms determines the destiny of the short circuit current in polycrystalline modules.

At higher working temperatures with irradiance greater than 1000 Wm^{-2}, an increase in the short circuit current and a reduction in open circuit voltage increases the efficiency of the polycrystalline solar cell. This result indicates that multicrystalline solar cells may be the better choice for power production in those regions of the globe having similar climatic conditions to equatorial region.

References

1. Bahaidarah H, Rehman S, Subban A, Gandhidasan P, Baig H (2015) Performance evaluation of a PV module under climatic conditions of Dhahran, Saudi Arabia. Energy Explor Exploit 33(6):909–930
2. Ciulla G, Brano VL, Franzitta V, Trapanese M (2014) Assessment of the operating temperature of crystalline PV modules based on real use conditions. Int J Photoenergy 2014:718315
3. Louwen A, de Waal AC, Schropp REI, Faaij APC, van Sark WGJHM (2017) Comprehensive characterisation and analysis of PV module performance under real operating conditions. Prog Photovolt Res Appl 25:218–232
4. Makrides G, Zinsser B, Norton M, Georghiou GE. Performance of photovoltaics under actual operating conditions. doi:https://doi.org/10.5772/27386 In book: Third generation photovoltaics, www.intechopen.com
5. Ye JY, Ding K, Reindl T, Aberle AG (2013) Outdoor PV module performance under fluctuating irradiance conditions in tropical climates. Energy Procedia 33:238–247
6. Malik AQL, Ming C, Sheng TK, Blundell M (2010) Influence of temperature on the performance of photovoltaic polycrystalline silicon modules in the Bruneian climate. AJSTD 26:61–72
7. Gaur A, Tiwari GN (2013) Performance of photovoltaic modules of different solar cells. J Solar Energy 2013:734581
8. Virtuani A, Mullejans H, Dunlop ED (2011) Comparison of indoor and outdoor performance measurements of recent commercially available solar modules. Prog Photovolt Res Appl 19(1):11–20
9. Zinsser, B.; Makrides, G.; Schubert, M.; Georghiou, G.E. & Werner, J.H. Temperature and irradiance effects on outdoor field performance. In: Proceedings of the 24th European photovoltaic solar energy conference, 2009, ISBN 3936338-25-6, 4083–4086, Hamburg, Germany
10. Huld T, Amillo AMG (2015) Estimating PV module performance over large geographical regions: the role of irradiance, air temperature, wind speed and solar spectrum. Energies 8(6):5159–5181
11. Aaditya G, Pillai R, Mani M (2013) An insight into real-time performance assessment of a building integrated photovoltaic (BIPV) installation in Bangalore (India). Energy Sustain Dev 17(5):431–437
12. Carr AJ, Pryor TL (2004) A comparison of the performance of different PV modules types in temperate climates. Sol Energy 76(1):285–294
13. Gxasheka AR, Van Dyk EE, Meyer EL (2005) Evaluation of performance parameters of PV modules deployed outdoors. Renew Energy 30(4):611–620
14. Biicher K (1997) Site dependence of energy collection of PV modules. Sol Energy Mater Sol Cells 47(1–4):85–94
15. Mayer EL, Mapuranga B (2005) The effect of temperature on the performance of photovoltaic modules at regulated voltage. In: Proceedings of 15th international photovoltaic science & engineering conference (PVSEC-15), Shanghai, China, p. 1256–1257
16. Bücher K (1996) Do we need site-dependent and climate-dependent module rating? In: Proceedings of 23rd IEEE photovoltaic specialists conference, Louisville, KY, p. 1056–1062
17. Emery K (1999) The rating of photovoltaic performance. IEEE Transact Electron Devices 46(10):1928–1931
18. Lorenzo E (1994) Solar electricity. Progensa, Sevilla
19. Kazmerski L, Photovoltaics L (1997) A review of cell and module technologies. Renew Sustain Energy Rev 1:71–170
20. Mayers DR, Emery K, Gueymard C (2002) Terrestrial solar spectral modeling tools and applications for photovoltaic devices. In: Proceedings of the 29th IEEE photovoltaic specialist conference, pp. 1683–1686, ISBN 0-7803-7471-1, New Orleans, USA
21. Gueymard C, Myers D, Emery K (2002) Proposed reference irradiance spectra for solar energy systems testing. Sol Energy 73(6):443–467
22. Huld T, Sample T, Dunlop ED (2009) A simple model for estimating the influence of spectrum variations on PV performance. In: Proceedings of the 24th European Photovoltaic Solar Energy Conference 2009, pp. 3385–3389, ISBN 3-936338-25-6, Hamburg, Germany
23. Gottschalg R, Betts TR, Hohl-Ebinger J, Herrmann W, Müllejans H. (2007) Effect of spectral mismatch on power rating measurements – a comparison on indoor and outdoor measurements for single and multi-junction devices. In: Proceedings of the 22nd European Photovoltaic Solar Energy Conference, 2007, pp. 2621–2624, Milan, Italy
24. Gottschalg R, Betts TR, Infield DG, Kearney MJ (2005) The effect of spectral variations on the performance parameters of single and double junction amorphous silicon solar cells. Sol Energy Mater Sol Cells 85(3):415–428
25. Beal RJ, Potter BG, Simmons JH (2014) Angle of incidence effects on external quantum efficiency in multicrystalline photovoltaics. IEEE J Photovoltaics 4(6):1459–1464
26. Malik AQ, Mohamad Fauzi Bin Hj Metali, Lim Chee Ming, Tan Kha Sheng. In: Proceedings of world renewable energy congress (WREN) 2009, Asia, p. 294–297
27. Malik AQ, Damit SJBH (2003) Outdoor testing of single silicon solar cells. Renew Energy 28:1433–1445

Standalone Photovoltaic System for Boys Hostel at Fiji National University, Natabua Campus, Fiji

Abdul Q. Malik and Host T. W. Karakaua

1 Introduction

Fiji is a tropical island nation located in Oceania in the South Pacific Ocean and comprises an archipelago of more than 332 islands, 110 of which are permanently inhabited. Rising sea levels and extreme weather conditions are big challenges for the Pacific island states. Despite contributing insignificantly to climate change, Fiji remains steadfast in its commitment to the Paris Agreement by continuing to incorporate low emissions and sustainable growth into its development agenda. For example, the 5-Year and 20-Year National Development Plan for Fiji is in line with Fiji's Nationally Determined Contributions (NDCs), seeking to achieve 99% renewable electricity generation by 2030 from a 2013 baseline of 60% and aiming to achieve 30% reduction in CO_2 emissions from the energy sector over the same period [1].

Among Pacific Island countries, Fiji is blessed with natural advantages that give the country remarkable natural energy potential. Fiji has a mountainous geography and powerful rivers that flow from the highlands to the sea. The country has wind potential and large tracks of land well suited for solar power generation. It is notable that the country in 2018 produced over half (54.86%) of its energy requirements from hydro-power, 0.25% from wind power, and 3.87% from Independent Power Producers (IPPs), namely Tropik Wood Industries Limited, Fiji Sugar Corporation, and Nabou Green Energy. Compared to last year, our energy mix was cleaner and more diversified, as our reliance on diesel fuel fell from 48.72% in 2017 to 41.02% in 2018 of our total energy generation mix [1].

The Government of Fiji is very keen to reduce the dependence of fossil fuel for the generation of power. The Department of Physics, Fiji National University, in line with the Government's commitment to achieve 99% of all energy requirements from renewable energy sources by 2030 has been working on different projects on renewable and sustainable energy. One of such projects is "Standalone Photovoltaic System for Boys Hostel at Natabua Campus, Lautoka."

2 Photovoltaic Generator/System

A photovoltaic generator/system consists of solar panel that absorbs and converts sunlight to electricity, battery charger to charge battery with the use of electricity generated by PV panels, solar batteries to store power that would be recalled when sun is not around, and an inverter that supplies power to AC load connected with the system. DC load connected with standalone PV system get power directly from the charge controller, cables, and accessories. Cables need to be ultraviolet resistant and suitable for outdoor applications. It is very important to keep power losses and voltage drop in the cable to a minimum. A typical PV generator is shown in Fig. 1.

Standalone PV system is meant to operate independent of the electric utility grid and is designed and sized to supply DC and/or AC electrical loads. It can provide DC and AC source of electrical power to homes in order to meet their energy requirement through the direct conversion of solar irradiance into electricity. Common applications are direct power to DC loads, water pumping, telecommunication, backup power, etc. PV systems range from small, rooftop-mounted, or

A. Q. Malik (✉) · H. T. W. Karakaua
Department of Physics, School of Pure Sciences, College of Engineering, Science and Technology, Fiji National University, Lautoka, Fiji
e-mail: Abdul.Malik@fnu.ac.fj

© The Author(s), under exclusive license to Springer Nature Switzerland AG 2022
A. Sayigh (ed.), *Sustainable Energy Development and Innovation*, Innovative Renewable Energy,
https://doi.org/10.1007/978-3-030-76221-6_11

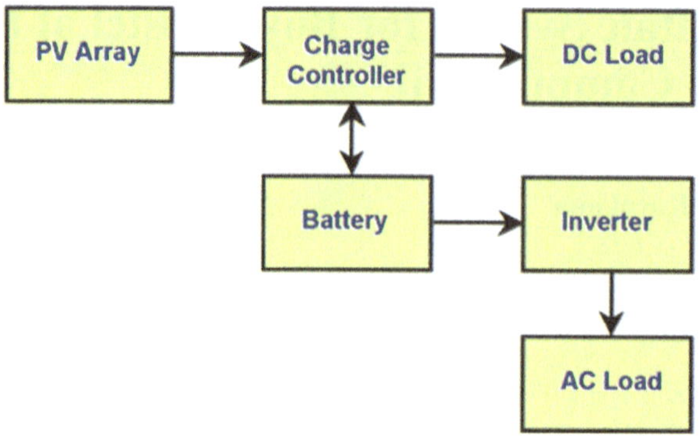

Fig. 1 Block diagram for a standalone PV system

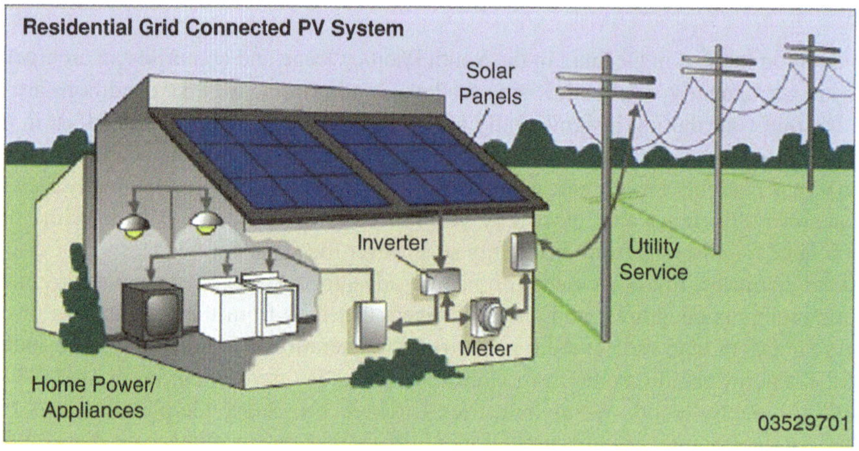

Fig. 2 Sketch of a grid-connected PV system. (Adapted from: www.energyeducation.tx.gov)

building-integrated systems with capacities from a few to several tens of kilowatts to large utility-scale power stations of hundreds of megawatts.

There are two types of PV systems: standalone (off-grid) and grid connected. The off-grid standalone system depicted in Fig. 1 involves the design, selection, and determination of parameters of the different components used in the system. The success of this process is based on factors such as geographical location of the off-grid solar system, weather condition, solar irradiance, and the load requirement [2–4].

The grid-connected systems are connected to the electricity grid. DC power generated by the PV system is to be converted to AC at the grid voltage through a specially designed inverter. The grid-connected system is shown in Fig. 2. Grid interactive systems can work with or without battery storage. The main advantage of this system is that the power can be fed into the grid or can be drawn from the grid as required [3, 4].

The depletion of conventional energy resources and increase in the energy demand over the globe lead to explore the alternative resources of energy. An extensive research has been conducted experimentally as well as simulation studies on the applications of PV systems as distributed energy resources to harness solar energy with low environmental impact. Khatib et al. [5] have presented their findings on optimization of size of various PV systems such as standalone PV systems, hybrid PV/diesel generator systems, hybrid PV/wind systems, hybrid PV/wind/diesel system, and grid-connected systems. Singh [6] presented a review on the progress made in solar power generation research and development since its inception. Eltawil and Zhao [7] emphasized on the importance of the grid-connected PV system regarding the intermittent nature of renewable energy generation and the characterization of PV generation with regard to the grid code compliance. Al-Shetwi and Sujod [8] picked up another important scenario on the grid-connected PV systems. They discussed how the high integration photovoltaic plants (PVPPs) can affect the operation, stability and security of utility grid. To minimize these issues, many countries have established new requirements for grid integration of solar photovoltaics. They discussed and summarized the most recent international regulation regarding photovoltaic integration and research findings on the compliance of these regulations, and they proposed recommendations for future research. Garg et al. [9] discussed the importance of system simulation to investigate the feasibility of PV system at a given location. Okwu et al. [10] described the principal criteria for the designing and installation of a PV system. A number of researchers presented their findings on the designing of a standalone PV system and discussed the reliability, life, and how to minimize the design cost [11–13].

3 PV Standalone System Description for Boys Hostel

The Boys Hostel is located in the Natabua Campus, Lautoka, Fiji National University. Lautoka is located at a latitude of -17.62° S and a longitude of 177.47° E in the tropical region of the globe. The bright sunshine hours and amount of global solar radiation for full calander year are tabulated in Table 1. These data were obtained from the Fiji Meteorological Services.

4 System Sizing

System sizing is the process that deals with evaluation of power demand in watt-hour per day that can be calculated with the list of electrical appliance and their hours of use each day known as load profile. This list is given in Table 2 along with the total electrical load of 169 KW h d^{-1}.

5 Size and Specification of Battery

The function of battery is to make sure the continuity of the energy supply and energy consumption to the electrical load as described in Table 2. The battery capacity is in ampere hours (Ah), and the energy consumption is in watt-hours (Wh). To relate these values with each other, the total daily load in Wh is divided with the voltage of the battery (48 V).

$$\text{Specific voltage of battery} = 48\ \text{V}$$

$$\text{Required bank capacity in Ah} = \frac{\left(\text{Average Whd}^{-1} \times \text{Day of autonomy}\right)}{\text{Battery voltage}}$$

$$\text{Required bank capacity in Ah} = \frac{143562 \times 1}{48} = 2991\ \text{Ahd}^{-1}$$

Table 1 Bright sunshine hours and global solar radiation over Lautoka

Month	Bright sunshine hours	Global solar radiation (MJ m^{-2})
January	10.5	22.47
February	6.4	20.51
March	8.6	20.40
April	6.4	17.37
May	10.5	15.88
June	10	15.82
July	9.9	17.10
August	9.8	22.80
September	4.6	20.60
October	4.6	13.15
November	10.3	27.60
December	11.5	27.50
Average	8.59	20.10

Table 2 Load profile for Boys Hostel located at Natabua Campus, FNU, Fiji

Serial no.	Device	Total no. of devices in Hall B	Rating	Total usage/day (24 h)
1	Fans	70	55 W	12 h = 46.2 KW
2	Room lights	105	54 W	7 h = 39.69 KW
3	Bathroom lights	9	54 W	12 h = 5.832 KW
4	Corridor lights	10	54 W	12 h = 6.48 KW
5	Music speaker/radios	15	32 W	7 h = 3.36 KW
6	Laptops	35	150 W	8 h = 42 KW
			Total load/day	= 143.562 KW h/day = 143,562 Wh day^{-1}

Table 3 Hoppecke battery specification

Battery type	Hoppecke 16 OPzS 2000
Normal voltage	2 V
Nominal capacity	2000 Ah

$$\text{Depth of discharge } (\text{DOD}) \cong 60\% = 0.6$$

$$\text{Final battery bank capacity in Ah} = \frac{\text{Required bank capacity in Ah}}{\text{Depth of discharge } (\text{DOD})}$$

$$= \frac{2991}{0.6} = 4985 \text{ Ah}$$

The recommended battery type used for PV solar system is deep cycle, and for the present project, Hoppecke battery is used and its specifications are given in Table 3.

$$\text{Batteries in series}: \frac{48 \text{ V}}{2 \text{ V}} = 24 \text{ batteries}$$

$$\text{Batteries in parallel}: \frac{\text{Final battery bank capacity } (\text{Ah})}{\text{Nominal voltage}} = \frac{4985 \text{ Ah}}{2000 \text{ Ah}} = 3$$

6 Specification of the PV Array

PV system sizing refers to the computation of PV components required to make the complete PV system meet the necessary load demand.

The efficiency of the battery is 80%, that is, 0.8.

$$\text{The average Ah to be supplied by the array} = \frac{\text{Daily average Ah taken from battery}}{\text{efficiency of the battery}}$$

$$= \frac{4985 \text{ Ah}}{0.8} = 6232 \text{ Ah}$$

$$\text{The array peaks amperes} = \frac{\text{Average Ah per day to be supplied by the array}}{\text{Peak sunshine hours per day}}$$

$$= \frac{4985}{8} = 623 \text{ A}$$

Table 4 I-V Characteristics of monocrystalline BSM500M-96 PV module

Type of PV cells	Monocrystalline
P_{max} (W)	500
V_{max} (V)	48.63
I_{max} (A)	10.28
V_{oc} (V)	59.01
I_{sc} (A)	10.87
Area per module (m^2)	2.56

The PV system designed is based on monocrystalline silicon solar module, and its I-V characteristics at standard test conditions (STCs) are tabulated in Table 4.

$$\text{Modules in parallel} = \frac{\text{Array peak ampere}}{\text{Peak ampere per module}} = \frac{623\ \text{Ah}}{10.28\ A}$$

$$\text{Modules in parallel} = 61$$

$$\text{Module in series} = \frac{\text{Battery bank voltage}}{\text{Nominal Module Voltage}} = \frac{48}{48.63} = 1$$

$$\text{Total number of modules require} = 61 \times 1 = 61$$

$$\text{Area required} = 61 \times 2.56\ \text{m}^2 = 157\ \text{m}^2$$

7 Specifying a Controller

The charge regulator ensures the optimization of the electricity generation by PV panels. It also protects the batteries against deep discharge and limits terminal charge, that is, protection against overcharges.

$$\text{Controller rating} = \text{Modules short circuit} \times \text{No of modules in parallel}$$
$$= 10.87 \times 61 = 663\ \text{A}$$

8 Specifying Inverter

Total watt of all appliance = 16,284 W.

For safety, the inverter should be 25–30% bigger size. The inverter required for present project would be of 200 kW, 48 V DC, 50–60 Hz frequency, and AC voltage 230 V.

9 Sizing the System Wiring

$$\text{Wiring Rating} = \frac{\text{Total AC connect load in (W)}}{\text{System DC voltage}} = \frac{16248}{48} = 339\ \text{A}$$

10 Conclusions

PV standalone system is designed based on the electrical load on watt-hour demand. This system generates green electricity without the emission of CO_2 and reduces the dependency on the fossil fuels. PV standalone system provides an ideal solution to electrify not only the scattered and/or remote locations within the country that cannot be connected with the National electric grid but also other residential and office buildings. Energy efficient houses could be designed based on either PV standalone or PV grid-connected generators.

Acknowledgments The authors record appreciations to the Fiji Meteorological Service for providing the meteorological data on solar radiation, sunshine hours, and ambient temperature.

References

1. 2018 Annual energy report. Fiji Energy Limited (efl.com.fj/about-us/company-information/company-reports)
2. Oladeji AS, Balogun OS, Aliyu SO (2017) Use of photovoltaic system for office building: the case study of National Centre for hydropower research and development, Nigeria. Niger J Technol 36(4):1208–1217
3. Sakiliba SK, Hassan AS, Wu J, Sanneh ES, Ademi S (2015) Assessment of stand-alone residential solar photovoltaic application in Sub-Saharan Africa: A case study of Gambia. J Renew Energy 2015:640327
4. Egbivwie EA, Ughwwmiakpor VO, Ayinuola AP (2017) Computer aided off-grid solar PV system design. Int J Sci Res Publ 7(9):310–317
5. Khatib T, Mohamed A, Sopian K (2013) A review of photovoltaic systems size optimization techniques. Renew Sust Energ Rev 22:454–465
6. Singh GK (2013) Solar power generation by PV (photovoltaic) technology: a review. Energy 53:1–13
7. Eltawil MA, Zhao Z (2010) Grid-connected photovoltaic power systems: technical potential problems – a review. Renew Sustain Energy 14:112–129
8. Al-Shetwi AQ, Sujos MZ (2018) Grid-connected photovoltaic power plants: a review of the recent integration requirements in modern grid codes. Int J Energy Res 42:1849–1865
9. Garg A, Bhatia A, Iqbal MS, Chauhan D Design of solar PV system. IRE J 1(9):43–48, ISSN: 2456-8880
10. Okwu PI, Okolo CC, Obinna O, Methew DO, Uche A (2017) Details pf photovoltaic solar system design calculations and the accessories. Int J Res Appl Sci Eng 5(VI):2933–2938
11. Ge H, Ni L, Asgarpoor S (2008) Reliability-based stand-alone photovoltaic system sizing design – a case study. Proceedings of the 10th international conference on probabilistic methods applied to power systems, pp. 1–8
12. Ibrahim UH, Aremu DA, Umwaha JI (2013) Design of stand-alone solar photovoltaic system for residential buildings. Int J Sci Technol Res 2(12):187–194, ISSN 2277-8616
13. Hasan MR, Islam MS, Hossain MA, Sarjana U, Hasan MR (2016) Design and development of a cost effective standalone PV system for residential buildings in Pabna, Bangladesh. Int J Sci Technol Res 7(9):1664–1669, ISSN 2229-5518

Part II
Solar Thermal, Heat Pumps and Geothermal Energy

Analysis of a Novel Thermochemical de-NO_x Cycle with Composite Multi-Halide Sorbent

C. Zhang, G. L. An, and L. W. Wang

1 Introduction

Exhaust gas emissions are severe pollution to the environment, and increasing emission standards have challenged NO_x removal technologies. There are severe challenges present to NO_x removal systems, including selective catalytic reduction (SCR) [1], diesel oxidation catalysts (DOCs) [2], three-way catalysts (TWCs) [3], and diesel particulate matter filters (DPFs) [4]. So far, the widely used de-NO_x system is the urea-SCR system [5].

For the urea-SCR system, harmful NO_x in flue gas is reduced to harmless N_2 by the atomized urea solution. Since the undecomposed urea and the deposition of by-products are at low temperature, and the lower the exhaust gas temperature, the easier the deposition, it is prone to slow on cold start and to produce crystal precipitation and block pipelines, especially the injector and the catalyst [6]. According to the research by Jacques et al., the SCR system is quite sensitive to hydrothermal aging, sulfur exposure, and N_2O production [7].

Hence, based on the above fact, the sorption-based ammonia storage technology with a sorption bed replacing the urea solution is promoted to solve the drawback of the urea-SCR system. Liu and Aika [8] studied the ammonia storage performance of $CaCl_2$-$CaBr_2$ mixture, which can be applied in de-NO_x cycle. Wang et al. [9] have proposed a composite sorbent-SCR system for ammonia storage/release, but it still troubles the inefficient conversion during cold start and the attenuation of sorption performance. Gao et al. and An et al. [10, 11] researched the composite multi-halide sorbent, which could change the sorption hysteresis of single halide and improve the sorption performance. Such a principle has not been applied in the de-NO_x system.

In this chapter, a novel NH_3-SCR system based on composite multi-halide sorbent is studied and analyzed.

2 The Principle of Thermochemical de-NO_x Cycle

In the thermochemical de-NO_x system, shown in Fig. 1, the composite multi-halide sorbent serves as a supply of ammonia, which will react with the NO_x in exhaust gas, whose reaction equation is shown in Eq. 1.

$$6NO_x + 4xNH_3 \rightarrow (2x+3)N_2 + 6xH_2O \tag{1}$$

Two kinds of halide, $CaCl_2$ and NH_4Cl, are always chosen. The reaction of the working pair of $CaCl_2$ and NH_4Cl is shown in Eqs. 2, 3, 4, and 5. Equation 2 is for the sorption/desorption process of NH_4Cl, and Eqs. 3, 4 and 5 are for the sorption/desorption process of $CaCl_2$. Since the reverse reaction of Eq. 3 always reacts at a temperature much higher than 120 °C, it is naturally impossible to react under this condition.

C. Zhang · G. L. An · L. W. Wang (✉)
Institute of Refrigeration and Cryogenics, Shanghai Jiao Tong University, Shanghai, China
e-mail: lwwang@sjtu.edu.cn

© The Author(s), under exclusive license to Springer Nature Switzerland AG 2022

A. Sayigh (ed.), *Sustainable Energy Development and Innovation*, Innovative Renewable Energy,
https://doi.org/10.1007/978-3-030-76221-6_12

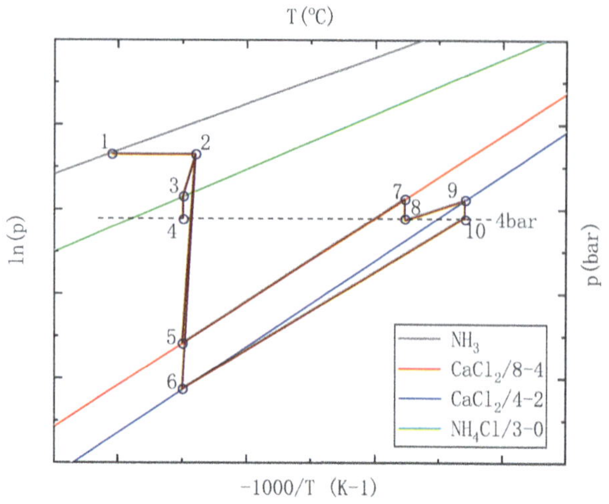

Fig. 1 Clapeyron diagram of $CaCl_2$ and NH_4Cl

$$NH_4Cl + 3NH_3 \Leftrightarrow [NH_4(NH_3)_3]Cl \tag{2}$$

$$CaCl_2 + 2NH_3 \Rightarrow [Ca(NH_3)_2]Cl_2 \tag{3}$$

$$[Ca(NH_3)_2]Cl_2 + 2NH_3 \Leftrightarrow [Ca(NH_3)_4]Cl_2 \tag{4}$$

$$[Ca(NH_3)_4]Cl_2 + 4NH_3 \Leftrightarrow [Ca(NH_3)_8]Cl_2 \tag{5}$$

For NH_4Cl, its starting temperature is low and the ammonia storage density is large, but the pressure is quite sensitive to temperature changes, which means that the system pressure is easily too high unless the temperature control technology is in quite a high level. Besides, for $CaCl_2$, its pressure is relatively low at the same temperature, and it can be maintained within 10 bar in a wide temperature range, but the ammonia storage density is smaller. Thus, $CaCl_2/NH_4Cl$-NH_3 working pair is chosen to combine the advantages of both. The Clapeyron diagram of the working pair is shown in Fig. 1.

As shown in Fig. 1, ammonium chloride corresponds to the line named NH_4Cl/3-0 (represents $[NH_4(NH_3)_3]Cl$ and NH_4Cl). Besides, the desorption process of calcium chloride is divided into two stages, which are respectively shown as $CaCl_2$/8-4 (represents $[Ca(NH_3)_8]Cl_2$ and $[Ca(NH_3)_4]Cl_2$ and $CaCl_2$/4-2 (represents $[Ca(NH_3)_4]Cl_2$ and $[Ca(NH_3)_2]Cl_2$).

From the perspective of de-NO_x cycling, there are two steps of the novel thermochemical de-NO_x cycle with the composite multi-halide sorbent, shown as follows:

1. Ammonia filling phase: The state of the ammonia tank is controlled at point 1. For the multi-halide reactor, the initial state of NH_4Cl is point 3, while that of $CaCl_2$ is point 6. Open the valve, since the system pressure is controlled by the ammonia tank, it is considered that the multi-halide reactor is switched to point 2 for sorption in a short time. After the reaction is completed, the multi-halide reaction is allowed to cool naturally, and NH_4Cl and $CaCl_2$ are restored to point 3 and point 5, respectively.
2. de-NO_x phase 1 (cold start phase): The temperature of the multi-halide reactor is controlled at point 3, and then open the valve. As the reaction is quite slow, assuming the temperature does not change, the NH_4Cl in the multi-halide reactor is switched to point 4 (30 °C, 4 bar), and the desorbed ammonia is injected into the catalytic reactor.
3. de-NO_x phase 2 (heat phase): After desorption of the ammonia complex of NH_4Cl is complete, heat the multi-halide reactor. At the same time, $CaCl_2$ is at point 7. When the valve is opened, only the $[Ca(NH_3)_8]Cl_2$ in the multi-halide reactor can desorb, and the state of $[Ca(NH_3)_8]Cl_2$ is switched to point 8. The desorbed ammonia flows into the catalytic reactor.
4. de-NO_x phase 3 (warning phase): When the amount of ammonia in the $[Ca(NH_3)_8]Cl_2$ is insufficient, it is difficult to maintain the output of ammonia at this temperature. Then, the temperature of the multi-halide reactor is raised until 76.9 °C so that the $[Ca(NH_3)_4]Cl_2$ therein reaches the point 9, desorption is performed at point 10, and spraying was continued through the catalytic reactor.

3 Experiments

3.1 Materials

The development of the composite sorbent is shown in Fig. 2, taking $CaCl_2/NH_4Cl$ as an example. First, the halide is dissolved in the water with a mass ratio of 1:1, and the halide solution is made. After that, the expanded nature graphite treated with sulfuric acid (ENG-TSA) is added to the solution. In order to improve the energy-storage density per unit mass/volume, the mass ratio of halide to ENG-TSA is optimized to be 5:1, which is equal to the mass ratio of $CaCl_2$, NH_4Cl and ENG-TSA which are optimized to be 5:5:2. After mixing, the material is put into the oven for drying. Then, put it into the designed mold and pressed it under the hydraulic pressure. At last, the consolidated material is transferred to the experiment platform of Rubotherm. The filling density of composite sorbent is 800 kg/m^3.

3.2 Experimental Test Unit

The resolution of the balance is 10 μg, and the mass of the adsorbent tested is about 0.2 g. The sample is placed in a steel basket suspended in a sealed steel chamber, and the temperature of the steel basket is controlled by the heat radiation of the circulating thermal oil. The oil temperature is controlled by a thermostatic bath (JULABO SE-6) with a temperature accuracy of 0.01 K. Below the basket is a PT100 temperature sensor with an accuracy of 0.1 K to monitor the temperature of the sorbent. The sealed steel chamber is connected to an ammonia tank used as a condenser/evaporator. The temperature of the ammonia tank is controlled by a thermostatic ethylene glycol water bath (JULABO F32-ME) with a temperature accuracy of 0.01 K. The working pressure of the measuring chamber is under the saturation pressure of ammonia and is detected by an absolute pressure sensor (Druck DPI-282) with an accuracy of 0.04% (Fig. 3).

3.3 Test Procedure

The equilibrium measurement is designed to study the relationship between sorption capacity and sorbent temperature under equilibrium conditions at constant pressure. For the desorption process, set the initial temperature as low as possible, and then advance to the next equilibrium state by increasing the temperature of the constant temperature oil bath (Toil) by 5 °C per step until the desorption is completed. For each step, the temperature can only be increased after the weight and temperature data have remained stable. In contrast, for the sorption process, the initial temperature should be set high enough, and Toil should also be decreased by 5 °C per step. The pressure is controlled within 1.0 MPa, and Toil remains at 20 °C and above.

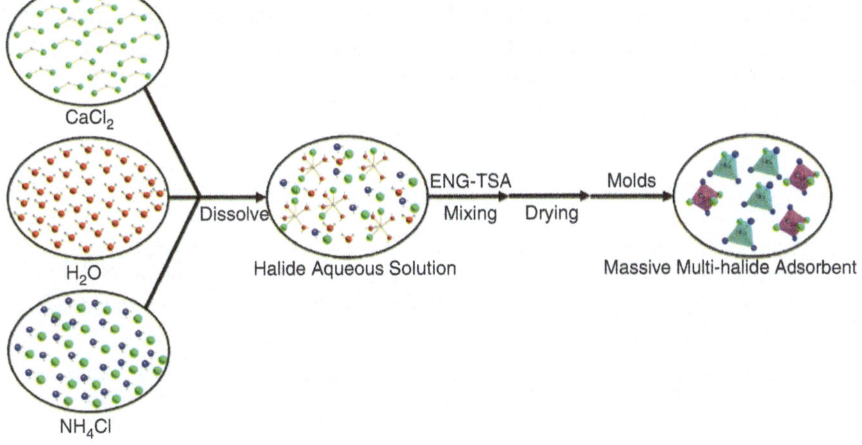

Fig. 2 Development of composite sorbent (taking $CaCl_2$ and NH_4Cl as an example)

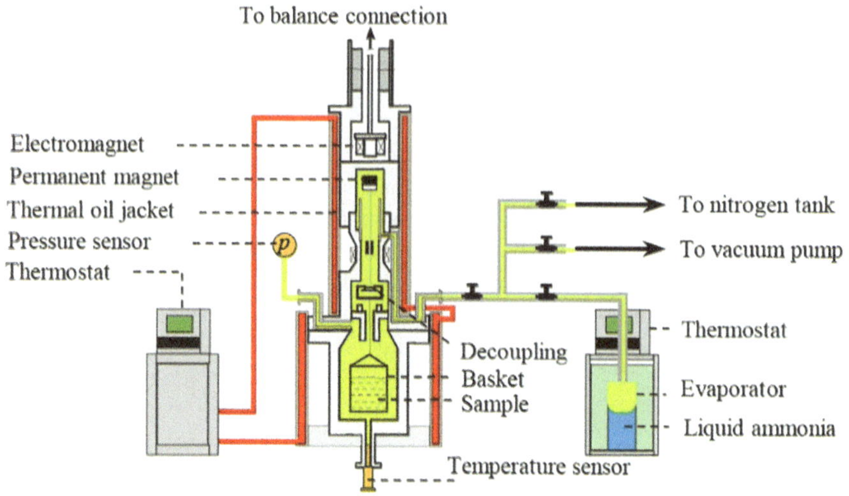

Fig. 3 Schematic diagram of Rubotherm balance for sorption analysis [11]

3.4 Test Results

3.4.1 Equilibrium for Thermodynamics (Sorption Capacity, Hysteresis)

Aiming to compare the sorption capacity and sorbent temperature between the composite sorbent of $CaCl_2$, which is chosen to be the best sorbent of single halide in the de-NO_x cycle [9], and that of $CaCl_2/NH_4Cl$ under equilibrium conditions at constant pressure, the temperature of the ammonia tank is controlled at 20.00 °C through the thermostatic ethylene glycol water bath for the equilibrium measurement. Therefore, their *x*-*T* graph is shown in Fig. 4, where *x* is the mass of the ammonia sorbed per gram of halide and *T* is the temperature of the sorbent.

3.4.2 Cycle Characteristics Comparison

When applying in the de-NO_x system, the composite sorbent is adjusted by three leading indicators, including the starting temperature of the desorption process, the thermal efficiency, and the available ammonia storage density.

Starting Temperature of the Desorption Process

When the de-NO_x system is operating, it needs to release the ammonia from the halide's ammonia complex. Hence, the starting temperature of the desorption process is quite essential. From Fig. 4, the starting temperature of the desorption process of $CaCl_2$ is 78.8 °C, while that of $CaCl_2/NH_4Cl$ is 20 °C, which is an approach to the temperature of the condenser/evaporator. Such an improvement of the starting temperature could provide a better performance of de-NO_x efficiency, which could better adapt to cold start conditions of the diesel engines.

Thermal Efficiency

The thermal efficiency is always equal to the incoming over the payment. Hence, in the de-NO_x cycle, its thermal efficiency is the amount of ammonia produced by desorption divided by the total input heat, shown in Eq. 8.

$$\eta = \frac{m_{NH_3}}{Q_{in}} = \frac{\Delta x_{max} m_{halide}}{Q_{sen} + \frac{\Delta x_{max} m_{halide} \Delta H}{M_{halide}}} = \frac{\Delta x_{max} m_{halide} M_{halide}}{M_{halide} Q_{sen} + \Delta x_{max} m_{halide} \Delta H} \quad (8)$$

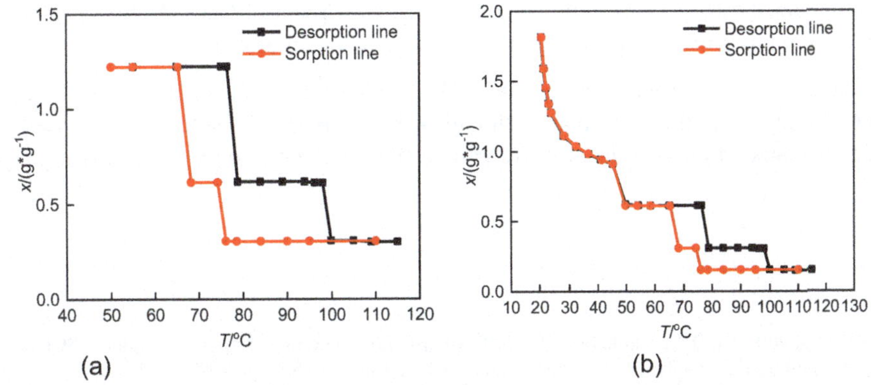

Fig. 4 The x-T graph of different halides: (**a**) $CaCl_2$ and (**b**) $CaCl_2/NH_4Cl$

Table 1 The thermal efficiency of the de-NO_x cycle with composite multi-halide sorbent

Type	Δx_{max}(g/g)	m_{halide}(g)	Q_{sen}(J)	ΔH(J/mol)	M_{halide} (g/mol)	η(g/kJ)
$CaCl_2$	0.9252	0.1667	483.24	41431.3	111	0.285
$CaCl_2/NH_4Cl$	1.6664	0.1667	483.24	36662.2	82.245	0.458

where η is the thermal efficiency of the de-NO_x cycle, m_{NH3} is the mass of ammonia produced by desorption, Q_{in} is the total input heat, Δx_{max} is the maximum mass of the ammonia sorbed per gram of halide, m_{halide} is the mass of the halide in the composite sorbent tested, Q_{sen} is the sensible heat, ΔH is the desorption enthalpy of the halide, and M_{halide} is the molar mass of the halide in the composite sorbent tested.

According to Fig. 4 and Eq. 8, the result is shown in Table 1.

According to Table 1, the thermal efficiency of the de-NO_x cycle with composite multi-halide sorbent of $CaCl_2/NH_4Cl$ is increased by 60.7% from that of $CaCl_2$.

Available Ammonia Storage Density

The higher available ammonia storage density is the key to the smaller mass of the composite sorbent. It can be defined as Eq. 9.

$$\rho = \Delta x_{max} \rho_a \frac{m_{halide}}{m_a} \tag{9}$$

where ρ is the available ammonia storage density of the composite sorbent, ρ_a is the filling density of composite sorbent in Rubotherm, and m_a is the mass of the composite sorbent.

Thus, the available ammonia storage density of composite sorbent of $CaCl_2$ is 0.617 gNH_3/cm^3, while that of $CaCl_2/NH_4Cl$ is 1.111 gNH_3/cm^3, whose increment is around 80.1%.

4 Conclusion

There are several advantages of the composite multi-halide sorbent applying in the thermochemical de-NO_x cycle, for it can serve as the ammonia supply of the de-NO_x cycle. The following conclusions can be drawn:

1. The thermochemical de-NO_x cycle based on the composite multi-halide sorbent can be driven at around 20 °C, which can adapt better to the cold starts of diesel engines than that based on the composite single-halide sorbent of $CaCl_2$.
2. The thermal efficiency of the de-NO_x cycle with composite multi-halide sorbent of $CaCl_2/NH_4Cl$ is increased by 60.7% from that of $CaCl_2$, which is the key to energy conservation and sustainable development.

3. The available ammonia storage density composite multi-halide sorbent of $CaCl_2/NH_4Cl$ is 1.8 times larger than that of $CaCl_2$, which is the core to achieving small volume with large-scale ammonia storage.

Furthermore, the overall performance of composite multi-halide sorbent is actually due to the existence of NH_4Cl. So if all NH_4Cl and no $CaCl_2$ are used, the performance should be better. In the de-NO_x cycle, however, NH_4Cl cannot be used alone because once the temperature rises, its pressure will be very high, bringing a potentially high risk of safety hazards.

References

1. Leskovjan M, Kočí P, Maunula T (2018) Simulation of diesel exhaust aftertreatment system DOC—pipe—SCR: the effects of Pt loading, PtO x formation and pipe configuration on the deNO x performance. Chem Eng Sci 189:179–190
2. Fayad M, Tsolakis A, Fernández-Rodríguez D, Herreros J, Martos F, Lapuerta M (2017) Manipulating modern diesel engine particulate emission characteristics through butanol fuel blending and fuel injection strategies for efficient diesel oxidation catalysts. Appl Energy 190:490–500
3. Zhang Q, Li M, Li G, Shao S, Li P (2017) Transient emission characteristics of a heavy-duty natural gas engine at stoichiometric operation with EGR and TWC. Energy 132:225–237
4. Jiang Y, Yang J, Cocker D, Karavalakis G, Johnson KC, Durbin TD (2018) Characterizing emission rates of regulated pollutants from model year 2012+ heavy-duty diesel vehicles equipped with DPF and SCR systems. Sci Total Environ 619:765–771
5. Balland J, Parmentier M, Schmitt J (2014) Control of a combined SCR on filter and under-floor SCR system for low emission passenger cars. SAE Int J Energy Inst 91(5):704–720. https://doi.org/10.4271/2014-01-1522
6. Praveena V, Martin MLJ (2017) A review on various after treatment techniques to reduce NOx emissions in a CI engine[J]. J Energy Inst:S1743967117300260
7. Jacques J, Pauly T, Zammit M, Ahari H, et al. (2016) Robust SCR design against environmental impacts. SAE Technical Paper 2016-01-0954, https://doi.org/10.4271/2016-01-0954
8. Liu CY, Aika KI (2004) Effect of the Cl/Br molar ratio of a CaC12−CaBr2 mixture used as an ammonia storage material[J]. Ind Eng Chem Res 43(22):6994–7000
9. Wang ZX et al (2018) Analysis of composite sorbents for ammonia storage to eliminate NOx emission at low temperatures[J]. Appl Therm Eng 128:1382–1390
10. Gao J et al (2017) Solution to the sorption hysteresis by novel compact composite multi-salt sorbents[J]. Appl Therm Eng 111:580–585
11. An, GL. Wang LW, Gao J. Wang RZ (2019) Mechanism of hysteresis for composite multihalide and its superior performance for low grade energy recovery. Scientific Reports, 9(1), 1563

Development of Innovative Technologies for Solar-Assisted Heat Pump for Residential Heat Supply

Ming Jun Huang, Sarah J. McCormack, Jian Yao, Yanjun Dai, and Neil J. Hewitt

1 Introduction

Energy-related carbon dioxide (CO_2) emissions have been seen to rise above 37% in 2018 when compared to 1990 level [1]. Furthermore, as stated, if this trend were to continue, the current rate of CO_2 emissions could increase approximately 21% by 2040 [2]. Approximately 30% of energy within the United Kingdom (UK) is consumed by buildings, where around 55% of this energy is required for heating, ventilation and air conditioning (HVAC) purposes [3]. These estimations are somewhat consistent across countries of the world. Moreover, space heating and hot water requirements make up approximately 40% of the total energy usage within UK buildings and are responsible for about 20% of the total greenhouse gas emissions [4]. Applying integrated Renewable Energys (RES) to buildings is crucial for achieving our binding EU targets of at least a 40% reduction in greenhouse gas emissions (GHGs) by 2030. The integration of highly efficient heating and cooling systems renewable technologies is a key strategy to achieve a low-energy-consuming building. Thus, the development of new sustainable and highly efficient heating technologies for buildings is imperative to meet current and future heating demand throughout the United Kingdom.

Solar energy technologies have been known for centuries although they still face various barriers with regard to their fast development. Solar thermal technologies may reduce conventional fossil fuels demand; however, they are often not able to reach potential due to the intermittent characteristic with ambient conditions. Recent studies have shown that the use of heat pump technologies as opposed to more conventional electrical heater and boiler systems can be used as an appropriate alternative for domestic hot water (DHW) applications as they are proven to be more viable to achieve low-energy consumption and carbon profile [5]. The overall idea of further improving the current conventional heat pump has been also brought forward in recent years in order to enhance system performance. The past number of years have provided contemporary research into the investigation of combining heat pumps with solar collectors in the globe to capture solar energy for water and space heating, also known as 'solar-assisted heat pump' (SAHP) systems [6].

Currently within the United Kingdom, there is 11 GW of installed heat pump capacity which accounts for approximately 38.2% of the nation's building heating market [7], while the residential sector represents a significantly smaller portion of existing heat pump installations. SAHP systems are combination of heat pumps and solar collector technologies which the solar thermal energy can be used to deliver heat directly or as a thermal energy source for a heat pump, while on the other hand, the electricity power produced by the solar PV can be used to drive the heat pump system [8]. According to Muhamed et al. [6], a SAHP system can be divided into two main categories: direct expansion (DX-SAHP) and indirect expansion (IX-SAHP). Their research has analysed the use of various types of SAHP systems, including PV/thermal collectors which are known as hybrid systems which convert solar radiation to electricity and thermal energy [9]. Incorporation of PVT

M. J. Huang (✉) · N. J. Hewitt
Belfast School of Architecture and the Built Environment, Ulster University, Coleraine, UK
e-mail: m.huang@ulster.ac.uk

S. J. McCormack
Trinity College Dublin, Civil, Structural and Environmental Engineering, Dublin, Ireland

J. Yao · Y. Dai
Institute of Refrigeration and Cryogenics, Shanghai Jiao Tong University, Shanghai, China

© The Author(s), under exclusive license to Springer Nature Switzerland AG 2022

A. Sayigh (ed.), *Sustainable Energy Development and Innovation*, Innovative Renewable Energy,
https://doi.org/10.1007/978-3-030-76221-6_13

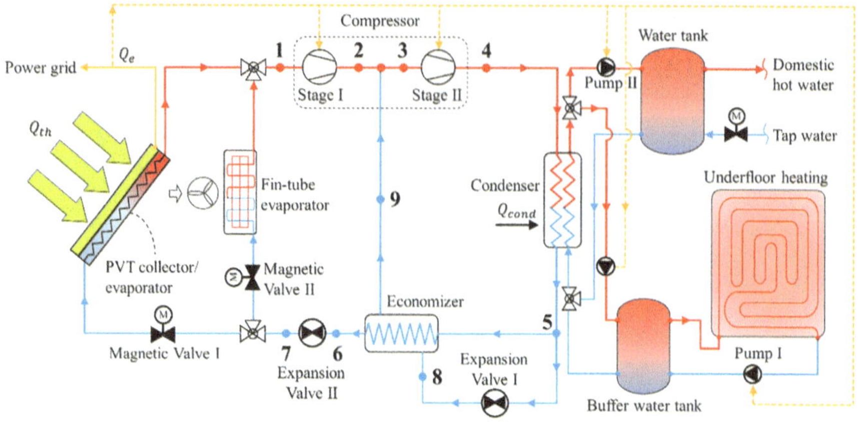

Fig. 1 Schematic diagram of the vapour-injection heat pump system employing PVT collector/evaporator for residential heating. (Adopted from [10])

modules into an air source heat pump (ASHP) could produce thermal energy and electricity simultaneously (see Fig. 1 [10]), and the PV cells could obtain higher electrical efficiency due to the lower operating temperature [11, 12].

The SAHP technologies have played a critical role in resolving potential issues/barriers that might occur with the application of geolocation, seasonal variation, reliability, material choice, operational strategies, storage, etc. It is important to consider the current and future trends in the subject area, including recent advances with the technology based on overall performance and the integration of thermal storage with enhanced heat transfer technologies into the SAHP system.

Improving low-energy application with heavy thermal mass and utilising renewable energy in buildings in the United Kingdom are vital in order to create a sustainable and dependable energy market as well as cutting CO_2 emissions. The flexibility in choosing heat sources, reduction of fuel consumption, increased indoor environmental quality and enhanced community energy management will also reduce costs for end users [13].

2 Heat Pump Development for SAHP

In nature, heat pump as high-energy efficiency systems is operated under vapour compression cycle which transfer energy from a low-temperature source like air (ASHP) or ground (GSHP) to a higher temperature using electricity as the input.

2.1 Solar-Assisted Air Source Heat Pump (SAASHP) Systems

Two major problems of the ASHP are the efficiency reduced in low ambient temperature when there is frost formed on the evaporators and the high compressor discharge temperatures [13]. The adoption of economised vapour injection (EVI) compressor could assist the ASHP to operate normally under low-temperature conditions [14, 15]. A SAASHP is a two-stage compressor with a variable inlet of solar energy captured for cold climates [16]. At an outside temperate of −25 °C, the SAASHP appeared to operate efficiently which was evidenced by a 55% increase in coefficient of performance (COP). Heat transfer enhancement was also analysed within this time period by Kong et al. [17] through the utilisation of new innovative components including an electronic expansion valve and micro-channel condenser. Similar values were obtained when compared to Qiu et al. [16], as it was found that a 55% COP enhancement was possible through a variation in ambient temperature from −2.9 °C to 12.8 °C, although it was later revealed that the micro-channel evaporator technology was not so efficient when compared to solar collector with respect to solar radiation intensity variation.

Wang et al. [18] developed the vapour injection technique based on flash tank with two-stage ASHP operation under cold climate and presented a novel control strategy to seek the optimised intermediate pressure for better performance. Additionally, Ko et al. [19] investigated the influence of volume ratio of a two-stage rotary compressor on ASHP experimentally, and it is found that the ASHP with VI cycle improves the heating power of 48% and the COP of 36% in the comparison with the ASHP at the ambient temperature of −15 °C. However, the low evaporating temperature in the fin-tube evaporator of the

ASHP system with VI two-stage cycle still limits the efficient operation of the HP system. Therefore, it is necessary to increase the evaporating operation temperature to achieve a higher system efficiency. On the other hand, solar thermal could be a widely available low-temperature heat source for heat pumps due to its characteristics [20, 21]. Solar-assisted heat pump (SAHP) has great potential in building energy saving [22–25]. The direct expansion solar-assisted heat pump (DX-SAHP) is one type of the SAHP, and it becomes state-of-the-art research due to its stability in heat production. Yao et al. [10] developed a detailed mathematical model of the vapour-injection heat pump system by incorporating PVT collector/evaporator and verified the effectiveness of the ASHP system with VI compressor. The influences of environmental conditions, mass flow rate ratio of injected vapour on system performance, performance comparison and economic performance have been then evaluated. The system can operate off-grid, and its COP can reach 4.0 which is 63.6% higher than the ASHP system with VI cycle at the ambient temperature of −10 °C and the solar irradiation of 500 W/m^2. The practical electrical efficiency of PV panels achieves at 15.1%, whereas the thermal efficiency of the system is 44.8%, and it is found that the evaporating and condensing temperatures are −6.1 °C and 51.5 °C, respectively.

2.2 Solar-Ground Source Heat Pump (SAGSHP)

The introduction of a solar-ground source heat pump (SAGSHP) technology coupled with a dual-coil ground source heat exchanger was developed about around 2004, which allowed for a much higher heat absorption rate per unit of surface area compared to that of a more conventional single-pipe vertical heat exchanger, and it resulted in a COP value improvement of approximately 21% [26]. Problems with high electricity demand in SAGSHP systems were identified in studies carried out by Ozgener and Hepbasli [27] and Kjellsson et al. [28], and they indicated that there was a lack of direct solar heating due to possibilities of solar collectors feeding the heat pump evaporator or potentially charging the ground.

Borehole heat exchanger (BHE) is a promising method for extracting heat from the deep geothermal energy which has been widely used for residential heating in high latitude areas. Yao et al. [29] have studied the solar-assisted photovoltaic/thermal (PVT) heat pump for buildings. The research results show that a larger mass flow rate could increase the BHE's heat extract capacity but also increase the flow resistance and pump power under nominal conditions. The maximum water temperature from this hybrid system could reach 40.8 °C while the solar fraction is 67.5% when the area of PVT module is 1000 m^2, solar irradiation is 600 W/m^2 and depth of the BHE is 2500 m. In the meantime, the heating COP of this hybrid system could reach 7.4, and the system could operate independently without power input from electrical grid.

3 Photovoltaic/Thermal (PVT) with Heat Pump

3.1 Development of PVT with HP

One of the most innovative solar technologies in recent years involves the enhancement of SAHP with photovoltaic and thermal (PVT) which dedicate many beneficial such as a reduction in space requirements and lower costs associated with the technology in terms of installation and operation. In an experimental numerical modelling study on PV-enhanced SAHP for hot water generation in the United Kingdom, it was concluded that the ambient temperature and low solar radiation values associated with the UK weather reduced the thermal efficiency although improved the electrical efficiency. An overall COP of 8.7 was concluded to which efficiency values of 50%, 40% and 10% were obtained [30]. PV/micro-channel enhanced DX-SAHPs were then introduced in 2016, which the studies found that parallel micro-channel evaporators could obtain higher power and heat generation as a result of higher refrigeration evaporation. This was the case with Zhou et al. [31] of which the experimental results show a COP of 4.7 and thermal efficiencies ranging between 56.6% and 69.7%.

3.2 SAHP Enhanced with PCM Integrated for Underfloor System and PV Regulation

Contemporary research has shown an increased interest into the possible integration of SAHP systems with storage capacities such as the use of phase change materials (PCMs). PCMs such as paraffin can be utilised in order to store excess solar energy during daylight hours while facilitating heat pump operation within non-daylight hours [32]. The solar energy incidents on the PV panel can be converted to electricity and the others will convert to thermal energy. The PV electricity

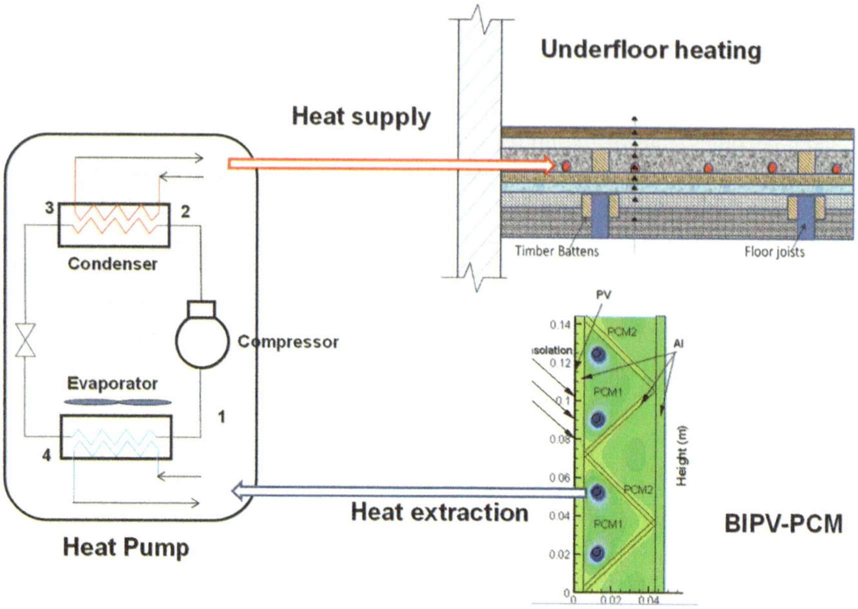

Fig. 2 A schematic diagram of using HP for supplying heat for PCM-layered underfloor heating system and extracting solar heat from BIPV-PCM through the cycling cupper pipes [13]

conversion efficiency will reduce when the temperature in PV increased. The cooling effect by the evapoerator of the ASHP can effective to remove the solar heat from the PV with thermal mass of PCM to enhance the thermal regulation capacity to keep the efficiency of the PV. The heat produced by the ASHP can be used to meet the heat demanding by the building ([13], see Fig. 2). Further study shows a 20-m^2 PVT panel module can output 21.4% of the electricity to power grid when the solar radiation intensity is 600 W/m^2 and meet the heat demand of a 100-m^2 room with PCM underfloor heating while maintaining the operation of the system. Meanwhile, the heating COP can reach 5.79 which is 70% higher than the conventional air conditioning system, and the electrical, thermal, overall efficiencies are 17.77%, 55.76% and 75.49%, respectively [33]. In addition, the investigation of micro-channel heat pipes and PCM storage has shown to be an emerging technology within the fourth quarter of the twenty-first century. Youssef et al. [34] have studied using a latent heat storage system in conjunction with a SAHP unit for DHW purposes. In their study, a 6–14% increase in COP value was obtainable under various operational modes. The effectiveness of storage capacity and ground heat exchangers were analysed in another study by Sakellariou et al. [35] through a modelling exercise of a SAGSHP validated along with a PVT system for DHW and space heating needs for a single-family residential home in the United Kingdom. The study found that the storage capacity and heat exchanger proved to be the most influential components with regard to the performance of the system.

A numerical simulation for a PV/T-enhanced SAHP coupled with PCM storage was carried out. The use of a PCM storage unit could facilitate hot water peak demand during morning periods while reducing the overall heating energy consumption by around 12.1% and 13.5% [3]. The study was based in the United Kingdom for a typical 24-hour residential demand profile in terms of DHW and space heating needs. Moreover, the study was conducted on using ASHP for the residential building with PCM underfloor heating [15, 36]. The thermal mass effect of the PCM with the HP on-off cycle operation has been studied along with the variable ambient influence.

4 Barriers to SAHP Integration

Moreover, it is known that the intensive electrification of the heating sector such as the introduction of heat pump systems may impact the electrical distribution grid. The widespread deployment of heat pump systems in the United Kingdom will overall increase peak demand by a substantial amount, which may require additional infrastructure investments in order to satisfy demand [37]. According to Love et al. [38], the peak electricity demand in the United Kingdom may increase as much as 14% as a result of the mass integration of heat pump technologies into the electricity network.

Further analysis is required in terms of the relationship between peak electricity demand and the penetration level of heat pump systems as the issue of voltage stability and overloading still remains. Thus, extensive research has been carried in order to discover new and innovative ways to reduce electricity demand and additional stress which may persist on the UK electricity grid as a result of heat pump integrations [39, 40]. However, the integration of solar technologies coupled with thermal energy storage systems may further enhance the viability of heat pump systems by alleviating the stress caused on the electricity grid through peak shaving and overall balancing demand. Therefore, artificial intelligence (AI) technology needs to be adopted for the comprehensive system operation based on the extensive field trial testing.

Other factors which may have a significant influence on SAHP performance include poorly insulated or excessively long pipework causing heat loss, electricity utilised by immersion element in order to top-up the production of DHW, and auxiliary heating in order to facilitate the heat pump system output in the event of low ambient temperatures.

5 Conclusions

The newest development of integrating renewable energy to buildings with innovative technologies has been discussed. Solar-assisted heat pump can significantly utilise the renewable energy for building energy supply. The PCM with heavy thermal mass can efficiently maintain the PV efficiency and reduce the energy consumption in the buildings.

Acknowledgements This research work is funded by the EU H2020 (CEC-H2020-LC-SC3-RES-4-2018; IDEAS, GA No 815271), EPSRC Lot-NET (EP/R045496/1) and Comfort Climate Box (EP/V011340/1).

References

1. IEA (2019) World energy outlook 2019. Available at: https://www.iea.org/reports/world-energy-outlook-2019. Accessed 7 Aug 2020
2. Badiei A, Golizadeh Akhlaghi Y, Zhao X, Shittu S, Xiao X, Li J, Fan Y, Li G (2020) A chronological review of advances in solar assisted heat pump technology in 21st century. Renew Sust Energ Rev 132(1):110–132
3. Kutlu C, Zhang Y, Elmer T, Su Y, Riffat S (2020) A simulation study on performance improvements of solar assisted heat pump hot water system by novel controllable crystallization of supercooled PCMs. Renew Energy 152(1):601–612
4. CCC (2016) Next steps for UK heat policy. The Committee on Climate Change, London
5. Bellos E, Tzivanidis C, Moschos K, Antonopoulos KA (2016) Energetic and financial evaluation of solar assisted heat pump heat pump space heating systems. Energy Conversion Management 120:306–319
6. Muhamed E, Riffat S, Omer S, Zeinelabdein R (2019) A comprehensive investigation of using mutual air and water heating in multi-functional DX-SAMHP for moderate cold climate. Renew Energy 130:582–600
7. EHPA (2014) European heat pump market and statistics report. Association, European Heat Pump
8. Poppi S (2017) Solar heat pump systems for heating applications – analysis of system performance and possible solutions for improving system performance system. KTH Industrial Engineering and Management, Stockholm
9. Fu Y (2014) Investigation of solar assisted heat pump system integrated with high-rise residential buildings. The University of Nottingham, London
10. Yao J, Zheng S, Chen D, Dai Y, Huang MJ (2020c) Performance improvement of vapor-injection heat pump system by employing PVT collector/evaporator for residential heating in cold climate region, submitted to the J. Energy
11. James A, Mohanraj M, Srinivas M, Jayaraj S (2021) Thermal analysis of heat pump systems using photovoltaic-thermal collectors: a review. J Therm Anal Calorim 144:1–39
12. Mohanraj M, Gunasekar N, Velmurugan V (2015) Comparison of energy performance of heat pumps using a photovoltaic–thermal evaporator with circular and triangular tube configurations. Build Simul 9(1):27–41
13. Huang MJ, Hewitt NJ (2018) Enhancing energy utilisation in building with combining building integarted PV and air source heat pump for underfloor heating using phase change materials, WREC2018, UK
14. Huang MJ, Hewitt NJ (2013) The experimental analysis of the effect of ambient factors on the defrosting of economised vapour injection compressor air source heat pump in marine climates. Int J Refrig 36(3):820–827
15. Huang MJ, Shah N, Wilson C, Hewitt NJ (2019) Analysis of the performance of developed EVI Air Source Heat Pump with Heat Demand During Seasonal Operations. ICR2019 : The 25th IIR International Congress of Refrigeration, Montreal, Canada, 24–30 Aug 2019. https://doi.org/10.18462/iir.icr.2019.1133
16. Qui G, Wei X, Xu Z, Cai W (2018) A novel integrated heating system of solar energy and air source heat pumps and its optimal working condition range in cold regions. Energy Conversion Management 174:922–931
17. Kong X, Yang Y, Zhang M, Li Y, Li J (2020) Experimental investigation on a direct-expansion solar-assisted heat pump water heater using R290 with micro-channel heat transfer technology during the winter period. Int J Refrig 113:38–48
18. Wang W, Li Y (2019) Intermediate pressure optimization for two-stage air-source heat pump with flash tank cycle vapor injection via extremum seeking. Appl Energy 238:612–626
19. Ko Y, Park S, Jin S, Kim B, Jeong JH (2013) The selection of volume ratio of two-stage rotary compressor and its effects on air-to-water heat pump with flash tank cycle. Appl Energy 104:187–196

20. Kuik O, Branger F, Quirion P (2019) Competitive advantage in the renewable energy industry: evidence from a gravity model. Renew Energy 131:472–481
21. Kumar R, Rosen MA (2011) A critical review of photovoltaic–thermal solar collectors for air heating. Appl Energy 88(11):3603–3614
22. Mohanraj M, Belyayev Y, Jayaraj S, Kaltayev A (2018) Research and developments on solar assisted compression heat pump systems – a comprehensive review (Part A: modeling and modifications). Renew Sust Energ Rev 83:90–123
23. Mohanraj M, Belyayev Y, Jayaraj S, Kaltayev A (2018) Research and developments on solar assisted compression heat pump systems – a comprehensive review (Part-B: applications). Renew Sust Energ Rev 83:124–155
24. Vaishak S, Bhale PV (2019) Photovoltaic/thermal-solar assisted heat pump system: current status and future prospects. Sol Energy 189:268–284
25. Esen H, Esen M, Ozsolak O (2015) Modelling and experimental performance analysis of solar-assisted ground source heat pump system. J Exp Theor Artif Intell 29(1):1–17
26. Bi Y, Guo T, Zhang L, Chen L (2004) Solar and ground source heat-pump system. Appl Energy 78(2):231–245
27. Ozgener O, Hepbasli A (2005) Performance analysis of a solar assisted ground source heat pump system for greenhouse heating: an experimental study. Build Environ 40(1):1040–1050
28. Kjellsson E, Hellstrom G, Perers B (2010) Optimization of systems with the combination of ground-source heat pump and solar collectors in dwellings. Energy 35:2667–2673
29. Yao J, Liu WJ, Zhang L, Tian B, Dai Y, Huang M (2020b) Performance analysis of a residential heating system using borehole heat exchanger coupled with solar assisted PV/T heat pump. Renew Energy 160:160–175. https://doi.org/10.1016/j.renene.2020.06.101
30. Zhang X, Zhao X, Xu J, Yu X (2013) Characteristics of a solar photovoltaic/loop-heat-pipe heat pump water heating system. Appl Energy 102:1229–1245
31. Zhou J et al (2016) Experimental investigation of a solar driven direct-expansion heat pump system employing the novel PV/micro-channels-evaporator modules. Appl Energy 178:484–495
32. Akbari H, Browne M, Ortega A, Huang MJ, Hewitt NJ, Norton B, McCormack S (2019) Efficient energy storage technologies for photovoltaic systems. Sol Energy 192:144–168
33. Yao J, Chen E, Dai Y, Huang M (2020) Theoretical analysis on efficiency factor of direct expansion PVT module for heat pump application. Sol Energy 206:677–694
34. Youssef W, Ge YT, Tassou SA (2017) Effects of latent heat storage and controls on stability and performance of a solar assisted heat pump system for domestic hot water production. Sol Energy 150:394–407
35. Sakellariou EI, Wright AJ, Axaopoulos P, Oyinlola MA (2019) PVT based solar assisted ground source heat pump system: modeling approach and sensitivity analysis. Sol Energy 193:37–50
36. Huang MJ, Hewitt NJ (2015) The energy conservation potential of using phase change materials as thermal mass material for air source heat pump driven underfloor heating system in a building, Book chapter, Progress in clean energy, volume 2: novel systems and applications, ISBN 978-3-319-17030-5, pp. 209–227
37. Fawcett, T., Eyre, N. and Layberry, R. (2015) Heat pumps and global residential heating., ECEEE summer study proceedings
38. Love J, Smith AZ, Watson S, Oikonomou E, Summerfield A, Gleeson C (2017) The addition of heat pump electricity load profiles to GB electricity demand: evidence from a heat pump field trial. Appl Energy 204:332–342
39. Le KX, Huang MJ, Wilson C, Shah N, Hewit NJ (2020) Tariff based load shifting for domestic cascade heat pump with enhanced system energy efficiency and reduced wind power curtailment. Appl Energy 257:113976
40. Le KX, Huang MJ, Shah N, Wilson C, Artain PM, Byrne R, Hewitt NJ (2019) Techno-economic assessment of cascade air-to-water heat pump retrofitted into residential buildings using experimentally validated simulations. Appl Energy 250:633–652

Bespoke Method for Installation of VIPs on Building Walls and Their Performance

Jose Tavares, Harjit Singh, and Maria Kolokotroni

1 Introduction

Vacuum insulated panels (VIPs) offer an excellent thermal resistance, up to five to eight times higher than traditional insulation materials. If performing without any loss of vacuum, a 30-mm thick VIP with a U value of 0.2 W/(m^2K) is comparable up to a 300-mm thick commonly used insulation materials, such as glass wool. VIPs help to create a homogeneous temperature distribution through building elements, reduce transmission of water vapour preventing condensation problems and meet minimum fire and sound regulations. The use of super insulation materials such as VIPs is one of the most promising solutions for the construction sector. Even including thermal loss due to the thermal bridge effect, they still perform five times better than typical insulation materials [1, 2] while keeping the total thickness of the insulation barrier to a minimum which is desirable for retrofit applications. Despite being a very efficient technology, only a small amount, 10%, is used in the building industry, and it is well known that the energy use of the EU's building stock accounts for about 40% of the overall energy use of which households make up about 25%. Additionally, the volume of new buildings per year is currently very low in many developed countries, a trend which seems it will continue in the future [3]. Therefore, retrofitting existing buildings will play an important role in order to meet the targets for decarbonisation of buildings as required by EU energy strategy for 2020, 2030 and 2050. Increasing the use of VIPs in the building industry could be an important solution to improve significantly the energy performance of buildings and fulfil the objective of delivering zero-energy buildings. However, there are a number of techno-economic challenges which remain in the way of their widespread adoption: the vulnerability of the VIPs at the stage of installation [4], the risk of surface condensation, the thermal bridging effect which can affect the panels performance in the long term [5, 6], predicting and ensuring service-life performance and the high initial cost compared to common insulating materials on average five times higher. Nonetheless, these have been shown to have a short payback period of 2.5 years in London when the cost of space savings is accounted for [7, 8]. Other barriers facing VIPs include the lack of long-term performance data from real case studies as well as the lack of accredited installers. This chapter outlines the construction details and challenges regarding the implementation of VIPs, their thermophysical performance and the instrumentation and monitoring regime used for long-term analysis of the thermal performance of a VIP-insulated wall system at Brunel University London.

2 Retrofit Approach and Case-Study Building

The installation was carried as part of the ReCO2ST project. The ReCO2ST Consortium consists of 17 partners from nine countries, 12 industrial partners and four non-profit research organisations with ongoing projects at four residential buildings: Brunel University London, Vevey – Switzerland, Cadiz – Spain, and Frederikshavn – Denmark [9]. In the

J. Tavares (✉) · M. Kolokotroni
Institute of Energy Futures, Brunel University London, Uxbridge, UK
e-mail: Jose.tavares@brunel.ac.uk

H. Singh
Institute of Energy Futures, College of Engineering, Design and Physical Sciences, Brunel University London, Uxbridge, UK

© The Author(s), under exclusive license to Springer Nature Switzerland AG 2022

A. Sayigh (ed.), *Sustainable Energy Development and Innovation*, Innovative Renewable Energy,
https://doi.org/10.1007/978-3-030-76221-6_14

Fig. 1 Façade student accommodation building

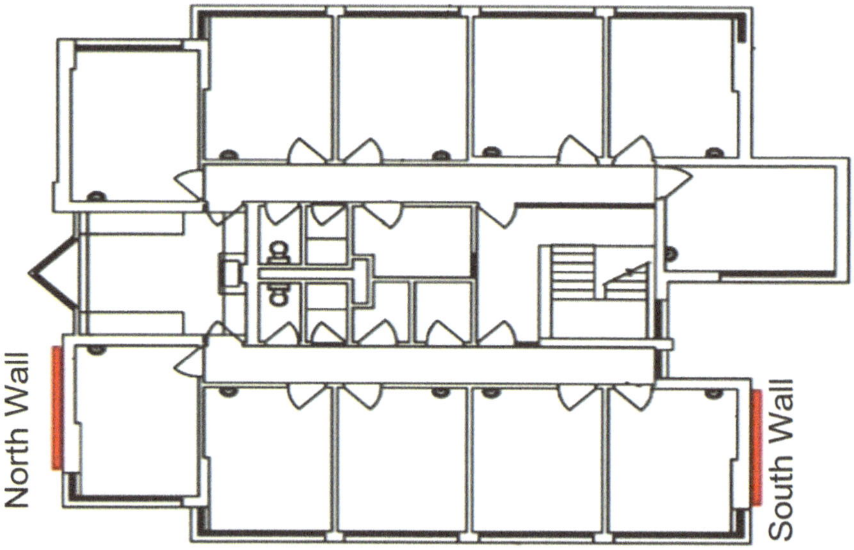

Fig. 2 Plan of the building – ground floor

demonstration buildings, the project aims to provide experimentally validated test results and case studies under real climatic conditions with the aim to increase the confidence of the building sector and homeowners in VIPs.

Apart from VIPs, the ReCO2ST project provides a range of low CO_2 technological and bespoke solutions to different retrofit scenarios: cool materials, PV arrays, smart windows, nature-based solutions, cooling evaporative ventilated façade and ventilation system with phase change materials.

This chapter reports on VIPs which were applied on the ground floor of a student accommodation building (Figs. 1 and 2) onto two walls. The north wall has limited exposure to direct solar radiation throughout the year, whereas the south wall has full exposure except early morning when is partially shaded by the adjacent building.

3 VIP's Properties, Assembling Method and Installation Challenges

The installed VIPs were supplied by va-Q-tec AG Germany. The panels, flat in shape of size 1000 × 600 × 30 mm, consist of a pressed powder board of fumed silica, non-flammable (class A1) and sealed under vacuum with a gas-vapour-tight film, covered on both sides by a 10-mm layer of EPS (Fig. 3a and b).

The aluminium-laminated envelop provides a good barrier for penetrating gases and water vapours. The EPS cover on both sides of the panel not only protects from haling damage but also protects it from surface condensation.

With a thermal conductivity value of 0.0070 W/(mK) at 20 mm thickness, VIPs are predicted to have a service life of up to 60 years by their manufacturer, va-Q-tec. In this project, nine VIP panels were installed on each wall supported by a wooden frame and covered by a water-proof cement board for external use. The bespoke installation procedure adopted, shown in Figs. 4 and 5, was developed by the project team after studying conventional methods and deliberations with the manufacturer.

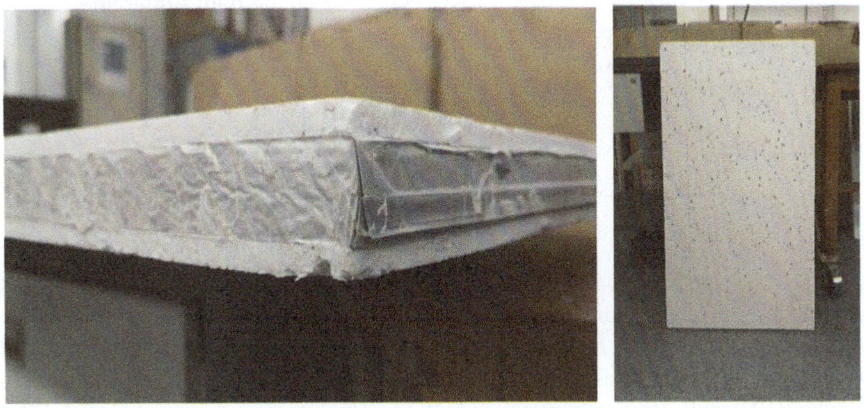

Fig. 3 (**a**) and (**b**) VIP panel with EPS cover on both sides

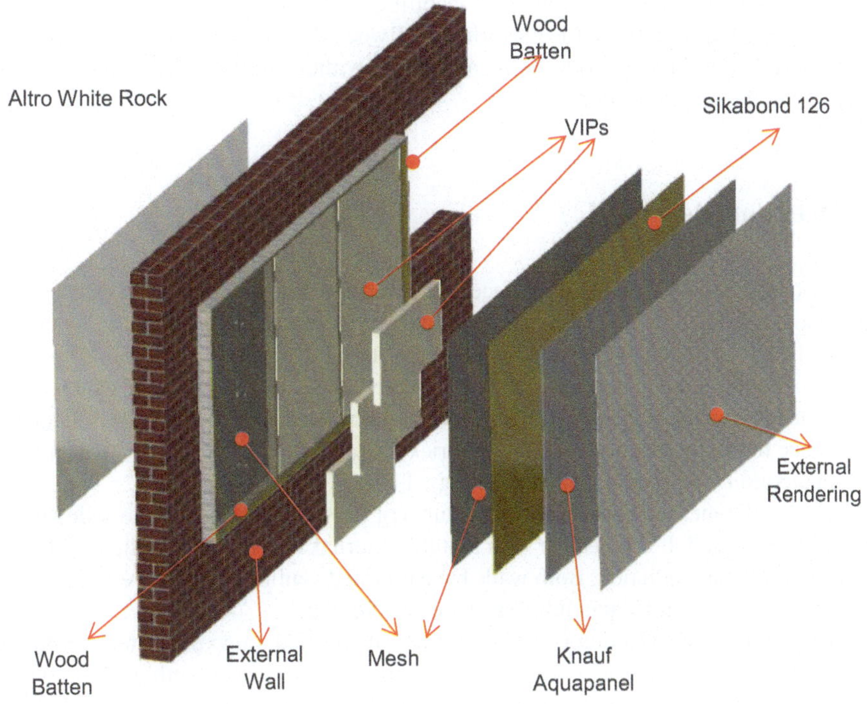

Fig. 4 VIP assembling method

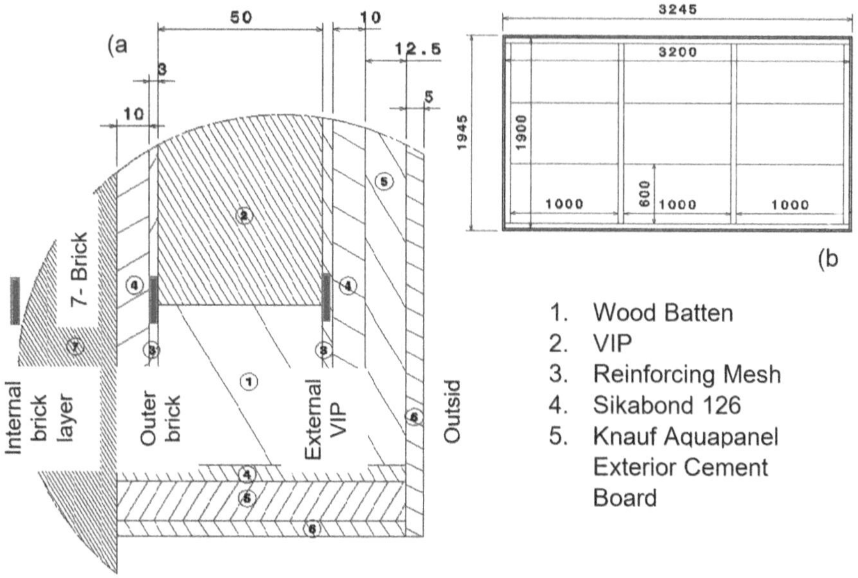

Fig. 5 (**a**) Side view – assembly components. (**b**) Assembly dimensions in mm

The foremost challenge for installation was the lack of accredited installers of VIPs in the United Kingdom. As a result, the project team developed a set of installation instructions which were implemented by a team of general builders who were constantly supervised by the project team. It has been ensured that the installation complies with UK's building regulations.

Some adaptations were made during the assembly, namely the introduction of self-adhesive neoprene strips 50 mm wide to protect the sides of the VIP panels against the wooden battens, which were not protected with EPS as shown in Fig. 3a. This simplified the handling process and, consequently, there was no damage to any of the VIP panels.

The second adaptation refers to the Sikabond 126 glue. The initial plan was to apply an overall layer of glue on top of all area of the VIP panels. Instead, only a smaller circular area of glue was applied at the centre and both sides of each panel, which was found to be sufficient to hold the panels in place. This reduced the overall weight of the system and reduced the working time and costs without affecting the insulation capabilities of the system.

Additionally, no extra steps were applied to the gap bridges between the panels, aiming to assess the efficiency of the VIPs as a product ready for installation directly from the manufacturer without further adaptations. The resultant outcome was very positive as described in the results section.

4 Operational Data and Results

Thermocouples were installed at (i) internal wall surface, (ii) external wall surface and (iii) the outmost surface of VIPs to monitor their performance, as shown in Fig. 5a. On the external surfaces, a total of 16 thermocouples (Fig. 6a and c) were employed at the centre of each VIP. On the internal layout, eight thermocouples were positioned parallel to the external layout (Fig. 6g). The thermocouples were connected to 12 data loggers of 4 channels each of which have been recording temperatures at 10-minute intervals since August 2019, and for the sake of economy of space, only January data are shown here. The temperature difference in the two sections (VIPs and brick) is presented in Figs. 9 and 10, and in more detail, the temperature difference on individual VIP panels is shown in Figs. 11 and 12.

Figures 7 and 8 show that the measured surface temperatures of the different layers have followed a similar pattern to the ambient air temperature as expected. It can also be seen that the external VIP surface resulted in slightly higher temperature on the south wall as a result of solar radiation. Both walls have returned similar temperature profiles because the rooms were kept at similar temperatures and the ambient conditions were also similar.

Figures 9 and 10 show that the introduction of VIPs on the walls created a maximum temperature difference of 14.2 °C between the external surface of the north wall and the ambient and 15.15 °C on the south wall implying a substantial amount of equivalent heating energy saving. Without the VIPs, this energy would be wasted in raising the temperature of the outer brick layer hence reducing thermal comfort for the building occupants, or additional energy would have been used to keep similar levels of thermal comfort.

Fig. 6 (**a**–**h**) Construction. (**a**) Frame and inner mesh with thermocouples. (**b**) VIPs with EPS covering. (**c**) Outer mesh with thermocouples. (**d**) Aquapanel cement board for exterior use. (**e**) External rendering. (**f**) Final external outcome after painting. (**g**) Layout of thermocouples on internal wall surface. (**h**) Final interior outcome with Altro White Rock

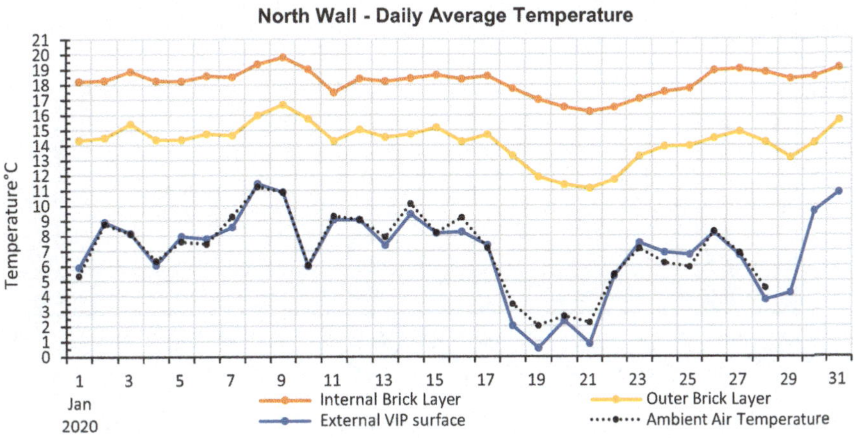

Fig. 7 Daily average temperature at three locations on the VIP north wall

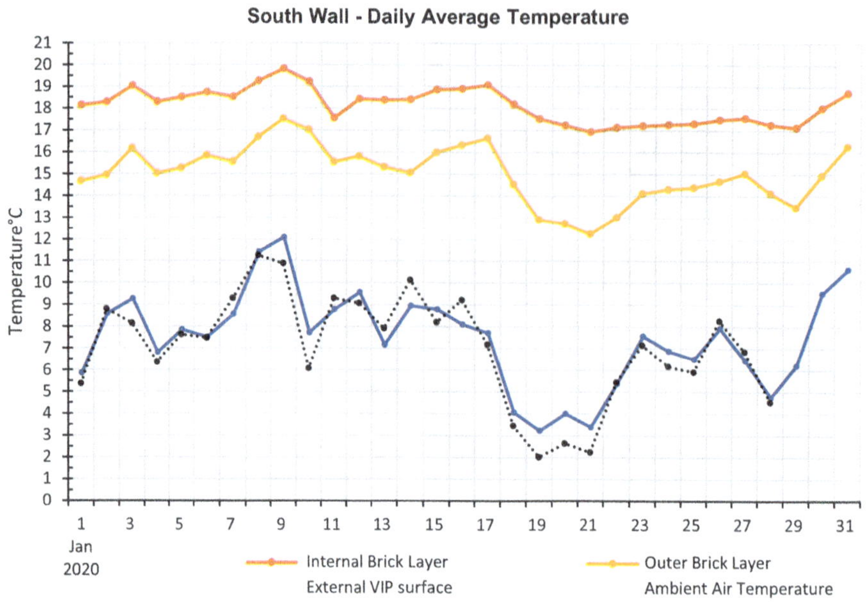

Fig. 8 Daily average temperature at three locations on the VIP south wall

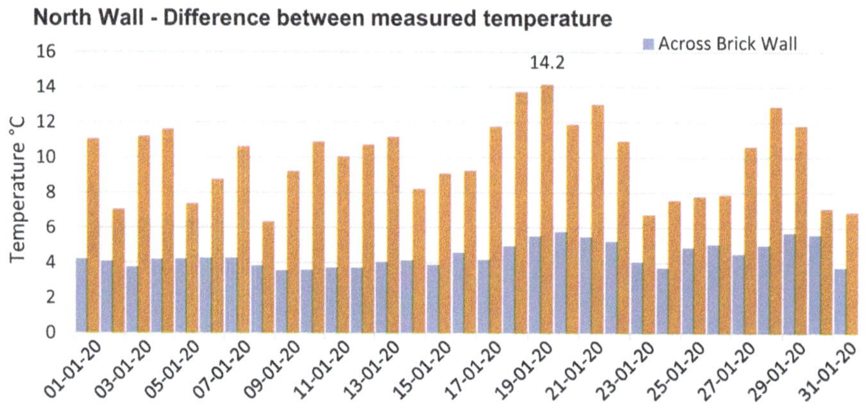

Fig. 9 North wall – Daily average temperature difference measured across double brick wall (blue)

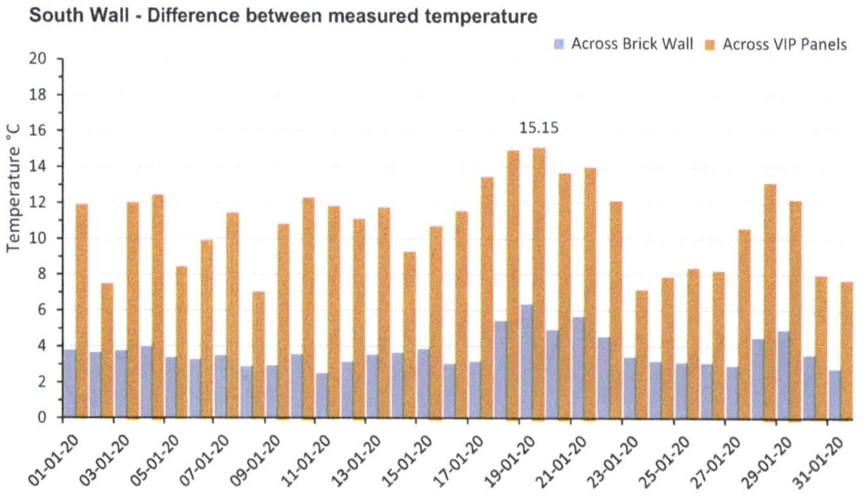

Fig. 10 South wall – Daily average temperature difference measured across double brick wall (blue) and across VIP centre panel (orange)

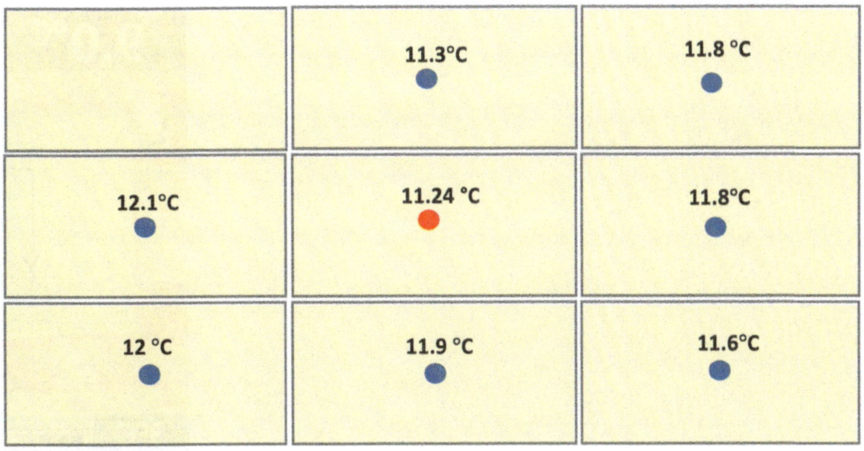

Fig. 11 North wall – January average temperature difference between internal wall surface and external VIP surface

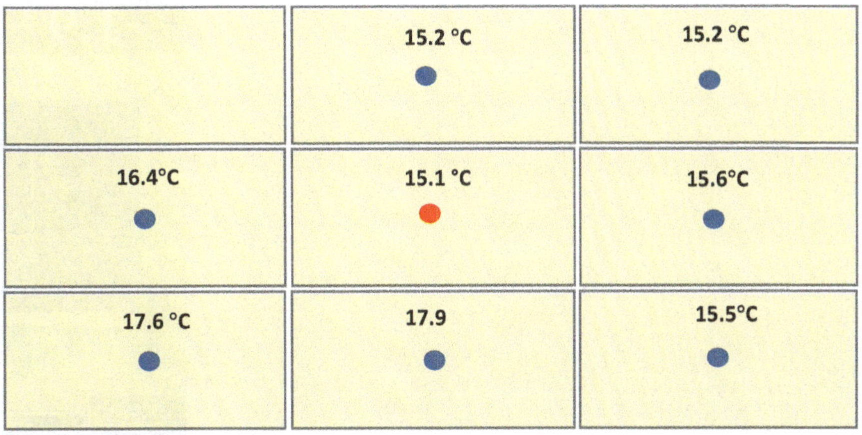

Fig. 12 South wall – January average temperature on external wall surface

Figure 11 shows that the VIPs performed consistently across the whole installation providing a stable temperature distribution to the wall. It is evident by the recorded temperatures that the installation method developed has significantly reduced thermal bridging effect between any two adjacent VIPs. A slight increase in temperature can be observed on the panels at the edge of the installation which reflects the effect of heat lost as a consequence of the uninsulated sections of the wall relative to the central panel which is surrounded by other VIPs. Similar temperature profiles were measured for the south wall, as shown in Fig. 12.

Figure 12 shows the average temperature for the external wall surface on the south wall installation as opposed to the temperature difference across all installation shown in Fig. 11. It can be seen that the temperatures recorded follow a similar pattern of higher temperatures on the VIPs at the edges of the installation relative to the central VIP. This can be as a result of the uninsulated areas of the wall; nonetheless, it is worth noticing that the south wall is partially exposed to direct sunlight for most of the day which implies the effect of solar radiation might account for the higher difference in temperature from the centre VIP as shown on Fig. 14. The January average temperature was 18.1 °C and 7.1 °C on the inside wall surface and on the external VIP surface respectively. With an average temperature of 15.1 °C at the central panel, the VIP installation proved to be an efficient barrier against heat loss.

Fig. 13 shows that the chosen installation method helped to reduce the effect of heat bridges which has been a major concern for building scientists and developers because these can lead to a complete loss of value that could be gained from VIP installations. Figure 14 shows the effect of solar radiation on the partially shaded installation on the south wall.

Fig. 14 Thermal image – south wall

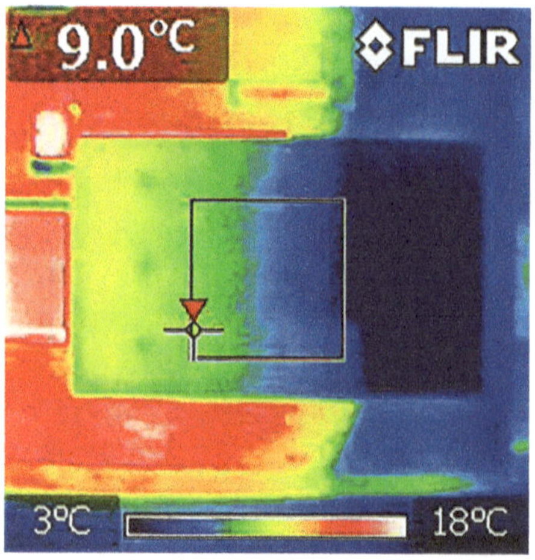

Fig. 13 Thermal image – north wall

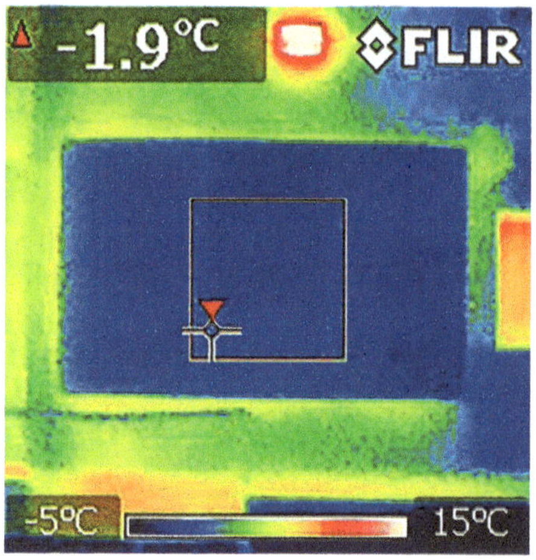

5 Conclusions

The installation method developed for the ReCO2ST project has significantly reduced thermal bridging effect between any two adjacent VIPs, hence reducing heat loss and thus improving the installation performance in the long term. Many installation challenges associated with the handling of the VIPs at the installation phase such as punctures were overcome through appropriate construction design methods. The choice of VIPs with EPS sheet covering provided by the ReCO2ST partner va-Q-tec also reduced the vulnerability of the VIPs while reducing the risk of surface condensation.

VIPs have been found to reduce heat transfer from the ambient with a fraction of the thickness compared to typically used insulation materials, leading to a significant reduction in the amount of heat energy required to maintain indoor thermal comfort conditions. Work continues to quantify the potential energy savings for this demonstration building.

Acknowledgement This project has received funding from the European Union's Horizon 2020 Research and Innovation program under Grant Agreement No 768576.

References

1. Brunner S, Simmler H (2008) In situ performance assessment of vacuum insulation panels in a flat roof construction. Vacuum 82:700–707
2. Ghazi-Wakili K, Bundi R, Binder B (2004) Effective thermal conductivity of vacuum insulation panels. Build Res Inf 32(4):293–299
3. European Commission (2010) EU energy and transport in figures – statistical pocketbook 2010. European Union, Luxemburg, ISBN 978-92-79-13815-7

4. Karami P (2015) Robust and durable vacuum insulation technology for buildings doctoral thesis in building technology. KTH University, Stockholm
5. Lorenzati A, Fantucci S, Capozzoli A, Perino M (2016) Experimental and numerical investigation of thermal bridging effects of jointed Vacuum Insulation Panels. Energ Buildings 111:164–175. https://doi.org/10.1016/j.enbuild.2015.11.026
6. Isaia F, Fantucci S, Capozzoli A, Perino M (2016) Thermal bridges in vacuum insulation panels at building scale. Eng Sustain:1–14. https://doi.org/10.1680/jensu.15.00057
7. Alam M, Singh H, Suresh S, Redpath DAG (2017) Economic analysis of Vacuum Insulation Panels (VIPs) used in non-domestic buildings. Appl Energy 188:1–8
8. Shukla N, Fallahi A, Kosny J (2013) VIP – technology review and cost study for building retrofits in Northern US locations. In: Thermal Performance of the Exterior Envelopes of Whole Buildings XII International Conference, ASHRAE, 2013, pp. 1–9
9. ReCO2ST. https://reco2st.eu/

The Value of CSP with Thermal Energy Storage in Providing Flexible Electric Power

Giancarlo Gentile and Giampaolo Manzolini

1 Introduction

Concentrated solar power (CSP) could play a major role in near future energy scenario, providing carbon-free, renewable, and dispatchable electric power to meet the increasing world energy demand. Even though the Levelized Cost Of Electricity of this technology is still non-competitive, several research programs are addressing its development and evolution [1]. CSP plants are typically coupled with thermal energy storage (TES) systems to store thermal power during sunlight hours and use it during the night.

This chapter aims to assess how TES makes CSP able to flexibly feed electric power into the grid, following the load profile to avoid curtailment and maximize the energy sale. TES is typically made of two tanks in which hot and cold heat transfer fluid (HTF) are stored and operates with thermal-to-thermal efficiencies above 97% [2]. This system is capable of storing large quantities of energy with a good energy density and specific energy (80–500 Wh/L, 80–250 Wh/kg), and it is economically viable with relatively low capital cost (20–30 $/kWh) [3, 4]. By means of an efficient communication with the grid, CSP plants with TES can follow the load profile reducing curtailment and operating at full load during peak demand hours. As already pointed out by J. Forrester [5], curtailment of other renewable sources such as photovoltaic energy could increase in the near future due to higher renewable penetrations, and there is the potential for CSP with TES to reduce overall renewable energy curtailment.

In the chapter, two algorithms are implemented in a Matlab suite to assess the capabilities of a CSP plant with TES to follow a residual load (RL) profile and to evaluate the economic profitability of a load-oriented control strategy. The RL is the difference between actual power demand and feed-in of non-dispatchable and inflexible generators. Several authors used the RL to address the effect of fluctuating renewable energy on electric power systems [6–8]. W. Schill [6], for instance, examined the effects of increasing amounts of fluctuating renewable energy on residual load to assess the storage requirements of a given power system. In this work, the RL is obtained as the difference between total electricity demand and the electric power produced by photovoltaic (PV) fields and wind farms.

The two algorithms are applied to the Gemasolar power plant, operating since 2011 in Sevilla, Spain, and with a nominal power output of 20 MW. The plant is a solar tower type CSP, adopting molten salts (solar salts, 60% $NaNO_3$ + 40% KNO3) as HTF and a steam Rankine cycle to feed a synchronous generator within the power block (PB) [9, 10]. The molten salts are heated up to 565 °C in the receiver and then either sent to the steam generator or stored in the TES hot tank. At the outlet of the steam generator, the molten salts have a temperature of around 290 °C and can be either sent to the receiver or stored in the TES cold tank [11]. This chapter is organised as follows: Section 2 describes the methodology applied to assess the CSP plant perfomances and the main energy flows in each analysed time step, Section 3 describes the two operating strategy algorithms implemented in Matlab, Section 4 reports the main results obtained with the two algorithms, and Section 5 draws the conclusions of this study and anticipates possible future works.

G. Gentile (✉) · G. Manzolini
Department of Energy, Politecnico di Milano, Milan, Italy
e-mail: giancarlo1.gentile@polimi.it

© The Author(s), under exclusive license to Springer Nature Switzerland AG 2022

A. Sayigh (ed.), *Sustainable Energy Development and Innovation*, Innovative Renewable Energy,
https://doi.org/10.1007/978-3-030-76221-6_15

2 Methodology

The Matlab suite computes, hour by hour and starting from real weather data, the thermal power transferred from the sun to the molten salts. The weather data adopted for the simulations are the IWEC (International Weather for Energy Calculations) direct normal irradiance (DNI) data provided by EnergyPlus, a program funded by the U.S. Department of Energy [12]. For each month of the year, one representative day is considered, and the annual calculations are carried out assuming that all the days of the same month have the same DNI profile. As stated by Rinaldi et al. [11], the overall efficiency of a CSP plant (solar-to-electric efficiency) can be defined as:

$$\eta_{\text{overall}} = \eta_{\text{optical}} \cdot \eta_{\text{thermal}} \cdot \eta_{\text{piping}} \cdot \eta_{\text{net,PB}} \tag{1}$$

where:

- η_{optical} compares the radiation hitting the receiver and the DNI arriving on the solar field.
- η_{thermal} takes into account the receiver thermal losses.
- η_{piping} considers the effect of piping thermal losses.
- $\eta_{\text{net, PB}} = \eta_{\text{PB}} \bullet \eta_{\text{aux}}$ accounts for the power block conversion efficiency and for the auxiliaries consumption, mainly circulating pumps and tracking systems.

Among these efficiencies, η_{optical} and η_{thermal} continuously change during the plant operation, depending on the sun position, the DNI, and the ambient conditions; the other two terms are instead almost constant.

Different methods are proposed in the literature to compute the optical efficiency of concentration systems: S. Polimeni [13] adopted the tool SolarPilot, developed by the National Renewable Energy Laboratory (NREL); another tool is Soltrace developed once again by NREL and capable of estimating the optical efficiency through Montecarlo analysis [14]. For this work, an accurate analysis of the optical efficiency during a whole operating year was beyond the research purposes. Therefore, the value of 58.31% is assumed, equal to the yearly average optical efficiency obtained by Rinaldi et al. [11] and related to the Gemasolar power plant. Analogously, the piping efficiency and the net steam cycle efficiency are assumed equal to the yearly average values reported in the same reference, respectively 99.87% and 38.55%. As a first approximation, these values are assumed constant throughout year.

The thermal efficiency of the receiver cannot be assumed constant throughout the year since its value significantly changes during off-design operation. Although, it is assumed that the receiver thermal losses ($\dot{Q}_{\text{loss,REC}}$) are constant and equal to the design value computed by Binotti et al. [15] for the Gemasolar power plant, 26.86 MW.

The thermal power available for the CSP plant can be computed as:

$$\dot{Q}_{\text{av}} = \left(A_{\text{SF}} . \text{DNI} . \eta_{\text{optical}} - \dot{Q}_{\text{loss,TES}}\right) . \eta_{\text{piping}} \tag{2}$$

where A_{SF} is the solar field area that in the case of Gemasolar is 304,750 m^2. This thermal power is used by the two algorithms described in Section 2, which define two different CSP plant operating strategies.

If the TES is full, the thermal power not used within the PB is lost through mirror defocusing. Hence, the thermal power transferred to the HTF is:

$$\dot{Q}_{\text{HTF}} = \dot{Q}_{\text{av}} - \dot{Q}_{\text{defoc}} \tag{3}$$

The thermal input of the power block can be lower, equal, or higher than $\dot{Q}_{\text{HTF}}$ depending on whether the TES is charged or discharged.

The thermal losses occurring in the two TES tanks depend on the molten salts' heat transfer properties, their temperature in each tank, the tanks surface and material, and the ambient conditions. In this study a simplified approach is adopted to compute the thermal flow exiting the two tanks according to U. Herrmann et al. [16]:

$$q_{\text{loss,TES}}\left[\frac{\text{kW}}{m^2}\right] = 0.00017 . T_{\text{HTF}}\,[\text{K}] + 0.012 \tag{4}$$

T_{HTF} is the average temperature of the heat transfer fluid in each of the two tanks, thus 290 °C and 565 °C. The thermal losses $\dot{Q}_{\text{loss,TES}}$ can be obtained multiplying, for each tank, $q_{\text{loss,TES}}$ to the tank surface. Each tank surface depends on the tank volume, which depends in turn on the mass of HTF that must be contained within the TES to guarantee the needed energy storage capabilities. In this respect, it is assumed that the tanks are characterized by equal height and diameter, and that the

HTF (solar salts) density ($q_{loss,TES}$) is function of the HTF temperature (T_{HTF}), through the experimental relation obtained by Bauer et al. [17]:

$$\rho_{HTF}\left[\frac{kg}{m^3}\right] = 2106 - 0.6679.T_{HTF}\left[^{\circ}C\right] \tag{5}$$

The comparison between the two algorithms described in Section 3 is carried out assuming that the daily load profile that must be covered by the CSP plant is equal to the 0.8% of the overall Spanish RL profile. The latter is equal to the difference between total power demand and the power produced by PV fields and wind farms in Spain. These data are reported by the Spain Electrical Grid Company [18].

For the daily simulations, the RL profiles related to July 20th and December 20th are considered for the summer and winter cases, respectively. For the annual calculations, the July RL data are used for the months of June, July, and August, the December RL data for December, January, and February, and the average between the two profiles for the remaining months. In each month, a sole daily simulation is carried out, using the DNI and RL data related to that month and a time step of 1 hour. The results in terms of energy produced, sold, curtailed, and lost through defocusing are then extended to the whole month. Lastly, it is assumed that, besides the TES, there are no other storage systems between the CSP plant and the loads; hence, all the energy produced by the plant and not consumed by the utilities is curtailed.

3 Operating Strategy Algorithms

A first algorithm, here referred to as Non-Matching Algorithm (NMA), is implemented in Matlab to simulate, starting from yearly DNI data of Sevilla, an operating strategy of the CSP plant that maximizes the power production. Hour by hour, the algorithm computes $\dot{Q}_{av}$ through Eq. 2 and, knowing the TES level and the PB capabilities, it decides whether to send the thermal power to the steam generator, to store it or, in case of full TES, to reduce it through mirror defocusing. The electric power produced within the power section is obtained multiplying the thermal power entering the steam generator and the net power block efficiency. The latter accounts also for auxiliary consumption and electromagnetic losses in the generator.

Secondly, another algorithm, hereby called Matching Algorithm (MA), is implemented to simulate a load-oriented plant control and assess the power plant capability to follow the RL profile. The algorithm receives as input the hourly energy required and exploits the TES to follow it as much as possible, adopting defocusing when necessary.

In both algorithms, a flexible operation of the PB is considered assuming 40% as minimum load calculated on thermal base (thermal input over thermal input at design conditions).

3.1 Non-Matching Algorithm

If no information are exchanged with the grid, the operating strategy maximizes the power production. Therefore, a linear programming (LP) problem is implemented in Matlab, aimed to maximize the electric energy produced in each hour, storing the remaining thermal energy in the TES and using defocusing when the TES is full.

The variables involved in this LP problem are:

- the electric power produced by the CSP plant (P_{prod});
- the thermal power exchanged with the TES, positive if the TES is discharged ($\dot{Q}_{TES}$);
- the thermal power entering the Power Block ($\dot{Q}_{in,PB}$);
- the thermal power lost through defocusing ($\dot{Q}_{defoc}$).

The parameters are as follows:

- the available thermal power ($\dot{Q}_{av}$) computed with Eq. 2;
- the net power block efficiency ($\eta_{net,PB}$);
- the TES maximum capacity ($E_{max,\,TES}$);
- the energy stored in the TES at the beginning of each time step ($E_{in,TES}$); this value is computed per each time step through Eq. 6 where Δt is 1 hour. For $t = 0$, $E_{in,TES}$ is set equal to zero but the simulations are carried out consecutively for three days to obtain a more realistic value of this parameter. For both the daily and the annual simulations, only the results related to the last of these three days are considered;

$$E_{\text{in,TES}}(t+1) = E_{\text{in,TES}}(t) - \dot{Q}_{\text{TES}}(t) \cdot \Delta t - \dot{Q}_{\text{loss,TES}}(t) \cdot \Delta t \tag{6}$$

- the thermal input required by the PB in nominal operating conditions ($\dot{Q}_{\text{nom,in,PB}}$);
- the minimum thermal input required by the PB to operate ($\dot{Q}_{\text{min,inPB}}$).

The objective function is:

$$\text{obj} = \max\left(P_{\text{prod}}\right) \tag{7}$$

The constraints are:

$$P_{\text{prod}} = \dot{Q}_{\text{in,PB}} \cdot \eta_{net,\text{PB}} \tag{8}$$

$$\dot{Q}_{\text{in,PB}} = \dot{Q}_{\text{av}} + \dot{Q}_{\text{TES}} - \dot{Q}_{\text{defoc}} \tag{9}$$

$$\dot{Q}_{\text{TES}} \leq E_{\text{in,TES}} \tag{10}$$

$$\dot{Q}_{\text{TES}} \geq E_{\text{in,TES}} - E_{\text{max,TES}} \tag{11}$$

$$\dot{Q}_{\text{in,PB}} \geq \dot{Q}_{\text{min,in,PB}} \tag{12}$$

$$\dot{Q}_{\text{in,PB}} \leq \dot{Q}_{\text{nom,in,PB}} \tag{13}$$

Please note that the constraint represented by Eq. 11 becomes meaningful when $\dot{Q}_{\text{TES}}$ is negative (the TES is charged) and it limits the thermal power sent to the TES when this is full.

The LP solver finds the values of the variables to maximize the objective function. When the LP problem is solved, the curtailed power can be computed as the difference between the power produced and the load.

3.2 *Matching Algorithm*

MA firstly computes the plant power production solving the same LP problem of NMA and then compares the results with the load. If the power production computed with NMA is lower than the residual load, the results of NMA are kept also for MA. In the other case, another LP problem is solved, this time aimed to minimize curtailment. This second LP problem has the same variables, parameters, and constraints of the NMA one, but the curtailed power (P_{curt})is considered as fifth variable Besides, the RL is introduced as parameter, and Eq. 14 is take into account as additional constraint.

$$P_{\text{curt}} = P_{\text{prod}} - \text{RL} \tag{14}$$

The objective function in this case is:

$$\text{obj} = \min\left(P_{\text{curt}}\right) \tag{15}$$

4 Results

The first target was to compute the power generation during the day starting from the DNI in Sevilla and using the TES size of the Gemasolar power plant, hence 13 hours of operation a nominal load [11]. In Fig. 1 they are reported the results obtained with NMA, using the DNI daily distribution representative of July.

The results highlight how the heat transfer fluid thermal power ($\dot{Q}_{HTF}$, equal to the available power ($\dot{Q}_{\text{av}}$) since defocusing does not occur) is remarkably lower than the sun power due to optical, thermal, and piping losses. The available power is sufficient to run the PB for 18 hours out of 24; the TES reaches the level of about 60% at 7 p.m. and then is discharged operating the PB until 2 a.m.

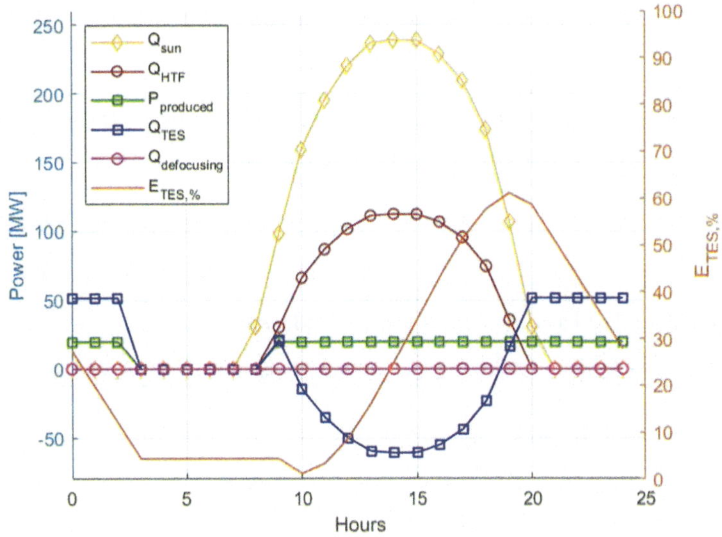

Fig. 1 Power flows and TES level on July (obtained with NMA)

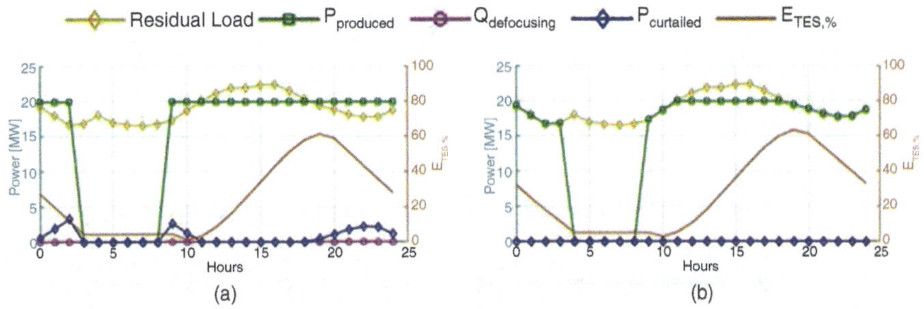

Fig. 2 Load, power flows, and TES level on July: (**a**) NMA results, (**b**) MA results

More interesting information about the plant flexibility are provided matching the power production with the RL profile. The results of this analysis obtained with NMA and MA, considering the summer operation, are reported in Fig. 2. A TES size of 13 hours, the weather data representative of July, and the summer RL profile are considered once again.

In Fig. 2a, it can be observed how the PB shut down occurs for 6 hours in the night and curtailment takes place for 11 hours. Fig. 2b points out that MA nullifies curtailment and reduces the shutdown hours to 5, as a result of the energy saved through partial load operation when the load is lower than the PB nominal power output.

A similar comparison can be carried out considering winter operation. Figure 3 shows the results obtained with both the algorithms considering the 20th of January.

In Fig. 3a, it can be noticed that curtailment takes place for 6 hours and that the power stored in the TES is not enough to cover the RL during the night. Differently, Fig. 3b highlights how the load-oriented control strategy is able to completely avoid curtailment, storing more energy in the TES, and covering the RL for more hours. A further confirmation of what already pointed out is given by the evaluation of the energy produced, sold, curtailed, or lost through defocusing on yearly basis. This calculation is carried out with the implemented Matlab suite, and the results are reported, for different TES sizes, in Fig. 4.

Firstly, it is possible to observe that the annual energy production is similar in the two cases NMA and MA, and for a TES size of 13 hours it is equal to around 80 GWh. This result is coherent with the Gemasolar power plant data reported in [10], according to which the yearly electricity generation is 80 GWh. This confirms the consistency of the assumptions made for the CSP plant efficiencies calculation.

Furthermore, the comparison between the yearly results obtained by means of the two algorithms shows that MA allows for a higher annual energy sale and hence for higher revenues. The curtailed energy is nullified by MA, but simultaneously defocusing takes place in a more remarkable manner. Precisely, with a TES of 13 hours, the annual energy sold is 67.2 GWh with NMA and 81.49 GWh with MA.

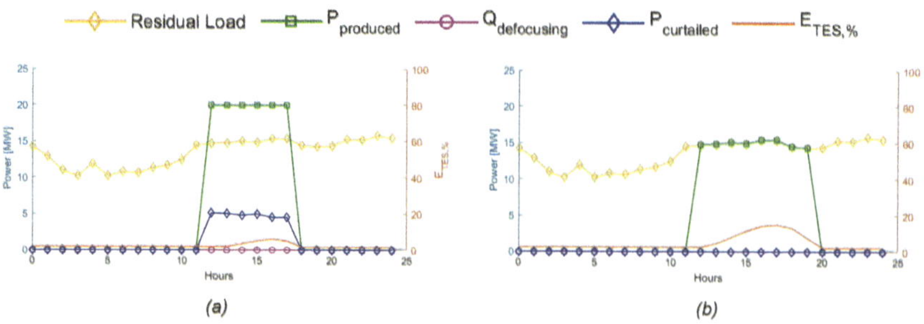

Fig. 3 Load, power flows, and TES level on January: (**a**) NMA results, (**b**) MA results

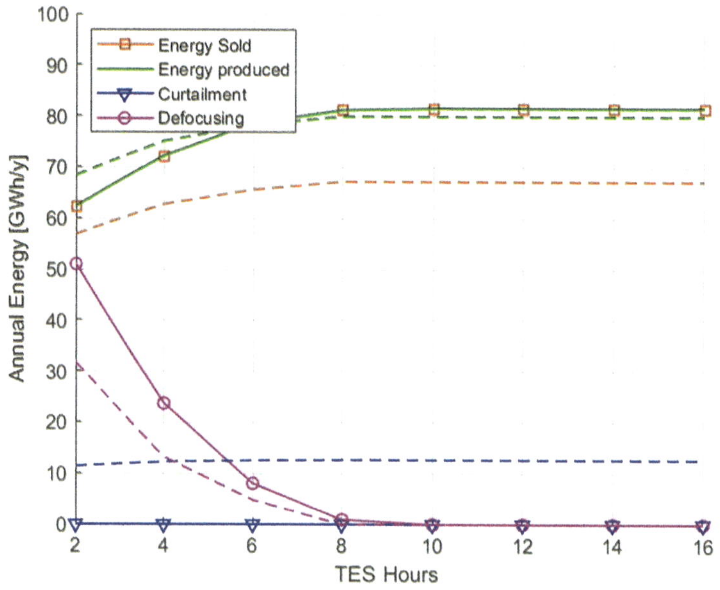

Fig. 4 Energy produced, sold, curtailed, and lost with defocusing, on yearly basis, computed with both NMA and MA; the dashed lines represent the results obtained with NMA. The TES capacity is expressed in terms of hours of nominal load operation

5 Conclusions

In this chapter, two algorithms are implemented in Matlab to simulate two control strategies of CSP plants with thermal energy storage. The comparison between the results obtained applying the algorithms to the Gemasolar power plant highlights how TES makes CSP able to follow a typical Spanish residual load profile, shifting the power production when needed by the utility. This confirms the value of CSP with thermal energy storage in providing a flexible and clean power production, aspect already highlighted by J. Forrester [5]. This feature typically belongs to peak load power plants such as gas turbines and not to renewable technologies, typically characterized by intermittent power generation.

Furthermore, the results show that a load-oriented control strategy not only minimizes curtailment, but increases the annual energy sold. As a matter of fact, the more effective usage of the TES enhances the plant availability during high-electricity demand hours leading to an increase of the annual energy sold. The results show that, under the assumptions above described, and considering the Gemasolar power plant as reference plant, a load-oriented control strategy would lead to about 21% increase in the annual revenue. Hence, the possibility to shift the power production following the load profile represents a value for the distribution grid and simultaneously enhances the economic profitability of the plant. Nevertheless, an adequate exchange of information between plant and utilities, along with effective control systems for the plant components are required to make a load-oriented operating strategy a feasible possibility. Moreover, the adoption of more compact turbomachinery, such as supercritical CO2 gas turbines, would facilitate procedures such as startup and shutdown, promoting the power generation flexibility [19].

Further works may involve a dynamic modeling of the transient undergone by the main plant components, such as receiver, generator, turbomachinery, and heat exchangers, to capture the complex dynamics involved and evaluate the time required by the plant to change its operating condition according to the utility.

Acknowledgments The authors wish to thank the Degree Program in Energy Engineering of Politecnico di Milano for its contribution.

References

1. Binotti M et al (2017) Preliminary assessment of sCO2 power cycles for application to CSP solar tower plants. Energy Procedia 105:1116–1122
2. Kuravi S et al (2013) Thermal energy storage technologies and systems for concentrating solar power plants. Prog Energy Combust Sci 39:285–319
3. Luo X et al (2015) Overview of current development in electrical energy storage technologies and the application potential in power system operation. Appl Energy 137:511–536
4. Glatzmaier G (2011) Developing a cost model and methodology to estimate capital costs for thermal energy storage. NREL technical report
5. Forrester J (2014) The value of CSP with thermal energy storage in providing grid stability. Energy Procedia 49:1632–1641
6. Schill W (2014) Residual load, renewable surplus generation and storage requirements in Germany. Energy Policy 73:65–79
7. Belderbos A, Delarue E (2015) Accounting for flexibility in power system planning with renewables. Int J Electr Power Energy Syst 71:33–41
8. Dallinger D et al (2013) Integration of intermittent renewable power supply using grid-connected vehicles – a 2030 case study for California and Germany. Appl Energy 104:666–682
9. Ushak S et al (2015) Using molten salts and other liquid sensible storage media in thermal energy storage (TES) systems. In: Advances in thermal energy storage systems. Woodhead Publishing, pp 49–63
10. NREL, Gemasolar thermosolar plant, [Online]. Availatble at: https://solarpaces.nrel.gov/gemasolar-thermosolar-plant
11. Rinaldi F et al (2014) Comparison of linear and point focus collectors in solar power plants. Energy Procedia 49:1491–1500
12. EnergyPlus, Weather data, [Online]. Available at: https://energyplus.net/weather
13. Polimeni S (2016) Techno-economic assessement of innovative high temperature solar receiver coupled with sCO2 cycles. M.Sc. thesis, Dept. of Energy, Politecnico di Milano, Milan, Italy
14. Wendelin T (2003) SolTRACE: a new optical modeling tool for concentrating solar optics. Solar energy, international solar energy conference, pp. 253–260
15. Binotti M et al (2017) Preliminary assessment of sCO2 cycles for power generation in CSP solar tower plants. Appl Energy 204:1007–1017
16. Herrmann U et al (2004) Two tank molten salt storage for parabolic trough solar power plants. Elsevier Energy 29:883–893
17. Bauer T et al (2013) Material aspects of solar salt for sensible heat storage. Appl Energy 111:1114–1119
18. Red Electica de Espana, Real time electrical energy demand, [Online]. Available at: https://demanda.ree.es/visiona/peninsula/demanda/total
19. Polimeni S et al (2018) Comparison of sodium and KCl-MgCl2 as heat transfer fluids in CSP solar tower with sCO2 power cycles. Sol Energy 162:510–524

Design, Realization, and Preliminary Tests of a Three-Dimensional Compound Parabolic Concentrator (CPC) Solar Cooker in Abidjan (Côte d'Ivoire)

Siaka Touré and Modibo Sidibé

1 Introduction

A rapid increasing in population is observed today, which leads to an increase in energy consumption everywhere in the world. As a result, there is an overexploitation of fossil fuel energy sources such as petroleum, wood fuel, and coal, which are responsible for serious environmental damages, such as deforestation. Solar cooking is a good solution to stop these environmental damages. In our country, the daily total radiation varies from 3 to 5 kWh/m^2, depending on the regions [1]. The use of sun radiation for cooking requires high temperatures, which are obtained by means of concentration systems. According to "Solar Cooker International" [2], foods start to cook between 82° and 91 °C. Different types of concentration systems are used to concentrate the solar radiation. Many works have been made about the parabolic concentrator solar cookers [3–5]. Many other works are about the hot box solar cookers [6, 7]. The conical solar cooker has been studied by several researchers [8, 9]. The ellipsoidal concentrator is also used for solar cooking [10] In Côte d'Ivoire, our studies about the technology of solar cookers include hot box, parabolic, spherical, conical, elliptic, and CPC types of solar cookers. A lot of studies have been made about the CPC solar energy collectors [11–17]. Most of them are related to the two-dimensional CPC concentrator. A three-dimensional CPC solar collector, designed and realized for solar cooking, is reported in this chapter.

2 Design and Realization of the Three-Dimensional CPC Collector

The compound parabolic concentrator (CPC) is a non-imaging collector. Its invention has been made since 1965 [18, 19]. The functioning principle of the CPC is shown in Fig. 1.

The CPC has two parabolic reflecting profiles AB and A′B′. The point B′ is the focal point of the parabolic arc AB while the point B is the focal point of the parabolic arc A′B′. Hence, all the solar rays arriving in the reflector aperture A′A that impinge the profile AB are concentrated in B′ as shown in Fig. 1. The two parabolas AB and A′B′ are identical. In Fig. 1, θ_c is the half acceptance angle. The principle of the construction of the two profiles AB and A′B′, starting from two identical parabolas A and parabola B, is shown in Fig. 2.

A solar ray comes into the CPC aperture at an angle θ. The rays for which $\theta < \theta_c$ reach the receiver. In a first step, the best half acceptance angle was investigated. It is assumed that the solar radiation is collected over a collecting time $\Delta h = t_f - t_i$, starting from t_i (before 12 h) to t_f (after 12 h). The hour angle is expressed as:

$$\omega = -15(t_i - 12) = 15(t_f - 12) \tag{1}$$

The corresponding collection time is Δh. From the calculation of the hour angle ω, the knowledge of the latitude and the declination δ, two angles θ_i (corresponding to δ) and θ_f (corresponding to - δ) were calculated as follows [20]:

$$\tan\theta_i = \frac{\cos\omega - \tan\varphi\tan\delta}{\cos\omega.\tan\varphi + \tan\delta} \tag{2}$$

S. Touré (✉) · M. Sidibé
Solar Energy Laboratory, FHB University Abidjan, Abidjan, Côte d'Ivoire

© The Author(s), under exclusive license to Springer Nature Switzerland AG 2022

A. Sayigh (ed.), *Sustainable Energy Development and Innovation*, Innovative Renewable Energy,
https://doi.org/10.1007/978-3-030-76221-6_16

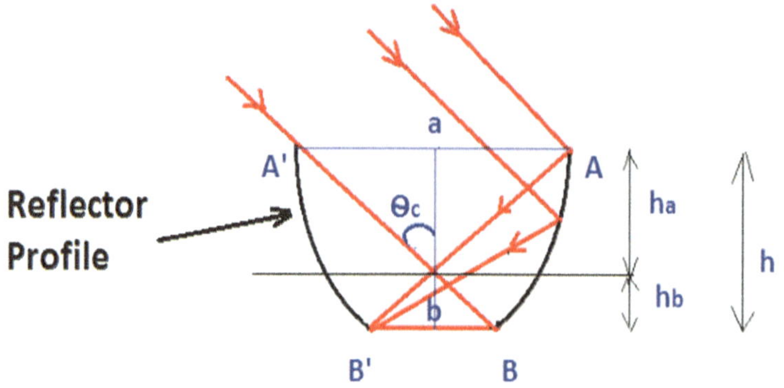

Fig. 1 Principle of the CPC reflector

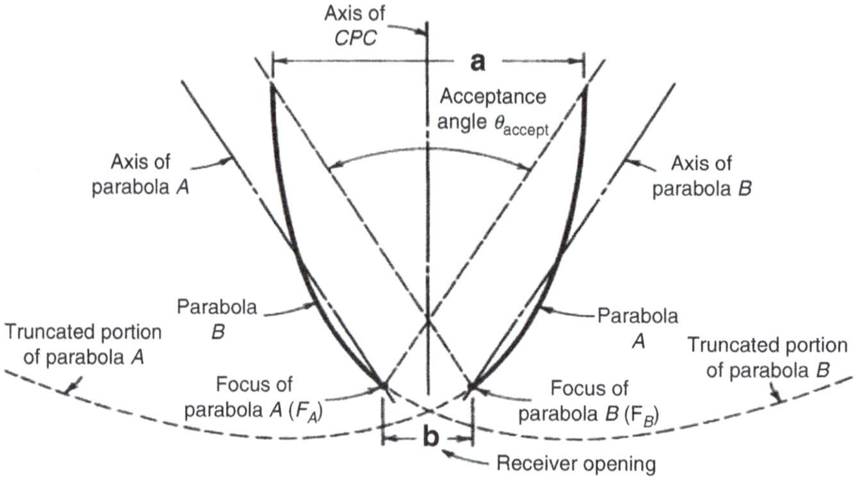

Fig. 2 Principle of the construction of CPC reflector profiles

$$\tan\theta_f = \frac{\cos\omega + \tan\varphi\tan\delta}{\cos\omega.\tan\varphi - \tan\delta} \tag{3}$$

That allowed the calculation of the half acceptance angle θ_c given by:

$$\theta_c = \frac{\theta_f - \theta_i}{2} \tag{4}$$

For $\delta = 23°27$, $\varphi = 5.3096$ (latitude of Abidjan), θ_c was calculated for different collection times. For the three-dimensional CPC, the geometric concentration ratio is given by Eq. (7). Table 1 gives θ_c and C for different periods of capitation of the solar radiation.

Figure 3 is a plotting of θ_c against Δh. A polynomial correlation was found, with a coefficient of determination $R^2 = 1$. It is expressed as:

$$\theta_c = 0.0257x3 + 0.0058x2 + 0.4551x + 23.045 \tag{5}$$

Equation (5) shows that the limiting value of θ_c, for small values of collection time, is 23°. The smaller is the half acceptance angle, the higher is the concentration ratio. This study shows that, for the design of our CPC reflector, it is better to choose a value of θ_c close to 23°. Practically, the value of θ_c chosen for the design of our CPC concentrator was $\theta_c = 20°$. Subsequently, using this value of θ_c, CPC was designed. Let b be the width of the circular receiver (absorber). The focal length of the parabola is expressed as [16, 21]:

Table 1 Half acceptance angles calculated for various collection times

Collection time Δh (h)	6	5	4	3	2
Collection period (h)	9–15	9:30–14:30	10–14	10:30–13:30	11–13
θ_c (°)	31.527	28.6685	26.605	25.151	24.184
C	3.657	4.344	C.9859	5.5362	5.9585

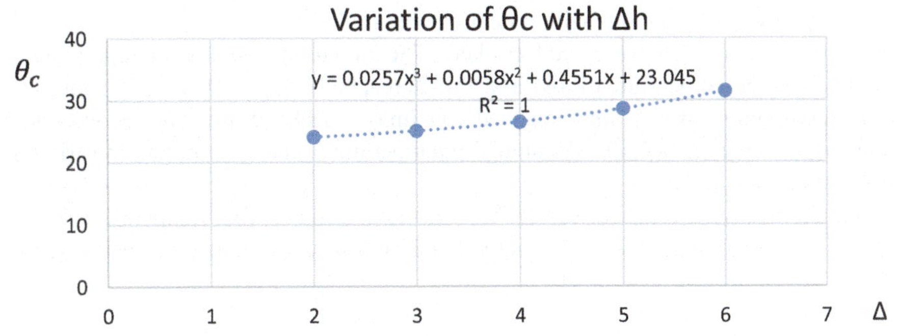

Fig. 3 Variation of the half acceptance angle with collection time

$$f = \frac{b}{2}\left(1+\sin\theta_c\right) \tag{6}$$

As for the width of the circular receiver, it was chosen as $b = 0.28$ m. Then from the value of θ_c, Eq. (6) yields $f = 0.188$ m. Then the width a of the CPC aperture was calculated. As it is assumed that the three-dimensional CPC is ideal, its concentration ratio is linked to a and b. It is expressed as:

$$\mathrm{C} = \frac{1}{\sin^2\theta_c} = \left(\frac{a}{b}\right)^2 \tag{7}$$

From Eq. (7) and the previous knowledge of θ_c and b, it was found $a = 0.8186$ m. As for the CPC length h, it is expressed as [21]:

$$h = \frac{a+b}{2\tan\theta_c} = \frac{f\cos\theta_c}{\sin^2\theta_c} \tag{8}$$

From the values of θ_c, a and b, Eq. (8) yields $h = 1.51$ m. For the practical realization of the non-imaging three-dimensional CPC, two parabolas were used to find the reflector profiles AB and A′B′ shown in Fig. 1. For that purpose, two parabolas A and B were designed with the same equation given below

$$y = \frac{1}{4f}x^2 \tag{9}$$

In Eq. (9), f is the focal length. A sheet of plywood was used to draw the x and y coordinates obtained from Eq. (9). Then, from the parabolic plywood, two parabolas A and B were made with steel rods. Then by using parabolas A and B, the CPC profiles AB and A′B′ were realized conformably to Fig. 2. Hence, the two-dimensional structure, made of steel rods, shown in Fig. 1, was obtained. As for the three-dimensional CPC, it is realized by rotating the two-dimensional profile around the axis of symmetry. For that purpose, the two-dimensional structure was reproduced 6 times to get 12 parabolic profiles. Then galvanized steel sheets were soldered on the CPC parabolic profiles to make the 12 facets of the reflective surface. Finally, square pieces of mirror, whose dimensions are 0.03×0.03 m^2, were set on the galvanized steel to increase the reflectance of the inner surface of the CPC. The CPC was glazed by a circular glass set at a distance $hg = 0.69$ m above the receiver. The receiver is made of stainless steel painted black. Its base is insulated by layers of air, plaster, plywood, and polystyrene. An opening allows the introduction of a cooking pan on the receiver. Figure 4 shows a schematic representation of the three-dimensional CPC solar cooker. The solar cooker is mounted on a support. A manual tracking system was also realized for the tracking of the sun. A photography of the three-dimensional CPC solar cooker, mounted on its support, is shown in Fig. 5.

3 Experimental Study of the CPC Solar Cooker

The cooking tests were performed on the roof of our Solar Energy Laboratory. During the experiments, the temperature data were measured by K-type thermocouples and recorded using Data loggers USB-TC08 connected to a computer. The global solar radiation I(W/m^2) was measured by an Eppley-type pyranometer which yields an electromotive force E (mV). The solar radiation is obtained from the following relationship: ($I = \frac{E}{C}$), where the constant C = 10.41 × 10^{-3} mV. W^{-1}m^2. Two cooking tests of eggs were carried out. The first one was made on March 26, 2020. It was a cooking test of two eggs in 0.5 L of water. The temperatures and solar radiation variations with time are given in Fig. 6.

It is worth noting that the two eggs were well cooked. The maximum cooking temperature (Tcook) reached was 83.44 °C. The maximum absorber temperature (Tabs1 and Tabs2) reached was 92.59 °C. The solar radiation conditions were rather moderate, as the maximum solar irradiance reached was Imax = 789.5 W/m^2. The second one was made on April 5, 2020. It was a cooking test of four eggs in 0.5 L of water. The temperatures and solar radiation variations with time are given in Fig. 7. The four eggs were well cooked.

The maximum cooking temperature reached was 86 °C. The maximum absorber temperature reached was 96.42 °C. The solar radiation conditions were also rather moderate, as the maximum solar irradiance reached was Imax = 757.6 W/m^2.

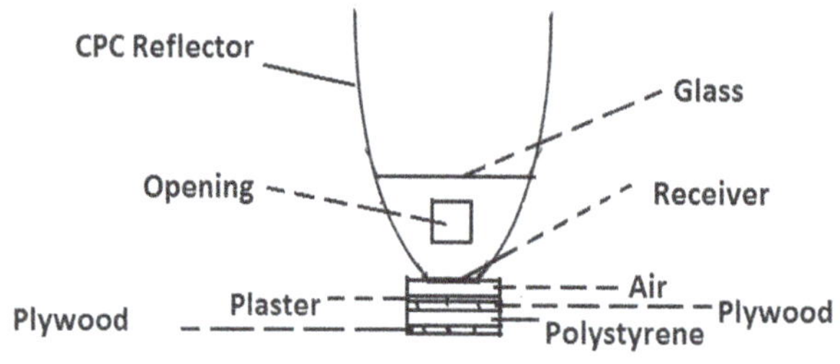

Fig. 4 Schematic representation of the CPC solar cooker

Fig. 5 Photography of the three-dimensional CPC and a cooking test

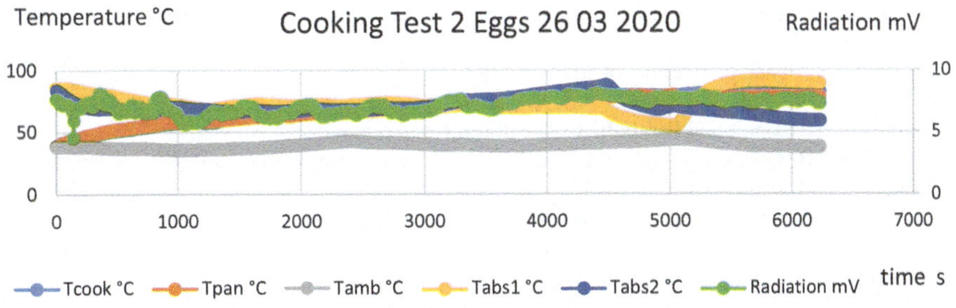

Fig. 6 Temperatures and irradiance evolutions with time on March 26, 2020

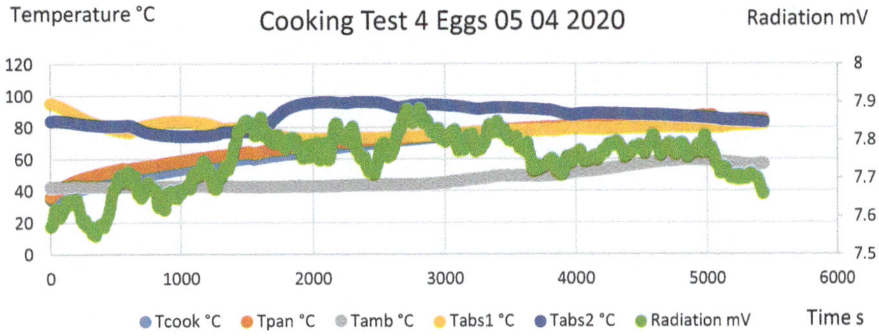

Fig. 7 Temperatures and irradiance evolutions with time on April 5, 2020

4 Conclusion

Despite the very moderate solar radiation conditions, cooking of eggs was made successfully with our three-dimensional CPC solar cooker. Higher temperatures may be obtained in high solar radiation conditions. The CPC reflector will be insulated in its lower part, to reduce the thermal losses. That will increase its efficiency.

References

1. Gnigbognima S, N'Cho P (2010) Forum Régional sur l'initiative énergie solaire de la CEDEAO. Présentation de la Côte d'Ivoire 1–12
2. Solar Cookers International (2004) Cuiseurs solaires: comment construire, employer et apprécier. www.solarcooking.org/francais/Plans-fr.pdf. Accessed 2016.12.25
3. Subbodh Kumar T, Kandpal C, Mullick SC (1993) Heat losses from a paraboloid concentrator solar cooker. Experimental investigations on effect reflector orientation. Renew Energy 3(8):871–876
4. Srinivasan M, Kulkarni LV, Pasupathy CS (1978) A simple technique of fabrication of parabolic concentrators. Sol Energy 22:463–465
5. Gavisiddesha SP, Revenkar PP, Gorawar MB (2013) Evaluation of thermal performance of paraboloid concentrator solar cooker. Int J Innov Res Technol Sci 1(3):58–65
6. Khalifa AMA, Taha MA, Akyurt M (1985) Solar cookers for outdoor and indoors. Energy 10(7):819–829
7. Toure S, Sidibe M, Fasinou W, Gbane A, Salami H (2004) A performance study of two solar cookers experimented in Abidjan (Ivory Coast). World Renewable Energy Congress VIII (WREC 2014), Denver, Colorado, USA, Aug 29–3 Sept 2004
8. Sharaf E (2002) A new design for an economical, highly efficient, conical solar cooker. Renew Energy 27(4):599–619
9. Toure S, Sidibe M (2017) Design and evaluation of a conical solar cooker in glazed and non-glazed configurations: cooking tests. IOP Conf Series Earth Environ Sci 93:012011. https://doi.org/10.1088/1755-1315/93/1/012011
10. Toure S, Sidibe M (2018) Design, realisation and experimentation of a solar cooker fitted with an ellipsoidal concentrator: preliminary results of cooking tests. IOP Conf Series Earth Environ Sci 188:012042. https://doi.org/10.1088/1755-1315/188/1/012042
11. Hsieh CK (1981) Thermal analysis of CPC collectors. Sol Energy 27:19–29
12. Norton B, Kothdiwala AF, Eames PC (1994) Effect of inclination on the performance of CPC solar energy collectors. Renew Energy 5(Part 1):357–367
13. Whang US, Yang YS (1998) A concentration enhanced CPC design. In: World renewable energy congress (WREC 1998). Elsevier Science Ltd, pp 2190–2193

14. Akiba S (2010) Optics of solar concentrators. Lecture at the First SFERA Summer School PROMES – CNRS Laboratory, Font Romeu – Odeillo, France, 10–12 June 2010
15. Johan N (2005) Optical design and characterization of a solar EBD – T – concentrator for photovoltaics. Lund University, Faculty of Engineering LTH Report 05/6
16. Figari A (2011) Some features of the Compound Parabolic Concentrator Conference Paper, May 2011
17. Tiruneh AT, Ndlela WN, Gadaga TH, Debesay T, Heikkilä J (2017) A four- wing compound parabolic concentrator Design for Heating and Sanitization of waste products. J Power Energy Eng 5:18–35
18. Winston R (1974) Principles of solar concentrators of a novel design. Sol Energy 16:89
19. Baranov VK (1975) Parabolocylindric reflecting unit and its properties. Geliotecknika 11(3–4):45–52. (English Translation, 36–41)
20. Chassériaux JM (1984) Conversion thermique du rayonnement solaire. (Dunod Paris)
21. Duffie JA, Beckman W (2013) Solar engineering of thermal processes. Chapter 7 Section 7.6 and 7.7, 337–349

Solar-to-Work Conversion from Meteorological Data

José Luis Duomarco

1 Introduction

Solar energy on top the surface of the sun (dia. =1.39 × 10^9 m) may be modeled by the undiluted Stefan-Boltzmann law of blackbody radiation sT_H^4 (BBR), with T_H = 5770 K and s = 5.6698 × 10^{-8} W/m²/K⁴. Radiation, after traveling from sun to earth (Distance = 1.49 × 10^{11} m), has a value $s\,T^4\,(d/2D)^2$ = 1370 W/m² on top of the atmosphere; known as the solar constant. After trespassing the atmosphere, solar energy on earth surface may be modeled as diluted BBR $f\,e\,sT_H^4$, and may be classified either as direct radiation, coming directly from the sun, or diffuse radiation, coming from different regions of the celestial vault after several scattering phenomena. The two additional constants are f the view factor and ε the dilution factor. The view factor of an assumed spherical sun seen from a horizontal slab S on earth's surface is $f = \text{sen}^2 d\cos q_o$, where $2d = 31'59''$ is sun's apparent diameter and q_0 the azimuthal angle, which varies from sunrise to sunset. Following are some examples of how to interpret these formulas, using meteorological data from selected places in Uruguay (BHI, DNI, DHI, GHI, T) [1].

1. *Direct normal irradiance (DNI)* – energy density flux (W/m²) of direct radiation from the sun, at normal incidence to the receiving surface, as measured with a pyrheliometer. It is usually projected on a horizontal surface with the factor $\cos q_o$ of the azimuthal angle of the sun, obtaining the *beam horizontal irradiance (BHI)*. The direct radiation dilution factor ε_{dr} and the view factor f adjust theory to experimental measurements.

$$\begin{aligned}
&\Phi_{\text{drh}} = f\ \varepsilon_{\text{dr}}\sigma T_{\text{H}}^4 = \varepsilon_{\text{dr}}\text{sen}^2\delta\cos\theta_0\sigma T_{\text{H}}^4 = \text{BHI} = \text{DNI}\times\cos\theta_0 \\
&2\delta = 31'\ 59'' \qquad T_{\text{H}} = 5773\ \text{K} \\
&f = \text{sen}^2\delta\cos\theta_0 = 2.1639\times 10^{-5}\times\cos\theta_0 \\
&\varepsilon_{\text{dr}} = \frac{\text{DNI}}{\text{sen}^2\delta\sigma T_{\text{H}}^4}
\end{aligned} \tag{1}$$

2. *Diffuse horizontal irradiance (DHI)* – energy density flux (W/m²) of radiation received from above by a horizontal surface, not arriving on a direct path from the sun, but has been scattered by molecules and particles in the atmosphere and comes equally from all directions, as measured with a pyranometer shaded from the beam radiation by a shade ring or disc. The diffuse radiation dilution factor ε_{df} adjusts theory to experimental measurements.

$$\begin{aligned}
&\dot{\Phi}_{\text{df}} = f\varepsilon_{\text{df}}\sigma T_{\text{H}}^4 = \varepsilon_{\text{df}}\sigma T_{\text{H}}^4 = \text{DHI} = \text{GHI} - \text{BHI} \\
&\delta = \frac{\pi}{2} \qquad \theta_0 = 0 \qquad T_{\text{H}} = 5773\ \text{K} \\
&f = \text{sen}^2\delta\cos\theta_0 = 1 \\
&\varepsilon_{\text{df}} = \frac{\text{DHI}}{\sigma T_{\text{H}}^4}
\end{aligned} \tag{2}$$

J. L. Duomarco (✉)
A.I.U., Montevideo, Uruguay

© The Author(s), under exclusive license to Springer Nature Switzerland AG 2022

A. Sayigh (ed.), *Sustainable Energy Development and Innovation*, Innovative Renewable Energy,
https://doi.org/10.1007/978-3-030-76221-6_17

3. *Global horizontal irradiance (GHI)* – energy density flux (W/m²) of radiation received from above by a horizontal surface as measured with a pyranometer. The GHI can thus be considered as the sum of BHI and DHI with a global dilution factor ε_g.

$$\dot{\Phi}_{\mathrm{drh}} + \dot{\Phi}_{\mathrm{df}} = \left(\varepsilon_{\mathrm{dr}} \mathrm{sen}\delta \cos\theta_0 + \varepsilon_{\mathrm{df}}\right)\sigma T_{\mathrm{H}}^4 = \varepsilon_{\mathrm{g}}\sigma T_{\mathrm{H}}^4 = \mathrm{GHI} = \mathrm{BHI} + \mathrm{DHI}$$
$$\varepsilon_{\mathrm{g}} = \frac{\mathrm{GHI}}{\sigma T_{\mathrm{H}}^4} \tag{3}$$

2 Maximum Work Out of Blackbody Isotropic Radiation

The radiation engine model of Fig. 1 consists of a radiation absorber (RA), two radiation reservoirs, direct (DR) and diffuse (DF), two constant temperature heat reservoirs ($\mathrm{HR_a}$, T_a) and ($\mathrm{HR_L}$, T_L) ($T_\mathrm{a} > T_\mathrm{L}$), and a work extractor (WE). The First and the Second Laws of Thermodynamics, applied to the RA in intervals of 1 hour, lead to ($i = 1 - 8760$) [3]:

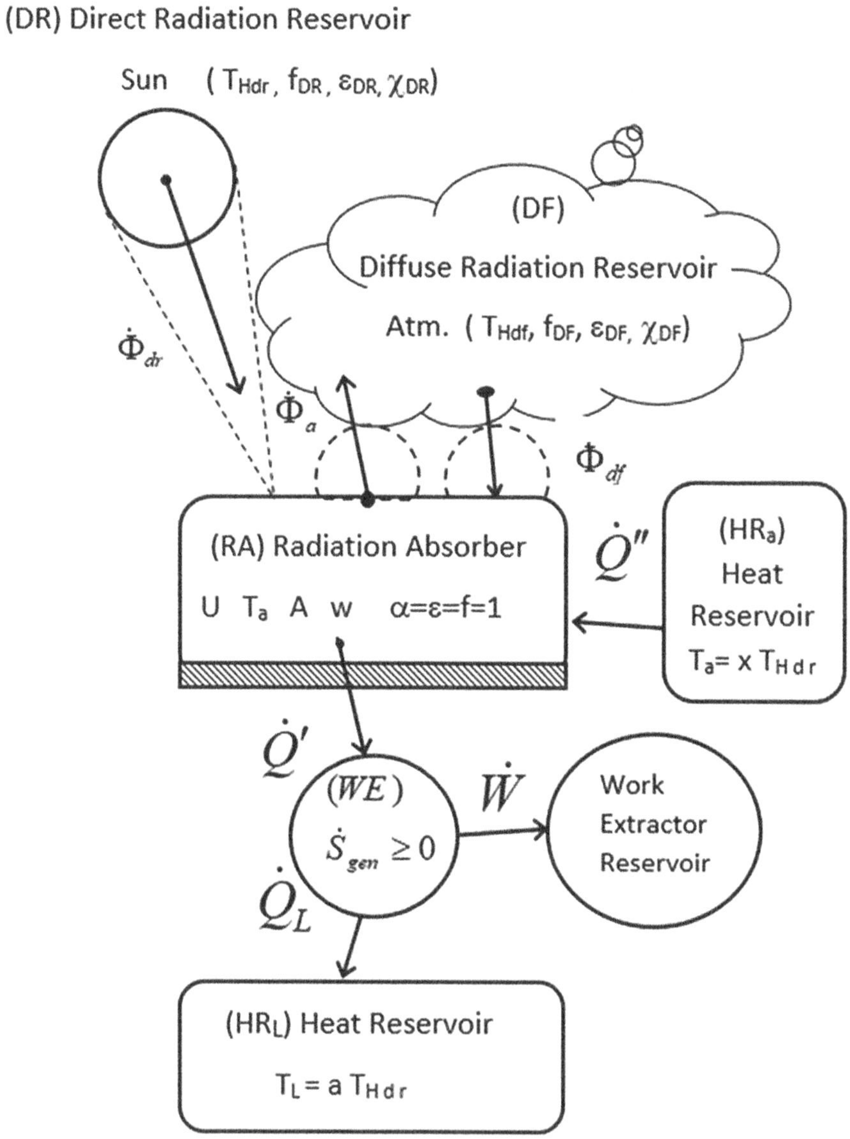

Fig. 1 Radiation energy converter

$$U_{i+1} = U_i + A\left(\Phi_{dr} + \Phi_{df} - \Phi_a\right)_i - Q_i' + Q_i'' \tag{4}$$

$$S_{i+1} = S_i + A\left(\Psi_{dr} + \Psi_{df} - \Psi_a\right)_i - \frac{Q_i'}{T_{ai}} + \frac{Q_i''}{T_{ai}} + S_{RAi} \tag{5}$$

Here the pairs (Φ_{dr}, Ψ_{dr}) and (Φ_{df}, Ψ_{df}) stand for energy and entropy from the sun, absorbed as direct and diffuse radiation by a horizontal surface, and the pair (Φ_a, Ψ_a) are energy and entropy emitted from the same surface to the atmosphere, all of them referred to unit area of this surface. In the operation of the RA, heat Q' is transferred to the WE and auxiliary heat Q'' is interchanged with the HR_a toward controlling the temperature T_a in a constant value during each hour. Both DU and DS are to be zero as the RA is considered to be in a steady state during each hour and S_{RA} is the entropy generation in the same interval.

Our model of RA is a slab with unit area $A = 1$ m^2 and width $w = 0.05$ m, made of different materials. Table 1 shows some of the materials used, with the properties required in defining a very simple hypothetical model; density ρ, melting temperature T_{melt} and specific heat capacity $c_{p.}$ The upper surface of the slab is nonselective, behaves as a Lambertian blackbody (radiation impinging is entirely transformed into internal energy), and is capable of receiving or emitting radiation over the whole hemisphere. The lower surface of the slab is assumed adiabatic. The WE is a thermal-cyclic motor converting heat Q' into mechanical work W, rejecting heat Q_L to the low temperature HR_L and is accompanied by an hourly entropy generation S_{WE}. It could also be interpreted as other device providing electrical work, or chemical work. A reversible Carnot motor could also be the case with $S_{WE} = 0$. The energy and entropy balances for the WE, assuming an integer number of cycles in the hour, are respectively:

$$W = Q' - Q_L \tag{6}$$

$$\frac{Q'}{T_a} - \frac{Q_L}{T_L} + \Delta S_M + S_{WE} = 0 \qquad \Delta S_M = 0 \Rightarrow \text{close cycle} \tag{7}$$

Using Eqs. 4, 6, and 7, we obtain an hourly conversion factor of solar radiation into work; incomplete as conditioning by Eq. 5 is missing; to be substituted by a maximization procedure:

$$W = \left[A_a\left(\Phi_{dr} + \Phi_{df} - \Phi_a\right) + Q''\right]\left(1 - \frac{T_L}{T_a}\right) - T_L S_{WE} \tag{8}$$

$$\eta = \frac{W}{A_a\left(\Phi_{dr} + \Phi_{df}\right)} = \left(1 - \frac{\Phi_a - Q''/A_a}{\Phi_{df} + \Phi_{dr}}\right)\left(1 - \frac{T_L}{T_a}\right) - \frac{T_L S_{WE}}{A_a\left(\Phi_{dr} + \Phi_{df}\right)} \tag{9}$$

3 Application to Actual Solar Radiation Data

Global Horizontal Irradiance (GHI) measures the incident solar radiation per unit area, on a horizontal surface, and is found by integration of irradiance over a specified time, in our case, 1 hour. Then the sum of direct and diffuse radiation can be expressed as:

$$\text{GHI} = \Phi_{drh} + \Phi_{df} \tag{10}$$

The typical daily GHI, taken as independent variable and shown in Fig. 3a for Montevideo, gives a particularly intrinsic insight of the functioning of the whole system. Substitution of GHI in the conversion factor, with a small modification, and

Table 1 Model of Radiation energy converter

Slab volume = 0.05 m^3	Density kg/m^3	Melting temperature °C	Specific heat capacity J/ (kg °C)	Heat capacity Wh/°C
Graphite	2090	3727	710	20.609
Diamond	3510	3823	509	24.814
Copper	8800	1083	380	46.444

assuming the WE is a reversible motor (SWE = 0), gives rise to a new upper bound η^+ that can be expressed as two (12) or three factors (13):

$$\eta = \left(1 - \frac{\sigma T_a^4 - Q''/A_a}{\mathrm{GHI}}\right)\left(1 - \frac{T_L}{T_a}\right) \le \eta^+ = \left(1 - \frac{\dfrac{\sigma T_a^4 - Q''/A_a}{\sigma T_H^4}}{\dfrac{\mathrm{GHI}}{\mathrm{sen}^2\delta\sigma T_H^4}}\right)\left(1 - \frac{a}{x}\right) \tag{11}$$

$$\eta^+ = \left(1 - \frac{a}{x}\right)\left(1 - \frac{x^4 - \zeta^4}{\iota^4}\right) \tag{12}$$

$$\eta^+ = \left(1 - \frac{a}{x}\right)\left(1 - \frac{x^4}{\iota^4}\right)\left(1 + \frac{\zeta^4}{\iota^4 - x^4}\right) = \eta_1\eta_2\eta_3 \tag{13}$$

where four nondimensional magnitudes are introduced:

$$x = \frac{T_a}{T_H}, \quad a = \frac{T_L}{T_H}, \quad \iota^4 = \frac{\mathrm{GHI}}{\mathrm{sen}^2\delta\sigma T_H^4}, \quad \zeta^4 = \frac{Q''/A}{\sigma T_H^4}$$
$$a_i \le x_i \le \iota_i \le 1 \qquad Q'' \ge 0 \tag{14}$$

But the thermal efficiency of a reversible cyclic motor operating between two heat sources of different temperatures, T_a and T_L ($T_a > T_L$), is precisely the first factor of (12) (Carnot motor efficiency). The second factor (12), then, should be equal to one, i.e., $x^4 = \zeta^4$ or $Q'' = A\ \sigma\ T_a^4$, and consequently none of Q'' is used in WE operation. It remains to be said consequently that the third factor (13) is not involved in the mechanism of direct transformation of radiation into work and should be excluded in the calculation of the exergy factor for the GHI.

Exergy, according to its maximum work definition, may be obtained by maximizing h^+ as function of $x = T_a/T_H$. For each individual hour-measurement of the GHI, the efficiency h^+ is a continuous function of the nondimensional magnitude $x = T_a/T_H$. This function, for x inside the interval, $a_j \le x \le \iota_j \le 1$, has a unique maximum ($x'$), that can be reached in a two-step iterative procedure. In the first step, the product of the first two parenthesis $h_1(x)h_2(x)$ is maximized by increasing x by small increments. In the second step, the resulting third parenthesis $h_3(x)$ value is used as a constant factor; a new third parenthesis results and the second step is repeated until the new third parenthesis converges to a single number. In Fig. 4, values of all three $h(x)$ are shown as function of GHI. In case x' exceeds $x'_{\mathrm{melt}} = T_{\mathrm{melt}}/T_H$ (rate of melting temperature of the RA material to sun's temperature), the procedure stops and x'_{melt} is selected.

Once the full equilibrium is accomplished, we are in a position to exclude the third factor. Finally summing all these values over the whole year gives the annual production of exergy from radiation:

$$\mathrm{GHIXa} = \sum_{j=1}^{8760}\left(1 - \frac{a}{x}\right)_j\left(1 - \frac{x^4}{\iota^4}\right)_j \times \mathrm{GHI}_j \tag{15}$$

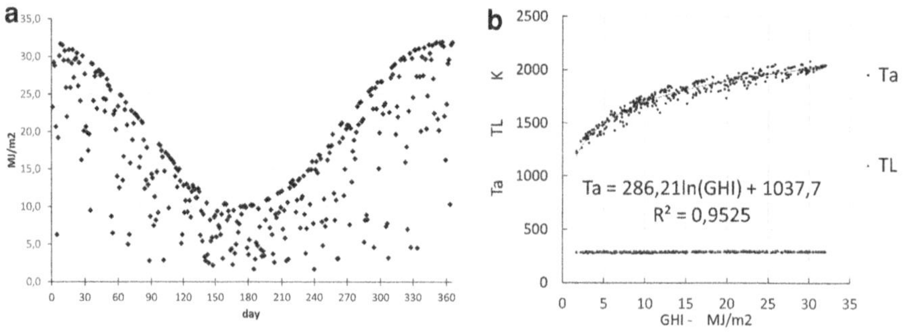

Fig. 3 (**a**) Annual typical daily Global Horizontal Irradiance (GHI) for Montevideo. (**b**) Slab's temperature (T_a), as function of GHI, between low ambient (T_L) and high Sun ($T_H = 5773$ K) temperatures with least squares adjustment

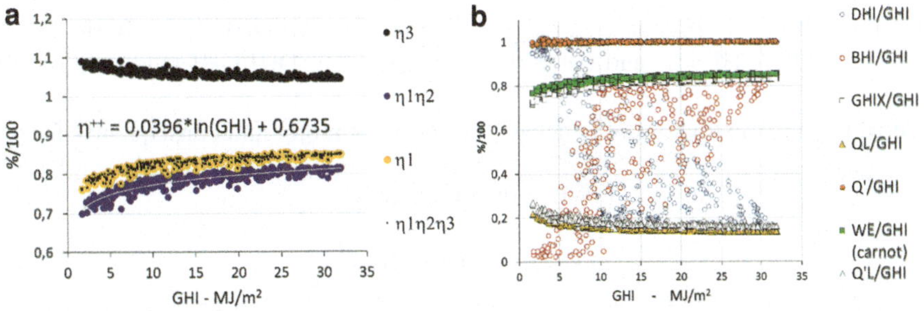

Fig. 4 (**a**) Three-factor expression of radiation-work conversion: (i) h1, Carnot factor; (ii) h3, complementary factor (>1); (iii) h1h2, exergy conversion factor GHI*h1*h2 = GHIX with least squares adjustment; (iv) h1h2h3 = h1, Carnot factor (h2h3 = 1). (**b**) Daily ratios DHI, BHI, GHIX, QL, Q', WE, Q'', and sT_a^4 to GHI versus GHI with typical Montevideo's meteorology data for RA

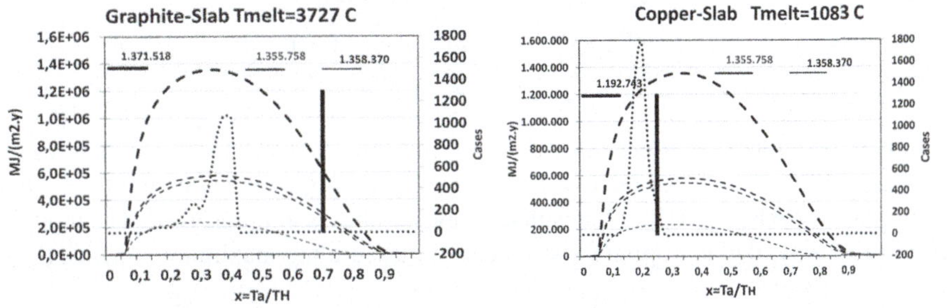

Fig. 2 Graphical representation of the annual global horizontal irradiance exergy determination (GHIX) with a graphite slab (melting $T = 3727$ °C), and with a copper slab (melting $T = 1083$ °C), in Montevideo. The following objects are shown: (i) frequency histogram of x corresponding to hourly maxima efficiencies; (ii) representation of the whole year function; (iii) representation of three 4-month functions

Table 2 Annual GHIX with copper and graphite slabs in Montevideo

W h / (m² y)	GHI	GHIXa	GHIXa / GHI	GHIXc	GHIXb
Copper slab	1,672,904	1,198,029	0.716137	Not applicable	Not applicable
Graphite slab	1,672,904	1,365,816	0.816454	1,358,370	1,355,758

Table 3 Annual GHIXa with graphite slab in different locations

Location	Lat. (deg.)	Lon. (deg.)	Altitude (m)	GHI (Wh/m²)	GHIX (Wh/m²)	GHIX/GHI
Montevideo	−34.83	−56.01	33	1,672,904	1,365,816	0.816454
Colonia	−34.45	−57.77	20	1,700,070	1,387,673	0.816244
Rivera	−30.90	−55.54	240	1,696,994	1,448,993	0.853858
Rocha	−34.49	−54.31	18	1,622,055	1,321,558	0.814743
Salto	−31.27	−57.88	50	1,754,052	1,432,393	0.816619

In Fig. 2a and b, the histograms of frequencies of $x = T_a/T_H$ that give maximum $h_1(x)h_2(x)$ values and accordingly maximum contribution to the GHIXa sums are shown. Instead of maximizing each hour and summing the results, the whole year maximum sum (GHIXb) or the sum of 12-month maxima values (GHIXc) could also be used as annual exergy value, where certainly GHIXa ≥ GHIXc ≥ GHIXb. Table 2 lists annual exergy results for copper and graphite in Montevideo. A low melting temperature limits the opportunities of reaching optimum values. In Table 3, GHIXa values for several sites in Uruguay are listed [2, 4].

Resulting daily slab's temperature T_a, jointly with Montevideo's typical ambient temperature T_L, is shown in Fig. 3b, with a least-squares fit.

Resulting daily ratios of DHI, BHI, GHIX, QL, Q', WE, Q'', and sT_a^4 to GHI versus GHI with typical Montevideo's meteorology data are shown in Fig. 4b. Interesting features are: (a) ratio Q'/GHI is equal to one, i.e., work extractor is fed exactly by the same GHI amount; (b).WE/GHI > GHIX/GHI, i.e., Carnot efficiency is greater than the exergy factor; (c)

$Q''/\mathrm{GHI} = \sigma\ T_a^4/\mathrm{GHI}$, i.e., none of Q'' is used in radiation conversion; (d) DHI/GHI + BHI/GHI = 1; (e) WE/GHI + QL/GHI = 1; (f) GHIX/GHI + QL′/GHI = 1; and (g) with increasing GHI, DHI/GHI tends to QL/GHI and BHI/GHI tends to WE/GHI.

All things considered the work extractor (WE) behaves as a Carnot motor and consequently for ($j = 1 - 8760$):

$$\begin{aligned} \mathrm{WE}_j &= \eta^+(x')A_a\mathrm{GHI} \\ Q_j' &= \eta^+(x')A_a\mathrm{GHI}\frac{T_a}{T_a - T_L} \\ Q_{Lj} &= \eta^+(x')A_a\mathrm{GHI}\frac{T_L}{T_a - T_L} \end{aligned} \tag{16}$$

Energy and entropy balance equations in RA are then:

$$\begin{aligned} & Q'' - Q' - A_a\sigma T_a^4 + A_a\mathrm{GHI} = 0 \\ & \frac{Q''}{T_a} - \frac{Q'}{T_a} - \frac{4}{3}A_a\frac{\sigma T_a^4}{T_a} + \frac{4}{3}A_a\frac{\mathrm{GHI}}{T_H} + S_{RA} = 0 \end{aligned} \tag{17}$$

An additional-alternative interpretation can be done for the case the work extractor behaves as a real-irreversible motor with a work or exergy production factor GHIXa less than WE. Heat QL increases to QL′ = QL+ (WE − GHIXa) and an entropy generation SWE = (WE − GHIXa)/T_L results, as reversibility is not preserved. The modified WE with exergy output GHIXa, becomes in this way, the appropriate instrument to measure the maximum work GHIXa, to be drawn out of GHI solar radiation.

Conclusion This chapter has shown how radiation exergy theory can be applied to the solar radiation meteorological data. Maximum work and thermodynamic exergy approaches are being used. In each of the annual 8760 hours, steady-state conditions are assumed. The auxiliary heat reservoir HRa that supplies heat Q'' is essential to stabilize the temperature of the RA in a value T_a. But it has only an auxiliary acting. The usefulness of the whole device lies in the fact that all happens as if only an amount GHI of energy at temperature T_a is fed into the work extractor. Two outputs are of interest, GHI $* \eta_{\mathrm{Carnot}}$ and GHI $* \eta_{\mathrm{exergy}}$. The first is the maximum work to be obtained with a reversible thermal cyclical motor between temperatures T_a and T_L. The second is the maximum work, or exergy, to be obtained from the incident GHI solar radiation.

References

1. Alonso-Suárez R, Bidegain M, Abal G, Modernell P. Memoria Técnica: Año Meteorológico Típico para aplicaciones de energía solar (AMTU). Series horarias típicas para 5 lugares del Uruguay. Laboratorio de Energía Solar del Uruguay (LES). Salto, Uruguay; 16 Apr 2016
2. Badescu V (2014) How much work can be extracted from a radiation reservoir? Physica A 410:110–119
3. Landsberg PT, Tonge G (1979) Thermodynamics of the conversion of diluted radiation. J Phys A Math Gen 12:551–562
4. Pons M (2012) Energy analysis of solar collectors, from incident radiation to dissipation. Renew Energy 47:194–202

Identifying Optimal Nanofluid Synthesis Conditions for Applications in Solar Thermal Concentrators

Anurag Pramanik, Harjit Singh, S. Suresh, Ram Chandra, and Virendra Kumar Vijay

1 Introduction

Climate change is one of the biggest existential threats to our society. Greenhouse gas emissions are leading to an increase in global temperatures and, in the long term, can induce catastrophic effects [1]. There is an immediate need to reduce the dependence on unsustainable resources such as coal and petroleum. Solar energy is a promising renewable and sustainable option with various applications including electricity generation, desalination, and thermal applications [2]. The total amount of solar energy on earth is very high but distributed over a large area and intermittent in nature. Hence, efficient solar radiation collection and storage are essential components for solar thermal applications.

The two main components of the solar collection include a heat transfer fluid and solar thermal concentrator. The heat transfer fluid can be innovated to directly absorb the incident solar radiation to completely substitute for selectivity coated metallic receivers. Significant advances have been made in developing novel designs of solar thermal concentrators which focus the solar radiation with high concentration ratio [2]. These use ethylene glycol, Therminol VP-1 (special formulation of diphenyl oxide and biphenyl), molten salts, etc., as working fluids passing through a selectively coated metallic receiver. However, they lack the absorption capacity for the solar radiation and therefore cannot be used directly for solar thermal applications [3]. Addition of NPs (1–100 nm) to these fluids to form a uniform suspension, also called "nanofluid," can significantly enhance (by up to 7 orders of magnitude) the fluids' solar energy absorption capability, and therefore, the use of nanofluids holds potential in solar thermal applications [3, 4]. Optimizing the nanofluid properties and stability so that it absorbs solar radiation across a broad spectrum of wavelength, especially in the visible and near infrared range (300–2500 nm), is an essential step in developing a new class of directly absorbing solar collectors [5, 6]. Developing robust protocols for nanofluid synthesis which will ensure long-term stability of such nanofluids is, therefore, a key step for its wider application in thermal applications [7–11].

In this study, we have used carbon NPs (<100 nm) and ethylene glycol-based nanofluid which show excellent absorption spectra in the visible and near infrared range [4]. The main objective of this study is to identify optimal conditions for maintaining stability of nanofluid over a prolonged duration of use. We have studied the effects of NP concentration and duration of sonication on nanofluid properties. Our results highlight the importance of these parameters when synthesizing carbon NP-based nanofluids for solar thermal applications.

A. Pramanik (✉) · H. Singh
Institute of Energy Futures, College of Engineering, Design and Physical Sciences, Brunel University London, Uxbridge, UK
e-mail: Anurag.Pramanik@brunel.ac.uk

S. Suresh
Department of Mechanical Engineering, National Institute of Technology,
Tiruchirappalli, Tamil Nadu, India

R. Chandra · V. K. Vijay
Centre for Rural Development and Technology, Indian Institute of Technology Delhi,
New Delhi, India

© The Author(s), under exclusive license to Springer Nature Switzerland AG 2022

A. Sayigh (ed.), *Sustainable Energy Development and Innovation*, Innovative Renewable Energy,
https://doi.org/10.1007/978-3-030-76221-6_18

2 Material and Methods

2.1 Synthesis of Nanofluid

Carbon NPs (Sigma-Aldrich) were weighed (40–300 mg/l) and added to ethylene glycol (Sigma-Aldrich), the base fluid. A nonionic surfactant, 0.1% (wt/vol) sodium dodecyl sulfate (Sigma-Aldrich), was also added to the mixture. The mixture was suspended using a bath-type sonicator for different durations (30 and 60 minutes).

2.2 UV–Vis Characterization

The nanofluid was characterized at various time points after synthesis using a UV–Vis spectrophotometer. The samples were enclosed in a Suprasil quartz cuvette with a path length of 1 cm. Measurements were taken over the 320–1000 nm range. Transmittance spectra of the pure base fluid (ethylene glycol) were also measured for reference.

3 Results and Discussions

The methodology for synthesis of nanofluid involves a two-step process [6]. First, the NPs are prepared as dry powders, and in the second step, they are dispersed in the base fluid with the aid of ultrasonic agitation. This is the most widely used and economical method for nanofluid synthesis. To further increase the stability of the nanofluid suspension, surfactant is added to improve the dispersibility of NPs [6]. Despite its simplicity, there are various parameters which can affect agglomeration of NPs. This can lead to precipitation of the NPs and can result in reduction in absorbance of the nanofluid as well as clogging of pipes, etc. Hence, to identify the important aspects of nanofluid synthesis and storage, we varied several parameters and measured nanofluid stability using spectral absorbance analysis. An outline of the overall experimental strategy and different variables is shown in Fig. 1.

Ethylene glycol is commonly used as a base fluid for solar thermal applications. The transmittance spectra for ethylene glycol are shown in Fig. 2.

The range of wavelength chosen covers the visible and near infrared region (NIR) which accounts for more than 80% of the solar energy. It is observed that ethylene glycol has very high transmittance, hence very low absorbance.

3.1 Effect of Nanoparticle Concentration on Nanofluid Spectral Transmittance

We next studied the effects of NP concentration on the absorption properties of nanofluid. We observed a concentration-dependent exponential decline in transmittance (and hence an increase in absorbance) of nanofluid (Fig. 3b). It is important to note that increasing the concentration of NP beyond 100 mg/L does not improve the absorption by a significant factor. Hence, we chose 100 mg/l concentration for the rest of the experiments in this study.

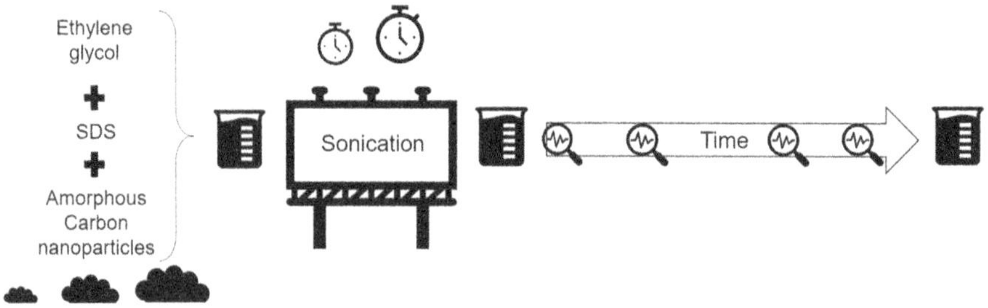

Fig. 1 Outline of experiments used to assess nanofluid stability for long-term storage

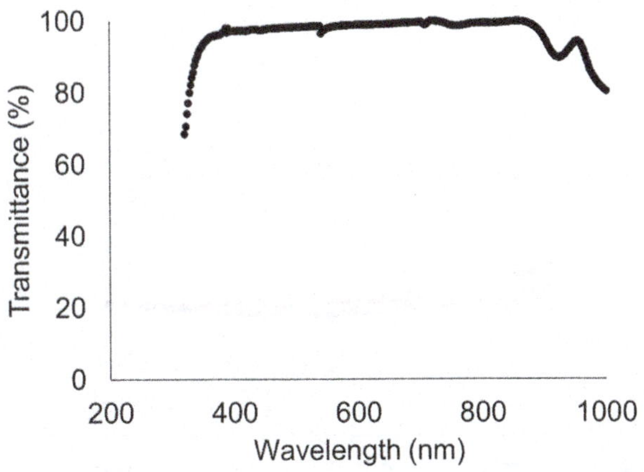

Fig. 2 Spectral analysis of pure ethylene glycol

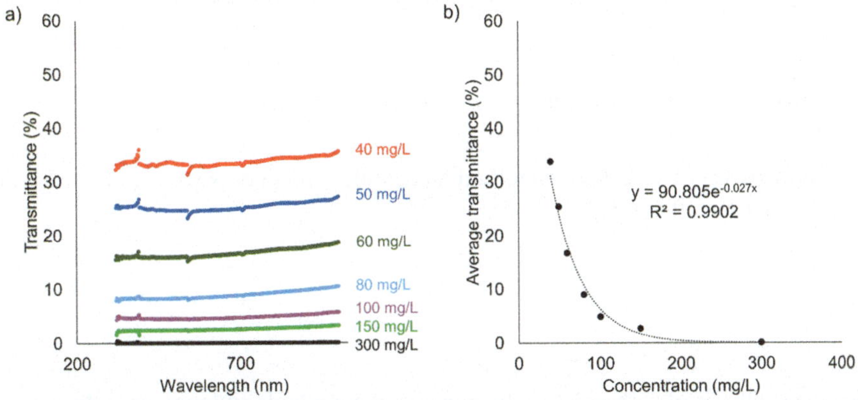

Fig. 3 Spectral properties of nanofluids synthesized. (**a**) Spectral transmittance measured over a spectral wavelength of 320–1000 nm. (**b**) Average transmittance measured over the spectral region against concentration

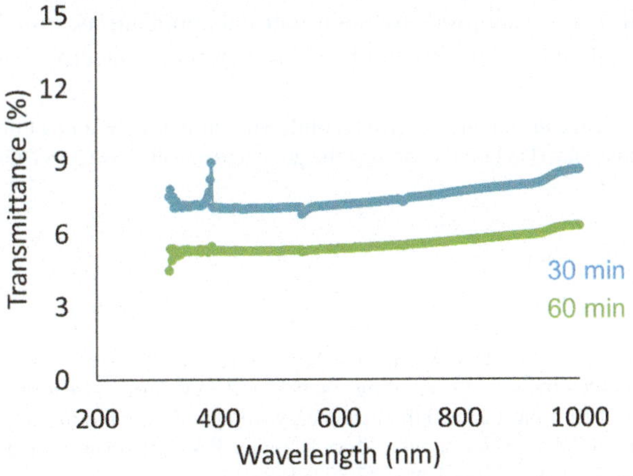

Fig. 4 Sonication duration improves nanofluid properties. Spectral transmittance of nanofluid after sonication for 30 minutes and 60 minutes

3.2 *Effect of Sonication Duration on Spectral Properties of Nanofluid*

The effect of sonication duration on nanofluid stability was studied. It was observed that 60 minutes of sonication led to a lower transmittance suggesting that increasing the sonication duration leads to more efficient dispersion of NPs (Fig. 4).

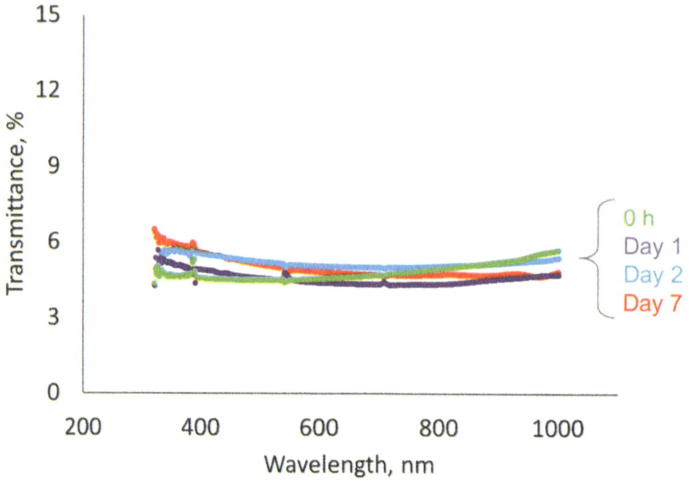

Fig. 5 Nanofluid storage for longer duration. Spectral transmittance curves for nanofluid stored for different duration is shown

3.3 *Long-Term Stability of Nanofluid with Continuous Stirring*

We next studied the long-term effects of storage of nanofluid on nanofluid properties with continuous stirring. We observed negligible effects due to storage even after 7 days (Fig. 5).

4 Conclusion

This study has described the synthesis of carbon NPs-ethylene glycol nanofluids and measured the role of various synthesis conditions on nanofluids' optical performance parameters. An exponential relationship between nanoparticle concentration and average transmittance was observed, which can be used to design nanofluid with desired spectral properties. A longer sonication duration was found to improve nanofluid properties (low transmittance). Finally, continuous stirring has been shown to aid in maintaining long-term stability of nanofluids. The optimal conditions identified in this study are being employed to synthesize carbon NP-based nanofluids for solar thermal applications. Future studies will aim at evaluating the effectiveness of the developed nanofluids as working fluids in solar thermal collectors delivering heat at up to 300 °C.

Acknowledgments The authors acknowledge the funding received for this work from the UK-India Education and Research Initiative (UKIERI) and the Department of Science and Technology (DST) India through the grant agreement IND/CONT/GA/18-19/16.

References

1. Nejat P, Jomehzadeh F, Taheri MM, Gohari M, Muhd MZ (2015) A global review of energy consumption, CO2 emissions and policy in the residential sector (with an overview of the top ten CO2 emitting countries). Renew Sustain Energy Rev 43:843–862
2. Tian Y, Zhao CY (2013) A review of solar collectors and thermal energy storage in solar thermal applications. Appl Energy 104:538–553
3. Khullar V, Tyagi H, Hordy N, Otanicar TP, Hewakuruppu Y, Modi P, Taylor RA (2014) Harvesting solar thermal energy through nanofluid-based volumetric absorption systems. Int J Heat Mass Transf 77:377–384
4. Hordy N, Rabilloud D, Meunier JL, Coulombe S (2014) High temperature and long-term stability of carbon nanotube nanofluids for direct absorption solar thermal collectors. Sol Energy 105:82–90
5. Rose BAJ, Singh H, Verma N, Tassou S, Suresh S, Anantharaman N, Mariotti D, Maguire P (2017) Investigations into nanofluids as direct solar radiation collectors. Sol Energy 147:426–431
6. Yu W, Xie H (2012) A review on nanofluids: preparation, stability mechanisms, and applications. J Nanomater 2012:1–17
7. Mahian O, Kianifar A, Kalogirou SA, Pop I, Wongwises S (2013) A review of the applications of nanofluids in solar energy. Int J Heat Mass Transf 57:582–594
8. Shah SNA, Shahabuddin S, Sabri MFM, Salleh MFM, Ali MA, Hayat N, Sidik NAC, Samykano M, Saidur R (2020) Experimental investigation on stability, thermal conductivity and rheological properties of rGO/ethylene glycol based nanofluids. Int J Heat Mass Transf 150:118981

9. Yu W, France DM, Routbort JL, Choi SUS (2008) Review and comparison of nanofluid thermal conductivity and heat transfer enhancements. Heat Transfer Eng 29:432–460
10. Mesgari S, Taylor RA, Hjerrild NE, Crisostomo F, Li Q, Scott J (2016) An investigation of thermal stability of carbon nanofluids for solar thermal applications. Sol Energy Mater Sol Cells 157:652–659
11. Taylor RA, Phelan PE, Otanicar TP, Adrian R, Prasher R (2011) Nanofluid optical property characterization: towards efficient direct absorption solar collectors. Nanoscale Res Lett 6:225

Thermal Energy Storage Technologies for Utilization of Renewable Energy Sources and EU Project HEART

Uroš Stritih

1 Introduction

Global development with its individual characteristics in each country will determine the future relevance of energy storages [1]:

- *Sensible heat* thermal storage is achieved by heating the storage medium (e.g., liquid sodium, molten salt, or pressurized water) and increasing its energy content but not changing state.
- On the other hand, *latent heat* is associated with changes of phase. Energy required during charging is used to convert a solid material in a liquid material (such as paraffin wax) or a liquid material into a gas. Phase change materials have the benefit of high thermal capacity. In order to use latent heat storage, the storage material should have a melting temperature within the range of the desired operation temperatures of the heat transfer fluid.
- Similarly *chemical heat* requires a specific temperature at which a chemical product is dissociated in a reversible chemical reaction and heat is retrieved when the synthesis reaction takes place. The development of such reactions is yet at a very early stage.

2 Sensible Storage Technologies

In sensible thermal energy storage systems, energy is stored by changing the temperature of a storage medium (water, air, oil, rock beds, bricks, concrete, or sand). The amount of energy introduced to the storage system is proportional to the temperature change, the mass of the storage medium, and the specific heat capacity of the storage medium.

Each medium or material has its own advantages and disadvantages, but usually its selection is based on the heat capacity and the available space for storage.

In Fig. 1, we can see sensible water storage which is used for solar energy systems.

Thermal storage needs to have the following properties:

- *Special Water Intake System*

Tanks need to have a special system of water stratification. Importantly, this system prevents the mixing of hot and cooled water returning to the tank. Unnecessary cooling, and thus heat loss, are significantly less in the described operating mode than in conventional tank filling.

- *Warm Hot Water Without Harmful Microorganisms*

In the heat sink, the hot water is heated by a flow system in a ribbed stainless steel heat exchanger. The hot water flow system is one of the most effective and safe, preventing the development of the harmful bacterium Legionella.

U. Stritih (✉)
University of Ljubljana, Faculty of Mechanical Engineering, Ljubljana, Slovenia
e-mail: Uros.Stritih@fs.uni-lj.si

© The Author(s), under exclusive license to Springer Nature Switzerland AG 2022

A. Sayigh (ed.), *Sustainable Energy Development and Innovation*, Innovative Renewable Energy,
https://doi.org/10.1007/978-3-030-76221-6_19

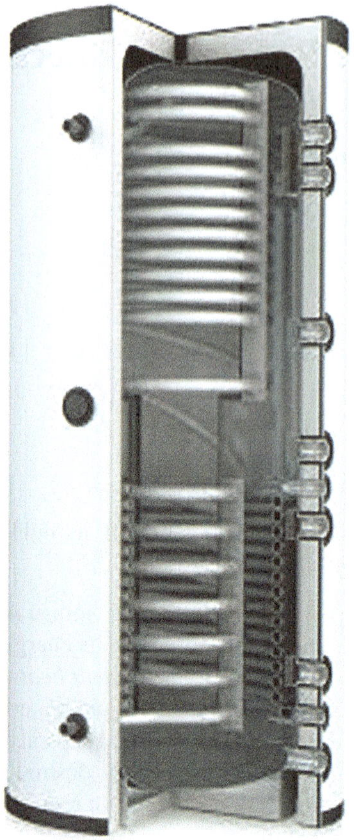

Fig. 1 Sensible (water) thermal energy storage [2]

- *Enough Hot Water in All Seasons*

Pavers allow the connection of different heating sources. During the heating season, the tank should be filled from the heating system, and during the transitional and summer time, the energy from the solar system is added. In the event that none of the above sources is available, the domestic hot water can be heated with an additional heating element.

- *Insulation Without Harmful Substances*

In order to maintain the heat generated, it is very important that the tank is properly insulated. All tanks should be equipped with modern insulations that are free of harmful substances. In the final stage of production, all the tanks should be sandblasted and painted with a protective paint that also protects the outside of the tank from corrosion.

Sensible storages can also be used in seasonal energy storage systems. An example can be seen in Denmark [2].

In this case, the heating system consists of two solar collector fields, a sensible heat storage tank, district heating pipes, a technical building in which are pumps and heat exchangers, and an area that is heated by district heating (Søndervang) with an absorption heat pump.

The heat storage unit presented in Fig. 2 contains 60,000 m^3 of water. Inlet and outlet pipes are driven along the bottom of the storage tank, which lies at the same height as the technical building. The feeder is covered with a 2.5 mm welded polyethylene lining. At the top of the substrate, there is also insulation and foil with vacuum fans that remove the moisture that comes out of the storage tank through the polyethylene lining. Inside and at the top of the lid, there are pipes with concrete that create a slope toward the center of the lid, where the collected rainwater drains away from the pumps.

Sensible tanks need certain space; therefore, latent thermal energy storage technologies are researched.

3 Latent Storage Technologies

Phase change materials (PCMs) can be used to store large amounts of thermal energy in small temperature ranges. In addition to sensible heat, PCMs store the latent heat of a phase transition. Most of the PCMs used in applications are solid-liquid PCMs storing heat in repeated melting and solidifying processes.

Fig. 2 Sensible (water) seasonal thermal storage [3]

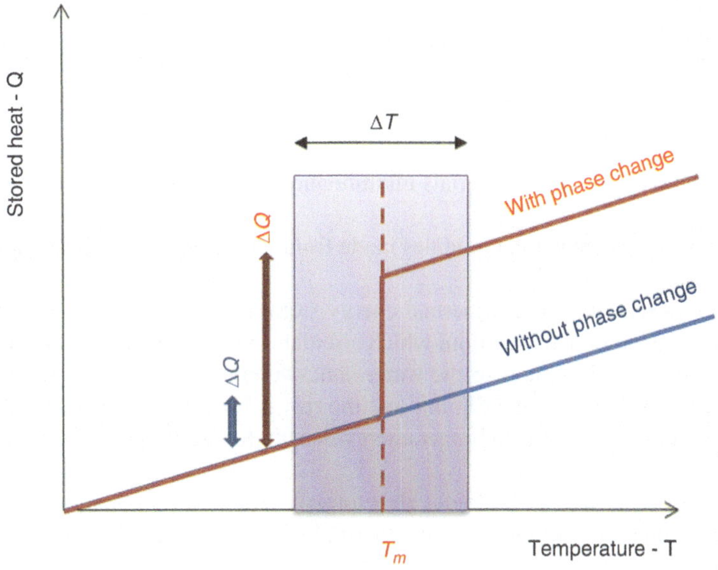

Fig. 3 Heat storage as sensible and latent TES

For efficient latent heat storage, the user has to identify a PCM with a phase change temperature that fits to the temperature level of the application. Figure 3 illustrates the difference between sensible and latent heat storage. Upon melting, while heat is transferred to the PCM, it keeps its temperature constant. Therefore, a larger amount of heat ΔQ can be stored within the temperature interval ΔT around the melting temperature T_m.

Storing of energy in PCMs takes place during the melting process while being used during the freezing process. The process of PCM melting and freezing can be explained by a change in the ambient temperature of the substance. As the ambient temperature rises above the PCM melting point, the chemical bonds in the substance begin to break and the PCM absorbs heat (melting state). When the ambient temperature drops below the melting point of the PCM, the substance begins to release heat and return to solid state.

The ability to store the latent heat of a substance is influenced by [4]:

- Specific heat in solid and liquid state
- Mass of the storage medium
- Temperature difference between start and melting temperature

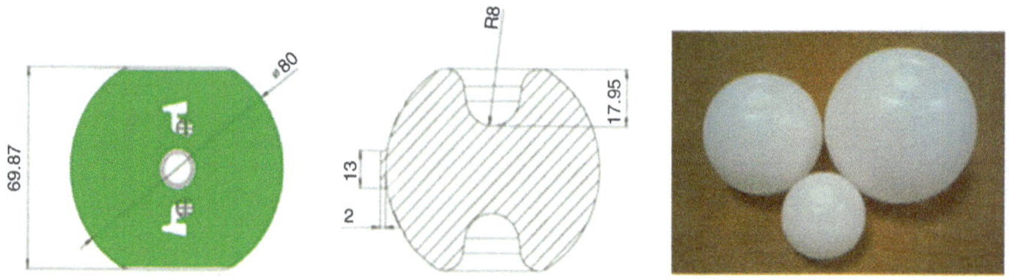

Fig. 4 Spherical encapsulation of PCM [8]

- Temperature difference between melting point and final temperature
- Specific melting heat

PCMs can be used for latent heat storage applications, building heating and cooling applications, photovoltaic module cooling, electronics cooling, and some other applications.

Crucially important for the development of heat storage systems are the media used by the storage tank for operation. The properties we want for the materials used are as follows:

- High energy storage density [Wh/kg or kWh/m^3]
- Low charge temperature
- Ability to high intake of sorbate [g sorbate / g sorbent]
- Proper transfer of heat and matter to provide predicted power
- Easy to handle—nontoxic
- Affordable (low cost per kWh of stored energy)
- Thermal stability—not to deteriorate

PCMs are made from organic or inorganic materials mixture and can be used for cold storage (as ice banks) or heat storage [5–7].

Macroencapsulation can be done in spherical modules made from HDPE. Technical drawing of encapsulation model and its picture is shown Fig. 4.

In Fig. 5, we can see a cross section of latent thermal energy storage. Spherical nodules filled with PCM are inserted in sensible TES unit with water as heat transfer medium which circulates in the unit and ensures heat exchange between PCM in nodules and water. By adding PCM nodules to the water tank, we obtain hybrid TES where PCM presents 40% of the volume according to the whole volume of the TES unit and the rest is filled by water which also acts as a thermal storage medium. Advantage of the proposed system is in higher energy density for thermal energy storage in comparison with water at the same temperature level.

Although latent thermal energy storage technologies are promising for many applications, such as solar systems, biomass systems, and heat and cold production from heat pumps, thermal losses occur. Therefore, chemical thermal energy storage technologies are researched.

4 Thermochemical Storage Technologies

Thermochemical heat storage is generally classified under chemical heat storage processes. Under the term thermochemical heat storage, we usually summarize sorption heat storage processes. Sorption can be defined as a phenomenon of fixation of a gas by a substance in solid or liquid phase. We distinguish between adsorption and absorption. The term absorption is used when the molecules of a substance in gas phase enter a liquid (usually) or solid thereby changing the composition of the liquid or solid. Adsorption is defined as binding of a gas on a surface of a solid or porous material. Adsorption is further divided in physical adsorption or physisorption and chemical adsorption or chemisorption [9].

Reversible sorption heat storage processes can be written in the following way:

$$AB + Q = A + B$$

where AB is a compound of components A and B and Q designates the heat supplied to dissociate AB into components A and B. Hence, the dissociation of compound AB represents an endothermic reaction. When components A and B are put in contact,

Fig. 5 Spherical encapsulation of PCM into tank [8]

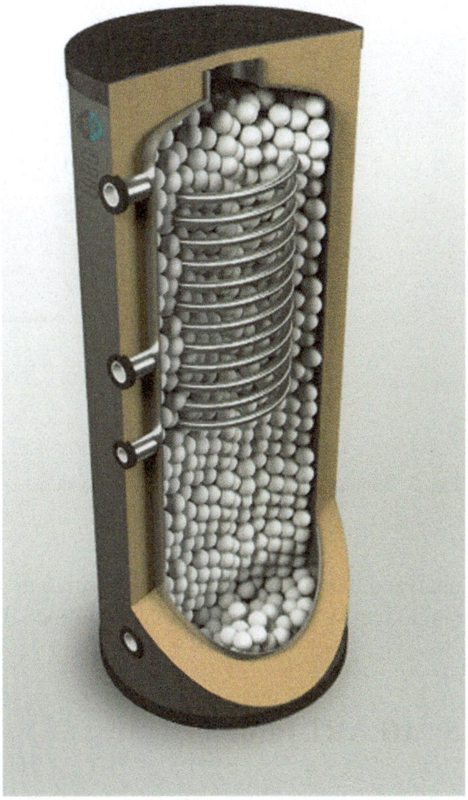

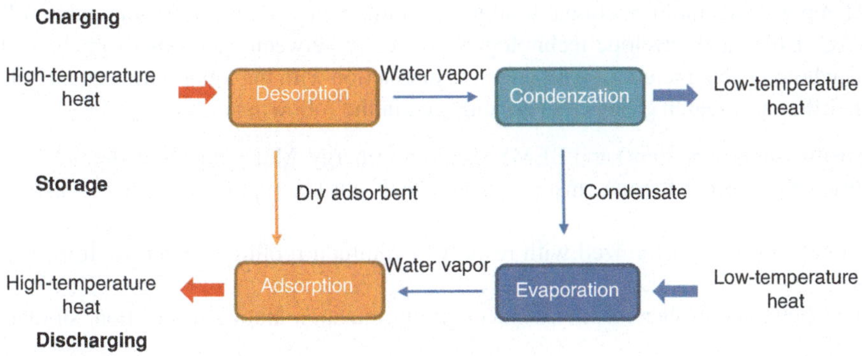

Fig. 6 Principle of operation of the sorption system for the energy storage

heat is released (exothermic reaction) during the forming of compound *AB*. The supplied heat Q can be stored with negligible heat loss as long the components *A* and B are separated. As a result of this, the dissociation of compound *AB* is called charging, while the forming of compound *AB* represents the discharging process of the thermochemical heat storage cycle. The component which is desorbed during charging is named adsorbate. The material capable of adsorbing (absorbing) the other component is called adsorbent (absorbent). The term adsorptive (absorptive) is also used for the adsorbate in the desorbed state, especially when adsorbate and adsorptive differ in chemical structure (not the case in physisorption). In heat storage applications, mainly water (vapor) is used as adsorbate because of its availability (i.e., cheap) and nontoxicity (Fig. 6).

Design of presented system in this chapter is similar to solar-driven sorption storage heating system as shown in Fig. 7. A real scale seasonal solar closed storage sorption system was analyzed for several months under realistic conditions in the heating period. Heat is provided in evacuated tube collectors in combination with buffer storage. Several sorption storage modules are included in the system that offers "charge boost" when vapor transfers between two of modules. This allows higher states of charge for a given charging temperature level. With validated developed numerical model, it has been

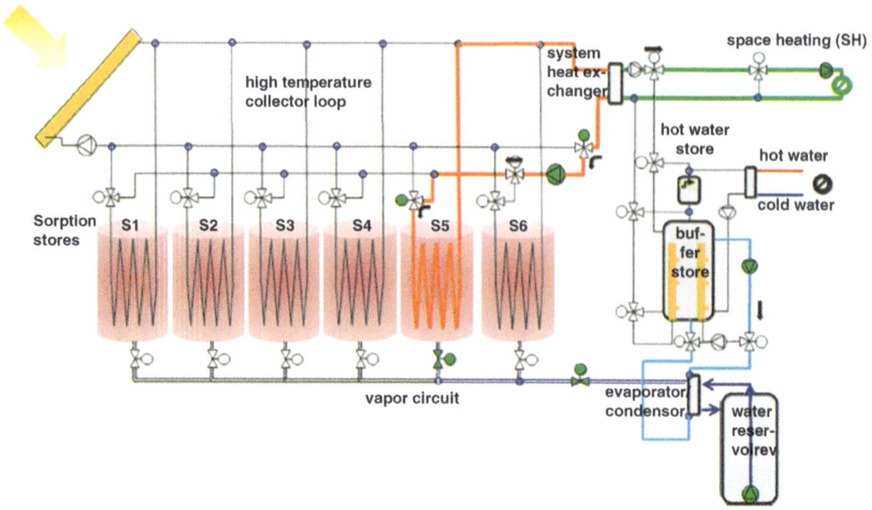

Fig. 7 Schematic of the sorption storage system

analyzed that for a storage system with 6 m^3 of material fractional energy, savings up to 75% has been achieved for standard single family house in the middle Europe [10].

5 HEART Project

Laboratory for Heating, Sanitary and Solar technology and Air-conditioning is currently involved in EU project HEART where we are in charge in testing of high-efficient storage tank.

The concept of HEART represents a multifunctional toolkit for building energy retrofit within which following subcomponents ICT, BEMS, HVAC, BIPV, and envelope technologies cooperate between each other to achieve higher levels of energy effectiveness of the whole system including with effective interaction with the Smart Grid. Those subcomponents are under development within different research groups that are engaged in the following areas:

- Cloud-based DSS (Decision Support System) and BEMS (Building Energy, Management Systems)
- Multifunctional prefabricated external thermal insulation in which wiring, pipes, vents, and sensing options can be integrated
- Universal PV tiles for pitched roofs, customized with respect to common roofing (for flat rooftops regular open racks systems would do)
- Multi-Input/Multi-Output power controller for interconnection optimization among production, storage, and consumption devices
- Hydronic air-to-water DC modular heat pump
- Advanced hot/cold storage tanks for thermal storage of heat pump that runs on PV
- Battery pack for PV electricity storage and power management (the only subcomponent not specifically developed in the project)
- DC smart fan-coils for heating/cooling of rooms (replacing existing radiators) [11]

A user who wants to use the HEART tool will register on the website to create a user account. He will then enter all relevant information about the building that he would like to renovate and also set goals (energy savings, return on investment, etc.), through which the tool will provide possible solutions for individual parts (subcomponents) of the renovation. The next step is to review the solutions in terms of energy, environment, economy, and manager, depending on boundary conditions set by the user. The solution you choose can serve as a reference when designing a renovation. The model can be updated at any moment, which allows us to check the various modifications that we would like to make in the planning and design phase, and to show us their effectiveness, giving us some feedback that can help us make a decision (Fig. 8).

HEART is able to transform an existing building in a highly efficient Smart Building. Its application allows not only to optimize the building's energy performance but also to make in extremely reliable, documented, and transparent. This latter

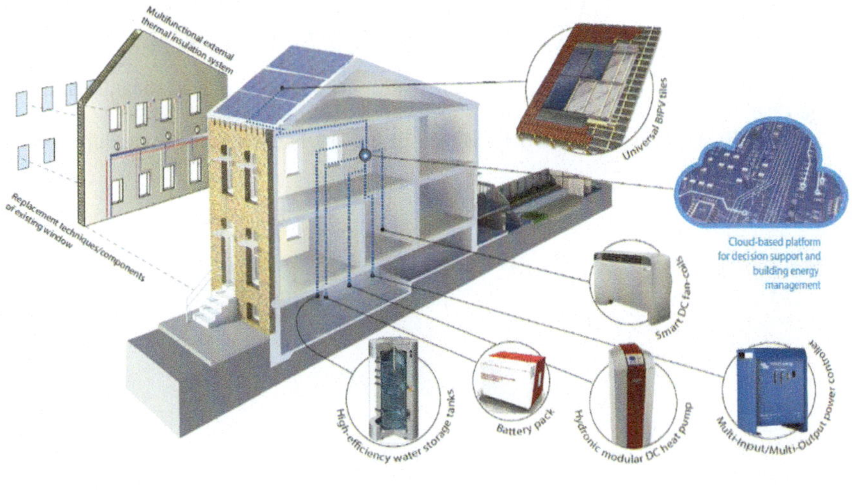

Fig. 8 Concept of HEART toolkit

aspect proves to be one of HEART's greatest strengths as it constitutes an instrument of control and guarantee for the simulation and promotion of incentives and (private or public) financial investments in building energy retrofit. Moreover, HEART is a multi-stakeholder tool, which can be used and applied from almost all actors involved in the renovation process (owners, inhabitants, designers, real estate developers and contractors, installers, building managers, financers, decision makers, etc.) in different phases of the building's life. HEART's contribution to the evolution of building retrofit can be briefly summarized through its main features:

- Retrofit planning and implementation optimization
- Reduction of energy consumption for heating, cooling, and DHW
- Reinforcement of RES exploitation
- Rationalization of energy flows (including renewables) inside the building and between building and smart grids

6 Conclusions

Storage technologies are nowadays a central component in energy-efficient buildings which implies also the substitution of fossil energy carriers by low carbon energy and closed carbon cycles, which means less CO_2 from fossil fuel power plants and a higher share of renewable generation. Renewable energy from solar and wind shows a high additional potential for electricity generation. By using heat pumps, electric cars, or synthetic fuels based on green hydrogen, methane or methanol (power-to-fuels), renewable energy will gain more and more importance and will contribute to the decarbonization of the heating, cooling, and transportation sector as well.

References

1. International Energy Agency. Energy conservation through energy storage. www.iea-eces.org
2. Stritih U, Paksoy H, Turgut B, Osterman E, Evliya H, Butala V (2015) Sustainable energy management: solar energy and thermal storage technologies in two Mediterranean countries. Manag Environ Qual 26:764–790
3. Dronninglund Solar thermal plant. www.drlund-fjernvarme.dk
4. Mehling H, Cabeza LF (2008) Heat and cold storage with PCM, Berlin. Springer, Berlin Heidelberg
5. Stritih U, Zavrl E, Paksoy H (2019) Energy analysis and carbon saving potential of a complex heating system with solar assisted heat pump and phase change material (PCM) thermal storage in different climatic conditions. European Journal of Sustainable Development Research 3(1):1–17
6. Stritih U, Charvát P, Koželj R, Klimeš L, Osterman E, Ostrý M, Butala V (2018) PCM thermal energy storage in solar heating of ventilation air: experimental and numerical investigations. Sustain Cities Soc 37:104–115
7. Osterman E, Butala V, Stritih U (2015) PCM thermal storage system for free heating and cooling of buildings. Energ Buildings 106:125–133

8. Global-E-Systems. GAIA energy storage ball – TES – thermal energy storage for heat and cold. www.global-e-systems.com
9. Mlakar U, Stropnik R, Koželj R, Medved S, Stritih U (2019) Experimental and numerical analysis of seasonal solar-energy storage in buildings. Int J Energy Res 43(12):6409–6418
10. Engel G, Asenbeck S, Köll R, Kerskes H, Wagner W, van Helden W (2017) Simulation of a seasonal, solar-driven sorption storage heating system. J Energy Storage 13:40–47
11. Grant Agreement No. 768921 – EU H2020 project HEART – H2020-EEB-2017. https://cordis.europa.eu/project/rcn/211645_en.html

Vacuum Insulation Panels for Thermal Energy Storage Systems

Sankarshan Verma and Harjit Singh

1 Introduction

The unavailability of solar irradiance during night or cloudy days has made the link between concentrated solar power (CSP) plants and thermal energy storage (TES) inextricable. Although TES technology at plants such as Andasol 3 (Spain), Diwakar (India), and Aurora Solar Energy Project (Australia) have been installed as a proof of concept, most of the CSP plants including Ivanpah Solar Electric Generating System (USA) and Waad Al Shamal ISCC Plant (Saudi Arabia) do not have a TES option [1]. One of the major reasons lie in the complexity involved in storage.

The thermal energy can be stored, as sensible heat either in the heat transfer fluid itself (direct storage) or in a different medium (in-direct storage), such as a molten salt, in two separate tanks: cold tank and hot tank. The molten salt from cold tanks is passed through a heat exchanger to the hot tank when solar irradiance is at its peak, thereby storing the excess thermal energy. While the irradiance is not strong enough, the salt is discharged from the hot tank and passed through the heat exchanger to extract the thermal energy which can be used to run the turbine. The most common molten salt used for this application, 60% $NaNO_3$ and 40% KNO_3, has a high freezing point (220 °C). This means that a drop in temperature of the salt at any point below 220 °C could cause crystallization of the salt, which could cause collapse of the entire plant [2]. Furthermore, the salt in hot tank must remain at a minimum specified temperature (around 393 °C for Andasol) so that enough heat can be extracted to run the turbine. At the same time, there is an upper limit on maximum temperature that the salt could have based on excess thermal energy available from the CSP plant. Thus, there is a clear ground to develop techniques to minimize heat losses from the TES tanks to maintain the temperature of the salt in both hot and cold TES tanks.

As Fourier law suggests, the rate of heat loss can be reduced by either increasing the thickness of insulation on the tank or by using an insulation with low thermal conductivity. Presently, due to high temperatures involved, thick layers of mineral wool or glass wool are used to insulate the TES tanks. Vacuum insulation panels (VIPs), which are increasingly being used in cold chain equipments like refrigerators, cold storage boxes, etc. [3, 4], could also be effective to suppress the heat losses from TES tanks, due to their extremely low thermal conductivity (0.004 W m^{-1} K^{-1} at room temperature) [5]. In the present work, we developed a computer model to compute the rate of heat loss from a hot TES tank throughout the day, with varying ambient temperature, including its charging and discharging processes. The rate of heat loss from walls, top, and bottom of the tank was separately calculated. The total rate of heat loss, which was the sum of these individual losses, was compared for various thicknesses of two types of insulation, that is, glass wool and VIPs. Also, based on extinction coefficients and $p_{1/2}$ values, a suitable VIP core material was chosen from the literature, and its thermal conductivity as a function of temperature was calculated.

S. Verma (✉) · H. Singh
Institute of Energy Futures, College of Engineering, Design and Physical Sciences, Brunel University London, Uxbridge, UK
e-mail: Sankarshan.Verma@brunel.ac.uk; harjit.singh@brunel.ac.uk

© The Author(s), under exclusive license to Springer Nature Switzerland AG 2022

A. Sayigh (ed.), *Sustainable Energy Development and Innovation*, Innovative Renewable Energy,
https://doi.org/10.1007/978-3-030-76221-6_20

2 Model

A fully parametric model of a cylindrical hot tank was developed in COMSOL Multiphysics 5.5 (see Fig. 1). The diameter and the height were set to 36 m and 14 m, respectively, as in the case of Andasol 1 [1]. The body of tank was made of 5-mm-thick steel.

The storage temperature in the hot tank was set at constant value of 393 °C. Taking advantage of the tank's geometry, the model was set as a 2D axis-symmetric geometry. The mesh in the tank's body and insulation was set as free triangular with element sizes ranging from 1.8 to 0.008 m. The mesh for regions inside tank and the ground was mapped accordingly. In total, the model consisted of around 89,727 domain elements and 7381 boundary elements.

2.1 Boundary Conditions

The storage tank loses heat through all three modes, that is, conduction, convection, and radiation. The heat escapes through walls and top lid to ambient atmosphere via natural convection and to the ground via conduction. The base of the tank is set as 1 m concrete, with a constant temperature of 25 °C at its bottom. The ambient temperature, which plays a significant role in convective heat transfer, was taken from a weather station located in Riyad, Saudi Arabia recorded on June 21, 2019 [6]. The profile of ambient temperature was slightly modified to make our model computationally less expensive (Fig. 2).

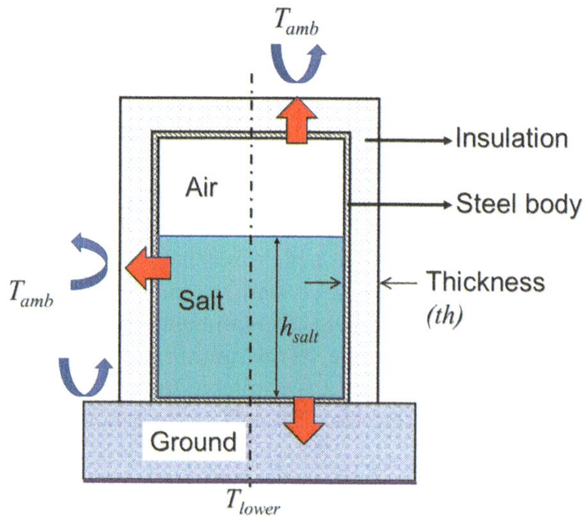

Fig. 1 2D schematic of the thermal storage tank. The red arrows represent the heat transferred due to conduction and radiation and the curved arrows represent convective heat transfer

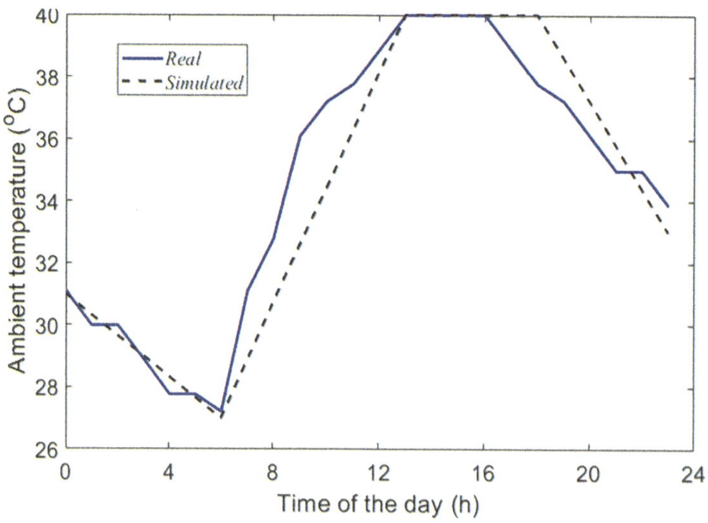

Fig. 2 Ambient temperature profile input to the model. "*Real*" denotes the actual temperature profile and "*Simulated*" denotes the one used as input in our model

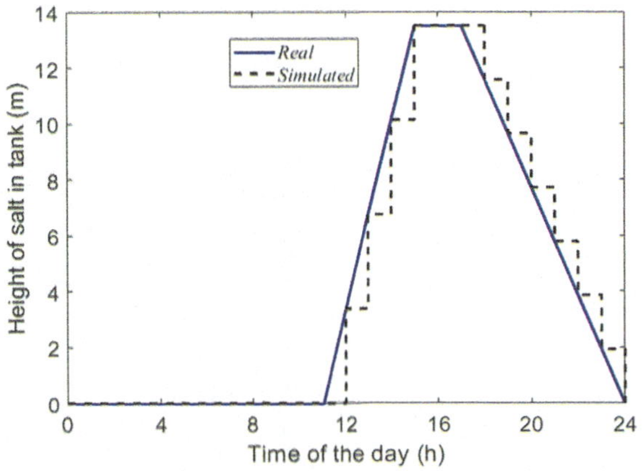

Fig. 3 Height of salt in the hot tank with time of the day

In this model, it was assumed that the peak hours in which the charging of hot tank occurs are from 11:00 to 15:00. During this time, the height of salt, which significantly affects heat flow rate through the tank, keeps increasing. After 15:00, when tank was fully filled, the salt stayed in the tank till sun's irradiation was strong enough to generate power, that is, 17:00, after which the hot tank began discharging the salt and the height of salt started falling. At any stage, it was assumed that the salt's temperature in the tank was 393 °C, as this was a nonnegotiable aspect for smooth functioning of the power plant. In our model, instead of considering a continuous rise or fall of height of the salt in the tank during charging and discharging, it was assumed that it occurs in steps of 1 hour, to make the model computationally less expensive (Fig. 3).

The model was run for two types of insulation materials, that is, glass wool and vacuum insulation panels, each with varying thickness. The model computes rate of heat loss for a full day at regular intervals, including filling.

and unfilling processes. A brief introduction to vacuum insulation panels and the heat transfer phenomenon through them is presented in Sect. 2.2.

2.2 *Vacuum Insulation Panels*

Vacuum insulation panels, or VIPs, are insulations which have the ability to provide a thermal conductivity value of 4 $mWm^{-1}\,K^{-1}$ at room temperatures, which is around 10 times lower than glass wool's thermal conductivity. VIPs consist of a porous core material made of powder/fibers/foam which is evacuated to pressures ranging from 1 to 0.1 mbar and sealed in a polymer-/ metal-based envelope [3, 4].

Generally, the heat transfer through VIPs is considered to be occurring parallelly, through gaseous conduction in pores, solid conduction through core's matrix and radiation, and thus the total thermal conductivity is a sum of the three conductivities (Eq. 1).

$$k_{\text{total}} = k_s + k_g + k_r \tag{1}$$

The evacuation at low pressure leads to suppression of gaseous conductivity (k_g) and convection in the core material, and highly porous nature of the core leads to low solid conductivity (k_s). This is the reason behind extremely low overall thermal conductivity offered by VIP. Various opacifying substances like silicon carbide and iron oxide are added in the core material to suppress the heat transfer through radiation, or radiative conductivity (k_r), which becomes a dominant mode at high temperatures.

As an input to our model, we chose a VIP sealed at 1 mbar with core made of 80% precipitated silica and 20% SiC. The temperature-dependent thermal conductivity of this VIP was calculated using data from literature [7] and Eqs. 1, 2, and 3 (see Fig. 4).

$$k_r(T) = \frac{16n^2\sigma T^3}{3\rho e} \tag{2}$$

$$k_g(T) = \frac{k_g^0(T)}{1 + {p_{1/2}}/{p}} \tag{3}$$

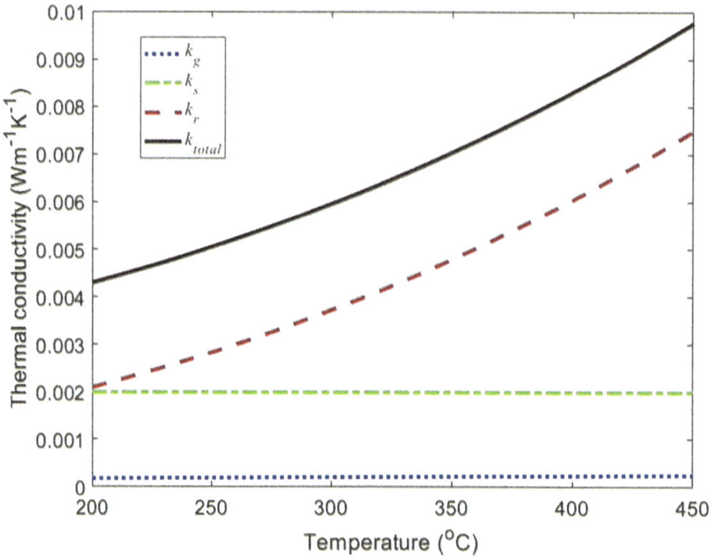

Fig. 4 Total thermal conductivity of the VIP and its constituents as a function of temperature

where n is the refractive index of core material which can be considered 1 because of its highly porous nature, σ the Stefan-Boltzman constant, T the mean operating temperature of the VIP, ρ the density of core material, e the specific extinction coefficient, k_g^0 the thermal conductivity of free air, $p_{1/2}$ a material specific constant, and p the pressure of air inside VIP.

$$k_{\text{glass wool}} = \frac{0.0318 - 0.0283}{30} \max\left(0,\ T - 263.15\right) + 0.0283 \tag{4}$$

3 Results and Discussions

The rate of heat loss was calculated for various thicknesses of glass wool and the VIP. The results are as shown below (Figs. 5 and 6).

The results clearly indicate how using VIPs can benefit the tank. The total rate of heat loss reached a peak of 1400 kW when the tank was insulated with 20-mm-thick glass wool which could be suppressed to around 420 kW by increasing the insulation thickness to 500 mm. On the other hand, a 100-mm-thick VIP is comparable in performance, with 500-mm-thick glass wool thus saving 5% of the tank's volume, which could be used to store more salt and thus more power. This is due to extremely low thermal conductivity of the VIP.

One notable characteristic of all heat loss graphs is the peak in the middle region. This peak occurs when the tank is fully filled with the salt and thus the heat loss through walls and top increases. As discussed earlier, the total rate of heat loss is a sum of that from top, bottom, and the walls (Fig. 6). It is interesting to note that with a tank-insulated with glass wool, the major loss occurred through the walls, whereas for a VIP-insulated tank, the major loss occurred through the bottom or the base. A solution to this could be to insulate the floor with VIPs as well.

4 Conclusion

A heat transfer model of a thermal storage tank containing the hot salt at 393 °C was developed. The rate of heat loss was calculated throughout the day which included the charging and discharging process for two types of insulation materials, that is, glass wool, which is conventionally used to insulate the tanks and a suitable VIP, which offered a thermal conductivity of around 8.5 $mWm^{-1}\,K^{-1}$ at 400 °C. The thickness of these materials was varied and the total rate of heat loss calculated at

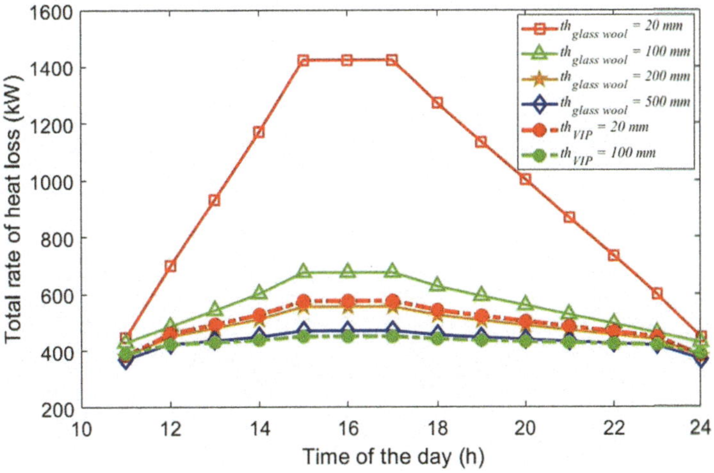

Fig. 5 Total rate of heat loss through the tank with varying insulation thicknesses of glass wool (solid) and VIP (dashed)

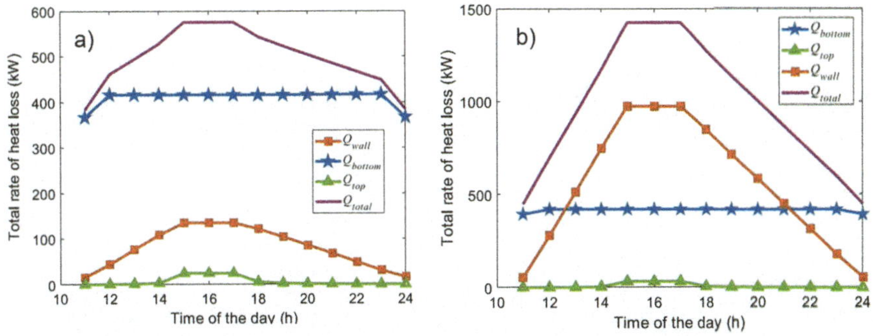

Fig. 6 Total rate of heat loss and its constituents for a hot tank insulated with 20 mm thick (**a**) VIP and (**b**) glass wool

every step. It was found that a 100-mm-thick VIP could be as efficient as 500-mm-thick conventional glass wool insulation through the day. Also, it was found out that the major heat loss in a VIP-insulated tank occurred through the base of the tank.

Acknowledgments The authors would like to acknowledge the funding received from EPSRC (UK) for the project SUNRISE (EP/P032591/1), which made this research possible.

References

1. NREL (n.d.) Concentrating solar power projects. https://solarpaces.nrel.gov/. Accessed 15 July 2020
2. Suárez C, Iranzo A, Pino FJ, Guerra J (2015) Transient analysis of the cooling process of molten salt thermal storage tanks due to standby heat loss. Appl Energy 142:56–65. https://doi.org/10.1016/j.apenergy.2014.12.082
3. Alam M, Singh H, Limbachiya MC (2011) Vacuum insulation panels (vips) for building construction industry - a review of the contemporary developments and future directions. Appl Energy 88:3592–3602. https://doi.org/10.1016/j.apenergy.2011.04.040
4. Verma S, Singh H (2020) Vacuum insulation panels for refrigerators. Int J Refrig 112:215–228. https://doi.org/10.1016/j.ijrefrig.2019.12.007
5. Fantucci S, Lorenzati A, Kazas G, Levchenko D, Serale G (2015) Thermal energy storage with super insulating materials: a parametrical analysis. Energy Procedia 78:441–446. https://doi.org/10.1016/j.egypro.2015.11.691
6. TWC Product and Technology LLC, Riyadh, Saudi Arabia Weather History. 2020. https://www.wunderground.com/history/daily/sa/riyadh/OERK
7. Caps R, Fricke J (2000) Thermal conductivity of opacified powder filler materials for vacuum insulations. Int J Thermophys 21:445–452. https://doi.org/10.1023/A:1006691731253

Modelling of Heat Pump System with Air and Solar Heat Sources Including Latent Thermal Energy Storage

Fran Torbarina, Kristian Lenic, Anica Trp, Mateo Kirincic, and Boris Delac

1 Introduction

As a part of the EU climate action, strategies and targets have been adopted by European Council with the long-term goal of EU becoming an economy with net-zero greenhouse gas emissions and become climate-neutral by 2050. In the following 10-year period, that is, by 2030, the key targets are at least 40% cuts in greenhouse gas emissions (from 1990 levels), at least 32% share of renewable energy, and at least 32.5% improvement in energy efficiency. According to the European Commission's data [1], space heating takes up a share of 68% in total end-use energy consumption which is why utilizing renewable energy sources in space heating applications has a great potential in meeting climate goals. Traditionally, heating energy has been obtained by burning of fossil fuels and biomass; however, heat pumps are being implemented more often in recent years due to their high efficiency and low ecology impact. Heating systems utilizing heat pumps with seasonal performance factor higher than 2.5 are considered as renewable energy sources [2].

Since the efficiency of heat pumps depends highly on the heat source and heat sink temperatures, utilizing solar energy for achieving higher evaporation temperature in heat pump systems has potential of additionally improving their performance. However, the mismatch in solar energy availability and highest heating demand in a day has to be taken care of. This is why special attention has to be paid to thermal energy storage design when utilizing solar energy in heat pump systems in order to maximize heat pump efficiency, prolong the daily operating time in favourable mode, and reduce the number of ON/OFF cycles. Sensible thermal energy storages (STES) with water as a storage medium are mostly used because of their low investment cost, but many researches have lately been done on latent thermal energy storages (LTES) with phase change materials (PCM) because of their ability to release thermal energy at constant or almost constant temperature which can be beneficial in heat pump systems.

Combining heat pumps with LTES has three main advantages: (1) energy stored in LTES can be used as a heat source for the heat pump enabling a near constant evaporating temperature which results in relatively constant coefficient of performance (COP), (2) defrosting cycles which are common with air source heat pumps and which deteriorate the average COP can be avoided, and (3) utilizing an LTES reduces required installed cooling capacity of the heat pump and simultaneously cuts down the investment cost [3]. The main disadvantages of LTES are transient heat transfer and low thermal conductivity of the PCM [4].

There are three main applications of systems utilizing both heat pump and LTES found in the literature: space heating, domestic hot water production, and space cooling. For space heating purposes, several designs concerning different placements of the LTES in heat pump system can be found in literature, but all fit well in one of the two characteristic designs: (1) LTES connected to condenser of the heat pump and (2) LTES connected to evaporator of the heat pump.

Heat pump systems in which LTES is connected with the condenser can be used for harmonization of the period of economically efficient operation with the period of heating demand. This way the system can operate during the low-cost electricity tariff storing the heat in the LTES and use it when needed during day when the electricity cost is higher. Several papers can be found in literature [5–7] where authors have investigated LTES connected to condenser of the heat pump. The main findings were that systems with LTES achieve lower number of heat pump ON/OFF cycles and lower energy losses

F. Torbarina (✉) · K. Lenic · A. Trp · M. Kirincic · B. Delac
University of Rijeka, Faculty of Engineering, Rijeka, Croatia
e-mail: fran.torbarina@riteh.hr

© The Author(s), under exclusive license to Springer Nature Switzerland AG 2022

A. Sayigh (ed.), *Sustainable Energy Development and Innovation*, Innovative Renewable Energy,
https://doi.org/10.1007/978-3-030-76221-6_21

compared to systems with sensible thermal energy storage. The authors pointed out that further research needs to be done to improve the heat transfer in LTES. By improving the heat transfer, the volume of the LTES could be significantly reduced [5]. Another application of these systems is eliminating peeks in heating loads. By storing the excess heat from the heat pump during the low-heat-need periods of the day, greater mean COP could be achieved [8]. The melting temperature of PCMs used in systems, where the LTES is connected to the condenser, was higher than 45 °C in all the papers that were used as reference.

The systems in which LTES is connected to the heat pump evaporator have been also investigated. Comakli et al. [9] investigated a system with heat pump, LTES, and solar collectors. They suggested a design in which during daytime, the heat pump operates as an air-source heat pump, while the heat from the solar collectors is used for charging the LTES and the excess heat is used directly for space heating. During nighttime, the heat pump uses the heat from the LTES as the heat source for the evaporator and operates with high efficiency. The authors stated that this design is suitable for locations with large number of cloudy days. Kaygusuz et al. [10] analyzed the economic feasibility and ecological impact of such a system.

Minglu et al. [11] investigated a system where both condenser and evaporator of the air-source heat pump are connected to the LTES. During heat pump operation, the excess heat from the condenser is stored in the LTES and used as a heat source for the evaporator during defrosting cycles. This way, the thermal comfort in the heated space is not deteriorated during defrosting cycles and the risk of working fluid leakage is reduced due to avoiding temperature straining which occurs when there are frequent defrosting cycles of the evaporator.

In this chapter, the benefits of integrating latent thermal energy storage in a system with solar thermal collectors and water-water heat pump are investigated. The heat pump and solar collector system with LTES is compared to a system utilizing a sensible thermal energy storage and to a system without thermal storage and without solar collectors. The systems have been compared in terms of thermodynamic performance and operating costs. As a result, heat pump COPs, total electricity consumption, electricity consumption of an auxiliary heater, and operating costs during a 3-week period for a southeast European location have been presented.

2 Systems Description

The schematic diagram of the investigated system with a latent thermal energy storage, System A, is shown in Fig. 1. The system consists of evacuated tube solar collectors (SC), latent thermal energy storage (LTES), water-water heat pump, water storage in solar loop (TES 1), two small inertial water storages (TES 2 and TES 3) connected to the heat pump evaporator and condenser side, plate heat exchangers (HE 1 and HE 2), auxiliary heater (Aux), dry coolers (DC 1 and DC 2), circulation pumps (P), and regulation valves (RV). Paraffin with melting temperature range from 27 °C to 28 °C, and 170 kJ/kg latent heat was used as the PCM for latent thermal energy storage.

The system operates in two daily regimes: (1) *Charging of the heat storage*: During the first part of the day, from 5:45 h to 16 h, the heat from solar collectors is used for charging the LTES. The heat pump operates using outside air as a heat source via dry cooler connected to the evaporator. In cases when the temperature of the outside air is below the minimum operating temperature (evaporator inlet temperature drops below −2 °C), heat pump stops and the auxiliary electric heater is turned on which heats the working fluid before it supplies the consumers. (2) *Discharging of the heat storage*: After 16 h, the heat source for the heat pump is switched from outside air to the LTES. The reason for that is an effort to achieve higher

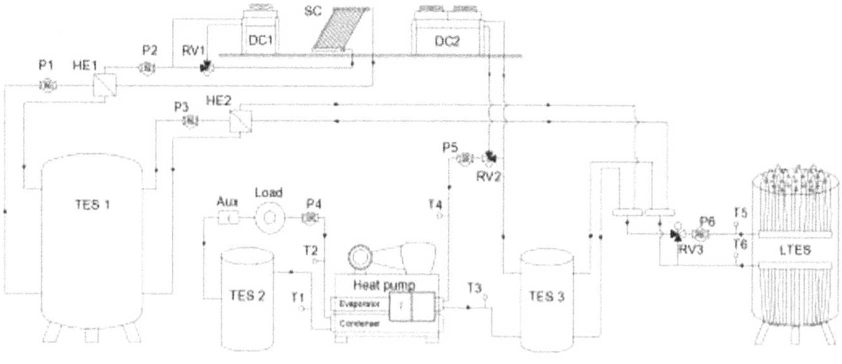

Fig. 1 Schematic diagram of the heating system utilizing heat pump, solar collectors, and latent thermal energy storage—System A

evaporation temperatures during the highest heating demand in late afternoon and evening in order to enable the heat pump to operate with higher COP. This would improve the efficiency of the system and reduce operating costs. Total daily operating time is from 5:45 h in the morning until 22 h in the evening.

The second considered system, System B, was equivalent to the one shown in Fig. 1, but instead of LTES it uses sensible thermal energy storage (STES)—a water storage with the same volume as LTES. The two operating modes were the same as well as in case of the System A. The third considered system, System C, consisted of the heat pump, two small inertial water storages (TES 2 and TES 3) connected to the heat pump evaporator and condenser, auxiliary heater (Aux), dry cooler (DC 2), and circulation pumps (P). The system operated using no thermal energy storages, in only one regime when the heat pump operates using outside air as a heat source during the whole day.

The systems were compared in terms of electricity consumption, operating costs, coefficient of performance (COP), and auxiliary thermal energy consumption. Daytime and nighttime electricity tariffs have been taken into account in a way that the electricity cost for the period from 7 h in the morning until 21 h in the evening was 0.11 EUR/kWh and from 21 h to 7 h the cost was 0.054 EUR/kWh [12].

3 System Modelling

Transient model of the system was developed in TRNSYS. It consists of more than 40 components where each one represents the mathematical model of an actual component such as water storage, heat pump, regulation valve, etc. (Fig. 2). The main used components (in TRNSYS referred to as Types) are briefly described in continuation.

For modelling of evacuated tube solar collectors, Type 538 was used [14]. The selected component calculates fin efficiency, loss coefficient, and transmittance-absorptance product at normal incidence based on the user-provided efficiency equation coefficients. The thermal capacitance of the collector is accounted for in this model as well. Sensible thermal energy storages (water storages) have been modelled using Type 534 [15]. This component models a fluid-filled, constant volume cylindrical storage tank with a vertical configuration. The tank is divided into isothermal temperature nodes where each node interacts thermally with the nodes above and below through several mechanisms: fluid conduction between nodes and through fluid movement (either forced movement from inlet flow streams or natural destratification, that is, mixing due to temperature inversions in the tank). For modelling of the latent thermal energy storage, Type 840 was used [16]. The model is based on the enthalpy approach in which the enthalpy is a continuous and invertible function of the temperature. It enables a detailed simulation of water tanks with integrated PCM modules or tanks filled with PCM. For modelling of a water-water heat pump, Type 927 was used [17], which models a single-stage heat pump. The user must provide a data file containing normalized capacity and power draw of the heat pump as functions of entering source fluid temperature and entering load fluid temperature on the basis of which the outlet evaporator and condenser temperatures are calculated.

Other used types for modelling of the proposed system were Type 654 (Single-speed circulation pump), Type 709 (Circular fluid-filled pipe), Type 511 (Dry fluid cooler), Type 91 (Heat exchanger), Type 647 and Type 649 (Mixing valve for fluid), Type 911 (Differential controller with lock-outs), Type 138 (Auxiliary heater), and Type 14 (General forcing function).

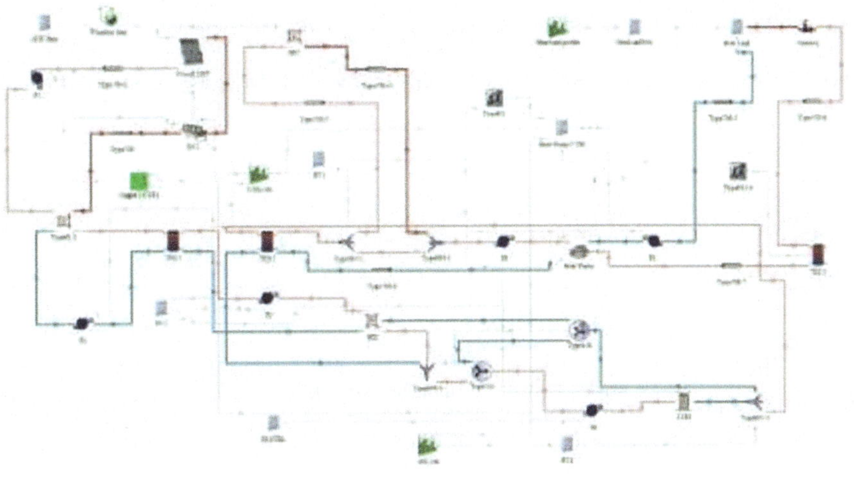

Fig. 2 System model diagram in TRNSYS

For reading of the weather data and radiation processing, Type 15 was used [13]. This component reads data from an external data file and interpolates it at time steps of less than one hour, making it available to other TRNSYS components. The simulations of the analysed systems have been performed for a 3-week heating period during February for location Zagreb, Croatia. Simulation time step was one minute.

4 Results and Discussion

Latent (LTES) and sensible (STES) thermal energy storage outlet temperatures during a 1-week period from February 7 to February 14 for System A and System B are shown in Fig. 3.

It can be seen that in the first part of the day, that is, during the charging process, the STES outlet temperatures are greater than the LTES outlet temperatures. During the second part of the day, that is, during discharging, the LTES outlet temperatures are higher, which have a positive effect on the efficiency of the heat pump.

Evaporator inlet temperatures for three compared systems are shown in Fig. 4 for the same 1-week period from February 7 to February 14.

Because the heat pump operates as an air-source heat pump during the first part of the day in all three cases, evaporator inlet temperatures are the same as well during those time periods. After 16 h in each day, however, the differences in evaporator inlet temperatures are evident. The System A with LTES achieves highest evaporator inlet temperatures, following with the System B with a water thermal energy storage, and the lowest evaporator inlet temperatures are achieved in case of System C when no thermal energy storage is utilized. Evaporator inlet temperature has a considerable effect on heat pump COP. The COP will be higher when the evaporator inlet temperature is higher as well. The COP profiles for three compared systems for the same period are shown in Fig. 5.

During the first part of the day, the COP values are quite similar for all three systems. It is because the heat pump operates in the same conditions during those periods. During the second part of the day, the benefit of integrating thermal energy storage and solar collectors in a system with heat pump can be observed because the values of COP in cases with LTES (System A) and STES (System B) are higher than the one of the systems without thermal energy storage (System C). During the two-day period from 942 h to 990 h, the heat pump in System C was off during late afternoons and evening which happened because the outside temperature was so low that the heat pump could not receive enough heat for evaporation. During those periods, the thermal energy for the consumers was supplied to the working medium by auxiliary electric heater. Comparing

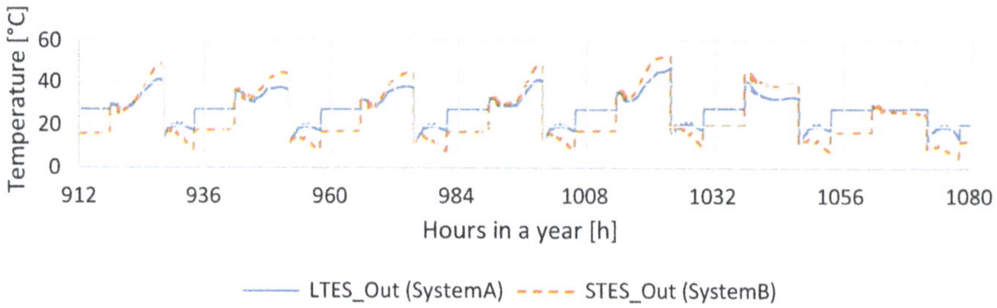

Fig. 3 Water outlet temperatures from thermal energy storages for System A and System B during a 1-week period (from February 7 to February 14)

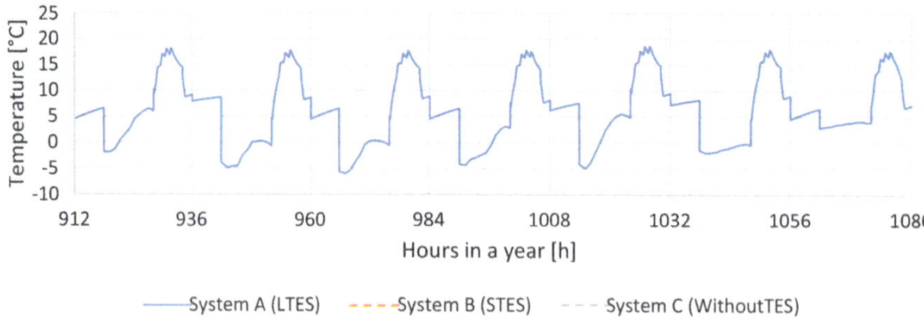

Fig. 4 Evaporator water inlet temperatures for all three compared systems during a 1-week period (from February 7 to February 14)

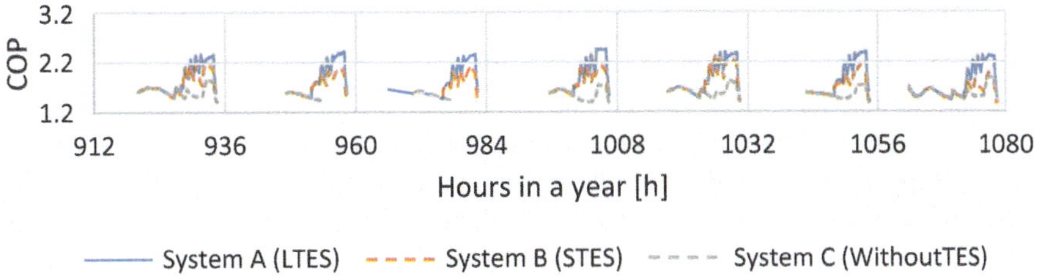

Fig. 5 COP profiles for three compared systems during a 1-week period (from February 7 to February 14)

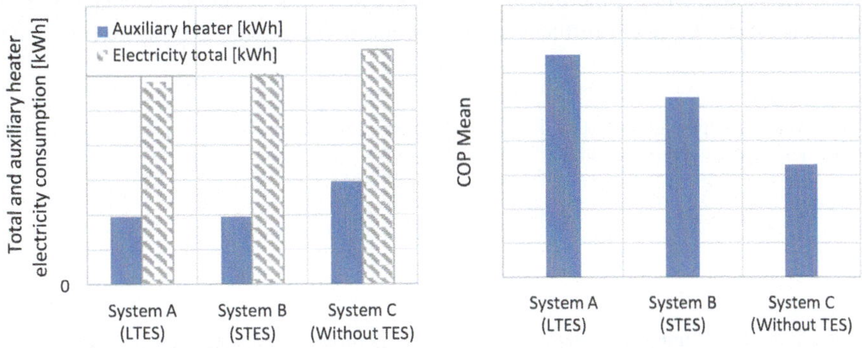

Fig. 6 Total electricity consumption, auxiliary heater consumption, and mean COP for three investigated systems during the analysed 3-week period

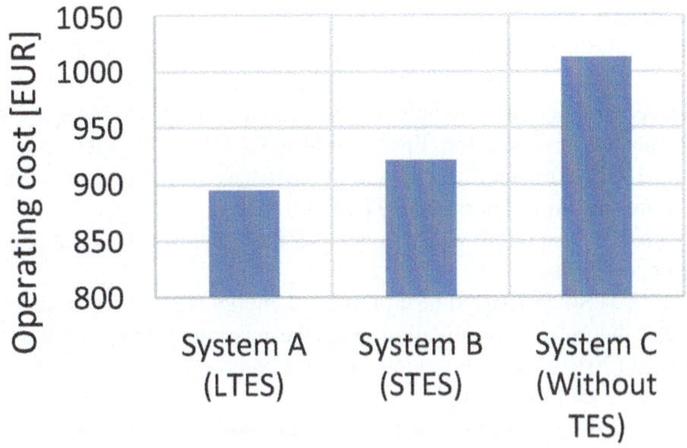

Fig. 7 Operating costs comparison for three considered systems during the analysed 3-week period

the system utilizing LTES (System A) with the one utilizing STES (System B), it can be observed that during the selected time period, the former one has a better thermodynamic performance because the COP is higher.

Total electricity consumption, auxiliary heater electricity consumption, as well as the heat pump COP for three investigated systems are show in Fig. 6.

System with LTES (System A) achieved lowest total electricity consumption and lowest thermal energy supplied by auxiliary heater followed by system with STES (System B) and the system without thermal energy storage (System C). On the other hand, the mean COP during the analysed 3-week period was highest for the system with LTES (System A), followed by system with STES (System B).The operating costs for three considered systems for a 3-week period are shown in Fig. 7.

As can be seen from Fig. 7, the operating costs were the lowest for the system utilizing LTES followed by the system with STES and with the system without thermal energy storage. Compared to the system with STES, the relative difference between operating costs was 2.92%, and compared to the system without thermal energy storage, the relative difference between operating costs was 13.1%.

5 Conclusion

In this work, unsteady simulations have been carried out with the goal to examine potential benefits of integrating latent thermal energy storage in systems with solar collectors and heat pump. Thermodynamic performance and operating costs have been analysed and compared for three different heat pump systems—the first one utilized solar collectors and LTES, the second one utilized solar collectors and STES, and the third one utilized no thermal energy storage, no solar collectors, and only worked as an air-source heat pump system. The system with LTES achieved higher average COP compared to the system with STES and to the system with no storage, due to the higher evaporator inlet temperatures which, in the end, resulted in lower electricity consumption and lower operating costs. It was found that utilizing latent thermal energy storage and solar collectors in a heat pump system can considerably improve mean COP compared to the system with sensible thermal energy storage and the system without thermal energy storage and solar collectors. Considering that the investment costs have not been accountant for in this work, further research would be necessary in order to prove the feasibility of the proposed system.

Acknowledgement This work was supported in part by the Croatian Science Foundation under the project HEXENER (IP-2016-06-4095) and in part by the University of Rijeka under the project number "uniri-tehnic-18-69."

References

1. European Commission, Energy consumption by end-use, 2013. Retrieved from ec.europa.eu/energy/content/energy-consumption-end-use_en
2. European Parliament and of the Council on the Energy Performance of Buildings, Commission Decision No. 2013/114/EU, 2013. O.J. L 62/27
3. Pardiñas AA, Justo AM, Diz R, Husevåg Kvalsvik K, Fernández-Seara J (2017) State-of-the-art for the use of phase-change materials in tanks coupled with heat pumps. Energ Buildings 140(1):28–41
4. Streicher W, Schultz J, Solé C, Cabeza L, Bony J, Citherlet S (2008) Final report of subtask C "phase change materials", Report C7 – IEA SHC Task 32, International Energy Association
5. Leonhardt C, Müller D (2009) Modelling of residential heating systems using a phase change material storage system. In: Proceedings 7th Modelica conference, Como
6. Agyenim F, Hewitt N (2010) The development of a finned phase change material (PCM) storage system to take advantage of off-peak electricity tariff for improvement in cost of heat pump operation. Energ Buildings 42(9):1552–1560
7. Kelly NJ, Tuohy PG, Hawkes AD (2014) Performance assessment of tariff-based air source heat pump load shifting in a UK detached dwelling featuring phase change-enhanced buffering. Appl Therm Eng 71(2):809–820
8. Maaraoui S, Clodic D, Dalicieux P (2012) Heat pump with a condenser including solid- liquid phase change material. In: International refrigeration and air conditioning conference, Purdue
9. Comakli O, Kaygusuz K, Ayhan T (1993) Solar-assisted heat pump and energy storage for residential heating. Sol Energy 51(5):357–366
10. Kaygusuz K (2000) Experimental and theoretical investigation of a solar heating system with heat pump. Renew Energy 21(1):79–102
11. Minglu Q, Liang X, Deng S, Yiqiang J (2010) Improved indoor thermal comfort during defrost with a novel reverse-cycle defrosting method for air source heat pumps. Build Environ 45(11):2354–2361
12. HEP Elektra, Tarrif Items, 2019. Retrieved from http://www.hep.hr/elektra/kucanstvo/tarifne- stavke-cijene/1547
13. TRNSYS 18 a TRaNsient SYstem Simulation Program Volume 4 Mathematical Reference
14. TESSLibs 18 Component Libraries for the TRNSYS Simulation Environment Volume 10 Solar Library Mathematical Reference
15. TESSLibs 18 Component Libraries for the TRNSYS Simulation Environment Volume 11 - Storage Tank Library Mathematical Reference
16. Schranzhofer H, Heinz A et al (2006) Validation of TRNSYS simulation model for PCM energy storages and PCM wall construction elements. In: Ecostock conference 31st May– 2nd June 2006, Pomona
17. TESSLibs 18 Component Libraries for the TRNSYS Simulation Environment Volume 11 - HVAC Library Mathematical Reference

Battery Storage: An Enabler for Utility and Grid Dynamics

Debmalya Sen

1 Indian Power Sector—Status Quo and Projections (Fig. 1)

The Indian power sectors' [1] present generating capacity is 366 GW, of which 54% is coal. But in just 2 years' time, the percentage of coal in the grid has come down from 59% to 54%, which in itself shows the way the sector is moving. The annual growth percentage of the sector as on date is around 6–8%. Renewable capacity as on date is 83 GW, which represents around 14% of the total generating capacity. Around 62 GWs of renewable capacity is under execution and bidding as on November 2019. Solar and wind both have rapidly increased their penetration into the Indian grid. In 2018 alone, renewables have pumped in 101.83 billion units into the grid. One sector which is highly unexplored due to infrastructure difficulties and cost is pumped hydro (85% unexplored). Analysis shows that India has potential of 96 GWs of pumped hydro, but at present only 4.8 GW is operational and additional 1 GW is under construction. As on date, India is 2.5% peak surplus and 4.6% energy surplus. When seen region wise, only South and East India experiences slight power deficits. With increase in demand and rise of urban population, the peaks being experienced across India has also increased considerably. The daily peaks experienced in 2019 have ranged as high as 150–170 GW. As indicated earlier, in order to cater to the increase in renewable generation, utilities have been bound to curtail generation from conventional plants. This can be inferred from the increase in reserve margin in the grid, which has increased from 51% in 2011 to as high as 72% in 2017.

The year 2019 promises to bring back the growth to track with around 50% more addition of RE capacity forecasted. 2018 was seen as a low investment year majorly due to currency fluctuations and antidumping duty which was imposed on imported solar panels. India has set a target to have 40% generation capacity through non-fossil fuel sources by 2030. Projections show that 75% of India will be powered by renewables by 2050. With this increase in renewables, therefore, what increases is the diversity in the grid, which is expected to be around 5% by 2022. This will also lead to considerable reduction in the spinning reserves present today. What this also signals that the ramping rates will be increasing as with massive solar ingress (75 GW by 2025), early morning and late evening ramps will get more steeper and for longer duration. As on date, the ramping rate is 150 MW/min for 40 minutes; this can go up to 340 MW/min for 3–4 hours by 2025.

As such the requirement of grid stability is well understood and appreciated; thus, there are various regulations and standards which have been regularized to make the grid responsible. The industry has also realized upon the importance of storage for grid reliability. IESA (India Energy Storage Alliance) [1] predicts that the storage market for India can be around 300 GW by 2025. The main markets include electric vehicles, UPS, renewables, and Telecom. There has been particularly huge focus on battery storage which has been a decrease of almost 85% at cell-level costs (Lithium Ion) over the last 8 years. This has mainly resulted from installations across the world of BESS projects. As on 2018, the total storage installation in India was 24 GWh; this is projected to reach 65 GWh by 2026. NITI Aayog is also about to launch the India Gigafactory document which predicts a 50 GWh market for India by 2025. Total potential is being predicted as 268 GWh (CAGR of 21% from 2018 to 2025). If we look into the application areas, around 68% of the demand comes from behind the meter applications, followed by grid applications.

D. Sen (✉)
Emerging Technologies, Wakad, Pune, Maharashtra, India
e-mail: dsen@ces-ltd.com

© The Author(s), under exclusive license to Springer Nature Switzerland AG 2022

A. Sayigh (ed.), *Sustainable Energy Development and Innovation*, Innovative Renewable Energy,
https://doi.org/10.1007/978-3-030-76221-6_22

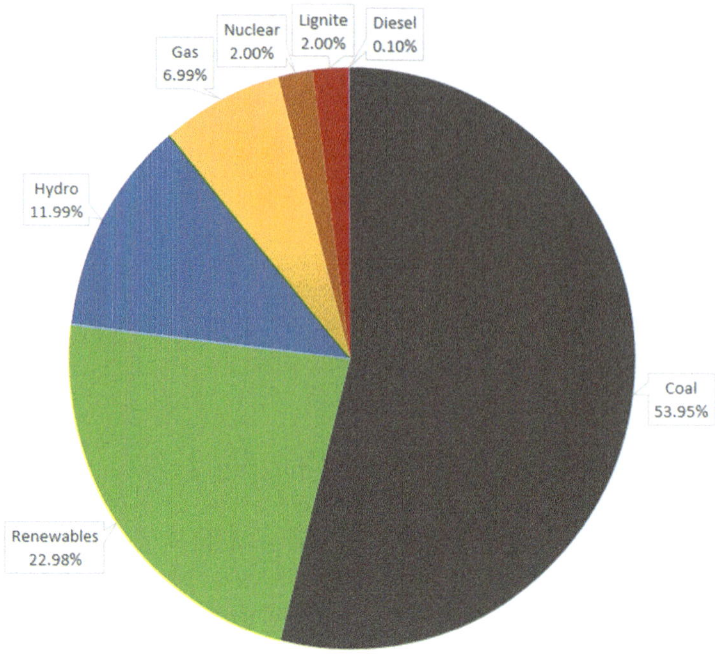

Fig. 1 Indian power generation capacity mix (Nov 2019)

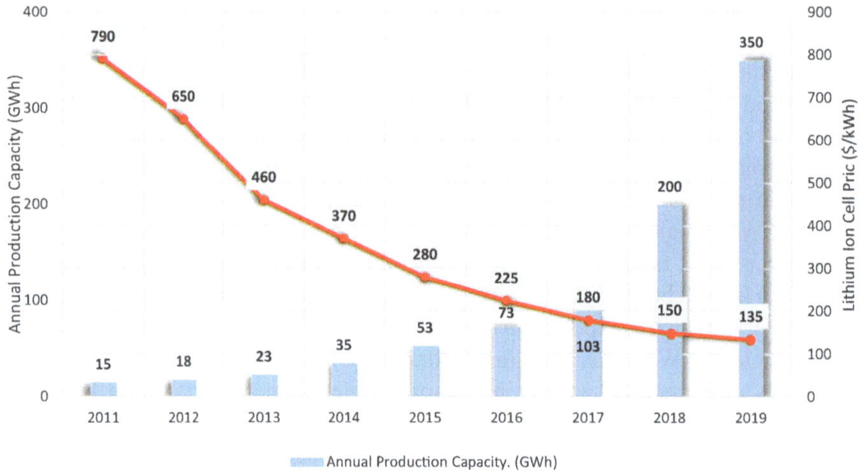

Fig. 2 Battery price and manufacturing capacity trend (CES Analysis)

The increasing percentage of renewables in the grid also will increase the ramping rate requirements. Studies predict 150 MW/min for 40 minutes at present to 340 MW/min for 3–4 hours by 2025 increase in ramping rates. Among usage and demand, AC usage is projected to increase significantly—24% of total energy consumption demand.

Potential of the storage market as per report is more than 300 GWh by 2030 growing at a CAGR of 21% from 2017 to 2025. The world has seen a steep decline in battery prices. Battery prices have decreased 85% from 2011 to 2018; at present battery prices are around USD150/ KWh, further 30% reduction expected in 2 years' time, almost reaching USD100/KWh by 2030.

Figure 2 shows how the price has fallen for battery packs from 2011 to 2018. The price drop mainly attributes to the economies of scale for each manufacturer crossing the 5 GWh mark by 2015.

2 Government Commitments

Government of India [2] has been actively promoting the greening of grid revolution. It has placed ambitious targets from time to time from 225 GWs of renewable capacity by 2022 (including hydro) to 30% of vehicles manufactured as EVs by 2030. India also made commitment in Paris Agreement to reduce GPD's emission intensity by 33% by 2030 from its level in 2005 and promotion of indigenous manufacturing and development of skills—Make in India. With policy frameworks like FAME I and now FAME II and also encouragement and incentives on local manufacturing, the government has made clear its intent. All these have given a clear indication to market on the way forward in the power sector. Recently, NITI Aayog has also been discussing upon the Gigafactory plan which is in the final stages of cabinet approval; the plan focusses on domestic manufacturing of batteries in India, so that the country does not have to rely on imports and becomes a market leader in the storage Industry.

3 Concerns and Utilities and DISCOM

The 2012 Indian Grid Blackout had given rise to many strict regulations to improve grid reliability. This includes stricter deviation settlement mechanism (DSM) protocols for both generators and DISCOMs with high penalties for violating schedules, narrower frequency bands, and more stringent rules for upgrades and maintenance cycles.

With the acceleration of renewable growth as discussed above, the grid variability also increases and over time, due to being run on power utilization factors mainly due to nonavailability of cheap domestic gas, the gas plants are slowly being shut down. Plants which are operational are facing trouble in extension of PPAs as the landed tariff is now being no more competitive with renewable tariff reaching Rs. 2.8/kWh. Thus, the spinning reserves which were prevalent in the grid is slowly depleting out thereby reducing the grid inertia. The concern for utilities and DISCOMs is twofold. One is the nonreliability of renewables; over time, forecasting and scheduling have been regularized for renewables, but the present forecasting software itself are not that accurate; this results in real-time variations which the utilities find hard to manage. These in-turn are subjected to penalties for full use or overdraws. The second concern is high solar generation. Unlike wind, solar is only available during day time; thus to cater for solar, the utilities back down other generation sources, but on the contrary, this is generally a low demand region. At a time when the demand actually starts rising (~5 pm), solar generation plummets. Thus, it is a situation where demand increases and supply decreases. To balance off this mismatch, utilities are faced with steep ramping blocks, and predictions show that such ramping will only become much steeper and longer going forward. IESA believes this can go to as high as 330 MW/min. The question arises, who will take care of such high ramping in a high RE grid. Thermal plants are not designed to cope with infrequent sudden changes, gas plants fit in well, but cheap gas availability is a concern, pumped hydro can, but we do not have enough installations of pumped hydro in India, this is where batteries can play a big role, as they have fast ramping capabilities.

Over and above this, the increase in rooftop solar in urban cities is creating problems of back feeding and reactive power support at the distribution transformer levels. The question which arises here is that is the existing infrastructure capable of handling this dynamic? One issue with DTs itself is its placement in fast-growing Indian cities, which makes upgrades a very exhaustive and expensive task. One should also not forget about the captive power generation, which also ends up relying on the grid for reliability and in turn increases uncertainty of operations for the grid. There has been talks of making the load management blocks from 15 minutes at present to 5 minutes, which means the level of load management dynamic will only increase going forward.

4 Storage as an Enabler: Focus—Utilities and DISCOMs (Fig. 3)

As illustrated, there are many solutions, but not many which makes technical and commercial viability. Battery storage is one solution which therefore is becoming more and more prominent for addressing these concerns. Till some years back, the mere price of batteries and their characteristics were making storage not a viable option for installation, but continuous developments of the storage sector and reduction in cost are slowly making storage deployments more and more attractive. The following section discusses on some of the potential applications of battery storage to address the current concerns being faced by utilities and DISCOMs.

Table 1 shows the various applications that battery storage can play a role in across the energy value chain.

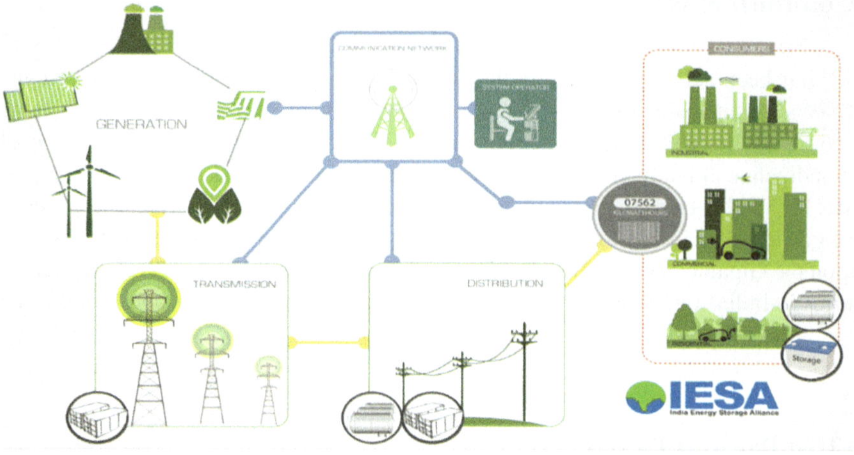

Fig. 3 An illustration of the connected ecosystem

Table 1 ESS applications across the energy value chain [3]

Transmission	National grid	RLDC/SLDC/IPP
Investment deferral	Delivering ramp rates	Frequency regulation
Voltage support	Output shifting	Firming of output
Congestion management	Acting as flexible reserve	Smoothing and deviation control
Reduction in losses	Saving investment in peakers and additional system capacity	

Table 2 Best case application of storage for various sectors [4]

Application	Grid/local	Hours
Avoiding of investments in new intermediate load plants through RE firming	Grid	4–8
Replacement of peaker plants	Grid	4–6
Ramping at grid scale level	Grid	0.5–1
Avoiding of wind and solar curtailment	Local	2–6
Frequency regulation	Grid	0.25–1
T&D congestion relief/upgrade investment deferral	Local	1–4
Deviation settlement for both inter- and intrastate	Local	0.5–2
Firming of output of wind and solar PV plants	Local	2–4

Some of the value proposition of energy storage mentioned can be served by central-level assets such as pumped hydro storage and compressed air storage. But these assets have limitations over the advanced storage technologies, which are their long gestation period in scale of 10–12 years and their inability to serve local transmission and integration issues.

A report on forecasting, Concept of Renewable Energy Management Centers and Grid Balancing by GIZ, states that from a long-term perspective, energy storage becomes a viable option. But the report recommends integration of RE through energy storage only after considering less costly balancing options like demand-side management, adaptation of conventional power plants, and regional balancing.

Energy storage assets can handle multiple applications and also lead to multiple cost benefits. However, co-owning and multiuse would require the regulations to evolve with a holistic view of benefits of energy storage (Table 2).

5 Case for ESS—Transmission Investment Deferral

The RE addition into the Indian power grid is expected to cross 175 GW by 2022, of which 160 GW comes from solar and wind energy resources and 120 GW is of utility scale projects that are connected directly to the transmission grid network. These two sources of power are variable in nature, and the typical total peak energy from solar + wind crosses 100 GW only

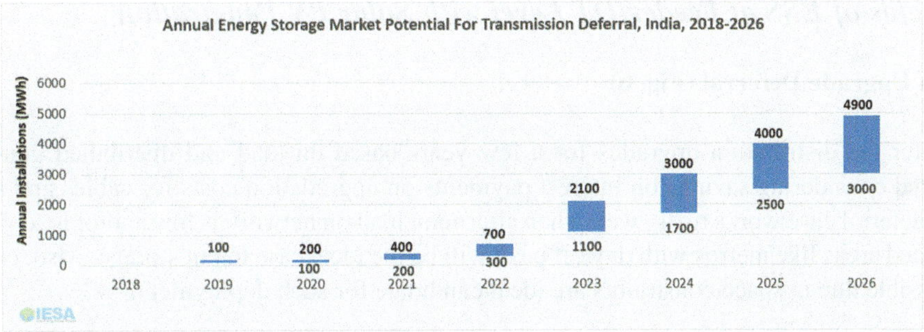

Fig. 4 Annual energy storage potential for transmission deferral [4]

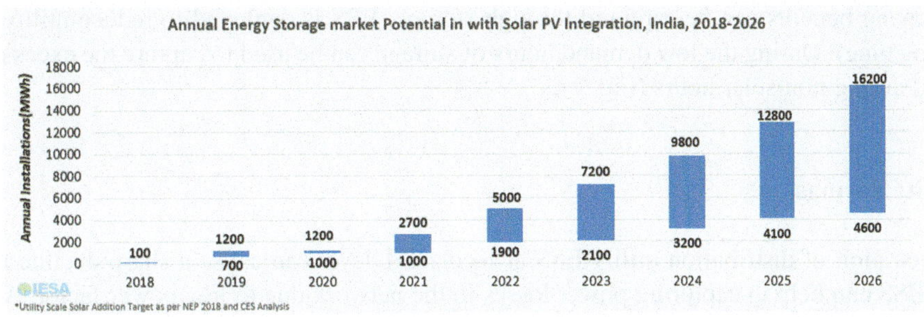

Fig. 5 Annual energy storage market potential (PV Integration)

for 29 days a year for 4 hours per day. This constitutes 1.3% of the total time during the year; the transmission line connected to the project is utilized for 120 GW. Energy storage can be used for deferring transmission capacity investments. Based on system modelling studies, optimum location for the addition of energy storage can be decided and integrated. This results in delivering the RE power to the consumers at the cheapest cost.

Most of these wind or solar farms require transmission capacity additions or transmission line expansions. As per the National Electricity Plan Vol II—Transmission, 47 GVA of transmission addition is necessary by 2022 with RE target hitting 175 GW. From CES analysis, it is observed that the peak RE generation crosses 100 GW only 1.3% of the time (in 2022) for which transmission investment can be saved with the installation of a BESS for 5 GW/15 GWh. While several wind and solar farms are farther than 50 kms away from the existing transmission structure, nearly 50% of the projects are well within this range (of 50 kms). By installing an ESS to store the peak generation, there can be significant savings on the transmission line investments. Hence by reducing 1.3% of the power from being transmitted across the lines, there can be savings on the transmission cost of constructing about 6 GVA of substations and transmission line capacities. These energy storage units can be also utilized for alternate application such as RE firming and RE output smoothening, thus the overall benefit is multiplied (Fig. 4).

The total potential capacity of ESS for green corridor investment deferral is around 5 GW/15 GWh as per CES analysis. The market becomes more attractive with the promotion of locally manufactured products which bring down the CAPEX. In India, RE curtailment is a major issue due to power evacuation challenges. Besides, the CAPEX involved in transmission capacity enhancement, obtaining right of way, land clearances for substations are all key challenges faced by developers. ESS is an immediate remedy for such cases, with a smaller gestation time.

6 Case for ESS at Distribution Side

Distribution grids are having a lot of issues in India, like low network reliability, power quality issues like low power factor and harmonics, and high technical and commercial losses. 415 V and 11 kV networks are considered not flexible for voltage and power quality fluctuations, as interventions like automated online tap changers (OLTC) are used mostly at 33 kV network and above. In such a scenario, higher penetration of rooftop solar PV, integration of EV chargers, and rising commercial loads can create a lot of difficulties for the operators in maintaining the network. ESS, as witnessed across the world, can provide a lot of flexibility at low- and medium-voltage electricity network (Fig. 5).

6.1 *Major Benefits of ESS at Feeder/DT Level with Solar PV Penetration*

6.1.1 Distribution Upgrade Deferral (Fig. 6)

ESS can help to defer the distribution upgrades for a few years based on load and distributed generation growth rates. Benefits are calculated considering savings on interest payments on upgradation costs for cables and substation equipment due to deferral. The deferral case works really well when afternoon load on network is low and/or load growth on the network is low. Hence, saturated areas like metros with day off peak will be the ideal case to put storage. Also, cities where substation upgrades are not possible due to space constraints are ideal candidate for such deployment.

6.1.2 System-Wide Peak Saving Benefits

System-wide peak saving benefits can be captured through storage. ESS is deployed to reduce utility loads during system peaks (mostly evening time). During the low demand hours of storage can be used to capture the excess solar production and the same can be used during non-solar hours (Fig. 7).

6.1.3 Power Quality Savings

Even a 40–50% penetration of distribution utility ESS at feeder/DT level can cause a sharp decline in PF and increase in voltage fluctuation. ESS can help in capturing power losses in the network due to low power factor. As per the data, feeder level load is having 0.9 PF, but when solar comes into the grid, then PF reduces to around 0.8. Reduction in PF happens

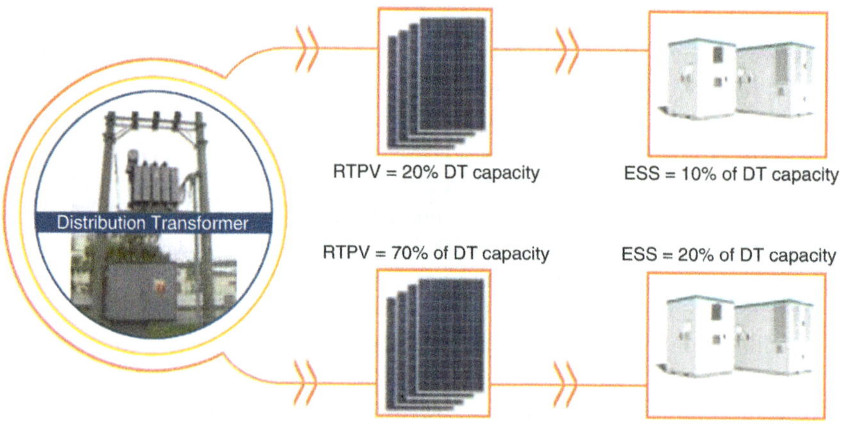

Fig. 6 DT applications for Storage

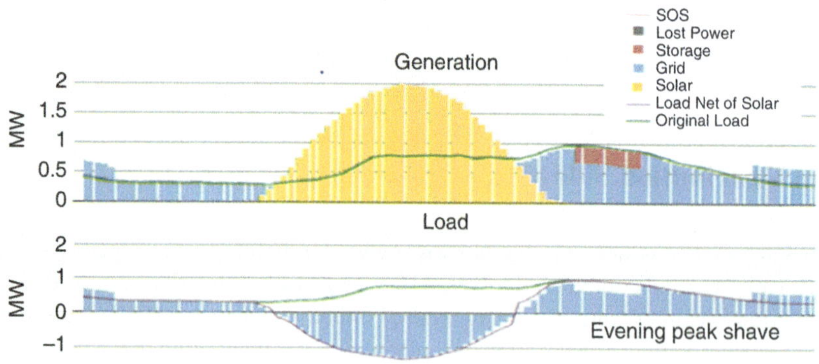

Fig. 7 Storage benefits as a peaker resource [4]

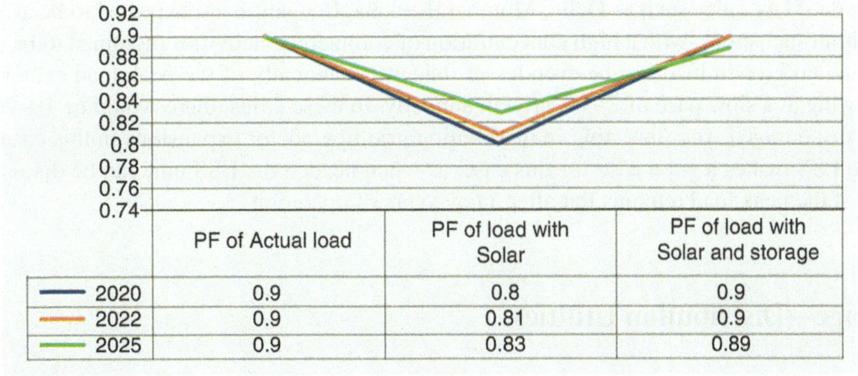

Fig. 8 PF Improvements with storage

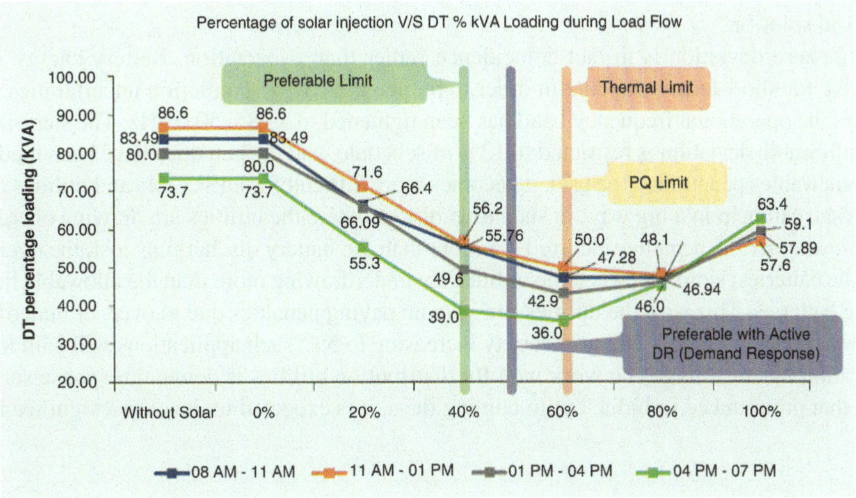

Fig. 9 CESC feeder load flow analysis (CES) [4]

because solar is giving only active power instead of active and reactive power; thus, ratio of apparent power to active power (i.e., PF) got reduced. With the help of storage, that ratio got increased back to its actual PF.

In the study conducted by CES for ISGF funded by MacArthur Foundation to develop an ESS Roadmap for Distribution Grid for Calcutta Electric Supply Corporation, few load centers in Kolkata were analyzed, where annual peak load growth has been 3% for the past 5 years. Introduction of Li-ion BESS at varying capacities was studied for 2020, 2022, and 2025. IRR was determined for each of these cases, and it was found that ESS capacities in the range of 10–20% of the distribution transformer capacity in these load centers was found to displace distribution deferral by up to 4 years at +IRR. Deferral apart from energy shifting and power factor improvement was one of the key applications. However, on case-to-case basis, different applications were found to drive the need for storage at distribution utility levels (Figs. 8 and 9).

The Study Yielded the Following Findings

1. 20% of DT capacity storage option is suitable for 2022 and 2025 to get maximum IRR. This storage option is also suitable for 2020 at a cost of negative IRR. But still there is high deferral benefit.
2. Solar penetration in the year 2022 and 2025 helps to reduce the original load curve, which may defer the transformer replacement up to certain year.
3. Maximum IRR is driven by low storage cost.
4. For higher penetration (year 2025 with 70% solar penetration), storage still provides economically viable results given appropriate size and technology.
5. The top metro cities in the country had peak load growth between 2% and 7% during 2018. During 2019–2026, various factors such as adoption of energy efficiency practices by residential and C&I customers in the metro cities and space saturation can lead to lower peak demand growth in these cities.

6. The peak load growth of key cities such as Delhi, Mumbai, Kolkata, Bengaluru are expected to be at 3–4% annually during 2019–2026. Within such cities, with a high concentration of commercial activities (assumed to be 10–20% of the total city load), the additional growth load can be even lesser than 1-2% annually of the peak load centers which have loads connected to it growing at a slow pace of 2–3% growth annually. In these cities, there would be 10–20% of load centers.
7. DISCOMs will be discouraged from investing in distribution grid upgrade or expansions in this case due to low returns and longer ROI. So ESS makes a good case for this case, as when needed the ESS unit can be disassembled and shifted to another location if the peak load remains flat after a few years of operation.

7 Case for Storage—Distribution Utilities

To make grid more reliable and attain energy security, deviation settlement mechanism was introduced in 2014. The same has been amended five times till date [1]. Over time penalties for not adhering to schedule has been made strict. Average price of electricity transacted through DSM is around Rs. 3/unit, but the maximum DSM rate is Rs. 8/unit. With new vector, the DSM penalty can go up to Rs.16/unit in case of violation of the band. The cost of DSM is unavoidable, but it can be optimized with Smart Grid solution.

As per the experts, zero deviation is in fact coincidence rather than expectation. Battery energy storage solution is an outstanding alternative for short-term balancing in order to reduce the cost of prediction uncertainties.

In its present form, the operational frequency band has been tightened to 49.85–50.05 Hz. The step size has been reduced to 0.01 Hz, maximum allowable deviation is restricted to 12% of schedule, and the penalties have been made much more defined. As such with more renewables penetrating the grid, it becomes more difficult for DISCOMs and utilities to adhere to schedules.

This is where storage can help in a big way; in such time blocks where the utilities are drawing energy more than the overdrawn limit, based on whether the penalties incurred are more than the battery discharging cost, the overdrawn amount can be met by discharging the batteries; similarly, when the utilities are underdrawing more than the allowable limit, the difference can be used to charge the batteries. This way, the utilities can save on paying penalties due to over- or underdraw of energy blocks. With increase of renewables in grid and with grid variety increasing to 5%, such applications will gain traction going forward.

One more application where storage can work well for distribution utilities is demand response services. The market for the same is still not that pronounced in India, but in coming days, it is expected to develop a significant market in India.

8 Conclusion

Energy storage is the way forward, even the toughest of critics have now acknowledged this fact. It is no longer an option but has become a necessity for the energy sector to remain sustainable going forward. Real on ground projects has further vindicated the fact that storage projects are viable not only technically but also on commercial grounds. The way forward will further see the mainstreaming of this technology for applications across the value chain in a more efficient and economic manner.

Looking across the world, there are many examples today, starting from the ESS for frequency response project in Chile (4 MWh) and Alamitos (400 MWh), the 129 MWh Hornsdale Tesla project, 10 MWh Hyundai Electric projects in Korea to manage energy shifts, and the world's biggest BESS project of 800 MWh in Rongke [5].

Overall, the future of technology looks promising with many options congregating in the power sector. It is a challenging and competing environment where the technology which remains agile, compatible, trailable, and convenient as well as customer oriented will lead and maintain the sustainability edge going forward.

References

1. Ministry of Power, Government of India (2019) Monthly Generation Report, Nov 2019. www.powermin.nic.in
2. "MNRE current notices and updates", Ministry of New and Renewable Energy. www.mnre.gov.in
3. "India stationary energy storage market overview report", IESA, June 2019, Accessed on Nov 2019
4. "Energy storage system – Roadmap for India 2019-2032" – ISGF, IESA. 6 Aug 2019. Accessed on Nov 2019
5. "Solving challenges in energy storage". Department of Energy. Storage success stories. Aug 2018. Accessed on Oct 2019

Performance, Modelling, Measurement and Simulation of Energy Efficiency for Heat Exchanger, Refrigeration and Air Conditioning

Abdeen Mustafa Omer

Nomenclatures

1	Condenser/heating
2	Evaporator
a	Air
ACH	Air changes per hour
b	Brine
Btu	British thermal unit
c	Coil
COP	Coefficient of performance (%)
cp	Specified heat at constant pressure (kJ/kg K)
DC	Direct current
Dp	Dew point
e	Electricity for other uses
f	Fan
g	Local acceleration of gravity (m/s^2)
GHP	Geothermal heat pump
GL	Ground loop
GSHP	Ground source heat pump
h	Enthalpy (kJ/kg)
hp	Heat pump
HP	Heat pump
hps	Heat pump system
hpsf	hps plus fan-coil heater
HRV	Heat recovery ventilator
HSPF	Heating season performance factor
i	Inlet
L	Latent
m	Mean, motor
N	Air change per hour (ACH) (h−1)
o	Outlet
o/d	Outdoor
P	Pressure (Pa) (kPa)
p	Pump
PRT	Platinum resistance thermometer
Q	Heat (thermal energy) (J)
Qc	Capacity (thermal power) (W)
R	Radius
S	Sensible
SEER	Seasonal energy efficiency ratio

A. M. Omer (✉)
Energy Research Institute (ERI), Nottingham, UK

© The Author(s), under exclusive license to Springer Nature Switzerland AG 2022
A. Sayigh (ed.), *Sustainable Energy Development and Innovation*, Innovative Renewable Energy,
https://doi.org/10.1007/978-3-030-76221-6_23

t	Temperature (Celsius) (°C)
T	Temperature (thermodynamic) (K)
tot	Total
V	Volume (m^3)
Vf	Volume flow (m^3/s)
W	Work (mechanical or electric) (J)
Wp	Power (mechanical or electric) (W)
η	Efficiency (%)
λ	Thermal conductivity (W/m/K)
ρ	Density (kg/m^3)
τ	Time (h.min.s)
φ	Relative vapour pressure (%)

1 Introduction

This section describes the different methods and techniques for providing energy for heating and cooling systems. It also covers the optimisation and improvement in the operation conditions of the heat cycles and the performance of the ground source heat pump systems (GSHPs).

With the improvement in people's living standards and the development of economy, heat pumps (HPs) have become widely used for air conditioning. The driver to this point was that environmental problems associated with the use of refrigeration equipment, the ozone layer depletion and global warming are increasingly becoming the main concerns in developed and developing countries alike. With development and enlargement of the cities in cold regions, the conventional heating methods can severely pollute the environment. In order to clean the cities, the governments drew many measures to restrict citizens from heating by burning coal and oil and encourage them to use electric or gas-burning heating. New approaches are being studied and solar-assisted reversible absorption heat pump for small power applications using water-ammonia is under development [1].

An air-source heat pump is convenient to use and so, it is a better method for electric heating. The ambient temperature in winter is comparatively high in most regions, so heat pumps with high efficiency can satisfy their heating requirement. On the other hand, a conventional heat pump is unable to meet the heating requirement in severely cold regions anyway, because its heating capacity decreases rapidly when the ambient temperature is below −10 °C. According to the weather data in cold regions, the air-source heat pump for heating applications must operate for long times with high efficiency and reliability when the ambient temperature is as low as −15 °C. Hence, many research works and developments have been conducted to enable heat pumps to operate steadily with high efficiency and reliability in low-temperature environments [2]. For example, the burner of a room air conditioner, which uses kerosene, was developed to improve the performance in low outside temperature [3]. Similarly, the packaged heat pump with a variable frequency scroll compressor was developed to realise high-temperature air supply and high capacity, even under the low ambient temperature of −10 to −20 °C [4]. Such a heat pump system can be conveniently used for heating in cold regions. However, the importance of targeting the low capacity range is clear if one has in mind that the air conditioning units below 10 kW cooling account for more than 90% of the total number of units installed in the European Union (EU) [5].

Conventional heating or cooling systems require energy from limited resources, for example, electricity and natural gas, which have become increasingly more expensive and are at times subject to shortages. Much attention has been given to sources subject to sources of energy that exist as natural phenomena. Such energy includes geothermal energy, solar energy, tidal energy and wind-generated energy. While all of these energy sources have advantages and disadvantages, geothermal energy, that is, the energy derived from the earth or ground, has been considered by many as the most reliable, readily available and most easily tapped of the natural phenomena.

Ground source-based geothermal systems have been used with heat pumps or air-handling units to satisfy building HVAC (heating, ventilation and air conditioning) loads. These systems are favoured because geothermal systems are environment friendly and have low greenhouse gas (GHG) emissions.

The installation and operation of a geothermal system of the present invention may be affected by various factors. These factors include, but are not limited to, the field size, the hydrology of the site, the thermal conductivity and thermal diffusivity of the rock formation, the number of wells, the distribution pattern of the wells, the drilled depth of each well and the building load profiles. Undersized field installations require higher duty cycles, which may result in more extreme water temperatures and lower HVAC performance in certain cases. Oversized field designs, on the other hand, require more wells, pumps and field plumbing and therefore will be more expensive, albeit adequate to handle almost any load circumstances. The detailed knowledge of the field rock (e.g. porosity, permeability, thermal diffusivity, heat capacity or other aquifer parameters) may facilitate the determination of the appropriate drilling depth for each well, as well as the number and position of such wells needed at that site. Some of this information may be obtained during the drilling operation.

2 Earth-Energy Systems (EESs)

The earth-energy systems, EESs, have two parts: a circuit of underground piping outside the house and a heat pump unit inside the house. And unlike the air-source heat pump, where one heat exchanger (and frequently the compressor) is located outside, the entire GSHP unit for the EES is located inside the house.

The outdoor piping system can be either an open system or a closed loop. An open system takes advantage of the heat retained in an underground body of water. The water is drawn up through a well directly to the heat exchanger, where its heat is extracted. The water is discharged either to an aboveground body of water, such as a stream or pond, or back to the underground water body through a separate well. Closed-loop systems, on the other hand, collect heat from the ground by means of a continuous loop of piping buried underground. An antifreeze solution (or refrigerant in the case of a direct expansion 'DX' earth-energy system), which has been chilled by the heat pump's refrigeration system to several degrees colder than the outside soil, circulates through the piping, absorbing heat from the surrounding soil.

In some EESs, a heat exchanger, sometimes called a 'desuperheater', takes heat from the hot refrigerant after it leaves the compressor. Water from the home's water heater is pumped through a coil ahead of the condenser coil, in order that some of the heat that would have been dissipated at the condenser is used to heat water. Excess heat is always available in the cooling mode, and is also available in the heating mode during mild weather, when the heat pump is above the balance point and not working to full capacity. Other EESs heat domestic hot water (DHW) on demand: the whole machine switches to heating DHW, when it is required.

Hot water heating is easy with EESs because the compressor is located inside. Because EESs have a relatively constant heating capacity, they generally have many more hours of surplus heating capacity than required for space heating. In fact, there are sources of energy all around in the form of stored solar energy, which, even if they have a low temperature, can provide the surroundings with enough energy to heat the soil, bedrock and ground water as a heat source for domestic dwellings as shown in Fig. 1, for example. Some emphasis has recently been put on the utilisation of the ambient energy from ground source and other renewable energy sources in order to stimulate alternative energy sources for heating and cooling of buildings. Exploitation of renewable energy sources and particularly ground heat in buildings can significantly contribute towards reducing dependency on fossil fuels.

Fig. 1 Using the soil, bedrock or groundwater as the heat source [6]

2.1 The Cooling Cycle

The cooling cycle is basically the reverse of the heating cycle. The reversing valve changes the direction of the refrigerant flow. The refrigerant picks up heat from the house air and transfers it directly in DX systems or to the ground water or an antifreeze mixture. The heat is then pumped outside, into a water body or return well (in the case of an open system), or into the underground piping (in the case of a closed-loop system). Once again, some of this excess heat can be used to preheat domestic hot water.

Unlike air-source heat pumps, EESs do not require a defrost cycle. Underground temperatures are much more stable than air temperature, and the heat pump unit itself is located inside; therefore, problems with frost do not arise.

2.2 Function of the GSHP Circuit

The collector liquid (cooling medium) is pumped up from the borehole in the tubing and passed to the heat pump. Another fluid, a refrigerant, circulates in the heat pump in a closed system with the most important characteristic of having a low boiling point. When the refrigerant reaches the evaporator, which has received energy from the borehole, the refrigerant evaporates. The vapour is fed to a compressor where it is compressed. This results in a high increase in temperature. The warm refrigerant is fed to the condenser, which is positioned in the boiler water. Here, the refrigerant gives off its energy to the boiler water, so that its temperature drops and the refrigerant changes state from gas to liquid. The refrigerant then goes via filters to an expansion valve, where the pressure and temperature are further reduced. The refrigerant has now completed its circuit and is once more fed into the evaporator where it is evaporated yet again due to the effect of the energy that the collector has carried from the energy source (Fig. 2).

Efficiencies of the GSHP systems are much greater than those of conventional air-source heat pump systems. A higher COP (coefficient of performance) can be achieved by a GSHP because the source/sink earth temperature is relatively constant compared to air temperatures. Additionally, the heat is absorbed and rejected through water, which is a more desirable heat transfer medium because of its relatively high heat capacity. The GSHP systems rely on the fact that, under normal geothermal gradients of about 0.5°F/100 ft (30 °C/km), the earth temperature is roughly constant in a zone extending from about 20 ft (6.1 m) deep to about 150 ft (45.7 m) deep. This constant temperature interval within the earth is the result of a complex interaction of heat fluxes from above (the sun and the atmosphere) and from below (the earth interior). As a result,

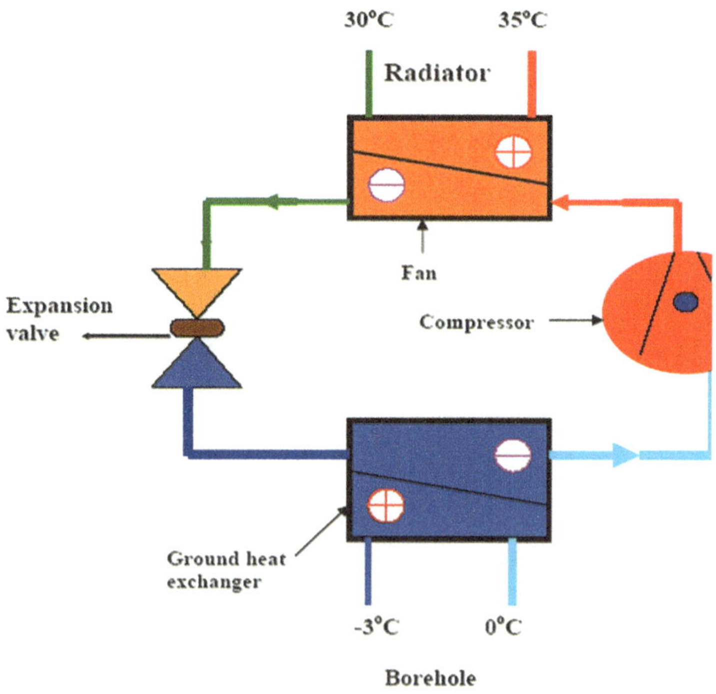

Fig. 2 Detail of the GSHP circuit

the temperature of this interval within the earth is approximately equal to the average annual air temperature [7]. Above this zone (less than about 20 ft (6.1 m) deep), the earth temperature is a damped version of the air temperature at the earth's surface. Below this zone (greater than about 150 ft (45.7 m) deep), the earth temperature begins to rise according to the natural geothermal gradient. The storage concept is based on a modular design that will facilitate active control and optimisation of thermal input/output, and it can be adapted for simultaneous heating and cooling often needed in large service and institutional buildings [8]. Loading of the core is done by diverting warm and cold air from the heat pump through the core during periods with excess capacity compared to the current need of the building [9–11]. The cool section of the core can also be loaded directly with air during the night, especially in spring and fall when nights are cold and days may be warm.

2.3 Free Cooling

The installation can additionally be fitted with fan convectors, for example, in order to allow connections for free cooling (Fig. 3). To avoid condensation, pipes and other cold surfaces must be insulated with a diffusion-proof material. Where the cooling demand is high, fan convectors with drip tray and drain connection are needed.

2.4 Refrigeration and Heat Pumps

The pressure (ps) is a function of how rapidly vapour can be removed through suction or formed through pressure. At equilibrium, the rate at which the vapour is formed (determined by Q) equals the rate at which it is removed. Therefore, both the heat transfer rate into the liquid (Q) and the vapour removal rate (suction pump capacity) determine the pressure and hence $T_{sat}(s)$ (Fig. 4). This is governed by the following set of equations.

$$Q = mh_{fg} \tag{1}$$

$$m = \rho g V \tag{2}$$

$$Q = \rho g V h_{fg} \tag{3}$$

$$Q = V h_{fg} / v_g \tag{4}$$

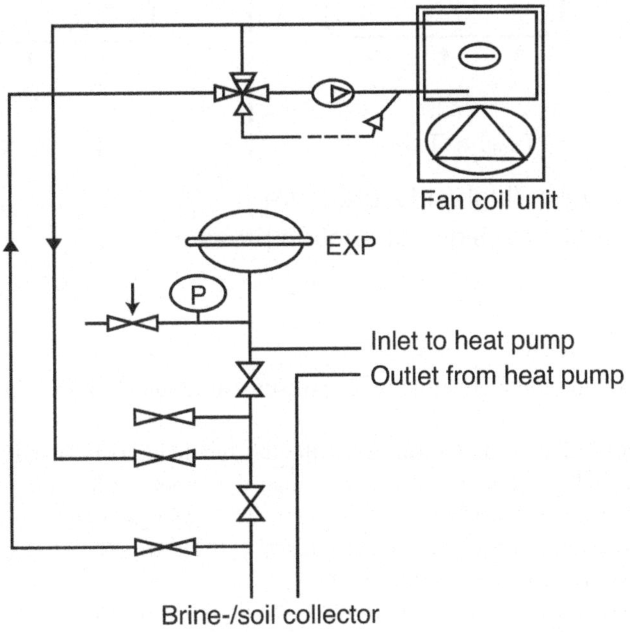

Fig. 3 Diagram of a cooling system

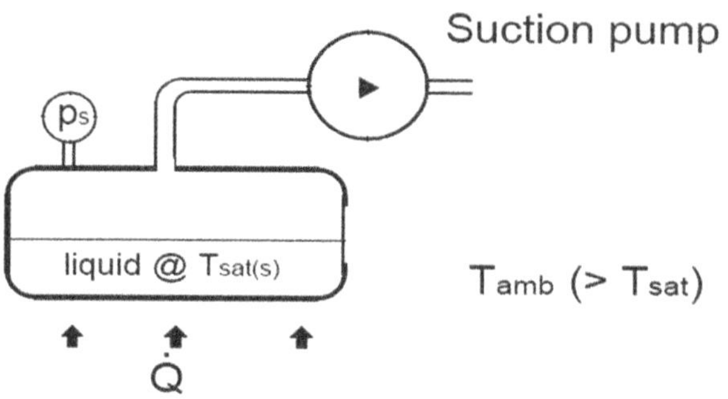

Fig. 4 Refrigeration cycle

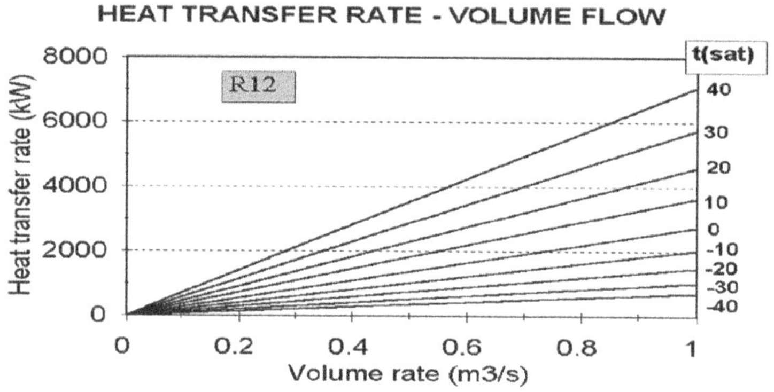

Fig. 5 Heat transfer rate versus volume rate. p_c = Condenser or 'high side' pressure. p_s = Evaporator, 'low side' or suction pressure.

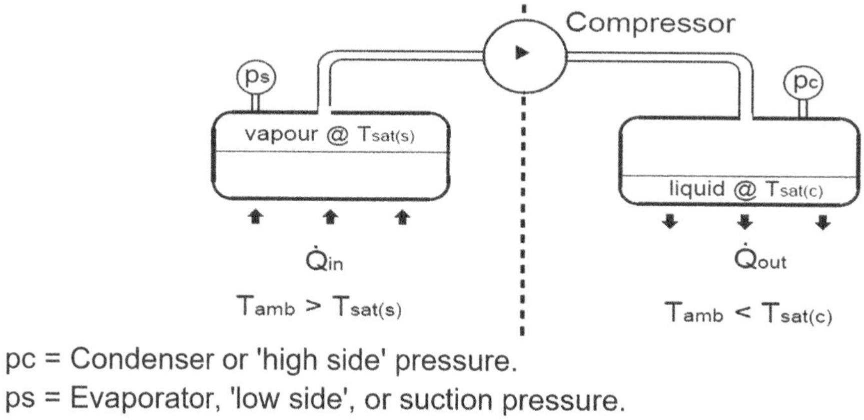

Fig. 6 Heat pumps

Both h_{fg} and v_g depend on the saturation temperature (or pressure) as assumed in Fig. 5, which describes the relationship represented by Eq. 4.

The right-hand side (RHS) of Fig. 6 is the 'converse' of the left-hand side (LHS) and constitutes a heat pump. Heat is 'pumped' from the LHS to the RHS. The main difference is that the vapour, after compression, will almost certainly be superheated and must cool to $T_{sat}(c)$ before condensing will occur. The same reasoning (in converse) applies to the RHS, as previously applied to LHS. Obviously, with the above system, the entire refrigerant would eventually end up on the RHS, and the heat pumping (and refrigeration) effect would cease.

Clearly, to ensure that the system can operate continuously, the liquid refrigerant needs to be fed from the RHS back to the LHS. This can be achieved by simply allowing it to flow back under its natural pressure difference. In this way, a continuous closed circuit refrigeration (or heat pump) system is obtained (Fig. 7).

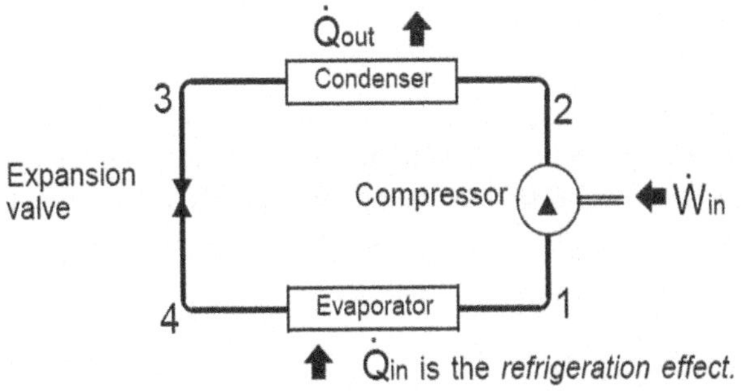

Fig. 7 Simplified refrigeration system diagram

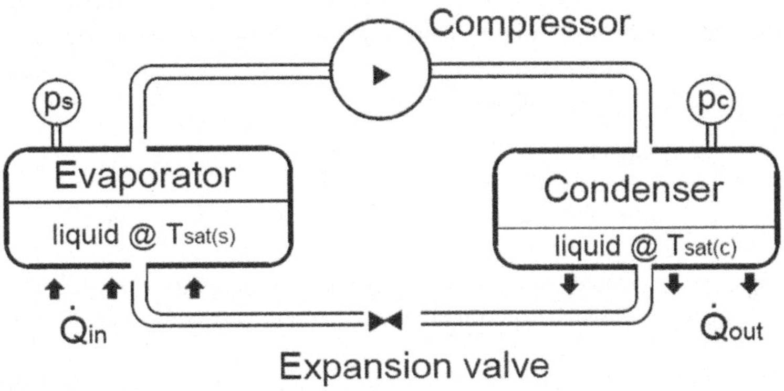

Fig. 8 Heat pump refrigeration cycle

Control of the liquid flow rate is needed to ensure that it equals the vapour formation rate, and an appropriate balance of liquid quantities in the evaporator and condenser is maintained. When the liquid passes through the expansion valve, it experiences a sudden drop in pressure, which causes instantaneous boiling (known as flashing). Vapour is formed using the liquid's sensible heat, which causes the liquid to drop in temperature to $T_{sat}(s)$. A saturated liquid/vapour mixture will enter the evaporator. Figure 8 explains this cycle in practice.

2.5 *System Performance*

The system balance requires the overall work done to be equivalent to the net energy used by the system. Hence,

$$Q_{out} - Q_{in} = W_{in} \tag{5}$$

For operation as a refrigerator, a measure of system performance is the amount of heat absorbed per unit work supplied to drive the system. This is known as the coefficient of performance [12].

$$COP_{ref} = Q_{in} / W_{in} \tag{6}$$

For operation as a heat pump, a measure of system performance is the amount of heat delivered per unit work supplied to drive the system. This is known as the coefficient of performance [13].

$$COP_{hp} = Q_{out} / W_{in} \tag{7}$$

It follows that (for the same system):

$$COP_{hp} = COP_{ref} + 1 \tag{8}$$

2.6 Vapour Compression Refrigeration

The term 'vapour compression refrigeration' is somewhat of a misnomer, it would be more accurately described as 'vapour suction refrigeration'. Vapour compression is used to reclaim the refrigerant and is more aptly applied to heat pumps. Vapour compression refrigeration exploits the fact that the boiling temperature of a liquid is intimately tied to its pressure. Generally, when the pressure on a liquid is raised, its boiling (and condensing) temperature rises and vice versa. This is known as the saturation pressure-temperature relationship.

2.7 Refrigerant Properties

In practice, the choice of a refrigerant is a compromise, for example, ammonia is good but toxic and flammable, while R12 is very good but detrimental to the ozone layer. Figure 9 shows some commonly used refrigerants and their typical ranges of usability.

Ideally, a refrigerant will have the following characteristics.

- Non-toxic – for health and safety reasons.
- Non-flammable – to avoid risks of fire or explosion.
- Operates at modest positive pressures – to minimise pipe and component weights (for strength) and avoids air leakage into the system.
- Has a high vapour density – to keep the compressor capacity to a minimum and pipe diameters relatively small.
- Easily transportable – because refrigerants are normally gases at standard sea-level (SSL) conditions, they are stored in pressurised containers.
- Environment friendly – non-polluting and non-detrimental to the atmosphere, water or ground.
- Easily recyclable and relatively inexpensive to produce.
- Compatible with the materials of the refrigeration system – non-corrosive, miscible with oil and chemically benign.

2.8 Cooling Mode

In the cooling mode, cool vapour arrives at the compressor after absorbing heat from the air in the building. The compressor compresses the cool vapour into a smaller volume, increasing its heat density. The refrigerant exits the compressor as a hot vapour, which then goes into the earth loop field. The loops act as a condenser condensing the vapour, until it is virtually all liquid. The refrigerant leaves the earth loops as a warm liquid. The flow control regulates the flow from the condenser, such that only the liquid refrigerant passes through the control. The refrigerant expands, as it exits the flow control unit and becomes a cold liquid.

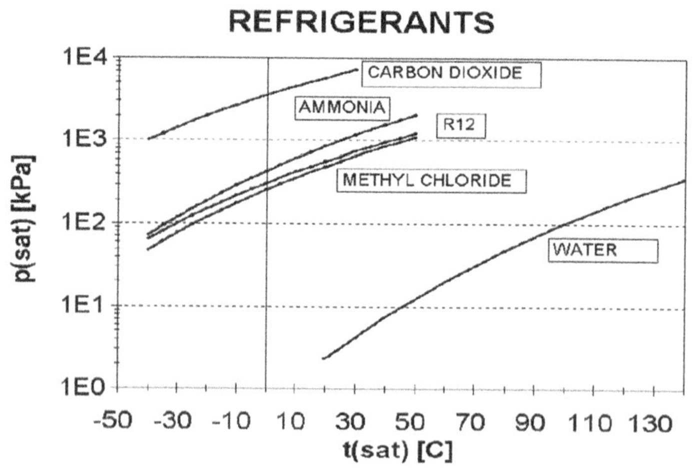

Fig. 9 Refrigerant chart

2.9 *Heat Pump Antifreeze*

A potential negative effect of all geothermal heat pumps (GHPs) is the release of antifreeze solutions to the environment. Antifreeze solutions are required in colder climates to prevent the circulating fluid from freezing. Antifreeze chemicals include methanol, ethanol, potassium acetate, propylene glycol, calcium magnesium acetate (CMA) and urea. These chemicals are generally mixed with water when used as a heat exchange fluid. These chemicals can be released to the environment via spills or corrosion of system components. Approved antifreezes include methanol, ethanol, propylene glycol, calcium chloride or ethylene glycol. These antifreezes must be mixed with water, at concentrations of 20% or less. Geothermal heat pumps for a single-family residence and the antifreezes for these units were evaluated by Heinonen et al. (1996) [14]. These authors evaluated the total energy consumption, corrosion due to the antifreeze, and the operational and environmental effects of six antifreeze solutions, namely methanol, ethanol, potassium acetate, propylene glycol, CMA and urea. However, the excluded salt solutions, such as sodium and calcium chloride, from their study because they pose serious potential corrosion problems. The differences in the total energy consumption for the studied antifreezes were considered minimal. Nevertheless, Heinonen et al. recommended that propylene glycol was a good choice based on its low health, fire and environmental risks (Table 1). Unfortunately, these authors did not assess the leak potential of these antifreezes in the plastic pipe (e.g. high-density polyethylene (HDPE) and CPVC SDR-11) commonly used for the ground loop (GL) [15].

However, the bond between the grout and borehole can be compromised by desiccation of the geological materials near the borehole, as the heat from the borehole lowers the moisture content of the geological materials and these materials contract. In areas with thick unsaturated zones, the bentonite grout may dry out over time, compromising the seal. To improve heat exchange, some advocate the use of spacers, which moves the heat conductor pipe to the side of the borehole, putting it in contact with the geological materials. However, the use of spacers appears to increase the environmental risk of antifreeze leaking into ground water, by reducing or removing the bentonite between the heat conductor pipe and geological materials.

3 Air Distribution

The air distribution system can make a big difference in both the cost and the effectiveness of geothermal heating and cooling. It also has an important effect on personal comfort and health. The air-handling component is either a separate cabinet or is part of the cabinet that houses the geothermal heat pump, and includes the blower assembly that forces air through the ductwork. The supply ductwork carries air from the air-handles to the rooms. Typically, each room has at least one supply duct and larger rooms may have several. The return ductwork moves air from the room back to the air handler. Most buildings have one or more main return ducts located in a central area. The cold liquid refrigerant is circulated through the air handler where it absorbs and removes the unwanted heat from the air and vaporises the refrigerant to a gas. The gas is compressed to increase its temperature and then the underground/underwater coils act as a condenser rather than an evaporator (as in the heating cycle) (Fig. 10). The heat in the refrigerant is transferred to the ground/water as the refrigerant condenses.

Refrigerants are present in the GSHP systems and so present the threat of the hydrochlorofluorocarbons (HCFCs) and toxicity. However, new types and blends of refrigerants (some using CO_2) with minimal negative impacts are approaching the market, as shown in Table 2. Because the GSHPs raise the temperature to around 40 °C they are most suitable for

Table 1 Cost and risk factors for heat pump antifreeze (Heinonen et al., 1996) [14]

	Antifreeze					
Factor	Methanol	Ethanol	Propylene glycol	Potassium acetate	CMA	Urea
Life cycle cost	3	3	2	2	2	3
Corrosion risk	2	2	3[a]	2	2	1
Leakage risk	3	2	2[a]	1[b]	1	1
Health risk	1	2	3	3	3	3
Fire risk	1[a]	1[c]	3	3	3	3
Environmental risk	2	2	3	2	2	3
Risk of future use	1	2	3	2	2	2

Notes: Ratings- 1 means potential problems and caution required, 2 means minor potential for problems and 3 means little or no potential problems

[a]DOWFROST HD

[b]GS-4

[c]Pure fluid only. Diluted antifreeze (25% solution) is rated 3

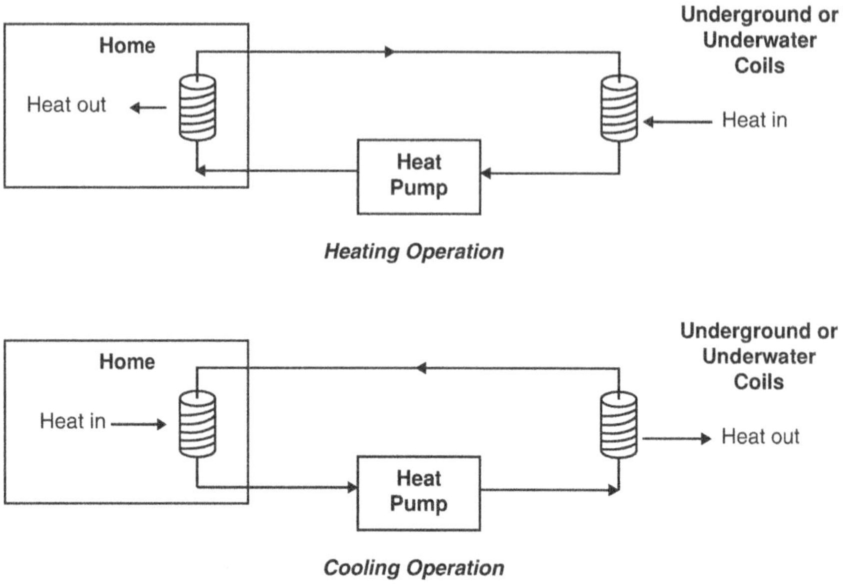

Fig. 10 Heating and cooling operations

Table 2 CO_2 emissions [16, 17]

System	Primary energy efficiency (%)	CO_2 emissions (kg CO_2/kWh heat)
Oil-fired boiler	60–65	0.45–0.48
Gas-fired boiler	70–80	0.26–0.31
Condensing gas boiler + low-temperature system	100	0.21
Electrical heating	36	0.9
Conventional electricity + GHSP	120–160	0.20–027
Green electricity + GHSP	300–400	0.00

underfloor heating systems or low-temperature radiators, which require temperatures of between 30° and 35 °C. Higher outputs, such as to conventional radiators requiring higher temperatures of around 60–80 °C, can be obtained through use of the GSHP in combination with a conventional boiler or immersion heater.

The GSHPs come in 15 models from 4 kW up to 30 kW (even up to 300 kW when connected in parallel). At least 65% of the heating and hot water energy consumption of a house can be saved (65–75% of heating costs with a heat pump) as a result of using such a system. However, sizing of the heat pump and the ground loops is essential for an efficient operation of the system. If sized correctly, a GSHP can be designed to meet 100% of space heating requirements. The sizing of the system is very sensitive to heat loads and should therefore be installed into properties with high-energy efficiency standards, particularly new build. It is a good and practical idea to explore ways of minimising space heating and hot water demand by incorporating energy efficiency measures (Fig. 11). This is known as the saturation pressure-temperature relationship (Fig. 12). The refrigerant exits the compressor as a hot vapour, which then goes into the earth loop field (Fig. 13).

3.1 Some Definitions

1. The word 'efficiency' is defined as the ratio of useful heat output to energy input, for example, if an open fireplace loses half its energy up the chimney it is said to be 50% efficient.
2. The COP or 'coefficient of performance' is found by dividing the useful heat output by the energy input, for example, a heat pump that produces 3 kWatts of heat for 1 kWatt of input power has a COP of 3. The open fireplace example with 50% efficiency would have a COP of 0.5 (1/2).

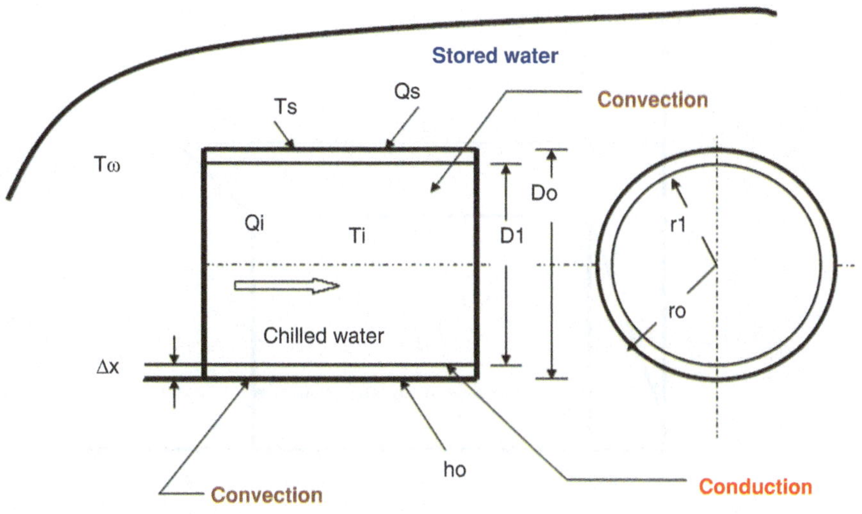

Fig. 11 Schematic of heat transfer through a circular tube heat exchanger

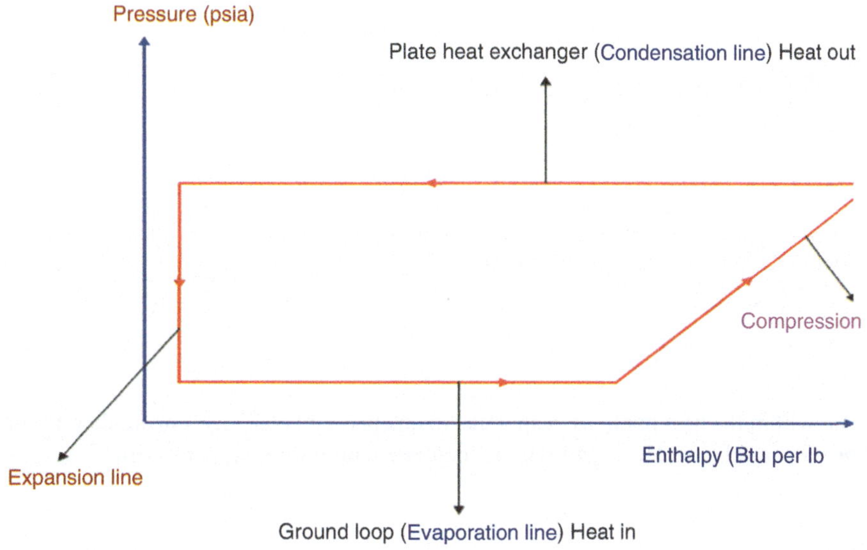

Fig. 12 The ideal cycle on pressure-enthalpy diagram

3. The heat 'source' is the outside air, river or ground, wherever the heat is being extracted from. Sometimes it is referred to as an ambient source.
4. 'Sink' is the name given to the part where the heat is usefully dissipated, such as radiators in the room, underfloor heating, hot water cylinder, etc.

3.1.1 Horizontal Collector

This can be either coiled 'slinky' or straight pipes that are buried 1.5–2 m deep in open ground (in gardens). The pipe is usually plastic and contains a glycol antifreeze solution.

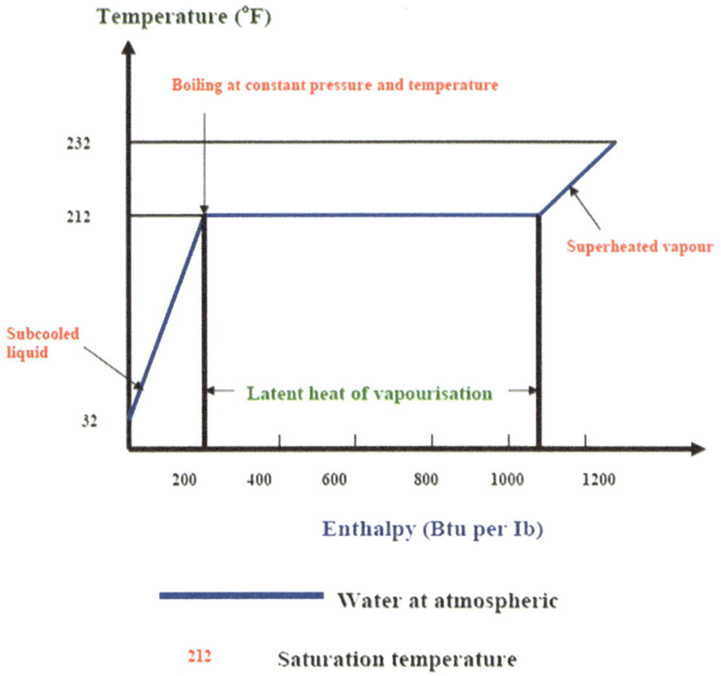

Fig. 13 Water undergoing a charge of state

3.1.2 Antifreeze

This is simply an additive to water that makes its freezing point lower. Common salt does the same thing, but ethylene or propylene glycol is more practical for heat pump systems.

3.1.3 Refrigerant

This is the working fluid within the heat pump. It evaporates in one part and condenses in another. By doing so, heat is transferred from cold to hot. This fluid is sealed in and will not degrade within the heat pumps life.

3.1.4 Heat Exchanger

This is a simple component that transfers heat from one fluid to another. It could be liquid-to-liquid, or liquid-to-air or air-to-air. Two heat exchangers are housed within the heat pump, one for the hot side (the condenser) and one for the cold side (the evaporator).

3.1.5 Slinky

This name is given to the way that ground collector pipes can be coiled before burying in a trench.

3.1.6 Passive Heat Exchange

When waste hot water preheats cold input water, it is said to be ‘passive’. This costs nothing to run. A heat pump is said to be ‘active’ if it can extract heat from cold waste water but requires a relatively small power input.

3.2 Some Refrigeration Characteristics

The seasonal energy efficiency ratio (SEER) may be applied to each of the components.

Assuming that kinetic energy (KE) and potential energy (PE) effects are negligible, that is, the SFEE (steady flow energy equation) is applicable; is

$$Q + W = \mathrm{m}\,\Delta h \tag{9}$$

Compressor:
Compression is assumed to be adiabatic:

$$\therefore Q = 0 \tag{10}$$

$$\mathrm{W12} = m\left(h_2 - h_1\right) \tag{11}$$

Or

$$W_{\mathrm{in}} = m\left(h_2 - h_1\right) \tag{12}$$

Condenser:

$$\mathrm{W23} = 0 \tag{13}$$

$$\therefore Q_{\mathrm{out}} = m\left(h_2 - h_3\right) \tag{14}$$

Expansion valve:

$$\mathrm{W34} = 0 \ \& \ \mathrm{Q34} = 0 \tag{15}$$

$$\therefore h_3 = h_4 \tag{16}$$

Evaporator:

$$\mathrm{W41} = 0 \tag{17}$$

$$\therefore Q_{\mathrm{in}} = m\left(h_1 - h_4\right) \tag{18}$$

Refrigeration effect
It follows that:

$$\mathrm{COP}_{\mathrm{ref}} = \left(h_1 - h_3\right) / \left(h_2 - h_1\right) \tag{19}$$

$$\mathrm{COP}_{\mathrm{hp}} = \left(h_2 - h_3\right) / \left(h_2 - h_1\right) \tag{20}$$

In order to determine the above equations, the specific enthalpy values will be needed. Because refrigerants work in the liquid/vapour phases, appropriate property charts or tables must be used.

3.3 The Ideal Refrigeration Cycle

- Isentropic compression (1→2)
- Constant pressure cooling/condensation (2→3)
- Throttling (3→4)
- Constant pressure vaporisation/heating (4→1)

The ideal refrigeration cycle is plotted in the p-h chart, as shown in Fig. 14.

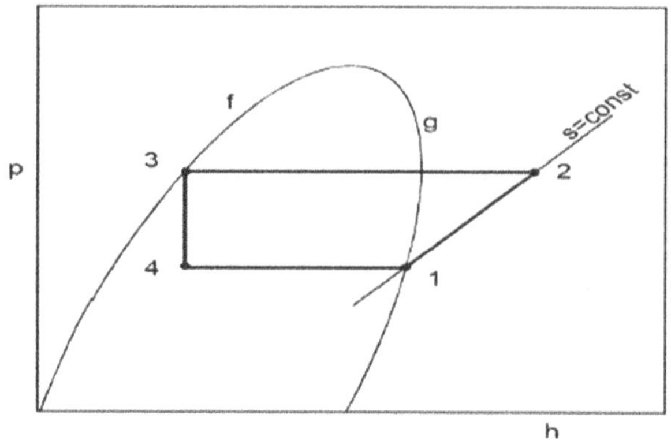

Fig. 14 Refrigeration cycle

3.4 Real Refrigeration Systems

Evaporator superheat

g→1 given in *K* above $T_{sat}(s)$

3.4.1 Isentropic Compressor Efficiency

$$\eta_{isen} = h_2' - h_1 / h_2 - h_1 \tag{21}$$

3.4.2 Condenser Subcooling

f→3 given in K below $T_{sat}(c)$

3.5 Refrigerant Properties (Charts and Tables)

Because refrigeration systems basically work between two pressures, and specific enthalpy is one of the most useful properties we need, refrigerant thermodynamic properties are normally presented in the form of a pressure-specific enthalpy (or p-h) chart.

This is done for convenience and is simply an alternative way of presenting property data, instead of, for example, p-V, or T-s, or h-s (Figs. 15, 16, and 17).

Other useful properties are also shown in the chart, vis: specific entropy, specific volume, temperature and quality. Regard these properties as 'contours'.

The pressure axis (*y*-axis) is typically logarithmic.

3.6 Pressure Drops in Evaporator and Condenser

Clearly, any or all of the above effects can be present, but the pressure drops are often small enough to be neglected (Figs. 18).

3.7 Refrigeration System Performance Improvement

Liquid-suction heat exchanger (Fig. 19 and 20).

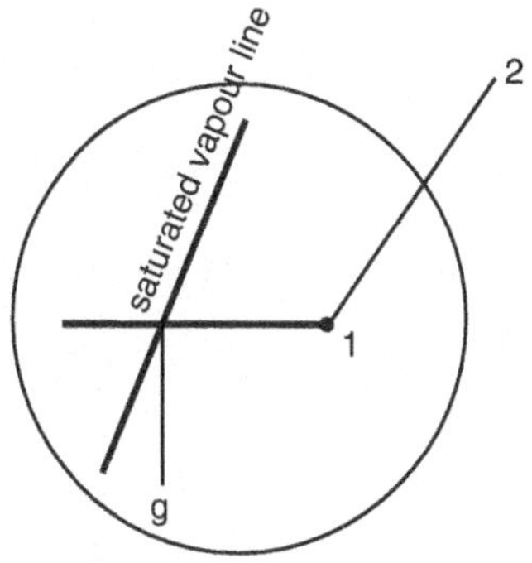

Fig. 15 Evaporator superheat

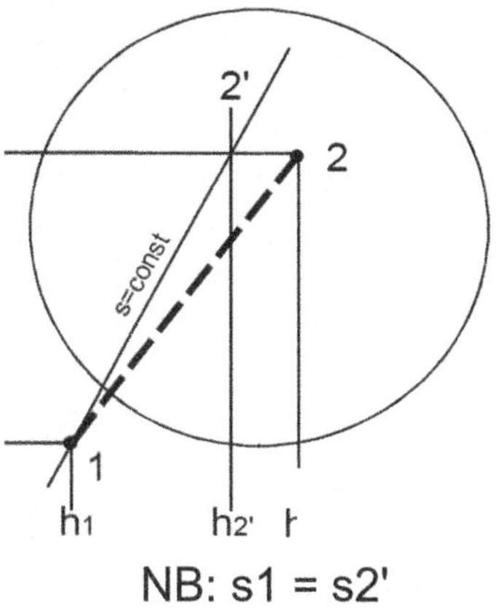

Fig. 16 Isentropic compressor

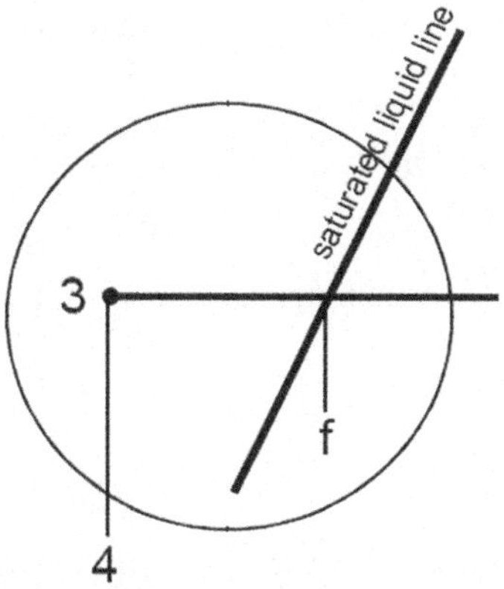

Fig. 17 Condenser subcooling

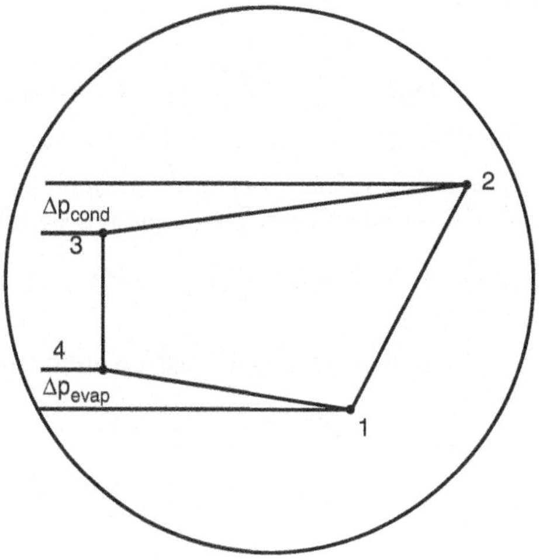

Fig. 18 Pressure drops in evaporator and condenser

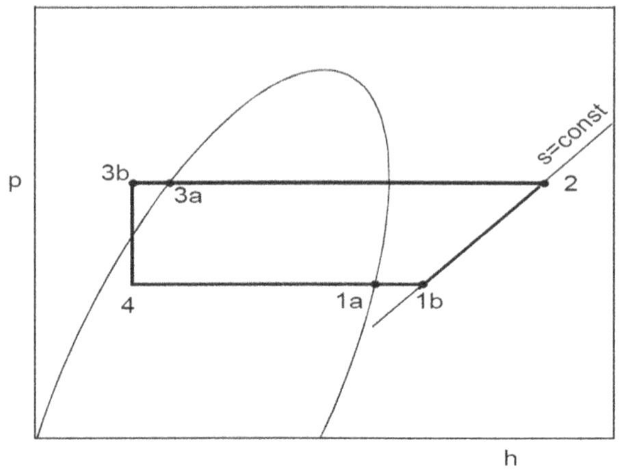

Fig. 19 Diagram of a liquid-suction heat exchanger

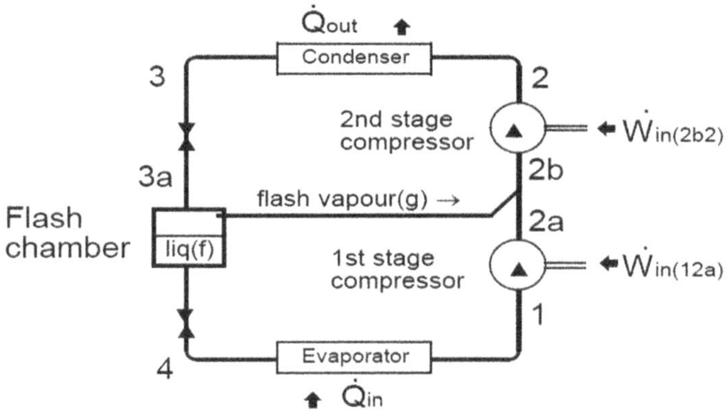

Fig. 20 Liquid-suction heat exchanger cycle

Assuming no losses:

$$H_{1b} - h_{1a} = h_{3a} - h_{3b} \tag{22}$$

3.8 Multiple Compression Using Flash Chambers

3.9 Diagram of Multiple Compressions Using Flash Chamber

At point 3a, we have a mixture of vapour and liquid, which is separated, in the flash chamber (Fig. 21). The proportion of the total mass flow that is liquid (and proceeds to the evaporator) is given by:

$$x_f = h_{3(g)} - h_3 / h_{3(g)} - h_{3(f)} \tag{23}$$

The remaining vapour mixes with the discharge from the first stage compressor to give different inlet conditions to the second stage.

Assuming adiabatic mixing:

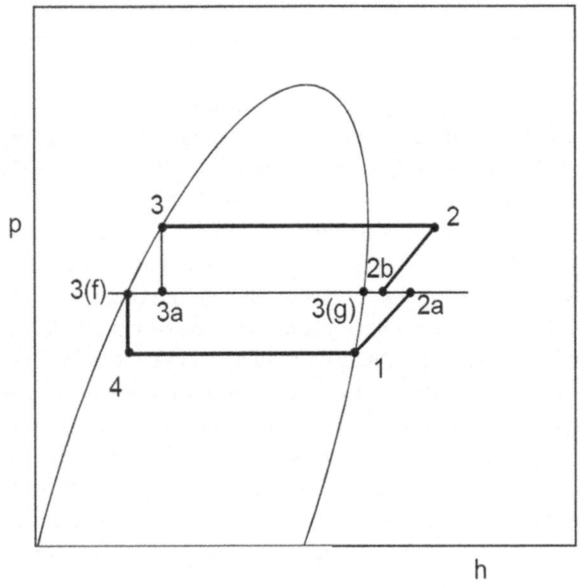

Fig. 21 Cycle of multiple compressions using flash chamber

$$1 * h_{2b} = x_f h_{2a} + (1 - x_f) h_{3(g)} \tag{24}$$

A similar equation can be used to find s_{2b}
Finally, the COP is given by:

$$\text{COP} = (x_f(h_1 - h_4)) / (x_f(h_{2a} - h_1) + (h_2 - h_{2b})) \tag{25}$$

4 Discussions

Thermal comfort is an important aspect of human life. Buildings where people work require more light than buildings where people live. In buildings where people live, the energy is used for maintaining both the temperature and lighting. Hence, natural ventilation is rapidly becoming a significant part in the design strategy for non-domestic buildings because of its potential to reduce the environmental impact of building operation, due to lower energy demand for cooling. A traditional, naturally ventilated building can readily provide a high ventilation rate. On the other hand, the mechanical ventilation systems are very expensive. However, a comprehensive ecological concept can be developed to achieve a reduction of electrical and heating energy consumption, optimise natural air condition and ventilation, improve the use of daylight and choose environmentally adequate building materials. Energy efficiency brings health, productivity, safety, comfort and savings to the homeowner, as well as local and global environmental benefits. The use of renewable energy resources could play an important role in this context, especially with regard to responsible and sustainable development. It represents an excellent opportunity to offer a higher standard of living to local people and will save local and regional resources. Implementation of greenhouses offers a chance for maintenance and repair services. It is expected that the pace of implementation will increase and the quality of work will improve in addition to building the capacity of the private and district staff in contracting procedures. The financial accountability is important and more transparent. Various passive techniques have been put in perspective, and energy-saving passive strategies can be seen to reduce interior temperature and increase thermal comfort, and reduce air conditioning loads [18–21]. The scheme can also be employed to analyse the marginal contribution of each specific passive measure working under realistic conditions in combination with the other housing elements. In regions where heating is important during winter months, the use of top-light solar passive strategies for spaces without an equator-facing façade can efficiently reduce energy consumption for heating, lighting and ventilation. The use of renewable energy resources could play an important role in this context, especially with regard to responsible and sustainable development. It presents an excellent opportunity to offer a higher standard of living to local people and will save local and regional resources.

Implementation of greenhouses offers a chance for conducting maintenance and repair services. Various passive techniques have been put in perspective, and energy-saving passive strategies can be seen to reduce interior temperature and increase thermal comfort, and reduce air conditioning loads.

Renewable energy is the term to describe a wide range of naturally occurring, and replenishing energy sources. The use of renewable energy sources and the rational use of energy are the fundamental inputs for a responsible energy policy. The energy sector is encountering difficulties because increased production and consumption levels entail higher levels of pollution and eventually climate changes, with possibly disastrous consequences. Moreover, it is important to secure energy at acceptable cost to avoid negative impacts on economic growth. On the technological side, renewables have an obvious role to play. In general, there is no problem in terms of the technical potential of renewables to deliver energy and there are very good opportunities for renewable energy technologies to play an important role in reducing emissions of greenhouse gases into the atmosphere – certainly far more than have been exploited so far. But there are still technical issues to be addressed to cope with the intermittency of some renewables, particularly wind and solar. However, the biggest problem with replying on renewables to deliver the necessary cuts in greenhouse gas emissions is more to do with politics and policy issues than with technical ones. The single most important step governments could take to promote and increase the use of renewables would be to improve access for renewables to the energy market. That access to the market would need to be under favourable conditions and possibly under favourable economic rates. One move that could help – or at least justify – better market access would be to acknowledge that there are environmental costs associated with other energy supply options, and that these costs are not currently internalised within the market price of electricity or fuels. It could make significant difference, particularly if, appropriate subsidies were applied to renewable energy in recognition of environmental benefits it offers. Cutting energy consumption through end-use efficiency is absolutely essential. And this suggests that issues of end-use consumption of energy will have to come onto the table in the foreseeable future.

The scientific consensus is clear – climate change is occurring. Existing renewable energy technologies could play a significant mitigating role, but the economic and political climate will have to change first. Climate change is real, it is happening now, and greenhouse gases produced by human activities are significantly contributing to it. The predicted global temperature changes of between 1.5 and 4.5 °C could lead to potentially catastrophic environmental impacts – including sea-level rise, increased frequency of extreme weather events, floods, droughts, disease migration from various places and possible stalling of the Gulf stream. This is why scientists argue that climate change issues are not ones that politicians can afford to ignore. And policymakers tend to agree, but reaching international agreements on climate change policies is no trivial task.

The most favourable orientation, which is due north, results in diminished excessive solar gains through the windows. However, most buildings cannot be oriented at will. If the only possible orientation is due south, and no external shade is used, the index reveals extra heat gains of some 0.26 over the value of a totally shaded window. Application of the model results from exploring the relative importance of the thermal inertia of walls, floor and ceiling. Heat stored in building materials, as proven in old, massive buildings, can be compensated during high insolation hours with thermal losses at night and early morning hours, when ambient temperatures are below 25 °C. Temperature variation will be lower for higher thermal capacities of building materials. However, it is known while thermal capacity increases the relative importance of individual heat flows change. For example, for lower wall temperatures, the contribution of radiative heat transfer will be reduced, and the relative importance of convective processes will increase, and thus the difficulty to calculate accurately the overall heat flows. The relevance of certain passive techniques is variable with prevailing weather.

Finally, the required temperature-dependent air transport properties were evaluated by the following expression, which are valid between 2 °C and 77 °C with temperature expressed in k:

$$\text{Thermal diffusivity}, \alpha = 1.534 \times 10^{-3}\ \text{T} - 0.2386\left(10^{-4}\,\text{m}^2\text{s}^{-1}\right)$$

$$\text{Kinematics viscosity}, \nu = 0.1016\text{T} - 14.8\left(10^{-6}\,\text{m}^2\text{s}^{-1}\right)$$

$$\text{Thermal conductivity}, k = 7.58 \times 10^{-5}\ \text{T} + 3.5 \times 10^{-3}\left(\text{Wm}^{-1}\text{K}^{-1}\right), \text{and}$$

$$\text{Thermal expansion coefficient}, \beta = \text{T}^{-1}\left(\text{K}^{-1}\right)$$

In order to depict the relative contribution of each of these techniques to inside temperature, a dimensionless index is defined as follows. When the interior temperature exceeds 25 °C, it will be considered as a temperature discomfort condition. This reference temperature is widely elements. Then the following expression:

$$F(t) = \max(T_t - 25.25) \quad (26)$$

I_s is a time function of truncated temperature and it will be able to estimate the overall discomfort by means of the integration along the day for each different scenario S:

$$A(S) = \int_S F(t)\mathrm{d}t \quad (27)$$

Then, for each passive technique, let:

$$A_{max} = \max\left[A(S) : \text{forallscenarios}\, S\right] \quad (28)$$

Finally, the normalised temperature index for each scenario S is:

$$I(S) = A(S) / A_{max} \quad (29)$$

Naturally, it would be preferred, for comfort reasons that this index would be small, preferably nil. It may be seen that the variable is directly related to temperature discomfort: the larger the value of the index, the farthest will inside conditions be from expected well-being. Also, the use of electricity-operated air conditioning systems will be more expensive the higher this variable is. Hence, the energy expenditure to offset discomfort will be higher when comparing two index values; the ratio of them is proportional to the expected energy savings. When the external shade blocks the windowpane completely, the excessive heat gains belong to the lowest values in the set, and the dimensionless index will be constant with orientation. For the climate conditions of the locality, it can be seen that a naked window can produce undesirable heat gains if the orientation is especially unfavourable, when the index can have an increase of up to 0.3 with respect to the totally shaded window.

5 Conclusion

With increasing worldwide awareness of the serious environmental problems due to fossil fuel consumption, efforts are being made to develop energy-efficient and environment-friendly systems by utilisation of non-polluting renewable energy sources, such as solar energy, industrial waste heat or geothermal water. The GSHPs are suitable for heating and cooling of buildings and so could play a significant role in reducing CO_2 emissions. Ground source or geothermal heat pumps are a highly efficient, renewable energy technology for space heating and cooling. This technology relies on the fact that, at depth, the Earth has a relatively constant temperature, warmer than the air in winter and cooler than the air in summer.

A geothermal heat pump can transfer heat stored in the earth into a building during the winter, and transfer heat out of the building during the summer. Furthermore, special geological conditions, such as hot springs, are not needed for successful application of geothermal heat pumps. The GSHPs are receiving increasing interest because of their potential to reduce primary energy consumption and thus reduce emissions of the GHGs. The GSHP is generally recognised to be one of the most outstanding technologies of heating and cooling in both residential and commercial buildings, because it provides high coefficient of performance (COP), up to 3–4 for an indirect heating system and 3.5–5 for a direct heating system.

The main benefit of using the GSHPs is that the temperature of the subsurface is not subject to large variations experienced by air. It is currently the most common thermal energy source for the heat pumps, and so would allow the construction of more efficient systems with superior performance. The GSHPs do not need large cooling towers and their running costs are lower than those of conventional heating and air conditioning systems. As a result, the GSHPs have increasingly been used for building heating and cooling with an annual rate of increase of 10% in recent years.

References

1. Luo L, Tondeur D, Le Gall H, Corbel S (2007) Constructal approach and multi-scale components. Appl Therm Eng 27:1708–1714
2. Luo L, Fan Y, Tondeur D (2007) Heat exchanger: from micro to multi- scale design optimisation. Int J Energy Res 31:1266–1274
3. Philappacopoulus AJ, Berndt ML (2001) Influence of de-bonding in ground heat exchangers used with geothermal heat pumps. Geothermics 30(5):527–545

4. Jo HY, Katsumi T, Benson CH, Edil TB (2001) Hydraulic conductivity and swelling of non-prehydrated GCLs permeated with single- species salt solutions. J Geotech Geoenviron Eng 127(7):557–567
5. Anandarajah A (2003) Mechanism controlling permeability changes in clays due to changes in pore fluids. J Geotech Geoenviron Eng 129(2):163–172
6. Omer A (2011) Performance analysis of direct expansion ground source heat pumps for buildings applications. PhD Thesis, Nottingham University, UK, May 2011, pp 135
7. Malin N, Alex W (2000. Building Green, Inc., 2000) Ground-source heat pumps: are they green? Environmental Building News 9(7/8):45
8. Omer AM (2008) Chapter 10: Development of integrated bioenergy for improvement of quality of life of poor people in developing countries. In: Magnusson FL, Bengtsson OW (eds) Energy in Europe: economics, policy and strategy- IB. 2008 NOVA Science Publishers, Inc., New York, pp 341–373
9. Brain G, Mark S (2007) Garbage in, energy out: landfill gas opportunities for CHP projects. Cogeneration and On-Site Power 8(5):37–45
10. Omer AM (2009) Environmental and socio-economic aspect of possible development in renewable energy use. In: Proceedings of the 4th international symposium on environment, Athens, 21–24 May 2009, pp 19–34
11. Omer AM (2009) Energy use, environment and sustainable development. In: Proceedings of the 3rd international conference on sustainable energy and environmental protection (SEEP 2009), Paper No.1011, Dublin, 12–15 Aug 2009, pp 45–54
12. Omer AM (2009) Energy use and environmental impacts: a general review. J Renew Sustain Energy 1(053101):1–29, United State of America
13. Omer AM (2009) Chapter 3: Energy use, environment and sustainable development. In: Mancuso RT (ed) Environmental cost management. 2009 NOVA Science Publishers, Inc., New York, pp 129–166
14. Heinonen EW, Tapscott RE, Wildin MW, Beall AN (1996) Assessment of antifreeze solutions for ground-source heat pumps systems. New Mexico Engineering Research Institute NMERI 96/15/32580, pp 156
15. Sims RH (2007) Not too late: IPCC identifies renewable energy as a key measure to limit climate change. Renew Energy World 10(4):31–39
16. Trevor T (2007) Fridge recycling: bringing agents in from the cold. Waste Manage World 5:43–47
17. Steele J (1997) Sustainable architecture: principles, paradigms, and case studies. McGraw-Hill Inc., New York, p 21
18. Omer AM (2012) Chapter 11: Ventilation and indoor air quality. In: Daniels JA (ed) Advances in environmental research, vol 24. 2012 NOVA Science Publishers, Inc., New York, pp 231–238
19. Omer AM (2013) Chapter 3, Ground-source heat pump technology advancements. In: Acosta MJ (ed) Advances in energy research, vol 13. 2013 NOVA Science Publishers, Inc., New York, pp 115–166
20. Omer AM (2014) Energy efficiency improvement utilising high technology: an assessment of energy use in industry, buildings development end environment, Social Science Research Network (SSRN), Canada, pp 1–36
21. Omer AM (2015) Low energy building design: heating, ventilation and air conditioning. Sci J Rev 4(2):30–46, England

Abdeen Mustafa Omer (BSc, MSc, PhD) is an Associate Researcher at the Energy Research Institute (ERI). He obtained both his PhD degree in the Built Environment and Master of Philosophy degree in Renewable Energy Technologies from the University of Nottingham. He is a qualified Mechanical Engineer with a proven track record within the water industry and renewable energy technologies. He has graduated from the University of El Menoufia, Egypt, with a BSc in Mechanical Engineering. His previous experience involved being a member of the research team at the National Council for Research/Energy Research Institute in Sudan and working director of research and development for National Water Equipment Manufacturing Co. Ltd., Sudan. He has been listed in the WHO'S WHO in the World 2005, 2006, 2007 and 2010. He has published over 200 papers in peer-reviewed journals, 50 review articles, 5 books and 50 chapters in books.

Effect of Natural and Forced Ventilation on Drying of Kiwi in an Indirect Solar Dryer

Hajar Essalhi, Mohammed Benchrifa, and Rachid Tadili

1 Introduction

Drying by solar energy is one of the deemed preservation methods for fruits and vegetables around the world for centuries. Open sun drying is the oldest and traditional method of drying where the food is spread over the ground or on trays. This method has a lot of disadvantages, such as the food being infected by insects, dirt, dust, pollution, and degradation through exposure to direct sun radiation. The introduction of the solar dryer as an energy-efficient technology in the drying processes can reduce losses of crops and improve the quality of the dried product. Several experimental studies reported the various methods used for drying of food products employing a solar dryer [1, 2] for grapes' drying, for red chili drying [3], and for black ginger drying [4]. Therefore, the objective of the present work is to study the effect of natural and forced ventilation on the drying of kiwi samples in an indirect solar dryer under the metrological conditions of Rabat, Morocco. World Renewable Energy Conference 2020, Lisbon 2020.

2 Materials and Methods

2.1 Description of the Solar Dryer

The indirect solar dryer consists of a solar air collector, a drying chamber, and a chimney. The solar air collector was composed of two corrugated aluminum sheets, painted in matt black paint. The two superimposed corrugated sheets were joined to form parallel cylinders, with a diameter of 9 cm. The absorber was of 1 m^2 area. A 6-mm-thick glazing was used as roofing. At the bottom of the solar air collector, a 5-cm-thick cork plate was added to minimize the heat loss. The heated air goes across the channels of the collector absorber to get at the drying chamber and throws upward across the samples. The drying chamber was fabricated from trespa wooden sheet and was insulated from all sides with glass wool of 5-mm thickness. The back of the drying chamber was equipped with a double-swing door of dimensions 95.5 cm by 48.2 cm. Four rectangular trays were made of galvanized iron wire of dimensions 95.5 cm by 26.2 cm. The upper part of the drying chamber was equipped with a vertical chimney made of aluminum of 70 cm height, as illustrated in Fig. 1.

To dry the kiwis, two measurement campaigns were carried out in May and July 2016: the first under natural convection and the second forced. Temperature measurements at different locations in the indirect solar dryer were realized using T-type thermocouples. Inclined solar radiation was measured, utilizing the Kipp & Zonen brand SP-Lite pyranometer. The measurements were collected and stored at the end of each hour using an acquisition system (CR10X). The solar air collector was placed facing south, inclined at an angle of 34° with respect to the horizontal corresponding to the latitude of Rabat. Fresh local kiwi fruits were purchased at the local market in Rabat, Morocco. Only good quality kiwis were utilized in the experiments. About 509 g of fresh kiwis cut into slices of 5 mm thickness, for each experiment, were dried, without any chemical pretreatment.

H. Essalhi (✉) · M. Benchrifa · R. Tadili
Laboratory of Solar Energy and Environment, Mohammed V University in Rabat, Faculty of Sciences, Rabat, Morocco

© The Author(s), under exclusive license to Springer Nature Switzerland AG 2022

A. Sayigh (ed.), *Sustainable Energy Development and Innovation*, Innovative Renewable Energy,
https://doi.org/10.1007/978-3-030-76221-6_24

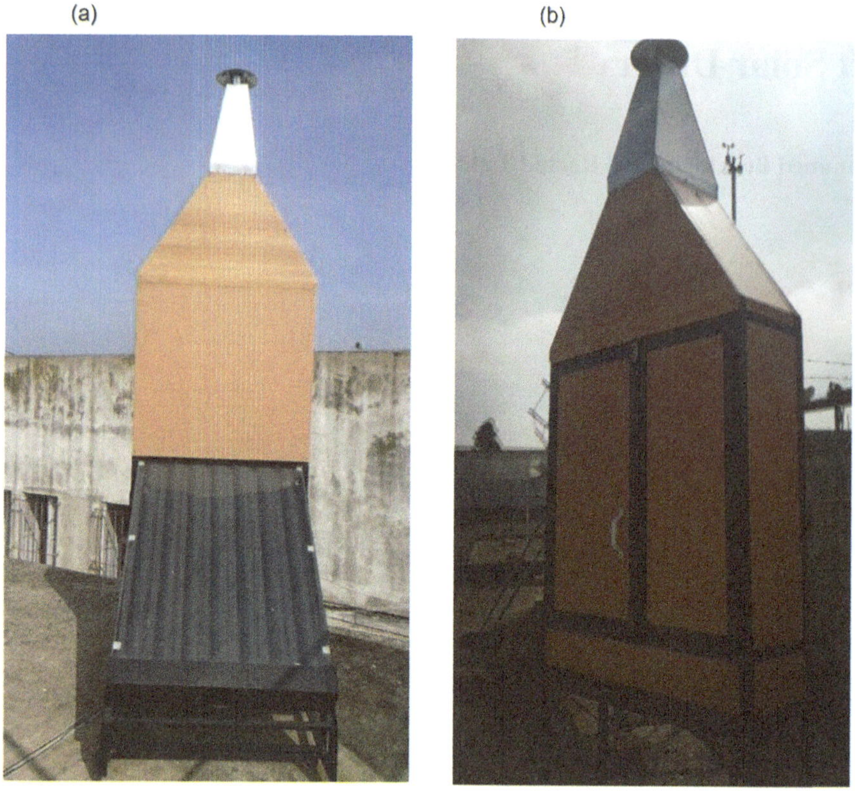

Fig. 1 (**a**) In direct solar dryer and (**b**) drying chamber

3 The Thermal Performance Results

The variations of air temperature at solar collector outlet, tray 1, tray 2, and solar radiation during the drying period of Kiwi under natural convection are shown in Fig. 2. The outlet air temperature of the solar air collector varied between 24 and 51 °C throughout the drying period. The air temperature at the two trays reached 48 and 45 °C at 13:00 h, respectively, on the second day of drying. An air temperature difference of 3 °C max observed for the temperatures was measured at the inlet, the first and second trays of the indirect solar dryer. The maximum solar radiation recorded was about 991 W /m^2 through peak sunshine hours.

Figure 3 shows the variation of air temperature at the solar collector outlet, tray 1, tray 2, and solar radiation through the drying experimentation under forced convection. An air temperature difference of 8 °C max was observed between the outlet of the solar collector and at the level of the two trays. During the entire drying period, the air temperature difference between the tray 1 and 2 is 1 °C max, except for the second drying day which is 3 °C max. According to these results, the use of a fan decreases the air temperature of the outlet of the solar collector, which significantly affects the other drying system temperatures. The absorber design allows good air circulation without resorting to the use of a fan [5].

Table 1 presents the comparison between solar drying of kiwi under natural and forced convection. Under natural convection, the mass of kiwi is reduced from 509 g to 87 g during 18 hours of drying, and within forced convection, it is reduced from 509 g to 87 g for 26 hours of drying. The drying time is not similar for the two convection modes, the drying time increases with the decrease in temperature, following the use of a high flow fan (150 m^3 / h). These results can be explained by the fairly wide absorber undulations diameter (9 cm), which allows moderate air circulation in natural ventilation. However, in forced circulation, the air circulates rather quickly, which prevents it from reaching high temperatures, and this leads to a slowing down of the drying process. In consequence, it takes more energy and more time to extract the same amount of water from the product by forced convection. We deduce from the results that the high-temperature parameter affects the drying process more then the flow parameter. The increase in the temperature of the drying air gives the last-mentioned a greater evaporating power, which creates a very short drying time. Therefore, we choose the natural convection mode for our dryer, which allows us to have a high drying air temperature and a suitable flow rate.

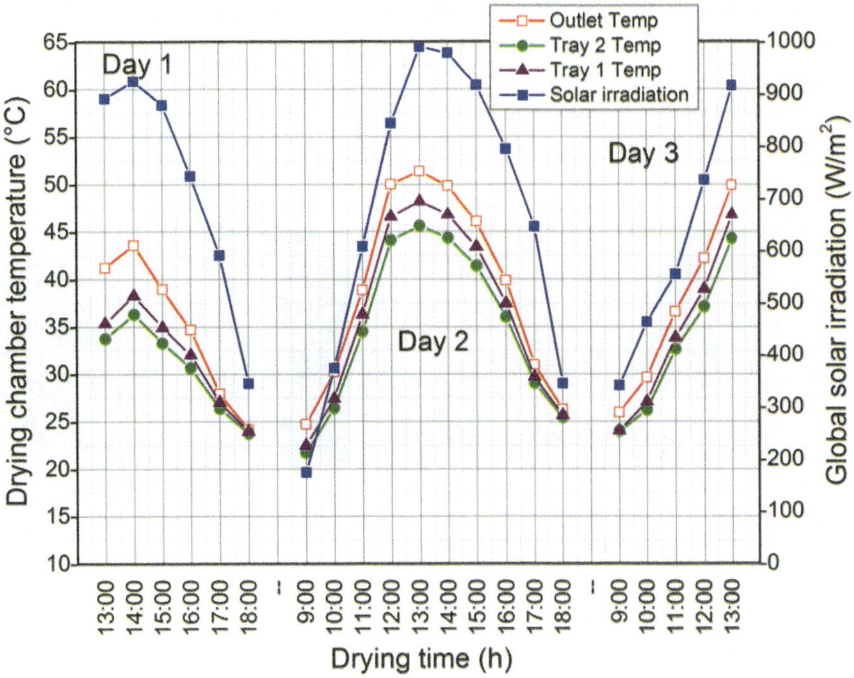

Fig. 2 Evolution of the air temperature of the drying chamber under natural convection

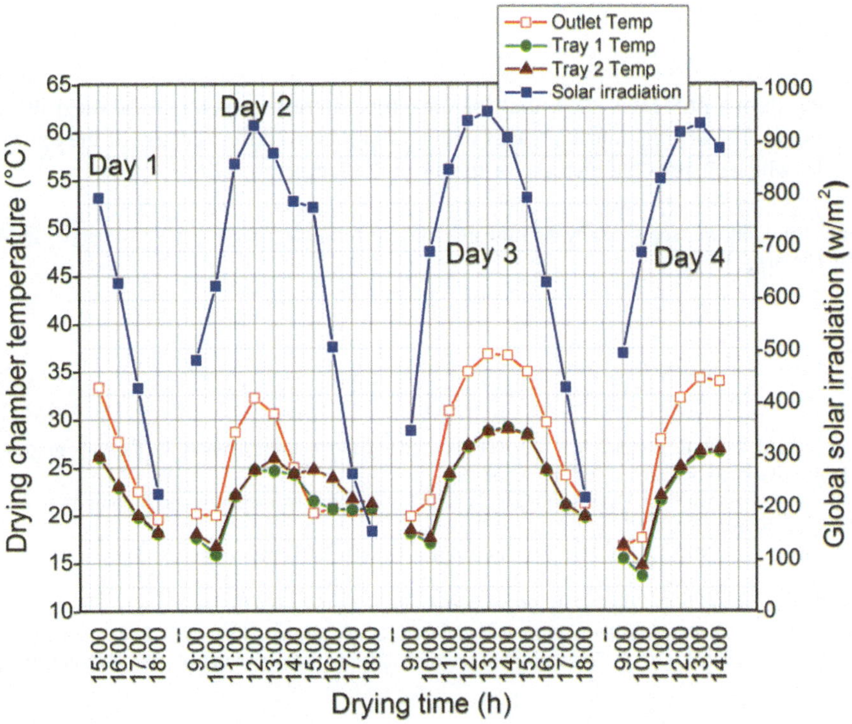

Fig. 3 Evolution of the air temperature of the drying chamber under forced convection

Table 1 Results of kiwi drying under two modes of convection

Product	Drying method	Initial mass (g)	Final mass (g)	Quantity of extracted water (%)	Drying time (h)
Kiwi	With ventilation	509	80	82	26
	Without ventilation	509	87	82	18

Fig. 4 (**a**) Kiwi samples before drying, and (**b**) kiwi samples after drying

We observe from Fig. 4 that, at the end of the drying process, the kiwis kept the same color for both modes of drying without using a chemical pretreatment.

4 Conclusion

The results of kiwi drying processes show that the air temperature parameter is the most influencing factor in the drying process of the product. The drying time decreases under natural convection because of the optimal flow entering through the absorber channels, which ensures the increase of the air outlet temperature.

Acknowledgment This work was supported by the Research Institute for Solar Energy and New Energies in Morocco (IRESEN) as part of the Inno Project Solar Thermal-13-FSS.

References

1. Essalhi H, Benchrifa M, Tadili R, Bargach MN (2018) Experimental and theoretical analysis of drying grapes under an indirect solar dryer and in open sun. Innovative Food Sci Emerg Technol 49:58–64
2. Bingol G, Roberts JS, Balaban MO, Devres YO (2012) Effect of dipping temperature and dipping time on drying rate and color change of grapes. Dry Technol 30(6):597–606
3. Fudholi A, Sopian K, Yazdi MH, Ruslan MH, Gabbasa M, Kazem HA (2014) Performance analysis of solar drying system for red chili. Sol Energy 99:47–54
4. Ekka JP, Bala K, Muthukumar P, Kanaujiya DK (2020) Performance analysis of a forced convection mixed mode horizontal solar cabinet dryer for drying of black ginger (Kaempferia Parviflora) using two successive air mass flow rates. Renew Energy 152:55–66
5. Essalhi H, Tadili R, Bargach MN (2018) Comparison of thermal performance between two solar air collectors for an indirect solar dryer. J Phys Sci 29(3):55–65

Energy and Exergy Performance Comparison of Three Different Solar Cookers Using Different Loads

Ashmore Mawire, Sibongiseni Simelane, and Patrick O. Abedigamba

1 Introduction

Fossil fuel energy resources are the main source of energy in domestic applications such as cooking of food in South Africa [1]. These fossil fuels used for cooking result in greenhouse gas emissions contributing to climate change. Solar cookers are environment-friendly devices which use the sun to cook food. The cookers are easy to use, have a low risk of fire being developed, are non-polluting and foods rarely burn [2]. They are basically three types of solar cookers which are direct-focusing (concentrating) solar cookers, box type of solar cookers and indirect solar cookers. Concentrating type of cookers focus solar radiation using reflectors onto a focal region where the pot is placed. Box type of solar cookers use the greenhouse effect to cook food in an insulated box (oven). A box cooker has a transparent glass or plastic top and may have one or more reflectors to enhance the performance of the cooker. The top is removable for placing cooking containers inside. The cooking containers and the bottom of the box cooker are dark-coloured or black for better absorption of solar radiation. Indirect solar cookers have the cooking pot physically displaced from the collector, and a heat-transferring medium is usually required to bring the collected energy into the cooking pot. These indirect cookers can be integrated with thermal energy storage (TES) systems for the possibility of indoor cookers; however, they are expensive to fabricate and maintain.

The two most viable options for solar cooking in developing countries are concentrating type of cookers and box type of solar cookers. A lot of recent research has been done on these two types of solar cookers [3–8]. A lot of work has been done on concentrating and box types of solar cookers to enhance their performance, but limited work has appeared in recent years evaluating the energetic and exergetic performance of solar cookers using different heating loads [9–11]. This justifies doing work on the energetic and exergetic performance of solar cookers with different cooking loads. In a bid to understand fully how variable loads affect the energetic and exergetic performance of different solar cookers, the aim of the study is to evaluate the energetic and exergetic thermal performance parameters of three different domestic solar cookers in Mahikeng, South Africa. The three domestic solar cookers evaluated are a parabolic dish solar cooker, a box solar cooker with a reflector and a box solar cooker without a reflector. The three cookers are compared experimentally with water heating experiments using 1.0, 1.5, 2.0, 2.5 and 3.0 kg loads on different days.

2 Experimental Methodology

The experimental tests of the different solar cookers were conducted in Mahikeng (25° 51′ 11.3796″ S latitude, 25° 38′ 24.6516″ E longitude) in the North-West province of South Africa. The three solar cookers were a parabolic dish solar cooker, a box solar cooker without a reflector, a box solar cooker with a single reflector. The parabolic dish solar cooker in the experiment had a diameter of 1.19 m. The box cooker without a reflector had a width of 0.55 m and a length of 0.75 m. The box cooker with a reflector had a width of 0.45 m and a length of 0.46 m. Each cooker was monitored for a minimum of two and half hours during water heating experiments to ensure sufficient time for a reasonable temperature rise of the water.

A. Mawire (✉) · S. Simelane · P. O. Abedigamba
Department of Physics and Electronics, North-West University (Mafikeng Campus), Mmabatho, South Africa
e-mail: ashmore.mawire@nwu.ac.za

© The Author(s), under exclusive license to Springer Nature Switzerland AG 2022

A. Sayigh (ed.), *Sustainable Energy Development and Innovation*, Innovative Renewable Energy,
https://doi.org/10.1007/978-3-030-76221-6_25

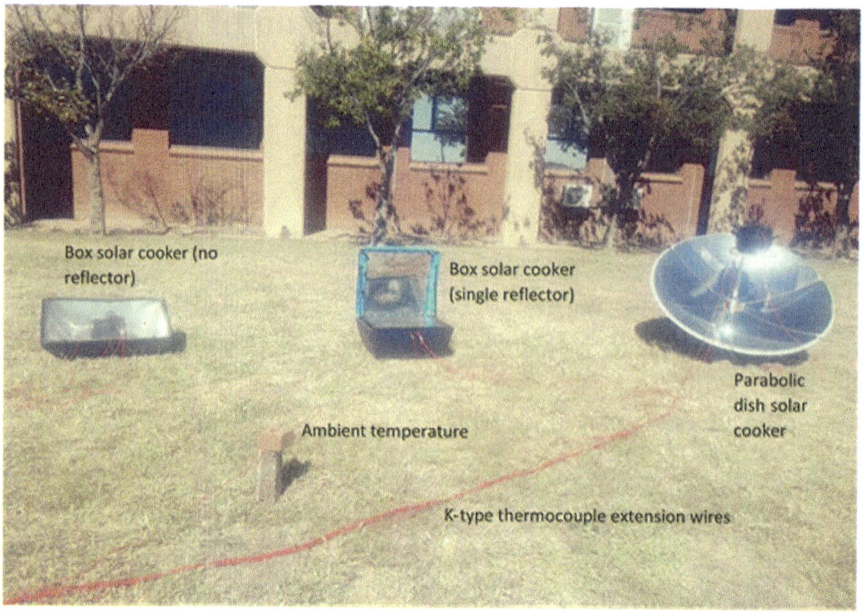

Fig. 1 A photograph of three different solar cookers connected to thermocouples

Temperatures in the solar cookers, the global solar radiation were monitored during the experimental tests. K-type thermocouples with an accuracy of ±2.2 °C were used to monitor the temperature. A CMP 11 (class 2) pyranometer was used to measure the global solar radiation [14]. Figure 1 shows the three solar cookers during the experimental tests. Thermocouple extension wires to the data-logger are shown in red. K-type thermocouples with an accuracy of ±2.2°C in the range of −200–1350 °C were used to measure the temperature at two different positions on each cooking pot, the water temperature in each pot and the ambient temperature. Black pots were used for better absorption of solar radiation. In each experimental test, equal masses of water were put into the pots that were placed onto the three solar cookers respectively. Five different masses of water are used, namely, 1.0, 1.5, 2.0, 2.5 and 3.0 kg in each of the five experimental tests to evaluate the effect of the load on the thermal performances of the cookers.

3 Thermal Analysis

The energy *input* E_i *to* the cooker is given by [12]:

$$E_i = GA \tag{1}$$

where G is the instantaneous solar radiation and A is the aperture area of the cooker. The exergy of solar radiation identical to the exergy input E_{xi} for the solar cookers is calculated from the Petela [13] expression as:

$$E_{xi} = G\left[1 + \frac{1}{3}\left(\frac{T_a}{T_s}\right)^4 - \frac{3T_a}{4T_s}\right]A, \tag{2}$$

where, T_a is the ambient temperature and T_s is the surface temperature of the sun given by Petela [13] as 5778 K.

The instantaneous output energy rate, E_o, of the cooker is determined through the water temperature inside the cooking pot as it rises due to the absorption of heat from the aperture area. It is expressed as [12]:

$$E_o = \frac{mc(T_f - T_i)}{\Delta t} \tag{3}$$

where c is the specific heat capacity of the water, m is the mass of the water, T_f is the final water temperature measured at each time interval and T_i is the initial water temperature inside each cooker. For the instantaneous output exergy analysis of the cooker, the ambient temperature is considered. The output exergy E_{xo} of the cooker is expressed as [12]:

$$E_{xo} = \frac{mc\left(T_f - T_i\right) - mcT_a \ln\left(\frac{T_f}{T_i}\right)}{\Delta t}. \tag{4}$$

The energy efficiency of the cooker is thus given as [12]:

$$\eta_e = \frac{E_o}{E_i}, \tag{5}$$

and the exergy efficiency of the cooker is given as [12]:

$$\eta_{ex} = \frac{E_{xo}}{E_{xi}}. \tag{6}$$

4 Results and Discussion

Average energy and exergy rates with respect to the water load are shown in the plots of Fig. 2. Higher average energy rates are seen for the parabolic solar cooker, which tend to increase with the water load possibly due to the higher initial energy rates. The increase in the energy rate for the parabolic dish solar cooker suggests that heating larger water loads is more efficient than heating small water loads. The more efficient solar box cooker with the reflector shows higher average energy rates when compared to the solar box cooker without the reflector. The average energy rate only increases marginally with an increase in the water heating load, suggesting that the water heating load has a small influence on the average energy rate during the experimental tests. In fact, for a load increase of 1.5–3.0 kg, the average energy rate only increases from around 70 W to around 78 W. The solar box cooker without the reflector with the least incident solar radiation also shows the lowest average energy rates, a marginal average energy rate variation with the water load especially from 2.0 kg to 3.0 kg.

The variation of the average exergy rate with respect to the water heating load for the three solar cookers is shown in Fig. 2b. As in the case of the energy rates, the parabolic dish solar cooker shows greater average exergy rates due to the greater initial exergy rate values compared to the other two cookers. The exergy rate also increases with an increase in the cooking load, suggesting that greater exergy is obtained with an increase in the cooking load due to more effective heat transfer. It is important to note that the average exergy rate increases marginally from about 8 W to 9 W for a load increase from 1.5 kg to 2.0 kg due to the highest wind speed which induces more heat losses for the 2.0 kg case. The average exergy rate values for the solar box cooker are almost constant with increasing load and therkkke is a slight drop for a load increase of 2.0–3.0 kg. The solar box cooker without the reflector shows the lowest average exergy rate values, these are almost constant and independent of the cooking load. The solar box cooker without the reflector also shows a slight drop of the average exergy rate values for a load increase from 2.0 kg to 3.0 kg. The low exergy rate values for solar box cookers are possibly due to lower concentration of solar radiation, and other researchers have also reported low exergy rate values for solar box cookers [9, 15].

Figure 3 shows the average energy and exergy efficiencies of the three solar cookers with different loads. The solar box cooker with the reflector shows the highest values of the energy efficiencies due to the reflector and proper effective insulation, which enables a greater absorption of solar radiation. This happens despite of the fact that it shows lower almost constant energy rates as compared to the parabolic solar cooker. The average efficiency of the box solar box cooker with the reflector varies from around 0.25 for the lowest load to around 0.52 for the largest load. The corresponding energy efficiencies of the solar box cooker without the reflector and the parabolic dish cooker increase from 0.11 to 0.18, and 0.08 to 0.28, respectively, for the same load increments. It is also important to note that the effect of the windy conditions for the 2.0 kg case results in a drop of the average efficiency for an increase in the load from 1.5 kg to 2.0 kg for all solar cookers. Generally, for all the solar cookers, the average energy efficiencies increase with the cooking load, suggesting more effective heat transfer and cooking with the larger load. The parabolic dish shows the lowest average energy efficiencies for cooking loads from 1.0 kg to 2.0 kg; however, the average energy efficiencies for the 2.5 kg and 3.0 kg loads are greater than those of the solar

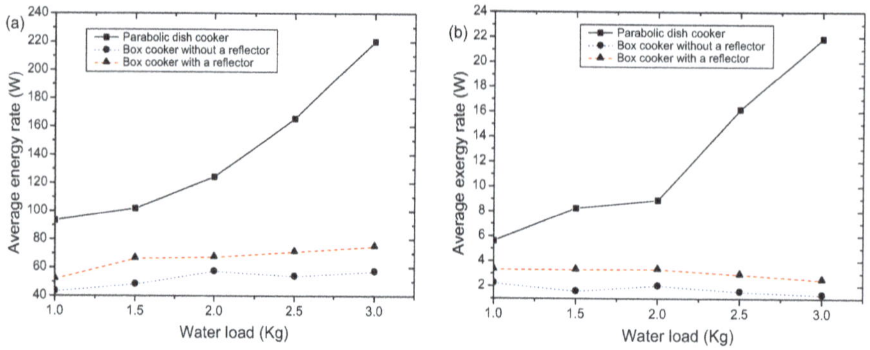

Fig. 2 (**a**) Variation of the average energy rate with increasing load for the different cookers, (**b**) the variation of the average exergy rate with increasing load for the different cookers

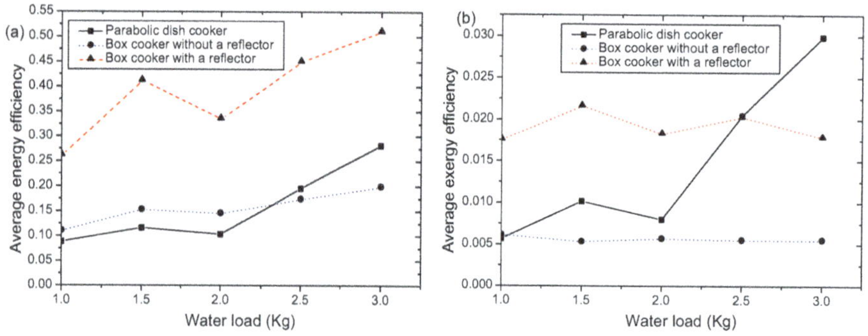

Fig. 3 (**a**) Variation of average energy efficiency with increasing load for the different cookers, (**b**) the variation of average exergy efficiency with increasing load for the different cookers

box cooker without a reflector. This again suggests that the efficiency of the parabolic dish solar cooker is more with the greater loads.

Energy efficiencies are a measure of only the quantity of energy and a true measure of the potential of the solar cookers is evaluated with exergy efficiencies which quantify the quality of the available energy. Average exergy efficiencies for the three solar cookers for different water loads are shown in Fig. 3b.

The solar box cooker without the reflector shows the lowest average exergy efficiencies which are almost constant (around 0.008), and independent of the cooking load possibly due to the very low values of cooking exergies obtained with this inefficient solar cooker. The solar box solar cooker with the reflector shows the highest exergy efficiency values fluctuating up and down around a mean value of around 0.02. There is no direct correlation between the cooking load and the average exergy efficiency for this cooker. On the other hand, the parabolic dish solar cooker shows a general increase of the average exergy efficiency with an increase of the heating load, except for the drop from the 1.5 kg to the 2.0 kg cases. The parabolic dish solar cooker shows an increase in the exergy efficiency from 0.007 to 0.03 with an increase in the load. The average exergy efficiency shows an almost linear and well-pronounced rise from the 2.0 kg load to the 3.0 kg, further supporting the idea that a large thermal mass for the parabolic solar cooker achieves better exergetic performance. The average exergy efficiency of the parabolic solar cooker is actually higher than that of the box solar cooker with the reflector for the highest cooking load of 3.0 kg. The average exergy efficiencies are low but these efficiencies are comparable to other efficiencies reported in recent literature with different types of solar cookers [9, 15].

5 Conclusion

Energetic and exergetic thermal performance parameters of three different domestic solar cookers in Mahikeng, South Africa were evaluated. The three domestic solar cookers evaluated were a parabolic dish solar cooker, a box solar cooker with a reflector and a box solar cooker without a reflector. The three cookers were compared experimentally with water heating experiments using 1.0, 1.5, 2.0, 2.5 and 3.0 kg loads Higher average energy and exergy rates were obtained with the parabolic dish solar cooker. The water heating load had a very small effect on the average energy and exergy rates of the solar box cookers. The parabolic dish solar cooker had the lowest average energy efficiencies at lower loads. At higher loads, the energy efficiencies of the parabolic dish solar cooker increased to be higher than those of the box cooker without a reflector. The average exergy efficiency of the parabolic dish solar cooker increases with the load. The solar box cooker with a reflector shows the highest average energy efficiency values which increased with the load. The energy efficiency values for the solar box cooker with the reflector increased from around 0.25 to around 0.52 with an increase in the load from 1.0 kg to 3.0 kg. The corresponding energy efficiencies of the solar box cooker without the reflector and the parabolic dish cooker increased from 0.11 to 0.18, and 0.08 to 0.28, respectively. The exergy efficiencies of the box cooker with a reflector and the box cooker without a reflector fluctuated up and down slightly at mean values of 0.02 and 0.008 with an increase of the load. On the other hand, the parabolic dish solar cooker showed an increase in the exergy efficiency from 0.007 to 0.03 with an increase in the load. The solar box cooker without a reflector shows the poorest results in all thermal performance parameters evaluated. The greatest cooking potential in terms of the cooking speed was shown by the parabolic dish solar cooker.

References

1. Wentzel M, Pouris A (2007) The development impact of solar cookers: a review of solar cooking impact research in South Africa. Energy Policy 35:1909–1919
2. Aramesh M, Ghalebani M, Kasaeian A, Zamani H, Lorenzini G, Mahian O, Wongwises S (2019) A review of recent advances in solar cooking technology. Renew Energy 2019(140):419–435
3. Kumar N, Agravat S, Chavda T, Mistry HN (2008) Design and development of efficient multipurpose domestic solar cookers/dryers. Renew Energy 33:2207–2211
4. Abu-Malouh R, Abdallah S, Muslih IM (2011) Design, construction and operation of spherical solar cooker with automatic sun tracking system. Energy Convers Manag 52:615–620
5. Mussard M, Gueno A, Nydal OJ (2013) Experimental study of solar cooking using heat storage in comparison with direct heating. Sol Energy 98:375–383
6. Sagade AA, Panja PS (2018) Experimental determination of effective concentration ratio for solar box cookers using thermal tests. Sol Energy 159:984–991
7. Sagade AA, Samdarshi SK, Panja PS (2018) Enabling rating of intermediate temperature solar cookers using different working fluids as test loads and its validation through a design change. Sol Energy 171:354–365
8. Pereira MC, Cavaco A, Tavares A (2018) Figures of merit and their relevance in the context of a standard testing and performance comparison methods for solar box – cookers. Sol Energy 166:21–27
9. Farooqui SZ (2014) A review of vacuum tube based solar cookers with the experimental determination of energy and exergy efficiencies of a single vacuum tube based prototype. Renew Sust Energ Rev 31:439–445
10. Mahavar S, Rajawat P, Punia RC, Sengar N, Dashora P (2015) Evaluating the optimum load range for box-type solar cookers. Renew Energy 74:187–194
11. Kumar N, Vishwanath G, Gupta A (2012) Effect of load variations on exergy performance of solar box type cooker. J Renew Sustain Energy 4:053125
12. Cuce E, Cuce PM (2013) A comprehensive review on solar cookers. Appl Energy 2013(102):1399–1421
13. Petela R (2005) Exergy analysis of the solar cylindrical–parabolic cooker. Sol Energy 79:221–233
14. Campbell Scientific (2019) High quality pyranometers double glass dome 2017. Available at: https://s.campbellsci.com/documents/au/product.../b_cmp6-cmp11-cmp21.pdf, Accessed 6 Mar 2019
15. Ozturk HH (2007) Comparison of energy and exergy efficiency for solar box and parabolic cookers. J Energy Eng 133:53–62

Experimental Investigation of Solar Still with Corrugated Absorber Surface for Domestic Desalination Use

A. Idrissi, M. Benchrifa, R. Tadili, M. Naciri, and S. El Madrab

1 Introduction

The main objective of desalination is removing dissolved minerals from seawater, groundwater, and treated wastewater. It becomes a proven technique in providing a solution to the problem of decreasing fresh water in rural and remote areas suffering from acute water crisis due to the contamination of groundwater and surface reserves either by natural or by anthropogenic reasons. Solar desalination is particularly important for locations where solar intensity is high.

Various configurations of a solar still are studied [1–5]. Through several techniques, the use of several configurations of a corrugated absorber plate is experimentally studied in the present work. Moreover, the plate is insulated in its bottom face by an air gap. Moreover, the performances of the solar still are presented via the measurement of the daily thermal efficiency and daily productivity of distilled water efficiency.

2 Description of the Experimental System and Experimental Plan

2.1 Components

The experimental study has as objective to investigate the performance of a solar still at the Laboratory of Solar Energy and Environment (LESE). The investigated solar still has a corrugated surface absorber. The realized solar still has several components as follows:

- Absorber:

  The absorber is used to receive and absorb the solar irradiation. This absorption leads to the temperature elevation in the base of the absorber. The base of the absorber is black and has as dimensions 810 mm × 1170 mm. Moreover, this surface is painted black to increase its absorption coefficient. During experimentation, three shapes of corrugated surface absorber are used:

  1. The corrugation shape is sinusoidal and the surface is in galvanized zinc (Fig. 1). The depth of this corrugation is 30 mm.
  2. The corrugation shape is sinusoidal with a great amplitude than the first absorber. The surface is in stainless aluminum (Fig. 2). In this case, the depth of the corrugation is 58 mm.
  3. The corrugation shape is triangular and the surface absorber is in stainless steel (Fig. 3).

- Glass cover:

  The absorber is covered by a glass cover which has 8 mm as thickness.

A. Idrissi (✉) · M. Benchrifa · R. Tadili · M. Naciri · S. El Madrab
Laboratory of Solar Energy and Environment, Mohammed V University in Rabat, Faculty of Sciences, Rabat, Morocco

© The Author(s), under exclusive license to Springer Nature Switzerland AG 2022

A. Sayigh (ed.), *Sustainable Energy Development and Innovation*, Innovative Renewable Energy,
https://doi.org/10.1007/978-3-030-76221-6_26

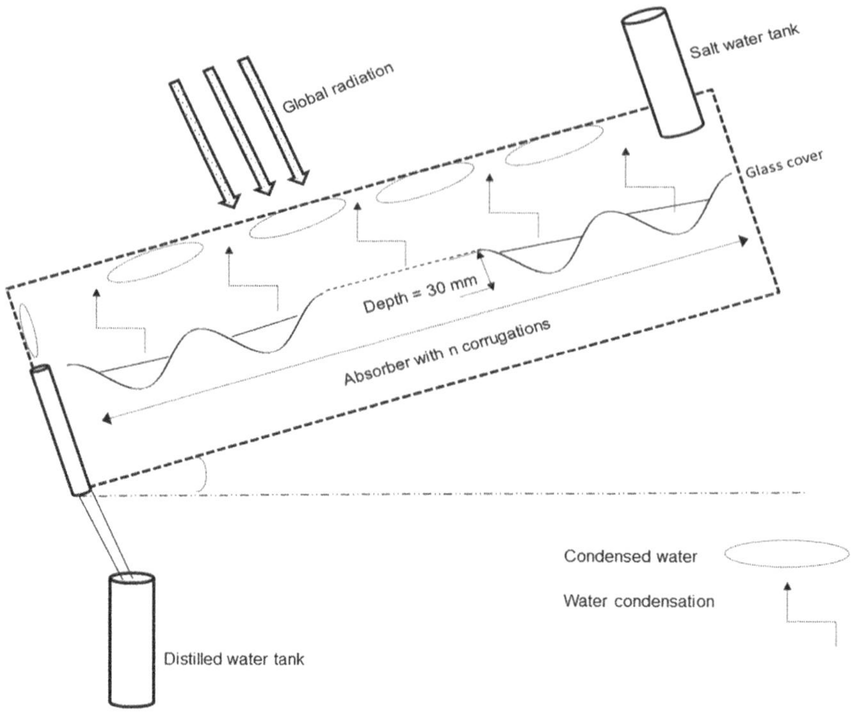

Fig. 1 Solar still with n sinusoidal corrugations

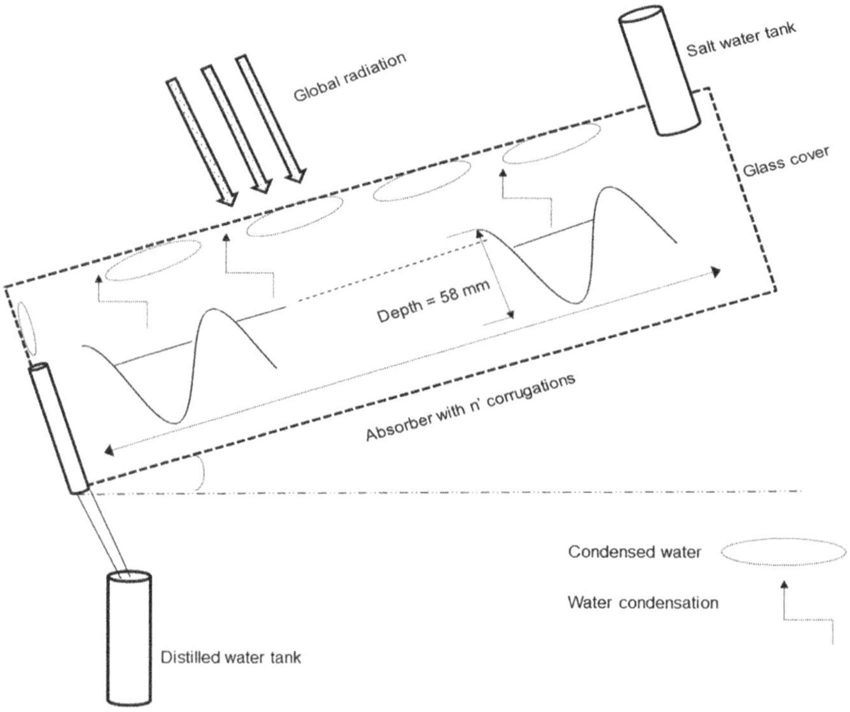

Fig. 2 Solar still with n' sinusoidal corrugations

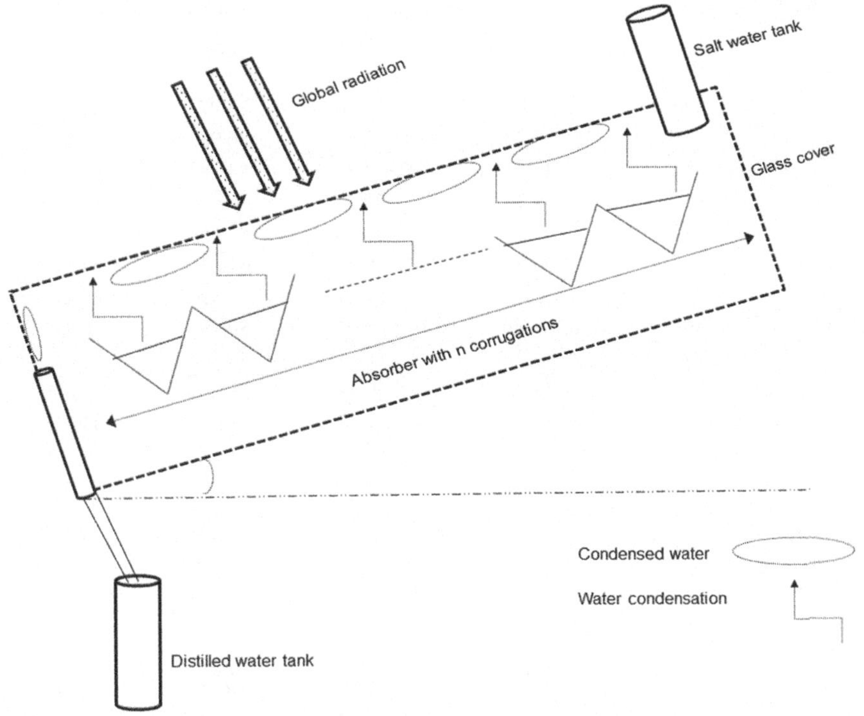

Fig. 3 Solar still with *n* triangular corrugations

- Angle of inclination:

  The solar still has an angle of inclination of 10°.
- Storage tanks:

  Two storage tanks for seawater and distilled water with a capacity of 5 L for each of them are used.
- Air gap:

  The corrugated absorber is maintained between two air gaps. The air gap in the bottom serves as insulation to reduce the energy losses.
- Gutter:

  A gutter is placed under the lower side of the absorber to collect the condensed water.

2.2 *Experimental Plan*

To measure the daily thermal efficiency and daily productivity of distilled water efficiency as indicators of the performance, a Kipp & Zonen brand SP-Lite pyranometer is used for global solar irradiation (Ig), graduated tanks and four thermocouples of type T to indicate glass cover temperature, air ambient temperature, water temperature, and absorber temperature. The presented measurements are taken for several days under different climate conditions.

3 Results and Discussion

For the three types of corrugated absorbers of a solar still, the tables below present the case when all of the solar stills give a high thermal performance. The hourly measures are taken between 8 h and 18 h.

Table 1 Temporal variations of measured indicators of performance for the first solar still

Hour	Ig (W/m^2)	Absorber (°C)	Air (°C)	Water (°C)	Glass (°C)
8	134.1	23.63	23.46	25.97	23.26
9	321.2	33.28	32.52	36.98	32.37
10	351	49.61	43.28	54.38	43.2
11	918	59.35	59.42	75.62*	58.22
12	973	68.12	53.18	67.61	49.04
13	977	60.72	55.62	64.22	34.89
14	956	61.85	57.24	61.32	48.41
15	862	65.77	63.94	59.93	50.83
16	735	59.99	58.65	56.5	48.01
17	558	51.82	49.38	50.4	43.02
18	356	36.79	37.37	38.32	35.02

Table 2 Temporal variations of measured indicators of performance for the second solar still

Hour	Ig (W/m^2)	Absorber (°C)	Air (°C)	Water (°C)	Glass (°C)
8	388.6	41.05	40.95	38.72	36.33
9	576.4	62.38	55.05	66.64	54.09
10	744	74.4	66.55	81.71	63.92
11	876	84.8	78.4	93.6	72.2
12	956	89.6	92.2	98.26	74.95
13	979	92	84.4	99.53	76.13
14	952	89.8	83.5	96.8	74.7
15	868	84.5	78.6	91.1	71.3
16	736	75.1	69.23	81.43	65.64
17	561.2	60.48	55.64	66.51	55.45
18	375.3	43.85	42.83	47.51	42.4

For the sinusoidal corrugations, Table 1 presents the measured Ig and obtained temperatures for absorber, air, water, and glass cover.

At the water temperature of 75.62*, the solar still produces 3 $L/m^2/day$. Moreover, during the process of distillation, some of the undulations become empty. That behavior leads us to fill the empty undulation by water. This decision makes losses in heat and in the produced distilled water.

As a solution to this problem, an absorber with great amplitude is used. The great amplitude reduces the number of sinusoidal corrugations. Thus, for the second type of sinusoidal corrugation, Table 2 illustrates the measured temperatures and global irradiation.

As seen in Table 2, the temperature of water does not exceed 100 °C and the temperature of the absorber becomes higher than that reached in the first type of corrugated absorber. The distilled water produced in this solar still is 5 $L/m^2/day$. On the other side, the disadvantage of this configuration is the duration of vaporization that it needs. This duration is equal to two successive clear days. To overcome this problem, manufacturing a corrugated absorber surface with medium amplitude and a greater number of corrugations becomes an optimal choice. In fact, the third corrugated surface of absorber has a triangular shape and the same number of corrugations as the first type of solar still. Table 3 shows the performance of a new solar still. The volume of the obtained distilled water is up of 8 $L/m^2/day$.

Moreover, a comparison between all the studied solar stills through the daily thermal efficiency $\boldsymbol{\tau}$ and daily productivity of distilled water efficiency ρ is presented in Table 4.

As shown in Table 4, the high thermal and productivity performances are attributed to the third solar still which has a triangular corrugated surface absorber.

Table 3 Temporal variations of measured indicators of the performance for the third solar still

Hour	Ig (W/m^2)	Absorber (°C)	Air (°C)	Water (°C)	Glass (°C)
8	355.9	39.9	36.78	41.3	48.06
9	549.5	52.36	47.97	59.23	51.5
10	718	64.32	58.65	70.26	61.74
11	849	71.86	66.74	79.29	67.41
12	927	77.47	73.77	87.67	72.76
13	955	79.52	76.02	89.82	73.92
14	925	78.47	74.32	88.17	72.27
15	851	72.86	70.55	83.56	68.16
16	595	61.49	59.71	62.75	57.96
17	407.6	45.32	45.18	48.89	43.39
18	307	33.66	32.58	31.36	31.57

Table 4 Thermal and productivity performances

Type of absorber	$\tau = \frac{\text{useful energy}}{\text{consumed energy}} (\%)$	$\rho = \frac{\text{volume of distilled water}}{\text{volume of salt water}} (\%)$
First solar still	39	85
Second solar still	46	81
Third solar still	50	86

4 Conclusion

In the present work, an experimental investigation of a solar still with corrugated absorber surface for domestic desalination use is carried out. This experimentation investigates the influence of shape corrugation on thermal efficiency and productivity of distilled water performance of a solar still. The higher thermal and productivity performances are obtained for a triangular corrugated surface absorber with 50% and 86%, respectively. In fact, the solar still with triangular corrugations produces 8 L/m^2/day of distilled water. Moreover, this solar still can be used in rural zones considering its low cost. A study containing the economic analysis and all working parameters will be carried out in future work.

References

1. Jani HK, Modi KV (2018) A review on numerous means of enhancing heat transfer rate in solar thermal based desalination devices. Renew Sust Energ Rev 93:302–317
2. Das D, Bordoloi U, Kalita P, Boehm RF, Kamble AD (2020) Solar still distillate enhancement techniques and recent developments. Groundw Sustain Dev 10:100360. https://doi.org/10.1016/j.gsd.2020.100360
3. Ayav Pİ, Atagündüz G (2007) Theoretical and experimental investigations on solar distillation of IZTECH campus area seawater. Desalination 208:169–180
4. Katekar VP, Deshmukh SS (2020) A review on research trends in solar still designs for domestic and industrial applications. J Clean Prod. https://doi.org/10.1016/j.jclepro.2020.120544
5. Sharon H, Reddy KS, Krithika D, Philip L (2020) Viability assessment of solar distillation for desalination in coastal locations of Indian sub-continent – thermodynamic, condensate quality and enviro-economic aspects. Sol Energy 197:84–98

Directly Absorbing Nanofluid-Based Solar Thermal Collectors for Cairo

Vishal Bhalla, Mohamed M. Sabry, Ahmed Ghitas, and Harjit Singh

1 Introduction

The nanofluid-based solar collectors [1] are being increasingly investigated for an efficient harnessing of solar energy [2] in which the nanoparticles absorb electromagnetic waves to generate heat [3–6]. In the present paper, a numerical model has been used to study the effect of parameters which influence the performance of the nanofluid-based solar collector. For the numerical modelling, the equations have been solved with finite difference technique by using forward difference implicit method. The most suitable nanoparticle material candidate and its optimum volume fraction in nanofluid has been identified. Further, the effect of working fluid flow rate and irradiation has been predicted. Finally, the contribution of three interfering parameters, volume fraction, mass flow rate and irradiation, towards the performance of nanofluid-based solar collectors has been evaluated using analysis of variance (ANOVA) technique [7].

1.1 Numerical Modelling of Nanofluid-Based Solar Collector

The schematic of a solar collector using nanofluid comprising of nanoparticles suspended in base fluid (Ethylene glycol) as working fluid is shown in Fig. 1. The length of the cylindrical receiver is 1800 mm and diameter 43 mm. For numerical modelling, following assumptions have been considered:

(a) The bottom surface of the receiver is transparent.
(b) The flow of fluid is laminar.
(c) The in-scattering from the nanoparticles is negligible.
(d) There is no emission from the nanoparticles.
(e) The scattering is independent scattering.

In order to simulate the intensity distribution within the receiver (along the diameter of the pipe) Eq. 1, which is a radiative transfer equation (RTE), has been used [8, 9].

$$\begin{aligned}\frac{dI_\lambda(\Omega)}{dy} &= -\left(K_{s_\lambda}+K_{a_\lambda}\right)I_\lambda(\Omega)+K_{a_\lambda}I_{b_\lambda}+\frac{K_{s_\lambda}}{4\pi}\int_{4\pi}I_\lambda(\Omega')p(\Omega'\to\Omega)d\Omega\frac{dI_\lambda(\Omega)}{dy}\\ &= -\left(K_{s_\lambda}+K_{a_\lambda}\right)I_\lambda(\Omega)+K_{a_\lambda}I_{b_\lambda}+\frac{K_{s_\lambda}}{4\pi}\int_{4\pi}I_\lambda(\Omega')p(\Omega'\to\Omega)d\Omega \qquad (1)\end{aligned}$$

V. Bhalla · H. Singh (✉)
Institute of Energy Futures, College of Engineering, Design and Physical Sciences, Brunel University London, Uxbridge, UK
e-mail: harjit.singh@brunel.ac.uk

M. M. Sabry · A. Ghitas
National Research Institute of Astronomy and Geophysics, Helwan, Cairo, Egypt

© The Author(s), under exclusive license to Springer Nature Switzerland AG 2022

A. Sayigh (ed.), *Sustainable Energy Development and Innovation*, Innovative Renewable Energy,
https://doi.org/10.1007/978-3-030-76221-6_27

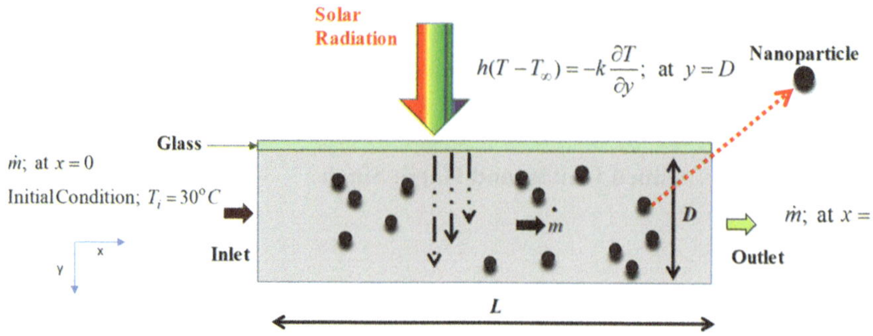

Fig. 1 Schematic of a directly absorbing nanofluid-based solar collector

where I_λ is the spectral incident intensity, I_b is the black body intensity, Ω is solid angle, $p(\Omega' \rightarrow \Omega)$ is phase function and $K_{a\lambda}$ and $K_{s\lambda}$ are, respectively, the spectral absorption and scattering coefficients.

In case of pure fluid (ethylene glycol), the attenuation of irradiation is due to absorption and the absorption coefficient ($K_{a\lambda,\text{EG}}$) for pure fluid is evaluated by Eq. 2.

$$K_{a\lambda,\text{EG}} = \frac{4\pi\kappa}{\lambda} \tag{2}$$

where κ is index of absorption and λ is the wavelength.

Hence, the RTE for pure fluid can be represented by Eq. 3, which is given as

$$\frac{dI_\lambda}{dy} = -\left(K_{a\lambda,\text{EG}}\right) I_\lambda \tag{3}$$

But with the addition of nanoparticles in the fluid, the absorption and scattering by nanoparticles take place, then absorption ($K_{a\lambda,\text{nanoparticle}}$), scattering coefficients ($K_{s\lambda,\text{nanoparticle}}$) and extinction coefficients ($Ke_{\lambda,\text{nanoparticle}}$) of nanoparticles are described by Eqs. 4, 5 and 6, respectively [10].

$$K_{a\lambda,\text{nanoparticle}} = \frac{6\pi f_v}{\lambda} Im\left\{\frac{m^2-1}{m^2+2}\left[1+\frac{\pi^2 d^2}{15\lambda^2}\left(\frac{m^2-1}{m^2+2}\right)\frac{m^4+27m^2+38}{2m^2+3}\right]\right\} \tag{4}$$

$$K_{s\lambda,\text{nanoparticle}} = \frac{4\pi^4 d^3 f_v}{\lambda^4}\left|\frac{m^2-1}{m^2+2}\right|^2 \tag{5}$$

and

$$K_{e\lambda,\text{nanoparticle}} = \frac{6\pi f_v}{\lambda} Im\left\{\frac{m^2-1}{m^2+2}\left[1+\frac{\pi^2 d^2}{15\lambda^2}\left(\frac{m^2-1}{m^2+2}\right)\frac{m^4+27m^2+38}{2m^2+3}\right]\right\} + \frac{4\pi^4 d^3 f_v}{\lambda^4}\left|\frac{m^2-1}{m^2+2}\right|^2 \tag{6}$$

where $K_{e\lambda}$ is extinction coefficient, which is sum of absorption and scattering coefficient, d is the diameter of the nanoparticle, m is complex refractive index $\left(m = \frac{m_{\text{particles}}}{n_{\text{fluid}}}\right)$, f_v is volume fraction of nanoparticles in the base fluid.

Finally, the attenuation of incident light for nanofluid can be evaluated by Eq. 7

$$\frac{\partial I_\lambda}{\partial y} = \left[-K_{a\lambda,\text{water}} + \left(-K_{e\lambda,\text{nanoparticles}}\right)\right] I_\lambda, \tag{7}$$

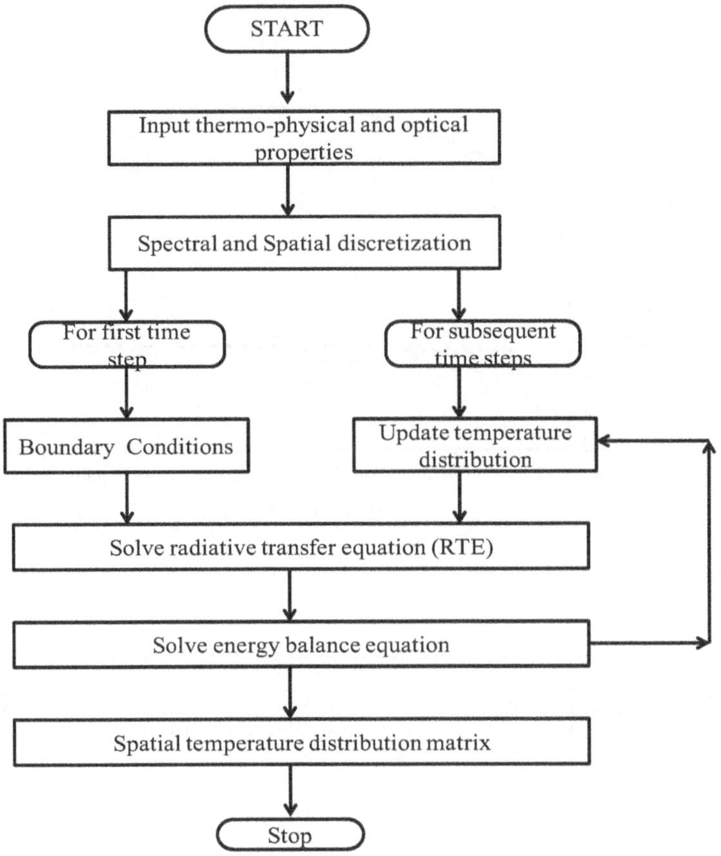

Fig. 2 Algorithm for spatial temperature distribution in nanofluid-based solar collector

Since in this model we assume independent scattering, hence we can add two intensities.

Finally, to find out the temperature gain in the nanofluid, RTE is solved simultaneously with steady-state two-dimensional heat transfer equation. The heat transfer equation is given in Eq. 8 [11].

$$k\frac{\partial^2 T}{\partial y^2} - \frac{\partial r_h}{\partial y} = \rho C_p \frac{\partial T}{\partial t}, \tag{8}$$

where k is the thermal conductivity of the nanofluid, r_h is radiative heat flux and C_p is the specific heat of the fluid.

The algorithm used to evaluate the spatial temperature distribution of nanofluid-based solar collector has been shown in Fig. 2.

2 Results and Discussion

2.1 Effect of the Nanoparticle Materials

In order to find the most suitable nanoparticle material able to deliver heat at highest outlet temperature of nanofluid, for a predetermined flow rate and other parameters fixed, with minimum volume fraction shown in Fig. 3, four candidate materials aluminium, silver, copper and graphite (diameter d = 25 nm) were chosen. The optical properties of these materials have been taken from Pallick et al. [12]. It can be seen in Fig. 3 that with the graphite nanoparticles, 70–75% higher temperatures can be achieved as compared to other nanoparticles. This is because the graphite has broadband solar spectrum as compared to other nanoparticles [13], due to which graphite is able to absorb more irradiation as compared to other nanoparticle materials. Due to higher temperature rise, graphite was chosen for further study.

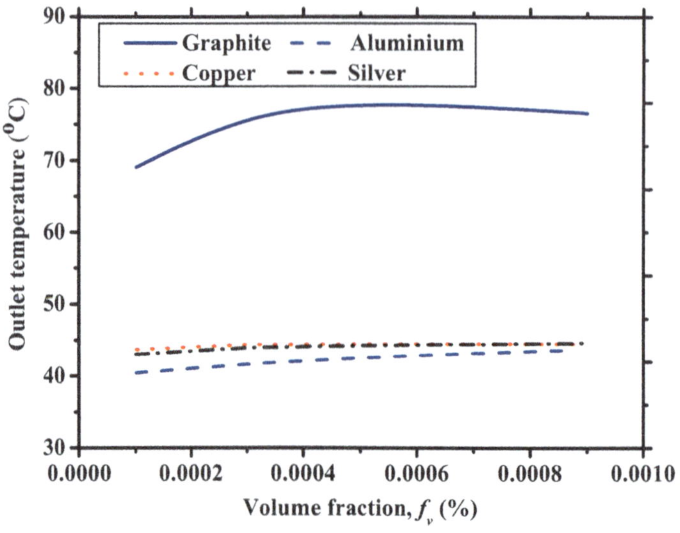

Fig. 3 Effect of different materials on the temperature rise

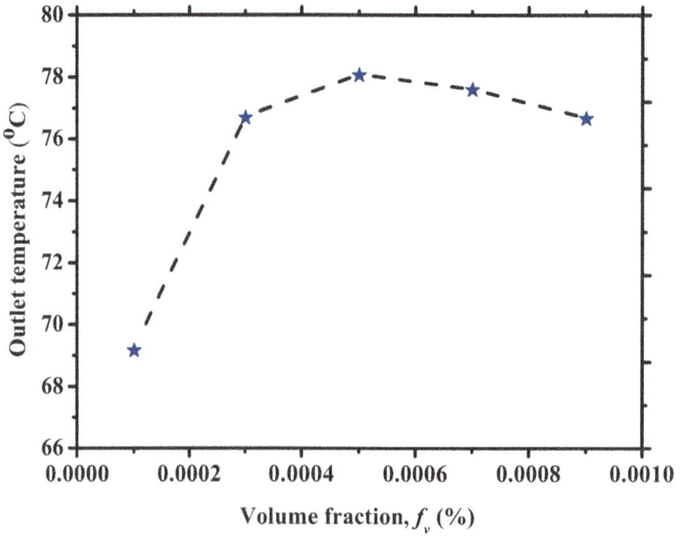

Fig. 4 Effect of volume fraction on nanofluid temperature

2.2 *Effect of Volume Fraction*

Five volume fractions (0.0001, 0.0003, 0.0005, 0.0007 and 0.0009%) of graphite nanoparticles have been considered and their effect on temperature predicted as shown in Fig. 4. The results show that 0.0005% is an optimum volume fraction on the basis of a highest temperature that could be achieved. The ability of a nanofluid-based solar collector in delivering heat at a desired temperature will obviously depend on the volume fraction of nanoparticles because a higher volume fraction will result into a larger attenuation of the sunlight. The attenuation varies exponentially with the extinction coefficient and hence also with the volume fraction of the nanoparticles. So, there has to be an optimum volume fraction of the nanoparticles corresponding to which the temperature gain is maximum.

2.3 *Effect of Mass Flow Rate*

In order to ascertain the effect of mass flow rate, different mass flow rates (0.001, 0.002, 0.03, 0.004, 0.005, 0.006 and 0.007 kg/s) have been considered. The results show that the outlet temperature decreases with increase of mass flow rate, see Fig. 5. This is due to fact that an increase in the fluid flow rate leads to reduction in the time for which the fluid is allowed to absorb sunlight, therefore leading to decrease the nanofluid temperature at the receiver outlet.

2.4 *Effect of Irradiation*

The effect of flux on the outlet temperature has been shown in Fig. 6. The outlet temperature was found to increase with rising flux. It can be concluded that in order to achieve high nanofluid temperature, the mass flow rate should be low and the irradiation should be high.

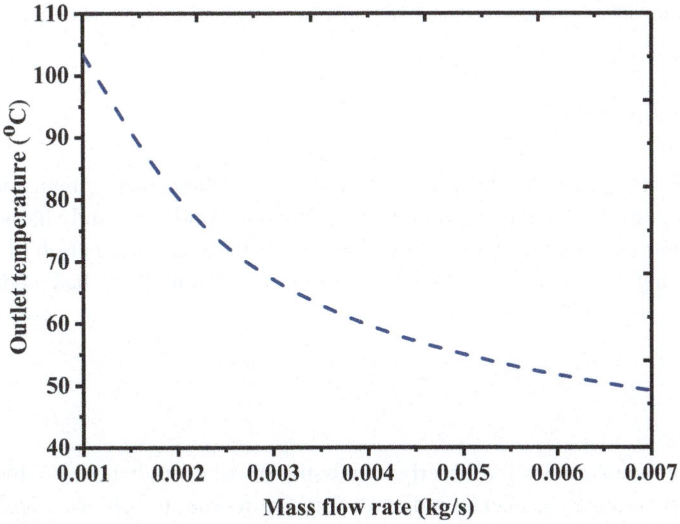

Fig. 5 Effect of mass flow rate on outlet temperature of the nanofluid

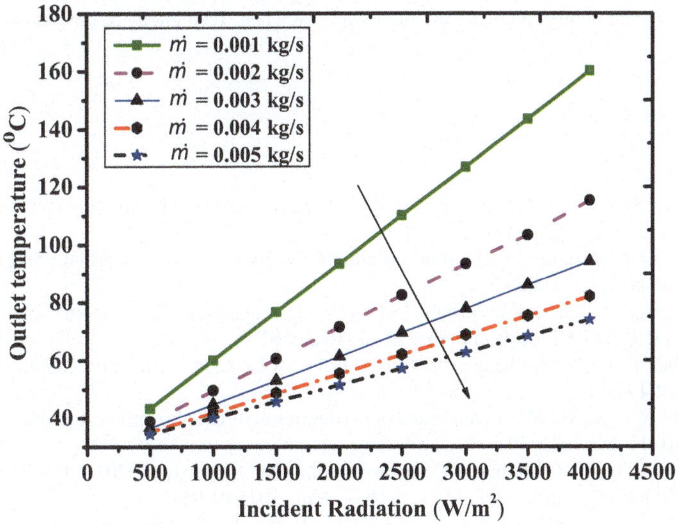

Fig. 6 Effect of irradiation on the temperature of nanofluid

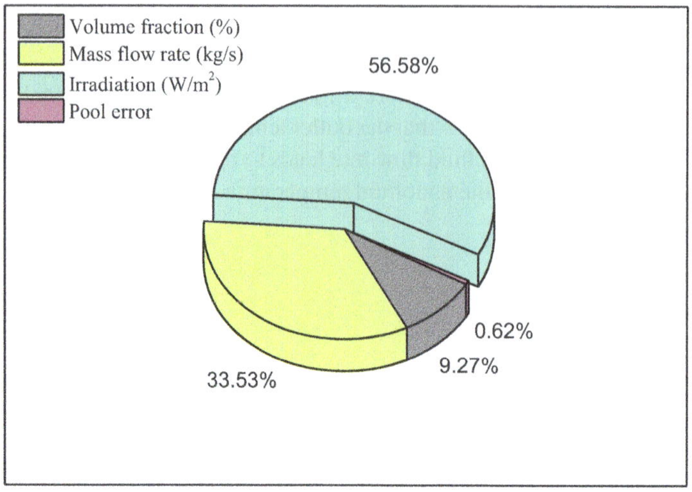

Fig. 7 ANOVA-predicted contributions of interfering parameters studied on the outlet temperature of nanofluid

2.5 Contribution of Each Factor

ANOVA has been performed to investigate the relative significance of three interfering parameters studied – volume fraction, mass flow rate and irradiation – on the outlet temperature of the nanofluid. The study has shown that incident flux and mass fraction have maximum individual contributions, see Fig. 7. A combined contribution of 90% was attributed to these two factors. The pooled error that accounts for combined factors or interaction effects had a 0.62% contribution.

3 Conclusion

The suitability of graphite nanoparticles for directly harnessing of solar irradiation at a low volume fraction of 0.0005% to achieve high nanofluid temperatures has been predicted. Further, in order to achieve a high temperature, the mass flow rate of the nanofluid should be proportionately low. The combined contribution of the incident flux and the mass flow rate on the temperature rise was found to be 90%, which shows that future studies should focus on the incident flux and mass flow rate of nanofluid to identify optimum design parameters for directly absorbing solar thermal collectors.

Acknowledgements Authors would like to thankfully acknowledge the joint funding received for the NoNSTOP project from UK's Newton Fund Institutional Links Programme (grant number 352029736) and Egypt's Science, Technology & Innovation Funding Authority (project ID 30870).

References

1. Muraleedharan M, Singh H, Suresh S, Udayakumar M (2016) Directly absorbing Therminol-Al2O3 nano heat transfer fluid for linear solar concentrating collectors. Sol Energy 137:134–142
2. Lenert A, Wang EN (2012) Optimization of nanofluid volumetric receivers for solar thermal energy conversion. Sol Energy 86:253–265. https://doi.org/10.1016/j.solener.2011.09.029
3. Taylor RA, Phelan PE, Otanicar T, Adrian RJ, Prasher RS (2009) Vapor generation in a nanoparticle liquid suspension using a focused, continuous laser. Appl Phys Lett 95:161907. https://doi.org/10.1063/1.3250174
4. Neumann AS, Urban J, Day S, Lal P, Nordlander NJ (2013) Halas, solar vapor generation enabled by nanoparticles. ACS Nano 7:42–49. https://doi.org/10.1021/nn304948h
5. Phelan P, Otanicar T, Taylor R, Tyagi H (2013) Trends and opportunities in direct-absorption solar thermal collectors. J. Therm. Sci. Eng. Appl. 5:021003. https://doi.org/10.1115/1.4023930
6. Hewakuruppu YL, Taylor RA, Tyagi H, Khullar V, Otanicar T, Coulombe S, Hordy N (2015) Limits of selectivity of direct volumetric solar absorption. Sol Energy 114:206–216. https://doi.org/10.1016/j.solener.2015.01.043
7. Jeffrey Kuo CF, Su TL, Jhang PR, Huang CY, Chiu CH (2011) Using the Taguchi method and grey relational analysis to optimize the flat-plate collector process with multiple quality characteristics in solar energy collector manufacturing. Energy 36:3554–3562. https://doi.org/10.1016/j.energy.2011.03.065

8. Brewster MQ (1992) Thermal radiative transfer and properties. Wiley
9. Hogan NJ, Urban AS, Ayala-Orozco C, Pimpinelli A, Nordlander P, Halas NJ (2014) Nanoparticles heat through light localization. Nano Lett 14:4640–4645. https://doi.org/10.1021/nl5016975
10. Veeraragavan A, Lenert A, Yilbas B, Al-Dini S, Wang EN (2012) Analytical model for the design of volumetric solar flow receivers. Int J Heat Mass Transf 55:556–564. https://doi.org/10.1016/j.ijheatmasstransfer.2011.11.001
11. Incropera FP, DeWitt DP, Bergman TL, Lavine AS (2007) Fundamentals of heat and mass transfer. Wiley. https://doi.org/10.1016/j.applthermaleng.2011.03.022
12. Palik ED (1998) Handbook of optical constants of solids, 1998th ed., Academic Press. http://gen.lib.rus.ec/book/index.php?md5=35C6AFB8DC7853DDDFA5ADD80D13FE5E
13. Khullar V, Bhalla V, Tyagi H (2018) Potential heat transfer fluids (nanofluids) for direct volumetric absorption-based solar thermal systems. J Therm Sci Eng Appl 10:1–13. https://doi.org/10.1115/1.4036795

The Impact of Efficient Insulation on Thermal Performance of Building Elements in Hot Arid Region

Ghanim Kadhem Abdulsada and Tawfeeq Wasmi Mohammed Salih

1 Introduction

Air-conditioning in Iraq, where the climate is extremely hot in the summer, consumes more than 60% of the electrical energy supplied to a residential building [1]. In such climate, where there is a high level of solar radiation and a large fluctuating of temperature values, the cooling load is exceeded the power capacity, and building elements suffer from undesired conditions lead to thermal stresses, and thus to crack and mold [2]. This issue could be treated by using efficient insulation system for exposed surfaces like external walls and roof. Thermal performance of an insulation system could be evaluated depending on the amount of heat transferred via it. In order to minimize the heat transfer, it is required to follow [3]:

- Low thermal conductivity materials to minimize the conduction
- Evacuated portions to minimize the convection
- Low emittance and high reflective surfaces to attenuate the radiation

Several studies were conducted in the development of efficient insulation for buildings in hot arid regions and the contribution of insulated walls and roof to the cooling load. Most of these researches investigated the optimum U-value required to ensure efficient insulation of building element.

A theoretical study carried out by Amjed et al. [4] has shown that the improvement of energy performance of Iraqi buildings can be achieved using insulation material, shades and air cooling. The U-value of the optimized wall must not be exceeding 0.12 $W/m^2.K$. For an efficient roof, it needs to satisfy U-value of 0.082 $W/m^2.K$ by using 15 cm of mineral wool.

A simulation program to figure out the contribution of insulation materials in hot climate in Spain was presented by ISOVER [5]. The simulation was carried out using the dynamic thermal building application DYNBIL. Comparisons with several measurements taken in inhabited buildings had shown a very good accordance. Excellent insulation of the roof and walls ($U < 0.2$ $W/m^2.K$) reduced the cooling load in summer. Insulation between the building and the ground was not required because the ground temperature was close to 20 °C. Hence, the ground can store excess heat in summer.

Another study carried out by Kalif A. et al. [6] tried to calculate the effect of packaging materials on the cooling load in Baghdad. The study conducted a survey of several materials and it also included testing and thermal analysis. A computer programming was done to estimate the amount of cooling load through the walls by using CLTD (cooling load temperature differences) method described by ASHRAE [7]. The U-value of materials was determined depending on the resistance method, where measured values of thermal conductivity of selected materials were taken. Results show that the percentage reduction in cooling load when using internal packaging materials ranges between 5% and 12%.

An experimental study carried out by Abdulsada and Salih [8] deals with the reduction of energy consumption in an Iraqi building by eliminating the effect of direct solar radiation in summer season using sufficient insulation system. The study has taken place in a terrace testing room using a new local thermal insulation to cover walls and roof with an air gap of 5 cm, taking into consideration several parameters like ambient temperature, solar radiation, orientation, and duration. The results

G. K. Abdulsada
Mechanical Engineering Department, Dijlah University, Baghdad, Iraq

T. W. M. Salih (✉)
Materials Engineering Department, Al-Mustansiriyah University, Baghdad, Iraq

© The Author(s), under exclusive license to Springer Nature Switzerland AG 2022

A. Sayigh (ed.), *Sustainable Energy Development and Innovation*, Innovative Renewable Energy,
https://doi.org/10.1007/978-3-030-76221-6_28

revealed that the proposed system with sufficient insulation can save the annual cooling load for the room by 6 MW and that the indoor temperature may down to 35 °C for outdoor temperature around 45 °C.

2 Procedure of the Work

The aim of this study is to satisfy efficient insulation for Iraqi buildings that can withstand extreme hot weather in the summer. The study has been done experimentally using two models built in Kirkuk mid-north of Iraq (35.5°N, 44.4°E). The first model was built in a traditional way without any insulation, while the second model was built with efficient insulation standards. The two models have the same dimensions (1.8 m × 1.8 m × 2 m). Each model has a window of 0.8 m × 0.5 m south orientation and a door of 1.9 m × 0.9 m north orientation, as shown in Fig. 1. The traditional model was built using a conventional procedure, where the walls are constructed from 20 cm common hollow block, 2 cm external cement plaster, and 1 cm internal gypsum plaster, while the roof is constructed from 15 cm reinforced concrete with 1 cm) internal gypsum plaster.

The efficient model has some innovative techniques. The outer face of the foundation was insulated by 4 cm cork to prevent heat transfer from the surrounding area to the interior space. Walls were built up to 2 m high using the standard 15 cm hollow block, and initially insulated by 2 cm cork. After that, walls were covered by bubbly wrap of 4 mm thickness made up from polyethylene as a condensation barrier. Finally, walls were insulated by 10 cm polyurethane panels and covered by reflected sheets. The overall U-value of the walls was 0.15 W/m^2·K. Roof was installed using standard polyurethane sandwich panel of 15 cm thickness, so the overall U-value for the roof was 0.19 W/m^2·K. Since the roof is exposed to high intensity of solar radiation, so it has been covered by a canopy of plastic layer. Double glazing closed-type window of plastic frame was used. The overall U-value of the window was 1.1 W/m^2·K. Door of UPVC frame and double insulated sheets was used with overall U-value of 1.8 W/m^2·K. Thermal bridges around the walls, roof, window, and door were avoided using EPS cork of 2 cm thickness. The airtightness was checked to decrease any leakage. In order to satisfy energy efficient standards, air has entered to the efficient model through a stream connected to an underground passage and extracted from an external diffuser, as shown in Fig. 2.

The study includes in-site measurements on the constructed models within 2 months (August and September). Several instruments have been used in the work. Lutron data logger system has been used to measure outdoor temperature, indoor temperature, and surfaces temperature for different locations each 15 min throughout the day. The global solar radiation was measured as well using solar power meter (Lutron-Model SPM-1116SD).

a. traditional model b. efficient model

Fig. 1 Constructed models used in the study. (**a**) Traditional model. (**b**) Efficient model

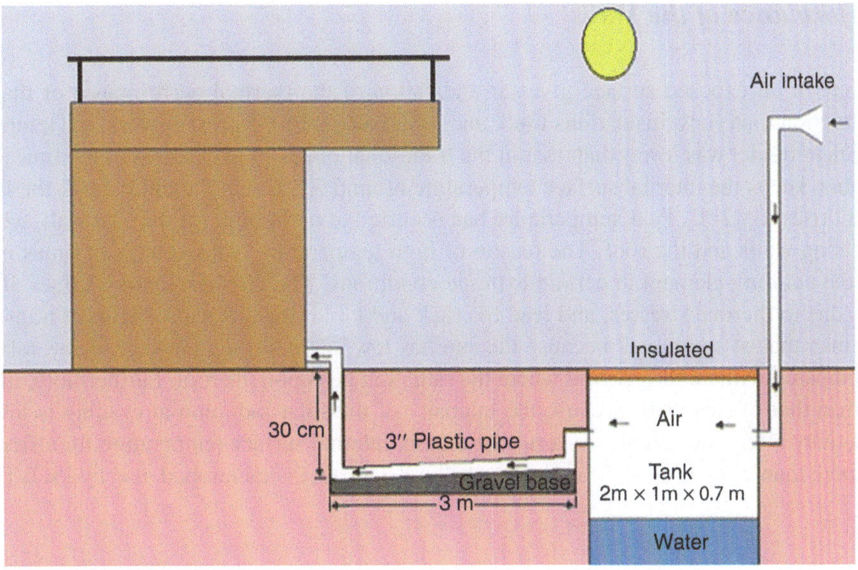

Fig. 2 Scheme of the efficient model

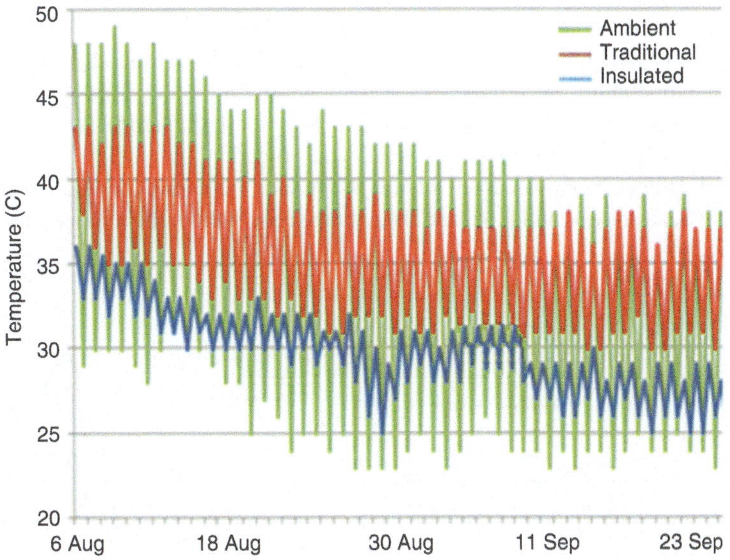

Fig. 3 Variation of indoor and outdoor temperatures during summer 2013

3 Results and Discussions

Data were collected in summer 2013 (August 06 to September 23), where there is high solar radiation and hot ambient temperature. Generally, results show valuable decreasing in the indoor temperature for the case of efficient insulated model compared to that of traditional one, as shown in Fig. 3. The average indoor temperature in efficient insulated model was 32 °C with maximum value of 34 °C. While, in the traditional one, it was 40 °C with maximum value of 44 °C. Results show clearly that there is a low fluctuation in indoor temperature of efficient model, about 4 °C, compared to that of traditional one which is fluctuated by 8 °C. However, maximum indoor temperature has occurred after 3 PM due to high accumulated heat. Furthermore, a local simulation program based on ASHRAE [7] relations has shown an energy saving up to 70% of the cooling load due to the efficient insulation system used.

3.1 Thermal Performance of the Walls

The internal temperature of an exposed surface gives an indication of the thermal performance of that element. It varies according to the construction materials, insulations used, and orientations. In the present work, the temperature of the internal surfaces in the efficient model was lower than that in the traditional one by 4–8 °C most of the time, as shown in Figs. 4 and 5. Efficient insulation keeps the internal surface temperature around 33 °C. On the other hand, the internal temperature of the un-insulated wall rises to 42 °C. Peak temperature has occurred at noon time for the east wall, while it is occurred at evening for other direction walls and the roof. The reason of high temperature rise in the peak times is due to the lack of thermal resistance of the building element in certain extreme conditions. The high fluctuating values affect the strength of construction materials due to thermal stresses, and lead to crack and mold. Results show that heat transmitted through east wall is actually more than that of south wall because the sun has low angle at the morning, so the solar radiation reaches directly to the walls compared with midday period where the radiation lies upon the roof. Furthermore, the heat transmission via west wall is less than that of east wall, because the intensity of the solar radiation at evening is lower than that at the morning due to dust activity afternoon usually. As a rule of thumb, internal surface temperature for efficient insulated building should not differ more than 1 °C to the indoor air temperature. This fact is mentioned also by ISOVER [5].

3.2 Thermal Performance of the Roof

Obviously, the roof is the most building element exposed to solar radiation. Therefore, it requires more processing besides the insulation to decrease the heat transfer. For example, the roof can be covered by shades of light fabrics or climbing plants. Shading can reduce indoor air temperature as much as 5 °C [4].

In the present work, the measured values of temperatures in the traditional model show extra rise of roof temperature, where the internal surface temperature was around 44°C, as shown in Fig. 4. So, a plastic layer was used to provide canopy shading to the roof of the efficient model beside the insulation. In general, the temperature of the internal surface of the efficient model with shaded roof has decreased to 32 °C, as shown in Fig. 5.

The effect of canopy shading is very valuable. Results show that the average reduction in the temperature of external surface of the roof was 10 °C, as shown in Fig. 6a, and the average reduction in the temperature of internal surface of the roof was 3 °C, as shown in Fig. 6b.

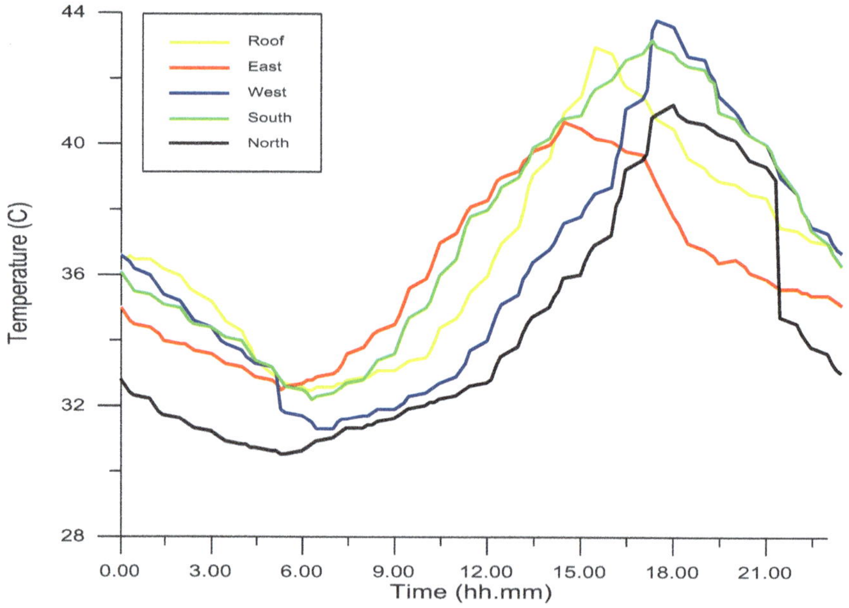

Fig. 4 Internal surface temperature in traditional model (15-08-2013)

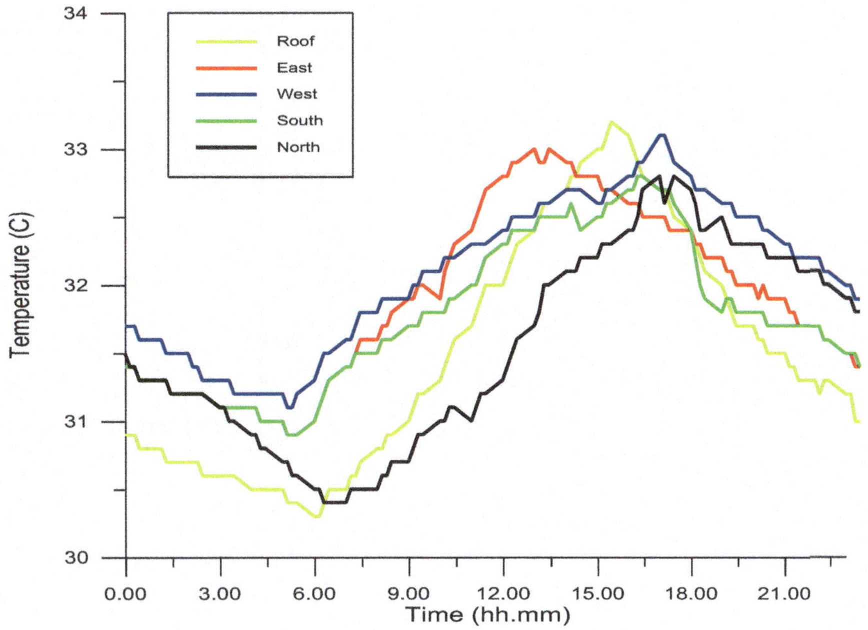

Fig. 5 Internal surface temperature in efficient insulated model (15-08-2013)

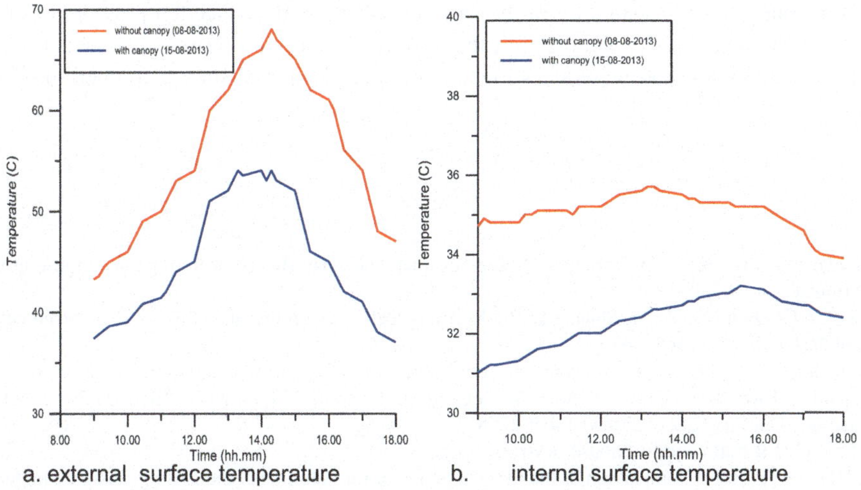

Fig. 6 Surface temperature of the roof in efficient insulated model. (**a**) External surface temperature. (**b**) Internal surface temperature

3.3 *Thermal Performance of the Foundation*

In order to recognize the effect of insulation on thermal performance of the foundation, earth temperature, as well as temperature of outer face of the foundation were measured for different depths, as shown in Fig. 7. It was found that earth temperature decreases gradually with depth far from the datum surface till a depth of 0.4 m where the temperature difference becomes negligible. Since the outer face temperature of the foundation is affected by the earth temperature, so it should be insulated for a certain depth to avoid thermal bridge and eliminate the heat transfer from the external surrounding to the inside. Hence, it is recommended to use cork of 4 cm thickness. Insulation between the floor and the ground is not required because the ground can absorb excess heat in summer.

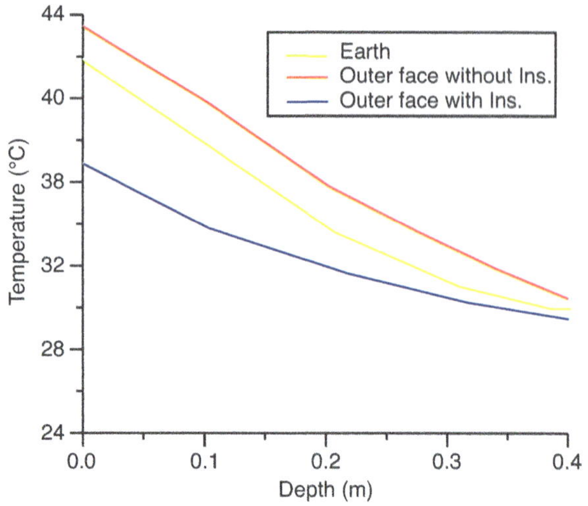

Fig. 7 Effect of insulation on foundation (August 06–10)

4 Conclusions

In hot summer and arid region, improving the thermal performance of building elements has many advantages in terms of energy consumption, economic impacts, and environmental protection. This goal could be done using efficient insulation. The results of the present experimental work have shown reduction in indoor temperature of efficient model by 8 °C comparing to that of traditional one, as well as stability in the variation of internal surface temperature of any element by less than 4 °C comparing with the high fluctuation occurred in the traditional one. Furthermore, results show that the effect of both efficient insulated roof and canopy shading reduces the temperature of internal surface temperature of the roof by 12 °C.

References

1. Abdulsada GK, Mohammed TW (2015) Experimental study to evaluate the performance of Iraqi passive house in summer season. Journal of Energy and Power Engineering 9:386–392
2. Yaohui M, Dezhi Z, Lin C, Zheng W, Yan Y, Shuang L (2018) Analytical study on thermal stress of autoclaved aerated concrete shear walls. Romanian Journal of Materials 48(2):162–167
3. Ecovolve (2007) Principles of thermal insulation: heat transfer via conduction, convection and radiation. http://beodom.com
4. Amjed A, Asaad K, Ali A. Towards a zero energy house strategy fitting for south Iraq climate. PLEA 2008 – 25th Conference on Passive and Low Energy Architecture, Dublin, 22nd to 24th October 2008
5. ISOVER (2008) The ISOVER multi comfort house. www.isover.net
6. Kalif A, Ismael A, Hussein N (2013) Comparison of the thermal behavior for natural and industrial packaging materials on the cooling load in baghdad. Journal of Babylon University (Engineering Sciences) 21(4):1450–1463
7. ASHRAE (1997) Chapter 27: residential cooling and heating load calculations. ASHRAE Handbook SI (Fundamentals)
8. Abdulsada GK, Salih TWM (2015) Experimental and theoretical study for the performance of new local thermal insulation in Iraqi building, Chapter 44 of the book: Renewable energy in the service of mankind, vol I. Springer, Cham

Part III
Sustainability and Low Energy Architecture

Walking with the Climate: Technologies and Methodology on Intermediate Urban Open Space Design

Alessandra Battisti

1 Introduction

The 2015 COP 21 conference in Paris – and those that followed, right up to the most recent meeting in Madrid – forcefully highlighted how, in recent years, the fight against climate change has become one of the most hotly debated environmental issues in the world and that an innovative vision for cities has never been as important as in this century, not to mention in the light of the COVID-19 pandemic and the problems caused by lockdown measures and social distancing. These observations become even more urgent and pressing if we consider that over half the world's inhabitants are currently living in cities and that this migratory trend will probably continue, to the point where over two-thirds of the world's population will be living in urban areas by 2050. India alone is set to double the numbers of city dwellers, adding 404 million people to its cities in the next 35 years [1]. In September 2019, the Secretary General of the United Nations, Antonio Guterres, hoped to foster a new vision of the environmental government of cities by asking world leaders to bring concrete, realistic plans for boosting their national contributions by 2020 to the 20th September meeting in New York (the very commitments to cutting greenhouse gases made during the 2015 Paris Agreement), in line with the aim of reducing greenhouse gas emissions by 45% over the next decade and achieving zero net emissions by 2050. Such a request had already been expressed in drastic terms in the IPCC's report of 2018, which indicated how the risks facing the human race, the biosphere and world economies were greater than previously anticipated and which encouraged action designed to reduce global temperatures [2], as well as reduce greenhouse gas levels [3], particularly focusing such action on two categories: the planet's great ecosystems (rainforests, great river systems, etc.) and urban areas, as these are exactly the places that are seriously hit by the effects of climate change, demonstrating a high degree of vulnerability and environmental, health and social risk.

All these forecasts are inspired by the same view of the built environment as the place, par excellence, where strategies designed to reduce climate change should be employed, strategies that could also contribute to solving other problems associated with urbanisation, such as poverty, asserting equality and the right to, and a guarantee of, a healthy life. As the greatest manifestation of man's presence on Earth, cities will be the ones to determine whether or not we will achieve inclusive economic growth that can produce employment without having to senselessly waste non-renewable resources thanks to an approach guided by environmental sustainability and resilience. That is why a systemic, integrated approach is necessary if we want to guarantee a sustainable form of urban development that takes into account climate change, factors of economic growth, social and environmental development and issues linked to urban health [3].

This is the direction in which global policies inspired by the UN's Sustainable Development Goals are moving [4]; clear, tangible goals, particularly as regards the possible achievement of Goal 3 – *Ensure healthy lives and promote well-being for all at all ages* – and Goal 11 – *Make cities and human settlements inclusive, safe, resilient and sustainable.*

Goal 3 aims to guarantee the health of all people of all ages and promote their well-being. This aim focuses on various areas of improvement, promoting mental health and well-being, the widespread availability of improved hygienic conditions and, in general, a greater awareness of environmental factors, particularly as the necessary condition for reducing infectious diseases.

A. Battisti (✉)
Dipartimento PDTA La Sapienza, Rome, Italy
e-mail: alessandra.battisti@uniroma1.it

© The Author(s), under exclusive license to Springer Nature Switzerland AG 2022

A. Sayigh (ed.), *Sustainable Energy Development and Innovation*, Innovative Renewable Energy,
https://doi.org/10.1007/978-3-030-76221-6_29

Goal 11 – 'Make cities and human settlements inclusive, safe, resilient and sustainable' – addresses cities, which are amongst the main culprits when it comes to energy consumption and carbon emissions, putting increasing pressure on the environment and thereby creating problems for public health. The safety of the people who live, work in or transit through cities should be guaranteed, both in terms of well-built public and private infrastructure and buildings and in terms of protection from forms of crime, domestic violence and/or harassment. When it comes to public health, the reduction of pollution, the improvement of air quality and the proper handling of waste, the water cycle and sewage are all key aspects. Socially vulnerable groups (such as people with disabilities, the elderly, children and the poor) should be guaranteed the same access to services, universal access to public gardens and parks that are safe and accessible, mutually supportive economic, social and environmental links between urban, periurban and rural areas, strengthening regional and national development planning, access to mobility and adequate, safe and low-cost housing, adopting and implementing integrated policies, territorial standards and plans that foster inclusion, the efficient use of resources, climate change adaptation and mitigation and resilience when faced with natural disasters [5].

To this end, we need to identify innovative solutions in order to help the inhabitants of cities behave in an increasingly informed way. Take, for example, the Swiss proposal of a '2000-watt society', where people are positively involved in influencing their own lifestyles and where they can proactively interact with their community.[1] It is only by identifying tangible, concrete and measureable goals that we can turn back the tide and have the power to make a difference as architects of our own planet.

2 Urban Heat Islands

In Europe, large-scale urbanisation is worsening the heat island phenomenon, year after year. This phenomenon contributes to increasing the thermal discomfort of city dwellers during the summer months and, as a result, increasing energy-consuming behaviours and practices that attempt to offset this phenomenon (such as the widespread use of artificial air-conditioning systems). Some cities, like Paris and London, have seen mortality rates rise exponentially after heat waves [6] and have adopted strategies to mitigate urban heat islands, attempting to react with initiatives that can protect the population, increasing their ability to adapt to the seriousness of the phenomenon.

Due to the increased incidence of heat waves, triggered by climate change, field literature has been focusing its attention for some years now on the important relationship between land sealing and mean radiant temperature and perceived temperature, particularly in urban areas with a higher number of artificial surfaces that absorb heat in summer and do not allow sufficient ground transpiration and evaporation. Furthermore, forecasts indicate that mean temperatures will continue to rise in the coming decades. As far as climate assessments and studies on thermal comfort are concerned, up until now they have often resorted to the use of simplistic instruments of analysis like the model adopted by global organisations such as the Intergovernmental Panel on Climate Change, which describes the effect of weather and climate on human beings using a simple index based on a combination of air temperature and relative humidity. The exclusion of important meteorological variables (such as wind speed and radiation flows), as well as thermo-physiological variables, significantly reduces the reliability of the results of the experiments carried out. Consequently, we can reasonably state that such discrepancies can hinder fairly 'obvious' and important microclimate considerations for many professionals working at a local level, such as architects and urban planners and designers [7].

The study of the relationship between city and climate, which was already an important issue when planning ancient Roman cities and buildings in the works of Vitruvius [8], was first systematically explored in the book *The Climate of London* written by Luke Howard in 1833, which is credited as the birth of urban climatology as an applied science. This work carefully describes the urban meteorological phenomena found in London at that time, and the way in which the city's geometric, physical and environmental parameters influenced the local climate by altering natural ventilation, humidity and temperature, making the London climate different to that of neighbouring rural areas [9]. Over a century later, Manley coined the phrase 'heat island' in 1958 in his treatise on the effect of artificial heat, which also focused on London. Today, this term is used to denote the urban environmental phenomenon where summer temperatures are higher than those of surrounding rural

[1] On 21 May 2017, 58% of Swiss citizens voted 'yes' in the referendum on the 2016 Energy Law, which aims to reduce the use of final energy by 43%, reduce the use of fossil fuels, dramatically increase the use of renewable fuels and close five nuclear power stations by 2035. This welcome reduction in consumption was founded on the '2000-watt society' study, carried out in the first decade of the 2000s by six federal scientific institutes, which joined forces to create the Swiss Federal Institutes of Technology Domain (ETH Domain). The study states the feasibility of reducing the flow of primary energy from 6000 to 2000 watts per capita (i.e. the equivalent of 4.5–1.5 tonnes of crude oil per year).

areas [10–13], which can be ascribed to the city's built environment, a phenomenon that sees variations in temperature from 2 °C up to 12 °C and can pose a serious threat to the health of the urban population [14, 15].

Experts often group heat islands into three types:

1. Surface urban heat islands, where the infrared radiation emitted and reflected by surfaces is measured, from which we can identify the parts of the built environment whose surfaces are hottest.
2. Urban canopy layer heat islands (UCL), which are found in the air layer occupied by human beings, from the ground up to the base of the tree canopy and building roof height.
3. Urban boundary layer heat islands, which start from roof and tree canopy height and extend upwards until the urban landscape no longer influences the atmosphere. This area normally does not extend more than a mile above the ground.

Surface and urban canopy layer heat islands, which are those most often studied in the field of architecture, are those mentioned in this document, where the built environment and its artificialisation has the most marked effect, as the flow of air and energy exchanges are checked on a microscale and the specific characteristics of a site are extremely important [16].

3 The Impact of Urban Heat Islands

The various forms of impact now universally attributed to the urban heat island phenomenon are the following:

- Increased energy consumption
- Increased emissions of air pollutants and greenhouse gases
- Deteriorating air quality
- Reduced health and comfort

3.1 Increased Energy Consumption

Over the past few decades, the unusual summer heat waves that have hit cities have increased energy consumption due to the almost continuous use of HVAC units with the resulting consumption of electricity, causing peaks in demand, a phenomenon that occurs on hot summer afternoons. Such peaks are extreme in densely built and populated cities. This often results in network overloads and associated blackouts and power cuts.

3.2 Increased Emissions of Air Pollutants and Greenhouse Gases

In order to cope with the greater demand for, and consumption of, primary energy during the summer, electricity companies rely on fossil fuel power stations that not only produce energy, but air pollution and greenhouse gases as well. The main pollutants produced by such power stations include:

- Sulphur dioxide (SO2)
- Nitrogen oxides (NOx)
- Particulates (PM)
- Carbon monoxide (CO) and
- Mercury (Hg).

3.3 Deteriorating Air Quality

The pollutants listed above are harmful to human health when released into the atmosphere and also contribute to complex problems affecting air quality such as ground-level ozone formation (smog), in addition to the problems associated with particulate pollution and acid rain. At an institutional level, the relevant authorities need to adopt an inclusive, multidisciplinary approach in order to implement effective action to reduce air pollution and improve well-being, involving territorial and transport planners, public health managers and environmental experts, as well as the participation of citizens.

3.4 Reduced Health and Comfort

Extreme temperatures have adverse effects on health, increasing mortality rates, hospitalisations and visits to A&E, due to the stress placed on cardiovascular and respiratory systems and the worsening of existing chronic illnesses. Amongst healthy individuals, increased daytime temperatures and reduced cooling at night can increase the risk of developing cerebrovascular diseases and arterial thrombosis due to increased blood thickness, plasma cholesterol and platelets.

The risk of suffering the effects of heat is greater amongst people with heart failure and ventricular tachycardia. Urban heat islands, combined with high levels of air pollution, can interfere with inflammatory mechanisms, increasing localised immunodeficiency and the risk of respiratory problems and infections.

Vulnerable groups, such as children, the elderly and those with chronic illnesses, are particularly at risk of suffering non-fatal heat stroke or increased mortality rates worsened by excessive heat or a sudden and dramatic increase in temperature. The Centres for Disease Control and Prevention estimate that excessive heat exposure contributed to over 8000 premature deaths in the United States from 1979 to 2003. That figure is higher than the number of deaths caused by hurricanes, lightning, tornadoes, floods and earthquakes combined.

4 The Conceptual Framework for Mitigation Measures: The Case Study of Prati, Rome

The study described in this article concentrates on the problem of how to mitigate UHIs in the Prati district of Rome (coordinates: 41° 54′ 41″ N, 12° 27′ 56″ E, 0 metres a.s.l.), located in Rome's first municipality, using climate walks that allowed us to create a georeferenced map of the open spaces we studied, which helped determine physiological comfort, that is, the balance between individuals and their surrounding environment, identified by considering the balance between energy coming in and going out. In order to identify this balance by inputting the phenomenon as researched by a number of field experts over the years into our numerical modelling work, a whole range of indices were proposed that could describe the well-being (comfort) of local inhabitants and could allow us to model their behaviour on the basis of a range of different planning options in order to come up with an instrument that could be used when planning public spaces.

Each of these indices can be used to describe this phenomenon [17], with varying degrees of accuracy.

They are divided into analytical/rational groups of indices (based on the principles of human thermal exchange), experimental groups (based on human response to various environmental factors) and comfort-based groups (measured through experiments conducted on humans) [18, 19].

The research takes into account the relationship between energy consumption and urban morphology, studies on intermediate open spaces, environmental and bioclimatic comfort, the interaction between biophysical and microclimatic factors and, last but not least, innovative monitoring systems, adopting a multidisciplinary approach that emphasises the contribution and overlapping of a number of different disciplines: the physical analysis of the microclimate and outdoor thermo-hygrometric comfort, the relationship between forms of open space redevelopment and design and comfort, the many aspects of comfort and health, the way surfaces covered with energy performing materials behave within a particular thermal and radiating environment, the effect that innovative technologies and materials could have on well-being, not only from an environmental and thermo-hygrometric point of view but from a social and accessible point of view as well, even contributing to the regeneration of the image and 'figurability' of open space [20].

The working method consisted of three phases:

- Phase 1 involved an examination of the site using maps and official climate data from weather stations, alongside a careful survey of the form of the built environment, the road network, the artificial and natural materials and the activities of local people.
- The Prati district (approximately 1.7 km^2) was chosen due to its cultural, historical, social and architectural importance in the urban space and its location: it is in the First Municipality of Rome, near the historic city centre. Overall, the area we analysed features a dense urban fabric, with buildings in rows dating from the twentieth century, a network of infrastructure with large roads where the only vegetation present are ornamental trees lining the avenues.

- Phase 2, which featured climate walks,[2] allowed us to draft a georeferenced map of the open spaces studied, which helped determine the level of physiological comfort, where the quantitative figures of the research were reached using the Universal Thermal Climate Index (UTCI) which aimed for a one-dimensional quantity adequately reflecting the human physiological reaction to the multi-dimensionally defined actual outdoor thermal environment. The human reaction was simulated by the UTCI-Fiala multi-node model of human thermoregulation, which was integrated with an adaptive clothing model. Following the concept of an equivalent temperature, UTCI for a given combination of wind speed, radiation, humidity and air temperature was defined as the air temperature of the reference environment, which according to the model produces an equivalent dynamic physiological response.
- Phase 3 was a scenario metaplanning phase where we redesigned the road network using what are generally experimental materials that offer greater permeability and high-performance colours, and implemented systems of green infrastructure and urban greenery that, when properly arranged, can cool and direct air flow appropriately, improving air quality thanks to the way they absorb pollutants.
- More specifically, highly reflective innovative materials with cold colours were used, particularly along urban pavements, in order to assess their energy and climate performance. Speaking of which, the research was based on studies concerning environmental regeneration practices and on trials using nanomaterials conducted by the author's research group in recent years. Whilst this work was also open to questioning their complex behaviour, we also considered it necessary to contemplate the use of plants and water when seeking comfort and well-being in open spaces. To this end, we consulted a number of different methodological approaches so as to guide our research during each separate phase of development: an interpretational approach designed to bring together theoretical contributions regarding open urban space and outdoor materials and organise them in a systematic manner; an abductive form of research, or rather a deductive, and therefore experimental, method of monitoring for simulating and verifying the comfort of particular technologies, considered to be environmentally and energetically efficient (Figs. 1 and 2).

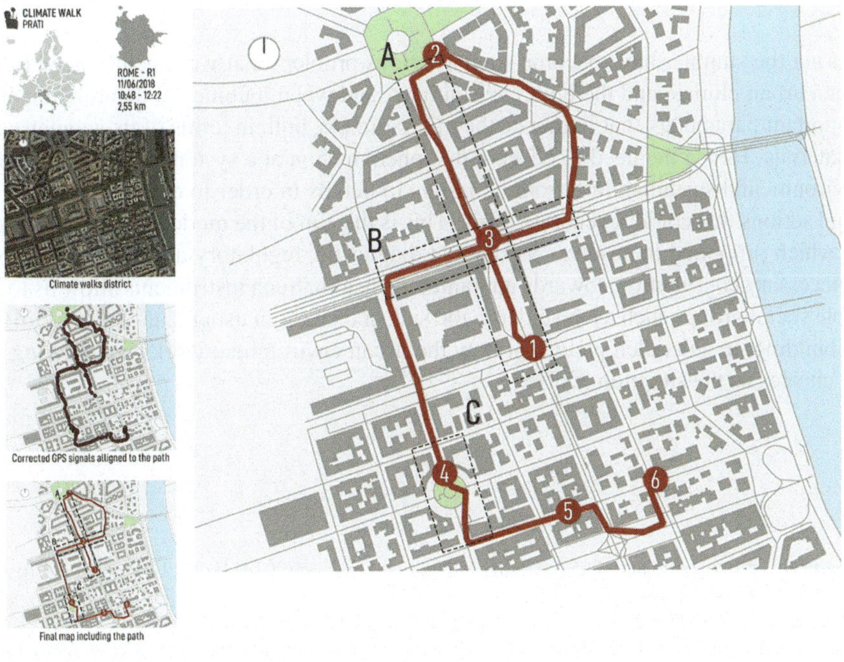

Fig. 1 Visualization of the route path and survey points

[2] The case study presented here was part of an experimental research project known as A.M.O.R. (Activating Munich Outdoor Resilience), carried out by the PDTA Department of Planning, Design and Technology of Architecture at Rome's Sapienza University and the Technical University of Munich's Faculty of Building Technology and Climate Responsive Design, funded by the MIUR (the Italian Ministry of Education, University and Research) and the DAAD (the German Academic Exchange Service, Ministry of Culture). Team Group: prof. Arch. Alessandra Battisti, prof. Ing. Thomas Auer - arch. Daniele Santucci; arch. Ata Chokhachian, arch. Simona Casciaro, Elisa Avancini - Elisa Cannarella - Cristoforo Cattivera

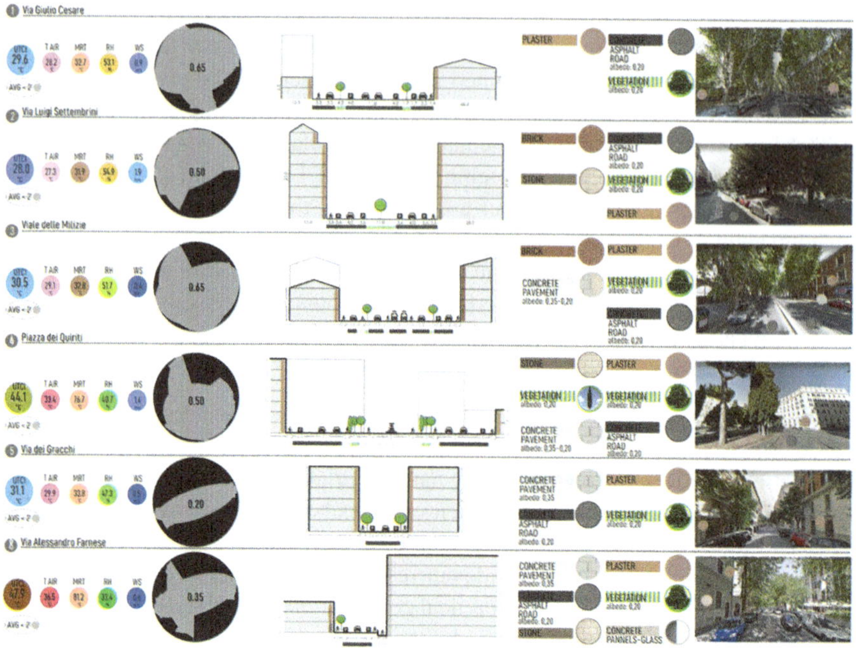

Fig. 2 Sky view factor and albedo analysis

5 Conclusions

UHIs in our cities during the summer have become such a serious problem that we urgently need to explore every facet of the interaction between urban climate and the built environment, and when it comes to this relationship, urban morphology is one of the most important parameters that influence the microclimate, both in terms of its geometry and space distribution and in terms of its materials. Hence the need to observe this phenomenon at a systemic level that includes everything from town planning to environmental sustainability, from buildings to details in order to develop the most complete view of the problems and potential actions that could regenerate cities. This is the aim of the modelling simulations of well-being at various planning scales, which propound themselves as suitable diagnostic, regulatory and monitoring tools that can make the right contribution, a necessary contribution, towards guaranteeing health. Such instruments allow us to support and safeguard the rights of vulnerable social groups such as the elderly, the sick and children using data, IoT and IoP, which have proved to be the best aides for building up a detailed picture of how the urban environment works and making it possible to estimate and correct planning choices as the situation evolves.

References

1. United Nations. Department of Economic and Social Affairs, Population Division (2014) World Urbanization Prospects: The 2014 Revision, Highlights (ST/ESA/SER.A/352), New York.
2. IPCC (2018) Global warming of 1.5°C. Intergovernmental Panel on Climate Change, Geneva
3. UFNCCC (2019) Climate action and support trends based on national reports submitted to the UNFCCC secretariat under the current reporting framework. United Nations Climate Change Secretariat, Bonn
4. United Nations. (2015). 17 goals for sustainable development. Available at: http://www.un.org/sustainabledevelopment/sustainable-development-goals/. United Nations. (2018) Global indicator framework for the Sustainable Development Goals and targets of the 2030 Agenda for Sustainable Development. Available at: https://unstats.un.org/sdgs/indicators/Global%20Indicator%20Framework%20after%202019%20refinement_Eng.pdf. Accessed Sept 2019
5. UN-HABITAT (2018) Tracking progress towards inclusive, safe, resilient and sustainable cities and human settlements: SDG 11 synthesis report - high level political forum 2018, UN, New York, https://doi.org/10.18356/36ff830e-en
6. Besancenot JP (2002) Vagues de chaleur et mortalité dans les grandes agglomérations urbaines. Environ Risques Santé 1:229–240
7. Matzarakis A, Fröhlich D, Bermon S, Adami PE (2018) Quantifying thermal stress for sport events—the case of the Olympic games 2020 in Tokyo. Atmos 9:479

8. Rykwert J (2002) L'idea di città Antropologia della forma urbana nel mondo antico. Adelphi, Milano
9. Howard L (1833) The climate of London, deduced from meteorological observations, made at different places in the neighbourhood of the metropolis, vol II. London, W. Philips
10. USEPA (2009a). Air quality, [Online] Available at: http://www.epa.gov/region4/airqualitytoolkit/3_sources/landuseurbandesignaffectairquality.pdf. Accessed 25 Apr 2011
11. Memon R, Leung D, Chunho L (2008) A review on the generation, determination and mitigation of urban heat island. J Environ Sci 20:120–128
12. Pearlmutter D (2000) Patterns of sustainability in desert architecture. *Arid Lands Newslett* 47:1–12, [online] Available at: http://ag.arizona.edu/oals/ALN/aln47/pearlmutter.html. Accessed 5 Mar 2019
13. USEPA (2009) Basic information, [online] Available at: http://www.epa.gov/heatisland/about/index.htm. Accessed 18 Dec 2018
14. Oke TR (1982) The energetic basis of the urban heat island. *Q J R Meteorol Soc* 108:1–24
15. AA,VV (2019) Piano Nazionale di Prevenzione degli effetti del caldo sulla salute Linee di Indirizzo per la Prevenzione, Ministero della Salute, Roma.
16. Golden J (2004) The built environment induced urban Heat Island effect in rapidly urbanizing arid regions – a sustainable urban engineering complexity. Environ Sci 1(4):321–249
17. Burton I, McGregor GR (2009) Biometeorology for adaptation to climate variability and change. Springer Netherlands, Dordrecht
18. Huizenga C, Hui Z, Arens E (2001) A model of human physiology and comfort for assessing complex thermal environments. Build Environ 36:691–699
19. Parsons KC (2002) The effects of gender, acclimation state, the opportunity to adjust clothing and physical disability on requirements for thermal comfort. Energ Buildings 34:593–599
20. Gehl J (1980) Life between buildings: using public space. Van Nostrand Reinhold, New York

Developing Sustainable Cities for Climate Change Mitigation

Despina Serghides, Stella Dimitriou, Ioanna Kyprianou, Elias Giannakis, and George Zittis

1 Introduction

Globalization is the unavoidable result of technological advancement in transport and communication, since the nineteenth century. Some parts of the world remain immune to it; however, one of its ramifications, climate change is a threat that knows no national boundaries. Climate change is affecting every part of the world in a variety of manifestations; however, these are most conspicuous in urban centres, where populations are concentrated. Higher traffic volume means more CO_2 emissions being released, further reinforcing adverse phenomena such as urban heat islands (UHIs). Effectively, urbanization means that within city boundaries, the built environment is altered without meaningful consideration to human needs.

Urban microclimates can be decisive factors in human behaviour, either enabling or disabling populations from spending time outside of their thermally conditioned comfort zones. Improved microclimates have cascading effects in numerous aspects: physical and mental well-being, energy conservation in buildings and in urban mobility, as well as reduced greenhouse gas (GHG) emissions. For example, UHIs have been linked to respiratory and cardiac diseases, increased mortality and overall decreased quality of life in urban centres [1–3]. UHIs are also linked to increased cooling demand during summer, burdening the electricity grids and leading to further anthropogenic heat and GHG emissions [4–6]. In this way, a vicious circle of augmented cooling demand and subsequent heat release is created, aggravating UHIs and hindering urban sustainability.

An array of research has been conducted to offer solutions and break the cycle of voracious energy consumption in cities. For instance, urban forests are proposed as alternative vegetation solutions in urban settings, offering multiple ecosystem services and fostering climate change mitigation [7, 8]. Proper design of urban spaces is also regarded as one of the pillars of sustainable cities, more so when it is joined by measures to sustain thermally comfortable conditions [9, 10]. These approaches fall within the spectrum of bioclimatic design and (other than urban greenery) include interventions such as cool paving materials reflecting solar radiation, thereby increasing albedo of the built environment and water surfaces cooling the surroundings through evaporative mechanisms [11–14].

Urban land use is also relevant to the various technological solutions available, since it allows to pinpoint the 'low-hanging' fruits. Roads act as ventilation corridors, impacting on the microclimate and energy consumption of the surrounding buildings [15]. Moreover, the amount of land devoted to road systems has been described as extensive [16], meaning that a lot of potential exists to mitigate urban heat by retrofitting them too. Data from the city of London suggest that approximately 9% of land is covered by residential and non-residential buildings, whereas 12% of London's land is covered by road [17]. In the metropolitan area of Barcelona, roads with a total of 2613.65 hectares correspond to 10% of the total area use. Although the demographics of these areas are vastly different than those of any city in Cyprus, land use does not take into consideration the height of buildings; thus, the road over total area percentage is expected to be similar.

A sustainable city should reflect local culture and society preferences and enable citizens to act in a sustainable manner. Satisfaction of urban dwellers relies on their mental health and well-being, with studies showing that increased pedestrian

D. Serghides (✉) · S. Dimitriou · I. Kyprianou · E. Giannakis
Energy, Environment and Water Research Center (EEWRC), The Cyprus Institute, Nicosia, Cyprus
e-mail: d.serghides@cyi.ac.cy

G. Zittis
Climate and Atmosphere Research Center, The Cyprus Institute, Nicosia, Cyprus

© The Author(s), under exclusive license to Springer Nature Switzerland AG 2022

A. Sayigh (ed.), *Sustainable Energy Development and Innovation*, Innovative Renewable Energy,
https://doi.org/10.1007/978-3-030-76221-6_30

activity, liveability and vitality can be enhanced through appropriate urban design practices [18, 19]. Moreover, urban design and microclimates facilitating outdoor activities is significant to the physical and mental health of residents and their behaviours in using outdoor urban spaces [19–22]. Other factors, such as building characteristics, orientation and neighbourhood location or the ability for users of the space to adapt it, have also been found significant towards the thermal responses of residents [23]. Although the present study is conducted in a warm Mediterranean country and primarily addresses summer discomfort, it has been shown that users of outdoor public spaces in winter cities also benefit from the spaces' adaptability in increasing the time spent outdoors [24].

This study aims to address the dimension of outdoor thermal comfort, by optimizing conditions in a typical neighbourhood of a coastal city in Cyprus (Limassol). Cyprus, located in the eastern Mediterranean climate change hotspot, is expected to be significantly impacted by global warming, mainly through strong warming during future summers [25]. The effect of urban vegetation was explored, in combination with changing the albedo of the materials used in outdoor spaces. Their impact on microclimate improvement was assessed through the improvement in a thermal comfort index (PMV). In particular, this study takes a closer look to the role of non-built – open areas in cities for GHG emissions reduction and concurrent climate change mitigation and adaptation.

2 Thermal Comfort Factors and Indices

The most commonly used indicator of thermal comfort is air temperature as it is easy to use and most people can relate to it [26]. However, air temperature should always be considered in relation to other environmental and personal factors, as by itself is not an accurate indicator of thermal comfort or thermal stress. These factors may be independent of each other, but together contribute to an individual's thermal comfort and they consist of four environmental factors (air temperature, radiant temperature, air velocity and humidity) and two personal-related conditions (clothing insulation and metabolic heat) [27].

The PMV (predicted mean vote) model is probably the best know human thermal comfort index, based on Fanger's comfort model [27]. It relates to the energy balance of the human body with the human's thermal feeling. PMV was originally developed for steady-state indoor situations, but through modifications, PMV can also be somewhat applied to outdoor situations [28].

The PMV scale is defined between −4 (very cold) and +4 (very hot), with 0 being the temperature of perfect comfort, as observed in Fig. 1. In applications, PMV reaches temperatures below and above these limits, as observed in the results of this study.

3 Methodology

3.1 *Identification of the Case Study Area*

The Municipality of Limassol has slightly over 100,000 inhabitants, according to the 2011 census, with the metropolitan area's population exceeding 192,000 in population. Different types of primarily residential (or mixed use) urban blocks were identified within the Limassol area (Fig. 2). The blocks under investigation in this study refer to primarily residential areas, largely depicted as orange in Fig. 2. Some parameters included in the simulations are based on the specific characteristics of the area, as stipulated by the Local Plan of Limassol [30]. The historical centre of the city – in yellow circle – was not considered in this study due to its irregular geometry.

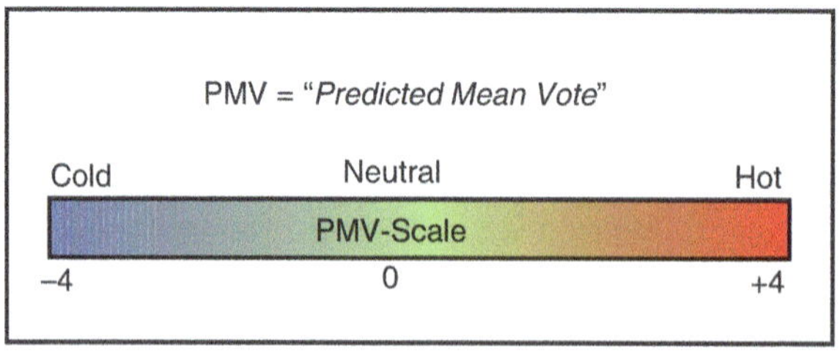

Fig. 1 PMV scale. (Source [29])

Fig. 2 Limassol planning zones. (Source [30])

The average length and width of various blocks within the K zones were measured. The tool used for the identification and measurement was Google Maps. Coverage and density coefficients of the residential area of Limassol municipality were obtained from the department of Land and Surveys, by selecting from the interactive map the plots within the area of interest [31]. The road and pavement dimensions were taken from a Government document from the Department of Town Planning and Housing [32]. The resulting plans of typical blocks were designed in AutoCAD and transferred to Envi-met spaces where the 3D model was created.

3.2 Current Situation Scenario

Typical materials used in Cyprus were applied in this case study. These refer to building volumes, roads and pavements. For the buildings, heavyweight concrete was the assigned material, whereas for the pavements, medium grey concrete of 0.5 albedo, and for the roads, asphalt of 0.2 albedo. Moreover, no trees were placed for the current situation simulation, to accurately reflect typical urban blocks in Cyprus.

3.3 Modified Situation Scenario

Trees (jacaranda mimosifolia) were placed equidistantly at 5 m around the central block, near the end of the pavement. These trees are simulated at a height of 15 m, have a foliage shortwave albedo of 0.6 and a foliage shortwave transmittance of 0.3. The pavement material was changed from grey to light grey concrete, which has an albedo value of 0.8, which is 0.3 higher than the initial material. Also, part of the car road on the northern and southern part of the block was substituted with a bicycle lane of 2 m width. The material used was red-coated asphalt of 0.5 albedo. The addition of bike lanes is based on the plan for the implementation of a connected and wide bike lane network throughout the city, according to the Limassol local plan (Fig. 3).

Following the interventions, a total of 45 trees were added and a 20% of the overall ground area (including pavements and the road) was modified, by changing their material.

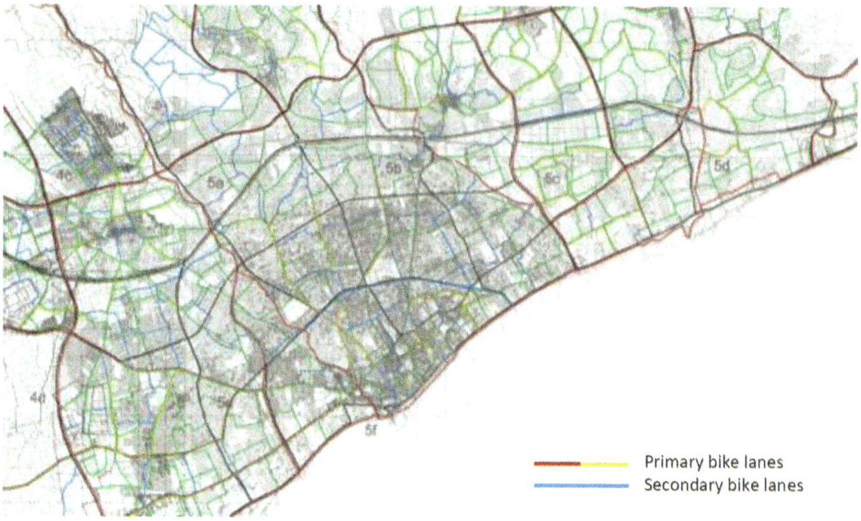

Fig. 3 Limassol planned primary and secondary bike lane network. (Source [33])

3.4 Weather Data

The weather data used correspond to values indicative for a typical summer day of the historical period 1986–2005, according to an ensemble of regional climate simulations made available from the EURO-CORDEX initiative [34]. Hourly values of meteorological parameters such as near-surface temperature, cloud cover, relative and absolute humidity, wind speed and direction were used as input for the high-resolution urban microclimate simulations.

3.5 Simulations

The simulations were performed using ENVI-met V4.4.5, running for 12 hours, from 8 AM to 8 PM, for a typical summer day. Sunset in June is at around 8 pm. The results from the simulations were processed with Biomet for a model of a 35-year-old male, of 80 kg, walking at a speed of 1.2 m/s. The clothing was set to 0.9 clo for winter and 0.5 clo for summer, in order to obtain the predicted mean vote (PMV). The outcome was visualized with Leonardo. The PMV was correlated with other parameters, such as the mean radiant temperature, the wind speed, the air temperature etc. in order to assess their impact on the thermal comfort. The same parameters were used to simulate the area after the measures.

3.6 Case Study Overview

A 100 m*100 m area was designed in Autocad and transferred to Envi-met environment. The area consists of a central 75 m*45 m block with six buildings of varying heights (1–3 storeys). The central block is surrounded by 2 m pavements and asphalt road of 6 m. Similar types of blocks complete the grid.

General information about the simulated block can be found in Table 1 and the 3D model of the area in Fig. 4 (current and modified scenario).

Table 1 General information of the typical block in the simulation

Average block type 1	
Width	45
Length	75
Ratio	1.7
Area	3375
Rotation	0
Coverage	0.5
Density	1.2
No of floors	Max 3
Height	Max 13.5 m

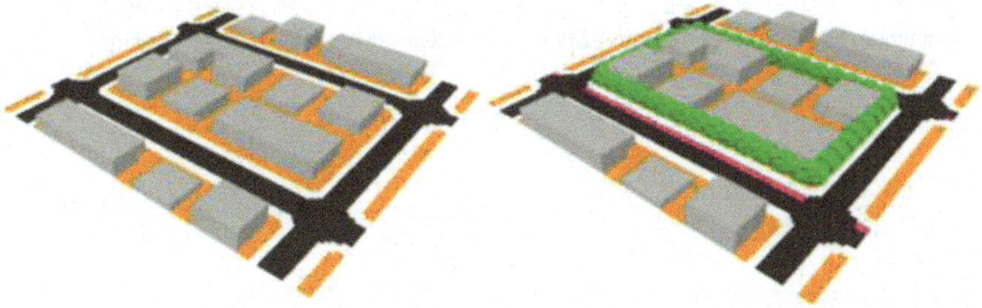

Fig. 4 3D model of the existing area (left) and the modified area (right)

4 Results

4.1 Current Situation Scenario

As expected, the outdoor thermal conditions during a summer day in Limassol are out of the comfort range. PMV exceeds the accepted comfort standards, with the average remaining above 2 up to 7 pm (see Fig. 5). The worse time to be outside is 2 o'clock in the afternoon.

A representation of the outdoor thermal comfort situation, at a height of 1.4 m from the pavement, at 2 o clock in the afternoon, is given in the form of PMV in Fig. 6. It is worth noting that only in the areas in shadow, the PMV is lower than 4.

4.2 Modified Situation Scenario

The block after the interventions was simulated for the same time frame, but only results from 9 am to 7 pm will be investigated, since after 7 pm, the median PMV is below 2 and by 8 pm even the maximum PMV is below 2.

The comparative results for the whole area, regarding the reduction in absolute maximum air temperature can be found in Fig. 7, for two different heights from the pavement ($k = 0$ and $k = 3$). The different heights were chosen based on the affected areas on the human body, 0.2 m ($k = 0$) being near the average person's calves and 1.40 m ($k = 3$) is near its neck.

A visual representation of the air temperature difference throughout the block at 2 o'clock can be observed in Fig. 8. The difference in ambient temperature ranges from 0.5 K to 1 K where only the pavement material is changed and between 1 K and 2 K, at the areas where trees are also present, with the median difference being 0.52 K.

The absolute difference in PMV at 2 pm, compared to the current state, does not exceed 0.3 where the only change was the pavements' material and was negligible where the road material was changed to red coated asphalt. More prominent reductions are observed, however, under the trees, reaching a maximum improvement of 1.86 (Fig. 9). Therefore, the change of the pavement material as a single intervention does not merit further investigation. As the combination of the measures has proven to be the most effective, only the central block up to and including the pavements and the first 3 m of the surrounding car road will be examined further. Therefore, the case study area is reduced from 100*100 m to a focus area of 80*60 m, which includes the central block, its surrounding pavements and the first 3 m of the car road. In the focus area, around 17% of the overall ground area (including pavements and the road) was modified, by changing the material.

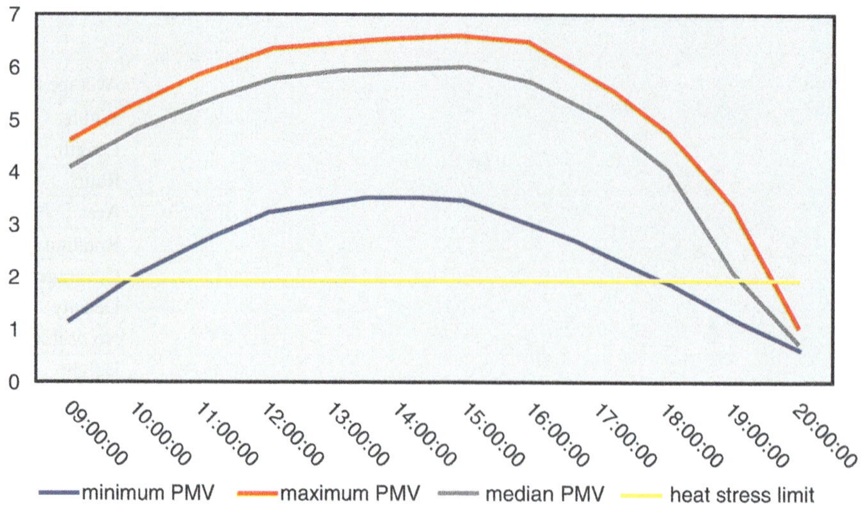

Fig. 5 PMV for a summer day in Cyprus, under current scenario

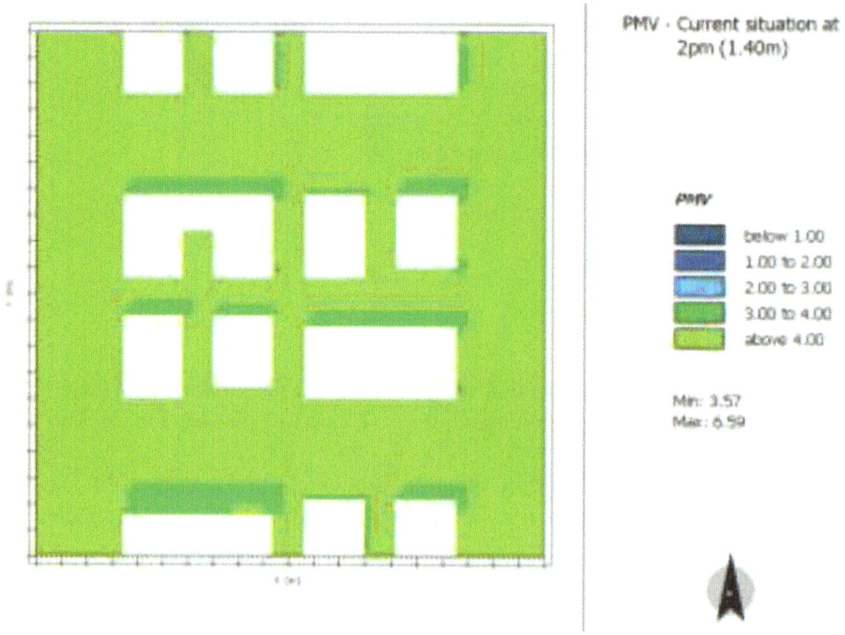

Fig. 6 PMV of the outdoor area simulated under current scenario

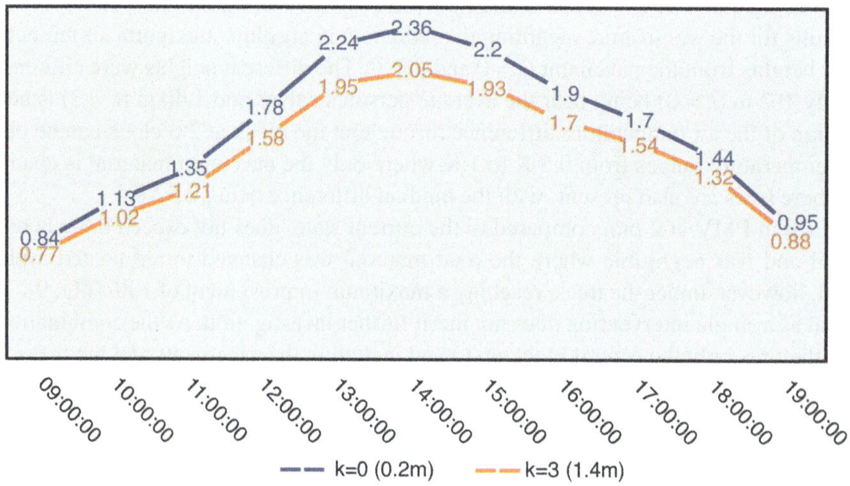

Fig. 7 Absolute maximum air temperature difference (°C)

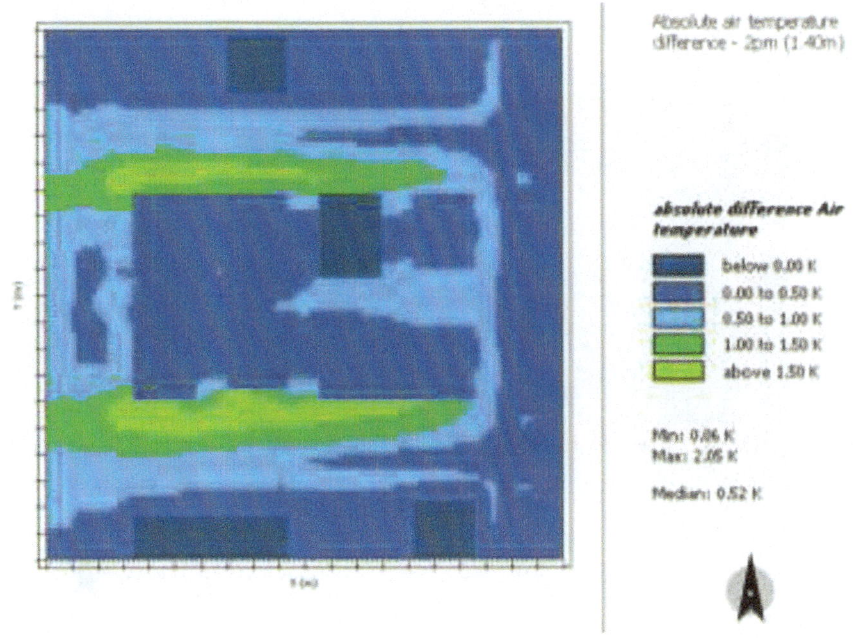

Fig. 8 Absolute air temperature difference at 2 o'clock

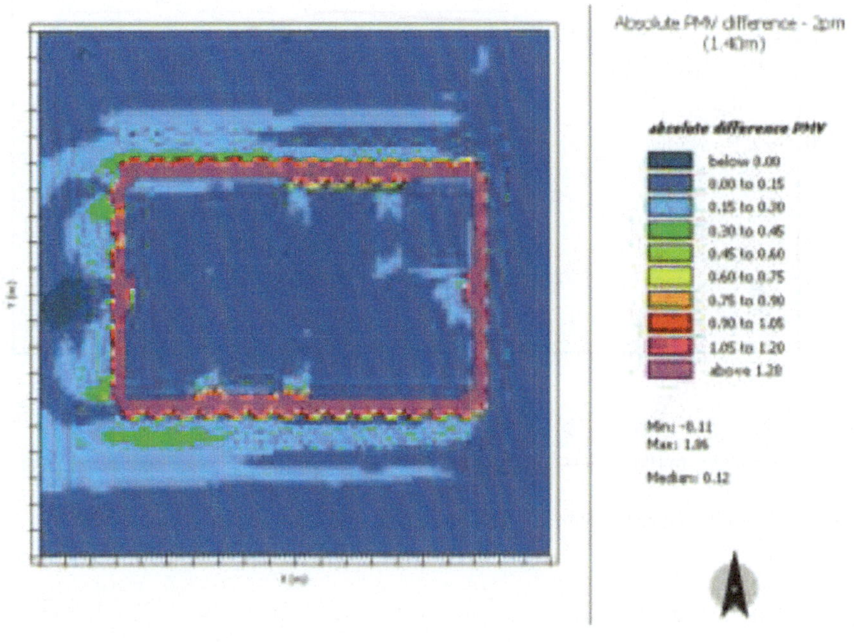

Fig. 9 Absolute PMV difference at 2 pm

4.3 *Modified Situation Scenario – Focus Area*

The median air temperature in the focus area, compared with the current situation of the same dimensions at 2 pm, is reduced by 0.63 K and the corresponding absolute median PMV difference is 0.19 (Fig. 10).

The PMV in the focus area after the interventions is improved overall by 0.2–0.35 but still remains above the accepted comfort range (Fig. 11). Only areas in shadow (Fig. 12), either from buildings or the trees result to PMVs lower than 2, therefore achieving thermal outdoor comfort, but only before 10 am and after 5:30 pm.

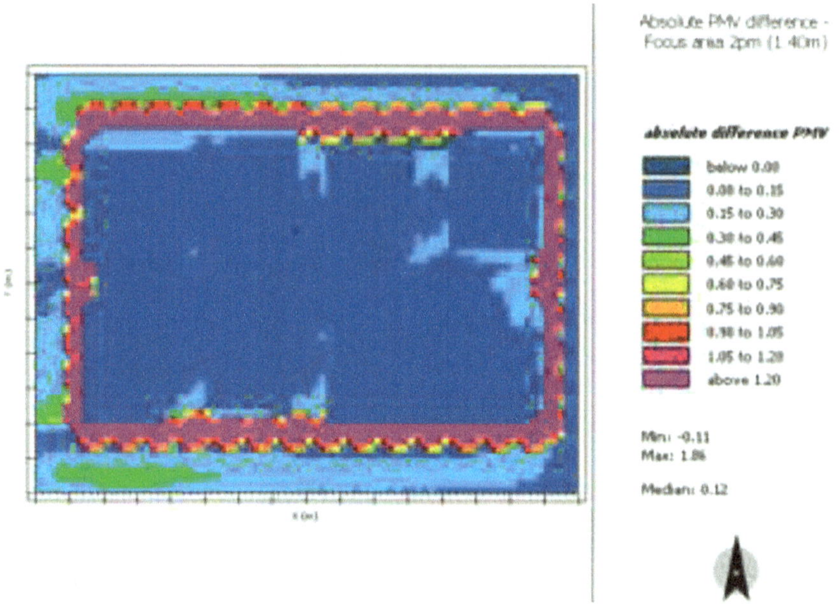

Fig. 10 Absolute PMV Difference – Focus Area at 2 pm

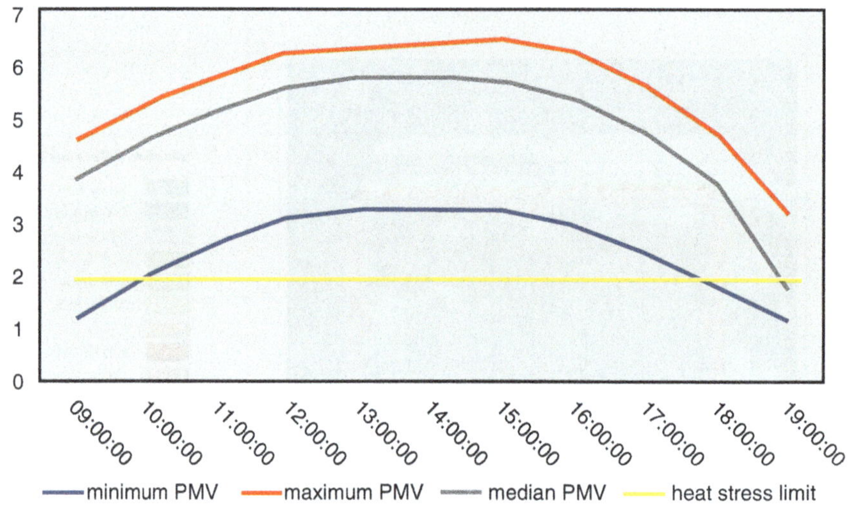

Fig. 11 PMV for a summer day in Cyprus, under modified scenario

5 Conclusions

This study detects extreme heat discomfort in an outdoor space by simulating a typical urban block of Limassol, Cyprus. The currently used materials and absence of vegetation render this outdoor space uninviting for pedestrian use during the summer, especially in the early afternoon.

The addition of trees on the perimeter of this typical city block combined with the modified pavement material has a positive impact on the ambient and radiant temperature. Specifically, the air temperature is reduced up to 0.54 K overall and up to 0.63 K in the focus area, reaching a maximum decrease of 2 K under tree shading.

On the other hand, the partial change of the asphalt road to a typical red-coated asphalt for the accommodation of bike lanes had no noticeable impact on the temperature or the outdoor comfort whereas the pavement material change had a reduced effect. More material alternatives should be studied in order to identify the most appropriate material for the bike lanes concerning thermal comfort, as their implementation will further help improve urban mobility and reduce the CO_2 emissions.

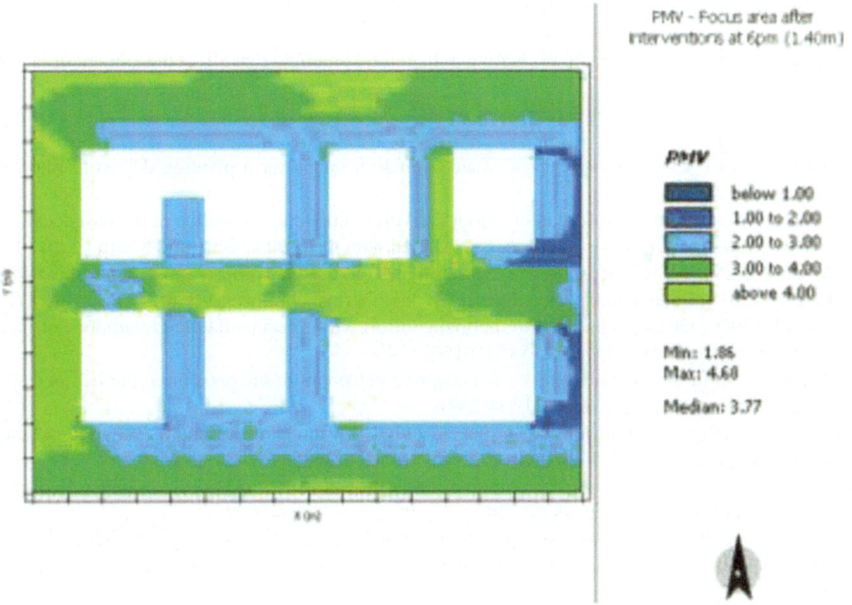

Fig. 12 PMV – Focus area at 6 pm

Overall, even the measures taken in this scenario are not sufficient to combat the extreme heat stress in this part of the world, with summertime PMV remaining above thermal comfort thresholds even after the interventions. Therefore, the study of additional – complementary – interventions is deemed necessary, including the integration of low vegetation, such as bushes around the pavements, and possibly the incorporation of shading devices above the pavements. Future urban microclimate conditions under a 'business-as-usual' pathway and efficiency of such mitigation/adaptation measures in a warmer climate regime have not been discussed in the present analysis but will be investigated in a follow-up study.

Bibliography

1. Wong KV, Paddon A, Jimenez A (2013) Review of world urban heat islands: many linked to increased mortality. J Energy Resour Technol 135:22101–22102
2. Sanchez-Guevara C, Núñez Peiró M, Taylor J, Mavrogianni A, Neila González J (2019) Assessing population vulnerability towards summer energy poverty: case studies of Madrid and London. Energ Buildings 190:132–143
3. Haritini T et al (2016) Health impact assessment for mortality associated with high temperatures in Cyprus. Proc 18th Mediterr Electrotech Conf Intell Effic Technol Serv Citizen. MELECON, p 1–4. https://doi.org/10.1109/MELCON.2016.7495465
4. Kolokotsa D, Psomas A, Karapidakis E (2009) Urban heat island in southern Europe: the case study of Hania, Crete. Sol Energy 83:1871–1883
5. Santamouris M (2016) Innovating to zero the building sector in Europe: minimising the energy consumption, eradication of the energy poverty and mitigating the local climate change. Sol Energy 128:61–94
6. Santamouris M, Cartalis C, Synnefa A, Kolokotsa D (2015) On the impact of urban heat island and global warming on the power demand and electricity consumption of buildings - a review. Energ Buildings 98:119–124
7. Amini Parsa V, Salehi E, Yavari AR, van Bodegom PM (2019) Evaluating the potential contribution of urban ecosystem service to climate change mitigation. Urban Ecosyst 22:989–1006
8. Amini Parsa V, Salehi E, Yavari A (2020) Improving the provision of ecosystem services from urban forest by integrating the species' potential environmental functions in tree selecting process. Landsc Ecol Eng 16:23–37
9. Loukaitou-Sideris A (2012) Addressing the challenges of urban landscapes: normative goals for urban design. J Urban Des 17:467–484
10. Chatzidimitriou A, Yannas S (2016) Microclimate design for open spaces: ranking urban design effects on pedestrian thermal comfort in summer. Sustain Cities Soc 26:27–47
11. Karakounos I, Dimoudi A, Zoras S (2018) The influence of bioclimatic urban redevelopment on outdoor thermal comfort. Energ Buildings 158:1266–1274
12. Manzano-Agugliaro F, Montoya FG, Sabio-Ortega A, García-Cruz A (2015) Review of bioclimatic architecture strategies for achieving thermal comfort. Renew Sustain Energy Rev 49:736–755
13. Serghides D, Dimitriou S, Kyprianou I, Papanicolas C (2019) The bioclimatic approach in developing smart urban isles for sustainable cities. Renew Energy Environ Sustain 4:2
14. Serghides D, Dimitriou S, Kyprianou I, Papanicolas C (2017) The adaptive comfort factor in evaluating the energy performance of office buildings in the mediterranean coastal cities. Energy Procedia 134:683–691

15. Leng H, Chen X, Ma Y, Wong NH, Ming T (2020) Urban morphology and building heating energy consumption: evidence from Harbin, a severe cold region city. Energ Buildings 224:110143
16. Litman T (2020) Transportation land valuation: evaluating policies and practices that affect the amount of land devoted to transportation facilities. Victoria Transport Policy Institute
17. Ministry of Housing Communities & Local Government (MHCLG) (2005) Land use by borough and ward. Available at: https://data.london.gov.uk/dataset/land-use-ward
18. Altunkasa C, Uslu C (2020) Use of outdoor microclimate simulation maps for a planting design to improve thermal comfort. https://doi.org/10.1016/j.scs.2020.102137
19. Chen L, Ng E (2012) Outdoor thermal comfort and outdoor activities: a review of research in the past decade. Cities 29:118–125
20. Akpinar A (2016) How is quality of urban green spaces associated with physical activity and health? Urban For Urban Green 16:76–83
21. Liu B, Lian Z, Brown RD (2019) Effect of landscape microclimates on thermal comfort and physiological wellbeing. Sustainability 11:5387
22. Elnabawi MH, Hamza N (2019) Behavioural perspectives of outdoor thermal comfort in urban areas: a critical review. Atmosphere (Basel) 11:51
23. Adunola AO, Ajibola K (2016) Factors significant to thermal comfort within residential neighborhoods of Ibadan metropolis and preferences in adult residents' use of spaces. SAGE Open 6:2158244015624949
24. Leng H, Liang S, Yuan Q. Outdoor thermal comfort and adaptive behaviors in the residential public open spaces of winter cities during the marginal season. https://doi.org/10.1007/s00484-019-01709-x
25. Zittis G, Hadjinicolaou P, Klangidou M, Proestos Y, Lelieveld J (2019) A multi-model, multi-scenario, and multi-domain analysis of regional climate projections for the Mediterranean. Reg Environ Chang 19:2621–2635
26. Gatto E et al (2020) Impact of urban vegetation on outdoor thermal comfort: comparison between a Mediterranean city (Lecce, Italy) and a northern European city (Lahti, Finland). Forests 11:228
27. Fanger PO (1970) Thermal comfort: analysis and applications in environmental engineering. Danish Technical Press, Copenhagen
28. Jendritzky G, Nübler W (1981) A model analysing the urban thermal environment in physiologically significant terms. Arch Meteorol Geophys Bioclimatol Ser B 29:313–326
29. Envi_Met. PMV/PPD (2020) Available at: https://envi-met.info/doku.php?id=apps:biomet_pmv. Accessed 21 Sept 2020
30. Department of Town Planning and Housing (2013) Local plan of Limassol - planning zones
31. Ministry of Interior. DLS portal. Available at: https://eservices.dls.moi.gov.cy/#/customer/application/dashboard. Accessed 21 Sept 2020
32. Department of Town Planning and Housing (2018) Traffic policy of local plans of Nicosia, Limassol, Larnaca and Paphos
33. Department of Town Planning and Housing (2013) Local plan of Limassol - secondary rad and bike lanes' network
34. Jacob D et al (2020) Regional climate downscaling over Europe: perspectives from the EURO-CORDEX community. Reg Environ Chang 20:51

Living Inspired by Nature: Two Scenarios of Green Buildings in Sri Lanka

Antonella Trombadore, Marco Paolini, and Matteo Torracchi

1 Introduction

Green building: "is the practice of increasing the efficiency of buildings and their use of energy, water, and materials, and reducing building impacts on human health and the environment, through better siting, design, construction, operation, maintenance, and removal—taking into account every aspect of the complete building life cycle."

In order to achieve a sustainable future in the building industry, we need to adopt a multidisciplinary approach covering a number of features such as energy saving, improved use of materials, material waste minimization, pollution and emissions control. To create a competitive advantage using environment-friendly construction practices, the whole life cycle of buildings should, therefore, be the context under which these practices are carried out. At international level, three general objectives have been identified which should shape the framework for implementing sustainable building design and construction, while keeping in mind the principles of sustainability issues (social, environmental and economic) identified previously. These objectives are resource conservation, cost efficiency and design for human adaptation [1].

2 Two Scenarios

Our starting point was the site analysis and the understanding of the client needs and requests; the client (a real estate company) asked for a new residential complex that can be adapted to this peculiar place in the middle of a subtropical wood.

Main physical constraints were given by the topography of the site with steep rocky slopes and very dense vegetation and the existing high-voltage line that divide the area into two parts. Our approach was based on merging the nature with the buildings [2] and we tried to mitigate the high slopes thank to *terraced buildings* that can smooth the transition between different heights and place. We developed a simple master plan that gives few but strong guidelines:

1. Buildings had to be located on two main lines/directions according to the topography of the place.
2. A main open/public area had to be located in the centre of the new development as a junction for all the surrounding buildings; according to our vision, all the shared and common amenities had to be located there.
3. The entire complex had to be surrounded by vegetation; this was done to enhance the relationship between the place and the new buildings and also to better fit with our vision about green constructions [3].

From these simple guidelines, we developed *two different scenarios*, both based on a site-specific approach, a strong visual images and a *basement* which house all the technical spaces, parking areas and shared amenities under a single roof that double as an extension of green outdoor area, that in our vision was not just plants or trees but an extension of the inside/private spaces into the common one (Fig. 1).

A. Trombadore (✉) · M. Paolini · M. Torracchi
Department of Architecture DIDA, University of Florence, Florence, Italy
e-mail: antonella.trombadore@unifi.it

© The Author(s), under exclusive license to Springer Nature Switzerland AG 2022

A. Sayigh (ed.), *Sustainable Energy Development and Innovation*, Innovative Renewable Energy,
https://doi.org/10.1007/978-3-030-76221-6_31

Fig. 1 General sustainability objectives applied to the project scenarios

Fig. 2 Scenario A: concept, master plan, section and 3D view

2.1 Scenario A

The idea is based on a simple repetitive and adaptive module that can be replicated and expanded with multiples and submultiples. With a system of this kind, we can get a diverse and multi-faceted image on one side by keeping low building costs thank to standard construction elements on the other side (Fig. 2).

As we can see in the site plan, the buildings follow master plan guidelines; they are arranged on *three main clusters*, one on north side with villas and bigger housing units with generous private terraces and common green courtyards to provide living space, light and air circulation, one on east side and one south side along the steep slopes. Units are arranged according to sun path to get maximum daylight and maximum benefits from *self-shading* and *cross-ventilation* to better cohabit and

Fig. 3 References: 1_module/grid combination, 2_steel frame construction system, 3_Bjarke Ingels Group – The Mountain, 4_v. Gimenez Arch. – Tree house, 5_Studio ninedots – REBEL, 6_Atelier du Pont – The doks, 7_Malka Architecture – Tangier Bay, 8_Precht – Botanical pavilion

mitigate the subtropical climate [4]. All housing units with private green spaces and terraces are facing towards the core open area where the clubhouse (with all common amenities are located) is located; this marks our approach, we did not want just to design buildings with plants and trees, we wanted to integrate greenery with the existing vegetation and with the future buildings into a coherent living environment plants grow around, near and over the constructions [5] (Fig. 3).

2.2 *Scenario B*

This scenario shows a more site-specific, less geometric and more natural approach although it all started from the same guidelines of scenario A. This layout is originated from the real topography of the site, the buildings follow and perfectly adapt to the different heights by merging the construction with the landscape (Fig. 4).

Seen from above, the buildings appear hidden under multiple layers of grass, trees and tropical vegetation, but if we go down to street level, you can see this strip that frames the housing units opened towards generous green terraces and patios.

As we can see in the site plan and 3D view, the general layout is still the same because of the same constraints, the site is still divided into three main clusters with villas on the north side and other units on east and south side, but the overall image is quite different except for the landscape integration (Fig. 5).

3 Comparison

Both scenarios come from the same conclusions about the context, few guidelines of the master plan generate these two different visions that present both a captivating architectural image by following a site-specific and sustainable approach (Fig. 6).

The general layout is almost the same while the aesthetic results are quite different; scenario A was intentionally thought as part of an adaptive, scalable and expandable construction system, while scenario B was conceived as two strong architectural signs that strongly characterize the project and perfectly fit with the site topography. Scenario B has less possibilities for future expansions or additions because of the more locked layout and because it is designed as a single gesture that cannot be modified without disturbing the entire composition.

4 Cost and Benefit

Green buildings do not need to cost more than conventional buildings. Sustainable building practices are becoming more mainstream, but there is still the assumption that using green methods costs more than traditional methods. But this is not the case if green methods are incorporated from the start. One of the biggest challenges to the cost-effective delivery of green buildings is the development of an understanding that green design is not opposite to conventional buildings, but is instead

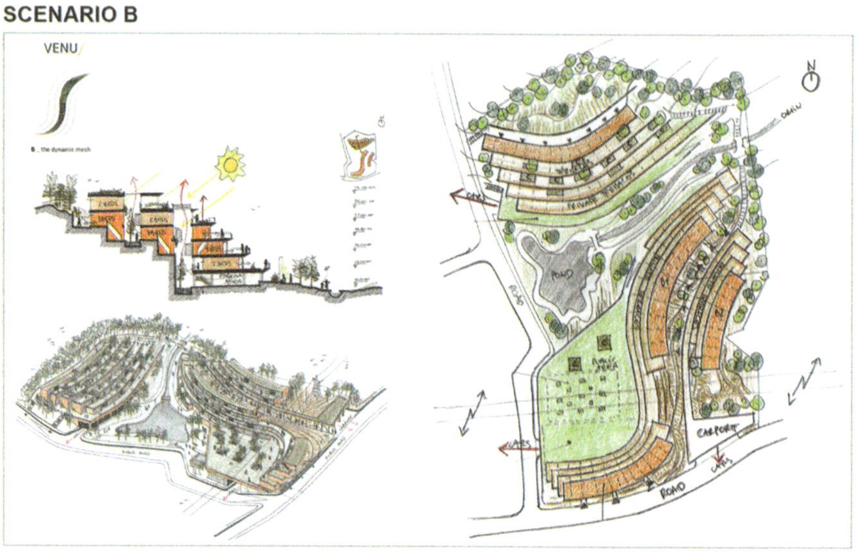

Fig. 4 Scenario B: concept, master plan, section and 3D view

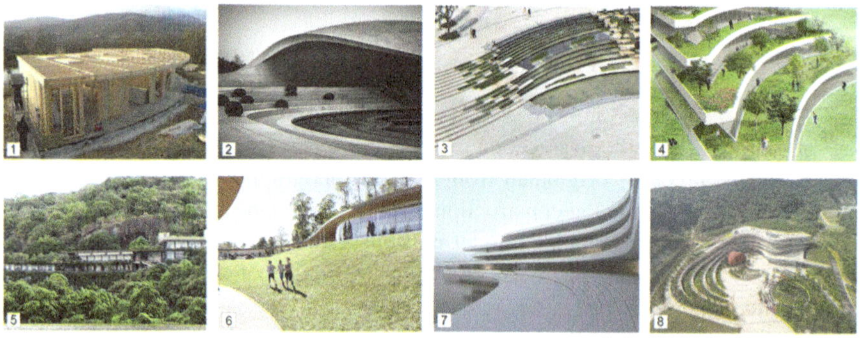

Fig. 5 References: 1_dry wood construction, 2_form sinuosity, 3,8_Studio Odile Decq. – Fanshan Geopark, 4_Vo Trong Nghia Architects – Green city hall, 5_Geoffrey Bawa – Kandalama Hotel, 6_SANAA – Grace farms,7_ V.Roman – villas concept, 8_Studio Odile Decq – Fangshan Geopark

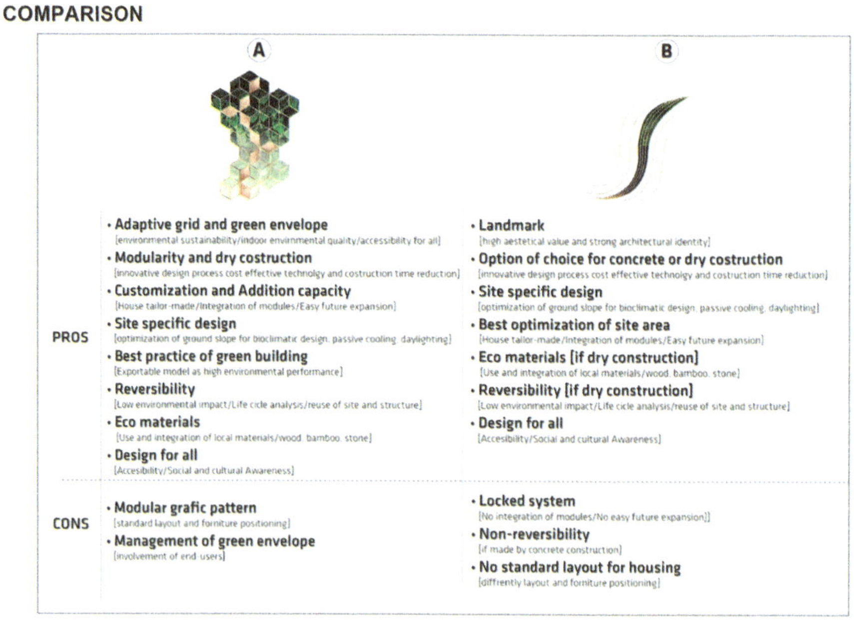

Fig. 6 Comparative scheme of the two project scenarios

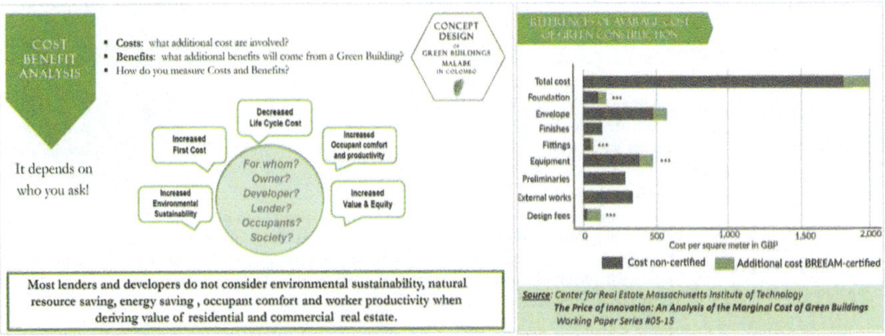

Fig. 7 (**a**) and (**b**) Focus on cost-benefit analysis of green building: design and construction

an integrated discipline for design that requires a different way of thinking. A developer's decision to build *green* does not necessarily hinge on whether or not it is going to cost more. It is more about a person's attitude. Studies show that certified green buildings, both commercial and residential, can command up to 30% higher prices than their traditional counterparts.

The higher the certification of green, the higher the price (which means that the LEED Platinum One Bryant Park is probably one of the more expensive office buildings in Manhattan). Occupancy rates have also been found to be up to 23% higher for LEED buildings. The world over, evidence is growing that green buildings bring multiple benefits. They provide some of the most effective means to achieving a range of global goals, such as addressing climate change, creating sustainable and thriving communities and driving economic growth. The benefits of green buildings can be grouped within three categories: environmental, economic and social. Green buildings harmonize with the local climate, traditions, culture and the surrounding environment. One of the most important types of benefit green buildings offer is to our climate and the natural environment. Green buildings can not only reduce or eliminate negative impacts on the environment, by using less water, energy or natural resources, but they can – in many cases – have a positive impact on the environment (at the building or city scales) by generating their own energy or increasing biodiversity. Green building benefits go beyond economics and the environment, and have been shown to bring positive social impacts too (Fig. 7).

Many of these benefits are around the health and well-being of people who live in green homes or work in green offices.

4.1 At a Global Level

The building sector has the largest potential for significantly reducing greenhouse gas emissions compared to other major emitting sectors [UNEP, 2009] such as energy efficiency, fuel switching and the use of renewable energy. The building sector has the potential to make energy savings of 50% or more in 2050, in support of limiting global temperature rises to 2 °C (above pre-industrial levels) [UNEP, 2016].

4.2 At a Building Level

Green buildings certified by the Indian Green Building Council (IGBC) result in energy savings of 40–50% and water savings of 20–30% compared to conventional buildings in India. Green buildings achieving the Green Star certification in South Africa have been shown to save on average between 30% and 40% energy and carbon emissions every year and between 20% and 30% potable water every year, when compared to the industry norm. Green buildings offer a number of economic or financial benefits, which are relevant to a range of different people or groups of people (Figs. 8, 9, 10, and 11). These include:

- Cost savings on utility bills for tenants or households (through energy and water efficiency)
- Lower construction costs and higher property value for building developers
- Increased occupancy rates or operating costs for building owners
- Job creation

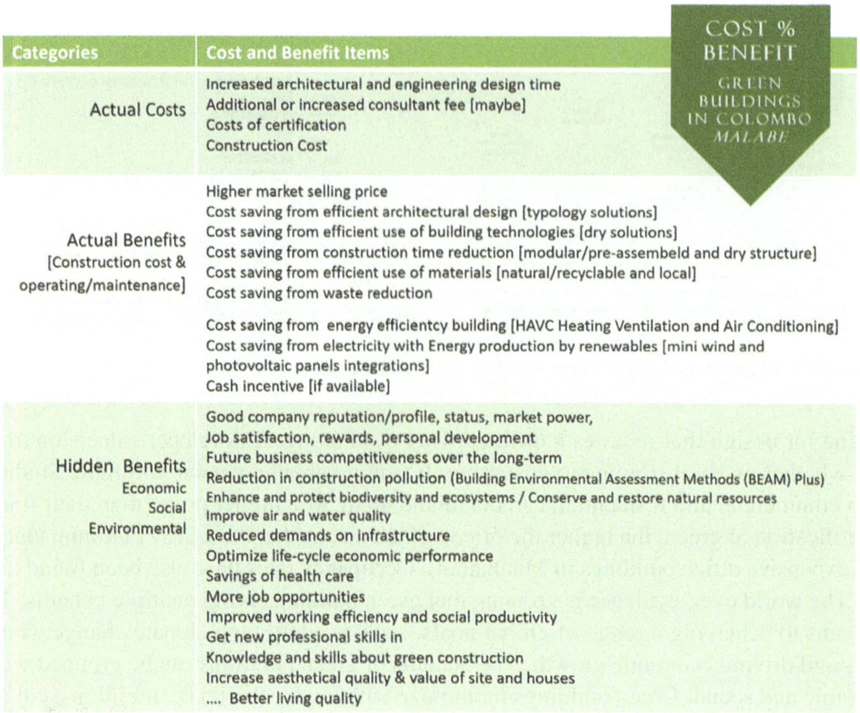

Categories	Cost and Benefit Items
Actual Costs	Increased architectural and engineering design time Additional or increased consultant fee [maybe] Costs of certification Construction Cost
Actual Benefits [Construction cost & operating/maintenance]	Higher market selling price Cost saving from efficient architectural design [typology solutions] Cost saving from efficient use of building technologies [dry solutions] Cost saving from construction time reduction [modular/pre-assembeld and dry structure] Cost saving from efficient use of materials [natural/recyclable and local] Cost saving from waste reduction Cost saving from energy efficientcy building [HAVC Heating Ventilation and Air Conditioning] Cost saving from electricity with Energy production by renewables [mini wind and photovoltaic panels integrations] Cash incentive [if available]
Hidden Benefits Economic Social Environmental	Good company reputation/profile, status, market power, Job satisfaction, rewards, personal development Future business competitiveness over the long-term Reduction in construction pollution (Building Environmental Assessment Methods (BEAM) Plus) Enhance and protect biodiversity and ecosystems / Conserve and restore natural resources Improve air and water quality Reduced demands on infrastructure Optimize life-cycle economic performance Savings of health care More job opportunities Improved working efficiency and social productivity Get new professional skills in Knowledge and skills about green construction Increase aesthetical quality & value of site and houses .... Better living quality

Fig. 8 The total mean construction and elemental costs per gross square metre for the BREEAM-certified and non-certified samples over the 2003–2014 (Q2) period. The light grey bars depict the non-certified samples' mean costs and the dark grey bars depict the BREEAM-certified samples' additional mean costs. Statistical significance at 1%, 5% and 10% levels denoted by *, ** and *** respectively

Fig. 9 (**a**) and (**b**) Examples of architectural solutions to implement the benefits of environmental sustainability

DESIGN PHASE and related costs

Categories	DESIGN PHASES	Standard DESIGN [% of the construction cost]	GREEN BUILDING DESIGN [% of the construction cost]
Actual Costs	Architectural Preliminar design Concept design	0 %	0 %
	Architectural and engineering modeling & design [Standard design process]	6%	-
	Architectural and engineering modeling & design [BIM applicaltion – parametric model]	-	5 %
	Energy performance evaluation [energy modeling]	0,5 % [if energy assessment is required]	1%
	Environmental strategies & technologies / Green building consultant fee for LEED or GBC assessment/certification	-	1 %
	Administrative procedures	0,5%	0,5%
	TOTAL COST of DESIGN PHASE	6 -7 %	7-8 %

Fig. 10 Comparative table of the relative costs between conventional construction and green building in design phase

Analysis of CONSTRUCTION PHASE

Categories % on total building cost	Standard Building costs	GREEN BUILDING cost - Scenarios A	GREEN BUILDING cost - Scenarios B
FOUNDATION 10%	TRADITIONAL FOUNDATION -Escavation -Earth moving -Concrete plinths	GROUND SCREW FOUNDATION Cost and time reduction Saving >25% [2.5%]	GROUND SCREW FOUNDATION Cost and time reduction Saving >25% [2,5%]
STRUCTURE 25%	CONCRETE STRUCTURE	STEEL FRAME /PREFABRICATED GRID Cost and time reduction Saving >15% [3,75%]	STEEL FRAME / WOOD Cost and time reduction Saving >10% [2,5 %]
ENVELOPE 20%	WINDOWS / EXTERNAL WALLS	WINDOWS /GREEN FACADES /TERRACE /SHADING DEVICES Cost implementation >20% [4%] Related Benfit of Energy reduction and indoor comfort	RIBBON WINDOWS TERRACE /SHADING DEVICES Cost implementation >25% [5%] Related Benfit of Energy reduction and indoor comfort
FINISHES 18%	INTERNATIONAL STYLE / HIGH LEVEL TARGET	USE OF NATURAL / LOCAL MATERIAL MODULAR GRID LAYOUT Cost and time reduction Saving >20% [3,6%]	USE OF NATURAL / LOCAL MATERIAL / TAILORED HOUSE LAYOUT Cost and time reduction Saving >10% [1.8 %]
FITTINGS/PLANS 12%	TRADITIONAL ELECTRICAL SYSTEMS OR AUTOMATION SYSTEMS	ELECTRICAL SYSTEMS / House Automation systems / BEAMs THERMAL PLAN /HVAC systems Cost implementation >15% [1,8%] Related Benfit of Energy reduction and indoor comfort	ELECTRICAL SYSTEMS / House Automation systems / BEAMs THERMAL PLAN /HVAC systems Cost implementation >15% [1,8%] Related Benfit of Energy reduction and indoor comfort
SERVICES & EQUIPMENTS 10%	TRADITIONAL CONCRETE POOL	GYM/POOL /BIO-LAKE / POND COMMON GREEN SPACE Cost reduction from integrated design Saving >20% [2%] Related Benfit of water management and environmental quality and indoor comfort	GYM/POOL /BIO-LAKE / POND COMMON GREEN SPACE Cost reduction from integrated design Saving >20% [2%] Related Benfit of water management and environmental quality and indoor comfort
EXTERNAL WORKS 5%	EARTH MOVING PARKING ROAD (Asphalted)	OPTIMIZATION SITE LAYOUT AND TOPOGRAPHY PARKING INTEGRATION SITE SPECIFIC ARCHITECTURAL DESIGN Cost reduction from integrated design Saving >10% [0,5%]	OPTIMIZATION SITE LAYOUT AND TOPOGRAPHY PARKING INTEGRATION SITE SPECIFIC ARCHITECTURAL DESIGN Cost reduction from integrated design Saving >10% [0,5%]
100 %	**Standard Building costs**	Total Cost Saving > 12,35% **6,55%** Total Cost implementation >5,8%	Total Cost Saving > 9,3% **2,5%** Total Cost implementation >6,8%

Fig. 11 Comparative table of the relative costs between conventional construction and green building in construction phase

5 Conclusion

The construction industry has historically lacked a collective line and a common idea of world development. The complex manoeuvres of change and the set improvement goals have long shown that the obstacles to a sustainable built environment are not really the technical solutions, but rather the way in which we collaborate effectively, ensuring that our collective efforts are truly aligned to achieve greater impact. The 2019 report dedicated to the SDGs (Sustainable Development Goals) shows that, despite progress in many critical areas of the planet, faster and more ambitious actions and policies are needed to achieve the necessary economic and social transformation.

This specific case of the project for Malabe demonstrates how the collaboration with local real estate agents, sensitive and attentive to the matter of sustainability, leads to real advantages both for an increasingly green building oriented towards greater respect for the environment, and at the same time increasingly calibrated on the liveability needs of the inhabitants. Not only the concept of absolute respect, which underlies this new way of building, brings with it a more cautious approach towards the exploitation of resources in their entirety but also concrete prospects for savings: buildings generated by a virtuous cycle are in fact less expensive by their nature, because they involve a reduction in waste, an optimization of activities and an increasingly marked recycling of elements and materials. The implication from the economic point of view is evident because the commercial value of the properties has increased and is therefore easily spendable even in the long term. Green buildings are considered increasingly attractive in the real estate market, thanks to the high living comfort and excellent energy efficiency that distinguishes them [6].

The natural environment is deteriorating at an alarming rate: Our idea of green buildings for Sri Lanka will be a great opportunity to reverse this course and engage a virtuous process.

The mission is to represent a real turning point that acts as a reminder of the surrounding area and the entire city to generate interest and transformation processes [7]. A positive example of how the synergy of several professional figures can give life to a case of local success between architecture, sustainability and collective life. The change of mentality is necessary, architecture can be a valid educator because it can sediment a careful and conscientious culture in individual people to bring back the feeling of belonging and gratitude towards the environment with which to communicate and collaborate in a smart way.

Our Malabe project focuses on creating buildings which are not only good for the environment but also support sporting area, green space, common outdoor facilities, for a healthier, happier and better *"living inspired by nature."*

References

1. U.S. Green Building Council (2011) Green building and LEED core concepts guide, U.S. Green Building Council
2. A+U Publishing (2019) Geoffrey Bawa – essence of Sri Lanka, Shinkenchiku
3. Bawa G (2007) Lunuganga, Times Editions
4. Trombadore A, Sala M, Ceccherini Nelli L (2007) Schermature solari, Alinea Editrice
5. Robson D (2016) In search of Bawa: master architect of Sri Lanka. Laurence King Pub
6. Trombadore A (2016) Mediterranean smart cities. Innovazione tecnologica ed ecoefficienza nella gestione dei processi di trasformazione urbana, Altralinea
7. Battisti A, Tucci F (2006) Qualità ed ecoefficienza delle trasformazioni urbane, Alinea

Buildings' Thermal Insulation Materials' Market and Drivers: Global Outlook and Assessment in the Context of SDGs

Sarah A. GadAllah and Mohsen M. Aboulnaga

1 Introduction

Buildings construction and operations correspond to 36 per cent of the global final energy use and nearly 40 per cent of energy-related carbon dioxide (CO_2) emission in 2017 [1]. In 2016, buildings in the EU are accountable for 41 per cent of the final energy consumption (456Mtoe) and 60 per cent of electrical energy use [2]. However, energy consumption of non-residential buildings in the EU-28 is 40 per cent more than residential buildings (250 kWh/m^2 compared to180 kWh/m^2) [3]. Climate change will force the global economy to shrink by 3 per cent by 2050 due to lack of climate resilience [4], but nevertheless the International Monetary Fund (IMF) stated that Corona virus (COVID-19) pandemic would drag the world into a same 3 per cent GDP contraction in 2020 [5], an earlier shrunk than expected.

The global strategies to lessen carbon emission through the improvement of energy efficiency raise the demand of thermal insulation materials. Buildings' thermal insulation materials play a vital role in reducing energy consumption, mitigating greenhouse gases emission, mainly CO_2 and sustainability of buildings.

The development of residential and commercial construction is the main driver of thermal insulation materials' global market. Their key utilization is mainly to restrain buildings' heat flow from one side of the insulation to inner leave of the external envelop in summer and the opposite direction in winter months [6]. In general, residential and tertiary construction manufacturers took over the thermal insulation materials application due to the rising demand that followed climate change impacts and global efforts to reduce energy use and mitigate CO_2 emissions. Nonetheless, sustainable development (SD) and the 17 essential interrelated goals (SDGs) that are meant to address the global changes cannot be overlooked when it comes to the analysis of the global thermal insulation market. However, various governments across the globe provide tax benefits for developing green buildings with minimum energy consumptions, hence directly boost the growth of building insulation market.

In fact, wall insulation became a dominant segment in the building insulation market in 2016, as a result of the incredible development in the housing sector, specifically in advanced countries such as the United States, Germany, and China. Generally, thermal insulation materials' application is controlled by residential and non-residential construction manufacturers, subsequent to its ascendant demand as a result of climate change, in order to reduce energy consumption. Research indicates that the building insulation market is divided into main zones comprising North America, Asia-Pacific including Australia and New Zealand, and Europe as well as Latin America, Middle East, and Africa as shown in Fig. 1 [7].

Asia-Pacific is looked upon as the main driver of evolution for the high-performance insulation materials' market, owing to the rise in the number of the recent manufacturing foundation and the expansion of the infrastructural investments. In addition, North America's growing investment in the construction industry and rigid regulations of energy efficiency are the stimulators of its thermal insulation materials' market demand. While in Europe, the renovation and restoration including re-insulation of existing structures are vital to the advancement of the insulation market [7]. The global thermal insulation market is mainly concerned with the commonly used thermal insulation materials represented in expanded polystyrene, extruded polystyrene, and polyurethane, as well as stone wool and glass wool.

S. A. GadAllah (✉) · M. M. Aboulnaga
Department of Architecture, Faculty of Engineering, Cairo University, Giza, Egypt
e-mail: maboulnaga@eng.cu.edu.eg

© The Author(s), under exclusive license to Springer Nature Switzerland AG 2022

A. Sayigh (ed.), *Sustainable Energy Development and Innovation*, Innovative Renewable Energy,
https://doi.org/10.1007/978-3-030-76221-6_32

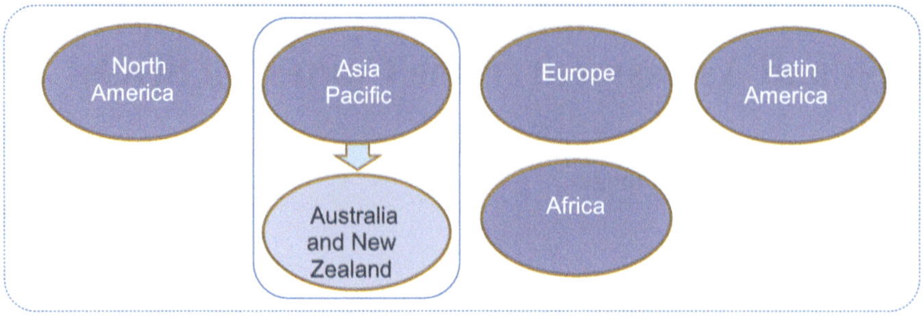

Fig. 1 Regions of buildings' thermal insulation materials' market. (*Source: Developed by Authors*)

Accordingly, this paper attempts to analyze the regions of the global thermal insulation market of conventional thermal insulation materials in order to determine each market's prevailing material. Along with emphasizing on thermal insulation materials' market drivers controlling the global and local market, presented in Egypt, through creating an exceptional relation with SDGs.

2 Objectives

This paper aims at investigating the thermal insulation materials' market globally and locally in a developing country as Egypt. The study is concerned with identifying the main drivers controlling the market through developing a relation between SDGs and these drivers; in order to recognize the most dominating one over the global and local market as well, that is represented in Egypt.

3 Methodology

The theoretical approach of this study is represented through the determination of the prevailing conventional thermal insulation material in the dominating countries through literature review. While, in order to examine in depth the controlling driver of the global and local thermal insulation markets; a methodology is developed which is based on analytical approach that utilizes a qualitative assessment to identify the weight of each driver by linking them to SDGs.

In order to pursue that, market drivers are weighed in relation to each goal of the SDGs based on a checklist of five indicators for the global market assessment as illustrated in Fig. 2. Four indicators were also developed to weigh the market 5 drivers in relation to SDG 13 "Climate Change" in the Egyptian market as shown in Fig. 3. These questions and indicators create a well-planned methodology to weigh the significance of each SDG relative to the addressed thermal insulation market driver. The criteria of weighing drivers is based on the SDGs applicability or implementation within the addressed market driver [8]. The driver that scores high average percentage regarding the 17 SDGs is considered the most important one controlling the thermal insulation market. The questions and indicators of thermal insulation market drivers used in the assessment globally and for Egypt in details are shown in Appendices 1 and 2.

4 Dominating Thermal Insulation Materials of the Global Market

This part reviews and assesses the dominating thermal insulation materials of the Global Market mainly in Asia, North America (the United States and Canada), and Europe:

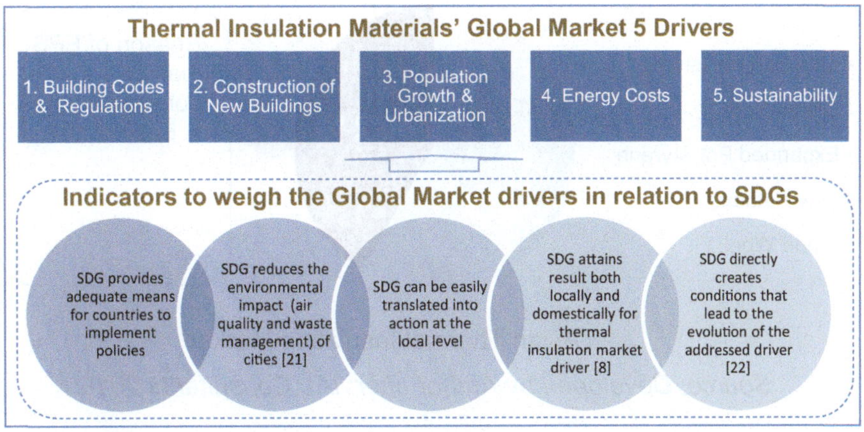

Fig. 2 Analytical approach—thermal insulation global market drivers/five indicators. (*Source: Developed by Authors*)

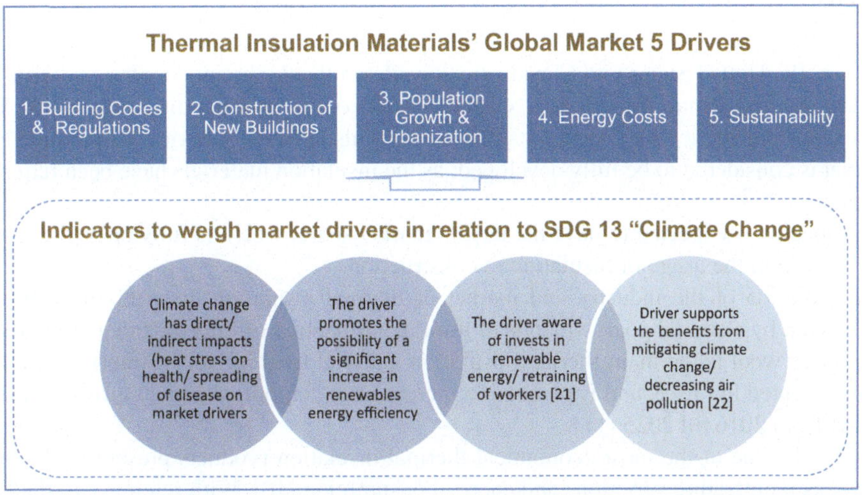

Fig. 3 Analytical approach—thermal insulation of Egypt market drivers/four indicators. (*Source: Developed by Authors*)

4.1 Asia

Asia is considered a large-scale continent consisting of a respectable number of developed countries including Japan, Singapore, China, and South Korea. Concerning the analysis of thermal insulation materials' market, a total of seven countries are studied, covering almost all of the commonly used insulation materials. They are represented in three major classifications: China, South Korea, and Japan, as well as the rising markets of India, Indonesia, Malaysia, and the Philippines.

The analysis focuses on the five commonly used insulation materials, owing to the predominance of China, which represents 83 per cent of the market demand [9]. Indeed, the less developed Asian countries have an undeniable insufficiency in the thermal insulation materials' utilization, due to the moderate climate that distinguishes the region, which aggravates the ignorance of the need of thermal insulation.

Nonetheless, in China, South Korea, and, to a subsidiary range, Japan, due to the maturity of its insulation market, the use of thermal insulation materials is more prevalent; owing to the increasing recognition of the benefits of insulation to realize energy efficient buildings and the rising energy prices, which increase the significance of the thermal insulation materials in construction. In fact, the dominant material of the thermal insulation market of Asia is expanded polystyrene that prevails with the largest proportion of 40.2 per cent as shown in Fig. 4 [9].

This is owing to the belief that expanded polystyrene is beneficial, in diverse applications such as floor, cavity walls, and roofs insulation, due to its flexibility and structural safety [10].

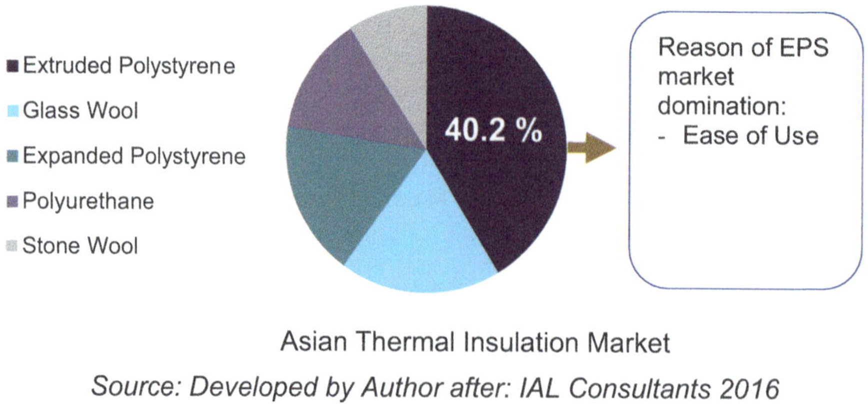

Fig. 4 Asian thermal insulation market (*Source: Developed by Author after: IAL Consultants 2016*)

4.2 North America

Regarding North America, the United States and Canada are considered the dominant countries in terms of thermal insulation uses. The rising demand of thermal insulation in residential and commercial buildings, due to the boosting of energy cost, is anticipated to lead the emergency for thermal insulation materials in North America. Consequently, North America's thermal insulation market is considered to be fully developed; as the insulation materials have been required for quite a prolonged time in this region.

The United States in specific is expected to have the largest market share, which is fully backed up by the realization of strict building codes that support the usage of insulation in construction.

As for Canada, the evolution of the industry and market of thermal insulation materials have been enforced by the Canadian government, driven by a crucial goal; which is to limit the utilization of materials that affect the ozone layer.

Generally speaking, glass wool is the main thermal insulation material used in North America with a share of 44.3 per cent of the market as represented in Fig. 5, and is planned to increase at a compound annual growth rate (CAGR) of around 3.7 per cent in the period from 2016 till 2025 [11].

Indeed, it is believed to be one of the most economical thermal insulation products present in the market, putting into consideration that it lessens temperature variations and energy consumption [10]. With reference to the United States, glass wool is truly proven to be the most exploited thermal insulation material for the years 2015 and 2016, and is estimated to hold its position till 2025; where it has a share of almost 45 per cent in the market [12]. This is also considered the case in Canada by around 50 per cent [13].

4.3 Europe

The European Union (EU) recently started advancing in thermal insulation through endorsing policies and rules to promote for the renewal of Europe's old buildings through insulation to prevent its heat gain or heat loss. These procedures are believed to enhance both the construction sector and EU market of buildings insulation materials in the period 2015–2027 [9].

Overall, a respected number of European companies can be considered as world leaders in the insulation products, that is, the present supply of insulation materials of the EU could be estimated a sustainable one.

The EU market of insulation materials actually depends on substantial countries that are considered its leading ones, as Germany, France, and Poland. Also, Western Europe countries cannot be underestimated; as they are accounted for almost 30 per cent of Europe's global market in 2015. Yet, this region's market is expected to have a slow development rate; owing to its maturity.

Insulation materials that are characterized by the best performance per unit cost, such as mineral (stone and glass) wool, polystyrene and polyurethane, possess the largest market potential, where glass and stone wool represent 58 per cent of the European thermal insulation market, thanks to their particularly well insulation properties [9].

On the other hand, expanded polystyrene (EPS) foam is still considered a common material when it comes to external wall insulation, thanks to its low price, especially in Eastern Europe countries. From the above analysis, it is clear that glass wool is deemed to be the best insulation material with 36 per cent as illustrated in Fig. 6 [14].

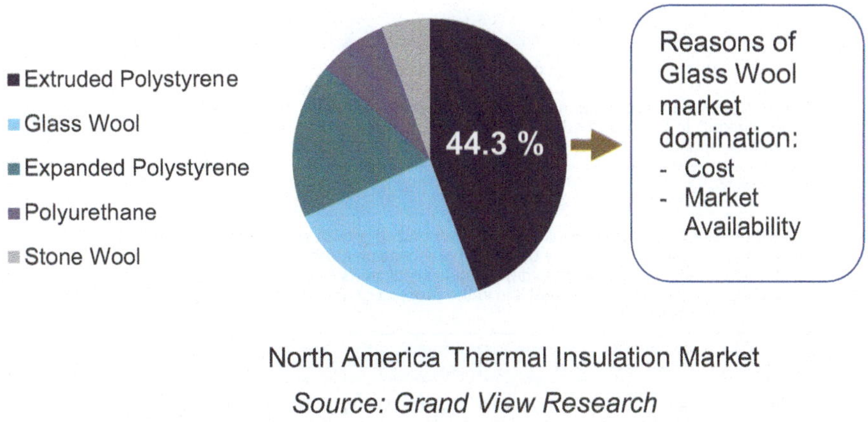

Fig. 5 North America thermal insulation market. (*Source: Grand View Research*)

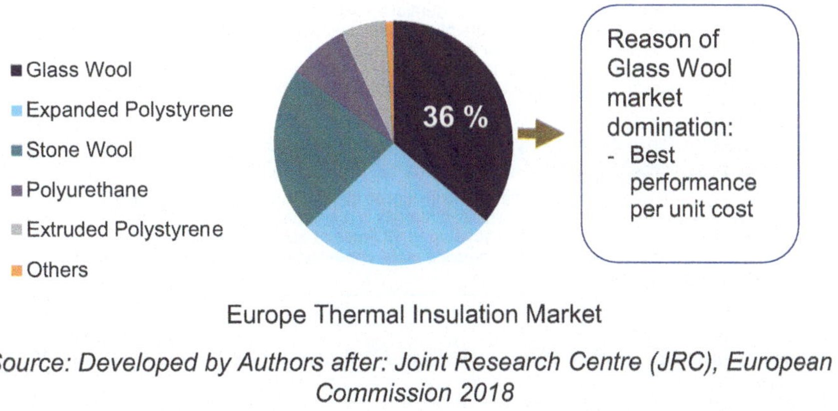

Fig. 6 Europe thermal insulation market. (*Source: Developed by Authors after: Joint Research Centre (JRC), European Commission 2018*)

5 Regions Dominating the Thermal Insulation Market

After analyzing Asia-Pacific, North America, and Europe markets, Asia-Pacific turns out to be the most profitable market in 2016 with more than 35 per cent share in the global market. This position is owed to the development of the region's construction industry as a result of industrialization and urbanization, along with the growing population which entails the enforcement of energy conservation practices by the governments. North America comes in third place after Europe in 2016 [15].

6 Main Drivers of the Global Thermal Insulation Market

Subsequent to the prior information about the three main regions that are dominating the thermal insulation market, the main drivers of the global thermal insulation market can be summarized in five main points as shown in Fig. 7 [16]. In relation to the drivers of the market, there is an undeniable connection between these drivers and the sustainable development goals (SDGs), where these drivers' significance determination directly depends on each goal.

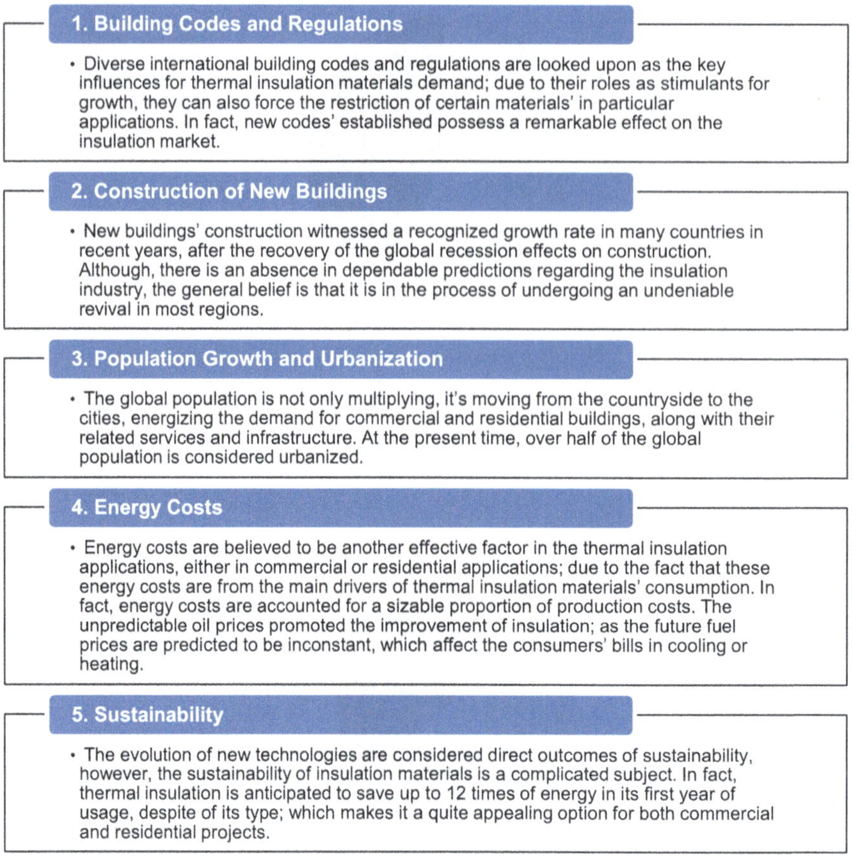

Fig. 7 Main drivers of global thermal insulation market. (*Source: Developed by Authors after: Insulation Outlook 2019*)

7 Main Drivers' Weight Based on Sustainable Development Goals—SDGs

From the literature review, it can be said that linking the thermal insulation materials to sustainable development has not been addressed. In attempt to carry out such linkage, market drivers are weighed in relation to each goal based on a checklist of five indicators as depicted in Appendix 1. It is clear from Table 1, which is based on the checklist, that almost 35 per cent of the SDGs have no relevance to the thermal insulation market drivers. Yet, the remaining SDGs crucially affect the determination of the market drivers' importance; where the table demonstrates that the "Sustainability" and "Energy costs" drivers are equally the most effective in the thermal insulation market, followed by "Building code and regulations" driver with minimal differences in the four drivers' percentages, down to the least important one; which is the Population Growth and Urbanization driver, as shown in Fig. 8.

Based on the addressed SDGs, it is revealed that SDG 14 "Life below Water" and SDG 15 "Life on Land" affect equally in the importance of market drivers. According to SDG 14 relevance to the thermal insulation market drivers' weight, it can be explained that, when thermal insulation materials are sustainable, they will not harm the environment or life below sea. Thermal insulation usage in buildings would mainly reduce energy consumption, hence, mitigate greenhouse gases. Therefore, limit the urban heat island effect and climate change impacts; accordingly, the impact of ecosystem on the environment and water will be lessened [8], whereas SDG 17 focuses on the global partnership needed across to achieve the mentioned 16 sustainable development goals, which explains that it affects all the market drivers.

Table 1 Thermal insulation market drivers' percentages based on SDGs. (*Source: Developed by Author*)

Thermal Insulation Market Drivers' percentages based on SDGs

Source: Developed by Author

Sustainable Development Goals (DGs)	Thermal Insulation Market drivers: 1. Building Codes and Regulations					2. Construction of New Buildings					3. Population Growth and Urbanization					4. Energy Costs					5. Sustainability				
Indicators	1	2	3	4	5	1	2	3	4	5	1	2	3	4	5	1	2	3	4	5	1	2	3	4	5
1. No poverty	X	X	X	X	X	√	X	√	X	√	√	X	X	X	X	X	X	X	X	X	X	X	X	X	X
2. Zero Hunger	X	X	X	X	X	X	X	X	X	X	X	X	X	X	X	X	X	X	X	X	X	X	X	X	X
3. Good health	√	√	√	√	√	√	√	√	√	X	X	X	X	X	X	√	√	√	√	X	√	√	√	√	X
4. Quality education	X	X	X	X	X	X	X	X	X	X	X	X	X	X	X	X	X	X	X	X	X	X	X	X	X
5. Gender equality	X	X	X	X	X	X	X	X	X	X	X	X	X	X	X	X	X	X	X	X	X	X	X	X	X
6. Clean water	X	X	X	X	X	X	X	X	X	X	X	X	X	X	X	X	X	X	X	X	X	X	X	X	X
7. Clean energy	√	√	X	√	X	√	√	√	√	√	X	X	X	X	X	√	√	√	√	√	√	√	√	√	√
8. Economic growth	X	X	X	X	X	√	X	√	X	√	X	X	X	X	X	X	X	X	X	X	X	X	X	X	X
9. Industry innovation	√	√	√	X	√	X	√	X	X	X	X	X	X	X	X	X	X	X	X	X	X	X	X	X	X
10. Reduced inequalities	X	X	X	X	X	X	X	X	X	X	X	X	X	X	X	X	X	X	X	X	X	X	X	X	X
11. Sustainable cities	√	√	√	X	X	X	X	X	X	√	X	X	√	√	√	X	X	√	X	√	X	X	√	X	√
12.Responsible consumption & production	X	√	X	X	X	√	X	√	X	X	X	X	X	X	X	√	X	√	√	√	X	√	√	√	√
13. Climate action	√	√	√	√	√	√	√	X	√	X	X	X	X	√	X	√	√	√	√	√	√	√	√	√	√
14. Life below water	X	X	X	X	X	X	X	X	X	X	X	X	X	X	X	√	X	√	X	X	√	X	√	X	X
15. Life on land	√	√	√	√	√	√	X	X	√	√	X	X	X	X	X	√	√	√	√	√	√	√	√	√	√
16. Peace & justice	X	X	X	X	X	X	X	X	X	X	X	X	X	X	X	X	X	X	X	X	X	X	X	X	X
17. Goals partnership	√	√	√	√	√	√	√	√	√	√	√	√	√	√	√	√	√	√	√	√	√	√	√	√	√
Percentage %	41	47	35	29	29	47	29	35	29	35	11	6	11	18	11	41	29	47	35	35	35	35	47	35	35
Average %	36					35					11					37					37				

(√) Yes (x) No

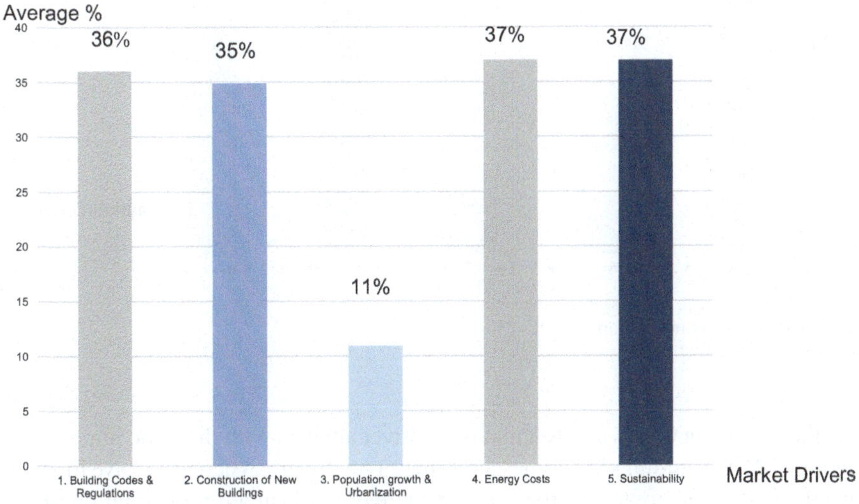

Fig. 8 Importance of market drivers in relation to SDGs

8 Egypt's Thermal Insulation Market Drivers' Weight

Indeed, climate change can be perceived as the most menacing challenge addressed in goal 13 that is climate action. It is truly influencing each country in each continent. National economies are disturbed and people's lives are affected due to the inconstant weather patterns and the highest recorded greenhouse gas emissions' levels in history. Although countries are leaping toward renewable energy to lessen emissions, climate change is a global challenge that is not affected by borders [17]. Speaking of the developing countries affected by climate change severe events, Egypt is looked upon as one of these countries that are tremendously influenced and most vulnerable by climate change impacts; therefore, the thermal insulation market drivers' significance weight in Egypt is based on climate action.

Egyptian market drivers are weighed according to a checklist of questions to determine the average percentage of the most important driver in relation to Egypt's status, as listed in Table 2. Figure 9 presents the importance of Egypt's market drivers in relation to SDG 13—Climate Action. Although these drivers' weight is based on climate action goal regarding Egypt, the thermal insulation driver in the Egyptian market is not dominated by "sustainability" driver, which is the key for climate action. However, it is controlled by "Energy costs" for the reason of the electricity tariff paid not for the purpose of mitigating

Table 2 Indicators of thermal insulation market drivers—Egypt. (*Source: Developed by Authors*)

Source: Developed by Authors

Indicator	Thermal Insulation Market Drivers				
	1.Building Codes & Regulations	2.Construction of New Buildings	3.Population Growth & Urbanization	4. Energy Costs	5.Sustainability
1	√	x	x	√	√
2	√	√	x	√	√
3	x	√	x	√	x
4	√	x	√	√	√
Ave. %	75	50	25	100	75

(√) Yes (x) No

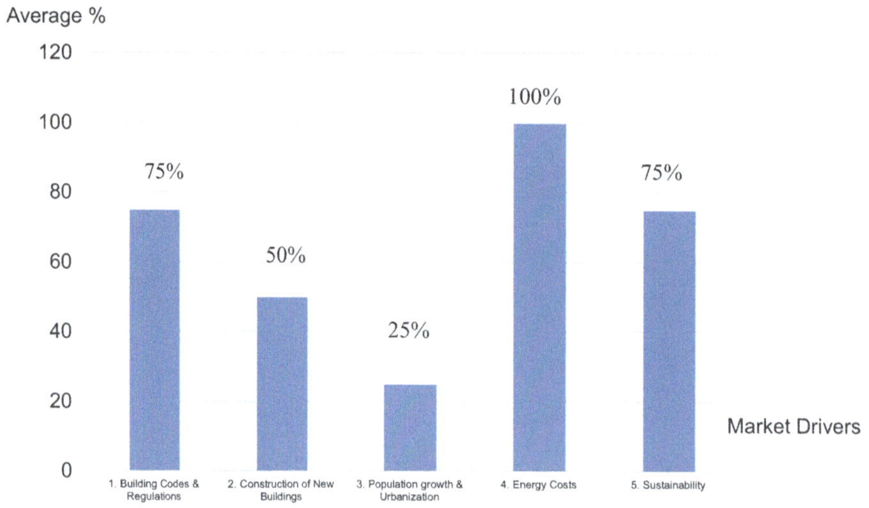

Fig. 9 Importance of Egypt's market drivers in relation to SDG 13

harmful emissions. Then, the "Building Codes and Regulations" driver appears with the same importance as "Sustainability" driver. As for, "Construction of New Buildings," it comes after these two drivers, whereas to have the population driver in the least position. Consequently, due to the control of the price of the thermal insulation material on the market, energy costs driver is seen as the main driver in Egypt's market.

9 Results and Discussions

By looking at the electricity tariff history in Egypt from 2008 till 2014, it was almost constant with no increase since the government was subsiding electricity. Nonetheless, the government decided in 2015 to liberate the energy prices and gradually increase the electricity tariff over a span of 4 years. Figures 10 and 11 illustrate the electricity tariff for residential and commercial buildings, respectively, between the years 2008 and 2018, where the figures are mainly concerned with the average usage for residential buildings, that is from 0 till 200 kilowatt hour (kWh) and the extreme usage of above 1000 kWh. The same goes for commercial buildings with average usage of 0.0 till 250 kWh and the extreme usage of above 1000 kWh [18].

It is clear from Fig. 10 that residential tariff from 0 till 200 kWh and above 1000 kWh is almost constant at 16 Piasters/kWh and 48 Piasters/kWh respectively, from 2008 till 2014, to begin rising in 2015 to 18.75 Piasters/kWh and 84 Piasters/kWh for more than 1000 kWh [19], the prices gradually increase in 2016 to 21.5 Piasters/kWh for 0 till 200 kWh, and in the same year, they arrive at 95 Piasters/kWh for the usage of above 1000 kWh, which is a notable escalation. Tariff prices

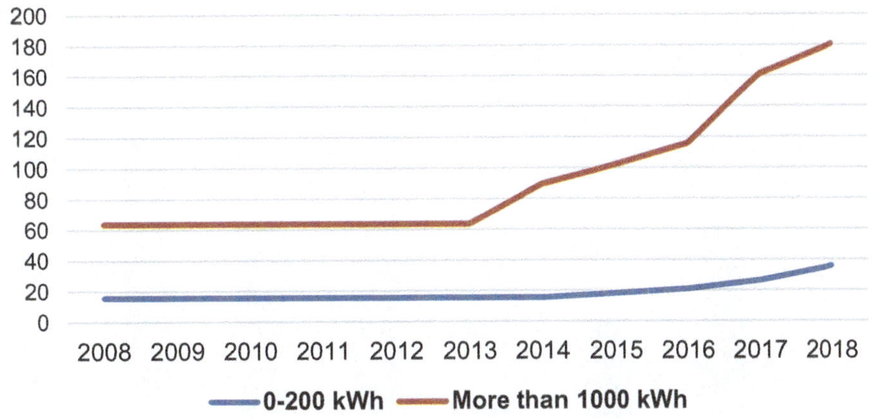

Fig. 10 Residential buildings electricity tariff from 2008 till 2018. (*Source: Developed by Authors after:* www.ceicdata.com & www.egyptera.org)

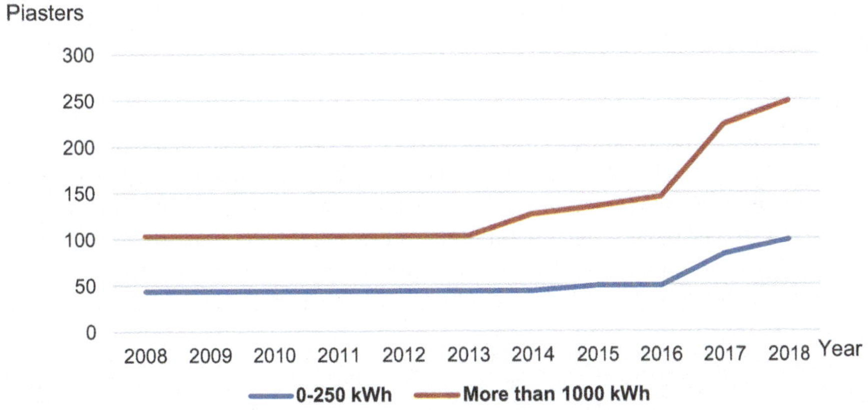

Fig. 11 Commercial building electricity tariff from 2008 till 2018. (*Source: Developed by Authors after:* www.ceicdata.com & www.egyptera.org)

obviously rise with a leap in 2017 to 27 Piasters/kWh and 135 Piasters/kWh, respectively, for the electricity use of 0 till 200 kWh and above 1000 kWh. In the year 2018, prices continue rising slightly to reach 36 Piasters/kWh for the utilization of 0 till 200 kWh and 145 piasters for the maximum use of above 1000 kWh [18].

According to Fig. 11, it is obvious that commercial tariff from 0 till 250 kWh and above 1000 kWh is almost constant at 44 Piasters/kWh and 60 Piasters/kWh respectively, from 2008 till 2014, the prices begin rising in 2015 to 50 Piasters/kWh for usage of 0 till 250 kWh and 86 Piasters/kWh for more than 1000 kWh [19]. In 2016, the prices remain constant at 50 Piasters/kWh for 0 till 200 kWh, and in the same year, they increase steadily to 96 Piasters/kWh for the usage of above 1000 kWh. Tariff prices clearly escalate with a leap in 2017 to 84 Piasters/kWh and 140 Piasters/kWh, respectively, for the electricity use of 0 till 250 kWh and above 1000 kWh, to keep ascending in the year 2018 and reach 100 Piasters/kWh for the use of 0 till 250 kWh and 150 Piasters/kWh for the maximum consumption of above 1000 kWh [18].

Despite the fact that an enormous number of residential units are needed to be constructed every year for the following 5 years, existing buildings are the massive percentage of the building stock, where Egyptian existing buildings include about 12 million units with almost 60 per cent residential; as they contribute with a giant share in the climate crisis [20]. Based on tariff increase and removing subsides, in addition to the discussion of the five drivers for thermal insulation market (Fig. 7), the thermal insulation market drivers clearly neglect the factor of the substantial magnitude of existing buildings, especially to developing countries. Therefore, it is strongly believed that existing buildings need modifications in view of the increasing electricity prices that put pressure on buildings' users and owners as shown in Figs. 10 and 11. Such modification can be displayed in the form of incorporating thermal insulation materials as a solution to control and minimize energy use for cooling, to directly reduce electricity bills. Hence, the modification of existing buildings is considered a new driver, which can clearly illustrate Egypt's status. Thus, Fig. 12 presents thermal insulation market drivers' weight in Egypt, after adding the new driver of "Enhancing Existing Buildings," in relation to SDG 13—Climate Action.

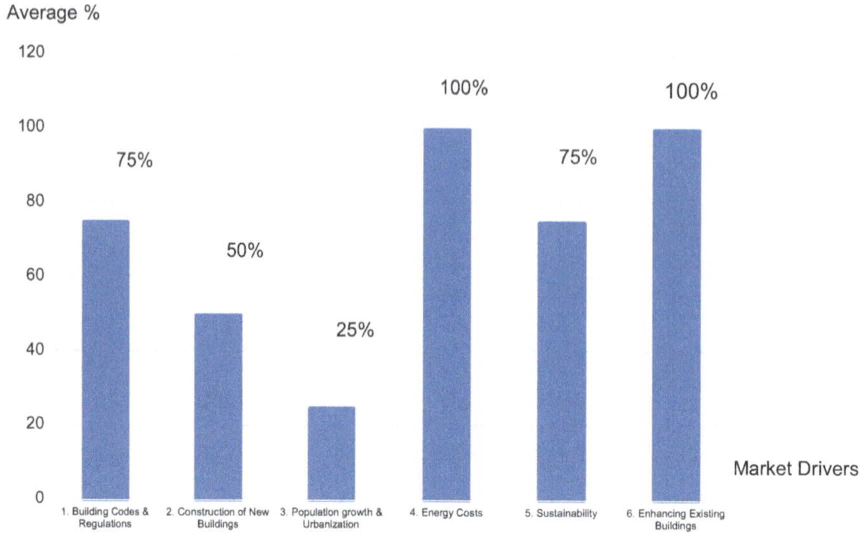

Fig. 12 Drivers' weight after addition, based on climate action (SDG 13) in Egypt. (*Source: Developed by Authors*)

Table 2 presents the results of the assessment and average percentage of thermal insulation market drivers in Egypt in connection to the four indicators to find the most significant driver for Egypt's market in relation to SDG 13—Climate Action.

In the Author's opinion, the new added driver can be looked upon as important as "Energy Costs driver" or more, when speaking of the Egyptian thermal insulation market and its tremendously rising electricity tariff. Whereas the rest of the drivers' weights are the same as listed in Table 2, where "Building Codes and Regulations" driver shares the same importance as "Sustainability" driver with average percentage of 75. Concerning "Construction of New Buildings," it possesses an average of 50 per cent. For the "Population Growth and Urbanization" driver, it is in the least position with an average of 25 per cent.

10 Conclusions

Based on the aforementioned global thermal insulation materials' market investigation presented in this paper, the global insulation market turned out to be divided on five regions: Asia-Pacific, North America, Europe, Latin America, Africa, and the Middle East. Yet, the most used thermal insulation material in each region's market is known only for the three prevailing regions, where extruded polystyrene is heavily used in Asia. Nonetheless, glass wool is the most used material in both North America and Europe. In fact, Asia-Pacific appeared to be dominating the global market due to its revenues, followed by Europe, and then North America. This basically left the Middle East and Africa with Latin America in the bottom places in relation to their markets' revenues.

On top of the global market's value, five important drivers are believed to control the global market. On the other hand, and based on the authors' opinion, these drivers' weights differ when related to the sustainable development dimensions and goals, where sustainability turned out to be the most important driver of the market, based on a checklist of five questions developed by the author. In fact, climate action is deduced to be the most important SDG globally, that mainly control the priority of the market's drivers of developing countries as Egypt accordingly.

In conclusion, the author reached an array of the drivers based on climate action in Egypt, in accordance with a specific checklist of indicators examining Egypt's market drivers in relation to SDG 13-Climate Action. The sustainability driver is believed to be from the bottom drivers prevailing Egyptian thermal insulation market, as it is claimed to be fully dominated by "Energy Costs." Such driver affects the insulation materials' prices, and somehow neglecting the climate.

After examining the critical case of electricity tariff rate in Egypt, the author thought that, it is fair enough to add a sixth driver, which is "Enhancing Existing buildings," that is believed to be the most important one of the Egyptian thermal insulation market drivers, that fulfills lowering consumed energy that leads to high electricity prices, along with reducing CO_2 emissions considering climate action.

Appendices

Appendix 1 – Indicators to weigh global market drivers in relation to SDGs:

1. Does this goal provide adequate means for countries to implement policies regarding the addressed market driver?
2. Does this goal reduce the environmental impact of cities by paying special attention to air quality and other waste management through the addressed market driver? [21]
3. Can this goal be easily translated into action at the local level through the addressed market driver?
4. Will the achievement of this goal result in more sustainable outcomes both locally and domestically regarding the addressed thermal insulation market driver? [8]
5. Does this goal directly create conditions that lead to the evolution of the addressed driver? [22]

Appendix 2 – Indicators to weigh market drivers in relation to SDG 13 "Climate Change" in the Egyptian market:

1. Does climate change have direct impacts (such as the effects of heat stress on health) and indirect impacts (through spreading of disease through climate conditions) on the addressed market driver?
2. Does this driver promote the possibility of a significant increase in renewables and increased energy efficiency to reduce global warming?
3. Does this driver invest in renewable energy and support for retraining of workers to start ending fossil fuel industries? [21]
4. Is this driver aware of the benefits from mitigating climate change and decreasing air pollution? [22]

References

1. The United Nations Environment Programme – UNEP. Global Status Report 2018, 7 12 2018. [Online]. Available at: https://www.unenvironment.org/resources/report/global-status-report-2018. Accessed 25 Feb 2020
2. Rousselot M. Energy efficiency trends in buildings - policy brief, ODYSSEE- MURE, June 2018. [Online]. Available at: https://www.odsse-mure.eu/publications/policy-brief/buildings-energy-efficiency-trends.html. Accessed 2 May 2020
3. Energy use in buildings – energy consumption, factsheets, European Commission, [Online]. Available at: https://ec.europa.eu/energy/eu-buildings-factsheets-topics-tree/energy-use-buildings_en/. Accessed 2 May 2020
4. The Economist Intelligent Unit (EIU), Global economy will be 3 percent smaller by 2050 due to lack of climate resilience, November 20, 2019. [Online]. Available at: https://www.eiu.com/m/global-economy-will-be-3-percent-smaller-by-2050-due-to-lack-of-climate-resilience/. Accessed: 21 Nov 2019
5. World Economic Forum, Agenda April 2020, International Monitory Fund, IMF- economy-Coronavirus-covid-19-recession. [Online]. Available at: https://www.weforum.org/agenda/2020/04/imf-economy-coronavirus-covid-19-recession/. Accessed 1 May 2020
6. MarketWatch. Thermal insulation materials market 2018 global trend, segmentation and opportunities, forecast to 2023, 4 10 2018. [Online]. Available at: https://www.marketwatch.com/press-release/thermal-insulation-materials-market-2018-global-trend-segmentation-and-opportunities-forecast-to-2023-2018-06-11/. Accessed 11 May 2019
7. Construction opportunities, "Insulation Materials," 8 10 2018. [Online]. Available at: http://www.constructionopportunities.in/IssueDetailPage?IssueMenuMasterId=16918&ParentMenuId=16918/. Accessed: 10 Nov 2019
8. SDGs Cities Guide. Practical tools for getting started with the SDGs [Online]. Available at: https://sdgcities.guide/chapter-2-practical-tools-for-getting-started-with-the-sdgs-4b7598170ebf/. Accessed 25 Feb 2020
9. The Asian market for thermal insulation products - IAL Consultants, 2016
10. Mordor Intelligence, Asia-Pacific expanded polystyrene market. 5 2018. [Online]. Available at: https://www.mordorintelligence.com/industry-reports/asia-pacific-expanded-polystyrene-eps-market/. Accessed 14 Jan 2019
11. Grand View Research, "Market Research Group," 8 10 2018. [Online]. Available at: https://www.grandviewresearch.com/industry-analysis/north-america-building-thermal-insulation-market/. Accessed 11 Nov 2019
12. The Statistics Portal, "U.S. building insulation market size by product 2014–2025," 8 10 2018. [Online]. Available at: https://www.statista.com/statistics/883385/market-size-building-insulation-us-by-product/. Accessed 23 Oct 2019
13. Kono J, Goto Y, Ostemeyer Y (2016) Factors for eco-efficiency improvement of thermal insulation materials. Chalmers Publication Library. Key Engineering Materials, 678
14. Joint Research Centre (JRC), European Commission (2018) Competitive landscape of the EU's insulation materials industry for energy-efficient buildings
15. Transparency Market Research. Transparency Market Research, 10 1 2019. [Online]. Available at: https://www.transparencymarketresearch.com/thermal-insulation-market.html/. Accessed 12 Oct 2019
16. Insulation Outlook. Factors affecting the Insulation Market, 10 1 2019, [Online]. Available at: https://insulation.org/io/articles/factors-affecting-the-insulation-market/. Accessed 12 Oct 2019
17. United Nations. Sustainable development goals, [Online]. Available at: https://www.un.org/sustainabledevelopment/climate-change-2/. Accessed 31 Jan 2019
18. Egyptian Electricity and Consumer's protection Institution. Electricity tariff, [Online]. Available at: http://egyptera.org/ar/t3reefa.aspx/. Accessed 10 June 2019

19. CEIC. Egypt electricity price, [Online]. Available at: https://www.ceicdata.com/en/egypt/electricity-price/. Accessed: 10 June 2019
20. ElGohary A, Khashaba S (2019) The challenge of greening the existing residential buildings in the Egyptian market base case. The Academic Research Community
21. Huan Y, Li H, Liang T (2019) A new method for the quantitative assessment of sustainable development goals (SDGs) and a case study on Central Asia
22. International Council for Science (2017) A guide to SDG interactions: from science to implementation

Innovative Urban Farming Techniques for Sustainability of Cities: from Farm to Fork – Case of the City Lanuvio, Rome, Italy

Mariam Shehata, Andrea D'Aprile, Giuseppe Calore, Lorianne Ginot, Maia Correrella, Mariam Elian, Nada Aly, Olakunle Sangodoyin, Salma Aboughanema, Sohaila Ali, and Mohsen Aboulnaga

1 Introduction

Cities generate over 80% of the global GDP and attract more people due to their economic growth [1]. Rapid growth in population leads to a huge increase in energy demands, CO_2 emissions and food requirements in cities [2, 3]. Climate change (CC) is considered as one of the biggest challenges nowadays. It affects the environment and cities negatively in various forms (severe events, storms, floods and health) [4–6]. The global temperature has increased by 1.4 °F in the last 140 years and the CO_2 level in the atmosphere has become 400.71 parts per billion.

Also, the cover of the world's forest has decreased by 1.5 million km^2 in 12 years [7]. In Italy, CC increased air pollution, floods, droughts, coastal erosion and caused reduction in agriculture. This affects food and water security in most regions [8].

Urban farms play an escalating and crucial role in food security and in generating clean and local food [9, 10]. Urban farms also create new job opportunities, boost the economic growth and contribute to the GDP, while eventually becoming green spots of the cities and acting as social catalysts [11, 12]. Urban farming is a key component in improving local economy, increasing social integration and achieving environmental sustainability [13]. Urban farming has been increasing in cities. For instance, more than 21,000 ha in Cagayan de Oro City, Philippines, 12% of urban land in Havana, Cuba and more than 11,000 ha in Jakarta, Indonesia, are dedicated for agriculture production [14–16]. Figure 1 presents some examples of urban farming in different cities.

According to the Food and Agriculture Organization of the United Nations (FAO), one quarter of the world's population is provided by food through urban and peri-urban farms and this number is expected to rise as shown in Box 1 [17]. With urban farms starting to be integrated to the city fabrics such as rooftop urban farms, their economic, social and environmental values increase as well. Urban farming provides the community with organic and clean food mainly vegetables, herbs and fruits [18]. Hence, it ensures food self-sufficiency and security, as well as it creates solutions to organic waste and improves the air quality mitigating CO_2 emissions. New technologies in agriculture have been developed to deal with areas with low soil fertility and water availability, cultivation in limited spaces and minimize the production impact to the environment and human health [13]. Some of these technologies are aquaponic and hydroponic systems, green walls, vertical farming (Fig. 2) as well as permaculture, energy from renewable sources, transforming organic wastes into fertilizers and rainwater harvesting [19–22].

M. Shehata (✉) · M. Elian · N. Aly · S. Aboughanema · S. Ali
Architectural and Technology Engineering Programme, Faculty of Engineering, Cairo University, Giza, Egypt

A. D'Aprile (✉) · G. Calore · L. Ginot · M. Correrella
Alma Mater Studiorum University of Bologna, Bologna, Italy
e-mail: lorianne.ginot@studio.unibo.it

O. Sangodoyin
Ladoke Akintola University of Technology, Ogbomoso, Oyo State, Nigeria

M. Aboulnaga
Department of Architecture, Faculty of Engineering, Cairo University, Giza, Egypt
e-mail: maboulnaga@eng.cu.edu.eg

© The Author(s), under exclusive license to Springer Nature Switzerland AG 2022

A. Sayigh (ed.), *Sustainable Energy Development and Innovation*, Innovative Renewable Energy,
https://doi.org/10.1007/978-3-030-76221-6_33

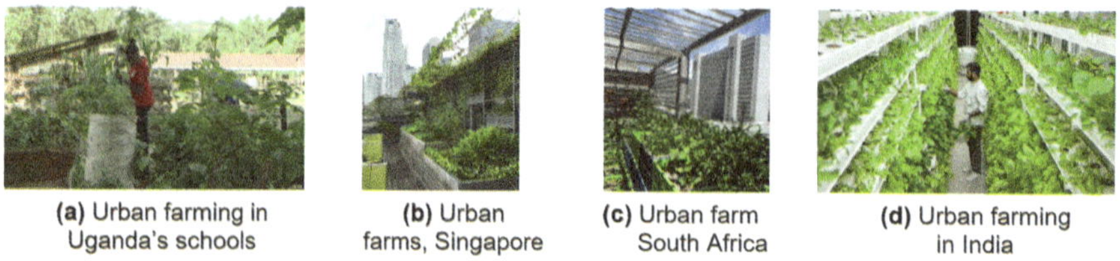

Fig. 1 Examples of urban farming in different cities [24–27]. (**a**) Urban farming in Uganda's schools. (**b**) Urban farming in Singapore. (**c**) Urban farming in South Africa. (**d**) Urban farming in India

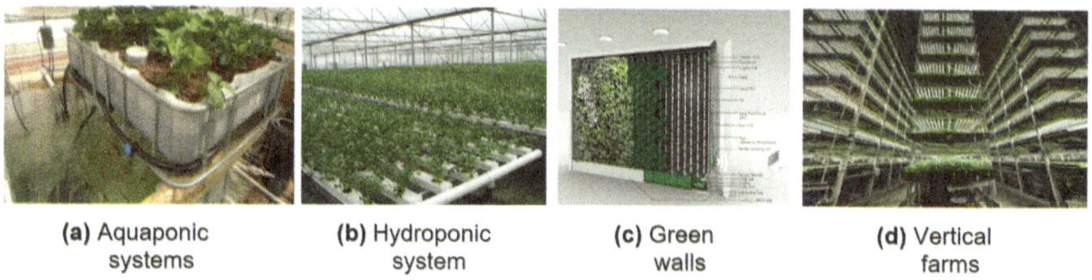

Fig. 2 Examples of urban farming types and technologies in cities [28–31]. (**a**) Aquaponic systems. (**b**) Hydroponic system. (**c**) Green walls. (**d**) Vertical Farms

Lanuvio is not far from the challenges of climate change (CC) and the need for food security. It is a village situated in 'Castelli Romani Regional Park' close to Rome, in the region of Lazio, Italy. The area has high cultural, historical, archaeological and naturalistic values that should be maximized and promoted. It is a popular and touristic place with festivals and social events promoting local food and traditionalism [23]. The site of ARES 118 (2500 m^2) was once used as a fire station. However, it is now a vacant land with no social or economic value. It is located next to the city centre, 800 m away from the main square and is reachable by train from other parts of Italy.

Box 1 Scale of Urban Farming in Providing Clean Food

'One quarter of the world's population is provided by food through urban and peri-urban farms and this number is expected to rise.'

Food and Agriculture Organization (FAO) – United Nations

This chapter presents a study conducted on an abundant site in Lanuvio, Rome, Italy, utilizing urban farm innovative technologies and integrating smart solutions to make this site a green hub in the city, generative and sustainable (economically, socially and environmentally viable). It also presents a conceptual project – named Green Symphony – that provides a social hub for the citizens of Lanuvio with smart and sustainable technologies focusing on agriculture.

2 Objectives

Focusing on the needs of the citizens and the necessity to restore the site (there is a train station 28 minutes away on foot next to the site as shown in Fig. 3), the objectives of Green Symphony project were to transform the abundant site of ARES 118 to be a sustainable node, an educational hub and a social innovation area to increase the liveability of the city. The conceptual project links arts, like music and agriculture creating a community focused farm. It reaches out to the community while using innovative environmentally friendly approaches, smart technologies to preserve energy and water and to produce pesticide-free food while focusing on the UN 2030 Agenda and SDGs, mainly SDG 3, 7, 9, 11, 12, 13 and 15 as illustrated in Fig. 4.

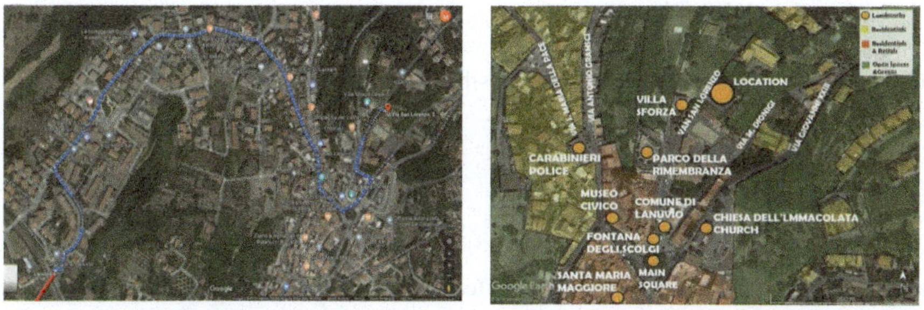

(a) Route to the site (b) Site's surroundings

Fig. 3 Accessibility to the site on foot and the surroundings. (*Image source: Authors based on Google Maps, 2020*). (**a**) Route to the site. (**b**) Site's surroundings

Fig. 4 Sustainable development goals for 2030 and the project targets. (*Image Source: UN Global compact, 2018*)

(a) Satellite footage of Lanuvio's location (b) Satellite footage of the site

Fig. 5 Satellite footage of Lanuvio and the site location in Rome metropolitan, Italy. (*Images source: Google Earth, 2020*). (**a**) Satellite footage of Lanuvio's location. (**b**) Satellite footage of the site

3 Issues and Challenges

With the increase in population and damages due to climate change, cities are lacking the resources to support their citizens. Therefore, cities are usually thought of as congested places lacking natural life and green sites. To change this situation, action is required that ensures the sustainability of the city and meets the needs of the community. The site of the study is located in the city of Lanuvio, Italy at coordinate 42°14′20.29″N 13°42′56.04″E as illustrated in Fig. 5. There are several challenges concerning the site of the study, which are as follows:

- The site being small (2500 m^2 only) with one usable building
- The land being abandoned and the building in a poor state (Fig. 5)

- Located along only one road and the only entrance
- Being a dull spot among the historic sites around it
- The main building being wall bearing with minimal possible changes

4 Methodology

This project study is a product of the participation of the team 'FENICE' in the international competition of UrbanFarm 2020. The methodology depends on two approaches: (a) a theoretical and analytical approach and (b) an applied approach. The analytical approach depends on the evaluation of the given site and its surroundings, and it produces a number of solutions to achieve its goals and objectives. These solutions include architectural, agricultural, social and ecological ones [32]. The applied approach is centred on integrating the environmental and energy systems as well as smart technologies including the urban farming systems. Figures 6 and 7 show the exterior state of the deserted building and the site shed.

5 SWOT Analysis

It is imperative to conduct a SWOT analysis of the study project to assess the strengths, weaknesses, opportunities and threats. Figure 8 summarizes the SWOT analysis that was carried out for per assessment. It is clear from the figure that project strengths and opportunities are of added value, but nevertheless the threats and weakness were addressed in the project assessment in the following parts.

Fig. 6 Building's exterior situation. (*Images source: UrbanFarm 2020* [32])

Fig. 7 Site and shed's situation. (*Images source: UrbanFarm 2020* [32])

Fig. 8 SWOT analysis of the study project. (*Source: Authors*)

6 The Study

The study's concept of architectural innovative and smart solution includes six fields: (a) environmental systems, (b) energy, (c) water, (d) materials, (e) emissions and (f) urban farming presented in Fig. 9 and discussed in the following section.

6.1 Concept

The name FENICE was inspired by the phoenix where in ancient Greek folklore, a phoenix is a long-lived bird that cyclically regenerated or is otherwise born again [33]. The study approach and slogan is basically to rebuild from ashes. We, therefore, rebuild the current land from its own ashes into an urban farm, which has a community focused approach. In order to create a farm that is fulfilling to the citizens' needs and making use of the abandoned site of Lanuvio, the project study focuses on linking arts (music) and agriculture through creating a community focused farm. Thus, the urban farm is called 'Green Symphony' to emphasize this link and create social synergy. The project has an area to host events and allow for social interaction. It also reaches out to the community while exploiting innovative environmentally friendly approaches, smart technologies to conserve energy, water use, yet produce pesticide-free food as well as focusing on 2030's sustainable development goals (SDGs).

6.2 Architectural Solutions

The project consists of four main spaces: workshops, a permaculture garden, outdoor recreational spaces and a kitchen. The two buildings are developed to be well insulated to save energy (heating/cooling). The remaining space will host a number of recreational activities with primary views of the historic sites and area dedicated to hosting events in the farm as shown in Fig. 10; and details are in Table 1.

6.3 Environmental Solutions

The restoration and valorization of degraded sites is part of the Regulation L. 29/97 that protects the regional park; thus, it is one of the main goals of the region of Lazio. Within this scope, Green Symphony could be a solution, as it provides local food and hosts events with economy of energy, water, the use of low impact materials and a negative carbon footprint (avoidance of impact).

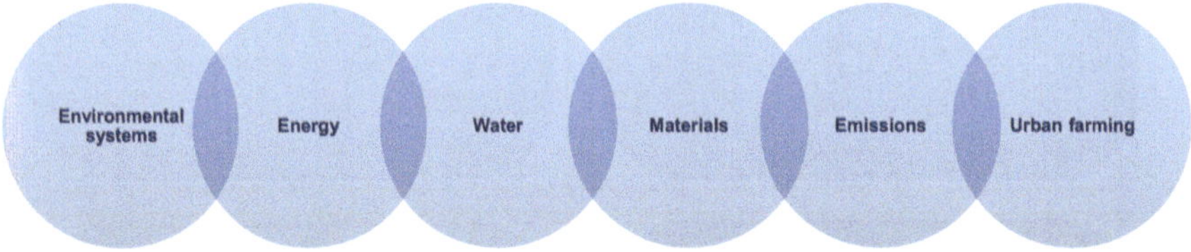

Fig. 9 The six sections examined in the study

Fig. 10 Elements of the study project planning in Lanuvio city, Italy (*source: authors*)

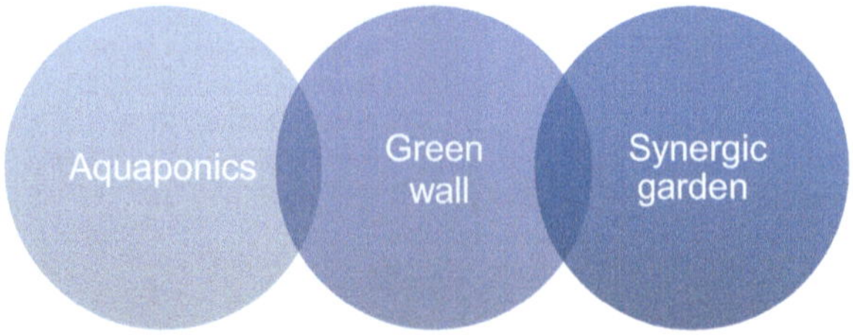

Table 1 Illustration for the architectural areas, locations and functions

Area	Location	Function
Workshop area	Under the old shed that was used formerly for ambulances.	Seminar area enclosed by glass walls, equipped with a wide screen and speakers.
Office	Main Building	Developed for the management of the farm's activities.
Green house	On the roof of the main building	Aquaponics system of agriculture is used where columns and trays are rented to the community.
Kitchen	The heart of the main building.	Sharing, collaborating, co-creating, and raising awareness are the main purposes for teaching the community the basics of using fresh produce to prepare traditional Italian dishes.
Eating area	Attached to the kitchen in the main building.	The community can order food according to the produce of the farm and the kitchen.
Storage room	Next to the kitchen in the main building.	The produce could be stored for a short time to be packaged and sent out to be sold.
Outdoor spaces	______________	An events area and interactive space in the landscape is designed to host different activities to boost the social dimension, in addition to a bike park.
External booths	Attached to the outer fence facing the main street.	For selling the excess products of fruit and vegetables.

Source: Authors

6.3.1 Energy

Energy production is one of the most critical factors that affect the environment. The building's envelop is insulated and the heat from the kitchen is harvested to decrease cooling loads. The pond is covered by a removable radiant barrier (TekFoil®) to preserve moderate temperatures during the winter. Dimmable LED lighting units combined with ABB i-bus® light and occupancy sensors are also used indoors to adjust light intensity and automatic turn off of the lights. LumiMotion® technology, in the landscape, is applied for turning the light on automatically when someone moves within a 10-m radius. These technologies reduce the energy use of lighting up to 60%. Moreover, renewable energy sources are chosen according to local abundance and 25 solar photovoltaic cells (315 kWp) are installed on top roofs. Each PV cell produces 320 kWh annually so the total solar energy produced will be 8 MW/year and it can be used to operate the two buildings and the pumps [34]. However, solar energy may not be sufficient to cover all the energy needs, hence, the farm is also supplied in certified 100% green electricity. Finally, piezoelectric Pavegen® tiles generate an average of 5 watts at 12–48 volts DC per footstep. This energy is used to load the batteries of humidity sensors in the garden.

6.3.2 Water

The aquaponic and green wall systems both use water from the fish pond to feed plants. This closed circuit reduces water consumption by 90% and prevents water waste. Likewise, saving water is a key aspect of the permaculture concept, so that mulching and a drip irrigation system in the garden are essential. Since Lanuvio has had substantial rates of rainy days [827 mm/m^2 between 2003 and 2018, according to the Regional Agency for Environmental Issues (ARPA)], the rainwater is collected through buildings, gutters and channels in the garden and stored in an underground tank size of 5 m^3 [35]. Such water can be used for farm irrigation instead of drinkable water. During drought periods, the needs in water are supplemented by the water network. In addition, the Internet of Things (IoT) technology can nowadays be applied in agriculture and save water up to 70% [36].

6.3.3 Materials

Green Symphony farm enrols in the 'Civiltà Contadina', which is non-profit association that is fostering biodiversity conservation by exchanging native seeds. Hence, it provides the first seeds that are restituted after 1 year, and then the remaining seeds are used for self-reproduction [37]. Materials used for renovations come from secondary raw materials; particularly, the columns of the aquaponics system and the irrigation system in the garden are built with recycled tubes. In the kitchen, no disposable tableware is used and food is stored into glass containers or jars. Wastes generated by the activities of the site are collected, carefully separated and sent to the municipal waste treatment.

6.3.4 Emissions

The carbon footprint (CF) of the life cycle of materials used in the study project has been computed based on data from the ADEME database (a French Agency for Environment) since no equivalent were found for Italian figures and similarity [38]. Table 8 (Appendix A) presents the carbon emissions' results after the project's implementation. The total CF generated is −77 tCO_2/year, mainly due to the good yields of both permaculture and aquaponics without any pesticides or fertilizers industrially produced. The CF of energy consumption is estimated at a maximum 8 tCO_2/year. This is due to heating, kitchen activities and pumps functioning, thus, the overall farm CF is −69 tCO_2/year. However, the renovation of the site has a CF estimated at 100 tCO_2 according to a computation suggested by the ADEME for housing renovation (2 tCO_2/year). In all, Green Symphony allows it to produce food and to enjoy recreational and didactical activities with a life cycle-negative carbon footprint (−67 tCO_2/year).

7 Urban Farm Design

7.1 Surroundings and Indoor Area

Green Symphony farm consists of a kitchen with an aquaponic farm on the rooftop, a workshop room, a synergic garden, a green wall and outdoor recreational spaces including a pond. The kitchen is situated in the main building offering cooking training using vegetables harvested from the farm and fish from aquaponics (refer to Table 9 in Appendix A for the number of sessions per year). Different innovative systems for agriculture are applied in the project (Fig. 11).

7.2 The Aquaponic System

The aquaponic system is a new agricultural technique used to grow plants, and it combines the hydroponic system (soilless agriculture on columns) with the aquaculture (fish breeding). An aquaponic closed cycle system that starts from an external pool containing fish and nutrients and arrives inside the rooftop greenhouse, where hydroponic columns are present with plants, is developed in this project. A vertical farm system that allows for good production in the project's space (about 65 m^2) is created.

7.2.1 The Pool

In order to guarantee a good amount of nutrients to vegetables production, a 50 m^3 pool is required; thus, it was decided to breed carps with a density of around 15 kg/m^3.

7.2.2 The Fishes

The study highlights that Cyprinus Carpio (common carp) originally from East Europe, resists a high range of water temperatures between 4 °C and 34 °C. Together, the common carp with mirror carp (same species) and the Amur carp (*Ctenopharyngodon idella*) are generated. All these fish are edible, and their breeding is widespread in aquaculture.

Fig. 11 Elements of the plan of the study project

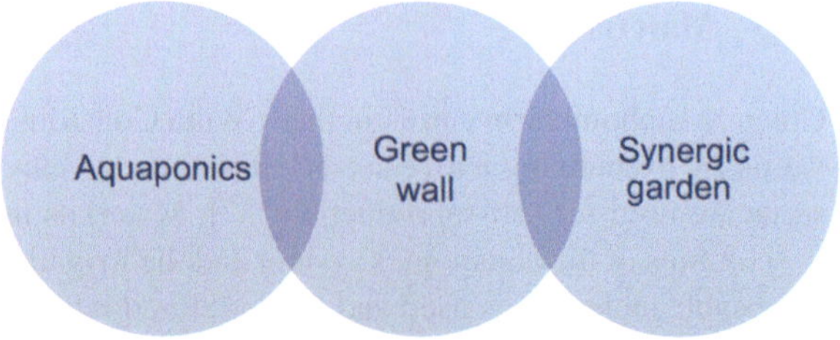

7.2.3 The Columns and Big Pots

The vertical hydroponic columns are pipes filled with an inert substrate like expanded clay. The typical plants that can be cultivated are presented in Fig. 12. Near the columns, some big pots are integrated and filled with the same inert substrate of the columns that are used to grow bigger vegetables like tomatoes or cucumbers. Humidity sensors are present in the greenhouse to regulate the water flow.

7.2.4 Estimated Production

An aquaponic system guarantees a faster plant growth and longer fructification time, with a reduction of water wasting of 90%. Based on a study conducted by a leading Italian company, it was found that there are some estimates which calculated a production between 6 and 11 kg/year of vegetables every 453 kg of fishes [39]. The estimated production can be found in Fig. 12. By considering the data and equally dividing the surface in four sections, it can obtain 16.25 m^2 for each type of vegetables. Table 2 lists the system's anticipated vegetables production around 5.3 tonnes/year.

7.3 Green Wall

At the city scale, green walls contribute to the recovery of degraded areas without occupying extra space. Thus, a green wall is developed in the project, not only for beauty, but also as an urban gardening, which increases yield of food production.

7.3.1 Green Wall System

The green wall will be fixed to the front side of the main building. The wall consists of 64 modular trays; pots will be placed on each modular tray in this system and secured by bars. In this case, an automatic irrigation system ensures good plant growth.

7.3.2 Green Wall Plants

Vegetable seeds and stems grow in a green wall making use of sterile potting mix. Most plants will be harvested continuously. The plants include 22 types (Fig. 13).

Fig. 12 Types and production of the vegetables and fruits used in the aquaponic system [39]

Table 2 Summary of the yield per year of the aquaponic system

Vegetables	Production			
	Kg/m²/year	*Surface*	*Kg/year*	*Tons/year*
Lettuce	95	16.25	1543.75	1,54375
Chard	75	16.25	1218.75	1,21875
Strawberries	58.80	16.25	955.50	0,9555
Basil	97.76	16.25	1588.60	1,5886

Source: Authors

Fig. 13 Types of vegetables, herbs and fruits cultivated in the green wall system [40–49]

7.3.3 Pests and Pathogens

The biological control and mechanical removal of pests, for example destroying cabbage moth eggs and larvae and leaf miners by hand, yet will control pests and pathogens.

7.3.4 Irrigation

Water is transported to the plants by small pipes made of second-hand plastics. Every tier of planter receives irrigation from four drip emitters (2/cell). A gutter system runs along each tier of planter, collecting excess water and draining it to the fish pond.

7.4 Synergic Garden

Synergic gardening is the Mediterranean climate version of natural farming and it is the application of permaculture philosophy. Vegetables will be associated with each other and with aromatic herbs in order to enable exchange of nutrients.

7.4.1 Garden Structure

The surface dedicated to synergic agriculture (745 m^2) will be split in two parts: one composed of 12 raised circular shrubberies with aromatic herbs combined with vegetables and one with 24 mounds of soil where crop rotation of vegetables is applied. The shrubberies will be 1-m high and 4-m diameter. Mounds will be 0.7-m high, 2-m wide and 8-m long. The total cultivated area will be 540 m^2, hence 205 m^2 are left for paths (large enough to host scholar groups).

7.4.2 Crops Combinations

Regarding the aromatic herbs combined with vegetable shrubberies, the combinations are listed below and Table 10 in Appendix A shows the benefits from those aromatic plants.

- Rosemary with carrots, salvia and beans
- Salvia with carrots, beans, rosemary and fennel
- Mint with tomato, broccoli, lettuce, carrots and potatoes
- Basil with fennel, lettuce, cucumber and tomatoes

Table 3 Summary of the yield per year

Vegetables	Yield (q/Ha)	Area cultivated (m^2)	q per 41.54 m^2
Carrots	400	41.54	1.7
Beans	40	41.54	0.2
Fennel	400	41.54	1.7
Tomato	800	41.54	3.3
Broccoli	200	41.54	0.8
Lettuce	300	41.54	1.2
Potatoes	250	41.54	1
Cucumber	250	41.54	1
Onion	300	41.54	1.2
Spinach	200	41.54	0.8
Pumpkin	500	41.54	2.1
Parsley	300	41.54	1.2
Chamomile	165	41.54	0.69

Source: Authors

- Chamomile with onion
- Thyme with spinach and pumpkins

7.4.3 Estimated Production

The cultivated area (540 m^2) is divided in 13 parts to grow different vegetables with an area of 41.54 m^2 for each species. Table 3 shows the calculations of the expected yield per year and results show that 1.7 tonnes per year could be produced from this area.

7.4.4 Irrigation System

The irrigation system used would drip irrigation linked with sensors for the humidity of the soil. A higher control of water waste is possible with this technological system:

- By raising shrubberies, it can take advantage of water drainage to irrigate plants.
- Part of water used for irrigation comes from rain collected in an underground water tank with 5000 L capacity (SIR5075 model, from ROTOTEC®). The choice of this capacity derives from the annual amount of rain and the estimated water needs for the garden that is 6.5 m^3 for 540 m^2.

7.4.5 Composting

The area will be provided with compost bins, with a comprehensive capacity of 3000 L. The structure will be handmade using pallets, while initial bacteria and microorganisms will be inoculated to make the composting process faster. An opening at the bottom of the bin will be necessary to harvest the mature compost, produced after 9–12 months, to fertilize the fields. The compost bin will be filled up with vegetable waste from the synergic garden, aquaponic and the kitchen. To cover the initial nutrient needs of plants, high-quality compost from local farmers is used.

8 Business Model

One of the project's goals is to be socially, environmentally and economically sustainable. Through the business model, it is possible to set the structure of the project when in operation and to calculate the costs and revenues. The project is summarized in a lean canvas as shown in Table 4.

Table 4 Lean canvas model of the project

Lean Canvas Model				
PROBLEM 1. Basic need of fresh healthy products 2. space and technology for small innovative production 3. Lack of popularity **EXISTING** AKTERNATIVES At a local level: 1. Cooperatives selling and transforming products 2. Didactic farms 3. Socials	**SOLUTION** 1. Product Selling 2. Rent of farming facilities 3. Events 4. Workshops and kitchen lessons 5. Advertisement on the app **KEY METRICS** The focus is on: - Events: at least 2 - Rent facilities: 16 columns and 18 trays	**UNIQUE VALUE PROPOSITION** innovative way to produce or to approach environmental sustainability, thanks to: -Vanguard techniques -Sustainable environment **HIGH-LEVEL CONCEPT**	**UNFAIR ADVANTAGE** site of Lanuvio: naturalistic values to promote, closeness to villa Sforza **CHANNELS** - Agricultural workers associations - Social media and events	**CUSTOMER SEGMENTS** 1. Consumers of local markets, typically families, elders 2. Local farmers, the younger ones from 35 to 50 years old 3. Young families with children 4. Groups: students from 6 to 13 years old with teachers 5. Other social sustainable producers
COST STRUCTURE Establishment costs for technologies Labor - voluntaries (0€)		**REVENUE STREAMS** 1. Weekly sales, based on seasons 2. Rent 3. Tickets 4. Fee on sales		

Source: Authors

8.1 *Products and Services Provided*

Our project operates on a multifunctional basis that provides the community with different services. Those services could be summarized as follows:

(a) Organic food sales: The farm sells organic food at the same time in the premises, at Lanuvio's market and through a 'mobile market' on bicycles.
(b) Rental of farming facilities: The locals can rent part of the farm's facilities annually where they can plant and harvest their own produce. While doing so, they are also revealed to farming techniques.
(c) Events and socializing area: Through the recreational and social events area, the locals can interact with each other and with the tourists.
(d) Workshops: The workshop room is where different classes and seminars are given to raise awareness about organic food planting, innovative farming techniques and permaculture, yet the kitchen gives lessons in basics of cooking.
(e) Yoga classes: Weekly courses hosted in our facilities, thanks to a specialized teacher, opened to anyone, subject to the purchase of a ticket.

8.2 *Targeted Customers*

Our farm targets a wide spectrum of users who interact together and exchange ideas. Different groups could be identified as follows:

(a) Local community: This is anyone who is living in Lanuvio and is looking out for consuming organic food or to learn about agriculture technologies to apply them in their backyard farms.
(b) Tourists: As the area is naturally full of historic sites, many tourists are visiting it.
(c) Volunteers: The farm welcomes all volunteers that could love new innovative experiences of work, whether they are from the direct community or international volunteers.
(d) School students: Since the farm spreads awareness and learning, students are especially targeted through periodic workshops and events, in which local schools may be directly involved.

8.3 *Organizational Structure*

The structure that best suits the mission of our project is the so-called 'social cooperative', which, from the regional law, 'is the one that promotes, supports and sustains human development and social integration with particular regarding of disadvantaged people', in particular the one related to the agricultural field (type B). As allowed by the law, our team will be made of volunteers (up to a maximum of half of the entire team) and specialized workers to handle the maintenance of the plantations.

Table 5 Sources of revenue with percentage

Item	Yearly Revenues in Euro	% of Total Revenues
Rent of the hydroponic columns	19,588 €	19,4%
Sales of products from permaculture	2,269 €	2,2%
Yoga classes	2,160 €	2,1%
Kitchen classes	74,850 €	74,1%
Sales of fishes	2,100 €	2,1%
Total	**100,967€**	**100,0%**

8.4 *Funding*

Starting from the region, already from last year some announcements were published offering subsidies for 'social cooperative of type B' as we are, with the aim of 'sustaining the social and working integration of disadvantaged people'. A fund comes from the MISE (Ministry of Economic Development) that offers specific public funding in favour of the birth of small cooperatives, with an allocated amount of 223 million €, which provides a feasible way to subscribe to the initial investment required. The Italian government also offers both non-repayable subsidies and easy credit term loans for the initial investment. The law provides up to a maximum of 40% of the expenses for 'market analysis, land, buildings, machineries and any other goods required for the production' as listed in detail in Table 11 (Appendix A).

8.5 *Cost Analysis*

This analysis has to follow the natural path of the project itself. Initially, a critical investment would be required to cover the renovation of the buildings, the technologies and so on, as shown in Table 12 (Appendix A). An initial investment of 380,000 € is expected, of which some expenses would be covered by the previously explained funding, and leave for an actually required initial investment of 250,000 € (refer to the operating margin below). Of course, every operating year, some expenses will be required for the maintenance and the activities pursued within the farm, as shown in Table 13 (Appendix A), for which we expect about 28,000 € of costs per year.

8.6 *Revenue Stream Analysis*

The biggest percentage of revenue is provided by the kitchen classes (74%) followed by the rent of the hydroponic columns (19%). The amount considered is the value of the yearly production from all the columns that can be guaranteed to those renting them. Details of this calculation are shown in Table 14 (Appendix A). For yoga classes, the revenue is fixed as 15% on the hypothetical fee requested by the teacher to its clients. An average yoga course could cost 60 €/month for two lessons per week, considering 20 clients per lesson. For details, refer to Table 15 (Appendix A). Moreover, tickets vary for the type of lessons proposed in the kitchen, as shown in Table 16 (Appendix A). All the different sources of revenue are shown in Table 5.

8.7 *Operation Margin and ROI*

The initial investment, funding, operating costs and revenues were summarized previously to calculate the operating margin and return on investments (ROI). The data produced refer to a fully operating year. A positive operating margin of about 73,000 € is expected, that is particularly favourable. In addition, the ROI is also in an optimum position (19%), due to the funding that allows us to reduce the 'weight' of the initial investments. Table 6 summarizes the calculation of the operating margin and ROI. Noteworthy is that even with the first years being less benefitting, the investments are paid off after a maximum of 5 years.

9 Results

The results are analysed by stating the features and the technologies implemented in the project and its impact on the economic, social and environmental sustainability. Table 7 shows a summary for all the interventions done in the project to analyse how sustainable the project is through extracting some percentages regarding the three pillars of sustainability. Also,

Table 6 Operating margin and ROI calculations

Feasibility calculation	
Initial investment - total	383,766 €
Granted initial funds	137,051 €
Costs uncovered by funds	246,714 €
Required initial investment (approximation)	250,000 €
Operation	
Operating revenues (per year)	100,967 €
Operating costs (per year)	27,954 €
Operating margin (per year)	73,013 €
ROI calculation	
Investment costs (funds included)	383,766 €
Profit of the year	**73,013 €**
ROI	**19%**

Table 7 Sustainable development pillars assessment of the project features

No.	Mission	Economic Sustainability	Social Sustainability	Environmental Sustainability
1	Presence of various functions indoors and outdoors	✓	✓	NA
2	keeping the old building structure	✓	NA	✓
3	Host music festivals	✓	✓	NA
4	Blend Music and Agriculture in one space	NA	✓	NA
5	Create workshop for raising awareness	✓	✓	NA
6	Create a space for restaurant & organic food market	✓	NA	NA
7	Integrate the production area with the selling, tasting and crafts area	✓	NA	NA
8	Create one universal space shared by all types of users	NA	✓	NA
9	Rental of farming facilities	✓	✓	✓
10	Architectural sustainable technologies	✓	NA	✓
11	Synergic garden (with permaculture technique)	✓	✓	✓
12	Green Walls	NA	NA	✓
13	Aquaponic system	✓	NA	✓
14	Greenhouse with hydroponic columns	✓	NA	✓
15	Redevelopment of the building	✓	✓	✓
16	Providing education for plant production, cooking & managerial skills	✓	✓	✓
	Total achieved	**13 /16**	**9/16**	**9/16**
		81%	**56%**	**56%**

the final results 3D visualizations are shown in Figs. 14, 15, 16, 17, 18, and 19. Figure 14 illustrates the whole site being developed as a bird's eye view. Figure 15 shows the view from the main entrance. In Fig. 16, the integration of greens in the project is shown where the main elevation includes green walls. Moreover, Fig. 17 shows the kitchen where people can learn cooking. Figure 18 presents the main section through the main building integrating different technologies in the indoors as the one space luminous, the aquaponic and the rainwater harvesting systems. Figure 19 shows the main elevation of the main building where different technologies are found as the information technology system, the green walls and the solar panels for saving energy.

10 Conclusions

Green Symphony project achieved its set of objectives by integrating the urban farm with its surrounding urban context. Also, this project covers all three pillars of sustainability: environmental, social and economic aspects to achieve a self-sustained site creating new jobs. In addition, it provides clean and safe vegetables and a productive hub for the city. Blending music, architecture and agriculture creates connectivity and social innovation through the variety of social spaces. Moreover, the project helps in creating liveable, healthy and sustainable urban environments in the commune of Lanuvio, in the metropolitan city of Rome, Italy. The project re-development contributes to attaining SDGs, particularly SDG 3, 7, 9, 11, 12, 13, and 15.

Fig. 14 Developed site after incorporating all urban farming technologies. (*Image credit: Authors, 2020*)

Fig. 15 Site entrance with various features. (*Image credit: Authors, 2020*)

Fig. 16 Shed and green wall in the restaurant. (*Image credit: Authors, 2020*)

Fig. 17 Interior of the kitchen in the building. (*Image credit: Authors, 2020*)

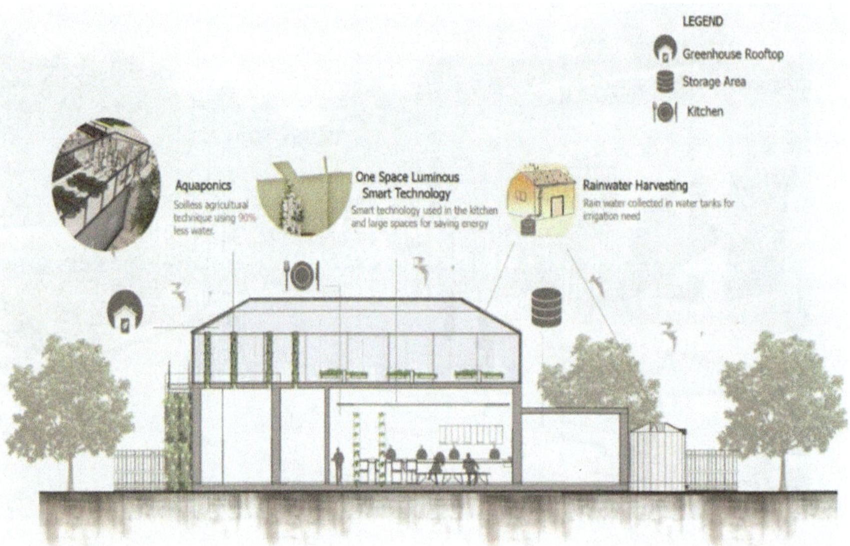

Fig. 18 Section through the main building. (*Image credit: Authors*)

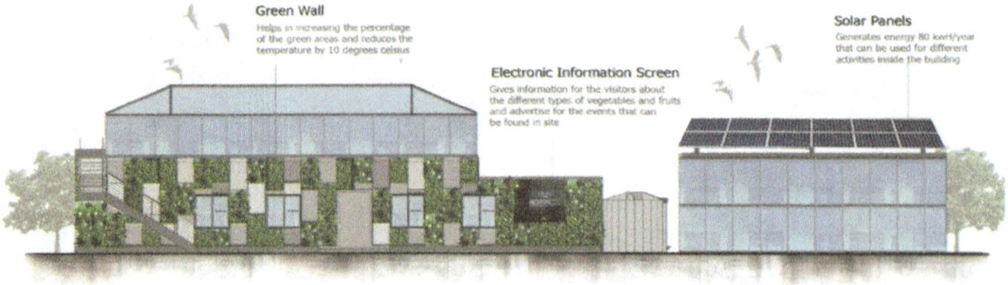

Fig. 19 The main elevation of the green walls, information boards and solar panels. (*Image credit: Authors*)

Appendix A

Table 8 Life cycle analysis and used materials' carbon footprint yearly by farms (without construction)

Kitchen		Database (ADEME)		
food purchase		**$kgCO_2/kg$**	**kg/yr**	**$kgCO_2/yr$**
directly from farmer	chicken	2.24	19.05	42.67
	eggs	1.48	76.20	112.78
	pork	2.42	19.05	46.10
	milk (cow)	0.94	26.21	24.64
from shop	fruits (average)	0.50	381.00	190.50
(contains packaging and distribution)	flour (wheat)	1.17	36.00	42.12
	olive oil	2.60	21.94	57.03
	white sugar	0.68	12.00	8.16
	vinegar	4. 17	6.30	26.27
	salt	0.54	1.00	0.54
	spices	2.19	1.00	2.19
	mustard	3.27	5.40	17.66
	butter	9.49	24.00	227.76
	cream	4. 17	20.56	85.75
	cheese (average)	4.30	127.00	546.10
total food purchase				1,430.27
		$kgCO_2/t$	t/yr	$kgCO_2/yr$
waste management	cardboard/paper	14.00	0.01	0.08
	metals	33.00	0.02	0.79
	glass	33.00	0.07	2.28
	plastic (average)	677.00	0.03	18.69
	domestic compost	0.01	0.50	0.00
total waste of kitchen				21.84
Permaculture	Database (ADEME)		Amount/yr	$kgCO_2/yr$
use of compost (direct emissions of nitrogen)	5.04	$kgCO_2$ /kgN spread	300 kgN/yr	**1,512.00**
draining/soil gestion (N leakage)	3,445.00	$kgCO_2/ha$	540 m2	**186.03**
conversion of soil (grassland to culture)	950.00	$kgCO_2/ha.yr$	540 m2/20yrs	**1,026.00**
pesticides (avoidance of impact)	-1,600.00	$kgCO_2$/k €	4,32€/yr	**-6.91**
fertilizers (avoidance of impact)	-4,795.00	$kgCO_2$/t nitrogen	300 kgN/yr	**-1,438.50**
vegetable consumption (avoidance of impact)	-0.50	$kgCO_2/kg$ (average)	1706 kg/yr	**-853.00**
total permaculture				**425.62**
Aquaponics	Database (ADEME)		Allocation	$kgCO_2/yr$
use of pellets	1,500.00	$kgCO_2$/k €	0.74	444.00
consumption of fish (avoidance of impact)	-3.08	kgCO2/kg (for trouts)	0.74	-854.70
consumption of vegetables (avoidance of impact)	-0.5	$kgCO_2/kg$ (average)		-2,500.00
pesticides (avoidance of impact)	-1,600.00	$kgCO_2$/k €		-0.51
fertilizers (avoidance of impact)	-4,795.00	$kgCO_2$/t nitrogen		-77,234.90
vegetable waste (at client's home)	48.1	$kgCO_2/t$		24.05
total aquaponics				-80,146.11
Green wall	Database (ADEME)		Allocation	$kgCO_2/yr$
use of pellets	1,500.00	$kgCO_2$/k€	0.26	156.00
consumption of fish (avoidance of impact)	-3.08	$kgCO_2/kg$ (for trouts)	0.26	-300.30
total green wall				-144.30
	Database (ADEME)		Amount/yr	$kgCO_2/yr$
maintenance of all machines and equipment	390.00	$kgCO_2$/k €	1500 €/yr	585.00
Festivals and workshops	Database (ADEME)		Amount/yr	$kgCO_2/yr$
new electronic devices	400.00	$kgCO_2$/k €	1000 €/yr	400.00
packaging recycled plastic (purchase)	202.00	$kgCO_2/t$	8 kg/yr	1.62
packaging recycled paper (purchase)	0.92	$kgCO_2/kg$	19 kg/yr	17.46
packaging recycled plastic (waste)	677.00	$kgCO_2/t$	8 kg/yr	5.42
packaging recycled paper (waste)	0.00	$kgCO_2/t$	19 kg/yr	0.00
total events				**424.49**

Table 9 Number of hosted kitchen lesson sessions per year

Type of training session	No. of sessions	No. of persons
One week	15/yr	10
Three days	10/yr	10
One day	6/yr	15

Table 10 Benefits for vegetables derived from aromatic herbs

Aromatic Herbs	Benefits
Chamomile	Improves flavor and growth.
Thyme	Rejects *Pieris rapae*.
Rosemary	Rejects flies. Rejects *Pieris rapae.*
Salvia	Rejects aphids, snails and caterpillars for its strong aroma.
Mint	Improves flavor and growth.
Basil	Rejects mosquitos and flies, improves flavor and growth.

Table 11 National funding for initial investments

Costs that Benefits from the National Fund for Cooperatives (NFC)	
Aquaponics	
Automatic Software	€ 1,500
Greenhouse Climate Control	€ 42,000
Hydroponics Columns	€ 25,000
Tray system	€ 30,000
Irrigation	
Water Tank	€ 3,400
Irrigation System	€ 2,500
Energy Supply	
Photovoltaic Panels	€ 15,000
Pavegon® Tiles	€ 20,000
Furnishings	€ 10,000
Building Works	
Excavation work for the water tank	€ 1,300
Excavation work for the fish pool	€ 31,427
Green walls	€ 10,500
Building Works - Workshop Building	
Total	€ 150,000
TOTAL	€ 342,628
%	40%
Part covered by funds	€ 137,051

Table 12 Initial investments of the project

Initial Investments			
Aquaponics		**Energy Supply**	
Automatic Software	€ 1,500	Photovoltaic Panels	€ 15,000
Pump	€ 300	Pavegon®Tiles	€ 20,000
Greenhouse Climate Control	€ 42,000	Administration	€ 25,000
Hydroponics Columns	€ 25,000	Furnishings	€ 10,000
Tray system	€ 30,000	Total	€ 70,000
Total	€ 98,800	**Music and workshop requirements**	
Permaculture		Speakers	€ 200
Membership with "Civiltà Contadina"	€ 20	Display	€ 1,000
Labor	€ 800	TOT	€ 1,200
Soil and Manure	€ 300	**Compost bin**	
Total	€ 1,120	Compost box	€ 400
Insect Box		**Building Works**	
Hive	€ 1,100	Excavation work for the water tank	€ 1,300
Bees	€ 660	Excavation work for the fish pool	€ 31,427
Other Expenses	€ 350	Landscaping tiling	€ 10,000
Total	€ 2,110	Greenhouse	€ 7.80
Irrigation		Total	€ 42,736
Water Tank	€ 3,400	**Building Works - Workshop Building**	
Irrigation System	€ 2,500	Surface	€ 62
Total	€ 5,900	Cost per m²	€ 2,419
Green wall		Total	€ 150,000
Initial Investment	€ 10,500	**TOTAL INITIAL COSTS**	**€ 383,766**

Table 13 Operating expenses

Operating costs per year	
Labor	
Wage (net, per employees)	€ 700
Month – total	€ 2,625
Total per year	€ 27,300
Energy Supply	
Needed electricity kWh	€ 12,000
kWh supplied by panels	€ 8,000
kWh supplied by electricity	€ 8,000
cost of electricity per kWh	€ 0.069
total cost electricity	€ 552
Seeds memberships	
Membership with Civiltà Contadina	€ 20
Water supply	
Kitchen + additional margin: assume 80 m^3 needed	
Cost per m^3	€ 1.02
Cube metres needed	80
Total cost of water supply	**81.6**
Total yearly operating cost	**27,953 €**

Table 14a,b Return of renting hydroponics columns Revenue from sale of fishes

	Monthly production (kg/m^2)	Average monthly prices on the market (€/kg)	Average monthly revenue (€)/m^2		
Lettuce	7.92	€ 0.75	€ 5.94	Monthly value of production/ m^2	**€ 20.09**
Chard	6.25	€ 0.80	€ 5.00	Total cultivable area (m^2)	**€ 65**
Strawberries	4.9	€ 1.10	€ 5.39	Total monthly value of production	**€ 1,306**
Basil	7.52	€ 0.50	€ 3.76	Total annual value of production	**€ 15,670**
Monthly value of production per m^2			€ 20.09	**Revenues**	**€ 19,588**

Table 15 Return from yoga classes

Yoga classes	Cost
Monthly revenue per client	€ 60
Monthly revenue	€ 1,200
Yearly revenue	€ 14,400
Yearly Revenue from Yoga	**€ 2,160**

Table 16 Return from kitchen classes

Kitchen sessions' type	Ticket per person	Number of persons/ class	Number of lessons/ yr	Total revenue
1 Day	€ 65	15	6	€ 5,850
3 Days	€ 165	10	10	€ 16,500
1 Week	€ 350	10	15	€ 52,500
Yearly Revenue from Kitchen Lessons				**€ 74,850**

References

1. McKinsey Global Institute (2011) Urban world: mapping the economic power of cities. [Online]. Available at: https://www.mckinsey.com/~/media/McKinsey/Featured%20Insights/Urbanization/Urban%20world/MGI_urban_world_mapping_economic_power_of_cities_full_report.pdf. Accessed 16 Aug 2020
2. Avtar R, Tripathi S, Aggarwal AK, Kumar P (2019) Population–urbanization–energy nexus: a review. Resources 8:136
3. Sims R, Schaeffer F et al (2014) Transport. In: Climate change 2014: mitigation of climate change. Contribution of Working Group III to the Fifth assessment report of the intergovernmental panel on climate change. Cambridge University Press, Cambridge/New York. Retrieved from: https://www.sciencedirect.com/science/article/pii/S2352146517308098./. Accessed 16 Aug 2020
4. Elasha B (2010) Mapping of climate change threats and human development impacts in the Arab region (Rep.). UNDP. Retrieved from: http://www.arabclimateinitiative.org/knowledge/background/Balgis_mapping%20CC%20threats%20and%20human%20dev%20impacts%20in%20Arab%20region.pdf Accessed 16 Aug 2020
5. Seneviratne S et al (2012) Changes in climate extremes and their impacts on the natural physical environment: an overview of the IPCC SREX report. Accessed 16 Aug 2020
6. United Nations (2018) Global Status Report 2018. Retrieved from: https://www.unenvironment.org/resources/report/global-status-report-2018. Accessed 16 Aug 2020
7. Sivaramanan S (2015) Global warming and climate change, causes, impacts and mitigation. https://doi.org/10.13140/RG.2.1.4889.7128 [MA1]. Accessed 16 Aug 2020
8. World Health Organization (2018) Climate change and health country profile: Italy. Retrieved from: https://apps.who.int/iris/handle/10665/260380. Accessed 16 Aug 2020
9. Hallett S, Hoagland L, Toner E (2016) Urban Agriculture: environmental, economic, and social perspectives. https://doi.org/10.1002/9781119281269.ch2
10. Binns T, Nel E (n.d.) The significance of urban agriculture in food security and sustainable livelihoods in response to economic restructuring in Zambia's copperbelt province (Rep.). Global education programme international development research fund
11. Kennard N, Bamford R (2020) Urban agriculture: opportunities and challenges for sustainable development. https://doi.org/10.1007/978-3-319-69626-3_102-1. Accessed 14 Aug 2020
12. Olivier D (2015) The physical and social benefits of urban agriculture projects run by non-governmental organisations in Cape Town (Unpublished master's thesis). Stellenbosch University. Retrieved from: https://core.ac.uk/download/pdf/37438493.pdf. Accessed 14 Aug 2020
13. Orsini F, Kahane R, Remi NW, Gianquinto G (2013) Urban agriculture in the developing world: a review. Agron Sustain Dev 33:695–720. https://doi.org/10.1007/s13593-013-0143-z. Accessed 14 Aug 2020
14. Potutan GE, Schnitzler WH, Arnado JM, Janubas LG, Holmer RJ (2000) Urban agriculture in Cagayan de Oro: a favourable response of city government and NGOs. In: Bakker N, Dubbeling M, Guendel S, Sabel Koschella U, de Zeeuw H (eds) Growing cities, growing food, urban agriculture on the policy agenda. DSE, Feldafing, pp 413–428
15. Cruz MC, Medina RS (2003) Agriculture in the city: a key to sustainability in Havana, Cuba. Ian Randle, Kingston. http://www.idrc.ca/openebooks/104-3/. Accessed 9Nov 2012
16. Purnomohadi N (2000) Jakarta: urban agriculture as an alternative strategy to face the economic crisis. In: Bakker N, Dubbeling M, Guendel S, Sabel KU, de Zeeuw H (eds) Growing cities, growing food, urban agriculture on the policy agenda. DSE, Feldafing, pp 453–466
17. Farming in urban areas can boost food security (2005) Retrieved from: https://www.fao.org/newsroom/en/news/2005/102877/index.html. Accessed 27 Apr 2020
18. Smit J, Nasr J, Ratta A (2001) Chapter 5: producing food and fuel in urban areas. In Urban agriculture: food, jobs, and sustainable cities. Published with permission from the United Nations Development Programme
19. Jena AK, Biswas P, Saha H (2017) Advanced farming systems in aquaculture: strategies to enhance the production. Innov Farming 2455–6521(2):84–89
20. Employing aeroponics as growing method (n.d.) Retrieved from: https://seattleurbanfoodsource.weebly.com/growth-research.html. Accessed 14 Aug 2020
21. Green Wall (n.d.) Retrieved from: https://verdeprofilo.com/en/vertical-gardening/green-wall. Accessed 14 Aug 2020
22. Beacham A, Vickers L, Monaghan J (2019) Vertical farming: a summary of approaches to growing skywards. J Hortic Sci Biotech 94:1–7. https://doi.org/10.1080/14620316.2019.1574214. Accessed 14 Aug 2020
23. Lanuvio (n.d.) Retrieved from: https://www.summerinitaly.com/guide/lanuvio. Accessed 14 Aug 2020
24. Deutsche Welle (DW) (2019) [Image]. Retrieved from https://www.dw.com/en/promoting-urban-farming-in-ugandas-schools/av-48108246. Accessed 17 Aug 2020
25. WOHA Architects. Raising the roof: cultivating Singapore's urban farming scene [Image]. Retrieved from https://cnaluxury.channelnewsasia.com/experiences/urban-farming-singapore-11925958. Accessed 17 Aug 2020
26. Kadapa-Bose S (2019) Soil-less & vertical, this style of farming will reduce water usage By 95%! [Image]. Retrieved from https://www.thebetterindia.com/199411/how-to-vertical-farming-hydroponics-fresh-produce-organic-india/. Accessed 17 Aug 2020
27. Barnett S (2013) Urban agriculture in South Africa [Image]. Retrieved from https://ozcf.co.za/2013/02/07/urban-agriculture-in-south-africa-food-in-the-city/on-the-roof-768x1024/. Accessed 17 Aug 2020
28. You Matter (2020) What is the aquaponics system? Definition, benefits, weaknesses [Image]. Retrieved from https://youmatter.world/en/definition/aquaponics-sustainable-benefits-system/. Accessed 17 Aug 2020
29. Huifa. PVC pipe flat hydroponic growing system [Image]. Retrieved from https://www.hfgreenhouse.com/products/hydroponic-system/. Accessed 17 Aug 2020
30. Outdoor Design Source (ODS) (2016) Vicinity green wall systems 14 June 2016, 3991 views [Image]. Retrieved from https://www.outdoordesign.com.au/news-info/vicinity-greenwall-systems/4492.htm. Accessed 17 Aug 2020
31. Markillie P. Vertical farms on the rise [Image]. Retrieved from https://worldin.economist.com/article/17522/edition2020vertical-farmsrise. Accessed 17 Aug 2020

32. UrbanFarm (2020) International challenge [Online]. Available at: https://site.unibo.it/urbanfarm/en/the-challenge/urbanfarm-2020. Accessed 25 Apr 2020
33. Phoenix mythological bird. Retrieved from: https://www.britannica.com/topic/phoenix-mythological-bir. Accessed 14 Aug 2020
34. PVWatts Calculator. Retrieved from: https://pvwatts.nrel.gov/. Accessed 14 Aug 2020
35. ARPA. Radar meteo: stima della pioggia I Servizio Idro-Meteo-Clima I ArpaER retrieved from: https://www.arpae.it/sim/?osservazioni_e_dati/radar. Accessed 17 Aug 2020
36. Ayaz M, Uddin A, Sharif Z, Mansour A, Aggoune El-Hadi (2019) Internet-of-Things (IoT)-based smart agriculture: toward making the fields talk. IEEE Access:1–1. https://doi.org/10.1109/ACCESS.2019.2932609
37. Orti G, Associazione A, Manifesto C. Civiltà Contadina Associazione per la salvaguardia della biodiversità. Retrieved from: https://www.civiltacontadina.it/. Accessed 14 Aug 2020
38. ADEME (2015) ADEME – Bilans GES site retrieved from: https://www.bilans-ges.ademe.fr/en/basecarbone/donnees-consulter/choix-categorie. Accessed 17 Aug 2020
39. Impatto Zero srl, Acquaponica, la redditività che non avete mai visto, Agricoltura 2.0, 2018 retrieved from: https://agricoltura2punto0.it/IT/ct/Mg/Azienda. Accessed 14 Aug 2020
40. Hunter C (2020) Cinnamon herbal properties: traditional herbal medicinal use [Image]. Retrieved from: https://www.thepracticalherbalist.com/advanced-herbalism/cinnamon-herbal-properties-traditional-herbal-medicinal-use. Accessed 17 Aug 2020
41. Burpee (2020) Tomato, sun gold hybrid [Image]. Retrieved from: https://www.burpee.com/vegetables/tomatoes/tomato-sun-gold%2D%2Dhybrid-prod001012.html. Accessed 17 Aug 2020
42. Zaden E (2020) [Image]. Retrieved from: https://www.enzazaden.com/uk/products-and-services/our-products/Herbs/Staro. Accessed 17 Aug 2020
43. Eden A (2020) Calypso Oleander [Image]. Retrieved from https://www.almostedenplants.com/shopping/products/348-calypso-oleander/. Accessed 17 Aug 2020
44. Plants with a purpose (2019) Dill, Bouquet Dill [Image]. Retrieved from: https://plantswithapurpose.net/product/dill-bouquet-dill/. 17 Aug 2020
45. Mizuna Gardens. [Image]. Retrieved from http://renee-oberdorf-34hn.squarespace.com/other-greens/. 17 Aug 2020
46. High Country Gardens. Standing ovation little bluestem grass [Image]. Retrieved from: https://www.highcountrygardens.com/perennial-plants/ornamental-grass/schizachyrium-standing-ovation-little-bluestem. 17 Aug 2020
47. MaskaRad (2018) Radish planting tips: how to plant radishes in the garden [Image]. Retrieved from: https://www.gardeningknowhow.com/edible/vegetables/radish/tips-how-plant-radish.htm. 17 Aug 2020
48. Seedway. Crunchy royale (Treated Seed) [Image]. Retrieved from: https://www.seedway.com/product/9457-crunchy-royale-treated-seed/. 17 Aug 2020
49. Caribbeangardenseed. Ethiopian kale seeds, Ethiopian mustard, Abyssinian mustard, African kale, highland kale [Image]. Retrieved from https://www.caribbeangardenseed.com/products/organic-lacinato-kale-seeds-also-know-as-dinosaur-kale-tuscan-kale-and-even-flat-black-cabbage-heirloom-cold-hardy-vegetable. 17 Aug 2020

Mediterranean University as Catalyst for Eco-Sustainable Renovation: The Experience of Med-EcoSuRe Cross-Border Living Lab

Antonella Trombadore and Gisella Calcagno

1 Introduction

Aligning to the global effort to reduce the climate change effects, the urgency to flatten the carbon curve is a pressing challenge for the construction sector: buildings are worldwide energivorous, responsible for nearly one third of the global energy consumption and relative CO_2 emissions.

In the Mediterranean area, where a large part of the built environment is consolidated, historicized and integrated in the cultural landscape, the challenge of energy efficiency lies in the renovation of existing buildings. Here, many gaps still exist between the experimentation of eco-sustainable models and prototypes and their actual and wide application in the real world, starting by the insufficiency of plans and strategies for cost-effective renovations in the Mediterranean climate and continuing with the inefficiency of tools to support an efficient decision making and management on the existing building.

To address them, this chapter presents the first results of the ongoing Med-EcoSuRe project, which is aimed to support eco-sustainable, cost-effective, high-performance and innovative energy renovations for public buildings in the Mediterranean area, targeting high educational buildings.

Following an exploration of the main issues and criticality related to the energy renovation of public buildings in the Mediterranean context, this chapter focuses on the Med-EcoSuRe project, and in particular on two of its innovative approaches: the Living Lab, with the cross-border implementation, to provide a shared platform for collaboration between stakeholders, and more experimentally the Digital Twin, to digitally drive a more informed decision making towards energy efficiency. In the conclusion, this chapter explains how outcomes will be exploited for the further development of the project and how they can contribute to an overall innovation in the field.

2 The Challenge of Public Buildings Renovation in the Mediterranean Area

At a global glance, buildings account for the largest share of both global final energy use (36%) and energy-related CO_2 emissions (39%) [1]. The decarbonization of the sector is both a critical and a strategic challenge to achieve the global commitments on climate change (Paris Agreement, 2015) and the ambitious UN 2030 Sustainable Development Goals (SDGs), which clearly draft the direction of transforming our world towards 'clean energy research and technology, including renewable energy, energy efficiency and advanced and cleaner fossil-fuel technology, and promote investment in energy infrastructure and clean energy technology' (SDG n. 7 on 'clean energy for all').

Looking at the Mediterranean basin, where energy use and emissions rate are in line with global trends, if countries in the southern shore are singularly implementing requirements for the energy performances of buildings, in the north the European Union is a unifier.

Considering the building sector very influent in the run towards Clean Energy for all Europeans [2], based on energy efficiency and renewable energy (with respective directives), the EU Energy Performance of Buildings Directive (2010/31/EU-EPBD) contains the concept of Nearly Zero Energy Building (NZEB), intended as a building with very high-energy

A. Trombadore (✉) · G. Calcagno
Department of Architecture (DIDA), University of Florence, Florence, Italy
e-mail: antonella.trombadore@unifi.it; gisella.calcagno@unifi.it

© The Author(s), under exclusive license to Springer Nature Switzerland AG 2022

A. Sayigh (ed.), *Sustainable Energy Development and Innovation*, Innovative Renewable Energy,
https://doi.org/10.1007/978-3-030-76221-6_34

performance and 'the nearly zero or very low amount of energy required (...) covered to a very significant extent by energy from renewable sources, including energy from renewable sources produced on-site or nearby' (art. 2). The Directive, adopted by Member States, poses a challenging timing for new constructions, requiring NZEB for public buildings from 2018 and for all new buildings from 2020, with a more generic reference on the need 'to stimulate the transformation of buildings that are refurbished into nearly zero-energy buildings' (art. 9). Since in the EU, 97% of the existing building stock is inefficient [3], that three quarters of buildings were built when energy efficiency requirements were limited or non-existent and that renovation rates are low (between 0.4% and 1.2%) [4], the latest legal provision on buildings energy performance is finally a turn on renovations. The amended EPBD (2018/844/EU) states the establishment of 'a long-term renovation strategy to support the renovation of the national stock of residential and non-residential buildings, both public and private, into a highly energy efficient and decarbonised building stock by 2050, facilitating the cost-effective transformation of existing buildings into nearly zero-energy buildings' (art. 1).

The emerging need to consider building renovation lies in the positive impact they can bring to society: not only energy and environmental benefits, such as reduced energy, poverty and greenhouse gas emissions, but also health improvements (better air quality and indoor comfort), jobs creation and financial savings.

In the overall process of buildings renovation, the public sector should play an exemplar and leading role by testing and showcasing the state of the art of renovation practices (processes and technologies) to stimulate the private sector, and the wider public, in addressing the clean energy objectives. Within the realm of public buildings, educational buildings (from schools to universities) represent a singular case, due to their congenital vocation towards education and awareness raising, which can also address energy and environmental issues [5]. According to Tascı, 'educational buildings can be seen as 3D textbooks written by architects: architecture and its products as a concrete material for sustainability education' [6], concept that can be extended to all the public buildings.

3 The Vision of the Med-EcoSuRe Project

The Med-EcoSuRe project (Mediterranean University as Catalyst for Eco-Sustainable Renovation) is financed under the ENICBC programme, which fosters the neighbourhood cooperation between the countries of the two shores of the Mediterranean Basin. The project addresses the challenges related to environmental protection and climate change adaptation and mitigation, supporting the energy rehabilitation of high educational buildings in the Med area.

The two basic pillars of the project are (1) universities can play a strategic role in the implementation of innovative strategies to stimulate efficient energy renovations of public buildings in the Med area as *hubs* to experiment technological innovation and energy-mix solutions and (2) university buildings can be renovated as *pilots/living labs* with an active involvement of researchers and students.

Networking a wide range of stakeholders (researchers, building and energy managers and technicians, companies, public organizations and end users), the project activated nine pilot universities/living labs engaged in the real-life energy renovation of the university building, and operating in the development of the specific knowledge and know-how on the design, construction and operation of high-performance energy renovations, as well as in the promotion and dissemination of EE best practices and behaviours in the Med area.

4 The Multilevel Living Lab Approach

In the last decades, Living Labs (LLs) are emerging worldwide as physical and virtual spaces in which people are trying to collectively solve a wide range of societal challenges; many of them pursue sustainable goals, a lot are placed in urban areas. Defined as 'user-centred open innovation ecosystems based on a systematic user co-creation approach, integrating research and innovation processes in real-life communities and settings' [7], LLs are characterized by common elements: multi-method approaches, user engagement, multi-stakeholder participation, real-life setting and co-creation [8].

For these features, the LL approach is congenial with the Med-EcoSuRe's objective to innovate the energy renovation process of the university building by introducing an active collaboration for the decision making. Adopting the LL principles, universities can act as catalysts for eco-sustainable renovations by implementing innovative methodologies and tools for the rehabilitation of public buildings, to support public authorities and future decision makers. Beyond the local level for the physical and real-life setting of the LL, the possibility to exploit the ICTs (Information and Communication Technologies) opens LLs to a larger scale for the cross-border implementation of several LLs, as experimented for energy efficiency pilots [9].

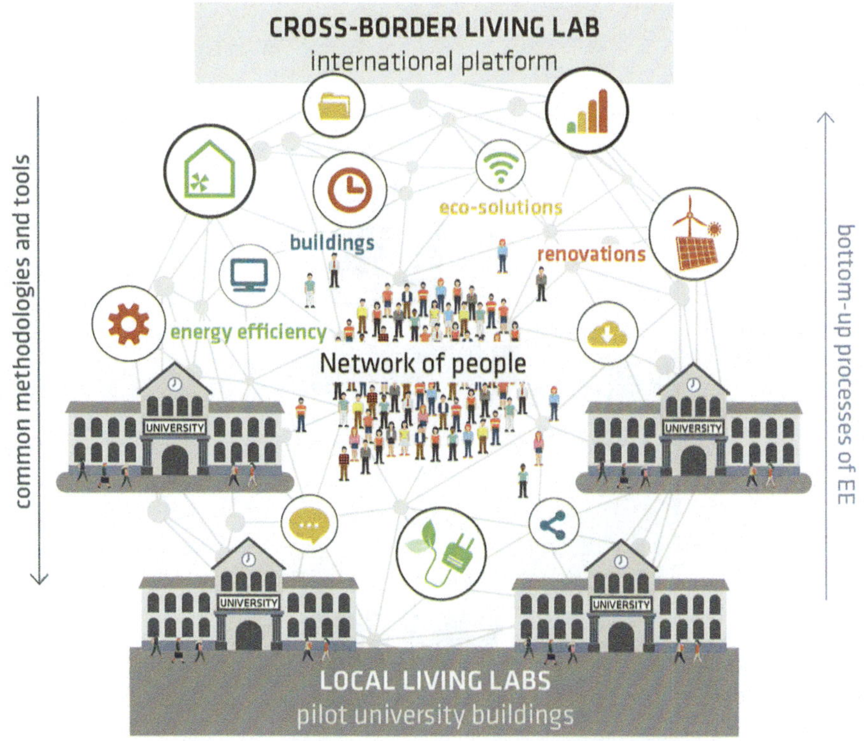

Fig. 1 The Med-EcoSuRe Living Lab structure

The project set-up a Cross-border LL, an international network of excellence, directed by partners and managed through a platform, that brings together the several local LLs/pilot university buildings in a common and integrated space, harmonized by common methodologies and tools, where to collaboratively develop innovation towards eco-sustainable energy renovation solutions for higher education institutions, within the specific climatic and cultural conditions of the Med area (Fig. 1).

At the strategic level, the Cross-border LL is managed in order (1) to connect partners with common visions, goals and approaches, (2) to plan the network and the development process, (3) to support collaborative testing on innovation and finally (4) to sustain in the assessments and enhancement of the achieved benefits and impacts [10]. The full implementation of Cross-border LLs requires three stages of LL networking: firstly, the exchange of best practices and lessons learned (learn from each other); secondly, the setting of tools, methods or infrastructure to exchange comparable information (harmonization and integration) and finally the performing of joint research activities [11].

At this stage of the project, the Cross-border LL is working on the collection of international and Mediterranean best practices on strategies, technologies and approaches which produced eco-sustainable solutions for high educational buildings (e.g. integration of renewables, smart envelopes, ventilation systems, involvement of end users, etc.) (Fig. 2).

The main foreseen output of the Cross-border LL is a toolkit, an innovative, interactive, bottom up and participative programme of training and education for technicians and students to implement a common knowledge and share cost-effective solutions for EE in university buildings in Med area. Exploiting the knowledge of the actors and experts community engaged in the network, the project collects and implements best practices, innovative scenarios and EE renovation measures, developing in parallel an international platform (ICT) combining physical and digital tools (Fig. 3). Due to the Covid-19 pandemic, the first seminars and workshops were webinars, but had the possibility to be enriched with the involvement of other ENI CBC projects addressing similar issues (BEEP, GreenBuilding, ESMES and BERLIN).

At the local and operational levels, local LLs adopt a quasi-experimental approach, including a pre-measurement, an intervention and a post-measurement, where the intervention is the real-life energy renovation of the existing university building.

Actually, local LLs are working on the planning and design of the respective pilot university building renovation, according to established, but in progress, action plans covering all the aspects of the renovation process (i.e. the selection and setting of the pilot university building/LL, the interaction with the stakeholders, energy audits, monitoring, end-user feedbacks, co-design, etc.). Sharing the common goal to reduce the buildings negative impact, university-based LLs contribute to the

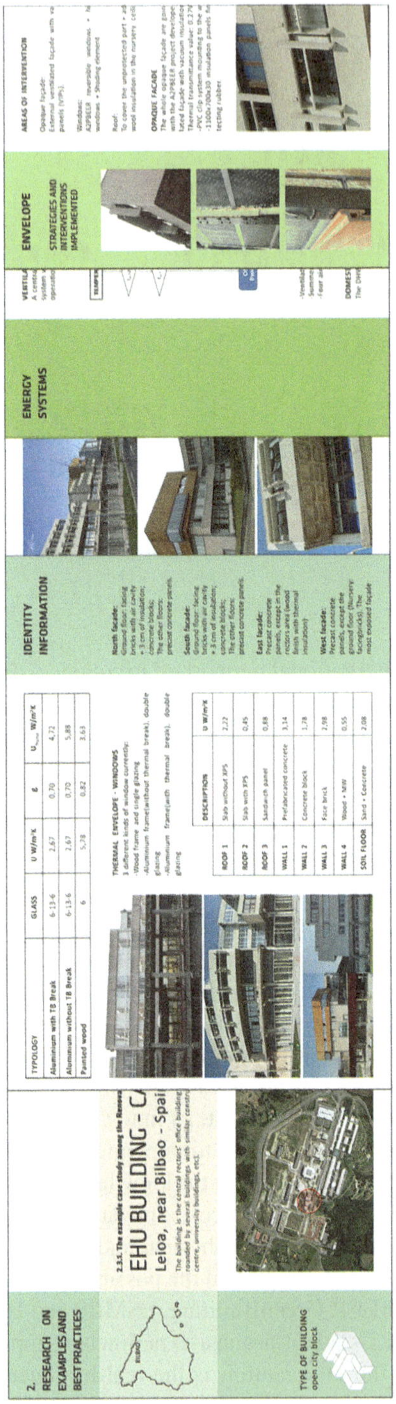

Fig. 2 Example of cataloguing EE renovation best practices

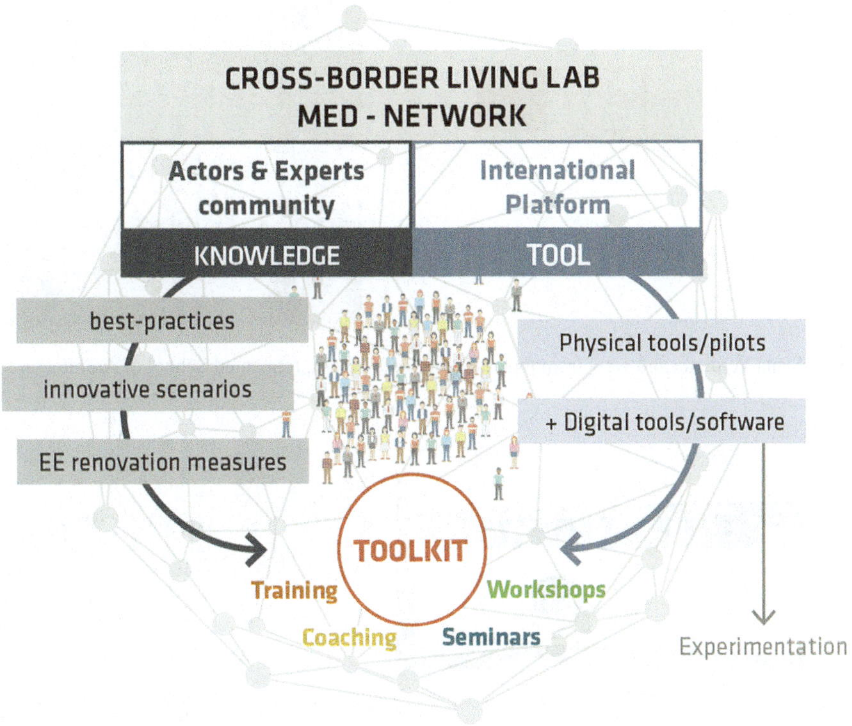

Fig. 3 Functioning of the Cross-border LL

diffusion of knowledge and know-how on EE building technologies and practices and, involving end users, educate and raise awareness about better energy behaviours and choices.

5 Experimenting the Digital Twin: The Advanced Living Lab

The Med-EcoSuRe Italian pilot university building/living lab is the School of Architecture hosted in the historical complex of Santa Verdiana (Fig. 4), placed in the quartiere of Santa Croce, in the UNESCO city centre of Florence.

As an experimentation within the project, the local team is setting up an advanced Living Lab in the pilot university building under renovation, where to combine virtual and physical data by exploiting the promising technology of the Industry 4.0: the Digital Twin (DT).

Born in the field of product lifecycle management, DT is based on the concept that physical objects (i.e. buildings) can be synchronized to virtual models with a bi-directional data connection in order to optimize the functioning of the physical twin (i.e. the building). Embracing the globally recognized digitalization opportunities for the construction and building sectors, that is, BIM, sensors and IoT [12], also supported by the EU policies [13], the interaction of the physical building with semantically enriched 3D models consents a real-time monitoring and data acquisition, processing and re-introduction of data in the buildings' lifecycle, as basis for decision making. The possibility to manage fully informed DT of buildings (real-time data) represents an evidence-based support for planning, design, construction, facility management, that is, for predictive maintenance, resource efficiency improvement, enhancement of tenants' comfort, what-if analysis for optimization of the building design [14]. Advancing the more definite BIM (Building Information Modelling), the DTs are exploiting the state of the art of digital technologies, such as sensors, Internet of Things (IoT) and, more experimentally, artificial intelligence and its advancements (big data analysis, semantic web, machine learning, deep learning) [15].

DTs have the potential to actively address the challenges related to the energy renovation of existing public buildings. The possibility to capture and elaborate a wide range of real-time EE data (e.g. consumptions or human comfort, through IoT), and to visualize them in virtual models, opens towards multiple uses: to simulate scenarios, to stimulate advanced participatory process, to support facility management and maintenance, to enhance the capability to plan and implement sustainable renovations, by analysing and evaluating technologies, performances and costs, and also to educate towards more responsible energy behaviours.

Fig. 4 First floor plan of Santa Verdiana and view from the adjacent Santa Teresa, renovated School of Architecture

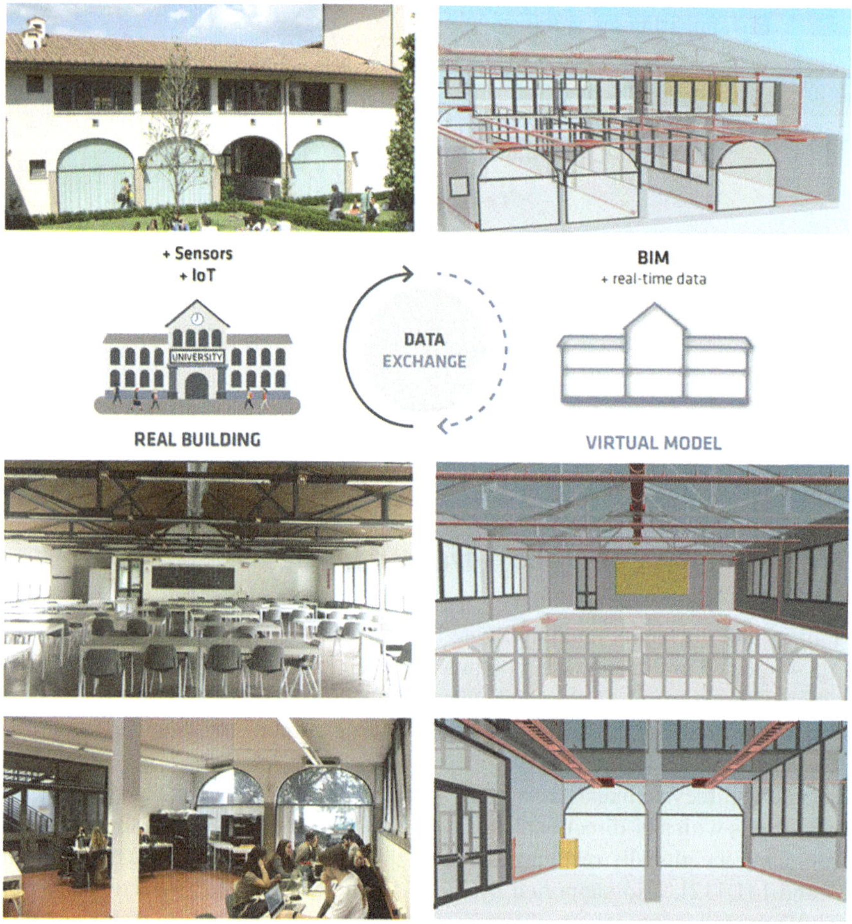

Fig. 5 Development of the Digital Twin of the pilot building

The LL is exploring the potentialities of DT for the decision making along the design, construction and operational phases of the university building renovation, with the scope to clearly evaluate the energy and environmental performances and optimize their efficiency through innovative solutions (technological and processual).

Actually, the data on the physical and technical assets of the pilot building were retrieved, collected and modelled in the 3D BIM model (IFC) (Fig. 2), to be implemented in the other BIM dimensions (time, costs, performances, facility management and sustainability) and connected to real-time data fed by sensors, in phase of installation, for IoT applications. It has to be noted that the history of the pilot building enriches the theme with considerations on heritage restoration, which can also be positively approached exploiting the information of digital models, such as in HBIM [16].

Combining with the Living Lab principles, the team is introducing a collaborative approach between the stakeholders involved in the building renovation (researchers, university building and energy managers and technicians, companies and end users) for the pre-measuring of the intervention (e.g. architectural data, audit, monitoring, sensors/IoT, end-users feedbacks and others). The idea is to integrate the data in BIM/DT environments as basis for the co-planning and design of the building renovation (works are foreseen in the next 2 years), and then for the management of the building, in particular for the energy aspects. The next phase, supported by the DT, is to evaluate different scenarios of integration of innovative technologies for the building energy renovation (e.g. the integration of renewable energy systems), also with companies, as evidence-based support for the decision making (Fig. 5).

6 Conclusion

The global need to reach major energy efficiency is compelling for the building sector, also in Mediterranean area where not only the complexity, beauty, but also fragility of the built environment matches with on-going socio-economic challenges.

In the overall process to renovate the existing building stock, universities can act as catalysts for innovative practices, technologies and behaviours related to energy efficiency, due to their nature of research and development (competences and network of stakeholders, e.g. companies and public administrations), and also education (not only students, but also public managers and technicians). The innovation on buildings renovation collectively developed in university-based *hubs/living labs* can be transferred firstly to the public sector, with methodologies and tools for a better EE management of the public buildings stock (e.g. schools), supporting the smart city's objectives [17], and then to the wider public, inspiring the private sector with innovative technologies and processes, and finally all citizens towards more sustainable behaviours (i.e. better use of energy).

The human-centred approach adopted by the Med-EcoSuRe project, manifested in the collaborative approach between stakeholders or in the optimization of the end-users' comfort and pro-active behaviour, assures the compliance with the Sustainable Development Goals to 'ensure access to affordable, sustainable and modern energy for all' (n. 7 – UN, 2015).

Moreover, the experimentation with the most innovative digital technologies in the building sector (BIM, sensors, IoT and Digital Twins) appears very promising to face the challenge to renovate, beyond buildings, the process of energy renovation, potentially augmentable with evidence and human-based digital data, for a more strategic and collaborative decision making along the design, the construction and the operational stages of the building. In particular, the architectural integration of renewable energy systems (e.g. PV in building envelope or site) can benefit from these approaches (Living Lab and Digital Twin) due to high-level amount of sharable data that can be used to advance scenarios or to connect with the smart city (e.g. smart grids).

Acknowledgements The article has been written by the Italian partner of the Med-EcoSuRe project: Department of Architecture (DIDA) – University of Florence, which is in charge to manage the Cross-border Living Lab and the Italian local pilot.

The project is financed under the ENICBC MED programme.

A special thanks to the other members of the partnership for their involvement in the Cross-border Living Lab: Mediterranean Renewable Energy Centre (Tunisia); University of Tunis El Manar (Tunisia); Thermothecnics Group at Thermal Energy Engineering Department – University of Seville (Spain); Energy Research Centre – An-Najah National University (Palestine); Naples Agency for Energy and Environment (Italy); Spanish Association for the Internationalization and Innovation of Solar Companies (Spain).

The implementation of the local LL/pilot in Florence was possible thanks to the co-joint work within the SPACEBeXLab – DIDALABS: Prof. Saverio Mecca, Prof. Antonella Trombadore, Prof. Carla Balocco, Prof. Maurizio De Lucia, Prof. Roberto Bologna, Prof. Elisabetta Cianfanelli, Prof. Debora Giorgi, Gisella Calcagno, PhD, Giacomo Pierucci, PhD, Arch. Juan Camilo Olano Salinas (BIM Specialist), Arch. Marco Paolini, Lorena Di Ianni and Cristina Micheloni.

References

1. IEA - International Energy Agency (2019) Global status report for buildings and construction. Towards a zero-emissions, efficient and resilient buildings and construction sector. Online. https://www.iea.org/reports/global-status-report-for-buildings-and-construction-2019. (08/20)
2. European Commission (2019) Clean energy for all Europeans. Online. https://op.europa.eu/en/publication-detail/-/publication/b4e46873-7528-11e9-9f05-01aa75ed71a1. (08/20)
3. BPIE - Buildings Performance Institute Europe (2019) Building renovation in the Clean Energy package: implication at local, national and EU levels. Online. http://bpie.eu/wp-content/uploads/2020/01/Build-Upon2_Summary-of-implications-of-the-Clean-Energy-Package-on-National-Renovation-Strategies_FINAL_2020.pdf. (08/20)

4. Energiewende A (2019) European energy transition 2030: the big picture. Ten priorities for the next European Commission to meet the EU's 2030 targets and accelerate towards 2050. Online. https://www.agora-energiewende.de/fileadmin2/Projekte/2019/EU_Big_Picture/153_EU-Big-Pic_WEB.pdf. (08/20)
5. Trombadore A, Alcamo G, Sala M (2015) TEENERGY experience: how to reduce energy demand in Mediterranean schools. In Scartezzini, JL (ed) CISBAT 2015 International Conference Future Building &District – sustainability from nano to urban scale. Lausanne Switzerland 9–11 September 2015, 1, p 131–136
6. Tascı B (2015) "Sustainability" education by sustainable school design. Procedia Soc Behav Sci 186:868–873
7. Hossain M, Leminen S, Westerlund M (2019) A systematic review of living lab literature. J Clean Prod 213:976–988
8. Malmberg K, Vaittinen I (2017) Living lab methodology. Handbook. Online. https://u4iot.eu/pdf/U4IoT_LivingLabMethodology_Handbook.pdf. (08/20)
9. Nina M, Launonen P, Oliveira ÁD (2014) APOLLON - using living lab methodologies in the cross-border context for energy efficiency pilots. In Helfert M, Krempels KH, Donellan B (eds) 3rd international conference on smart grids and green IT systems (Smartgreens 2014). Barcelona, Spain, 3–4 April, 2014, p 336–342
10. Schaffers H, Turkama P (2012) Living labs for cross-border systemic innovation. Technology innovation management review, 2, p 25–30. Online. https://timreview.ca/article/605. (08/20)
11. Lievens B, Schaffers H, Turkama P, Ståhlbröst A, Ballon P (2011) Cross border living labs networks to support SMEs accessing new markets. In Cunningham P, Cunningham M (eds) eChallenges e-2011 conference proceedings. Online. https://www.academia.edu/28066081/Cross_border_living_labs_network_to_support_SMEs_accesing_new_markets. (08/20)
12. IET and ATKINS (2019) Digital Twins for the built environment. An introduction to the opportunities, benefits, challenges and risks. Online. https://www.theiet.org/media/4719/digital-twins-for-the-built-environment.pdf. (08/20)
13. European Commission (2019) Supporting digitalisation of the construction sector and SMEs. Including building information modelling. Online. https://ec.europa.eu/growth/content/report-supporting-digitalisation-construction-sector-and-smes_en. (08/20)
14. Khajavi S, Motlagh N, Jaribion A, Werner L, Holmstrom J (2019) Digital twin: vision, benefits, boundaries, and creation for buildings. IEEE Access 7:147406–147419
15. Boje C, Guerriero A, Kubicki S, Rezgui Y (2020) Towards a semantic construction digital twin: directions for future research. Autom Constr 114:1–16
16. Gigliarelli E, Calcerano F, Cessari L (2017) Heritage Bim, numerical simulation and decision support systems: an integrated approach for historical buildings retrofit. Climamed 2017 – Mediterranean conference of HVAC; historical buildings retrofit in the Mediterranean area, 12–13 May 2017, Matera, Italy. Energy Procedia 133:135–144
17. Trombadore A (2015) Mediterranean smart cities. Innovazione tecnologica ed ecoefficienza nella gestione dei processi di trasformazione urbana. Altralinea Edizioni s.r.l., Firenze

Application of Renewable Energies in a University Building with Complex Topology

Eusébio Z. E. Conceição

1 Introduction

In this study, a Building Dynamic Software, developed by the authors in the last decade, is applied. This software was developed to be used mainly in complex topology buildings, with high occupation level and with new Heating, Ventilating, and Air-Conditioning (HVAC) system. The main objective is the development of more comfortable occupied spaces, with HVAC systems more efficient and with buildings with low energy consumption levels. The software is used to analyze the buildings, vehicles, and other systems' thermal response. Conceição and Lúcio [19] and Conceição and Lúcio [16] analyzed the solar radiation influence in the buildings, while Conceição et al. [3], Conceição et al. [5], Conceição and Lúcio [15], and Conceição et al. [22] analyzed the thermal comfort and air quality. In both examples, the application and development of more comfortable occupied spaces is used.

The thermal comfort level, developed by Ole Fanger [27] and presented in ASHRAE-55 [1] and ISO 7730 [25], using Predicted Mean Vote (PMV) and the Predicted Percentage of Dissatisfied (PPD) indexes, is used in spaces equipped with HVAC system. These indexes consider the environmental variables as indoor air temperature, indoor air velocity, indoor air relative humidity, mean radiant temperature, clothing level, and activity level.

In Conceição and Lúcio [15], Conceição and Lúcio [19], and Conceição and Lúcio [16], using the Building Dynamic Software, the integral comfort level to evaluate the thermal comfort level in occupied was used. However, in Conceição and Lúcio [17], Conceição [7], and Conceição et al. [11], the Human Thermal Response is used to evaluate the thermal comfort level in detail. In Conceição et al. [23], as example, the Computational Fluids Dynamics are used. Finally, in Conceição et al. [21] and Conceição and Lúcio [20], in a coupling of three software, the Human Thermal Comfort, Computational Fluids Dynamics, and Building Dynamic Software are also used.

The adaptive thermal comfort level, developed in spaces not equipped with HVAC system, was developed, as example, in Yang et al. [30], Yao et al. [31], and Conceição et al. [10]. Ole Fanger and Toftum [28], also for spaces not equipped with HVAC system, developed an extension of the PMV model to be used in warm climates. Other models, developed to be used in spaces equipped with HVAC system, are presented in Conceição et al. [9], using a preferred air temperature, and in Conceição et al. [13], using a control based in PMV index.

To evaluate the indoor air quality, the carbon dioxide concentration is used (see ASHRAE-62 [2] and Ministério das Obras Públicas [26]. The airflow rate or the air exchange rate, used in the Building Dynamic Software, can be seen in Conceição et al. [24], Conceição et al. [8], and Conceição et al. [12].

In this work, a numerical model, that simulates the Building Dynamic Software and evaluates the indoor environment comfort, in transient conditions, is used. The greenhouse application, in winter conditions, and the underground spaces, in summer conditions, are applied. The first one to warm and the second one to cold internal spaces, using renewable energies, are applied.

E. Z. E. Conceição (✉)
FCT-Universidade do Algarve, Faro, Portugal
e-mail: econcei@ualg.pt

© The Author(s), under exclusive license to Springer Nature Switzerland AG 2022

A. Sayigh (ed.), *Sustainable Energy Development and Innovation*, Innovative Renewable Energy,
https://doi.org/10.1007/978-3-030-76221-6_35

2 Numerical Model

The numerical model used in this work considers the Buildings Dynamic Software which simulates buildings with complex topology and works in transient conditions. More details of this numerical model can be analyzed in Conceição [6], Conceicao et al. [5], Conceição and Lúcio [18], and others.

The model is based on the energy balance integral equations and mass balance integral equations:

- The energy balance integral equations are developed for the air, the different transparent bodies, the interior bodies, and the opaque bodies of the building. These equations consider the accumulated sensible heat and the heat flux due to the conduction, convection, radiation, evaporation, and others.
- The mass balance integral equations are developed for the water vapor and the air contaminants, considering the accumulated mass and the mass flux due to the convection, diffusion, and others.

In the resolution of the equations system, the Runge–Kutta–Felberg method with error control is used.

The model was validated for school buildings under winter conditions in Conceição et al. [4] and under summer conditions in Conceição and Lúcio [14].

3 Numerical Methodology

The Building Dynamic Software was applied in a University building, mainly consisting of classrooms, laboratories, offices, and other spaces. The building is divided into 319 spaces, distributed by four floors, and is composed by 329 transparent surfaces (windows) and 3585 opaque surfaces (internal and external walls, doors, and others).

The perspective of the University building considered in this study can be seen in Fig. 1. In this figure, the external and internal details are presented. In this figure, in black are represented the interior and exterior walls, in blue are represented the windows, in green are represented the doors, and in orange and red are represented different roofs.

The occupied cycle considers people with an activity level of 1.2 met. The clothing level considered was 1 clo for a typical winter day and 0.5 clo for a typical summer day.

In the ventilation cycle is considered an airflow rate changing between compartments calculated in accordance with the Portuguese standard RECS [29].

In winter conditions, internal warm greenhouses spaces and building passive strategies are used, while in summer conditions, internal cold underground spaces and building passive strategies are considered. In both, simulations are used mixing ventilation inside each compartment. In this numerical simulation, winter and summer conditions are used. In the first one, the warm internal greenhouses are used, while in the second one the cold underground spaces are used.

3.1 Warm Internal Greenhouses

In this study, two greenhouses in each floor (3 floors) are considered. However, in this work, only results of the second floor are considered. Two spaces with a large window area turned to South are considered, namely:

- One to the building South-East side. This space is used in the morning and the greenhouse, number 102, to heat twelve spaces with windows turned to West. In this study, only the results of the spaces 115, 116, and 123 are considered.
- Another to the building South-West side. This space is used in the afternoon and the greenhouse, number 64, to heat ten spaces with windows turned to East. In this study, the spaces 65, 70, 73, and 79 are considered.

3.2 Cold Underground Spaces

In this study, only one underground space, located under the ground floor, is considered. However, in this work, only results of the second floor are presented. Two air flow topologies are considered, namely:

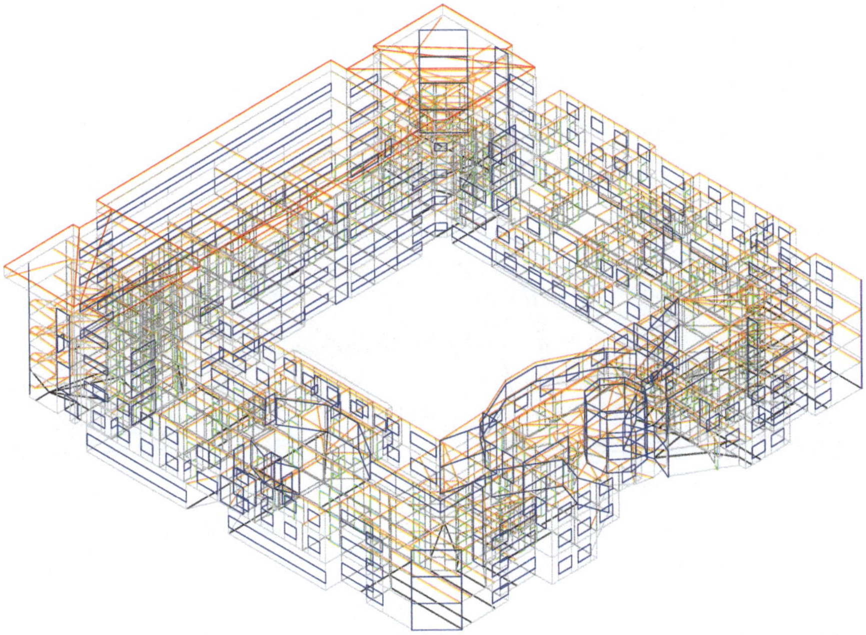

Fig. 1 Perspective of the University building considered in this study

- In the morning, for the second floor, and the underground space, number 1, is used to cold seven spaces with windows turned to East. In this study, the spaces 65, 70, 73, 103, 104, 105, and 113 are considered.
- In the afternoon, for the second floor, and the underground space, number 1, is used to cold eight spaces with windows turned to West. In this study, the spaces 53, 55, 56, 58, 59, 61, 62, and 63 are considered.

4 Results and Discussion

In winter and summer thermal simulations, the evolution of internal air temperature and occupants' thermal comfort level are presented. In the thermal comfort level, the Predicted Mean Vote is used.

4.1 Winter Conditions

Figure 2 shows the evolution of air temperature, in the second floor level, in the West side building spaces, in winter conditions, when is used an internal greenhouse located in South-East side building spaces. The evolution of air temperature in the second floor level, in the East side building spaces, in winter conditions, when is used an internal greenhouse located in South-West side building spaces is presented in Fig. 3.

In these figures, and in the following figures, the W is associated with the situation without greenhouse space, while the G is associated to the situation with greenhouse.

In accordance with the obtained results, it is possible to conclude the following:

- When the internal greenhouse is considered, the air temperature in each occupied space increases 2–3 °C, while the non-occupied internal greenhouse space decreases 4–5 °C.
- When the non-occupied internal greenhouse space is located in South-East side and the occupied building spaces are located in the West side, the increase in the air temperature is verified mainly in the morning.
- When the non-occupied internal greenhouse space is located in South-West side and the occupied building spaces are located in the East side the increase in the air temperature is verified mainly in the afternoon.

Figure 4 presents the evolution of Predicted Mean Vote, in the second floor level, in the West side building spaces, in winter conditions, when is used an internal greenhouse located in South-East side building spaces. The evolution of Predicted

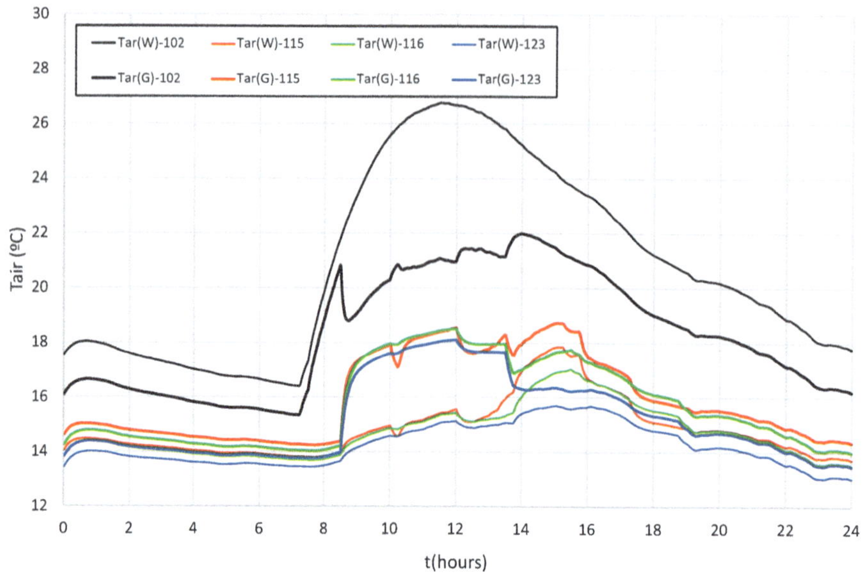

Fig. 2 Evolution of air temperature (Tair) in the second floor level, in the West side building spaces, in winter conditions, when is used an internal greenhouse located in South-East side building spaces

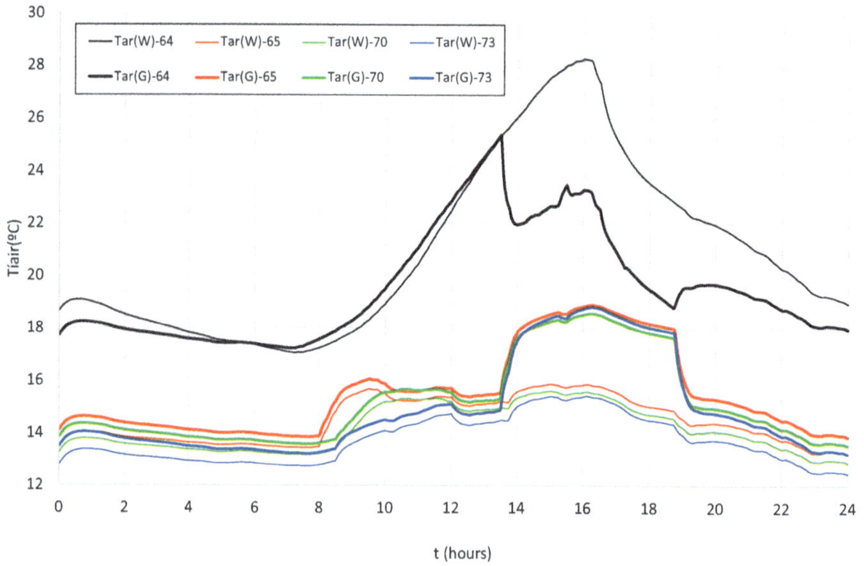

Fig. 3 Evolution of air temperature (Tair) in the second floor level, in the East side building spaces, in winter conditions, when is used an internal greenhouse located in South-West side building spaces

Mean Vote in the second floor level, in the East side building spaces, in winter conditions, when is used an internal greenhouse located in South-West side building spaces is shown in Fig. 5.

When the internal greenhouse is not considered, the thermal comfort level in each occupied space is not acceptable. However,

- when the non-occupied internal greenhouse space is located in South-East side and the occupied building spaces are located in the West side, the thermal comfort level is near the acceptable level in the afternoon;
- when the non-occupied internal greenhouse space is located in South-West side and the occupied building spaces are located in the East side, the thermal comfort level is near the acceptable level in the morning and in the afternoon.

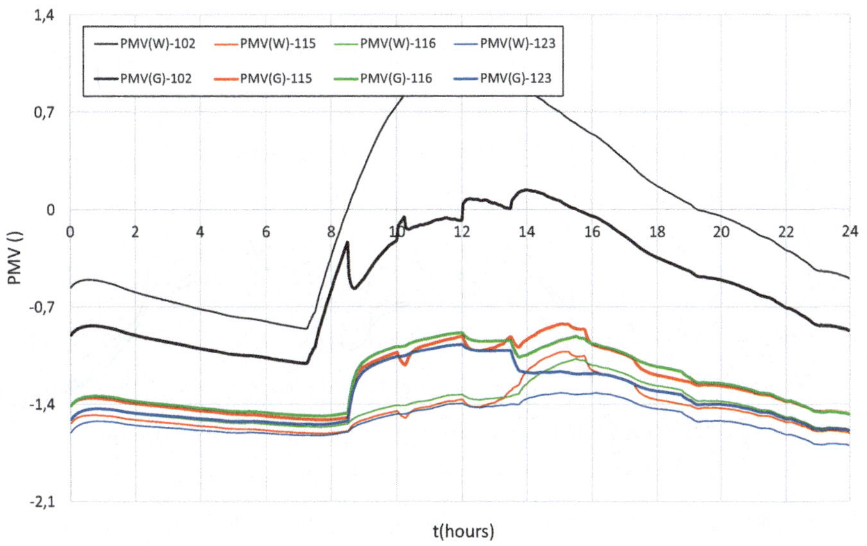

Fig. 4 Evolution of Predicted Mean Vote (PMV) in the second floor level, in the West side building spaces, in winter conditions, when is used an internal greenhouse located in South-East side building spaces

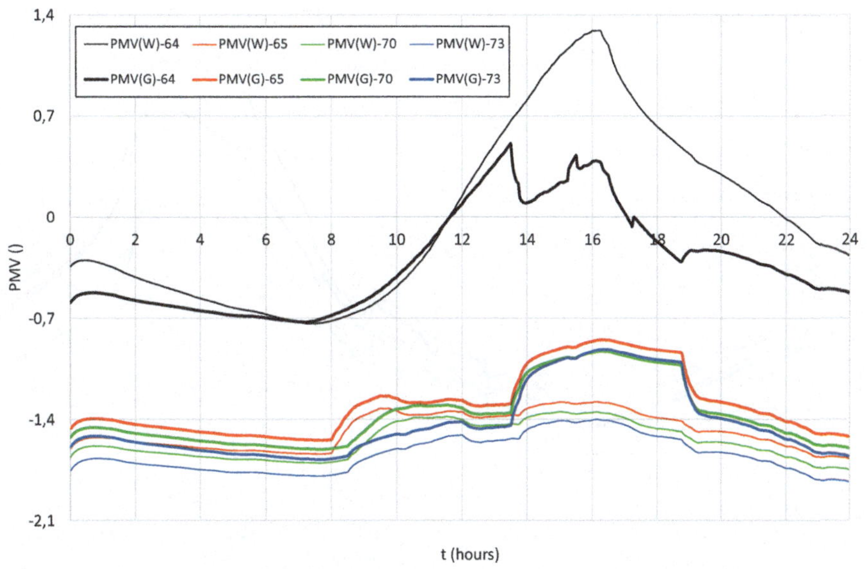

Fig. 5 Evolution of Predicted Mean Vote (PMV) in the second floor level, in the East side building spaces, in winter conditions, when is used an internal greenhouse located in South-West side building spaces

4.2 Summer Conditions

Figure 6 shows the evolution of air temperature in the second floor level, in the East side building spaces, in summer conditions, when is used an internal cold underground space. The evolution of air temperature in the second floor level, in the West side building spaces, in summer conditions, when is used an internal cold underground space is presented in Fig. 7.

In these figures, and in the following figures, the W is associated to the situation without internal cold underground space, while the U is associated to the situation with internal cold underground space.

In accordance with the obtained results, it is possible to conclude the following:

- The internal air temperature is higher in spaces with windows turned East than in spaces turned to West.

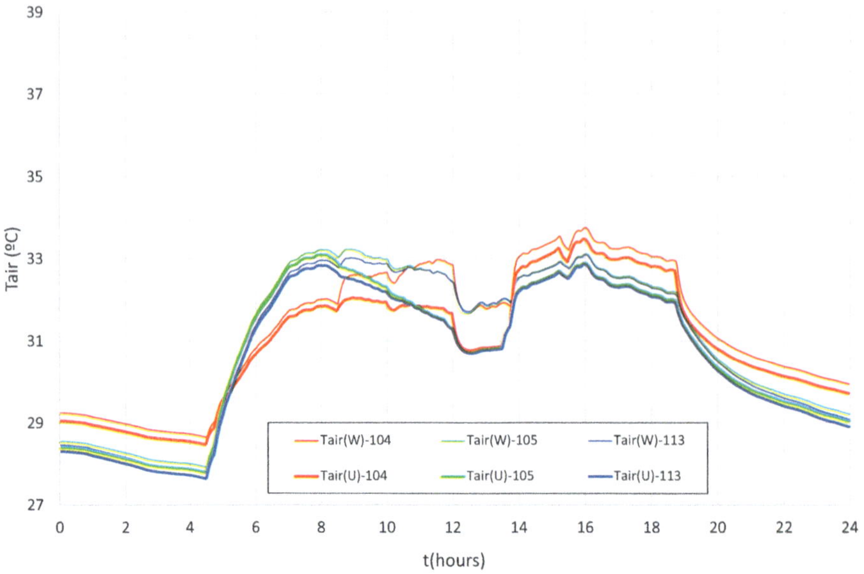

Fig. 6 Evolution of air temperature (T air) in the second floor level, in the East side building spaces, in summer conditions, when is used an internal cold underground space

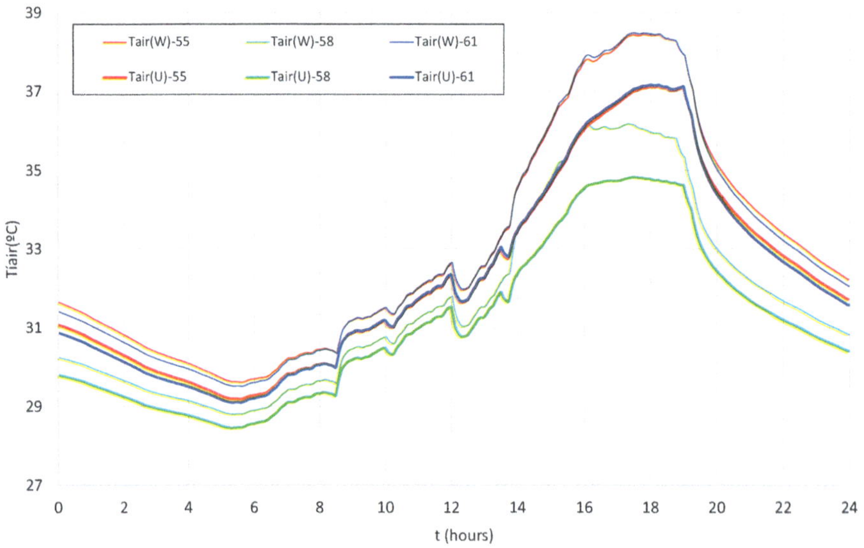

Fig. 7 Evolution of air temperature (T air) in the second floor level, in the West side building spaces, in summer conditions, when is used an internal cold underground space

- The air temperature in spaces with windows turned to East only decreases the internal air temperature in the morning, when the influence of internal cold underground space is verified. In the afternoon (without the influence of internal cold underground space) the internal air temperature increases slightly due to the high levels of external air temperature.
- The air temperature in spaces with windows turned to West only decreases the internal air temperature in the afternoon, when the influence of internal cold underground space is verified.
- When the internal cold underground space is considered, in general, the air temperature in each occupied space reduces more than 1 °C.

Figure 8 shows the evolution of Predicted Mean Vote in the second floor, in the East side building spaces, in summer conditions, when is used an internal cold underground space. The evolution of Predicted Mean Vote in the second floor, in the West side building spaces, in summer conditions, when is used an internal cold underground space is presented in Fig. 9.

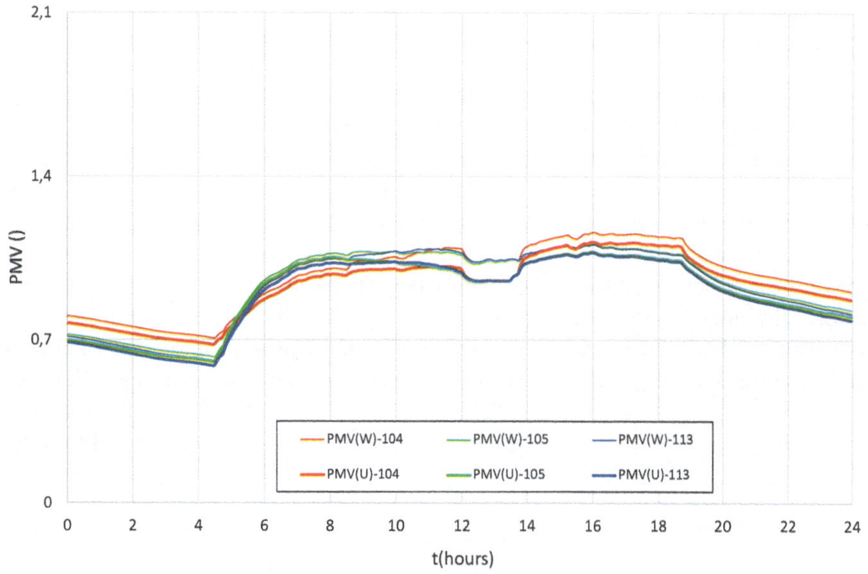

Fig. 8 Evolution of Predicted Mean Vote (PMV) in the second floor, in the East side building spaces, in summer conditions, when is used an internal cold underground space

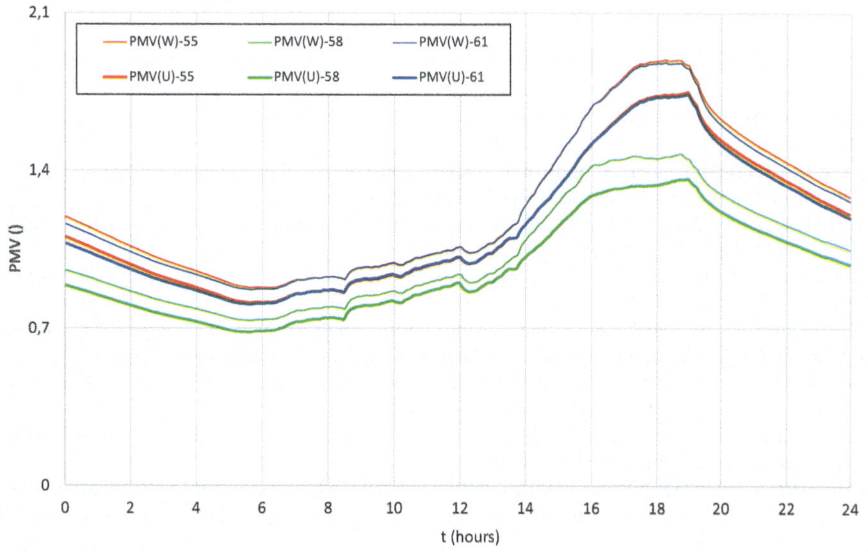

Fig. 9 Evolution of Predicted Mean Vote (PMV) in the second floor, in the West side building spaces, in summer conditions, when is used an internal cold underground space

When the internal cold underground space is not considered, the thermal comfort level in each occupied space is not acceptable. However,

- when the occupied building spaces are located in the East side, the thermal comfort level is near the acceptable level in the morning and in the afternoon;
- when the occupied building spaces are located in the West side, the thermal comfort level is more near the acceptable level in the morning than in the afternoon.

5 Conclusions

In this numerical study, an application of renewable energies in a University building with complex topology is presented. In winter conditions, internal warm greenhouse spaces and building passive strategies are used, while in summer conditions, cold underground spaces and building passive strategies are also used.

The internal warm greenhouse spaces, in winter conditions, in the present work, increase the air temperature inside the occupied spaces and improve the thermal comfort levels that the occupants are subjected. In the West side building spaces, the thermal comfort indexes are near the acceptable values during all day, and in the East side building spaces, the thermal comfort indexes are near the acceptable values during mainly the afternoon.

In summer conditions, the internal cold underground space decreases the air temperature in the occupied spaces. The application of the extension of the PMV model, to be used in warm climates, promotes thermal comfort indexes near the acceptable values, mainly in spaces with windows turned East, during all day, and in spaces with windows turned West, during the morning.

In future works, the increase of airflow rate from the internal warm greenhouse spaces and from the internal cold underground spaces to the occupied spaces will be made.

Acknowledgments The author would like to acknowledge to the project (SAICT-ALG/39586/2018) from Algarve Regional Operational Program (CRESC Algarve 2020), under the PORTUGAL 2020 Partnership Agreement, through the European Regional Development Fund (ERDF) and the National Science and Technology Foundation (FCT).

References

1. ASHRAE-55 (2017) Thermal environmental conditions for human occupancy, ANSI/ASHRAE Standard - 55
2. ASHRAE (2016) Ventilation for acceptable indoor air quality, J Phys A Math Theor. https://doi.org/10.1088/1751-8113/44/8/085201
3. Conceição E, Gomes J, Awbi H (2019) Influence of the airflow in a solar passive building on the indoor air quality and thermal comfort levels. Atmos 10(12):766. https://doi.org/10.3390/ATMOS10120766
4. Conceição E, Silva A, Lúcio M (2004) Numerical study of thermal response of school buildings in winter conditions. In Proceedings of the 9th Conference on Air Distribution in Rooms (Roomvent 2004), Coimbra, Portugal, p 5–8
5. Conceicao EZE, Silva MCG, André JCS, Viegas DX (2000) Thermal behaviour simulation of the passenger compartment of vehicles. Int J Vehicle Des. Inderscience Publishers 24(4):372–387
6. Conceição EZE (1996) Estudo Aero-Térmico do Comportamento Térmico de Veículos de Transporte de passageiros. Ph. D., DEM-FCT University of Coimbra
7. Conceição EZE (2000) Evaluation of thermal comfort and local discomfort conditions using the numerical modelling of the human and clothing thermal system. In RoomVent'2000-7th International Conference on Air Distribution in Rooms, p 131–136
8. Conceição EZE, Lúcio MMJR, Vicente VDSR, Rosão VCT (2008) Evaluation of local thermal discomfort in a classroom equipped with cross flow ventilation. Int J Vent. https://doi.org/10.1080/14733315.2008.11683817
9. Conceição EZE, Lúcio MMJR, Ruano AEB, Crispim EM (2009) Development of a temperature control model used in HVAC systems in school spaces in Mediterranean climate. Build Environ 44(5). https://doi.org/10.1016/j.buildenv.2008.06.015
10. Conceição EZE, Nunes ARL, Gomes JMM, Lúcio MMJR (2010) Application of a school building thermal response numerical model in the evolution of the adaptive thermal comfort level in the Mediterranean environment. Int J Vent 9(3):287–304. https://doi.org/10.1080/14733315.2010.11683887
11. Conceição EZE, Rosa SP, Custódio ALV, Andrade RL, Meira MJPA, Lúcio MMJR (2010) Study of airflow around occupants seated in desks equipped with upper and lower air terminal devices for slightly warm environments. HVAC and R Research 16(4):401–412. https://doi.org/10.1080/10789669.2010.10390912
12. Conceição EZE, Farinho JP, Lúcio MMJR (2012) Evaluation of indoor air quality in classrooms equipped with cross-flow ventilation. Int J Vent 11(1):53–68
13. Conceição EZE, Gomes JMM, Ruano AE (2018) Application of HVAC systems with control based on PMV index in university buildings with complex topology. IFAC-PapersOnLine 51(10):20–25. https://doi.org/10.1016/j.ifacol.2018.06.230
14. Conceição EZE, Lúcio MMJR (2006) 'Numerical study of thermal response of school buildings in summer conditions. In HB 2006 - healthy buildings: creating a healthy indoor environment for people, proceedings
15. Conceição EZE, Lúcio MMJR (2008) Thermal study of school buildings in winter conditions. Build Environ 43(5):782–792. https://doi.org/10.1016/j.buildenv.2007.01.029
16. Conceição EZE, Lúcio MMJR (2009) Numerical study of the thermal efficiency of a school building with complex topology for different orientations. Indoor Built Environ 18(1):41–51. https://doi.org/10.1177/1420326X08099550
17. Conceição EZE, Lúcio MMJR (2010) Evaluation of thermal comfort conditions in a localized radiant system placed in front and behind two students seated nearby warmed curtains. Build Environ 45(10):2100–2110. https://doi.org/10.1016/j.buildenv.2010.03.006
18. Conceição EZE, Lúcio MMJR (2010) Numerical simulation of passive and active solar strategies in buildings with complex topology. Build Simul 3(3):245–261. https://doi.org/10.1007/s12273-010-0010-7

19. Conceição EZE, Lúcio MMJR (2010) Numerical study of the influence of opaque external trees with pyramidal shape on the thermal behaviour of a school building in summer conditions. Indoor and Built Environment 19(6):657–667. https://doi.org/10.1177/1420326X10377546
20. Conceição EZE, Lúcio MMJR (2016) Numerical simulation of the application of solar radiant systems, internal airflow and occupants' presence in the improvement of comfort in winter conditions. Buildings 6(3):38. https://doi.org/10.3390/buildings6030038
21. Conceição EZE, Lúcio MMJR, Awbi HB (2013) Comfort and airflow evaluation in spaces equipped with mixing ventilation and cold radiant floor. Build Simul 6(1):51–67. https://doi.org/10.1007/s12273-012-0093-4
22. Conceição EZE, Lúcio MMJR, Lopes MC (2008) Application of an indoor greenhouse in the energy and thermal comfort performance in a kindergarten school building in the South of Portugal in winter conditions. WSEAS Trans Environ Dev 4(8):644–654
23. Conceição EZE, Vicente VDSR, Lúcio MMJR (2008) Airflow inside school building office compartments with moderate environments. HVAC R Res 14(2):195–207. https://doi.org/10.1080/10789669.2008.10391003
24. Conceicao EZ, Silva MC, Viegas DX (1997) Air quality inside the passenger compartment of a bus. J Expo Anal Environ Epidemiol 7(4):521–534
25. ISO (2005) ISO 7730: ergonomics of the thermal environment analytical determination and interpretation of thermal comfort using calculation of the PMV and PPD indices and local thermal comfort criteria. Management. https://doi.org/10.1016/j.soildyn.2004.11.005
26. Ministério das Obras Públicas (2006) O Regulamento dos Sistemas Energéticos de Climatização em Edifícios (RSECE)-Decreto-Lei n.o 79/2006, Diário da República
27. Ole Fanger P (1970) Thermal comfort. Analysis and applications in environmental engineering. Danish Technical Press, Copenhagen
28. Ole Fanger P, Toftum J (2002) Extension of the PMV model to non-air-conditioned buildings in warm climates. In Energy and buildings. https://doi.org/10.1016/S0378-7788(02)00003-8
29. RECS (2013) Regulamento de Desempenho Energético dos Edifícios de Comércio e Serviços (RECS) - Requisitos de Ventilação e Qualidade do Ar Interior - Portatia nº 353-A/2013 de 4 de Dezembro, Diário da República
30. Yang L, Yan H, Lam JC (2014) Thermal comfort and building energy consumption implications - a review. Appl Energy. https://doi.org/10.1016/j.apenergy.2013.10.062
31. Yao R, Li B, Liu J (2009) A theoretical adaptive model of thermal comfort - adaptive predicted mean vote (aPMV). Build Environ. https://doi.org/10.1016/j.buildenv.2009.02.014

Impacts of Ecological and Psychological Human Behaviors: Assessment of Selected Buildings in New Cairo, Egypt to Attain Livability and Mitigate Climate Change

Mai Barakat and Mohsen Aboulnaga

1 Introduction

Mitigation and adaptation of climate change (CC) is considered the center of the world's attention in the last decade and still, where more than half of CO_2 emissions are caused by human activities [1]. Nonetheless, it has been decreased due to the COVID-19 pandemic in the past 4 months of lockdown worldwide. Humans are the first affected by CC severe events, such as heat waves, floods, and disasters, in addition to coronavirus (COVID-19) pandemic; hence, causing many health and physiological problems, for example, migration. The United States Global Change Research Program (USGCRP), also known as the Climate Change Science Program (CCSP) 2001–2008, indicated that human activities are the main reason for CC and causing significant risks for public health and welfare [2]. High temperature and increased pollution due to urban heat island effect (UHIE), and increased stress, anxiety, and aggressive behaviors threaten the ecosystems [3–6].

This issue is summarized through the latest social isolation and psychological problems that have occurred to the majority of vulnerable users in most of the upgraded areas worldwide, where 17 million people and 42 million people were forced to leave their homes due to natural disasters in 2009 and 2010 respectively [7]. To address such discord, architects ought to understand the physiological behavior of buildings' users to anticipate the least damage impacts to the surrounding environments. Since city's inhabitants are key players in mitigating CC, understanding their needs, behaviors for adapting the design of their desire and surroundings (urban areas and buildings) is vital. In fact, without innovative planning for context and long-term change, cities will be not resilient. Therefore, a new approach for promoting well-being and achieve psychological comfort for users should be introduced as a main element in urban areas/buildings' development to ensure sustainability and livability.

A complex system that cannot differentiate between physical, social, and psychological dimensions concerning human, social development, and welfare is what should be reached for attaining a sustainable society. Beyond the level of satisfaction, the issue of environmental pollution, urbanization, and consequently the high rate of GHG emissions, CC impacts, and rising diseases, notably the deadly coronavirus (COVID-19) are of prime importance [8]. Different shapes, colors, textures, or even furniture define the comfort of the space, leading to balanced living conditions and public health enhancement, and that's why sustainable design is the key element for a better operation under such environmental conditions. Environmental psychology needs a high awareness of sustainable development and its goals (SDGs) to protect the present and future generations [9]. There is a conjunction between architecture, health, and technology when assessing the rate of emissions, hence livability (quality of life), but the greatest challenge is the concerns about public authorities and policies that require restrict permissions for a positive environmental change. On the other hand, responsibility linked to each individual separately to control his/her activities and its consequents [10].

To mitigate or limit the impacts of human activities on the global and to find a new way to improve the well-being and attitude of the buildings' users, by presenting a well-integrated design to the psychologically sustainable buildings, an innovative and sustainable approach is needed.

M. Barakat (✉)
Department of Architecture, Faculty of Engineering, Misr University for Science and Technology – MUST, Giza, Egypt

M. Aboulnaga
Department of Architecture, Faculty of Engineering, Cairo University, Giza, Egypt
e-mail: maboulnaga@eng.cu.edu.eg

© The Author(s), under exclusive license to Springer Nature Switzerland AG 2022

A. Sayigh (ed.), *Sustainable Energy Development and Innovation*, Innovative Renewable Energy,
https://doi.org/10.1007/978-3-030-76221-6_36

2 Objectives

The objectives of the study are mainly to: (a) create an approach for mitigating CC by enhancing human behaviors and activates; and (b) minimize human stresses through sustainability behaviors and livability framework. Figure 1 illustrates other sub-objectives for the study.

3 Environmental Psychologies

Environmental psychology deals with the relationship between the individual and his/her surroundings (life-space). This includes not only the environment to provide us with all that we need to survive, but also the spaces in which to appreciate, understand, and act to fulfil other needs and aspirations. Strengthening the urban ecosystem and improvement of economic-social balances constitute the essence of a sustainable city model. Urban farming such as vertical green walls/garden practices, for example, contribute to mitigating UHIE, lowering air pollution and improving air quality as well as increasing energy efficiency through zero-kilometers clean food and reducing noise pollution, in addition to a lessening stress caused by the vibrant urban life in megacities [11–17]. Vertical green walls/gardens also contribute to enhancing city aesthetics and creating new job opportunities and boosting the economy [18, 19].

New researches that support this view looked at the relationship between well-being and ecologically responsible behavior. Two studies found out that happier people are living in more ecologically sustainable ways and that there is not a straightforward trade-off between well-being and environmentally responsible behavior. Figure 2 describes the consequents of human activities [20, 21].

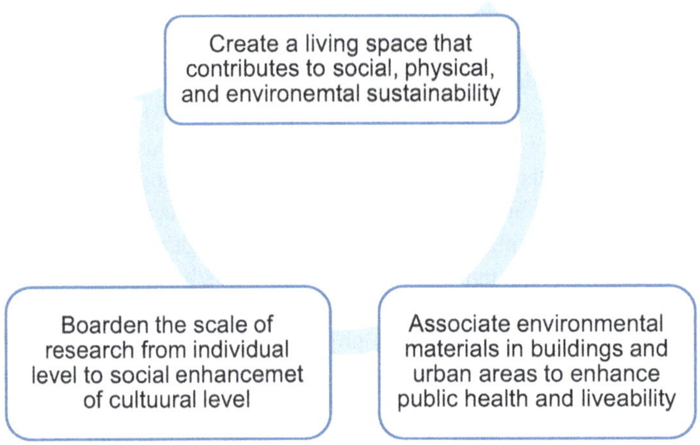

Fig. 1 Sub-objectives of well-integrated sustainable design of psychological architecture

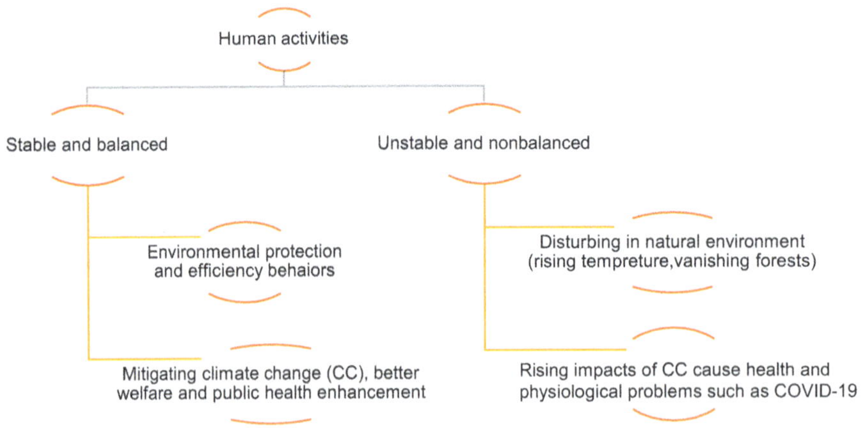

Fig. 2 Relationship between human activities and climate change

4 Methodology

In this study, the adopted methodology is centered on an environment-based approach to enhance sustainable behavior. This approach involves a review part followed by data collection and processing, and analysis as well as setting a qualitative environmental assessment targeted to present a definite solution for the sustainable design to minimize stress, enhance well-being, and achieve livability. Another part is the comparative method between different case studies along different scales expressing human comfort and better environmental psychology in urban areas and buildings. To achieve this, a questionnaire was developed, which includes 18 indicators to measure the livability categories (human well-being, transportation, aesthetic quality, and economic impacts, as well as environmental impacts and social impacts). There are 18 dimensions that relate to livability categories. To measure the satisfaction degree of users against each indicator, a scale is set which includes three degrees of satisfactions: very satisfying, suitable, and poor. Appendix 1 shows the questionnaire used for the study including these indicators.

4.1 Understanding Livability

Livability is an ensemble concept that its' factors relate to several complex characteristics or states, including sustainability, quality of both life and place, and healthy communities [22]. It is also the most immediate appearance of sustainability dimensions, which refers to the ability of an environment (a place or a community) to meet the social needs of citizens without compromising the ability of future generations to meet their full range of human requirements [23]. Livability is linked to territoriality [24].

Territoriality is defined as the boundaries and property lines which are marked. This can be the placement of barriers, shrub boarders, and the use of different colors of walkways to mark an area. Psychological deterrents to trespass are developed to establish territoriality [24]. With the massive impacts of coronavirus (COVID-19) worldwide, this condition is even more applicable. The coronavirus (COVID-19) has forced Governments worldwide to impose strict measures and policies' confinements for health purposes and protect humans and changed the social life, while visiting public spaces, workplaces, hotels, restaurants, and cafés after almost more than 100 days of lockdown. Examples in restaurants can be seen in The Netherlands (Fig. 3a) and office spaces (Fig. 3b) and elevators (Fig. 3c). All these images illustrate the marked boundaries and property lines to protect humans and well-being.

4.2 Livability in Cities

A livable city is a confident place to live in and is more resilient as well, with competitive social, economic, and environmental advantages. A livable city is also a healthy one that is promoting health, equity, and well-being. International agendas such as the sustainable development goals (SDGs), the New Urban Agenda, and the Healthy Cities Movement (HCM) increasingly call for urban settings to promote health and environmental resilience. The prioritization of creating healthy, livable,

(a) Social distance at restaurants in Amsterdam

(b) Personal protection office's spaces

(c) New measures in elevators in Indonesia

Fig. 3 Territoriality and personal spaces in/outside buildings due to COVID-19 impacts. (**a**) Social distance at restaurants in Amsterdam. (*Image source:* https://www.sciencedirect.com/topics/computer-science/environmental-design. *Photo Credit: Anne Lakeman*). (**b**) Personal protection office's spaces. (*Image source:* https://www.mediamatic.net/en/page/377822/testing-serres-separees-at-mediamaric-eten). (**c**) New measures in elevators in Indonesia. (*Image source:* https://www.huffingtonpost.ca/entry/elevators-coronavirus_ca_5e7b89d9c6b6b7d80959c210)

and sustainable cities responds to an established evidence base supporting the link between cities and health, and well-being outcomes as well as responding to the global trends of rapid population growth and urbanization. For example, already half of the world's population lives in cities, and an estimated two-thirds of people will be living in urban settlements by 2050, with the most rapid rates of urbanization occurring in low-to-middle-income countries (LMICs) [25, 26].

4.3 Examples of Livable Cities

Sustainable cities programs, designed to promote the creation of high-quality jobs, affordable housing units, and environmental quality, while reducing energy use and mitigating CO_2 emissions are urgently needed for crafting livable cities. Often these plans represent a cooperative effort on the part of the public and private sectors to reach shared goals. In Moline, Illinois, for example, a public-private partnership, Renew Moline, was created to attract business to the decaying downtown riverfront and to develop a major new waterfront "commons" to draw jobs and visitors downtown. In Tucson, Arizona, the city and its private and public sector partners built a large-scale, pedestrian-oriented, mixed-use development whose homes use half the energy consumed by the typical Tucson area house. Santa Monica, California, has launched "sustainable cities" programs to create high job qualities, better housing units, and suitable environment quality, moreover energy use, and toxic emissions reduction [27]. In addition to enhancing the livability of Times Square in Manhattan in New York after implementing the PlaNYC plan, it has been transformed to a pedestrian friendly urban space.

Since March 2020, the livability concept has changed with new parameters as preventive measures due to the horrible and deadly impacts of coronavirus (COVID-19). Figure 4 illustrates these examples in the United States. Examples in Canada, Europe, and Asia are shown in Figs. 5, 6, 7, and 8.

(a) Renewed Moline waterfronts in Illinois | **(b)** Mixed land use in Tuscon, Arizona | **(c)** Liveable Times Square after PlaNYC in New York

Fig. 4 Examples of livable cities – United States. (**a**) Renewed Moline waterfronts in Illinois. (*Image Source:* http://www.vandewalle.com/monona-waterfront-redevelopment/). (**b**) Mixed land use in Tucson, Arizona. (*Image Source:* https://azbex.com/nearly-400ksf-new-mixed-use-proposed-in-tucson/). (**c**) Livable Times Square after PlaNYC in New York. (*Image source:* http://www.880cities.org/doablecity/making-change/)

(a) Livable urban space, French old town in Montréal | **(b)** Ottawa's city centre | **(c)** Liveable center of Ottawa

Fig. 5 Examples of livable cities in Canada – North America. (**a**) Livable urban space, French old town in Montréal. (*Image credit: Mohsen Aboulnaga*). (**b**) Ottawa's city center. (*Image credit: Mohsen Aboulnaga*). (**c**) Livable center of Ottawa. (*Image credit: Mohsen Aboulnaga*)

Fig. 6 Examples of livable cities in Japan and Singapore – Asia. (**a**) Fujisawa city's social area in Yokohama. (*Image credit: Mohsen Aboulnaga*). (**b**) Fujisawa smart and city in Yokohama. (*Image credit: Mohsen Aboulnaga*). (**c**) Livable Robinson towers garden in Singapore. (*Image source:* https://www.designboom.com/architecture/kpf-robinson-tower-singapore-sustainable-urbanism-07-31-2019/)

Fig. 7 Examples of livable cities in South Korea – Asia. (**a**) Livable EWH University spaces, Seoul. (*Image credit: Mohsen Aboulnaga*). (**b**) Livable urban spaces in Songdo smart city. (*Image credit: Mohsen Aboulnaga*). (**c**) Livable Incheon National University – Songdo. (*Image credit: Mohsen Aboulnaga*)

Fig. 8 Examples of livable urban spaces in cities – Europe. (**a**) Hafen city waterfront in Berlin, Germany. (*Image credit: Mohsen Aboulnaga*). (**b**) Livable urban space in Vienna, Austria. (*Image credit: Mohsen Aboulnaga*). (**c**) Livable urban space in Delft, The Netherlands. (*Image source:* https://land8.com/the-building-that-wants-to-be-a-landscape-at-delft-university-of-technology/). (**d**) Livable urban space in Paris, France. (*Image credit: Mohsen Aboulnaga*). (**e**) Livable waterfront in Ghent, Belgium. (*Image credit: Mohsen Aboulnaga*). (**f**) Livable urban space in Brussels, Belgium. (*Image credit: Mohsen Aboulnaga*)

5 Case Studies and Assessment

In this section, two case studies were selected for the analysis. The area, location, and characteristics of the two buildings considered for the assessment are the ABC Bank and Bank Misr in New Cairo, Egypt. These case studies were selected for the availability of data and information, access, and physical entry to meet with the bank staff and employees in these two organizations, knowing that such exercise is not easy in such corporations.

5.1 Case Study 1 – First ABC Bank

The first case is the international bank ABC in New Cairo which was constructed in 2015 and officially opened in 2018. The First ABC Bank has acquired a skeleton building that was originally designed to function as an administrative building. Thus, the original building design was developed to accommodate bank requirements. With a total built-up area of around 12,700 m^2, the newly designed building consists of a 7-storey building as illustrated in Fig. 9a [28].

5.2 Case Study 2 – Bank Misr

The second case study is Second Bank Misr (BM), which is a 34-meter high building and is built on a plot area of 11,245 m^2 with a footprint area of 4322 m^2. Floors from 1st to 7th floors are allocated to BM's office areas, which is located in the gateway to the Money and Business District as presented in Fig. 9b [28].

5.3 The Conceptual Livability Indicators

Livability indicators are vital in assessing the level of quality of life both at the city and the building level. To conduct the assessment, it is important to review and understand these indicators. In the context of sustainability, there are three main sustainability dimensions to be considered notably: social, economic, and environmental dimensions as well as other three demotions such as viability, livability, and equity.

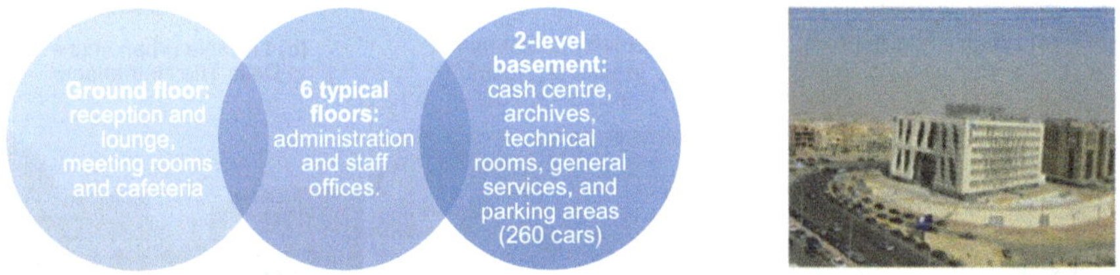

(a) The First ABC Bank floors' description and general view

(b) Bank Misr new office building's description and general view

Fig. 9 The office buildings selected for the study and assessment. (*Images' Source:* http://www.ecgsa.com/project/10361/1/1/4034). (**a**) The First ABC Bank floors' description and general view. (**b**) Bank Misr new office building's description and general view

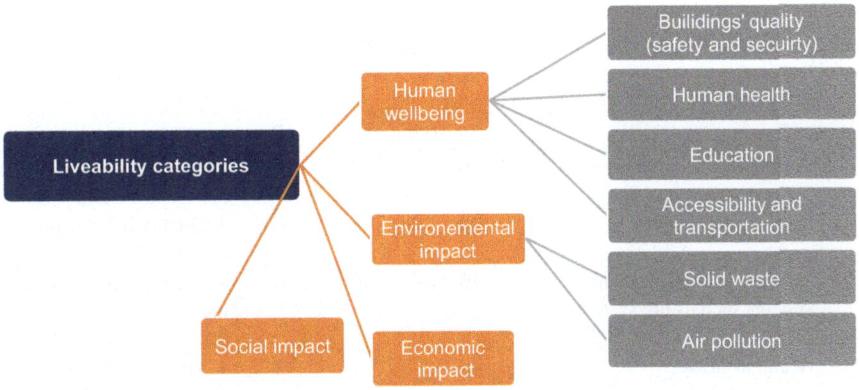

Fig. 10 Sum up of livability main categories

For the fourth dimension of sustainability – livability; there are subcategories notably: accessibility, walkability and transportation, building quality, human health, economic development, education and culture, and leisure. These are representing the human well-being component, whereas solid waste and air pollutants are representing the environmental impact component [29].

These dimensions encompass the main aspects of livability that are applied in the selected case studies. An assessment is created by integrating most of the livability dimensions and indicators that could suit the selected cases in Egypt. Figure 10 illustrates most of the livability categories used in assessing the different buildings and urban areas in the city.

The conduced assessment for the study is a qualitative assessment and not a quantitative one, due to some unavailable and inaccessible data, and security of organizations. Livability dimensions in this study, includes: promoting health and well-being, selecting the transportation category as it has a huge impact on pollution, safety, accessibility, and well-being. In addition to the employment rate, the workspace and management of income affect economic dimensions. All are the result of analyzing the assessment to gain the developed consensus regarding livability [29].

It is imperative to mention that although these dimensions are globally recognized, but not all of them can suit other countries. This is mainly due to their dependence on the human psychology, climatic condition, cultural issues, and comfort of the selected country as well as the weight given is not accurately estimated and is deduced according to the country's needs and priorities. Nevertheless, livability is rather defined as "suitability for human living" [30]. It refers to the extent in which the attributes of a particular place can, as they interact with one another and with activities in other places, satisfy residents by meeting their economic, social, and cultural needs, yet promoting their health and well-being in addition to protecting natural resources and ecosystem functions [30].

To understanding livability, it is in fact a sustainability principle, which encompasses features or relate to several other complex characteristics or states, including quality of both life and place, and healthy environments or communities. Livability, as a crosscutting concept, also contributes to the assessment of the cumulative impacts of public and private actions and failures to act and helps capture some of the externalities ignored or inaccurately valued by market mechanisms. These mechanisms include lending and investment policies, risk or reward assessments, and consumer, business, and government purchasing decisions [31].

5.4 Assessment of Livability

The assessment is based on six main categories: (a) human well-being, (b) transportation, (c) aesthetic quality, (d) economic impacts, (e) environmental impacts, and (f) social impacts. All were selected as they are the integration of the three main variables: social, economic, and environmental. In this study, a questionnaire was developed to assess livability in the selected two cases. This includes the livability category in response to the related 18 dimensions and degree of users' satisfaction as well as measured by 18 indicators. Three satisfaction levels (poor, suitable, and very confined) were created in the questionnaire to obtain the response of targeted employees' for the livability assessment as shown in Table 1.

In the assessment, meetings and interviews using the questionnaire were conducted targeting a population sample of the banks' users (50 employees), but due to availability, it was answered by only 30 employees of different positions in both case studies (First ABC bank and Bank Misr). These levels present the total weight of assessment as well as the related dimension and indicator. Table 2 shows the questionnaire used in the assessment including each category, dimensions, users' degree of satisfaction, and each related indicator that was answered by both banks' employees.

Table 1 Categories, dimensions, and indicators of the livability assessment – the questionnaire

Livability category	Dimension	Satisfaction degree of users			Indicator
		Very Satisfying	Suitable	Poor	
(a) Human Well-being	Feeling safe with security system	10	20	0	1.Existing of security system [Full day and night (24/7)]
	Building quality	10	20	0	2. Quality of structure, finishing, and maintenance are all Suitable for users
	Maintenance for mechanical systems such as fire systems & AC	10	20	0	3. Specified time and periods for the usual maintenance with safety measures for ensuring occupants during Maintenance period
	Health insurance	10	18	2	4. The organization ensure a good health insurance companies for the employees and staff
(b) Transportation	Private transportation	10	10	10	5. Specified transportation (small buses) for employees in addition to a well-designed large parking that occupy all employees and visitors' private cars
	Distance from public transport to the destination	4	13	13	6. Nearby Metro station or bus stop or inaccessible area for taxies or UBER transport
	Existing of main services around the area	12	14	4	7. Essential services (hospital/police station) in the area to ensure safety and care
(c) Aesthetic quality	Fine views and good details	8	12	10	8. Existing of green areas and wide streets
	Create a sense of place	8	20	2	9. The surrounding built environment inside/outside the building help in achieving comfort
	Interesting Facades	8	18	4	10. The external façade appearance for employees visual acuity
(d) Economic impacts	Hours paid compared to salary	6	24	0	11. Detected by users, occupants or employees
	Effect of organization over local & international income	6	15	9	12. Does this organization annual income affect the national and International income?
	Job growth rate	6	12	12	13. Justice in job growth rate compared to effort paid
(e) Environmental impacts	Use of renewable or low consumption devices/systems	9	21	0	14. Existing of LED lighting or sustainable AC systems or Double-skin façades or sun shades or double glazing?
	Existing of noise	9	15	6	15. Crowded area or nearby schools or and disturbing sources?
	Occupants density inside the space	8	18	4	16. Number of occupants/m^2 In case of pandemic disease as COVID-19, it is 1.2/m^2
(f) Social impacts	Existing of a social network	18	7	5	17. A connecting network as a Facebook or What's App., so that all members can get to know each other for better communications.
	Existing of a gathering safe space inside the community	0	3	27	18. Existing of a well-prepared restaurant or café that gather employees inside the organization.

Note: The numbers presented in the table refer to the exact number of 30 employees who had responded to the questionnaire to measure the 18 indicators.

6 Results and Discussions

Figures 11 and 12 present the results of the conducted assessment using questionnaire and data processing and outcomes. The employees' satisfactions of livability four categories – human well-being, transportation, aesthetic quality, and economic impacts – are depicted in Fig. 11, whereas Fig. 12 presents the employees' satisfactions for the livability category – environmental impact. It is clear from Fig. 11 that human well-being category has the degree of stratification of the employees about 37.4% (62.6% suitable), whereas that of transportation, it is higher (55%) as indicated in Fig. 11b. For the aesthetic quality, Fig. 11c shows that the employees' degree of stratification is 38.5%, 37.5%, and 25% and responded poor, suitable, and much very satisfied (confident), respectively. In terms of economic impacts, Fig. 11d indicated that 62.5% of the employees are feeling for such livability category – suitable, whereas 24% responded very satisfied (confident) and 25% stated poor.

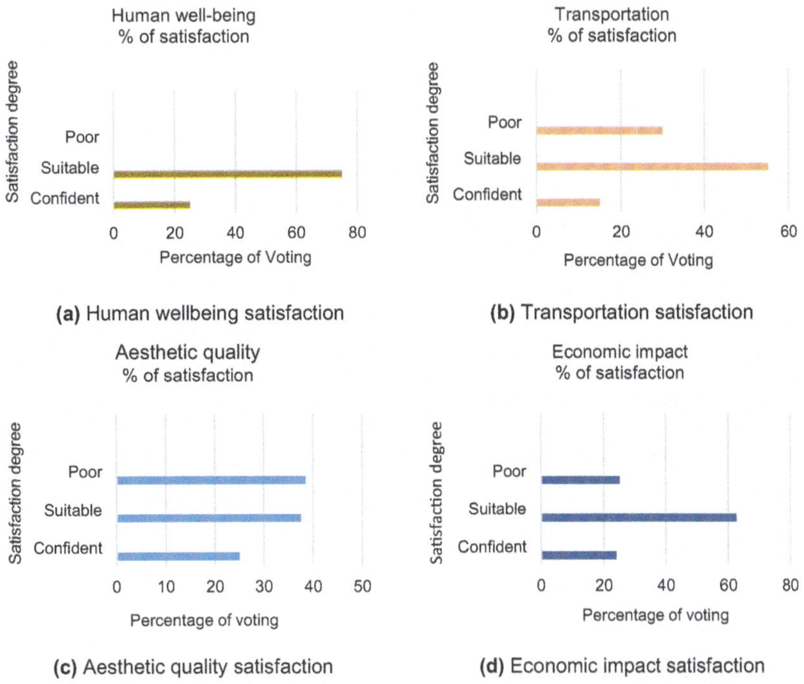

Fig. 11 Employees' satisfaction of Livability categor. (**a**) Human well-being satisfaction. (**b**) Transportation satisfaction. (**c**) Aesthetic quality satisfaction. (**d**) Economic impact satisfaction

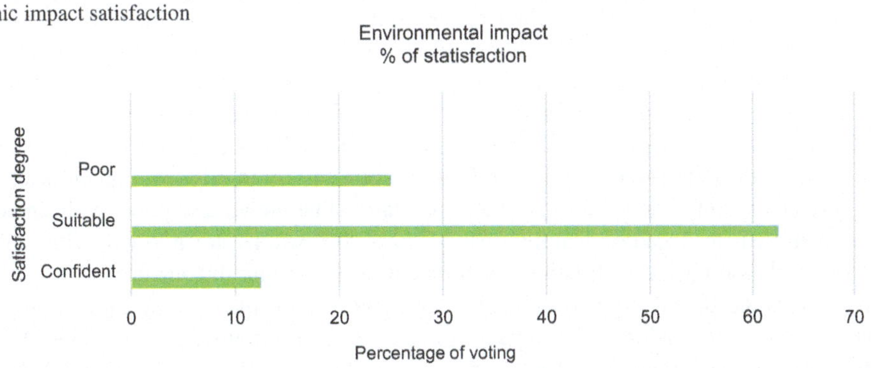

Fig. 12 Employees' satisfaction of livability category – Environmental impacts

For the livability category – environmental impact, it is clear from Fig. 12 that 62.5% of the employees see this category as suitable, whereas 25% and 12.5% portray this poor and very satisfied (confident) respectively.

By analyzing the results of the assessment, it was clear that concerning the human well-being category, employees are very satisfied (confident) concerning its dimensions and indicators as both organizations (banks) pay effort in building maintenance, employees' insurance, and infrastructure.

Most of the employees are also not satisfied with transportation means, which consequently affects their health and energy. Hence, better sustainable transport is needed and recommended in this case. Therefore, it indicates that the livability category is not attained and needs further improvement. Also, the aesthetic quality is not the best in their opinion as they do not completely feel satisfied (confident) with the views or the external elevation. It was indicated by the employees that "static is always boring." Concerning the economic impacts category, most employees are very satisfied with their salaries and benefits although they do not feel the best toward justice in their positions compared to their qualifications, however.

Regarding the environmental impacts category, employees are confident since they are satisfied with renewable systems, saving lighting, and efficient AC systems beside they agree with the sound insulation system that separates the building from external noise. Finally, concerning the social impacts category, employees have a gathering WHAT's APP group for most of them and they share their important events, but they miss having a gathering space inside the organization. This created repulsion between employees as most of them do not know each other in person (Fig. 13).

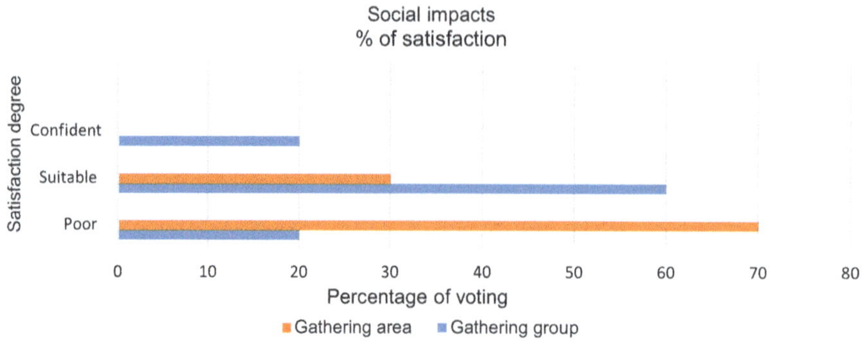

Fig. 13 Employees' voting for livability category – Social impacts

7 Conclusions

The livability in cities and buildings was reviewed. Examples of livability in various cities worldwide are presented. Selected case studies (office buildings – First ABC Bank and Bank Misr) in New Cairo in Cairo, Egypt were examined and assessed in terms of livability level and attainment. The qualitative assessment and surveys' results were descriptively and thoroughly analyzed and spatially represented using a developed questionnaire encompassing livability categories, dimensions to identify the user degree of satisfactions for each livability indicator. Based on the results, employees' ambitions exceed just issues related to infrastructure, water supply, and electricity, to more determining issues related to social and cultural needs.

It can be concluded that the livability, in terms of the six major categories, was almost satisfied except for transportation, social, and cultural issues. For transportation, this is related to connectivity to city-level amenities as this would facilitate accessibility. Also, aesthetic quality is related to aesthetic standards, fine views, and parks with fine details, as well as site configuration, mass, and pleasing façades' features. All of these dimensions are not satisfying enough and truly affect employees' secondary needs for a better well-being environment, hence ensure livability.

Climate change (CC) should be integrated into local planning processes, plans, policies, and programs. There is a need for city-level climate action plans to address the impacts, threats, and opportunities presented by CC. Some tools and principles have been developed to support cities in this process.

Thus, it is imperative to state some specific recommendations as a framework that may be further required for better livable buildings and cities. This framework includes the following:

1. Ensure the safety of employees
2. Adopt new technologies for employees' work records and consequently ensure economic and social justice
3. Ensure accommodation or creative space for more activities during the workday to allow strong social relationships and enhancement in the work rate
4. Create internal and external green spaces for better workspace views
5. Renovate for parking spaces and surrounding streets design for different transportation methods
6. Incorporate urban farming and renewable energy (solar PV panels) on the roof to mitigate CO_2 emissions and clime change.
7. Water and waste management services to ensure better facilities and suitable maintenance system for users
8. Manage users' density inside the selected space amid the coronavirus (COVID-19) pandemic to ensure better air quality and fewer diseases or infections

Appendix 1

Table 2 Developed questionnaire for categories, dimensions, and indicators of livability assessment

Livability category	Dimension	Satisfaction degree of users			Indicator
		Very Satisfying	Suitable	Poor	
(a) Human Well-being	Feeling safe with security system				1.Existing of security system [Full day and night (24/7)]
	Building quality				2. Quality of structure, finishing, and maintenance are all suitable for users
	Maintenance for mechanical systems such as fire systems & AC				3. Specified time and periods for the usual maintenance with safety measures for ensuring occupants during Maintenance period
	Health insurance				4. The organization ensure a good health insurance companies for the employees and staff
(b) Transportation	Private transportation				5. Specified transportation (small buses) for employees in addition to a well-designed large parking that occupy all employees and visitors' private cars
	Distance from public transport to the destination				6. Nearby Metro station or bus stop or inaccessible area for taxies or UBER transport
	Existing of main services around the area				7. Essential services (hospital/police station) in the area to ensure safety and care
(c) Aesthetic quality	Fine views and good details				8. Existing of green areas and wide streets
	Create a sense of place				9. The surrounding built environment inside/outside the building help in achieving comfort
	Interesting Facades				10. The external façade appearance for employees visual acuity
(d) Economic impacts	Hours paid compared to salary				11. Detected by users or occupants and employees
	Effect of organization over local & international income				12. Does this organization annual income affect the national and International income?
	Job growth rate				13. Justice in job growth rate compared to effort paid
(e) Environmental impacts	Use of renewable or low consumption devices/systems				14. Existing of LED lighting or sustainable AC systems or Double-skin façades or sun shades or double glazing?
	Existing of noise				15. Crowded area or nearby schools or and disturbing sources?
	Occupants density inside the space				16. Number of occupants/m^2 In case of pandemic disease as COVID-19, it is 1.2/m^2
(f) Social impacts	Existing of a social network				17. A connecting network as a Facebook or What's App., so that all members can get to know each other for better communications.
	Existing of a gathering safe space inside the community				18. Existing of a well-prepared restaurant or café that gather employees inside the organization.

References

1. Science for understanding and responding for climate change (2019) Available at: https://www.nap.edu/read/12782/chapter/5. Accessed 21 June 2020
2. Cianconi P, Betrò S, Janiri L (2020) The impact of climate change on mental health: a systematic descriptive review, PMCID: PMC7068211. Available at: https://www.ncbi.nlm.nih.gov/pmc/articles/PMC7068211. Accessed 21 June 2020
3. Berdahl P, Bretz S (1997) Preliminary survey of the solar reflectance of cool roofing materials. Energ Buildings 25:149–158. Accessed 21 June 2020
4. Akbari H (2005) Energy saving potentials and air quality benefits of urban heat island mitigation (PDF), 19, p 251K, Lawrence Berkeley National Laboratory. Accessed 21 June 2020
5. Center for Disease Control and Prevention (2006) Extreme Heat: A Prevention Guide To Promote Your Personal Health And Safety. Accessed at 21 June 2020
6. James W (2002) Green roads: research into permeable pavers. Stormwater 3(2):48–40. Accessed: 21 June 2020
7. United States Environmental Protection Agency, EPA. Available at: https://www.epa.gov/heat-islands/heat-island-impacts. Accessed: 21 June 2020
8. Mensah J (2019) Sustainable development: meaning, history, principles, pillars, and implications for human action: literature review, 5:1. https://doi.org/10.1080/23311886.2019.1653531. Available at: https://www.tandfonline.com/doi/full/10.1080/23311886.2019.1653531. Accessed at 21 June 2020
9. United Nations (2017) World health organization, sustainable development goals. Available at: https://sustainabledevelopment.un.org/index.php?page=view&type=30022&nr=192&menu=3170. Accessed: 21 June 2020
10. Fleury-Bahi G, Pol E, Navarro O (2017) Introduction: environmental psychology and quality of life, chapter from book handbook of environmental psychology and quality of life research, p .1–8. Available at: https://www.researchgate.net/publication/307513038_Introduction_Environmental_Psychology_and_Quality_of_Life/. Accessed 25 Apr 2020
11. Achieving Sustainable Development and Promoting Development Cooperation (2002) Available at: https://www.un.org/en/ecosoc/docs/pdfs/fina_08-45773.pdf. Accessed: April 25, 2020
12. Giorgos Goniadis, International Hellenic University (2015) Introduction to sustainable development, p 80. Available at: https://www.ihu.edu.gr/icsd/docs/introduction-to-sustainable-development.pdf. Accessed: 21 June 2020
13. Duhl LJ, Sanchez AK (1999) Healthy cities and the city planning process. Available at: https://www.euro.who.int/__data/assets/pdf_file/0009/101610/E67843.pdf. Accessed 21 June 2020
14. Aboulnaga M, Mostafa M (2019) Mitigation of heat island effect in megacities through districts' prioritization for urban green coverage applications: Cairo – as a case study. Renew Energy Environ Sustain, 4(5). Available at: https://www.rees-journal.org/. Accessed 22 July 2019
15. Duxbury N, Hosagrahar J, Pascual J (2015) Culture 21, Why must culture be at the heart of sustainable urban development. Available at: http://www.agenda21culture.net/sites/default/files/files/documents/en/culture_sd_cities_web.pdf. Accessed 21 June 2020
16. Pascual J (2006) Exploring local policies for cultural diversity. In Local policies for cultural diversity, UCLG-agenda 21 for culture. Available at: http://www.agenda21culture.net/index.php/docman/agenda21/17-report1exenpdf/file. Accessed 21 June 2020
17. Shaheed F (2014) Reflections on culture, sustainable development, and cultural rights. International Award UCLG - City of Mexico – Culture 21, UCLG. http://www.agenda21culture.net/images/a21c/awards/winners/art_FS2_ENG.pdf/. Accessed 21 June 2020
18. United Nations (1999) The universal declaration of human rights. Available at: http://wwwunorg/en/documents/udhr/. Accessed 21 June 2020
19. UNESCO (2014) Culture for development indicators: methodology manual. Paris: UNESCO
20. New economics foundation (2005) Well-being and the environment, chapter 4 from one planet living book. Available at: https://neweconomics.org/uploads/files/88cb2c4314731314d0_gxm6bq37b.pdf. Accessed 25 Apr 2020
21. Moser G (2007) Environmental psychology, chapter from Encyclopedia of Social Psychology. Available at: https://www.researchgate.net/publication/234167386_Environmental_Psychology. Accessed 25 Apr 2020
22. Quality of life (2002) Community and quality of life: data needs for community informed decision making, p 15–19. Available at: https://www.nap.edu/catalog/10262/community-and-quality-of-life-data-needs-for-informed-decision. Accessed 21 June 2020
23. Tom W (2015) Sustainable development: a bird's eye view. Available at: https://www.mdpi.com/2071-1050/3/10/1637/htm
24. Stachowski J (2016) Effective physical security, chapter 3 - Influences of physical design. Available at: https://www.sciencedirect.com/topics/computer-science/environmental-design. Accessed 25 Apr 2020
25. Davern M. World economic forum, 2020, entitled: we must address these 3 factors, to make our cities more vibrant and 'liveable'. Available at: https://www.weforum.org/agenda/2019/09/this-is-what-makes-a-city-liveable/. Accessed 21 June 2020
26. Alderton A, Davern M, Nitvimol K et al (2019) What is the meaning of urban liveability for a city in a low-to-middle-income country? Contextualizing liveability for Bangkok, Thailand, Global Health, 15:51. Available at: https://doi.org/10.1186/s12992-019-0484-8. Accessed 21 June 2020
27. Stein KE. Community and quality of life, 2007, chapter 1: concept of livability and indicators. Available at: https://www.nap.edu/read/10262/chapter/4#45. Accessed 25 Apr 2020
28. ECG (2017) ABC Bank New Headquarters, projects. Available at: http://www.ecgsa.com/project/10361/1/1/4034. Accessed 25 Apr 2020
29. Thanoon M, & Haykal H (2020) Influences of the Accessibility and Availability of Green Spaces on the Liveability of Residential Complexes in Erbil City. Pubs.sciepub.com. Retrieved 17 July 2021, from http://pubs.sciepub.com/ajcea/8/2/1/index.html.
30. Zanella A, Camanho A & Galvão T (2014) The assessment of cities' livability integrating human wellbeing and environmental impact. Research gate. Retrieved 17 July 2021, from https://www.researchgate.net/publication/271922686_The_assessment_of_cities%27_livability_integrating_human_wellbeing_and_environmental_impact.
31. Elsawy AA, Ayad HM, Saadallah D (2019) Assessing liveability of residential streets – case study: el-attarin, Alexandria, Egypt. Alex Eng J 58(2):745–755. Available at: https://www.sciencedirect.com/science/article/pii/s1110016819300547. Accessed 25 Apr 2020

The Visual Representation of Urban Contexts for the Education to the Sustainability: The App VISOSMappING

Paola Puma

1 Introduction

Often the theme of sustainability finds its more traditional declination mainly centered on the architectural scale of the built heritage but, as the ongoing global health emergency has highlighted, even cities have problems of fragility on which our daily behaviors can have a greater or lesser impact, and we need a new vision applying the notion of sustainability not only to the single building but also in relation to the entire urban environment.

Therefore, also to achieve an economic and social recovery that does not restore but improve the previous conditions, it is necessary to start from different assumptions that favor the development of a more resilient vision of the future of cities.

Education for Sustainable Development (ESD) therefore represents the first and necessary investment pivot on the future to initiate and consolidate the transition toward local and global sustainability.

The ESD shapes the culture of those who will be tomorrow's citizens, who will have the power to decide and influence how the future will be, then especially early childhood is increasingly recognized in the broader education community as an essential part of ESD and life-long learning.

We can outline a synthetic reference framework on education for sustainability, which puts the 2030 Agenda in the foreground, and the concerted action of UNESCO.

In the Agenda 2030, each SDG has specific educational objectives associated, described in the document "Education for the Sustainable Development Goals: Learning Objectives" edited in 2017 by UNESCO.

This framework intends to provide a general strategy and specific lines of action associated to the 2030 Agenda and to contribute to develop a growing time engagement of different agents.

Moreover, SDG 4 "Quality education" is directly involved in the ESD: the 2019 EU's performance on the SDG reports "education, vocational training and lifelong learning have a central role in building a sustainable, resilient, competitive and cohesive Europe for the future, as they enable people to reach their full potential. Throughout all life stages, education and training are crucial aspects of human development and key drivers for growth, jobs and social cohesion" [1].

UNESCO is the other major organization which, due to its institutional mission, is strongly committed to ESD.

The Learning Objectives of Education for Sustainable Development Goals by UNESCO are focusing on foreseeing, planning, and acting the necessary knowledge, skills, attitude, and behavior to accomplish the SDGs to be generally spread around the world and to transmit fundamental ideas practically applicable to all levels of education, from early childhood to universities: "ESD has to be understood as an integral part of quality education, inherent in the concept of lifelong learning: All educational institutions – from preschool to tertiary education and in non-formal and informal education – can and should consider it their responsibility to deal intensively with matters of sustainable development and to foster the development of sustainability competencies. ESD provides an education that matters and is truly relevant to every learner in the light of today's challenges" [2].

In its current strategy plan "UNESCO Operational Strategy on Youth 2014–2021" is declared: "As part of the post-2014 Education for Sustainable Development (ESD) programme framework, UNESCO will support the integration of ESD into education plans and curricula, including in non-formal education and life-long learning, with a focus on the key sustainable development challenges of climate change, disaster risk reduction and biodiversity. Youth will be addressed both as

P. Puma (✉)
Department of Architecture, University of Florence, Florence, Italy
e-mail: paola.puma@unifi.it

© The Author(s), under exclusive license to Springer Nature Switzerland AG 2022

A. Sayigh (ed.), *Sustainable Energy Development and Innovation*, Innovative Renewable Energy,
https://doi.org/10.1007/978-3-030-76221-6_37

beneficiaries and drivers of sustainable development throughout these activities" [3]. And moreover "Public institutions, social organizations, schools, and families may all contribute to the development of citizens who are more aware of the importance of working toward sustainable development goals. Among these, higher education institutions play a fundamental role in educating competent professionals who can work to achieve sustainability... Therefore, it is necessary to promote educational practices that help students become aware of the importance of exercising active and responsible citizenship that responds to the sustainability challenges of the 21st century" [4].

Many and increasing number of topics of early childhood responses to ESD have been focused in the UN Decade of Education for Sustainable Development (DESD, 2005–2014) [5] and produced documents, reports, and successful examples of programs addressing ESD in early childhood settings and practices [6].

2 Approaches and Methods

ESD increasingly confront researchers and teachers with new contexts and perspectives.

The complexity of today's world in fact requires teachers and students to acquire adequate access keys based on the fusion of different types of formal, non-formal, and informal teaching. Non-formal learning usually takes place outside the program provided by formal school education and is aimed at promoting the personal and social development of participants. Informal learning, on the other hand, takes place during daily activities and consists of "learning by doing," which allows children to spontaneously increase their maturity and to implement, with spontaneity and naturalness, thanks to participation "in first person," their knowledge [7].

The importance of this multidimensional learning dynamic is recognized at all levels and enhances the didactic experiences conducted outside the traditional training areas. From this point of view, the importance of knowledge acquired in extra-curricular settings (both real and virtual) is enhanced and the traditional teaching lessons now have the obligation to interact with other "new" forms of education, also to establish the accustomed to the fluidity of learning places (lifelong learning).

Innovation in teaching techniques and the development of alternative learning methods and contexts must therefore be enhanced with special attention to active learning characterized by permeable borders between the different themes and training levels.

In this perspective, the training offer increasingly moves knowledge acquisition from school to environments of daily life, shopping centers, libraries and museums, parks, and public squares [7].

The VISOSMappING educational app we have designed is part of this line of integration between learning environments to prepare a didactic support that favors the active participation of the student, brings the experiences of everyday life closer to the didactic setting, and uses participatory methods that overturn the centrality of the educational process from teacher to student, making the teacher play the role of facilitator of the discovery process.

Especially working in the primary school, the scientific, cultural, and social targets are part of a complex learning environment needing of cross-disciplinary contents and methodologies.

The idea to conceive a tool targeting pupils of primary schools focusing on sustainability and on the educational meaning of consumption was born working in a collaborative way in a joint group consisting of researchers of two disciplines, representation of architecture and researchers of the training processes.

To intertwine the formal education with the current communication technologies, the concept has chosen the educational app as the preferred tool and chosen the field application in the case study as small area in the historical center of Florence.

The concept is focusing on the idea of making children build the visual map of the places of their everyday life observed under the lens of the impact of consumption on the city, thus underlining how even the built environment is an important area for practicing virtuous behavior from a sustainable perspective.

The VISOSMappING structure is therefore centered on the representation of urban contexts and is articulated on various levels consisting of mapping, sharing of images collected, and organization of images in an interactive timeline by "surfing" those places through time [8].

2.1 *The Survey and Representation Workflow*

One of the goals of the project has been the will to make the users able to build visual storytelling stories, then the operational scheme is conceived in a visual setting both for the teachers and the pupils following the traditional architectural survey and representation workflow [9].

To apply this principle all the steps are designed using visual and intuitive tools:

- The data acquisition is obtained by mapping, a process available on smartphones or tablets.
- The data representation is obtained using a drag and drop editor.
- The visual communication is obtained by posting the contents on visual repositories.

3 Discussion

Adding to some research developed in the last years that have analyzed the effectiveness of learning by digital tools, also the recent experience held during the COVID-19 emergency's lockdown has shown how simply being connected to the network is not a successful learning condition as is the relationship between the children.

The project concept integrates the contents of education to sustainability into a design that recalls the framework referring to SDG, and UNESCO, and UE strategies above mentioned together with the Century skills, the goals set in 2016 by the New Skill Agenda Europe [10] (Fig. 1).

To stimulate the educational path toward the culture of sustainability, the app assumes two content's axes as thematic assumptions and scientific background:

1. *Sustainable consumption is a daily training process* that can spontaneously become virtuous starting from children only by facilitating their awareness and familiarity with the virtuous buying habits that children observe every day in the places of the city they pass through and experience during all their activities; here the direct reference is to the Cognitive learning objectives, Socio-emotional learning objectives, Behavioral learning Objectives in Learning objectives for SDG 12 "Responsible Consumption and Production" [2].
2. *The city's visual storytelling* allows to make understandable to children the places of everyday life told in a collaborative way by digital educational tools, aiming to make children think how much individual consumption transform the collective habitat.

3.1 Sustainable Consumption as a Training Process

"The gradual decentralization of the production society has allowed a closer and freer understanding of the universe of consumption, which is free from moral prejudice to consumption interpreted only as waste or as a source and consequence of alienation. At the same time, and as a consequence, consumption has become increasingly central in society and their value-added load has gradually reached the traditional weight of productive work in the formation of human beings [11]. In short, continue understanding consumption only as the result of historical processes would be misleading or at least would only give a very partial view of the past. 'We need to break', Trentmann insists, 'with a tradition that has treated the material culture as a separate sphere of everyday life' [12]. So if the objects that surround us and consumption, material and immaterial, help determine what we are, consuming is a training process. ... The bottom line of the platform therefore insists on the opportunity of a serious comparison with the educational value of consumption as the first step of every consumption education study. To this end, it is necessary to develop an Educational App that enables to virtually meet the consumption places that have followed from the first steps of the affluent society in Italy (era, as mentioned, the influence of consumption was particularly evident On people's lives and on society), up to now" [8].

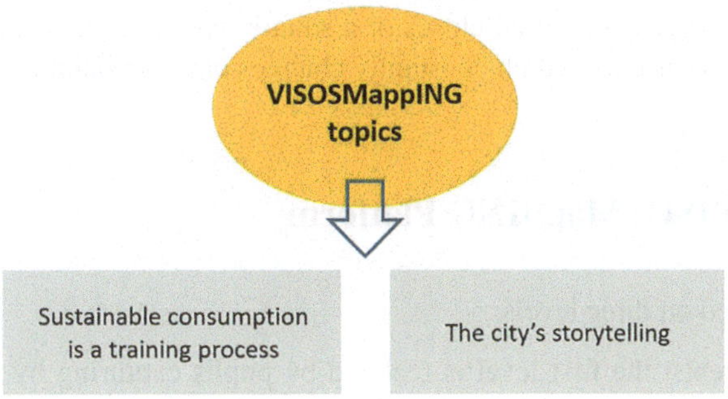

Fig. 1 The app topics

Fig. 2 The Florentine historical midtown

Fig. 3 The unbalances from mass tourism in the Florentine historical midtown: before and after the lockdown

3.2 The city's Visual Storytelling

How the situation created in some cities after the lockdown due to the COVID-19 emergency is brutally highlighting, the cities that have based their economy on mass tourism (and that in many European cities had made the historical centers unlivable) are today in the difficulties deriving from consumption models unsustainable for these fragile environments, because they are based on the exploitation of artistic and monumental heritage and an unbalanced commercial offer on the sole b&b and food [13].

In particular, the art cities such as Venice or Florence suffer greatly from the weakness of an economy that considers food only as an economic resource and not as a resource and common good (Fig. 2).

The test area is represented by the historical center in Florence and particularly by the portion around "Piazza San Pier Maggiore" and the neighboring zones close to the "Arco di San Pierino."

In this area, we can register the impact of the mass tourism in the last 5 years on the Florentine historical center, where the old shops frequented by the residents have been gradually replaced by shops and services devoted to tourists and where the increasing demand of food has modified the image of the city (Fig. 3).

Therefore, ViSosMapping has been tested in this area to apply the potentials of the platform in a portion of city where perfectly clear are the signs, the consequences, and the contradictions of material consumerism models which provoked the unsustainable livability of this place we can highlight by visual storytelling [14].

ViSosMapping allows the children to represent these places by observing them through the lens of the small daily transformations that a child can notice (like a shop that closes or a square that is being restored and open as a playground) and living the city as a place of daily experience which constantly changes our own habitat (Fig. 4).

4 The Case Study: The VISOSMappING Platform

The design of the app is articulated on three levels:

1. Sharing their own visual contents: the first level is realized by pupils capturing by smartphones and tablet the place's pictures and geotagging it with the geographical position the geo-tagging. To create personalized maps allows the chil-

Fig. 4 The area study in the Florentine historical midtown

dren to express their own description and image (new imagination?) of the city and to integrate also multimedia contents such as images, videos, and tags which highlight visited places, favorite places, etc.

2. Building the timeline: the second level is represented by laboratory activities, where the teacher and the children share and upload the collected visual data to build the interactive timeline; "The logical/mathematics functions typical of the computer language are so introduced in a graphic way as if they were parts of a puzzle or a Lego to freely assemble. Colours and shapes then lead the user into *the* creational process that becomes, therefore, an assemblage for pre-packed parts" [8, 15].
3. Exploring the platform: in the third level, the children can use the platform single or in group by consulting the contents (but posted contents can be modified by different participants only if the processing takes place in a contextual presence so as to encourage participants' physical presence and interaction in a real social place).

The sharing of the experience and the open availability of the final contents is allowed from the upload on the social networks or the hosting video services, such as YouTube or Tik Tok.

VISOSMappING has been designed taking attention to make the interface intuitive and to "user friendly" procedures: taking street pictures, the drag and drop editing, and the app concept in progressive web app format, assuring the adaptability between web and mobile app.

5 Results and Conclusions

The experience to map their neighborhood using daily tools allows pupils to be aware by simple tools on how apparently distant aspects as food consumption in their habitual streets and squares are social actions related in a far more evident way than it is thought with the sustainable life in and of the city.

The training value of consumption becomes hidden content and then is translated by the app language in visual-audio expressions in line with the EDS goals.

"Part of the necessary value system is for teachers to realise that they are part of the larger world and, in turn, have an obligation to influence each child to seek a more sustainable world for the future. This generation of children, as well the next, and the next, will have to overcome the unsustainability caused by previous generations. We cannot embrace ideas of protecting children and childhood from reality. Children are part of the social reality and have to be prepared early on to advocate for sustainability and work actively towards solutions in cooperation with adults. Adults and teachers have the responsibility to lay the foundations for young children's sustainable learning and for making sure that the next generation is interested in creating a sustainable world." [16].

Excessive consumption of resources poses a threat to our planet every day, so it's necessary to educate children on what they can do to minimize it to develop a future more sustainable urban livability. The educational institutions widely work on many multidimensional initiatives on environment education and urban and local environmental problems. In this setting, environment means a complex way of permanently interacting with eco-systems and social and cultural systems, including economic and political aspects all contributing to build a robust environment citizenship (Figs. 5 and 6).

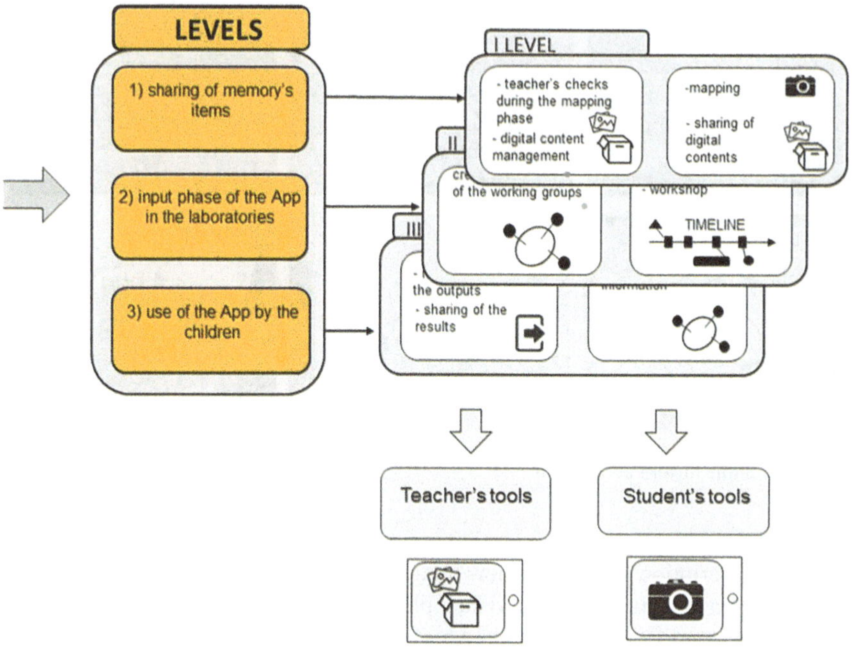

Fig. 5 The app concept

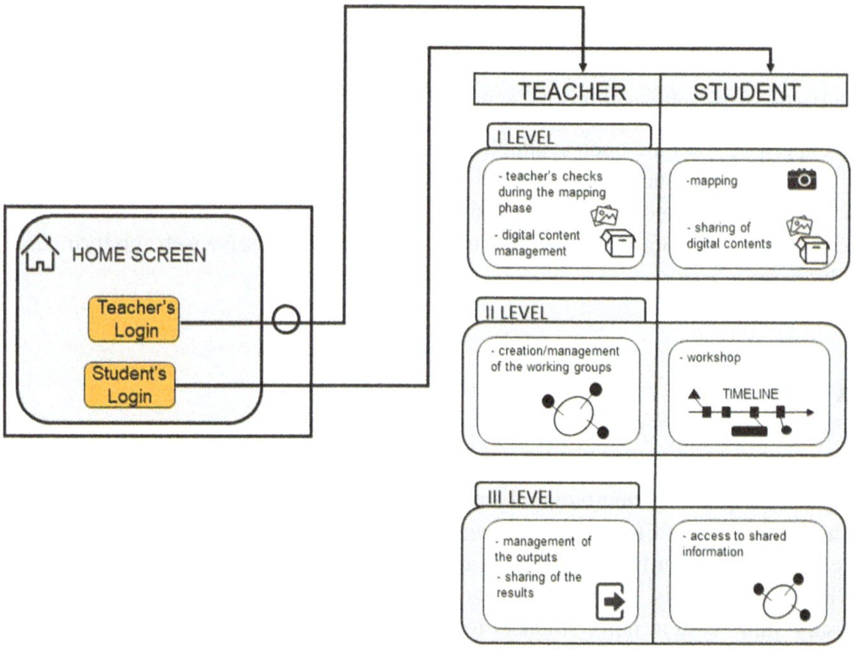

Fig. 6 The app scheme

References

1. The eu's performance on the sustainable development goals (2019) Annex II, https://ec.europa.eu/commission/sites/betapolitical/files/reflection_paper_sustainable_annexii_en.pdf
2. UNESCO (2017) Education for the sustainable development goals: learning objectives; UNESCO: París, France, p 1–68. https://unesdoc.unesco.org/ark:/48223/pf0000247444?posInSet=1&queryId=bc2e9a9d-6733-4f79-9ee5-cfe22da4e4c6
3. UNESCO (2014) UNESCO operational strategy on youth 2014–2021. https://unesdoc.unesco.org/ark:/48223/pf0000227150

4. Jordi CF (2020) Reflective learning in higher education, sustainability special issue, MDPI
5. UN Decade of ESD. https://en.unesco.org/themes/education-sustainable-development/what-is-esd/un-decade-of-esd
6. UNESCO (2012) Education for sustainable development good practices in early childhood, United Nations decade of education for sustainable development (2005–2014), education for sustainable development in action, good practices N°4, 2012, education for sustainable development section, UNESCO Education Sector. https://unesdoc.unesco.org/ark:/48223/pf0000217413
7. Commissione delle comunità europee, Direzione Generale Istruzione e Cultura della Commissione europea (2000) Memorandum sull'istruzione e la formazione permanente. https://archivio.pubblica.istruzione.it/dg_postsecondaria/memorandum.pdf
8. Nicastro G, Oliviero S, Puma P (2017) Forming conscious consumption: visual memories from the 1950s Up to Date in the Multimedia Representation of the VISOSMappING Platform, Proceedings Immagini? 1:892; MDPI; ISBN: 978-3-03842-681-3, https://www.mdpi.com/2504-3900/1/9/892
9. Bini M, Bertocci S (2012) Manuale di rilievo architettonico e urbano, CittàStudi, Milano
10. www.ec.europa.eu. A new skills agenda for Europe. Available online: http://ec.europa.eu/social/main.jsp?catId=1223
11. Oliviero S (2016) Lavoro, consumo, formazione: una prospettiva storico-educativa. Rivista di Storia Dell'educazione, 1:125–140, ISSN 2384-8294
12. Trentmann F (2016) Empire of things: how we became a world of consumers, from the fifteenth century to the twenty-first. Penguin, London, ISBN 978-0713999624
13. D'Eramo Marco (2017) La maledizione del Tourist District. In Puma Paola (ed), Firenze, la trasformazione del centro antico, Edifir, Firenze, p 15–19
14. Puma P. Mapping esperenziale del centro storico di Firenze: Le trasformazioni della scena urbana, dell'immagine e dell'immaginario. In Belli G, Capano F, Pascariello I (eds) Proceedings of the VIII Congresso AISU La città, il viaggio, il turismo, Napoli, Italy, 7–9 September 2017. http://www.iconografiacittaeuropea.unina.it/index.php/2-non-categorizzato/46-atti-aisu
15. Chen N-S, Hwang G-J (2014) Transforming the classrooms: innovative digital game-based learning designs and applications. Educ Technol Res Dev 62:125–128. https://doi.org/10.1007/s11423-014-9332-y
16. Ingrid PS, Eunhye P (2017) How to educate children for sustainable learning and for a sustainable world. Int J Early Child 49:273–285. https://doi.org/10.1007/s13158-017-0197-1

Thermal Performance of Roof Solutions for Social Housing in Portoviejo, Ecuador

Dania González Couret and José Fabián Véliz Párraga

1 Introduction

There is a trend to increase research works focused on improving the quality of social and popular housing, and particularly, their indoor thermal environment. Then, numerous precedent studies can serve as reference to social housing for Portoviejo's warm and humid climate [1–3]. The present paper's authors carried out an exploratory study of the indoor thermal environment in social housing located in the warm and humid Portoviejo's climate [4]. It demonstrated the necessity to look for alternative solutions to reduce thermal load towards interior spaces through the roofs, using locally available resources. This paper intends to expose and discuss the results of an experimental research carried out in lab models to evaluate the relative thermal behaviour of different housing solutions in Portoviejo city, Ecuador.

2 Materials and Methods

The results presented in this paper are part of a major research, in which a theoretical framework of the variables influencing housing indoor environment (particularly, those referred to as thermal transference through the roofs) was identified. The second step of the research consisted on an exploratory study of the indoor thermal environment in housing usually built by The Ministry of Urbanism and Housing (MIDUVI) (Fig. 1). An intentional sample of 92 housing was selected, representing diverse conditions of the design identified variables in order to compare resulting measured indoor temperatures.

Measurements of temperature and relative humidity were carried out by using a Kestrel 4200 (Pocket Air Flow Tracker) with precision of ±1 °C, located inside the housing (living room) at 1.50 m over the floor and programmed to measure each hour. In every case, measures were also made in an outdoor reference point located less than 10 m from the studied housing and using the same equipment, inside a small meteorological cabinet at 2.00 m height under a tree. Each case was measured during 24 hours. Results were compared to simultaneous measurements taken at the local meteorological station.

The third step of the research, which constitutes the essential objective of the present paper, consisted on an experimental study with unitary lab models (1 m^2 interior surface and 2.60 m height) in order to compare the relative performance of different roof solutions under Portoviejo's conditions, for what an inventory of locally available resources was made (natural or recycled).

Three minimum spatial unitary modules were built by concrete block walls, 10 cm width, in an experimental polygon, separated 3.50 m between them, to avoid shadows. Weather conditions (sunny, cloudy, or rainy) were observed and recorded each day, since temperature differences (interior–exterior) vary according to that. Indoor air temperature measurements were carried out each hour, covering the experimental modules with the different studied roof solutions, and simultaneously to measurements of outdoor temperatures (Fig. 2). Finally, a statistical processing of the temperatures measured in real zinc roof housing and in the experimental modules covered by a zinc roof was made, in order to find their correlation.

D. G. Couret (✉)
Technological University of Havana, Havana, Cuba

J. F. V. Párraga
Technical University of Manabí, Manabí, Ecuador

© The Author(s), under exclusive license to Springer Nature Switzerland AG 2022

A. Sayigh (ed.), *Sustainable Energy Development and Innovation*, Innovative Renewable Energy,
https://doi.org/10.1007/978-3-030-76221-6_38

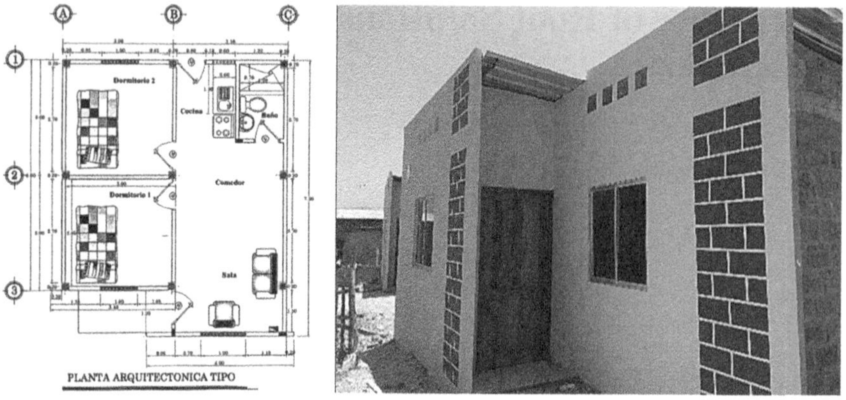

Fig. 1 Social housing built by MIDUVI in Ecuador

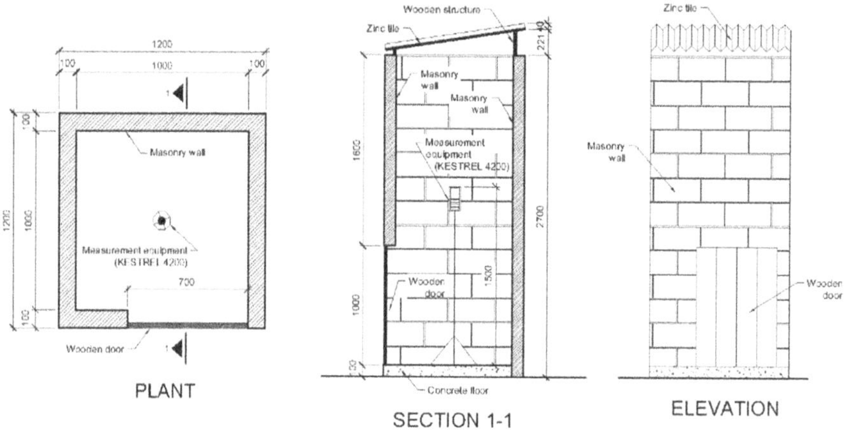

Fig. 2 Unitary experimental modules

3 Results and Discussion

3.1 Indoor Temperature in Zinc Roof Housings

In the previous exploratory study [4], it was possible to conclude that specific conditions in peripheral contexts do not induce significant differences on indoor air temperature. Notable differences according to architectural typologies neither exist. It is also possible to appreciate that popular auto-constructed housing offers better conditions than the MIDUVI ones. According to orientation, differences neither are significant. Then, since typologies and contexts are very similar, building materials (roofs and walls) are the most influencing variable on indoor temperature.

Interviewed people almost unanimously considered their houses are hot or very hot, during the day, mainly in the afternoon, when the sun is shining, except the inhabitants of housings with bamboo walls, who say they are relative fresh during the day. On the contrary, those who live in concrete roof housing ensure they are very hot, even at night.

3.2 Indoor Temperature in Experimental Zinc Roof Modules

Taking into account that zinc tile is the most used roof building material, because of its availability and cost as well as a relative better behaviour, its influence on indoor temperature was verified, adding over it mortars elaborated with different local materials (natural or recycled), which could improve its performance.

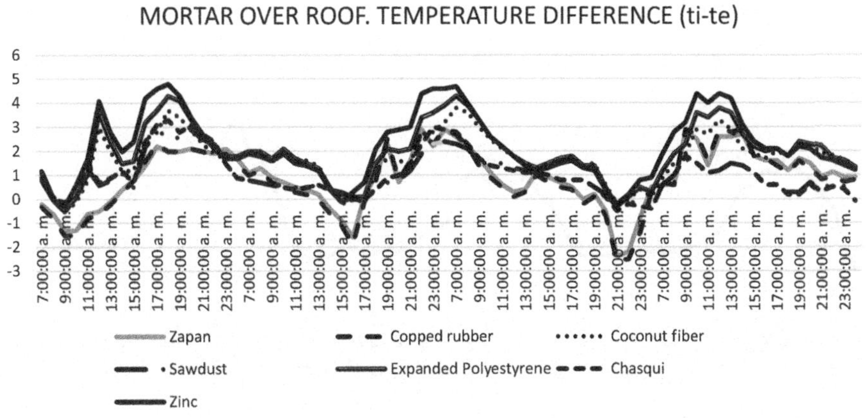

Fig. 3 Hourly difference of temperature between inside the experimental module with mortar made by different materials over the zinc roof and the outdoors reference point

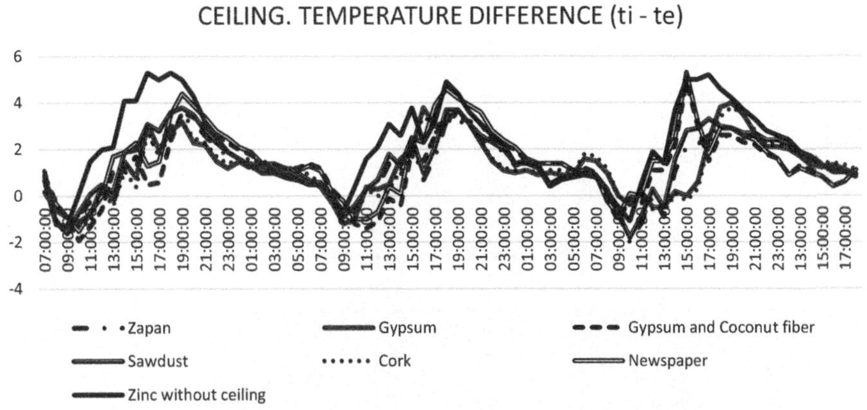

Fig. 4 Hourly difference of temperature between inside the experimental module with ceiling of different materials added to zinc roof and the outdoors reference point

An added mortar with zapan (a fibre coming from the trunk of the banana tree) produces an indoor air temperature reduction of near to 2 °C at 4.00 pm. Using in the added mortar recycled rubber coming from car tyres, indoor air temperature in the experimental module is reduced between 1.5 °C and 2 °C, while with coconut fibre added to the mortar, reduction achieves nearly 2 °C in the afternoon. A mortar with sawdust also provoke a similar effect in the afternoon, but less than the former cases (slightly higher than 1 °C), while employing expanded polystyrene (espuma-flex) allow to achieve maximum reduction of only 1 °C at 4.00 pm (Fig. 3).

To add a ceiling below the tilted tiles zinc roof is a tradition in this region, not to improve the indoor thermal environment, but because of aesthetic reasons. Taking this into account, measurements were carried out in the experimental modules, adding to the zinc roof, diverse ceiling solutions elaborated with different low thermal transfer materials locally available (natural or recyclable).

Gypsum is a material traditionally used as ceiling in Portoviejo, and according to measurements carried out in the experimental modules, it produces a reduction of indoor temperature up to 2 °C between 3.00 pm and 4.00 pm, despite the sky conditions (cloudy or sunny). Cork modular panels are industrially produced elements with low conductivity and available on the market at a relatively low price. Despite its use is not traditional in housing, the experimental research showed a reduction of indoor air temperature of more than 2 °C at 3.00 pm, including rainy days.

The rest of the tests developed in the experimental modules were made adding to the zinc roofs different ceiling solutions using alternative materials locally available. A sawdust ceiling over a textile network allowed to reduce the indoor air temperature in the afternoon up to 2 °C between 2.00 pm and 4.00 pm. Newspaper, recognized as highly insulating material which constitutes an available and recyclable waste, was also tested as a ceiling for the zinc roof, generating an indoor air temperature reduction of up to 2 °C between 3.00 pm and 4.00 pm (Fig. 4). The major temperature reduction is achieved by adding a zapan ceiling to the zinc roof, which is maintained during the whole afternoon and achieve up to 3 °C (Fig. 5).

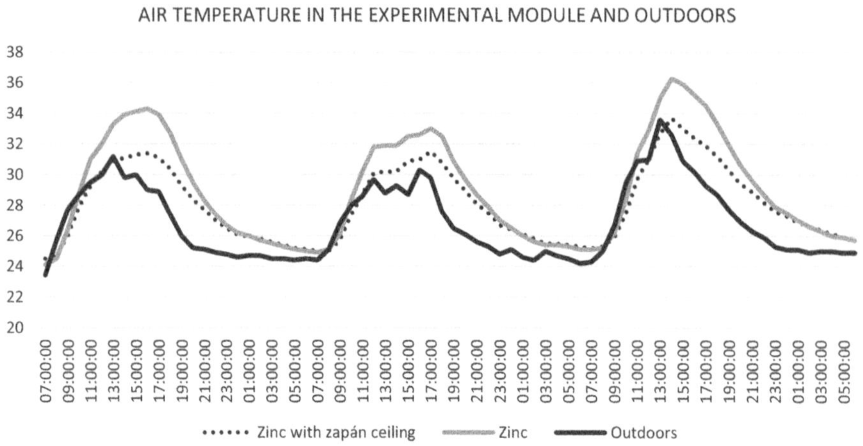

Fig. 5 Outdoor and indoor temperatures registered in experimental modules covered by a "zinc" roof and "zinc" with a "zapan" ceiling

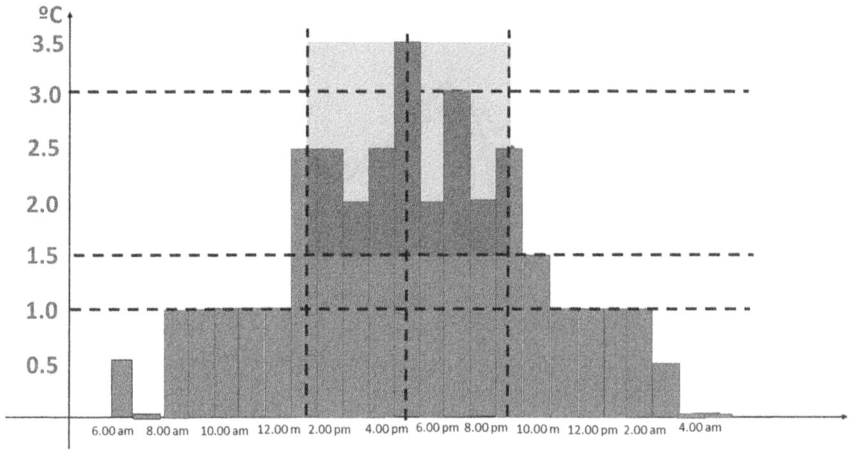

Fig. 6 Combined histograms of hourly indoor temperature differences in monitored housing and in unitary experimental modules with zinc roof

3.3 Correlation Between Housing and Experimental Modules

Between 1.00 pm and 9.00 pm, indoor temperatures in modules are higher than in housing with respect to outdoors, with Δt differences between 1.5 °C and 2.5 °C, being possible to achieve 3 °C at 7.00 pm (Fig. 6). However, results are comparable in equal conditions, despite the verified scale effect could means real reductions slightly inferior to those registered in the test.

4 Conclusion

In the peripheral urban areas where social housing is predominant, building materials, particularly in roof, constitute the variable most influencing on indoor air temperature.

Zinc roof housing, which is the predominant solution, is hot in the afternoon, but the addition of mortars elaborated with natural fibres or other components over the zinc roof improves its thermal behaviour. The best performances correspond to zapan, recycled rubber and coconut fibre (Δt up to 2 °C).

The inclusion of alternative and locally available materials as ceiling in the traditional zinc roof is even more effective in reducing indoor temperature. Such is the case of the newspaper, or zapan, which can contribute with reductions up to 3 °C.

References

1. Triana MA, Lamberts R, Sassi P (2015) Characterisation of representative building typologies for social housing projects in Brazil and its energy performance. Energy Policy 87:524–541
2. Mesquita L, Kos JR (2017) Towards more resilient and energy efficient social housing in Brazil. Energy Procedia 121:65–70
3. Naicker N, Teare J, Balakrishna Y, Wright C, Mathee A (2017) Indoor temperatures in low cost housing in Johannesburg, South Africa. Int J Environ Res Public Health. https://doi.org/10.3390/ijerph14111410
4. González Couret D, Veliz Párraga JF (2019) Vivienda de interés social en Portoviejo. Ambiente Térmico Interior. Revista AUS. 26/35–41/ segundo semestre 2019. https://doi.org/10.4206/aus.2019.n26-07

Mass Transit: An Integrated Approach to Sustainable Urban Development – The Case of Greater Cairo Region

Khalid Z. El Adli and Abdullau Al-Attar

1 Introduction

Cities around the globe have, in recent years, been facing numerous challenges including among others, rapid urbanization, climate change, scarcity of natural resources, and shrinking public budgets, ultimately leading to unstainable urban futures. Yet, it is anticipated that by 2050 around 67% of the world's population will be living in cities [1]. This move toward urbanization further exacerbates threats posed by development as cities continue to consume scarce natural resources. On the one hand, consumption of natural resources including water, food, energy, and fertile land pollutes the urban environment and thereby affects the health and quality of life. On the other hand, urban sprawl and unplanned growth destroys natural resources and leads to massive agriculture land loss, consequently reducing food resources and resulting in economic and social losses [2].

Yet, cities are the economic engines, they are the major consumers and distributors of goods and services responsible for most of the world's energy consumption and 50% of the world's economic output [3]. City growth should thus be kept-up with supporting public services and infrastructure. To sustain scarce natural resources and overcome challenges and threats posed by rapid urbanization, the UN Habitat introduced the "Sustainable Development Goals (SDG's)" including "Goal 11 – Sustainable Cities and Communities" [4]. This is in prospect of recognizing an approach to better plan, build, and manage our cities and urban places. Sustainable cities should thus embrace environmental, economic, and socially responsible solutions and strategies. Sustainable land-use planning strategies entail adopting innovative housing, green urban development, energy conservation and renewable energy, sustainable building practices, green economy and governance, waste management, and recycling as well as sustainable green mobility and transport [5].

The essence of sustainable transport and mobility is to reduce car dependence without compromising mobility and accessibility in urban areas. It is thus essential that sustainable transport systems incorporate comprehensive and integrated transport solutions that are entwined with the complex characteristics of urban form and socioeconomic development – this remains one of the most delicate tasks of transportation planning.

2 Sustainable Mobility and Transport

Sustainable transport encompasses the planning and design of urban transport systems with the objective of providing safe mobility and accessibility of people and goods through effective systems in terms of cost, management, and technology while responding to social, economic, and environmental challenges. Sustainable transport strives to minimize redundant trips, encourage a change in behavior and improve the operational efficiency of motorized transport [6]. It is assessed in terms of accessibility, health and safety, cost-effectiveness, consumption of natural resources, and reduction of pollutants. A proclaimed approach to achieve sustainable mobility and transport is via Transit-Oriented Development (TOD). TOD evolves around mass transit and multimodal transport hubs supported by high-density residential and mixed-use development along corridors and activity nodes. It pursues low-carbon transport solutions emphasizing nonmotorized transport and as a result

K. Z. El Adli (✉) · A. Al-Attar
Faculty of Urban and Regional Planning, Cairo University, Cairo, Egypt
e-mail: khaled.eladli@cu.edu.eg; abdallahfarouk@cu.edu.eg

© The Author(s), under exclusive license to Springer Nature Switzerland AG 2022

A. Sayigh (ed.), *Sustainable Energy Development and Innovation*, Innovative Renewable Energy,
https://doi.org/10.1007/978-3-030-76221-6_39

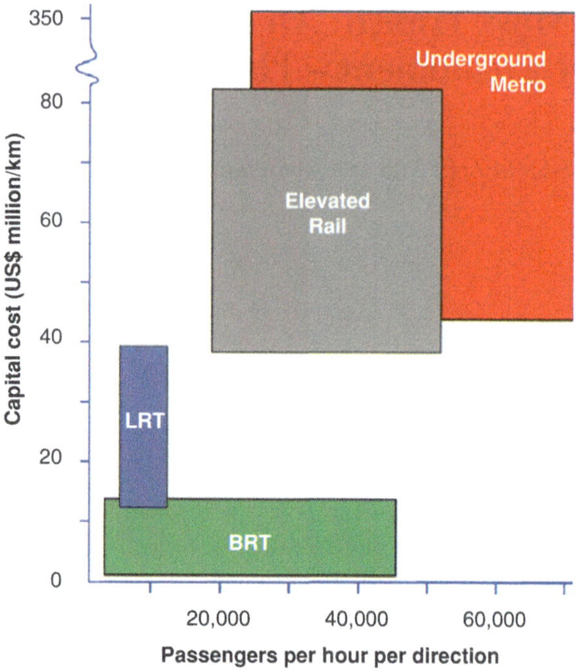

Fig. 1 Transit modes cost per km. (Source: [9])

reduces dependency on petroleum-based modes [6]. Increasing the efficiency of transport systems in cities thus benefits the environment and contributes to equitable access, it contributes to the city's attractiveness to investment, human capital, and economic growth [7, 8].

Sustainable mobility and transport are thus both a goal toward greening the built environment and a common target to be achieved in city planning at variable scales of development: regional, urban, and local. Integrating sustainable green transport in urban development requires an understanding of the classification and characteristics of the different modes of transport (see Fig. 1). Transport modes can possibly be classified into four main categories [10, 11]:

- *Paratransit*: This mode of transit includes car sharing, taxi, Jitneys as well as Dial and Ride (DAR).
- *Street transit*: Street transit mode is considered a low capacity public transport mode. It includes such transport modes as Regular Bus (RB), Express Bus Service, Trolleybus (TB) and Street Cars (SCR).
- *Semi-rapid transit*: This public transport mode could be characterized as a medium capacity mode and includes: Bus Rapid Transit (BRT), Light Rail Transit (LRT) and Automated Guided Transit (AGT).
- *Rapid transit*: Characterized as High-Performance Mode. This mode of transport includes Light Rail Rapid Transit (LRRT), Monorails as well as Regional Commuter Rail (RGR).

Cities with good transit planning and operations generally adopt an integrated transport network making use of a variety of modes than do cities in which transit is neglected [10, 12]. Planning integrated public transport systems contributes to the success of cities. Sustainable public transport systems should thus warrant the following key preferences of modal characteristics [13]:

- Ensuring the diversity of modes of public transport as a multimodal system
- Linking properties of transit modes with the urban character of served areas
- Supporting a progression of transit systems with various characteristics
- Minimizing transit gaps at various urban scales to avoid use of private modes

It is thus safe to conclude that investing in green mobility and urban transport systems is both a key indicator and contributor to the development of sustainable cities.

3 Transport Trends and Approaches

3.1 North American Cities

Differences in urban fabric, forms, and morphologies affect transit solutions and trends. In North America, following World War II, many cities witnessed vast suburbanization resulting in loss of residential and business activities in city centers. This move toward suburbia had a strong effect on emerging private mobility and was to a large extent responsible for public transit losing its ridership. Car-dependent settlement patterns developed with almost no investment in public transport as congestion was tackled with further car-based infrastructure. This approach was regrettably embraced and adopted by several cities around the globe, mainly in developing nations.

However, the failure of this approach to offer sustainable solutions prompted many North American cities to reconsider its transport and mobility strategies. Toward the end of the 20th century cities such as Edmonton, Calgary, and Portland among others pursued further investment in public transport. Rail transit including heavy; interurban; regional and commuter rail projects, light rail transit, as well as street transit projects including streetcars; busses and trolley busses were among the varied modes of public transport adopted [14, 15]. However, the new expansions of mass rapid transit systems represented a costly endeavor and were thus generally less abundant, usually linking the CBD with suburbs and other city centers [16].

Portland's LRT was planned with the object of making the central part of the city a traffic-free zone while minimizing parking within the city center. The network was planned at two levels (see Fig. 2) [17]. The first level serves the city center via a street tram connecting the northwestern part of the city center with the southwestern part. The second level links the suburbs and the city center through an integrated LRT network. Both networks contributed to making the city center a distinct area [17].

3.2 Western European Cities

Public transport modes in Western European cities are the main responsibility of local governments. Hence, as a consequence of reconstruction movements following World War II and the raising of environmental awareness, funds were channeled into public transport. Medium-sized cities largely adopted an integrated transport paradigm.

Many cities sought to revive the streetcar infrastructure, light rail systems as well as bus networks while introducing pedestrian corridors. Additionally, many cities invested heavily in underground and elevated rapid rail systems thus offering an integrated public transport network [15]. Mass rapid transit systems were considered as the ideal model in such cities as London and Paris. These systems having the optimal catchment areas, offer to date, rapid service not only to the old city centers but also to new central business districts thus affording a high degree of mobility and accessibility [16].

To support interurban and intraurban transit services, the city of Karlsruhe in Germany offers various transit modes including heavy rail, light rail, street trams as well as metro (see Fig. 3). This integrated transit network links suburbs and rural areas with the city and city center. Strategies for integration include using urban regional light rail transit with new and exclusive tracks; sharing urban and regional light rail with heavy rails on existing rail network for interurban service; and connecting medium and large cities with shared use of the national railway network to support transit services in city centers [17].

Fig. 2 Portland city embraced an LRT network at two levels. (Source: [17])

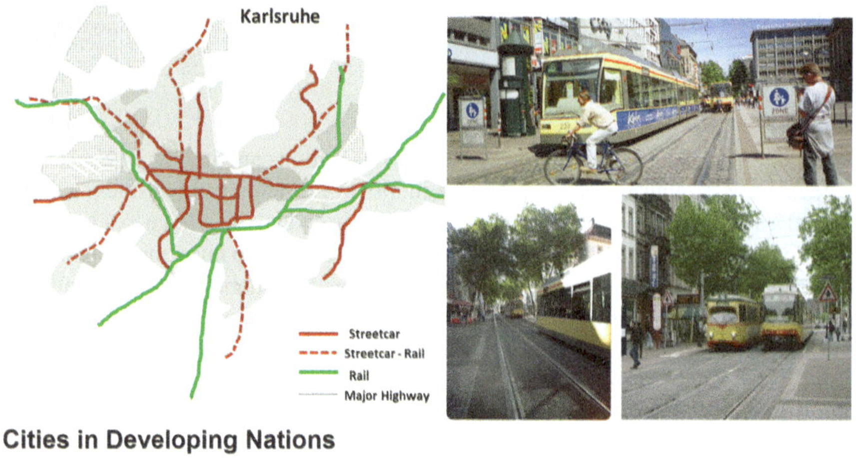

Fig. 3 The city of Karlsruhe utilizes multi-modal transit to link suburbs and rural areas with the city. (Source: [17])

3.3 Cities in Developing Nations

Urban characteristics, land use regulations and zoning ordinances including densities and FAR play a significant role in the selection of transit modes [18]. In Africa, Asia, and South America, most large cities tend to enjoy higher average densities than most North American and European cities. Ahmedabad boasts a density of 18,430 ppl./km^2, Bogota 21,276 ppl./km^2, Cairo 19,376 ppl./km^2, Jakarta 14,464 ppl./km^2 and Seoul 16,589 ppl./km^2 [19]. Hence, in an attempt to support higher density, cities such as Ahmedabad, Bogota, and Curitiba embraced and implemented with great success the BRT, a Semi-Rapid Transit system), as an upgrade from street transit. The BRT's success stories are attributed to its relatively low investment cost compared to other transit systems, its ability to attract riders and promote high density in urban areas. The BRT has since received increased attention due to these advantages as well as to its short implementation period.

Furthermore, by exploring modal splits among cities worldwide, it became apparent that the percentage of travelers using a particular type of transport varies substantially as a consequence of development trends, land use patterns, and densities. On the one hand, Asian and European cities largely show more transit shares as compared to less dense cities in North America and Australia which tend to be more car-dependent. On the other hand, cycling and walking seem to constitute the prevalent mode of transport in many developing nations due to their low costs and lack of modern transport infrastructure (see Fig. 4).

4 Urban Mobility Trends in the GCR

4.1 Public Transport in the GCR

In Egypt, around 43% of the nation's population currently live in cities, 56% of whom reside in the Greater Cairo Region (GCR) and Alexandria Governorate [20]. Nevertheless, over the past four decades, urban planning, infrastructure and service delivery in the GCR have not been able to cope with rapid urbanization.

Consequently, the GCR has a relative high share of public – collective transport. These include a variety of formal and informal modes including the Cairo Metro Network spanning over 76 kms (over three lines), the CTA and GCR bus networks in addition to informal microbus routes [21]. These modes account for 12.436 million daily motorized trips during a typical 2001 weekday [22]. Figure 5 illustrates the distribution of transit type shares of daily trips as follows:

- 52% of daily-motorized public transport trips representing 6.5 million trips by shared taxis and microbuses.
- 28% of daily public transport trips representing 3.5 million trips by public buses.

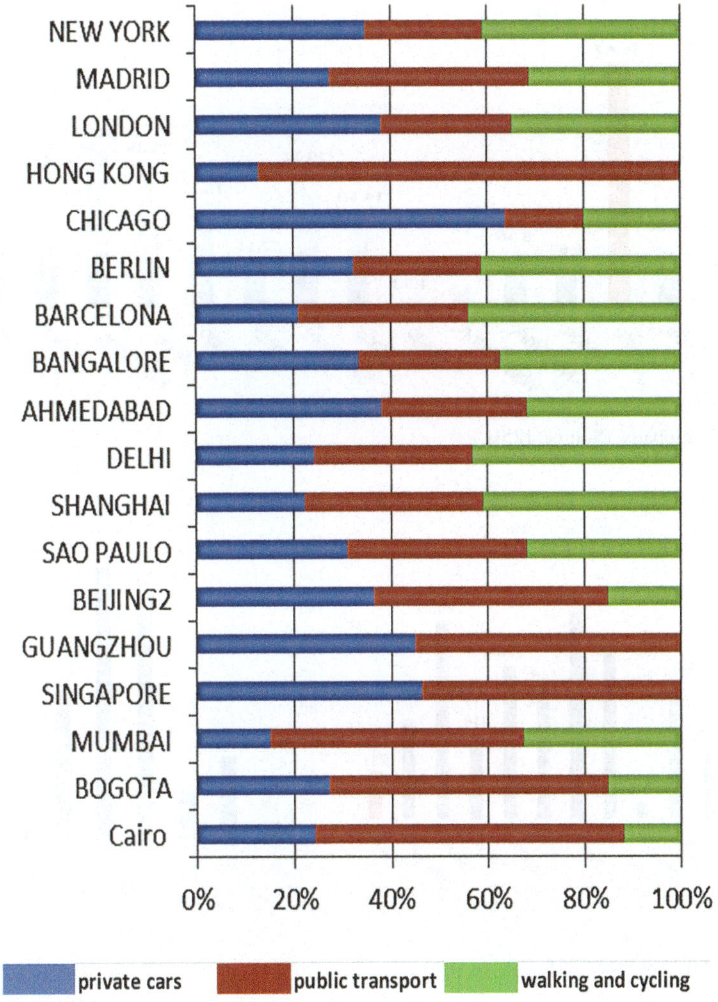

Fig. 4 Modal splits among global cities. (Source: [19])

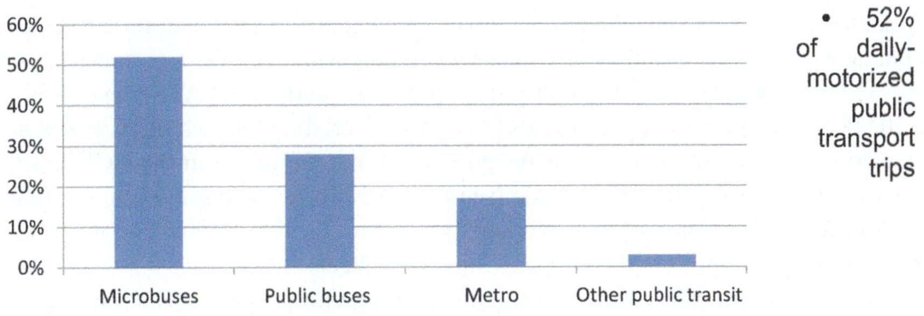

Fig. 5 Shares of daily transit trips by type in the GCR. (Source: [22])

- 17% of public transport daily transit trips representing over 2 million trips by metro. However, in 2008, the daily transit trips by metro reached 3 million. It has also been reported that the metro in the GCR has the highest number of boarding's per kilometer of any metro system worldwide [23].
- 3% of daily public trips were undertaken by other public transit.

It is thus but evident that the fragmented solutions of informal public transport represented in microbuses play a primary role in daily trips and have a significant impact on the high public transport shares in the GCR.

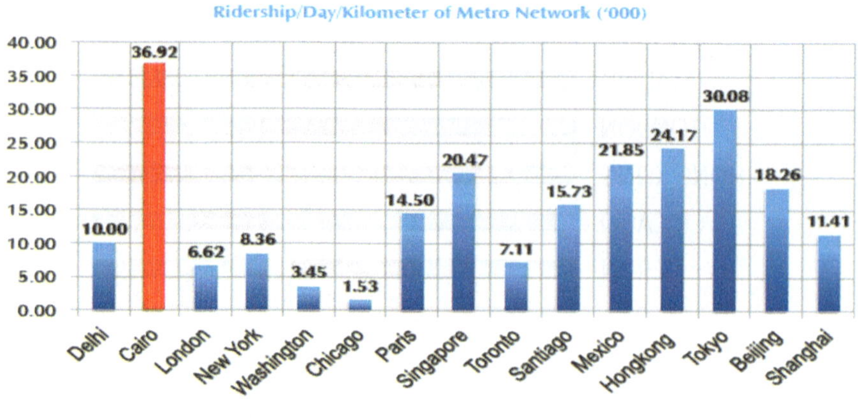

Fig. 6 Ridership for metro systems globally. (Source: [25])

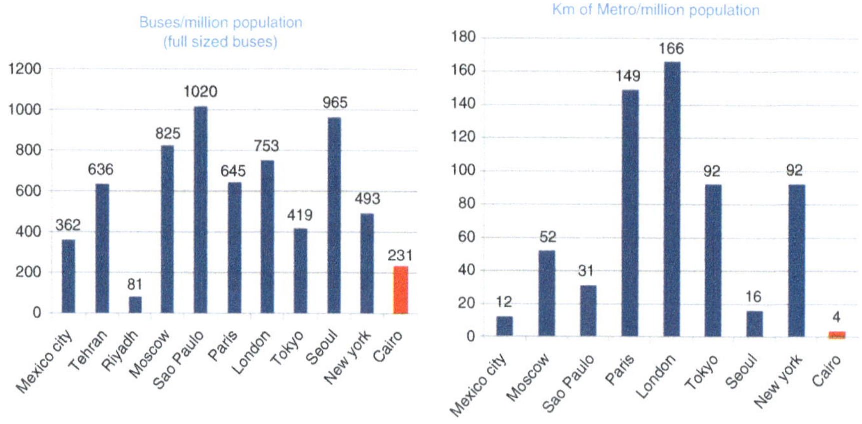

Fig. 7 Transport supply ratio globally. (Source: [25])

4.2 *Transport Supply in the GCR*

Comparing public transport supplies in GCR with other cities around the globe is central to determining the transport infrastructure characteristics, identify gaps and challenges, and arrive at an appropriate model for sustainable transport.

In 2016, the GCR public bus network spanned more than 13,500 kms with over 473 bus routes operated by 4900 standard busses and minibuses [21]. According to Cairo Regional Area Transport Study (CREATS) developed by JICA, shared taxis (microbuses) operate 20,000 vehicles covering 503 routes [24]. Moreover, the Cairo Metro Network negotiates over 76 kms (over three lines) and is currently growing [21]. It has been reported that the metro in the GCR has the highest number of boarding's per kilometer of any metro system worldwide (see Fig. 6) [23]. A comparative study of GCR's public transport supplies with global cities reveals that Cairo's Metro Network ranks low, lacking kilometers of sufficient transit service per million urban population. The same applies to the Greater Cairo's bus fleets (see Fig. 7).

Nonetheless, despite ranking low on public transport supplies, public transport in GCR has higher values of average daily trips among many global cities, this is due to informal transport such as microbuses and shared taxis [25]. The shortage of formal public transport supply coupled by vast rates of informal urban developments at both the city core and fringes, paved the way for fragmented solutions of demanded informal public transport to support interurban and intraurban trips.

4.3 *Challenges of Public Transit Planning in GCR*

Based upon a study undertaken by the Development Research and Technological Planning Center (DRTPC) at Cairo University and CREATS, public transport in the GCR faces a number of challenges. These have been identified to include [22, 24]:

- Disaggregate planning
- Inappropriate faring policies as well as controlled fares
- Shortage of funding
- Absence of dedicated lanes and /or bus priority at junctures
- Minimal modal coordination
- Dominance of informal transport

4.4 BRT Planning in Giza City

In view of recognized challenges, the Governorate of Giza under the direction of the then Governor H.E. Dr. Khalid Z. El Adli acted on developing a coordinated and comprehensive development plan for the city of Giza. The plan that was to transform the city of Giza within 7 years, links historic and cultural sites with the city core though an integrated transit network. The coordinated action plan entailed introducing a sustainable mobility and transport network featuring a BRT system, water taxis, bike sharing, and cycle routes as well as linking all transit modes with existing and planned metro lines (see Fig. 8).

Phase One of the action plan involved working closely with The Institute for Transportation and Development Policy (ITDP), UN Habitat and the UNDP in coordination with the International Association of Public Transport (UITP) to identify potential BRT corridors and phasing within the city of Giza.

The planning processes for the BRT corridor was commenced by ITDP through a series of sequences and progressions including: data collection, corridor selection, planning BRT services, proposing physical designs and alignments, advising on station design criteria, signal design and intersection, designing pedestrian accesses, supporting bicycle infrastructure and sustaining multimodal integration solutions. However, due to the complexity of administrative and financial systems within the Governorate coupled by the implementation intricacy, stages of the BRT project processes were not necessarily completed sequentially. Instead, due to local pressures, the Governorate embarked on planning and aligning phase one of the main corridor and surrounding areas.

In 2015, alignment of the proposed BRT was set to run from Remaya Square to Gameat Al Dewal Al Arabeya via King Faisal Road running through Cairo University Road and Al Dokki Street with an overall length of 16.1 kms with the main transport hub being at Gameat Al Dewal Al Arabeya Street. The transit hub was to be connected with metro lines, regional rail and buses, water taxi, bike sharing stations, and to be serviced with multistory parking and commercial initiatives. The BRT corridor was redesigned by the first author to accommodate curbside stations, ensure safety of both pedestrian and cyclists, integrate transit modes, and provide services and amenities. Phase one of the scheme involved planning and constructing of four vehicular underpasses and related intersections where high passenger volumes have been observed. Unfortunately, only one of the four underpasses spanning over 1.8 kms, has since been constructed and is currently operational. However, the Governorate has recently secured necessary financial credits to implement the proposed BRT and support projects. The authors are thus hopeful that the BRT along with a planned rapid transit monorail connecting 6 October City with Giza City will soon come into existence and become functional (Fig. 9).

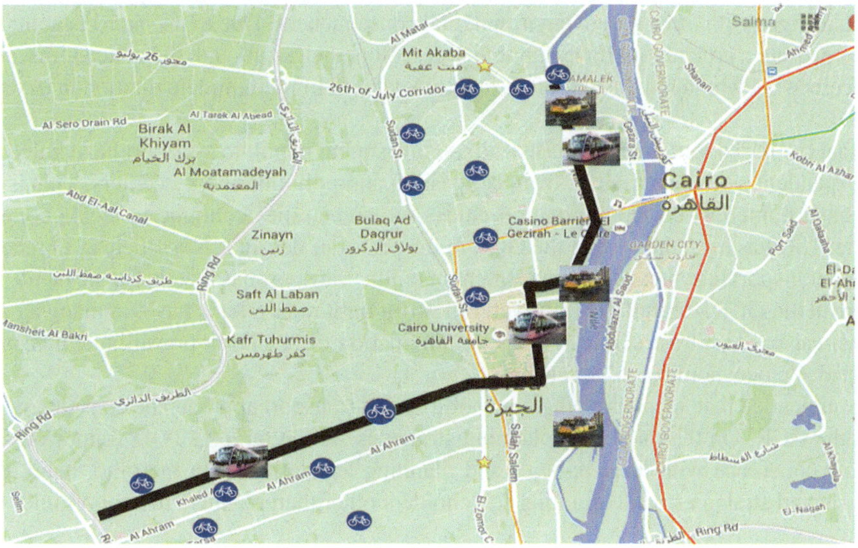

Fig. 8 Proposed BRT corridor for the city of Giza. (Source: [9])

Fig. 9 Proposed alignment for BRT stations at King Faisal Road. (Source: [23])

Fig. 10 Proposed solutions for intersections alongside corridor routes. (Source: [9])

5 Conclusion

In view of the above discussions, it is but evident that an integrated multimodal transit approach constitutes the way forward to sustainable urban futures. Not only does sustainable mobility and transport boost the quality of life in cities but it also helps mitigate urban challenges with often severe environmental consequences. Due to its varied benefits, sustainable mobility and transport thus constitute a crucial tool for sustaining cities. Nonetheless, though the concept is well acknowledged by both planners and concerned stakeholders, it is too often overlooked with limited implementation in developing nation cities (Fig. 10).

The authors argue that car-based settlement patterns with no investment in public transport have failed to offer sustainable urban solutions. Car-dependent settlements suffer from congestion and pollution in city centers. This is apparent in several North American cities and numerous developing nation cities. Yet, some cities continue to tackle congestion with further car-based infrastructure. This has been the case in the GCR in the past four decades. Nonetheless, in recent years, interest in sustainable mobility and transport has intensified. Investment in public transport has recently gone hand-in-hand with further car-based infrastructure in the GCR. Regretfully however, timeworn light rail systems have, in many a case, been torn down to make way for further car-based infrastructure. This is in anticipation of resolving prevailing congestion while failing to notice environmental and socioeconomic consequences.

It is thus highly recommended that an integrated green mobility and transport approach be considered to foster sustainable settlement patterns in the GCR. On the one hand, planning strategies should target a TOD framework addressing specific challenges and ensuring coherence with local policies and planning legislations. On the other hand, detailed local plans should reflect transit-oriented strategies while granting stakeholders other than planners the opportunity to participate in the planning process.

Furthermore, it is crucial to endorse sustainable financing options and diversify local funding sources. An innovative property tax and development fee could help raise money for funding sustainable mobility and transit. Real estate developers could be charged to link their new projects to existing transit networks. Local governments could also cash-in part of the value increases of real estate dues benefiting from transit networks for up-keep purposes. Public-private partnerships (PPPs), should also be considered [5]. Additionally, green transit strategies should capitalize on the huge potential for investors to change the current status of transport networks in the GCR.

To conclude, further efforts are required to create a pertinent vision for sustainable green mobility and transport in the GCR with huge prospects for enhancing the quality of life.

References

1. Höjer M, Wangel J (2014) Smart sustainable cities: definition and challenges. Adv Intell Syst Comput 310:333–349. https://doi.org/10.1007/978-3-319-09228-7_20
2. Five big challenges facing big cities of the future | World Economic Forum. https://www.weforum.org/agenda/2018/10/the-5-biggest-challenges-cities-will-face-in-the-future/. Accessed 05 May 2020
3. UN-HABITAT (2006) “State of the World’s Cities 2006/2007. Sustainable Development Knowledge Platform,” State of the World’s Cities 2006/2007. https://sustainabledevelopment.un.org/index.php?page=view&type=400&nr=1129&menu=35. Accessed 02 July 2020
4. Goal 11: sustainable cities and communities | UNDP. https://www.undp.org/content/undp/en/home/sustainable-development-goals/goal-11-sustainable-cities-and-communities.html. Accessed 08 May 2020
5. El Adli K, Abd El Aziz N (2020) Greening the urban environment: an integrated approach to planning sustainable cities. In Greening the urban environment, Springer- In print
6. Lewis EA. Development Bank (2015) Green city development tool kit, Mandaluyong - Philippines. Accessed 02 July 2020. [Online]. Available: www.adb.org
7. Assadian A, Nejati M (2011) Challenges faced by megacities in the future. Inf Resour Manag J 24(2):76–88. https://doi.org/10.4018/irmj.2011040106
8. OECD (2013) Green growth in cities
9. El Adli K. Mass transit: an integrated approach to sustainable urban development, 19–20 November, 2019 | CEDARE, The future of E-mobility and urban planning in Egypt seminar (CEDARE-FES), Sep. http://web.cedare.org/sgp/events/2019/11/21/the-future-of-e-mobility-and-urban-planning-in-egypt-1920-november-2019/. Accessed 03 July 2020
10. Vuchic VR (2007) Urban transit systems and technology, vol 1, 1st edn. John Wiley & Sons, Inc., Hoboken
11. Marshall S (2004) Streets and patterns. Taylor & Francis. https://www.routledge.com/Streets-and-Patterns/Marshall/p/book/9780415317504
12. Bashandy S (1990) The impact of transportation technology on city form. Cairo University
13. Nielsen G, Lange T, Mulley C, Nelson JD. Network planning and design for public transport success-and some pitfalls
14. Post RC (2007) Urban mass transit: the life story of a technology. Greenwood Press, London
15. Samuels JM et al (2001) Making transit work, TRB Special Report 257, Washington, D.C. Accessed 27 May 2020. [Online]. Available: http://onlinepubs.trb.org/onlinepubs/sr/sr257.pdf
16. Kennedy C, Miller E, Shalaby A, MacLean H, Coleman J (2005) The four pillars of sustainable urban transportation. Transplant Rev 25(4):393–414. https://doi.org/10.1080/01441640500115835
17. Klassen P (2009) Investigation into the potential of modern streetcars in Halifax Nova Scotia | Tram | Light rail. Dalhousie University, Halifax Nova Scotia
18. Suzuki H, Cervero R, Iuchi K (2013) Transforming cities with transit, 1st edn. The World Bank, Washington, DC
19. Rodrigue J-P et al (2017) Modal split, journey to work trips, selected cities | The geography of transport systems. In: The geography of transport systems, vol 2020. Routledge, New York. https://transportgeography.org/?page_id=5055. Accessed 30 May 2020
20. World Population Review (2020) Cairo population 2020 (Demographics, Maps, Graphs), Cairo population 2020. https://worldpopulationreview.com/world-cities/cairo-population/. Accessed 03 July 2020
21. CAPMAS (2017) CAPMASS Public Transport report 2015–2016, CAPMAS, CAIRO. [Online]. Available: https://www.capmas.gov.eg/Pages/SearchGeneral.aspx?Search_id=النقل+العام
22. DRTPC (Feb. 2009) Research study on urban mobility in greater Cairo; trends and prospects final report, Giza. Accessed 31 May 2020. [Online]. Available: https://planbleu.org/sites/default/files/publications/greater_cairo_final_report_0.pdf
23. ITDP (2015) Bus rapid transit for Greater Cairo: prefeasibility assessment, no. June, p 12–20. [Online]. Available: https://itdpdotorg.wpengine.com/wp-content/uploads/2016/06/BRT-for-Greater-Cairo-151123-1.pdf
24. UNDP (2008) Sustainable transport: full sized project | UNDP transparency portal. https://open.undp.org/projects/00045900. Accessed 06 June 2020
25. Nakat Z, Herrera S, Cherkaoui Y (2014) Cairo congestion traffic study, World Bank Gr., no. Cairo Congestion Traffic study, p 0–12. [Online]. Available: https://www.worldbank.org/en/country/egypt/publication/cairo-traffic-congestion-study-executive-note#:~:text=The studywas intended to,one-fifth of Egypt’s population

Energy Management and Savings in the Domestic Situation: A Case Study

Tony Book

Description of a house that was recently refurbished to a high specification of insulation and windows is given.

In addition, a 5.4-kWp PV system was installed on the south-east facing roof, an underfloor heating system and various controls which can be monitored by smart phone apps.

The hot water comes from an immersion heater (3 kW) primarily from the roof through a smart switching system. This detects when the house electricity load is exporting and instead diverts this energy to the hot water cylinder (210 Litre).

The underfloor system which is zoned in every room is heated by gas. There is also a 7-kw log burner in the main living area, which due to the efficiency of all the above is hardly ever used. There is also a 10-kW battery storage system (3200 mAh).

This chapter demonstrates several points. The base line of about 7000 kWh per annum is reduced by about 80% with the use of the PV running in tandem with the battery system. The dwelling is virtually self-sufficient most of the summer months.

The efficiencies of energy diversion for hot water, battery charging and running an electric vehicle are demonstrated, which reduce the export of PV to less than 10%.

The financials of the above are discussed, together with discussion of using smart variable tariffs that are now becoming increasingly available.

The house shown in Fig. 1 was recently refurbished to a high specification of insulation and K glass windows. Additional installations were a 5.4-kWp PV array on the south-east facing roof, underfloor heating and various system controls which can be monitored by smart phone apps.

The hot water comes from an immersion heater (3.2 kW) powered primarily from the roof through a smart switching system. This detects when the PV generated electricity is exporting and instead diverts this energy to the hot water cylinder (210 Litre).

The underfloor system, which is zoned in every room, is heated by gas. There is also a 7kw log burner in the main living area, which due to the efficiency of all the above is hardly ever lit. When it has been used the living area, which is quite large (7 m × 6 m), overheats to about 25 °C even if the outside temperature is around 0 °C (Fig. 2).

New electrical distribution boards were installed, future proofing for installation of battery storage.

Appliances such as oven and hob that require high power are connected to one set of circuits and those that require lower power such as TV, fridge, lighting, phone hubs, and computers are connected to another set of circuits (Figs. 3 and 4).

1 Electricity Meters

There are two electricity meters – one is the smart meter for power readings recorded half hourly and the other is the generation meter for recording the harvesting of energy from the PV system. The former is for billing from the utility company and the latter is for collecting the feed in tariff from the energy produced from the PV generation.

T. Book (✉)
Consultant, Brighton, UK

© The Author(s), under exclusive license to Springer Nature Switzerland AG 2022

A. Sayigh (ed.), *Sustainable Energy Development and Innovation*, Innovative Renewable Energy,
https://doi.org/10.1007/978-3-030-76221-6_40

Fig. 1 5.4 kWp PV system

Fig. 2 Distribution boards and control

Fig. 3 Upper Distribution Board and 100 Amp Supply

Fig. 4 Lower distribution board (Lower —right-named eddi (DWH diverter))

2 Power Diverter and Hub

The white and grey box on the lower left is the 'smart' controller which diverts the exportable energy. The system is connected via a hub which balances the excess load between the hot water tank and the car charger (Figs. 5 and 6).

This readout on the simple screen of the power diverter shows that about 5 kW have been diverted to the hot water tank by 2 pm on a mid-March day saving £1 of energy – average saving over a year is about £400.

The data can also be seen on an app (see later) (Fig. 7).

This example shows, towards the end of a day in May, that 7.75 kWh has been diverted into a 210 litre tank.

Figure 8 shows a hybrid car being charged. About 75% of the energy over the year came from the PV panels and the remainder from the grid, usually off peak.

The car had a battery range of 30 miles in the winter months and up to 50–55 miles in the summer months. Additionally, there was a 35 litre petrol tank which provided energy to a small generator which powered the electric motor, extending the total range by another 300 miles.

The car was driven about 10,000 miles per year.

The hybrid was exchanged for an electric vehicle (EV) (Fig. 9) which averages 5 miles (8 km) per kW in the summer and 3 miles (4.8 km) per kW in the winter.

The 50 kW battery has a range of about 225 miles in the summer.

At the time of publication, the car has travelled about 5000 miles and consumed 1306 kW, 589 kW from the grid at 5p/kWh and 546 kW from the PV system on the roof. In addition, 150 KW @25p was purchased on journeys (The car battery during the journey was freely charged at various stations). This works out at about 1.3 pence per mile or 1 euro-cent per kilometre.

Although there is a price premium on EV purchase when compared to ICE cars (internal combustion engines), EVs are cost-effective, that is, they require low maintenance and running costs. The financial break-even is about 30,000 miles when compared with an equivalent ICE car (Fig. 10).

The hot water tank and underfloor heating manifold on the left.

The hot water tank is heated by PV-generated energy for about 300 days a year. There are coils in the tank which can be heated by the gas boiler, but this system has been used on less than a dozen occasions in 4 years (Fig. 11).

Each room has its own control and is a controllable zone by itself.

In terms of energy management, not a lot of thought is given to water usage other than heating it.

It pays to be mindful of how water volume is used. There are two people in this household, who mainly shower. The average volume of water usage is about 240 litres per day in the winter months and 400 litres per day during the summer period. This extra amount in the summer is due to watering the garden, including irrigation for growing vegetables.

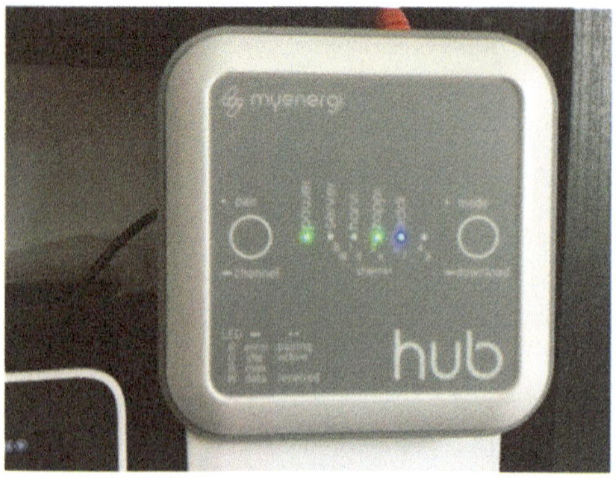

Fig. 5 Power diverter. Blue light shows that the temperature of the tank is 60°C

Fig. 6 Diverter screen

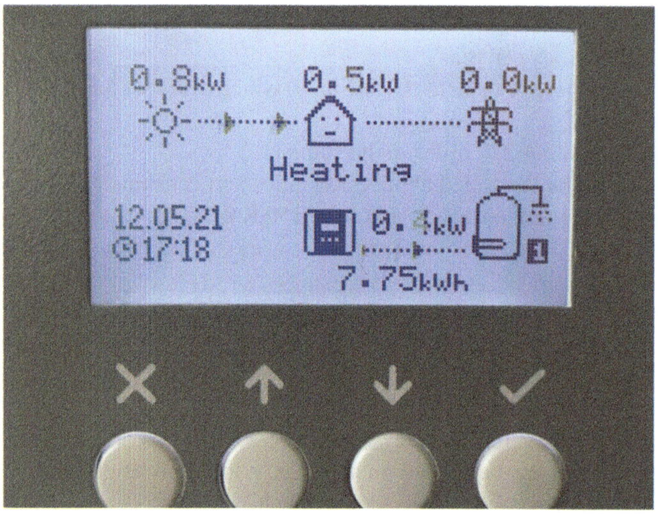

Fig. 7 Hybrid Car

Fig. 8 Hybrid car plugged in

Fig. 9 EV being charged

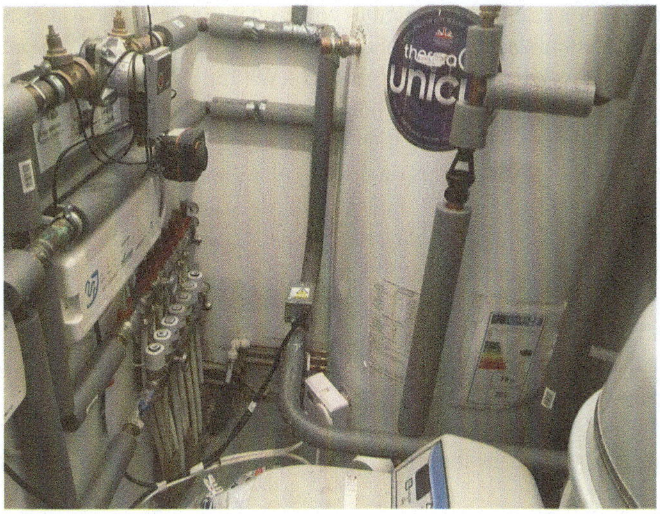

Fig. 10 Plant room

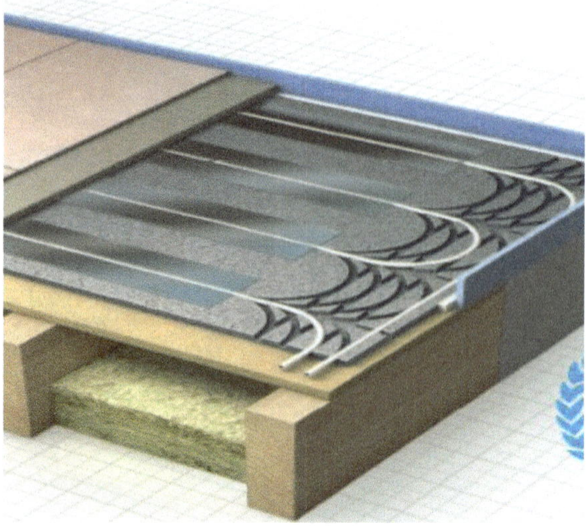

Fig. 11 Underfloor heating system

3 Water Usage

In order to ameliorate the water use in summer (4 × 250 litre), water butts filled via down pipes from the roof have been added to reduce the water volume purchased (Figs. 12 and 13).

There were further improvements after 3 years:

The incorporation of a charge inverter and battery pack to "time shift" electricity consumption throughout the year.
The purchase of an all-electric car (EV).
Acquisition of a smart car charger which 'fuels' the car only when there is surplus on the house. This works in tandem with the smart hot water diverter and is very efficient.

There is the opportunity to obtain split plunge pricing of electricity from a utility company and pay for power by the half hour and even get paid for using it in the off peak times. This is dependent on the carbon intensity which is difficult to predict from week to week. By definition, the carbon intensity tends to be higher in the summer as the grid sources more 'clean' energy from PV, wind and other renewables (Figs. 14, 15, and 16).

The storage system comprises 16 × Calb 2000 mAh lithium-ion phosphate batteries controlled by a set of 123SmartBMS micro-processors. These cards calibrate the batteries to ensure that they are neither overcharged nor overdischarged. The parameters are set at 90% and 20%, respectively. In addition, the BMS prevents the batteries from overvoltage and over/undertemperature. The charge-inverter was sized to match the PV array (5.4 kWp) with charge and discharge rates of 4 kW from the grid or the PV system.

In the winter months, the charge-inverter is scheduled to come on during off peak times at 2.30 am. To charge from 20% to 90%, it generally takes about 2.5 hours and uses about 10 kW and costs 50p (5p/kWh on the Octopus GO-Faster Tariff). At 5.00 am, the hot water is boosted for 90 minutes. This uses 5 kW (15p). The charge-inverter is 'held all day before being switched back on. This stops the batteries supplying the hot water and using up some of the battery storage. It also allows the BMS time for absorbtion and to float the cell units, thereby conditioning them and increasing their life. The EV can be charged during this time if required, but that depends on driving plans and current car range levels.

During the day in the winter, the property runs on off peak electricity stored overnight and released during the day. Without any added PV energy, on a typical winter's day, the batteries will last until early evening and may or may not cover the whole of the evening meal cooking.

Gradually as the PV energy increases in late Januaury into February, the battery supply goes through to at least 2.30 am, with a combination of off peak grid and solar energy.

As the spring comes round, the battery state of charge (SOC) is reduced to 60–80%, allowing the increasing solar radiaton to make up the difference during the day.

Similarly, the grid-sourced water heating is cut back too. By late March, the batteries and solar generation are capable of meeting all the system charging, hot water heating, house requirements of cooking, washing, drying and the EV, around the clock.

This continues until late September when the grid starts to be used more frequently.

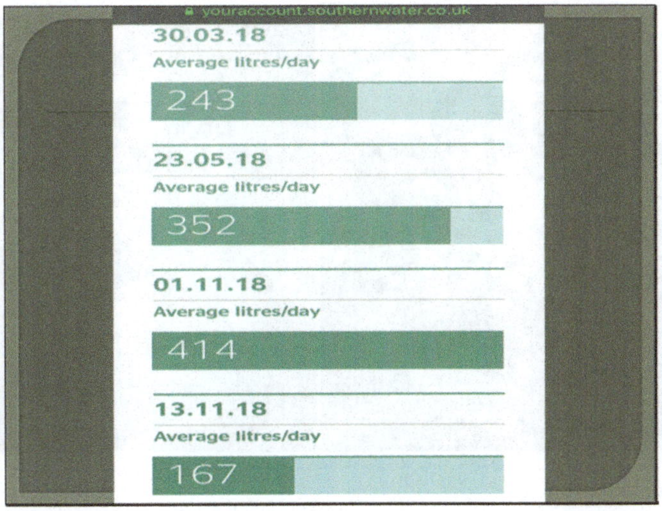

Fig. 12 Water consumption readings

Fig. 13 Three of the four 250 litre water butts

Fig. 14 *Two inverters*: on the left is the PV inverter and on the right is a charge-inverter which controls the 10 kW battery system

Fig. 15 The CALB 16 × 2000 mAh battery set

Fig. 16 A BMS smart card ensures that the batteries run between 20% and 90% charge and do not overvoltage or overheat

4 Examples of Spring and Summer Energy Usage

Note: The base line when away on holiday is 2kWh. This is the amount required just to keep the tank hot at 60 °C from one day to the next (Fig. 17).

27 kWh consumed, of which 20 kWh generated.

The hot water was boosted for an hour around 6 am as there had been no sunshine the previous day.

The car was charged during the morning and more water was heated around the middle of the day and early afternoon.

Only 13 kW was imported (purchased). The evening consumption shows a small increase around supper time and the rest for lighting and evening TV (Fig. 18).

A small car is charged first then rest of the house followed by a lot of water heating on a sunny day.

Only 4 kW imported all day and 7.5 kW actually used (Fig. 19).

The car was charged overnight with off-peak electricity and there was washing machine usage/cooking in the morning (Fig. 20).

A good summer's day produces over 30 kW. (The annual amount is 5500 kWh.)

Once again, plenty of hot water was produced. There is 84% green contribution on this day which is typical for the summer months (twice that gained in the winter). Additional saving was made due to energy management using the diverter (Figs. 21 and 22).

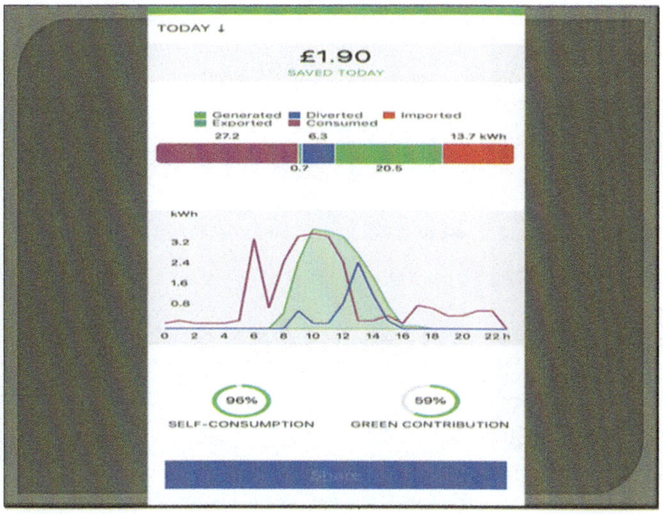

Fig. 17 App screen of performance

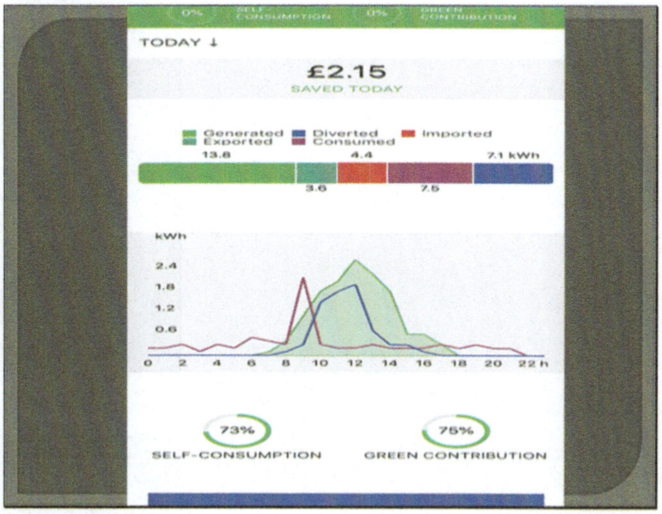

Fig. 18 Typical spring day

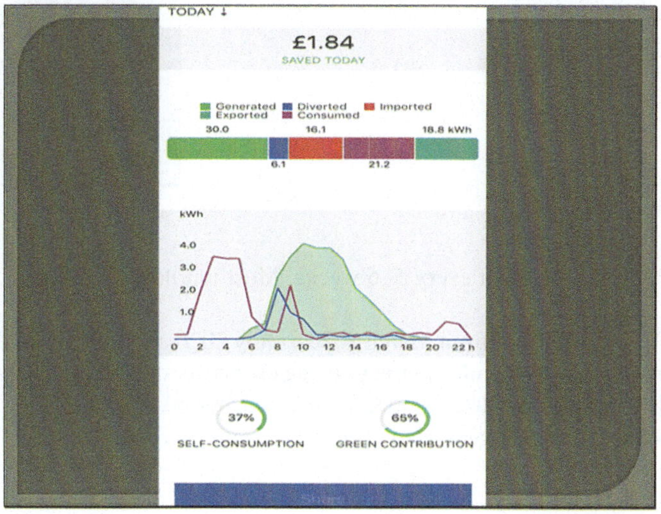

Fig. 19 A summer's day

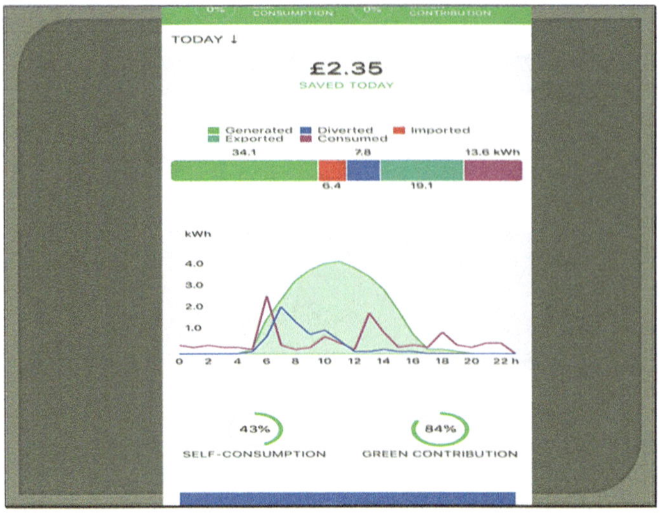

Fig. 20 More activity

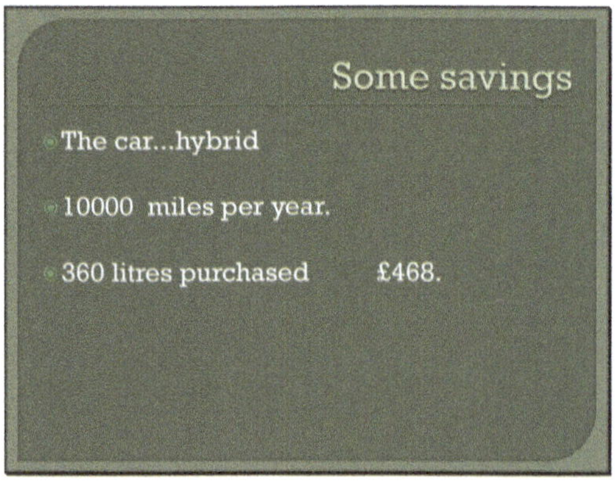

Fig. 21 Car financials

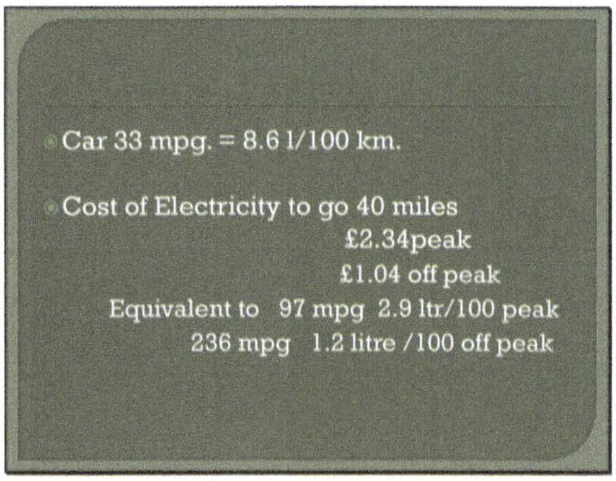

Fig. 22 Hybrid car costs

The running costs are very attractive.

The hybrid car was filled with petrol about every 5–6 weeks. Most journeys were local except for occasional 400 mile round trips.

During the recent Corona Virus lockdown, from March 2020 to July 2020, the car was used for short journeys only. About 900 miles was covered in short trips up to 40 miles journeys using electricity only. NO petrol was purchased for five months. Indeed one has to be mindful of the gasoline deteriorating in that time period (Fig. 23).

It is estimated that about a third of the PV generated off the roof is used to charge the car. This produces a very economical result.

Figure 24 is self-explanatory.

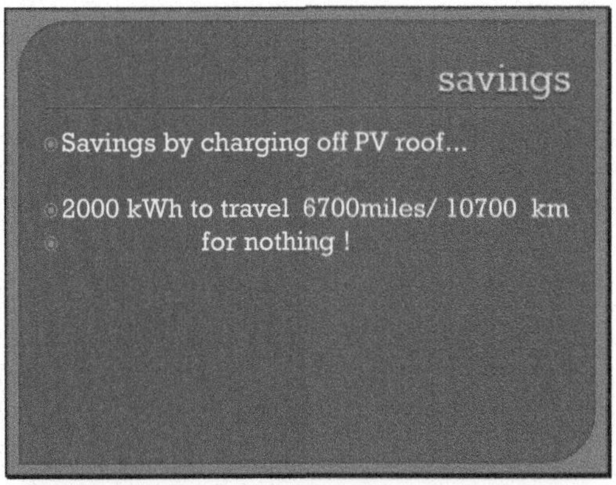

Fig. 23 Saving in hot water, lighting and cooking

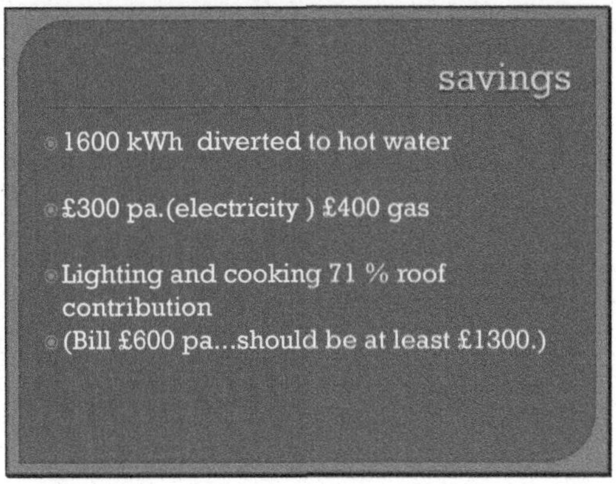

Fig. 24 Bill reduction by £700 per year

About a quarter of the PV output heats the hot water and the 'roof' contributes significant savings to the utility bills and running the car

Figure 25 is self-explanatory.

About a quarter of the PV output heats the hot water and the 'roof' contributes significant savings to the utility bills and running the car (Fig. 26).

On a slightly cloudy day, the PV system generates 2.6 kW, the batteries float and the hub balances the car and hot water (Fig. 27).

A winter's day (off peak 00.30–04.30). Eddi = DHW, Zappi = Car, Home = Battery Charger. At 00.30 am the batteries get charged at 2.30 am, the car receives 5.9 kW and the hot water about 5 kW. The PV heats the water further 1 kW around mid day from the winters sun, which produces 5.3 kWh over the day. It is a heavy day of use; overall 27 kWh is imported.

Off peak was changed to 02.30–0630 after six months to ensure that the DHW was at its hottest at late as possible (Fig. 28).

From 6 am, the PV starts generating and the batteries (home) get charged up first.

The car gets a good 15.6 kW, which is the equivalent of about 62 miles (100 km), and the hot water takes nearly 6 kW. In sum, 34.5 kW was generated by the PV and 2.2kw imported. Only 0.9 kW was exported, so that 95% of the PV-generated energy was used up. Once the sun goes down, the batteries cover evening cooking, TV, lighting and all consumption through the night.

In the winter, the batteries do not get much charge from the winter sunshine and need topping up overnight to get through the day. On the night tariff (Octopus Go) at 5p per unit, it only costs about 50p, which is a third of the day-time tariff. The savings are considerable.

The household consumes 7500 kWh per year excluding space heating (gas).

5400 kWh comes from the PV system. About half the year, the property is self-sufficient, defined as importing of energy of 1Kwh or less per day. It takes 1 kW per day to run the 16 battery control cards. The winter average consumption is 11 kWh per day, most of that from the batteries. There is a lot of debate about the economics of the PV and/or batteries. It depends a lot on life style choice. These days, and it is improving all the time, PV will pay back in 5–7 years and batteries in 7–9 years.

Buying this technology is equivalent to buying energy forward which stays at today's price, whilst the actual tariffs rise with inflation and governments' energy policies, not only in this country but also throughout the world.

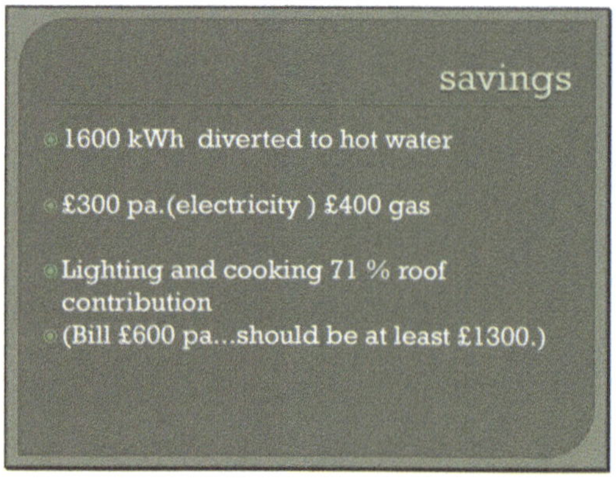

Fig. 25 Home energy status

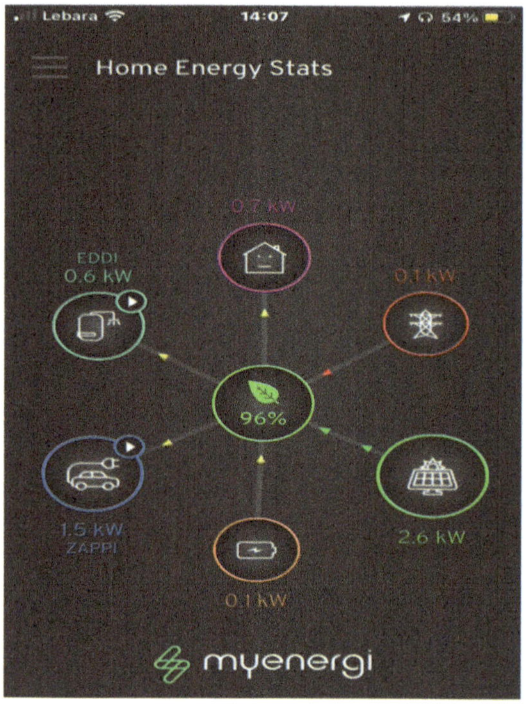

Fig. 26 Readings of A winter's day (off peak 00.30-04.30)

It is important to choose the most economical electricity tariffs, and this is best researched from the price comparison websites. The best prices can be found from products that use Smart Meters where supplies are priced by the half hour and can be monitored by an IDH (in-home display).

For space heating, which is still using a gas boiler (at least for another decade), shopping around for the most economic tariff is important. A total of 10,000–20,000 kWhs of gas are being purchased per year, so even small variations in tariff can make a difference to the budget, also taking into consideration the daily standing charge.

The monthly consumption of the household is 600 kWh/month throughout the year.

Apart from space heating in using municipality gas, all cooking, washing, drying, TV, etc. are electric.

All lighting is by LED.

The consumption of grid electricity in the winter months:

Per month

About 200 kWh comes from the PV system, the rest from the batteries.

The time shift of supply from off peak to peak times has changed the ratio of grid consumption from 60:40 to 10:90 with commensurate savings.

Summer months:

Per month.

Fig. 27 Shows the PV system starts generating electricity from (06:00 AM)

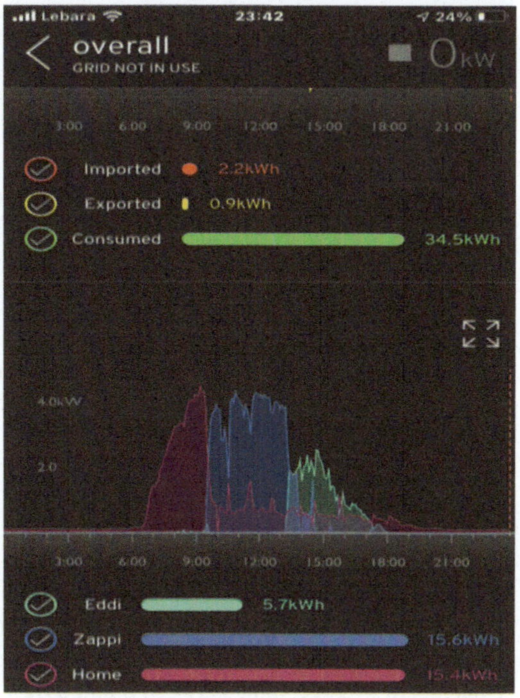

Fig. 28 Summer time

700 kWh is supplied by the PV. On average, the only 35 kW is imported and monthly bill is £12–£14, of which £7.50 is the standing charge.

Sample reports from a smart phone app analysing the smart meter data.

5 Typical Winter Consumption

10 Jan 2021 to 9 February 2021 (31 days)

Your total electricity cost is £47.77 inc 5% VAT.

The unit consumption cost is £40.02 inc VAT.

The standing charge cost is £7.75 inc VAT.

On Go, you would have paid an estimated £47.77 inc 5% VAT – est. 6.81p/kWh.

Your electricity usage is STABLE during this period.

Your electricity cost is STABLE during this period.

You have a total consumption of 587.5kWh.

Your average unit rate is 6.81p/kWh.

Your average factual unit rate is 8.13p/kWh (includes your standing charge).

Your average standing charge is 25.00p/day.

Your national carbon footprint is 98.68 kg CO_2 or 170 g CO_2/kWh.

Your regional carbon footprint is 88.55 kg CO_2 or 151 g CO_2/kWh.

Both the national carbon footprint (across Great Britain) and your regional footprint (for your distribution area) are calculated. An average value between 0 and 59 g CO_2/kWh is very low, 60–159 g CO_2/kWh is low, 160–259 g CO_2/kWh is average,

260–359 g CO_2/kWh is high and 360 g CO_2/kWh or more is very high. You can improve your carbon footprint by avoiding peak hours, reducing your base load and planning consumption when carbon intensity on the grid is the lowest.

Your estimated base load is 52 W, totalling 39.1kWh.

This is an estimate of the amount of electricity used when you are not actively using anything. This includes devices that are kept running all the time such as routers and fridges, smart lights, standby power for plugged in appliances, etc. An average home has a base load of 105–120 W. Lowering your base load can make a big impact on your electricity cost and can be as simple as turning a device off at the socket when not in use.

6 Typical Spring Consumption

14 April 2021 to 14 May 2021 (31 days)

Your total electricity cost is £15.53 inc 5% VAT.

The unit consumption cost is £8.42 inc VAT.

The standing charge cost is £7.11 inc VAT.

On Go, you would have paid an estimated £14.87 inc 5% VAT – est. 9.92p/kWh.

This is an average import unit of electricity cost rate in p/kWh.

Your electricity usage is STABLE during this period.

Your electricity cost is STABLE during this period.

You have a total consumption of 71.8kWh.

Your average unit rate is 11.72p/kWh.

Your average factual unit rate is 21.62p/kWh (includes your standing charge).

Your average standing charge is 22.94p/day.

Your national carbon footprint is 14.83 kg CO_2 or 204 g CO_2/kWh.*

Your regional carbon footprint is 10.86 kg CO_2 or 163 g CO_2/kWh.

* Some of the CO_2 data was not available. Instead, estimates were calculated based on the available data.

Both the national carbon footprint (across Great Britain) and your regional footprint (for your distribution area) are calculated. An average value between 0 and 59 g CO_2/kWh is very low, 60–159 g CO_2/kWh is low, 160–259 g CO_2/kWh is average, 260–359 g CO_2/kWh is high, and 360 g CO_2/kWh or more is very high. You can improve your carbon footprint by avoiding peak hours, reducing your base load and planning consumption when carbon intensity on the grid is the lowest.

Your estimated base load is 6 W, totalling 4.5kWh.

This is an estimate of the amount of electricity used when you are not actively using anything. This includes devices that are kept running all the time such as routers and fridges, smart lights, standby power for plugged in appliances, etc. An average home has a base load of 105–120 W. Lowering your base load can make a big impact on your electricity cost and can be as simple as turning a device off at the socket when not in use.

7 Conclusions

The system discussed in this chapter has been operating for four years and has been constantly monitored, using smart phone apps. It is proving to be very efficient in making significant savings of both energy and running costs. The return on investment is about 20% and the residence uses about half the normal fossil fuel energy compared to a similar sized house without such a system.

The system saves about 6 tons of CO_2 per year. The gas boiler uses about 3 tons of CO_2 per year. If it was realistic to equip half the homes in the UK similarly over the next 10–15 years, then the savings (of CO_2) would be over 30 million tons per year.

The Scope of Advanced Smart Vacuum Insulation Technologies for Net Zero Energy Buildings

Saim Memon

1 Introduction

The perception at which, currently, net zero energy building (NZEB) cannot be imagined without glass is because glazed windows play an imperious role of allowing natural daylighting but with repercussion of space-heating energy loss in cold-arid and space-cooling energy loss in hot-arid areas [1]. As such, evitable energy losses through glazed windows of buildings indirectly contribute to carbon emissions, thus impelling climate change. This is particularly due to an inadequate thermal transmittance value (U-value) of a glazing. In this chapter, brief discussions of the scope of advanced smart vacuum insulation technologies developed by the author of this chapter's collaborative contributions are presented. This chapter presents the new and existing published findings by the author on (i) Vacuum Glazing (VG), (ii) Triple Vacuum Glazing (TVG), (iii) Electrochromic Vacuum Glazing (EVG), (iv) Semi-transparent Photovoltaic Electrochromic Vacuum Quadruple Glazing (STPV-EVQG), (v) Translucent Vacuum Insulation Panel (TVIP), and (vi) Fusion-sealed Vacuum Glazing (FVG).

2 Vacuum Glazing (VG)

A VG encompasses evacuated space between two sheets of low e-coated glass supported by pillars and edges sealed with hermetic (vacuum-tight) material [2]. Figure 1a illustrates a design proposed in Memon et al. [3] distinguished in terms of its composite edge seal made of $Sn_{56}Pb_{39}Zn_3Sb_1$-$AlTiSiCu_1$ wt%) (CS-186), it was considered as an alternative to indium-sealed VG. This composite hermetically sealed the edges of glass sheets, as exhibited in Fig. 1b, and obtained the vacuum pressure of 0.0042 Pa, as presented in Fig. 1c. The pump-out hole was also sealed with this composite as shown in Fig. 1d. Thus, predicted the U-value of 0.91 $Wm^{-2}\,K^{-1}$, as shown in the hot-side isotherm in Fig. 1e, and experimentally achieved 0.97 $Wm^{-2}\,K^{-1}$, as demonstrated in Fig. 1f. This method overcomes the cost issue, without compromising the scope of using tempered glass and soft coatings, but its composite edge seal owns 39% of Pb, complete details are published elsewhere [3].

3 Triple Vacuum Glazing (TVG)

An evacuated cavity of a typical VG will not achieve ultra-high vacuum pressure under laboratory conditions. Thus, an extra layer of evacuated space allows an improvement to the thermal performance (U-value) <0.5 $Wm^{-2}\,K^{-1}$. The concept of TVG introduced in [4] and a new development of TVG, with the same composite edge seal utilised for VG, reported by Memon et al. [5]. Figure 2a illustrates a design diagram of TVG and Fig. 2b displays the constructed TVG in which an evacuated pressure attained was 0.48 Pa as demonstrated in Fig. 2c. A 3D FEM model of the constructed TVG was developed and the hot-side isotherm is shown in Fig. 2d. The U-value obtained was concluded to be 0.33 $Wm^{-2}\,K^{-1}$. It was predicted using the

S. Memon (✉)
Solar Thermal Vacuum Engineering Research Group, London Centre for Energy Engineering, School of Engineering, London South Bank University, London, UK
e-mail: S.Memon@lsbu.ac.uk; https://saimmemon.com

© The Author(s), under exclusive license to Springer Nature Switzerland AG 2022

A. Sayigh (ed.), *Sustainable Energy Development and Innovation*, Innovative Renewable Energy,
https://doi.org/10.1007/978-3-030-76221-6_41

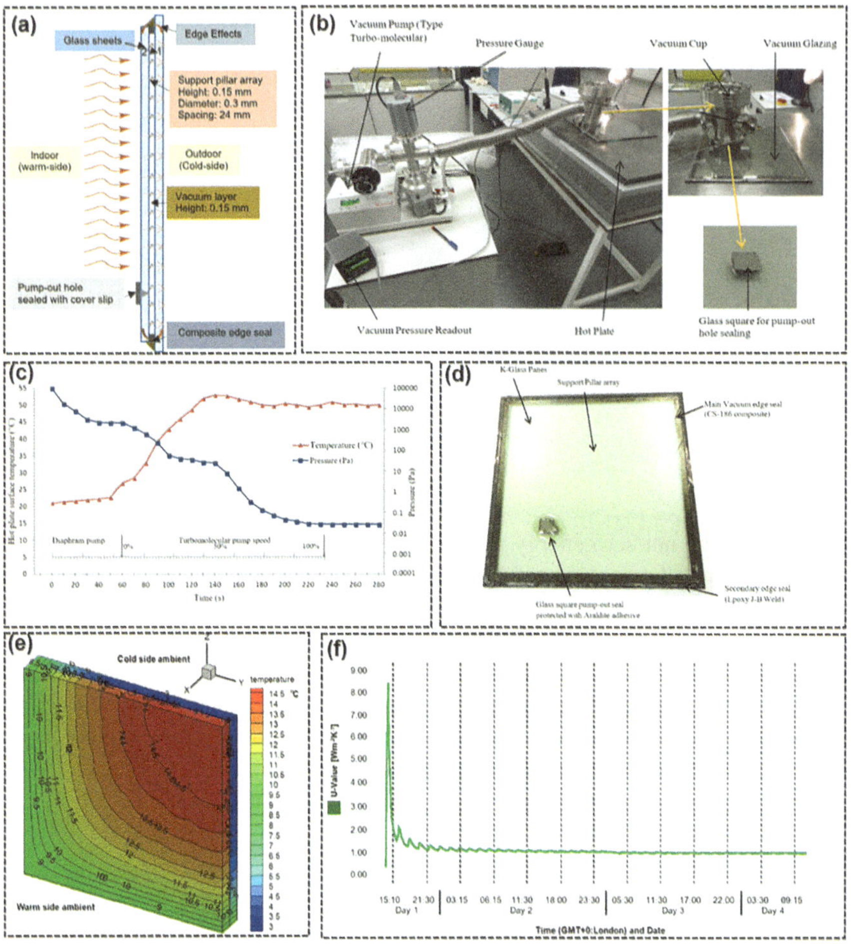

Fig. 1 Illustration of VG. (**a**) Design diagram, (**b**) experimental setup for evacuation and pump-out hole sealing, (**c**) temperature/pressure profiles, (**d**) constructed VG, (**e**) FEM model for thermal performance predictions, and (**e**) experimental measurement of U value

developed model that the thermal performance could be improved by shortening the width of the composite edge seal as the U-value was compromised due to the edge effects [6]. This method was the first successful construction of TVG, but it is complex-to-construct, mainly due to the need of precision involved in ultrasonically soldering the edges of glass sheets. Long-lasting endurance and gradation of vacuum pressure are mainly reliant on the edge-seal [7]. Thus, such stepped arrangement of glass sheets, as exhibited in Fig. 2a and b, and the level of precision required to ultrasonically solder the edges are developmental challenges. If cavity-vacuum pressure leaks, then the thermal performance could deteriorate to 2.4 Wm^{-2} K^{-1}. Full details of TVG are reported elsewhere [5–7].

4 Electrochromic Vacuum Glazing (EVG)

An idea of EVG was first reported in [8]. In this study, EVG is an integration of electrochromic film, made of WO_3 (Tungsten anhydride), with solder glass edge-sealed VG, made by SPACIA [9], to form a single component capable of performing the task of repelling heat escape from glazed windows and exhibit the smart characteristics of an electrochromic film, as illustrated in Fig. 3a, which is a reversible change from opaque to transparent by the application of low-electric field from power supply, as illustrated in the in Fig. 3b. This electrochromic film required electrical power of <0.4 Wm^{-2} that allowed the reaction of WO_3 with the electrons, and charge balancing ions enabled its reverse transformation. Using the hot-guarded box thermal testing method, following ISO 9869 standards, experimental measurements of U-value were performed for minimum 72 h maintaining a minimum of 5 °C. EVG achieved U-value of 0.82 Wm^{-2} K^{-1}, as shown in Fig. 3c, which is slightly better than VG. Although degradation occurs in electrochromic films caused by a variety of environmental factors, such as

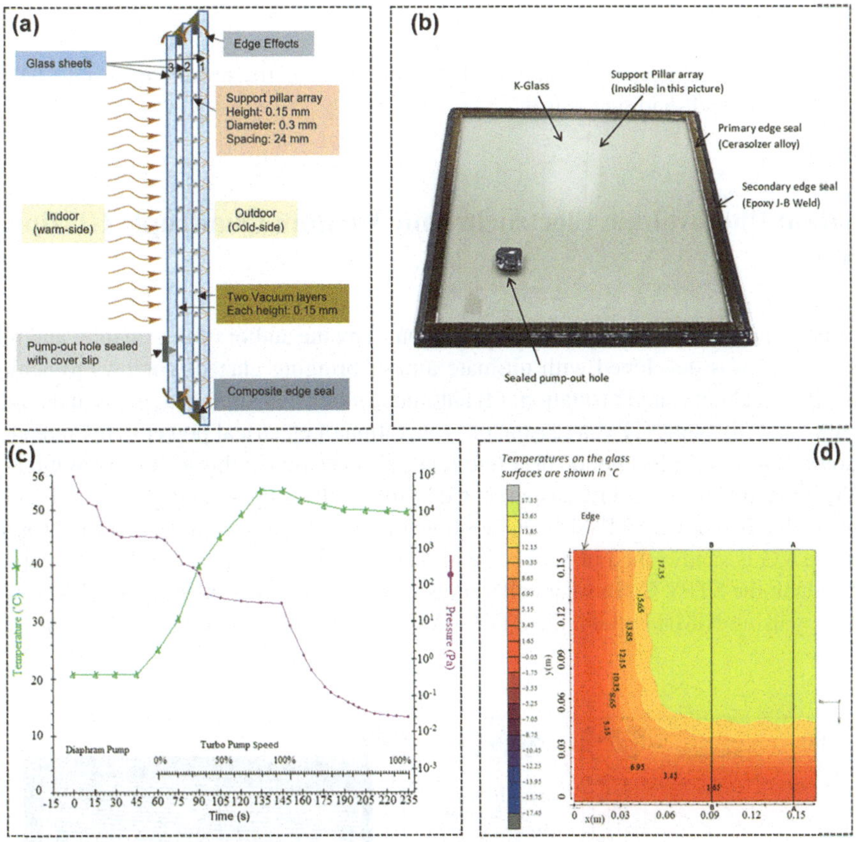

Fig. 2 Elucidation of TVG. (**a**) Design diagram, (**b**) constructed sample, (**c**) experimental temperature/pressure profiles, and (**d**) 3D FEM model for thermal performance prediction

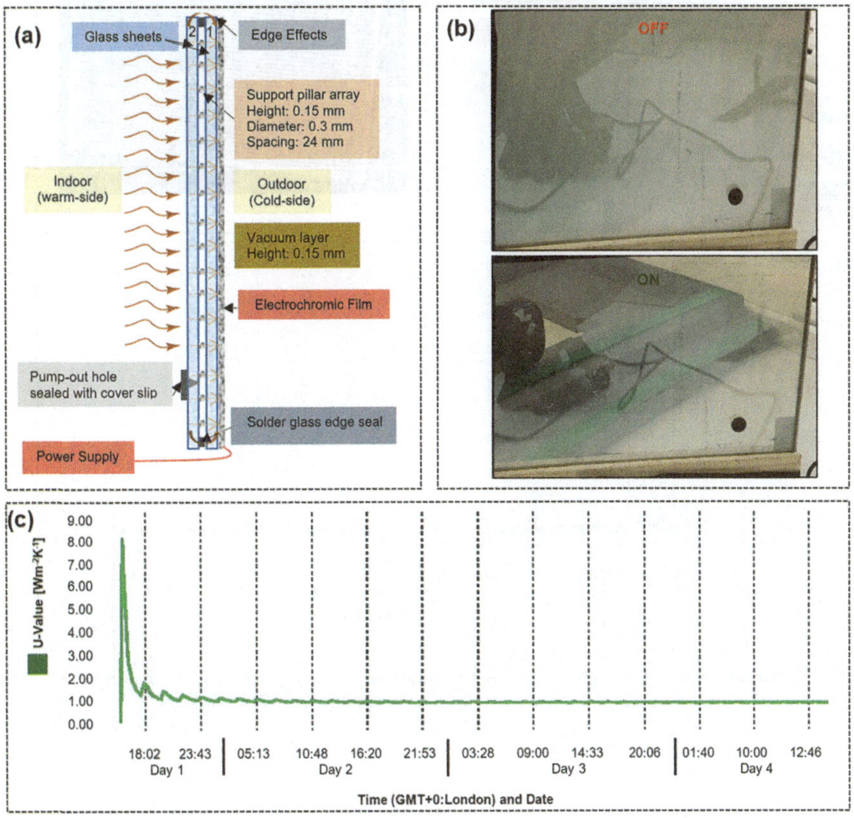

Fig. 3 (**a**) Design diagram of EVG, (**b**) reversible change from opaque to transparent by the application of low-electric field from power supply, and (**c**) experimental measurement of U value of EVG with heat flux meter

water, it hinges on ways of chemical deposition and the advancement of electrochromic materials. EVG is a promising solution for adaptive solar heat gain and shading control.

5 Semi-Transparent Photovoltaic Electrochromic Vacuum Quadruple Glazing (STPV-EVQG)

A STPV-EVQG conception is developed and patented (provisional) by the author of this chapter and initial results published elsewhere [10]. This concept was developed with ultimate aim of bringing glazed windows to achieve (a) lower thermal transmittance U-value, (b) excellent sound insulation, (c) automatic control of solar heat gains and control over transparency with electrochromic film, and (d) With STPV, a reasonable generation of electrical power to operate electrochromic film and electronic device to control over WIFI for mobile user interface. The research in this area is ongoing. Figure 4a shows design diagram of STPV-EVQG that integrated semi-transparent CdTe solar cell strings-based glazing to EVG. Initial investigations are presented here that is the assembled STPV-EVQG as shown in Fig. 4b, in which the electrochromic film layer faces the internal environment and VG is sandwiched in between and STPV on the outer part to allow the capture of the solar radiations. To make it self-reliant, the STPV CdTe solar cell strings-based glazing was integrated to EVG to supply the power and store the generated energy using NiMH battery, the STPV-EVQG achieved U-value of 0.7 $Wm^{-2} K^{-1}$ as shown in Fig. 4c.

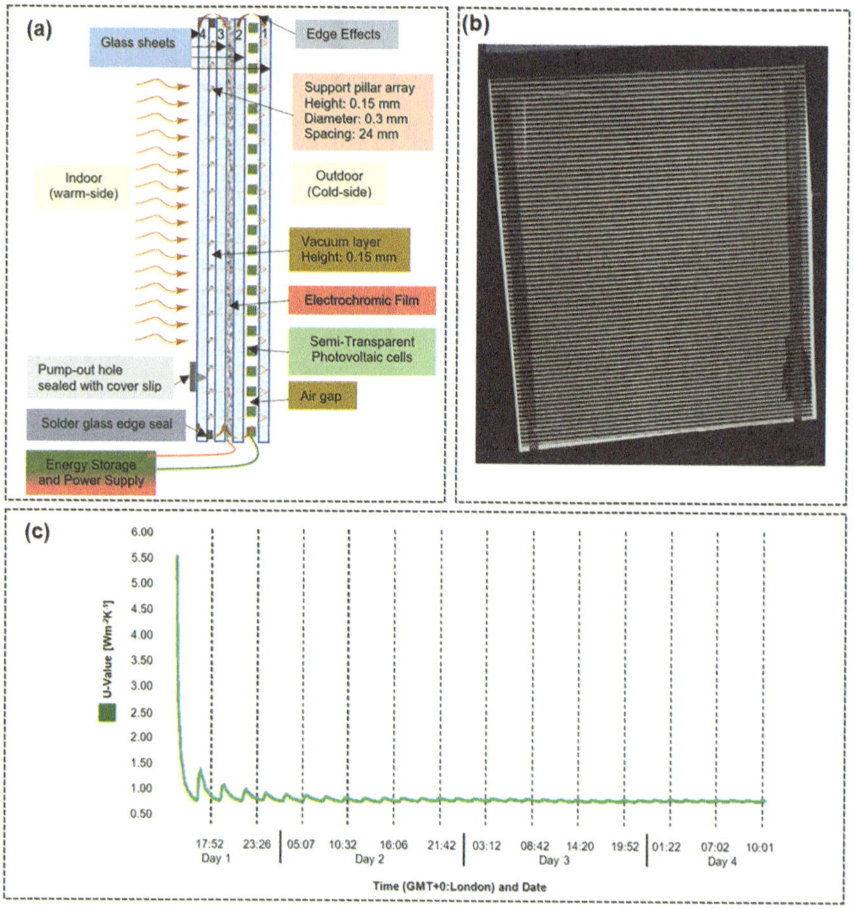

Fig. 4 Illustration of STPV-EVQG. (**a**) Design diagram, (**b**) assembly for initial experimentations, and (**c**) experimental measurement of U-value with heat flux meter

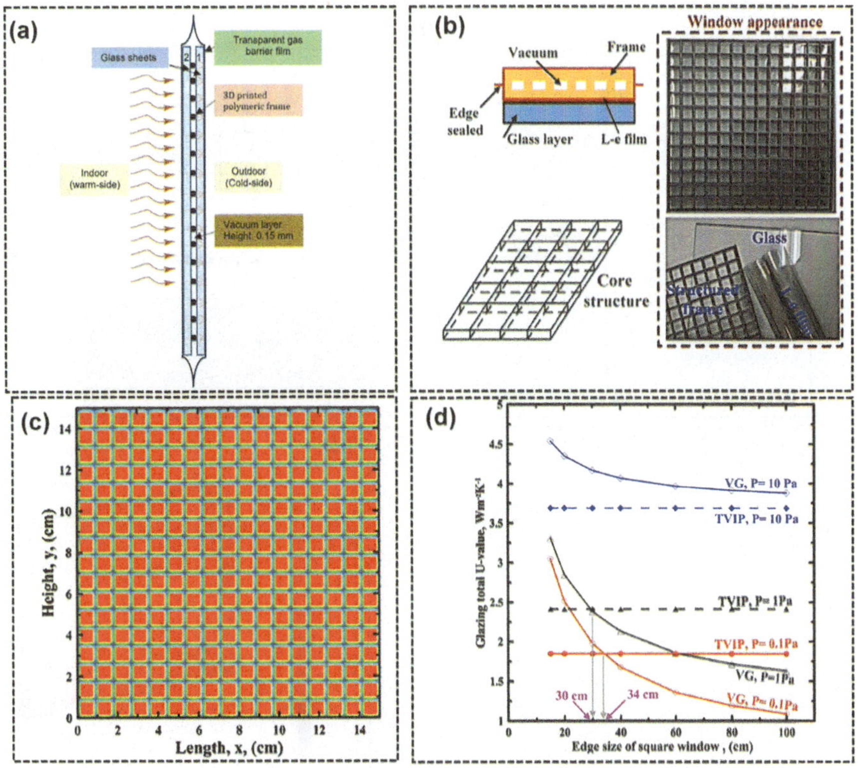

Fig. 5 Illustration of TVIP. (**a**) Design diagram, (**b**) hollow-frame 3D printed structured-core material encapsulated in a transparent multi-layered polymeric envelope, (**c**) hot-side isotherm of TVIP as a result of the 3D-validated FEM simulation, and (**d**) comparative U-value

6 Translucent Vacuum Insulation Panel (TVIP)

Traditionally, VG suffers with the complex-to-construct and subsequent edge effects issues. A concept of TVIP [11] is eminent in terms of it has no edge effects, easier to construct, its cost is lower than VG and has lower thermal conductivity as compared to traditional opaque vacuum insulation panels (VIP) as shown in Fig. 5a. The TVIP has been a progressive development with a scope to replace existing insulated windows. This TVIP consists of a hollow-frame structured-core material encapsulated in a transparent multi-layered polymeric envelope and the spacers are 3D-printed, as shown in Fig. 5b [11]. The TVIP with frame and mesh spacers achieved thermal conductivity of $7{\cdot}10^{-3}$ Wm^{-1} K^{-1} at a vacuum pressure of 1 Pa and the light transmittance of 0.88. Figure 5c shows hot-side isotherm of TVIP as a result of the 3D-validated FEM simulation. On the one hand, TVIP has no edge effects, and on the other hand, 3D-printed polymeric structure increases the thermal short-circuit around the central TVIP area. Thus, TVIP achieved 1.7 Wm^{-1} K^{-1}, which is slightly higher than VG as shown in Fig. 5d. Full details of TVIP are reported elsewhere [11]. However, the TVIP with the transparent gas barrier envelope which is made of PET (Polyethylene terephthalate), SE-PET (Silica Evaporated PET), PA (Polyamide) and CPP (Cast Poly Polypropylene) layers. It means the problems of future gas leaks, outgassing and absorption of moisture challenges are remained that limit the practicality of this TVIP design.

7 Fusion-Sealed Vacuum Glazing (FVG)

Advancement in hermetic (vacuum-tight) edge-sealing materials has been one of the challenges since decades because of the existing cost, use of hazardous substance and complexity-to-construct issues in VG. A novel concept of glass-metallic textured layer for hermetic fusion edge-sealed VG invented, as shown in Fig. 6a, and reported in Memon and Eames [12] when Sn62-$B_2O_3$38 wt% fused with Sn90-In10 wt% alloy, as shown in Fig. 6b. The pump-out hole seal of FVG achieved at the hot-plate surface heat induction of 50 ± 5 °C with the cavity-vacuum pressure of $8.2{\cdot}10^{-4}$ Pa, as shown in Fig. 6c. The FVG is cost-effective (at least 70% reduction to edge-sealing materials such as indium, CS-186 and solder-glass cost), energy efficient (ultrasonic soldering-free), and Pd-free (hazardous substance-free) solution and has a potential for mass production.

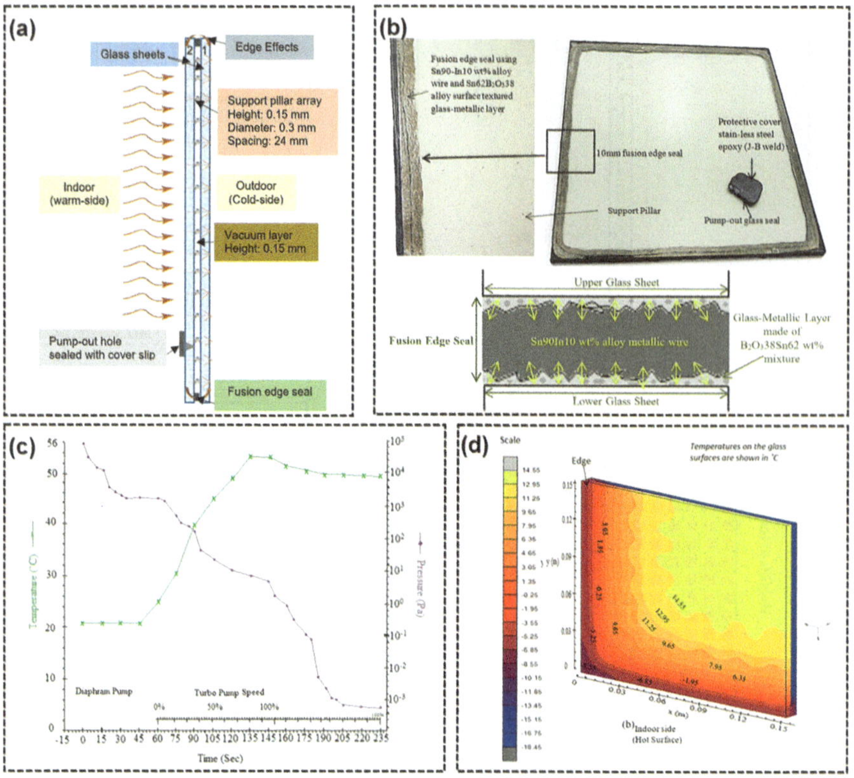

Fig. 6 Illustration of FVG. (**a**) Design diagram, (**b**) constructed photograph showing the fusion edge-seal composition, (**c**) experimental temperature/pressure profiles, and (**d**) 3D FEM simulated model for thermal performance prediction

A validated 3D FEM predicted central and overall U-value of FVG to be 1.03 and 1.24 $Wm^{-2} K^{-1}$, respectively. Figure 6d shows the influence of edge effects due to the use of a 10 mm wide fusion seal, the edge effects can be seen to extend to a distance of up to 72 mm from the edge on the warm side surfaces. Overall, it can be seen that the FVG is implied to be the most low-cost solution for mass production. Full details are reported elsewhere [12].

8 Conclusion

This chapter briefly discussed the scope of advanced smart vacuum insulation technologies for NZEB. The results implicate that VG achieved U-value of 0.97 $Wm^{-2} K^{-1}$, it was further reduced with TVG that achieved U-value of 0.33 $Wm^{-2} K^{-1}$. To control the solar heat gains from 0.41 (transparent mode) to 0.13 (opaque mode), ECG is constructed that achieved U-value of 0.82 $Wm^{-2} K^{-1}$, it required electrical power of <0.4 Wm^{-2}. To make it self-reliant, the semi-transparent PV glazing was integrated to EVG to supply the power and store the generated energy using NiMH battery, the STPV-EVQG achieved U-value of 0.7 $Wm^{-2} K^{-1}$. Due to the cost/edge-effects of hermetic glass edge-sealing materials, a new structured core transparent vacuum insulation panel (TVIP) is briefly discussed that achieved U-value of 1.72 $Wm^{-2} K^{-1}$ but suffered the vacuum pressure degradation. This chapter briefly discussed novel concept of glass-metallic textured layer for hermetic fusion edge-sealed vacuum glazing (FVG) invented when Sn62-$B_2O_3$38 wt% fused with Sn90-In10 wt% alloy achieved vacuum pressure of $8.2 \cdot 10^{-4}$ Pa that achieved the U-value of 1.24 $Wm^{-2} K^{-1}$.

Acknowledgements This work was supported by Daiwa Anglo-Japanese Foundation (12549/13360), EPSRC of the UK (EP/G000387/1), LSBU NSS Project grants.

References

1. Memon S, Katsura T, Radwan A, Zhang S, Serageldin AA, Abo-Zahhad EM, Sergey S et al (2020) Modern eminence and concise critique of solar thermal energy and vacuum insulation technologies for sustainable low-carbon infrastructure. Int J Solar Thermal Vacuum Eng 1(1):52–71. https://doi.org/10.37934/stve.1.1.5271. ISSN online (2716-6953)
2. Memon S (2013) Design, fabrication and performance analysis of vacuum glazing units fabricated with low and high temperature hermetic glass edge sealing materials. PhD Thesis. Loughborough University, UK. https://dspace.lboro.ac.uk/2134/14562
3. Memon S, Fang Y, Eames PC (2019) The influence of low-temperature surface induction on evacuation, pump-out hole sealing and thermal performance of composite edge-sealed vacuum insulated glazing. Renew Energy 135:450–464
4. Manz H, Brunner S, Wullschleger L (2006) Triple vacuum glazing: heat transfer and basic mechanical design constraints. Sol Energy 80(12):1632–1642
5. Memon S, Farukh F, Eames PC, Silberschmidt VV (2015) A new low-temperature hermetic composite edge seal for the fabrication of triple vacuum glazing. Vacuum 120:73–82
6. Memon S (2017) Experimental measurement of hermetic edge seal's thermal conductivity for the thermal transmittance prediction of triple vacuum glazing. Case Stud Ther Eng 10:169–178. ISSN 2214-157X. https://doi.org/10.1016/j.csite.2017.06.002
7. Memon S, Farukh F, Kandan K (2018) Effect of cavity vacuum pressure diminution on thermal performance of triple vacuum glazing. Appl Sci 8(9):1705
8. Fang Y, Eames PC (2006) Thermal performance of an electrochromic vacuum glazing. Energy Convers Manag 47(20):3602–3610
9. Memon S (2014) Analysing the potential of retrofitting ultra-low heat loss triple vacuum glazed windows to an existing UK solid wall dwelling. Int J Renew Energy Dev 3(3):161–174
10. Radwan A, Katsura T, Memon S, Serageldin AA, Nakamura M, Nagano K (2020) Thermal and electrical performances of semi-transparent photovoltaic glazing integrated with translucent vacuum insulation panel and vacuum glazing. Energy Convers Manag 215:112920
11. Katsura T, Memon S, Radwan A, Nakamura M, Nagano K (2020) Thermal performance analysis of a new structured-core translucent vacuum insulation panel in comparison to vacuum glazing: experimental and theoretically validated analyses. Sol Energy 199:326–346
12. Memon S, Eames PC (2020) Design and development of lead-free glass-metallic vacuum materials for the construction and thermal performance of smart fusion edge-sealed vacuum glazing. Energ Buildings 227:110430

Studies of the Effect of Glazed Facades and Vegetation on Urban Heat Islands in Brasilia

Gustavo Cantuaria

1 Introduction

Urban centres are usually places with the highest temperature indexes due to the heat islands, and it is possible to notice a great difference of temperature in relation to areas farthest from the centre. Each city has its characteristics and therefore the intensity and timing of the occurrence of heat islands may vary. The phenomenon can be aggravated by several reasons. In this research, the focus is on the use of glass on façades, where, especially the more reflective, can be one of the contributors to the occurrence of diurnal urban heat islands. Increase in energy consumption, mainly by the use of air conditioning, is another interference, in order to compensate the excessive heat and create climates more pleasant to the internal environments, generating a greater waste of money. Furthermore, thermal discomfort is imposed on the population that has to live or pass through the place with buildings of glazed facades, due to excessive heat and glare.

2 Climatic Context of Brasilia

In arid and semi-arid regions, like Brasilia, characterized by the subtropical climate of altitude, the annual climatic cycle presents wide variations between the seasons, with particular significance in the case of ambient temperature and humidity, not only yearly but also on a daily basis. Due to the usually high percentage of clear days, the quantity and intensity of solar radiation is also relevant. The climatic features described coupled with increased urban construction, result in intensified heat islands during the warm and dry season, accentuating the strain on urban microclimates, and causing environmental discomfort [1]..

3 Methodology

This research focuses on urban heat islands. The method used in this study was remote sensing, where measurements of surface temperatures are taken. The measurements are of the amount of energy reflected and emitted by different surfaces, by different facades, in the same urban area. Urban fragments were photographed in the south wing and the north wing areas of the gregarious sector of Brasilia. The areas chosen to measure the heat islands were the commercial and hotel sectors, located on the north and south wings, as they have a large number of buildings with varying facades, some with almost 100% of the facade covered by glass and others with a lower percentage. Differences are due to the year of construction and renovation of the buildings. The more recent buildings tend to have a greater glazed facade surface area as it eludes a "wealth and trendiness" status.

G. Cantuaria (✉)
University Centre of Brasilia (UniCEUB), Brasilia, Brazil
e-mail: gustavo.cantuaria@ceub.edu.br

© The Author(s), under exclusive license to Springer Nature Switzerland AG 2022

A. Sayigh (ed.), *Sustainable Energy Development and Innovation*, Innovative Renewable Energy,
https://doi.org/10.1007/978-3-030-76221-6_42

During the fieldwork, photographs were taken using the FLIR C2 photo camera, which provides both thermal photos with the measurement of the temperatures emitted by the surfaces at the chosen points, as well as the original photos of the site. Photos were taken of the west-facing facades and of the ground below the chosen buildings, at two distinct times: first in the beginning of the afternoon, from 12:00 to 15:00, when the sun is more intense, and second at night, from 19:00 to 20:00, when it is possible to note the existence or non-existence of the islands of heat. Comparisons were made between the two photos taken at the selected locations and at their respective times. In addition to the photos, the THDL-400 Thermo-hygrometer was used to measure temperature and humidity of the general sectors at the same time.

Brasilia being a planned city ensures certain prerequisites for its constructions, such as maximum height of the buildings, however, these requirements do not influence the materials that will be used on its façades. The earlier, modernist architecture of the South commercial and hotel sectors, is different from the later contemporary architecture of the North commercial and hotel sectors. The South Sector presents façades with lower percentage of glass, use of solar protections and balconies, which does not exclude the existence of buildings with a large part of its facade covered by glass, but shaded. Furthermore, the South commercial sector contains more shaded locations, compared to other sectors, in both wings. These shadows are provided by vegetation, by buildings themselves and space configurations, having a much larger number and proximity to the buildings. The relative air humidity of the city centre, which is already low, is aggravated by all these factors, especially in the dry season, which was the time chosen to illustrate this research.

4 Study Cases and Results

The areas selected for research were SCS (South Commercial Sector) and SCN (North Commercial Sector), as well as SHS (South Hotel Sector), and SHN (North Hotel Sector). These sites were chosen because they are part of the gregarious scale of Brasilia, places of meeting, socializing, and tours. They also form the central core of the plane-like form of Brasilia. Figures 1-6 present temperature recordings and thermal images of some of the urban fragments recorded, measured, and analysed.

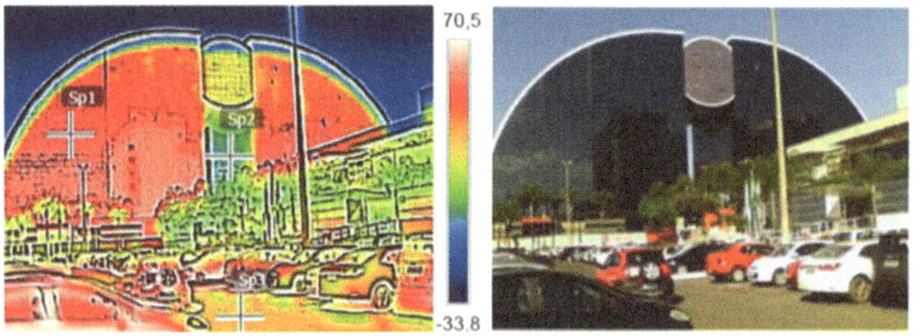

Fig. 1 Brasilia Shopping, daytime

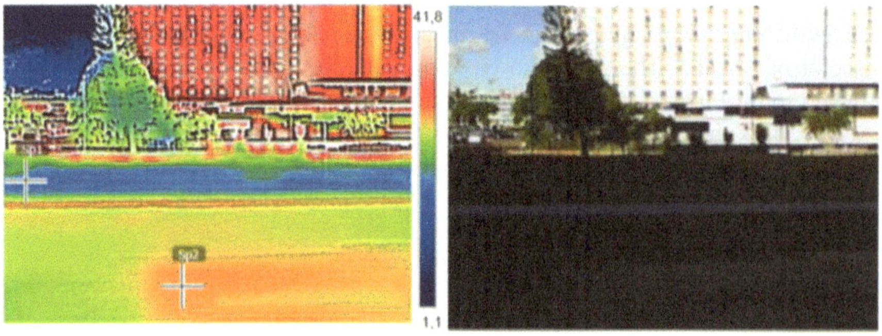

Fig. 2 Ground microclimate close to the National Hotel, daytime

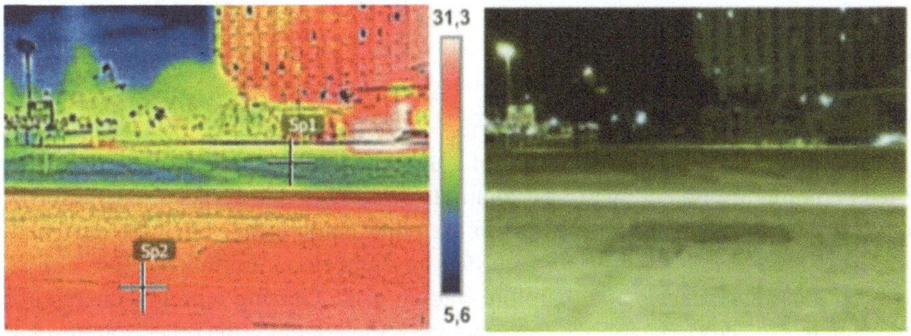

Fig. 3 Ground microclimate close to the National Hotel, night time

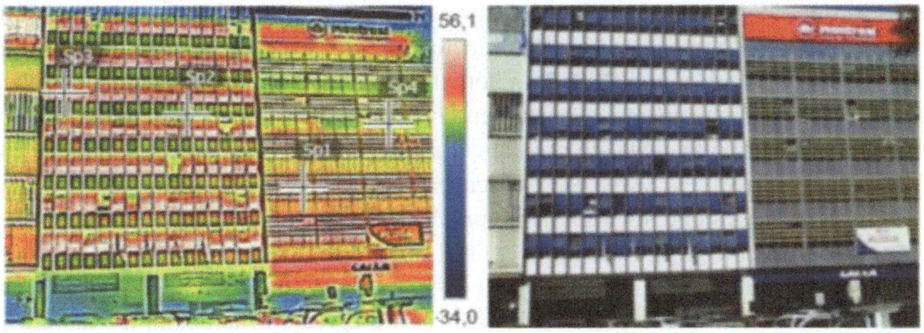

Fig. 4 Newton Rossi and Montreal buildings, respectively, daytime

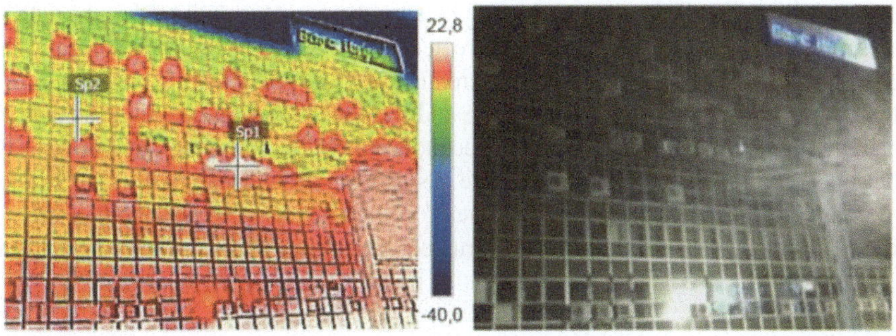

Fig. 5 Bancorbrás building, night time

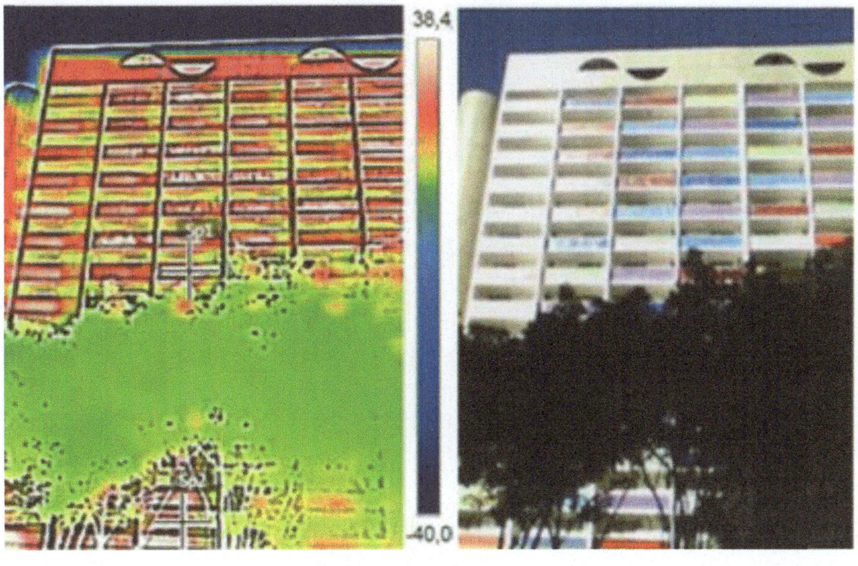

Fig. 6 St. Paul's daytime

5 Conclusion

This study suggests that during the day, the extensive use of glass on facades affects the microclimate thermally, increasing surrounding surface temperatures by reflecting solar radiation. They also cause visual discomfort for pedestrians due to excessive glare from reflection of direct solar radiation. On the other hand, facades that are designed with solar protection elements like balconies and brise soleils, or use concrete for their envelope, minimize glare effects and solar reflection. Energy efficient glazings are more expensive and therefore seldom used. Glazing with low energy efficiency stimulates the use of air conditioning, which also adds to heating the air externally. On hot days, the reflection of solar radiation is increased by the use of glass. The heat comes from increased absorption of solar radiation by surfaces that retain it like pavements and asphalt. These urban surfaces are also dry, aggravated by the lack of green areas and water bodies, generating energy accumulation, further heating the air.

The increase in temperature of the local microclimate during the day, suggests the occurrence of an unusual phenomenon of diurnal heat islands. At night, the microclimate is not affected directly by the glazed facades, as the radiation is quickly dissipated. The "traditional" heat island, at night, exists in both the North and South Sectors. This is due to the space configuration, with excess built surfaces, of pavements, asphalts, and buildings, aggravated by the amount of cars on site, and the sparse vegetation, which does not provide great sun protection or increased humidity during the day. The south wing is older than the north wing, containing more trees, fewer glazed facades, and greater use of architectural elements to block the sun, therefore avoiding visual discomfort due to glare and providing more shading, ameliorating the thermal sensation. Furthermore, the use of shading elements and radiation control, intercepts the energy in the correct place before it reaches the building, causing the heat to be obstructed, reflected and dissipated into the external air. In addition to acting efficiently, the control elements bring expression to the building, as can be observed in the examples of modernist architecture, as opposed to contemporary "glass boxes". It is possible to minimize the losses or gains of heat by the energy balance of glazing on facades.

Research results in the north wing suggest the occurrence of what can be called diurnal urban heat islands, caused by the high reflectance of the solar rays by the glazed and mirrored facades that are absorbed by the ground, generating an increase in surface temperatures, humidity reduction, and causing greater microclimatic thermal discomfort during the day. There is a clear indication that vegetation makes a difference in the creation of more comfortable microclimates and in the mitigation of urban heat islands. In order to improve the microclimate and to reduce the heat islands, it is important the use of urban vegetation, which shades the ground and blocks it from direct solar radiation, lowering the absorption and emissivity of the radiation, while increasing the humidity of the microclimate. As shown in the research, the sites studied have a higher temperature where there is no vegetation shading density. Therefore, results from the researched urban fragments, strongly indicate the presence of heat islands due to dense constructed areas with sparse vegetation and significant amount of paved areas.

Reference

1. Cantuaria GAC (2001) Trees and microclimatic comfort: with special reference to Brasilia, Brazil. PhD. Architectural Association School of Architecture

Buildings' Morphology, Solar Rights and Zero Energy in High Density Urban Areas

Isaac Guedi Capeluto

1 Introduction

Solar exposure during winter and shading during the hot season greatly influences the quality of open areas between buildings providing proper conditions to sustain urban life. Furthermore, common zoning regulations include rules for the location of solar energy systems in buildings and open areas and therefore sun access should be ensured in order to guarantee their efficient operation. Mutual shadowing between buildings may be an obstacle to install photovoltaic arrays integrated in the building envelope and open spaces around buildings compromising the feasibility of the zero energy building (ZEB) equation. Hence, the zero energy concept should be considered at the urban scale investigating different morphologies. To help reducing energy consumption and greenhouse gas emissions generated by buildings, they must become from energy consumers to energy producers. In a zero energy (or net zero) building/complex (similar to zero emissions), its annual energy consumption is considerably reduced by energy efficiency measures and is supplied by renewable energy sources on the site [1]. During the measurement period (1 year), most of the zero energy buildings are connected to the urban grid and use its energy when on-site energy production is minor to building's loads. The energy is returned to the grid when on-site energy production is excess.

There are several definitions for zero energy building/complex. The main difference between them is the limits of measurement – the location of renewable energy sources. In accordance to different definitions, the energy production source can be found within the building footprint or inside the site boundary, however, the building owner can purchase "green" energy off-site or import renewable energy source (biomass, ethanol, etc.) for on-site energy production.

Most of the existing examples that implement zero energy projects deal with relatively low buildings located in low urban density areas, and then with a very high potential for solar exposure and collection in PV systems located in their envelope or surroundings areas. In this work, we will challenge the relationship between urban density, environmental quality, and energy production by examining the energy generation potential of several morphologies for renewal of a urban zone undergoing a densification program in Tel Aviv.

2 Densification Programs in Israel

The land is one of the most precious natural resource in Israel. Despite the importance of protecting the green open spaces and natural areas, the reality shows a constant reduction of these areas. According to the "Sustainability Plan for Israel 2030" [2] a recommended design strategy to cope with this threat, is concentrated dispersion and revitalization of existing city centers. To minimize damage to open spaces, in the early 2000s, an urban renewal plan was proposed. Within the framework of the plan, the planning authorities in Israel proposed two main tracks for the promotion of municipal urbanization processes [3]:

I. G. Capeluto (✉)
Faculty of Architecture and Town Planning, Technion – Israel Institute of Technology, Haifa, Israel
e-mail: arrguedi@technion.ac.il

© The Author(s), under exclusive license to Springer Nature Switzerland AG 2022

A. Sayigh (ed.), *Sustainable Energy Development and Innovation*, Innovative Renewable Energy,
https://doi.org/10.1007/978-3-030-76221-6_43

1. Urban renewal (increased building rights)
2. Urban redesign (clearance and construction)

The aims of both tracks are to achieve an efficient land utilization within built-up areas, the creation of a new housing inventory in cities, and their improvement by upgrading existing buildings and infrastructure. In the context of the "Urban Renewal" track, the committee promotes the intensification of built-up areas without demolition activities and maximal use of existing infrastructures. In contrast, the "Urban Redesign" route requires fully re-planning of an existing area with significant plot densification and related infrastructure renewal. The case study complex presented in this work is facing an approved "Urban Redesign" densification program. In this case, the density will increase by more than 200% compared to the existing situation. The densification plan and general design layout was already approved by District Committee in 2015.

3 Design for Solar Rights and Solar Access

The consideration of solar rights and solar access in urban design is fundamental in order to allow passive heating of buildings in winter and to improve the comfort conditions of people in streets, sidewalks, and open spaces. It can reduce the energy consumption of the building if used indoors, while insolation of exterior spaces may create climatically comfortable areas, which can be used for outside activities in winter. On the other hand, proper shading should be provided in order to avoid overheating of buildings and create pleasant spaces during summer.

Knowles [4] suggested a method for ensuring solar access to each residential unit in a community. De Kay [5] made a comparative analysis of various envelopes allowing daylight access. Schiler and Uen-Fang [6] developed a computer program for the generation of solar envelopes for flat-rectangular sites based on Knowles work, and Koester [7] presented energy armatures using passive resources like winds and rain water for urban sustainable development. The model SustArc expanded upon the term "solar envelope" and offered the creation of two different types of envelopes [8]: Solar Rights Envelope – SRE: A 3D constraint which defines the maximum height of a built volume which would not violate the solar access for the defined surroundings on specific hours, days, and months. Solar Collection Envelope – SCE: A complementary 3D constraint which defines the lowest heights from which a new built volume would receive solar access, given a built context and on specific hours, days, and months. The volume included between both envelopes is called the Solar Volume – SV.

Solar envelopes have been traditionally advocated as a means to achieve sustainable and energy-efficient development and the use of SCE for self-shading building envelopes was shown to have energy performance improvements [9]. However, a recent study indicates that at least under some climate conditions and building types, the use of traditional solar envelopes has a negative effect on total energy use including transportation as well as larger negative impact on developable density [10]. This is partially dependent on lower densities leading to increased distance traveled which leads to increased energy use. The study focuses on single family detached residential homes in the United States, but raises broader questions regarding the overall usefulness of solar envelopes as a design strategy.

Capeluto and Plotnikov [11] proposed a new approach for the generation of solar envelopes which is based on specific requirements throughout the year as opposed to uniform general requirements for the entire immediate environment. This is especially important for achieving high density since the more refined and specific the solar envelopes are, the bigger their obtained volume will be. Applying different parameters to the surrounding context such as buildings and parks and combining the resulting envelopes to one: The definition of solar access is further refined by first acknowledging that the context is not uniform and requires a combination of solar access requirements. One example of that would be the need to ensure solar access for different facing facades on different times of the day. Another example may take place in a mixed-use context – an office building does not have the same quantitative and qualitative solar access requirements as a residential building, a school, a public garden, and so on.

Therefore, by providing adequate solar access to the surroundings based on climate, orientation, and use, the envelope reacts to the environment needs more accurately while maintaining a maximum achievable volume, while ensuring solar access for desired conditions for the entire year and as a result maximizing the land use and the project's feasibility.

4 Urban Densification Alternatives

The studied plot is located in Tel Aviv – Jaffa (32.5N, 35E), one of the densest cities in Israel.

Fig. 1 Aerial view of the existing situation

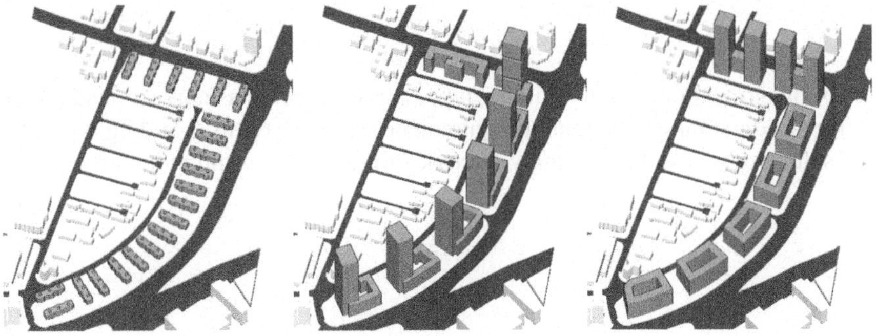

Fig. 2 Isometric view of studied alternatives: existing (left), approved (center), and proposed (right)

The existing buildings in the plot are elongated with linear typology. They are located following the nearby road geometry; therefore, the same building typology appears with varied orientations (Fig. 1). The plot contains twenty-one residential buildings; all of them are four-floor constructions with 22 apartments in each building [12, 13].

In this chapter, we analyze several alternatives with increased urban density (Fig. 2):

0. Existing – Current Plot Condition, Including the Addition of Two Floors in Top of each existing buildings.
1. Approved – redesign as approved by District Committee, as designed by Yashar Architects Ltd.
2. Proposed – redesign to maximize usable roof area and electricity production, keeping same density as the previous alternative.

The aim of this study is to check how near in the way to zero energy can we get within plot borders with future plot density conditions. Can we improve the approved design in terms of energy consumption and electricity generation within plot borders?

The existing condition density is 10.7 units/dunam. It does not match to the minimal density requirements by in force National Building Plan 35 [14], which is 12 units/dunam. With the addition of two floors to existing buildings, we achieve a density of 16.6 units/dunam. The approved plot redesign will increase the density by about 3 times. Both densification design alternatives, (1) approved and (2) proposed, both with improved insulation, were designed to achieve the same density, namely, 32.4 housing units/dunam (Table 1).

The approved option shows the design alternative by Yashar Architects Ltd. This alternative contains 3 different building block types: an 8-story long type building, a 30-floor tower with 8-floor base on the lower levels, and a 6-story building connected to the 25-floor tower which appears 5 times at plots number 2, 3, 4, 5, and 6. In this case, the potential surfaces for PV installation are reduced and unclear because of mutual shading.

The proposed alternative was designed to increase on plot solar yield potential. The strategy was to maximize roof surfaces fully exposed to solar radiation to increase on-site electricity production. It also contains 3 basic building block types: 2 similar blocks of 8-story buildings connected to two 20-floor towers above (with minor geometry differences due to actual plot borders) which appear only once at the northern plot 1 to maximize façade exposure and prevent mutual shading, and a European (courtyard) type block which appears 5 times. This block is a nine-story building with flat roof fully exposed to sun radiation.

Table 1 Comparison of density by plot number and design alternatives

		Units amount by design alternatives			
Case study plot with new site division	Plot number	0. Existing	Existing + 2 floors	1. Approved	2. Proposed
	Plot Density (unit/dunam)	10.7	16.6	32.4	32.4
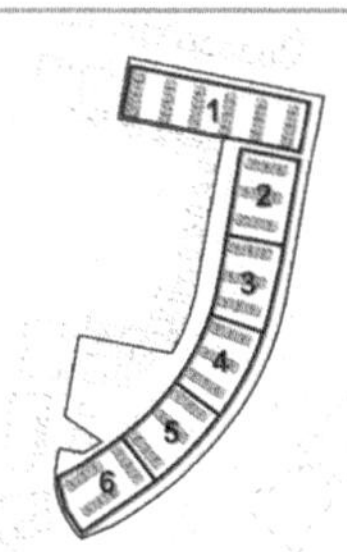	1	132	204 (132 + 72)	391	680
	2	66	102 (66 + 36)	201	143
	3	66	102 (66 + 36)	201	143
	4	66	102 (66 + 36)	201	143
	5	66	102 (66 + 36)	201	143
	6	66	102 (66 + 36)	201	144
	Total	**462**	**714 (462 + 252)**	**1396**	**1396**

In both design options, the ground floor of all building types was considered as non-residential, public floor, as approved in the original plan. The consideration of differential requirements for solar rights (building type, facades, and open areas) can heavily affect the resulting design and accordingly increase or reduce the resulting density.

5 Results

As mentioned earlier, a zero energy project must first reduce its energy use by energy efficiency measures, and then use renewable energy sources on the site to balance the demand. In order to determine the overall energy consumption, two conditions were checked: existing condition with no envelope insulation, and improved insulation according to Israeli Energy Rating of Buildings Standard [15].

According to the results, there is a direct relationship between urban density, energy consumption, and area needed and available for electricity production. The study showed that even near zero energy goal is difficult to achieve using low efficiency (12%) PV systems implemented on the roofs, 50% energy generation should be a realistic target. Existing buildings' protrusions and balconies, limits PV array installation on main facades. However, assuming more effective solar cells, it is theoretically possible to achieve zero energy goal for the existing plot situation with improved envelope using solar panels roof installation with about 30% efficiency. When plot density increased (addition of 2 floors), more efficient panels are needed. Zero energy for this condition will be possible to achieve with 40% efficient PV panels (Fig. 3).

Figure 4 presents the total energy consumption of the entire complex for alternatives 1 – approved and 2 – proposed, comparing it with the electricity generation capacity using PV in roof and facades assuming different panel's efficiencies. It was found that about two-thirds of roof surfaces of the approved design (alternative 1) are fully exposed or partly shaded, while another third is shaded most of the year. Additionally, the upper levels of south-oriented facades can also be considered for energy generation. Due to mutual shading and small roof area of the several towers, in this alternative, electricity generation is mainly based on vertical arrays, while roof electricity generation is about 20% only of the total production.

On the other hand, the proposed alternative (alternative 2) is composed mainly by lower courtyard buildings, which offer larger, fully exposed roof areas for maximum electricity generation. In this case, PV roof array which is easy to install and maintain, assuming 20% efficiency can provide more than 85% of the total energy used for HVAC and lighting, depending less on complex vertical façade installation and maintenance. In this case using facades as well, it is possible to achieve almost a zero energy complex, a task that will be easier when more efficient panels become available.

6 Conclusions

According to the study, there is a clear relationship between urban density, energy consumption, and area needed and available for electricity production. However, architectural design plays an important role in this equation. There are density limitations for developing nearly zero energy urban complexes and ensure basic environmental qualities. However, there is a challenging opportunity to push the limit and achieve nearly zero energy complexes if "good architectural design" is implemented, and more efficient systems developed.

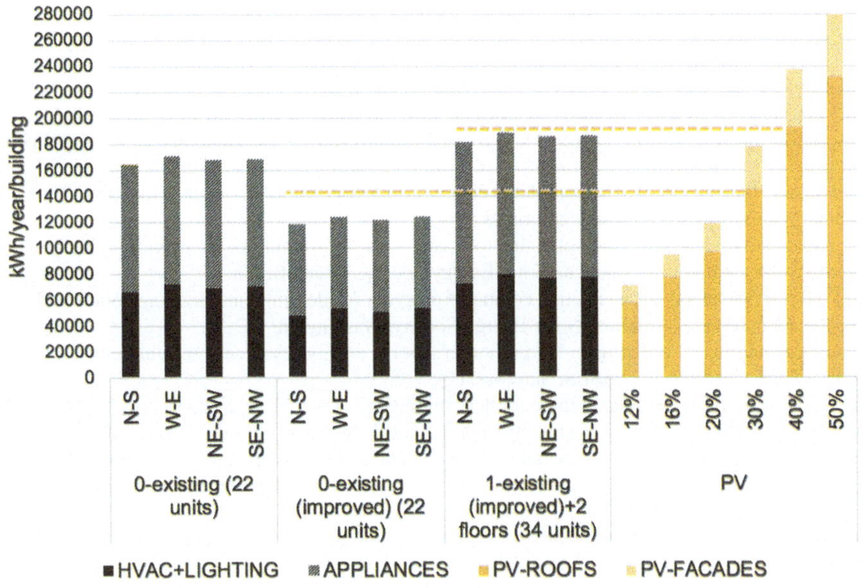

Fig. 3 Comparison of energy consumption of a single building with various orientations in relation to electricity production by solar panels with varying efficiency (kWh/year/building)

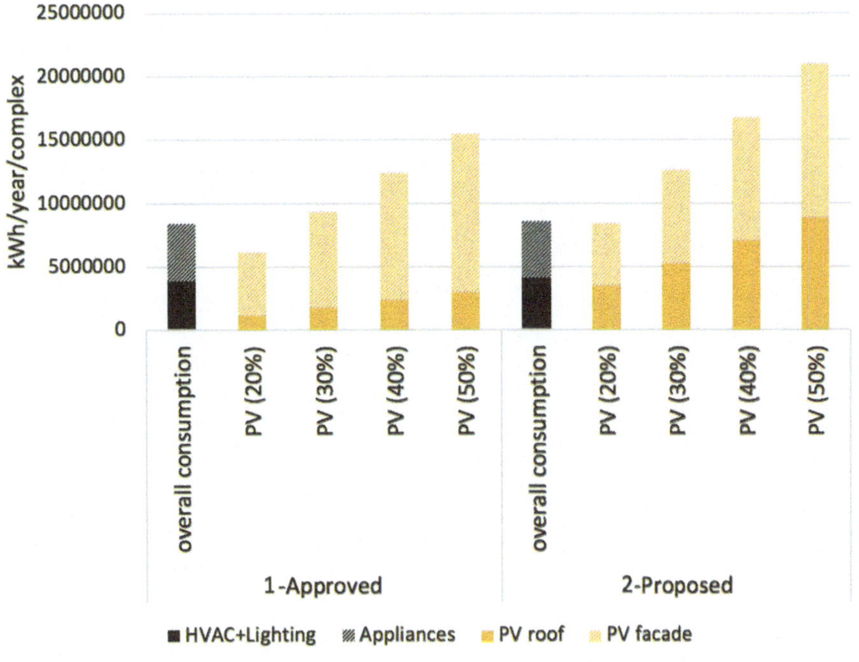

Fig. 4 Case study. Approved and proposed design alternatives' overall electricity consumption compared to possible energy production by PV roof and facade installation with various efficiencies (kWh/year/complex)

Horizontal roof PV installations are easier to install and maintain and have less impact on building appearance, being suitable for any building orientation. Façade surfaces covered by PV require highly exposed facades design. Adjacent buildings, shading elements, or balconies will decrease facade energy production potential.

The study shows that zero energy or near zero energy goals in high-density conditions is difficult to accomplish if urban and building design relies mainly on façade installation as in the proposed option. It is important to implement solar-oriented design on the urban level as much as on building scale in order to reduce energy consumption and maximize energy on-site production. In the future, with increased efficiency of PV panels and proper design at urban and building scales, zero energy should be an attainable goal for high-density developments.

References

1. Torcellini P, Pless S, Deru M, Crawley D (2006) Zero energy buildings: a critical look at the definition. ACEEE Summer Study Pacific Grove, NREL
2. Ministry of Environmental Protection, Jerusalem Institute for Israel Studies (2012) Sustainability 2030, Sustainability Plan for Israel 2030, (in Hebrew)
3. Ministry of Construction and Housing (1999) The complete version of Increased Building Rights Guidelines (in Hebrew)
4. Knowles RL (1981) Sun rhythm form. MIT Press, Cambridge
5. DeKay M (1992) A comparative review of daylight planning tools and a rule–of–thumb for street width to building height. In: Proceedings of the 17th National Passive Solar Conference. American Solar Energy Society, Cocoa Beach
6. Schiler M, UenFang P (1993) Solvelope: an interactive computer program for defining and drawing solar envelopes. In: Proceedings of the 18th National Passive Solar Conference. American Solar Energy Society, Washington, DC
7. Koester RJ (1994) Energy armatures — ordering an integration of passive energy resources for community sustainability. In Proceedings of the 19th National Passive Solar Conference — ASES, San Jose, CA
8. Capeluto IG, Shaviv E (2001) On the use of solar volume for determining the urban fabric. Solar Energy 70(3):275–280. Elsevier Science Ltd.
9. Capeluto IG (2003) Energy performance of the self-shading building envelope. Energy Build 35:327–336
10. Niemasz J, Sargent J, Reinhart CF (2013) Solar zoning and energy in detached dwellings. Environ Planning B: Planning Des 40:801–813
11. Capeluto IG, Plotnikov B (2017) A method for the generation of Climate-based, context dependent parametric solar envelopes, architectural science review. Taylor & Francis, pp 395–407
12. Kolodiy O, Capeluto IG (2016) Towards near zero–energy neighbourhoods in high density conditions: achievable target or utopian dream? PLEA 2016, 1108–1113, Los Angeles, California, ISBN 978-0-692-74961-6
13. Kolodiy O, Capeluto IG (2017) Towards near zero energy in high density residential areas. PLEA 2017, Edinburgh, III, 3563–3570, ISBN 9780992895754
14. Ministry of Interior (2005) Planning and building regulations, Tama 35 – Integrated National Master Plan for Construction, Development and Preservation, (in Hebrew)
15. Standards Institute of Israel (SI) (2011) Standard of Israel 5282, part 1, energy rating of buildings, Residential Buildings (in Hebrew)

Integrating Building Information Modeling Tool for Enhancing Energy and Daylight Performance in buildings' retrofit and Attaining Sustainability

Maryam Elsharkawy, Mohsen Aboulnaga, and Ayman Mahmoud

1 Introduction

Earthwide temperature increases and global environmental changes and challenges have been turning out to be prevalent issues, ranging from urban heat island effects (UHIEs) to climate change (CC) impacts [1, 2]. Such impacts and challenges attract global attention on the importance of enacting robust policies and planned actions for preserving the environment. In the European Union 27 member states, energy use and CO_2 emissions of buildings account for 40% and 36%, respectively [3]. About 35% of buildings in the EU are old (over 50 years), but nearly 75% of these buildings is energy inefficient [4]. In the MENA (Middle East and North Africa) region, buildings are responsible for more than 50 percent of the total energy use, yet are inefficient [5].

In Egypt, the building sector (residential, commercial, government, and public utilities) accounts for 67 percent of the 2016 total electrical energy [6]. Educational buildings (universities, research centers, and schools) worldwide are classified as the highest institutions in terms of energy use and greenhouse gas (GHG) emissions, yet the third of the total utilized areas [7]. Hence, sustainable transformation of university buildings is imperative for governments, policymakers, and owners/ developers due to high-energy use intensity/m^2 according to the American National Energy Audit by CBECS (Commercial Buildings Energy Consumption Survey) [8].

There are real challenges that Egypt may face to meet future energy demands due to the rapid population increase [9], if no retrofitting measures are enacted and effectively implemented in the existing building stock. Due to global CC manifestation, buildings are increasingly in demand of energy use for cooling in summer months and heating in winter months, in response to the huge use of air-conditioning or heating. Such impacts increased the dependence on active solutions rather than on passive ones. However, in the context of low-energy cost, conventional concerns of effective envelope and green building materials were uprooted to a second level of significance and resulted in buildings that include leakages into windows and poor thermal insulations permitted the envelope to become viable to heat transmittance, thus high-energy use rates. Nevertheless, this is no longer the case since the Government of Egypt introduced new legislatives on the electrical energy tariff in 2014, 2016, 2018, and the last in June 2020, which led to lifting the energy subsidies and increased the electricity cost for all buildings' types [10–13].

Existing educational buildings form a large portion of the total building heritage that should be preserved and protected against deterioration worldwide [14]. For both occupants' well-being and performance enhancement, old buildings should never be neglected, but be retrofitted and renovated to increase their energy performance, comfort, and assets' value. The current building stock comprises mainly a high percentage of old buildings with high carbon footprints, and even the newly constructed buildings still have lofty carbon footprints. According to a study review conducted in 2016, the existing building stock forms 86% of the total buildings [15]. However, the negative effect of the building sector on the environment could be reduced by improving the efficiency of the current buildings rather than building new ones [16–18]. A fascinating case

M. Elsharkawy (✉)
Department of Architecture, Faculty of Engineering, Cairo University, Cairo, Egypt
e-mail: mariem670@eng.cu.edu.eg

M. Aboulnaga · A. Mahmoud
Department of Architecture, Faculty of Engineering, Cairo University, Cairo, Egypt
e-mail: ayman.mahmoud@eng.cu.edu.eg

© The Author(s), under exclusive license to Springer Nature Switzerland AG 2022
A. Sayigh (ed.), *Sustainable Energy Development and Innovation*, Innovative Renewable Energy,
https://doi.org/10.1007/978-3-030-76221-6_44

emerged to serve as an educational and moralistic approach; it aims at demonstrating the environmental supervision in schools to teach the children for the first time what it means to be in a world that holds stewardship for both the society and the environment [19].

The building stock constitutes a large share for old school construction, where more than 50% of these buildings are close to reaching the end of their life span, which gives priority to focus on sustainable retrofitting of the existing educational buildings' stock [20]. If schools are built according to sustainable principles and practices, a significant effect on energy and other resources' utilization would be better management and saved. Although the advantages of the sustainable design are highly noted in the recent 5 years and since the Paris Climate Agreement – 2015 [21], they are not considered primarily for the building sector. Nonetheless, the educational mission itself aims at cherishing the sustainability in design, construction, and how these principles and measures are executed.

A better environment for learning is available in sustainable schools [22]. The school designs are usually comprised of single corridor sided classrooms, where the opposing hallway combines expansive openings, which give adequate sunlight or a two-sided passageway that does not have adequate natural airflow, and these areas are being enhanced with electrical ventilation sources [23]. Regardless of the absence of ordinary air flow in the passages and hallways, usually the installation for mechanical ventilation within those spaces does not appear to be frequent, thus, thermal and visual conditions within the spaces are poorly provided [24].

The building efficiency and energy performance are highly affected by the building's cover-surface (envelope), since the energy consumed for cooling, and heating, or providing other electricity needs, is affected by the efficiency of the envelope system [25]. A building envelope compromises both opaque and translucent components, which the sun heat and light that impinges on these surfaces contribute to and affect the interior thermal conditions. The capacity of the envelope components to convey the thermal heat depends on the thermal properties of materials. According to Huang et al. (2014), mechanical systems are sometimes utilized to adapt the interior conditions to comfort levels [26].

Multiple studies explored the performance of the building envelope's different components for the sake of both the initial design decisions and the retrofitting options [27]. Fan and Xia investigated the components of the building envelope including walls, windows, and roofs for optimizing their energy performance with high economic value in a retrofitting approach [28]. They formulated a multi-objective optimization model to reach for optimum solutions.

Based on a study review conducted by Gan et al. on building performance optimization actions, it aimed at enhancing the building performance through their envelopes, HVAC systems, and other systems [29]. In this review, multiple performance assessment methods were used to realize that building thermal properties in addition to occupants' behavior are the most significant factors that account for high-energy consumption and carbon emissions through the building life cycle [29]. Also, Soler et al. studied the performance of the opaque components in the building envelope, and they assessed all the materials involved in the construction process. In this study, they optimized the integration of layers' properties that shaped the opaque components [30]. Using a mathematical procedure, it was possible to provide an efficient material integration scheme with low carbon emissions [30]. In addition, Sartori and Calmon explored the retrofit of two existing buildings in terms of building performance enhancement, analyzing their embodied, operational, and wasted energy values [31]. They simulated different retrofit scenarios that included different building envelope measures and concluded that insulating materials, such as mineral wool, sheep wall, and expanded polystyrene (EPS), are highlighted significantly to improve the external walls and pitched roofs' performance [31]. Also, EPS as a building thermal insulating material is assessed under variable conditions and is proven to be highly efficient [32].

Many studies emphasized the significance of the building envelope in improving the performance of the whole building. However, building façades particularly are the most influential factors in the performance of the building envelope in general and specifically, these façades are highly influencing on the thermal and visual performance of educational building spaces [33, 34]. In Egypt, educational buildings are highly significant, as they enclose young Egyptians for a long period of time. Students enrolled in precollege academic schools reach 17 million in Egypt within 40,000 schools (public and private) and incorporate 1.6 million employees [35]. The old educational structures in Egypt are conceived as heritage that can never be neglected or undeveloped.

Owners, developers, and government should pay more attention to the retrofitting of an existing school building through optimizing building energy consumption to reduce CO_2 emissions and buildings' footprint, yet preserving our heritage. It is significant to optimize the performance of old educational buildings in Egypt that are constructed and characterized with high-energy consumption and inefficient energy performance levels. One of the main parameters that influence buildings' energy consumption is the building envelope, mainly the façade. Inefficient façades are considered the main building elements contributing to high-energy consumption and utility costs, despite the direct effect on the occupants' discomfort. Thus, there is an urgent need to adopt an efficient retrofitting approach for existing buildings' façades to enhance the performance of these buildings and the educational experience, and above all to allow for sustainable usage of the total energy.

This research highlights the main energy consumption problem related to the building façade elements in old educational buildings in Egypt. The study presents an assessment of a case study in order to reduce the total energy consumption and improve the building performance level through analyzing two retrofit schemes emphasizing on the main façade of an educational building.

2 Objectives

The objective of this study aims at investigating how buildings' façade retrofit represents an energy-efficiency alternative for improving the energy performance of existing buildings. The study focuses on exploring and assessing buildings' envelopes and their energy efficiency through retrofitting schemes of existing educational buildings. In this research, an old public-school building is examined and analyzed for potential sustainable reuse and higher energy performance. Considering a typical educational building, it is possible to approach improvements for this type of building based on the Cairo climatic conditions. Part of this attempt is to explore the energy reductions due to several improvements in the façade system and conductive materials in order to identify the most efficient façades deduced from the retrofit process in the selected educational building.

3 Methodology

The research method depends on analytical and measurable (applied) approaches. The analytical approach includes an assessment of the current building and its envelope as well as energy performance. The applied (simulation) of different façade retrofit schemes is performed in an existing educational building case study, located in Cairo, Egypt, as shown in Fig. 1. The analysis of the total building's energy performance, using the DOE-2.2 engine, is primarily applied to analyze the existing energy performance of the case study. A measurable procedure and defined set point for the building parameters are shown in Table 1, where the building baseline material properties are identified. Possible façade retrofit schemes are suggested for the transparent and opaque components of the main façade. They are evaluated for their operational energy consumption savings percentages (Fig. 1). The educational building refrains a typical single-sided corridor layout. Since the building main windows are oriented south-west as shown in the sun-path diagram of the building provided in Fig. 2, they are mostly subject to direct daylight. The energy consumed in the building is thus affected by the envelope thermal performance. In addition, the energy consumed depends on the occupancy level, and the behavior related to the school activities, schedules of classes, and operational habits. Understanding if the building has a seasonal operational schedule is crucial, and the level of occupancy of the different classes or floors throughout the building. The occupancy schedules' profile in the classrooms shows the percentage of occupancy in the classrooms during the day, where the peak hours are during the school day between 08:00 and 16:00 and only a drop in occupancy percentage during the break hours.

3.1 *Envelope Thermal Profile*

The core of this research is the building façades' performance. Hence, it is a fundamental part of this study to establish the relationship between the properties of the envelope and the building operational energy, which is partially wasted due to the building's current inefficient performance. The R-value refers to the capacity of the façade system to transfer thermal energy

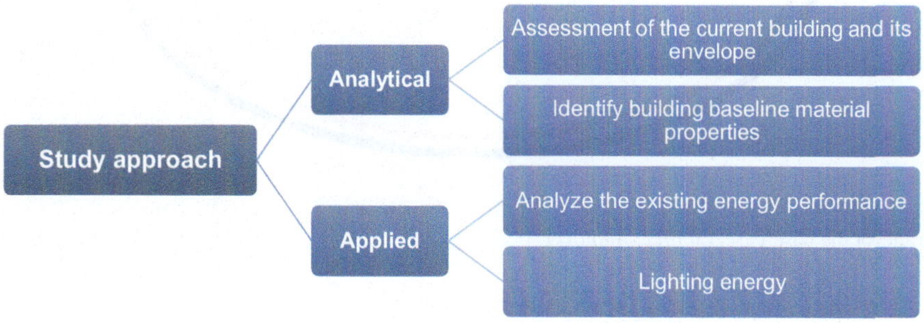

Fig. 1 Approaches conducted for the assessment of an educational building. (*Source: Authors*)

Table 1 The examined school building data

Building model dimensions		
Shape	Rectangular	
Volume	143 m^3	
Foot print	470 m^2	
External wall total area	758 m^2	
North east (NE) façade South west (SW) façade	390 m^2(434–18-30) 212 (434 m^2–222)	
NE and SW façade	156 (2 × 90 m^2–20-4)	
Windows total area	126 m^2 (W 1.2 m × H 1.6 m)	
Voids opening area	272	
Window wall ratio (WWR)	26%	
Building envelope specifications		
Exterior wall	120-mm thick wall with plaster, and paint Both sides with U-value 2.89 W/m^2 K	
Roof materials	Sloped concrete over reinforced concrete, and 50-mm thermal insulation, 20-mm water proofing, sand, and mortar Cement tiles with U-value 0.23 W/m^2K	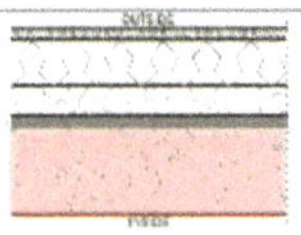
Glazing	Single glazing	Single glazing
Thermal properties		
Exterior walls	U-value	R-value
	2.89 W/m^2.K	0.34 m^2K/W
Roofing	0.23 W/m^2.K	4.35 m^2K/W
Glazing	5.44 W/m^2.K	0.18 m^2K/W
Glazing frame	2.88 W/m^2.K	0.34 m^2K/W
Glazing	Solar heat gain coefficient	0.42

Source: Authors

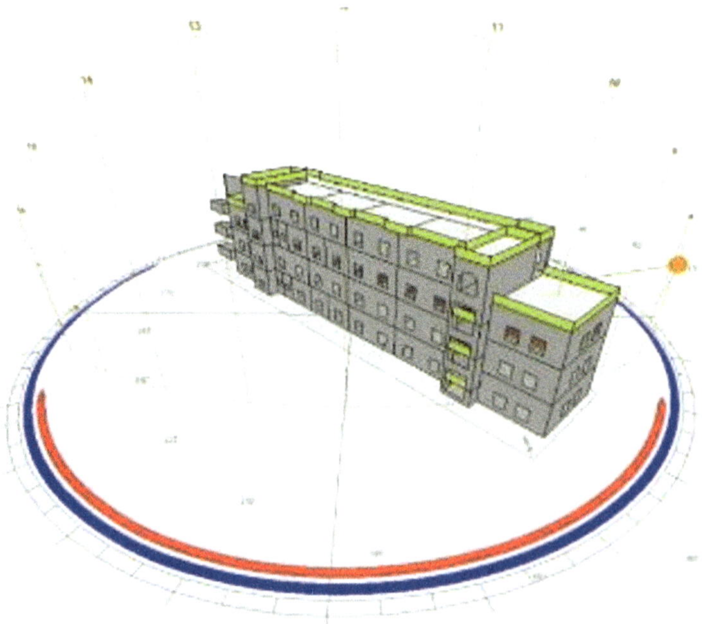

Fig. 2 Sun-path diagram for the base-case. (*Source: Authors*)

through façade components. After estimating the thermal resistance of the wall, the load through the façade is calculated. Also, both the summer and winter conditions are analyzed, despite cooling being the predominant load in the building. Two calculations have been carried out: (a) instantaneous heat transfer (for peak high and low temperatures) and (b) cumulative heat transfer.

In order to pursue such an assessment, the energy simulation programs employed in simulating the current energy consumption before and after the retrofit of the building façades are Ecotect and Insight 360 [36–38]. In addition, the monthly energy consumption for the different combinations of opaque and translucent parts of the façade is analyzed in two retrofit scenarios.

4 Retrofit Schemes

Based on the typology developed, two different schemes were chosen and simulated for comparison against the baseline case of the building. Energy reductions could be visualized from these different schemes identified in Fig. 3 and Table 2.

4.1 The Simulation

In the applied assessment and simulation, the outer wall composition and insulation for the single-/double-skin façade typologies are set by using five types of wall systems as the following (1A-5A); while the translucent component of the façade is simulated using four types of glazing retrofit schemes as the following (1B-4B). These settings are as follows:

- 1A- Standard Egyptian wall 120-mm thickness with U-value 2.89 $W/m^2.K$;
- 2A- Standard Egyptian wall 250-mm thickness with U-value 2.31 $W/m^2.K$;

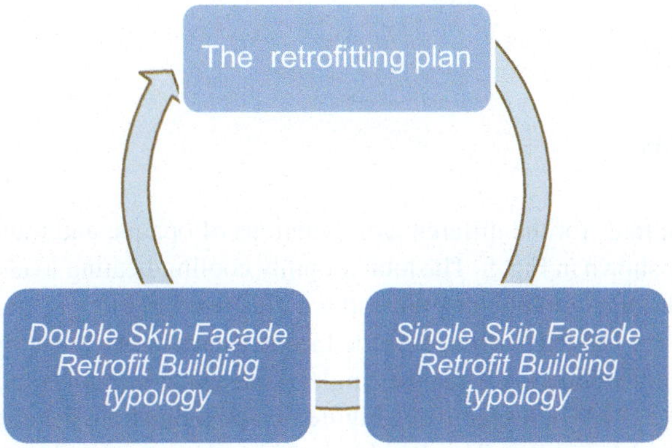

Fig. 3 The retrofitting plan of the case study. (*Source: Authors*)

Table 2 Simulated retrofit schemes

South-west façade		WWR	Description	Façades' interventions
Single skin/ regular window		26%	The original façade is maintained, but some components or all of them are upgraded or replaced	Window replacement Different walls' compositions Shadings added
Double skin/ fully glazed		90%	An additional glazed skin is added to the original façade, allowed to receive ventilation. Both the original façade and the additional skin are modified.	Glazing systems Different walls' compositions Shadings added Full-height double skin with catwalk

Source: Authors

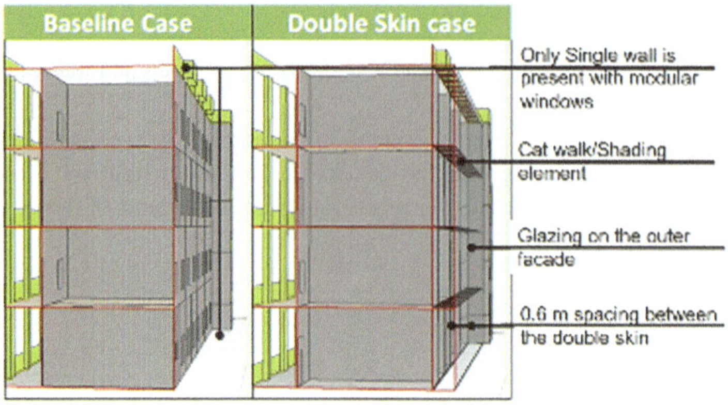

Fig. 4 Single- and double-skin approaches for the examined educational building. (Source: Developed by authors)

- 3A- ASHRAE 90.1-2013 American standard with U-value 0.86 W/m^2.K [38];
- 4A- Autoclaved aerated concrete (ACC) with U-value 0.44 W/m^2.K; and
- 5A- Exterior insulation finishing system (EIFS) with U-value 0.34 W/m^2.K.

- Scenario 1 B- Single glazing with U-value 5.44 W/m^{2}K
- Scenario 2 B- ASHRAE 90.1-2013 glazing 2.71 W/m^{2}K
- Scenario 3 B- Double glazing with air gap 1.5 W/m^{2}K
- Scenario 4 B- Double glazing with argon filled 1.1 W/m^{2}K

In all the above, 1B, 2B, 3B, and 4B, these are considered with the main properties for this glazing type, which are defined through the simulated model which shows that the SHGC is 0.81, VT is 0.639, and glazing thickness is 6 mm with total thickness 24 mm (Fig. 4).

5 Results and Discussions

The monthly energy consumption rates for the different combinations of opaque and translucent parts of the analyzed façades in the two retrofitting schemes are shown in Fig. 5. The total monthly cooling/heating loads are compared with the different wall compositions for the single-skin façade typology in kWh and are illustrated in Fig. 5a, and for the double-skin façade typology in kWh they are presented in Fig. 5c. The EIFS applied on the façade shows the most energy saving results of the different walls applied scenarios. The total heating/cooling loads for the single-skin/regular windows façade building and double-skin façade building are 134,941 kWh and 84,312 kWh, respectively. Figure 5b and d illustrate the energy performance of different glazing systems applied for the building façade in kWh. The double glazing with argon-filled system applied on the single-skin façade shows the most energy saving results of the different systems applied. The total annual energy consumption (heating/cooling loads) for the building was found for single-skin and the double-skin façade typology to be 12,742 kWh and 14,502 kWh, respectively. However, the energy consumption of all systems shows their maximum values during the summer months (June, July, and August) to encounter for the electricity used by cooling loads in these hot months as shown in Fig. 5.

The energy saving is compared with the different wall compositions for both scenarios – the single-skin façade and the double-skin façade. Table 3 presents the energy saving of such comparative analysis. Using different wall compositions within each of the skin typologies allows different energy saving percentages. Considering the 120-mm wall as the baseline model wall composition; with comparison to the other wall compositions of the EIFS gives the highest energy saving percentage in both typologies. But when compared with the regular window and the double-skin typologies, the first typology results show higher energy saving percentages, higher than 42% of the total energy saving (Table 3). This very high percentage is related to the higher opaque content in the simulated regular window typology. It is clear that in the single-skin/regular windows, the façade typology that incorporates wooden windows with 26% WWR, as shown in the baseline case, reveals better thermal performance over the applied double-skin/fully glazed for this building. This is due to the fact that the direction of the south-west orientation of the façade of the simulated building shows lower performance with the increased glazing area represented in the double skin, especially with the absence of any additional shading device. Nevertheless, different glazing types are compared for their total energy savings within the simulated single-skin building and double-skin building scenarios.

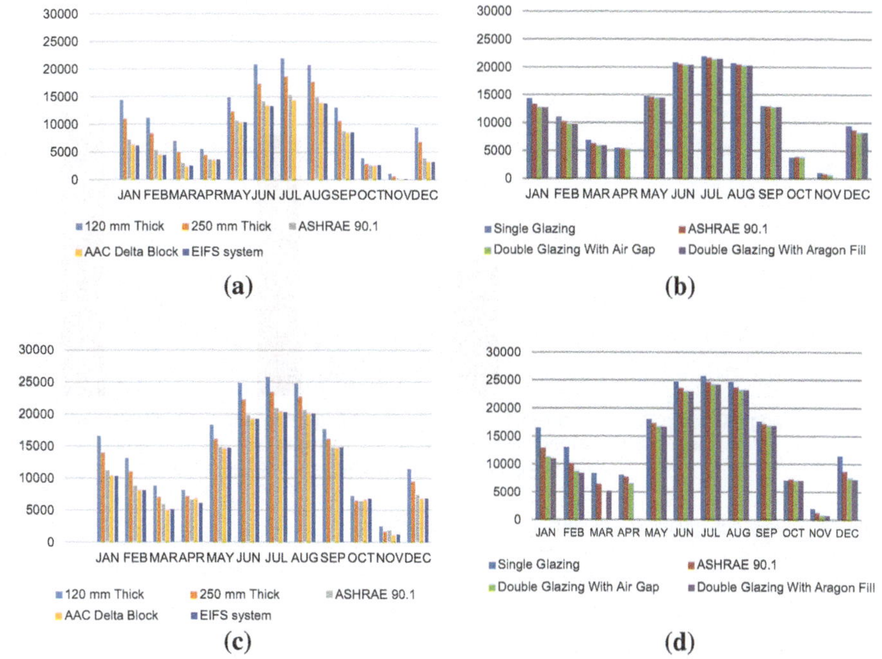

Fig. 5 Monthly energy consumption for the different retrofit façade retrofit schemes. (**a**) Opaque in a regular window façade. (**b**) Translucent in a regular window façade. (**c**) Opaque in double-skin façade. (**d**) Translucent in double-skin façade

Table 3 Energy saving percentage for the wall compositions applied within the façade typologies

Energy saving %	120 mm thick baseline	250 mm thick	ASHRAE 90.1 American standard	AAC delta block	EIFS
Single-skin façade	%0	19.46%	38.80%	41.33%	42.03%
Double-skin façade	%0	%12.11	%23.00	%24.66	25.23%

Table 4 Energy saving percentage for the glazing types applied within the façade typologies

Energy saving %	Single glazing baseline	ASHRAE 90.1 American standard	Double glazing with air gap	Double glazing with argon fill
Single-skin façade	0%	%3.3	%4.87	12.40%
Double-skin façade	0%	9.43%	17.35%	19.08%

The simulation readings for the building's operational energy consumption are compared with the base-case values and percentages as listed in Tables 3 and 4. Considering the single glazing as the baseline model glazing property; with comparison to the other glazing models of both retrofit typologies, the double glazing with argon fill gives the highest energy saving percentage in both typologies, as shown in Table 4.

However, in comparison between the single-skin and the double-skin typologies, the double-skin shows higher energy saving percentages concluding better skin efficiency. Though, it is concluded that using the double-skin façade with more efficient glazing systems that allow more light transmission due to the higher window-to-wall ratio results in higher levels of natural lighting as provided in Fig. 7, and higher energy savings in kW/h as shown in Fig. 6. Such modification resulted in lowering electricity bills' costs.

When applying the single glazing system without any upgrade in the façade's glazing, but altering the exterior wall system with different wall compositions, higher efficiency for the single-skin/regular window façade typology of design is provided. Only with higher glazing standards, the double-skin façade results in a more efficient retrofit scheme. An efficient retrofit scheme is the one that considers all the comfort-related variable conditions, that is, suitable interior temperature range, higher natural lighting levels lying within the comfort range for reading (300–400) Lux, good view to the outside, and more environmental indoor air quality.

Comparing the operational energy loads and the total saving percentages for the optimum selection of façade materials (opaque and translucent) within both the double-skin façade and single-skin façade shows that a maximum of 72% energy

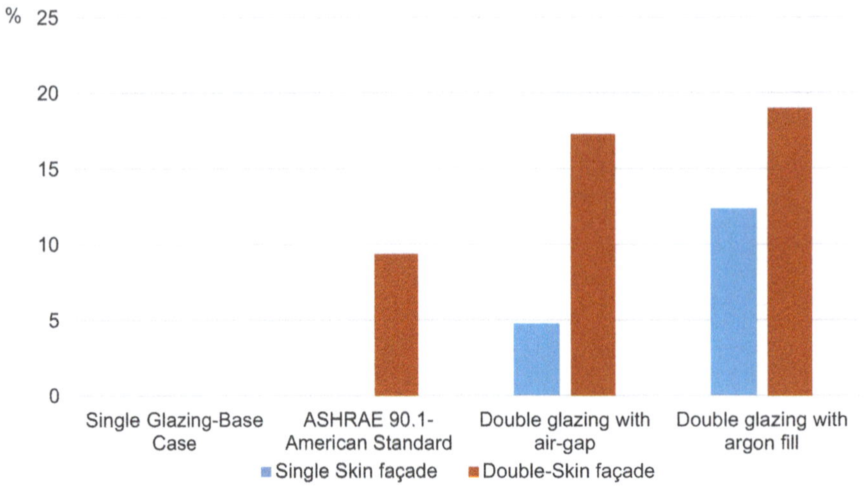

Fig. 6 Comparing the energy saving percentage for the different glazing systems

Table 5 Energy load reduction and saving percentage for optimum façade design systems

Energy savings	Baseline single skin	Baseline double skin	Optimum material selection (single skin)	Optimum material selection (double skin)
Total loads (kW/h/m^2)	74	92	21	58
Energy saving %	0%	0%	72%	37%

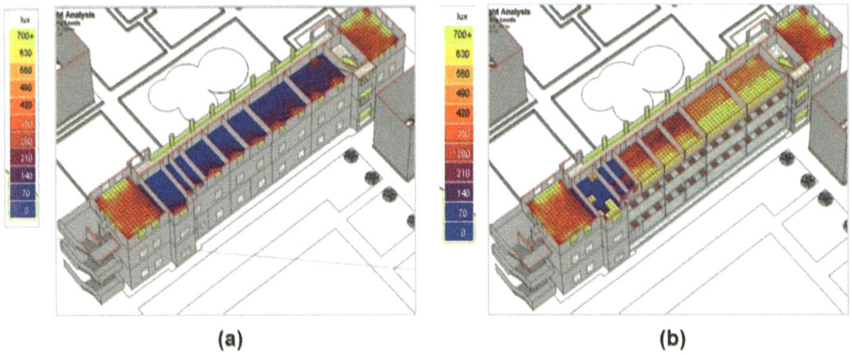

Fig. 7 Daylight distribution for the upper floor of the retrofitted school building – 21 March. (**a**) Single-skin (wall/window) façade. (b) Double-skin full glazing façade

saving could be the outcome of a regular window/single-skin façade retrofit (Table 5), which proposes the most efficient combination of the external façade wall composition and glazing type in a simulated building optimum façade retrofit, to be able to compare their total energy consumption profiles and their related energy saving percentages which highlighted the capability of the regular window façade in retaining higher thermal performance levels. In contrast, Fig. 7 presents the positive effect of the double-skin façade in enhancing interior daylight conditions. Finally, a comparative evaluation approach is provided in Table 6, combining the simulation results of the two building typologies for the façades. The comparison provides optimum estimated energy savings based on the studied criteria.

Table 6 Proposed checklist for achieving low-energy performance in educational buildings

Performance criteria	After retrofitting Single-skin façade	Double-skin façade
Retrofitting the opaque system only		
Opaque system	EIFS	EIFS
Opaque U-value	0.58	0.58
Energy consumption (kW/h/m^2)	43	68.8
Energy reduction	42%	25%
Retrofitting the translucent system only		
Translucent system	Double-glazing (argon filled)	Double-glazing (argon filled)
Translucent U-value	1.85	1.85
Energy consumption (kW/h/m^2)	65	74
	12.4%	19%
Combined retrofit of optimum opaque and translucent façade systems		
Energy consumption (kW/h/m^2)	**31**	52.4
Air shaft	None	500 mm shaft
Air flow	Normal	**Stack-cooling effect**
Illuminance %	26%	**66%**
Costs	**Moderate** **Lower construction and maintenance costs**	**High**

6 Conclusions

This study is a compilation of building retrofits with special emphasis on the façade system. In this regard, the current concern of global warming and the essential role of buildings urge the need for more environmental studies. Currently, the tangible benefits of retrofitting the façade in existing buildings are not broadly studied. Therefore, the focus is to comprehend the building retrofit as an energy-efficient tool for decreasing the operational energy consumption in buildings, and one particular building example gives an idea of the possible savings in similar hot-climatic conditions.

The building total energy consumption is 145,459 kWh and the total energy consumption per square meter 74 kWh/m^2. The simulated energy consumption does not take into consideration the air gaps or the thermal bridges or some other heat leakages through the façade systems. The surveyed energy use intensities in the existing school buildings range from 25 to 50 kWh/m^2. Therefore, it was found that the building is relatively inefficient in the overall energy consumption, when compared to buildings of the same vintage.

The façade modifications that include exterior wall replacements and insulation upgrade affect the building's total energy consumption by maximum 42% in a single-skin building typology and 25% in a double-skin typology. Also, the high-efficiency glazed system within building façades could reduce the energy consumption by 12.4% and 19% in a single-skin building and double-skin building typologies, respectively. Illumination levels are highly enhanced with a proper design for the façade openings. However, the thermal analysis results are not highly accurate for rigorous quantitative analysis of a detailed building assessment and are accounted for further calibration values using building energy meters.

All the previously indicated percentages of reduction are expressed as a function of the total heating/cooling-related energy consumption in the building, which is considered two-thirds of the total energy, and the other one-third energy consumption is not affected by the façade retrofit, but, affected by the internal loads, which are mainly not affected by the façade alteration. However, if all other loads are encountered for in the analysis, then the total savings compute 28.9% of that universe. It also seems a challenging task to achieve the remaining energy reduction toward net zero energy (NZE) in this particular existing building.

Further exploration should consider other retrofit aspects such as lighting and mechanical retrofit. This study should contribute to the emerging literature on retrofitting investments by proposing a guideline approach to façade retrofits and implementing an integrated analytic framework to examine different façade retrofit typologies and reach an effective building performance-related decision. This participates in creating retrofit decision framework to integrate the façade performance dimension and support decision-making.

References

1. Aboulnaga M, Mostafa M (2019) Mitigating heat island effect in mega cities through districts' prioritisation for urban green coverage application: Cairo – Egypt as a case study. Renew Energy Environ. Sustain 4(5) EDP Sciences. https://doi.org/10.1051/rees/2019002
2. Zhang J (2018) Impacts of heat mitigation strategies and pollutant transport on climate and air quality from urban to global scales (Doctoral dissertation, University of Southern California)
3. Tettey UYA, Gustavsson L (2019) Primary energy and CO2 emissions implications of different insulation, cladding and frame materials for residential buildings. IOP Conference Series: Earth and Environmental Science 297(1):1, 2020. IOP Publishing
4. Ferrante A, Mochi G, Predari G, Badini L, Fotopoulou A, Gulli R, Semprini G (2018) A European project for safer and energy efficient buildings: pro-get-one (proactive synergy of integrated efficient technologies on buildings' envelopes). Sustainability 10(3):812
5. Franco S, Mandla VR, Rao KRM (2017) Urbanization, energy consumption and emissions in the Indian context a review. Renew Sust Energ Rev 71:898–907
6. Aliyu AK, Modu B, Tan CW (2018) A review of renewable energy development in Africa: a focus in South Africa, Egypt and Nigeria. Renew Sust Energ Rev 81:2502–2518
7. Leal Filho W, Salvia AL, Do Paco A, Anholon R, Quelhas OLG, Rampasso IS, Brandli LL (2019) A comparative study of approaches towards energy efficiency and renewable energy use at higher education institutions. J Clean Prod 237:117728
8. Griffith B, Long N, Torcellini P, Judkoff R, Crawley D, Ryan J (2008) Methodology for modeling building energy performance across the commercial sector (No. NREL/TP-550-41956). National Renewable Energy Lab (NREL), Golden, CO (United States)
9. The Population Division of the Department of Economic and Social Affairs UNPESA or World Population Prospects 2019. https://www.un.org/development/desa/pd
10. ElSobki MS Jr (2016) Egypt's renewable energy. Ministry of Electricity and Renewable Energy (MoERE)
11. Imam R, Yassin M (2018) The potential for integrating solar thermal energy in both centralized and decentralized systems in Egypt. In: Energy sustainability, vol 51418. American Society of Mechanical Engineers, p V001T06A005
12. Igrioglu G, Ostry A, Allam M (2020) Integrated governance for coherent implementation of the SDGs in Egypt
13. El Gindi SAEM (2020) Integration of energy efficiency measures and renewable energy technologies as an approach to sustainable development in Egypt. J Urban Res 36(1):102–117
14. Webb AL (2017) Energy retrofits in historic and traditional buildings: a review of problems and methods. Renew Sust Energ Rev 77:748–759
15. Ruparathna R, Hewage K, Sadiq R (2016) Improving the energy efficiency of the existing building stock: a critical review of commercial and institutional buildings. Renew Sust Energ Rev 53:1032–1045
16. Liu G, Li X, Tan Y, Zhang G (2020) Building green retrofit in China: policies, barriers and recommendations. Energy Policy 139:111356
17. Saikia P, Pancholi M, Sood D, Rakshit D (2020) Dynamic optimization of multi-retrofit building envelope for enhanced energy performance with a case study in hot Indian climate. Energy:117263
18. Pisello AL, Petrozzi A, Castaldo VL, Cotana F (2016) On an innovative integrated technique for energy refurbishment of historical buildings: thermal-energy, economic and environmental analyses of a case study. Appl Energy 162:1313–1322
19. Hart RA (2013) Children's participation: the theory and practice of involving young citizens in community development and environmental care. Routledge
20. Hu M (2019) Cost-effective options for the renovation of an existing education building toward the nearly net-zero energy goal—life-cycle cost analysis. Sustainability 11(8):2444
21. Falkner R (2016) The Paris agreement and the new logic of international climate politics. Int Aff 92(5):1107–1125
22. Biedenweg K, Monroe MC, Oxarart A (2014) The importance of teaching ethics of sustainability. Int J Sustain Higher Educ 14(1): 6–14
23. Brown, D. (2017). The performance of light: exploring the impact of natural lighting in the New UMass School of Performance
24. Becker R, Goldberger I, Paciuk M (2007) Improving energy performance of school buildings while ensuring indoor air quality ventilation. Build Environ 42(9):3261–3276
25. Ali H, Hashlamun R (2019) Envelope retrofitting strategies for public school buildings in Jordan. J Build Eng 25:100819
26. Huang Y, Niu JL, Chung TM (2014) Comprehensive analysis on thermal and daylighting performance of glazing and shading designs on office building envelope in cooling-dominant climates. Appl Energy 134:215–228
27. Kheiri F (2018) A review on optimization methods applied in energy-efficient building geometry and envelope design. Renew Sust Energ Rev 92:897–920
28. Fan Y, Xia X (2017) A multi-objective optimization model for energy-efficiency building envelope retrofitting plan with rooftop PV system installation and maintenance. Appl Energy 189(3):327–335
29. Gan VJ, Lo IM, Ma J, Tse KT, Cheng JC, Chan CM (2020) Simulation optimisation towards energy efficient green buildings: current status and future trends. J Clean Prod 254:120012
30. Soler D, Salandin A, Bevivino M (2020) Using integer Linear Programming to minimize the embodied CO2 emissions of the opaque part of a façade. Build Environ 177(106883)
31. Sartori T, Calmon JL (2019) Analysis of the impacts of retrofit actions on the life cycle energy consumption of typical neighbourhood dwellings. J Build Eng 21:158–172
32. GadAllah S, Aboulnaga M (2020) Climate action and SDGs' attainment: insulation materials' impacts assessment. SMC J 11:76–80. Available at: http://www.sustainablemediterraneanconstruction.eu/SMC/The_Magazine_n.11_files/111.pdf
33. Mohelníková J, Novotný M, Mocová P (2020) Evaluation of school building energy performance and classroom indoor environment. Energies 13(10):2489
34. Aboulnaga M, Wanas A, Hammad M, Hussein M (2017) Sustainability of higher educational buildings: retrofitting measures to enhance energy performance—the case of AASTMT business management school building, Egypt. In: Mediterranean green buildings & renewable energy. Springer, Cham, pp 117–150
35. Pouezevara S, Mekhael S, Darcy N (2014) Planning and evaluating ICT in education programs using the four dimensions of sustainability: a program evaluation from Egypt. Int J Educ Dev ICT 10(2):120–141

36. Ecotect (2011) [Online]. Available https://openei.org/wiki/Ecotect. Accessed 1 Apr 2015
37. Autodesk Insight 360 (2019) [online]. Available https://insight.autodesk.com. Accessed 11 July 2020
38. Halverson MA, Rosenberg MI, Hart PR, Richman EE, Athalye RA, Winiarski DW (2014) ANSI/ASHRAE/IES standard 90.1-2013 determination of energy savings: qualitative analysis (no. PNNL-23481). Pacific Northwest National Lab. (PNNL), Richland, WA, USA

Climate Change Adaptation: Renewable Energy and Climate Actions in Cities to Mitigate Climate Change and Enhance Liveability – A Diagnostic and Strategic Study

Mohsen Aboulnaga, Naguib Amin, and Bruno Rebelle

1 Introduction

Climate change (CC) has no borders or boundaries since it has been manifested globally in most of the cities [1, 2]. Many cities across the world suffer from CC risks and impacts; it is clear that no single city is safe from CC risks [3–5]. The increase in urbanisation and CC impacts has a direct effect on the increase of urban temperature [6]. The urbanisation sprawl in cities coupled with some impacts of climate change (CC) is leading to the emergence of urban microclimates [7–10]. According to the Climate Change Resilience Index (CCRI), a new framework was established by the Economist Intelligence Unit (The EIU). Based on a study of the World's 82 largest economies and measure their ability to conform CC, the global economy will be 3 per cent smaller by 2050 due to CC impacts [11].

Between 2011 and 2018, the European Union (EU) initiated a major policy design project "Cleaner Energy Saving Mediterranean Cities" (CES-MED), under the European Neighbourhood and Partnership Instrument (ENPI), to support South Mediterranean countries to develop sustainable energy and climate action plans (SECAPs) [12]. A recent study on the concepts and methodologies for measuring the sustainability of cities reviewed synthesises the recent development in methods being exploited to the impacts of cities on environmental sustainability [13]. Incidentally, better data and methods are becoming available for understanding the complex and woven functioning of cities and their impact on sustainability [14].

COP 21 and Paris Agreement on Climate Change have added impetus to the necessity for pragmatic approaches in boosting the climate change adaptation (CCA), in order to increase measures and actions towards strengthening climate resilience [15, 16]. The evolving know-how of experts from European financing institutions and the European Commission (EC) offers insights into how these climate issues can be best integrated into projects' development and implementation [17]. The CES-MED project is a clear manifestation of such development that is aiming to assist governorates in developing sustainable energy action plans (SEAPs) (SECAPs) and adapting to CC risks [13]. Within this framework, the "Integrating Climate Change Information and Adaptation in Project Development: Emerging Experience from Practitioners" has been developed to provide assistance to aid practitioners assess climate change risks and vulnerabilities and integrate adaptation measures into project planning, design, and implementation [18, 19].

M. Aboulnaga (✉)
Department of Architecture, Faculty of Engineering, Cairo University, Cairo, Egypt
e-mail: maboulnaga@eng.cu.edu.eg

N. Amin
EU Funded Project CES-MED, Beirut, Lebanon

EU Funded Project CES-MED, Rabat, Morocco
e-mail: namin@climamed.eu

B. Rebelle
Transitions, CES-MED SECAP Consultant, Paris, France
e-mail: bruno.rebelle@transitions-dd.com

© The Author(s), under exclusive license to Springer Nature Switzerland AG 2022

A. Sayigh (ed.), *Sustainable Energy Development and Innovation*, Innovative Renewable Energy,
https://doi.org/10.1007/978-3-030-76221-6_45

2 Objective

The objective of this work focuses on the strategic renewable energy and climate action plans with emphasis on two seaside cities in Egypt where the SEAPs and SECAPs were conducted, applied, and implemented based on each city's strategy in 2030, and its clean energy capacity. The SECAPs for both the city of Hurghada and the city of Luxor were developed as part of the CES-MED project, funded by the European Union between 2011 and 2018. In this paper, the SECAP for the city of Hurghada is highlighted and discussed.

3 Methodology

The SECAP is considered as a strategic manifesto and an operational tool [12, 13]. It defines a global framework, with quantifiable objectives to be reached by 2030, based on the emissions' reference inventory (baseline emission inventory (BEI)) and the detailed assessment of energy consumption [12, 20–23]. Diagnostic studies were carried out to identify the gap in each city in terms of climate change and renewable energy. In this paper, the focus is given to SECAP of the city of Hurghada, prior to providing a detailed account of the concrete measures undertaken to reduce GHG emissions and promote the development of sustainable energy. The national and regional strategies on climate change adaptation (CCA), the renewable energy capacity, and future needs were assessed to draw the gap at the local government level. In terms of vulnerability, climate change risks (CCRs) by sectors have been assessed and presented. To conduct the SEACP, it was vital to understand and address the four main issues, respectively, to achieve the set mission (Fig. 1).

3.1 The Steps of the Sustainable Energy and Climate Action Plan (SECAP)

The SECAP for the city of Hurghada encompasses five main phases. Figure 2 illustrates these five phases, including the SECAP report, CCA and analysis, and priority action plans [12, 14, 21].

3.2 National and Regional Strategies on Climate Change Adaptation

To craft and develop the sustainable energy and Climate Action Plan – SECAP for the city of Hurghada, it was imperative to get a solid and comprehensive understanding of the national and regional strategies on climate change adaptation (CCA) as well as sectoral policies. Figures 3 and 4 illustrate the regional CCA strategies and sectoral policies in brief. Nevertheless,

Fig. 1 Main sections of the SECAP methodology

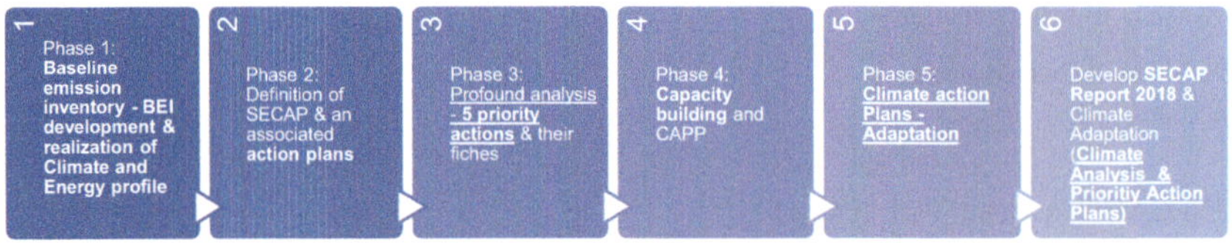

Fig. 2 Steps for sustainable energy and climate action plans. (Source: http://publications.jrc.ec.europa.eu/repository/handle/JRC93697)

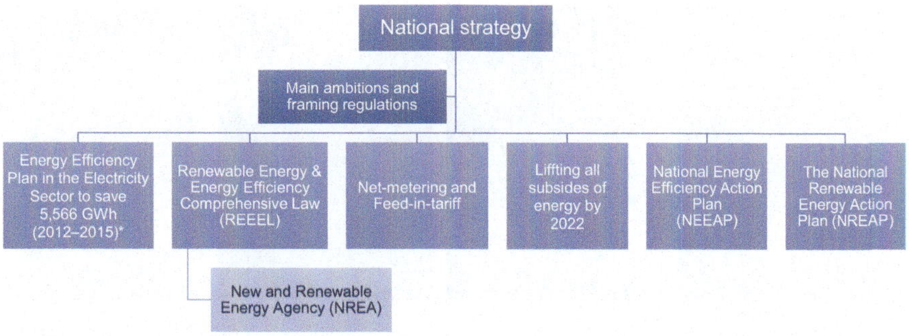

Fig.3 Nationalstrategiesonclimatechangeadaptation,Egypt.(Source:http://www.ces-med.eu/publications/egypt-municipality-hurghada-sustainable-energy-and-climate-action-plan-secap/)

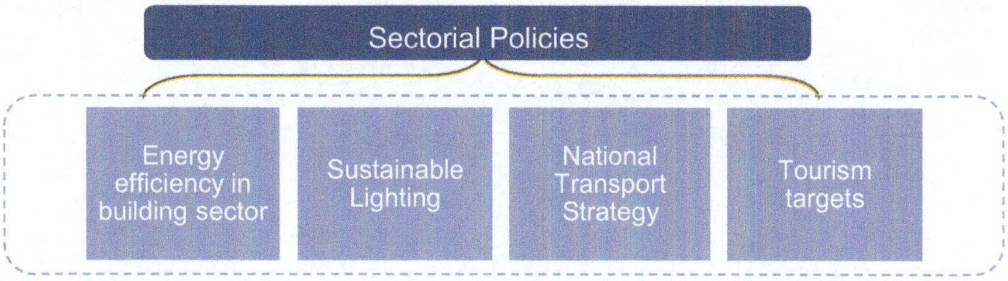

Fig. 4 National sectoral policies on climate change adaptation – CCA. (Source: http://www.ces-med.eu/publications/egypt-municipality-hurghada-sustainable-energy-and-climate-action-plan-secap)

since the completion of the CES-MED project, more strategies on CCA have been developed and executed [24–26]. The most recent CCA strategies were developed and implemented in March 2020, during and after the devastating severe storm that hit Egypt between 9 and 12 March 2020 [27].

3.3 The Case Study and Considered Scope

In developing the SEACP, it was importantly needed to determine the administrative areas of the city of Hurghada, where the SECAP and its sections will be implemented. Thus, the following areas are presented below after consultation with the senior officials of the Governorate of Red Sea. Figure 5 presents the administrative areas of the City of Hurghada for climate risks' assessment, climate adaptation action (CAA), and vulnerability analysis.

4 Sustainable Energy and Climate Objectives: City of Hurghada

Keeping in mind its specific values and the local context, the City of Hurghada develops a strategy consistent with Egypt's Vision 2030, energy strategy 2035, and the Sustainable Development Strategy – SDS 2030 [28–31]. This strategy is structured around two levels, as shown in Fig. 6. In 2019, the Governorate of Red Sea, nonetheless, reported that three green buildings were developed in the city of Hurghada, where green roofs were implemented, and renewable energy was integrated [32]. However; the SECAP study indicated that more measures were to be taken for climate change adaption (CCA) [20, 21].

The city of Hurghada plan encompasses short- and medium- to long-term strategic objectives:

Short-term objectives in 2020:

- Reduce energy consumption across the board by around 10 per cent to 15 per cent
- Mitigate GHG emissions by 10 per cent in 2020, compared to the business as usual (BaU) scenario of the 2015 baseline [21]

Fig. 5 The administrative areas of the City of Hurghada for climate adaptation and vulnerability analysis. (Image source: https://satellites.pro/Hurghada_map.Egypt)

Fig. 6 Actions taken by the City of Hurghada (municipality) to reduce energy use in all sectors. (Image source: https://www.egypttoday.com/Article/1/52769/In-pics-Egypt%E2%80%99s-rooftops-going-green)

Medium-term and long-term objectives in 2030:

- Boost efforts to reduce energy consumption and improve efficiency, resulting in a continuous trend of improvement in energy intensity (energy consumption compared to the gross domestic product (GDP)).
- Continue reducing the GHG emission with the objective of reaching at least a reduction of 27 per cent by 2030 and if possible, going to a 30 per cent reduction, compared to the BaU scenario. Such an effort will place the city of Hurghada in the appropriate trajectory to match the collective target agreed at COP 21 end of 2015 and that of Paris Agreement.
- Improve the quality of life in Hurghada and optimise service delivery to inhabitants and all stakeholders in order to speed up the energy transition towards sustainable development,
- Develop renewable energy production capacities in and around the city as well as in the southern region of the Governorate, as presented in Fig. 7 [21].

Prior to the climate adaptation actions (CAAs), it was necessary to calculate the baseline emission inventory (BEI), which is part of the preparation for SECAP of the city of Hurghada. The BEI is based on the calculation of greenhouse gas emissions (not a measure) and covers eight sectors: (a) residential buildings; (b) tertiary buildings and public lighting; (c) industry; (d) transport; (e) waste and water management; (f) agriculture (crops, animal production, and fishing); (g) tourism; and (h) renewable energy [20, 21]. The scope and methodological principles of the GHG emissions are based on the data collection and BaU scenario forecasts for the city can be found in the CES-MED publications of the SECAP documents [20, 21].

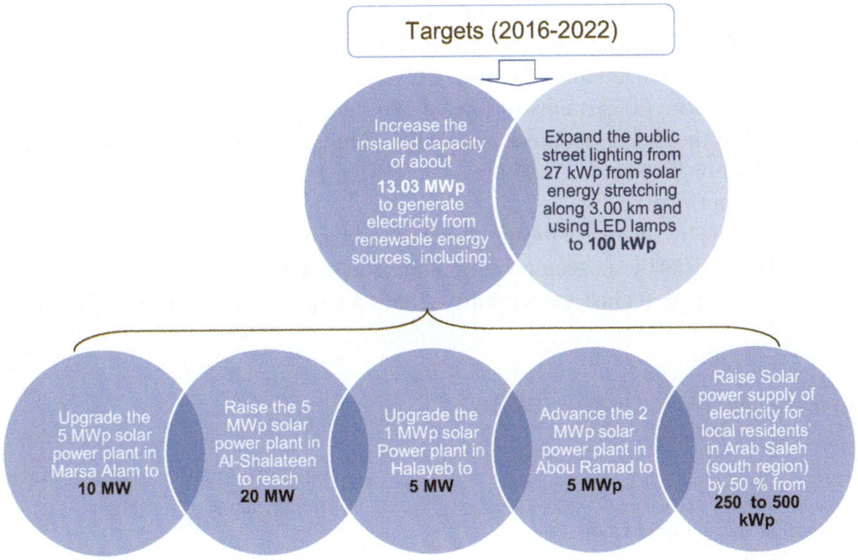

Fig. 7 Anticipated renewable energy needed capacities in the city of Hurghada and south Region of the Governorate of Red Sea. (Source: Developed by authors based on official meetings with Hurghada's municipality senior staff)

5 SECAP Development: Case Study

5.1 *Adaptation Cycle-Specific Steps (ACSS)*

It is imperative to develop the adaptation scoreboard of the Adaptation Cycle-Specific Steps for the city of Hurghada. This scoreboard was based on the European Commission Joint Research Centre (EC-JRC) guidelines that include six steps utilised to assess the city's climate actions [33, 34].

5.2 *Risk Assessment and Vulnerability Analysis*

The risk assessment and vulnerability analysis, which are based on the Future Cities Adaptation Compass Tool (FCACT) [35, 36], Governors' (Mayors) Adapt [37], and European Climate Adaptation Platform (ECAP) [38, 39], were conducted using a set of parameters and interviews with local government officials.

5.3 *Climate Adaptation*

In SECAP planning, a whole study and methodology of climate adaptation action (CAA) of the city of Hurghada was conducted based on the city's vulnerability to CC, climate risk assessment, and CAA adopted. The structure of this section has been developed based on the extensive literature review. Also, introduction to CC impacts, particularly in Mediterranean countries with emphasis on Egypt, mainly urban areas, costal zones, agriculture, water and ecosystems, health, and tourism is presented in [21]. This section focuses on the climate overview in Hurghada; highlighting climate trends [21]. The adopted national and regional strategies on CCA are followed by a subsection dedicated to climate data and projections feeding in estimations of the climate change impacts in the future presented in the SECAP documents [21]. The following section addresses the adaptation scoreboard, including a self-assessment from the Governorate of Red Sea against the standard adaptation scoreboard in the SECAP. The CCRs by sectors in the city of Hurghada and the Governorate's score in the adaptation cycle-specific steps are highlighted. The risk analysis, vulnerability assessment, and risks' assessment template are based on the Future Cities Adaptation Compass tool. They also depict the national CCA, and mitigation measures, including the climate change action plan (CCAP), and adopted measures within the framework of the INDCs (Intended Nationally Determined Contributions) based on adaptation challenges in the studied four sectors. Furthermore, this section sheds lights on the national adaptation action plan, mainly in coastal zones, water resources, irrigation as well as agricultural, health, tourism, building, and energy sectors. Finally, it highlights the adaptation actions in the city of Hurghada in terms of strategic actions [20, 21].

In this study, the ranking of climate change vulnerabilities (CCVs) in Egypt, based on that of the OECD, was exploited to identify the risk in terms of three criteria: (a) certainty of impact; (b) severity of impact; and (c) the importance of resources [21]. These factors were applied to the city's resources and risk ranking, including: costal, water, agriculture, and energy resources. The assessment was set on five sets of ranking scale (High, High-medium, Medium, Medium-low, and Low). Table 1 shows the results of CCV in Egypt.

For the climate change risk (CCR) by sectors in the city, a set of indicators (High exposure, Exposed, and Low exposure) were developed. The climate change risks by six sectors (agriculture and food security; water stress and drought; ecosystems; tourism; urban; and health) was mapped against vulnerability. Table 3 (Appendix 1) presents the vulnerability of city sectors by CCR.

Table 1 Ranking of climate change vulnerabilities in Egypt (OECD)

Resources/ risk ranking	Risk	Certainty of impact	Severity of impact	Importance of resources
Coastal resources	SLR (sea level rise) Coastal erosion and soil salinisation	High-medium	High	High
Water resources	Decreased Nile flow Low precipitation / salt water intrusion / decreased Nile flow Low precipitation Salt water intrusion	Medium	High	High
Agriculture	SLR Soil salinisation High temperature	Medium-low	Medium-low	High-medium
Energy resources	High temperature Decreased Nile flow	Medium-low	Medium-low	Medium-low

Table 2 Governorate's (Municipality's) score in the adaptation cycle-specific steps (SECAP template and JRC guidelines) – City of Hurghada, Governorate of Red Sea

Adaptation cycle steps	Actions	Grades (A–D)
Step 1: Preparing the ground for adaptation	Adaptation commitments defined/integrated into the local climate policy	A
	Human, technical, and financial resources identified	B
	Adaptation team (officer) appointed within the municipal administration and clear responsibilities assigned	D
	Horizontal (e.g. across departments) coordination mechanisms in place	A
	Vertical (e.g. across governance levels) coordination mechanisms in place	A–B
	Consultative and participatory mechanisms set up, fostering the multistakeholder engagement in the adaptation process	B
	Continuous communication process in place	A
Step 2: Assessing risks and vulnerabilities to climate change	Mapping of the possible methods and data sources for carrying out a risk and vulnerability assessment	A
	Assessment of climate risks and vulnerabilities undertaken	D
	Possible sectors of actions identified and prioritised	C
	Available knowledge periodically reviewed and new findings integrated	B
Steps 3 and 4: Identifying, assessing, and selecting adaptation options	Full portfolio of adaptation actions compiled, documented, and assessed	D
	Possibilities of mainstreaming adaptation in existing policies and plans assessed, possible synergies and conflicts identified	B
	Adaptation actions developed and adopted	C–B
Step 5: Implementation	Implementation framework set up with a clear milestones	C
	Adaptation actions implemented and mainstreamed as defined in the SECAP document*	N/A
	Coordinated actions between adaptation and mitigation set of measures	C
Step 6: Monitoring and evaluation	Monitoring framework in place for adaptation actions	C–D
	Appropriate monitoring and evaluation indicators identified	C–B
	Regular monitoring of the progress and reporting to the relevant decision-makers	A
	Adaptation strategy and/or action plan updated, revised, and readjusted according to the findings of the monitoring and evaluation procedure	D

Table 3 Climate change risks by sectors – City of Hurghada, Governorate of Red Sea

Climate change risks by sectors	Vulnerability
Agriculture and food security	
Global and regional studies indicate generally deficits and decline in crops' yields for wheat, rice, and maize.	Not applicable
National-scale studies agree with each other that crop yields in Egypt could decline with climate change impacts.	Not applicable
Increasing pressures on food security as a result of climate change.	Low exposure
Water stress and drought	
Water stress could increase with climate change.	Exposed
Mean precipitation patterns could decrease with climate change.	Exposed
Groundwater could experience increase of salinity due to SLR or droughts.	Low exposure
Ecosystems	
Coral reefs' growth suffered from underwater activities – a decline in growth by 40% during 1987–2000. A drastic decline of ~49% and 45% between 2000 and 2013, respectively. Also, coral reef bleaching in the sea areas will increase 80% by 2060.	Exposed
Water acidification, which affects and treats mangroves in sea biodiversity, could include over 1000 species.	Exposed
Tourism	
Heat waves will affect attractiveness and tourism movement in Egypt.	Exposed
Loss in beaches' tourism due to coral reefs' bleaching in the Red Sea area.	Exposed
Urban	
Increases thermal discomfort and heat strokes, particularly in urban areas due to high temperatures and pollutions.	Exposed
Infrastructure failure due to storms and resulted floods.	Exposed
Air quality in cities decreases due to climate change.	Exposed
Health	
Higher temperatures, water stress, and malnutrition increase rift valley fever, avian influenza, and diarrhoea.	Exposed
Dust and sand storms affect patients with respiratory history and problems.	Exposed
Heat waves and higher temperatures increase cases of heat strokes and death in elderly citizens.	Exposed

5.4 Adaptation Scoreboard

The adaptation scoreboard is part of the SECAP developed by the Joint Research Center (JRC) of the European Commission [13, 23]. The city of Hurghada has realised a self-assessment of its adaptation status, setting a grade from A to D, in line with its progress regarding the Adaptation Cycle Steps, based on meetings and interviews with the Coordinator of Planning Department, Governorate of Red Sea (July and August 2017). More specifically:

- "A", corresponds to completion level of 75–100%
- "B", corresponds to completion level of 50–75%
- "C", corresponds to completion level of 25–50%
- "D", corresponds to completion level of 0–25%

The city of Hurghada (Municipality) has developed a score based on grades ranging from A to D, according to the above four grades to each one of the adaptation cycle-specific steps, as presented in the following Table 2. Nonetheless, in the age of coronavirus (COVID-19), there are parameters in some sectors that might be reviewed, such as urban sector (air quality in cities decreases due to CC), as well as health sector (dust and sand storms affect patients with respiratory problems). Since the social distancing and confinement measures reduced air pollution in most of the cities globally, that may help the patients with respiratory problems [40, 41]. However, this may all be temporarily depending on governments' policies, regulations, and measures developed/or to be developed during and post COVID-19.

6 Mapping Climate Change Risks by Sectors Against Vulnerability

6.1 Risk Assessment and Vulnerability Analysis

In order to conduct a risk assessment and vulnerability analysis, as a first step, the climate hazard types should be identified. These hazard types in general for the Maghreb and Mashreq countries, in particular including Egypt, are presented in Table 4 (Appendix 1). The Governorate of Red Sea (municipality) was called in to assess the impact that each climate hazard type

Table 4 Climate hazard types

General climate hazard types	Applicable for Maghreb and Mashreq regions, Governorate of the Red Sea, Egypt
Extreme heat	√
Extreme cold	
Landslides	
Storms[a]	√
Droughts	√
Sea level rise[b]	√
Floods	√
Extreme precipitation	
Forest fires	
Ice and snow	Took place in Cairo, Egypt in December 2013 after 112 years and in Alexandria in 2015 and 2016

[a]Took place in the Governorate of the Red Sea, mainly in Ras Ghareb in October 2016
[b]The Red Sea has very little waves and SLR is not comparable to that of Mediterranean

has on a series of vulnerable/impacted sectors, such as: (a) population (public health), (b) infrastructure (transport, energy, water, and social), (c) built environment (building stock and materials), (d) economy (tourist and agriculture), and (e) biodiversity (coastal zone ecosystems and green zones/forests).

These sectors have been identified as the most relevant for the Maghreb/Mashreq region, utilising information from the Future Cities Adaptation Compass Tool, Governors' (Mayors) Adapt, as well as the European Climate Adaptation Platform website and Covenant of Mayors' (CoM's) recommendations for the SECAPs [35–39]. Table 5 in Appendix 2 presents the results of the risk assessment analysis which were used to conduct the vulnerability analysis formulated by the City of Hurghada (municipality).

This exercise is based on sources such as the Future Cities Adaptation Compass Tool [35] and the United Nations Framework Convention on Climate Change (UNFCCC). Hence, the vulnerability analysis was carried out. In order to elaborate on the risk assessment, all date and information in Tables 4 and 5 were developed and realised based on six meetings' interviews with the Coordinator of Governorate of Red Sea. The impact of each risk is identified in the scale of High/Medium/Low; where specific climate projections are available, a risk analysis combining probability and impact was realised; additional information can be found in [21]. This is given in Table 5 (Appendix 2).

7 Results and Discussions

The ranking of CC vulnerabilities in Egypt was measured by five indicators (Low, Medium-low, Medium, Medium-high, and High). It is clear from Table 1 that the ranking of CC vulnerabilities indicates that the coastal resources, in terms of certainty of impact, severity of impact, and importance of resources, are High-medium and High. For water resources, these are medium and high for the three criteria. Regarding agriculture and energy resources, these are Medium-low, except for the importance of resources; it is High-medium for energy resources. It can be deduced from Table 3 that for the CCRs of the city of Hurghada in terms of Vulnerability; Ecosystems, Tourism, Urban, and Health sectors are exposed to CCR but agriculture and food security, and water stress are at low exposure to exposed. The adaptation actions by sectors, energy supply and renewable energy development, and project's fiche can be found in the SECAP document, pages 106–109 and 160–170 [21].

8 Conclusions

The SECAP for the City of Hurghada, Governorate of Red Sea was developed as part of the CES-MED project, funded by the European Union. The CES-MED project provided policy design support for the Municipality and the City of Hurghada in developing the SECAP, which has been pursued and coordinated with the National Committee of the Ministry of Foreign Affairs (National Focal Point) in collaboration with the Ministry of Local Development and in close coordination with the Governorate of Red Sea. The SECAP structure has been crafted, including four sections: (i) Governorate Climate and Energy

Table 5 Risk assessment of the city of Hurghada

	Receptors	Weather sensitivity	Future risk	Impact/ risk	Impact
Population	Public health	Extreme heat	Increased number of deaths	Low	Low-medium
			Reinforcement of heat stress	Medium	
			Increased infectious diseases	Low	
			Altered allergic patterns	Low	
		Droughts	Increased allergic incidents	Low	Low-medium
			Decreased air quality	Low	
			More respiratory problems	Medium	
		Sea level rise (significantly fewer waves due to coral reef)	Increased incidents of asthma and pneumonia	Low	Low
			Increased waterborne diseases	Low	
			Limitations to the healthcare access	Low	
		Storms	Limitations to the healthcare access	Low	Low
			Increased numbers of injuries and deaths	Low	
		Floods	Limitations to the healthcare access	N/A	N/A
			Increased numbers of injuries and deaths	N/A	
Infrastructure	Transport	Extreme heat	Damages on roads	Low	Low
			Modification of transport frequency and means	Low	
			Air quality problems	Low	
			Higher maintenance costs	Low	
		Droughts	Difficult transport of bulk material	Low	Low
		Sea level rise (Significantly fewer waves due to coral reef)	Damages	N/A	N/A
		Storms	Damages	Low	Low
			Mobility problems	Low	
		Floods	Damages	Low	Low
			Mobility problems	Low	
	Energy	Extreme heat	Blackouts and inability to cover demand load	N/A	N/A
			Damages, especially in the thermal power plants	N/A	
		Droughts	Blackouts and inability to cover demand load	N/A	N/A
			Higher maintenance costs	N/A	
			Cooling problems in power plants	N/A	
		Sea level rise (Significantly fewer waves due to coral reef)	Damages	Low	Low
			Shutdown of power plants near rivers, etc.	Low	
			Operational difficulties	Low	
			Higher maintenance cost	Low	
		Storms	Damages/failures in the production facilities and distribution grid/power cuts	Low	Low
		Floods	Damages/power cuts	Medium	Medium
	Water	Extreme heat	Water scarcity (desalinated water)	Low	Low
			Water quality issues (bottled water)	Low	
		Droughts	Water scarcity (desalinated water)	N/A	N/A

(continued)

Table 5 (continued)

	Receptors	Weather sensitivity	Future risk	Impact/risk	Impact
			Water quality issues (desalinated water)	N/A	
		Sea level rise (Significantly fewer waves due to coral reef)	Increased underground water salinity	Low	Low
			Water management issues	Low	
			Damages	Low	
			Water quality issues	Low	
			Higher maintenance costs	Low	
		Storms	Increased damages and related maintenance costs	Low	Medium
			Water management issues	Medium	
		Floods	Increased damages and related maintenance costs	Low	Low
			Water management issues	Low	
			Water quality issues	Low	
	Social	Extreme heat	Increased need for air-conditioned public spaces	Medium	Medium
		Droughts	Increased numbers of people presenting respiratory problems and burdening the healthcare facilities	Low	Low
			Inability to cover the water demand	Low	
			Difficulties in the operation of certain facilities due to lack of water (e.g. swimming pools)	Low	
		Sea level rise (Significantly fewer waves due to coral reef)	Potential damages in the coastal area facilities	Low	Low
			Loss of coastal public spaces (beaches)	Low	
		Storms	Damages	Low	Low
			Increased maintenance costs	Low	
		Floods	Damages	Low	Low
			Increased maintenance costs	Low	Low
			Flooding at the city level of the afflicted public building infrastructure (schools, hospitals, etc.) – Difficulties in providing the envisaged services	Low	
Built environment	Buildings' stock and materials	Extreme heat	Concrete's damages	High	High
			Increased cooling demands	High	
			Higher maintenance costs	High	
			Urban heat island effect	High	
		Droughts	Higher water demand	Medium	Medium
		Sea level rise (Significantly fewer waves due to coral reef)	Sinkholes collapse	Low	Low
			Extensive damages/loss of property	Low	
			Impact on coastal zone economy	Low	
		Storms	Damages	Low	Low
			Increased maintenance costs	Low	
		Floods	Damages	Low	Low
			Increased maintenance costs	Low	

	Receptors	Weather sensitivity	Future risk	Impact/ risk	Impact
Economy	Tourism	Extreme heat	Change of the tourism season – lower touristic flows	Medium	Medium
			Reduction of the tourism-related economy	Medium	
		Droughts	Increased water supply costs	Low	Low
			Potential increase of indirect costs for the tourists (infrastructure related) and reduction of touristic flows	Low	
		Sea level rise (Significantly fewer waves due to coral reef)	Damages and even complete destruction of touristic infrastructure, nearby coastal areas and deltas	Medium	Medium
		Storms	Damages to touristic facilities	Medium	Medium
		Floods	Damages to touristic facilities	Medium	Medium
			Potential effects on the touristic flows, in areas with flooding history	Medium	
	Agriculture	Extreme heat	Changes in growth cycle	N/A	N/A
			Damages/loss of harvest	N/A	
			Livestock loss and impacts on health	N/A	
			Lower crop yields	N/A	
			Increased fire risks	N/A	
		Droughts	Damages/loss of harvest	N/A	N/A
			Lower crop yields	N/A	
			Livestock loss and impacts on health	N/A	
			Land degradation	N/A	
			Increased fire risks	N/A	N/A
		Sea level rise (Significantly fewer waves due to coral reef)	Damages/loss of harvest in areas near delta, sea etc.	N/A	N/A
			Increased water salinity will result in existing crops' long-term destruction	N/A	
		Storms	Damages/loss of harvest in afflicted areas	N/A	N/A
			Surface soil erosion	N/A	
		Floods	Damages/loss of harvest in afflicted areas	N/A	N/A
			Livestock loss	N/A	
			Surface soil erosion	N/A	
Biodiversity	Coastal zone eco-systems	Extreme heat	Loss of specific species (fish, etc.)	Medium	Medium
		Droughts	Increase of coastal water salinity	Low	Low
		Sea level rise (Significantly fewer waves due to coral reef)	Loss of specific species (fish, etc.)	Medium	Medium
			Soil erosion	Medium	
			Water salinisation	Medium	
		Storms	Soil erosion	Medium	Medium
		Floods	Soil erosion	Medium	Medium
	Green zones/ forests	Extreme heat	Fires and destruction of the ecosystem, flora and fauna	N/A	N/A
		Droughts	Fires and destruction of the ecosystem, flora and fauna	N/A	N/A
		Sea level rise (Very less waves due to coral reef)	Increase of underground water salinity and destruction of the ecosystem	Low	Low
		Storms	Destruction of trees and other damages	N/A	N/A
		Floods	Destruction of trees and other damages	N/A	N/A

Strategy [21]; (ii) BEI [20, 21]; (iii) Actions Planned Section [20, 21]; and (iv) Climate Adaptation. However, the CCA suggested that more actions are needed, especially in the time of COVID-19.

Acknowledgments Authors would like to thank the European Union; ENPI-South, EuropeAid/132360/C/SER/Multi; Contract No.: ENPI 2012/309-311 for funding the CES-MED Project. We also thank the Directorate-General for Neighbourhood and Enlargement Negotiations (DG-NEAR) for supervising this project and to CES-MED Team Leader, Transitions and Energies Demain teams for developing the SECAP with me and for their hard work, support, and cooperation throughout of this ongoing project. We express our sincere thanks to H.E. General Ahmed Abdullah, former Governor of Red Sea and General Abdelfattah Tamam, Secretary General – Governorate of Red Sea, and Dr. Mohamed Badr, former Governor of Luxor for their support and cooperation. Thanks go to Eng. Ayman Sultan Planning Department – Governorate of Red Sea and Dr. Ramdan Seddik, Director of Environmental Affairs – Governorate of Luxor for their assistance and collaboration in project. We are thankful to Ambassador Abu Bakr Hefny Mahmoud and Ambassador Adel Ibrahim, former Deputy Assistant Minister for EU Affairs and Counsellor Mohamed El-gammal, Ministry of Foreign Affairs – Egypt for their endless support and follow-up during the period from May 2016 to April 2018. Last but not least, we are grateful to Dr. Ahmed Azzam for facilitating the CES-MED project (2015–2018).

Appendices

Appendix 1

Appendix 2

References

1. The United Nations: Sustainable Development Goals (2020) Climate change knows no national borders, UN Chief says. [Online], available at: https://www.un.org/sustainabledevelopment/blog/2015/11/climate-change-knows-no-national-borders-un-chief-says/. Accessed 12 Sept 2020
2. ACTIONAID (2016.) Climate change knows no borders – an analysis of climate-induced migration, protection gaps and the need for solidarity in South Asia, Report, 8 December 2016, [Online], available at: https://actionaid.org/publications/2016/climate-change-knows-no-borders/. Accessed 12 Sept 2020
3. Mohsen M. Aboulnaga, Amr F. Elwan, and Mohamed R. Elsharouny (2019) Urban climate change adaptation in developing countries – policies, projects, and scenarios, Ch.2, doi https://doi.org/10.1007/978-3-030-05405, ISBN 9783030054052 [Online], Springer. Available at: https://rd.springer.com/book/10.1007/978-3-030-05405-2#siteedition-corporate-link/. Accessed 28 Jan 2019
4. Nicholls RJ, et al. (2007) Ranking of the world's cities most exposed to costal flooding today and in the future, OECD, [Online], available at: https://www.oecd.org/environment/climte-change-risks/. Accessed 12 Sept 2020
5. CDP (2020) Cities at risk: dealing with the pressures of climate change, [Online], available at: https://www.cdp.net/en/research/global-reports/cities-at-risk/. Accessed 12 Sept 2020
6. Chapman, S., et al., (2017). The impact of urbanization and climate change on urban temperature: a systematic review. Landscape Ecol, Aug 2017, Springer Sciences & Business Media B.V. 2017, [Online], available at: https://doi.org/10.1007/s10980-017-0561-4/. Accessed 12 Sept 2020
7. Tahir AA, et al. (2014) Impact of rapid urbanization on microclimate of urban areas of Pakistan. Air Quality Atmosphere & Health 8(3), July 2014, [Online], available at: https://doi.org/10.1007/s11869-014-0288-1/. Accessed 12 Sept 2020
8. Paranunzio R, et al. (2019). Evaluating the effects of urbanization evolution on air temperature trends using nightlight satellite data, Atmosphere 10, 117; doi: https://doi.org/10.3390/atmos10030117, MDPI, [Online], available at: https://www.mdpi.com/journal/atmosphere/. Accessed 12 Sept 2020
9. Revi A, Satterthwaite DE, Aragón-Durand F, Corfee-Morlot J, Kiunsi RBR, Pelling M, Roberts DC, Solecki W (2014) Urban areas. In: Field CB, Barros VR, Dokken DJ, Mach KJ, Mastrandrea MD, Bilir TE, Chatterjee M, Ebi KL, Estrada YO, Genova RC, Girma B, Kissel ES, Levy AN, MacCracken S, Mastrandrea PR, White LL (eds) Climate change 2014: impacts, adaptation, and vulnerability. Part A: global and sectoral aspects. Contribution of working group II to the fifth assessment report of the Intergovernmental Panel on Climate Change. Cambridge University Press, Cambridge and New York, pp 535–612.10. Available at: https://www.ipcc.ch/site/assets/uploads/2018/02/WGIIAR5-Chap8_FINAL.pdf. Accessed 12 Sept 2020
10. Rosenzweig C, Solecki W, Romero-Lankao P, Mehrotra S, Dhakal S, Bowman T, Ibrahim SA (2015) ARC3.2 summary for city leaders. Urban Climate Change Research Network, Columbia University, New York. [Online], available at: https://pubs.giss.nasa.gov/docs/2015/2015_Rosenzweig_ro02510w.pdf. Accessed: September 10, 2020

11. Resilience to Climate change (2019) Global economy will be 3 percent smaller by 2050 due to lack of climate resilience. The Economist Intelligence Unit [Online], available at: https://www.eiu.com/n/globale- economy-will-be-3-percent-smaller-by-2050-due-to-lack-of-climate-resilience/. Accessed 21 Nov 2019
12. The Cleaner Energy Saving Mediterranean cities, CES-MED, [Online], available at: https://www.ces-med.eu/. Accessed 4 Apr 2018
13. Integrating climate change adaptation into project development: emerging experience. Available at: http://www.afd.fr/lang/en/home/projets_afd/changement_climatique/. Accessed 15.08.2016
14. Guidebook – 'How to Develop a Sustainable Energy Action Plan (SEAP) in South Mediterranean Cities', 2014. European Commission, EUR 27016 EN – Joint Research Centre – Institute for Energy and Transport, ISBN 978-92-79-44693-1; Doi: https://doi.org/10.2790/392701, [Online], Available at: http://publications.jrc.ec.europa.eu/repository/handle/JRC93697/. Accessed 1 Apr 2019
15. The United Nations: Climate Change (2020) What is the Paris Agreement? Art.7, [Online], available at: https://unfccc.int/process-and-meetings/the-paris-agreement/what-is-the-paris-agreement/. Accessed 13 Sept 2020
16. The United Nations: Climate Change, (2020). What do adaptation to climate change and climate resilience mean? Framework Convention on Climate Change, UNFCCC, [Online], available at: https://unfccc.int/topics/adapatation-and-resilience/the-big-picture/what-do-adaptation-to-climate-change-and-climate-resilience-mean/. Accessed 13 Sept 2020
17. European Commission EC (2018), Europe is ready for climate impacts: Commission evaluates its strategy, [Online], available at: https://ec.europa.eu/clima/news/europe-ready-climate-impacts-commission-evaluates-its-strategy_en/. Accessed 13 Sept 2020
18. The Cleaner Energy Saving Mediterranean cities, CES-MED [Online], available at: https://www.ces-med.eu/. Accessed 4 Apr 2018
19. Aboulnaga, M. 2016. Recommended national sustainable urban and energy savings actions for Egypt, cleaner energy saving mediterranean cities (CES-MED), ENPI 2012/309-311, EU funded project, EuropeAid/ 132630 /C/ SER/Multi, Hulla & Co. Human Dynamics – KG, October 10, 2016, Available at: http://ces-med.eu/. Accessed 27 Feb 2019
20. Aboulnaga M, Amin N, Rebelle B (2020) Climate adaptation action: the role of clean energy and strategic action plans of south Mediterranean cities. In: Sayigh A (ed) Green buildings and renewable energy – innovative renewable energy. Springer, Cham, pp 11–33. Available at: https://doi.org/10.1007/978-3-030-30841-4_2/. Accessed 13 Sept 2020
21. Sustainable energy & climate adaptation plan (SECAP), City of Hurghada, Governorate of the Red Sea – Egypt, cleaner energy saving and Mediterranean cities CES-MED, [Online], available at: http://www.ces-med.eu/publications/egypt-municipality-hurghada-sustainable-energy-and-climate-action-plan-secap/. Accessed 22 July 2020
22. Sustainable energy & climate adaptation plan (SECAP), City of Luxor, Governorate of Luxor – Egypt, cleaner energy saving and Mediterranean cities CES-MED, [Online], available at http://www.ces-med.eu/publications/egypt-municipality-luxor-sustainable-energy-and-climate-action-plan-secap/. Accessed 22 July 2020
23. Bertoldi P, Cayuela DB, Monni S, de Raveschoot RP (2010) JRC scientific and technical report: existing methodologies and tools for the development and implementation of sustainable energy action plans (SEAP), Italy, [Online], available at: https://publications.jrc.ec.europa.eu/repository/bitstream/JRC56513/reqno_jrc56513_existing_methodologies_and_tools_for_the_development_and_implementation_of_sustainable_energy_action.pdf/. Accessed 15 Oct 2016
24. Egypt's national strategy for adaptation to climate change and disaster risk reduction (NSACCDRR), 2011. Available at: http://cairoclimatetalks.net/sites/default/files/Adaptation%20Strategy%20-%20Final%20-%20E.pdf/. Accessed 24 Aug 2016
25. EgyptToday (2109) Egypt, UK tackle resilience, adaptation at UN climate change summit, [Online], available at: https://www.egypttoday.com/Article/1/75087/Egypt-UK-tackle-resilience-adaptation-at-UN-climate-change-summit/. Accessed 13 Sept 2020
26. The United Nations Development Programme - UNDP (2019). INDC Project Actions and Impacts: Egypt (2019), [Online], available at: https://www.ndcs.undp.org/content/ndc-support-programme/en/home/impact-and-learning/library/indc-project-actions-and-impacts-egypt.html/. Accessed 13 Sept 2020
27. Baldi M, et al. (2020) Climatology and dynamical evolution of extreme rainfall events in the Sinai Peninsula – Egypt. Sustainability 2020, 12, 6186; doi.org/10.3390/su12156186/ [Online], available at: https://www.mdpi.com/journal/sustainability/12-06186-1/pdf/. Accessed 13 Sept 2020
28. Egypt's Vision and SDS 2030 (2016), [Online], available at: https://arabdevelopmentportal.com/sites/default/files/publication/sds_egypt_vision_2030.pdf/. Accessed 20 Mar 2019
29. Sustainable Development Goals Knowledge Platform - Egypt, United Nations (2020). Voluntary national review 2018: key massage of Egypt VNR 2018, [Online], available at: https://sustainabledevelopment.un.org/memberstates/egypt/. Accessed 13 Sept 2020
30. Ministry of Planning, Monitory, and Administrative Reforms (2018) Egypt's voluntary national review 2018, [Online], available at: https://sustainabledevelopment.un.org/content/documents/27736Egypt_VNR_Presentation_Script.pdf. Accessed 13 Sept 2020
31. State Information Service (2020).Egypt to update 2030 vision amid coronavirus crisis, 2 June 2020, [Online], available at: https://www.sis.gov.eg/Story/147408/Egypt-to-update-2030-Vision-amid-Coronavirus-crisis?lang=en-us/. Accessed 14 Sept 2020
32. Egypt Today (2019). [Online], available at: https://www.egypttoday.com/Article/1/52769/In-pics-Egypt%E2%80%99s-rooftops-going-green/. Accessed 8 Sept 2020
33. European Commission (2020) Adaptation preparedness scoreboard: country fiche for The Netherlands. [Online], available at: https://ec.europa.eu/clima/sites/clima/files/adaptation/what/docs/country_fiche_nl_en.pdf. Accessed 14 Sept 2020
34. Shakou LM, Wybo JL, Reniers G, Boustras G (2019) Developing an innovative framework for enhancing the resilience of critical infrastructure to climate change. Saf Sci 118:364–378. [Online], available at: https://doi.org/10.1016/j.ssci.2019.05.019/. Accessed 14 Sept 2020
35. Future Cities Urban Networks to face climate change (2016) The future cities adaptation compass: a guidance tool for developing climate-proof city Regions, [Online], available at: http://www.future-cities.eu/fileadmin/user_upload/pdf/FUTURE-CITIES_Adaptation-Compass_Guidance.pdf. Accessed 14 Sept 2020
36. Solecki W, Seto KC, Balk D, Bigio A, Boone CG, Creutzig F, Zwickel T (2015) A conceptual framework for an urban areas typology to integrate climate change mitigation and adaptation. Urban Clim 14:116–137. [Online], available at: https://doi.org/10.1016/j.uclim.2015.07.001/. Accessed 14 Sept 2020
37. Covenant of Mayors - COM, [Online], available at: https://www.covenantofmayors.eu/en/. Accessed 5 Oct 2016
38. European Commission - Climate Adapt (2020) [Online], available at: https://climat-adapt-eea.europa.eu/about/. Accessed 14 Sept 2020

39. Wrobel M, Bisaro A, Reusser D, Kropp JP (2013) Novel approaches for web-based access to climate change adaptation information – MEDIATION adaptation platform and ci: grasp-2. In: Hřebíček J, Schimak G, Kubásek M, Rizzoli AE (eds) Environmental software systems. Fostering information sharing. ISESS 2013. IFIP advances in information and communication technology, vol 413. Springer, Berlin, Heidelberg. [Online], available at: https://doi.org/10.1007/978-3-642-41151-9_45/. Accessed 14 Sept 2020
40. Startoulias D et al (2020) Air quality development during the COVID-19 pandemic over a medium-sized urban area in Thailand. Sci Total Environ 746:141320. https://doi.org/10.1016/j.scitotenv.2020.141320/. Accessed 14 Sept 2020
41. Broomandi P, Karaca F, Nikfal A, Jahanbakhshi A, Tamjidi M, Kim JR (2020) Impact of COVID-19 event on the air quality in Iran. Aerosol Air Qual Res 20(8):1793–1804

An Approach for Healthier, Smart and Sustainable Cities in the Time of COVID-19 Crisis: Mapping SSC Indicators with Quality of Life and SDGs

Sara Tarek and Mohsen Aboulnaga

1 Introduction

Cities worldwide are not only facing challenges such as rapid population increase, high energy demand, air pollution and climate change, but also the new deadly coronavirus (COVID-19) that has been rapidly sweeping the globe since early December 2019 till date [1–5]. The impact of COVID-19 resulted in more than 21 million infected cases and nearly one million death cases [6]. In controlling the spread of the deadly coronavirus, governments were forced to impose confinement policies to regulate movement and transportation in and out of cities worldwide [7].

Hence, it is vital to assess the role of smart cities and smart buildings in overcoming such crisis and provide rapid solutions to the deadly coronavirus pandemic in improving a better operation of the urban areas and cities. This paper presents a study that focuses on smart sustainable cities – SSCs' dimensions, characteristics and indicators as well as highlights the SSC contribution to attain SDGs. It presents the results of mapping SSC indicators to assist in creating healthier cities with a better quality of life in the age of COVID-19.

2 Objective

The objective of the study is to identify and develop the main dimensions for applying smart sustainable city solutions in urban areas in the time of COVID-19. The research investigates the main definitions, solutions and indicators for smart sustainable cities that help in coping with global issues and crises to achieve greener and healthier urban environments in cities. It also maps the SSC indicators with Quality of Life (QoL) dimensions as well as maps selected SSC indicators with SDGs.

3 Methodology

To conduct this research work, the methodology comprises of two main approaches: the first approach presents a concise review for the selected literature about what and how pandemics do impact cities through history, and in response to achieving healthier environments it reviews the Quality of Life (QoL) dimensions. It also highlights the United Nations Agenda 2030 and the challenges it faces due to COVID-19 crisis. This approach also introduces the SSC as an effective solution for cities in response to the pandemic by recalling the definitions and previous researches that identify the smart cities and sustainable cities together with their indicators. The second approach aims at identifying a framework for smart sustainable cities by pointing out its integration with SDGs as a solution for cities in the time of COVID-19. This follows a descriptive-analytical approach to recognize the indicators for SSC implementation. This approach also highlights the selected standards and indicators covering the different dimensions and indicators for SSC implementation and mapping to develop a solution for cities' recovery from the COVID-19 crisis.

S. Tarek (✉) · M. Aboulnaga
Department of Architecture, Faculty of Engineering, Cairo University, Cairo, Egypt
e-mail: sarat@eng1.cu.edu.eg; maboulnaga@eng.cu.edu.eg

© The Author(s), under exclusive license to Springer Nature Switzerland AG 2022

A. Sayigh (ed.), *Sustainable Energy Development and Innovation*, Innovative Renewable Energy,
https://doi.org/10.1007/978-3-030-76221-6_46

4 Smart Sustainable Cities' Approach as an Effective Solution for COVID-19

Since the COVID-19 outbreak in December 2019, cities are shifting towards healthy and digital urban life to minimize the spread of the coronavirus. This has been manifested in most of the cities globally, for example, New York, London, Paris, Amsterdam, Singapore, China, Rome and Berlin, as shown in Fig. 1 [8–16]. Apart from improving air quality, ecosystem and lower CO2 emissions, apparently the Coronavirus has already boosted the efforts for digital, smart and greener urban environments. During the lockdown, cities witnessed some changes like clean air, virtually empty streets, fewer airplanes in the skies, global carbon emissions dropping by 17% [17] due to the imposed policies and confinement measures (complete lockdown) that forced people globally to stay home and practise their daily activities from home for more than 100 days. Thus, they were forced to use digital technologies in most of their activities, for example, working, attending classes and attending events online. It has become clearer now that there is a vibrant combination between people, technology and data/information to find out innovative solutions for the multiple influences that resulted from the COVID-19 crisis. Figures 1 and 2 show examples of the changes in urban life in various cities due to the COVID-19 impacts [18].

To achieve a healthier urban environment, nevertheless, different aspects shall be identified regarding the connections between the urban environment itself, health and environmentally sustainable cities. Figure 3 presents the strategies of health and well-being and its mutual effect on environmental ones. Shifting towards greener smarter solutions speeds up cities' movements towards achieving smart sustainable goals. However, the definitions of both smart and sustainable cities (SSCs) should be differentiated and clarified, so that it will be easier to identify the criteria a city must have to be classified as a smart and sustainable one. The next section identifies both smart and sustainability concepts, then it presents combining both smartness and sustainability together to identify the SSC approach and indicators.

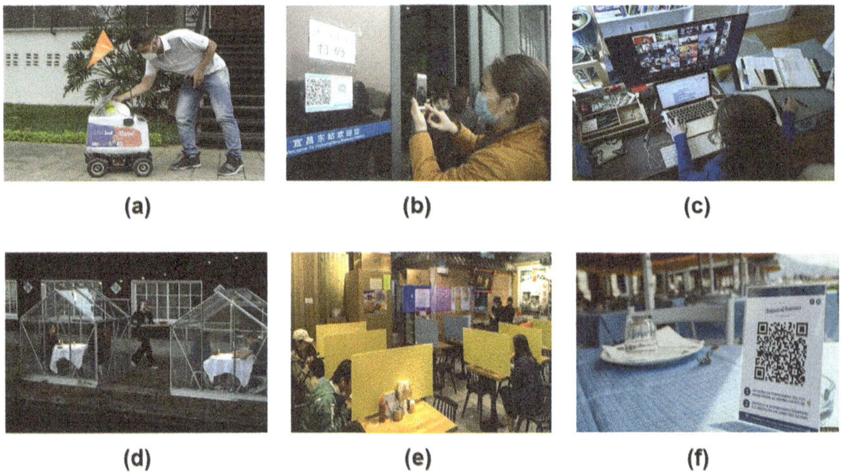

Fig. 1 Measures and actions taken in cities that redesigned urban life due to COVID-19. (**a**) Using robots for delivery in Bogota, Colombia. (Credit: Fredy Builes) [19]. (**b**) A traveller scans a health QR code after buying a train ticket in Yichang. (Credit: Wu Yanling) [20]. (**c**) Digital transformation of education through intensified Zoom sessions online. (Credit: Olivier Douliery) [21]. (**d**) Art centre uses isolated greenhouses for Coronavirus, Amsterdam measures. (Credit: Peter Dejong) [22]. (**e**) Barrier sheets at a restaurant, Hong Kong, China. (Credit: Alamy) [14]. (**f**) Restaurants shift to QR codes touch-free ordering, Naples, Italy. (Credit: Reuters) [14]

Fig. 2 Improved air quality and ecosystem, and empty streets resulted from COVID-19 [18]. (**a**) Clear skies and clean air in Los Angeles – California, United States. (Credit: Laura J. Nelson). (**b**) Improved ecosystem, mountain goats in Wales, UK. (Credit: P. Byrne/PA). (**c**) Improved air quality in street of Barcelona, Spain. (Credit: Reuters/Nach Doce. (**d**) Clean air in an empty square in Rome, Italy. (Credit: Reuters/A. Lingria)

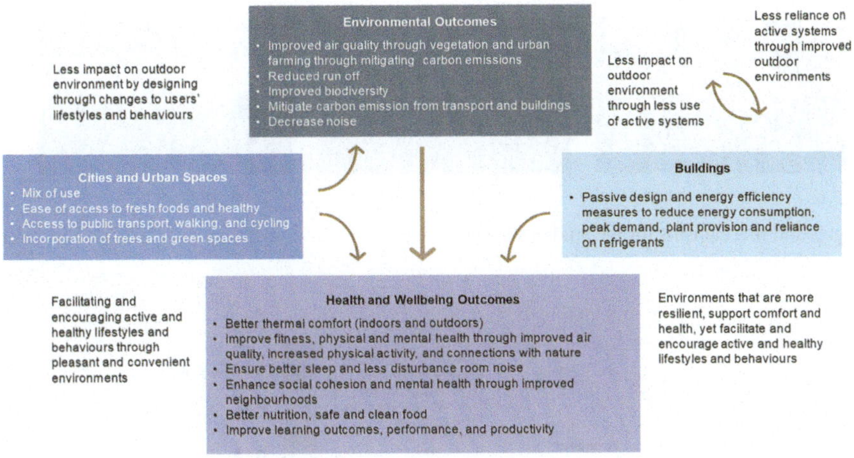

Fig. 3 Framework of the built environment influence on both health and well-being. (Source: Developed by authors after Capelli and Conserva [23, 24])

5 Smart City and Sustainable City

It is imperative to review and discuss smart city and sustainable city in order to understand and highlight their definitions and indicators for a better insight into the study.

6 Procedures

Different models and frameworks were seemingly addressed to identify SSC indicators. However, the concept is still inconsistent; hence, appropriate and specific indicators are needed to be developed on different levels to ensure SSC achievement at local levels. Therefore, this research study adopts a descriptive-analytical approach to identify the indicators for SSC development/design and implementation. Since the research aims at developing SSC as a solution for a better city life in the time of and post COVID-19, this could be achieved by identifying the main indicators for implementing SSC in order to achieve a healthy environment with a better quality of life.

After reviewing different approaches and methods to identify SSC dimensions and indicators, and since it is very broad and systematically updated, and in order to attain the research objective, the research selected SSC methods that are targeting cities, communities scale with main focus on SDGs and the development of SSC strategies and mechanisms, namely: U4SSC – ITU-T Y.4903/l.1603, CITYKEYS and European Smart City Ranking, as presented in Fig. 4 [25–27].

7 Results and Discussions

Based on the above assessment, a framework illustrating the main contribution to this study in which the research question was put forward – can smart sustainable cities SSC assist in overcoming the crisis of COVID-19 pandemic and provide rapid solutions to it in operating urban areas? Evidence of the research results proposes that implementing SSC strategies, through the integration between four dimensions (economic, environmental, technological and socio-cultural), together are essential key to achieve healthier, cleaner and greener cities to face the consequences of the pandemic. Figure 5 presents the proposed framework of the SSC implementation to achieve the above-mentioned objective. The SSC definition merges between traditional urban sustainability and novel needs for cities related to ICT (Information Communication Technology) and smart applications and services.

The implementation of SSC depends on the integration between indicators from many fields, in particular economic, social, environment as well as urban environment. It also depends on the balance between sustainability and smartness and broad usage of SSC indicators. Nevertheless, to achieve a healthier urban environment and to enhance the quality of life to control consequences of the spread of Coronavirus impacts, besides identifying its contribution to SDGs to attain a better

Fig. 4 Selected SSC strategies and mechanisms for the study [80–81]

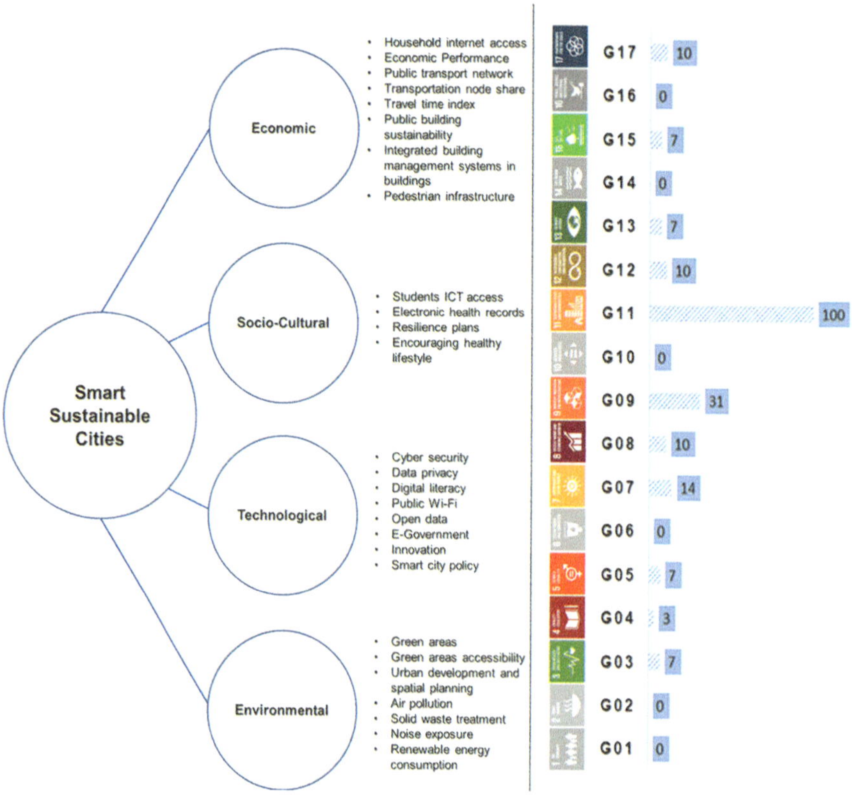

Fig. 5 Framework on the implementation of SSC as a solution for cities amidst COVID-19, showing the indicators' contribution to SDGs. (Source: Developed by authors)

sustainable future, the results show that SDGs applying SSC approaches in urban areas with the new development measures can assist in post COVID-19 adoption and attain SDGs, mainly SDG 3, 4, 5, 7, 8, 11, 12, 13, 15 and 17.

It is clear that all the indicators are directly related to almost 11 of the United Nations' SDGs, particularly SDG 3, 4, 5, 7, 8, 9, 11, 12, 13, 15 and 17. The percentage of the indicators contributing to the SDGs is as follows: 100 per cent of the indicators contribute to the SDG 11; 31 per cent of the indicators contribute to the SDG 9; 14 per cent of the indicators contribute to the SDG 7; 10 per cent of the indicators contribute to SDG 8, 12 and 17; and 7 per cent of the indicators contribute to SDG 3, 5, 13 and 15 as well as 3 per cent of the indicators contribute to the SDG 4.

The results also highlight how there should be an integration between different systems and processes for enhancing the role of ICT that support economy, people and collaborative governance, besides focusing on security, privacy, information consistency as well as financial aspects for the growing needs for technological updates in the time of COVID-19 crisis.

8 Conclusions

A smart sustainable city is a new concept that has substantially appeared since 2010. Policymakers, decision makers, researchers and academics have drawn attention to this new concept, but its understanding and implementation need deeper understanding, work and research. The present research work identified the likely indicators needed to assess Smart Sustainable Cities, through identifying the definitions, dimensions and indicators for achieving both sustainable and smart cities. It investigated the promising contribution of the SSC concept in creating healthier and cleaner cities as an emerging need amidst the COVID-19 pandemic. It highlights smart and sustainable solutions for cities regarding achieving better liveability, so it reviewed the Quality of Life (QoL) dimensions developed by the OECD to ensure achieving the research main goal. It also recalled the strategies, methods and indicators for SSC implementation established by the international organization and focused on those established by U4SSC – ITU-T Y.4903/l.1603, CITYKEYS and European Smart City Ranking.

The research work synthesized the key indicators with the reviewed QoL dimensions and identified the likely relationship between them for a better deployment of these indicators in the SSC development and implementation process. The indicators were classified into four categories: Economic, Socio-cultural, Technological and Environmental and were mapped against the UN SDGs to highlight its role in the sustainable development, and how it will enhance the achievement of the UN target by 2030 for cities. The various aspects of the presented research work supported the notion that 'Adopting SSC approaches in cities policies is a prompt and flexible solution to face the impacts resulted from COVID-19 pandemic and to keep the track to achieve the UN SDGs – as specified – by 2030'. Furthermore, it points out the key indicators for achieving SSC, which may serve in the decision-making process of adopting the concept/approach.

The study points out that in the light of making cities and sustainably classified cities smart and more sustainable, decision makers should focus more on smart solutions that might have substantial contribution relevant to sustainability. These solutions should be responding to environmental pressures, climatic change, changes in socio-economic needs, enhancing urban living, safety, health and well-being. This could be achieved through the integration of smart models and sustainable approaches.

Regarding the study limitations, reviewing and analysing SSC reference literature showed that there are important gaps in this field of research. Besides, there is a lack of the existence of comprehensive frameworks and methodologies that could be used as tools for evaluating SSC or smart dimensions' contribution to sustainability, which could add a systematic perspective for SSC. Also, qualitative approach is a good method to indicate smart and sustainable assessment, variables and indicators, but quantitative methods are essential for better reading and further investigation.

The findings and conclusions point out directives for future research, namely: further development of the proposed framework that incorporates SSC indicators and UN SDGs and supporting it with a quantitative approach to investigate its reliability, developing a decision-making tool to achieve SSC integrating the mentioned indicators and QoL dimensions together.

References

1. Kammen DM, Sunter DA (2016) City-integrated renewable energy for urban sustainability. Science 352(6288):922–928, New York, NY. https://doi.org/10.1126/science.aad9302
2. Bierbaum R, Smith JB, Lee A, Blair M, Carter L, Stuart F, Ch F, Fleming P, Ruffo S, Stults M, McNeeley SH, Wasley E, Verduzco L (2013) A comprehensive review of climate adaptation in the United States: more than before, but less than needed. Mitig Adapt Strateg Glob Change 18:361–406. Available at: https://doi.org/10.1007/s11027-012-9423-1. Accessed: 17 Aug 2020
3. Cheval, S.; Mihai Adamescu, C.; Georgiadis, T.; Herrnegger, M.; Piticar, A. & Legates, D.R. (2020) Observed and potential impacts of the COVID-19 pandemic on the environment. Int J Environ Res 17, 4140. Available at: https://doi.org/10.3390/ijerph17114140. Accessed 10 Aug 2020
4. Rosenzweig C, Solecki WD, Hammer SA, Mehrotra S (2011) Climate change and cities: first assessment. Report of the Urban Climate Change Research Network. Available at: https://www.adaptationclearinghouse.org/resources/climate-change-and-cities- first-assessment-report-of-the-urban-climate-change-research-net.html. Accessed 17 Aug 2020
5. Scientists probe link between Covid-19 and deadly air pollution (2020) https://www.ccacoalition.org/en/news/scientists-probe-link-between-covid-19-and-deadly-air-pollution. Accessed 16 Aug 2020
6. World Health Organization (WHO) (2020) Coronavirus disease COVID-19 situation report 209. Available at: https://www.who.int/emergencies/diseases/novel-coronavirus-2019/situation-reports. Accessed 16 Aug 2020
7. OECD (2020) COVID-19 crisis response in MENA countries. Available at: https://www.oecd.org/coronavirus/policy-responses/covid-19-crisis-response-in-mena-countries-4b366396/. Accessed 16 Aug 2020
8. United Nations (2020) COVID-19 in an Urban World. Available at: https://www.un.org/sites/un2.un.org/files/sg_policy_brief_covid_urban_world_july_2020.pdf. Accessed 17 Aug 2020
9. The death of the city. (2020) [online], Available at: https://www.politico.eu/article/the-death-of-the-city-coronavirus-towns-cities-retail-transport-pollution-economic-crisis/. Accessed 17 Aug 2020

10. International Labour Organization (2020) Social protection responses to the COVID-19 pandemic in developing countries: Strengthening resilience by building universal social protection. Available at: https://www.ilo.org/wcmsp5/groups/public/%2D%2D-ed_protect/%2D%2D-soc_sec/documents/publication/wcms_744612.pdf. Accessed 17 Aug 2020
11. Have UK cities been hotbeds of the Covid-19 pandemic? (2020) [Online], Available at: https://www.centreforcities.org/blog/have-uk-cities-been-hotbeds-of-covid-19-pandemic/. Accessed 17 Aug 2020
12. Ing W (2020) Chancellor to unveil £3 billion retrofit funding. Architect J. Available at: https://www.architectsjournal.co.uk/news/chancellor-to-unveil-3-billion- retrofit-funding/10047507.article. Accessed: August 17, 2020
13. The World Bank In Egypt (2020) [Online], Available at: https://www.worldbank.org/en/country/egypt/overview/. Accessed 17 Aug 2020
14. Saadat S, Rawtani D, Hussain CM (2020) Environmental perspective of COVID-19. Sci Total Environ 728:1. [Online], available at: https://doi.org/10.1016/j.scitotenv.2020.138870. Accessed 17 Aug 2020
15. How Covid-19 could redesign our world (2020) BBC. [online], Available at: https://www.bbc.com/future/article/20200527-coronavirus-how-covid-19-could-redesign-our-world/. Accessed 15 Aug 2020
16. OECD (2020) Cities policy responses. OECD policy responses to coronavirus (COVID-19). Available at: http://www.oecd.org/coronavirus/policy-responses/cities-policy-responses-fd1053ff/. Accessed 17 Aug 2020
17. Le Quéré C, Jackson RB, Jones MW et al (2020) Temporary reduction in daily global CO2 emissions during the COVID-19 forced confinement. Nat Clim Chang 10:647–652. Available at: https://doi.org/10.1038/s41558-020-0797-x. Accessed 15 Aug 2020
18. Coronavirus in Scotland: 100 days of life under lockdown (2020). [Online], Available at: https://www.bbc.com/news/uk-scotland-53162959. Accessed 17 Aug 2020
19. Business insider (2020) Softbank-backed delivery startup Rappi is testing out robots for contactless delivery – take a look. Available at https://www.businessinsider.com/delivery-startup-using-robots-in-colombia-to-avoid-coronavirus-2020-4. Accessed 16 Aug 2020
20. Transform News (2020) Hubei recovers railway services, Wuhan not included. Available at: http://www.transformcn.com/en/2020-03/26/content_239675.htm. Accessed 18 Aug 2020
21. The Guardian (2020) 'Zoom is malware': why experts worry about the video conferencing platform. Available at: https://www.theguardian.com/technology/2020/apr/02/zoom-technology-security-coronavirus-video-conferencing. Accessed 18 Aug 2020
22. Insider (2020). A restaurant in Amsterdam built small greenhouses for diners to enjoy a meal in while social distancing. Available at: https://www.insider.com/amsterdam-restaurant-mini-greenhouses-cubicles-social-distance-2020-5.Accessed 18 Aug 2020
23. Capelli AM, Conserva BF (2020) The environmental sustainability at the time of COVID-19 pandemy: an holistic approach through LEED ® v4 BD+C rating system, doi:https://doi.org/10.13140/RG.2.2.24299.90407. Accessed 20 July 2020
24. World Health Organisation – WHO (2019) WHO global strategy on health, environment and climate change: the transformation needed to improve lives and well-being sustainably through healthy environments. Health, environment and climate change. Report by the Director-General. Available at: https://www.who.int/docs/default-source/climate-change/who-global-strategy-on-health-environment-and-climate-change-a72-15.pdf?sfvrsn=20e72548_2. Accessed 17 Aug 2020
25. ITU (2016) ITU-T Y.4903/L.1603 Key performance indicators related to the sustainability impacts of information and communication technology in smart sustainable cities. Available at: https://www.itu.int/rec/T-REC-Y.4903/en. Accessed 17 Aug 2020
26. ITU (2014) ITU-T FG-SSC Smart sustainable cities: An analysis of definitions. Available at: https://www.itu.int/en/ITU-T/focusgroups/ssc/Documents/Approved_Deliverables/TR-Definitions.docx. Accessed 17 Aug 2020
27. European Telecommunications Standards Institute ETSI TS 103 463 – Key performance indicators for sustainable digital multiservice cities. (2017) Technical specification. https://www.etsi.org/deliver/etsi_ts/103400_103499/103463/01.01.01_60/ts_103463v010101p.pdf. Accessed 17 Aug 2020

Urban Farming as an Innovative Tool for Generative and Sustainable Cities: The Case of Galliera City in Bologna, Italy

Yasmina Ragab, Emanuele Durante, Brando Di Giovanni, Luca Gialli, Patricia González Mariscal, Laura Gabrielli, Michele Lapomarda, Mohamed Shahwan, Sherifa El-Haggan, and Mohsen Aboulnaga

1 Introduction

Cities emit about 60% of the global CO_2 emissions, and about 75% of the global primary energy (buildings and transport). Cities are responsible for around 65% of global energy demand [1] and 75% of natural resources consumption [2]. Urban areas today host around 55% of the world's population, which is expected to reach 68% by 2050. Such urbanization rate will increase pressure on urban infrastructure due to high demand and scarce resources and climate change (CC) [3]. The Ex-SIAPA industrial site is an emblematic example of that where a city centre can be redesigned in order to foster sustainability and food production in order to create a new economic source. The city is located in the Metropolitan City of Bologna, in the municipality of Galliera, the site after being a dichlorodiphenyltrichloroethane (DDT) and chemical production city engine in the 1990s became today an abandoned wound in the heart of the city with no added value to the community [4].

Urban strategies are suggested for the area's requalification in order to return it to its past glory, creating a new economic reproductive centre [4]. Urban agriculture (UA) and resilient food system have been addressed in the past few years, as they are the solution to create resilient and sustainable cities as they tackle food security, field's remediation and new business' interventions along with community development all at once; urban farms are considered to have significant social and economic impacts [5, 6]. Urban farms enhance access to fresh fruits and vegetables with low carbon footprint with the minimization of transport and packaging material and cost, as it provides multidimensional benefits [7].

In view of the confinement policies imposed by governments worldwide resulting from the spread of the coronavirus (COVID-19) that hit the world and the limitation on transportation and movement of citizens, rooftop urban framing can be a perfect solution to help in social distances since it provides zero-kilometre clean food in cities [8–11].

Urban farming (UF) has been already implemented in many cities around the world such as Tokyo, Havana, Milan and Rome as well as Singapore and others, causing significant benefits in terms of producing clean and fresh food [12–14]. Besides the vertical forest project at the Porta Nuoavo area in Milan, Italy produces vegetables and fruits equivalent to 30,000 square meters of woodland [15]. Hence, urban farms as a localized form of food production and consumption produce food, feed and fibres to meet the needs of rising populations, protect the environment and the expansion of the natural resources supply, counterbalance CC, sustain the economic viability of agriculture system and enhance air quality, engagement in agricultural activities as well as liveability inside communities [16]. Therefore, UF could be a strategic solution or a mean to limit the spread of coronavirus by adopting a self-consumption routine of cleaner vegetables and fruits without the need to go to buy them from the weekly markets or supermarkets.

Y. Ragab (✉) · M. Shahwan · S. El-Haggan
Architectural Engineering and Technology Programme, Faculty of Engineering, Cairo University, Cairo, Egypt

E. Durante · B. Di Giovanni · L. Gialli · L. Gabrielli · M. Lapomarda
Alma Mater Studiorum University of Bologna, Bologna, Italy

P. G. Mariscal
Humboldt University of Berlin, Berlin, Germany

M. Aboulnaga
Department of Architecture, Faculty of Engineering, Cairo University, Cairo, Egypt
e-mail: maboulnaga@eng.cu.edu.eg

© The Author(s), under exclusive license to Springer Nature Switzerland AG 2022

A. Sayigh (ed.), *Sustainable Energy Development and Innovation*, Innovative Renewable Energy,
https://doi.org/10.1007/978-3-030-76221-6_47

This chapter presents a study conducted on a site in the city of Galliera, Bologna, Italy, to transform it to be green and sustainable through integrating innovative and smart solutions of UA, yet make the project a social hub for the citizens with educational experience and provide economic sustainability of the city.

2 Objectives

The aim of this work is to revitalize the abandoned Ex. SIAPA industrial site, in the metropolitan city of Bologna, in the municipality of Galliera, to become a smart and sustainable city engine by focusing on urban farming as a key factor to create a generative and resilient city. The project aims at implementing smart solutions and urban farm technologies for a sustainable, energy-efficient, resilient, socially engaging and liveable site. It is also aimed at applying an interactive and impact-orientated approach to create a positive value in meeting the social, environmental and economic needs of the local community by creating an innovative educational, social and agriculture-based centre in the midst of the urban setting.

3 Issues and Challenges

In this extremely globalized context that we are living in, our consumerism and disgorged life consciousness, with no regards to crops seasonality or sustainability, lead to the abandonment of rural as well as urban areas. The SIAPA area is an emblematic example of that, and it is located in the metropolitan city of Bologna, in the municipality of Galliera, North of Italy. There are many challenges associated with both the city and the site (Fig. 1), as follows:

- The site is considered a wound in the heart of the city after being its centre.
- The site is abandoned with no regards for its potential to be retrofitted.
- The adaptive reuse potential in the site is discarded and the soil is contaminated.
- Vegetables and fruits are transported from surrounding areas increasing carbon and water footprint of the city.
- The site area is huge, ranging around 194,000 m^2 [17].

4 Methodology

The methodology depends on two approaches: theoretical and practical. The theoretical approach includes data collection and general information regarding the site's location, patterns of current use and activities. The practical approach encompasses site assessment through several visits and observations of the location and surrounding urban areas; this was also accompanied by practical data gathering, processing and analysis of the site and the surrounding context to identify the gaps and mapping the linkage of the site to transportation network in terms of accessibility and connectivity. In addition, the analytical approach includes how smart and green, architectural, urban and technical interventions are then proposed and intergraded in the existing building and site to attain sustainable development and SDGs. This work was part of 'Hop-E' team project in the UrbanFarm 2020 International Challenge [18].

(a) The municipality of Galliera is 55 km away from the center of Bologna [17]

(b) The site at the heart of Galliera dividing the city and creating an obstacle for its development [17]

(c) Entrance, boundaries and accessibility analysis to the site [17]

Fig. 1 The abundant site in the city of Galliera, Bologna, Italy. (**a**) The municipality of Galliera is 55 km away from the centre of Bologna [17]. (**b**) The site at the heart of Galliera dividing the city and creating an obstacle for its development [17]. (**c**) Entrance, boundaries and accessibility analysis to the site [17]

5 Concept and the Process of Project Development

To pursue these approaches, it was imperative to understand the word 'SIEPE' in Italian which means bush. It is similar to SIAPA the old name, which was related to an industrial destination. The project's main concept is centred on restoring the site to its past glory by exploiting and integrating a smart, sustainable and innovative approach and technological solutions that create a socially inclusive agriculture-based centre. One of the key aspects of the success of the project development is the integration of UF strategies into the buildings and related site, along with other sustainable measures (environmental, economic, social, liveable and viable and financial). The transformation of the project includes many aspects, which are architectural, social, agricultural and economical aspects. The buildings are integrated together through different types of gardens, which serve the community, the students, the production, and gives a green identity to the new site.

The site area (194,000 m^2) is capable of incorporating various typologies within its borders such as: educational, residential, communal and industrial. These buildings' design revolves around five main activities: (a) producing water, (b) electricity, (c) cultivating food, (d) recycling water and waste, (e) teaching about sustainability, UF and innovation in the field of UA, (f) engaging the community and (g) selling the products as depicted in Fig. 2.

There are four main buildings (educational buildings, residential buildings, technical buildings and communal areas). The educational buildings include a research centre, a library, seminars and workshop areas. For the residential buildings, these encompass the students' dorms and the lodge, whereas the technical buildings consist of aquaponics and hydroponics greenhouses, the brewery, the biogas plant, the nexus of Quattro lakes for fish repopulation and the wastewater treatment plant. Finally, the communal areas comprise the market, the restaurants, the lounges, the museum and the day-care centre.

The whole project is integrated by adding a set of gardens having recreational, productive and educational purposes which complement all our functions but in the outdoor space. Figure 5 in Appendix A illustrates the project's layout and buildings.

5.1 *Sustainable Architectural and Innovative Solutions*

The architectural concept is based on two pillars: (a) creating a contrasting experience in every part of the journey and (b) delivering a remarkable identity to the new project. A contrast between opposites is presented through the experience exhibited in the following: new against old, waste versus food, consumption versus production, curvilinear opposed to rectilinear, demolition versus restoration and teaching versus learning. It is aimed to design a vibrant, sustainable and resilient learning hub that tackles producing food in clean urban environments. Starting with the exposure and the accessibility to the site, a

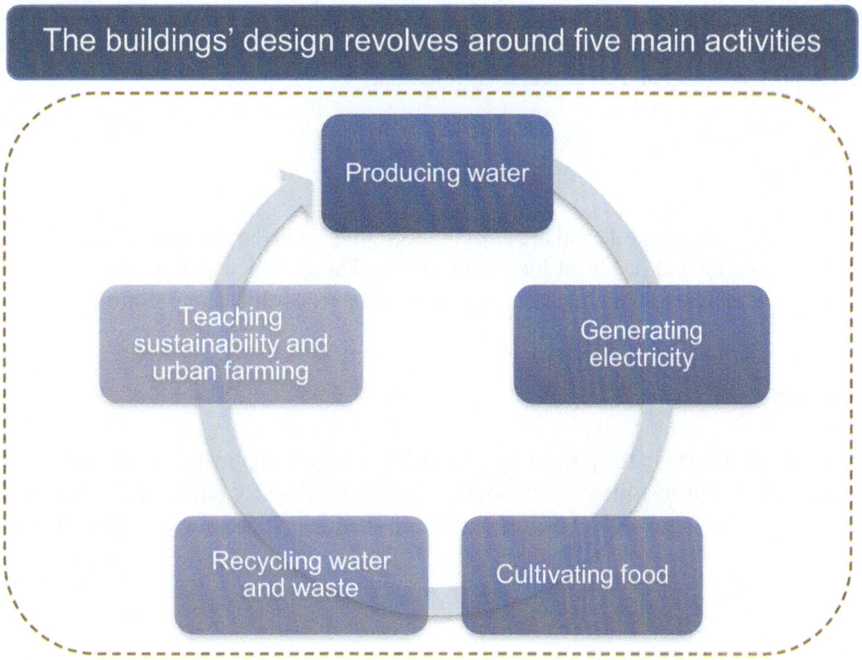

Fig. 2 Buildings' design and activities of the project development

main spine is designed to link the two parts of the city along Bologna-Ferrara axis, taking the visitor through a journey from the main street to the canal end, where there will be a harbour that allows for transportation to the other part of the city creating the perfect link. The main spine passes along various activities and always presents the visitor to something new and interesting along the journey as shown in Fig. 5 in Appendix A.

5.1.1 Art and Science Museum

The science museum, located at the start of the journey of the spine, is reflecting the contrast between art and science. It is also composed of two main exhibition halls. The first is a hall that holds the arts and crafts created by the kids and students, which embodies the contrast between food production and waste recycling, and the second is a hall that exhibits information and work regarding urban farming and sustainability' pillars.

5.1.2 Research Centre

A research centre's main aim is to innovate and search for what's new and better. It is fit in an archaeological building with a historical value and roots to the city, to hold the research centre. The outer skin preserves the building's old bricks and value to ensure sustainability, whereas the interior is incorporating high-tech and modern facades' treatments creating another type of contrast. Moreover, the research centre has labs, seminars and lecture halls (Fig. 6 in Appendix B).

5.1.3 Column of Hop

The project's whole identity is based on hop (*Humulus lupulus*) which is a dioecious plant member of the Cannabaceae family, whose female flowers are used in the brewing industry [19]. Furthermore, a landmark with three columns of hop marks the beginning of the journey; moreover, along the way, columns of hop accompany the visitors throughout their urban experience and create a green identity to the site.

5.1.4 Curvilinear Shed

Two curvilinear and modern sheds create two gateways marking the start and the end of the journey. The sheds' modern design contrasts with the site's old and rectilinear identity. One gateway marks the market location that is part of the centre of the project. The other gateway symbolizes the beginning of the nexus of Quattro lakes zone, which is a significant zone that merges between the project social and technical dual function.

5.1.5 Nature Fusion Cells

These cells are designed in all gardens on site and are closed glass cubicles in the outdoor, as the climate in Bologna is mostly cold, and the gardens are a serene and beautiful location to be at. These cubicles allow for the people to feel the outdoor nature, while sitting in a comfortable environment and can be opened from all sides in summer.

5.1.6 Hop-E Mobile App

The Internet of Things (IoT) is applied in the project through using a mobile application to market all events on site including planned lectures, conferences, seminars, different workshops dates, yet allow for online subscription for the different events. Moreover, it can be used to facilitate the rent of dorms and lodge available via a simple mobile application.

(a) Nature Fusion Cells Exterior (b) Aerial view of lakes and garde

Fig. 3 Developed layout of the project in Galliera, Bologna, Italy. (**a**) Nature fusion cells exterior. (**b**) Aerial view of lakes and gardens. (*Images source: Designed by authors*)

5.2 Main Zones' Development

The students' dorms are developed with a hydroponic greenhouse located on the roof, this greenhouse is added and designed based on the social survey and the community needs. A set of gardens is also developed with various functions including educational gardens, students' gardens and therapeutic gardens with aromatic plants, as well as community gardens and international spices gardens. The market place, which is located in the centre of the project, stands as the main social core of the project, where people gather, engage and sell organically and locally produced products in addition to the restaurant; the market creates the communal core. This restaurant is juxtaposed by an amphitheatre and a set of cafes and lounges. In addition, a learning hub formed where the seminars, the library and the workshops are located, students and kids can work on making art and crafts from waste and also learn about urban farming. The journey then ends with the nexus of Quattro lakes that have a dual function as a social and technical complex, used for fish repopulation to be used in aquaponic farming, and is used to create a social gathering space keeping the site far end alive and interactive (Fig. 3).

5.3 Environmental Engineering Approach

In the project development, avoiding water and energy waste through the implementation of energy and water efficiency measures and renewable technologies were considered. Focusing on the implementation of water and waste recovery, along with the reclamation and the upgrading of the polluted areas, the project tackles many aspects such as: (a) sustainable drainage systems, (b) green roof integrated photovoltaic systems, (c) microbial fuel cell, (d) anaerobic digestor for electricity production and (e) phytoremediation as shown in details and Fig. 6 in Appendix B in Table 2 in Appendix C.

6 Urban Agriculture Interventions

Urban farming (UF) is a key factor for sustainability and city's resilience, and it plays a central role in the revitalization of the area. To grow safe food in a polluted site such as the Ex. SIAPA, the only solution is given by the use of soilless technologies like hydroponics and aquaponics systems. From a high-tech production system to purely educational and recreational gardens, an impact-orientated approach is utilized to create a positive value in meeting the needs of the local community. The old warehouses have been retrofitted into high-tech greenhouses for the hydroponics and aquaponics production of several fruits and vegetables. Adopting these technologies will allow water savings from 70% to 90% compared to the open-field cultivations.

Environmental control is the prerequisite for a stable production over the time. Even though the initial investment could be high, the benefits of indoors growth will prevent the occurrence of problems such as bad weather conditions and plant diseases, distinguishing the products over the industrial competitors by producing fruits and vegetables, 100% nickel and pesticides free and allowing a fair compensation to the workers. The food production and the subsequent commercialization of the products is what will make the requalification process possible, and even if beer cannot be considered as food. It has been thought that adding a social craft brewery to the many activities developed in the process can be interesting under many aspects.

The beer production is using only zero-kilometre ingredients, and every single cereal used will come from the nearby local farmer. A special ingredient for beer production and also for the entire requalification of the ex SIAPA area is the hop flower. Hop for us represents a needed agricultural innovation; even if this plant grows spontaneously along banks of rivers in central and northern Italian regions, we import 3500 tons of hops flowers per year from Germany, England and the United States that equals to the 98% of the Italian demand [21]. Growing these flowers inside a Mediterranean greenhouse demonstrates that producing such hops locally is not only a more sustainable choice, but also a strategic added value in the creation of a new product. Moreover, the life cycle of the exhausted hop will not end after the brewing process. In fact, it will be stored and then reused as a substrate for the mushroom production that will take place inside a dedicated room of the microbrewery. Our farming concept turns around wastes as an opportunity to generate, reducing environmental burden. All CO_2 emissions from the biogas plant until the beer fermentation is stocked into the mushrooms and plants, improving the growth performances and reducing our carbon footprint.

7 Business Model

It is vital to ensure the project's economic and financial sustainability; therefore, a draft cost analysis was carried out to envisage how the project outcomes would lead to the set targets. Some indicated results from the Draft Van Analysis are illustrated in Fig. 4. The project is developed to be profitable from the early stages using an old building to host 80 students' rooms from the University of Bologna. Post this stage, three different production lanes and the energy and water management parts were developed. Consequently, the social activities were commenced with two objectives: a) increase the biodiversity in Galliera, destroyed by the DDT production, and b) provide a new social life to Galliera, thus attracting workers, people with different disabilities, students and less fortunate people to help in starting the intensive production of hops and different vegetables. The main objective is to build an integrated farming production, reduce waste, and reuse all the resources in an efficient way. In this project, it is important to ensure sustainability and constant selling to our products in two different methods: stakeholders constantly sell our products. Two different ways were applied: (a) zero-kilometre restaurants and markets and (b) large-scale organized distribution, in the middle of Italy, which starts from an intermodal and logistics centre in Bologna, just 20 km far from Galliera.

In terms of investment, this project's total cost is estimated to be around six million €, yet from different European funds, we can reach around four million €. This is gained due to waste management, an increase of the biodiversity in the river, social inclusive activities and rehabilitation of people with a different type of problems (Leonardo da Vinci funds, FEASR funds and LIFE projects).

Nevertheless, the production activities have an ROI between 12% and 15%. From this project, we could also have different types of intangible value and profit:

- Offering job opportunities to immigrants or people with disabilities or low competitive skills
- Rehabilitation of people with problems as addiction or ex-prisoners
- Creating a farming start-up hub, with the help of Alma-Cube the incubator of bologna, to give the students a space to develop their own idea and business

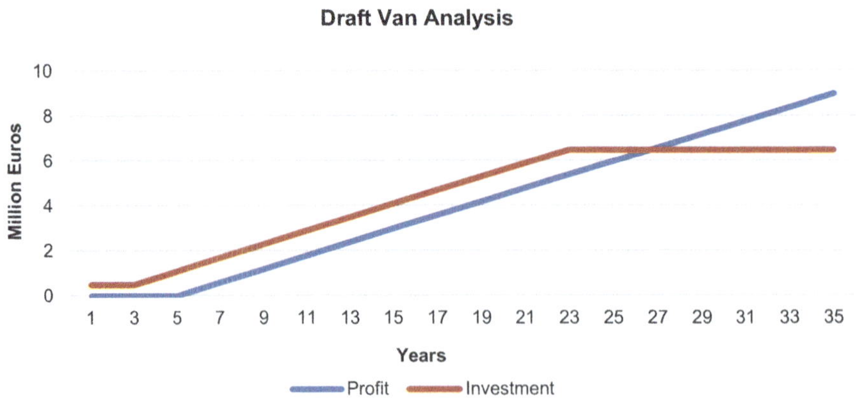

Fig. 4 Elements of the environmental engineering system used in the project

Our target is to have a three-year span to start all the activities, and to avoid the complexity of starting all these different activities all together; it has been intended to build an oriented project management and applying a plan-do-check-act logic and have a continuous improvement method to circularly increase the optimization of the project like developing new type of seeds, hop or technologies to increase the production of the bio-digester. Finally, concerning the main production stream of hops, the focus is centred on entering the supply chain in order to develop a 100% Italian beer which could be useful as a marketing strategy to help other people develop their own production that will make other producers become part of the ecosystem. Furthermore, to produce hops, we will need a greenhouse production system and Artigiafer Greenhouse Technology, nonetheless with an estimation of total costs and revenues.

8 Results and Discussions

The assessment of the project revealed the following results and its relation to the three pillars of sustainability. Table 1 presents the sustainability assessments. By the end of this research project, the following were accomplished: (a) revitalize the whole site with only 5% demolishing of the total footprint; (b) generate renewable energy using photovoltaic cells over roofs with a total area of 4000 m^2; (c) reduce overall energy consumption and carbon footprint by the implementation of green walls over roofs and walls, double skin facades and radiant cooling materials; (d) produce annually more than 100 tons of local vegetables and fruits and produce more than 70 tons of hops and (e) provide accommodation for 200 students.

9 Conclusions

Hop-E research work is a regenerative project that implemented urban farming in the city of Galliera, in order to return the Ex-SIAPA site to its past glory being the city engine, yet the new heart of the city is an innovative agriculture- and education-based centre that addresses the three aspects of sustainability attaining the SDGs, mainly SDG 4, 6, 7, 9, 11, 12, 13 and 15.

The envisioned new retrofitted and revitalized site creates a spiral that integrates all community members while adopting circular economy process, creating a liveable, resilient and sustainable site. The aim to create a generation of innovative, socially, economically and environmentally aware community members has been achieved. In this project, sustainability principles, to produce, recycle, teach and engage, were developed, adopted and materialized.

Acknowledgements The authors would like to thank Cairo University and Faculty of Engineering, Egypt for encouraging the team to conduct this research and fund the travel to present this project at the Finale of the UrbanFarm 2020 Challenge in Pordenone, Italy. Also, the authors sincerely thank UrbanFarm 2020 for the golden opportunity to participate in such international challenge. Finally, the authors express their utmost appreciation to Alma Mater Studiorum University of Bologna and Alma Mater Foundation in Italy for the support and providing support for the research work and competition, particularly Professor Francisco Orsini and Dr. Giuseppina Pennisi at Alma Mater Studiorum University of Bologna, Italy. Finally, thanks to Professor Mohsen Aboulnaga at Cairo University and Arch. Mona Mostafa for supporting and supervising the project.

Appendices

Appendix A

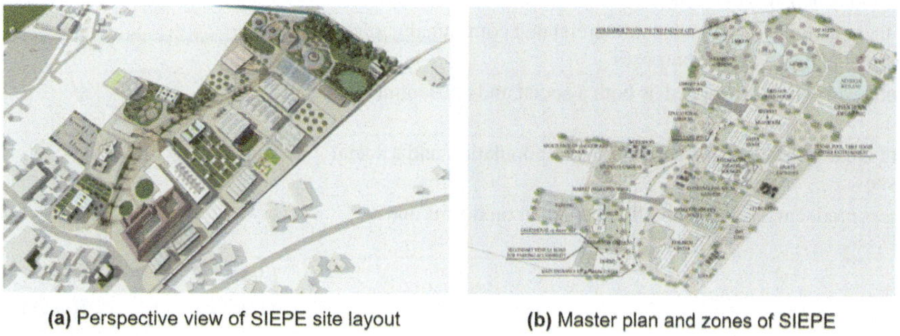

(a) Perspective view of SIEPE site layout **(b)** Master plan and zones of SIEPE

Fig. 5 Developed layout of the project in Galliera, Bologna, Italy. (**a**) Perspective view of SIEPE site layout. (**b**) Master plan and zones of SIEPE. (*Image source: Designed by authors*)

Appendix B

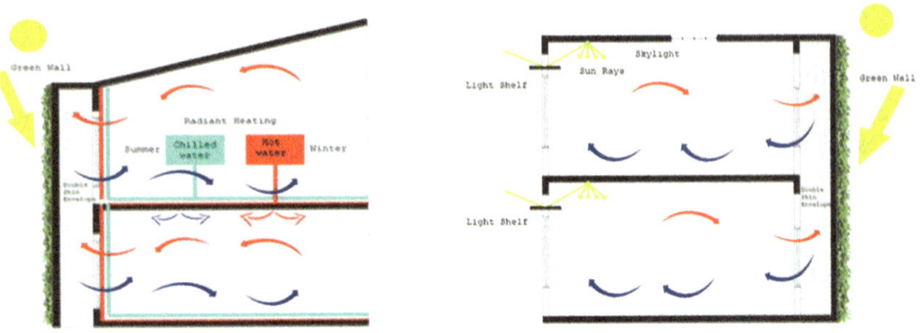

(a) A section through the research centre depicting green walls and radiant cooling

(b) A section through the dorms' centre showing green walls, light shelves, and radiant cooling

Fig. 6 Applied environmental solutions in the research centre and dorms centre in Galliera, Italy. (**a**) A section through the research centre depicting green walls and radiant cooling. (**b**) A section through the dorms' centre showing green walls, light shelves and radiant cooling. (*Source: Designed by authors*)

Appendix C

Table 1 Sustainability pillars assessment of the project interventions

No.	Mission	Environmental sustainability indicator	Social sustainability indicator	Economic sustainability indicator
1	Revitalizing the Ex-industrial site with minimum demolition and maximum retrofitting of existing industrial warehouses	√	x	√
2	Providing an educational and agriculture-based hub in the city centre	√	√	√
3	Creating a social hub and spiral that target all community members: elderly, children, research facilities, disadvantaged people, etc.	x	√	x
4	Developing a main agriculture concept based on waste as a resource	√	x	√
5	Merging between hydroponics, aquaponics and aquaculture farming to be included in social activities and educational courses	√	√	√
6	Providing zero-km leafy crops and vegetables grown on site	√	x	√
7	Ensuring energy saving and energy production using different environmental strategies	√	x	√
8	Harvesting rainwater and soil reclamation	√	x	√
9	Integration of building management system and Internet of Things (IoT) for ease of operation and control	√	x	x
10	Designing an art and science museum promoting sustainable lifestyle, urban farming and waste recycling	√	√	x
11	Designing a research centre for urban farming development	√	√	x
12	Designing library, co-working spaces and workshops along with a market, restaurant, students' dorms and sport facilities	x	√	x
13	Developing a set of therapeutic, educational, commercial and communal gardens used for both educational and agriculture purposes	√	√	x
14	Establishing a microbrewery that is designed as both a social and agronomical research lab	x	√	x
15	Creating the nexus of Quattro lakes that is used for fish repopulation and a social gathering space on site	√	√	x
16	Establishing a business model and revenue totally dependent on dorms and production lanes	x	x	√
Total		**13/16**	**9/16**	**8/16**
Percentage		**81.25%**	**56.25%**	**50%**

√ Indicates that the sustainability pillar exists
x Indicates that the sustainability pillar does not exist

Table 2 A five-elements environmental engineering approach

Environmental engineering elements		Items applied in the project
1	Sustainable drainage systems	Intercept rain water before it reaches the ground Storing rain water in underground water-proofing tanks Reusing water for cleaning machines, cooling and irrigation
2	Green roof-integrated photovoltaic systems	Mono-crystalline SI with an efficiency of 20% on extensive green rooftops that boost panels' efficiency and durability Benefits: The interception of storm water runoff Reduction of the urban heat island effect (UHIF) Provision of habitat for wild birds, flora and fauna Prevents high rates of evapotranspiration enabling the usage of plants that increases CO_2 sequestration
3	Microbial fuel cell (MFC) [20]	The ability of microorganisms to transfer electrons outside the cell to produce electrical power from wastewater Usage of wastewater coming from our microbrewery through a cell of anode, cathode separated by an exchange membrane (*Anode*): Organic substrate is oxidized generating electrons and protons (*Electrons*): Reach the cathode (generating an electric current) (*Protons and electrons*): They are reduced in cathode, combining with oxygen, to form water
4	Anaerobic digestor for electricity production	Biogas from an anaerobic digestor used to produce electrical power Heat loss from the engine will then be used to heat the digester when needed
5	Phytoremediation	Three aquifers, a deep one and two shallow ones, currently contaminated, mainly by organo-chlorides Treating contaminations by an anaerobic/aerobic ponds for the deep aquifer and in site treatment with poplars Enhance pollutant degradation processes, naturally occurring in aquatic ecosystems Constitutes an upgrade for the naturalistic value of the area

Source: Developed by authors

References

1. IRENA, Renewable energy in cities, 2016. [Online]. Available: https://www.irena.org/-/media/Files/IRENA/Agency/Publication/2016/IRENA_Renewable_Energy_in_Cities_2016.pdf. Accessed 6 May 2020
2. Chowdhury PKR, Weaver JE, Weber EM, Lunga D, LeDoux STM, Rose AN, Bhaduri BL (2020) Electricity consumption patterns within cities: application of a data-driven settlement characterization method. International Journal of Digital Earth 1(13):119–135
3. United Nations Environment Programme, Resource efficiency & green economy, [Online]. Available: https://www.unenvironment.org/explore-topics/resource-efficiency/what-we-do/cities/resource-efficiency-green-economy. Accessed 6 May 2020
4. ALMA MATER STUDIORUM - Università di Bologna, UNIBO / Galliera - Former Siapa Area, 2020. [Online]. Available: https://site.unibo.it/urban-farm/en/cities-and-locations-2020/galliera-former-siapa-area. Accessed 6 May 2020
5. RUAF, Urban Agriculture And City Region Food Systems, [Online]. Available: https://ruaf.org/urban-agriculture-and-city-region-food-systems/. Accessed 6 May 2020
6. Hallett S, Hoagland L, Toner E (2016) Urban agriculture: environmental, economic, and social perspectives. Hortic Rev:65–120
7. Little N, What is urban agriculture?, University of Maryland, 3 September 2019. [Online]. Available: https://extension.umd.edu/learn/what-urban-agriculture. Accessed 6 May 2020
8. Policy responses to COVID-19, 14 Aug 2020. [Online]. Available: https://www.imf.org/en/Topics/imf-and-covid19/Policy-Responses-to-COVID-19. Accessed 8 Aug 2020
9. Ritchie H, Ortiz-Ospina E, Beltekian D, Mathieu E, Hasell J, Macdonald B, Giattino C, Roser M, Policy responses to the coronavirus pandemic, 3 Aug 2020. [Online]. Available: https://ourworldindata.org/policy-responses-covid. Accessed 8 Aug 2020
10. Martinez S, 0KM Food: a revolution that benefits everyone, The PRISMA, 5 Nov 2018. [Online]. Available: http://theprisma.co.uk/2018/11/05/0km-food-a-revolution-that-benefits-everyone/. Accessed 8 Aug 2020
11. Pancorbo O, Zero km food products start to take Spain by storm, EFEAgro, December 2016. [Online]. Available: https://www.euractiv.com/section/agriculture-food/news/sr-agri-local-zero-kilometre-products-start-to-take-spain-by-storm/. Accessed 8 Aug 2020
12. Burton P, Lyons K, Richards C, Amati M, Rose N, Fours LD, Pires V, Barclay R, Urban food security, urban resilience and climate change, 2013. [Online]. Available: https://www.nccarf.edu.au/sites/default/files/attached_files_publications/Burton_2013_Urban_food_security.pdf. Accessed 6 May 2020
13. Thorpe D, Urban agriculture – a next big thing for cities, Smart Cities Dive, 2017. [Online]. Available: https://www.smartcitiesdive.com/ex/sustainablecitiescollective/urban-agriculture-next-big-thing-cities/338111/. Accessed 6 May 2020
14. Zimmer L, The world's first commercial vertical farm opens in Singapore, InHabitat, 26 Oct 2012. [Online]. Available: https://inhabitat.com/the-worlds-first-commercial-vertical-farm-opens-in-singapore/. Accessed 6 May 2020
15. Stefano Boeri Architetti, "Vertical Forest," Stefano Boeri Architetti, [Online]. Available: https://www.stefanoboeriarchitetti.net/en/project/vertical-forest/. Accessed 6 May 2020

16. National Institute of Food and Agriculture, "Urban Agriculture," United States Department of Agriculture, 2015. [Online]. Available: https://foodtank.com/news/2015/07/urban-farms-and-gardens-are-feeding-cities-around-the-world/. Accessed 6 May 2020
17. UrbanFarm 2020 International Challenge, [Online]. Available: https://site.unibo.it/urban-farm/en/cities-and-locations-2020/galliera-former-siapa-area. Accessed 25 April 2020
18. Orsini F, Pennisi G, D'Alessandro A, Kratochvilova D, Steffan G, Paoletti M, Sabbatini G, D'Ostuni M, Trombadore A, Gianquinto G (2020). Bridging interdisciplinary knowledge for sustainable urban landscapes: results from the international student competition UrbanFarm2019. Acta Hortic 1298:7–106. https://doi.org/10.17660/ActaHortic.2020.1298.15
19. Snyder R, Conway S, Humulus lupulus - Hops, "Food for Thought", 2008. [Online]. Available: https://academics.hamilton.edu/foodforthought/Our_Research_files/hops.pdf. Accessed 6 Aug 2020
20. Logan BE, Hamelers B, Rozendal R, Microbial fuel cells 2006. [Online]. Available: https://web.mit.edu/pweigele/www/SoBEI/Info_files/Logan%202006%20Environ%20Sci%20Technol.pdf. Accessed 6 Aug 2020
21. Cricca L, AgroNotzie, 21 Sept 2018. [Online]. Available: https://agronotizie.imagelinenetwork.com/vivaismo-e-sementi/2018/09/21/luppolo-una-nuova-eccellenza-del-made-in-italy/60016. Accessed 5 Aug 2020

The Need for Renewable Energy Towards Sustainable, Green and Smart Buildings: Case of Emirate of Ajman

Abdul Salam Darwish

1 Introduction

The energy use per capita for developing countries is escalating rapidly, while the energy demands for developed nations is rising inevitably. Energy use is increasingly correlated to life quality, causing an increase in energy demand with an increase in population. As a result, as forecasted by the German Advisory Council on Global Change (WBGU), the use of fossil fuels would continue for more than a decade, peaking in 2030, then progressively declining towards 2100, Fig. 1 [1]. The Emirate of Ajman is one of the seven states constituting the United Arab Emirates. With an area of just 259 square kilometres, Ajman is the smallest of the emirates by area [2]. The Emirate of Ajman is bordered on its north, south and east by the Emirate of Sharjah. It has a population of some 463,000 [3]. It is located in the Northern Emirates where FEWA (Federal Electricity & Water Authority) [2, 4] caters to the needs of electricity and potable water for Northern Emirates of Ajman, Ras Al Khaima, Fujairah and Umm Al Qwain. Climate change, global warming, environmental degradation and the increase in natural disasters are the consequences of relying on conventional energy resources. UAE is one of the pioneers who has recognised this effect where sustainable practices and conservation are considered as the proper pathways to sustain life in a pleasant environment.

In this aspect, UAE is leading countries to deploy clean, renewable energy systems by harnessing advanced technologies. Activities and efforts are practiced all over the Emirates in many aspects in order to achieve the overall energy targets. Over the last several decades, there has been an increasing awareness of the disparaging connection between traditional economic development and the over-exploitation of the world's natural resources. Pollutants, harmful residues from land stripping and the production of building materials increasingly stress the environment. According to some estimates, the worldwide construction industry consumes about 40% of total raw materials at a rate of about three billion tons per year. In the United States alone, buildings consume about 65% of the total energy consumption and cause (30%) greenhouse emissions [5]. Since built environments comprise such a large and ever-increasing portion of the world's total greenhouse emissions and since the majority of those emissions are derived from a building's life cycle, this is the area of urban development where meaningful and long-term change must immediately be implemented. The power demand in Ajman is met by burning an increasing amount of natural resources domestically. Its rapid socio-economic growth, with increasing population, expansion of urbanisation and growing industrial activities, has made the region big energy consumers. Electricity demand has dramatically doubled over recent years. It can be seen from the annual analysis report published by [6] that residential buildings are responsible for 40% of energy consumption, industrial 22%, commercial 18% and others 20%, as shown in Fig. 2.

'Green building' is a term utilised all over the world that has become synonymous with sustainability. This prevalent connection, however, does not designate that all green building is, in fact, sustainable. Green building is a very consequential step towards sustainable construction and is defined as the practical application of viable methods that take into account not only the building site's environmental conditions but withal strives to make the best utilisation of resources throughout the building's phases. These phases include site selection, design, construction, operation, maintenance and demolition. In order to neutralise the energy use in this country, renewable energy resources were the key to going forward [7]. However, a proper and organised management framework is needed for implementation and investment supported by smart technology. This chapter presents the most significant factors to construct an efficient management framework.

A. S. Darwish (✉)
School of Civil Engineering, University of Bolton, Bolton, UK
e-mail: a.darwish@bolton.ac.uk

© The Author(s), under exclusive license to Springer Nature Switzerland AG 2022

A. Sayigh (ed.), *Sustainable Energy Development and Innovation*, Innovative Renewable Energy,
https://doi.org/10.1007/978-3-030-76221-6_48

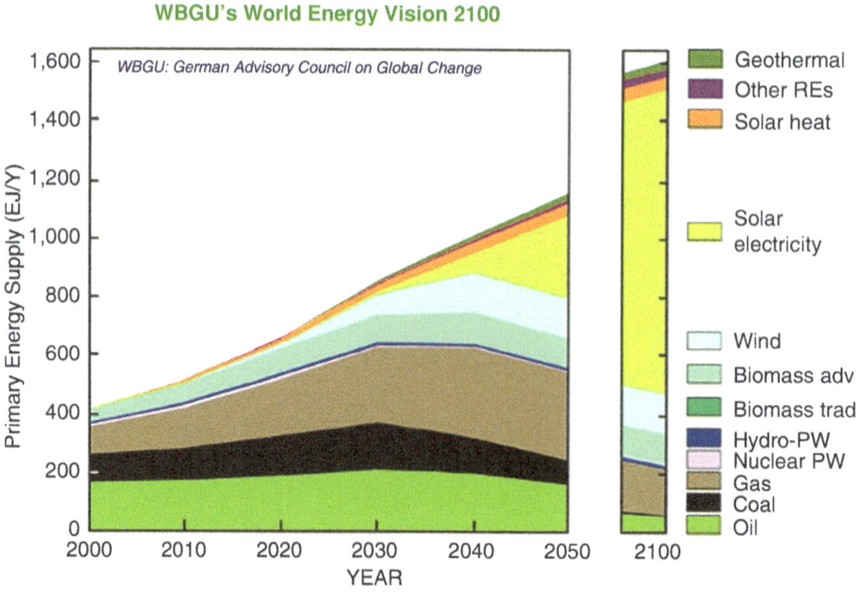

Fig. 1 A forecast of the global energy supply until year 2100 [3]

Fig. 2 Energy consumption sectors

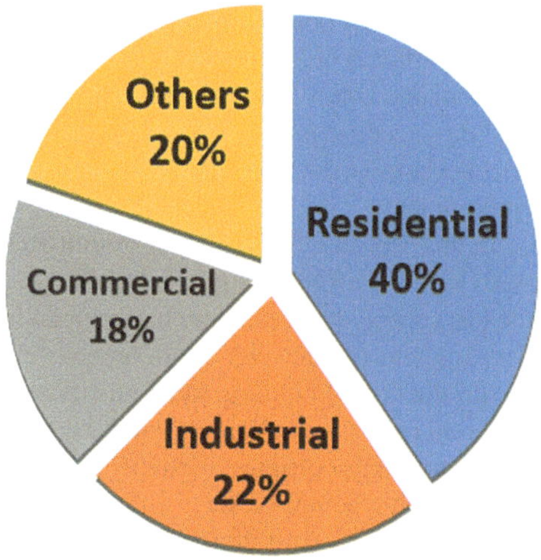

2 Electricity Status and Demand

The most basic needs for electricity in this area are air-conditioning, water desalination, lighting and sub-industrial activities. The shortage of electricity supply has made many firms use alternatives to the grid supply that is the generators run by diesel, adding more pollutions to the environment. Electricity demand is high compared to the electricity generated locally at the Emirate of Ajman. The real figures considering the northern Emirates within the Federal Electricity & Water Authority (FEWA) (Fig. 3) show the difference between the generated and the consumed amount of electricity detected from data given in [8].

The northern Emirates region has experienced an increase in the number of consumers, which is also expected to be doubled in a few years (Fig. 4). This increase will add more pressure on the national grid for electricity demand producing high carbon dioxide emissions.

Figure 5 [9] represents the annual per capita carbon dioxide emissions (tonnes) for several countries, clearly showing that UAE is amongst the highest in CO2 emissions.

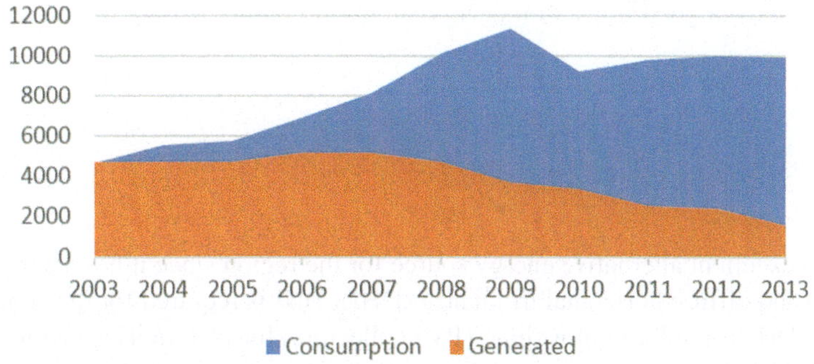

Fig. 3 FEWA electricity generation compared with consumption [GWh]

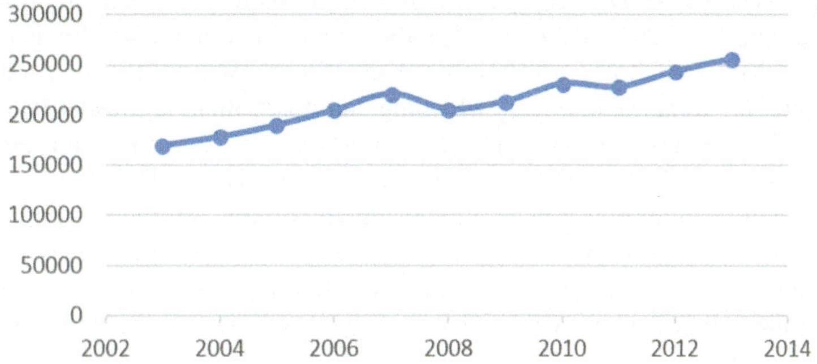

Fig. 4 Growth in the number of consumers – electricity – FEWA

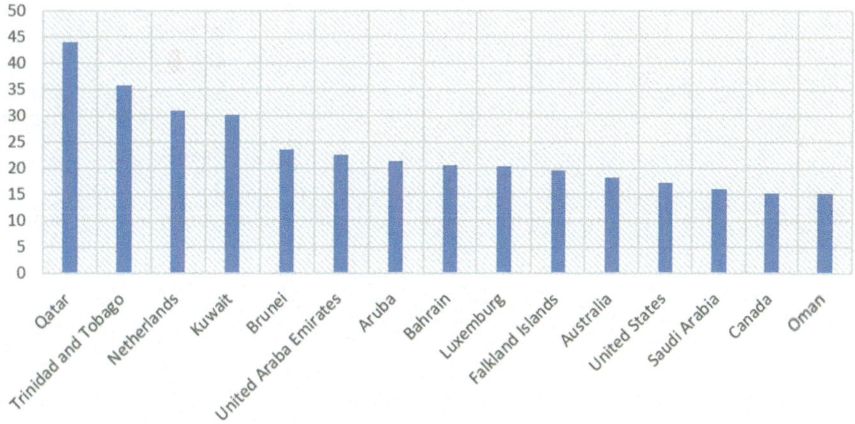

Fig. 5 Annual per capita carbon dioxide emissions (tonnes)

In order to overcome all future expected increase in energy demand, the Emirate of Ajman stated its 2021 Vision in February 2014 as 'its objective is to create a happy society through an active economic movement contributing in building a Green Economy and enhancing sustainable development, aiming to achieve prosperity in the various fields' [10].

3 Potential of Renewable Energy Resources

There are no specific studies found to evaluate the available renewable energy resources at Ajman as individuals. However, many studies and reports were published for the whole country that showed Ajman has good potential in solar energy and biomass. Other resources have not studied thoroughly to reflect their high potential, such as wind, wave, tidal and geothermal, which would have a good impact on energy supply if comprehensive research work is carried out using measured data.

4 Solar Energy

Solar energy could be an excellent alternative energy source for the region since it has a considerable high amount of direct normal irradiance (DNI) and diffuse horizontal irradiance (DHI). DNI is required for the concentrated solar plant collectors (CSP), and both DNI & DHI for solar photovoltaic (PV) collectors, for electricity generation. However, climate and cost challenges might be added, such as temperature and dust, with the current struggle to compete economically with fossil fuel. High temperature and dust reduce solar collectors' efficiency leading to lower power output and high power loss [11]. Masdar Institute and Vaisala provided a comprehensive view of the UAE's wind resource through the creation of wind and solar atlas [12] and made available freely to support investment in this respect. Data are available at multiple heights. The atlas was constructed from meteorological data and ensemble modelling approach calibrated and tested for an accurate fit. In addition, the atlas provides the UAE solar irradiance maps using a robust satellite-based solar mapping tool developed and validated at Research Center for Renewable Energy Mapping and Assessment at Masdar Institute [13]. A proposed site was selected by the present work in Ajman, shown in Fig. 6, to demonstrate the solar and wind energy potential. For example, the amount of direct normal irradiance (DNI) and diffuse horizontal irradiance (DHI), Table 1, indicates a well-promising solar energy potential that can be utilised for power generation using all possible affordable systems.

As the cooling load is the highest portion of the total energy demand, the possible technology that could be used is vapour compression chillers by thin-film PV modules or Fresnel concentrators, which are the most economically viable combination. Other developed systems may also be considered.

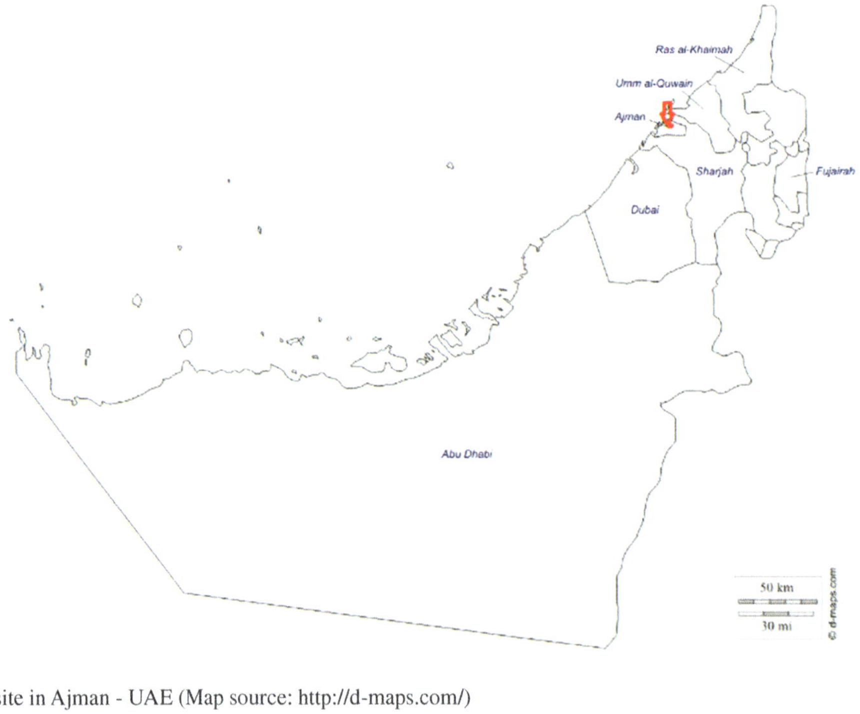

Fig. 6 The selected site in Ajman - UAE (Map source: http://d-maps.com/)

Table 1 The amount of direct normal irradiance (DNI) and diffuse horizontal irradiance (DHI) for selected position in Ajman [13]

Month	Diffuse horizontal irradiance (DHI) kW/m^2	Direct normal irradiance (DNI) kW/m^2
January	41.3	163
February	60	132.8
March	86.2	136.7
April	99.8	113.5
May	101.7	148
June	98.2	149.3
July	102.7	140.2
August	99.8	129.6
September	71.1	156.8
October	54	172.2
November	45.1	138
December	45	133.2
Total	904.9	1713.3
Average	75.41	142.775

5 Suitable Solar Energy Technologies

Solar energy utilisation technologies suitable for Ajman and other Emirates are as follows:

- Photovoltaic technologies
  - Crystalline (C-Si, mc-Si, UMG-Si, ribbon)
  - Concentrators (high concentration photovoltaic (HCPV)) and (low concentration photovoltaic (LCPV))
  - Third-generation solar cells
    - Dye-sensitised solar cells (DSSC)
    - Organic photovoltaic (OPV)
    - Nanotechnology
    - Hybrid systems
  - Thin film
    - (CIGS) Copper indium gallium selenide solar cells
- Thermal technologies
  - Solar thermal heat (domestic, hot water, cooling, process heat, desalination)
  - Solar thermal power (linear concentrating Fresnel, parabolic dish, parabolic trough and power tower).

6 Wind Energy

It is believed that wind energy can play an important role in Ajman, in increasing the range of the available renewable energy sources and securing the generated power supply at the time when solar energy falls short at night time or through any unexpected climate change periods when no enough sunshine hours are available. The right wind generator system, proper site selection and suitably planned spreading of wind energy systems are the keys to a strategic implementation to overcome this challenge and assure maximum power extraction from the available wind energy. The size of installed wind turbines in this region depends on the potential of the available wind power at the proper hub height for each selected area with a cost-effective impact. It will require comprehensive energy policies and plans for technology infrastructure. For this purpose, detailed long-time wind speed data characteristics at short periods are needed. All probability distributions and vertical velocity profiles with related curves are essential for investment and decision-making process. According to [14], at least 1400 full load hours, capacity factor 16%, needed from a wind turbine to be economically viable. Figure 7 shows the annual mean wind speeds at 80 m height for the same site shown in Fig. 6. A possible appearance of wind speeds higher than 5 m/s

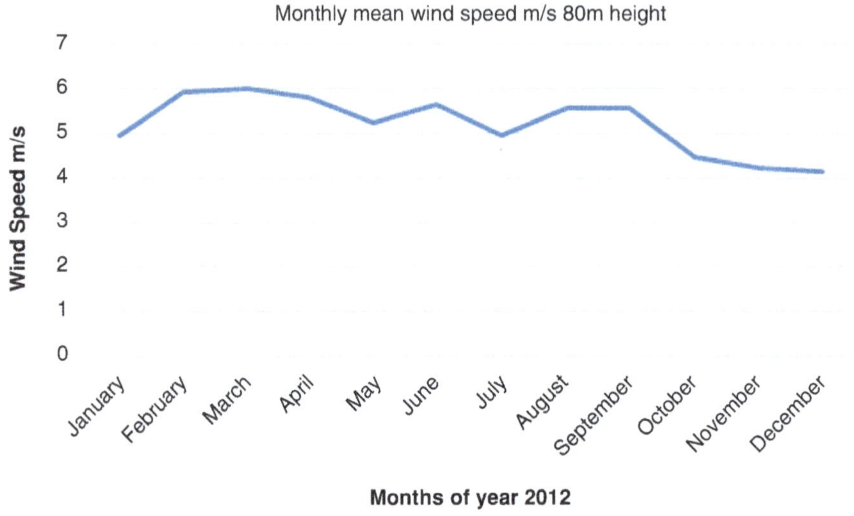

Fig. 7 Annual mean wind speeds at 80 m height (estimated data at [13])

is monitored [15, 16], indicating a possible energy utilisation from the wind for Masdar City as an example of a similar site could reach to 3307 MWh of power annually. This is by using a 500 kW wind turbine to achieve a capacity factor of 7.5% at 657 full load hours and a 3.5 kW turbine would achieve 832.2 full load hours with a capacity factor of 9.5% generating an annual output of 29 MWh [16]. Average wind speeds of 5.1 m/s found to achieve 1200–1300 full load hours with a capacity factor of 13.7–14.8%. Ajman wind resources potential is much less compared to solar, but they are reasonable, especially on-shore and off-shore, for a specific application and local grids. Mean wind speeds are shown in Fig. 7. However, the present research proposes that the most suitable technology to start with is low-speed vertical axis wind turbine 10–15 KW installed on the rooftop of the buildings and adjacent to the parking areas as they are widely available in this region.

7 Waste to Energy

There is a considerable amount of per capita waste generation disposed of in the area. The non-recyclable waste can provide a limited energy source and contribute to the baseload power. Ajman is looking for options to generate electricity from waste to cover a reasonable percentage of Ajman's daily needs and support the criteria of green, renewable and sustainable energy sources, in line with the country's objective of treating 75 per cent of its wastes by 2021.

8 Geothermal

The type of energy is still not been investigated to the point that the country starts utilising its potential. The country is not located in the high geothermal gradient areas. However, it might be suitable for hot water for thermal applications such as cooling.

9 Hydro/Ocean

The area's geomorphology of high mountains next to the coast could utilise the seawater pumped storage hydropower systems. This needs a full investigation of proper investment [17]. Wave and tidal energy potential need appropriate studies. Small innovated systems could be worthy if suitably introduced to contribute to the overall energy resources.

10 Renewable Energy and Sustainable Smart Buildings: The Right Solution

In order for the Emirate of Ajman to approach their vision, mission and energy target [18], the Department of Municipality and Planning in Ajman and Emirates Council for Green Buildings signed a plan of cooperation and coordination agreement between them. This has the aim to set down unified frameworks and environmental policies to generalise and spread the green buildings' special conditions and criteria, to gain environmental benefits and carbon footprint reduction for the UAE generally and the Emirate particularly [19]. In this respect, activities and events are put forward to provide qualification, training, educational programmes, actions to reinforce the green buildings and raising public awareness to reduce energy and water consumption. The sustainable building offers the following benefits:

- Costs for lower life cycle
- Less insurance costs
- Increased property value
- Increased productivity
- Better image
- Less risks
- Increase health for occupants
- Better infrastructure consequences
- Improved environment and economy

It is consequential that all incipient constructions strike a meticulous balance between economic, social and environmental considerations. Sustainable building innovation engenders an economic steadiness that in turn engenders high grades of economic magnification and service. These factors improve project delivery and increase effectiveness and productivity. Environmental sustainability defends the environment by avoiding contamination, enhancing biodiversity and emerging a successful eco-friendly transport plan. It also manages natural resources by retaining technology and education and predicting to increase energy efficiency and guarantee the proper use of resources. This sustainability is attained through a social process that distinguishes the needs of everyone and creates a partnership with local societies to ensure that everyone's needs are satisfied. The main objectives and strategies of sustainable buildings are [20]:

- Resource conservation
  - Energy conservation
    - Selection of resources and building methods
    - Energy-efficient deconstruction and recycling design
    - Intensive transportation design for low energy
    - Energy-efficient technological process development
  - Water conservation
    - Well-designed and efficient water plumbing fixtures
    - New techniques for rainwater collection
    - Implementing recirculating systems
  - Land conservation
    - Introduction of methodology to re-use the building materials
    - Projects near exiting infrastructure
    - Organising and planning projects that are not far from the existing projects
  - Materials conservation
    - Implement the latest waste management systems
    - Interrelate the natural and normal materials
    - Take into considerations all designs that serve the prevention of pollution
- Cost efficiency
  - Purchase cost (initial)
  - Main recovery cost
- Design for human adaption
  - Health and comfort protection
  - Physical resources protection

11 Smart, Sustainable Buildings

Smart, sustainable buildings will implement the most advanced technological instrumentation in order to achieve energy efficiency goals and long-term social sustainability. Mapping out smart building's design requires strategic arrangements and depends on the utilisation of multiple factors. In order to achieve a perfectly integrated, smart building, construction should not only be smart but must be green and consists of intelligent systems in different design and operation prospects. There are three different approaches to the definition of intelligent buildings [21]:

- Performance-based approach
- Services-based approach
- System-based approach

The performance-predicated approach replicates the ways that the building is acting according to its users' environment and demands, considering the efficient utilisation of resources in an efficient cost manner and according to international standards and regulations. The accommodations-predicated approach perceives the quality of comfort that the buildings are providing, including smart functions, communications, computerisations, etc. The system-based approach refers to all of the available, highly designed technology that is contained within the building's design and construction. By integrating the right construction and organisation of these three approaches, an ideal composition of structural design, system and service will result in a highly efficient automated building, communication network systems, safety and comfort. This combination will adhere to all regulations, to fulfil the smart building philosophy [22] (Fig. 8).

High-performance buildings' systems and services include:

- HVAC control
- Fire controller
- Transmission control
- Camera/video streamers and security controller
- Data communications
- Lighting control and power monitoring
- Automation integrated system
- Advanced intelligent integrated system

Smart buildings will need to have the capability of integrating into a smart city with an advanced electrical grid system, which will also be able to utilise potential renewable energy options. This capacity will require an intelligent power system for power delivery with the latest two-way communication and control systems for the smart grid, as well as the use of 'smart metres' for commercial and residential buildings.

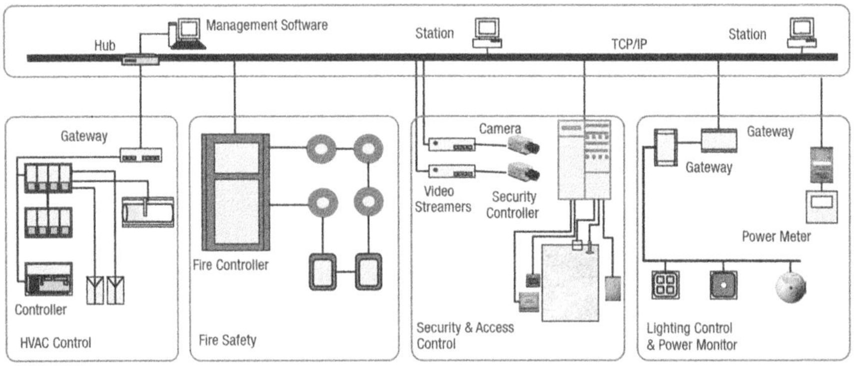

Fig. 8 An integrated intelligent building system [23]

12 Impact of Smart Grids on Intelligent Buildings and Cities

It is essential to incorporate a smart grid for the following reasons:

1. A smart grid is a more reliable grid, with fairer pricing and optimum energy use.
2. To bring new businesses and consumers to the location to support the city economy.
3. The new generation is very tight; IT experienced that could smoothly be integrated with the smart technology results in successful overall integration.
4. To control the usage peaks for optimum energy use.
5. Grid stability and reliability with improved efficiency for energy consumption.
6. It decreases energy costs.

13 Smart Building Elements

Smart building initiates several elements that are likely to be included within the initial design procedure. The followings are the most recommended elements:

- On-site electricity generation technology
- Smart lighting controlled and integrated systems and increasing daylight entrance
- Implementing the right DC grid for a DC loads
- Balancing load/generation
- Designing and installing an energy management system
- Street lighting
- Making available the most efficient electrical and thermal energy storage.
  - Energy efficiency building materials such as active/passive techniques for heating and cooling, proper glazing and insulation
  - Include the principles of increasing shading criteria
  - High-efficiency heat-recovery ventilation systems

14 An On-Going Project Proposal in the Emirate of Ajman – UAE

The present research has finished a proposal for a commercial building with a zero-energy footprint with no grid support. Energy self-sufficient and carbon-neutral green sustainable smart building with photovoltaic wall-mounted design to match the climate, weather conditions and energy resources over the year at that area. Roof- and park-mounted vertical axis wind turbines are to be installed to support electricity demand at night-time hours. An efficient innovated solar thermal system to satisfy the building air-conditioning requirements discarding all old conventional systems. Moreover, this research recommends a proposal to the Government of Ajman to adopt a building regulation to consist of 10% of solar power added to new buildings planned in the Emirate of Ajman. This could be a starting point for the near future investment to achieve a considerable reduction in CO2 emission.

15 Conclusion

The energy use is increasingly correlated to life quality, causing an increase in the energy demand with the increase in populations resulting in a considerable amount of carbon footprint damaging the environment and risking health issues. Ajman is one of the United Arab Emirates that has stated its 2021 Vision to engender happiness to the society through a proper, efficient economic movement contributing to building a green economy and energising sustainable development. In this aspect, new buildings should be designed with features that meet the future challenges of technological, environmental and societal progress. Integrating the recent developed sophisticated communications and control systems into green sustainable construction methodologies and energy consumption for a protected environment. A more comfortable, built environment can be engendered while instantaneously reducing carbon footprint sites. The utilisation of eco-cordial materials and innovative procedures will result in optimised energy performance and a perpetual carbon dioxide monitoring with the latest commissioning quantifications and verification processes. There are no specific studies found to evaluate the available renewable energy resources at Ajman as individuals. However, many attempts were found for the whole country that showed good potential in solar energy and biomass and recommended further research and evaluations of other renewable resources. Ajman is enjoying a high potential of solar and biomass that can satisfy all the Emirate energy demand and support the rest of the United Arab Emirates if these resources started to be utilised integrating the green-smart-sustainable concept. As the cooling load is the highest portion of the total energy demand, the possible technology that could be used is vapour compression chillers by thin-film PV modules or Fresnel concentrators, which are the most economically viable combination. Other developed systems may also be considered. Wind levels vary within the scope of a local intensity as it can vary significantly between locations not far from each other, only 40 metres depending on the roughness and the effects of altitude and direction as well as other influential factors. Roof- and park-mounted vertical axis wind turbines are recommended to be installed to support electricity demand at night-time hours.

References

1. Kurochans.net (2020) WBGU World Energy Vision 2100. [online] Available at: http://www.kurochans.net/kenkyu/energy_problem/wbgu.html. Accessed 4 June 2020
2. Wahid J, Basim S, Arar M, Awad J, Samarhan S (2020) Urbanisation in Ajman: pushing by housing development. Int Trans J Eng Manag Appl Sci Technol:1–11
3. Leung I (2011) Australia's solar industry: threats and opportunities. Available at: http://www.electronicsnews.com.au/Features/Australia-s-solar-industry-threats-and-opportuniti. Accessed 9 Dec 2015
4. Quora.com (2020) [online] Available at: https://www.quora.com/Is-Dubai-a-country-or-a-city. Accessed 3 June 2020
5. Uqba F, Ahmed R (2012) The employment of technical development to the direction of high-technology buildings within the framework of sustainable design. Damascus Univ J Sci Eng:226–246
6. UAE Ministry of Energy (2014) Statistical annual report, electricity and water 2011–2013. Available at: https://www.moenr.gov.ae/en/knowledge-center/reports.aspx. Accessed 13 Dec 2015
7. Emirates 24|7 (2020) Renewable energy shaping the future of sustainability. [online] Available at: https://www.emirates247.com/news/emirates/renewable-energy-shaping-the-future-of-sustainability-2020-01-05-1.691242. Accessed 4 June 2020
8. FEWA (2015) Home. Retrieved from Federal Electricity & Water Authority. Available at: http://www.fewa.gov.ae/. Accessed 12 Dec 2015
9. Asif M (2016) Growth and sustainability trends in the buildings sector in the GCC region with particular reference to the KSA and UAE. Renew Sust Energ Rev 55:1267–1273
10. MoEW (UAE Ministry of Environment and Water) (2015) United Arab Emirates state of green economy report 2014. MoEW, Dubai
11. Mondoc B, Pop F. Factors influencing the performance of photovoltaic power plant. 3rd International conference on modern power systems MPS 2010, 18–21 May 2010, CLUJ-NAPOCA, ROMANIA, pp. 271–276. Available at: http://ie.utcluj.ro/files/acta/2010/Number5/MPS_2010_Mondoc.pdf. Accessed 9 Dec 2015
12. VAISALA (2015) UAE Wind Atlas informs middle eastern wind development. Available at: http://www.vaisala.com/. Accessed 11 Nov 2015
13. Assesment R (2020) Home – research center for renewable energy mapping and assesment. [online] Research Center for Renewable Energy Mapping and Assesment. Available at: https://recrema.masdar.ac.ae/. Accessed 4 June 2020
14. Alnaser WE, Eliagoubi B, Al-Kalak A, Trabelsi H, Al-Maalej M, El-Sayed HM, Alloush M (2004) First solar radiation atlas for the Arab world. Renew Energy 29(7):1085–1107. ISSN 0960-1481
15. Alaeddine M, Mona AA, Mahieddine E (2013) Solar energy in the United Arab Emirates: a review. Renew Sust Energ Rev 28:340–375
16. Janajreh I, Taleb I (2010) Wind data collection and analyses at Masdar City for wind turbine assessment. Int J Thermal Env Eng 1(1):43–50
17. Masdar Institute/IRENA (2015) Renewable energy prospects: United Arab Emirates, REmap 2030. IRENA, Abu Dhabi. Available at: www.irena.org/remap. Accessed 12 Oct 2015
18. Ajman DO (December 2017) Media centre – News. Retrieved from Government of Ajman: http://www.am.gov.ae
19. Ajman MA (2013) Annual report 2013. Ajman - UAE, Ajman, Municipality and Planning Department. Available at: www.am.gov.ae/en/OurDept/Documents/Annual_Report_2013_EN.pdf. Accessed 6 Oct 2015
20. Akadiri PO, Chinyio EA, Olomolaiye PO (2012) Design of a sustainable building: a conceptual framework for implementing sustainability in the building sector. Buildings 2012(2):126–152
21. Darwish A (2016) Sustainable green smart buildings: future energy survivor. ISESCO. J Sci Technol 12(21)
22. (GORD) GO (2014) Global sustainability assessment system (GSAS). Gulf Operations for Research & Development (GORD), Qatar. Available at: http://www.gord.qa/uploads/formsnew/GSAS_Overview_07_for_web.pdf. Accessed 7 Sept 2015
23. Wang S (2010) Intelligent buildings and building automation. Spon Press, Oxon & New York

Sustainable Buildings for Sustainable Community in the Emirate of Ajman – UAE

Yaser Omar Rajab Kaied, Abdul Salam Darwish, and Peter Farrell

1 Introduction

The United Arab Emirates, among many countries, is one of the highest energy consumers per capita in the world where environmental pollution and carbon emission has been a significant challenge [1]. Growing economic, population growth and modernisation in lifestyle increased the demand for energy in GCC countries and specifically UAE. UAE is amongst the highest countries in the world in terms of per capita energy consumption and environmental emissions with the building sector having a very significant contribution. UAE has the second highest share in the GCC's construction market, accounting for 29% of the total, in the region, as indicated in Fig. 1, and Table 1 lists the ten largest construction projects under completion [2]. Ajman is the smallest of the seven emirates. Its area is 259 km^2 and placed 10 miles between the emirates of Umm Al-Quinn (see Fig. 2) [3, 4]. Three regions are distinguished: Masfut, agricultural area and town centre.

The increase in population and migration over the years put additional demand for all wellbeing resources (see Table 2) [4], which shows Ajman is the highest in population compared with Ras Al Khaima (RAK) and Uum Al Quin.

Government officials at the Emirate of Ajman have expressed their great interest and enthusiasm for developing sustainable buildings that are built with all green building standards indicating a need for a proper rating system and energy performance certification [5].

The concept of sustainable development (SD) has been considered in many institutions and new policy, and academic voices in the field of education put forward the importance of a holistic approach to the concept of sustainable development [6]. It is realised that SD must cover environmental and human wellbeing sustainability. In this respect, SD has matured and is defined as 'meeting the needs of the current generation without compromising the ability of future generations to meet their own needs' [7, 8]. Residential buildings are responsible for 40% of energy consumption, industrial 22%, commercial 18% and others 20% (see Fig. 3) [9].

Buildings' design prospects should include the two main issues: flexibility for the people who lived in them and surroundings to select the way they would like to work and live with. Moreover, mixed use of buildings that contain use for work or living. This requires a proper design that includes all community benefits and wellbeing. This is to ensure that those buildings can be adapted and altered to meet the requirements of the workforce while also resolving community instabilities [10]. Energy-saving principles must be at the heart of any buildings' design concepts that need the latest energy-saving technologies, including heating, cooling with proper management software. Moreover, technology can facilitate the interrelationship between the sustainable buildings-community combinations. The lack of information is a significant issue that affects building evaluation and raising awareness for consumers. Modelling, in this case, is essential to fill all information gaps and support rating and assessment processes. The role of building information modelling (BIM) in reducing the fragmentation among professionals at each and across building delivery stage, using evidence from the literature, is discussed in [11] and building information modelling system introduced. This is to provide a platform for professionals to work in an integrated environment at any stage of the building delivery process.

Y. O. R. Kaied (✉) · A. S. Darwish · P. Farrell
School of Civil Engineering, University of Bolton, Bolton, UK
e-mail: yk1res@bolton.ac.uk; a.darwish@bolton.ac.uk; P.Farrell@bolton.ac.uk

© The Author(s), under exclusive license to Springer Nature Switzerland AG 2022

A. Sayigh (ed.), *Sustainable Energy Development and Innovation*, Innovative Renewable Energy,
https://doi.org/10.1007/978-3-030-76221-6_49

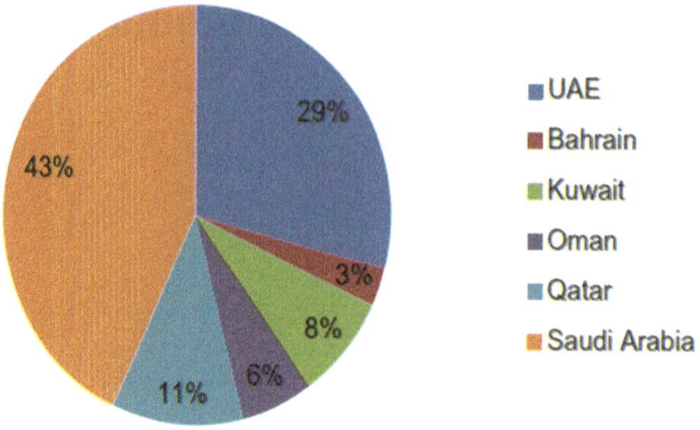

Fig. 1 Construction projects in the GCC region by investment [2]

Table 1 Top ten most significant projects in the UAE [2]

Project	Value (US$ Billion)
Barakah nuclear power 3	4.0
Upper Zakum full field: early production facility: offshore: EPC 2	3.8
Ruwais refinery expansion project: process package 2 (RFCC)	3.1
Barakah nuclear power plant 1	3.1
Abu Dhabi airport expansion: midfield terminal complex	3.0
National housing scheme project	2.9
Ruwais refinery expansion project: offsites & utilities (package 3)	2.7
Barakah nuclear power plant 22.5	2.5
Ruwais carbon black and delayed 2.5 Coker project	2.5
Integrated gas development: Ruwais 2.54th NGL train package	2.5

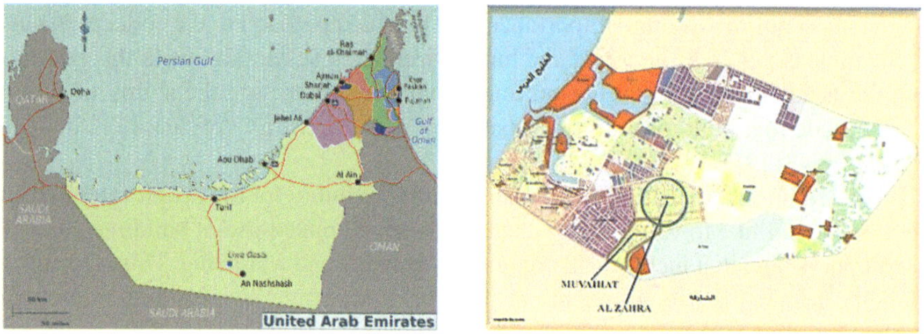

Fig. 2 Emirate of Ajman location within the UAE [3, 4]

Table 2 The three emirates at the North of the UAE [4]

		Population (thousands)					
Year/Emirate	Area (km^2)	2005	2010	2015	2020	2025	2030
Ajman	259	179,000	460,000	441,000	463,000	490,000	577,000
R.A.K	1684	102,000	250,000	235,000	245,000	258,000	272,000
Uum Al Quin	777	30,000	73,000	67,000	70,000	73,000	77,000

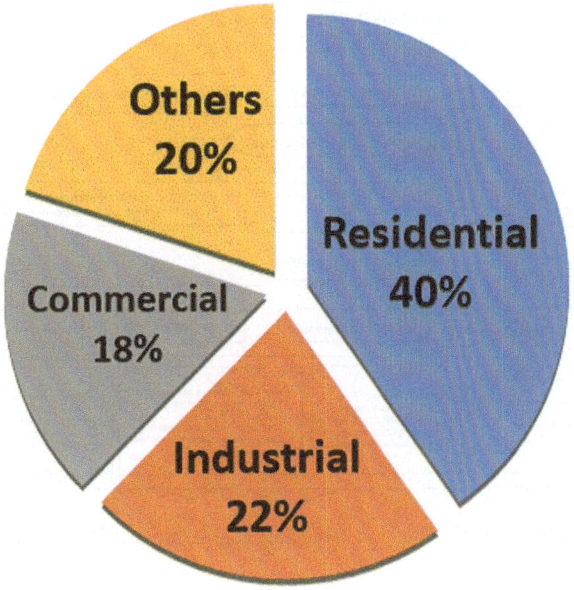

Fig. 3 Energy consumption sectors [9]

2 Sustainable and Green Buildings: Ajman's Energy and Environmental Solution

There exists a strong interrelation between the development of green buildings' principles towards a sustainable building with a resulting consequence of energy performance efficiency. Ajman has managed to achieve more than 10% electricity saving [9] and is still in the process of attaining double this saving with sustainable buildings' progress achievements at this Emirate.

The benefits of sustainable buildings are [12]:

Lower life cycle costs

- Lower insurance fees
- Higher property value
- Higher productivity
- Improved image
- Reduced risks
- Healthier for tenants and visitors
- Reduced effects on infrastructure
- Better for the environment and local economy

The concept of the sustainable (or green) building is a recent response to address environmental and health problems that stem from buildings and reduce impacts of the building sector on the natural environment, as well as on people. Although definitions vary, sustainable building refers to the application of sustainability principles to design, construct and manage buildings to mitigate environmental footprints of the building sector and its surroundings, and consequently on humans [13].

Many challenges delay the use and implementation of green building construction and renovation. Some of the challenges are related to fundamental factors of building construction and renovation like cost, maintenance and operation [14]. Sustainable building innovation creates economic stability that, in turn, generates high grades of economic growth and employment. These factors improve project delivery and increase profitability and productivity [12]. Figure 4 [5, 12] shows the practical framework for sustainability implementation in building construction, strategies and methods to achieve resource conservation.

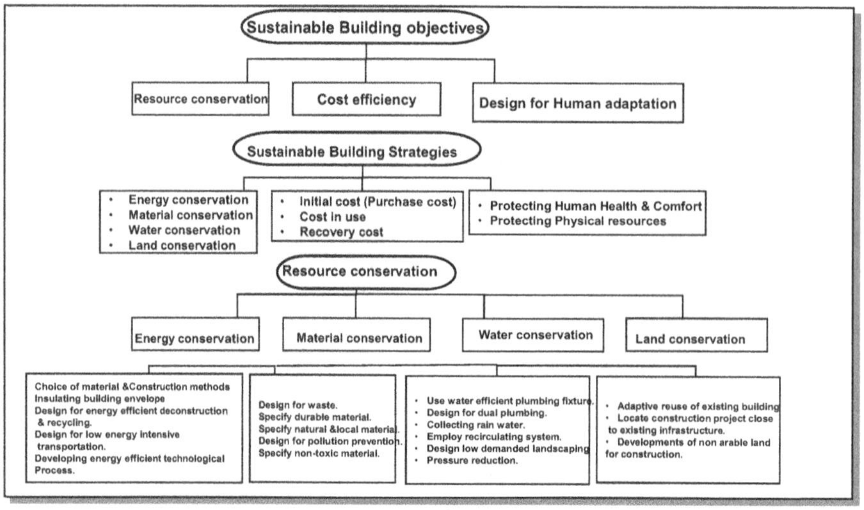

Fig. 4 A practical framework for sustainability implementation in building construction, strategies and methods to achieve resource conservation [5, 12–15]

3 Sustainability of Mixed-Use Buildings in Ajman

As a result of COVID-19 crisis, there is a need for buildings, venues and facilities that have the rapid-response characteristics, prefabricated, and highly adaptable buildings. These buildings should be with facilities that are tailor-made for modular construction and mixed use with the amplification of pop-up, flexible and adaptable spaces. As a result, several key performance factors are to be considered in the near future when designing sustainable mixed-use as follows [15]:

- Healthy building standards will come to the forefront, reducing development risk.
- The introduction of outside air will become a more significant part of new-builds.
- Smaller, horizontal distribution systems will likely become more prevalent.
- Foyers and building lobbies can provide a safer and cleaner shared environment.
- Self-sustaining districts and connected communities will experience a resurgence.
- Microgrids provide efficiencies.
- Integrated districts forge human connections.

Ajman government has started to invest and open wide for investment in green buildings considering green building system standards [5]. The use of green concrete and thermal brick with other environmentally-friendly materials, as well as the sustainable building procedure, is a priority issue. To have these, buildings are sustainable in a way to support the community, workforce and economy; mixed-use buildings are the key to achieving government targets for sustainability development. The energy performance of the buildings and used facilities are to be considered with the aim of attaining energy-saving.

4 Sustainable Energy Performance and Certification Solution

Sustainable energy is the energy that consumed at insignificant rates compared to its supply and with manageable collateral effects, especially environmental effects [16]. Increased levels of energy consumption and conservation are highly correlated with technological innovations, technology adoptions, user's lifestyle, economic growth, etc. 20–40% of the entire energy consumption in developed countries (40% in Hong Kong, 37% in the United States, 39% in the UK and 31% in Japan) refers to the energy usage of buildings [17].

Regarding the importance of environmental assessment and energy performance of buildings, it is essential to implement the most practical theoretical perspectives, trends, applications and constraints towards the development of green, environmentally sustainable buildings. In this respect, building energy efficiency, thermal performance and material efficiency is the significant parameter of sustainable energy performance indicators. They should take into consideration during building energy performance evaluations [18].

5 Sustainable Community Needs

Sustainability for the community is, in principle, meant to contribute significantly to the community which cannot be achieved without having all to work together with stakeholders, including local government, utility providers, developers and individuals themselves. Several needs are associated with a community for sustainable development such as [19]:

- Protecting and enhancing the environment (minimise waste and limit pollution)
- Meeting social needs (creating sustainable buildings, grant settlement, protect diversity and local distinctiveness, health, food, water, housing, fuel and education, skills empowerment)
- Promoting economic success (creating a local economy, value unpaid work, access to facilities and opportunities)

Moreover, employment opportunities are continuously improving, resulting in a more rewarding working life.

A critical task is also to educate users on the importance of sustainable design for reducing the energy consumption levels of buildings, and their harmful influence on natural environments. Ultimately, it highlighted that sustainable buildings must be developed based on a clear understanding of the main targets of sustainable developments for a low-carbon future and applicable smart technology, as elaborated in Table 3 [18]. Low awareness of green buildings, confusion over implementation processes and low coordination among key agencies have proved to limit green building activity [20].

6 Sustainable Materials

It is vital to express the severe necessity for analysing the energy consumption level of buildings and to innovative new solutions for achieving sustainability in the built environment. Building materials play an essential role in increasing the sustainability of buildings and contributing to economic prosperity. The ecological impact of construction materials can be significantly reduced by encouraging the use of the best eco-innovation in production plants and substituting the use of finite natural resources for waste generated in other manufacturing processes, thus resulting in increased competition to produce more eco-efficient products [21]. This is illustrated in Fig. 5.

Table 3 Main targets of sustainable developments [18]

1	Resource efficiency
2	Energy efficiency (including greenhouse gas emissions reduction)
3	Pollution prevention (including indoor air quality and noise abatement)
4	Harmonisation with environment
5	Integrated and systemic approaches

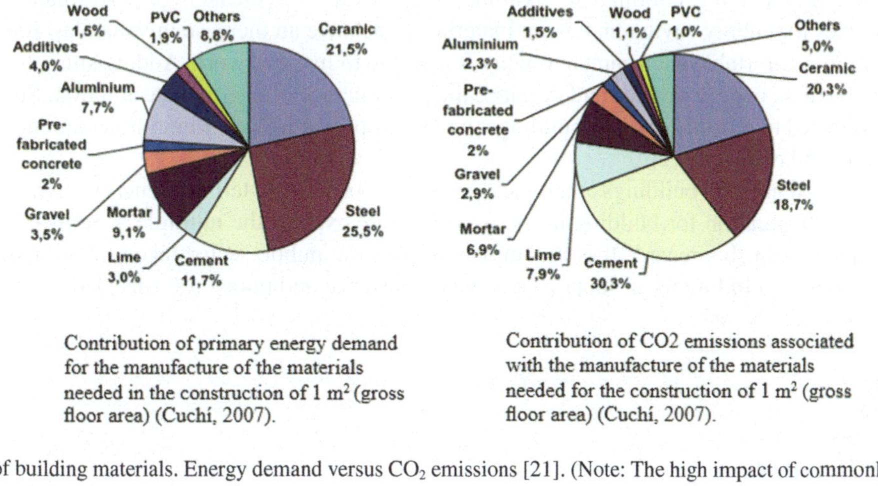

Fig. 5 A comparison of building materials. Energy demand versus CO_2 emissions [21]. (Note: The high impact of commonly used materials such as steel, cement and ceramics is notable)

The use of building materials has a substantial impact on the environment, mainly because of the large quantity of non-renewable resources with the potential for depriving future generations of their use [22, 23]. Building materials have a substantial impact on the environment during their life cycle. They create different types of pollutions over that life cycle, which harm the environment starting from the extraction of raw materials to the disposal of demolition waste.

Referring to the role of materials, the use of kenaf-fibres insulation boards is highly recommended for application in sustainable green buildings [24]. Kenaf-fibres have excellent thermal insulation properties, which absorb a high percentage of produced CO_2 compared with the other alternatives [18]. In order to ensure the reduction of energy consumption, zero energy building concepts should be implemented as an essential factor when developments need to achieve sustainable outcomes [25].

7 Sustainable Energy-Community Framework Model

A sustainable energy community developed through organising and introducing all required regulations to ensure clean energy resources are used with energy performance buildings and facilities used for the community wellbeing [21].

8 Role of Renewable Energy in Sustainable Green Buildings

Improving energy efficiency in the buildings and integrating renewable energy technologies into it are essential practices that require in-depth assessment and data collection in order to produce strategic procedures for decision-makers. Renewable energy has been providing the most efficient, sustainable source. Integrating this source of energy with the existing building energy systems can lead to zero energy building in many forms as far as it is economically and technically viable [9]. Two systems are mainly applicable for integration: solar and wind systems. The systems that have usually employed in buildings are PVs and solar thermal collectors. Solar thermal systems can supply thermal energy for space heating, cooling and the provision of hot water for the needs of houses or other buildings. Solar systems can be installed in different ways, such as a roof- or wall-mounted. New technology is available that implements thin-film panels and wall tile converted panels. Battery systems are also being developed to store energy and use it when no solar irradiation is available. Wind technology is also possible to use in the form of vertical axis turbines that can be installed on rooftops or in spaces surrounding buildings.

9 Renovated Buildings

Retrofitting existing buildings can frequently be more cost-effective than building a new facility since existing buildings comprise the largest segment of the built environment. It is essential to initiate energy conservation retrofits to reduce energy consumption and the cost of heating, cooling, and lighting [26]. The aim is to create high-performance buildings. By applying sustainable practices for energy-saving and green materials to achieve an increase in value, last longer and contribute to a better, healthier, more comfortable environment quality for people to live and work. Adding smart technology will control the usage of energy, increase energy savings and organise the performance of energy consumption. Sustainability initiatives are needed to be introduced to retrofits existing buildings to reduce operating costs. Environmental impacts increase building adaptability, durability and resiliency [26].

Green and sustainably renovated buildings could yield significant benefits in terms of energy and CO_2 reduction, cost savings and improved health situation for building users; making more explicit the multiple benefits of sustainable buildings needs further consideration in this regard. It is recommended that the public sector take critical actions to accelerate the number of green buildings, including fiscal support, technical assistance and policy reforms [13].

10 Case Study

10.1 Building Description

A non-residential building was selected for analysis. The building layout plan is shown in Fig. 6 [27]. The building includes offices, information technology facilities, laboratories, facilities areas, meeting rooms and storage. The structure of the building is designed in a way such that it responds to its surroundings, creating an interactive environment for users and visitors.

Fig. 6 Building layout [27]

The outdoor environmental effect on the building, which has many rooms with at least one wall exposed to extreme weather conditions, will have a high possibility of energy loss with increased airconditioning demand. The building consists of two floors; thus, rooms on the ground floor have no roof and, therefore, no direct sunshine to their ceilings. However, first floor rooms have a ceiling that has direct sunshine, which adds additional air-conditioning demand load. External doors are operating most of the working hours, allowing a considerable amount of energy loss, with no protective sealing available for energy conservation. The windows are mainly single glazing with an accumulated area of 71 m^2 [27]. Internal finishing parameters are shown in (Table 4).

10.2 Building Energy Assessment

Energy assessment (Fig. 7) shows that the building is with rating E, which indicates the size of this building consumes much of energy. However, at the same time, it shows that the building is retrofit and has more energy-efficient practices.

10.3 Converting to Sustainable Building

Different retrofits can be applied to the building envelope to increase the energy performance of the building and convert it to be sustainable. Options are such as adding insulation to improve the wall resistance or replacing the windows with lower U-Value glazed ones with proper airtightness. This will help to reduce the external heat gained. The most gained energy saving can well achieve through the reduction in the cooling load. This can be achieved by using a more efficient air-conditioning system and by reducing the heat gain with proper attention to space cooling load corresponding to sensible and latent cooling loads (keeping with comfortable temperature 24 °C and humidity to about 50% relative). Roof insulation should be enhanced, and sunshine breakers must be installed on the roof. Smart technology also must be implemented in the building [13].

Table 4 Information for internal finishing

Description	Room no.	Floor	Walls (up to the building ceiling)	Skirting	Ceiling
Lobby entrance	G1	Granite/ marble floor and step for (trade and riser) 60*60 cm granite tiles for flooring	Smooth plaster +primer + Fenomastic skill paint	10 cm height ceramic skirting	Epson board false ceiling
Lobby	G2	60*60 cm granite tiles	Smooth plaster +primer + Fenomastic skill paint	10 cm height ceramic skirting	Epson board false ceiling
Room 01	G3	60*60 cm granite tiles	Smooth plaster +primer + Fenomastic skill paint	10 cm height ceramic skirting	Epson board false ceiling
Room 02	G4	60*60 cm granite tiles	Smooth plaster +primer + Fenomastic skill paint	10 cm height ceramic skirting	Epson board false ceiling
Pantry	G5	No-slip ceramic tiles	Glazed ceramic tiles	–	60*60 aluminium false ceiling TYP
Wash area	G6	No-slip ceramic tiles	Glazed ceramic tiles	–	60*60 aluminium false ceiling TYP
Wash area	G7	No-slip ceramic tiles	Glazed ceramic tiles	–	60*60 aluminium false ceiling TYP
Toilet	G8	No-slip ceramic tiles	Glazed ceramic tiles	–	60*60 aluminium false ceiling TYP
Toilet	G9	No-slip ceramic tiles	Glazed ceramic tiles	–	60*60 aluminium false ceiling TYP
Entrance lobby	M1	Granite/marble floor and step for (trade and riser) 60*60 cm granite tiles for flooring	Smooth plaster +primer + Fenomastic skill paint	10 cm height ceramic skirting	Epson board false ceiling
Meeting room	M2	60*60 cm granite tiles	Smooth plaster +primer + Fenomastic skill paint	10 cm height ceramic skirting	Epson board false ceiling
Entrance lobby	S1	Granite/ marble floor & step for (trade &riser) 60*60 cm granite tiles for flooring	Smooth plaster +primer + Fenomastic skill paint	10 cm height ceramic skirting	Epson board false ceiling
Store 1	S2	60*60 cm granite tiles	Smooth plaster +primer + Fenomastic skill paint	10 cm height ceramic skirting	Epson board false ceiling
Store 2	S3	60*60 cm granite tiles	Smooth plaster +primer + Fenomastic skill paint	10 cm height ceramic skirting	Epson board false ceiling

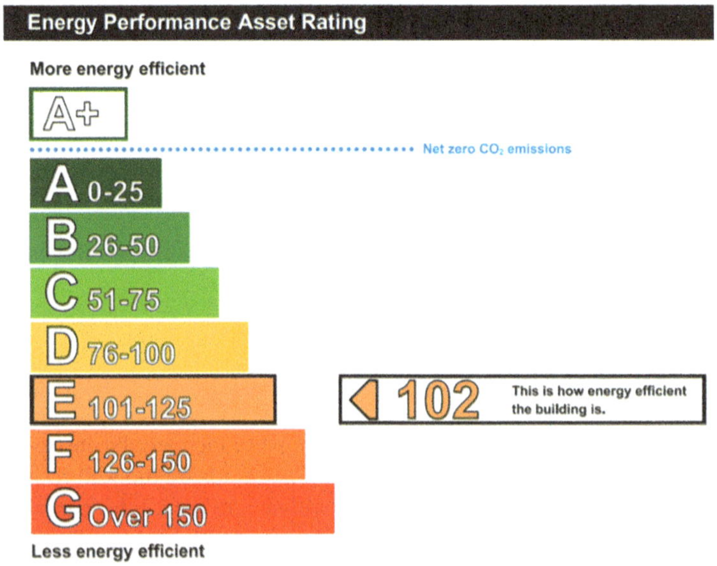

Fig. 7 Energy performance rating system building [28]

Renewable energy systems have a significant impact on energy savings and building sustainability. It is recommended that this building can have renewable energy systems added to generate around 100 kW power self-powered, with a pilot project that consists of 60 kW solar PV and solar thermal roof - garages shed mounted, and 40 kW wind energy system.

11 Conclusions

This paper analyses and emphasises the significance of governments developing public and industrial awareness in order to limit and avoid the effects of climate change, with the goal of achieving the energy sustainability target. Sustainability has become our life need through which many aspects are involving, such as buildings. Energy performance has to assist continuously and thoroughly to achieve sustainable buildings. The structure can then be converted to obtain sustainable green and smart characteristics. The non-residential building was selected as a case study. Energy assessment for the building showed that the way such buildings are operating results in low energy ratings due to a visible design and construction practices. It showed that by adding energy-saving options and technologies, the building could well become more energy efficient. Renewable energy systems are recommended for use to give the building a percentage of self-power. Practices such as develop and implement an integrated energy efficiency strategy are one of the critical factors for achievement.

References

1. Jamil M, Ahmad F, Jeon YJ (2016) Renewable energy technologies adopted by the UAE: prospects and challenges – a comprehensive overview. Renew Sust Energ Rev 55:1181–1194
2. Asif M (2016) Growth and sustainability trends in the buildings sector in the GCC region with particular reference to the KSA and UAE. Renew Sust Energ Rev 55:1267–1273
3. Quora.com (2020) Dubai is a city. [online] Available at: https://www.quora.com/Is-Dubai-a-country-or-a-city. Accessed 3 June 2020
4. Wahid J, Basim S, Arar M, Awad J, Samarhan S (2020) Urbanisation in Ajman: pushing by housing development. Int Trans J Eng Manag Appl Sci Technol 11:1–11
5. Pradeep A (2020) Ajman municipality starts work on green building. [online] Middle East Construction News. Available at: https://webcache.googleusercontent.com/search?q=cache:T0voIhwJY-YJ:https://meconstructionnews.com/33129/ajman-municipality-starts-work-on-green-building+&cd=2&hl=en&ct=clnk&gl=uk. Accessed 3 June 2020
6. Sinakou E, Boeve-de Pauw J, Goossens M, Van Petegem P (2018) Academics in the field of education for sustainable development: their conceptions of sustainable development. J Clean Prod 184:321–332
7. Darwish AS (2016) Sustainable green smart buildings: future energy survivor. J Sci Technol ISESCO 12(21):35–42. Available Online: https://www.isesco.org.ma/ISESCO_Technology_Vision/NUM21/doc/5.pdf
8. United Nations (1987) Report of the World Commission on Environment and Development, General Assembly Resolution 42/187, 11 Dec 1987. Retrieved: 2007-11-14
9. Darwish AS (2018). Renewable energy utilization - aspects and innovations: sustainable, green, and smart buildings in the Emirate of Ajman. *5th Ajman International Environment Conference*, Ajman – UAE, 6–7 Mar 2018
10. Scott C (2020) Why Sustainable Buildings Need To Focus On Community And Collaboration. [online] the Guardian. Available at: <https://www.theguardian.com/sustainable-business/sustainable-buildings-focus-community-collaboration. Accessed 1 June 2020
11. Fadeyi MO (2017) The role of building information modeling (BIM) in delivering the sustainable building value. Int J Sustain Built Environ 6(2):711–722
12. Darwish AS (2017) Green, smart, sustainable building aspects and innovations. In: Sayigh A (ed) Mediterranean green buildings & renewable energy. Springer, Cham
13. Balaban O, Puppim de Oliveira JA (2017) Sustainable buildings for healthier cities: assessing the co-benefits of green buildings in Japan. J Clean Prod 163:S68–S78
14. GhaffarianHoseini AH, Dahlana ND, Berardi U, GhaffarianHoseini A, Makaremi N, GhaffarianHoseini M (2013) Sustainable energy performances of green buildings: a review of current theories, implementations and challenges. Renew Sust Energ Rev 25(2013)
15. Finn F, Paterson D (2020) Redesigning mixed-use environments for a post-pandemic world | Dialogue blog | Research & insight | Gensler. [online] Gensler. Available at: https://www.gensler.com/research-insight/blog/redesigning-mixed-use-environments-for-a-post-pandemic-world. Accessed 5 June 2020
16. Renewable Energy & Efficiency Partnership (August 2004) Glossary of terms in sustainable energy regulation (PDF). Retrieved 2008-12-19. Available Online: www.reeep.org/sites/default/files/Glossary%20of%20Terms%20in%20Sustainable%20Energy%20Regulation.pdf
17. Juan Y, Gao P, Wang J (2010) A hybrid decision support system for sustainable office building renovation and energy performance improvement. Energ Buildings 42:290–297
18. GhaffarianHoseini A et al (2013) Sustainable energy performances of green buildings: a review of current theories, implementations and challenges. Renew Sust Energ Rev 25:1–17
19. Corporate Business and Partnerships (2020) Characteristics of a Sustainable Community Stafford Borough Council. [online] Staffordbc.gov.uk. Available at: https://www.staffordbc.gov.uk/characteristics-of-a-sustainable-community#:~:text=Sustainable%20communities%20can%20be%20described,to%20the%20environment%20is%20minimised. Accessed 5 June 2020

20. Van Schaack C, BenDor T (2011) A comparative study of green building in urban and transitioning rural North Carolina. J Environ Plan Manag 54(8)
21. Darwish A (2014) Eco-friendly buildings: the central factor in transitioning to a green economy. Int J Environ Sustain 3(1):54–62
22. Godfaurd J, Clements-Croome D, Jeronimidis G (2005) Sustainable building solutions: a review of lessons from the natural world. Build Environ 40(3):319–328
23. Ofori G (2002) Singapore construction: moving towards a knowledge-based industry. Build Res Inf 30(6):401–412
24. Ardente F, Beccali M, Cellura M, Mistretta M (2008) Building energy performance: a LCA case study of kenaf-fibres insulation board. Energ Buildings 40:1–10
25. Marszal AJ, Heiselberg P, Bourrelle JS, Musall E, Voss K, Sartori I, Napolitano A (2011) Zero energy building – a review of definitions and calculation methodologies. Energ Buildings 43:9719
26. Paradis R (2018, March 26) Retrofitting existing buildings to improve sustainability and energy performance. Retrieved from National Institute of Building Sciences: www.wbdg.org/resources/retrofitting-existing-buildings-improve-sustainability-and-energy-performance
27. Municipality & Planning Department (2019) City of Ajman population and waste management. Municipality & Planning Department, Ajman. (Data requested by official communications)
28. Koo C et al (2014) Development of a new energy efficiency rating system for existing residential buildings. Energy Policy Elsevier 68(2014):218–231. https://doi.org/10.1016/j.enpol.2013.12.068

Mediterranean Hospital Energy Performance Mapping: The Energy Auditing as a Tool Towards Zero Energy Healthcare Facilities

Despina Serghides, Stella Dimitriou, and Ioanna Kyprianou

1 Introduction

The alarming trends of energy consumption and subsequent pollution emissions during the past century have called for energy efficiency standards in all areas, from appliances to the built environment. In the European Union (EU), regulation is achieved through directives addressed to all member states (MSs), and in the construction sector, the Energy Performance of Buildings Directive (EPBD) constitutes the main set of guidelines. The concept of near-zero energy buildings (nZEB) was introduced in the 2010 recast of the EPBD, stating that all new buildings must comply with national regulations by 31/12/2020, in order to have a very high energy performance and obtain the energy they need from renewable sources [1]. These measures apply to the residential and commercial sectors alike, with some facilities bearing a higher load of energy consumption than others.

Healthcare facilities are one of the highest energy consumers, since they require constant supply of life-guaranteeing energy for patients. They consume vast amounts of energy not only for equipment, but also for proper ventilation, heating, cooling and lighting of the buildings themselves. The smooth operation of these facilities is fundamental to provide patients with the appropriate recovery conditions; therefore, hospitals and clinics should be exemplars of best practices, instead of derelict or inefficient facilities, as is sometimes the case [2]. Nowadays more than ever, the necessity of efficient and reliant healthcare facilities becomes evident, in the face of a global pandemic.

Ensuring the energy-efficient operation of these complex buildings is therefore focal to the wellbeing of populations attending them; however, it is also a significant element of sustainability goals. Energy and emission reduction are amplified when the scale moves from a simple dwelling to an intricate facility. In Cyprus, healthcare facilities are responsible for approximately 8% of total energy consumption in the tertiary sector [3]. Moreover, hospitals and clinics in Cyprus take up more area (in m^2) than any other category within the service building stock [4]. Improving the energy consumption of these complex buildings will therefore have a significant impact on the services sector as a whole.

Ameliorating the pollution profile and energy performance of healthcare facilities is not uncommon in the field of zero energy buildings. For instance, a Spanish study assessed the amounts of CO_2 emitted as a result of energy consumption in 14 hospitals of the region [5]. Additionally, a case study based in Kosovo showed that cost-efficient energy improvements of hospitals can approximately halve the energy consumption and CO_2 emissions of those buildings [6]. Other studies have shown that 60–75% of the energy consumed in this type of buildings relates to the buildings themselves and the services they provide to meet required conditions according to occupant needs [7]. Hospitals and clinics in Cyprus consume mostly electricity (with generation stations fuelled by oil) and oil for space thermal conditioning, and sanitary hot water is largely covered by solar collectors (by 85%). Nevertheless, the remaining percentage for water heating still amounts to one of the largest energy needs, following space cooling [4].

In Cyprus, healthcare facilities include private (app. 30%) and public health centres (app. 60%) [4]. In the public sector, there are six general hospitals, three specialized centres, two small rural hospitals and 39 health centres, as well as many sub-centres for provision of primary health care services. In the private sector in 2016, there was a total of 74 hospitals, polyclinics and diagnostic centres. At the moment, private healthcare facilities consume up to 75% more energy than public counterparts [4, 8]. It has been estimated that primary energy consumption can be reduced by more than 50% if a class B energy refurbishment is implemented, with further improvements if the refurbishment is based nearly zero energy standards [8].

D. Serghides (✉) · S. Dimitriou · I. Kyprianou
Energy, Environment and Water Research Centre (EEWRC), The Cyprus Institute, Aglantzia, Cyprus
e-mail: d.serghides@cyi.ac.cy

© The Author(s), under exclusive license to Springer Nature Switzerland AG 2022

A. Sayigh (ed.), *Sustainable Energy Development and Innovation*, Innovative Renewable Energy,
https://doi.org/10.1007/978-3-030-76221-6_50

The present study introduces the competency of energy auditing as in identifying the correct opportunities and maximising the potential for energy savings and reduction of pollutant emissions in selected buildings offering healthcare services in Cyprus. The audits were part of the ZenH Balkan project (zero energy hospitals), which aims to improve energy efficiency in the building sector by producing benchmarks and designing guidelines for zero energy hospitals. The project also aspires to improve the technical capacity of professional groups and government officials. Findings of this analysis offer cost-effectiveness estimates which will ultimately be used to seek financial tools and implement energy efficiency improvements in hospitals across the Mediterranean region.

2 Methodology

Three clinics were selected from a pool of five which had previously been energy audited, in the framework of ZenH Balkan and corresponded to a 22% of the private clinics in the Capital District. The selection of the original five clinics was made based on their size, all having a conditioned area of around 1000 m^2 and a minimum number of 10 beds. The three clinics, in which the standard audits were performed, were chosen based on the accessibility provided by the clinics and the availability of requested information.

The energy auditing of the clinics was conducted in accordance to the European Standards: EN 16247-1:2012 [9] and EN 16247-2:2014 [10]. The team of accredited energy auditors, which performed the audits, was supervised by the Cyprus Institute's project research team (authors of this chapter), who provided the Energy Audit team with the contact details of the representatives of the five clinics and the questionnaire template to be used for the energy assessment of the buildings.

The hospital representatives provided the following information to the energy audit team:

- Architectural drawings: Plans, sections and elevations
- Electromechanical drawings: Information on the lighting system, HVAC system and the hot water production of the building
- Energy Consumption information: Electricity, fuel and water bills for the past 3 years

The energy auditors conducted at least two visits at each of the buildings. The visits included discussions with hospital representatives regarding the energy management of the building and inspection of the buildings' services (e.g. boilers, chillers, lighting) as well as measurement taking, including placement of electricity metres for a week.

The outcome of the audits will be summarized under the following sections:

- Overview of the Three Clinic Buildings
- Envelope of the Clinics
- Energy Systems of the Clinics
- Energy Consumption of the Clinics
- Energy Performance Comparison of the Three Clinics
- Potential Improvements
- Conclusion

3 Overview of the Three Clinic Buildings

3.1 Climate Zone

All hospitals are located in Nicosia and thus fall within the same climate zone, specifically a hot semi-arid climate (Köppen climate classification BSh). The characteristics of the climate zone are given in Table 1.

3.2 Floor Areas of All Clinics

A summary of the total floor areas of the three clinics is presented in Table 2. Clinic A is located in a building constructed in 1987. It is developed in three floors (i.e. basement, ground floor and first floor) and the floor plan is overall L shaped in outline. The entrance is located on the North-East corner of the building.

Table 1 Climate zone for all clinics

Hospital climatic zone	
Altitude:	162 m
Latitude:	35° 09′
Longitude:	33° 24′
Degree days (heating):	867
Degree days (cooling):	320
Temperature max:	43.4 °C (July, August)
Temperature min:	−2.9 °C (December)

Table 2 Floor areas of all clinics, per floor and air-conditioning status

Clinic ID	Floor type	Floor area (m²)		
		Non-conditioned	Conditioned	Total
Clinic A	Basement	154.8	873.4	1028.2
	Ground floor	50.7	1166.2	1216.8
	First floor	50.7	1026.1	1076.7
	Total	**256.1**	**3065.7**	**3321.7**
Clinic B	Ground floor	14.2	408.6	422.9
	First floor	10.1	327.6	337.8
	Second floor	14.0	301.1	315.0
	Roof	0.0	21.7	21.7
	Total	**38.3**	**1059.0**	**1097.3**
Clinic C	Basement	269.5	27.1	296.6
	Ground floor	33.1	252.1	285.2
	First floor	33.1	264.5	297.6
	Second floor	33.1	266.8	299.9
	Total	**368.9**	**810.4**	**1179.4**

Table 3 All primary spaces and their functions, along with the occupied area in m² and as a percentage of the total floor area of the whole building

Space	No of rooms			Floor area (m²)			% of total building floor area		
Clinic ID	A	B	C	A	B	C	A	B	C
Patients rooms	21	14	11	384.37	215.17	150.73	13%	20%	19%
Offices	32	8	8	845.22	178.05	145.29	28%	17%	18%
Bathrooms-toilet	38	21	21	175.36	79	69.68	6%	7%	9%
Corridors	8	2	4	544.72	19.5	128.89	18%	2%	16%
Operating theatre	3	–	2	124.92	–	41.17	4%	–	5%
Storage	9	–	5	120.02	–	38.86	4%	–	5%
Living room-waiting	4	–	5	318.5	–	139.37	10%	–	16%
Laboratory	8	–	2	237.08	–	24.88	8%	–	3%
Computer, X-ray room, other	–	2		–	43.32		–	4%	
Stairs	–	2	–	–	19.37	–	–	2%	
Other (specify)	19	6	21	315.48	138.43	71.57	10%	13%	9%
Total area (m²)				3065.67	1060.61	810.44			

Clinic B was built in 1977 and refurbished in 2007. It consists of four floors which are divided into zones (i.e. ground floor, first floor, second floor and roof). The building's floor plan is overall rectangular with a length to width ratio of 1.8:1 and the entrance being on the North-West side of the ground floor. Clinic C was constructed in 2009, and it consists of four floors (i.e. basement, ground floor, first and second floor) and the building's floor plan is overall rectangular, highly compact with a length to width ratio of 1.6:1 with the entrance being on the South-East side of the ground floor. Table 3 presents the primary spaces of each clinic, indicating their function and the occupied space in the total floor area of the respective buildings.

4 Envelope of the Clinics

Clinic A has double glazed windows and 5 cm polystyrene on the roof. The other envelope elements are not insulated. Clinic B has Ytong thermal insulating brick on the external walls and outer thermal insulating coating on both external walls and columns-beams, as well as double glazed windows. Clinic C, same as Clinic A, has double glazed windows and 5 cm polystyrene on the roof. The other envelope elements are not insulated (Table 4).

Regarding the shading of their windows, all clinics have internal shading devices that are used mostly during the summer months. On the contrary, external shading is found only in Clinic C, and partially in Clinic B (on the North-West orientation), with Clinic A having none at all (Table 5).

5 Energy Systems of the Clinics

5.1 HVAC

Characteristics of HVAC operation for each clinic are presented in Table 6. Starting with the HVAC of Clinic A, the system operates 24 hours a day throughout the year. The main air-conditioning (heating/cooling) system consists of chillers and boilers that feed the fan coil units (FCUs) installed in each room. In the building, there are a total of 112 FCUs (water-air). On each floor, there are heat recovery units, positioned above the corridors.

The central HVAC system controller is generally set to 22.0 °C for heating, whereas for cooling the chillers are set to 7.0–12.0 °C. There are two air-changes per hour. Rooms are usually maintained at a temperature of 22.8–23.9 °C and relative humidity of 43.1%, and corridors are usually maintained at 22.9 °C and a relative humidity of 32.8%. Operating rooms are usually maintained at an air-temperature of 19 °C and 48–50% relative humidity. There are also some independent A/C units throughout the building.

Clinic B's HVAC system operates 24 hours a day throughout the year. The central HVAC system controller is generally set to 22.0 °C for heating, and the chillers are set to 7.0–12.0 °C. There are two air-changes per hour. Rooms and corridors are usually maintained at 22.5 °C (winter and summer). Operating rooms are usually maintained at an air-temperature of

Table 4 Building components actual construction and U-values for all clinics

Component	U-value ($Wm^{-2} K^{-1}$)	U-value ($Wm^{-2} K^{-1}$)	U-value ($Wm^{-2} K^{-1}$)
	Clinic A	Clinic B	Clinic C
External wall	1.031	0.412	1.388
External column/beam	2.941	2.941	2.941
Horizontal roof	0.529	0.529	0.529
Windows	4.10	4.10	4.10

Table 5 Usage of shading system in the three case study clinics

		External shading (% use)		Internal shading (% use)	
Clinic ID	Orientation	Winter	Summer	Winter	Summer
Clinic A	North-East			20	80
	South-East			20	80
	South-West			20	80
	North-West			20	80
Clinic B	North-East			20	80
	South-East			20	80
	South-West			20	80
	North-West	100	100	20	80
Clinic C	North-East	100	100	20	80
	South-East	100	100	20	80
	South-West	100	100	20	80
	North-West			20	80

Table 6 HVAC features for all clinics

Clinic ID	Space of HVAC	Temperature (°C)	Relative humidity (%)	Air velocity (ms^{-1})
Clinic A	Hospital rooms	22.8–23.9	43.1	–
	Hospital corridors/other	22.9	32.8	–
	Operation room	19.0	48–50	100% fresh air
Clinic B	Hospital rooms	22.5	42.9	–
	Hospital corridors/other	22.5	42.9	–
	Operation room	19.0	–	100% fresh air
Clinic C	Hospital rooms	22.0	–	–
	Hospital corridors/other	22.0	–	–
	Operation room	19.0	50	100% fresh air

19 °C (Table 6) (winter and summer). For the operating rooms, there is 100% fresh air intake. The main units are usually serviced two times per year. The main air-conditioning (heating/cooling) system of the building is comprised of chiller machines and oil boilers that feed the FCU installed in each room. The building has a total of 41 FCUs. The exhaust and fresh air of the building are satisfied by using heat recovery units in the second floor and air handling units (AHUs) in the other floors. Also, independent AHUs are used in operating theatre, positioned in the roof. AHU units use 100% fresh air. There are also independent A/Cs.

As for Clinic C, the HVAC system also operates 24 hours a day throughout the year. The central HVAC system controller is generally set to 26.0 °C for cooling and 21.0 °C for heating. There are two air-changes per hour. Rooms and corridors are usually maintained at 22.0 °C. Operating rooms are usually maintained at an air-temperature of 19 °C and a relative humidity of 50% (Table 6). For the operating rooms, there is 100% fresh air intake. The main units are usually serviced two times per year. There also 75 PV panels installed on the roof of the clinic.

Moreover, Clinic C has a variable refrigerant volume (VRV) inverter-driven unit system. It has 17 ceiling-type cassettes and 26 VRV concealed in the roof. The set indoor temperature is achieved via electronic controllers which communicate with outdoor units to provide the required amount of cooling at any given time. The electronic controller also has a programed timer for switching on and off the units if required. This system is economical and versatile. On the second floor, where most patient rooms are located, there is a central heating with radiators in each room. In total there are 29 radiators on the second floor. Heat recovery units are used on each floor, to control the amount of fresh air intake. In the operating theatres, there are independent A/C units.

5.2 *Hot Water*

Clinic A has an average of 2000 litres daily water consumption. Hot water is provided by 2 boilers and 30 solar panels (2 × 1 m), at 58 °C. The maintenance schedule for the hot water is once a year.

Similarly, Clinic B also has an average daily water consumption of 2000 litres. Hot water is provided by two boilers and six vacuum panels (16 tube), at 60 °C. The maintenance schedule for the hot water is once a year. Most pipes are insulated; however, improving the insulation of the pipes is expected to result in increased energy savings.

Finally, Clinic C consumes on average 1000 litres on a daily basis. Hot water is provided by one boiler and six solar heating panels (2 × 1m), boiled at 60 °C. The maintenance schedule for the hot water is once a year and all pipes are well insulated.

5.3 *Lighting System*

In all the clinics, the lighting is comprised of LED lighting (a mixture of LED panels, spotlights and T5) installed throughout the buildings. In Clinic A, they operate in most areas for 6 hours per day and in Clinics B & C, between 6 and 8 hours. The hours of operation in the corridors is 8 and 10, respectively.

5.4 *Electrical Energy Monitoring*

The monitoring was performed during winter; thus, the increase in electricity energy consumption in all the case study buildings corresponds to possible increase in lighting, medical equipment and laundry, with the exception of the second clinic which partly uses electricity for its heating needs. The instrument used is the power quality analyser, of type Kyoritsu, previously calibrated model KEW 6315.

5.4.1 Clinic A Monitoring

The hospital uses electricity for cooling, lighting, medical equipment and laundry. During weekdays (16–18/01/2019 and 21–22/01/2019), the electricity consumption presents a step increase from 8:30 to 9:30 am and remains at elevated levels (with fluctuations) until 6:30 pm. After 6:30 pm, the consumption is reduced steadily until midnight and then remains stable throughout the morning hours up until 8:30 am. During Saturday (19/01/2019), there is an increase in electricity consumption from 8:30 am to 14:30 pm, whereas on Sunday (20/01/2019), the electricity consumption remains constant throughout the day (Fig. 1).

5.4.2 Clinic B Monitoring

The clinic uses electricity for all its needs, with the exception of hot water provision. The electricity needs are very high and there is no photovoltaic system. There are two electricity metres (Meter A and Meter B) in the building. The energy audit team installed electricity recording equipment on both counters for 1 week (Figs. 2 and 3). The first metre recorded week 06–11/12/2018 and the second metre recorded week 06–12/02/2019.

Electricity consumption during weekdays in both counters is higher than at weekends because doctors do not examine patients on Saturday and Sunday. On weekdays, the hours with the highest consumption are observed between 8 am and 3 pm. There are, of course, exceptions, because in a clinic anything may occur at any moment.

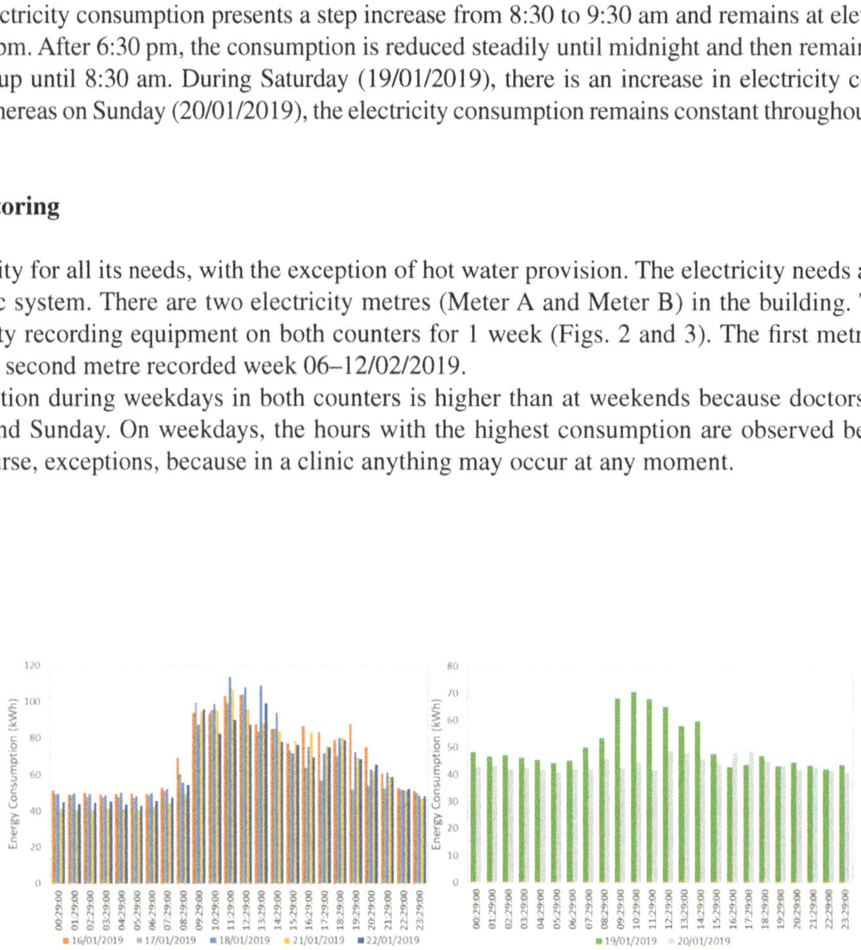

Fig. 1 Electricity consumption for Clinic A on weekdays (left) and weekend (right)

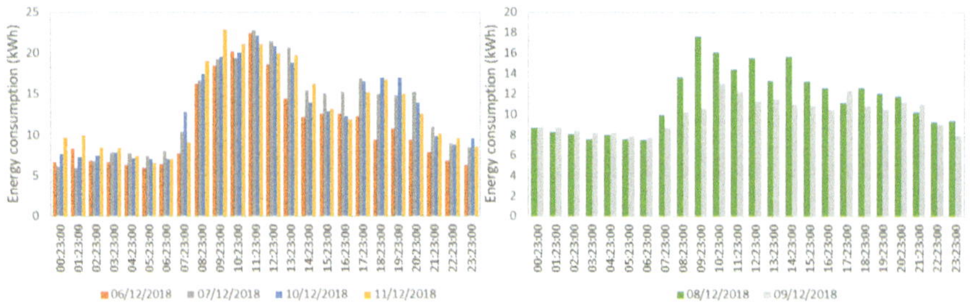

Fig. 2 Electricity consumption for Clinic B, for weekdays (left) and weekend (right) for electricity metre A

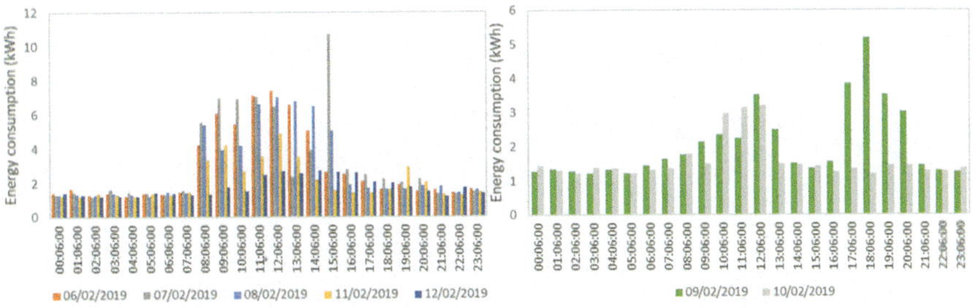

Fig. 3 Electricity consumption for Clinic B, for weekdays (left) and weekend (right) for electricity metre B

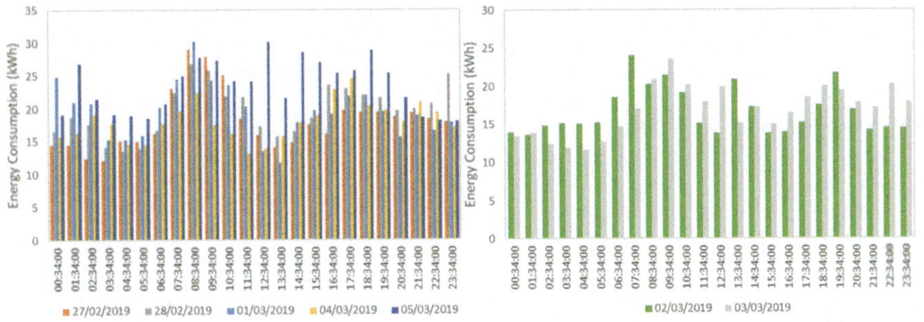

Fig. 4 Electricity consumption for Clinic C on weekdays 27/02/2019–01/03/2019 (left) and weekend 02–03/03/2019 (right)

5.4.3 Clinic C Monitoring

The clinic uses electricity for all its equipment (HVAC System (cooling), lighting, medical equipment, hot water, and other equipment). The audit team installed electricity recording equipment for 1 week, from 27/02/2019 to 05/03/2019 (Fig. 4).

From Fig. 4, it is clear that the electricity consumption during weekdays is higher than at weekends, similarly to clinic B. On weekdays, the hours with the highest consumption are observed between 8 am and 3 pm.

6 Energy Consumption of the Clinics

6.1 Clinic A

6.1.1 Electricity

Based on the monthly electricity bill data, the hospital's total annual electricity consumption is approximately 742,698 kWh (in 2017; 2016: ECTotal ≈ 779,110 kWh). The highest electricity consumption is during the summer months and the lowest during the winter months (Fig. 5). Specifically, the highest energy consumption is in July–August (2016: ECaverage ≈ 89,930 kWh; 2017: ECaverage ≈ 92,695 kWh; 2018: ECaverage ≈ 88,247 Wh) and the lowest is usually in February (2016: ECaverage ≈ 48,120 Wh; 2016: ECaverage ≈ 41,970 Wh).

6.1.2 Diesel Oil

The month of the highest diesel oil refill varies depending on the year. For example, the three months with the highest diesel oil refills:

- 2016: January (DO ≈ 5400 Lt), March (DO ≈ 5060 Lt) and December (DO ≈ 6850 Lt)
- 2017: February (DO ≈ 7350 Lt), December (DO ≈ 6740 Lt) and March (DO ≈ 6900 Lt)
- 2018: January (DO ≈ 7000 Lt), February (DO ≈ 4600 Lt), October (DO = 4390Lt)

Fig. 5 Monthly electricity consumption for years 2016–2018, for Clinic A

6.2 *Clinic B*

6.2.1 Electricity

Based on the electricity bill data issued every 2 months (i.e. electricity metre A), the hospital's total annual electricity consumption is approximately 134,289 kWh (in 2017; 2016: ECTotal ≈ 139,827 kWh). The highest electricity consumption is on average during the summer months of July–August and the lowest during the winter months of January–February (Fig. 6, left).

Based on the monthly electricity bill data (i.e. electricity meter B), the hospital's total annual electricity consumption is approximately 75,133 kWh (in 2017; 2016: ECTotal ≈ 55,350 kWh). The highest electricity consumption is during the summer months and the lowest during the winter months (Fig. 6, right). Specifically, the highest energy consumption is in July and August (2016: ECaverage ≈ 10,400 kWh; 2017: ECaverage ≈ 17,600 kWh; 2018: ECaverage ≈ 11,600 Wh) and the lowest is usually in January–February (2016: ECaverage ≈ 180 kWh; 2018: ECaverage ≈ 2240 kWh). The difference in electricity consumption between the different months within the same year could be attributed to the high need for cooling. Any difference in the energy consumption for the same season but different years could be due to the weather conditions. The very low electricity consumption in the winter of 2018 is due to failure of a large chiller, which has since been replaced. Until its replacement, however, only the small chiller in the operating room was working.

6.2.2 Diesel Oil

The month of the highest diesel oil refill varies depending on the year, due to different heating demand in each year. For example, the three months with the highest diesel oil refills are:

- 2016: January (DO ≈ 2700 Lt), February (DO ≈ 2100 Lt) and November (DO ≈ 1985 Lt)
- 2017: January (DO ≈ 2080 Lt), March (DO ≈ 2490 Lt) and December (DO ≈ 2810 Lt)
- 2018: March (DO ≈ 1910 Lt), February (DO ≈ 1840 Lt), October (DO ≈ 1525 Lt)

6.3 *Clinic C*

6.3.1 Electricity

Based on the monthly electricity bill data, the hospital's total annual electricity consumption is higher in 2016 (in 2018: ECTTotal=195,993kWh, 2017: ECTotal ≈ 202,832kWh, 2016: ECTotal ≈ 204,045 kWh). The highest electricity consumption is during the summer months and the lowest during the winter months (Fig. 7). Specifically, the highest energy

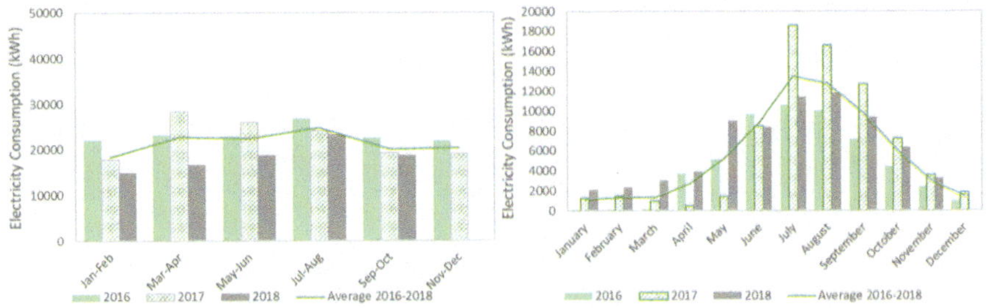

Fig. 6 Bimonthly electricity consumption for electricity metre A (left), and monthly electricity consumption for electricity metre B (right), in Clinic B

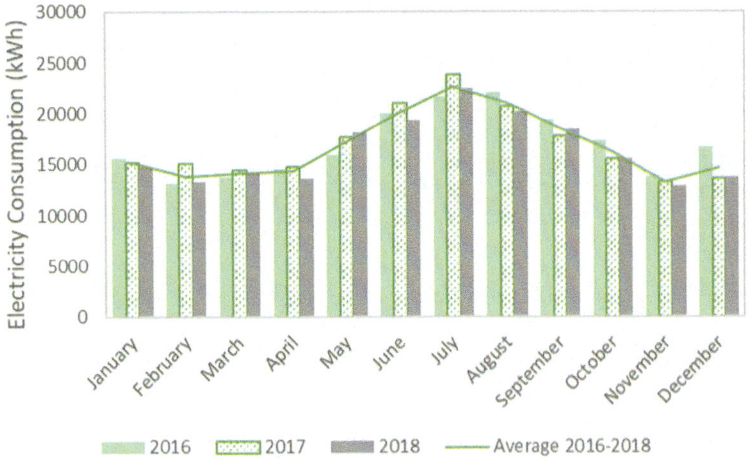

Fig. 7 Monthly electricity consumption for years 2016–2018, in Clinic C

consumption is in July–August (2016: ECaverage ≈ 21,872 kWh; 2017:ECaverage ≈ 22,259 kWh; 2018: ECaverage ≈ 21,263 kWh) and the lowest is usually in February.

6.3.2 Diesel Oil

The month of the highest diesel oil refill varies depending on the year. For example, the three months with the highest diesel oil refills:

- 2016: March (DO ≈ 1392 Lt), September (DO ≈ 1422 Lt) and December (DO ≈ 1023 Lt)
- 2017: February (DO ≈ 1400 Lt), May (DO ≈ 1210 Lt) and September (DO ≈ 1291 Lt)
- 2018: January (DO ≈ 1178 Lt), June (DO ≈ 1085 Lt), August (DO ≈ 786 Lt)

7 Energy Performance Comparison of the Three Clinics

The energy monitoring revealed that during weekends, the consumption is both lower and more stable throughout the day in all clinics, compared with weekdays. The consumption is especially stable during Sundays, since there are no outpatients. The highest energy demand is detected from around 8:30 am to 2:30 pm, since most of the surgeries are performed during the morning and at the same time consulting rooms are in operation.

The clinic with the lowest energy consumption per unit area – based on the electricity and diesel oil bills of 2017 – is clinic C, with a consumption of 303 kWh/m^2. Clinic B has similar energy consumption profile (336 kWh/m^2), whereas Clinic A has the highest energy consumption of the three, at 401 kWh/m^2. The comparatively lower energy consumption of Clinics B and C is attributed to their mechanical systems that are newer and the properties of their envelope elements. In particular, although all clinics have insulated roof and double glazing, Clinic B uses thermally insulated bricks for its external walls and Clinic C has external shading in all the orientations (Fig. 8).

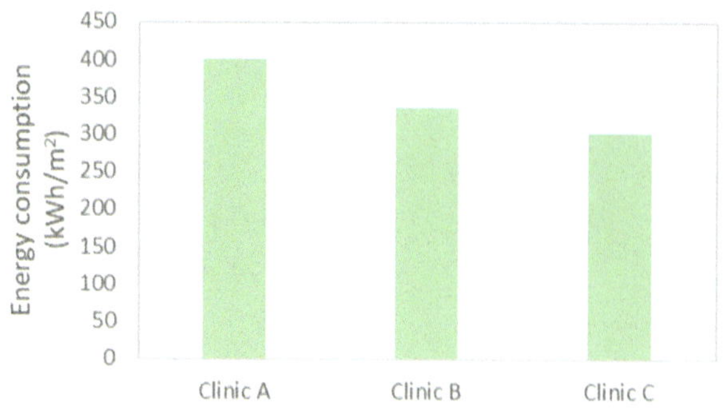

Fig. 8 Energy consumption per area, for the three case study clinics

Table 7 Proposed improvements per clinic

Improvement/upgrading	*Clinic A*	*Clinic B*	*Clinic C*
Motion sensors	√	√	√
BMS system	√	√	√
Wall insulation	√	√	√
External shading	√	√	
Mechanical systems	√		
Onsite RES	√	√	

8 Potential Improvements

Suggested improvements, based on the audits' outcome, can be found in Table 7.

Overall, in the consulting rooms of all the clinics, motion sensors could be installed to ensure that if the rooms are not occupied lights are automatically switched off. Also, installing a BMS system in the buildings could result in significant energy savings for all the clinics. Improvement of the building envelope thermal properties is suggested as well for all clinics, with the addition of insulation on the external walls. More specific improvements, such as upgrading the mechanical systems and increasing the onsite RES, are recommended to some of the audited clinics.

9 Conclusions

Deep energy audits are necessary for developing a comprehensive profile of the energy performance of hospital buildings in Cyprus, where information is scarce. Through performed audits in three private clinics in Nicosia, various deficiencies were identified such as in the thermal insulation of the envelope and the window shading. Also, the overall lack of energy monitoring and smart systems was observed and differences in the onsite RES incorporation were noted. It can be deducted that there is ample potential for improvements almost in every aspect of the energy performance of the clinics.

Further energy monitoring, during the summer months, is deemed necessary so that a holistic depiction of the energy performance of the clinics will be produced. Additionally, energy simulations, an essential tool for modelling the thermal performance of the building, should be used in unison with onsite monitoring and measurements.

Deep energy audits in hospitals should be extended to more healthcare institutions in Cyprus, both public and private, in order to create a complete energy profile of the sector and be able to propose suitable refurbishment solutions.

The study indicates the significance of auditing for identifying the energy deficiencies of buildings and consequently for determining the appropriate interventions for achieving energy efficiency and in effect improving the energy performance of the buildings.

Bibliography

1. Economidou M, Todeschi V, Bertoldi P, D'agostino D, Zangheri P, Castellazzi L (2020) Review of 50 years of EU energy efficiency policies for buildings. https://doi.org/10.1016/j.enbuild.2020.110322
2. Prada M, Prada IF, Cristea M, Popescu DE, Bungău C, Aleya L, Bungău CC (2020) New solutions to reduce greenhouse gas emissions through energy efficiency of buildings of special importance-hospitals. https://doi.org/10.1016/j.scitotenv.2020.137446
3. Serghides D, Dimitriou S, Kyprianou I (2020) Paving the way towards zero energy hospitals in the Mediterranean region:159–167. https://doi.org/10.1007/978-3-030-30841-4_11
4. Economidou M, Zangheri P, Paci D (2017) Final Report Long-term strategy for mobilizing investments for renovating Cyprus national building stock (D1.8). https://ec.europa.eu/jrc. Accessed 9 July 2020
5. García-Sanz-Calcedo J (2019) Study of CO_2 emissions from energy consumption in Spanish hospitals. 26. https://doi.org/10.21595/vp.2019.20965
6. Hyseni D, Tashevski DJ, Filkoski RV, Shesho IK (2020) Energy efficiency in complex buildings. IOP Conf Ser Earth Environ Sci 410:012024. https://doi.org/10.1088/1755-1315/410/1/012024
7. Maassen W (2017) (nearly) Zero Energy Hospital Buildings
8. MECIT (2015) Εναλλακτική προσέγγιση για την βελτίωση της ενεργειακής απόδοσης των κτιρίων που ανήκουν και χρησιμοποιούνται από κεντρικές κυβερνητικές αρχές σύμφωνα με την παράγραφο 6 του άρθρου 5 της Οδηγίας 2012/27/ΕΕ για την ενεργειακή απόδοση (Υποδειγματικός ρόλος.
9. CEN, Energy audits - Part 1: General requirements, EN 16247-1:2012. https://standards.cen.eu/dyn/www/f?p=204:110:0::::FSP_PROJECT:35014&cs=149DEEE262998D2F15F2B882FFD9DDEE2. Accessed 9 Sept 2020
10. CEN, Energy audits - Part 2: Buildings, EN 16247-2:2014. https://standards.cen.eu/dyn/www/f?p=204:110:0::::FSP_PROJECT:35015&cs=161FB10C3795A65B1B289B17D0A75C3FE. Accessed 9 Sept 2020

Study of the Influence of Solar Shading Devices in the Solar Inputs in Buildings

Andrés Vilaboa Díaz and Pastora M. Bello Bugallo

1 Introduction

The European Union is committed to establishing a decarbonised, sustainable and competitive energy system by 2050. One of the main objectives is the decarbonisation of the current real housing stock, responsible for 36% of all EU CO_2 emissions. The promotion of passive-type solutions (urban vegetation, green roofs, thermal insulation, solar shadow devices, ...) contributes to reducing air conditioning and cooling needs in a sustainable way with the environment [1].

In Spain, most of the energy consumption in buildings is destined for heating. According to Eurostats, in 2017 the final energy consumption for heating accounted for 43% of the total compared to 1% for cooling. In addition, 63% of the energy used for heating has been obtained from non-renewables energies [2] (Fig. 1).

The use of solar radiation can be an ally when it comes to reducing CO_2 emissions from the EU, reducing our heating demand in the cold months. However, poor control of solar radiation can become unwanted contributions in the summer months, increasing energy demand in air conditioning. Therefore, the solar shading devices become indispensable for the control of the thermal comfort of the building. In addition to the solar shading devices, the optimal design of the type of window is one of the important points to consider in the initial design of the building, and several parameters for the selection of the optimal glazing must be taken into account.

The orientation of the facade will be decisive for the optimal selection of the size of the glazed surface. Thus, the optimum proportion of glazing for south facades for moderate climates is between 38% and 42%, reducing this proportion with east and west orientations [3]. Comparing a glazed surface of 30, 60 and 100%, for the surface of 60%, there is an energy increase of 23%, and for an area of 100%, it produces an increase of up to 47%. In addition, this increase in the glazed area does not necessarily reduce the energy consumption in lighting that would represent 10% of the total energy consumed [4].

The thermal transmittance (U-value) and the solar factor of the window (g-value) are important parameters to take into account when selecting the optimal window. A window with low U-value and a high g-value allows reducing the energy demand in heating; however, this type of configuration would increase the energy demand in air conditioning. According to studies obtained through mathematical models, depending on g-value, a reduction in U-value from 1.2 to 0.8 W/(m^2 K) can reduce the demand for air conditioning by 5–15% [5]. The standard triple and quadruple glazing window models result in lower heating and cooling needs, while in the case of quintuple windows, the results were opposite [6].

The importance of selecting a good glazing will have an impact on the building's air conditioning costs and therefore, it is necessary to find a balance between the U-value and the g-value to reduce the energy demand in the buildings. In climates where solar contributions reduce heating needs but increase cooling demand, the introduction of solar control systems helps to achieve more sustainable buildings.

When establishing the type of solar shadow device, not only the influence on the comfort conditions must be evaluated, but also the light quality will be altered. Nielsen et al. compared the annual energy demand in air conditioning and lighting for offices in Denmark for three cases: without solar shading, with fixed solar shading and with dynamic solar shading. The solar shading device used were the venetian blinds, and the simulation showed that with the dynamic solar shading, the total energy demand is reduced with respect to the fixed sunscreens, obtaining a reduction of up to 16% [7]. In addition, the use of venetian blinds as solar control systems based on the angle of solar incidence allows to avoid the entry of direct solar radiation but improving lighting [8].

A. Vilaboa Díaz (✉) · P. M. Bello Bugallo
Department of Chemical Engineering, School of Engineering, Universidade de Santiago de Compostela, Santiago de Compostela, Spain

© The Author(s), under exclusive license to Springer Nature Switzerland AG 2022

A. Sayigh (ed.), *Sustainable Energy Development and Innovation*, Innovative Renewable Energy,
https://doi.org/10.1007/978-3-030-76221-6_51

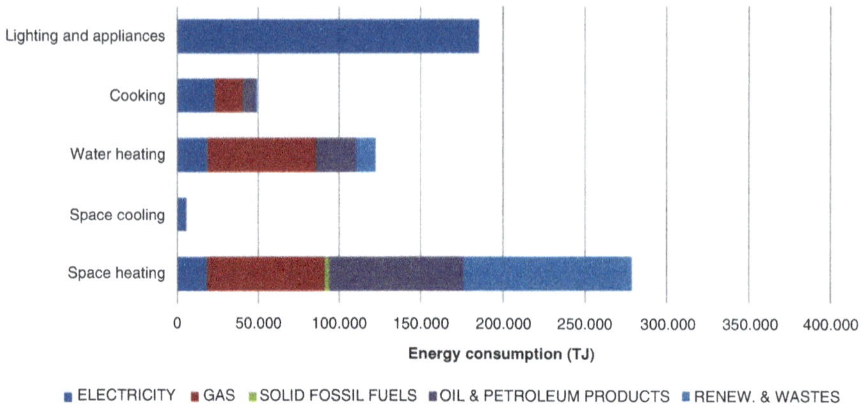

Fig. 1 Final energy consumption in households in Spain in 2017 [2]

2 Objective

The objective of this work is to evaluate the influence of solar device for the control of the energy obtained from the sun in two cities of Spain. The influence of the overhangs for solar control and the potential of dynamic solar shading devices will be analysed. In this way, the importance of the optimal dimensioning of the overhangs and awnings and the selection of the *g*-value of the window will be analysed.

3 Case Study

The simulations will be carried out in different climatic zones of Spain. The north and south of Spain will be taken as a reference. The location taken as a reference will be Ourense (42° latitude), which has a mild climate. The area of southern Spain taken as a reference will be Murcia (38° latitude) with a dry climate [9]. The following table shows the main characteristics of the selected locations (Table 1):

The study will take as reference the days of maximum and minimum solar declination. The maximum solar declination (23.45°) occurs in midsummer (June 20–21), reaching the Tropic of Cancer, while the minimum decline (−23.45°) occurs on the winter solstice (December 21–22), moment in which the sunlight reaches the Tropic of Capricorn.

The glazed surface taken as a reference will consist of a height of 1.5 m. The particular conditions analysed at each point will be described individually.

4 Methodology

The study will be carried out in the following stages:

- The efficiency of the windows properties be studied, analysing the *g*-value and *U*-value of windows.
- The influence of the overhang will be analysed according to the time of year including different windows.
- The potential of dynamic solar control devices will be also analysed, showing the results that could be achieved with an awning.

5 Development and Presentation of Results

5.1 Study of the Influence of the Windows Properties in Solar Contributions

In this first part of the study, the influence of the *g*-value of the window on the two climatic zones of Spain previously defined (Ourense and Murcia) will be analysed. For this, three windows with different *g*-value will be taken as reference. This modification of the *g*-value will allow greater or lesser passage of radiation inwards.

Table 1 Summary of the case studies analysed

	Ourense	Murcia
Latitude	42° 19′	38° 0′
Average annual temperature [°C]	14.9	18.6
Lowest monthly average temperature [°C]	3.4 (January)	4.7 (January)
Highest monthly average temperature [°C]	30.6 (August)	34.2 (August)
Annual hours of sunshine [hours]	2054	2967

Table 2 Type of glasses used for simulations

Case	Type of glass	*U*-value (W/m²•K)	*g*-value
1.	Single glass	5.7	0.83
2.	Double glass – 4(12)4	1.7	0.50
3.	Double glass with solar control – 6(12)6	1.6	0.10

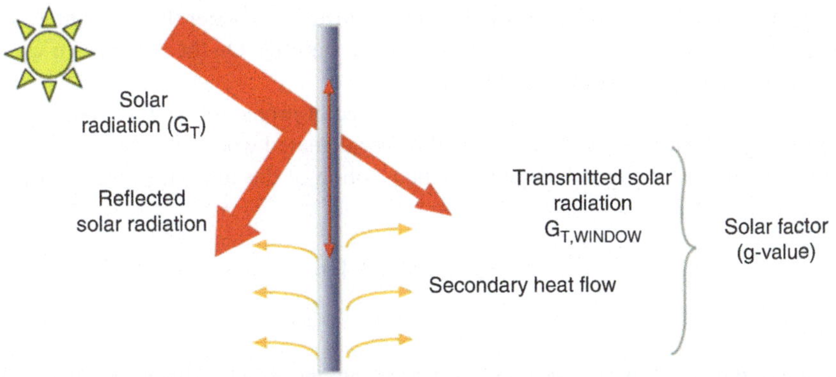

Fig. 2 Global transmission of energy through glass. (Adapted from IDAE [10])

The glasses used for this study are summarised in Table 2.

The g-value is the overall energy transmission rate or ratio between the fraction of solar radiation transmitted by the translucent surfaces ($G_{T,WINDOW}$) and the external solar radiation (G_T) (Fig. 2).

Therefore, if the *g*-value is defined as the percentage of solar radiation that enters the interior:

$$G_{T,\text{window}} = G_T \cdot g_{\text{value}}$$

The incident total solar radiation (G_T) depends on the direct radiation ($G_{b,t}$), the diffuse radiation ($G_{d,t}$) and the shortwave radiation that is reflected in the Earth's surface or by nearby surfaces ($G_{r,t}$) [11].

$$G_T = G_b \cdot \cos\theta + G_{d,t} + G_{r,t}$$

This equation can be shown as a function of the geometry of solar radiation:

$$G_T = A \cdot e^{-\frac{B}{\sin\alpha_S}} \cdot \cos\theta + C \cdot G_d \cdot \left(\frac{1+\cos\beta}{2}\right) + G \cdot \rho \cdot \left(\frac{1+\cos\beta}{2}\right)$$

where:

- G_T is the incident total solar radiation [W/m²].
- G_d is the diffuse radiation on a horizontal surface [W/m²].
- G is the global radiation on a horizontal surface [W/m²].
- A is the apparent extraterrestrial irradiance [W/m²].
- B is the atmospheric extinction coefficient.

- C is the proportion of diffuse radiation on the horizontal surface.
- α_s is the solar height [°].
- θ is the angle of the sun rays on the surface to study [°].
- β is the angle between the surface plane and the horizontal [°].
- ρ is the terrestrial albedo.

Combining these expressions, the total energy transmitted inwards is obtained. The results obtained for the different windows properties are shown in Table 3.

The g-value of the window reduces proportionally the entrance of solar radiation inwards. G-value of 0.1 only allows 10% of solar radiation both in summer and winter. Therefore, it may be necessary to prioritize the entry of solar radiation in winter and avoid it in summer. For this, it seems essential to introduce solar control devices that prevent the entry of solar radiation in summer and allow it during winter.

5.2 *Study of the Influence of Fixed Solar Shading Device in Solar Contributions*

In summer, the angle of incidence of the sun's rays on the earth's surface is greater than in winter; this causes a greater contribution of solar energy and causing an increase in cooling needs in summer. On the contrary, in winter, incident solar radiation is a great ally to minimise energy consumption in winter.

To study the optimal design of solar shading device, the influence of protection on the most unfavourable day of the year will be studied. The projection of the sun's rays on the window is calculated when it has a fixed solar shadow device (Fig. 3).

The shadow generated by the overhang will depend on the position of the sun. Thus, the solar projection on the window will be:

$$h_{\text{shadow}} = tg(\alpha_s) \cdot L$$

where:

- h_{shadow} is the projection of the sun's rays on the window, taking into account the distance between the window (h_{window}) and the overhang (h_1).
- L: is the length of the overhang.
- Finally, if the orientation of the facade (γ) is taken into account, the solar projection on the window will be described as follows.

The two analysed climatic zones (Ourense and Murcia) will be taken as a reference for the day when the sun reaches the highest and lowest position. In addition, different lengths of the overhang will be analysed. The following cases will be assumed:

- CASE A: Overhang with a length of 0.5 metres.
- CASE B: Overhang with a length of 1 metre.
- CASE C: Overhang with a length of 1.5 metres.
- CASE D: Overhang with a length of 2 metres.

The influence of the entrance of solar radiation towards the interior has been analysed by modifying the g-value of the window and the length of the overhang. The results of combining both sun protection systems will be shown below. To do this, the three different proposed windows (CASE 1, 2 and 3) will be evaluated with the four overhang options (CASE A, B,

Table 3 Summary of the solar radiation transmitted depending on the type of glass

	Ourense		Murcia	
Reference day (dd/mm)	20/06	21/12	20/06	21/12
Solar radiation on horizontal surface [kWh/m²·d]	9.60	2.86	9.61	3.56
Solar radiation on perpendicular surface south orientation [kWh/m²]	3.07	5.38	2.64	5.81
CASE 1: Single glass [kWh/m²·d]	2.55	4.46	2.19	4.83
CASE 2: Double glass – 4(12)4 [kWh/m²·d]	1.54	2.69	1.32	2.91
CASE 3: Double glass with solar control – 6(12)6 [kWh/m²·d]	0.31	0.54	0.26	0.58

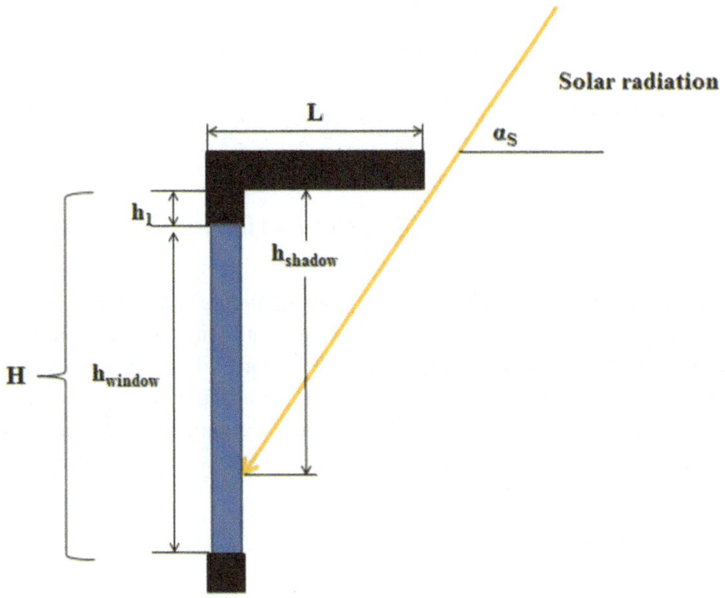

Fig. 3 Solar projection on a window with an overhang

Table 4 Summary of the solar radiation transmitted depending on the type of glass and length of overhang

	Ourense		Murcia	
Reference day (dd/mm)	20/06	21/12	20/06	21/12
Solar radiation on horizontal surface [kWh/m^2•d]	9.60	2.86	9.61	3.56
Solar radiation on perpendicular surface south orientation [kWh/m^2•d]	3.07	5.38	2.64	5.81
CASE 1-A: Single glass + Overhang of 0.5 m [kWh/m^2•d]	1.82	4.11	1.44	4.24
CASE 1-B: Single glass + Overhang of 1 m [kWh/m^2•d]	1.45	3.65	1.21	3.65
CASE 1-C: Single glass + Overhang of 1.5 m [kWh/m^2•d]	1.34	3.19	1.12	3.07
CASE 1-D: Single glass + Overhang of 2 m [kWh/m^2•d]	1.28	2.73	1.06	2.48
CASE 2-A: Double glass – 4(12)4 + Overhang of 0.5 m [kWh/m^2•d]	1.00	2.41	0.78	2.38
CASE 2-B: Double glass – 4(12)4 + Overhang of 1 m [kWh/m^2•d]	0.77	2.14	0.68	1.85
CASE 2-C: Double glass – 4(12)4 + Overhang of 1.5 m [kWh/m^2•d]	0.70	1.86	0.62	1.32
CASE 2-D: Double glass – 4(12)4 + Overhang of 2 m [kWh/m^2•d]	0.67	1.58	0.58	0.84
CASE 3-A: Double glass with solar control – 6(12)6 + Overhang of 0.5 m [kWh/m^2•d]	0.20	0.48	0.17	0.51
CASE 3-B: Double glass with solar control – 6(12)6 + Overhang of 1 m [kWh/m^2•d]	0.15	0.43	0.15	0.44
CASE 3-C: Double glass with solar control – 6(12)6 + Overhang of 1.5 m [kWh/m^2•d]	0.14	0.37	0.14	0.37
CASE 3-D: Double glass with solar control – 6(12)6 + Overhang of 2 m [kWh/m^2•d]	0.13	0.32	0.13	0.30

C and D). Therefore, a total of 12 cases will be analysed for the two climatic locations (Ourense and Murcia) and at two different times of the year (06/20 and 12/21), as tabulated in Table 4.

By combining the *g*-value of the glass and the length of the overhang, a reduction of the solar radiation input greater than 95% is achieved when a 1 metre length overhang is added (CASE 3-B, 3-C and 3 -D).

With respect to CASE-1, where a *g*-value of 0.87 is used, the overhangs reduce the solar radiation between 40.6 and 58.3% solar radiation in Ourense and between 45.4 and 59.8% in Murcia. The introduction of overhangs greatly reduces the entry of solar radiation in Murcia; this reduction is greater because the solar radiation has a greater inclination angle.

Finally, with respect to case B, the entrance of solar radiation is avoided by 50% without using overhangs. Comparing the two climatic zones, in Ourense a reduction between 67.6% and 78.2% is reached and in Murcia between 70.5% and 78.8%.

During winter, solar radiation must be used to reduce the energy used for heating. Therefore, analysing only this point of view, the option that allows a greater entry of solar radiation would be the cases in which the *g*-value has a value of 0.87 (CASE-1). In these cases, the solar radiation input is reduced to a greater extent in Murcia by increasing the length of the overhang. Thus, in Ourense for CASE-1, adding an overhang leads to a reduction between 23.6% and 49.1% and in Murcia between 27.1% and 57.3%.

The control of solar radiation by means of fixed solar shading device is complex since it does not allow taking advantage of solar energy in winter and avoiding it in summer. Therefore, the introduction of systems that allow regulating solar contributions takes on a fundamental role in the construction of sustainable housing, so dynamic solar control systems will have a very important weight in new buildings.

5.3 *Study of the Influence of Dynamic Solar Shading Device in Solar Contributions*

Unlike fixed solar shading device, dynamic solar shading devices allow regulating solar energy contributions in the building; this allows taking advantage of solar contributions in winter and minimizing solar contributions in summer. Therefore, these types of technologies maximise energy efficiency in buildings. This versatility allowed by dynamic solar shading device is the main advantage over fixed solar shading device. In this part, the efficiency of one of the dynamic solar shadow devices is analysed. In this case, the awnings will be selected, because they are devices that can be automated and can produce a complete sun protection. In this case, the angle of inclination (β_a) formed between the awning and the surface of the window will be analysed (Fig. 4).

The shade produced by the awning will be formed by two variables:

- L_y: the projection of the awning itself on the surface of the glass.
- $h_{s,y}$: the projection of solar radiation on the surface of the glass

$$h_{\text{shadow}} = L_y + h_{s,y}$$

The influence of the awning angle at the inlet of solar radiation inward will be analysed. Continuing with the hypothesis assumed in the previous point, a window with a height (H) of 1.5 metres will be analysed taking as reference the day 06/20, day in which the sun reaches its maximum solar height. In Table 5, the inward energy input is shown starting from the length described in Cases A, B, C and D.

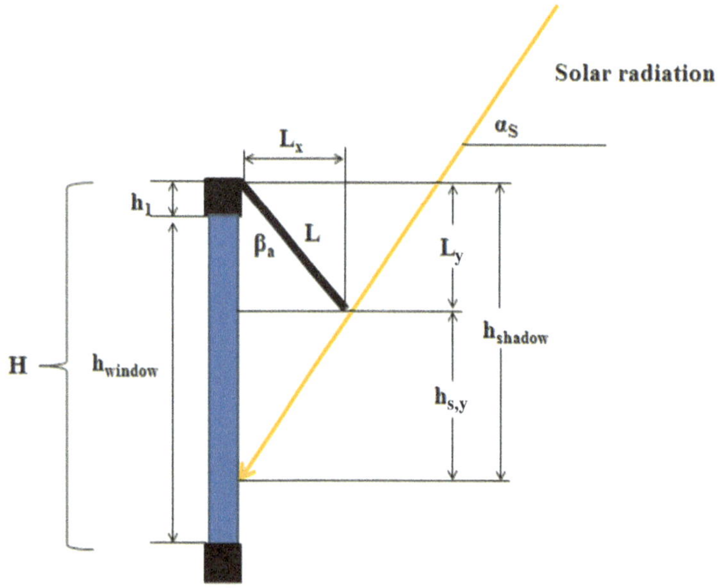

Fig. 4 Solar projection on a window with a dynamic awning

Table 5 Solar energy input for a 1.5 m window (g-value=1) depending on the awning angle

	Ourense				Murcia			
Awning angle [°]	90°	70°	50°	30°	90°	70°	50°	30°
CASE A: Awning of 0.5 m [kWh/m²•d]	1.99	1.83	1.80	1.93	1.74	1.62	1.58	1.67
CASE B: Awning of 1 m [kWh/m²•d]	1.53	1.36	1.25	1.20	1.46	1.30	1.20	1.15
CASE C: Awning of 1.5 m [kWh/m²•d]	1.41	1.21	1.07	1.01	1.35	1.16	1.02	0.97
CASE D: Awning of 2 m [kWh/m²•d]	1.34	1.10	1.00	1.00	1.28	1.05	0.96	0.96

As can be seen, varying the angle of the awning allows greater control of solar radiation, allowing direct solar radiation to be prevented. Therefore, being able to adjust the angle of the awning would reduce the entry of solar radiation in summer, and being a mobile solar control system could allow the entry of radiation in winter. However, it can be seen that the length is also of particular importance; in the case of using a 0.5 m awning, it would not be enough to prevent the entry of solar radiation even if the angle between the awning and the window is reduced.

Therefore, mobile solar control systems must be automatic systems, which allow varying the length and angle depending on the solar position. The analysed system can be installed after the construction, so they are technologies to take into account in modifications in the current buildings.

6 Conclusions

Solar shading devices play an important role in the energy consumption of buildings. Currently, a wide variety of technologies for solar control are available on the market, from the simplest to developing technologies that not only prevent solar radiation, but also allow the energy from the sun to be reused. It is the job of engineers and architects to select the most appropriate technology from the initial phase of building design, in order to build energy-efficient buildings.

The selection of the *g*-value and the length of the overhang in the design stages of the building are important. Thus, longer lengths of the overhang allow minimizing solar radiation in summer and allowing the contribution of winter energy. On the other hand, low values of the *g*-value of the window will prevent the entry of solar radiation that benefits the reduction of cooling in summer, but increases the cooling needs in summer. Therefore, it is essential to find a balance between both parameters.

Dynamic solar control device has the versatility that the fixed system does not have. In the case analysed, the awnings can be an efficient and low-cost system to reduce the entry of solar radiation in summer.

In the case analysed in Ourense, an awning with a length of 1.5 metres, placed perpendicular to the facade, will allow the passage of 1.41 kWh/m^2•day; however, with a mobile solar control system, the solar contribution would be 1.01 kWh/m^2•day, that is, improves by 29% to fixed systems.

Finally, the potential of fixed solar shading device is highlighted, capable of preventing the entry of direct solar radiation but allowing the passage of diffuse solar radiation, so that they will maintain the lighting conditions inside the buildings.

References

1. EC, Directive 2018/844 of the European Parliament and of the Council of 30 May 2018 amending Directive 2010/31/EU on the energy performance of buildings and Directive 2012/27/EU on energy efficiency
2. Eurostats Database. https://ec.europa.eu/eurostat/data/database. Accessed Dec 2019
3. Obrecht TP, Premrov M, Leskovar VZ (2019) Influence of the orientation on the optimal glazing size for passive houses in different European climates (for non-cardinal directions). Sol Energy 189:15–25
4. Poirazis H, Blomsterberg A, Wall M (2008) Energy simulations for glazed office buildings in Sweden. Energ Buildings 40(7):1161–1170
5. Grynning S, Gustavsen A, Time B, Jelle BP (2013) Windows in the buildings of tomorrow: energy losers or energy gainers? Energ Buildings 61:185–192
6. Thalfeldt M, Kurnitski J, Voll H (2015) Comparison of simplified and detailed window models in office building energy simulations. Energy Procedia 78:2076–2081
7. Nielsen MV, Svendsen S, Bjerregaard JL (2011) Quantifying the potential of automated dynamic solar shading in office buildings through integrated simulations of energy and daylight. Sol Energy 85(5):757–768
8. Gomes MG, Santos AJ, Rodrigues AM (2014) Solar and visible optical properties of glazing systems with venetian blinds: numerical, experimental and blind control study. Build Environ 71:47–59
9. Agencia Estatal de Meteorología. http://www.aemet.es/es/portada. Accesed Dec 2019
10. IDAE, Instituto para la Diversificación y Ahorro de la Energía (2008) Soluciones de Acristalamiento y Cerramiento Acristalado. Madrid
11. American Society of Heating, Refrigerating and Air-Conditioning Engineers, Inc (2011) ASHRAE handbook: heating, ventilating, and air-conditioning applications, Atlanta

The Effect of Colored Transparent Filters in Transmission of Solar Radiation of Building Facades

Bahar Sultan Qurraie and Maryam Singery

1 Introduction

Transparent facades of buildings are the main transmitters of radiant energy into the building, which can be minimized with great tricks, especially in areas with hot climates. These include a variety of glasses and transparent polymers or advanced types of shading. In terms of solar control, the properties of the glass vary depending on the value of radiation conductivity. Even most glasses do not pass most of the ultraviolet and infrared waves. The radiation conductivity value can be changed by applying the coating layers obtained from some precious metals and / or metal oxides onto the glass. Such coatings directly affect the rate and intensity of the radiation conductivity [1]. The US Department of Energy (DOE) suggests that having spectrally selective capabilities of glasses causes a reduction in the solar load of the building [2] (Fig. 1).

2 Stained Glass: Orosi Glass

Color in stained glass is in conjunction with psychology and emotions. The spiritual impression of everyone is being revealed by crossing the threshold of historical place filled by color lighting. Inspissated reds and majestic yellows are colors that were commonly putted into stained glass because of the sensations they pointed [4]. The glass produced from natural materials (e.g., sand, limestone, and plant ashes) so the color changing appeared by adding the heavy metals. The staining was attained during the fusion process by the addition of variable amounts of different metal oxides (e.g., Co, Mn, Cu, Fe, etc.) [5].

Historical building technologies have taught us a lot about how to design. Orosi glasses are one of these stained glass technologies. Orosi glasses are aesthetically, climatically, traditionally, and symbolically used and integrated with daylight. It is a technique used to reduce sunlight in Iranian traditional architecture. Through years of use, it has been able to control the solar energy in depth and is preferred over the different types of structures found in the extreme hot climates of Iran. Orosi are used in wooden cages in pure four colors of blue, green, yellow, and red, and they are mounted together in natural ways without using metal as can be seen in Fig. 2.

According to their color, they showed variability in reducing the damage rates. At the same time, Orosi glasses attract attention with their beauty and they are popular by the users by affecting the human soul. The preferred absorber spectrum in the range of visible wavelength causes the formation of color in the glass. According to the results of the research in the case of Orosies by Haghshenas et al. [7] in UV-VIS wavelength range, as shown in Fig. 3, the transmission specimen does not match the human susceptibility curve in glass specimens. Blue and Red glasses are the colors with the highest human susceptibility. But yellow glasses provide the lightest; they pass the yellow-orange spectrum and the maximum spectrum is between 62 and 75 nm which does not include the visual sensitivity of people. The highest transmission spectrum for green glasses is also suitable for people's visual senses and is considered as the most suitable color according to the optimum glass light transmission [3].

B. S. Qurraie · M. Singery (✉)
Department of Architecture, Karabuk University, Karabuk, Turkey

Department of Architecture, Tabriz Branch, Islamic Azad University, Tabriz, Iran

© The Author(s), under exclusive license to Springer Nature Switzerland AG 2022

A. Sayigh (ed.), *Sustainable Energy Development and Innovation*, Innovative Renewable Energy,
https://doi.org/10.1007/978-3-030-76221-6_52

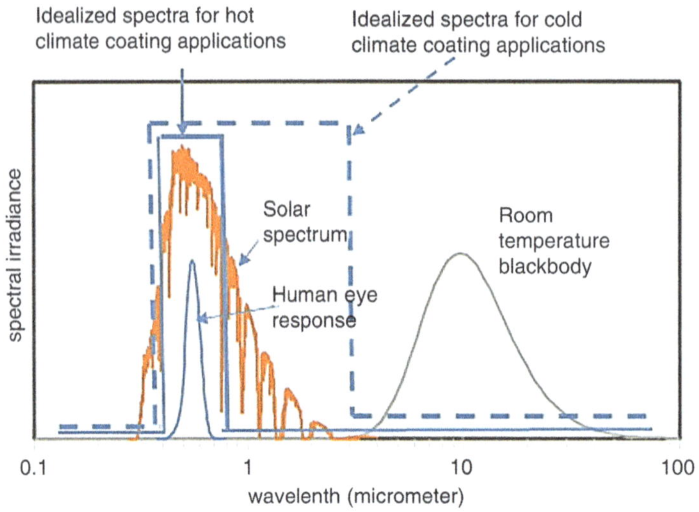

Fig. 1 Transmission spectra by the US Department of Energy for hot and cold climates [3]

Fig. 2 Chinese knotting in Orosi windows [6]

3 Thermal Radiation

Different glass types have different properties for the amount of solar radiation reflected, transmitted, absorbed, and re-radiated [8]. In this study, the amount of energy passing through the transparent colored specimens (colored transparent filters [CTFs]) is examined. To do this, and to harmonize the laboratory environment with real windows, color filters were placed between the two ordinary glasses and the third glass with spacing inside. The graph of the spectra transmitted percentage from both devices is put together and completed for each sample. Then, according to the corresponding formulas, the amount of energy passing through each sample color can be calculated.

In laboratory tests:

- The type of glasses is "Soda-lime Optical Glass" with 2 mm thickness.
- The colored transparent filters (CTFs) are in six colors of "green, yellow, blue, red, orange, and purple," and the pieces are in type of "Fomito Flash Speedlight Color Gel Kit Filter 30HP W/gel band and Reflector for Canon, Nikon, Olympus, Pentax, Yongnuo, Neewer and Godox."

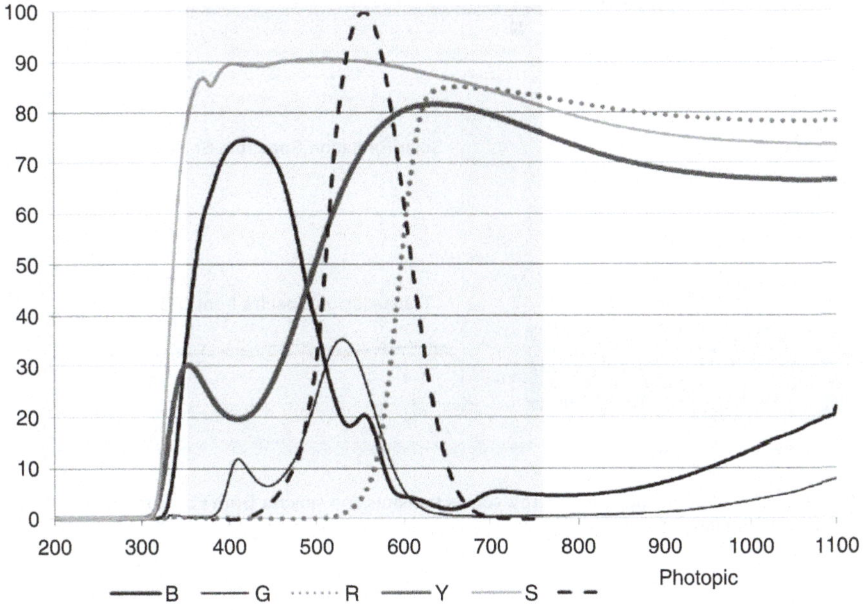

Fig. 3 Test results of the transmission spectra were obtained from the colored glass [7]

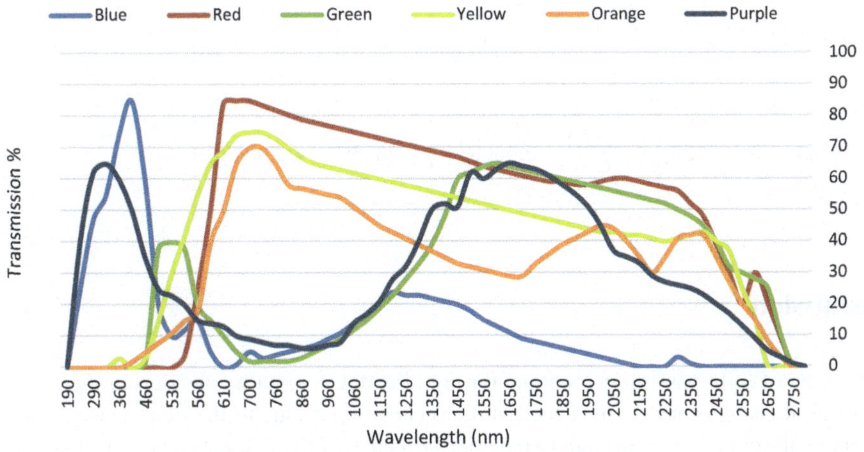

Fig. 4 Transmission spectra by CTF's color in the wavelength range between 190 nm and 3000 nm

- Transmission percent of spectrum through the samples between 190 and 1100 nm had been achieved by BECKMEN DU 530 Spectrophotometer and near Infrared wavelengths transmittance calculated by GL-6021 Galaxy Series FT-IR Spectrometer, with Infrared MIO465-0053-03 in 1100–3000 nm wavelengths.

The transition percentage from the two devices was combined in the range of UV-VIS-IR wavelengths and the calculations were made via MATLAB. These two devices cover wavelengths of 190–3000 nm. For longer ultraviolet wavelengths, different spectrophotometers should be used, which is not necessary with the reduction of solar energy. All colors have different transmittance amount at different infrared wavelengths (Fig. 4).

Transmittance irradiation by colors can be seen in Fig. 5 which shows the collection of each color's transmittance and solar radiation in the wavelength range of 190–3000 nm. According to multiplying these two graphs, the transmittance irradiation of CTFs at right angle 90° has been calculated as Table 1.

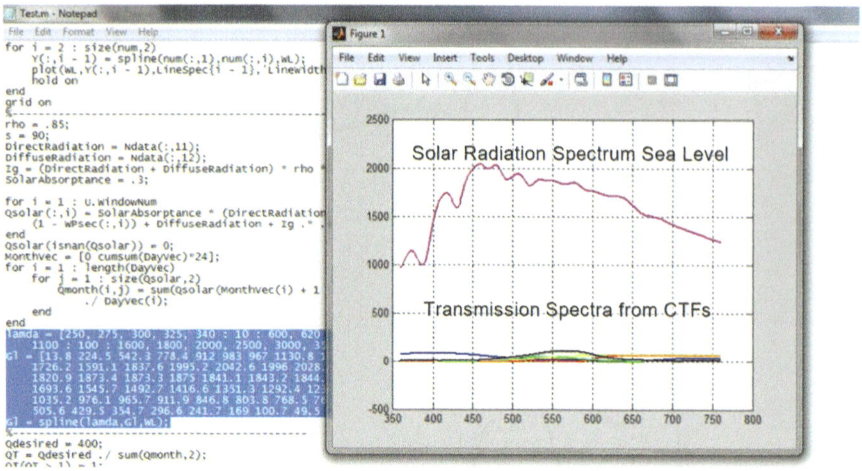

Fig. 5 Solar radiation spectrum energy versus transmission spectra from CTFs in the wavelength range 360–760 nm

Table 1 Transmittance irradiation of CTFs and transmittance irradiation of CTFs per solar irradiance

	Blue	Red	Green	Yellow	Orange	Purple
Transmittance Irradiation	296042.6	662585.8	296847.7	616126.4	474173.4	356262.5
Transmittance Irradiation/Solar Irradiance	0.221444	0.495623	0.222046	0.460871	0.354688	0.266489

4 Light Transmission

According to the diagram of light passing percentage through the devices, with considering the corresponding equation, the light transmission level of each color becomes accessible to each other. Due to the obtained graphs of transmission spectra of CTFs, the percentage of light passing through two devices for each of the colors has been investigated. "The light transmittance of the composite was measured at a wavelength from 200 to 1100 nm using a transmission optical spectrometer" [9].

Transmission spectra by CTF's to wavelengths drawn by MATLAB between 360 and 710 nm can be seen in Fig. 6, where the black graph is the sensitivity of the human eye. So, to calculate the ratio of light transmission of CTFs to each other Eq. 3.22 can be used: (where λi is the wavelength of CTFs, λp is the wavelength of the sensitivity of the human eye, and τ_λ is the spectral transmission).

$$\text{Lum Lamda}(i) = \frac{\int_{\min \lambda i, \lambda p}^{\max \lambda i, \lambda p} \tau_\lambda}{\int \tau_{\lambda p}} \tag{1}$$

Luminance transmittance of CTFs can be seen in second row of Table 2 as a proportion ratio to the sensitivity of human eye. As shown in Table 2, yellow transparent filter has the maximum lighting transmission with 45% considering the sensitivity of the human eye.

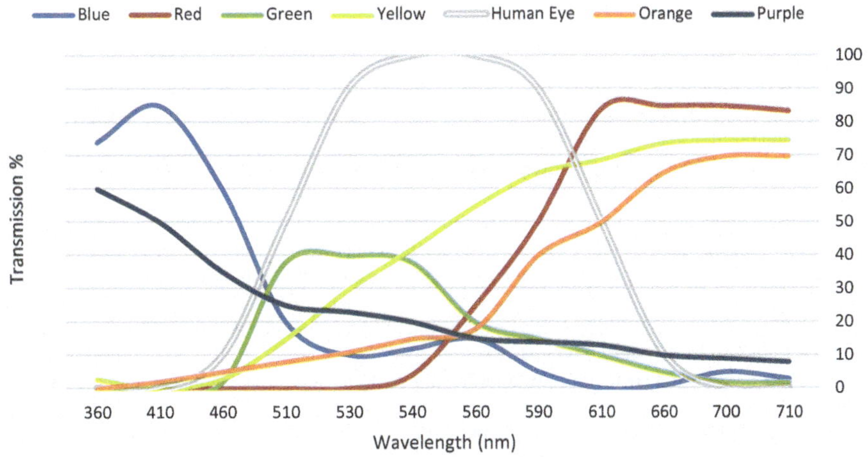

Fig. 6 Transmission spectra ranged from CTFs to wavelengths between 360 and 710 nm

Table 2 Light transmission of CTFs

	Blue	Red	Green	Yellow	Orang e	Purple
$\int_{400}^{700} \tau_\lambda$	4055	330 0	3775	532 0	317 0	495 0
Lum Lamda (400-700)	0.34732 3	0.2826 55	0.3233 4	0.45567 5	0.2715 2	0.42398 3

5 Conclusion

With the advancement of technology, many advances have been made in the field of reformers, including glass and transparent polymeric materials. Various filters and covers on glass and smart glass are among them. In the study, without considering the change in the material of the glass and the change in g value, the amount of passing spectra was investigated by applying transparent colored filters. In order to be closer to the actual conditions in the lab work, colored filters were placed in the bathtub with glass. And thus, the sample was prevented from moving in the bathtub and the spectrum was obtained in a vertical manner. Normally, certain conditions occur due to the fact that liquid is put into the bathtub but since the samples are solid, it was very important to pay attention to how they are placed, keep them clean and suitable for the full bathtub size.

According to the data obtained from the spectrum passing through the tested CTFs (By two Spectrophotometers), transmittance irradiation and light transmission energy were obtained in separate graphs. By reviewing the results gained, in comparison with other colors, red and yellow samples transmitted the highest amount of solar irradiation and blue, purple, and green ones have the lowest amount of irradiation transmittance. In the case of lighting transmission; yellow and purple have the most lighting transmittance in comparison with other colors. According to what was obtained in hot climates, to reduce the radiant energy received from the transparent facades of the building, transparent colored filters can be used, regardless of the type of glass applied. Blue and green can be the best options in this selection according to the minimum amount of transmitted radiant energy and the maximum amount of transmitted light energy. Other colors can be used to a lesser extent.

References

1. Sev A, Gür V, Özgen A (2004) Cephenin Vazgeçilmez Saydam Malzemesi Cam, 2. Ulusal Yapı Malzemesi Kongresi:6–8
2. Alvarez G et al (2005) Spectrally selective laminated glazing consisting of solar control and heat mirror coated glass: preparation, characterization and modelling of heat transfer. Sol Energy 78(1):113–124
3. Haghshenas M, Ghiyabaklu Z. (2006) The effect of colored glass on the light and energy transmission in the visible region. http://dsspace.gazi.edu.tr (Thesis)
4. Zenner GM (2008) Nanoparticle stained glass. In: Wisconsin UO (ed) Research science and engineering center. Madison Materials, Madison
5. Sterpenich J, Libourel G (2001) Using stained glass windows to understand the durability of toxic waste matrices. Chem Geol 174(1):181–193
6. Sarbangoli HS, Pourlamar SH (2014) Chinese knot ornaments in the house of village buildings in Masouleh. Fine Arts Magazine Tehran University 19(4):55–66
7. Haghshenas M, Bemanian MR, Ghiabaklou Z (2015) Analysis of the photo–damaging performance of persian "orosi" in carpeted and non–carpeted spaces. İran Bilim ve Teknoloji Üniversitesi 25(2):94–99
8. Mitchell R, Kohler C, Arasteh D (2006) THERM 5.2/WINDOW 5.2 NFRC simulation manual, Lawrence Berkeley National Laboratory. University of California, Berkeley
9. Sato H et al (2002) Effects of the difference between the refractive indices of constituent materials on the light transmittance of glass-particle-dispersed epoxy-matrix optical composites. Philos Mag B 82(13):1369–1386

Part IV
Bio-Energy, Biofuels and Waste-to-Energy

Sustainable-Business Waste Management: *A Case the Emirate of Ajman – UAE*

Khaled M. AlHosani and Pouria Liravi

1 Background

Waste is an unavoidable product of society, and it is a challenge to realise how to manage large quantities of different types of waste in a way that has benefits for society, the economy and the environment and governments are facing a formidable challenge in trying to find solutions [1]. There are different types of opportunities for new business models to be applied to waste-related markets [2]. Greater cooperation between the authorities and businesses, involving joint teams from different sectors of municipal departments and industries are to be within the ways of going forward. As the worldwide economy revenues to grow, the United Arab Emirates' businesses must have a proper strength in order to take advantage of the emerged opportunities in all aspects. Procurement, trade and benefit will play an essential role in developing markets and generate demand for new products, systems and facilities. For this study and to approach a framework model to support, authorities set up initiatives to commercialise waste management and introduce opportunities for local and global business, the Emirate of Ajman is to be selected for two reasons: the first reason is that this Emirate is intending to start an intensive effort to develop waste-business correlations which give this work originality and the second reason is primary and secondary data are available and accessible.

The importance of this study is that it will further benefit the future planning and development of the United Arab Emirates in the waste management and associated business sector.

This research work aimed to Empower industries to manage waste which have been argued as the most innovative and practical approach to waste management strategy set to get businesses to place sustainability at the top of their priority list [3, 4]. This does not only help extend producer responsibility but creates a circular economy that retains the value of materials within the economy. Renowned benefits, therefore, exist for the Emirate of Ajman if industries are empowered to manage waste; these benefits go beyond the reduced level of waste sent to the landfills and incineration sites but has leveraged benefits to the economy as a whole.

2 Methodology

The main aim of this research work is to empower industries and private sectors to support waste management. This is for the benefits to go beyond the reduced level of waste sent to the landfills and incineration sites and consequently has leveraged benefits to the economy. It is, therefore, essential to implement a regulatory framework and policies to support businesses in expanding the evidence base across a variety of consumer sectors, taking into account the repair, maintenance and service opportunities [5]. The research methodology includes collecting, analysing and integrating quantitative (surveying, questionnaires and experimenting) and qualitative (interviews) research work. The reason behind this initial thought of hybrid methodology is that it will enable the researcher to collect and analyse both numerical and non-numerical data [6]. The research methodology will, therefore, use a sequential exploratory research design to establish evidence in the area of study. This

K. M. AlHosani (✉) · P. Liravi
University of Derby, Derby, UK
e-mail: k.alhosani1@unimail.derby.ac.uk; p.liravi@derby.ac.uk

© The Author(s), under exclusive license to Springer Nature Switzerland AG 2022

A. Sayigh (ed.), *Sustainable Energy Development and Innovation*, Innovative Renewable Energy,
https://doi.org/10.1007/978-3-030-76221-6_53

variant of the mixed methods ensures that qualitative data are first collected with the help of interviews. The resulting outcome of both the qualitative and quantitative triangulated ad integrated to arrive at key study conclusions.

3 Waste Status in the Emirate of Ajman

Due to the continuous growth in population and the standard of life as well as the industrial development in the United Arab Emirates (Fig. 1) [7], a sharp increase in the municipal and industrial waste have resulted, see Fig. 2 [7]. Figure 3 [7] shows the ongoing waste streams (2010–2020). Figure 4 [7] shows the forecasting of all waste in the Emirate of Ajman. All these issues are forming a big challenge to the Government at present and in future. Each Emirate of the UAE must develop its own waste management. Consequently, a new approach to this development is needed.

The primary waste streams in the Emirate of Ajman are characterised as follows [7]:

- Municipal solid waste (MSW) – waste generated by household residences, commerce, offices and public institutions
- C&D (Construction & demolishing buildings)
- Industrial
- Tyre
- Agricultural/green

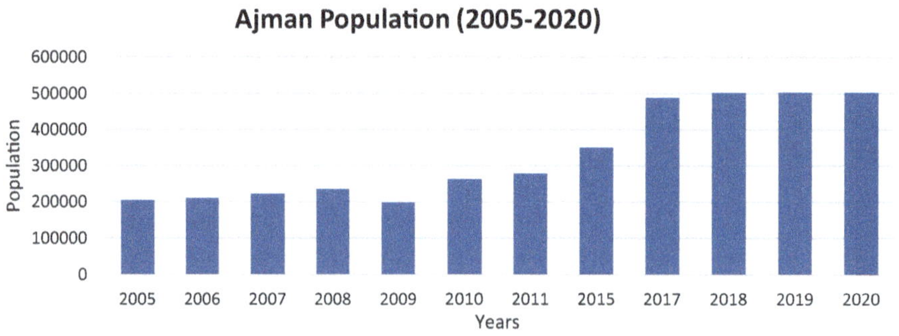

Fig. 1 Ajman population from 2005 to 2020 [7]

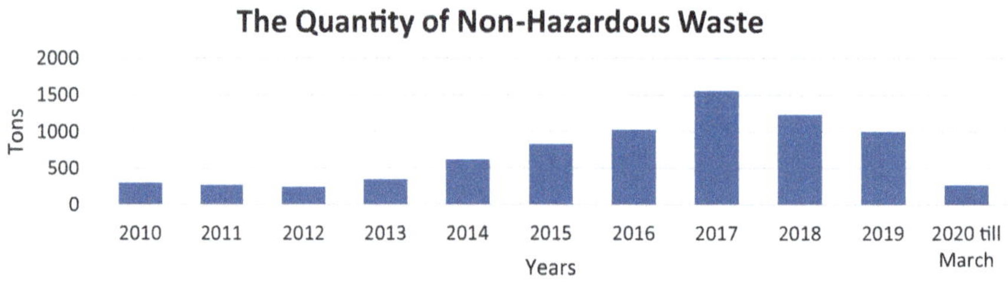

Fig. 2 The quantity of non-hazardous waste [7]

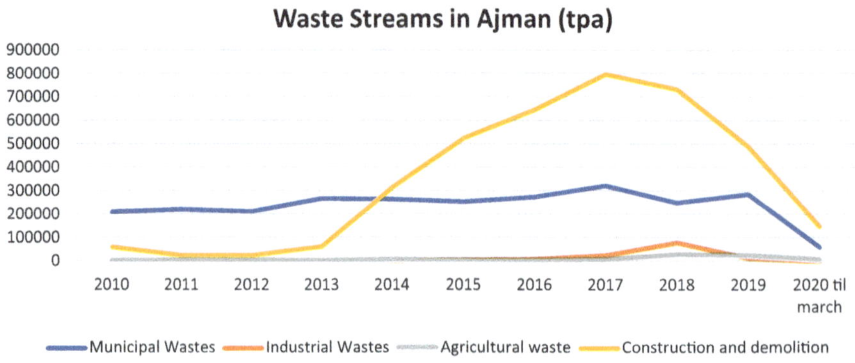

Fig. 3 Waste streams (2010–2020) [7]

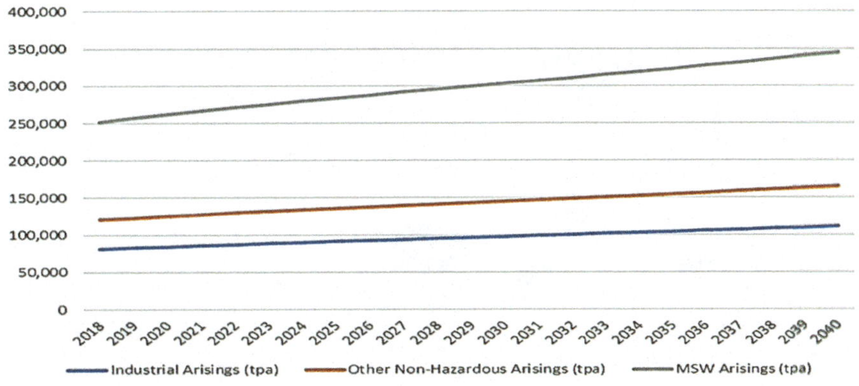

Fig. 4 The growth of three waste streams in Ajman until the year 2040 [7]

- Sludge
- Hazardous – residue or ash resulting from activities and operations containing hazardous
- Materials or properties of hazardous substances
- Medical/clinical
- Other

There exists a lack of waste management correlated with business opportunities policies and practices in the UAE. Therefore, this research seeks to understand the present waste situation for a selected city, quantifying its waste and types, then modelling solutions to create business opportunities in order to minimise the waste and improve the business sectors economic aspects, which will have positive consequences on the society and results in a better environment. In addition, the research aiming to promote an appreciation that both the economy and environment can benefit from carefully considered waste management practices, which will, in turn, lead to the sustainability of resources and the preservation of the environment, particularly considering the existential threats the global environment faces today. Moreover, municipal waste management companies compete and function in institutional environments that occasionally let them act on markets. At the same time, in other cases prohibit them from using the full capabilities of their expertise in resource management and infrastructure in an economically competitive manner [8]. This has limited some opportunities for them to expand with strategies for waste management and business marketing aspects. A proposal for a centralised waste management system with a theoretical framework was introduced, which implements an intelligent sensor-based substructure for waste separation and collection at the right time [9]. This proposal can fill in the present gap that the waste management system has in many cities and Ajman is one of those cities.

Converting the practices for waste management in a way that can attract the private and the public sectors to invest and create initiatives for business opportunities could well help sustain the waste management process and reduce the associated problem with liquid/solid waste accumulation, resulting in a more favourable impact on the environment. The UAE needs to introduce different strategical practices and ways in which the authorities shape the markets in which the local and international businesses operate. Moreover, the use of material resources and waste needs to be looked at carefully within the operation of business factors to help understand the type of barriers and required motivations for more efficient resource use is to be focused on with best practices and methods to overcome the obstacles [10]. In addition, there is a need for identifying business opportunities throughout the supply chain, from product design and manufacturing to materials management, recovery and reuse, which formed another objective of this research [11]. All these are never studied thoroughly and create a gap to be filled by this research study.

4 Emirate of Ajman Location and Industrial Waste

Ajman is the smallest Emirate in the north of UAE. Figure 5 shows the use of land which is distributed in a way that totals 3076 plots that has two plots for multi-use, 28 commercial, 2617 residential, 7 of recreational & touristic, 391 industrial and 31 agricultural plots. Due to the continuous increase in population and the improved lifestyle, these figures are increasing, indicating the spread of waste-associated problems and challenges. This has to be monitored, controlled and has a proper sustainable management framework model.

Despite having the Emirate of Ajman's area is small, the Emirate enjoys a significant amount of industrial activities which consequently add more challenges to the waste management. At the same time, it explores many opportunities as the waste association business needs proper regulations, policies and business organising to achieve sustainable waste management. Table 1 (produced by the author) shows a list of the available industries with their waste-related activities 1 being the high productive capacity of waste and 0 being the low productive capacity of waste.

Fig. 5 Emirate of Ajman [9]

Table 1 Industrial activities firms in Emirate Ajman [7]

No. of industries	Activities (English)	Waste production
52	Plastic bags industry	1
42	Building metal works industry	1
42	Lubricant industry	1
39	Grease industry	1
36	Home furniture industry	0
28	Manufacturing of kitchen furniture and fixtures industry	1
28	Manufacturing of kitchen furniture and fixtures	0
23	Manufacturing of wooden doors and windows industry	1
22	Household linen industry	0
19	Metal products coating industry	1
17	Manufacturing of tissue and paper towels	1
16	Metal construction industry	1
16	Industry of bottles and plastic containers	1
14	Manufacturing of fire-resistant wooden doors industry	1
14	Detergents and disinfectants industry	1
13	The women's clothing industry	0
13	Office furniture industry	0
13	Making containers and boxes from cardboard	1
13	Making containers and boxes from cardboard	0
12	Men's clothing industry	0
7	Central and retail gas filling industry	0
7	The manufacturing of fiberglass boats industry	1
7	Children's clothing industry	0
7	Mirrors and glass cutting industry	1
6	Steel tubes industry	1
6	Aluminium foil and coil industry	1
5	Plastic sheeting industry	1
5	Sand brick industry	0
5	Tire recycling industry	0
4	Installing pipelines and specialised installations	1
4	Steel pipe fittings industry	1
3	Gases industry, liquefaction and filling	0
3	Maintenance of fiberglass products	1
3	Car upholstery industry	1
2	Petroleum gas filling industry	1
2	Metal ships industry	1
2	Repairing of boats	1
2	Metal boxes industry	1

(continued)

Table 1 (continued)

No. of industries	Activities (English)	Waste production
2	Manufacturing of marble products for decoration	1
2	Calcium carbonate powder industry	1
2	Artificial marble industry	1
1	Manufacturing offshore platforms and drilling towers	1
1	Ship repair and maintenance industry	1
1	Repair and maintenance of ship engines	0
1	Manufacturing of wallpaper and wall coverings	1
1	Manufacturing of iron castings industry	1
1	Manufacturing of rolled stainless steel sheets and coils	1
1	Manufacturing of vials industry	1
1	Mosaic tiles industry	1
1	Nets making industry	0
1	Stone cutting and grinding industry	1
568		

5 Ajman Municipality Waste and Public Health Department

The organisational framework of the Public Health and Environment Sector is demonstrated within Fig. 6 and the organisational chart of the Waste Management and Public Health Department provided in Fig. 6.

6 Ajman Municipality's Strategic Goals and Most Important Material Topics and the SDGs

United Nations Sustainable Goals (SDGs) have been seriously considered by the UAE aiming to achieve economic growth, environmental benefits and social wellbeing working for the best of the present and future generations. Waste management also has substantial attention, facilitated by the implementation of smart advanced technology with the required regulations and proper initiatives to achieve sustainable waste management. Therefore, empowering the right business for overall topics associated with the waste corresponding practices is an essential goal for the Emirate of Ajman in specific. All the factors mentioned above are necessary for an effective waste-business interrelation framework model. Table 2 shows the guiding topics and corresponding goals to support this argument [12].

Environmental topics are recently intensively considered by many companies aiming sustainable development achievements. Therefore all customers, shareholders, employees and owners are inquiring on the practices and methodologies that can be implemented in order to deliver sustainability. Waste business over the world are placing strategies to be competitive and governments are also setting up initiatives and regulations to support this development and empower the waste business aiming to reach a sustainable waste management [10]. Ajman authorities are considering many initiatives such as organisational strategy which is essential to create long-term value for our ecological, social and economic environment. Moreover, initiatives are to achieve a significant waste reduction and encourage reuse and reduction in food waste. Other practices are avoiding single use packaging, recycling as many items of waste as possible and even the residues from energy recovery can be used as aggregates in road building reaching zero landfill consequently allowing waste business associated companies to examine the actual business benefits of shifting to a more sustainable model [13].

7 Sustainable Waste Management, Sustainable Business – Key Factors and Smart Waste for Ajman Municipality

The increased growth in lifestyle and urbanisation leads to a growth in the economy and consequently increasing opportunities for growing waste-associated business market. The main factors behind the sustainable development of this market would be the minimum impact on the environment, creative ideas for the implementation of smart technology, waste disposing efficiency, garbage monitoring, transportation, treatment and transportation. Many challenges are arising; important ones

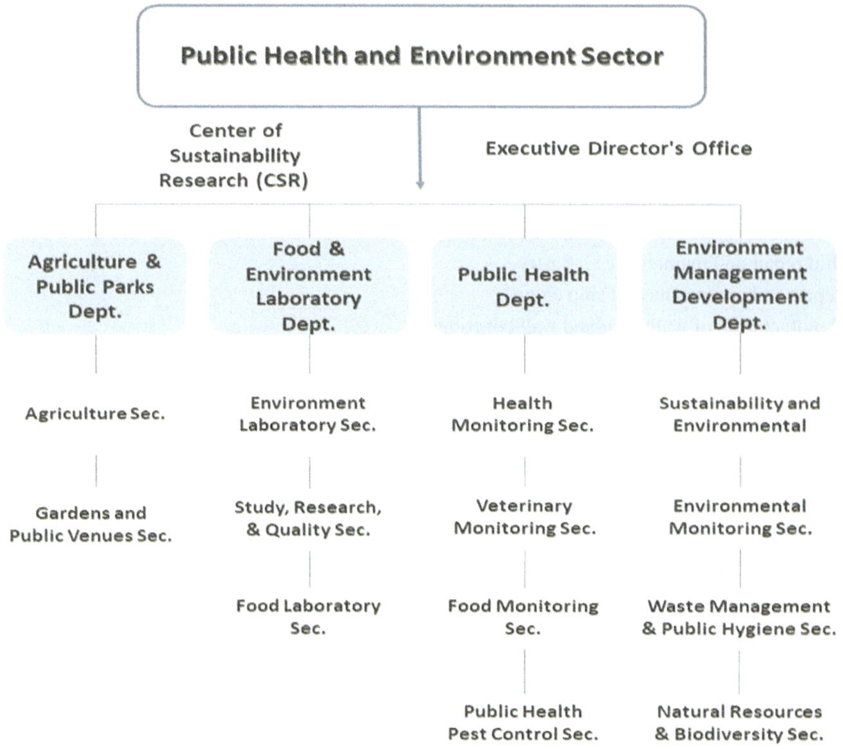

Fig. 6 Ajman municipality public health and environment sector organisational framework

Table 2 Ajman municipality's strategic goals and most important material topics and the SDGs [12]

Ajman Municipality's strategic goals and the SDGs

The Municipality's Strategic GoalsDescription	Relevant SDG
G1- Achieve integration between infrastructure and residential elementsSustainable environment while developing a modern infrastructure in Ajman.	**SDG 7:** Affordable and clean energy **SDG 11:** Sustainable cities and communities
G2- Qualify and protect the environment	**SDG 2:** Zero Hunger **SDG 13:** Climate action
G3- Manage the wastes efficiently and effectivelyEmpowerment	**SDG 11:** Sustainable cities and communities **SDG 12:** Responsible consumption and production
G4- Develop the Biological diversity Transparency and Integrity	**SDG 11:** Sustainable cities and communities **SDG 15:** Life on landSDG 16: Peace, justice and strong institutions
G5- Ensure best use of resourcesTeamwork	**SDG 6:** Clean Water and Sanitation **SDG 11:** Sustainable cities and communities
E1- Develop human resources Loyalty and Belonging	**SDG 5:** Gender Equality and Women's empowerment

Ajman Municipality's Most Important Material Topics and the SDGs

Aspect	Description	Relevant SDG
Most Important Material Topics	Infrastructure and Roads (Reduce Traffic Congestion)	**SDG 9:** Industry, innovation and infrastructure **SDG 11:** Sustainable cities and communities
	Waste Management Practices	**SDG 13:** Climate action **SDG 14:** Life below water **SDG 15:** Life on land
	Community Health and Well-Being (i.e. Food Safety, Air Quality)	**SDG 3:** Good health and well-being **SDG 13:** Climate action
	Climate Change and Energy Efficiency	**SDG 7:** Affordable and clean energy **SDG 13:** Climate action

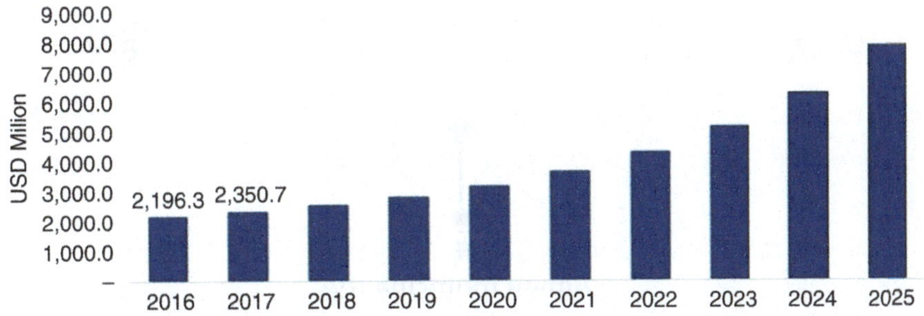

Fig. 7 Global smart waste management market expectation [14]

are comprising a need for advanced infrastructure, and advanced smart technology needs to be implemented to support this advanced. In order to achieve the improved operational efficiency of waste management lifecycle. As global smart waste management is sharply expanding and expected to reach a market value of 7877 Million in 2025, see Fig. 7 [14]. This increased expansion opens wide the available facilitated technological aspects that support the advanced waste market and enable the Ajman municipality to use the latest advancement in smart waste management.

The management of waste in critical spaces requires that this is done sustainably to render embedded economic, social and ecological rewards [15, 16]. This requires that waste is first seen as an unavoidable product of the industrial activity, and managing waste remains a formidable challenge that requires a strong alliance between private and public partners to find amicable solutions [17]. In Ajman, key initiatives have been introduced to make the area of waste management attractive to private investors [18]. Nonetheless, to ensure that waste management is integrated into the very industry systems and structures that currently exist, these industries must be empowered enterprises to manage waste through the reduction, reuse, recycling and repurposing of waste in a manner that meets all the three dimensions of economic, social and ecologic sustainability towards overall reduced environmental impact. This analogy remains a central aspect of the Ajman Municipality Sustainability model, as suggested in Fig. 8. The model shows a flow of roles and practices starting with the governmental activities such as the required practices, the incentives that are provided to empower business, regulations to control these practices with factors that support practical cooperation and the export and import activities. It follows that the municipality will take all the required data and analyse to practically perform the regulations set up by the government according to standards referred to each activity for quality assurance to cover the private business sector and industrial sector. This shows how well could the waste management be performed in a sustainable way when applying this model that specifies the right practices for each sector and explores in specific what would the private business provide to achieve successful management in sustainable performance. Therefore, it is recommended that a particular sequence of activities should be followed, starting from data collection and reviewing all possible waste management activities and setting up the required plan for the practical policies taking into consideration the continuous challenges in the Emirate's context. This is to be followed by formulation, implementation and monitoring of the action plan in view of the challenges pertinent to the Emirate.

Continuous evaluation of the process should have the primary goal of the environmental benefits and risks assuring that the business sector is aware of the regulations that are followed and performed by setting third part auditions.

However, the subject of sustainable waste management must be first explored on the grounds of the tri-dimensional framework of economic, social and ecological sustainability [19]. The author is, therefore, suggesting a conceptual framework presented in Fig. 9, which constitute of mixed methodology of qualitative and quantitative research work.

8 Strategic Risks for Business – Waste Management

Strategic risks have been identified by the municipality [12] that may affect the vision and objectives that have been set up between the period 2017 and 2018. Some of the essential risks related to the waste management are as follows:

- The administrative and financial independence of the department is hindering the speed of its adoption to cope with the urban development witnessed by the world, and hindering the speed of its completion of the basic projects in the Emirate and development projects due to the limited annual budgets.
- Pollution from facilities outside the Emirate's borders cannot be controlled.
- Lack of technical staff to control the facilities producing liquid waste.

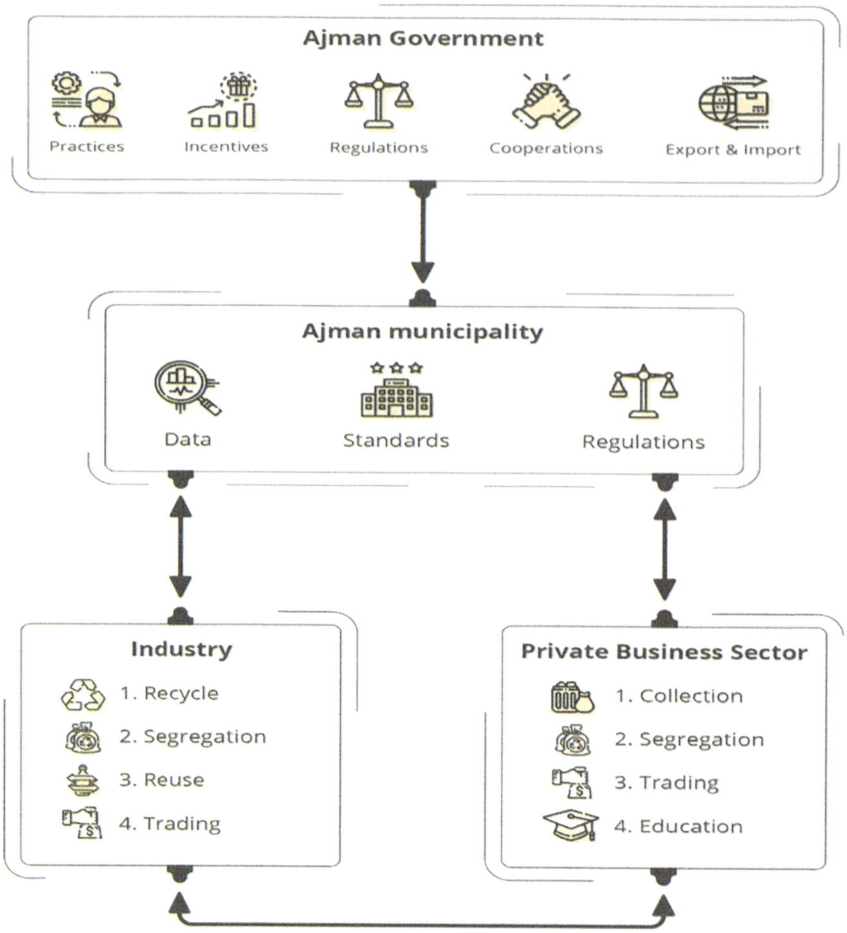

Fig. 8 Model for waste management sustainability. (Source: Author)

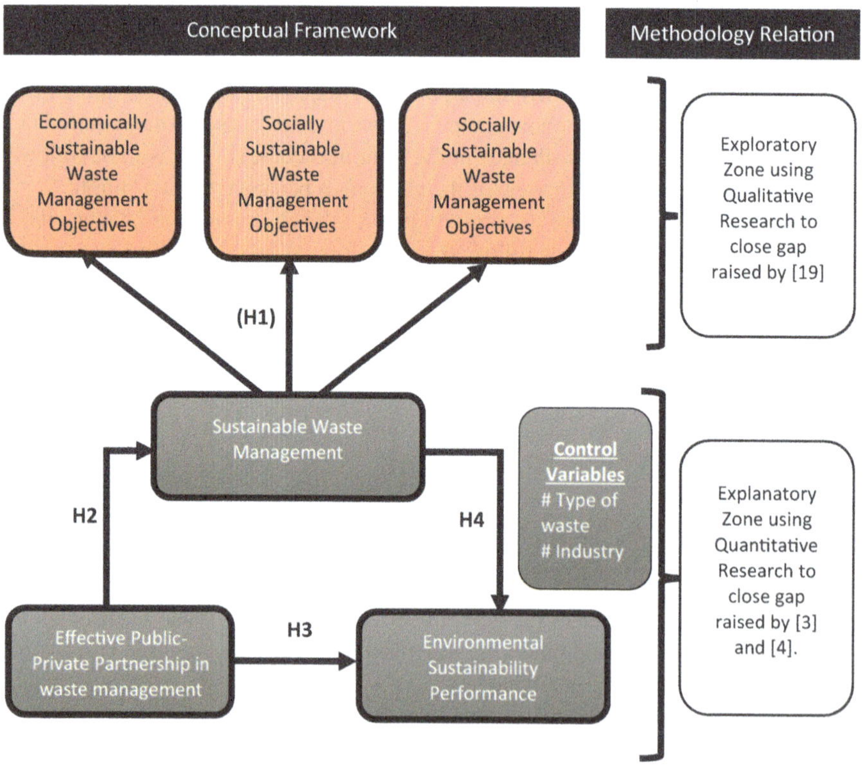

Fig. 9 Conceptual framework and methodology relation (H is Hypothesis). (Source: Author)

- Lack of a market for companies deal with waste management for specialised purposes in the Emirate.
- Lack of adequate waste sorting and recycling facilities in the Emirate.
- Lack of adequate sorting of wastes from the source and weak culture from the community to adopt waste sorting from the source.
- Lack of markets for some types of recycled materials such as recycled plastic bags.
- Violations of construction work and excavation in the Emirate's reserves without obtaining any approvals from environmental management and development, failure to maintain biodiversity in the Emirate, failure to comply with international agreements.

These risks have been carefully considered by the Municipality and a plan must, therefore, be managed to reduce their impact. A need for a regulated framework is to enhance this action. Empowering the waste business concepts and private organisations could well contribute to achievement for the sustainable waste management.

9 Positive Environmental Impacts Achieved Through Digitalisation Initiatives

One of the advanced practices that recently developed at the Emirate of Ajman is the digital master plan for Ajman 2017–2022 aimed to create a road map to transform the government services as well as their digitally. Five objectives are included for 30 priority government services every 100 days leading to an expected achievement of financial efficiency. For example, Fig. 10 shows indicators of reduction in environmental impact through digitalisation by reducing paper use in 2018 as a positive environmental impacts achieved through digitalisation initiatives [12].

10 Business Advantages for Waste Minimisation – Initiatives

Currently, approximately 77% of solid waste generated in the UAE is disposed of at landfill facilities with approximately 20% diverted from landfill and approximately 4% is recycled as reported by Ajman Municipality. In specific the Emirate of Ajman is targeting to recycle 75% by 2021 achieving a reduction to 1.5 kg/day per capita waste as shown in Fig. 11 [7]. A

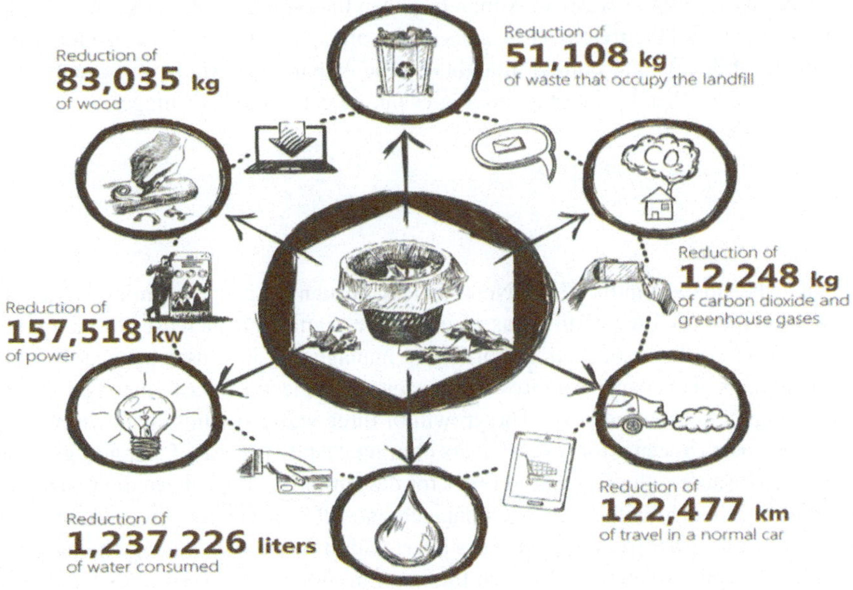

Fig. 10 Indicators of reduction in environmental impact through digitalisation by reducing paper use

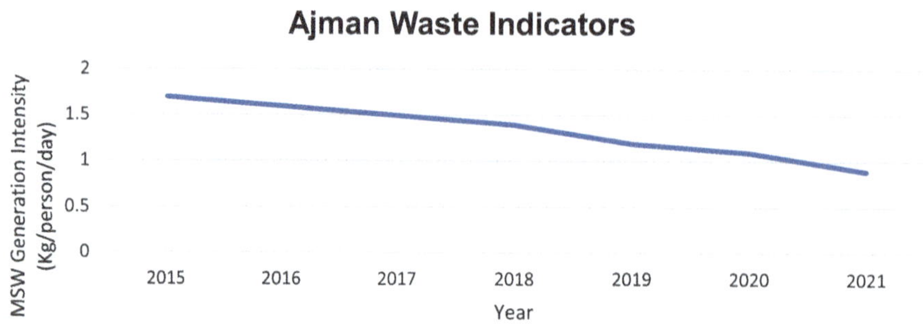

Fig. 11 Ajman municipal solid waste indicators. (Produced by author, data from [12])

need for different framework of practices to achieve this target. Initiatives such as proper technique and methodology to calculate the national waste indicators, collecting accurate data about the municipal solid waste (MSW) periodically, develop education and awareness programs and campaigns in order to reduce the waste generation rate and increase the segregation of municipal solid waste at source and create an integrated system within other municipalities within the UAE to achieve an efficient road map for the initiatives that lead to the management framework.

11 Waste Treatment Technologies Business Opportunities

The Emirate of Ajman has a total of three materials recovery facility (MRF) sorting facilities. Mixed municipal solid wastes (MSW) arrive at the three facilities where the recyclable materials are removed using a mixture of manual and automated picking and separation methods. The recyclable items that are segregated by the Ajman MRFs include:

- All kinds of plastic materials
- Paper
- Organic waste
- Metals
- Cardboard

Initiatives to deal with the amount of waste in Ajman has been thought through by the Government. Many aspects are to contain the waste management and the treatment is an essential practice that must be considered within the proposed management framework model. Table 3 shows initial thoughts by the Ajman [12]. Many of these thoughts were considered and implemented with improvements that use latest technologies taking care of disadvantages.

12 Conclusions

The Emirate of Ajman is the smallest in the UAE. Nevertheless, Ajman is an active Emirate and constitute many industrial, tourist and other business. Increased population has resulted in a significant increase in the amount of all types of waste which have added challenges to the associated authorities to manage this increase sustainably and to achieve the Emirate sustainable development goals. This paper explores the existence of a lack of waste management correlated with business opportunities policies and practices in the UAE. The growth of three waste streams is shown, which indicates a continuous increase in the amount of waste expected for several years coming despite the use of technological advances and other modern practices. The research discusses the fact that in order for the authorities to achieve their sustainable development goals, and in specific the goals interrelated with the waste management, different factors are introduced and discussed. The proposed factors are to construct an effective waste-business interrelation framework model which consists of policies, practices and regulations and modelled in a framework to form the main initiatives for an advanced and effective sustainable waste management at this Emirate. The model is to empower the waste-business relationship by converting the practices for waste management in a way that can attract the private and the public sectors to invest. Creating initiatives for business opportunities could well help sustain the waste management process and reduce the associated problem with liquid/solid waste accumulation, resulting in a more favourable impact on the environment. The importance of this study is that it will further benefit the future planning and development of the United Arab Emirates in the waste management and associated business sector.

Table 3 Advantages and disadvantages of treatment technologies

Technology	Advantages	Disadvantages
Mechanical biological treatment	Combines MRF sorting with composting, can produce refuse-derived fuel	Low-quality compost, waste may require further treatment in some markets
Unsorted/residual waste materials recovery facility (dirty MRF)	Cheap to implement, can be used with existing waste collection regimes	Gives poorer quality recyclates than clean MRF
Energy from waste (EfW) – Incineration	Reduces mass disposed greatly, proven technology. Can treat most waste streams	May be unpopular in some markets due to pollution concerns and visual impact
Fluidised bed combustion technology	Reduces mass disposed greatly, proven technology. Very low emissions. Can treat most waste streams	May be unpopular in some markets due to pollution concerns and visual impact. More expensive than EfW above
Anaerobic digestion	Comparatively simple technology. (cf. EfW, etc.). Can produce green energy via methane capture. Produces soil enhancer	Expensive compared with other organic treatments such as composting
Composting	Simple technology, cheap to buy. Produces soil enhancer	Can only treat biodegradable wastes. Needs special facilities (IVC) for managing some pathogen risks
Materials recovery facility (clean MRF)	Roughly same cost of machinery as dirty MRF. Good quality of materials produced	Needs dedicated collection, so can be costly to implement
Pyrolysis and gasification	Reduces mass disposed greatly, proven technology. Very low emissions. Can treat most waste streams	May be unpopular in some markets due to pollution concerns and visual impact. More expensive than other thermal technologies above. Unproven in most markets with chequered past so far economically
Co-incineration	Reduces mass disposed greatly, proven technology. Uses existing facilities	May cause additional pollution due to inconsistencies in the input waste streams. Can cause corrosion in boilers unless combustion chamber modified

References

1. Singh J, Laurenti R, Sinha R, Frostell B (2014) Progress and challenges to the global waste management system. Waste Manag Res 32(9):800–812
2. Michael J (2019) Smells like money: the business of waste management. [online] Bplans Blog. Available at: https://articles.bplans.com/smells-like-money-the-business-of-waste-management/. Accessed 10 Oct 2019
3. Cameron S (2019) How the resource and waste strategy can empower industry. https://sctimes.io/news/article/7/6476
4. Arranza JL (2019) Empowering industries for zero waste economy. https://businessmirror.com.ph/2019/08/05/empowering-industries-for-zero-waste-economy/
5. DEFRA (2018) Our waste, our resources: a strategy for England. [online]. Crown, London. Available at https://assets.publishing.service.gov.uk/government/uploads/system/uploads/attachment_data/file/765914/resources-waste-strategy-dec-2018.pdf. Accessed 21 Oct 2019
6. Creswell JW, Clark VLP (2017) Designing and conducting mixed methods research. Sage publications, California
7. Municipality & Planning Department (2020) City of Ajman population and waste management. Municipality & Planning Department. (Data requested by official communications), Ajman
8. Hultman J, Corvellec H (2012) The European waste hierarchy: from the sociomateriality of waste to a politics of consumption. Environ Plan A Econ Space 44(10):2413–2427
9. Esmaeilian B, Wang B, Lewis K, Duarte F, Ratti C, Behdad S (2018) The future of waste management in smart and sustainable cities: a review and concept paper. Waste Manag 81:177–195
10. Ravindra K, Kaur K, Mor SD (2015) System analysis of municipal solid waste management in Chandigarh and minimisation practices for cleaner emissions. J Clean Prod 89:251–256
11. Visvanathan C (2011) Business and employment opportunities in waste management and recycling in Asia. Waste Manag 31(6):1083–1084
12. Ajman Municipality and Planning Department (2018) Sustainability report for years 2017 – 2018. Ajman Municipality, Ajman
13. Commercial Waste (2020) Waste management and sustainability - commercial waste. [online] Available at: https://cleanstreets.westminster.gov.uk/waste-management-sustainability/. Accessed 10 June 2020
14. Market research future, 2020. Smart waste management market by component, growth and analysis – 2025 | MRFR. [online] Marketresearchfuture.com. Available at: https://www.marketresearchfuture.com/reports/smart-waste-management-market-994. Accessed 9 June 2020
15. Seadon JK (2010) Sustainable waste management systems. J Clean Prod 18(16–17):1639–1651
16. Silva A, Rosano M, Stocker L, Gorissen L (2017) From waste to sustainable materials management: three case studies of the transition journey. Waste Manag 61:547–557
17. Yeomans J (2019) Waste management using multicriteria population-based simulation-optimisation algorithms. J Waste Manag Disposal 2:1–8
18. Abdullah A (2017) Private firms to manage Ajman's waste. Khaleej Times. Retrieved from https://www.khaleejtimes.com/nation/ajman/private-firms-to-manage-ajmans-waste
19. Allaoui H, Choudhary A, Elsaid S, Aghezzaf EH (2015) A framework for sustainable waste management: challenges and opportunities. Manag Res Rev 38(10):1086–1097

Energy Recovery in Wastewater Treatment Plants: Utilization of Hydro and Biogas Energy

Serhat Kucukali, Irfan Ar, Asiye Günal, and Mümtaz Ak

1 Background

Concerning the development of small hydropower in existing wet infrastructures such as drinking water, wastewater, and irrigation, water networks represent a convenient potential of energy recovery [1]. The benefits and simplifications of these facilities compared to river-type hydropower plants could be summarized as follows: (i) all infrastructure access are present, which will reduce the investment cost and risk considerably, (ii) the facility has a guaranteed discharge through the year, (iii) the generated electricity will be used at the infrastructure system and the excess electricity can be sold to the government, (iv) there is no land acquisition and significant operating costs (Fig. 1), and (v) since the hydropower potential is a function of the available discharge and head, the required data (i.e., discharge, head, flow duration curve) are usually well-known [2, 3]. For instance, Samra wastewater treatment plant (WWTP) in Jordan is an example for electricity production from wastewaters before and after the wastewater treatment plant. The four hydro turbines installed at Samra WWTP will generate 21 GWh/year electric energy which has an important environmental benefit as Jordan generates its electricity mainly from oil. This production is estimated to avoid the emission of more than 17,000 tons CO_2 per year. Also, the facility covers the 98% of Amman city electricity consumption by using the biogas and hydropower (both at inlet and outlet) energies together [4].

Wastewater treatment plants consume large amounts of electric energy. Chen and Chen [6] reported that wastewater treatment is estimated to use around 3–5% of the total electrical energy load in developed and developing countries around the world. Moreover, Cakir [7] reported that artificial aeration process is alone responsible for half of the total energy cost in a wastewater plant.

The installation of a hydropower plant at the outlet of the wastewater treatment plant will cover the electric consumption considerably and it will produce clean and feasible energy. In this context, this study aims to show the possible benefits of the installation of a hydro turbine at the outlet of a waste-water treatment plant. This could be an alternative clean energy solution to reduce the consumption of energy supplied by the national electric grid mostly fed by fossil fuels and to induce the minimization of CO_2 emissions to the atmosphere. In the case of hydro energy, electric generation, E, is proportional to the discharge, Q, and the head, H, as follows:

$$E = \rho g Q H \eta t \tag{1}$$

where ρ is the density of fluid ($\rho_{\text{wastewater}} = 1080\ \text{kg/m}^3$), g is acceleration due to gravity, η is the efficiency of the hydro turbine for the corresponding discharge, and t is the working time of the turbine for the relevant time period. Ak et al. [8] used a fuzzy logic decision analysis tool to select the most appropriate low-head hydro-turbine alternative at the outlet of WWTPS and they found the Archimedean Screw as the most suitable technology for the hydropower development at the outlet of a WWTP

S. Kucukali (✉)
Hacettepe University, Civil Engineering Department, Ankara, Turkey

I. Ar · A. Günal
Gazi University, Ankara, Turkey
e-mail: irfanar@gazi.edu.tr

M. Ak
Netsu Engineering, Ankara, Turkey

© The Author(s), under exclusive license to Springer Nature Switzerland AG 2022
A. Sayigh (ed.), *Sustainable Energy Development and Innovation*, Innovative Renewable Energy,
https://doi.org/10.1007/978-3-030-76221-6_54

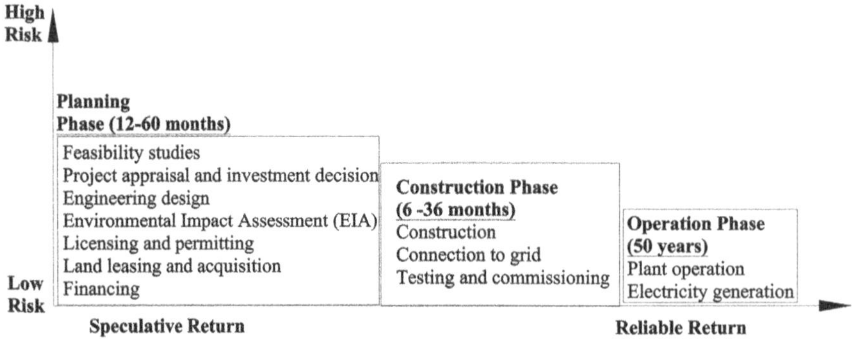

Fig. 1 The life cycle phases of renewable energy projects. (Adapted from Lawrence and Dickson [5])

due to its superior performance in economic and environmental criteria. In this context, this study aims to show the possible benefits of the installation of a hydro turbine at the outlet of a waste-water treatment plant. This could be an alternative clean energy solution to reduce the consumption of energy supplied by the national electric grid mostly fed by fossil fuels and to induce the minimization of CO_2 emissions to the atmosphere. Energy Recovery efficiency, ER, is calculated from:

$$\mathrm{ER} = \frac{E}{E_c} \tag{2}$$

where E is the generated electricity and E_c is the consumed electricity of the facility.

2 Case Study: Tatlar WWTP

Tatlar WWTP which is the one of the biggest wastewater treatment plants of the Europe has been operated by the Municipality of Ankara since 1993. The combined sewer system carries domestic and industrial wastewater and rainwater by gravity and there is no need for pumping. Because of its high population and the industrial activities, Ankara produces enormous amount of wastewater. Wastewater of Ankara metropolitan area is discharged into Ankara Creek. The plant was built as an active sludge project with 765,000 m^3/day wastewater treatment capacity per day [8]. Process of the treatment is anaerobic sludge stabilization, active sludge technique, and removal of sludge water by using strip filter technique. By the year 2025, the capacity of the plant is planned to be increased to ensure treatment of 6.3 million people's wastewater which will be about 1.377.000 m^3/day [8]. The function of wastewater treatment plant is to enhance the water quality of the receiving water body from class IV to class II by applying: (i) physical treatment (screenings/grit disposal), primary settling, activated sludge aeration, and secondary settling (Fig. 2).

Then, the treated wastewater effluent is discharged into the Ankara Creek (Fig. 3). The discharge is measured at the outlet of the plant at the Parshall flume by an ultrasonic displacement meter. Assessing reliable and sufficient discharge data during the planning phase of a hydropower development is necessary to accurately estimate the energy yield. Accordingly, long-term (at least 1 year) on-site discharge measurements and the analysis of the measured data by an expert are key steps in reducing the uncertainty related to the energy yield assessments. Uncertainty analysis is becoming increasingly important tool in measurements. The uncertainty in the discharge measurements in the Parshall flume is expected to be less than 3%, while the error in the level measurements at the project site is expected to be below 0.1%.

In this study, to calculate the annual hydroelectric generation, daily discharge data of Tatlar WPPT for the year 2013 (Fig. 4) is used. From those data, monthly hydroelectric generation is computed. The construction duration for the Archimedean screw is significantly shorter (i.e., 9 months) than that of the conventional hydro turbine which creates an opportunity cost due to the early electric generation of the screw. Also it has lower O&M costs compared to other conventional low-head hydro turbines [8]. The supply and installation of the Archimedean screw requires less than 10 months due to the fact that the screw is a compact technology and does not require assembly of many parts (Fig. 5). This characteristic of the screw also reduces the supply risk as well [9, 10]. Also, Archimedean screw is the only low-head hydropower technology that causes aeration by creating turbulence and mixing at the free-surface of the flow. Hence, the screw enhances the dissolved oxygen content of the passing water. Also, there is 3 MW installed capacity biogas power plant which has been operation in Tatlar WWTP.

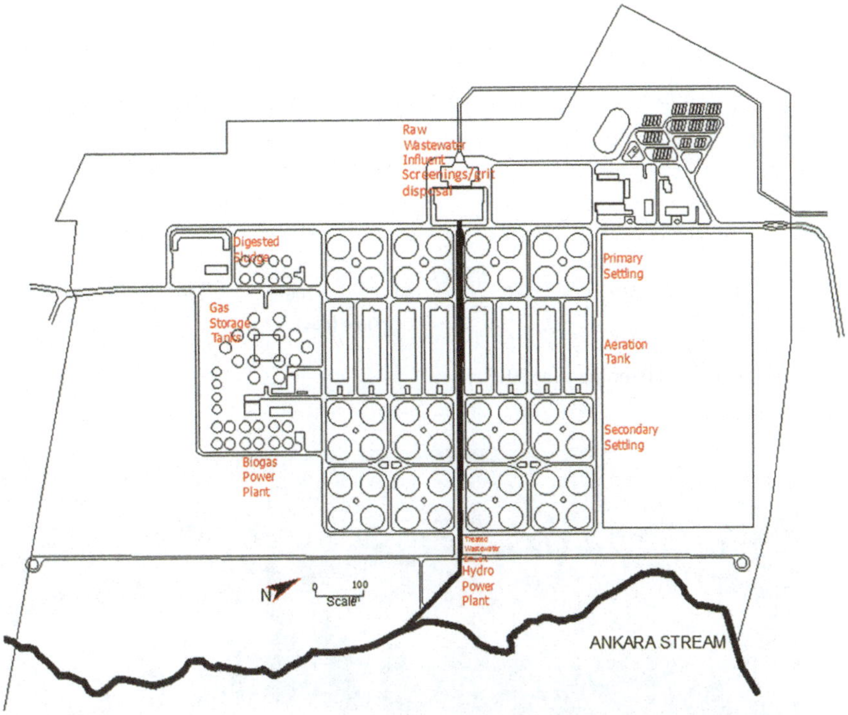

Fig. 2 Layout plan of the Tatlar WWTP

Fig. 3 Outlet of the Tatlar WWTP

The monthly electric generation from biogas power plant is shown in Fig. 6. Biogas power plants avoid greenhouse gas (GHG) emissions considerably relative to other power plants. The biogas power plants have nearly four times higher GHG emission reduction potential compared to other renewable power plants in terms of CO_2e tons/MWh. This is due to the fact that in biogas power plants, methane is tapped and converted into useful energy. Methane is a greenhouse gas and it has a global warming potential nearly 23 times greater than CO_2, hence 1 ton of methane equals about 23 tons pf CO_2e [11]. Monthly hydroelectric and biogas generations based on the measured values and their comparison with the monthly electric consumption of the Tatlar WWTP for the year 2013 is shown in Fig. 6. The facility recovers over 100% its electric consumption by harnessing the biogas and hydro energies. Moreover, this combined electric production is estimated to avoid the emission of more than 60,000 tons CO_2 per year.

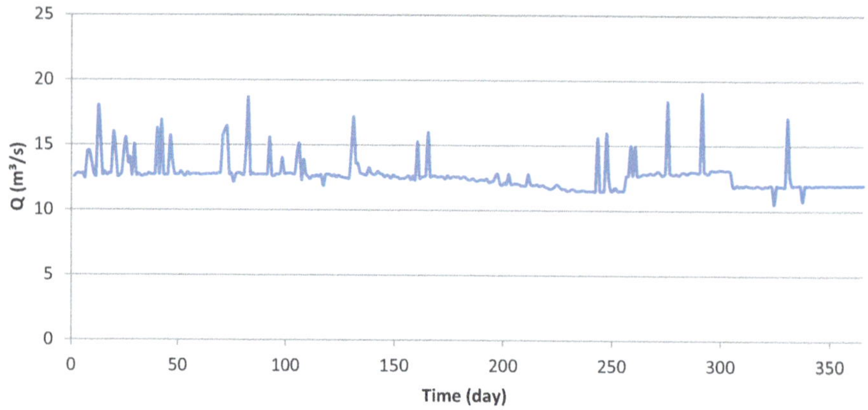

Fig. 4 Daily discharge data of Tatlar WWTP for the year 2013

Fig. 5 Photo of the archimedean screws at the outlet of Tatlar. Project data: $Q = 15\ \text{m}^3/\text{s}$, $H_g = 9.25$ m, $H_n = 8.76$ m, $P = 4 \times 250\ \text{kW} = 1000$ kW

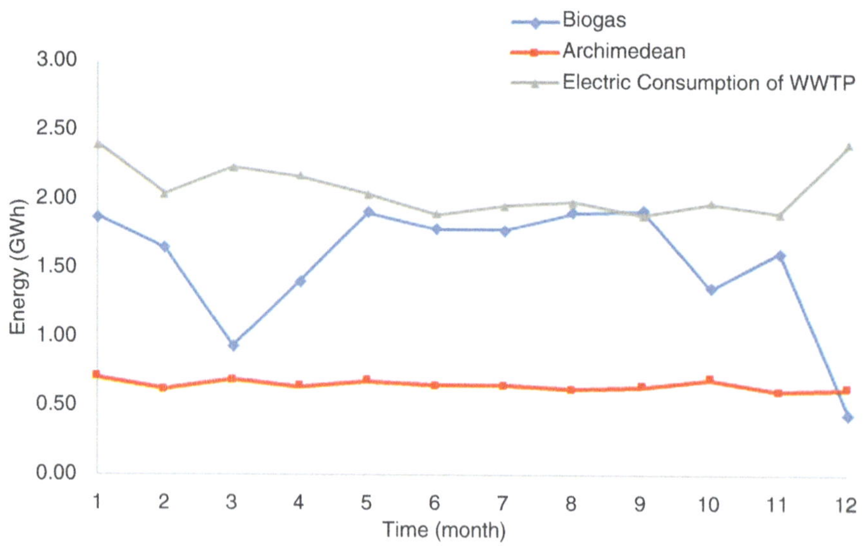

Fig. 6 Monthly hydroelectric and biogas generations based on the measured values and their comparison with the monthly electric consumption of the Tatlar WWTP for the year 2013

3 Conclusions

This study focuses on the energy aspect of wastewater treatment plants. Utilization of the existing biogas and hydropower energy potential in wastewater treatment plants have been analyzed. The new energy laws and the market conditions of the most of the countries around the world create opportunity to harness this potential. For a case study, Tatlar WWTP in Ankara has been investigated in a detailed manner. The annual electricity consumption of Tatlar WWTP was 26.5 GWh in 2013. The installation of an Archimedean screw at the outlet of the wastewater treatment plant has a payback period of less than 3 year and hydroelectric generation covers the electric consumption about 30% producing clean and feasible energy. The finding of the case study demonstrated that by using biogas and hydro energy together, the energy recovery efficiency of wastewater treatment plant can exceed 100% which is consistent with the operational data of Samra wastewater treatment plant in Jordan. Those energy-related technological solutions can make considerable contributions to sustainable wastewater treatment plant development.

References

1. ESHA-European Small Hydropower Association (2010) Energy recovery in existing infrastructures with small hydropower plants: multipurpose schemes-overview and examples
2. Kucukali S (2010) Hydropower potential of municipal water supply dams in Turkey: a case study in Ulutan dam. Energy Policy 38(11):6534–6539
3. Cottin C, Choulot A, Denis V (2012) Multipurpose schemes: feedback on electromechanical equipments set within drinking water networks. In: Proc., 2012 HidroEnergia Conf., Wroclaw
4. Denis V (2008) Wastewater turbining before and after treatment: an optimal use of existing infrastructures. In: Proc., 2008 HidroEnergia Conf., Bled
5. Lawrence S, Dickson P (2010) Clean energy infrastructure. In: Underhill MD (ed) The handbook of infrastructure investing. Wiley, New York
6. Chen S, Chen B (2013) Net energy production and emissions mitigation of domestic wastewater treatment system: a comparison of different biogas-sludge use alternatives. Bioresour Technol 144:296–303
7. Çakır C (2014) Atıksu aritma tesislerinde hava dağilimi optimizasyonu ile enerji tasarrufu, 4. Turkish-Ge. Water Partnership-Day, Antalya, Turkey (in Turkish)
8. Ak M, Kentel E, Kucukali S (2016) A fuzzy logic tool to evaluate low-head hydropower technologies at the outlet of wastewater treatment plants. Renew Sust Energ Rev 68:727–737
9. Steel E, Lilleland O (2015) Using supplier performance data to improve project performance. In: Proc., Hydro 2015 Conf., Bordeaux, 10.02, USB
10. Williamson SJ, Stark BH, Booker JD (2011) Low head pico hydro turbine selection using a multi-criteria analysis. World Renewable Energy Congress 2011, Linköping, CD
11. Bayon R, Hawn A, Hamilton K (2009) Voluntary carbon markets: an international business guide to what they are and how they work. Earthscan, Sterling

Mechanical Biological Treatment Facility: Energy Recovery Assessment

Sani Muazu, Hossein Mirzaii, T. Andy Augousti, and Redha Benhadj-Djilali

1 Introduction

Many governments across the world have set out targets for reducing greenhouse gas (GHG) emissions as a strategy for tackling global warming (GW) that causes climate change. One of the strategies for tackling GW is decarbonisation of energy supply from fossil fuels. A popular globally promoted strategy for decarbonisation of energy supply from fossil fuels is by increasing the penetration of renewable energy sources. One of the popular policy adaptations for tackling fossil fuel energy mix decarbonisation is the European Energy Roadmap 2050, which seeks to reduce GHG emissions by 80–90% by the year 2050 by increasing the share of electricity generation from biomass and waste by 7.3 and 10.8%, respectively. The dry and wet parts of municipal solid waste (MSW) can be converted into refuse-derived fuel (RDF) which can be used for energy production in dedicated facilities (waste incineration), including industrial processes, such as cement production and coal-based power production [1]. This highlights the critical role the energy derived from waste can play in achieving energy supply mix decarbonisation [1, 2].

Mechanical biological treatment (MBT) technology has been identified as the most preferred waste treatment technology that can effectively recover recyclable materials and high calorific value RDF from MSW. RDF has been promoted as a viable alternative energy substitute in conventional fossil fuel energy generation sources, and as such it can play an important role towards decarbonisation of energy supply from fossil fuels and increasing energy security [3, 4]. Energy security is a crucial element of economic posterity, and different countries have developed energy management policies that promote energy efficiency and security of supply. One of such policies is the Portuguese energy program called Intensive Energy Consumption Management system – SGCIE. The program was developed in 2008 to promote energy efficiency and monitoring in facilities with annual energy consumption higher than 500 ton (11.63 MWh) per year [5].

Even though energy recovery from RDF can help towards decarbonisation of energy mix, a recent study has shown that its potential for decarbonisation is limited to the type of energy used for its recovery from MSW [2] and the type of avoided referenced primary energy it will substitute in the lifetime assessment study [6]. Another study by Liikanen et al. [7] also shows that RDF utilisation for electricity generation in Brazil produces more GHG emissions, as the avoided referenced primary energy it will have substituted comes from hydropower renewable energy. Hydropower constitutes 75% of electricity generation in the country [7]. Hence, its potential environmental and other socio-economic benefits must be examined in a life cycle perspective, considering all induced burdens and gains from upstream and downstream processes in order to assess its sustainability towards increasing decarbonisation of energy mix [2]. This call for the need of further environmental and socio- economic research studies on MBT plant that will consider the local operating conditions and characteristics of MSW in order to assess and evaluate other useful performance parameters that can offer comprehensive understanding of the technology for better decision support making [6].

Majority of MBT plant published performance studies focused mainly on environmental assessment based on life cycle assessment (LCA) methodology neglecting other crucial parameters that are very important to local decision-making such as demand for space, energy load burden to the local energy security, and local capacity for utilisation of MBT plant output

S. Muazu (✉) · H. Mirzaii
Renewable Energy Engineering, Kingston University, London, UK
e-mail: K1560411@kingston.ac.uk

T. Andy Augousti · R. Benhadj-Djilali
School of Engineering and the Environment, Kingston University, London, UK

© The Author(s), under exclusive license to Springer Nature Switzerland AG 2022

A. Sayigh (ed.), *Sustainable Energy Development and Innovation*, Innovative Renewable Energy,
https://doi.org/10.1007/978-3-030-76221-6_55

products. LCA is a structured, comprehensive, and internationally standardized methodology (ISO 14040, 14044: 2006) that can be used to evaluate the environmental effect of a product or processes throughout its life cycle or lifetime, which is known as a "from cradle to grave" analysis [8]. The entire life cycle is included in the assessment, which means that the product under study is followed from the initial extracting and processing of raw materials through manufacturing, distribution, and use, up to final disposal, including all transportation involved [9]. LCA is an important part of the strategy to increase the transparency of technical and scientific data about plant functioning and consequences of the proposed development.

Even though LCA is mostly dedicated to environmental impact studies, it can be modified to include other parameters such as energy sustainability performance indicator, which can be used to account for further gains that are more beneficial to local planning and this will improve the benefits of LCA as a valuable tool for modelling waste management at both local and national scenarios [10].

In this study, the energy sustainability performance potential of MBT plant is assessed using life cycle direct net energy production balance assessment (NEP) methodology, which is a performance indicator for energy recovery rate of MBT plants.

The aim of the study discussed in this chapter is to assess the performance of MBT plant in Nigeria using life cycle direct net energy production balance assessment (NEP) methodology. The NEP assessment provides decision-makers with an important input for supporting decision-making on waste treatment technologies to local energy policies that promote security of supply.

2 Methodology

2.1 Net Energy Production (NEP) Assessment

NEP is an energy performance indicator that represents energy recovery efficiency at operation level of a project [11]. The NEP methodology used in this study is based on the work of Fei et al. [11] and Cimpan and Wenzel [6]. The NEP methodology considers all the input and output flows of all the systems and variables involved in energy and material recovery from MSW in order to assess the positive and negative benefits of MBT plant when compared with the existing MSW management in Nigeria.

$$\mathrm{NEP} = E_{\text{savings}} - E_{\text{consumption}} \tag{1}$$

$$E_{\text{consumption}} = E_{\text{MSW trans}} + E_{\text{MBT}} + E_{\text{prod trans}} \tag{2}$$

$$E_{\text{savings}} = E_{\text{RDF}} + E_{\text{ener sav}} \tag{3}$$

Where,

- NEP is the net energy production $\left(\frac{\mathrm{MJ}}{\mathrm{ton}}\right)$
- E_{savings} is total energy savings from utilising MBT products in energy recovery and substitution $\left(\frac{\mathrm{MJ}}{\mathrm{ton}}\right)$
- $E_{\text{consumption}}$ is total energy consumption involved in each assess scenario of MBT plant operation $\left(\frac{\mathrm{MJ}}{\mathrm{ton}}\right)$
- E_{MBT} is the energy consumed by MBT plant to recover output products $\left(\frac{\mathrm{MJ}}{\mathrm{ton}}\right)$
- $E_{\text{MSW-trans}}$ is total diesel fuel consumed for MSW transport to the MBT Plant $\left(\frac{\mathrm{MJ}}{\mathrm{ton}}\right)$
- $E_{\text{prod-trans}}$ is total diesel fuel consumed for MBT plant output products to the utilisation points $\left(\frac{\mathrm{MJ}}{\mathrm{ton}}\right)$
- E_{RDF} is Energy savings from RDF utilisation in energy recovery processes $\left(\frac{\mathrm{MJ}}{\mathrm{ton}}\right)$
- $E_{\text{ener-sav}}$ is Energy savings from Fe, Al, PET and HDPE recovery and subsitution $\left(\frac{\mathrm{MJ}}{\mathrm{ton}}\right)$

The lower heating value of RDF is determined using Dulong's empirical equation [4]:

$$\mathrm{HHV} = 0.336C + 1.418H - 0.145O + 0.0941S,$$

$$\mathrm{LHV} = \mathrm{HHV} - 0.0244(W + 9H),$$

$$W = \left(\frac{\mathrm{WW} - \mathrm{DW}}{\mathrm{WW}}\right) * 100$$

HHV higher heating value $\left(\frac{\text{MJ}}{\text{ton}}\right)$, LHV lower heating value $\left(\frac{\text{MJ}}{\text{ton}}\right)$ *C, H, O, S* element content (%); *W* moisture content (%) in wet weight (WW) and and dry (DW) weight basis

2.2 Goal and Scope Definition

The goal of this study is to assess the energy sustainability performance potential of MBT plant compared with the existing MSW management system in Kano metropolis located in north western Nigeria.

2.3 Scenarios and Boundary

Three MSW management scenarios are compared in this study using NEP performance assessment indicator.

2.3.1 Scenario 1

The first scenario is the current MSW management scenario in Kano metropolis. Kano State is the second most populous state in Nigeria with an estimated state-wide and metropolitan capital population of 13 and 4 million people, respectively [12]. In 2014, the annual generation of MSW in the metropolitan capital was estimated to be 1.8 million tons [13]. Available public record on collection and disposal shows that in 2008 an estimated 83,000 tons of MSW was collected across Kano metropolis by the state waste management body (REMASAB) and disposed of in landfill [12]. By conservative estimate of 8% annual growth in collection, the current MSW collection is projected to be 186,931 tons/year. The average composition of MSW across Kano metropolis, based on characterisation studies conducted in 2010, is shown in Table 1 [14]. Empirical data derived from analytical experiments for different components of MSW [15, 16] were used to calculate its physical and chemical properties in this study.

MSW management in Kano metropolis is entirely by landfilling, and collected MSW from the city centres is mostly shipped to four major and other small uncontrolled landfills within the city. A recent study shows that there are over 300 government's authorised and non-authorised MSW disposal landfills that are scattered within the metropolis [17]. Compactor trucks and other mini vans are mostly employed to transport waste from collection points to government-authorised disposal landfills. The boundary of current MSW management scenario in Kano metropolis is shown in Fig. 1.

2.3.2 Scenario 2

The second scenario is the MBT plant with process flow sheet shown in Fig. 2. The MBT plant model used in this study was developed in a spreadsheet using Microsoft® Excel visual basic for applications (VBA) programming language in order to generate life cycle inventory results that are used as inputs for conducting the NEP assessment in this study. The plant accepts and treats 176,000 tons of MSW for a period of 4000 hours per year. The MBT plant is assumed to be located at 50 km distance from waste collection point, residue landfill and 100 km away from energy recovery and recycling plants. The annual MBT plant electricity use is 43.01 kWh/ton and 1.36 L/ton diesel consumption.

2.3.3 Scenario 3

The third scenario is the same as the second scenario but considers a situation where there is no capacity to utilise the MBT plant output products.

Table 1 Composition of 1 ton of MSW in Kano metropolis [14]

MSW components	(%)	Wet (tons)	Dry (tons)	Carbon	Hydrogen	Oxygen	Nitrogen	Sulphur	Ash	Total
Food scrap	27.67	276.67	83.00	39.84	5.31	31.21	2.16	0.33	4.15	83.00
Vegetable	10.33	103.33	41.33	19.01	2.48	15.71	1.41	0.12	2.60	41.33
Paper and cardboard	15.00	150.00	141.00	61.19	8.18	62.46	0.42	0.28	8.46	141.00
Textile and rubber	7.00	70.00	63.00	30.24	4.03	25.20	1.39	0.13	2.02	63.00
Plastic materials	11.00	110.00	107.80	64.68	7.76	24.58	–	–	10.78	107.80
Ash/dirt	13.00	130.00	119.60	29.06	3.59	4.78	0.60	0.24	81.33	119.60
Metal	9.33	93.33	90.53	–	–	–	–	–	90.53	90.53
Glass	6.67	66.67	65.33	–	–	–	–	–	65.33	65.33

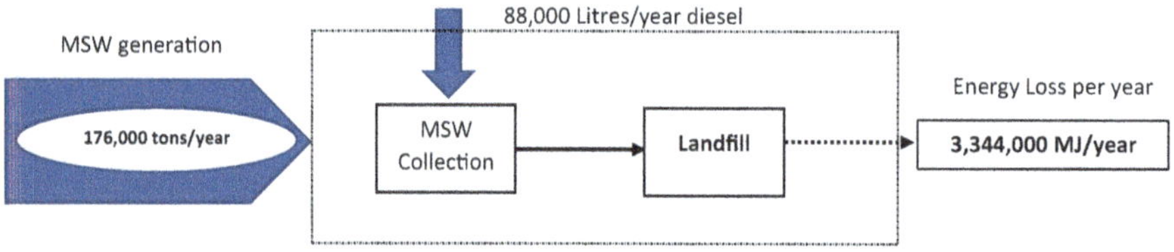

Fig. 1 Energy use flow for Kano metropolis MSW scenario 1 assessment for year 2020

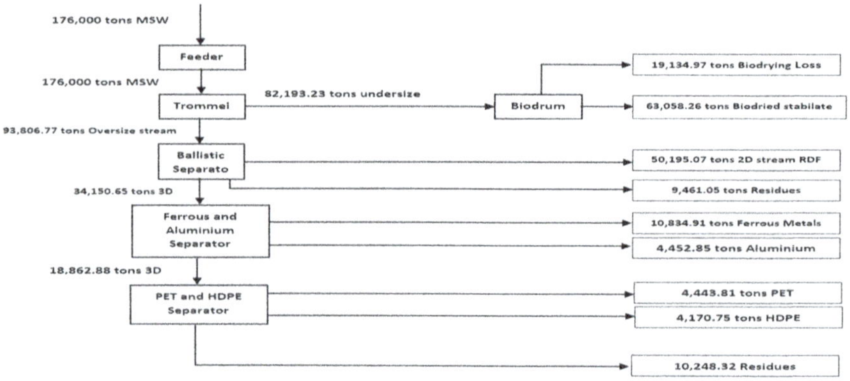

Fig. 2 MBT Plant Process Flow Chart and Mass Balance

2.4 Life Cycle Inventory Energy Assessment

To establish common basis of assessment, all 3 scenarios involve treatment of 176,000 tons of MSW per year.

For scenario 1, the assessment boundary starts at the point of collection and disposal at landfill located at 50 km. Energy consumption ($E_{\text{consumption}}$) is estimated based on the volume of diesel consumed by MSW collection trucks that travel a total round distance of 100 km at a rate of 0.05 Litres/ton-km [6].

For scenario 2, the assessment boundary energy consumption ($E_{\text{consumption}}$) occurs at three stages. The first stage is the same as scenario 1. The second stage is the annual sum of electricity and diesel consumed by the MBT plant to treat the input MSW from scenario 1 to generate output products in Fig. 2. The third stage involves diesel consumption by long haul truck for transport of MBT plant output products: RDF to incineration facility, ferrous metal, aluminium, polyethylene terephthalate (PET) and High-density polyethylene (HDPE) plastics to reprocessing facilities, and biodried stabilate and residues to landfill. The diesel consumption of long haul truck is assumed to be 0.03 L/ton-km [6]. All electricity and diesel energy consumption units are converted to energy equivalent (1 L = 38 MJ).

Table 2 Avoided energy from reprocessing of recyclable materials (per tons of secondary material produced) [18]

Recyclable materials	Recycling efficiency (%)	Energy savings (MJ_{eq})
PET	81	−15,940
HDPE	81	15,940
Ferrous metal	100	174,634
Aluminium	100	146,76

Energy savings ($E_{savings}$) are estimated based on the following: the RDF recovered is assumed to be used in electricity generation plant with net electrical conversion efficiency of 19%. Ferrous metal and aluminium are assumed to be processed in the steel and aluminium manufacturing plants to produce new metallic materials. PET and HDPE are assumed to be granulated and reprocessed to produce new plastic materials. The recycling of all recyclable materials is assumed to substitute the production of the same materials from virgin raw materials with the following substitution rate: ferrous and aluminium 0.9 and PET and HDPE 0.81. Energy savings from recycling of the recyclables are shown in Table 2. Energy consumption ($C_{onsumption}$) is estimated based on the treatment of input MSW Lower Heating Value (LHV) means Lower Heating Value of 10.14 MJ/kg and moisture content of 28.49%w/w [18].

3 Results and Discussion

3.1 *Results*

The net energy production (NEP) assessment for three scenarios is presented below. Total energy consumption ($E_{consumption}$) for scenario 1 shown in Table 3 is estimated to be 190 MJ/ton and energy savings ($E_{savings}$) is zero, since there is no recovery of recyclable materials or RDF.

Total energy consumption for scenario 2 ($E_{consumption}$) is estimated to be 754.35 MJ/ton as presented in Tables 4 and 5. The estimated modelled electricity consumption of the MBT plant is 43.01 kWh/ton and this value is within the range of typical MBT plant electricity consumption estimates (4–73 kWh) reported by Ecodeco Company MBT plants data which shows an electricity consumption of 48 kWh/ton [19]. A study on MBT plant by McDougal et al. [19] and Reza et al. [4] shows closely related electricity consumption figures of 55.5 kWh/ton and 48.5 kWh/ton, respectively. This proves that the results of the energy consumption of the MBT plant are within acceptable limit of what is reported in the literature. Results for all the energy consumption processes in the second scenario assessment boundary show that the transport of MBT plant output products has the highest energy consumption of 357.98 MJ/ton (47.45%) followed by MBT plant with 206.37 MJ/ton (27.36%). MSW transport has the lowest energy consumption of 190 MJ/ton (25.19%).

The total energy savings ($E_{savings}$) for scenario 2 are given in Table 6. As it can be seen, aluminium recovery and recycling processes have the highest energy savings of 4418.29 MJ/ton (73.35%) followed by ferrous metal with 903.48 MJ/ton (15%).

The total energy consumptions ($E_{consumption}$) for scenario 3 is the same as for scenario 2 at 754.35 MJ/ton and the total energy ($E_{savings}$) savings is zero, as there is no local capacity to utilise the MBT plant output products.

3.2 *Discussion*

The net energy production (NEP) for scenario 1 is negative, and it can be expressed that for every 1 ton of MSW disposed in landfill without any recovery of materials or energy, an estimated 190 MJ is lost under the existing MSW management system in Kano.

The net energy production (NEP) for scenario 2 is 5269.27 MJ/ton, and this means that for every 1 ton of MSW treated at the MBT plant, an estimated 5269.27 MJ is saved based on the current MSW composition for Kano metropolis in Table 1.

The net energy production (NEP) for scenario 3 shown in Table 7 is negative at −754.35 MJ/ton, and it can be expressed that implementation of scenario 3 will result in more energy loss than scenario 1. This is because there is no capacity for utilisation of outputs.

The NEP assessment for the three scenarios assessed in Fig. 3 indicates that scenario 2 is better than scenarios 1 and 2 in terms of NEP efficiency.

Table 3 Energy consumptions ($E_{consumption}$) for scenario 1

MSW treatment process	Quantity (tons)	Total diesel consumption (L/year)	Energy consumptions (MJ/ton)
MSW landfill	176,000.00	880,000.00	190.00

Table 4 MBT plant electricity and diesel energy consumptions for scenario 2

MBT processes	Type of fuel	Unit	Energy consumptions	Energy consumptions (MJ/ton)
Feeder	Electricity	kWh	26,022.22	0.53
Trommel screen	Electricity	kWh	47,049.61	0.96
Ballistic separator	Electricity	kWh	239,466.19	4.90
Metal separator	Electricity	kWh	14,991.67	0.31
Plastic separator	Electricity	kWh	8,335.89	0.17
Biodrum	Electricity	kWh	7,043,191.58	144.07
Baler	Electricity	kWh	72,862.44	1.49
Building areas and dedusting	Electricity	kWh	64,391.98	1.32
Biodrying air treatment	Electricity	kWh	54,035.83	1.11
	Electricity	kWh	**7,570,347.42**	**154.85**
Front-end loader	Diesel	Litres/year	146,666.67	31.67
Skidloader	Diesel	Litres/year	91,967.42	19.86
	Diesel	Litres/year	**238,634.09**	**51.52**
			Total	**206.37**

Table 5 Transport Energy Consumption for MSW and MBT Plant Output Products in Scenario 2

MSW & MBT plant outputs	Recovered quantity (tons)	Recovered percentage (%)	Total diesel consumptions (L/Year)	Energy consumptions (MJ/ton)
Raw MSW	176,000.00	0	880,000.00	190.00
PET plastics	4,443.81	2.52	111,184.11	24.01
HDPE plastics	4,170.75	2.37	104,352.23	22.53
Ferrous metals	10,834.91	6.16	271,089.55	58.53
Aluminium	4,452.85	2.53	111,410.32	24.05
RDF	50,195.07	28.52	811,654.24	175.24
Residues	19,709.37	11.20	59,128.12	12.77
Stabilate	63,058.26	35.83	189,174.77	40.84
Biodrying loss	19,134.97	10.87		
	176,000.00	**100.00**	**Total**	**547.98**

Table 6 Net energy production (NEP) for scenario 2

MBT plant output products	Recovered quantity (tons)	Energy savings (MJ/ton)
PET plastics	4,443.81	300.80
HDPE plastics	4,170.75	282.32
Ferrous metals	10,834.91	903.48
Aluminium	4,452.85	4,418.29
RDF	50,195.07	118.73
Residues	19,709.37	0
Stabilate	63,058.26	0
Biodrying loss	19,134.97	0
Total	**176,000.00**	**6,023.62**

Table 7 Net energy production (NEP) for scenario 3

Processes	Energy outflows (MJ/ton)	Energy inflows (MJ/ton)	Net energy production (MJ/ton)
MSW transport	0	190.00	−190.00
MBT plant	0	206.37	−206.37
MBT products	0	357.98	−357.98
		Total	**−754.35**

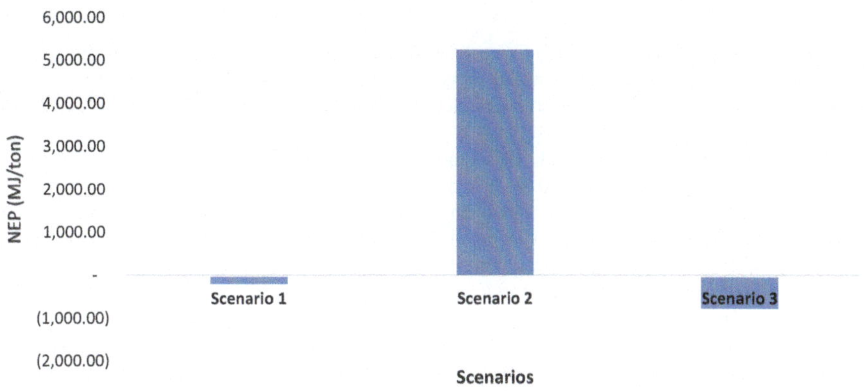

Fig. 3 NEP plot for three scenarios

4 Conclusion

This chapter presents a study on the role of MBT technology and its benefit in local waste management system in Kano metropolis city located in north western Nigeria. The NEP assessment results for existing MSW management (scenario 1) in this study show that projected disposal of 176,000 tons of MSW in Kano metropolis for the year 2020 will lead to a massive energy loss of 880,000 litres of diesel per year. This quantity of diesel is the equivalent of daily diesel consumption demands of 22,420 vehicles in Kano. This is based on estimates of 39.25 litres/day-vehicle reported in a study on the daily average fuel consumption of highway vehicles for some cities in Nigeria [20].

Based on current average annual composition of MSW in Kano metropolis, it can be said that implementation of MBT plant in the local MSW management system will bring about positive benefits of energy savings of 5269.27 MJ per every ton of MSW treated. A scenario 3, where there is no local capacity to utilise the outputs of the MBT plant, will result in more energy loss of 754.35 MJ/ton and this is much worse than the current system (scenario 1) of MSW treatment in Kano metropolis with energy loss of 190 MJ/ton.

The NEP assessment results shed more light on energy consumptions ($E_{consumption}$) and energy savings ($E_{savings}$) along the boundary of assessment for the three scenarios. The NEP for scenario 1 can be improved by introducing MSW collection trucks with less diesel fuel consumption or electric vehicles in order to improve the local energy security.

The NEP assessment results for scenario 2 provides clearer picture on energy consumption and savings processes and potential for optimisation to improve the NEP savings. The first stage can be improved with implementation of suggestions for scenario 1 above. Improving the composition of the input MSW to the MBT plant in the second stage will result in more recovery of recyclable materials and RDF, which will translate into improved energy savings from recycling and energy utilisation and lesser quantity of residues and stabilate for transport in stage 3. Reducing the quantity of residues and stabilate will reduce the volume of diesel consumption and improve the NEP of scenario 2. Similarly, utilisation of long-haul trucks or electric vehicles for MBT plant output products will significantly increase the NEP of scenario 2.

The results of NEP assessment for scenario 3 indicate that implementation of MBT plant alone will not bring about any benefits to the local MSW management system if there is no capacity for utilisation of the output products.

The results of NEP assessments presented above are only valid for the average MSW composition and moisture content shown in Table 1. A future change in composition may result in better or worse energy savings. Hence, more research works need to be done to test the sensitivity of changing composition of input MSW to NEP indicator for both scenarios assessed.

Based on new data provided in this study, it can be concluded that existing MSW management in Kano metropolis by landfill disposal is a serious threat to the energy security of the city. There is need for new MSW management solution for Kano metropolis.

Acknowledgement This work was supported by the Petroleum Technology Development Fund (PTDF) of the Nigerian Government under the scholarship number PTDF/ED/PHD/MS/1234/17.

References

1. Cimpan C, Maul A, Jansen M, Pretz T, Wenzel H (2015) Central sorting and recovery of MSW recyclable materials: a review of technological state-of-the-art, cases, practice and implications for materials recycling. J Environ Manag 156(2015):181–199
2. Longo S, Cellura M, Girardi P (2020) Life cycle assessment of electricity production from refuse derived fuel: a case study in Italy. Sci Total Environ 738(2020):139719
3. Álvaro R (2018) Design and modelling of a large-scale PV plant. Available at: https://upcommons.upc.edu/bitstream/handle/2117/120416/master-thesis-alvaro-roca-rubi.pdf. Accessed 22 July 2020
4. Reza B, Soltani A, Ruparathna R, Sadiq R, Hewage K (2013) Environmental and economic aspects of production and utilization of RDF as alternative fuel in cement plants: a case study of Metro Vancouver Waste Management. Resour Conserv Recycl 81(2013):105–114
5. Bernardo H, Oliveira F, Quintal E (2014) Energy management in municipal solid waste treatment: a case study of a mechanical biological treatment facility. Available at: https://www.eceee.org/library/conference_proceedings/eceee_Industrial_Summer_Study/2014/4-undertaking-high-impact-actions-the-role-of-technology-and-systems-optimisation/energy-management-in-municipal-solid-waste-treatment-a-case-study-of-a-mechanical-biological-treatment-facility/2014/4-093-14_Bernardo_PR.pdf/. Accessed 23 Mar 2020
6. Cimpan C, Wenzel H (2013) Energy implications of mechanical and mechanical-biological treatment compared to direct waste-to-energy. Waste Manag (New York, NY) 33(2013):1648–1658
7. Liikanen M, Havukainen J, Viana E, Horttanainen M (2018) Steps towards more environmentally sustainable municipal solid waste management — a life cycle assessment study of São Paulo, Brazil. J Clean Prod 196(2018):150–162
8. Roy P (2009) A review of life cycle assessment (LCA) on some food products. J Food Eng 90:1–10
9. Berlin J (2002) Environmental life cycle assesssment (LCA) of Swedish semi-hard cheese. Int Dairy J 12(11):939–953
10. Nondek L (2010) Optimization model of integrated MSW management. Available at: http://www.wasteconsult.net/files/downloads/2010-12%20-%20Nondek%20-%20Optimization%20Model%20MSW%20management.pdf. Accessed 12 Feb July 2019
11. Fei F, Wen Z, Huang S, Clercq DD (2018) Mechanical biological treatment of municipal solid waste: energy efficiency, environmental impact, and economic feasibility analysis. J Clean Prod 178(2018):731–739
12. Ali F (2012) Groundwater pollution threats of solid waste disposal in Urban Kano, Nigeria: Evaluation and protection strategies. Available at: https://www.escholar.manchester.ac.uk/api/datastream?publicationPid=uk-ac-man-scw:198714&datastreamId=FULL-TEXT.PDF. Accessed 15 June 2020
13. Daura LA (2016) Determination of greenhouse gas emission from incineration of municipal solid waste in Kano, Nigeria. IJSER 7(7):912–915
14. Nabegu AB (2010) An analysis of municipal solid waste in Kano metropolis, Nigeria. J Hum Ecol 31(2):111–119
15. Anders L, Holger E (2011) Waste characterization: approaches and methods. In: Thomas HC (ed) Solid waste technology & management, 1 & 2, First edn. Blackwell Publishing Ltd, pp 62–84
16. Yusuff AS, John W, Okoro O, Ajibade A (2014) Physico-chemical composition and energy content analysis of solid waste. A case study of Castlereagh District, Northern Ireland. Am J Eng Sci Technol Res 2(1):1–9
17. Naibbi AI, Umar UM (2017) An appraisal of spatial distribution of solid waste disposal sites in Kano Metropolis, Nigeria. J Geoscience Environ Prot 5(11):24
18. Cremiato R, Mastellone M, Tagliaferri C, Zaccariello L, Lettieri P (2018) Environmental impact of municipal solid waste management using Life Cycle Assessment: the effect of anaerobic digestion, materials recovery and secondary fuels production. Renew Energy 124(2018):180–188
19. EPEM (2020) Mechanical biological treatment facilities. Available at: http://www.epem.gr/waste-c-control/database/html/MBT-01.htm. Accessed 25 Mar 2020
20. Maxwell UN, Harold CM (2012) Analysis of consumption pattern of highway transportation fuel in Nigeria. OPEC Energy Rev, Organization of the Petroleum Exporting Countries 36(1):25–34

Climate Change and Renewable Energy-2020

Ali Sayigh

1 Introduction

The Covid-19 pandemic has led to the reduction in demand for energy especially in transport: air, land and sea, reducing carbon emissions, and resulted in cleaner air. Now we have seen how life looks with fewer planes, cars and less pollution. Some are hopeful that we won't want to revert back to old-style transport and energy systems. This forced action due to the pandemic is a fraction of what could be achieved if there was a real commitment to reduction in the use of fossil fuels. Notwithstanding this, there is still the urgent need to cut back massively on all activities that lead to carbon emissions [1].

Climate change and the resulting thermal warming are happening but they can be slowed down by taking aggressive measures as those we are taking with the present pandemic. In other words if there is a will there is a way.

We live on a small planet populated by 7.7 billion people which in 2050 will have reached be 10 billion. Resources are limited, water is limited, air is limited and climate change is a global problem not a regional problem.

As the pandemic decreases we will emerge into a very different world from a climate perspective. Emissions of greenhouse gases in 2020 have been reduced which has provided a thinking period. But the falling emissions driven by the pandemic are not sustainable, and easily reversed. Now is the time to realize that the warning from millions of scientists and climate experts about climate change should not be ignored. We should use this Covid-19 sudden shock to rethink our ultimate purpose as human beings. Dr. Antonio Guterres, UN Secretary General, said it clearly, 'We will not fight climate change with a virus' [2].

The CO_2 concentration in the atmosphere, in May 2020, reached 417.2 ppm. This is higher than last year by 2.4 ppm. Therefore, now the fear is that global temperature could reach 2 °C more than it is now. If Covid-19 caused 850,000 deaths (August 2020) (referring to: Our World in Data), global warming could cause more death and financial calamity than the Covid-19 (see Fig. 1)

The effect of global warming will result in millions of deaths and billions of destruction worldwide. Global warming does not respect place nor status, all countries rich or poor, will face catastrophe.

It is obvious that for those who do not have the wealth and the manpower as well as the technology, global warming will be more devastating for them. Figure 2 shows melting snow in the North Pole and the Arctic destroying many of the animals and causing more rain and floods in other parts of the world.

Desertification is another major consequence of global warming in many countries. According to the world Meteorological Organization (WMO), the 20 warmest years on record have been in the past 22 years, with the period 2015–2018 making up the top four. The global average temperature for the first 10 months of 2018 year was nearly 1 °C above the levels of between years 1850 and 1900. If this trend did continue, the WMO says temperatures may rise by *3–5 °C by 2100* (*see* Fig. 3).

If the temperature reaches over 50 °C with dry grass or hay or wooded forest, there is a major risk of self-ignited wild fires. According to the World Resources Institute:

Global warming happening since 1970 has accelerated due to increased burning of more fossil fuels liberating excessive CO_2 and other green gases. All the excessive CO_2 in the atmosphere which has not been absorbed by the sea or plants will create a blanket over the Earth which will trap the heat which has come from the Sun within the Earth's atmosphere. This will make life unbearable and eventually unsustainable. The oceans are already saturated and cannot absorb any more CO_2, while the maximum quantity which plants can absorb is 30%. Furthermore, the cutting of trees and forest fires have resulted in reducing the carbon dioxide sink to about 20%. Emitted CO_2 lasts 20 years [8] (Fig. 4).

A. Sayigh (✉)
Chairman of World Renewable Energy Congress, DG of WREN, FRSA, Brighton, UK
e-mail: asayigh@wrenuk.co.uk; http://www.wrenuk.co.uk/

© The Author(s), under exclusive license to Springer Nature Switzerland AG 2022

A. Sayigh (ed.), *Sustainable Energy Development and Innovation*, Innovative Renewable Energy,
https://doi.org/10.1007/978-3-030-76221-6_56

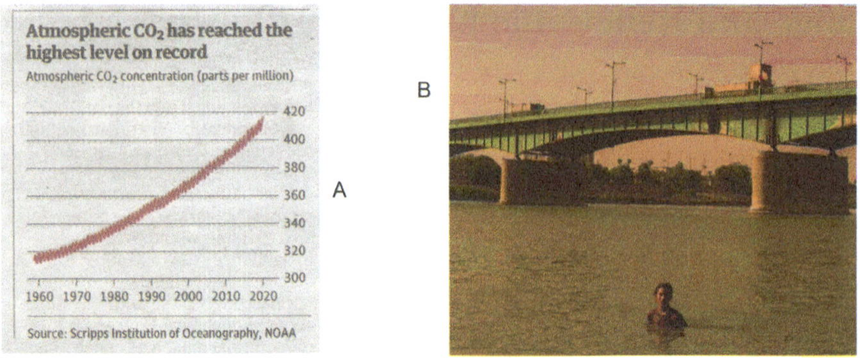

Fig. 1 (**a**) Scripps Institute of Oceanography, NOAA [3]. (**b**) A boy in Baghdad cooling himself in the river Tigris on 5 August 2020; temperature reached over 50 °C [4]. In Kuwait, temperature reached 54 °C in the summer [5]

Fig. 2 (**a**) North Pole. (**b**) Vince flood, Italy [6]

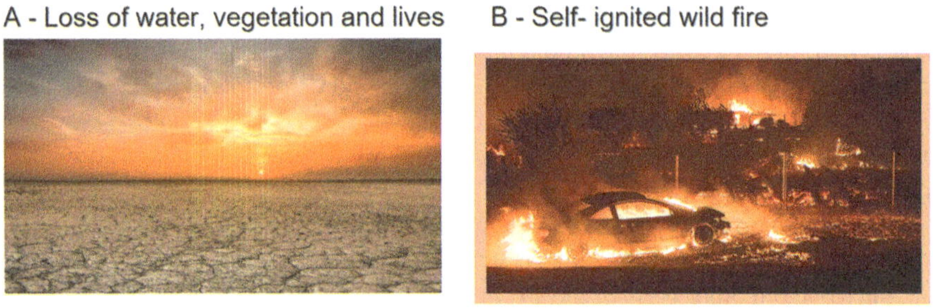

Fig. 3 Effects of climate change [7]. (**a**) Loss of water, vegetation and lives. (**b**) Self-ignited wild fire

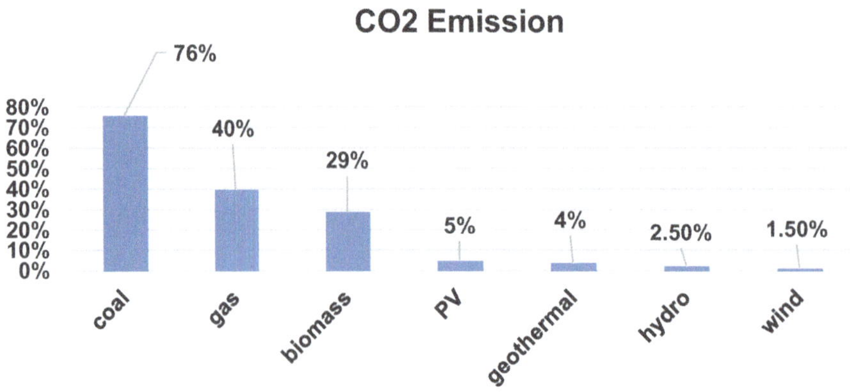

Fig. 4 Percentage (%) of carbon emission from energy sources

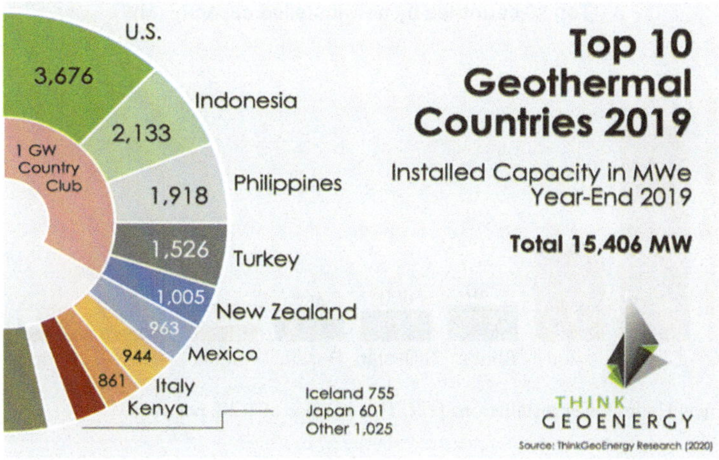

Fig. 5 The top 10 geothermal countries 2019

2 Geothermal Energy

Geothermal energy is the energy extracted from tapping the Earth heat at different depths. The deeper you drill the hotter layers become. Heat has been radiating from the centre of the Earth for some 4.5 billion years. At 6437.4 km (4000 miles) deep, the centre of the Earth has around the same temperature as the Sun's surface, 9932 °F (5,500 °C) [9]. At the end of 2019, the Global Power generated from geothermal applications was 15.406 GW [10] (Fig. 5).

3 Hydropower

Hydropower capacity at the end of 2019 is 1307 GW [11]. It is expected to remain the world's largest source of renewable electricity generation and supplies electricity to many parts of the world. Hydropower capacity increases led by China, India and Brazil. In the sustainable development scenario (SDS), hydropower generation will expand 3% per annum until 2030. However, capacity expansion recently is in the downward trend and expected to continue due mainly to less large-project development in China and Brazil, where concerns over social and environmental impacts [11] have questioned the overall benefits of large-scale dams. Additionally, more than 50 countries have mini-hydropower resulting from rainfall or small rivers and streams. The average cost of electricity from hydropower is about €0.06 per kWh (Figs. 6 and 7).

4 Tidal Energy

Tidal energy is a renewable source of energy produced by the surge of ocean waters during the rise and fall of tides. The first was located in La Rance, France. The largest facility is the Sihwa Lake Tidal Power Station in South Korea. China, France, England, Canada and Russia have much more tide potential yet to be utilized [14], and this huge potential is too great to be ignored and could be crucial in offsetting the use of fossil fuel.

5 Biomass and Biogas and Waste to Energy

Biomass is one of the main tool in reducing carbon emissions, and importantly, it is one form of renewable energy which is available in every country.

According to Svetlana Ladanai and Johan Vinterbäck – Report 013, ISSN 1654-9406, Uppsala, Sweden [15]: 'It is the fourth largest energy source after coal, oil and natural gas – it can produce different forms of energy. Renewable energy options, capable of providing all the energy services required, both locally and globally. The advantages of biomass as energy

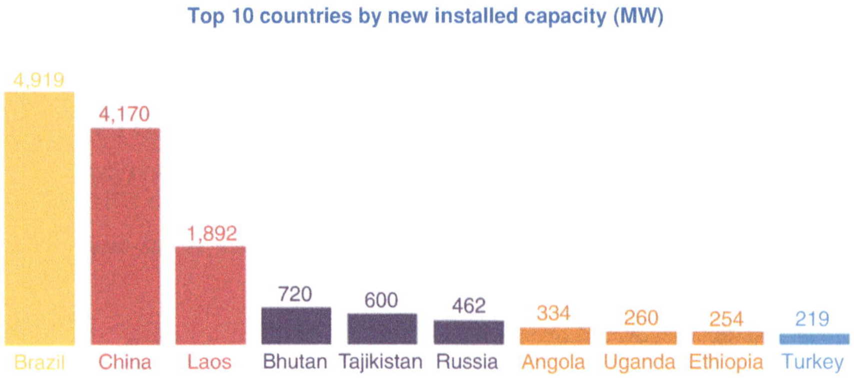

Fig. 6 Top 10 countries added new Hydropower installations [12]. This is a rise of 1.2% per year

Fig. 7 Rance Tidal Station, Brittany, France [13], 500 GWh. Its length is 700 m. and the tide range is 8 metre

resource is common and widespread across the globe. For example, the annual global primary production of biomass is equivalent to the 125×10^{10} GW of solar energy captured each year. If one takes 5% of this power, or 6.25×10^{10} GW, then this should cover more than 50% of the world's total primary power demand. The potential for energy from biomass depends in part on land availability. Assuming'. According to IEA [16], power generation from biomass end of 2019 was 68.5 GW, and expected to reach 137 GW by 2030.

In June 2020, IEA produced a report of bioenergy generation analysis which has regional biofuel and waste to energy (W-t-E) combined in kilo-ton oil equivalent (ktoe), it is (1,328,839) [17].

The World's largest W-t-E plants are in Shenzhen, China and in United Arab Emirates – Dubai commenced operation in 2020. Each plant can process 5500 tonnes of waste per day.

The power capacity of Dubai plant is 185 MW, producing 2% of Dubai's annual energy consumption, while Shenzhen East has a capacity of 165 MW, handling one-third of the city's waste (Fig. 8).

By the end of 2019, there were 2450 plants around the world using 368 million tonnes of waste a year. Each tonne of waste produced 550 kWh. Therefore, in 2019, W-t-E was equivalent to 23.1 GW. Therefore, the combined capacity for biomass and W-t-E is 91.6 GW

Fig. 8 Dubai's W-t-E largest plant and Shenzhen East's largest plant with PV on top [18]

6 Anaerobic Digestion (AD)

This is a process by which organic material is broken down by microorganisms in the absence of oxygen, producing biogas, a methane-rich gas used as a fuel, digestant, or a source of nutrients as fertilizer. The process normally takes 15–30 days for the organic waste to be processed by an AD plant. The biogas created is used to generate electricity or heat with a combined heat and power unit (CHP). The biofertilizer is pasteurized to make it pathogen-free and can be applied on farmland successfully. The technology is widely used to treat wastewater and can be employed to treat organic wastes from domestic and commercial food into manures and biofuel crops [19]. It is also used in the fermentation process of beer and lager production. While in developing countries, it is used on small farms and in buildings to provide power for cooking and lighting [20].

There are 473 operational AD plants in the UK, including 80 bio-methane-to-grid plants, and a further 327 projects are under development [20, 21].

7 Solar Thermal

Looking at solar radiation intensity globally, the area of the Middle East and North Africa are the richest in the world (see Fig. 9). In 2019, 652.4 GW_{th} cumulative were installed globally, which if converted to electrical power at 30% efficiency will produce 196 GW [21] (Figs. 10, 11, and 12).

Solar water heaters have improved extensively since selective coatings were introduced which increased their efficiency. Figure 13 shows some modern collectors. Figure 14 shows global concentrating solar power.

In 2019, solar thermal systems had the power of producing 479 $GW_{th.}$ This is equal to saving of 43 million tonnes of oil saved and 138 million tonnes of CO_2 emissionavoided [24], while Concentrated Solar Power (CSP) has generated 6.451GW_e.

8 Photovoltaic Technology

Since 1839, when Becquerel discovered the photovoltaic effect, many types of solar cells have been developed around the world. Their efficiency has been significantly improved [14], and their cost dropped. There is no country which does not have photovoltaic installation, whether megawatt in solar farms or small-scale kilowatt applications [25] (see Table 1).

The dominant solar cells in the market are crystalline and polycrystalline cells. Thin-film cells, such as CdTe, CIGS, a-Si and TF-Si cells, are much cheaper because they use one-tenths of the silicon used by crystalline cells but have less efficiencies. Figure 15 shows the best research cell efficiencies [26].

Another is Kurnool Ultra Mega Solar Park, in India: This has a capacity of 1000 MW.

It is worth noting that more than 35 countries have more than 1 GW, PV capacity.

Photovoltaic cells and usage can cover every possible shape and application. PV can be incorporated in building facades whether in situ or in cladding or attached to bridge structures (see Fig. 16). PV can play an important role in the advent and can increase the production and sale of electric cars (EV) providing widespread charging stations.

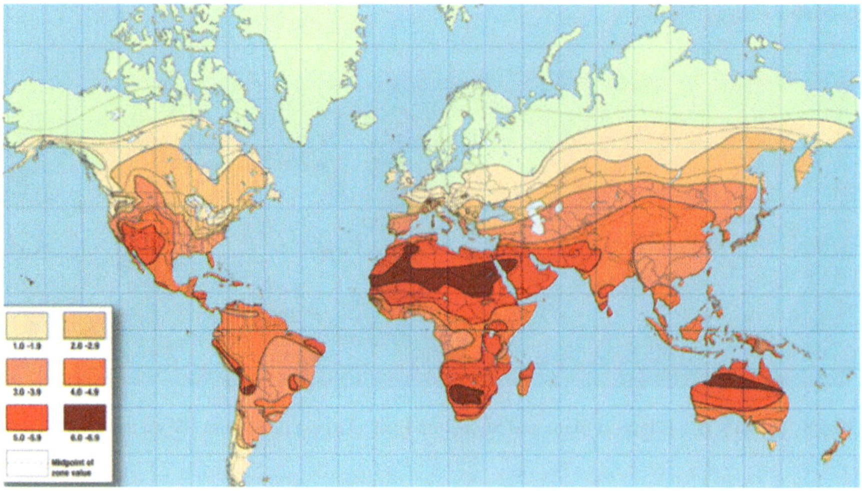

Fig. 9 Solar radiation intensity globally – kWh/m^2

Fig. 10 Some of the recent flat plate collectors made by an Austrian company [22]

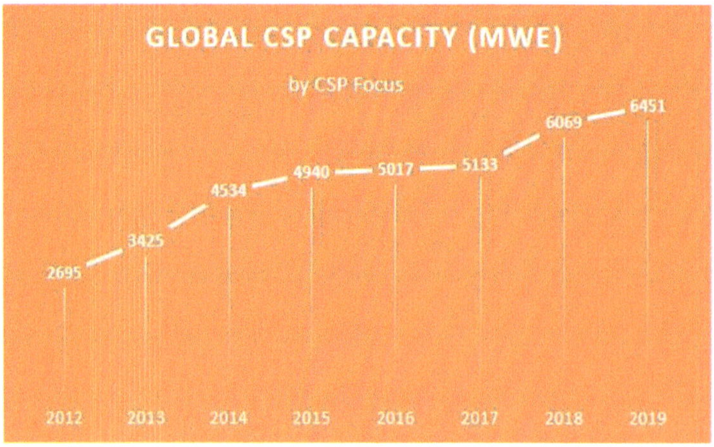

Fig. 11 Global CSP in MWe [23]

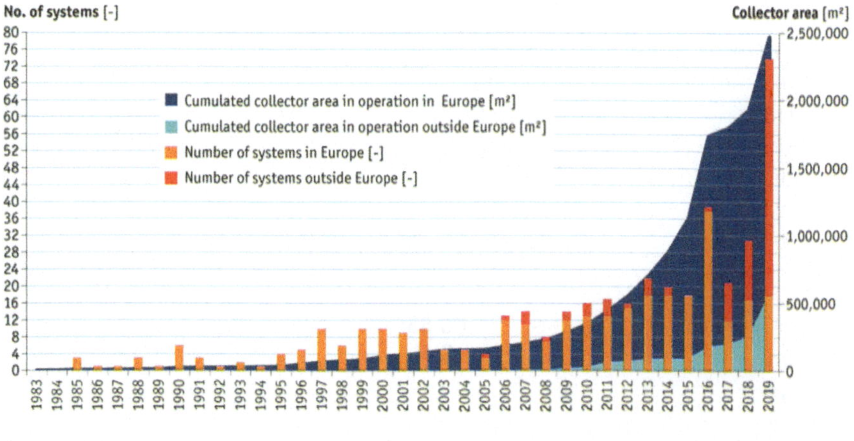

Fig. 12 Most large solar thermal systems globally at the end of 2019

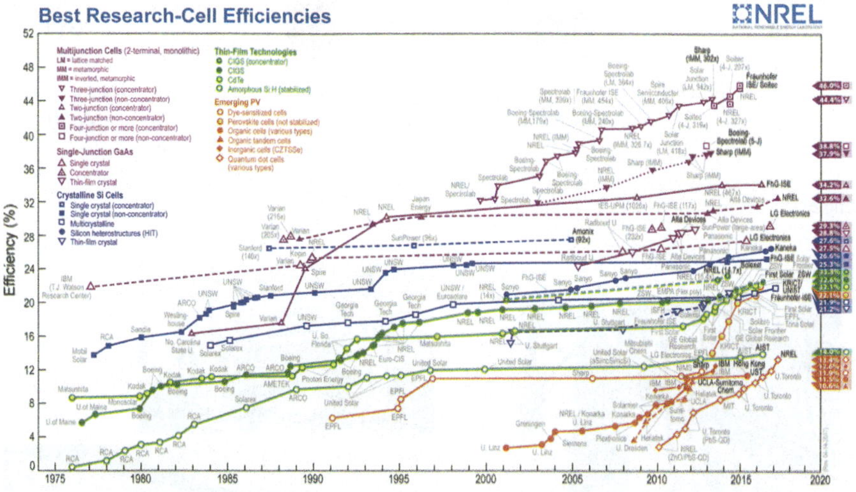

Fig. 13 World's largest PV farm, Longyangxia Dam Solar Park in Qinghai province – China, 850 MW [27]

Fig. 14 World's largest PV farm, Longyangxia Dam Solar Park in Qinghai province – China, 850 MW [27].

Table 1 Selected countries with PV capacity in 2019

No.	Country	PV-GW	No.	Country	PV-GW
1	China	200	7	Australia	14.6
2	USA	115	8	United Kingdom	13.3
3	Japan	54	9	South Korea	11.2
4	Germany	49.2	10	France	9.9
5	India	42.8	11	Spain	8.4
6	Italy	20.8	12	Rest of the world	113.2

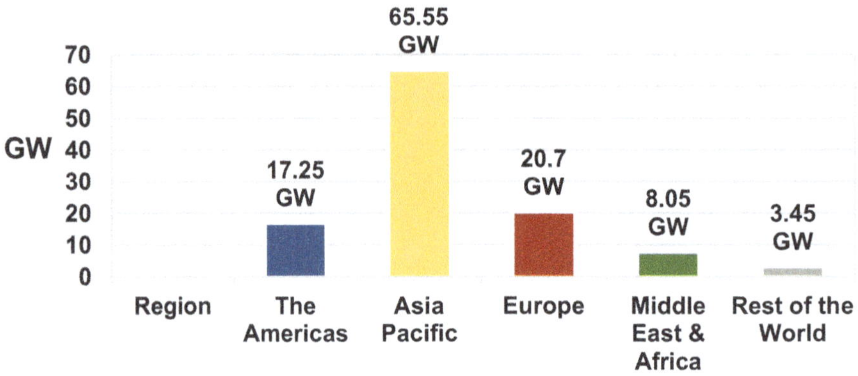

Fig. 15 Best cell efficiencies

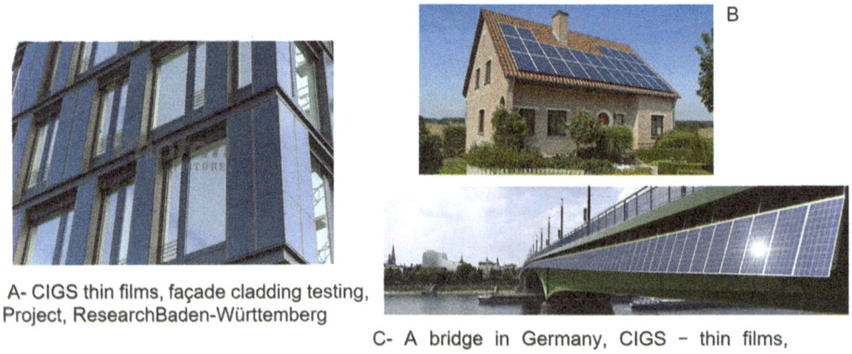

Fig. 16 Both the house and the bridge are using passivated – emitter rear contact solar cells – PERC, Cologne, Germany. (**a**) CIGS thin-film, façade cladding testing project, Research Baden-Württemberg. (**c**) A bridge in Germany, CIGS – thin films

In 2020, crystalline single solar cell efficiency has reached 26%. Nowadays it is possible to buy solar panels for large-scale projects with all their associated inverters, connections and stands for US$ 0.37 per Watt. For smaller PV systems (<1 kW) the cost is US$ 0.50.

PV can also be floated on a lake or sea. It is normally more expensive than installing it on land, but it has the advantage of not occupying needed land and it will have less dust on the covers. One of the biggest floating PV farm is in China (see Fig. 17).

9 Wind Energy Technology

Great advances have been made in wind energy, in the design of the structures, of the turbines, which have led to a greatly increased efficiency and lowered costs so that nowadays it is a major producer of renewable electricity [29].

The development of wind energy became more active during the late 1970s and 1980s due to the impetus given by the urgent necessity of climate change. As it was stated in Fig. 4, it has the lowest carbon footprint of all methods of power generation.

Fig. 17 The 40 MW floating solar power plant at Huainan, China, is able to power a small town [28]

Modern wind technology has been developed and is based in two sections, onshore and offshore. Other developments such as floating windmills and vertical axis windmills and single-blade mills are not commonly used.

In principle, the higher the turbine the better is the performance. Also 90% of all turbines are three-bladed with a perfect aero foil shapes. Figure 18 shows the future height of hubs in comparison with the existing structures in the USA, in particular, the Haliade-X, GE new 12-MW machine which will be ready in 2021 [30].

One of the largest wind farms on-shore in the UK (see Fig. 19).

Another one of the largest on-shore wind farm is Alta Wind Energy Centre (AWEC) in Tehachapi, Kern County, California (see Fig. 20).

10 Off-Shore Wind Farms

The United Kingdom has the most off-shore wind power globally. It has nearly 8 GW. In total the UK has combined wind power capacity of 60.0 GW. The largest wind farm off-shore in the UK is the Walney Extension; it has a capacity of 659 MW (Figs. 21 and 22).

In 2018, EON completed a 432 MW off-shore wind farm in Brighton, UK supplying electricity of 1400 GWh per year to more than 350,000 homes, or half of the homes in Sussex and saving 600,000 tonnes of CO_2 a year. The farm consisted of 116 MHI Vestas 3.45 MW turbines, their height is 140 metre and the blade length is 55 metre (see Fig. 23).

The World off-shore wind capacity is 29 GW at the end of 2019 (see Fig. 24).

The cost of electricity from wind energy is much cheaper than that generated from fossil fuels. In Brazil, it is USD 20.8 per MWh, while in Denmark, it is USD 22.8 per MWh. In Saudi Arabia's first commercial wind farm, the 400 MW Dumat Al Jandal project, was contracted in July, 2019, at USD 19.9 per MWh.

11 Transport Progress

Recently, EV sales have soared. Only about 17,000 electric cars were on the world's roads in 2010. By 2019, that number had swelled to 7.2 million, 47% of which were in China. Nine countries have more than 100,000 electric cars on the road, and in 20 countries, they have reached market shares above 1% (see Fig. 25) [32].

If an EV car is charged from main power station, then its carbon emission is 50% compared to the emissions of a car using a fossil fuel. However, if an EV car is charged from PV station, then its carbon emission is less than 10%.

It is a fact that air travel contributes 15% carbon emission. Figure 26 shows a 9-seater electric plane which is being used in the USA for short flight, less than 100 mile. Figure 27 shows a PV charging station for EVs.

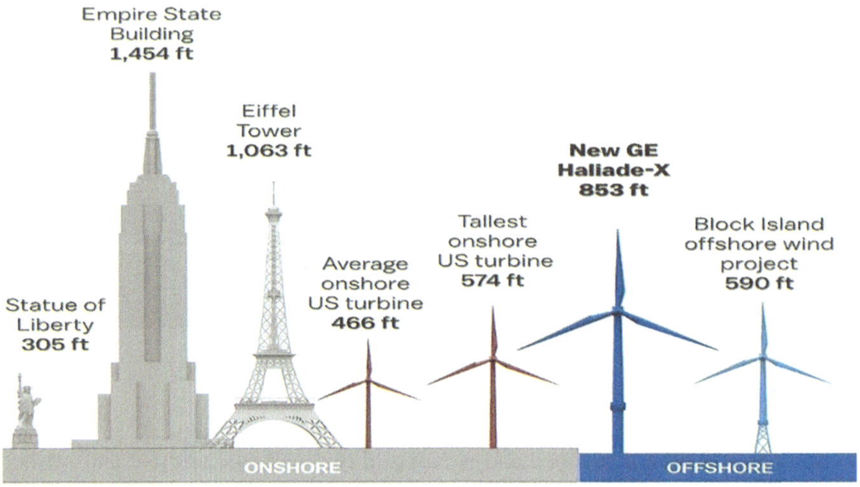

Fig. 18 How the expected Haliade-X, GE machine compares with others

Fig. 19 UK Whitelee Wind Farm. It has 215 Siemens and Alstom wind turbines, total capacity 539 (MW), with the average of 2.5 MW per turbine

Fig. 20 AWEC wind farm. Capacity 1550 MW, 586 machines in (1.5–3.0) MW

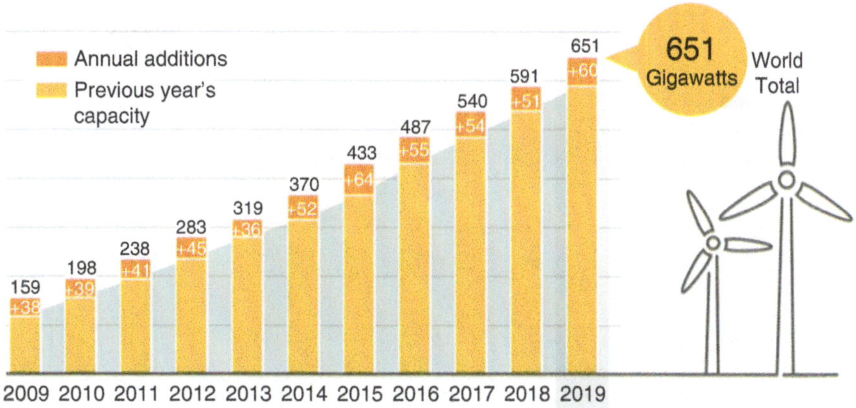

Fig. 21 Global wind energy capacity 2009–2019, in GW [31]

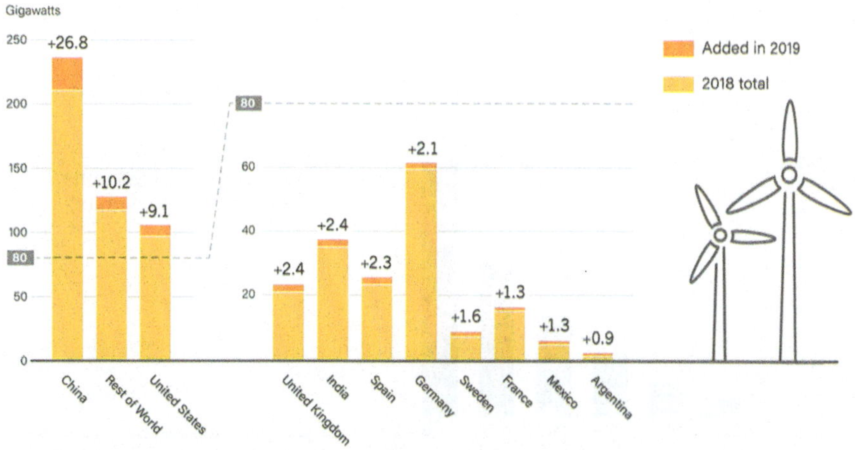

Fig. 22 Top 10 countries' wind power capacity globally

Fig. 23 Brighton off-shore wind farm

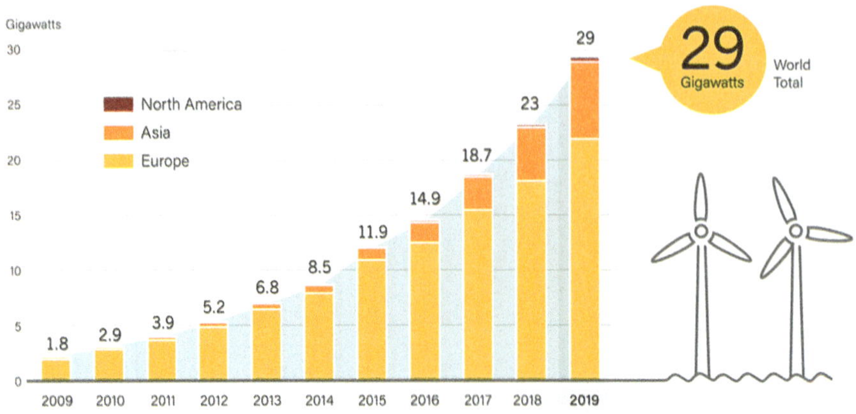

Fig. 24 Global off-shore wind capacity in three regions 2009–2019

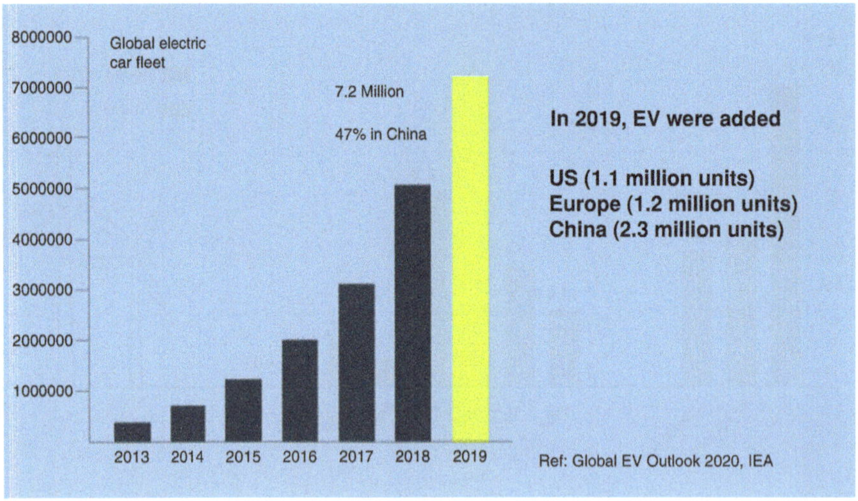

Fig. 25 EV progress globally by the end of 2019

Fig. 26 The Cessna Caravan magni X – nine-seater electric airplane has a speed of 114 mile/hour [33]

Fig. 27 The age of electric travel

12 Conclusion

It is a paramount importance to tackle climate change by reducing or eliminating all fossil fuels in our daily life. We must reshape our thinking and the ways we function. We must actively involve in energy conservation, efficiency and management. We should recycle materials, use waste to energy, construct our buildings using materials such as wood and timber, use natural ventilation, plant vegetation for shading and utilize natural lighting. In addition, we must improve public transport and the introduction of EV and hydrogen fuel.

Always use renewable energy for electricity, heating, cooling, agriculture, transport and industry. Rationalize the use of food and water, since world population presently is 7.8 billion and will be 10 billion in 2050. The pandemic Covid-19 has proved that the world is like one city.

If we convert all renewable energy forms into power as shown below, we have so far 2912.906 GW, noting that at present, the global power generation is 3082.200 GW [34]. We must not be complacent. We must actively press on to achieve 100% global power generation.

Geothermal = 15.406
Hydropower = 1307.000
Biomass and W-t-E, wood fire = 91.100
Solar thermal applications = 196.000
Photovoltaic systems = 652.400
Wind energy applications = 651.000
Total = 2912.906, this is 94.5% of global need of electricity

References

1. Elliott D. Renewable energy: can it deliver? 10th July 2020 by Polity
2. Porritt J (2020) Hope in hell, a decade to confront the climate emergency. Simon & Schuster UK Ltd., 222 Gray's Inn Road, London WC1X 8HB
3. Scripps Institute of Oceanography, NOAA
4. The Washington Post: https://apple.news/A4vGHzoYwQCyemGOEszhbow
5. The Guardian News Paper 2 April 2018
6. Telegraph, World News 2018
7. WMO Report 2018
8. The Guardian Newspaper, 3 July 2017

9. Ref: Geothermal Energy Association. http://geo-energy.org/Basics.aspx
10. Richter A, Think geoenergy, 27 January 2020
11. IEA Hydropower, Tracking report 2020
12. 11 Hydropower Status Report, International Hydropower Association
13. https://en.wikipedia.org/wiki/Rance_Tidal_Power_Station
14. https://www.nationalgeographic.org/encyclopedia/tidal-energy/
15. Ladanai S, Vinterbäck J. Report 013, ISSN 1654-9406, Uppsala, Sweden
16. Bioenergy power generation – tracking report 2020
17. Bioenergy power generation – Analysis, IEA, June 2020
18. Schmidt HL (2019) United Arab Emirates Government 2019
19. World energy resources, waste to energy 2016
20. https://www.renewableenergyhub.co.uk/blog/anaerobic-digesters-future-renewable-energy/
21. https://www.nnfcc.co.uk/publications/report-anaerobic-digestion-deployment-in-the-uk, published 16.4.2018. NNFCC
22. Solrico, Green one Tech – Solar Collectors, – 1, Austria, established 1991
23. Global solar water heaters market size, share & trends analysis report 2018, Dublin September 3, 2018, CISION
24. https://webstore.iea.org/download/direct/793
25. Sayigh A (2017) Photovoltaics for sustainable electricity and buildings. Springer, Cham
26. National Renewable Energy Laboratory, NREL, 6 April, 2020
27. 26 large solar farms in the world, the largest published in Origin by Justine Summers, 24 October 2018
28. The Guardian 4 November 2017
29. https://www.powermag.com/changing-winds-the-evolving-wind-turbine/
30. GE, Vox Research
31. https://www.evwind.es/2020/07/05/wind-energy-expanded-19-in-2019-with-around-60- w-of- w-capacity/75563
32. IEA, EV global outlook 2020
33. The Guardian News Paper 28 May 2020
34. IEA, World Energy Outlook 2019

Part V
Policy, Finance, Social Impact, Education and Sustainability

The European Green Deal: Unlocking the EU's Potentials of Renewable Energy?

Rainer Hinrichs-Rahlwes

1 Introduction and Context

When the new European Commission (EC) took office on 1 December 2019, the *European Green Deal* was announced to be one of the major lighthouse projects of the upcoming legislative period. In her Mission Letter Commission President Ursula von der Leyen entrusted Frans Timmermans with the role of Executive Vice President for the *European Green Deal* and responsible for the climate action portfolio [1]. He became responsible for leading a group of Commissioners tasked with developing and implementing what she hopes to become *"Europe's hallmark"*. Main elements of the European Green Deal are outlined in the letter: *"At the heart of it is our commitment to becoming the world's first climate-neutral continent. It will require collective ambition, political leadership and a just transition for the most affected"*. A number of legislative and non-legislative elements should be presented in the first 100 days of the new Commission: a *European Climate Law* should enshrine the commitment to become climate neutral by 2050 and therefore increase the EU's greenhouse gas reduction target for 2030 from 40% to *"at least 50%"*. Negotiating with other countries, an increase to 55% was considered. A *Just Transition Fund* shall protect regions and people which are most affected by the transformation towards carbon neutrality. And *"tax policies enable us to deliver on our climate ambitions. This will include the work on the Carbon Border Tax, as well as the review of the Energy Taxation Directive"*. Implementation of existing instruments like the Emissions Trading System (ETS) [2] and the Effort Sharing Directive (ESD) [3] should be assured and enhanced. A *European Climate Pact* shall be developed *"bringing together regions, local communities, civil society, industry and schools. Together they will design and commit to a set of pledges to change behaviours across our society"*.

When this paper was written in May 2020, some details of the *European Green Deal* had been published with the *Commission Work Programme 2020* [4]. First proposals had been subject to public consultations and feedback loops, and some were officially presented to the European Council and the European Parliament for consideration and – where appropriate – adoption.

2 Objectives of the European Green Deal

From the beginning, the new EC underlined the high priority of the *European Green Deal*. Building on an EC Communication *The European Green Deal* [5], which was published only 10 days after the new Commission had taken office, the initial *Commission Work Programme* highlighted the *Green Deal* as the first of *"six headline ambitions"*, branding it (in **bold letters**) as the solution for the global challenges of climate change, biodiversity loss, security and prosperity. *"**The European Green Deal is our new growth strategy**. It will help create jobs and make Europe more competitive globally. Our new industrial strategy will be essential in making this happen as an enabler of both the ecological and digital transitions"*. The ambition is high: *"The European Green Deal provides a roadmap with policies and measures to deliver the transformative change*

R. Hinrichs-Rahlwes (✉)
European Renewable Energies Federation (EREF), Brussels, Belgium

German Renewable Energy Federation (BEE), Berlin, Germany

© The Author(s), under exclusive license to Springer Nature Switzerland AG 2022

A. Sayigh (ed.), *Sustainable Energy Development and Innovation*, Innovative Renewable Energy,
https://doi.org/10.1007/978-3-030-76221-6_57

we need across all sectors". As a first major deliverable, the *European Climate Law* shall enshrine the objective of climate neutrality by 2050 and it will be accompanied by proposals for higher *"EU ambition to reduce greenhouse gas emissions by 2030"*.

There will be a *"Strategy for Smart Sector Integration and a Renovation Wave"* and proposals for better exploiting *"Europe's offshore renewable energy"* and a *"Strategy for Sustainable and Smart Mobility"*. A *"new EU Biodiversity Strategy for 2030"* and a *"'Farm to Fork' Strategy"*. To provide sufficient financing, the EC also proposed a *"European Green Deal Investment Plan"* and a *"Renewed Sustainable Finance Strategy"*, which – together with a *"Just Transition Mechanism"* and a *"Just Transition Fund"* – shall help the most affected regions and sectors. In addition, a *"European Climate Pact"* will be developed. It *"will bring together all of these efforts, involving regions, local communities, civil society, schools, industry and individuals"*. A detailed timetable is provided as an Annex to the *Work Plan*.

Strategic considerations and context are provided in the Communication [5]. The European Green Deal *"is a new growth strategy that aims to transform the EU into a fair and prosperous society, with a modern, resource-efficient and competitive economy where there are no net emissions of greenhouse gases in 2050 and where economic growth is decoupled from resource use"*. It aims at including a broad range of policy areas, such as health, environment, social justice, energy, climate change and sustainable development. It is designed to be *"an integral part of this Commission's strategy to implement the United Nation's 2030 Agenda and the sustainable development goals"* [6].

Figure 1 provides an illustration of the context and the various elements. The most important ones for this paper are *"Increasing the EU's climate ambition for 2030 and 2050"*, *"Supplying clean, affordable and secure energy"*, *"Mobilising industry for a clean and circular economy"*, *"Building and renovating in an energy and resource efficient way"*, *"Accelerating the shift to sustainable and smart mobility"*, and the overarching approach of *"Mainstreaming sustainability in all EU policies"*.

The EC announces a *"plan to increase the EU's greenhouse gas emission reductions target for 2030 to at least 50% and towards 55% compared with 1990 levels in a responsible way"*. This is significantly more than the existing 40% target, but far below a 65% reduction, which NGOs, Greens and several other Members of the European Parliament are suggesting to be in line with the Paris Agreement. Given that energy accounts for more than 75% of the EU's greenhouse gas emissions prioritising energy efficiency is highlighted. For the supply side, a *"power sector must be developed that is based largely on renewable sources, complemented by the rapid phasing out of coal and decarbonising gas"* – a strong, but ambiguous statement, which leaves room for Member States to pursue their own priorities, not all of which are sustainable.

The new *EU industrial strategy* was adopted in March 2020 [7]. For significantly accelerating sustainable building and renovation of buildings, the EC plans to *"engage in a 'renovation wave' of public and private buildings"* and to *"rigorously enforce the legislation related to the energy performance of buildings"*. A *"strategy for sustainable and smart mobility"* is planned to be adopted in 2020. Policies and strategies to *"ramp-up the production and deployment of sustainable alternative transport fuels"* and *"striving for drastically less polluting"* transport, especially in cities are key objectives of the new EC.

"Mainstreaming sustainability in all EU policies" is focusing on financial aspects. The EC estimates *"that achieving the current 2030 climate and energy targets will require €260 billion of additional annual investment, about 1.5% of 2018*

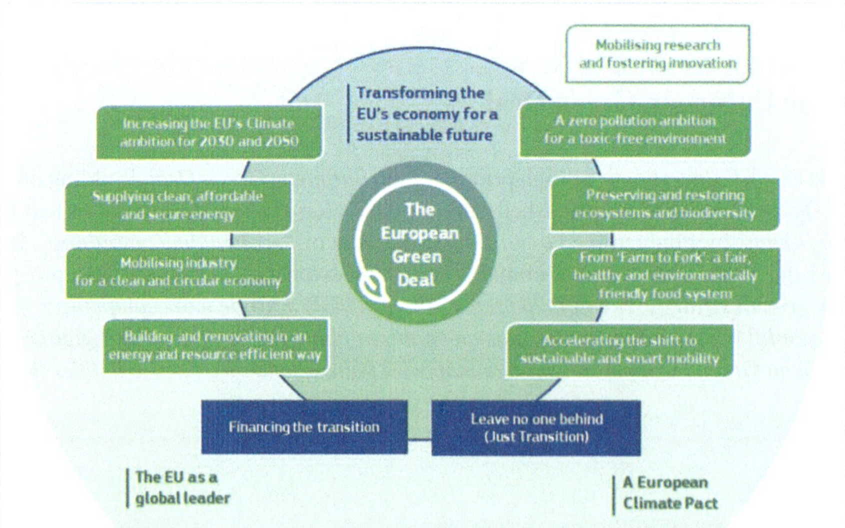

Fig. 1 The European Green Deal – Source [5]

GDP". A *Sustainable Europe Investment Plan* will be presented, as well as a *Just Transition Mechanism,* including a *Just Transition Fund.* Well-designed tax reforms will help boost sustainable growth and ensure that national budgets follow the priorities of the *European Green Deal.* A *European Climate Pact* shall involve citizens and stakeholders in discussing, developing and strengthening activities. A public consultation [10] about preferred activities for the *Climate Pact* took place.

3 Components of the European Green Deal

Not all building blocks of the *European Green Deal* are fully developed yet, but the EC started working on it at high speed – and is trying to keep the momentum going despite the COVID-19 crisis. The draft *European Climate Law,* the *Just Transition Mechanism* and *Fund,* the *European Climate Pact*, and the *Industry Strategy* were already presented, or public consultations were opened to involve citizens and stakeholders. Some elements may be delayed due to COVID-19. In the following, the status of various building blocks will be described and analysed.

The proposal for a ***European Climate Law*** [11] was presented just before the end of the new EC's first 100 days in office. Building on the headline target to become a climate-neutral continent by 2050, and on a Communication from the predecessor Commission of November 2018 [12], the *"proposal aims to complement the existing policy framework by setting the long-term direction of travel and enshrining the 2050 climate-neutrality objective in EU law, enhancing adaptation efforts, establishing a process to set out and review a trajectory until 2050, regular assessment and a process in case of insufficient progress or inconsistencies*".

Article 1 sets out *"a binding objective of climate neutrality in the Union by 2050". Article 2* defines the process towards increasing the EU's 2020 reduction target: *"By September 2020, the Commission shall review the Union's 2030 target for climate [...] in light of the climate-neutrality objective [...], and explore options for a new 2030 target of 50 to 55% emission reductions compared to 1990. Where the Commission considers that it is necessary to amend that target, it shall make proposals to the European Parliament and to the Council as appropriate ... By 30 June 2021, the Commission shall assess how the Union legislation [...] would need to be amended [...]".*

Article 3 introduces a potentially strong instrument for the EC to steer towards climate neutrality, which was seldom used in climate and energy policy so far. *"1. The Commission is empowered to adopt delegated acts [...] by setting out a trajectory at Union level to achieve the climate-neutrality objective [...] until 2050. At the latest within six months after each global stocktake [...] of the Paris Agreement, the Commission shall review the trajectory. 2. The trajectory shall start from the Union's 2030 target for climate [...]".* According to *Article 9*, the delegated acts will *"enter into force only if no objection has been expressed either by the European Parliament or the Council [...]".* This would enable the EC to move forward on the basis of majority votes, preventing individual Member States from blocking ambitious trajectories, which may be needed to eventually achieve the 2050 target.

Article 4 deals with *"Adaptations to climate change"* and *Article 5* sets 5-year periods for assessing progress achieved, starting by 30 September 2023, mandating the EC to assess existing frameworks and propose or take additional measures, if needed.

Article 6 mandates the EC to regularly assess national measures by *"30 September 2023, and every 5 years, thereafter".* The EC can issue recommendations, but without the right to impose penalties. *"Where the Commission finds [...] that a Member State's measures are inconsistent with that objective [...], it may issue recommendations to that Member State. The Commission shall make such recommendations publicly available".* The Member State has to take the recommendations into consideration, but *"If the Member State concerned decides not to address a recommendation or a substantial part thereof, that Member State shall provide the Commission its reasoning".* Public blame and shame are the strongest tools the EC can apply according to the proposal.

In March 2019, the European Parliament endorsed the climate-neutrality target [13], and in November 2019, it underlined that the EU should strive to reach the target *"by 2050 at the latest"*, and it had strengthened the call by declaring a climate and environment emergency [14]. The European Council had a more difficult discussion because some Member States were not willing to endorse the greenhouse gas neutrality objective by 2050 and therefore tried to build a blocking minority against the endorsement. It took until 12 December 2019 that the Council endorsed the objective [15], but Poland was not yet willing and insisted on another round of negotiations – or as the Council Conclusions have it: *"One Member State, at this stage, cannot commit to implement this objective as far as it is concerned, and the European Council will come back to this in June 2020".*

Based on the *European Climate Law* and as a background for an amendment regarding the 2030 target, the objective of the ***2030 Climate Target Plan*** [16] is to analyse *"the economic, social and environmental impacts"* and *"possible policy measures".* Following an *Inception impact assessment,* a public consultation was open until 23 June 2020. Adoption was planned for the second half of 2020. For the ***European Climate Pact***, the ***New EU Strategy on Adaptation to Climate Change*** and for some other elements delays until 2021 were announced due to the Covid-19 pandemic.

Accelerating development and deployment of energy efficiency and renewable energy is at the heart of the green transformation. Therefore, Member States need a common understanding with the EC which financial incentives could be offered. This is why EC regularly provides guidelines about admissible state aid and thus accelerate clearance for new or modified legislation. In this context, the ***Guidelines on State aid for environmental protection and energy 2014–2020*** [17] are an important tool, because they outline what the EC considers admissible state aid. As the existing guidelines are based on legislation up to 2020, many of the articles will soon expire. The assessment of necessary adaptions to new legislation, such as the revised *Renewable Energies Directive* [18] or the revised *Energy Efficiency Directive* [19] and other parts of the *Clean Energy for all European Package* [20] has already begun. A new draft was expected around the end of 2020. For the revised guidelines to facilitate the uptake of renewable energy and energy efficiency measures it will be important for the EC to develop a proposal which helps Member States to design their energy markets and – where needed – support systems in a way that large- and small-scale, centralised and decentralised, utility-scale and prosumer-driven projects have a fair chance on a level playing field.

On 10 March 2020, the EC presented ***A new Industrial Strategy for Europe*** [7] comprising a chapeau document [8], a communication titled ***An SME Strategy for a sustainable and digital Europe*** [9] and explanatory factsheets. The ambition is high: ***"We now need a new industrial way for Europe, fit for the ambitions of today and the realities of tomorrow.*** *At the heart of this is the ability of Europe's industry to lead the twin transitions and drive our competitiveness. It cannot afford to simply adapt – it must now become the accelerator and enabler of change and innovation"*. It lays out *"what we want to achieve by 2030 and beyond"* and this means that *"Europe needs an industry that becomes greener and more digital"*. The proposal highlights a broader approach: *"All industrial value chains, including energy-intensive sectors, will have a key role to play"*. Europe must *"create lead markets in clean technologies"* and not only for a few big players: *"Regulatory policies, public procurement, fair competition and the full involvement of SMEs will be essential to make this happen"*. This should be supported by *"policies and financial instruments at EU and national level"*.

The EC underlines that the *"next five years will be decisive to set the right enabling conditions for this transition"*. The communication describes key actions for *"Making it happen"* in a number of policy areas. There is the need for *"certainty for industry. A deeper and more digital single market"*, and there is the need for *"Upholding a global level playing field"*, and *"Supporting industry towards climate neutrality"*. This comprises the proposal for a new *Just Transition Mechanism* and a *Just Transition Fund* [21] which should *"mobilise €100 billion to ensure a fair transition for carbon intensive regions as they continue to transform their industries and economies"*.

The EC highlights the need for *"a more strategic approach to renewable energy industries"*. This approach, however, mentions the examples *"offshore energy"*, and *"clean hydrogen"* in the context of a *"new strategy for smart sector integration"*. Other renewable sources are not mentioned. And *"clean hydrogen"* is not limited to **green hydrogen**, which leaves the door open for fossil-based hydrogen. The strategy includes a *"focus on sustainable and smart mobility"*, and the announcement that the EC *"will propose a* ***Carbon Border Adjustment Mechanism*** *in 2021 to reduce the risk of carbon leakage, in full compatibility with WTO rules"*. To generate support for clean technologies in Europe by pricing carbon and other pollutants, much will depend on how this mechanism will be designed.

The *New Industrial Strategy* is linked to the need of *"leveraging and blending EU money with Member States and other institutional partners, as well as crowding in private investors"*. This shall be achieved by an *InvestEU* and a *Green Deal Investment Plan.* To mobilise private money, market failures shall be mitigated, among others by reviewing State Aid Guidelines including those for *Important Projects of Common European Interest (IPCEI),* together with the *European Climate Law* and the recently agreed [22, 23] criteria for sustainable investment in the context of a proposal from the EC for a *Regulation on the establishment of a framework to facilitate sustainable investment* (**EU Green Taxonomy**) [24].

Another part of the *European Green Deal* could be the proposal for a revision of the ***Energy Taxation Directive*** [25], which was announced for June 2021. Taxation in the EU, however, requires unanimity among Member States. Therefore, it may be difficult to reach a meaningful result, which would accelerate Europe's shift towards climate-neutrality.

4 Conclusions and Outlook

Since the IPCC's *Special Report on 1.5 °C Global Warming* [26] underpinned the urgency of climate action, the concept of Greenhouse Gas Neutrality within the planetary boundaries is supported by serious climate scientists and by an increasing number of governments, politicians, stakeholders and civil society organisations. Evidence is growing that the remaining carbon budget is too rapidly decreasing. With more extreme weather events and ice caps melting faster than expected, strong efforts to fully implement the Paris Agreement are needed more than ever. It is becoming ever more likely that net zero emissions by 2050 may be too late to avoid the most dangerous impact of the climate crisis, and that the industrialised countries of the Global North have to reduce emissions more quickly than the developing countries of the Global South. At the same time, with rapidly decreasing costs of renewable energies making them the cheapest option for energy supply around the world for most end-uses, there may be a chance to reduce greenhouse gas emissions quickly enough to avoid more tipping points to be reached. With renewable energy available everywhere, emissions-free processes are clearly the better choice compared to emissions removal from the atmosphere in future.

The EC's emphasis on the *European Green Deal* and the commitment to Greenhouse Gas Neutrality is widely applauded – with some advocating for an earlier target year than 2050, and others criticising too high ambition. The same applies for the 50 to 55% reduction milestone proposed for 2030, which some consider too ambitious, whereas others all calling for at least 65% reduction by 2030 to be in line with the Paris Agreement. The EC's intention to present trajectories and impact assessments only in September 2020 has been criticised for being too late to constructively feed in to the preparation of the next global climate conference, which was planned to take place in November 2020, but is now postponed due to the COVID-19 crisis.

The European Parliament and the European Council will discuss and modify the EC's proposals. They will suggest improvements for strategies, and they will have to agree on the various pieces of legislation, including the ambition level, the target year and the instruments and policies to achieve the objectives. Despite the well-received *Green Deal* commitments, the aspiration alone will not suffice to guarantee successful implementation. Milestones and regular monitoring on the way to (net) zero emissions by 2050 (or earlier if the global responsibility of the industrialised countries is duly accepted) every 5 or at least 10 years will have to be set. Specific targets and milestones for rapidly increasing renewable shares towards 100% and to achieve efficiency gains in all end-uses will have to be included. Binding responsibilities for Member States' fair shares to the Union's energy transformation and penalties in case of non-compliance will have to be agreed and enacted. State Aid rules, support systems and market designs will have to be adapted to unleash the full potential of renewable energy. Administrative and regulatory barriers, which are still slowing down the growth of renewable energy, will have to be removed. Direct and indirect subsidies for unsustainable energy will have to be rapidly phased out to create and ensure a level playing field, where clean, safe and cheap renewable energy can eventually outcompete dangerous and dirty fossil and nuclear energy sources.

With the COVID-19 crisis, the EC is struggling not to lose the momentum for the *European Green Deal.* Some delays seem inevitable already with attention shifting to recovery packages. A revised work plan will be presented around the end of May. Some Member States suggest postponing the green deal until the crisis is over. An increasing number of politicians, stakeholders, NGOs and individuals as well as major industries and their associations [27] are calling for a ***Green Recovery*** plan as the cheapest, safest and most sustainable way back to normality. On the global level, a broad coalition of more than 100 companies, associations, NGOs and other stakeholders launched a *Call to Action in Response to COVID-19: Renewable Energy is a Key Part of the Solution* [28]. The call concludes: *"The actions outlined above would support a sustainable economic recovery, make societies more resilient, and help to fulfil the objectives of the Paris Agreement and the 2030 Agenda for Sustainable Development. In responding to the crisis, governments are currently in a position to recognise the benefits of renewables, continue building broad public support for the transformative decarbonisation of societies, and pave the way for a clean, low-carbon economy".*

References

1. von der Leyen U, President of the European Commission, Mission Letter to Frans Timmermans, Brussels, 1 December 2019. https://ec.europa.eu/commission/commissioners/sites/comm-cwt2019/files/commissioner_mission_letters/mission-letter-frans-timmermans-2019_en.pdf
2. Directive 2003/87/EC of the European Parliament and of the Council Of 13 October 2003 establishing a system for greenhouse gas emission allowance trading within the Union and amending Council directive 96/61/EC, OJ L275, 25 October 2003, consolidated version as of 8 April 2018. https://eur-lex.europa.eu/legal-content/EN/TXT/PDF/?uri=CELEX:02003L0087-20180408&from=EN
3. Regulation (EU) 2018/842 of the European Parliament and of the Council of 30 May 2018 on binding annual greenhouse gas emission reductions by Member States from 2021 to 2030 contributing to climate action to meet commitments under the Paris Agreement and amending Regulation (EU) No 525/2013, OJ L156/26, 19 June 2018. https://eur-lex.europa.eu/legal-content/EN/TXT/PDF/?uri=CELEX:32018R0842&from=EN
4. European Commission, A union that strives for more, COM(2020) 37 final, Brussels, 29 January 2020. https://ec.europa.eu/info/sites/info/files/cwp-2020-publication_en.pdf
5. European Commission, Communication from the Commission to the European Parliament, the European Council, the Council, the European Economic and Social Committee and the Committee of the Regions – The European Green Deal, COM(2019) 640 final, Brussels, 11 December 2019. https://ec.europa.eu/info/publications/communication-european-green-deal_en

6. Transforming our world: the 2030 agenda for sustainable development. Website accessed 04 May 2020, 17:30 CEST. https://sustainabledevelopment.un.org/post2015/transformingourworld
7. European Industrial Strategy. Website accessed 04 May 2020, 18:15 CEST. https://ec.europa.eu/info/strategy/priorities-2019-2024/europe-fit-digital-age/european-industrial-strategy_en
8. European Commission, Communication from the Commission to the European Parliament, the European Council, the Council, the European Economic and Social Committee and the Committee of the Regions – a New Industrial Strategy for Europe, COM(2020) 102 final, Brussels, 10 March 2020. https://ec.europa.eu/info/sites/info/files/communication-eu-industrial-strategy-march-2020_en.pdf
9. European Commission, Communication from the Commission to the European Parliament, the Council, the European Economic and Social Committee and the Committee of the Regions – an SME Strategy for a sustainable and digital Europe, COM(2020) 103 final, Brussels, 10 March 2020. https://ec.europa.eu/info/sites/info/files/communication-sme-strategy-march-2020_en.pdf
10. European Climate Pact, Public consultation. Website accessed 05 March 2020, 18:45 CEST. https://ec.europa.eu/info/law/better-regulation/have-your-say/initiatives/12219-European-Climate-Pact
11. European Commission, Proposal for a Regulation of the European Parliament and of the Council establishing the framework for achieving climate neutrality and amending Regulation (EU) 2018/1999 (European Climate Law), COM(2020) 80 final, Brussels, 4 March 2020. https://ec.europa.eu/info/sites/info/files/commission-proposal-regulation-european-climate-law-march-2020_en.pdf
12. European Commission, Communication from the Commission to the European Parliament, the European Council, the Council, the European Economic and Social Committee and the Committee of the Regions and the European Investment Bank – a clean planet for all - a European strategic long-term vision for a prosperous, modern, competitive and climate neutral economy, COM(2018) 773 final, Brussels, 28 November 2018. https://eur-lex.europa.eu/legal-content/EN/TXT/PDF/?uri=CELEX:52018DC0773&from=EN
13. European Parliament resolution of 14 March 2019 on climate change – a European strategic long-term vision for a prosperous, modern, competitive and climate neutral economy in accordance with the Paris Agreement (2019/2582(RSP)), 14 March 2019. https://www.europarl.europa.eu/doceo/document/TA-8-2019-0217_EN.pdf
14. European Parliament resolution of 28 November 2019 on the climate and environment emergency (2019/2930(RSP)), 28 November 2019. https://www.europarl.europa.eu/doceo/document/TA-9-2019-0078_EN.pdf
15. European Council, European Council meeting (12 December 2019) – conclusions, EUCO 29/19, Brussels, 12 December 2019. https://data.consilium.europa.eu/doc/document/ST-29-2019-INIT/en/pdf
16. European Commission, 2030 Climate target plan. Website accessed 06 May 2020 17:30 CEST. https://ec.europa.eu/info/law/better-regulation/have-your-say/initiatives/12265-2030-Climate-Target-Plan
17. European Commission, Communication from the Commission, Guidelines on State aid for environmental protection and energy 2014–2020, Official Journal of the European Union 2014/C 200/01, 28 June 2014 https://eur-lex.europa.eu/legal-content/EN/TXT/PDF/?uri=OJ:C:2014:200:FULL&from=EN
18. Directive (EU) 2018/2001 of the European Parliament of the Council of 11 December 2018 on the promotion of the use of energy from renewable sources (recast), Official Journal of the European Union L 328/82, 21 December 2018. https://eur-lex.europa.eu/legal-content/EN/TXT/PDF/?uri=CELEX:32018L2001&from=EN
19. Directive (EU) 2018/2002 of the European Parliament of the Council of 11 December 2018 amending Directive 2012/27/EU on energy efficiency, Official Journal of the European Union L 328/210, 21 December 2018. https://eur-lex.europa.eu/legal-content/EN/TXT/PDF/?uri=CELEX:32018L2002&from=EN
20. European Commission, Clean energy for all Europeans package, published 20 October 2017, last update 12 March 2020. Website accessed 08 May 2020 17:45 h CEST. https://ec.europa.eu/energy/topics/energy-strategy/clean-energy-all-europeans_en
21. European Commission, Proposal for a regulation of the European Parliament and of the Council establishing the Just Transition Fund, COM(2020) 22 final, Brussels, 14 January 2020. https://ec.europa.eu/info/law/better-regulation/ and Annex. https://ec.europa.eu/info/law/better-regulation/
22. energypost.eu, Luca Bonaccorsi, Defining green investments, ending greenwash: the EU's new Taxonomy Regulation, 13 February 2020. Website Accessed 6 May 2020 18:30 h CEST. https://energypost.eu/defining-green-investments-ending-greenwash-the-eus-new-taxonomy-regulation/
23. Council of the European Union, Press Release 15 April 2020, Sustainable finance: Council adopts a unified EU classification system. Website Accessed 6 May 2020. https://www.consilium.europa.eu/en/press/press-releases/2020/04/15/sustainable-finance-council-adopts-a-unified-eu-classification-system/
24. Proposal for a Regulation of the European Parliament and of the Council on the establishment of a framework to facilitate sustainable investment, COM(2018) 353 final, Brussels, 24 May 2018. https://eur-lex.europa.eu/legal-content/EN/TXT/PDF/?uri=CELEX:52018PC0353&from=EN
25. European Commission, Proposal for a Directive, EU Green Deal – Revision of the Energy Taxation Directive. Website accessed 6 May 2020 18:45 h CEST. https://ec.europa.eu/info/law/better-regulation/have-your-say/initiatives/12227-Revision-of-the-Energy-Tax-Directive
26. IPCC: global warming of 1.5°C. An IPCC Special Report on the impacts of global warming of 1.5°C above pre-industrial levels and related global greenhouse gas emission pathways, in the context of strengthening the global response to the threat of climate change, sustainable development, and efforts to eradicate poverty. https://www.ipcc.ch/site/assets/uploads/sites/2/2019/06/SR15_Full_Report_Low_Res.pdf from https://www.ipcc.ch/sr15/. Accessed 10 May 2020
27. AVERE et al. Call to EU leaders: #EUGreenRecovery to restart Europe, Brussels, 31 March 2020. https://windeurope.org/wp-content/uploads/files/policy/position-papers/20200331-Joint-letter-call-to-EU-leaders-green-recovery-to-restart-europe.pdf
28. IRENA coalition for action, call to action in response to COVID-19: renewable energy is a key part of the solution, Abu Dhabi, 28 April 2020. https://coalition.irena.org/-/media/Files/IRENA/Coalition-for-Action/CfA_flyer_2020.pdf

The Opportunities and Challenges for 100% Renewable Energy

David S. Renné

1 Introduction and Background

We, as global citizens, are facing unique and urgent challenges in how to address the impacts of climate change on our lifestyles and livelihoods. We are faced with two basic choices: adapting to the effects of climate change or mitigating its causes. This paper focuses on two key points in the mitigation argument: (1) we must reduce our global greenhouse gas emissions to levels that ensure that global temperature increases do not exceed 1.5 °C by the end of this century and (2) the wide-scale adaptation and plummeting cost of renewable energy technologies can help us meet these emission reduction targets by achieving a rapid and complete decarbonization of all of our end-use energy needs.

Currently, our energy consumption results in nearly three-fourths of all anthropogenic carbon emissions to the atmosphere. The remaining sources are associated with land use practices, such as deforestation and agricultural practices, and release of greenhouse gases from melting of permafrost in the arctic regions. Focusing on the energy sector, the International Energy Agency [1] shows that over the past two decades, energy-related carbon dioxide emissions grew from about 21 GT/yr to over 33 GT/yr (Fig. 1). Recent indications that the growth rates may finally have leveled off were contradicted by additional growth in 2018–2019. Even though developed economies show improved efficiency and lower CO_2 emission rates due to the conversion of coal plants to natural gas and to the vast increase in the use of renewable technologies (Fig. 2), Asia shows continued high demand for coal, which has resulted in a net global increase in CO_2 emissions recently. Of course, we now know that emission levels dropped considerably in 2020 due to the COVID-19 crisis.

Despite the improved efficiencies in the power sector, especially with the widespread application of renewable energy technologies, the challenges to reduce carbon emissions to levels required to cap global warming at 1.5 °C are immense and urgent. This urgency is clearly demonstrated in a series of graphics produced by the New Climate Institute, in a presentation by Höhne [2] at the SB-50 in Bonn in June 2019. Over 10 years ago, the Intergovernmental Panel on Climate Change (IPCC) recommended that global warming must be capped at 2 °C above preindustrial levels to avoid catastrophic climate change [3]. Figure 3, based on [3], shows the emissions reduction curve required to achieve this cap by the end of this century. Historical emissions from all sources are also shown. Under this scenario, starting at an emissions level of around 28 GT/yr, Fig. 3 shows we must reduce all carbon emissions to zero by around 2070–2080. But 10 years later, the IPCC [4] came out with new findings that show the urgency of capping global warming at 1.5 °C. However, between 2007 and 2018, total carbon emissions (including energy-related emissions) have risen to over 41 GT/yr, and we now face a challenge of bringing all emissions down to zero by around 2050 (Fig. 4).

The Paris Climate Agreement of 2015 [5] also established an aspirational target of capping global warming at 1.5 °C, and signatory countries submitted Nationally Determined Contributions (NDCs) that consist of pledges each country will make to reduce its own emissions over a certain time frame. This was an important step in giving the Paris Agreement actual commitments from the signatory countries to work with; however, in reality, as shown in Fig. 5, these national contributions fall far short of achieving the emission reduction goals required under the 2018 IPCC assessment [4].

These figures show that we must urgently transform our energy system to clean renewable energy technologies in order to be on a pathway to limit global warming to no more than 1.5 °C above preindustrial levels and therefore mitigate the dire impacts of climate change as forecast by the IPCC.

D. S. Renné (✉)
Dave Renné Renewables, Boulder, CO, USA

© The Author(s), under exclusive license to Springer Nature Switzerland AG 2022

A. Sayigh (ed.), *Sustainable Energy Development and Innovation*, Innovative Renewable Energy,
https://doi.org/10.1007/978-3-030-76221-6_58

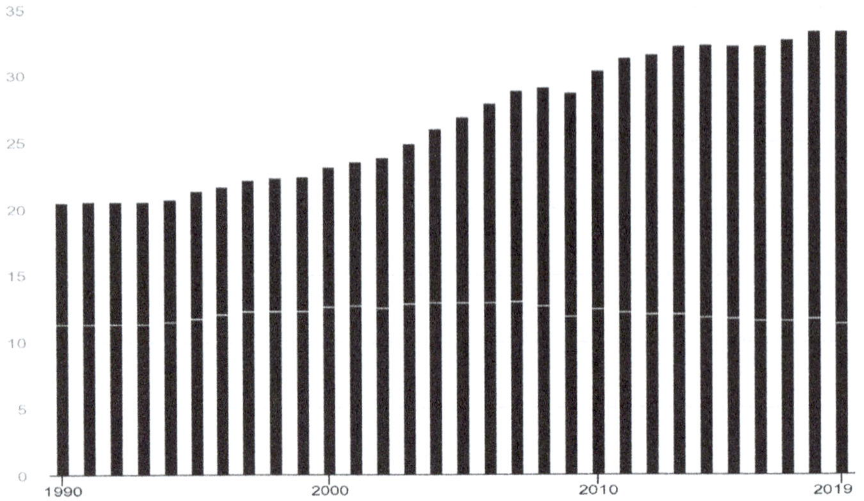

Fig. 1 Annual global energy-related CO_2 emissions. Bottom portion: developed countries. Top portion: developing countries. (Source: International Energy Agency, 2019: https://www.iea.org/articles/global-co2-emissions-in-2019)

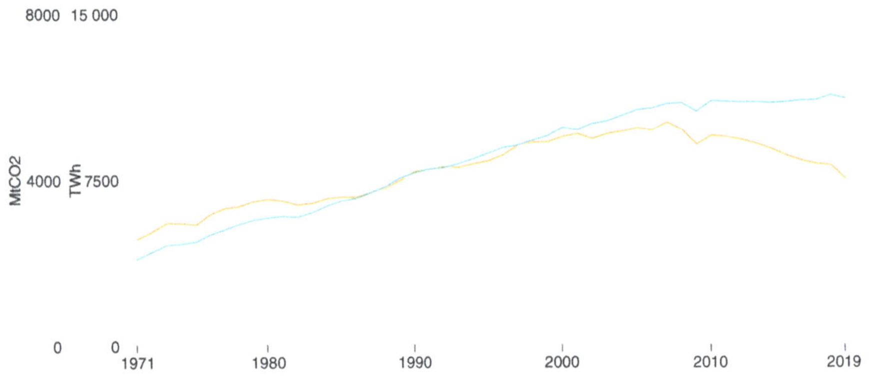

Fig. 2 Global electricity generation (blue line) and power sector CO_2 emissions (orange line). (Source: International Energy Agency, 2019: https://www.iea.org/articles/global-co2-emissions-in-2019)

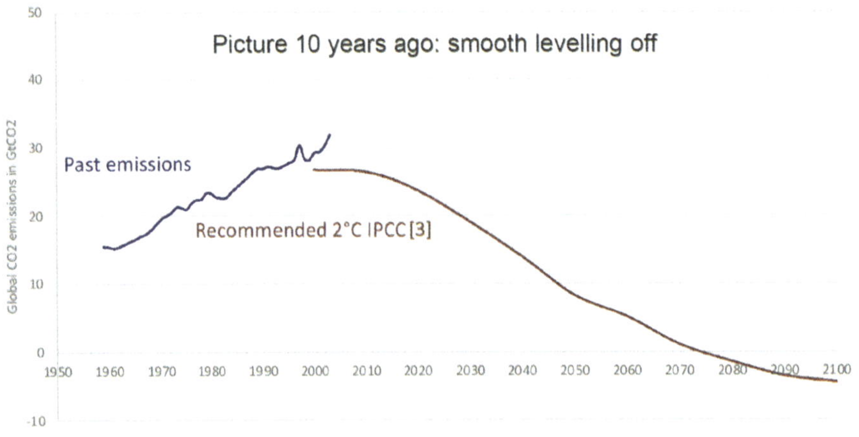

Fig. 3 Actual total anthropogenic carbon dioxide emissions, and the pathway for emissions reduction recommended by the IPCC [3]

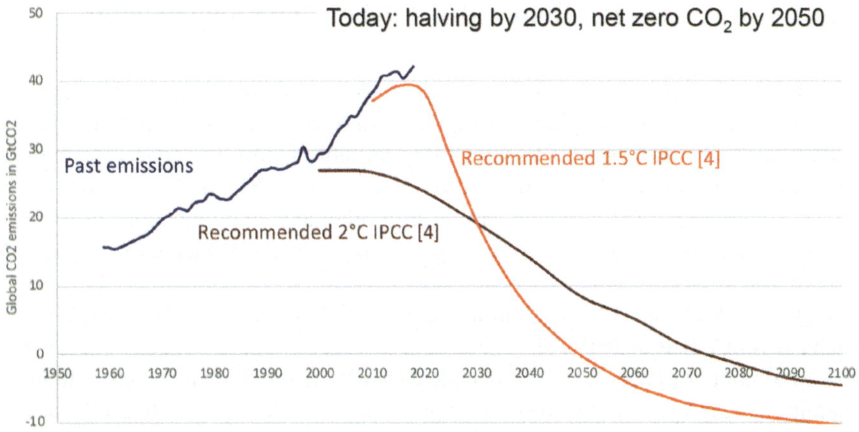

Fig. 4 Current total global carbon dioxide emissions and the revised emission reduction pathway recommended by the most recent IPCC assessment [4] to cap global warming at 1.5 °C

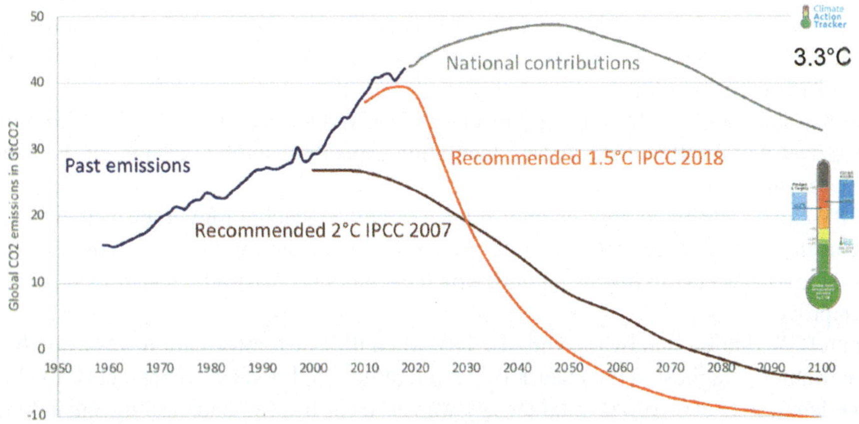

Fig. 5 Same as Fig. 4, with the addition of the emissions reduction pathway that would be achieved if all countries followed their NDCs as specified in the 2015 Paris Climate Agreement

2 The Concept of 100% Renewable Energy

The concept of transforming our energy system to 100% renewables is increasingly suggested as a strategy to achieving the required decarbonization. In Sect. 3 we explore the progress made thus far in transforming our energy system to renewables, and in Sect. 4 we investigate various scenarios that have been published recently where renewables represent a significant, if not exclusive, pathway for a decarbonized energy supply.

IRENA's Coalition for Action has recently published two white papers that address the concept of the transition to 100% renewables: the first on status, trends, and lessons learned from the perspective of regional and national governments [6], and the second from the perspective of utilities that have established a target of 100% renewables in the power sector [7]. Our working definition of 100% renewables is derived from [7] as follows: *100% Renewables is achieved when all sources of energy to meet all end-use energy needs (power, transport, and heating and cooling) in a certain location, region or country are derived from renewable energy resources (including bioenergy, geothermal, hydropower, ocean, solar, and wind) 24 hours per day, every day of the year. Renewable energy can either be produced locally to meet these end-use energy needs or can be imported from outside the region using supportive policies and installations, such as electrical grids, hydrogen, or heated water. Any storage facilities to help balance the energy supply must also use energy derived only from renewable resources.*

3 Current Status and Projections for Renewable Energy: The Solar Sector

Throughout the decade of the 2010s worldwide renewable energy deployments maintained a remarkable and steady upward climb. This is due to a variety of factors such as plummeting technology costs, favorable policies, access to financing, and technology improvements…especially with respect to integrating variable renewable energy technologies into the power grid. Much of this progress has been thoroughly reported in the annual Global Status Reports published by the Renewable Policy Network for the 21st Century (REN21), although a number of other annual publications providing reliable and useful data are also available. Most of the information in the following paragraphs is derived from the REN21 (2020) Global Status Report [8].

Globally, by the end of 2019, global solar PV installations reached 627 GW capacity (Fig. 6). This figure reflects that an additional 115 GW capacity was installed in 2019 [9].

Figure 6 is remarkable in a number of ways; for the decade as a whole, the annual growth rate averaged 39%, and the annual market growth (additional capacity each year) is over 30%. In 2019 alone, excluding China, the global PV market grew about 44%. Much of the growth as well as the world's current total capacity is due primarily to strong market conditions in a few countries. In 2019, for example, the largest PV markets were in China (30.1 GW), India (9.9 GW), the USA (13.3 GW), Japan (7.0 GW), Vietnam and Spain (4.8 GW each), Germany (3.8 GW), and Australia (3.7 GW). However, throughout the decade, a few other countries, primarily in Europe, but also in southeast Asia, have played leading roles in PV installations.

Growth in deployments of technologies in the other solar sectors (heating and cooling, and Concentrating Solar Power, or CSP) has also been impressive throughout the decade, although recently solar PV has truly become the dominant solar technology of choice [8]. Global solar heating and cooling capacity was 479 $GW_{thermal}$ by the end of 2019, and CSP capacity was at 6.2 GW, with more projects in the pipeline.

By way of comparison, total installed wind capacity reached 651 GW by the end of 2019 [8]. Combined, solar and wind deployments represented over half of all new electricity generation capacity installed throughout the world in 2019, a trend that has been growing in recent years.

The growth in solar PV can be attributed to three inter-related trends: (1) technology improvements, reliability, and durability, which increase consumer confidence and lower life cycle costs; (2) enabling policy incentives, which encourage further use of these technologies; (3) access to private capital, which becomes increasingly available as project risks decline and returns on investment improve.

Figure 7 (taken from ref. [8]) shows that by the end of 2016, when the latest information was available at the time of this writing, the share of renewables is highest in the electricity sector, although this sector represents only 20% of our total end-use energy. On the other hand, over 50% of our total energy was used for heating and cooling (including cooking and high temperature industrial process heat), and over 30% was used in the transport sector. However, the contribution by renewable technologies to these latter two sectors is much smaller than in the power sector. While the electricity sector shows over 25% of all energy provided coming from renewables (albeit a majority of this is hydro), only 3.3% of the transport sector is renewable energy (primarily biofuels) and less than 10% of the heating and cooling sector is renewable (primarily modern bioenergy, solar thermal and geothermal).

Therefore, while Fig. 6 shows very positive trends in solar PV, Fig. 7 highlights that significant challenges remain, especially if society wants to rely on renewable energy technologies as the primary means of addressing the urgent climate

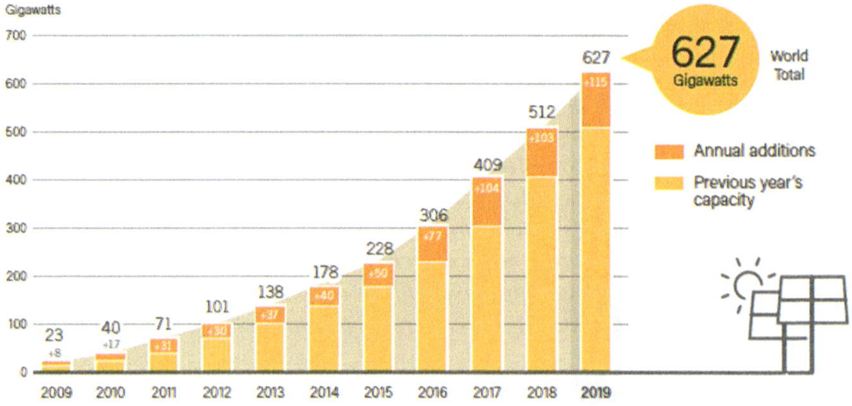

Fig. 6 The growth of global PV capacity over the past decade. (From [8])

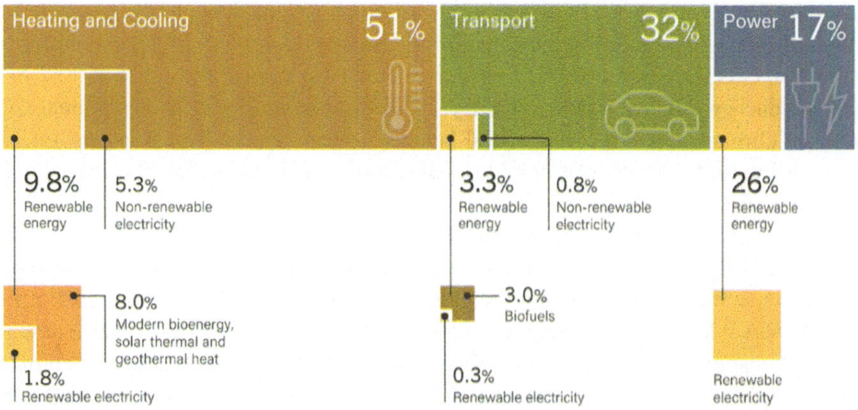

Fig. 7 Renewable energy in total energy consumption, by sector, 2016. (From [8])

challenge outlined earlier. To a large extent newly installed renewable capacity is addressing primarily the power sector which also happens to be the smallest of the three sectors. Although there are current trends that the power sector is growing significantly, including the electrification of both the transport and the heating and cooling sectors, as a society, we are not yet well-positioned to address the urgent climate challenge without a further significant transformation in all of our end-use energy consumption choices.

In the next section we will examine a few recently published pathways that result in significant decarbonization of our total end-use energy consumption by mid-century, with a particular focus in the solar PV sector. Further massive electrification of all end-use energy sectors, combined with extensive new energy efficiency measures, is inherent in all of these energy transformation scenarios.

4 Pathways for Decarbonizing Energy: The Role of Renewables

A number of studies have been conducted over the past decade that explore various pathways for achieving climate mitigation targets, with particular emphasis on the 2.0 °C or 1.5 °C global warming caps. Although the studies generally focus on decarbonization of the energy system, some go further and examine how to mitigate CO_2 emissions from non-energy sources as well, and a few focus specifically on achieving energy-related decarbonization using 100% renewable energy technologies. Summaries of some outcomes of a few recent studies are covered in IRENA's Global Renewables Outlook 2020 [10].

Three studies are analyzed here. Specific attention is given to the contribution of solar PV to the energy system decarbonization and toward achieving 100% renewable energy supply. Even in the more conservative studies we find that solar PV can make a substantial contribution, providing very strong market opportunities for these technologies in the years to come.

4.1 IRENA's Sustainable Energy and Climate Change Analysis

Since 2014, even before the Paris Climate Agreement was signed, IRENA has been undertaking studies to determine how the world can achieve sustainable energy and climate change, known as the Global Energy Roadmap, or REMap. Initially their analyses went out to 2030, using a pathway that doubles the share of renewable energy. In 2018 they extended the outlook to 2050 with their publication of "Global Energy Transformation: A Roadmap to 2050 (2018 Edition)" [11], updated in 2019 [12], and further elaborated upon in their first Global Renewables Outlook 2020 [10].

The Roadmap examines in detail a "Transforming Energy Scenario," which is described as an ambitious, yet realistic transformation pathway based largely on renewable resources and energy efficiency measures aimed to keep the rise of global warming to well below 2 °C and toward 1.5 °C by the end of this century. This scenario contrasts to the "Planned Energy Scenario," which is their reference case and is based on governments' current energy plans, targets, and policies, including the NDCs established by individual countries (as exemplified in Fig. 5). An even lesser-ambitious scenario, the Baseline Energy Scenario, is predicated on policies and commitments in place just prior to the Paris Agreement.

Even the Transforming Energy Scenario does not bring carbon emissions down to the level of the most recent IPCC recommendations, which call for a fully decarbonized energy system by mid-century [4]. For example, ref. [10] shows that this scenario, albeit ambitious, cuts 70% of the world's energy-related CO_2 emissions by 2050, 90% of which is accomplished through renewables and energy efficiency measures. By 2050, this scenario still shows annual CO_2 emissions of 9.5 Gt. Nevertheless, significant climate, economic, and social benefits arise from the Transforming Energy Scenario beyond the Planned Energy Scenario, and ref. [10] does offer additional recommendations to cut CO_2 emissions beyond 2050 to zero or net-zero.

The Transforming Energy Scenario is achieved through a number of factors. First and foremost, the costs of solar PV and other renewables will continue to decline, and policies that currently support subsidies to the fossil industry will shift to renewables. The scenario specifically assumes that the global-weighted average Levelized Cost of Energy (LCOE) for solar PV will drop to USD 0.040/kWh by 2050, a decrease of 58% from 2018. Overall, the scenario assumes a major shift toward the electrification of the transport and heating and cooling sectors, resulting in more than a doubling of the current contribution of the power sector, so that electricity supply represents 49% of total energy demand. Furthermore, the percentage of renewables providing this electricity supply will grow from 25% (2017 levels) to 86% by 2050, with solar and wind representing the majority of this supply due in large part to their falling prices. In fact, solar PV will provide 25% of total electricity demand (currently the contribution of solar PV to the total electricity supply is 2.8%). As shown in Fig. 8, this will require a much greater capacity expansion (8519 GW) than wind.

Of the three scenarios evaluated, the Transforming Energy Scenario is the most conservative. We now examine two other studies that show a more aggressive use of renewables and electrification of end-use energy demands that result in even higher capacity scenarios for solar PV.

4.2 Achieving the Paris Climate Agreement Goals – UTS (Springer)

The University of Technology Sydney (UTS) has recently published a Springer book titled "Achieving the Paris Climate Agreement Goals" [13] that directly addresses the IPCC 2018 recommendations of completely decarbonizing the energy sector by mid-century. The approach of this comprehensive volume is to examine mitigation strategies for three scenarios of future global warming by the end of this century: 5.0 °C (the "reference scenario"), 2.0 °C (the Paris Agreement target), and 1.5 °C, or the level urgently recommended by the IPCC 2018 [4]. The authors approach their analysis using a variety of computer model and statistical approaches. The analysis goes beyond the energy sector to examine mitigation strategies for non-energy-related CO_2 emissions.

For this paper we focus on the 1.5 °C strategy, and the specific role that solar PV plays in achieving this strategy. In contrast to the 5.0 °C scenario, which is based on current policies and the International Energy Agency's World Energy Outlook

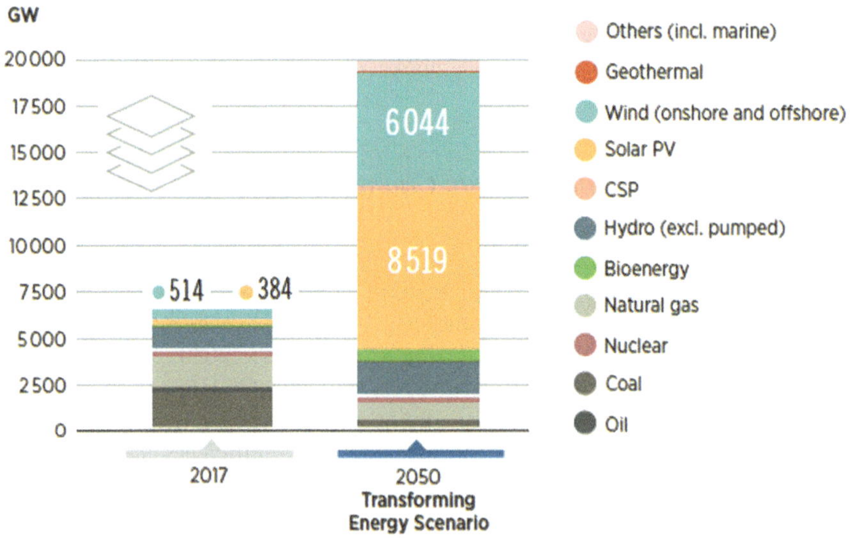

Fig. 8 The growth in capacity levels required to achieve expected electricity supply by 2050 under the Transforming Energy Scenario. Under these capacity levels, solar PV will contribute 25% of total electricity supply in 2050 (see ref. [10])

2017 [14], the 1.5 °C scenario is achieved by allowing for no more than 450 Gt of total CO_2 emissions to occur between 2015 and 2050. To achieve such a reduction in the energy sector, the study shows that the heating and cooling sector and transport sectors must be heavily electrified, and very aggressive energy efficiency measures must be undertaken. By 2050, based on the assumptions for this scenario, total global energy demand must be reduced to 412 EJ/yr (114,454 TW-hr),[1] or 26%, from 2015 levels of 556 EJ/yr (154,457 TW-hr).[2] Global electricity production will be 65,300 TW-hr (235 EJ/yr) supplying about 57% of all end-use demand (IRENA's Transforming Energy Scenario has a 49% contribution from the power sector). The 1.5 °C scenario requires a 100% supply of renewable electricity (compared with 86% in IRENA's transforming energy scenario), or a capacity of 25,700 GW.

Achieving the 1.5 °C scenario requires not only a substantial reduction in end-use energy consumption due to efficiency measures, but also a significant application of low-cost variable renewable energy technologies. Costs are a major driving factor here, and as a result, solar PV's share of the final energy supply grows to 12,684 GW, or about 49% of total electricity capacity. This compares with 2019 levels of 627 GW, and a nearly 50% higher solar PV capacity in 2050 than IRENA's Transforming Energy Scenario.

4.3 LUT University and Energy Watch Group (EWG) Study on 100% Renewables

The LUT/EWG study, conducted jointly by the Finnish Lappeenranta University of Technology (LUT) and the Energy Watch Group (Berlin) [15] examines least-cost options for achieving a cap of 1.5 °C global warming by the end of this century based on significant electrification of all end-use energy demand and a 100% renewable energy system. The model simulates a transition to 100% renewables across nine major regions and 145 subregions at 5-year time increments from 2015 to 2050. The LUT/EWG study projects that electrification will grow by a factor of four to five by 2050 from 2015 levels, due primarily to substantial electrification of the transport and the heating and cooling sectors. By 2050, 90% of all end-use energy supply will come from the power sector (Fig. 9).

Cost considerations result in the scenario favoring wind and solar technologies as the most predominate sources of electricity supply in the 2050 scenario. Specifically, solar PV provides 69% of all primary energy, or 76% of electricity, by 2050, as shown in Fig. 10. This translates to a solar PV supply of 103,713 TW-hr/yr (373 EJ). Unlike the other two studies, this study does not project what this means in terms of installed PV capacity; however, if we consider that the average global solar resource where all PV systems are installed is 4.5 kW-h/m^2-day, and that the performance ratio of these systems averages 0.8, this translates to a total PV capacity of 79 TW by 2050, substantially higher capacity than forecast by either the IRENA Transforming Scenario or the UST 1.5 °C Scenario.

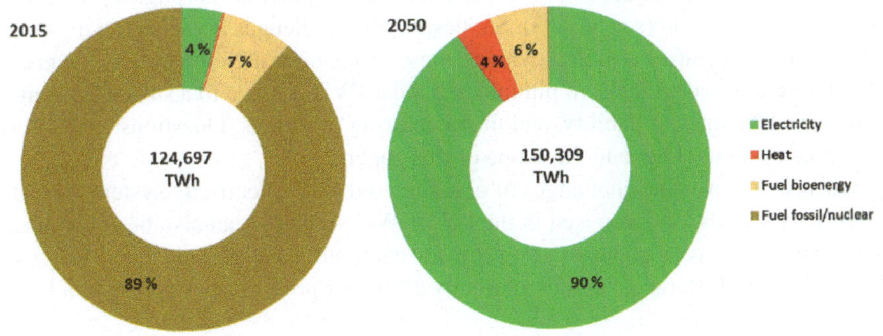

Fig. 9 Share of main fuels in the total primary energy demand globally, 2015 and 2050, in the LUT analysis [15]

[1] As compared with 538 EJ/yr in IRENA's Transforming Energy Scenario.

[2] This number is slightly lower than the 571 EJ/yr reported in [10].

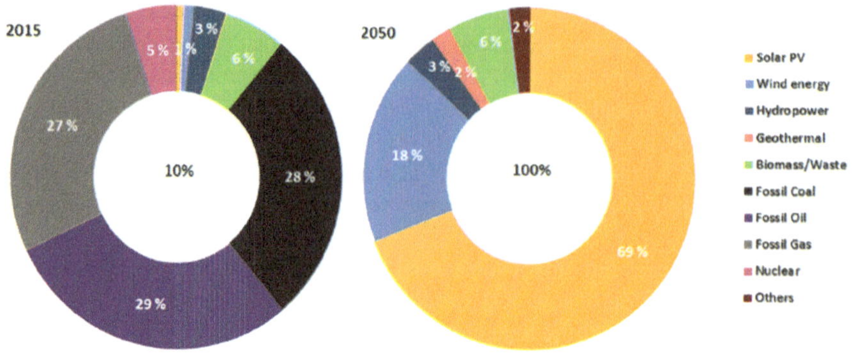

Fig. 10 Shares of primary energy supply in 2015 and 2050 in the LUT/EWG analysis [15]

5 Summary

This paper has shown that the consensus of the scientific community as represented by the IPCC [4] for achieving the Paris Climate Agreement targets requires all carbon-related emissions to fall to zero by mid-century. As depicted in Fig. 5, this timeline for decarbonizing the world's energy system will require an immediate and sharp reversal of global CO_2 emission trends.

Renewable energy, and especially, as of late, solar and wind technologies, are rapidly making strong contributions to the decarbonization of the energy sector, primarily in the power sector. The most recently available Global Status Report [8] shows that by the end of 2019, renewables contributed over 27% of our total power supply, and this contribution grows every year. Nevertheless, much more needs to be done to replace fossil fuel usage with renewables in both the transport and the heating and cooling sectors, where only a small fraction of renewables is currently meeting the needs of these sectors.

In this paper, three different studies were reviewed that lay out various pathways to decarbonize the energy sector by mid-century, primarily through rapid expansion of renewable energy deployments coupled with more efficient use of energy and electrification of the transport and heating and cooling sectors. The results of these three studies, focusing specifically on the contribution from solar PV to address the most aggressive carbon-reduction scenarios, are provided in Table 1.

The information in this table strongly suggests that as governments step up efforts to mitigate climate change, the opportunities for substantial growth in PV markets are immense. Even in the more conservative scenarios the annual additions of PV must increase by at least a factor of four over current levels (~115 GW/yr in 2019), which translates to an annual capacity growth rate of 9–10%. A full-scale effort to supply all of our end-use energy with renewables by 2050, as called for in the LUT/EWG analysis, translates into an annual solar PV capacity growth rate that is roughly 17%, compared with the 39% growth rate experienced this past decade (see Sect. 3). So, despite the challenges associated with such a long-term and sustained scale-up of solar PV manufacturing and deployment, these projections are clearly feasible given the capacity growth we have experienced in the past decade, thanks to plummeting solar PV costs due to a strong learning curve combined with technology breakthroughs in efficiency, reliability, and manufacturing processes. Questions such as critical materials availability and recycling of used systems is beyond the scope of this paper.

There is growing evidence that even the challenges of operating a reliable electricity system consisting almost entirely of variable renewable energy, as specifically proposed in the LUT/EWG scenario, can also be overcome. Although such a discussion is beyond the scope of this paper, Perez et al. [16], for example, have demonstrated that a very high, and in fact, over-capacity of PV combined with storage can meet virtually all of our power requirements at a lower cost than with conventional power systems.

Clearly, addressing climate mitigation goals makes for a very bright future for the solar PV industry. The message to governments is that they must seize on the technological and economic growth opportunities provided by renewables through enabling policies and regulations to achieve the ambitious transformation of our energy system to a fully carbon-free system within the next few decades.

Table 1 Summary of outcomes from the three carbon emission reduction scenarios examined in this Chapter

Name of study	Name of scenario	Electricity in total end-use energy consumption	Electricity supplied by PV	Projected PV capacity by 2050, GW
IRENA [10]	Transforming energy scenario	49%	25%	8519
UTS [13]	1.5 °C Target increase scenario	57%	49%	12,684
LUT/EWG [15]	100% RE by 2050	90%	76%	~79,000

References

1. International Energy Agency (IEA) (2019) Global energy and CO_2 status report. https://wwwieaorg/reports/global-energy-co2-status-report-2019. Updated March 2019
2. Höhne N (2019) Mitigation ambition. Presented at the Side Event: "Nationally driven ambition perspectives," SB 50, Bonn, 21 June 2019. New Climate Institute
3. IPCC (2007) Climate change 2007: synthesis report. Contribution of working groups I, II and III to the fourth assessment report of the intergovernmental panel on climate change. In Pachauri RK, Reisinger A (eds) Core writing team. IPCC, Geneva, Switzerland, p 104. https://www.ipcc.ch/report/ar4/syr/
4. IPCC (2018) Summary for policymakers. In Masson-Delmotte VP, Zhai HO, Pörtner D, Roberts J, Skea PR, Shukla A, Pirani W, Moufouma-Okia CP, Pidcock R, Connors S, Matthews JBR, Chen Y, Zhou X, Gomis MI, Lonnoy E, Maycock T, Tignor M, Waterfield T (eds) Global warming of 1.5°C. An IPCC Special Report on the impacts of global warming of 1.5°C above pre-industrial levels and related global greenhouse gas emission pathways, in the context of strengthening the global response to the threat of climate change, sustainable development, and efforts to eradicate poverty. https://www.ipcc.ch/sr15/
5. United Nations (2015) The Paris agreement. https://unfccc.int/files/essential_background/convention/application/pdf/english_paris_agreement.pdf
6. IRENA Coalition for Action (2019) Towards 100% renewable energy: status, trends, and lessons learned. International Renewable Energy Agency. https://coalition.irena.org/-/media/Files/IRENA/Coalition-for-Action/IRENA_Coalition_100percentRE_2019.pdf
7. IRENA Coalition for Action (2020) Towards 100% renewable energy: utilities in transition. International Renewable Energy Agency. https://coalition.irena.org/-/media/Files/IRENA/Coalition-for-Action/IRENA_Coalition_utilities_2020.pdf
8. REN21 (2020) Renewables 2020 global status report. Renewable energy policy network for the 21st century, Paris. https://www.ren21.net/wp-content/uploads/2019/05/gsr_2019_full_report_en.pdf
9. IEA PVPS Task 1 (2020) Snapshot of global PV markets 2020. International Energy Agency Photovoltaic Power Systems Technology Cooperation Program Task 1 "Strategic PV Analysis and Outreach," Report IEA-PVPS T1-37:2020April 2020. https://iea-pvps.org/wp-content/uploads/2020/04/IEA_PVPS_Snapshot_2020.pdf
10. IRENA (2020) Global renewables outlook edition: 2020. Energy transformation 2050. International Renewable Energy Agency, April 2020. https://www.irena.org/-/media/Files/IRENA/Agency/Publication/2020/Apr/IRENA_Global_Renewables_Outlook_2020.pdf
11. IRENA (2018) Global energy transformation, 2018 edition. A roadmap to 2050. International Renewable Energy Agency. https://www.irena.org/-/media/Files/IRENA/Agency/Publication/2018/Apr/IRENA_Report_GET_2018.pdf
12. IRENA (2019) Global energy transformation, 2019 edition. A roadmap to 2050. International Renewable Energy Agency. https://www.irena.org/-/media/Files/IRENA/Agency/Publication/2019/Apr/IRENA_Global_Energy_Transformation_2019.pdf
13. Teske S (ed) 2019: achieving the Paris climate agreement goals. Springer-Open. https://doi.org/10.1007/978-3-030-05843-2
14. IEA (2017) Word energy outlook 2017: a world in transformation. International Energy Agency Flagship Report, November 2017. Paris. https://www.iea.org/reports/world-energy-outlook-2017
15. Ram M., Bogdanov D., Aghahosseini A., Gulagi A., Oyewo A.S., Child M., Caldera U., Sadovskaia K., Farfan J., Barbosa LSNS., Fasihi M., Khalili S., Dalheimer B., Gruber G., Traber T., De Caluwe F., Fell H.-J., Breyer C., 2019: Global Energy System based on 100% renewable energy – power, heat, transport and desalination sectors. Study by Lappeenranta University of Technology and Energy Watch Group, Lappeenranta, Berlin, March 2019
16. Perez M, Perez R, Rábago KR, Putnam M (2019) Overbuilding and curtailment: the cost-effective enablers of firm PV generation. Sol Energy 180:412–422. https://doi.org/10.1016/j.solener.2018.12.074

"Zero-Carbon World": Opportunities and Threats for New and Old National Economies Because of Energy and Industry Transition

Winfried Hoffmann

1 Introduction

According to the Paris agreement the world has to move as quickly as possible toward a Zero-Carbon world. This can be done expressly and least costly with renewables, notably photovoltaics (PV) and wind with appropriate energy storage. This necessary transition together with the needed infrastructure implies a new and huge production industry for those new products. In addition the Zero-Carbon world has also to end transportation fueled by oil. This will also cause a transition from the existing combustion engines for all transportation sectors toward electric motors (batteries and fuel cells) and green fuels for aviation. Another important sector for today's CO_2 – emissions is the heating (oil and gas), ventilation and air conditioning (electricity) of houses which can be overcome by proper insulation. The introduction of green hydrogen, produced by hydrolysis of water with renewable electricity, is also needed urgently to suppress greenhouse gas emissions. This energy transformation will cause a fundamental industry transition: not only will existing power plants become obsolete and be replaced by a fleet of renewable power systems, but also many traditional machines and processes will be replaced by new ones. One example is the replacement of the old steel making process using coal by usage of green hydrogen.

2 Why "Zero-Carbon World" and Energy Transition Toward 100% Renewables

The energy situation for 2015 is shown in Fig. 1 (left). The primary energy of about 160 PWh (Peta = 10^{15}) is still dominated by fossil fuels (~80.5%), seconded by biomass (mainly firewood, ~10%), nuclear (~5%), hydro (~2.5%) and other renewables (~2%). The secondary energy of about 100 PWh contains ~22% electricity and the rest is petrol, oil, gas and (process) heat [1–3].

The losses associated with the transformation from secondary energy into what we really need, the end-user energy, are still ~50% due to a number of machines and devices, which need to be replaced by more efficient ones. One important aspect in today's energy scenario is the fact of inequality: only a quarter of mankind (~2 bn) consumes three quarters of energy (~75 PWh SE). Energy efficiency increase could realistically decrease this number by a factor of 2.5, resulting in 15 PWh SE usage for 1 bn people with today's quality of life as we know it from our privileged countries [4, 5]. Assuming this energy efficiency increase and a future population of 10 bn people [6], we can anticipate a future SE level of 150 PWh by allowing everyone the same quality of life as we have today. The dominance of fossil fuels today is the main cause for the increasing levels of greenhouse gases due to CO_2 emissions.

In order to meet the important Paris goal to limit the greenhouse gas emissions at a level to keep the temperature increase well below 2 °C, it is important to realize that not only the CO_2 – emissions from burning coal, gas, and oil must be taken into account, but also the contribution of other greenhouse gases like methane, N_2O and F-gas as well as the increasing land use change (burning of jungle for pasture and agriculture). Normally the effect of these additional contributions is calculated in CO_2 – equivalents. As seen in Fig. 2 (left), the global CO_2 – emissions from fossil sources amounted in 2017 to ~37 Gt,

W. Hoffmann (✉)
Hanau, Germany
e-mail: winfried@hoffmann-ase.de

© The Author(s), under exclusive license to Springer Nature Switzerland AG 2022
A. Sayigh (ed.), *Sustainable Energy Development and Innovation*, Innovative Renewable Energy,
https://doi.org/10.1007/978-3-030-76221-6_59

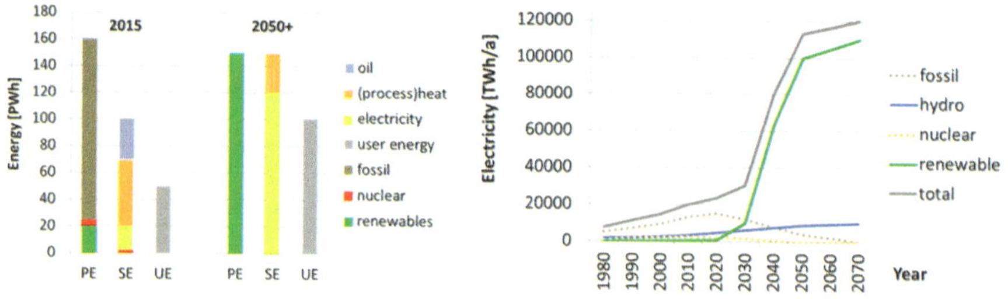

Fig. 1 *Left picture*: Prime, secondary, and (end) user energy (PE, SE, and UE) for 2015 [1–3] and 2050+ [4, 5]; *right picture*: implication for development of electricity production

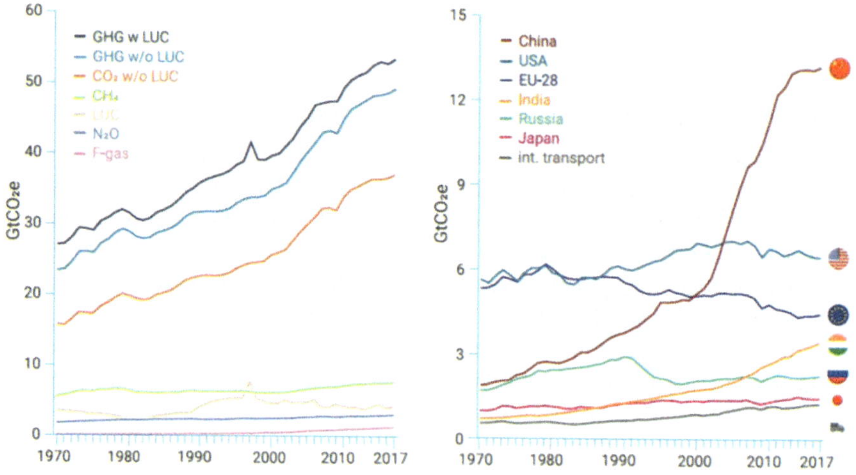

Fig. 2 *Left*: Global greenhouse gas emissions ([7], including LUC = land use change), *right*: development of greenhouse gas emissions in major industrial countries

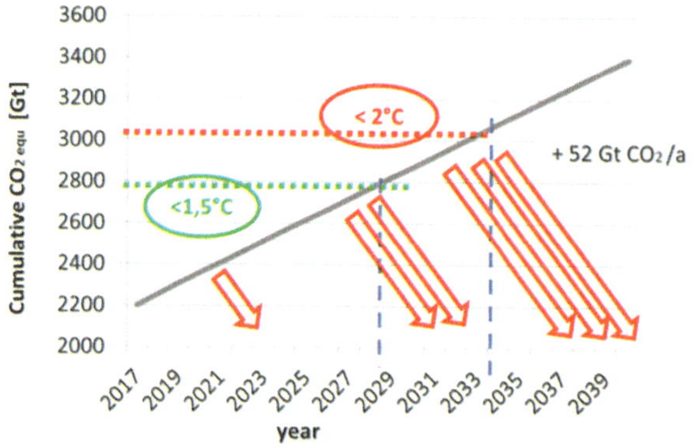

Fig. 3 Cumulative equivalent of atmospheric CO_2 and two scenarios for keeping the global temperature increase below 1.5 and 2 °C, respectively, at a constant annual increase of 52 Gt CO_2 (2017 level)

while another equivalent of almost 50%, namely ~16 Gt, was contributed by these additional processes. The above-mentioned temperature increase of below 2 °C will only allow a cumulative equivalent of ~800 Gt CO_2, which is reached in only 13 years if we continue as in 2017 (Fig. 3). Therefore we must drastically change and not only allow no further increase but to reverse the situation and emit quickly less toward the "Zero-Carbon economy." Looking to the most important contributors as seen in Fig. 2 (right), it is obvious that this will be a major challenge: only Europe and less US decreased the emissions over time, Russia, Japan, and also China, after a sharp increase due to its industrialization, stayed ~constant. The big question is with India, Latin America, and Africa. If their industrialization were to follow the same pathway as China, the additional 800 Gt would be reached very quickly in less than 10 years.

Table 1 Today's prices in $ct/kWh for various electricity generating technologies including storage cost; (+) indicates future increase and (−) future decrease of prices

Electricity generating technology	Price [$ct/kWh]
PPA (Power Purchase Agreement) new nuclear (Hinkley Point, UK)	20(+)
"clean coal/gas" (CSS)	>10(+)
PPA new wind parks (on- and off-shore)	~4(-)
PPA new PV power stations (>10 MW)	
northern regions (Ger, UK)	~4(-)
southern regions (Arabia)	~2(-)
Electricity storage with large scale batteries	~5(-)

Very fortunately the world has seen in the last 10 years an abrupt drop in prices for renewable technologies, in particular PV and wind (Table 1). There is now widespread acceptance in the fact that already today PV and wind are the lowest cost electricity generating technologies with Levelized Cost of Electricity (LCOE) in the range of 2–5 $ct/kWh, compared to >20 for new nuclear and >10 for clean fossil [4]. Even adding storage cost (<5 $ct/kWh) makes renewables plus storage cheaper than traditional technologies [8].

The annual energy offered by renewable energies in a sustainable way has been estimated by [9], which concludes that 90% is attributed to solar, 9% to wind, and only 1% to all other renewables (including hydro, biomass, geothermal, wave, and tidal) and is shown in Fig. 4. The total amount was calculated to >3000 PWh, which exceeds the future needs described above by a factor of 20. In order to allow for the local offerings from geothermal, hydro, biomass/waste, wave/tidal, and also wind, a simplified contribution from the various renewables was suggested [4, 5] to be 60% solar (with PV adding most), 20% wind, and 20% all other renewables.

The consequences by changing (1) all primary energies to renewables, (2) to assume an energy efficiency factor of 2.5 and (3) to use mainly electricity producing renewables like wind and solar are significant compared to today's situation as seen in Fig. 1: (i) by definition the primary energy is equal to the secondary energy, (ii) the relative contribution of electricity to the SE increases from ~20% to 80%, possibly even 90% and (iii) the end-user energy increases significantly, which allows all of the future 10 bn people to have a similar quality of life as we, in today's industrialized world, have now.

3 Industry Transition as Consequence of a Zero-Carbon World

3.1 Electricity Generation

One important consequence of the projections described above is a complete change of the globally existing fleet of electricity generating power plants: all fossil and nuclear plants will be replaced by predominantly wind and solar power plants as seen in Fig. 5. A study by [10] has shown that such a global change can be done not only environmentally safe but also cost-effective.

3.2 Mobility

Another important contributor to the global CO_2 – emissions is mobility, which accounts ~one quarter of today's secondary energy use, mostly fossil fuels (petrol and diesel for cars, trucks and buses, kerosene for planes and heavy crude oil for ships). A Zero-Carbon world needs a complete change away from these fossil fuels. This can be done in two different pathways:

1. **Replace fossil oil by liquid biomass.** This would allow to use the existing infrastructure to store and distribute the green fuel and also to continue the use of combustion engines. The first would please the international oil and gas companies and the second the international car manufacturers. However, this has a tremendous drawback: the areas needed to grow

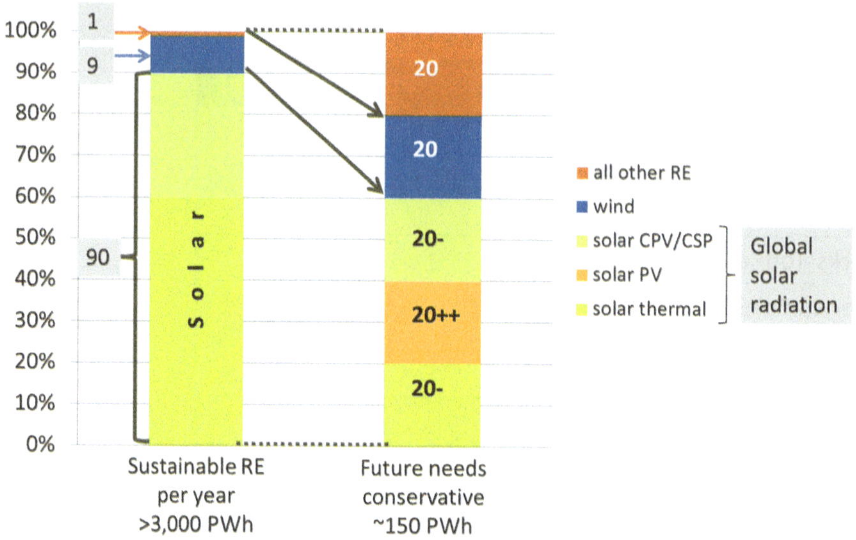

Fig. 4 Globally available sustainable renewable energy (ref.: WBGU Germany [8]) and the future needs together with the contribution from the various energy generating technologies (ref.: W. Hoffmann [4, 5])

Fig. 5 Exit from traditional nonrenewable sources (fossil and nuclear power stations) toward mainly to wind and solar

all this necessary biomass would compete with the urgently necessary areas for food production. It also has been demonstrated that the same area of land can deliver 50–100 times more energy in the form of electricity from PV and/or wind compared with the energy content of biomass [4]. Therefore we suggest a second pathway:

2. **Replace fossil oil by green electricity.** This would allow to use electricity directly with batteries in all those applications, where recharging must not happen too often within a day, typically cars, which drive in 80% of normal use less than 100 km per day. For those travelling >1000 km per day, buses and trucks one could use fuel cells driven by hydrogen derived from the hydrolysis of water with green electricity. Ships could best use liquid natural gas (LNG), but produced from green hydrogen plus CO_2 to obtain methane (CH_4 = LNG). For the foreseeable future the long-range planes would best be fueled by green fuels made from methane and processed further to bio-kerosene. This is shown in Fig. 6.

A simple reason why this differentiated pathway (2) described above will realistically continue to use three different technologies is shown in Fig. 7: for all three cases, we start with 100 kWh green electricity and calculate the distance traveled. The e-car with a battery can travel ~600 km by using the stated efficiencies. In the case of fuel cells, we have to take the two efficiency drops electrolysis (~75%) and fuel cell (~75%) into account, resulting in 56% (in comparison with 90% for the battery). Hence this fuel-cell driven car can only travel 333 km. In the third case of green fuel we have to take into account the efficiency of methanization (~50%) and the lower efficiency (~1/3) for the combustion (+gear) engine compared with the electric motor. This will only allow for a distance of 75 km.

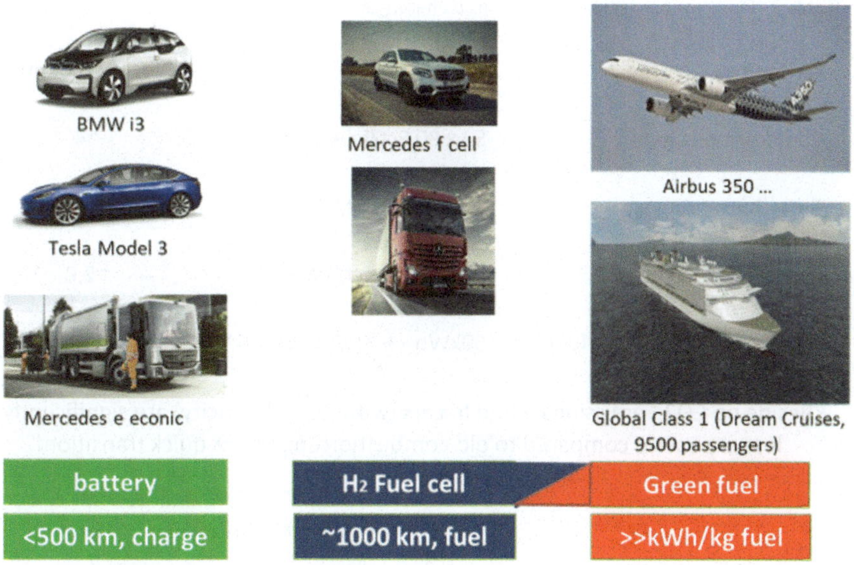

Fig. 6 The future of mobility – cars, lorries, ships, and planes

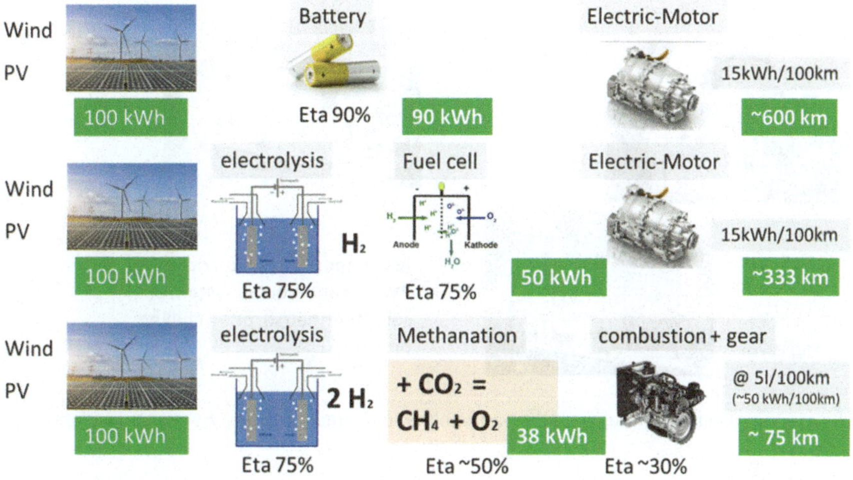

Fig. 7 Mobility: distance covered for the same amount of RE for different technologies

The fuel cost (only) for the three discussed cases is shown in Table 2 together with today's fossil fuel cost in a typical combustion engine. The following simplifying assumptions have been made: (a) for today's petrol price we have excluded the tax, (b) for the PV electricity cost we have used the Northern Europe amount which would bisect in sunny regions in the south and we have added a margin for distribution and selling the PV electricity of 50%, (c) the cost for chemical transformation from hydrogen to methane/green fuel was not taken into account and would increase the stated number. The comparison shows that the very distinct cost ranges for the three cases will allow these different application routes in the future. If we include the simplified technology and lower cost for the electric motor compared to the combustion engine will further trigger the change away from today's combustion engine. Only for the application where the high energy density fuel kerosene is a must – as for planes – the high price for this green fuel is justified and will be paid.

3.3 Houses

The heat requirement for houses changed significantly from the past with typical ~75 W/m^2 toward ~50 W/m^2 and is further reduced with low energy and passive houses down to 15–30 W/m^2 (Fig. 8). With these lower requirements we can reduce the amount of oil and gas accordingly and the remaining needs for heating and cooling can be done by solar thermal and PV.

Table 2 Fuel cost per 100 km for various technologies

technology	calculation	Cost [$/100km]
Combustion engine	5 l x 80 $ct/l (w/o tax)	~4
E-car (battery)	15kWh x 5 $ct/kWh x 1,5*	~1,1
Fuel cell car (hydrogen)	27kWh x 5 $ct/kWh x 1,5*	~2,0
Combustion with green fuel	50kWh x 5 $ct/kWh x 1,5*	~>3.8**

Beside no CO2 –emissions e-and fc-cars (with RE – electricity) are significantly less expensive compared to old combustion engine → quick transition!

* Includes distribution & selling of RE-electricity

** excludes the cost of chemical transformation

Fig. 8 Development of heat requirement for houses from the past into the future mainly driven by proper insolation

3.4 Chemistry

As a last example of how the industrial landscape will change due to the required Zero-Carbon world we take a look into the global chemical industry. In Fig. 9 is summarized how green hydrogen can be used in a variety of important chemical reactions.

As a specific example we take a look into the global production of steel as shown in Fig. 10. Today iron ore is reduced by coal which amounts to 2.7 Gt CO_2 representing an astonishing 7% of the global CO_2 – emissions. These emissions can be omitted when green hydrogen is used as shown in the lower part of Fig. 10.

4 "Old" and "New" National Economies

It is interesting to analyze how different national economies have tackled the findings described before. We will take a closer look what happened in Germany – representing also some other countries in Europe – as an example of an "old" economy and China representing a "new" economy. Before this, we will shortly analyze the two countries in the years of 2000–2010.

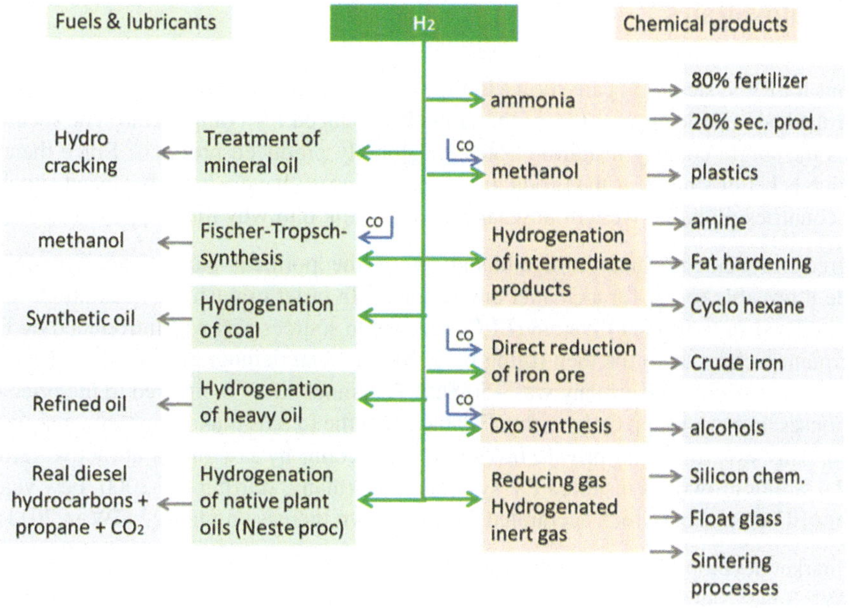

Fig. 9 The importance of green hydrogen for the chemical industry

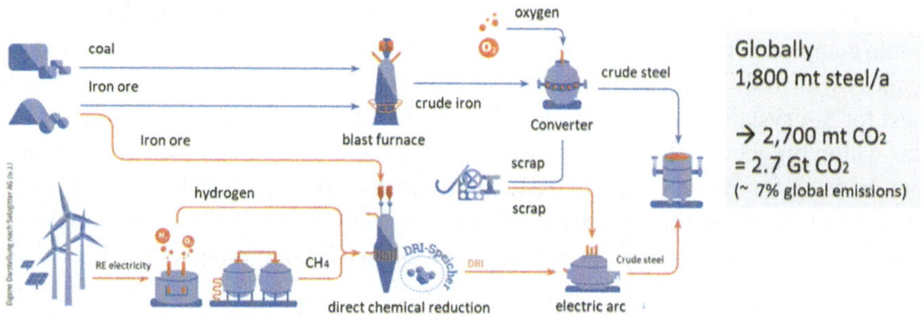

Fig. 10 Impact of green hydrogen to reduce significantly the CO_2 emissions for steel production (lower part: iron ore plus hydrogen) in comparison to the traditional process (upper part: iron ore plus coal)

The industrial economy in Germany in these years can be characterized by:

- Well-functioning electricity production and infrastructure since many decades run by well-established oligopolistic national utilities (e.g., RWE, EON, and Vattenfall).
- Many SME's as "hidden champions" serving global and national big industrial conglomerates in the machine building sector.
- First of its class automobile production industry with best combustion engines (e.g., Mercedes, BMW, and Audi).
- Big chemical industries (e.g., BASF, Bayer, and Merck)

The situation in China was quite different:

- Big efforts to industrialize the country by investing many GW coal-fired power stations (compare the increase of CO_2 shown in Fig. 2).
- Manufacturing for many international companies (toys, but also semiconductors).
- Cheap work force.
- Clear goal to not only have production as a service, but also to establish an own production industry in areas where globally needed products evolve.
- A population of more than a billion people where most had a low living standard. Private companies at that time did not exist.
- Last, but not least, a strictly organized country with a one-party state.

5 Case Study "Photovoltaics"

The annual installations for PV systems [11] are shown in Fig. 11. They can be clearly divided into three phases: a **first phase** until 2011, where Europe had installed up to 80% of the globally produced PV components. The **second phase** reaching out to 2017 with China as the single country installing >50% of globally produced products. Since then we have in the **third phase** a situation, which is better suited for the production industry, as we have a more balanced situation for the PV installations in the various countries globally. We will now take a closer look into why this happened as described.

First phase The "green movement" and the establishment of the political party "Die Grünen" in Germany in the 1990/2000 years made it possible to ask for a cleaner environment. In order to push the installation of green technologies – PV and wind – a special Feed-in tariff (FiT) system (EEG, renewable sources act) was introduced by Hans-Josef Fell which was supported by Hermann Scheer from the then ruling party SPD (social democratic party). It should be remembered that in 2000, the LCOE of PV electricity in Germany was ~50 $ct/kWh, much more compared to the price of ~20 $ct/kWh from the grid to domestic houses in these years. Only with a FiT based on the LCOE plus reasonable margin private people would invest in a PV system. This was done and private investors could count by law on an attractive return on investment for 20 years. This caused a strong growth of installed PV systems in Germany, starting in ~2000. Few years later, a similar FiT was introduced in other EU-countries, which accounted for the shown increase up to ~22 GW in 2011.

In parallel to this market development a number of European companies invested in the manufacturing of PV solar cells and modules (e.g., RWE Solar, Siemens Solar, Bosch Solar, Q-Cells, Solarworld, BP Solar, and Shell Solar), feedstock silicon (Wacker), materials like silver paste (Heraeus), glass and encapsulation foils, inverters (e.g., SMA, Fronius, ...), production equipment (e.g., Schmid, Manz, Baccini, Centrotherm, ...) and BOS-components (e.g., Schletter, ...). There was also an increasing number of companies to plan and install the PV systems (EPC's [engineering, procurement, and construction] and whole-salers, for example, Phoenix Solar, IBC, Densys/PV5, Solnet group, MHH Solartechnik, ...). While most of the non-PV cell and module companies named above are still active in this field, all of the PV cell and module producers from above exited the business with only small niche module producers in place today. Beside feedstock silicon, the specific investment was highest for Si-crystals, wafer-, cell- and module production, which was one of the reasons why until the late 1990s only companies within bigger conglomerates survived in this field (like RWE, Siemens, BP, and Shell). This changed when after the market increase for PV in Germany/EU entrepreneurs (like the founders of Solarworld and Q-cells) could successfully place an IPO and could use the money for investments. Unfortunately, the subsidiaries of the big conglomerates were either ended or sold to other companies as the needed investments did not fit into the portfolio of the respective mother company.

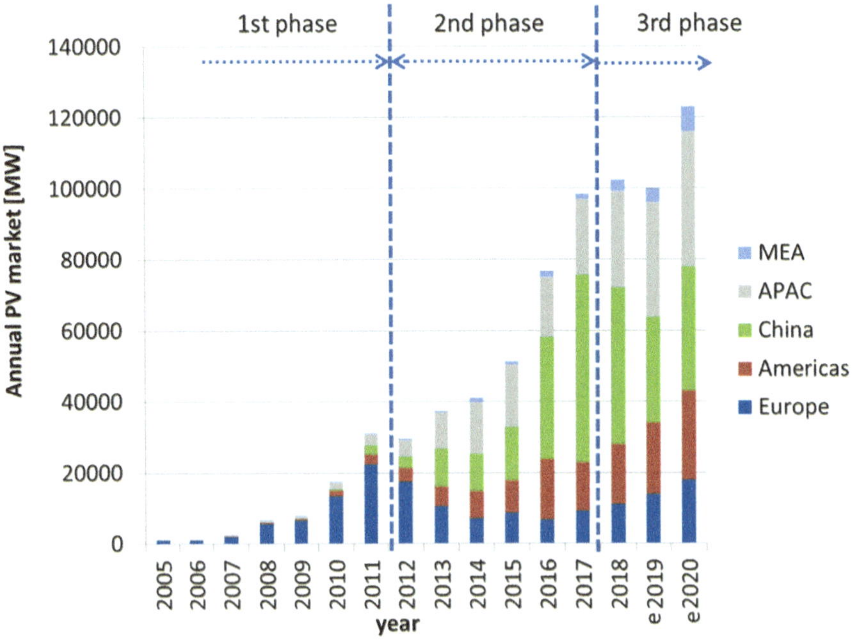

Fig. 11 Three distinct phases in the development of the PV development [11]

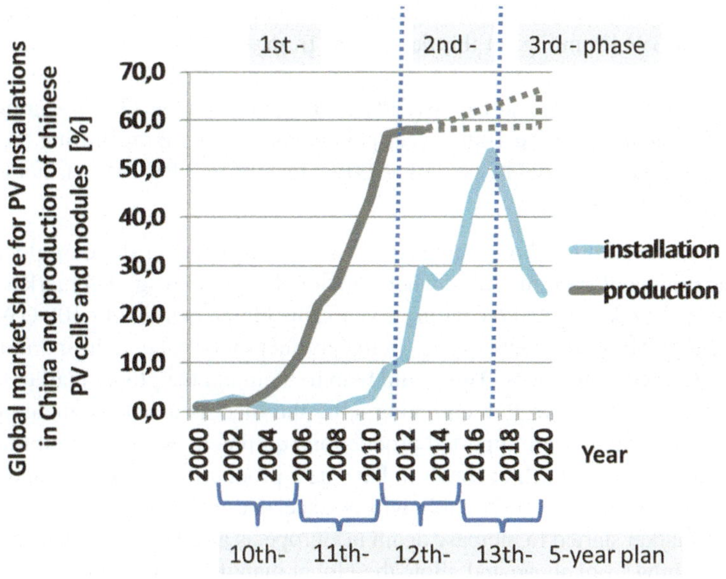

Fig. 12 Market share for production and installation of PV solar modules in China [11, 12]

The situation in China was quite different. When in the late 1990s the content for the 9th 5-year plan (1995–2000) was discussed, politicians together with industrially experienced managers and scientific institutions analyzed the landscape for future products. Photovoltaics was found to be one of the sweet spots. The market increase in Europe, paid by the millions of electricity consumers, was taken as a good market entry for solar cells and modules produced in China and exported to Europe. The Chinese government put aside a big investment sum (a report from the "Deutsche Bank Research" concluded a sum of ~50 bn $) which was foreseen via the Chinese state bank for investments in PV manufacturing plants. At the same time the government also allowed to start privately owned companies. Hence some entrepreneurs started their PV manufacturing, asked low interest money from the Chinese State bank and also went to the New York Stock Exchange to get additional money. The author remembers Mr. Shi Zhengrong from the company Suntech who very successfully built up the then biggest solar cell and module producer and became the first billionaire in the PV community and one of the first in China. Manufacturing of solar cells and modules kick-started in ~2004 as seen in Fig. 12 and reached from almost zero within 7 years a global market share of ~60%. The further development is difficult as indicated, as Chinese companies have in the last years started subsidiaries in other countries. Two important facts contributed to the quick increase of market share:

1. The companies for production equipment in Europe profited from the joint development together with European cell and module producers (named above) and the excellent research community (e.g., Fraunhofer-ISE, ISFH) to not only build hardware to process wafers into efficient solar cells but also to know the process parameters for the individual process steps. Chinese manufacturers were able to purchase the latest state of the art technology for their production and translated it toward economy of scale.
2. Most of the EPC-companies in Germany/EU took advantage of lower priced modules from China to maximize their profit. Major reason for the lower prices was the fact that the quick ramp-up caused a severe overproduction globally with subsequent prices below production cost and with the bigger economy of scale and new equipment Chinese manufacturers could produce at less cost compared to much smaller German/EU-based companies.

Second phase The flip-side of the nice increase of PV installations in Germany/Europe was the recognition by the utilities that a measurable amount of electricity was produced by former customers and taking business away from them. This ended in political decisions which drastically reduced the FiT, thereby reducing the potential return on investment and resulting in a quick decrease of annual PV installations in Germany and also in other EU-countries as seen in Fig. 11.

When China realized that all of a sudden their major markets in Europe collapsed with no other market in place, they had only one choice: create a local market to absorb all the PV components from the existing Chinese production industry. In addition in Europe and US, antidumping duties and anti-subsidy countervailing duties on Chinese PV products were enforced which made exports to EU and US difficult (minimum import prices for solar cells and modules in EU). As seen in Fig. 11 China showed a similar astonishing development for the local market: starting in 2010, it took again only 7 years to grow the market from just a few MW toward ~52,000 MW in 2017, which represented a global market share of >50% (see Fig. 12).

Third phase As seen in Fig. 12 there is a dramatic decrease in the market share of Chinese PV installations caused by a significant decrease of installed GW. From Fig. 11 this decrease is from ~52 GW in 2017 down to 44 and 30 GW in 2018 and 2019, respectively. The official statements from the Chinese government can be summarized as follows: (i) the cumulative installations in 2017 with ~130 GW exceeded already by 20% of the goal from the 13th 5-year plan for the year 2020, which was set at 105 GW; (ii) the preparation for the 14th 5-year plan is concentrating on shifting the ~100% subsidy-driven PV installations, in particular, the large green fields, toward ~100% subsidy-free PV market, notably to increase the share of domestic and decentralized PV systems [13].

It is tempting to speculate on an additional reasoning (iii): with the 50% market share of global installations it is possible to influence significantly the pricing of solar modules, when a major change of the global market is initiated. Hence, with the decision to decrease the Chinese market by 23 GW within 2 years, the global market of ~100 GW was affected negatively by >20%. Knowing that we have globally still overcapacity in the production of solar cells and modules, the consequence is a fierce market competition with unhealthy prices. This could help to maintain the global market share of Chinese producers.

There is another development in China worth mentioning: after the successful achievement of >60% global market share for producing wafers, cells and modules with imported production equipment and important materials, the goal formulated in the 13th 5-year plan (2016–20) was the achievement to have new production equipment and materials to >80% made in China. Again, as often in China, this goal most probably will be exceeded in 2020.

As seen in Fig. 11 the PV market started to increase again in Europe as also in many other regions in the world. This may mitigate a little the #(iii)-reasoning from above and allow the global manufacturing industry to export into a more balanced portfolio of countries with a sizable PV market. Also, it may trigger at last moment the buildup of manufacturing along the value chain in Germany/EU, India, and other countries.

6 Case Study "Li-ion Batteries"

The most influencing industry in the development of Lithium-ion batteries is the automotive industry. The pioneer Tesla who cooperated with Panasonic, one of the leaders for Li-ion batteries, paved the way toward e-mobility. It was China again which pushed this new mobility concept into economy of scale. It was clear that Chinese car manufacturers would have a hard time to compete with German/EU car manufacturers on the basis of combustion engines. Hence, there was a window of opportunity to enter into the car manufacturing arena with a new concept. New manufacturers, like BYD (build your dream), quickly ramped up production of electric cars and also batteries. The company Geely has the clear goal to produce by 2020 90% e-cars from their production portfolio. Remarkable is the strategic purchase of international companies like Volvo and London Taxi as well as the investment as largest single stockholder of Daimler AG by Geely. Among the three biggest producers in 2019 for Li-ion batteries are two Chinese: After LG Chem (51 GWh capacity) from Korea are CATL (Contemporary Amperex Technology Ltd. with 40 GWh) and BYD (25 GWh) from China. The pace of Chinese companies to conquer top market positions in future industry areas on a global scale is admirable.

In contrast, we had in Germany a completely different attitude in this field: being the most advanced car manufacturing country with highly sophisticated and best-of-its-class combustion engines and smooth automatic gearboxes the old CEO's like Mr. Winterkorn from VW and Mr. Zetsche from Mercedes neglected the change toward electric cars. Very arrogantly they argued publicly that for many more years to come e-cars will not be able to compete. When the EU started to install clear goals and limits for the average number of CO_2 emissions for each fleet of the respective car manufacturers it became quickly clear that even the best combustion engine will not be able to reach this. However, unfortunately, the management tried to meet these goals by introducing defeat devices which let into what is known as the Dieselgate scandal starting in September 2015. By today, the new management of the before-mentioned companies have completely changed. Late, but hopefully not too late they will be able to make this turn-around. Recognizing Germany and EU as relevant e-car market there are also two foreign investments worth to mention: Li-ion battery production by the Chinese company CATL in Thuringia/Germany and Tesla (car and battery) in Brandenburg/Germany. In addition, German carmakers are planning with consortia to build also battery factories in the near term. In this context it is worth mentioning that these consortia in Germany are offered substantial support from the ministry of economics both in terms of money and arguments in media. This willingness was not seen 15 years before when the PV industry asked for similar support. One could speculate why this is the case: batteries are an essential part of future e-cars and the automotive industry is seen in politics as a very important part of the economic welfare in Germany; 15 years ago in the case of PV there was no supporting industry behind this – in those days expensive – new way of electricity production, even more, utilities were successfully fighting against this decentralized technology.

7 Conclusion

What are the lessons learned from these exercises? We will discuss five supporting measures which are seen as decisive for the establishment of new products and services in a country: **First**, it is obviously sometimes obstructive to have a long running successful industry for important services in a country which hinders the introduction of new products and services. **Second**, an open-minded analysis including politics, industry, and science can help a lot to identify future important products and services. **Third**, once a product is chosen which only by economy of scale will be able to compete against the old existing products, an appropriate market support must be defined, which enables the end customer to decide for the new product. **Fourth**, if the production industry for this new product is seen important for the economy well-aligned financial measures have to be in place, and last but not least, **Fifth**, companies and highly motivated managers must be available with an attitude of reaching shortest "time to market" to produce at economy of scale quality products and sell these in a competitive market environment.

For this, we have analyzed in two countries – Germany as an old economy and China as a new economy – the two areas: technology for producing electricity (PV) and the mobility sector (batteries). Tables 3 and 4 summarize our analysis:

As seen in Fig. 13 the electrification in China started parallel to the industrialization program. Today China has the highest electricity consumption globally and this with a gigantic fleet of coal-fired plants. The pollution in most cities is harmful to the population which is a major reason to install increasingly renewable technologies like wind and solar PV.

The mentioned "Made in China 2025" program sets the clear goal that national champion companies should be developed in 10 high-tech industries, including robotics, semiconductors, electric vehicles, and renewable industries like wind and PV.

Looking to the qualitative assessment for the five supporting measures, it is obvious that Germany with only one "+" and one "0", but three "–" is far behind China which shows five "+" for this new industry. As a result it is not surprising that in Germany, but also in the other European countries there is no longer a production company for PV wafers, cells, and modules compared to the >60% market share globally in China.

Table 4 summarizes the situation for batteries for e-cars

Again the analysis shows that in Germany with only one "+", one "0" and 3 "–" compared to the 5 "+" evaluation in China, it is not surprising that today, only one Chinese company (CATL) is producing in Germany for the European e-car

Table 3 The five supporting measures as seen in Germany and China for PV; + and – indicate a positive and negative attitude to each of the 5 supporting measures, respectively, 0 means neither/nor

PV	Germany (old economy)		China (new economy)	
(1st)	Since many decades established utilities with strong ties to politics ➔ no interest to change	–	Electrification started in the 1990s due to the industrialization program with steep increase – open-minded	+
(2nd)	Fragmented groups with very different interests	–	One state party defines 5-year programs with all groups	+
(3rd)	The green party together with social democrats established Feed-in tariff	+	A Feed-in tariff was defined to grow the market	+
(4th)	Because of strong old industries there was no need to push for new ones	–	Strong interest to conquer new and future industries, embedded in "Made in China 2025" program	+
(5th)	Available, but short "time to market" is often hindered	0	Available with also short "time to market" philosophy	+

Table 4 The five supporting measures as seen in Germany and China for Li-ion batteries; + and – indicate a positive and negative attitude to each of the five supporting measures, respectively, 0 means neither/nor

Li-ion battery	Germany (old economy)		China (new economy)	
(1st)	Since many decades many big and powerful car makers with best-of-its-class combustion engines ➔ no interest for e-cars and batteries	–	Although biggest market there is strong dependence to cooperate with foreign car makers in Joint Ventures ➔ strong interest to have own technology	+
(2nd)	Strong interest to keep the old technology	–	Very open-minded to develop new technologies like e-cars	+
(3rd)	A buyer's premium is demanded but not yet approved from politics	+	High subsidies have been given to kickstart the sale of e-cars	+
(4th)	Because of strong old industries there was no need to push for new ones	–	Strong interest to conquer new and future industries, embedded in "Made in China 2025" program	+
(5th)	Available, but short "time to market" is often hindered	0	Available with also short "time to market" philosophy	+

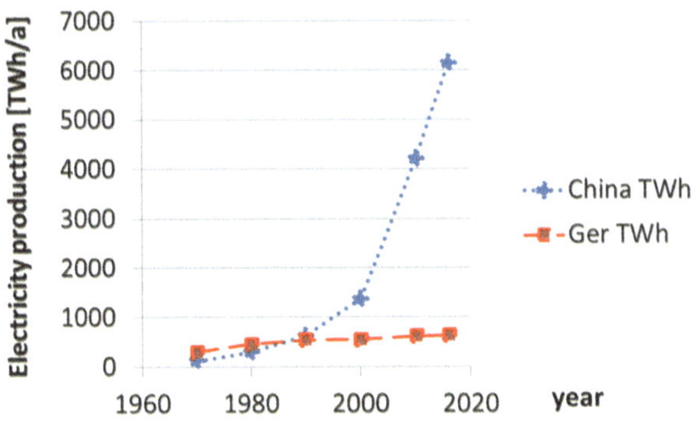

Fig. 13 Electricity Production in Germany and China

market, but no single German company. Just recently (2021) VW has decided to form a joint venture with Northvolt to also produce Li-ion battery cells and modules in Germany.

One important result from this analysis is obvious: when old economies rest on their well-established old technologies with even reluctance to start new ones, new economies have an easy time to conquer the production leadership for the future with new and better products. Of course, Germany and the other European countries cannot push decisions as the state party in China – but we must find appropriate measures and financing instruments if Europe as an old economy would like to keep its quality of life and welfare. The European Green Deal, the Carbon Boarder Adjustment Mechanism and hopefully some financial instruments (e.g. via National Recovery Plans) will enable European companies to compete globally in important new industries - PV modules and new batteries are just two examples among many others.

Acknowledgments I would like to thank many of my former colleagues from industry, scientific institutes, and associations for many enlightening discussions and great friendship.

References

1. IEA (International Energy Agency) (2017) World energy outlook
2. BP Energy Outlook (2018)
3. REN 21 (2019)
4. Hoffmann W (2014) The economic competitiveness of renewable energy. Scrivener Publishing/Wiley, ISBN 978-1-118-23790-8
5. Hoffmann W (2020) In Petrova-Koch V, Hezel R , Götzberger A (eds) PV as a major contributor to the ~100% renewably powered world and solving the climate battle. Springer Series in Optical Sciences 140, p. 9-30, ISBN 978-3-030-22863-7
6. UN (United Nations) (2013) World population prospects: the 2012 revision, highlights and advance tables, working paper No. ESA/WP.228
7. UNEP (2018)
8. Hoffmann W (2014) PV as one of the major contributors to a future 100% renewably powered world – importance and evidence for cost effective electricity storage. PVSEC Amsterdam
9. WBGU (2011) Advisory Board to the German Government on global change, Flagship report 2011, World in transition: a social contract for sustainability, ISBN 978-3-936-191-37-0
10. Ram M, Bogdanov D, Aghahosseini A, Oyewo AS, Gulagi A, Child M, Fell H-J, Breyer C (2017) Global energy system based on 100% renewable energy-power sector, ISBN: 978-952-335-171-4
11. SolarPower Europe (2019) Global market outlook
12. Own research (2020)
13. Personal communication Frank Haugwitz (2019, 2020) AECEA (Asia Europe Clean Energy Advisory Co Ltd.

Renewable Energy Technologies: Barriers and Policy Implications

Jyoti Prasad Painuly and Norbert Wohlgemuth

1 Introduction

Between 1990 and 2017, total worldwide energy demand increased at an average annual rate of 1.7%, hardly outpaced by renewables which increased by 2% annually on average [1]. Solar PV and wind energy were the fastest growing forms of renewable energy, with average annual growth rates of 37.0% and 23.4%, respectively (Fig. 1). The use of solid biofuels and charcoal increased by a mere 1.0%, dragging – as a result of their weight within the renewables category – the annual average growth rate of all renewables to 2.0%.

With these impressive growth rates, can it still be said there are barriers to the growth of renewables? First, the share of renewable energy in total primary energy supply is still as low as 10.2% in the OECD (in 2017). This share ranges from 0.4% in the Middle East to 9.0% in China to 47.3% in Africa. Second, there is a huge potential for the development of renewables and increasing its share in the total primary energy supply. Third, even the current pace of renewable development owes to support from various quarters, including governments and donors, absence of which may throttle its development, and finally, the growth is uneven, with a few countries primarily contributing to renewable energy development and potential remains untapped in others due to a variety of barriers.

Though barriers exist, these may vary in their intensity (importance) requiring major efforts where intensity is high. For example, high subsidies to fossil fuels in a country can make economic viability a major barrier, and coverage, for example, some barriers are specific to a technology or a country/region. Also, barriers have interdependencies, and addressing the key barriers can help others getting addressed more easily. For example, high subsidies to fossil fuels in a country can make economic viability a major barrier. Barriers can also be specific to a technology or a country/region.

2 Literature on Barriers to Renewable Energy

The European Union has an ambitious "Green Deal" that requires almost complete decarbonisation of the Union's energy system by 2050. The study in Ref. [2] provides a meta-analysis of projections for the shares of renewables in selected EU states. They find that "a favourable investment climate also requires the absence of severe barriers in the non-economic environment". Their paper provides a detailed overview of the main barriers and enablers for individual EU countries. Not surprisingly, countries face similar barriers and can achieve deployment through common enablers. Consequently, there is significant potential to learn from each other. The study referred here [3] presents a comprehensive literature review of barriers to large-scale market integration of intermittent renewables in the EU's electricity market. The study cited herein [4] provides an overview of public policy initiatives to promote renewable energy. They stress five pivotal actions for a sustainable energy future: strengthening policy commitment to renewable energy; mobilising investment; institutional, technical,

J. P. Painuly (✉)
UNEP DTU Partnership, Technical University of Denmark, Copenhagen, Denmark
e-mail: jypa@dtu.dk

N. Wohlgemuth
Department of Economics, University of Klagenfurt, Klagenfurt, Austria
e-mail: norbert.wohlgemuth@aau.at

© The Author(s), under exclusive license to Springer Nature Switzerland AG 2022

A. Sayigh (ed.), *Sustainable Energy Development and Innovation*, Innovative Renewable Energy,
https://doi.org/10.1007/978-3-030-76221-6_60

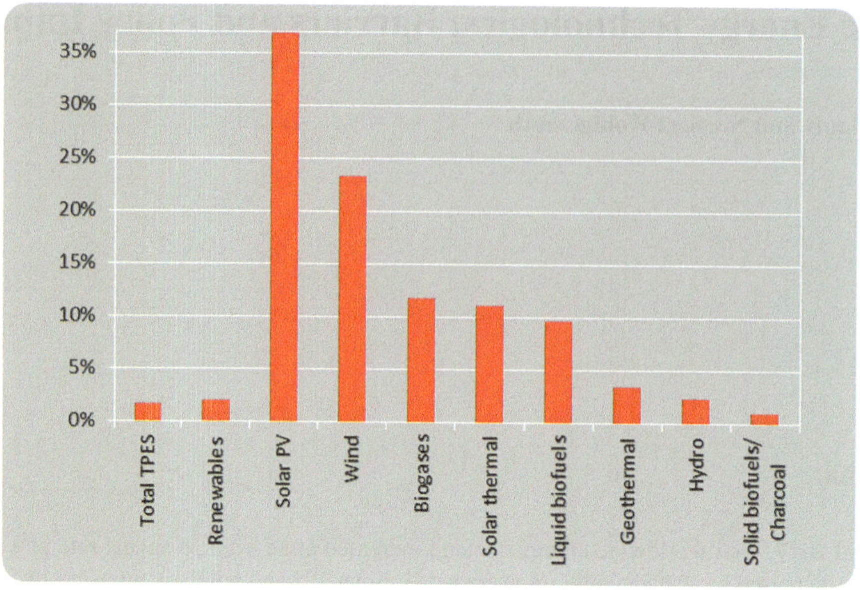

Fig. 1 Average annual growth rates of world renewables supply, 1990–2017. (Source: [1])

and human capacity building; harnessing cross-cutting impacts of renewable energy on other sustainable development goals; and enhancing regional engagement and international cooperation. Government policies that complement each other (e.g. in the areas of land use, transportation, employment, agricultural, food security, water, trade concerns, infrastructure) are more likely to be effective. Sen and Ganguly conclude that a shift towards renewables might begin with a prominent role for energy efficiency, the "hidden fuel". Conserving energy by increasing efficiency in its use increases, ceteris paribus, the share of renewables [5].

The integration of (low cost) renewables in the electricity system causes organisational problems in many cases, for example, concerning utility resource planning methods and processes that ensure the least cost service provision. The study referred in here [6] concludes that, to overcome these barriers, regulators and utilities "will need to expand and upgrade their analytical capabilities".

In regions with little growth in electricity demand, such as Europe, increasing generation by renewables depresses wholesale electricity prices, making renewables somehow victims of their own success, indicating the limits of marginal cost pricing. Wholesale electricity prices slumped from about € 80/MWh in 2008 to € 30–40. In an industry where marginal costs are of great importance to the overall economics (merit order), the economics of renewables affects electricity markets to an extreme extent, leading electric utilities to separate their renewables (and grid) businesses from (loss-making) conventional generation. The study referred herein [7] claims that the world is caught in a vicious circle: renewable energy subsidies increase their deployment, which depresses prices, thereby further increasing the need for financial support. In the extreme case of 100 percent (non-dispatchable) renewables, the marginal cost-driven market price would fall to zero, deterring any investment that is not fully subsidised. Therefore, the more successful policies to support renewables are, the more expensive and less effective policies become. The utility model of generating electricity is in many cases broken, as are markets. The "zero marginal cost society" (Jeremy Rifkin) may not be as easy to achieve as commonly thought, but lessons can be learned from the communications industry [8].

The study cited herein [9] analyses the effect of higher penetration of renewables on the price of fossil fuels. Usually, fossil fuel prices are assumed to be exogenous; however, a massive increase in the use of renewables is likely to depress fossil fuel prices, keeping all other factors constant. Lower prices for fossil fuels, as a result of the increasing use of renewables, are, therefore, likely to inhibit the economic competitiveness of renewables. They fall victim to their own success. This "green paradox" and carbon leakage are examples of this development. On the other hand, the crude oil price crash of March 2020 also makes renewables less cost-competitive as a result of changing relative prices between fossil fuels and renewables.

The economics of individual fuels is to a large extent affected by market failure – the lack of correct pricing. The study under Ref. [10] shows that internalisation of negative environmental spillovers of fossil fuels "would serve to hasten this transition process" towards renewables. This internalisation could take place, for example, by fuel and carbon taxes and provides an important contribution to the much-cited levelling of the playing field between different types of energy.

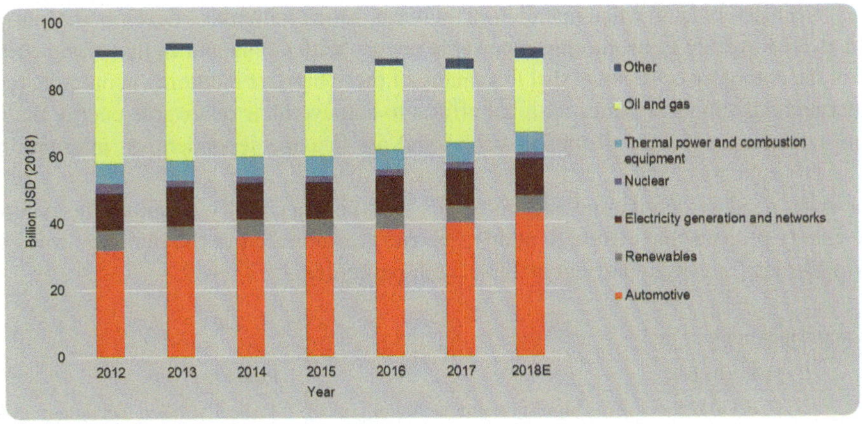

Fig. 2 Global reported corporate R&D spending in energy-related sectors, 2012–2019. (Source: [5])

A sub-optimal level of R&D expenditure can also be considered a case of market failure. The composition of public energy-related R&D expenditure changed substantially over time. In 1974 most of public energy-related R&D in OECD countries had been spent on nuclear energy research. Had this amount been spent on renewables, they would probably already have achieved full cost competitiveness. Concerning the private sector, Fig. 2 shows worldwide corporate R&D spending in energy-related sectors between 2012 and 2018. In contrast to public-funded energy R&D activities, private R&D activities account only for a modest share in total R&D expenditure.

A literature review reveals a general consensus that technical and economic barriers are no longer the principal barriers preventing some renewables from achieving a greater role in energy systems in many countries. The study under Ref. [11], for example, by collecting data through an online questionnaire sent to energy professionals, finds that social, technological, and regulatory barriers strongly affect the deployment of renewable energy all over the globe, while economic barriers strongly influence it indirectly.

For a renewable energy policy to be successful, The study referred in here [12] concludes that policymakers have to prioritise their objectives: large deployment of renewables, speed of adoption, and cost of the policy. All three objectives cannot be achieved by a single policy. The study referred in here [13] argues in favour of the possibility of large-scale electricity systems based on 100 percent predominantly intermittent renewables that meet key requirements of reliability, affordability, and energy security. In addition, such an energy transition can be implemented much faster than historical energy transitions.

In summary, in the early years, most of the barriers were common across countries and many even across various types of renewables. The list of the barriers included awareness, economic and financial, technical, political and regulatory, institutional, capacity, market, and social and cultural. However, over a period of time, many of these barriers have been addressed for some renewables, particularly in developed countries and emerging economies, which the scale of investment in the renewables reflects. In many developing countries, most or all of these barriers still remain, but the nature of barriers in countries that are leaders in renewables has changed.

Technology-specific barriers have emerged that impede large-scale deployment. Increased use of bioenergy requires a sustainable framework that also considers environmental issues. For geothermal, economic viability and sustainability of the enhanced geothermal systems on a large scale is a challenge. Ecological and social impacts in the case of new hydropower projects can have ecological and social issues, while ocean energy development may require testing infrastructure, enabling policies and regulations. Environmental concerns and public acceptance in the case of wind energy can be an issue. In many countries, issues include complicated licensing procedures and difficulty with land acquisition and permissions.

3 Barriers Identification and Policy Frameworks

Table 1 presents a summary of the barriers indicating their dimensions and impact on renewable energy projects.

Several types of barriers and categorisations can be found in the literature beyond this general categorisation. Also, the presence and intensity of barriers may vary across countries. However, in general, though many early barriers had been addressed in most developed countries through governmental policy support, new barriers such as variable renewable energy (VRE) absorption in the grid, end-use barriers in some sectors, and similar have emerged with expansion of RETs, requiring further governmental support. Financial barriers also remain in some cases due to upfront invest requirement, and environmental issues may also come up when installation increase. Along with many developed economies, emerging economies have also been at the forefront of promoting renewable energy with China and India among countries with large-scale installations. Some of the emerging economies still face many of these barriers including economic, technical, financial, etc., and may require international support besides national efforts to achieve their renewable energy ambition. Finally, most of the other developing countries still face all these barriers and are heavily dependent on international support to address the same.

The policy frameworks need to identify and address issues specific to a country / region. The experience from implementation of renewable energy projects and efforts to address barriers as indicated in the literature review indicates the need to consider the following issues (related to barriers in Table 1) in particular:

Table 1 Barriers and their dimensions

	Dimensions	Remarks
Economic barriers	Cost of technology is high (investment and/or cost of power in case of power production)	Typically, when technology is in the initial stage of development, ocean thermal for example
	Market failure – competing technologies are subsidised or their costs not internalised	Subsidy to fossil fuels, external costs of fossil fuels not considered
	High import duties/taxes	Sometimes levied to promote local industries
Technical and infrastructure barriers	Technology availability	Technology patents, especially for new technologies
	A lack of knowledge about technology and skill to maintain	Need for training
	Access to the grid (for renewable power)	Due to distance, it can be expensive
	Grid capacity (to carry renewable power)	Strengthening of the grid may be needed
	System capability to absorb variable renewable energy (VRE)	The system may not be flexible
	A lack of skilled personnel and training facilities	
	Infrastructure barriers for some end-use	
Awareness and information barriers	A lack of information about technology, its economics, and its features	A lack of appropriate institution to promote RE
	A lack of information about benefits compared to the existing technology	Need for capacity building
	A lack of capacity to understand the technology and its benefits	
Financial barriers	High-risk perception of renewable energy technology (RET) projects	This leads to high financing cost and also makes access to finance difficult
	Difficult to obtain project finance	This is due to a lack of familiarity with RETs with mainstream finance institutions as well as the high-risk perception of RET projects
	High upfront cost	Increases financial risk and cost of finance; can make project unviable
	May require additional investment in the system (grid strengthening) by distribution companies	Distribution company may not be interested or financially sound to invest
Regulatory and policy barriers	Absence of enabling policies/regulations	RETs may need regulations such as feed-in-tariffs, quotas, etc. to compete with a traditional technology (fossil fuels) since external costs not charged to the latter
	Uncertain policies	Policy uncertainty can lead to a lack of confidence and consequently affect investment in RE
Institutional and administrative barriers	A lack of institutional capacity to promote RETs	Leads to a lack of awareness about REs, their true costs, and benefits
	Inadequate support – for example for land acquisition	Can increase the cost of RE
	A lack of involvement of stakeholders in policymaking	This can lead to misplaced priorities and policies that do not work
	Complicated and bureaucratic licensing and other procedure	These lead to delays and increased cost
Social and environmental	A lack of public acceptance due to social reasons	It can lead to increased cost and smaller market
	Environmental barriers	Noise pollution, zoning regulations for example
	Aesthetic considerations	Visibility issues leading to public opposition in case of wind turbines for example
End-use barriers	A lack of infrastructure and high cost of its development for renewable energy heating, charging stations for electric vehicles for use of renewable electricity	These are usually technology and end-use specific barriers

- A lack of awareness/information about technology can lead to its non-acceptance. For example, the high upfront cost of a solar hot water system can dissuade users not knowing that it has a short payback period, and its life cycle costs are small, particularly in regions where there is good solar radiation and alternate source of hot water is high tariffs electricity. It is therefore important to identify and focus on the removal of such barriers.
- Some core barriers lead to the presence of other barriers, and therefore, once the core barriers are addressed, other related barriers may simply disappear. For example, the economic barrier can be on account of the high cost of technology (cost barrier), market failure (barrier due to direct or indirect subsidy to competing technologies), or high import tariffs (trade barriers). Addressing any one of these three barriers or a breakthrough in technology leading to lower cost of technology may help address not only the economic barrier but make other barriers also irrelevant.
- With maturing renewable technologies and their increased penetration, new barriers have emerged in countries with significant contributions from renewables. From "no more wind turbines in my backyard" to "no more of this intermittent power in my grid", new challenges have emerged, some requiring sophisticated approaches, complexities of which are themselves a barrier. The integration of VRE in the system requires increased flexibility in the system as well as better forecasting techniques for VRE generation. Better system operations and market designs that have incentives to absorb VRE may be other measures. Increased flexibility may require integration with other regional grids that have more flexibility, energy storage, and innovative measures such as vehicles to a grid (V2G) system that can provide flexibility. Policies that promote these measures will be needed. Danish integration of higher wind energy in the grid, for example, has been possible due to flexibility in conventional production, strong transmission and distribution networks, and a larger exchange of power with neighbouring countries. Replicating Danish experience in a different setting can be very challenging. Similarly, battery storage and vehicle to grid (V2G) technologies, which have been suggested as opportunities to address the barrier related to the absorption of intermittent renewable power in the grid, are still new and in the pilot stage with their own barriers in implementation.
- The current state of renewable energy development indicates the success of efforts made by various countries to address various barriers that renewable energy faced. The results, however, are mixed in countries depending on the availability of resources and their capacity to handle the challenges faced by them. As a result, the development of renewable energy is dissimilar across countries. Several countries where renewable energy has made good inroads have now reached to stage where besides problem related to absorption of VRE in the grid, they face new challenges that include increased application of renewable energy to heating and cooling, and transport sector, two crucial sectors to increase its penetration. In the transport sector, there is a need for integrated policies to address three issues: use of renewable fuel, vehicles that use renewable fuels, and infrastructure development for renewable fuel distribution [14]. Similarly, for greater penetration to heating and cooling, integrated policies for renewable fuel use and infrastructure development may be needed.
- Some policy areas that have good potential to push renewable energy production and use but difficult to implement include a carbon tax, energy subsidy reform, and behavioural changes. The first two send price signals to the producers as well as consumers influencing their behaviour. Often, these measures are also cited to ensure that renewables are provided a "level playing field". Energy subsidy reform and carbon tax are meant to compensate for the external costs that producers and users of fossil fuels put on the society. Behavioural changes can also be brought about through higher awareness and information, in addition to taxes and penalties. Due to difficulty in implementation, these policies have been substituted by a variety of promotional measures including incentives, tax rebates, feed-in tariffs, etc. in several countries to promote renewables successfully.

4 Conclusions

From the available literature, the following conclusions can be drawn:

Despite the growth of renewable and support by governments and other stakeholders, there are barriers to its further development. Though broadly the category of barriers has not changed, they vary across regions, technologies, end uses, and in their dimension in terms of intensity and coverage. It is therefore important to examine the barriers for technology not only for their existence but also importance, intensity, and similar other features.

Several types of barriers and categorisations can be found in the literature. A lack of awareness/information about technology still remains an important barrier, particularly in developing countries. Though the list of barriers may sometimes look long in a country, it is important to identify core barriers, as removal of such barriers may also eliminate other related barriers.

New barriers have emerged with increased penetration of renewable technologies, which in many cases are technology or end-use specific. These include absorption of VRE in the grid, and end-use barriers such as increased application of renewable energy to heating and cooling, and transport sector, two crucial sectors to increase its penetration. The solutions to address these barriers are emerging slowly and may take time to prove their efficacy. Battery storage and vehicle to grid (V2G) technologies, for example, have been suggested as opportunities to address the barrier related to the absorption of intermittent renewable power in the grid. These are still new and in the pilot stage with their own barriers in implementation.

Policies such as carbon tax and energy subsidy reform that provide level playing field and address externality issues related to use of fossil fuels have long been suggested to increase penetration of renewable technologies. These policies induce behavioural changes through price signals by shifting costs of fossil fuel production and use to their producers and end users but have been found difficult to implement due to a variety of reasons. These have been substituted by various other measures including incentives, tax rebates, feed-in tariffs etc. in several countries to promote renewables successfully. The challenge however remains to address the new barriers, particularly in developed and emerging economies where renewables' penetration is significant.

References

1. IEA (2019) Renewables 2019. International Energy Agency, Paris
2. Shivakumar A, Dobbins A, Fahl U, Singh A (2019) Drivers of renewable energy deployment in the EU: an analysis of past trends and projections. Energ Strat Rev 26:100402
3. Hu J, Harmsen R, Crijns-Graus W, Worrell E, van den Broek M (2018) Identifying barriers to large-scale integration of variable renewable electricity into the electricity market: a literature review of market design. Renew Sustain Energy Rev 81:2181–2195
4. Sen S, Ganguly S (2017) Opportunities, barriers, and issues with renewable energy development – a discussion. Renew Sustain Energy Rev 69:1170–1181
5. IEA (2019) World energy investment 2020. International Energy Agency, Paris
6. Go R, Kahrl F, Kolster C (2020) Planning for low-cost renewable energy. Electr J 33:106698
7. Blazquez J, Fuentes-Bracamontes R, Bollino CA, Nezamuddin N (2018) The renewable energy policy Paradox. Renew Sustain Energy Rev 82:1–5
8. Lo H, Blumsack S, Hines P, Meyn S (2019) Electricity rates for the zero marginal cost grid. Electr J 32:39–43
9. Foster E, Contestabile M, Blazquez J, Manzano B, Workman M, Shah N (2017) The unstudied barriers to widespread renewable energy deployment: fossil fuel price responses. Energy Policy 103:258–264
10. Owen AD (2006) Renewable energy: externality costs as market barriers. Energy Policy 34(2006):632–642
11. Seetharaman MK, Patwa N, Saravanan GY (2019) Breaking barriers in deployment of renewable energy. Heliyon 5(2019):e01166
12. Blazquez J, Nezamuddin N, Zamrik T (2018) Economic policy instruments and market uncertainty: exploring the impact on renewables adoption. Renew Sustain Energy Rev 94:224–233
13. Diesendorf M, Elliston B (2018) The feasibility of 100% renewable electricity systems: a response to critics. Renew Sustain Energy Rev 93:318–330
14. IRENA, IEA, and REN21 (2018) Renewable energy policies in a time of transition. IRENA, OECD/IEA, and REN21

Energy Poverty and Policy Implications in Two Mediterranean Countries

Ioanna Kyprianou, Anaïs Varo, Salvador Martí I. Puig, and Despina Serghides

1 Introduction

According to the latest Intergovernmental Panel for Climate Change (IPCC) reports, the Mediterranean is expected to confront elevated temperatures and lower precipitation rates, while it is already in higher risks of drought and temperature extremes [1]. The Mediterranean region has also become a hotspot for research in the field of energy [2–10]. This chapter is focused on one of the lesser known research topics, the phenomenon of energy poverty. At the moment there is no consensus for an official definition; however, a commonly accepted one is that energy poverty occurs when a household is unable to secure a level and quality of domestic energy services, that is sufficient for its social and material needs [11].

It is estimated that 50–125 million people are affected at some degree by energy poverty in Europe [12]. Poor energy performance of the residential sector means that energy-deprived individuals are either living in thermally uncomfortable conditions or consuming too much energy to attain an acceptable level of thermal comfort. Different countries and even regions within the same country are affected by different elements of energy poverty in a variety of degrees. For example, Mediterranean countries exhibit high excess winter mortality rates, albeit the mild winters of the region, whereas Scandinavian countries present some of the highest indications of hidden energy poverty [12–14]. The main indicators suggested by the European Energy Poverty Observatory (EPOV) are presented in Table 1, for the two countries participating in this research and the EU average [14].

Overall, Cyprus seems to be significantly different than Spain and the EU average when it comes to the two consensual indicators, while Spain also shows a worse performance than the EU average. Generally, Spain maintains percentages close to the EU average; however, in the two expenditure-based indicators, Cyprus seems to be doing slightly better than Spain (and both countries are better than the EU average). Conclusively, the worst indicator seems to be the ability to keep homes warm in Cyprus, despite the country's mild climate, indicating a poor energy performance of the housing stock.

Nevertheless, it needs to be noted that typical energy poverty indicators have been designed for cooler countries, with special emphasis on heating services, disregarding non-heating energy uses [15, 16]. Indoor overheating problems are prominent both for the Spanish and the Cypriot populations, something which is not represented in large-scale investigations.

Aside from the measurement and monitoring of energy poverty, the EU has assigned member states with the responsibility of dealing with the issue within their own territories, through a range of measures [17–21]. To this moment, no concrete action has been taken at the EU regulation level; however, the new EU Green Deal appears to be promising an array of effective energy poverty mitigating actions.

Research is also often country specific and can be shaped by political agendas. As a result, some countries have more advanced research and implementation frameworks in place. For example, research in Spain is much more prolific than in Cyprus, when it comes to energy poverty [22]. However, there has been limited research on juxtaposition of energy poverty situations and mitigating measures in different countries – a gap which is partly filled by this study. This chapter aims to provide an overview of main measures in energy poverty alleviation in two Mediterranean countries that have significantly

I. Kyprianou (✉) · D. Serghides
Energy, Environment, Water Research Centre, The Cyprus Institute, Nicosia, Cyprus
e-mail: i.kyprianou@cyi.ac.cy

A. Varo · S. M. I. Puig
Political Science Area, University of Girona, Girona, Spain

© The Author(s), under exclusive license to Springer Nature Switzerland AG 2022

A. Sayigh (ed.), *Sustainable Energy Development and Innovation*, Innovative Renewable Energy,
https://doi.org/10.1007/978-3-030-76221-6_61

Table 1 Indicators of primary energy poverty for Cyprus, Spain and EU average [14]

Type of indicator	Indicator	% population in Cyprus	% population in Spain	% population for EU average	Source
Consensual	Inability to keep the house adequately warm	21.9%	9.1%	7.3%	EU-SILC (2018)
	Arrears on utility bills	12.2%	7.2%	6.6%	EU-SILC (2016)
Income/ expenditure-based	High share of energy expenditure in income (2 M) indicator	12.0%	14.2%	16.2%	HBS (2015)
	Low absolute energy expenditure (M/2)	13.2%	13.0%	14.6%	HBS (2015)

different approaches in research, energy poverty detection and measure implementation (Cyprus and Spain). This study highlights that similarities can be found in countries with distinct scales and modus operandi; it also indicates that there is no specific recipe to mitigate energy poverty.

2 Methods

A set of common indicators for the two countries has been employed, obtained from the EPOV, as it gathers databases from EU-SILC and Household budget surveys (HBS). EPOV is a framework that manages a suite of indicators which should be viewed and used in combination, differentiating between primary and secondary indicators [14]. For purposes of this study, the following primary indicators are presented: inability to keep home adequately warm, arrears on utility bills, high share of energy expenditure in income (2 M) and low absolute energy expenditure (M/2). It should be noted that a significant indicator highly relevant to Mediterranean climates is the ability to keep dwellings comfortably cool in summer. This indicator has been collected via ad hoc modules only in 2007 and 2012, and its collection will soon stop [23]. For this reason, however pertinent it is, it is not included in our study.

In addition, the main EP-mitigating measures will be discussed, through a literature review of scientific publications, as well as documents drafted by government bodies and regional authorities. A subsequent analysis of the approaches followed in the two member states highlights the similarities and disparities.

3 Energy Poverty in Cyprus

Research and published material examining energy poverty in Cyprus are very limited. Several departments from the Ministries of Labour and Energy, Commerce, Industry and Tourism coordinate to create monthly databases of individuals that have successfully applied to be considered as vulnerable consumers. The decision-making process is based on certain criteria for each group of vulnerable consumers (e.g. one set of criteria for people suffering from multiple sclerosis and a different set of criteria for beneficiaries of social welfare aid). The categories of vulnerable consumers can be found in Cyprus' Integrated National Energy and Climate Plan, pp. 93–94 [24].

Measures aiming at protection of vulnerable consumers in Cyprus do not differentiate between the urban and rural population in Cyprus. Aids for heating oil were once provided to residents of mountainous and remote areas; however, with the onset of the economic crisis, this subsidy was ceased. In April 2019 and following a particularly wet and intense winter season for the island, government officials announced that a financial aid would be provided once more to remote regions, although it is not directly targeting energy poverty mitigation and is not subject to energy consumption patterns or type of fuel consumed [25].

The main mitigating measures include protection from electricity disconnection (subject to approval by a medical board) and better terms of energy refurbishment and renewable technologies (RES) schemes (see [26] for full list). As for numerical indications of beneficiaries, these were up to now anecdotally reported, but an effort was made to report and explain the reasoning of calculating vulnerable populations in the recent NECP report. Although there were 52,645 beneficiaries (5.8% of population), only 23,593 beneficiaries are actually considered energy poor (2.62% of population), and 13.370 (1.5% of population) have applied and benefited from the measures since 2013 [24].

The provided measures are constantly evolving to facilitate increased uptake of RES and energy efficiency measures, with favourable terms for the populations enlisted as vulnerable [22]. However, the lack of accuracy in the detection of the energy poor is a major hindrance in the effectiveness of all measures. Firstly, there are no limitations to the contracted power that energy poor households consume. Therefore, there is uncontrollable electricity consumption at lower rates for some groups of consumers. Moreover, the cost of the special tariff is covered by public service obligations (PSOs), imposed on all non-vulnerable consumers in Cyprus [27]. This possibly adds additional burdens in the form of PSOs for energy poor populations that are not accurately detected by the current methodology, thereby making a hard situation harder. Secondly, at the moment there are no economic criteria for groups such as blind individuals or those with physical disabilities. Finally, there have been no studies on the effectiveness of the established measures, in order to either confirm their validity and relevance or suggest substantiated adaptations, towards mitigating energy poverty in Cyprus. In essence, a holistic approach is lacking.

Overall, there has been no research or national reports on energy poverty in Cyprus; however, relevant legislation and a framework of detection exist on a national level. Through a range of measures, the Republic of Cyprus has managed to partially address drivers of energy poverty, although this is done sporadically. Because of the lack of correlation between the current national mechanisms and the widely used indicators of energy poverty, it is difficult to assess whether the state is providing sufficient and efficient support to the energy poor. Furthermore, it is unclear whether individuals living in energy poor conditions are correctly identified. The low absorption rate for measure uptake is further evidence of possible misinterpretation of the energy poor consumer profile in Cyprus, the lack of public awareness and heavily bureaucratic procedures. In spite of the shortcomings of research and approaches followed, this member state is creditable to continuous efforts (albeit misguided ones at times), including the recovery of financial aids for remote rural areas.

4 Energy Poverty in Spain

Until March 2019 there was no official specific interpretation of energy poverty in Spain. The new National Strategy on Energy Poverty (2019–2024) [28] provides, for the first time, an official definition on energy poverty and vulnerable consumer. The new definition establishes that '*energy poverty refers to a situation where a household cannot satisfy their basic needs in terms of energy as a consequence of a non-sufficient income level and, in some cases, can be aggravated further by living in a non-energy-efficient house*' [28]. It can be highlighted that the definition proposed includes, besides the economic dimension, the (poor) state of buildings in terms of energy efficiency. The second official definition proposed in the National Strategy is the one for vulnerable consumers, understood as an '*electricity or other thermic energy consumer in energy poverty situation who is eligible for support measures established by public administrations*' [28]. This definition is more ambiguous than the previous one, and it delegates its concretion to the specific targeting method of the energy poverty policies in place. Moreover, the only measure implemented nationally, the social tariff, includes a definition on vulnerable electricity consumers (excluding other vulnerable consumers or a general definition of energy poor population).

Spain is a decentralised country with different spheres of decision, and a large and diverse number of actors, who interact and collide. Energy poverty is no exclusion to this; it is a multifactorial phenomenon that involves different jurisdictions that, in turn, are distributed among the different government levels. Recent decisions of the Spanish Constitutional Court [29–33] have clarified the jurisdiction distribution of policy areas connected to energy poverty.

The new National Strategy proposes 19 measures, most of whom – 11 out of 19 policies – refer to energy poverty diagnosis and information provision improvement. Amongst the set of initiatives, not implemented yet, some of them can be regarded as significant steps. An example is a new Social Tariff, which would be the fifth new version of the current Social Tariff explained below, based on the universalisation to all basic supplies, its automatic application and the continuous coordination among multilevel public administrations in its implementation. Other relevant initiatives are the establishment of a minimum vital supply for vulnerable consumers that cannot be restricted, even in cases of non-payment. Finally, there are a set of measures focused on energy efficiency improvement of buildings, including short-term and long-term measures.

Of all the measures included in the new National Strategy, the most relevant one currently in implementation in Spain is the social tariff or social bond. It was first approved in 2009 and has been modified four times since then, thrice due to Court decisions related to its funding system [34] and the last reform because of political reasons. The main goal of the measure is to implement a discount on the last resort tariff in the electricity bill for vulnerable households and to protect severely vulnerable households from disconnection.

The social bond was created as a reaction to the fast increase of the electricity rates after the energy sector liberalisation in Spain. The measure asserted on freezing the electricity rates at the time of the measure approval (i.e. in 2009). This resulted in a remarkable discount, due to the important price increase that came in the following years. The second version of the Social Bond changed from the frozen rate to a 25% discount on the electricity bill. In 2017 another version of the Social Bond version was approved, allowing a 25% or 40% discount on the regulated rates[1] (VPSC) for vulnerable households and prohibiting the disconnection in severe cases when other administrations (regional or local) cover the remaining 50% of the bills. The social bond was modified, for the last time, in October 2018 by the new Spanish government ruled by the Socialist Party,[2] as discussed below.

The measure scope has changed in its different versions. Currently, it is applied to all national territory, and there is a strict criterion to select and classify eligible consumers. Before the last reform, criteria were simpler and less restrictive: beneficiaries were all population with a low contracted power (<3 kW) in their main residences, plus some specific personal criteria like the condition of pensioners, large family status or being part of a household with all its members unemployed. It was criticised for several reasons, but the most important one was its lack of specific attention to vulnerable consumers and vulnerable groups. In the current last version new eligible population criteria were established. The targeting scheme now identifies three types of vulnerable consumers, depending on the requirements covered. A common requisite for all eligible consumers is having a contracted power equivalent to or below 10 kW, provided it is contracted in their usual residence and they fulfil certain criteria. In addition, there is a new category of vulnerable consumers in a situation of social exclusion (as accredited by the local social services). In this last case, there is a prohibition of disconnection (but only if another public administration, regional or local, agree to disburse 50% of bills). The limited disconnection protection provided by the new social tariff regulation is complemented with a prohibition of disconnection for medical reasons, eligible only to consumers requiring electricity-powered life support equipment.

The last reform in October 2018 [35] lightly expanded the scope of the measure and its targeting, including single-parent families as vulnerable households and increasing the upper limit of energy consumption by 15%. Moreover, the new reform has forbidden disconnection of energy supply for all eligible households with children or dependent people. Finally, with these modifications, a new measure with a national scope was announced: a 'thermal voucher'. The measure consists on a direct subsidy for heating and hot water, with the subsidy amount varying, according to household income and climate zone.[3]

This study only considered the most relevant Spanish energy poverty measure with a national scope. Despite that, as already outlined, Spain is a significantly decentralised country, meaning that regions can develop their own energy poverty policies (although in some cases, that regional initiative can collide with the national government approach). To illustrate this diversity on policy initiatives tackling energy poverty in each of the territories, it is underscored here that half of the regions with high impact of energy poverty have specific measures to combat the situation, and only five of the total have not developed any territorially particular measure.[4] These regions, at the same time, are the regions considered 'rural' or with a low population density. Moreover, most of the measures implemented have a financial nature, aiming to cover energy expenses, but there is a significant lack of policies to tackle other drivers of energy poverty.

5 Energy Poverty Policy Approaches: Similarities and Differences

A number of issues were identified in the two cases studied, allowing the identification of similarities and differences. These have been disaggregated in different dimensions: the definition, the policy approach and main measures, the energy drivers addressed through measure implementation and, finally, the targeting method.

The main differences of the first dimension (definition) can be observed in Table 2. In general, the two definitions are similar in that they are both now legislated and categorise vulnerable consumers based on criteria. Nevertheless, the differences in Table 2 highlight the better-balanced nature of the Spanish definition, which accounts for consumption limitations, among other criteria. On the other hand, the definition in Cyprus has based the classification on specific circumstances and sometimes (but not always) income criteria. Although the Republic of Cyprus has offered legislation a decade ago and has

[1] The Voluntary Price for Small Customer (VPSC) is the system for establishing the price of electricity introduced by the government for customers in the regulated market with a contracted power that does not exceed 10 kW.

[2] In June 2018 a Motion of censure against the former government ruled by the Popular Party was brought to the Parliament. After that, the Socialist Party formed a government with the support of the rest of the left-wing parties.

[3] The measure rules identify six different climate zones and distribute the eligible consumers according to the height over the sea and the location of the capital of the region where the household is located.

[4] To see a complete overview of the territorial energy poverty policies in Spain, see Annex 1.

Table 2 Comparison between Cypriot and Spanish energy poverty definitions

	Cyprus	Spain
Energy poverty definition	Energy poverty may relate to the situation of customers who may be in a difficult position because of their low income as indicated by their tax statements in conjunction with their professional status, marital status and specific health conditions and, therefore, are unable to respond to the costs for the reasonable needs of the supply of electricity, as these costs represent a significant proportion of their disposable income [36]	Energy poverty refers to a situation where a household cannot satisfy their basic needs in terms of energy as a consequence of a non-sufficient income level and, in some cases, can be aggravated further by living in a non-energy-efficient house [28]
Subject of the definition	Consumers	Households
Drivers addressed by definition	Low income level Personal circumstances: Professional status, marital status, health conditions	Low income level Low energy-efficient buildings
Point of reference within definition	Costs for the reasonable needs of the supply of electricity	Satisfaction of basic needs in terms of energy

reviewed the criteria for vulnerable consumer groups several times, these revisions do not seem to be driven by research. Conversely, research in the EU, which offers ample evidence for 'easy fixes' (e.g. a definition reliant on energy consumption patterns), has not impacted on the existing situation. Nonetheless, it has been argued that even the Spanish approach is not as well-rounded as it seems, being heavily reliant on economic situations. Moreover, the definition for Cyprus is legislated specific to the use of electricity (as there is no natural gas market at the moment), while the Spanish definition does not restrict fuel type and considers the energy performance of dwellings.

The second dimension identified is the regional/national policy approach and the relevant implementation of measures. Cyprus is highly centralised, and all measures are equally distributed across different regions, significantly eliminating the rural/urban dichotomy. A centralised interposition on energy poverty situations could be useful in terms of coordinated, homogeneous implementation of the measures and to ensure equality among citizens. Nevertheless, there is a notable absence of deep knowledge on the specific situations of energy poverty, the variations of the phenomenon depending on the specific contextual factors or the available resources to generate operative and effective policy structures.

On the other hand, in Spain there is no national measure implementation other than the social tariff. Instead, local authorities are acting autonomously and addressing energy poverty however they deem suitable. The decentralised character of Spanish regulation was intensified after the crisis, with municipalities taking initiatives to cover a significant number of competences in front of the lack of intervention of the centralised government in emergency situations. Nevertheless, multi-level policies to tackle complex problems like energy poverty had already been incorporated in the DNA of the administrative scheme's model. Of course, the freedom to act as they deem appropriate allows lack of action, justified by 'non-significant' levels of energy poverty within the communities' territories.

It is interesting to note that while in Cyprus the first interventions addressing energy poverty dated from before the crisis, since 2006 [22], energy poverty only appeared on the Spanish political agenda at the beginning of the recession, after the first consequences of the economic crisis and the first version of the social tariff (which was approved in 2009).

As for the drivers addressed through energy poverty mitigating measures, both countries have achieved consumer protection from disconnections and reduced energy tariffs for the vulnerable, nationally. Therefore, the low income and high energy price drivers are addressed to some degree – although in Spain consumers are more comprehensively protected from disconnection, in Cyprus they are only kept connected in life and death situations. Even so, the lack of synergy among government levels in Spain causes disturbances: incongruent national and regional measures can generate inequalities between energy poor citizens of different areas. Additionally, in the special tariff measure, Cyprus is found to be lagging behind Spain, as it doesn't impose any limitations to the energy consumption of beneficiaries. In Spain, if the household's benefit recipients spend more energy than stipulated, they must pay the surplus at regular rates. A possible criticism for Spain is that specific vulnerable groups, beyond the households with children, are not considered in the establishment of these limits. The final driver of energy poverty is the inefficient building stock. In Cyprus, nationally implemented measures promote energy refurbishments with a higher level of support for vulnerable consumers. However, there are no initiatives to support the low-income populations in attaining the initial investment required, even if the amount is reduced by 75%. Conversely, in Spain, regional governments are offering tailor-made solutions to their territories, often promoting energy efficiency. Moreover, the definition of energy poverty refers to energy performance of buildings, thereby inherently ensuring that the correct populations are targeted.

The final aspect of comparison involves the targeting method, where the Spanish methodology is superior, as it separates vulnerable consumers in multiple tiers and incorporates the dimension of energy inefficient housing stock. Despite its superiority, Spain's targeting method may still be considered insufficient. Firstly, the economic requisite is the basis of the eligible criteria, becoming decisive to select the beneficiaries (even if the criteria also include references to specific groups). Secondly, the economic criterion only considers the income before taxes, without taking into consideration other factors like housing costs or other expenses. For example, a household from Madrid, the most populated and expensive city in Spain, will have the same discount than a household from a rural region, where the housing costs can be significantly reduced.

Finally, the last reform on the special tariff in Spain has introduced an interesting, but yet undeveloped, measure that could impact the detection of energy poor households: the heating voucher. The most suggestive element of this new measure is the consideration of different climatic zones of the country to determine the final amount of the voucher in each specific case.

6 Conclusions

Climate change studies indicate the Mediterranean region as a hotspot, and the area's countries should have national action plans in place to mitigate global warming effects, more frequent natural hazard episodes (e.g. heat waves) and social inequalities. Energy poverty encompasses these dimensions and has being recognised as an impediment to society in both countries examined here. EPOV primary indicators indicate that both countries have to deal with higher than the EU average difficulties, augmented additionally by economic crisis repercussions.

The obvious findings of this study are the variances between the two countries in several aspects of energy poverty detection and mitigation (Table 3). Nevertheless, some similarities can be found in the otherwise so different approaches. Both have classified groups of vulnerable consumers according to sets of criteria and have developed measures which are available only to certain groups of those consumers.

Cyprus appears to be a cohesively ruled nation that has complied with EU regulations in providing official definitions of vulnerable consumers and energy poverty but has not furthered research into the topic. In this way, although various nationally regulated measures exist, there has been no evaluation of their effectiveness. Judging from the low absorption rates of

Table 3 Main findings of this study, including similarities, differences and recommendations for each country

		Cyprus	Spain
Differences	Consumption limitations	No limitations	Limited consumption under the special tariff regime – varies for different VC groups
	Regulatory framework	Highly centralised – very limited distinction between rural and urban regions	Decentralised but with national obligations – autonomous communities adjust measures on their own
	Definition of energy poverty	Based on personal circumstances and economic criteria Limited to electricity use	Based on personal circumstances, economic criteria and technical characteristics of buildings Not limited by fuel types
Similarities	VC definition	Several groups defined	Several groups defined
	Measure implementation	Beneficiaries selected based on different criteria	Beneficiaries selected based on different criteria
Recommendations	Energy policy research is reinforced and encapsulated in the process of strategic policy-making. Adapt the existing operational frameworks to allow for policy assessment Provide distinction between energy poverty and vulnerable consumers and effectively improve the profiles of energy vulnerable consumers Examine more closely the urban-rural dichotomy Allow decentralised action Aim for long-term structural change rather than temporary solutions		Continue developing the measures proposed in the National Strategy at the same time, rather than focusing on regional and local initiatives Examine more closely the urban-rural dichotomy Maintain basic level of national regulation Aim for long-term structural change rather than temporary solutions

measures, this is deemed necessary. Moreover, although at one time there was a distinction between urban and remote rural areas, the onset of the economic crisis resulted in the abolishment of such measures for several years. Recent efforts to revive the financial aids towards remote and mountainous regions are now observed in Cyprus; however, this is not being done within the context of energy poverty.

On the other hand, Spain operates in a very dissimilar way, with autonomous communities taking over control of major measures aiming to support the energy poor population locally. Until the approval of the new National Strategy on Energy Poverty, one nationally regulated measure exists, the special tariff, which incorporates disconnection protection measures as well. It needs to be highlighted that the new National Strategy proposes diverse measures addressing energy poverty to be developed in the next 5 years. One of the main measures included in the strategy is the creation of a new social bond and several complementary policies, such as the establishment of a minimum vital supply. These advances are optimistic and obliges the future governments to engage with solutions tackling energy poverty.

As for the regional approach, centralised and decentralised avenues have both advantages and disadvantages. While unified systems allow for even monitoring, evaluation of measures at equal grounds and international comparisons, they also mean that specificities are lost between the urban-rural transition. Moreover, decentralised methodologies highlight the ability to adapt measures to localities and promote regional initiatives. Despite that, they can also portend uncontrollable measure implementation or even lack of it, as a result of the absence of obligations from a central authority. Local authorities in Cyprus should be given the freedom to investigate and act upon the energy poor populations within their own territories, as they see fit. On the other hand, for Spain it is recommended to retain a basic level of national control to avoid misuse of power or lack of it. The current situation in Spain carries optimism, in that the scientific community, the public and the relevant stakeholders are all engaged in the discussion.

The recent approval of the National Strategy on Energy Poverty in Spain has signified a big step towards a comprehensive and holistic approach to the energy poverty phenomenon. Despite its focus on diagnosis and information provision measures, the strategy provides a solid starting point to tackle the phenomenon properly, taking into account various drivers. It is advised for Spain to continue developing the measures proposed in the National Strategy at the same time, rather than focusing on regional and local initiatives. For Cyprus, it is recommended that energy policy research is reinforced and encapsulated in the process of strategic policy-making. The existing operational frameworks seem to be flexible to change and should enable adaptations that allow for policy assessment. An important aspect that should be improved is the distinction between two terms that have been combined: energy poverty and vulnerable consumers. The profiles of *energy vulnerable consumers* in Cyprus are far from ideal, and emphasis should be placed upon this, before any more alterations are considered. Finally, both countries should examine more closely the urban-rural dichotomy. Additional support is required for both countries in remote/low-density localities, and emphasis should be placed on action plans to mitigate energy poverty on the long term.

On the whole, this study highlights the robust mechanisms encountered in Spain, activated by national and regional governments and fuelled by citizen action. Cyprus may be lagging behind in terms of research, accurate detection of energy poverty and stakeholder engagement; however, the country's regulation is far less convoluted, and there are many 'low hanging fruit' solutions. Spain offers an elaborate framework, with more effective detection approaches, but measures are reliant on the different levels of government, at times unable to work synergistically.

In a nutshell, Cyprus presents an over-simplified image, whereas Spain seems to be found on the other end of the spectrum, presenting a much more intricate research and regulatory framework. It can be concluded that there is no specific recipe that will instantly obliterate energy poverty. It is rather a matter of producing research and evidence of the effectiveness of policies or the need to adjust them. Therefore, while there is heavy critique on the Spanish strategy, the country leads a contemplative pathway and encourages innovative actions. This should be encouraged for Cyprus as well, instead of simple adherence to EU regulations and persistence to austerity measures. Ultimately, balanced strategies that are well thought and also enable pan-European approaches should be the goal for member states that suffer from energy poverty.

Annex 1 Specific regional measures addressing energy poverty in the 17 Spanish autonomous communities

Autonomous community	Energy poverty impact[a]	Specific measures and type of measures
Andalucía	High	There are no specific measures to tackle energy poverty
Aragón	Low	Direct subsidy to pay electricity and gas bills. Energy Efficiency Plan (2016–2019) [37]
Asturias	Low	There are no specific measures to tackle energy poverty

Autonomous community	Energy poverty impact[a]	Specific measures and type of measures
Islas Baleares	Low	Creation of a work group against energy poverty Direct subsidies to third sector entities involved in projects addressing energy poverty [38]
Canarias	High	There are no specific measures to tackle energy poverty
Cantabria	High	There are no specific measures to tackle energy poverty
Castilla-La Mancha	High	Regional plan against poverty, which includes emergency aids and the prohibition of disconnection during the winter period [39]
Castilla y León	Low	Collaboration agreements within the regional government and the supply companies to avoid basic supplies disconnections [40]
Cataluña	Low	Catalan act 24/2015, July 29th, on urgent measures to address the housing and energy poverty emergency. This law introduces the precautionary principle (general prohibition of disconnection) and the basic supplies guarantee for vulnerable households [41]
Extremadura	Medium	Royal Decree, 1/2016, may 10th, of extraordinary measures against social exclusion. It includes measures to guarantee the access to electricity, gas and water minimum supply [42]
Galicia	Medium	Law 7/2017, December 14th, of energy efficiency measures and guarantee of electricity access It regulates the regional procedure to complement the National Social Tariff [43]
Comunidad de Madrid	Medium	Direct subsidy to pay electricity and gas bills [44]
Región de Murcia	High	Direct subsidies to municipalities and municipal associations of social services to intervene in energy poverty situations [45]
Navarra	Medium	There are no specific measures to tackle energy poverty
País Vasco	Low	Rent to guarantee a minimum income, as a periodic aid to cover basic necessities [46] Social emergency support, as a non-periodic subsidy to cover specific expenses [47]
La Rioja	Medium	Economic support to municipalities to avoid disconnections [48]
	High	Minimum income in social exclusion situations, which provides an extension of the aim for basic supplies (electricity, gas and water) [49]

Source: Elaborated by the authors

[a]The Energy Poverty Impact is determined by the data from the Household Budget Survey and the Survey on Living Conditions on five different indicators (2 M, M/2, Inability to keep home adequately warm, Arrears on utility bills and supply disconnections). Three types of impact with regard to the national energy poverty average have been defined: high, medium and low impact. Regions with a High Impact of energy poverty are those with two or more indicators above the national average. Regions with a medium impact are the areas with all the indicators within the average or only one indicator above or below the national average. Regions with a low impact of energy poverty are those with two or more indicators below the national average

References

1. IPCC (2018) IPCC SR1.5 – Chapter 3: Impacts of 1.5°C global warming on natural and human systems Notes: TSU compiled version. 1–246. https://doi.org/10.1002/ejoc.201200111
2. Simoes SG, Gregório V, Seixas J (2016) Mapping fuel poverty in Portugal. Energy Procedia 106:155–165. https://doi.org/10.1016/j.egypro.2016.12.112
3. Gaspar R, Antunes D (2016) Energy efficiency and appliance's characteristics considered prior to purchase: differences and similarities between the United Kingdom, Germany, Portugal, Greece, Poland, Spain and Italy. Int J Green Energy 13:546–555. https://doi.org/10.1080/15435075.2014.962033
4. Ramos A, Labandeira X, Löschel A (2016) Pro-environmental households and energy efficiency in Spain. Environ Resour Econ 63:367–393. https://doi.org/10.1007/s10640-015-9899-8
5. Semprini G, Gulli R, Ferrante A (2017) Deep regeneration vs shallow renovation to achieve nearly zero energy in existing buildings: energy saving and economic impact of design solutions in the housing stock of Bologna. Energ Buildings 156:327–342. https://doi.org/10.1016/j.enbuild.2017.09.044
6. Martín-Consuegra F, de Frutos F, Oteiza I, Agustín HA (2018) Use of cadastral data to assess urban scale building energy loss. Application to a deprived quarter in Madrid. Energy Build 171:50–63. https://doi.org/10.1016/j.enbuild.2018.04.007
7. Fabbri K (2015) Building and fuel poverty, an index to measure fuel poverty: An Italian case study. Energy 89:244–258. https://doi.org/10.1016/j.energy.2015.07.073
8. Kalogirou SA, Pashiardis S, Pashiardi A (2017) Statistical analysis and inter-comparison of the global solar radiation at two sites in Cyprus. Renew Energy 101:1102–1123. https://doi.org/10.1016/j.renene.2016.09.027
9. Phimister E, Vera-Toscano E, Roberts D (2015) The dynamics of energy poverty: evidence from Spain, Econ. Energy Environ. Policy 4:153–166. https://doi.org/10.5547/2160-5890.4.1.ephi

10. Kyprianou I, Serghides D (2018) Energy poverty in Cyprus and the use of geographic information systems, in: Sixth Int Conf Remote Sens Geoinf Environ. https://doi.org/10.1117/12.2326172
11. Bouzarovski S (2017) Energy poverty: (dis)assembling europe's infrastructural divide. https://doi.org/10.1007/978-3-319-69299-9
12. Atanasiu B, Kontonasiou E, Mariottini F (2014) Alleviating fuel poverty in the EU: investing in home renovation, a sustainable and inclusive solution. https://doi.org/10.1007/978-3-540-69738-1_14
13. Thomson H, Snell C, Bouzarovski S (2017) Health, well-being and energy poverty in Europe: A comparative study of 32 European countries. Int J Environ Res Public Health 14:584. https://doi.org/10.3390/ijerph14060584
14. EPOV (2018) Indicators & Data
15. Herrero ST (2017) Energy poverty indicators: A critical review of methods. Indoor Built Environ 26:1018–1031. https://doi.org/10.1177/1420326X17718054
16. Simcock N, Walker G, Day R (2016) Fuel poverty in the UK: beyond heating? People Place Policy Online 10:25–41. https://doi.org/10.3351/ppp.0010.0001.0003
17. European Union (2009) Directive of 2009/72/EC of the European Parliament and of the Council of 13 July 2009 Concerning Common Rules for the Internal Market in Electricity and Repealing Directive 2003/54/EC. Off J Eur Union L211:L 211/55–L 211/93. https://doi.org/10.1126/science.202.4366.409
18. European Union (2009) Directive 2009/73/EC of the european parliament and of the council of 13 July 2009 concerning common rules for the internal market in natural gas and repealing directive 2003/55/EC. Off J Eur Union L211:94–136. https://doi.org/10.1126/science.202.4366.409
19. European Commission (2015) Energy Union Package
20. European Commission (2016) Clean energy for all
21. The European Parliament and the Council of the European Union (2018) Directive (EU) 2018/844 of the European Parliament and of the Council amending Directive 2010/31/EU on the energy performance of buildings and Directive 2012/27/EU on energy efficiency. Off J Eur Union 2018:75–91
22. Kyprianou I, Serghides DK, Varo A, Gouveia JP, Kopeva D, Murauskaite L (2019) Energy poverty policies and measures in 5 EU countries: A comparative study, Energy Build. https://doi.org/10.1016/j.enbuild.2019.05.003
23. Thomson H, Bouzarovski S (2018) Addressing energy poverty in the European Union: state of play and action
24. T. Mesimeris, N. Kythreotou, M. Menelaou, C. Rousos, C. Karapitta-Zachariadou, G. Partasides, T. Antoniou, N. Hadjinikolaou, K. Piripitsi, A. Kalaika, A. Christophidou, M. Chandriotis, G. Papageorgiou, M. Chrysaphis, C. Ellinopoulos, E. Stougiannis, A. Sotirelis, M. Lambrinos, D. Psyllides, A. Christodoulou, A. Karaolis, A. Kleanthous, Cyprus' Integrated National Energy and Climate Plan, 2020
25. Press and Information Office (n.d.) Press Release: scheme for provision of financial assistance to support mountain residents from the Ministry of Interior
26. Kyprianou I, Serghides D (2019) Dealing with energy poverty in Cyprus–an overview. Int J Sustain Energy:1–13. https://doi.org/10.1080/14786451.2019.1699560
27. EAC (2018) Public Service Obligations (P.S.O.)
28. Ministerio para la Transición Ecológica (2019) Estrategia Nacional contra la Pobreza Energética 2019–2024. 177
29. Constitutional Court Decision (STC) (n.d.) 17TH March 2016 (Num. 62)
30. Constitutional Court Decision (STC) (n.d.) 11th January 2018 (Num. 2)
31. Constitutional Court Decision (STC) (n.d.) 24th May 2018 (Num. 54)
32. Constitutional Court Decision (STC) (n.d.) 4th October 2018 (Num. 102)
33. Constitutional Court Decision (STC) (n.d.) 17th January 2019 (Num. 5)
34. Supreme Court (2012) Supreme Court decision 419/2012, February 7th of 2012
35. Spanish Parliament (2018) Royal Decree-Law 15/2018 of 5th October, on urgent measures to promote the energy transition and consumer protection
36. MECIT (2015) Ministerial Order for Vulnerable Electricity Consumers 289/2015
37. Aragon Parliament (2016) Law 9/2016, 3rd of November, on energy poverty reduction in Aragón
38. Balearic Islands Government, Social Services and Cooperation Regional Ministry decision, 20th of September 2017, to open a call for grants targeted at private non-profit organisations which manage aids to reduce energy poverty, (2017)
39. Castilla La Mancha Government (2015) Plan regional contra la Pobreza Energética de Castilla-La Mancha
40. Castilla y Leon Parliament, Law 4/2018, 2nd July, on planning and operating rules of the Network for the protection and inclusion of social or economic vulnerable people and families in Castilla y León, (2018) 563
41. Catalan Parliament, Catalan Act 24/2015, July 29th, on urgent measures to address the housing and energy poverty emergency, (2015) 4–7. https://doi.org/BOE-A-2012-5403
42. Extremadura Government, Royal Decree, 1/2016, May 10th, of extraordinary measures against the social exclusion, 89 (2016). https://doi.org/10.1002/ajt.12205
43. Galicia Parliament (2017) Law 7/2017, December 14th, on energy efficiency measures and guarantee of the electric power access in Galicia. 58153–58162
44. Madrid Government (2017) Government Council decision, 19th of September 2017, on regulation and procedure rules of direct granting of individual aids for paying electricity and/or gas bills
45. Murcia Government (2018) Decree 154/2018, July 27th, of direct subsidies to municipalities and municipal associations of social services of the Region of Murcia for actions against energy poverty
46. Basque Country Parliament (2008) Law 18/2008, December 23rd, on the Income guarantee and Social Inclusion from the Basque Country
47. Basque Country Government (2011) Decree 4/2011, January 18th, on Social Emergency Aids in the Basque Country
48. Rioja Government (n.d.) Subsidies to Local Entities to avoid energy supply disconnection to families in social exclusion situation
49. Valencian Parliament (2017) Valencian Community Law 19/2017, December 20th, of the Valencian minimum income for inclusion

Enhancing Higher Education Buildings' Performance Through Energy Audit and Retrofitting: Climate Action to Mitigate CO_2 Emissions and Attain SDGs

Mohsen Aboulnaga, Abdulrahman Amer, and Radwa Ahmed

1 Introduction

The world noticeably confronts colossal energy demands resulting from rapidly increasing population (70% of the world's population will be living in cities by 2050) and the scarcity of natural resources [1]. Cities consume huge amounts of energy and natural resources due to urban activities resulting from enormous numbers of inhabitants in 38 mega cities [2, 3]. The United Nations (UN) has reported that urban areas are expected to accommodate 6.3 billion people by 2050 in comparison to 3.4 billion people in 2009 with a huge rise by around 85 per cent [4]. In addition, the world is currently suffering from the Coronavirus (COVID-19) pandemic deadly impact since the end of 2019 and during 2020. This pandemic instantly causes new challenges to all governments, mainly educational (university and schools) and health institutions [5, 6].

Most cities are quite vulnerable to the impacts of climate change severe events, where these cities are not planned to bear such shocks and challenges, that is, resilient. The existing built environment would necessarily need a substantial intervention to upgrade infrastructure of these cosmopolitans in order to be able to remain viable, bearable and equitable in the environmental, economic and social dimensions [7]. Hence, upgrading cities and urban environment, adopting sustainability principles and transforming buildings to be low energy would have a huge influence on reducing energy use, mitigating CO_2 emissions and offsetting climate change (CC) as shown in Fig. 1. Higher education buildings contribute significantly to electrical energy use in cities, mainly due to activities, high occupancy rates and effectiveness of behavioural patterns of users on electrical energy consumption within these buildings.

The World Bank reported that cities account for about 70 per cent of global energy use [8]. It is estimated that about 80 per cent of global carbon emissions are produced from cities and urban areas each year [9]. One-third of global energy end use is happening within buildings, which are responsible for about 33 per cent of CO_2 emissions [10]. According to the International Energy Agency (IEA) statistics of 2018, the building and construction sector is consuming more than one-third of the global energy end use and is responsible for about 39 per cent of the total carbon emissions worldwide as referred in the Global Status Report for Buildings and Construction 2019 [11]. Most of these buildings are not efficient enough to cope with sustainability and CC adaptation measures. Also, traditional methods of designing buildings enormously deplete the natural resources, pollute air and impact global warming.

1.1 Commercial and Educational Buildings

Tertiary buildings, especially office and educational buildings, are considered highly intensive energy use. Educational buildings are even consuming higher rate of energy due to the fact that they generally operate for long periods, contain large quantities of users and host diverse variety of activities [12, 13]. The business sector in the United Kingdom (UK) is

M. Aboulnaga (✉)
Department of Architecture, Faculty of Engineering, Cairo University, Giza, Egypt
e-mail: maboulnaga@eng.cu.edu.eg

A. Amer · R. Ahmed
Researcher - Environmental Design and Energy Efficiency Programme, Department of Architecture, Faculty of Engineering, Cairo University, Giza, Egypt

© The Author(s), under exclusive license to Springer Nature Switzerland AG 2022

A. Sayigh (ed.), *Sustainable Energy Development and Innovation*, Innovative Renewable Energy,
https://doi.org/10.1007/978-3-030-76221-6_62

Fig. 1 Climate change severe events and impacts in Cairo, Egypt (2019–2020). (**a**) Severe floods in Heliopolis, March 12, 2020. (Source: www.alahramdaily.com). (**b**) Severe floods in New Cairo, March 12, 2020. (Source: Weather channel IBM 2020) (**c**) Severe floods in Heliopolis, March 12, 2020. (**d**) Severe floods in Cairo, October 2019 2020

contributing to approximately 18 per cent of the total carbon emissions produced by the country in 2017 [14]. Governments globally are focusing nowadays on retrofitting existing buildings of this type to attain a valuable reduction in energy use and GHG emission [15]. Public buildings only including educational, health care, office and commercial facilities are sharing 13 per cent of the total electrical energy in Egypt [16]. Improving the energy performance of educational buildings in Egypt is going to lessen the load on the national electricity grid and shrink the carbon footprint of the country [17].

Mehreen et al. conducted a pilot study on a higher education multi-purpose building at Heriot-Watt University in the UK, where most higher education buildings exploit building energy management systems (BEMS). Nevertheless, a huge amount of energy waste can be avoided through understanding the building users' behaviour, utilizing occupancy patterns and activities to redesign control strategies to reach an optimal building energy performance [18]. Careless behaviour patterns of occupants in higher education buildings have an influencing negative impact on the building energy demand in terms of heating, ventilation and air conditioning (HVAC), artificial lighting and space appliances as well. One-third of designed building energy demand could be added to the default building energy need because of users' attitude [19]. Such energy waste in buildings can be averted if energy audit is used to underpin the areas with poor energy performance and to identify inefficient systems, hence enhancing energy efficiency.

This chapter presents an assessment of a higher education building located in Giza, Greater Cairo, Egypt to transform it through energy audits to be a low-energy building.

2 Objective and Scope

This work aims to enhance the energy performance of a case study of an educational building in order to reduce the total energy consumed by non-domestic buildings sector through following a retrofitting strategy that applies energy efficiency/saving and renewable energy techniques. The study addresses a process of upgrading the environmental qualities of a high education and high-occupancy buildings to be sustainable and low-energy building. It focuses on reducing energy consumption, lowering CO_2 emissions of the building under assessment to counterbalance climate change. The research work examines the sustainability of energy use in educational buildings with a focus on the case study of The Chemical Engineering Building, Faculty of Engineering at Cairo University [20]. It also analyses the total and specified energy use and GHG emissions statistics in terms of time of use, floor area, space uses and occupancy rate, activities and energy services to suggest solutions to enhance its performance.

3 Methodology

The methodology of the research work is based on analytical and applied approaches. The analytical approach is centred on pursuing a baseline assessment of the building's total energy use and cost, in addition to total CO_2 emissions, and then comparing it with its energy performance after retrofitting measures and simulation. The comparison can be achieved through conducting an energy audit and retrofitting simulation of the case study of a higher educational building. The analytical approach also focuses on an energy audit, assess energy use, and recommends actions to reduce energy use and/or increase production. The verification, monitoring and analysis of use of energy, including submission of technical report containing recommendations for improving energy efficiency with cost benefit analysis and an action plan to reduce energy consumption [21]. The applied approach concentrates on incorporating innovative solutions and suggesting energy improvements through simulation of the retrofitted case.

In higher education buildings, the annual electrical energy consumption relies on time of use and influx of building users such as: students, staff members and visitors. The retrofitting strategy is based on two main pillars: space uses and energy services [21]. The energy audit is composed of five key phases, including: (a) site visit and observation, (b) gathering preliminary data, (c) analysing data and information (simulation of baseline case), (d) applying innovation and suggested improvements and e) Simulation of the retrofitted case. [22] The methodology's analytical approach (Energy audit and assessment) is illustrated in Fig. 2.

3.1 *The Case Study*

The baseline assessment is conducted on the Chemical Engineering building, Faculty of Engineering at Cairo University. Such building, which was established in 1952, is a multi-story building including four storeys with a plot area of 2170 m^2 and a total built-up area of 6853 m^2 [20].

3.2 *Building's Observation*

The first phase is conducted by visiting and observing the building's status, in which an introductory investigation is proceeded to be familiar with the case study. It begins with defining the exact location of the building, and then taking a walk through its spaces in four floors for identifying the different zones of activities, facilities and energy modes [22]. Figure 3 demonstrates the exact location of the examined building within the campus of the Faculty of Engineering at Cairo University. In this stage, thorough investigations are carried out on the building's layout, floor plans and elevations after modelling using computer software. Figure 4 presents the four elevations of the building under study. The examined building's floor plans and envelops including horizontal and vertical components as well as opaque and transparent elements are shown in Appendix 1.

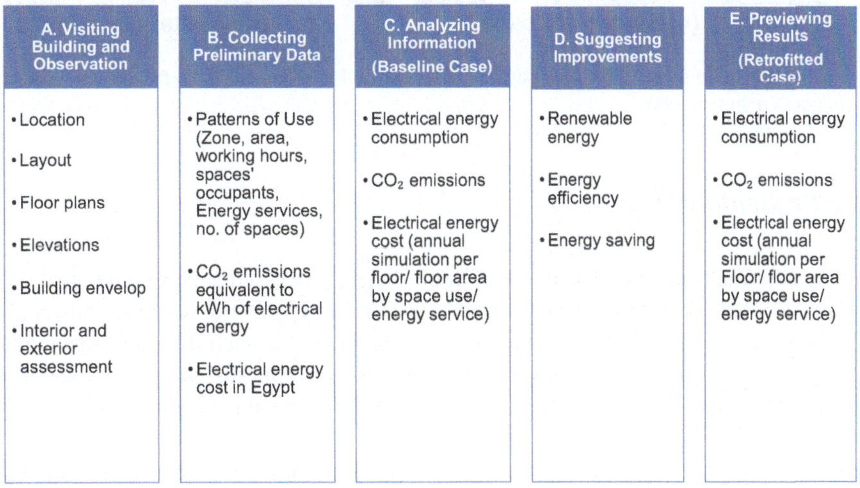

Fig. 2 The methodology's analytical approach – energy audit

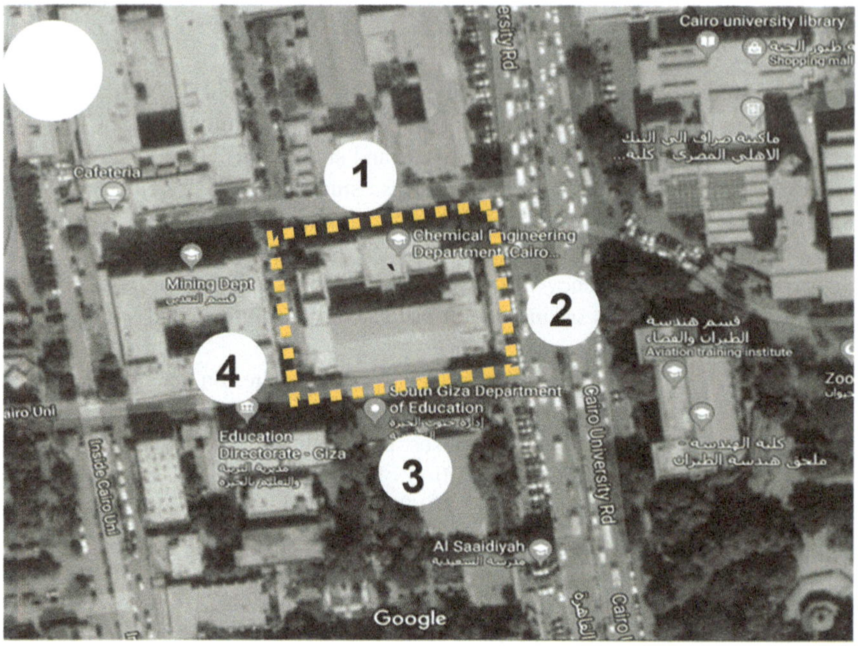

Fig. 3 The case study building location at Cairo University campus in Giza, Egypt

Fig. 4 Façades of the case study building at Cairo University campus in Giza, Egypt. (**a**) Northern façade. (**b**) Eastern façade. (**c**) Southern façade. (**d**) Western façade. (Images credit: authors)

3.3 Collecting Preliminary Data

This phase considers gathering necessary data that will be classified, processed and evaluated to deliver a clear understanding of the current conditions of the building energy performance and to be able to suggest and recommend solutions for the baseline case (Fig. 4). A set of surveys and interviews were conducted for each floor and whole building during December 2019 to identify the patterns of use and activities in the various zones of the case study. This involves actual working hours, spaces' areas and occupancy and the energy services provided [21, 22]. Survey and energy audit tables of each floor are depicted in Appendix 2.

The amount of CO_2 emissions equivalent to generating 1 kWh of electrical energy depends mainly on the source of energy used in electricity generation in the power plants. The carbon emissions intensity by source (kg CO_2/kWh) is derived from both the operational emissions from the generation source and the emissions generated during the generating sources construction. Table 1 lists the intensity of carbon emissions based on different generation source types [23]. The International

Table 1 Intensity of carbon emissions by source

Power Plant Generation Source	Intensity of Carbon Emissions (kg CO_2/kWh)
Gas	0.5
Oil	0.65
Coal	0.90
Nuclear	0.005
Hydro	0.005
Solar	0.058
Wind (on-shore)	0.00464
Wind (off-shore)	0.00525

Source: Hammond, GP & Jones, CI, 2008

Table 2 Electricity tariff, commercial buildings, in Egypt

Monthly Consumption Segments No.	Monthly Consumption Segments (kWh)	Price (L.E.)
1	0 - 100	0.55
2	100 - 250	1.00
3	250 - 600	1.15
4	600 - 1000	1.45
5	**More than 1000**	**1.50**

Source: Egyptian Electric Utility and Consumer Protection Regulatory Agency, 2018

Energy Agency (IEA) reported in 2018 that the majority of power plants in Egypt (around 80%) are using natural gas for electricity generation, 11 per cent for oil, 7 per cent for hydro-electric sources and only 2 per cent for renewables. The amount of CO_2 emissions released from generating 1 kWh of electricity in Egypt is estimated to be an average of 0.4725 kg/kWh [24]. In the study, these data are used to calculate the emission from the buildings due to energy use. The cost of electrical power for commercial buildings in Egypt is categorized upon monthly consumption segments that rely on the actual use electricity in one month [25]. Table 2 shows different monthly consumption segments and the electrical tariff for each segment.

4 Energy Audit: Baseline Assessment of the Case Study

Electrical energy savings can be significantly optimized in higher education buildings on the long run through carrying out an energy audit, which aims at enhancing the building design, upgrading energy management system, applying renewable energy technologies in addition to replacing traditional electrical appliances with energy efficient products to shorten the annual electrical energy demand [18]. An energy audit is defined as a thorough investigation process of the energy consumption of an existing building in order to optimize energy use through detecting operating problems, improving building systems as well as applying energy saving/efficiency and renewable energy techniques [26, 27]. In this study, an energy audit is carried out to enhance the building's energy performance by assessing the existing conditions of patterns of use, then utilizing appropriate energy conservation potentials. It is conducted in the whole building (Fig. 4), including all building zones

Fig. 5 Levels of energy audit conducted on the higher education building. (Image Credit: authors)

Table 3 The baseline case appliances' electrical power in watts

Energy Services								
Lighting			HVAC			Appliances		
Symbol	Type	Power (W)	Symbol	Type	Power (W)	Symbol	Type	Power (W)
Floor 1	2 x Florescent Lamp (20W)	40	AC 1.5	AC (1.5 hp)	1125	PC	PC	100
Floor 2	2 x Florescent Lamp (40W)	80	AC 3.0	AC (3.0 hp)	2250	LT	Laptop	75
Floor 3	3 x Florescent Lamp 40W)	120	CF	Ceiling Fan	100	PJ	Projector	450
Floor 4	4 x Florescent Lamp (40W)	160	WF	Wall Fan	75	SP	Speaker	50
			SF	Stand Fan	125	PT	Printer	75
						SC	Scanner	25
						PCM	Photocopying Machine	1450
						KT	Kettle	1800
						BL	Blender	350
						EC	Electrical Cooker	750
						FG	Fridge	1200

Source: Authors based on energy audit visits

throughout two main levels: (a) Level 1, walk through assessment and (b) Level 2, survey and data analysis [28], where the energy services in the building are shown in Fig. 5. The energy audit covered all electrical appliances and equipment used in all spaces, including lighting bulbs, cooling, ventilation and other apparatus. Table 3 refers to all baseline case appliances found in the building and their electrical power in watts; they are categorized according to the three types of energy service provided in the building [21, 22].

5 Analysis of the Baseline Case

In this stage, the gathered information about the patterns of use in the previous step was analysed and tackled. Data have been classified for each floor into the three main energy services provided in the building: (a) Lighting, (b) HVAC and (c) appliances as listed in Table 3. It was found that the annual electrical energy consumption in the ground floor is 54,525 kWh/ year (23.54 kWh/m^2/year/ floor), and the carbon emissions are approximately 25.76 ton/year, which costs EGP 81,788 yearly (an equivalent of US$ 5,065 at an exchange rate of Central Bank of Egypt (CBE) as of June 5, 2020). In the ground floor, pantries have the highest amount of electricity consumption (20,069 kWh of power per year) based on space use as presented in Fig. 6. Pantries, corridors and study rooms are the main contributors to energy use in the ground floor and are responsible for 76.4 per cent of annual power use (Fig. 7). Lighting has also the highest amount of electricity consumption in the ground floor (29,327 kWh/ year) as shown in Fig. 8. Lighting and appliances are the main contributors to energy use in the ground floor and form 92.6 per cent of the annual power use (Fig. 9).

In the first floor, the annual electrical energy consumption is 55,889 kWh/ year (27.36 kWh/m^2/year/floor area), and the carbon emissions are approximately 26.41 ton/year, which costs EGP 83,834 yearly (an equivalent of US$ 5,191.31 at an exchange rate of CBE as of June 5, 2020). In this floor, pantries have the highest amount of electricity consumption in the first floor based on space use (23,917 kWh/ year) as illustrated in Fig. 10. Pantries, staff offices and corridors are the main contributors to energy use in the first floor and form 85 per cent of annual power use (Fig. 11). Lighting has the highest amount of electricity consumption based on energy service (25,365 kWh/ year) as shown in Fig. 12. Appliances and lighting are the main contributors to energy use and are responsible for 88.1% of the annual power use (Fig. 13).

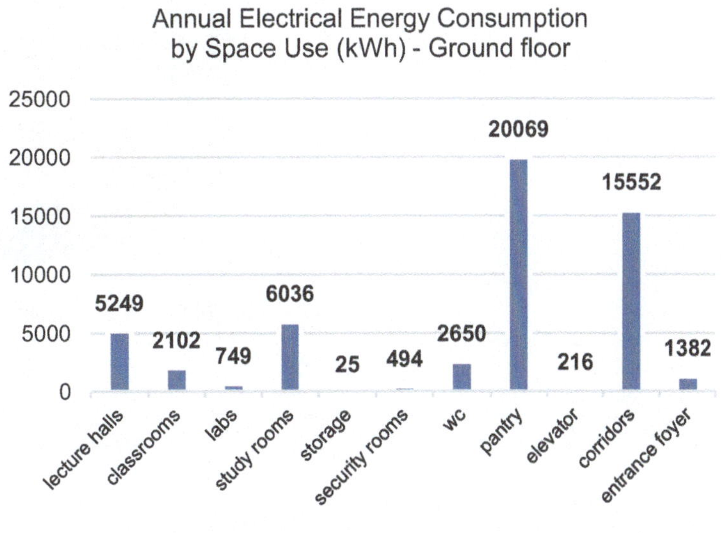

Fig. 6 Annual electrical energy use by space (kWh)

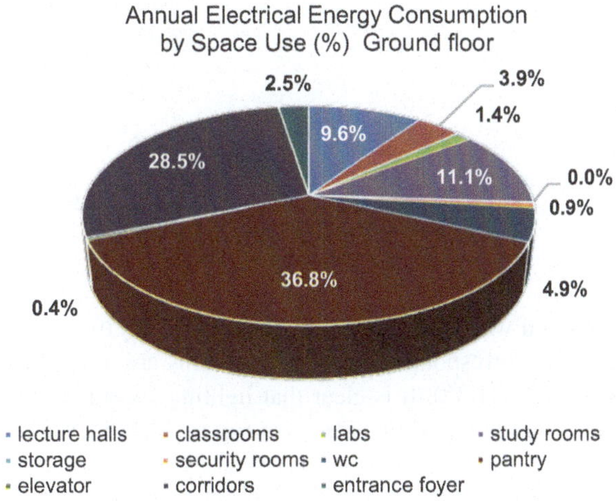

Fig. 7 Annual energy use by space (%)

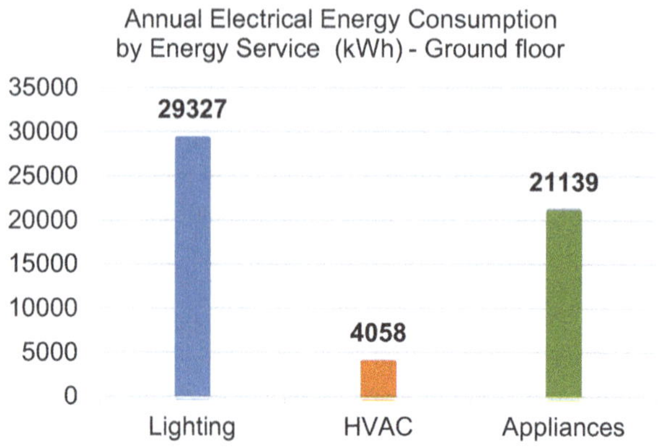

Fig. 8 Annual electrical energy use by service, ground floor

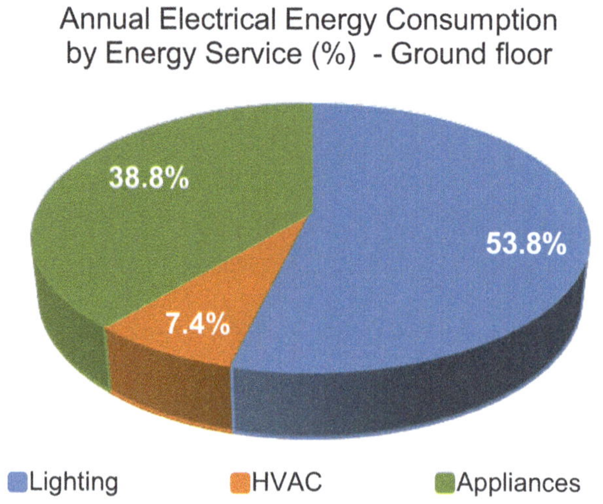

Fig. 9 Annual energy use by service (%)

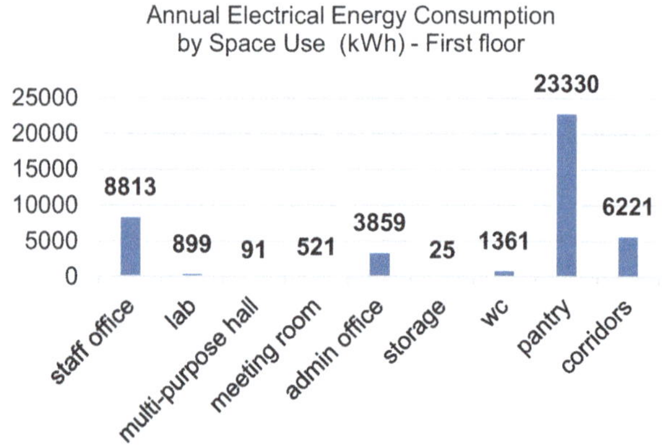

Fig. 10 Annual electrical energy use by space, ground floor

For the second floor, the same trend was observed where the annual electrical energy consumption is 41,895 kWh/ year (29.19 kWh/m^2/year/ floor area). The corresponding carbon emissions are approximately 19.79 ton/year at a cost of EGP 62,841 yearly (an equivalent of US$ 3,891.35). It is clear that lighting consumes 14,252 kWh/ year, and pantries have the highest amount of electricity consumption based on space use (up to 20,105 kWh of power per year as shown in Fig. 14). Pantries, staff offices and corridors are the main contributors to energy use in the second floor and accounts for 83.8 per cent of annual power use (Fig. 15). Appliances have the highest amount of electricity consumption based on energy service

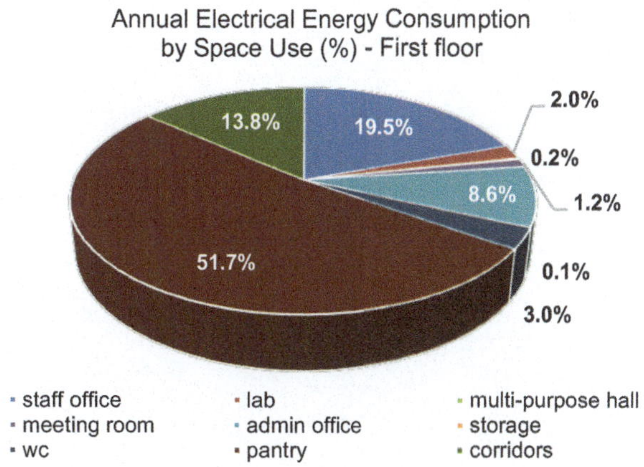

Fig. 11 Annual energy use by space (%)

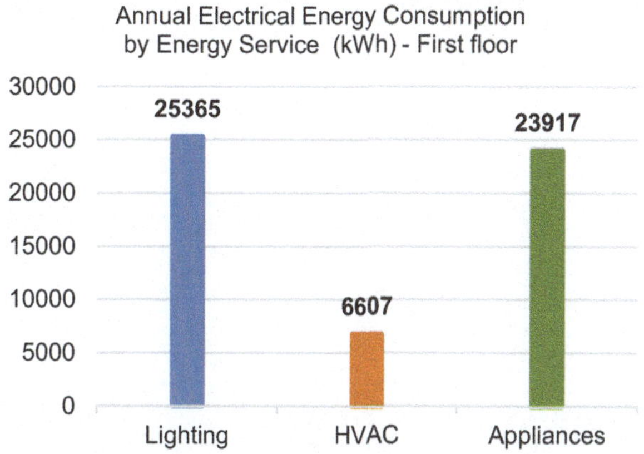

Fig. 12 Annual electrical energy use by service, first floor

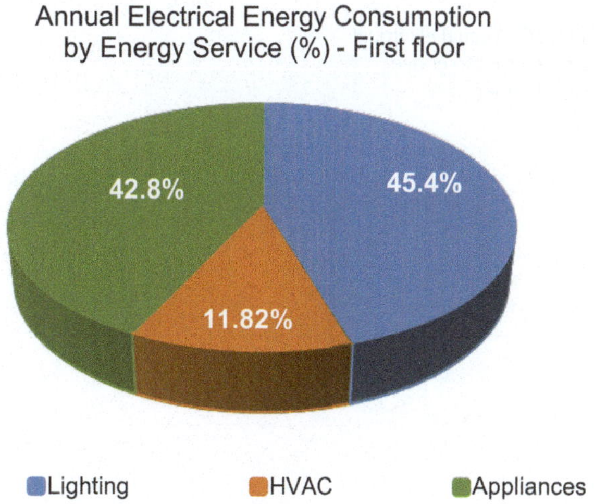

Fig. 13 Annual energy use by service (%)

(20,846 kWh/ year) as presented in Fig. 16. Appliances and Lighting are the main contributors to energy use in the second floor and are responsible for 83.8 per cent of annual power use (Fig. 17).

For the third floor, the annual electrical energy consumption is 1,260 kWh/ year (1.19 kWh/m^2/year/floor area), and the related carbon emissions are approximately 0.59 ton/year, at a cost of EGP 1,890 yearly (an equivalent of US$ 120.38). It is obvious that the examination hall has the highest amount of electricity consumption in the third floor based on space use (846 kWh of power per year) as presented in Fig. 18. Examination hall is the main contributor to energy use in the third floor

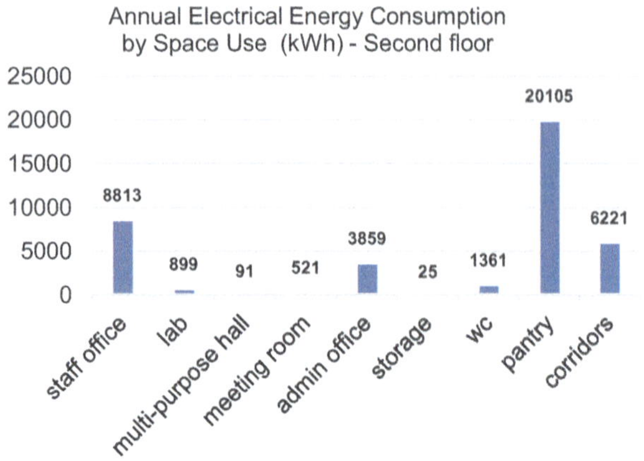

Fig. 14 Annual electrical energy use by space, second floor

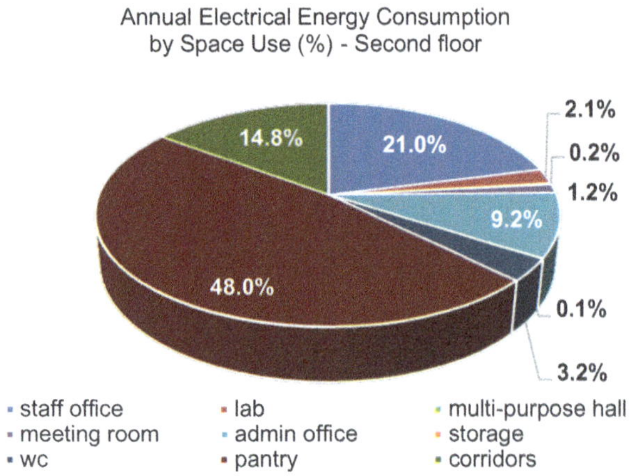

Fig. 15 Annual energy use by space (%)

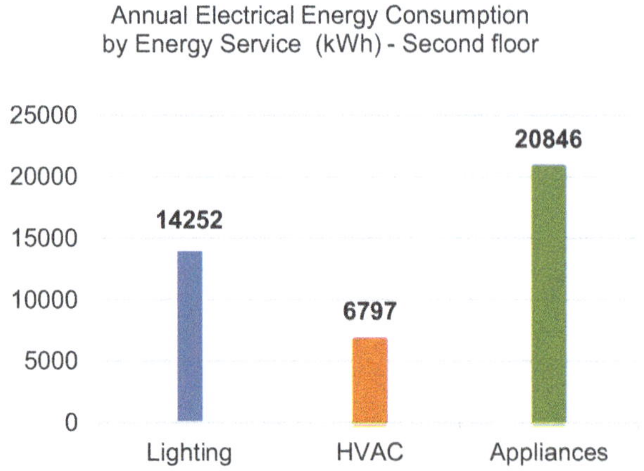

Fig. 16 Annual electrical energy use by service, second floor

accounting for 67 per cent of annual power use (Fig. 19). Lighting still has the highest amount of electricity consumption based on energy service (589 kWh/ year) as illustrated in Fig. 20. Lighting is also the main contributor to energy use (46.7 per cent of annual power use) as highlighted in Fig. 21.

Based on the above energy audit, the annual electrical energy consumption in the whole building is estimated 153,569 kWh/ year (22.41 kWh/ m^2/ year) as presented in Fig. 22. This indicates that pantries have the highest amount of electricity consumption in the whole building based on space use (up to 63,504 kWh of power/ year) where pantries, corridors and staff offices are the main contributors to energy use in the whole building and responsible for 73.6 per cent of annual power use

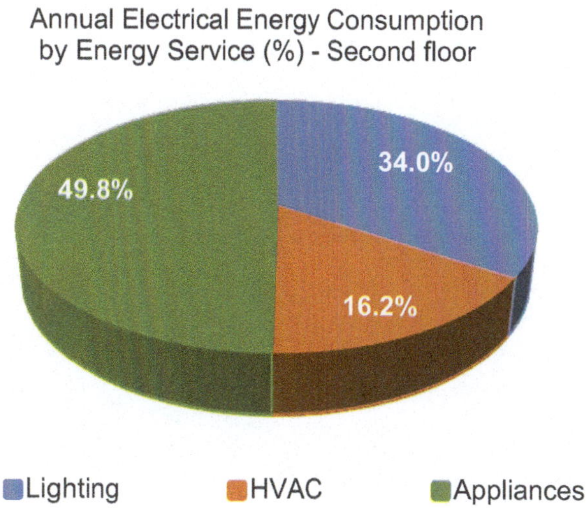

Fig. 17 Annual energy use by service (%)

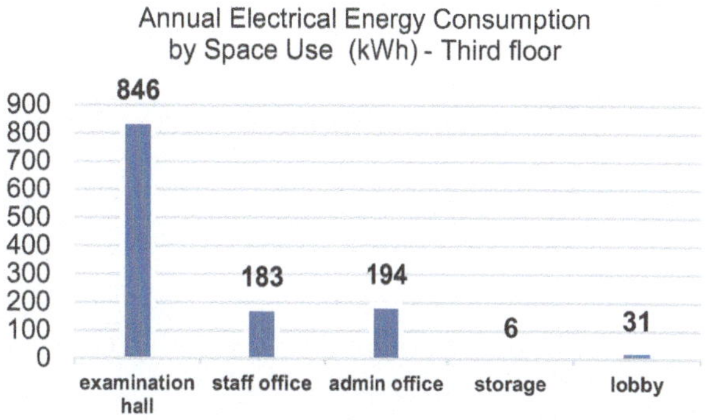

Fig. 18 Annual electrical energy use by space, third floor

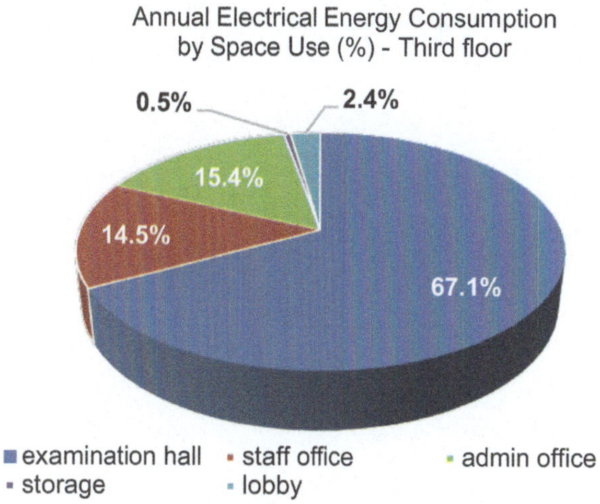

Fig. 19 Annual energy use by space (%)

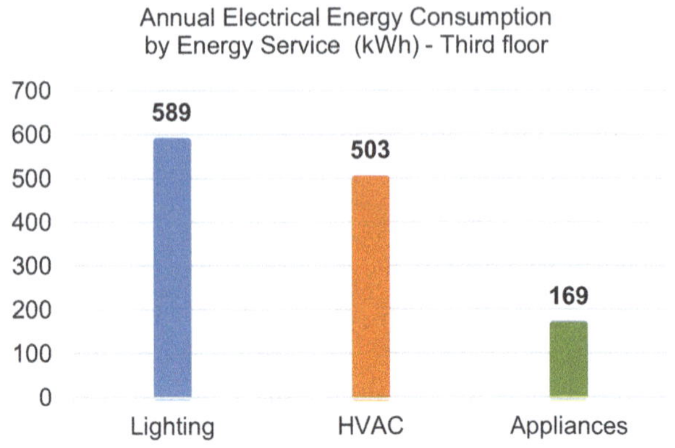

Fig. 20 Annual electrical energy use by service, third floor

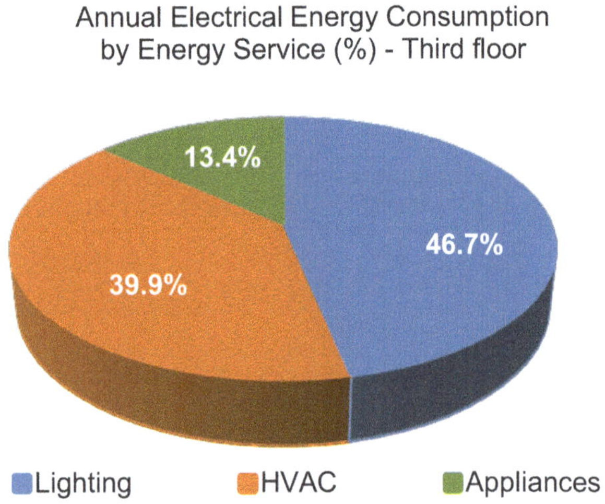

Fig. 21 Annual energy use by service (%)

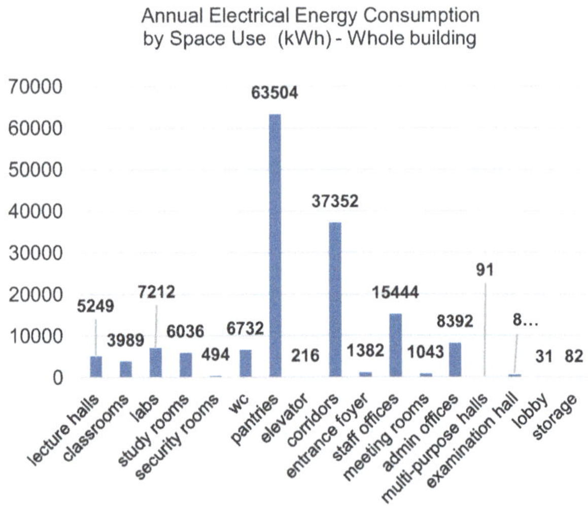

Fig. 22 Annual electrical energy use by space (kWh), the whole building

as shown in Figs. 23, 24, and 25. Based on the energy audit, it clear that the whole building annual electrical energy consumption, especially that per floor area, is high (22.41 kWh/ m^2/ year) and based non-renewable energy sources and not sustainable. By looking at Unilever headquarters' office building in Hafen city – Hamburg, Germany, it depicts one of most efficient buildings in Europe with an annual electrical energy use of less than 100 kWh/ m^2 [26]. Hence, the next section of the study focuses on the building's retrofitting and enhancing energy use.

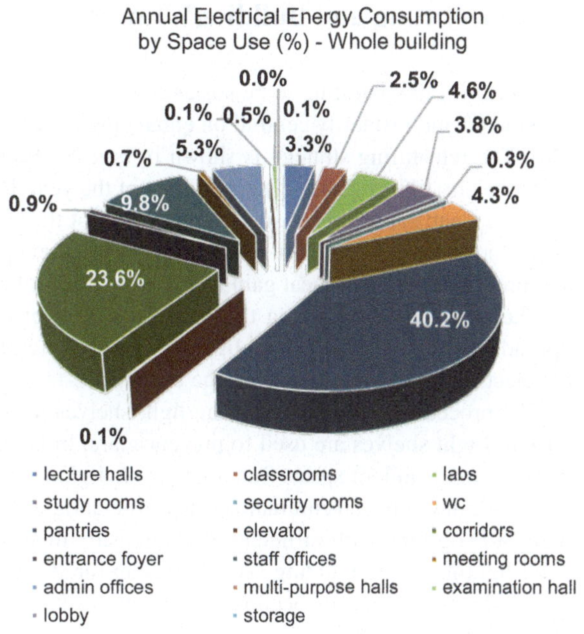

Fig. 23 Annual energy use by space (%) the whole building

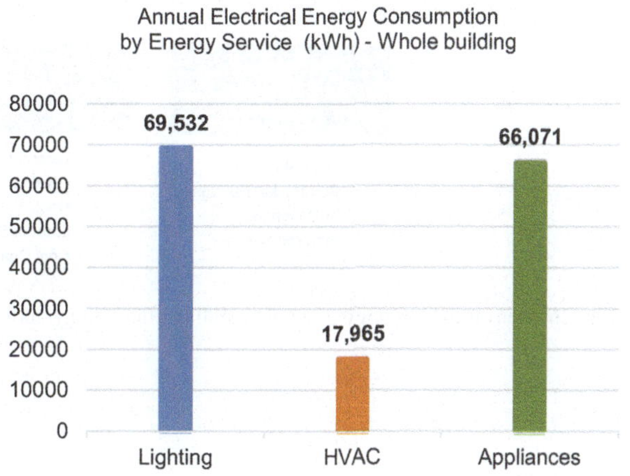

Fig. 24 Annual electrical energy use by service (kWh), the whole building

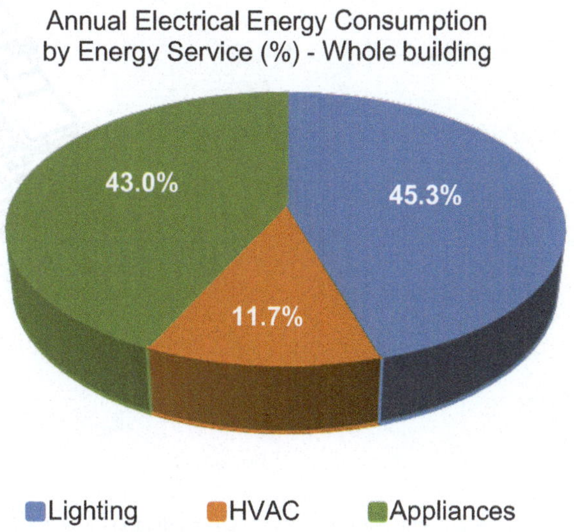

Fig. 25 Annual energy use by service (%), the whole building

6 Retrofitting and Building's Energy Improvements

To transform this building to be sustainable and energy efficient higher education building, a retrofitting strategy has been considered and primarily centred on enhancing the energy services in the building in addition to utilizing renewable energy [20]. The retrofitting strategy is shown in Fig. 26. Renewable energy alternative chosen for the case study is solar energy since the city like Giza has clear sky most of the year. Based on the calculation, 34 solar photovoltaic panels are exploited to generate clean energy and installed on the roof of the building. These PV panels would generate 18.7 MWh of electricity per year according to the highest power output per panel up to 250 W (Fig. 27). The solar PV panels also shade the roof surface and prevent direct solar heat gain that adds to the cooling loads in the indoor spaces.

The strategy of enhancing the lighting system of the building is following the procedures of promoting daylight and upgrading the existing artificial lighting system. Integration of daylight and artificial lighting has a significant influence on the electricity used for lighting in the building.

The procedures include installing light shelves through the openings of laboratory spaces that are open to the southern façade. Light shelves are used to prevent glare, block direct solar heat gain on the windows and distribute daylight evenly throughout the indoor space. Sun tracking skylights are installed on the roofs of the third and some spaces of the second floor to provide daylight and eliminate using artificial lighting during day time. All existing florescent lighting lamps to be replaced with LED lighting with sufficient light intensity; this can save up to 70–80 per cent of electrical energy consumption of artificial lighting in the building [26]. Fixing infrared motion sensors is essential for energy efficiency of artificial lighting in spaces that have unsteady occupancy frequency like corridors, lobbies and entrance foyers and will save much wasted energy allocated for illumination.

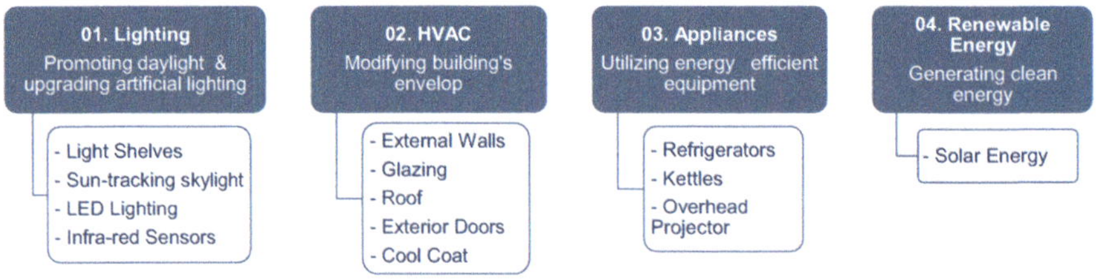

Fig. 26 Retrofitting strategy of whole building after energy audit results

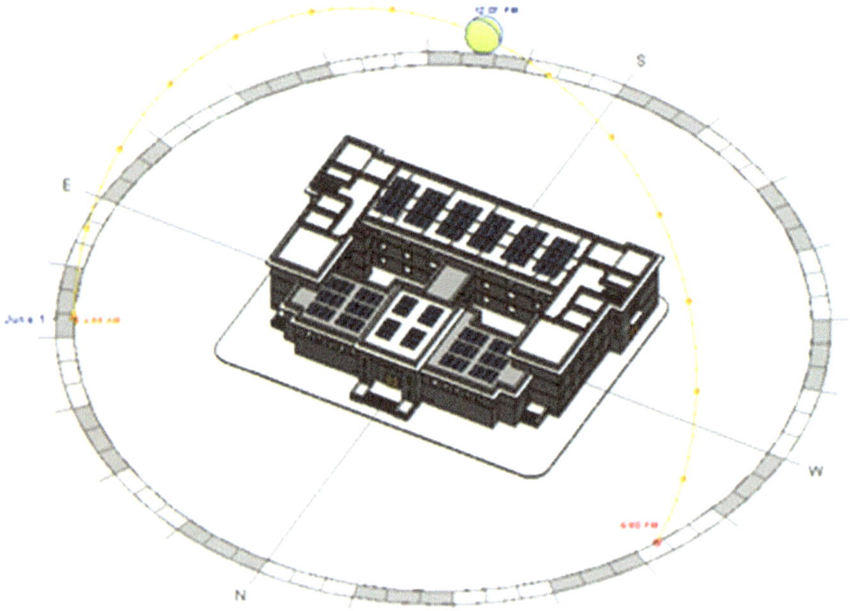

Fig. 27 Arrangement of solar PV panels on the top of the roof of the building

Also, incorporating a 50-mm-thick thermal insulation layer of polyurethane over the existing external wall profile will lower the U-value of the external wall by 1.12 W/m^2 C° to reach a value of 0.41 W/m^2 C° in the retrofitted case compared with 1.51 W/m^2 C° in the baseline case. The current ordinary single glazing openings are going to be replaced with double pane, low-e coated, argon filled, high-performance glazing to upgrade the U-value of transparent elements of the external building envelop from 4.75 W/m^{2}C° to 1.15 W/m^2 C°.

Moreover, the retrofitting actions involve coating the roof surface of the building with light coloured, cool with high solar reflective coat. Cool roofs have an influential role in lowering that warming impact of urban heat island effect where they are capable to reflect undesired direct solar heat gain from the sun. All proposed modifications to the building envelop of the case study are intended to increase the insulation potential of building, thus minimizing the cooling load in summer and heating loads in winter yet reduce energy consummation dramatically. Due to the fact that pantries consume about 40 per cent of the total annual energy demand of the whole building refrigerators, kettles and electric cookers have to be exchanged with energy efficient alternatives. An LED overhead projector was suggested to save energy. Using energy-star appliances in the different spaces of the building could save up to 43.6 MWh yearly.

Energy consumption was estimated using calculations of users' patterns and working hours during different times throughout the academic year. Lighting, HVAC and different appliances energy consumption have been calculated after retrofitting by each floor as shown in the below figures and tables of the patterns of use after modifications to measure the impact of each improvement on energy consumption and the related building's CO_2 emissions (Figs. 28 and 29).

Figure 30 presents the electrical energy use by service of the whole building after retrofitting and modification process including the lighting, HVAC and appliances energy use, which are dropped to 14,926 kWh/year, 14,077 kWh/year and 31,468 kWh/year for lighting, HVAC and appliances, respectively. This reduction in energy service is reflected in percentage in Fig. 31.

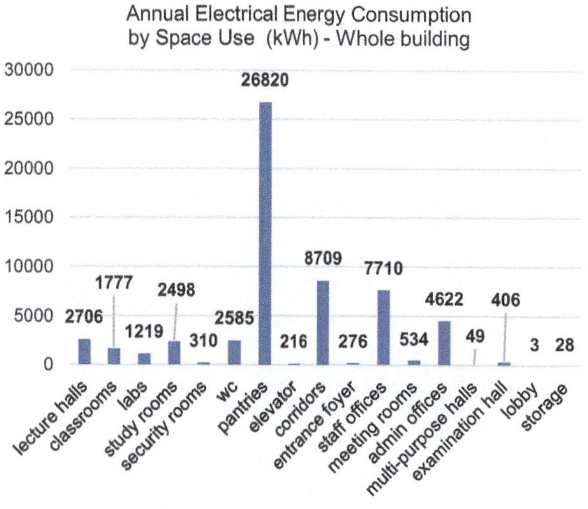

Fig. 28 Annual electrical energy use by service, whole building after retrofitting process

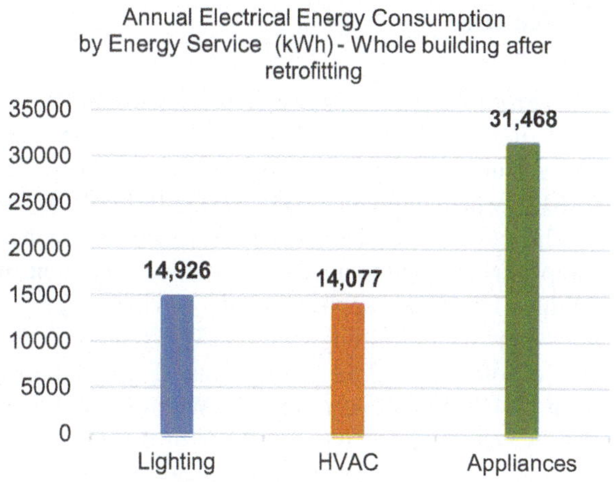

Fig. 29 Annual energy use by service (%), whole building after retrofitting

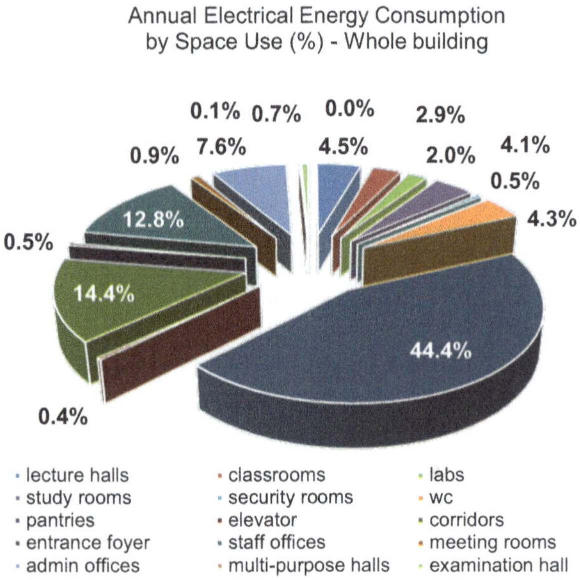

Fig. 30 Annual electrical energy use by service - whole building after modification process

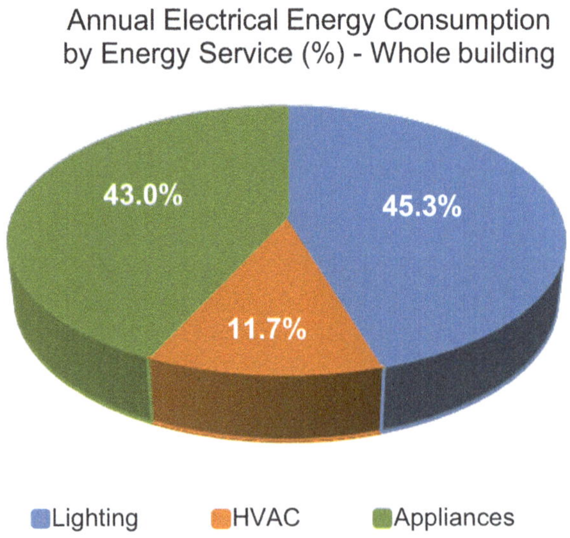

Fig. 31 Annual energy use by service (%) whole building after modification

7 Results

The retrofitting process, based on previewing all floors energy use and enhancing of the whole building energy performance; it was found that the annual electrical energy consumption is estimated 60,470 kWh/ year (8.82 kWh/m^2/ year/ floor area). The calculated carbon emissions are approximately 28.57 ton/year. It costs USD 3877.29 yearly (an equivalent to EGP 62,614), where more than USD 10387.10 (EGP 167,740) is being saved each year.

The retrofitting and improvements carried out on the case study of one of the higher education buildings at Cairo University campus in Giza led to a reduction in the energy consumption of the whole building all over the year. This enhancement in lighting, HVAC and appliances led to saving of energy use of more than 60 per cent (60.470 kWh/ year) when compared of that before the retrofitting 153,569 kWh/ year. Also, the annual electrical energy savings in the whole building per floor area after retrofitting is 13.59 kWh/m^2/ year (Fig. 32). It dropped by 39.4 per cent due to retrofitting from the original energy use from 22.41 kWh/m^2/ year to 13.59 kWh/m^2/ year. Figure 32 illustrates the reduction in the annual electrical energy consumption by space use (17 types) in kWh of the whole building before and after retrofitting. Figure 33 shows the reduction in lighting, HVAC and appliance before and after retrofitting.

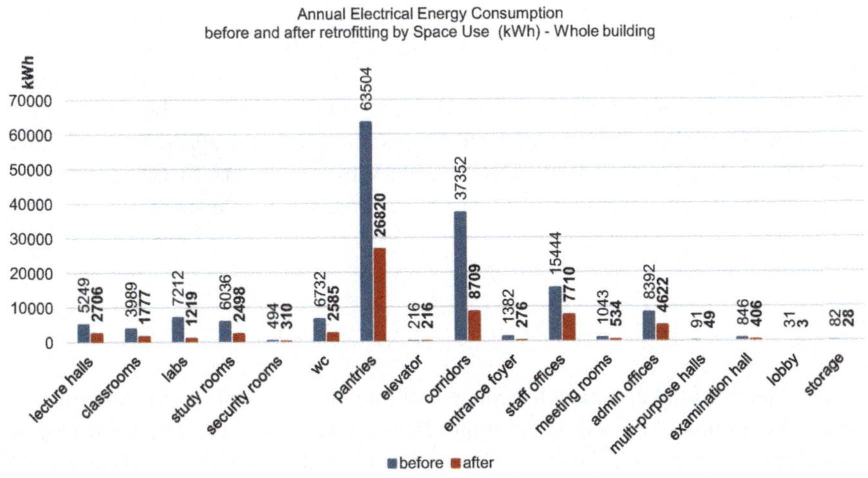

Fig. 32 Annual electrical energy consumption by space (kWh) of the whole building before and after retrofitting

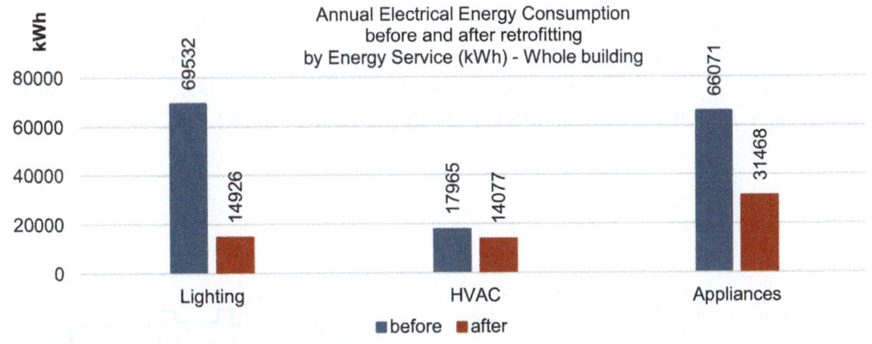

Fig. 33 Annual electrical energy consumption by service of the whole building before and after retrofitting

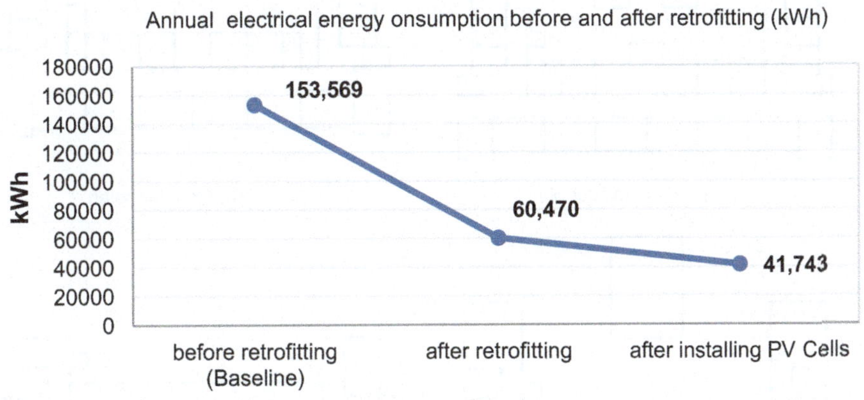

Fig. 34 Comparison between baseline case energy consumption after retrofitting and installing PV cells (kWh) of the whole building

In terms of annual electrical energy consumption by energy service (lighting and appliance), it dropped from 69,532 kWh and 66,071 kWh to 14,926 kWh and 31,468 kWh respectively, a reduction of 78.2 and 52.4 per cent. In contrast, the reduction in HVAC found to be 17,965 to 14,077 kWh (21.64 %) as illustrated in Fig. 33.

By adding the 34 solar PV panels on the building's rooftop, these led to generating clean energy from renewable resources at 18,727 kWh yearly. Using clean energy saves up to 12 per cent of electrical energy. Figure 34 illustrates how total energy will be saved after retrofitting techniques used and generating energy to be 72.8 per cent for the whole building, as the annual energy consumption before any improvements was 153,569 kWh/year, and then it becomes 41,743 kWh/year after retrofitting and generating energy form PV cells as shown in Fig. 34. In addition to saving energy, CO_2 emissions will be mitigated by 28.57 ton/year after retrofitting process. It was dropped by 60.63 per cent from 72.56 ton/year before retrofitting to 28.57 ton/year after enhancement. Thus, the energy cost decreases to EGP 62,614 yearly (an equivalent to US$ 3,988.15 yearly) after retrofitting. This cost dropped from EGP 230,354 (US$ 14,672.23) in the base line case by about 72.8 per cent of cost saving. It is clear from Fig. 34 that the annual energy consumption has dropped from 153,569 kWh/ year to 60,470 kWh/ year due to the retrofitting, but it further dropped to 41,743 kWh/ year, that is, by 60.62 per cent from the baseline case and almost 31 per cent according to the generated clean electricity from renewable sources.

8 Conclusions

The retrofitting process was an essential approach to reduce building energy consumption and mitigate CO_2 emissions. The annual electrical energy consumption in the whole building before retrofitting is 153,569 kWh/year and found to be 41,743 kWh/year and after retrofitting resulting of a saving of 72.8 percent (111,826 kWh/year). Also, the annual electrical energy consumption in the whole building per floor area before retrofitting is calculated to be 22.41 kWh/ m^2/ year and is reduced to 6.10 kWh/ m^2/ year after retrofitting (saving 16.31 kWh/ m^2/ year. The equivalent CO_2 emissions based on the annual electrical energy consumption in the whole building before retrofitting is 107.50 ton/ year. The retrofitting process led to mitigating the annual CO_2 emissions in the whole building to be 29.22 ton/year (saving 78.28 ton/year). In addition, the integration of solar PV panels generated electrical energy in the whole building of 18,727 kWh/ year, which is 31 per cent of the total energy demand after retrofitting. The study revealed that annual energy use due to improvement in lighting, AC and appliances are reduced by 78.5, 78.35 and 52.4 per cent, respectively. This is translated to saving of US$10,387 (equivalent to EGP 167,740 at an exchange rate of 1 US$ = EGP 16.1489 as per June 5, 2020).

Appendix 1

Fig. 35 Building's floor plans (ground, first, second and third floors) used for the energy audit and assessment. (**a**) Ground floor plan. (**b**) First floor plan. (**c**) Second floor plan. (**d**) Third floor plan

Appendix 2

Table 4 Energy consumption of the ground floor (baseline case)

Floor	Zone	No.	Energy consumption: Lighting								HVAC									Appliances								
			Type	No.	Power (W)	hrs/d	days/w	months/y (4 Wks/)	hrs/y	Energy (kWh)	No.	Type	No.	Power (W)	hrs/d	days/w	months/y	hrs/y	Energy (kWh)	No.	Type	No.	Power (W)	hrs/d	days/w	months/y	hrs/y	Energy (kWh)
Ground	Lecture hall	2	FL01	16	40	6	5	6	720	921.6	2	CF	16	100	6	5	3	360	1152	2	SP	4	50	6	5	6	720	288
			FL04	8	160	6	5	6	720	1843.2		WF	2	75	6	5	3	360	108		PJ	1	450	6	5	6	720	648
												SF	2	125	6	5	3	360	180		LT	1	75	6	5	6	720	108
	Classroom	4	FL03	4	120	6	5	6	720	1382.4	4	CF	2	100	6	5	3	360	288	4	LT	1	75	6	5	6	720	216
												WF	2	75	6	5	3	360	216									
	Lab.	5	FL04	4	160	3	2	6	144	460.8	5	CF	4	100	3	2	3	72	144	5	PC	1	100	3	2	3	72	36
												WF	4	75	3	2	3	72	108									
	Study room	2	FL04	12	160	6	6	8	1152	4423.68	2	CF	4	100	6	6	4	576	460.8	2	LT	4	75	6	6	8	1152	691.2
												WF	2	75	6	6	4	576	172.8									
												SF	2	125	6	6	4	576	288									
	Storage	8	FL02	4	80	1	1	2	8	20.48	8	WF	2	75	1	1	1	4	4.8	8								
	Security room	2	FL03	2	80	3	6	10	720	230.4	2	WF	1	75	4	6	5	480	72	2	PC	1	100	4	6	10	960	192
	WC	4	FL03	2	120	9	6	10	2160	2073.6	4	WF	2	75	8	6	5	960	576	4								
	Pantry	2	FL03	2	120	9	6	10	2160	1036.8	2	WF	2	75	8	6	5	960	288	2	KT	1	1800	2	6	10	480	1728
																					BL	1	350	1	6	10	240	168
																					EC	1	750	2	6	10	480	720
																					FG	1	1200	24	7	10	6720	16128
	Elevator	1									1									1	EL	1	100	9	6	10	2160	216
	Corridors	3	FL02	30	80	9	6	10	2160	15552	3									3								
	Entrance foyer	1	FL01	16	40	9	6	10	2160	1382.4	1									1								
									12824	**29327.4**								**6076**	**4058.4**								**15144**	**21139.2**

***No AC in Ground floor only ceiling, wall or stand fans**

Table 5 Energy consumption of the ground floor (retrofitted case) – lighting

Floor	Zone	No.	Type	No.	Power (W)	hrs/d	days/w	months/y	hrs/y	Energy (kWh)	No.	Type	No.	Power (W)	hrs/d	days/w	months/y	hrs/y	Energy (kWh)	No.	Type	No.	Power (W)	hrs/d	days/w	months/y	hrs/y	Energy (kWh)
			Energy consumption																									
			Lighting									HVAC (air									Appliances							
Ground	Lecture hall	2	FL01	16	8	6	5	6	720	184.32	2	CF	16	100	6	5	3	360	1152	2	SP	4	50	6	5	6	720	288
			FL04	8	32	6	5	6	720	368.64		WF	2	75	6	5	3	360	108		PJ	1	450	6	5	6	720	648
												SF	2	125	6	5	3	360	180		LT	1	75	6	5	6	720	108
	Classroom	4	FL03	4	24	6	5	6	720	276.48	4	CF	2	100	6	5	3	360	288	4	LT	1	75	6	5	6	720	216
												WF	2	75	6	5	3	360	216									
	Lab.	5	FL04	2	32	3	2	6	144	46.08	5	CF	4	100	3	2	3	72	144	5	PC	1	100	3	2	3	72	36
												WF	4	75	3	2	3	72	108									
	Study room	2	FL04	12	32	6	6	8	1152	884.736	2	CF	4	100	6	6	4	576	460.8	2	LT	4	75	6	6	8	1152	691.2
												WF	2	75	6	6	4	576	172.8									
												SF	2	125	6	6	4	576	288									
	Storage	8	FL02	4	16	1	1	2	8	4.096	8	WF	2	75	1	1	1	4	4.8	8								
	Security room	2	FL03	2	16	3	6	10	720	46.08	2	WF	1	75	4	6	5	480	72	2	PC	1	100	4	6	10	960	192
	W.C.	4	FL03	2	24	9	6	10	2160	414.72	4	WF	2	75	8	6	5	960	576	4								
	Pantry	2	FL03	2	24	9	6	10	2160	207.36	2	WF	2	75	8	6	5	960	288	2	KT	1	1800	2	6	10	480	1728
																					BL	1	350	1	6	10	240	168
																					EC	1	750	2	6	10	480	720
																					FG	1	1200	24	7	10	6720	16128
	Elevator	1									1									1	EL	1	100	9	6	10	2160	216
	corridors	3	FL02	30	16	9	6	10	2160	3110.4	3									3								
	entrance foyer	1	FL01	16	8	9	6	10	2160	276.48	1									1								
									12824	**5819.4**								**6076**	**4058.4**								**15144**	**21139.2**

Table 6 Energy consumption of the first floor (baseline case)

Floor	Energy consumption (kWh)																											
	Zone	No.	Lighting								No.	HVAC								No.	Appliances							
			Type	No.	Power (W)	hrs/d	days/w	months/y	hrs/y	Energy (kWh)		Type	No.	Power (W)	hrs/d	days/w	months/y	hrs/y	Energy (kWh)		Type	No.	Power (W)	hrs/d	days/w	months/y	hrs/y	Energy (kWh)
First	Staff office	6	FL03	4	120	6	5	6	720	2073.6	6	AC3	1	2250	4	5	3	240	3240	6	LT	3	75	2	5	6	240	324
												CF	2	100	6	5	3	360	432		PT	1	75	1	5	6	120	54
												WF	2	75	6	5	3	360	324									
	Classroom	4	FL03	4	120	6	5	6	720	1382.4	4	CF	2	100	6	5	3	360	288	4								
												WF	2	75	6	5	3	360	216									
	Lab	7	FL04	4	160	3	2	6	144	645.12	7	CF	4	100	3	2	3	72	201.6	7	PC	1	100	2	2	6	96	67.2
												WF	4	75	3	2	3	72	151.2									
	Meeting room	1	FL04	12	160	3	2	6	144	276.48	1	CF	4	100	3	2	3	72	28.8	1	LT	10	75	3	2	6	144	108
												WF	2	75	3	2	3	72	10.8		PJ	1	450	3	2	6	144	64.8
												SF	2	125	3	2	3	72	18		SP	2	50	3	2	6	144	14.4
	Admin. office	1	FL04	12	160	6	5	10	1200	2304	1	CF	4	100	9	5	5	900	360	1	PC	4	100	9	5	10	1800	720
												WF	2	75	9	5	5	900	135		PCM	1	1450	2	5	10	400	580
												SF	2	125	9	5	5	900	225		PT	1	75	1	5	10	200	15
	Storage	8	FL02	4	80	1	1	2	8	20.48	8	WF	2	75	1	1	1	4	4.8	8								
	W.C.	4	FL03	2	120	9	6	10	2160	2073.6	4	WF	2	75	9	6	5	1080	648	4								
	Pantry	2	FL03	2	120	9	6	10	2160	1036.8	2	WF	2	75	9	6	5	1080	324	2	KT	1	1800	2	6	10	480	1728
																					BL	1	350	1	6	10	240	168
																					EC	1	750	2	6	10	480	720
																					FG	1	1200	24	7	12	8064	19353.6
	Corridors	3	FL02	30	80	9	6	10	2160	15552	3									3								
									9416	**25364.5**								**6904**	**6607.2**								**12552**	**23917**

Table 7 Energy consumption of the first floor (retrofitted case) - all

Floor	Energy consumption (kWh)																											
	Zone	No.	Lighting								No.	HVAC								No.	Appliances							
			Type	No.	Power (W)	hrs/d	days/w	months/y	hrs/y	Energy (kWh)		Type	No.	Power (W)	hrs/d	days/w	months/y	hrs/y	Energy (kWh)		Type	No.	Power (W)	hrs/d	days/w	months/y	hrs/y	Energy (kWh)
First	Staff office	6	FL03	4	24	6	5	6	720	414.72	6	AC3	1	2250	2	5	3	120	1620	6	LT	3	75	2	5	6	240	324
												CF	2	100	6	5	3	360	432		PT	1	75	1	5	6	120	54
												WF	2	75	6	5	3	360	324									
	Classroom	4	FL03	4	24	6	5	6	720	276.48	4	CF	2	100	6	5	3	360	288	4								
												WF	2	75	6	5	3	360	216									
	Lab.	7	FL04	2	32	3	2	6	144	64.512	7	CF	4	100	3	2	3	72	201.6	7	PC	1	100	2	2	6	96	67.2
												WF	4	75	3	2	3	72	151.2									
	Meeting room	1	FL04	12	32	3	2	6	144	55.296	1	CF	4	100	3	2	3	72	28.8	1	LT	10	75	3	2	6	144	108
												WF	2	75	3	2	3	72	10.8		PJ	1	220	3	2	6	144	31.68
												SF	2	125	3	2	3	72	18		SP	2	50	3	2	6	144	14.4
	Admin. office	1	FL04	12	32	6	5	10	1200	460.8	1	CF	4	100	9	5	5	900	360	1	PC	4	100	9	5	10	1800	720
												WF	2	75	9	5	5	900	135		PCM	1	1450	2	5	10	400	580
												SF	2	125	9	5	5	900	225		PT	1	75	1	5	10	200	15
	Storage	8	FL02	4	16	1	1	2	8	4.096	8	WF	2	75	1	1	1	4	4.8	8								
	WC	4	FL03	2	24	9	6	10	2160	414.72	4	WF	2	75	9	6	5	1080	648	4								
	Pantry	2	FL03	2	24	9	6	10	2160	207.36	2	WF	2	75	9	6	5	1080	324	2	KT	1	1500	2	6	10	480	1440
																					BL	1	350	1	6	10	240	168
																					EC	1	750	2	6	10	480	720
																					FG	1	425	24	7	12	8064	6854.4
	Corridors	3	FL02	30	16	9	6	10	2160	3110.4	3									3								
									9416	**5008**								**6784**	**4987**								**12552**	**11097**

Table 8 Energy consumption of the second floor (baseline case)

Floor	Zone	No.	Lighting Type	No.	Power (W)	hrs/d	days/w	months/y	hrs/y	Energy (kWh)	No.	HVAC Type	No.	Power (W)	hrs/d	days/w	months/y	hrs/y	Energy (kWh)	No.	Appliances Type	No.	Power (W)	hrs/d	days/w	months/y	hrs/y	Energy (kWh)
Second	Staff office	8	FL03	4	120	6	5	6	720	2764.8	8	AC3	1	2250	4	5	3	240	4320	8	LT	3	75	3	5	6	360	648
												CF	2	100	6	5	3	360	576		PT	1	75	1	5	6	120	72
												WF	2	75	6	5	3	360	432									
	Lab.	6	FL04	4	160	3	2	6	144	552.96	6	CF	4	100	3	2	3	72	172.8	6	PC	1	100	3	2	6	144	86.4
												WF	4	75	2	2	3	48	86.4									
	Multi-purpose hall	2	FL01	20	40	3	1	2	24	38.4	2	CF	4	100	3	1	1	12	9.6	2	LT	2	75	3	1	2	24	7.2
												WF	2	75	3	1	1	12	3.6		PJ	1	450	3	1	2	24	21.6
												SF	2	125	3	1	1	12	6		SP	2	50	3	1	2	24	4.8
	Meeting room	1	FL04	12	160	3	2	6	144	276.48	1	CF	4	100	3	2	3	72	28.8	1	LT	10	75	3	2	6	144	108
												WF	2	75	3	2	3	72	10.8		PJ	1	450	3	2	6	144	64.8
												SF	2	125	3	2	3	72	18		SP	2	50	3	2	6	144	14.4
	Admin. office	1	FL04	12	160	6	5	10	1200	2304	1	CF	4	100	6	5	5	600	240	1	PC	4	100	6	5	10	1200	480
												WF	2	75	6	5	5	600	90		PCM	1	1450	2	5	10	400	580
												SF	2	125	6	5	5	600	150		PT	1	75	1	5	10	200	15
	Storage	8	FL02	4	80	1	1	2	8	20.48	8	WF	2	75	1	1	1	4	4.8	8								
	W.C.	2	FL03	2	120	9	6	10	2160	1036.8	2	WF	2	75	9	6	5	1080	324	2								
	Pantry	2	FL03	2	120	9	6	10	2160	1036.8	2	WF	2	75	9	6	5	1080	324	2	KT	1	1800	2	6	10	480	1728
																					BL	1	350	1	6	10	240	168
																					EC	1	750	2	6	10	480	720
																					FG	1	1200	24	7	10	6720	16128
	Corridors	3	FL02	12	80	9	6	10	2160	6220.8	3									3								
									8720	**14252**								**5296**	**6797**								**10848**	**20846**

Table 9 Energy consumption of the second floor (retrofitted case)

Floor	Energy consumption (kWh)																											
	Zone	No.	Lighting								No.	HVAC								No.	Appliances							
			Type	No.	Power (W)	hrs/d	days/w	months/y	hrs/y	Energy (kWh)		Type	No.	Power (W)	hrs/d	days/w	months/y	hrs/y	Energy (kWh)		Type	No.	Power (W)	hrs/d	days/w	months/y	hrs/y	Energy (kWh)
Second	staff office	8	FL03	4	24	6	5	6	720	552.96	8	AC3	1	2250	2	5	3	120	2160	8	LT	3	75	3	5	6	360	648
												CF	2	100	6	5	3	360	576		PT	1	75	1	5	6	120	72
												WF	2	75	6	5	3	360	432									
	Lab.	6	FL04	2	32	3	2	6	144	55.296	6	CF	4	100	3	2	3	72	172.8	6	PC	1	100	3	2	6	144	86.4
												WF	4	75	2	2	3	48	86.4									
	multi-purpose hall	2	FL01	20	8	3	1	2	24	7.68	2	CF	4	100	3	1	1	12	9.6	2	LT	2	75	3	1	2	24	7.2
												WF	2	75	3	1	1	12	3.6		PJ	1	220	3	1	2	24	10.56
												SF	2	125	3	1	1	12	6		SP	2	50	3	1	2	24	4.8
	meeting room	1	FL04	12	32	3	2	6	144	55.296	1	CF	4	100	3	2	3	72	28.8	1	LT	10	75	3	2	6	144	108
												WF	2	75	3	2	3	72	10.8		PJ	1	220	3	2	6	144	31.68
												SF	2	125	3	2	3	72	18		SP	2	50	3	2	6	144	14.4
	Admin. office	1	FL04	12	32	6	5	10	1200	460.8	1	CF	4	100	6	5	5	600	240	1	PC	4	100	6	5	10	1200	480
												WF	2	75	6	5	5	600	90		PCM	1	1450	2	5	10	400	580
												SF	2	125	6	5	5	600	150		PT	1	75	1	5	10	200	15
	storage	8	FL02	4	16	1	1	2	8	4.096	8	WF	2	75	1	1	1	4	4.8	8								
	W.C.	2	FL03	2	24	9	6	10	2160	207.36	2	WF	2	75	9	6	5	1080	324	2								
	pantry	2	FL03	2	24	9	6	10	2160	207.36	2	WF	2	75	9	6	5	1080	324	2	KT	1	1500	2	6	10	480	1440
																					BL	1	350	1	6	10	240	168
																					EC	1	750	2	6	10	480	720
																					FG	1	425	24	7	10	6720	5712
	corridors	3	FL02	12	32	9	6	10	2160	2488.32	3									3								
									8720	**4039**								**517 6**	**4637**								**10848**	**10098**

Table 10 Energy consumption of the third floor (baseline case)

Floor	Zone	No.	Lighting Type	No.	Power (W)	hrs/d	days/w	months/y	hrs/y	Energy (kWh)	No.	HVAC Type	No.	Power (W)	hrs/d	days/w	months/y	hrs/y	Energy (kWh)	No.	Appliances Type	No.	Power (W)	hrs/d	days/w	months/y	hrs/y	Energy (kWh)
Third	Examination hall	1	FL01	96	40	6	2	2	96	368.64	1	CF	21	100	6	2	1	48	100.8	1	LT	2	75	6	2	2	96	14.4
												WF	21	75	6	2	1	48	75.6		SP	6	50	6	2	2	96	28.8
												SF	7	125	6	2	1	48	42									
												AC1.5	4	1125	6	2	1	48	216									
	Staff office	2	FL03	4	120	6	2	2	96	92.16	2	CF	2	100	6	2	1	48	19.2	2	LT	4	75	6	2	2	96	57.6
												WF	2	75	6	2	1	48	14.4									
	Admin office	2	FL03	4	120	6	2	2	96	92.16	2	CF	2	100	6	2	1	48	19.2	2	PC	1	100	6	2	2	96	19.2
												WF	2	75	6	2	1	48	14.4		PCM	1	1450	1	2	2	16	46.4
																					PT	1	75	1	2	2	16	2.4
	Storage	2	FL02	4	80	1	1	2	8	5.12	2	WF	2	75	1	1	1	4	1.2	2								
	Lobby	2	FL01	4	40	6	2	2	96	30.72										2								
									392	**589**								**388**	**503**								**416**	**169**

Table 11 Energy consumption of the third floor (retrofitted case)

Floor	Energy consumption (kWh)																											
	Zone	No.	Lighting								No.	HVAC								No.	Appliances							
			Type	No.	Power (W)	hrs/d	days/w	months/y	hrs/y	Energy (kWh)		Type	No.	Power (W)	hrs/d	days/w	months/y	hrs/y	Energy (kWh)		Type	No.	Power (W)	hrs/d	days/w	months/y	hrs/y	Energy (kWh)
Third	Examination hall	1	FL01	48	8	6	2	2	96	36.864	1	CF	21	100	6	2	1	48	100.8	1	LT	2	75	6	2	2	96	14.4
												WF	21	75	6	2	1	48	75.6		SP	6	50	6	2	2	96	28.8
												SF	7	125	6	2	1	48	42									
												AC1.5	4	1125	3	2	1	24	108									
	Staff office	2	FL03	2	24	6	2	2	96	9.216	2	CF	2	100	6	2	1	48	19.2	2	LT	4	75	6	2	2	96	57.6
												WF	2	75	6	2	1	48	14.4									
	Admin. office	2	FL03	2	24	6	2	2	96	9.216	2	CF	2	100	6	2	1	48	19.2	2	PC	1	100	6	2	2	96	19.2
												WF	2	75	6	2	1	48	14.4		PCM	1	1450	1	2	2	16	46.4
																					PT	1	75	1	2	2	16	2.4
	Storage	2	FL02	2	16	1	1	2	8	0.512	2	WF	2	75	1	1	1	4	1.2	2								
	Lobby	2	FL01	2	8	6	2	2	96	3.072										2								
									392	**59**								**364**	**395**								**416**	**169**

References

1. Mother Nature Network, Healthy Spaces, [online]. Available at: https://www.mnn.com/health/healthy-spaces/blogs/majority-world-population-will-live-cities-2050-UNreport#:~:text=Almost%2070%25%20of%20population%20will%20live%20in%20cities%20by%202050,should%20get%20better%20at%20it)&text=May%2021%2C%202018%2C%201%3A37%20p.m.&text=The%20dizzying%20pace%20of%20human,according%20to%20a%20new%20report. Accessed 28 Nov 2019
2. While A, Whitehead M (2013) Cities, urbanisation and climate change. Urban Studies 50(7):1325–1331
3. Demographia World Urban Areas (2019) Population, land area and urban densities; demographia world urban areas, 15th Annual Edition: 291904. [Online]. Available at: https://www.newgeography.com/content/006272-demographia-world-urban-areas-2019-population-land-area-urban-densities/. Accessed 7 May 2020
4. United Nations 'UN' (2010) World urbanization prospects, the 2009 Revision: Highlights. United Nations, New York
5. International Association of Universities (2020). Covide-19: Higher education challenges and responses, [online]. Available at: https://www.iau.net/. Accessed 5 July 2020
6. International Association of Universities (2020) Impact of COVID-19 on higher education worldwide – Resources for higher education institutions, updated 24 April 2020, [online], https://www.covid-19_and_he_resources.pdf. Available at: https://www.iau.net/. Accessed 5 July 2020
7. Satterthwaite D, Huq S, Reid H, Pelling M, Romero-Lankao P (2009) Adapting to climate change in urban areas: the possibilities and constraints in low- and middle-income nations. In: Bicknell J, Dodman D, Satterthwaite D (eds) Adapting cities to climate change: understanding and addressing the development challenges. Earthscan, London, pp 3–50
8. The World Bank Urban Development, [Online]. Available at: https://www.worldbank.org/en/topic/urbandevelopment/overview/. Accessed 28 Nov 2019
9. United Nations Environment Programme 'UNEP' / UN-HABITAT (2005) Climate change: the role of cities: involvement, influence, implementation. UNEP/ UN, Nairobi
10. International Institute of Applied Systems Analysis, Energy End-Use: Buildings, [Online]. Available at: https://iiasa.ac.at/web/home/research/Flagship-Projects/Global-Energy-Assessment/GEA_CHapter10_buildings_lowres.pdf. Accessed 28 Nov 2019
11. Global Status Report for Buildings and Construction (2019) International Energy Agency. Available at: https://www.iea.org/reports/global-status-report-for-buildings-and- construction-2019/, Accessed 29 Nov 2019
12. Department of Energy and Climate Change 'DECC', [Online]. Available at: https://www.gov.uk/2050-pathways-analysis/. Accessed: November 30, 2019
13. Chung HM, Rhee KE (2014) Potential opportunities for energy conservation in existing buildings on university campus: a field survey in Korea. Energy Build 78:176–182
14. Department of Business, Energy and Industrial Strategy, National Statistics (2017) UK greenhouse gas emissions, Provisional Figures, [Online]. Available at: https://assets.publishing.service.gov.uk/government/uploads/system/uploads/attachment_data/file/695930/2017_Provisional_Emissions_statistics_2.pdf. Accessed 30 May 2020
15. Azar E, Menassa CC (2012) A comprehensive analysis of the impact of occupancy parameters in energy simulation of office buildings. Energy Build 55:841–853
16. Gallachoir BPO, Keane M, Morrissey E, O'Donnell J (2007) Using indicators to profile energy consumption & to inform energy policy in a university: a case study in Ireland. Energy Build 39:913–922
17. Bahar B, Berrin SD, Cüneyt D (2009) Energy efficient retrofit methods at the building envelopes of the school buildings, http://www.salford.ac.uk/__data/assets/pdf_file/0015/142422/063-Basarir.pdf. Accessed 30 Nov 2019
18. Gul MS, Patidar S (2015) Understanding the energy consumption and occupancy of a multi-purpose academic building. Energy Build, Elsevier B.V., 87:155–165
19. Nguyen TA, Aiello M (2013) Energy intelligent buildings based on user activity: a survey, Energy Build 56, p. 244–2571
20. Chemical Engineering Department, Faculty of Engineering, Cairo University, [Online]. Available at: https://che.eng.cu.edu.eg/en/ Accessed 29 Nov 2019
21. Aboulnaga M, Wanas A, Hammad M, Hussein M (2017) Sustainability of higher educational buildings: retrofitting measures to enhance energy performance – the case of aastmt business management school building, egypt, mediterranean green buildings and renewable energy. Selected Papers from the World Renewable Energy Network's Med Green Forum, Chapter 9, p. 117–150, Springer Cham, Available at https://doi.org/10.1007/987-3-319-30746-6/. Accessed 20 Nov 2019
22. Ma Z, Cooper P, Daly D, Ledo L (2012) Existing building retrofits: methodology and state-of- the-art. Energy Build 55:889–896
23. Parliamentary Office of Science and Technology (2006) Carbon footprint of electricity generation, Number 268, 7 Millbank, London
24. International Energy Agency, [online]. Available at: https://www.iea.org/data-and statistics?country=EGYPT&fuel=Energy%20supply&indicator=Electricity%20generation%20by%20source. Accessed 4 Dec 2019
25. The Egyptian Electric Utility and Consumer Protection Regulatory Agency, [online]. Available at: http://egyptera.org/ar/Tarrif2018.aspx#, Accessed 5 Dec 2019
26. Sterling E, Collett C, Turner S, Downing C (1994) Commissioning to avoid indoor air quality problems, In Proceedings of the ASHRAE Transactions, p. 867, Symposia
27. Rahman MM (2009) Building energy conservation and indoor air quality assessment in a subtropical climate, Master of engineering thesis. Central Queensland University, Rockhampton
28. Alajmi A (2012) Energy audit of an educational building in a hot summer climate. Energy Build, Elsevier B.V., 47:122–130

JANUS: A Comprehensive National Modelling Tool to Support Energy-Emissions Planning in Portugal

Ricardo Aguiar, R. P. Borges, Paulo Martins, Ana Andrade, and Isabel Cabrita

1 Introduction

The activities of energy planning by the governments have increased in complexity as well as time frame. Increased complexity arises mainly from the availability of more detailed statistical and survey data for the energy demand sectors, and from a broader range of technologies at the energy supply side, in particular, those relying on intermittent renewable energy sources. The time horizon for planning has also been increasing, as response to the planning needs regarding deployment of new energy technologies that must consider learning curves, as well as the appearance of medium- and long-term public policy targets, mostly related to climate change mitigation. In this context short-term projections of historical data – the main tool up to a near past – are of little value because they implicitly assume slow innovation rhythms. Another aspect that has changed the landscape of energy planning is the acceleration of planning cycles, trying to keep up with the swift innovation pace of technologies and policy views that occur at the current fast change phase of the energy paradigm transition. Such rapid change in structure and technological mix of the energy sector invalidates simple projections. For all these reasons, numerical modelling of the energy sector has become essential nowadays.

We briefly review the use of energy models for supporting public policies in Portugal and describe the JANUS model, the first in-house national energy-emissions model developed and run by a body of the Portuguese Public Administration, the Directorate-General for Energy and Geology (DGEG).

2 Background

In the European Union, different member states have implemented their own energy sector models, and the European Commission (EC) itself has sponsored the construction of Europe-wide models that are used in energy planning at the European level, as well as at national level for member states that do not yet have that kind of modelling capacities. Such is the case of the well-known PRIMES/GEM-E3 models [1] and more recently of the POTEnCIA model under development at the Joint Research Center at Seville [2].

In Portugal, energy planning by the government has long since been supported by the model Valorágua, developed by EDP, the most important electricity utility; nowadays the model is run by REN, the TSO utility. Valorágua still supports the annual cycle of security of supply planning, managed by the Portuguese Directorate-General for Energy and Geology (DGEG). In recent years, DGEG has also been able to make projections by introducing simulated inputs in the official EC SHARES tool [3] that is used to declare national energy statistics to Eurostat.

For national planning in the medium and long term, during the last 10 years, the Ministry in charge of the Environment has relied on external consultancy, mainly by the Nova University of Lisbon, that owns the TIMES_PT model, or by spin-off organisations of this university. TIMES ("The Integrated MARKAL-EFOM System") was developed as part of the IEA Energy Technology Systems Analysis Program [4] and TIMES_PT is an adaptation to the Portuguese context. Meanwhile

R. Aguiar (✉) · R. P. Borges · P. Martins · A. Andrade · I. Cabrita
Division of Studies, Research and Renewables, Directorate-General for Energy and Geology, Lisboa, Portugal
e-mail: ricardo.aguiar@dgeg.gov.pt; rui.borges@dgeg.gov.pt; paulo.martins@dgeg.gov.pt; ana.andrade@dgeg.gov.pt; isabel.cabrita@dgeg.gov.pt

© The Author(s), under exclusive license to Springer Nature Switzerland AG 2022

A. Sayigh (ed.), *Sustainable Energy Development and Innovation*, Innovative Renewable Energy,
https://doi.org/10.1007/978-3-030-76221-6_63

the ministry in charge of energy has relied also on external consultancy, but in this case mainly from a spin-off of the University of Lisbon; these consultants used a proprietary model mounted over the LEAP platform [5].

In summary, the Portuguese Governments have been depending on energy modelling results at European level that could neither be adjusted nor support analysis of different strategic options, and on models run by utilities and external consultants that provide only indirect control and limited opportunities for testing policy alternatives.

In this context, and as required by a new and more ambitious policy cycle of the EC, there was a serious and urgent call for a national energy-emission model owned and managed by the Portuguese Public Administration by the time the elaboration of the National Energy-Climate Plan 2021-2030 (NECP) [6] started in mid-2016 and to be ready by the end of 2019. The Ministry in charge of energy decided then to promote the development of such an in-house model at DGEG, to be up and running in time for the Portuguese NECP elaboration: the JANUS model was born and named after the ancient Roman god of beginnings, transitions and endings.

3 Modelling Approach

Alternatives for mounting a national energy-emissions model were considered. At this first stage, the main choice was between building a new model or adapting and improving an existing model. A survey of existing models for the Portuguese energy system was performed and it was concluded that many already existed; however, apart from the TIMES_PT model, they addressed only the electricity system. As regards the available modelling platforms, and again apart from the TIMES_PT, most options were focused on the electricity system, not representing the entire range and complexity of demand, transmission, storage, supply and import-export, and/or had a rigid structure that could not be easily altered at will to cope with technological and policy innovations.

Thus, the DGEG team decided to build a new national energy-emissions model almost from scratch. The LEAP platform was selected for the task, as it displayed several advantages or desired characteristics, as follows. The main use of the JANUS model was (at first) to help set targets and trajectories, which are policy decisions and not purely the results of a technical process. Therefore, a model was needed that could work easily in back casting, setting first targets and the set of specific technologies and measures available, and then working out what policy packages were needed to get there – such as new energy efficiency regulations, support for new installed renewable power, etc.; which meant a fast-running model was desirable so that many options could be easily explored. As LEAP is based on energy balances, the running times are minutes rather than hours (except in optimization mode and at very high time resolution). Incidentally, very long running times and the difficulty in performing back-casting exercises were reasons for not selecting the TIMES framework instead. The other two main reasons were the difficulty to control the adoption of technologies, as the economic optimization process tends to instantaneously and unrealistically jump to the best available technology at a given moment, and the strategic interest of having in Portugal two national models of different technical approaches whose results could be cross checked.

Other features of LEAP that were appreciated included the possibility to increase the complexity of the model by steps, the ability to implement innovations that are not yet mature (such as hydrogen technologies) and the capacity to represent different regions of a territory, a feature that was used to build sub-models for mainland Portugal and for the Autonomous Regions of Madeira and Azores.

4 Model Structure and Processes

Having set up some key assumptions, the computations in JANUS proceed sequentially downwards along a chain of processes. This starts with the computation of energy demand at the economic sectors, the transmission of fuels (for convenience, electricity is also referred as fuel), the supply of fuels and finally the extraction, import and export of resources. Reporting results, getting indicators and drawing graphs are very important features in practice, therefore additional external tools were built at DGEG to complement the software's own tools.

Population is a key driver of energy demand, thus a satellite demographic model for Portugal was built to feed population data and forecasts into JANUS. This demographic model provides data very similar to the central scenario of the official model of the National Statistics Institute [7] but allows to explore other scenarios for sensitivity analysis.

Macroeconomic parameters are also key drivers of energy demand. Another satellite model provides the evolution of added value shares by economic sector; in conjunction with a GDP scenario, this enables to compute the absolute added values, which are linked to energy demand in each sector. Typically, the first three GDP projection years rely on the latest governmental perspectives, and the following rely on the latest EU Reference Scenario [8].

The more important other key assumptions include values at every 10 years (referred as technological "curves") of efficiencies of the various devices and of their availability by time slot of the year. Curves for imported and exported energy prices, and for CO2 emission licenses, are also set up but they are not key assumptions, as they are only used for post-processing economic calculations.

The linkages between sectorial final energy demand *E* and the population or macroeconomic parameter *K* selected are represented by means of elasticities, *e*, as a function of time, *t*. The following general form is used:

$$E(t) = A(K,t)I(t) \tag{1}$$

where *A* is an activity variable that is considered to show more correlations with the sector energy demand such as GDP, added value, passenger-km, etc. and *I* is the final energy intensity. *A* and *I* are projected separately. Typically,

$$A(K,t)/A(K,t_0) = (1+a)(K/K(t_0))^e \tag{2}$$

where *a* is a constant and t_0 is the last historical data year, currently 2015 at JANUS (version 5). For instance, in the case of international aviation, *A* is value added, *K* is GDP of the European Union, $a = 4\%$ and $e = 1.0$, but for other sectors, in general, *K* is either population, national GDP or GDP per capita, $a = 0$ and *e* is between 0.6 and 0.9.

The trend of final energy intensity with respect to the activity, *I*(*t*), is prescribed rather than modelled as it depends on energy efficiency policies and obligations, for example, EU Directives. Nevertheless, the historical trend and the technological innovations are also taken into account when projecting the trend. Logistic curves were preferred for the trends that are undergoing transitions, while logarithmic curves are used by processes moving towards saturation.

The evolution of the fuel mix in each sector is also prescribed in a similar manner, namely considering historical trends, existing and foreseen innovations as well as public policies, aims and impacts.

For coherence with the National Energy Balances, the economic sectors were defined as following: agriculture, livestock and forestry; fisheries; public works and civil construction; extractive industry (mining); manufacturing industry; residences; services; transportation by road and by rail, domestic and international navigation and domestic and international aviation.

The road transport and industry sectors are especially important in the Portuguese context as each represents around 25% of final energy demand; therefore, they deserve special treatment. The manufacturing industry was represented at the level of 14 sub-sectors, with separate drivers (added value for the sub-sector) and energy mixes. Road transportation is handled via a detailed stock turnover model of vehicles, currently named HERMES. This model addresses passenger and freight transportation separately, and represents various stock types: trucks, buses, vans, cars and motorcycles.

Considering the expected large role of hydrogen in the future energy mix, JANUS also includes a module to handle demand of fuels used as non-energy feedstocks for the chemical industry.

Electricity transport (high and very high voltage AC power lines) and distribution (medium and low voltage grids) are modelled as losses. Some power production devices, for example, residential photovoltaic systems (PV), are connected to the distribution but not to transport power lines. Losses in natural gas reception at ports and pipelines are also modelled as losses.

The technologies modelled for electricity supply are as follows: small PV systems at buildings, medium-size PV systems, cogeneration and a set of power plants supporting the national electricity system. Those power plants based on renewables include the technologies of biogas burning, biomass burning, waste burning, PV, PV with batteries associated, concentrated solar PV, concentrated solar thermal power (CSP), ocean waves, wind onshore, wind onshore with energy storage associated, wind offshore (floating turbine type), small hydro systems, large hydro systems (run of river and dams with or without pumped storage), enhanced deep geothermal and hydrogen turbines. Those based on fossil fuels include fuel oil and gasoil generators and coal power plants. Finally, there are the combined cycle gas power plants: nowadays relying on natural gas, but in the future fed by a gas mix with increasing shares of hydrogen and other renewable gases.

There are also dedicated wind and PV power plants that provide energy for electrolysers and synthetic fuel production devices but are not (or seldom are) connected to the national transport electricity grid.

The supply of fuels is the most complex chain of processes in the model. It includes: fossil fuel refineries (including natural gas reforming processes); production of methane from biogas cleaning, biomass gasification and hydrogen methanation; first-generation and advanced biofuels (gasoline and

diesel components, but also aviation kerosene) and hydrogen production via electrolysis, biomass gasification and thermoelectrochemical processes. The high complexity comes from the interlinkages between the various fuel and electricity technologies, and from handling, in addition to energy, the emissions and sinks of methane, water, hydrogen, oxygen and carbon dioxide.

Still, only a fraction of the possible value chains of fuels is considered. The value chains that have been addressed, were selected on the basis of the studies performed by DGEG for proposing a roadmap for hydrogen in Portugal [9].

Although LEAP offers several optimization procedures for the energy dispatch, for example, by energy cost or by merit order, it was concluded that a certain computational order of technologies represented best the actual power dispatch.

Gas storage is not explicitly modelled. Electricity storage is difficult to handle explicitly with LEAP as the energy balances are done at each time step without dependence in the previous one. However, storage operation could be represented in two ways: at device level (e.g. PV with batteries), the daily availability curves are smoothed out by the effect of storage at the cost of some loss in the average daily value; and at large scale, a fraction of the produced electricity is exported, then imported at another point of the model to produce electricity at some energy loss by means of hydro dams, batteries and fuel cells.

Energy resources are not difficult to handle as there are no significant reserves and no extraction of fossil fuels in Portugal. Thus, it is only necessary to check that demand of renewable energy resources is within their respective yields. In practice, only biomass and biogas availability present constraints for the next decade.

Greenhouse gas emissions are computed using the classic method of emission factors. In most cases, these factors were derived from the national communications to the UNFCCC.

5 Data and Calibration

The data loaded in JANUS consist of:

(i) Historical data 2007–2016 of energy demand, production, transmission losses, imports and exports, by fuel type, were obtained from the official National Energy Balances. This required consistency checks and sometimes corrections, mainly when obtaining the mainland energy balance by subtraction of the autonomous regions' energy balances from the national energy balance; as well as basic demographic and macroeconomic data, namely population, GDP and added value of each economic sector.

(ii) Nominal power, average conversion efficiency of each category of energy production devices and availability of these devices in two daily time slots and four seasons, including for electricity production, cogeneration, transformation of fossil fuels (refineries) and biofuels, energy losses in storage (mainly reverse hydro dams), transport and distribution networks; this was far from being a trivial problem as the reporting categories of the energy balances are not aligned with those of installed nominal power, etc., requiring the use of multiple statistical sources.

(iii) Economic data for each category of devices (CAPEX and OPEX at 10-year intervals), often using European surveys like ASSET [10] or even international surveys by IRENA [11] as specific Portuguese data for some categories simply were never collected.

Verification was performed mainly by adjusting the availability of energy production devices to obtain the recorded energy demand. Calibration was performed by deriving correlations between sectorial energy demand and macroeconomic variables during 2007–2014, then trying to predict 2015–2016. It is relevant to remark that while verification is an essential step, insisting on obtaining very high calibration levels of the correlations between energy demand and economy is not so important, and can even be harmful for a good modelling. Indeed, the interannual fluctuations of energy supply and demand result from specific circumstances of each year, such as weather, economic cycles, fluctuations of energy prices in the international market, introduction of innovations, etc. It is not only impracticable to include all these parameters in the correlations but also irrelevant because they are unrepeatable, one-of-a-kind shocks (as the introduction of natural gas in 1997, or the impact of the Covid-19 pandemics in 2020), and thus will be akin to noise superimposed at the signal of the energy demand and supply. Furthermore, as the energy panorama has been changing rapidly in the past decade, only the most recent couple of years is meaningful as a starting point for projections and really matter for calibration.

6 Using the Model

The model requires a reference scenario that might be either a "business-as-usual" or a "current policies and measures" scenario. Then, "new policies" scenarios are built by modifying parts of this reference scenario. Sensitivity analysis is explored by modifying the key assumptions.

We refer two major instances where JANUS was already used. The first was at the use of version 4 for the preparation of the Portuguese NECP. It required exploring about 200 different scenarios until the final NECP renewable energy targets and the cocktail of technologies and energy efficiency measures to be applied were stabilized. It is remarked that the NECP targets are political; the JANUS scenarios are used as one more piece of information for decision (although a very important one), not as the outcome of a technical procedure.

The second instance was the national strategy for hydrogen, EN-H2 [12]. Shortly after delivery of the final NECP to the EC, there was a growing perception that the NECP strategy of massive electrification in all sectors was neither fully effective nor cost optimal. For instance, some industrial processes require high temperatures that are easy to obtain by burning fuels but very expensive when achieved by electrical means. Another problem is related to the numerous stranded assets left by swift electrification, and the ability to raise high investment amounts in short time, for promoting the required technological changes. Meanwhile, as the technological and business case around hydrogen had been developing fast, enriching the NECP vision with hydrogen technologies presented a useful path to improve the energy transition strategy, while keeping the same policy targets. In this context, to help building the EN-H2, JANUS version 5 was assembled during the first trimester of 2020. This was done essentially by adding more hydrogen value chains to the version 4 structure, as well as by updating the performance of technologies and refreshing the macroeconomic scenarios in the wake of the Covid-19 pandemic. JANUS time range was also extended to 2050, so that the EN-H2 scenarios could confirm other long-term views, such as those of the carbon neutrality roadmap [13], that Portugal is well placed to achieve carbon neutrality by 2050.

7 Conclusions

The aim of this paper is to introduce for the first description in the scientific literature of the JANUS model developed and adopted by the Portuguese Directorate-General of Energy and Geology, in order to support the national energy planning. This paper also provides evidence on how an in-house national model offering high quality could be assembled nearly from scratch in a few years' time, providing the public administration of the country with a very flexible and promptly available tool to assist the energy planning by the governments – as in fact it already happened with the Portuguese NECP 2021-2030 and the Portuguese Hydrogen Strategy.

Considering that other modelling and forecasting tools are currently being developed at DGEG, namely based on EnergyPlan [14], there is the purpose that JANUS will continue to be developed as an umbrella model for the other tools and satellite models. Priority areas for JANUS development include

higher time resolution, more extensive economic data (in particular at the energy demand side) and detailed satellite models for more energy demand sectors. Better assessment of the weather variability on security of supply, and impacts on climate change, public health and jobs are also foreseen.

References

1. Modelling tools for EU analysis. Official EU website. https://ec.europa.eu/clima/policies/strategies/analysis/models
2. POTEnCIA (Policy Oriented Tool for Energy and Climate Change Impact Assessment. Official EU website. https://ec.europa.eu/jrc/en/potencia
3. Eurostat, SHARES (Renewables). Official EU website. https://ec.europa.eu/eurostat/web/energy/data/shares
4. IEA (2020) Overview of TIMES modelling tool. Official website of the International Energy Agency. https://iea-etsap.org/index.php/etsap-tools/model-generators/times
5. Introduction to LEAP, Low emissions analysis platform. Stockholm Environment Institute. https://leap.sei.org
6. PNEC, Plano Nacional Energia-Clima 2021–2030 (National Energy-Climate Plan). Ministerial Resolution no. 53/2020, from July 10. Official Journal of the Portuguese Republic. https://dre.pt/
7. Statistics Portugal. Official website at https://www.ine.pt
8. Energy modelling - EU Reference Scenario 2016. EU Open Data Portal. https://data.europa.eu/euodp/en/data/dataset/energy-modelling
9. Hydrogen integration in value chains (in Portuguese) (2019) Report of Project POSEUR-01-1001-FC-000004 Evaluation of potential and impact of hydrogen in Portugal – strategy towards sustainability. Hidrogénio em Portugal – Estratégia para a Sustentabilidade. DGEG. ISBN 978–972-8268-48-0
10. Technology pathways in decarbonisation scenarios (2018) ASSET – Advanced system studies for energy transition project report. European Commission, Directorate-General for Energy, Contract no. ENER/C2/2016-489/SI2.742171
11. Renewable Power Generation Costs in 2017 International renewable energy agency, Abu Dhabi. ISBN 978-92-9260-040-2
12. National Strategy for Hydrogen, Ministerial Resolution no. 63/2020, from August 14. Official Journal of the Portuguese Republic. https://dre.pt/

13. Carbon Neutrality Roadmap - Long Term Strategy for Carbon Neutrality of the Portuguese Economy by 2050. https://descarbonizar2050.pt/en/ and Ministerial Resolution no. 107/2019, from July 1, Official Journal of the Portuguese Republic, https://dre.pt
14. EnergyPlan - Advanced energy system analysis computer model. Department of Development and Planning, Aalborg University. https://www.energyplan.eu/

Energy Transition in Morocco and Covid-19 Crisis: Opportunities and Challenges

Hassan Nfaoui and Ali Sayigh

In 2009, Morocco defined a new energy strategy. It has set itself the objective of bringing the contribution of renewable energy (RE) to mixed electrical power installed to 52% by 2030, so the energy consumption reduction due to the energy efficiency (EE) to 25%, ensuring the same service and remaining competitive. The RE contribution to the national electricity consumption is increasing, but Morocco uses more coal to meet its electricity demand. During the COP22 on climate change (CC) held in Marrakech in 2016, Morocco committed to reducing its greenhouse gas (GHG) emissions by 32% before 2030.

Morocco is characterized by an unequal distribution of water resources over time. It has built more than 145 large dams with a storage capacity of 18.67 billion m^3. It benefits from a strategic location in the north of Africa, with varied relief and 3,500 km of coastline. Fortunately, Morocco has significant solar and wind potential. In Ouarzazate, for example, the average annual daily sunshine ranges from 7.9 to 11.2 hours/day. Concerning the wind potential, the average annual wind speed is at 10 meter height; for Tétaoun in the north, it could reach 11 m/s; but for the Dakhla in the south, it has an average of 8 m/s.

The contribution of the installed hydroelectric power is 1800 MW, which is linked to rainfall. Hydropower represents 5.1%–13.7% of the net electricity consumption. In 2020, solar energy is 962 MW. Morocco has become one of the first African investors in RE and the first user for solar energy in Africa. In 2016, wind power installed in Morocco generated 899 MW, or 11% of the national installed power, generating 2859 GWh, or 8% of the national electricity consumption, while solar represents only 1%. In 2020, regarding wind energy, Morocco (1360 MW) is the second in Africa after South Africa (2295 MW).

RE is a promising alternative economic sector. The share of renewable energies in the installed electricity mix is 35% in 2019. Many international organizations regard Morocco as a pioneering African reference, thanks to its experience and achievements of RE sector.

Oil is one of the local and international challenges, especially for non-oil-producing countries, like in the case of Morocco. Energy is a key element in both the competitiveness and the sovereignty of a country. It is far that RE offers as an alternative to fossil energy. This energy imbalance is one of the reasons for the weakness of the Moroccan economy. The energy transition (ET) is necessary and profitable, not only for environmental reasons but also for economic reasons.

From the geopolitical dimension of energy in Morocco, if energy constitutes 3% of the Moroccan gross domestic product (GDP), then 97% depends entirely on it. This means that without stable energy security, the problem risk of the Morocco's economy fragility remains. The energy model to be adopted by Morocco must first aim at reducing its energy dependence without giving up the fight against CC.

In consideration of Covid-19, the fight against CC will slow down and investments in green technologies will decrease. Much of the green funding will go to those that are already functioning and profitable such as solar or wind.

The cost of RE has become competitive with fossil fuels. Locally produced alternative energies are more resistant to the worldwide supplies disruption. The green economy and the integration of the smart technology sector create more new jobs.

Because of Covid-19 and according to the International Energy Agency (IAE), in 2020, the world energy demand should drop by 6%, the biggest oil shock since the 1973 one. In the other hand, IAE estimates that the renewable electricity production will increase by 5% and the global CO_2 emissions are expected to fall by 8%, the largest decline in history.

H. Nfaoui (✉)
Solar Energy & Environment Team, Energy Centre, Mohammed-V University, Rabat, Morocco

A. Sayigh
WREC, WREN, Brighton, UK
e-mail: h.nfaoui@um5r.ac.ma

© The Author(s), under exclusive license to Springer Nature Switzerland AG 2022

A. Sayigh (ed.), *Sustainable Energy Development and Innovation*, Innovative Renewable Energy,
https://doi.org/10.1007/978-3-030-76221-6_64

Guterres, the Secretary General of the UN, has suggested that the huge sums of money that will be spent to overcome the pandemic must create new businesses as part of a clean and ecological transition to a green and sustainable economy. GHGs, like the virus, do not respect national borders. He stressed that the solution lies in building a healthy and resilient future for humanity as well as for the planet.

According to Rifkin, a prospective specialist and ardent defender of smart electric grids, in the 20th century, the second revolution took place in the United States of America. It is due to the discovery of the telephone which converged with a new cheaper source of energy, the oil, and the internal combustible engine. In his opinion, the second revolution is in decline. We are now dealing with the emergence of a third industrial revolution generated by the digitalization of communications, energy, and transports, and this is what we call the internet of objects. When communication internet integrates with the one of digitalized renewable energies (solar and wind) and the one of automated and digitalized transport, we can start to produce and share all kinds of physical objects of energy, and products originated from 3D printing with a marginal cost close to zero. For him, the ET as part of the third industrialized revolution has already started in Germany and China.

He said as well: "if we set up the energy Internet, it means that all humanity will produce its own energy locally from the solar and wind and shared through all the continents by giving up fossil fuels. In addition, the establishment of the transport internet means that we could eventually share electric vehicles and reduce our ecological imprint. The new society at zero marginal cost is the only economic plan, I know, capable of responding in time to global warming, whose consequences would be terrible for humanity. This ET must be accomplished quickly, with skill and determination if we want to provide a viable future for generations."

The ET is being done in the context of what is called 3D (decentralization, decarbonization and digitalization), three dimensions that define the world of energy, within the framework of an integrated and prospective vision

Once Covid-19 started to spread rapidly in Europe and the World Health Organization (WHO) declared the alert on the virus, Morocco also announced a state of health emergency for 3 months which caused major damage for certain economic sectors, particularly tourism and transport, and led to an economic recession. Morocco could bounce back better from this pandemic because it has made renewable energies the cornerstone of its ET.

Today, Morocco is faced with a decisive historical challenge which is the necessity of winning the bet of modernization as well as scientific and intellectual progress. It is required to double its efforts to reach it, so that it could fill the gap that separates it from the West in this domain. It has no other choice than to be up to date to keep rhythm with the scientific revolution of the West and the profound technological transformations that the modern world is experiencing in terms of digital culture and computer sciences.

Faced with these challenges, there are key factors to boost research and development, such as revising national priorities (R&D roadmap with an orientation law) and setting up the infrastructure for a national smart grid.

In the eyes of the political and economic security challenges facing Africa, it is appropriate and in the interest of Morocco to bet on scientific research in renewable energies and investment in the green economy.

Let us hope that the world after the Covid-19 crisis will be resolutely more ecological, Morocco is committed to it.

1 Introduction

The challenges of the energy transition (TE) are of several natures: strategic (security of energy supply continuously, stably, and at low cost), social (access to energy for all), economic (rational and green economy), and environmental climate change (CC) and environmental protection. It is for all these reasons that Morocco, like most countries in the world, was forced to engage in the ET process. It is necessary and profitable, not only for environmental but also for economic reasons because Morocco is a non-oil-producing country.

Energy is a main factor for economic activity. The inconvenience is how to guarantee sustainable energy security that reassures and maintains growth as well as the standard of living, while ensuring access to energy at affordable costs. On the other hand, we must respond to the growing demand for energy around the world while limiting its ecological impact

But among the consequences of the health crisis due to the Covid-19 epidemic on economic and social activity, the world demand of energy should fall by 6% in 2020, that is, 7 times the decline after the 2008 crisis, the most big energy shock since that of the 1973 one [1].

The more expensive and stable the oil, the more we try to work on the savings of energy. On the other hand, for a low price, it is an opportunity for manufacturers to take advantage of this favorable occasion to adapt and invest in the green economy. Fossil products are limited in time and even if the price would not increase in the medium term we must avoid wasting these natural resources and preserving them for future generations.

Currently, solar, wind, and geothermal energy are at almost zero marginal cost and are competitive with fossil energy. The variation in the price of solar technology is a decreasing exponential curve close to that which is experienced by computer integrated circuits a few years ago. Their exploitation to produce electricity requires setting up technological infrastructure such as the smart grid [2, 3].

The electricity production can be adapted to consumption demand instantly, but the production is intermittent and irregular when it comes from solar or wind resources because these sources by nature depend on weather conditions. Due to the digital revolution, this disadvantage is no longer up to date thanks to the regularization of distribution by smart grids. Storing the produced energy to ensure supply security is no longer a major problem, the consumer can become a player in the energy sector (consumption optimization) by installing solar panels or wind turbines. It can produce energy and distribute it as needed through the national smart grid, avoiding waste as much as possible.

The ET is being done in the context of what is called 3D (decentralization, decarbonization and digitalization), three dimensions that define the world of energy, within the framework of an integrated and prospective vision. However, we note that the digitization of energy sector is lagging behind that of communication one, while the digital revolution is not new. The digitalization of ET is an industrial and social transition accompanied by a transition of the paradigm.

The evolution of energy sector requires the transition to the 3D vision:

- Decentralization.
  - Centralized fossil production without any form of electricity storage, implicitly energy dependence, and technical losses by the transport of the national electric grid.
  - Paradigm change requires the transition to the more flexible decentralized model and the possibility of engaging in the fight against CC.
- Digitization.
  - Best route for industrialists concerned with optimizing their economic and ecological performance.
  - Maximum energy input from renewable sources.
- Decarbonization.
  - Financially attractive resource.
  - Reduction in greenhouse gas (GHG) emissions with a lower cost.
  - Management according to a decentralized model by a smart digital infrastructure.
  - Consumer also producer, making this configuration specific to the needs and the local resources.

The treatment of the whole reverts to optimization of energy sources available locally and better control of their use. This requires building bridges between the different actors (producers-consumers).

The fossil fuel resources are limited in time, the price of renewable energy (RE) is falling, and none of the countries in the world is spared from the damage of CC. All these subjects converge to the climate problem. According to Ségolène Royale, the president of COP21, ET is an economic and environmental opportunity for the world. The industrialized countries carried out their industrial revolution in the last century using fossil fuels. They are the cause of the global warming. She urged them to ratify the Paris Agreement and commit to its realization [4].

The big swing over of COP21 is investing in the green economy and the attribution of a pollution cost (carbon price), which will be reinvested and reinjected into clean production and RE. The ET is essential so that we can collectively fight against CC. Each country decides its energy model according to its potentials, its initiatives, its capacity for innovation, and the time limit of the ET, so that it is as fast as possible. There is an opportunity to achieve this extraordinary common goal, unique in history, on which for the first time the international community agrees not only to build the world of today but to take an interest in the world of tomorrow that we will not live [4].

There are three human activities that have destroyed the planet: global warming, pollution of all kinds, and overexploitation of natural resources. These three bad behaviors must be stopped to restore harmony between human beings and our natural environment. Human beings cannot live without nature and when we try to live against it by destroying it, it turns violently against us. These challenges can be met by the exchange of knowhow, the technology transfer, and the collective intelligence of the international community [4].

In accordance with Rifkin, an American prospective specialist and the architect of the third industrial revolution, the first industrial revolution was born in Great Britain in the 19th century. It is a result of the convergence of communications (steam power printing for the telegram), energy (coal), and transport (steam engine on rail locomotives). In the twentieth century, the second revolution took place in the United States of America. It is due to the discovery of the telephone which converged with a new cheaper source of energy, the Texas oil, and the internal combustible engine. According to him, the second revolution has reached its peak. It is in decline. We are now dealing with the emergence of a third industrial revolution generated by the digitalization of communications, energy, and transports, and this is what we call the internet of objects. When communication internet integrates with the one of digitalized renewable energies (solar and wind) and the one of automated and digitalized transport, we can start to produce and share all kinds of physical objects of energy, and products originate from 3D printing with a marginal cost close to zero. For him, the ET as part of the third industrialized revolution has already started in Germany and China [2, 3].

The energy internet like the communication one costs money to set up the infrastructure to communicate like the internet that we are currently using. But once it is in place, the real cost to maintain, manage, and use this system is almost free to send information and communicate. If we apply this technology to electrical lines, the entire electrical network will be transformed into a real energy internet. So it will cost money of course. The faster we put this in place, the faster we will have energy at almost zero marginal cost. It will completely change the situation. It is a question of survival. It is a new step in history and it is inevitable [2, 3].

Stéta is she, him for Rifkin, an economist, is not of this opinion and opposes rigorous arguments put to her. For her, the new marginal cost society is not on the agenda. As far as energy is concerned, the digital world is increasingly deregulated. Regarding the rhythm of growth (slow, moderate, etc.) and the transition to the third industrialized revolution, most economists have no answer for this topic. Thanks to advances in technology, people consume less and they are more and more efficient. Digitization leads us to the decentralization even faster. People want to take ownership of things and, therefore, we see toward local RE that people want to exploit [5].

What he said about the German economy which made more and more use of solar, wind energy, etc. is absolutely true. However, the German economy is currently heavily dependent on coal. It has chosen to renounce nuclear and switch to higher carbon energy for the moment than it was in the past. So that will pose a whole bunch of economic, social, and geopolitical problems for Germany. It is not at all an easy and neutral decision [5].

For the choices that the Chinese authorities have made to develop clean energies, in reality, it is a perfectly rational and perfectly economic choice for them. It was not at all an enthusiasm which was born from the wish that China wanted to contribute to a necessarily cleaner world. First of all because the Chinese economy and the whole China suffers terribly from pollution. The objective of China is also to become the pioneer in a number of industries and clean energy is one of them. This is particularly the case for the solar panel industry, where Chinese solar panels are increasingly winning global markets [5].

Energy is at the top of the priority pyramid in Africa. It is an essential lever for the development of the African continent. It is a prerequisite for the development of transport, infrastructures, education, and health. Energy is also more important for Morocco as a non-producer of fossil fuels.

Morocco imports more than 90% of its energy needs (Fig. 1) and its electricity consumption increases by about 6% per year. For these reasons, to guarantee sustainable energy security which reassures and maintains growth so the standard of living, while ensuring access to energy at affordable costs and of course preserving the environment, in 1980, Morocco adopted the Rural Electrification National Program (RENP). However, for the Rural Electrification Global Program (REGP) launched in 1995, more than 30% of this program was achieved through the use of decentralized electrification technologies

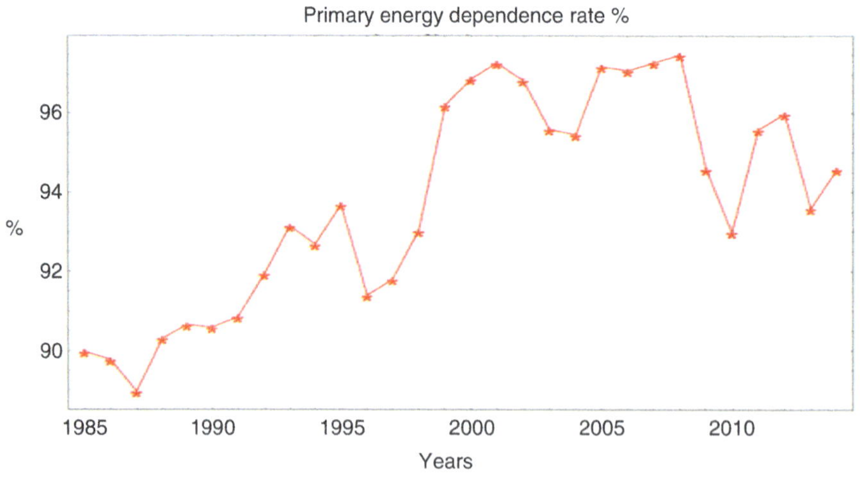

Fig. 1 Primary energy dependency ratio [6–9]

by RE (solar, wind, and hydro). In 2009, a new and solid national energy strategy was developed, particularly in sustainable energy domain, for the needs of its national economy and to support its energy independence. It is based on the development of green economy and increase in the renewable energy contribution to the national mix electricity installed at 52% by 2030. This will enable Morocco to reduce its GHG emissions and to fight against CC, thus contributing to the preservation of the environment. It has also developed a national energy efficacy (EE) strategy aimed at achieving a 25% less consumption of energy. This strategy targets the most energy-intensive sectors, in particular transport, building, industry, as well as agriculture and public lighting [6–9].

The situation and characteristics of energy consumption since 1998 have shown that Morocco is highly dependent on imported fossil fuels to meet its energy needs. In 2016, petroleum products represented 54%, and more than a quarter came from coal, the most polluting fossil product. Consumption of natural gas increased significantly from 2005. It represented 6% in 2016 [6–9].

Morocco is characterized by unequal distribution of water resources over time. It has built more than 145 large dams with a storage capacity of 18.67 billion m^3. It has developed a great expertise in the field of studies and construction of dams, which is recognized worldwide. It has a strategic location in the North West of Africa, only 14 kilometers from Europe, with varied relief and 3500 km of coastline. It has available an important solar and wind potential. In the Ouarzazate region, for example, the annual average daily global irradiation on a horizontal surface varies from 3.86 to 7.85 kWh/m^2, which means an annual average of 7.9–11.2 hours of sunshine per day [10, 13]. Concerning the wind potential, the annual average wind speed, at 10 meter high, for the Tetaoun zone in the north could reach 11m/s, but for the Dakhla zone, it could exceed 8 m/s [7–9].

In 2009, a large-scale solar energy use strategy was launched by the creation of the Moroccan Agency for Solar Energy (MASEN), which has become the Moroccan Agency of Sustainable Energy (MASE), to develop renewable energies, particularly hydraulic, solar, and wind. In 2015, MASEN has so far managed only development in the solar power sector, and has been tasked to do the overall strategic planning of renewable energy projects, including wind and hydro. In concordance with the Electricity National Office (ONE), it will take charge of all projects for electricity generation from renewable sources, including feasibility studies, planning, financing, development, and construction, as well as operation and maintenance [7].

Under the impetus of the state, with laws, tax, and regulations given to the RE sector,

Morocco managed to carry out several large solar and wind energy projects. They are already operating in a practical environment, entirely perfect from the technical and economical point of view. The results are tangible. RE is a promising alternative economic sector. The share of renewable energies in the installed electricity mix is 35% in 2019. Many international organizations regard Morocco as a pioneering African reference, thanks to its experience and achievements in the RE sector.

At the United Nations meeting on CC held in Marrakech in 2016, Morocco committed to reduce its GHG emissions by 32% by 2030, with the integration of smart digital equipment that are efficient in energy consumption. The energy efficiency (EE) is one of the keys to achieve these objectives. It is considered a national bet that requires collective awareness to reduce and consume energy efficiently [7].

The EE is one of the tasks of the Moroccan Agency for Energy Efficiency (AMEE). Plus the implementation of government policy, which aims at energy independence through the promotion of EE, helps the government to formulate its strategies and regulatory directives in the form of draft laws with regard to EE, such as thermal regulation of housing. In this context, the AMEE has identified an energy saving potential of 25%, all economic sectors combined, by 2030 [7]. According to the World Economic Forum, Morocco is at the top of the list for the Arab and African energy transition index for the year 2020, thanks to its energy strategy and setting the accompanying legislative measures up to generate clean electricity[18].

Once Covid-19 began to spread rapidly in Europe and World Health Organization (WHO) declared the alert on the virus; Morocco also announced a state of health emergency for 3 months, which caused major damage to certain economic sectors including energy, tourism, transport, etc. and led to an economic recession [19].

The Coronavirus crisis has imposed the need for digital transformation in Moroccan public administrations because the adoption of digital solutions is an essential means of ensuring the continuity of administrative work and reducing the exchange of correspondence and paper documents, especially during the sanitary confinement period.

One of the repercussions of the changes associated with the national and international economic situation owing to the Covid-19 crisis is that the Moroccan government had to establish an amending finance law for the year 2020, imposed by the conditions of the law on health emergency, and the repercussions it had on the national economy within 3 months of sanitary confinement. One of the most important clauses aimed to finance the digital administration so that the Moroccan administration transits from the paper era to the digital one. For the first time, it has placed digital transformation among the priorities of the finance law, considering it as a lever for development, in the same way as the sectors which require a proper advancement due to the pandemic impact, such as education, scientific research, health, employment, and social protection [19].

The amending finance law coincided with the adoption of the digital management law of administrative services. Through which the government aimed to establish a legislative framework to legalize and accelerate the digital transformation of the Moroccan administration, allowing the activation of the digitization [19].

The Covid-19 health crisis is now impacting all countries and has consequences for all of us. During the confining, the health crisis was having priority on economy crisis. Guterres, the Secretary of the UN, warned that the crisis caused by Covid-19 could cause a loss of 8.5 trillion dollars for many sectors of the economy worldwide, particularly energy, transport, and tourism [20].

Guterres suggested also that the huge sums that are being spent to overcome the pandemic must create new businesses in the framework of a clean and ecological transition to a green and sustainable economy. The greenhouse gas (GHGs), like the virus, do not respect national borders. He insisted that the solution is in the construction of a healthy and resilient future for humanity as well as for the planet [20].

The World after the Covid-19 crisis will be resolutely more ecological. For this reason and in the light of the security, political, and economic challenges facing Africa, it is appropriate and in the interest of Morocco to bet on research and development in RE, investment in the green economy and the energy internet by setting up the infrastructure of a national smart electrical grid.

2 Energy Sector Characteristics

2.1 Energy Consumption

The Moroccan energy balance remains characterized by the predominance of petroleum products which represents, in 2016, 54% of the national energy consumption, evaluated at 19.58 Mote. Coal ranks second with 26%, a most polluting source of energy (Fig. 2).

2.2 Electricity

Until 1996, of thermal and hydraulic origin, almost all electricity production is provided by the Electricity National Office (ONE) [7]. The gradual liberalization of the electricity sector began in 1997, either by deprivation of refining companies or by concessional electricity production. The electrical power installed by ONE and the autoproducers is 10938MW in 2018 (Table 1, Fig. 3). Figure 3 shows that the hydraulic power has not changed much since 2005. Thermal and wind power have increased significantly since 2008, especially for thermal power.

Figures 4a,b, and 5 show that the evolution of the net electrical energy demand has increased significantly, from 19.52TWh in 2005 to 35.87TWh in 2016, an increase of 84%. The contribution of hydroelectricity depends on rainfall. For the same

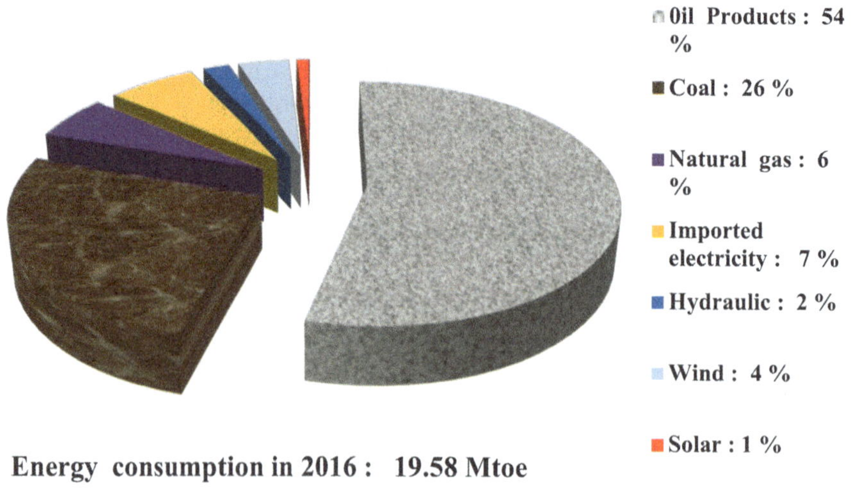

Fig. 2 Distribution of national nest energy consumption in 2016 [6–9]

Table 1 Comparison of the installed electric power in Morocco in 2018 to that of 2017 [6–9]

	Installed electrical power (MW)		
	2018	2017	Var:18/17 (%)
Hydraulic	***1770***	***1770***	***0***
Thermal	*7237*	*5851*	*23.7*
Steam coal	4281	2895	47.9
Fuel vapor	600	600	0
Gas Turbine	1230	1230	0
Diesel	263.7	263.7	0
Combined cycle	834	834	0
Gas oil	28.3	28.3	0
Wind	***1220***	***1018.4***	***19.8***
Wind ONEE	204.5	204.5	0
Wind private	37.3	37.3	0
Wind IPP	351.7	351.7	0
Wind law 13-09	626.5	424.9	47.4
Solar	***710.8***	***180.8***	***293.1***
Solar ONEE	20.8	20.8	0
Solar MASAN	690	160	331.3
Total	***10937.8***	***8820.2***	***24***

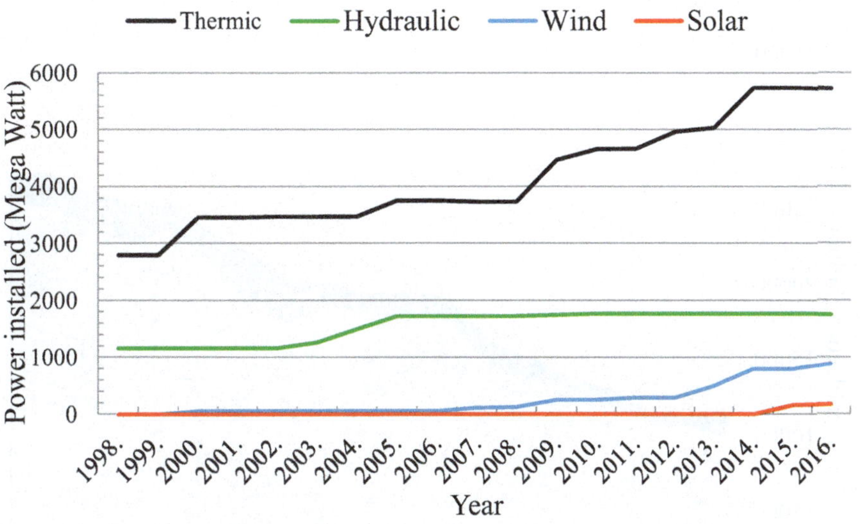

Fig. 3 Evolution of the national electric power [6–9]

installed capacity, 1800MW, hydroelectric power generation was 3631 TWh in 2010 and it dropped to 1816TWh in 2012 [6–9]. The contribution of wind energy to electricity consumption became significant since 2012 (Fig. 4a). On the other hand, the first kWh of electricity from solar energy is injected into the national electricity grid only in 2015 (Figs. 4a). Figure 4b shows that the contribution of renewable energy (RE) to the national electricity consumption increases and Morocco uses more coal to meet its electricity demand. Since 1999, the consumption of primary energy and electricity have increased annually on average by 4.5% and 6%, respectively (Fig. 6). RE is a promising alternative economic sector in Morocco. The share of renewable energies in the installed electricity mix is 35% in 2019 [6–9].

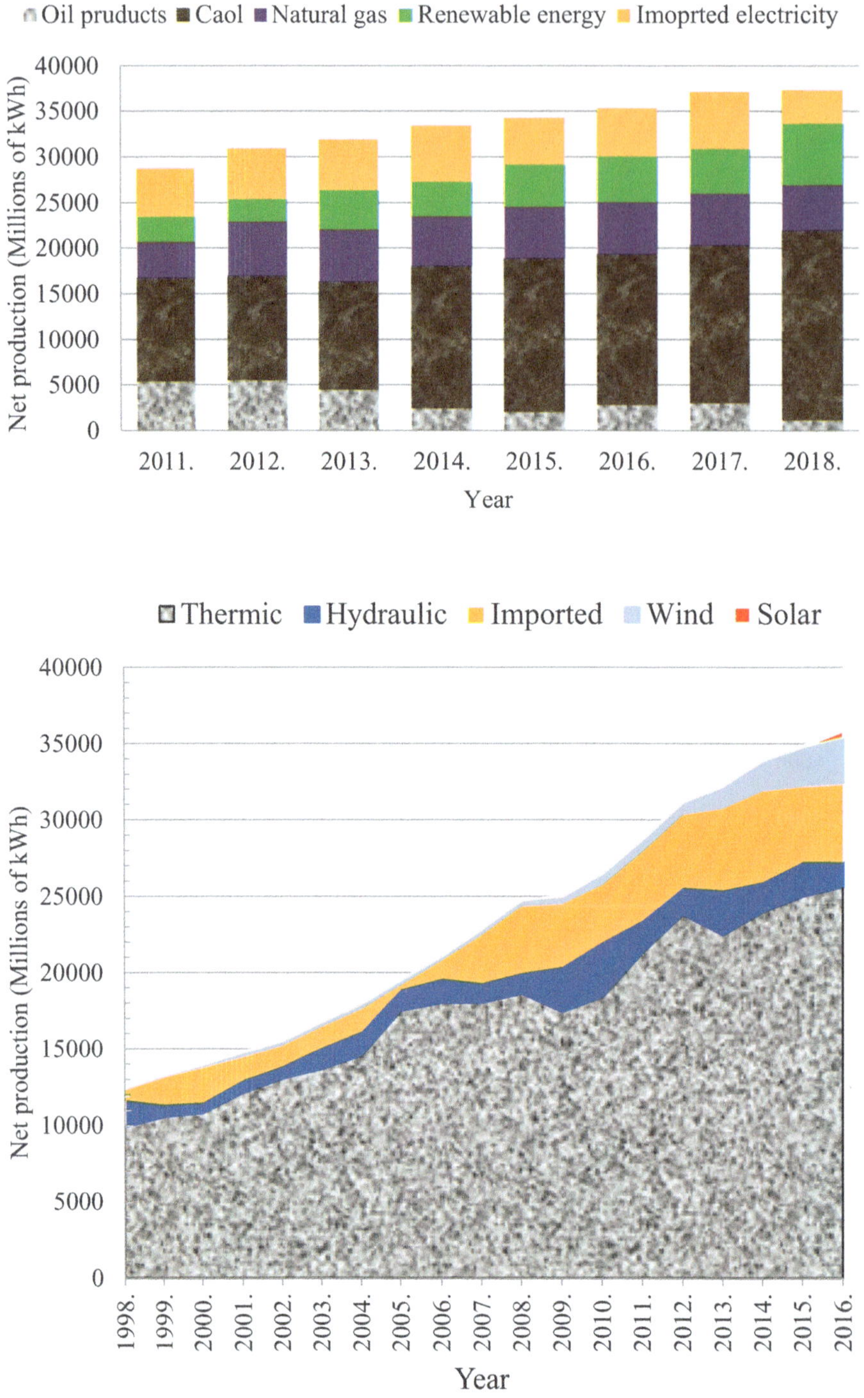

Fig. 4 (**a**) Evolution of the production of national net electricity called from 2011[6–9]. (**b**) Evolution of the production of national net electricity called from 1998[6–9]

3 Renewable Energy (RE)

3.1 Hydraulic

The policy of dam construction in Morocco has been a top priority since 1960. In 2018, Morocco has 145 large dams with a storage capacity of 18.67 billion m^3 available and 14 others under construction (3.21 billion m^3). They constitute a pillar of agricultural development and the electricity production. Figure 7 shows that the surface water resources are in average year

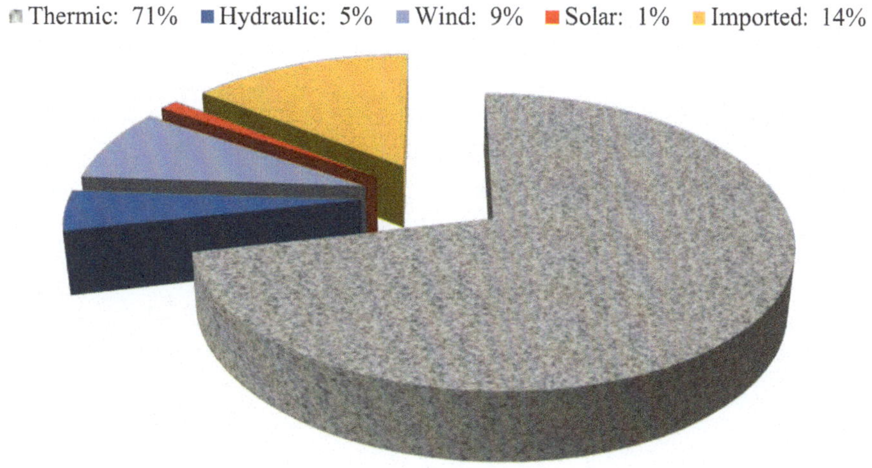

Fig. 5 Distribution of national net electricity consumption in 2016 [6–9]

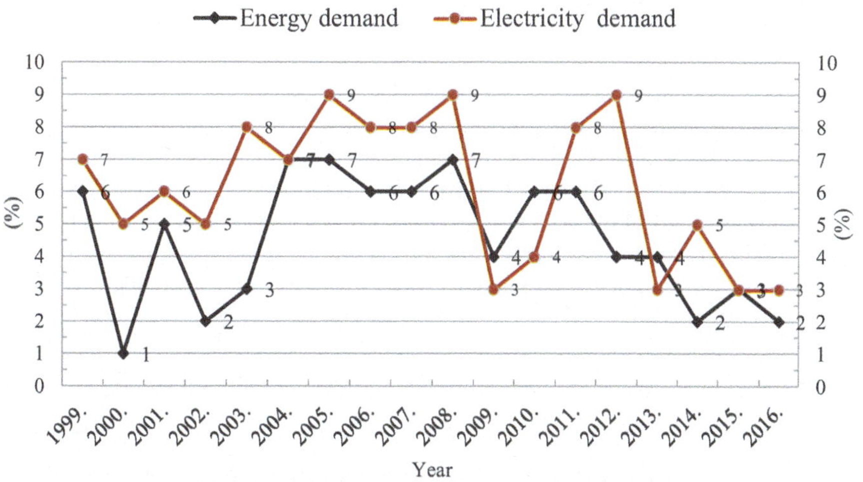

Fig. 6 Evolution of energy and electricity demand growth [6–9]

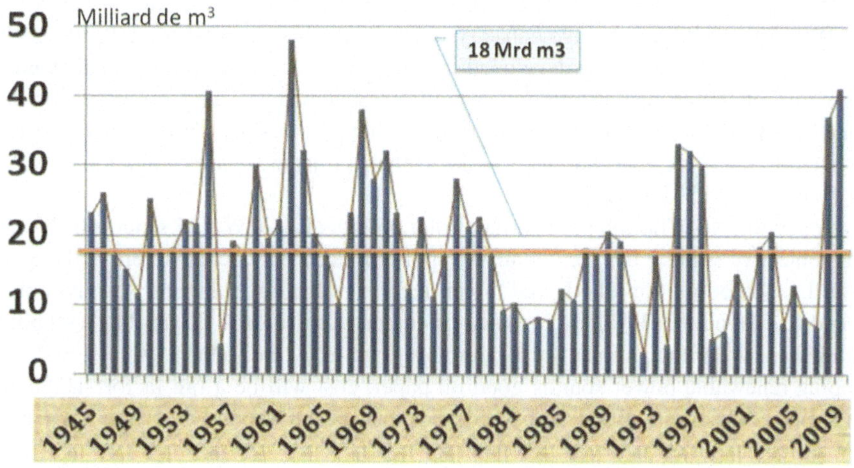

Fig. 7 Unequal distribution of water in time [9]

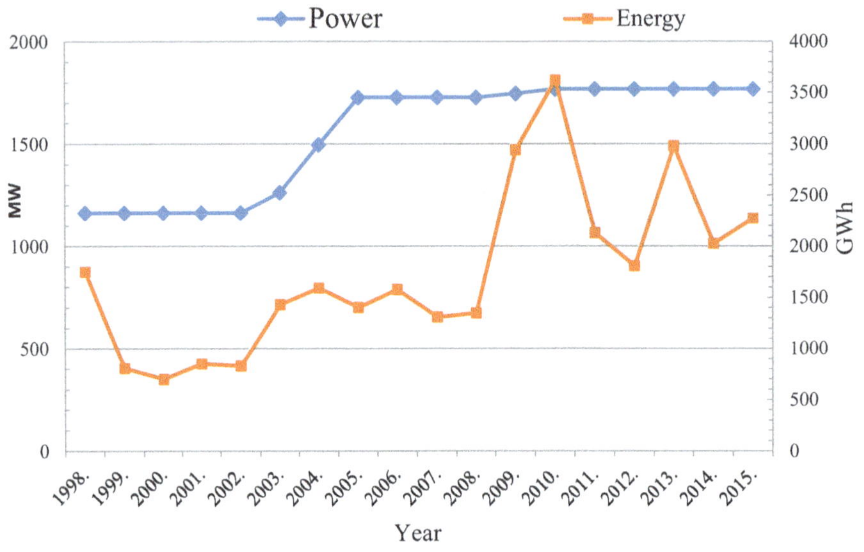

Fig. 8 Evolution of hydraulic power and electricity [6–9]

at 18 billion m^3. This unequal distribution of water resources over time makes it necessary to store supplies from wet years for use during periods of drought [6, 7, 21].

The hydroelectric sector has also been particularly important since 1954, the date of commissioning of the Bin El Ouidane dam (135 MW). It is the first dam built in Morocco. Plus irrigation, it produces around 287 GWh/year [7]. The contribution of installed hydroelectric power (1800 MW), which depends on rainfall, is between 5.1% and 13.7% of the net electricity consumption (Fig. 8). The dam Al Wahda (240 MW) is the largest one in Morocco. It was inaugurated in 1977. Apart from supplying water for drinking and irrigation, it generated hydroelectric power of around 400 GWh/year [6–9, 21].

3.2 *Solar*

In 2020, solar in Morocco is as follows: 962 MW in operation through 7 projects and more than 2700 MW committed through 20 projects [6–9]. The first hybrid power plant "Ain Beni Mathar" (470 MW), incorporating 20 MW of solar power, was built in 2010 by ONE. It is a Solar-Natural Gas Hybrid Plant. The second giant solar power station "Noor-Ouarzazate" (582 MW), one of the largest solar energy complexes in the world, was installed between 2016 and 2018 by MASEN. It uses concentrated solar technology (CSP) (360MW), heliostats technology with a solar thermal tower (150 MW), and photovoltaic technology (72 MW). The electricity storage capacity is 7 hours, which will be used during peak hours. It consists of four multitechnology solar subplants, which is developed in full compliance with international standards, both technologically and environmentally, including a platform for scientific research, tests, and training. For the solar plant "Noor-Ouarzazate," the price of KWh reached $ 0.19, comparable to that of the solar power plant Redstone Arsenal (USA), a competitive price with traditional sources of energy [7, 9]. Two others projects are in exploitation by MASEN: Noor Laâyoune (185 MW) and Noor Boujdour (120 MW) (Fig. 9).

In 2019, ONE developed the Noor Tafilalt photovoltaic solar project (120 MW) (Missour (40MW); Erfoud (40MW); and Zagoura (40MW)). Its objective is to secure the country's supply of electrical energy, improve the quality of service for the isolated regions, as well as the promotion of RE and the protection of the environment. Also two other solar power plants (Boujdour (20MW) and Laâyoune (85MW)) are installed by the private sector in 2018. Plus a significant capacity of solar electric power is installed as part of the national solar pump program or by individuals [7].

But, while developing renewable power sounds good on paper, cost will be a big factor; The launch of the Noor CSP project has helped the price of electricity produced by CSP to come down to around $0.16 per kWh, but that looks expensive compared to solar PV which has fallen as low as $0.03 per kWh [22].

It remains to be seen whether the cost of CSP will fall low enough to be globally commercially competitive, and deliver cost-effective renewable power for Moroccan consumers. CSP also uses large amounts of water to keep the mirrors clean – a real problem in water-stressed Morocco. At the same time, Morocco has not totally kicked the fossil fuel habit – coal still makes up the biggest part of energy production today (35%) and is set to be expanded over the next 5 years. The new energy mix will include at least 3900 MW of energy from natural gas, and the search for hydrocarbon deposits on Morocco soil continues [22].

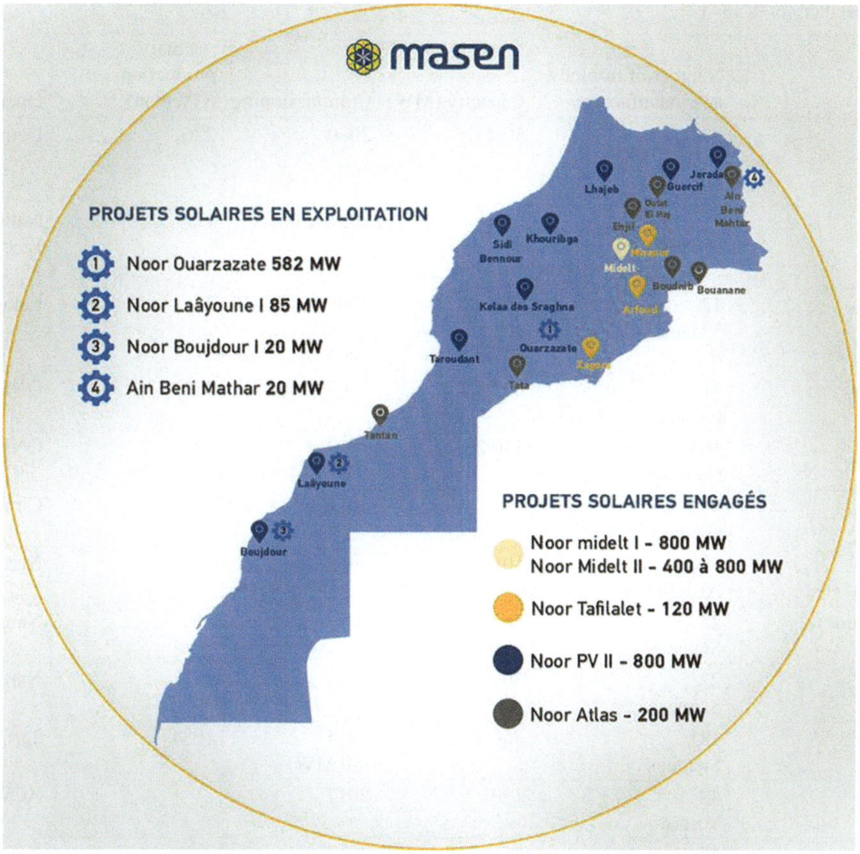

Fig. 9 Solar projects in exploitation or undertaken by MASAN [7]

3.3 Wind

The first wind farm, installed in Morocco, is Koudia Al Baida (Tetouan), with a capacity of 50 MW, put into service in 2000. However, the largest wind farm is Tarfaya in southern Morocco (300.1 MW), put into service in 2014 [6–9, 14–17].

In 2016, wind power installed in Morocco is 899 MW, or 11% of the national installed power, generating 2859 GWh, or 8% of the national electricity consumption. In 2020, it reached 1360 MW (Table 2, Fig. 9). For wind energy, in Africa, Morocco ranks third after South Africa (2295 MW) [9] and Egypt (1388 GW). The largest wind farm is the Tarfaya one (301 MW), installed in 2004. As an example, wind energy generated 2858 GWh in 2018 (8% of Moroccan annual electricity consumption) [6–9, 14–17].

Morocco is also investing in wind. A consortium of Enel Green Power, Nerefa, and Siemens won a bid to build five new wind farms at different sites across Morocco. Their combined capacity will be 850MW, a huge increase taking Morocco closer to its aim of producing 14% of electricity from wind by 2020. The unit cost in the tender documents was one of the lowest in the world, at just $0.03 per kWh [22].

4 Energy Efficiency (EE)

4.1 Building Sector

Buildings are an essential subject for the energy transition (ET). The building sector is the second energy consumer, representing 27% of the national energy consumption, including 18% in the residential sector and 9% in the tertiary sector [7]. The national EE program aims to exploit, first and foremost, the significant source of energy savings in the building sector, by applying the following measures:

Table 2 Wind farms installed [6–9, 14–17]

Sites	Number of turbines and manufacturer	Capacity (MW)	Commissioning	Annual production (GWh/an)	Operator/ owner
Koudia Al Baida (Tétouan)	84 Vestas	50.4	2000	226	Detroit Energy Company
Koudia Al Baida (Tétouan)	7 Enercon	3.5	2001	12	National Office of Electricity, ONE
Tétouan	12 5 Gamesa 6	10.2 10 12	2005 2008 2009	38	Lafarge Ciments
Amogdoul (Essaouira)	71 Gamesa	60.35	2007	210	ONE
Tanger	165 Gamesa	140.25	2009 (107.1 MW)	526.5	ONE
Laâyoune	6 Gamesa	5.1	2011	16	Ciments du Maroc
Akhfenir (Tan Tan)	61 Alstom/GE	101.87	2013		Nareva Holding
Foum El Oued (Laâyoune)	22 Siemens	50.6	2013		Nareva Holding
Haouma (Tétouan)	22 Siemens	50.6	2013		Nareva Holding
Tarfaya	131 Siemens	*300.1*	2014 (50 MW)	1084	Tarfaya Energy Company
JbelKhalladi (Tanger)	40 Vestas	120	2017		ACWA Power
Akhfenir (Tan Tan)	56 GE	95	2017		Nareva Holding
Aftissat I Boujdour	87 SGRE	200	2018	87	Nareva Holding
Taza	50 GE	150	2020		Public-private
Total		***1359.97***			

4.1.1 Regionalization Program

The regionalization program was launched in 2012. It aims to develop RE programs and strengthen EE in the municipalities. Its purpose, in particular, is to train and support elected members, engineers, and municipal technicians for the implementation of municipal energy planning, to improve their knowhow on good practices to adopt for carrying out economic demonstration projects in energy [7]. In addition to the regionalization program, there were two important projects: "Green Cities" Project and Ouarzazate "Eco-district" Project.

4.1.2 Public Lighting

In Morocco, street lighting, in general, constitutes the second budget item, after the salaried bill, and the first item of energy expenditure for the municipalities. Public lighting alone represents about 35% of the energy consumption of a municipality. To offset the continuous increase in the energy consumption of public lighting, the municipalities have started to give more and more importance to the thematic of EE in their planning, and to take measures to optimize their energy consumption [7].

For its part, ONE launched the "INARA" operation in 2007. It is, in particular, the introduction of LED technology for the needs of residential and public lighting for municipalities where the ONE distributes electricity. This program has resulted in cumulative electrical energy savings of approximately 3,370 GWh since the start of the operation, as well as a reduction in CO_2 emissions of 427,731 tons per year [7, 24].

4.1.3 Implementation of the EE Code in the Building

This is the adoption of the Thermal Construction Regulation in Morocco (RTCM) and Decree No. 2-136-874 in 2014 for the application of Law No. 47-09 relating to EE, as well as the realization of the software "Binayate," the first software of energy performance diagnostic. It is a preliminary thermal evaluation of the construction projects and control of the conformity with the RTCM in the Building [7, 24].

4.1.4 Green Mosques

As part of the strengthening of EE, the first phase of the green mosques' energy upgrade program (15,000 mosques) was launched in 2015.

This program meets a double objective:

- Raising citizens' awareness of environmental protection and EE techniques for the use of renewable energy.
- Reducing the energy consumption of mosques by introducing economical lighting, solar water heaters, thermal management of mosques, as well as the use of photovoltaic solutions for the electricity production.

4.1.5 Solar Water Heaters

Morocco enjoys a significant solar potential which amounts to more than 3000 hours per year of sunshine, or a daily irradiation of 5.6 kWh/m^2 [10–13]. In order to promote the solar thermal sector, the development program of solar water heaters aims at increasing mobilization of the national energy resources and the preservation of the environment.

4.2 Transport

In Morocco, the transport sector is the largest consumer of energy with nearly 38% of the national energy consumption. According to the AMEE, the transport sector, which depends almost exclusively on petroleum products, also contributes to more than 23% of the country's GHG emissions. The annual growth in fuel consumption in this sector is more than 5% per year [7, 24].

In fact, several measures aimed at reducing the fuel consumption of vehicles are already being implemented in Morocco; in particular, the promotion of low-consumption vehicles, scrapping programs for old trucks and taxis, integration of electric vehicles into the transport fleet of urban municipalities, implementation of the obligation to submit energy-consuming transport fleets to an energy audit, and promotion of collective travel (tram, bus, etc.).

4.3 Industry

In the framework of an experimental project and in order to establish an EE culture in the industrial sector, MEME initiated several actions through the realization of voluntary energy audits for several industrial companies. This will push energy managers to implement ISO 5001 with the aim of reducing energy consumption, in order to improve their knowhow in terms of the implementation of the energy management system in industry.

With the aim of instituting a compulsory and periodic energy audit system to rationalize the energy use in companies, a decree intended for audit organizations for enforcing the Law No. 47-09 relating to EE was put into action in 2019 [7].

4.4 Agricultural Sector

Energy consumption in the agricultural sector represents around 7% of national energy consumption. The program to optimize the energy performance for farming concern was launched by AMEE in partnership with the "Crédit Agricole du Maroc" Group (GCAM), it is intended to support producers in actions aimed at reducing their energy bill, optimizing their consumption, and encouraging the RE use. This will improve the competitiveness of the agricultural sector and the contribution to environmental protection, through the reduction in GHG emissions [7, 24].

Solar pumping is an alternative for existing diesel irrigation systems. In 2013, AMEE, in partnership with the Global Environment Facility (GEF), developed a project to support the national solar pump program for irrigation purposes in order to replace existing pumps using subsidized diesel and butane.

5 Energy Transition (ET)

The main objective of the Moroccan solar plan is to increase its energy independence and to make renewable energies the cornerstone of the Moroccan national energy policy. Morocco made the choice of an energy transition (ET) toward sustainable by integrating renewable energy sources into its energy mix.

Morocco is very dependent on energy imports. The energy bill has recorded an annual growth rate of 18.8% on average since 2002, essentially reflecting the rise in prices of fossil fuel and the growth in demand [7].

Since 2009, Morocco has adapted a national energy strategy for large-scale electricity production from solar and wind energy and the strengthening of EE. Plus its efficiency in favor of the environment, this solution contributes to its energy independence. In this context, renewable sources have become an essential element for ET and must be accompanied by political decisions and public investments.

This approach was developed on the direct supervision of the sovereign, the King of Morocco, Mohammed VI, so that it is operational. The government has taken a new legal plan of action accompanying this strategy: the law no. 13-09 on RE liberalizes the electricity production from sources renewable, which is no longer the monopoly of ONE. The law no. 58-15 opens up the possibility for auto electricity producers generated from renewable sources to connect to the national low-voltage electricity grid, while the law No. 48-15 establishes a national electricity regulator [7].

5.1 Opportunities

Morocco faces imminent danger of shrinking water tables (aquifers) and dam drainages. Despite the Moroccan policy to rationalize the use of water, the consequences of global warming will be greater than those of other countries, for example, Spain, due to its location in a geographical area that has suffered for years from the decline of the rains and the acceleration of the desert climate [10, 11, 23]. This has pushed Morocco to rethink its water policy and take into account the climatic specificities of Morocco by developing national skills and expertise in the field of feasibility studies of dam constructions, recognized worldwide. Today Morocco could export it to other countries wishing to take advantage of the Moroccan experience.

The efficient way of dealing with the exploitation of RE has made it a model to follow for electricity production from renewable sources and to fight against CC. Note that investment in RE projects is estimated at $30 billion between 2016 and 2030. The model concerned is the subject of international admiration and praise. Morocco is considered as a model for RE use for African countries. The result today is an ET that benefits the country's economy while putting it among the nations respecting their climate commitments [7].

MASEN has signed several cooperation agreements with African countries in order to support them in the production of electricity from RE. It also appointed partnerships with donors, in particular with the African Development Bank (ADB), to support African countries, in particular those in the Sahel region targeted by "Desert to Power" in development of RE technologies best suited to their climatic conditions and their specific energy needs. It also created a coalition for access to sustainable energy, jointly led by Morocco and Ethiopia, focusing on universal access to energy through the massive use of clean energy [7].

In addition to (on top of) the economic and environmental interest of the solar pumping, there is the industrial component which will make emerge a niche of local industry specialized in the manufacture and the installation of solar pumping equipments. It will also lead to the creation of energy service companies operating in the sector in order to perpetuate these techniques. On top of that, the job creation potential of the green economy is enormous [7].

The realization of wind farms will participate in the economic and social development through the involvement of the Moroccan firms (feasibility studies, electrical works, civil works, etc.), industrial integration (parts manufacturing of wind turbines, etc.), and creation of indirect jobs during the construction and direct jobs during the operational phase. Furthermore, the wind power generation will contribute to the Moroccan energy independence, reduce the purchase of fossil fuels, and consequently the currency savings.

Last but not least, the solar water heater development program will also contribute to the advancement stage of a local production sector, the creation of new activities, the jobs, thus allowing the emergence of a national market. Its objective is to install 1.7 million m^2 of solar panels by 2020, that is, a cumulative energy saving of around 100 Ktoe [7, 24].

5.2 Perspectives

5.2.1 Hydraulics

In Morocco, the gap between precipitation and the level of water collected is very large. The rainfall varies on average between 50 and 250 billion m^3. Morocco aims to reach a storage capacity of 30 billion rainwater and to build 35 dams by 2030 [1à].

In order to improve the technical and economic operation conditions of the electric production equipment and the transport national grid, ONE launched the project of the second energy transfer station by pumping (ETSP). It will be established at the level of Abdelmoumen dam located approximately 70 km northeast of Agadir [è].

With a power of 350 MW, this station will participate in meeting the country's electricity demand during peak periods as well as in optimizing the exploitation of the means of electric production, in particular those based on renewable energy. The project will be developed by ONE in the framework with immediate possession [7].

This project will also develop the region socially and economically, reduce CO_2, and conserve water resources as much as the STEP operates in a closed circuit and consumes practically no water.

5.2.2 Solar

MASEN is currently carrying out another gigantic solar power plant in Midelt with a capacity of 800MW and will be commissioned in 2022. It is a hybrid power plant: CSP and photovoltaic. The price per kWh is estimated at 0.68 Dh and the amount of CO_2 avoided is around 675 360 tons/year. In addition, there are also four other projects undertaken: Noor Midelt II (400 to 800 MW), Noor Tafilalet (120 MW), Noor PV (800 MW), and Noor Atlas (200 MW) [7].

Regarding ONE, it is carrying out the Noor Atlas photovoltaic project (120 MW). This project consists of seven photovoltaic power plants which will be located in the regions of Tan Tan, Tata, Boudnib, Ain beni Mathar, Outad el Haj, and Enjil. The year 2018 was marked, in particular, by the establishment of the environmental and social impact study and the launch of the prequalification relating to construction of these projects [7].

5.2.3 Wind

Under the stated objective by Morocco to reach 20% of its electric power installed from wind by 2030. Table 3 shows five other wind farms in progress with a total capacity of 1300MW. Three of them will be installed in south of Morocco. The concerned sites were selected in different windy regions across Morocco in order to provide better production and stability of the national electric grid. In addition, other projects are being considered for electricity production or for supplying sea water desalination [6–8].

Major cities in south of Morocco such as Tarfaya, Tan Tan, Laâyoune, etc. are connected to the national grid. Unfortunately, in 2020, Dakhla city is still electrified by a diesel power plant (37.5MW). To cope with the significant increase in electricity consumption for Dakhla region, a new diesel generator of 16.5 MW capacity is committed to be installed by ONE [6–8].

Increasing wind electricity production will contribute to the reduction in the share of coal and petroleum products imported, the two most polluting energy products. In addition, the wind energy use helps environmental protection by reducing CO_2 emission due to the use of petroleum products and coal, which limits CC such as floods, droughts, etc.

The Dakhla region that has not been connected to the national grid until now is supplied with electric power generated only by a power diesel plant. So, wind power can contribute to the electrification of this region by coupling it to the diesel plant.

Compared to solar energy, the electricity from wind and hydro power is twice less expensive. Installing more wind farms, the objective set by the Moroccan national energy strategy, namely the wind power share will be established at 14% of the installed national electric power capacity, is likely to be realized in 2030 [7].

Table 3 Wind farms under construction [6–8]

Sites	Number of turbines and manufacturer	Capacity (MW)	Commissioning	Operator/Owner
Oualidia I	11 WinWind	50.4	2020	Private
Safi	61 TBD	200	2021	Private
Aftissat I Boujdour	87 SGRE	200	2021	Private
(**)	370 SGRE	850	2020-2022	Public–Private
Total		***1300.4***		

(**) Midelt (180MW); Boujdour (300MW); Jbel Lahdid (200MW); Tiskrad (100MW); and Tangier II (70MW)

5.2.4 Electricity

The European electricity grid is the largest interconnected network in the world in terms of power transmitted. Since 1997, the Moroccan power grid is the only African network connected to the European one. Morocco's planning is to be a regional center for the export of competitive clean electricity to its neighbors, Spain and Algeria (Fig. 11). In addition to the project of electrical connection with Mauritania through the cities of Laâyoune and Dakhla, in 2018, Morocco and Portugal have agreed about the construction of a submarine electrical cable connecting the two countries of 1000MW capacity. Thus, technoeconomic feasibility studies are undertaken in the ground. In 2019, Spain and Morocco decided to increase the electric energy capacity exchanged between them. RE transition can then enable a regional renewable electricity sharing and electrical systems can be utilized on the regional level [4].

On the top of that, Mauritania shares with Morocco a similar wind resource mainly for the coastal area. So, there is possibility to have some interesting social and economic development in this region once the electric interconnection is carried out.

5.3 *Challenges*

Due to climate change (CC), Morocco is suffering from extreme phenomenon: water stress, recurrent droughts, and floods. To secure its water resources, Morocco must have a strong will so that the construction rate of dams accelerates.

According to the World Resources Institute (WRI), an American think tank specializing in environmental issues, the disruption of water capacity in Morocco is due to increased demand, mismanagement, and CC. The most marked decline is that of the second largest dam, Al Massira (Fig. 10), whose capacity decreased by 60% in 3 years due to recurrent drought, increasing irrigation, and water requirements for the nearby big cities like Casablanca. Despite recent rain, the WRI said water is now at the lowest level in a decade [23] (Fig. 12).

Electricity demand is soaring in Morocco due to economic development and rising living standard. With few resource options, mainly coal use, it has lofty ambitions for developing RE to offset acute power shortages, especially in summer months. After all power generation projects and structural reforms to support them are pushing forward in morocco, in many cases RE is the best solution, particularly tapping into solar and wind resources. Figure 13 shows that despite the RE contribution increased by 100% between 2001 and 2018, the coal contribution also increased by more than 40%.

The Moroccan Conjuncture Center (MCC) believes that several technical, economic, and social challenges will have to be mastered in order to succeed ET. Especially since today, the energy supply relies mainly on fossil fuels which represent more than 80% of the primary energy consumed in the world and put increasing pressure on the environment [25].

The MCC observes that the institutional energetic architecture adopted in Morocco remains very concentrated. It is criticized by several actors who claim a decentralization of energy skills in order to reproduce the German success. The German model includes more easily participatory initiatives and strengthens the participation of municipalities and cities. But, also to fight against climate change is to take action in eliminating fossil fuels and committing to a 100% RE future [25].

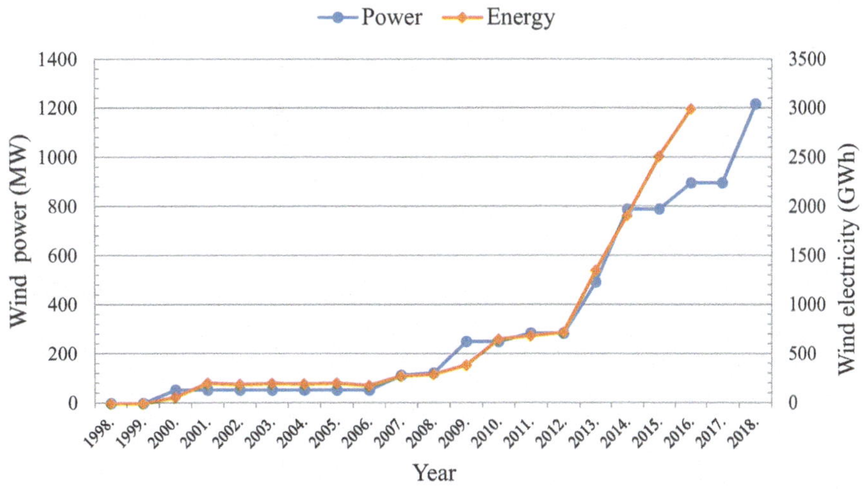

Fig. 10 Evolution of the wind power and energy [6–9, 15, 16]

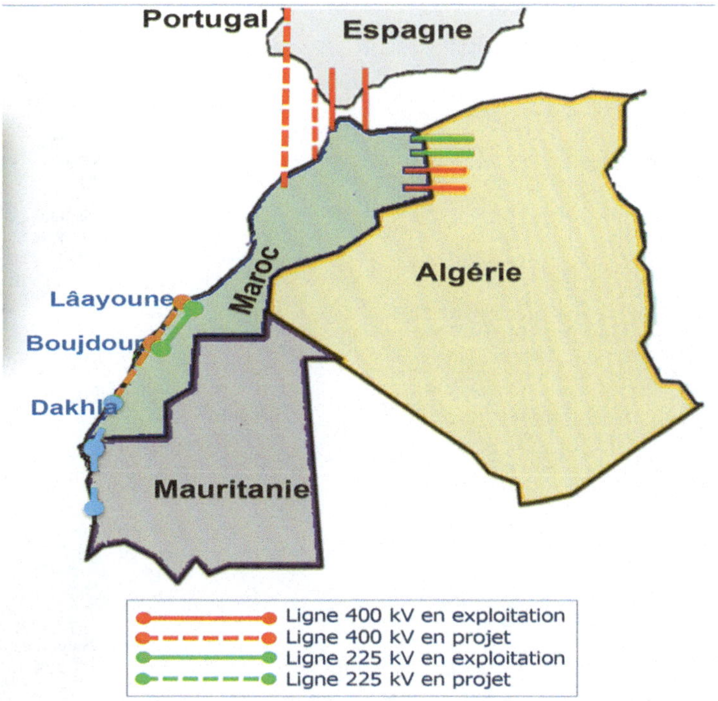

Fig. 11 Electrical interconnection [9]

6 Energy Transition (ET) and Covid-19

6.1 Health, Economic, and Social Crises

During the confining, the health crisis was the number one priority. But during the lock in period, the countries affected by the epidemic began to take a serious look at the economic and social crisis resulting from the health crisis.

The countries of the world are seeking to restore economic activity through austerity measures whose severity varies from country to country. However, the objective is to get out of sanitary containment procedures and boost economic activities with the adoption of precautionary measures to avoid the epidemic virus recession.

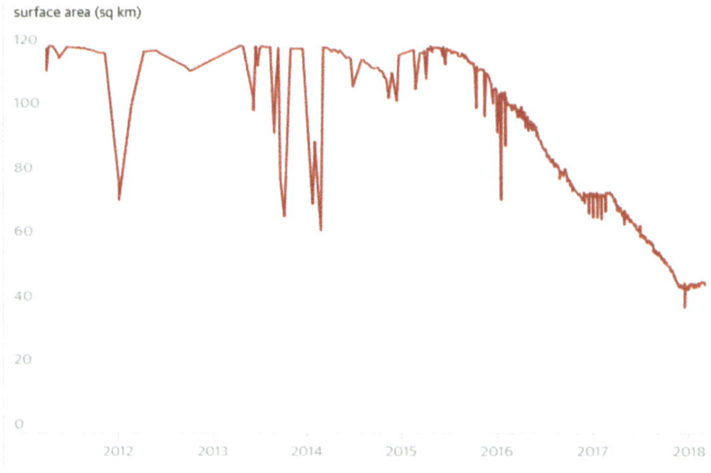

Fig. 12 Starkest decline of AL Massira dam [23]

Fig. 13 Net electricity consumption in 2011 (**a**) and 2018 (**b**) [7]

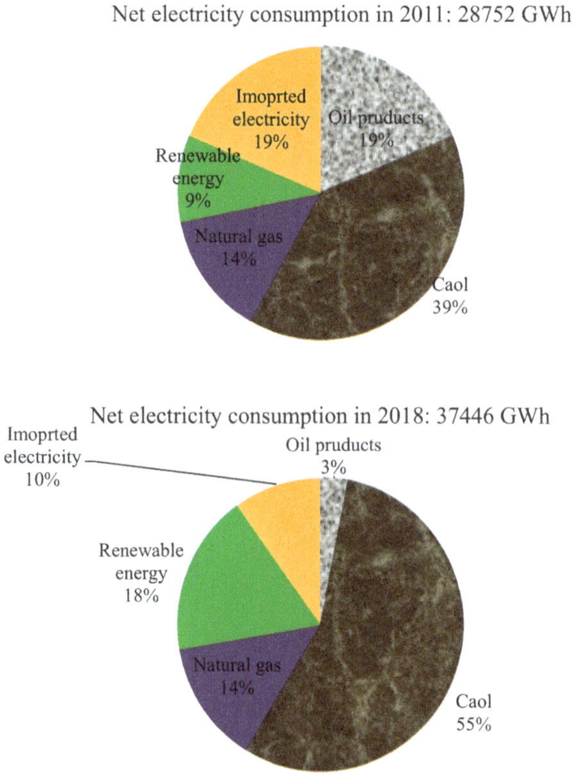

However, this restoration faces many challenges given the continued spread of the virus, even at a low rate, and the lack of a vaccine or efficacious treatment.

The decision to shut down certain economic sectors and the restraint of others depends on the contribution of these sectors to gross domestic product (GDP) and on the extent of the direct impact of such sector on the deterioration of economic indicators significantly, in order to avoid to reach the stage of depression.

The disruption or shutdown of a sector that employs a large workforce can accelerate the transition from recession to depression, which will affect other economic sectors, including the healthy one, and what all countries are trying to avoid

For the American philosopher Noam Chomsky, despite the gravity of this epidemic, the greatest danger is CC. We will come out of the epidemic at a very high price, but we will never recover from the melting of the polar ice caps at the poles, rising sea levels, and other negative effects due to CC.

Guterres, the Secretary of the United Nations (UN), in a videoconference on the financing for sustainable development, warned that the crisis caused by the emergence of the Corona virus could cause a loss of 8.5 trillion dollars for the worldwide production. It is the biggest economic recession since the Great Depression in the thirties of the last century [20].

According to the World Health Organization (WHO), since January 2020, the world has been facing a deteriorate crisis due to the emerging Corona virus epidemic. It has caused the collapse of world trade, the accelerated decline in energy markets, and huge losses in many sectors of the economy, particularly transport, tourism, and spare time activities [26].

6.2 *Impacts of the Covid-19 Crisis on Energy*

Because of Covid-19 and according to the IAE, the world energy demand should drop by 6% in 2020, the biggest shock since the oil shock of 1973. On the other hand, IAE estimates that the total global use of RE will rise by about 1% and the renewable electricity generation will rise by nearly 5% in 2020. Global CO_2 emissions are expected to plunge by almost 8% this year, the largest decline in history. The energy industry emerging from the crisis will be different from what happened before, the governments can direct the world toward a more resilient and sustainable energy, the key to the economic recovery plans [1].

According to Taherian, clearly, the CC will have headwinds and the Covid-19 will slow some green activities. Low oil prices make green energy less competitive, some investors will defer their investments from green technology. Still a question of funding for multiplying new green projects. To achieve a true alternative green economy, many of the new green technologies need to scale up. Much of the green funding is for increasing scaling up technologies that are already working and being profitable like solar or wind [27].

Despite the economic recession and the drop in global energy demand, its consumption, in particular, electricity, will gradually increase in the medium term, and in the short term depending on the rhythm of the development of the epidemic and the resumption of economic growth.

6.3 *Green and Sustainable Economy in the Service of Healthy and Resilient Future*

Countries affected by the epidemic will change their annual budgets to austerity ones, thereby reducing public investment to boost their national economy. However, the epidemic crisis offers a precious opportunity for political decision makers to initiate important and structural reforms aimed at reviving the economy. It is also a favorable occasion to accelerate the transition to a green economy, by investing in profitable economic activities, such as digital technologies, renewable energies, etc., creating more new jobs.

Guterres, the President of the UN, during a videoconference on the financing for sustainable development, underlined the need for unity and solidarity worldwide to face the health and economic crises He emphasized that all our efforts must go toward building sustainable and flexible ways that allow us not only to overcome Covid-19 but also to face the climate crisis, reduce inequalities, and eliminate poverty and hunger [20]

Also, on the occasion of the International Nourishment Earth Day, he mentioned that the Covid-19 pandemic is the greatest ordeal that the world has known since the Second World War. We must join forces to save lives and soften the terrible economic and social consequences of this crisis. But there is another problem that requires our urgent attention, the environmental crisis affecting our planet. Biodiversity is declining sharply. CC is such that we are approaching the point of no return. We must take a decisive action not only to protect our planet from the Corona virus but also from CC that threatens the existence of humanity itself. The following are among the climate-related actions he proposed [28]:

- The huge sums that are being spent to overcome the pandemic must also create green jobs and new businesses in the framework of a clean and ecological transition to ensure sustainable growth and favor the transition from a gray economy to a green one.
- Public funds must be invested in the future, not in the past, and be allocated to sustainable sectors and projects which contribute to the protection of the environment. Polluters must start paying for the pollution for which they are responsible.

These principles, he said, can help better overcome this crisis together. The GHGs, like the virus, do not respect national borders. He insisted that the solution is in the union of efforts to build a healthy and resilient future for humanity as well as for the planet.

In a video statement, in the opinion of Von Der Leyen, the European Commission President, "European Green Deal" could be for Europe a motor for its recovery. She said: “As we now plan to slowly go back to work and to invest billions of Euros to restart our economy, we should avoid falling back in old, polluting habits,” “We can make our society and our planet healthier by investing in renewable energy, by driving clean cars, by renovating our houses and making them energy efficient, by buying sustainable food, reusing materials rather than throwing them away or producing low carbon steel," and she added “That is the essence of the European Green Deal” [21].

Besides, Timmermans, the executive Vice-President of the commission, on the European Climate Law, believes that the "European Green Deal" would be a lifeline to get out of the corona virus crisis and it would be crucial as Europe would lose out twice if it mobilized investment to restore the old economy before it made it green and sustainable [20].

6.4 Energy Transition in Morocco and Covid-19

Oil is a part of both local and international stakes, especially for non-oil-producing countries, such as the Morocco case. We know that the energy bill represents the big Moroccan trade deficit. In addition, energy is a key element both of competitiveness and of the sovereignty of a country.

For this reason, most of the developed countries in transition to a postepidemic world will try to be advantageously positioned in Africa. In addition, being a large consumer market, Africa is rich in primary resources, plus its huge investment needs to achieve its economic and social growth and development. For this justification and in the light of the security, political, and economic challenges facing the African continent, it is appropriate and in the interest of Morocco to bet on scientific research in renewable energy and the investment in the green economy, to ensure sustainable development, by creating more opportunities for new jobs linked to industrialization and exploitation.

ET in Morocco is necessary and profitable, not only for environmental but also for economic reasons because it is a non-oil-producing country. According to the World Economic Forum, Morocco is at the top of the list for the Arab and African TE index for the year 2020, thanks to its energy strategy adopted in 2009 and setting up of the accompanying legislative measures to generate clean electricity on a large scale.

Due to its geographic proximity to the European Union, which adopts an energy policy based on alternative energies, Morocco can benefit from it by improving its experience in this domain so that it becomes one of the main sources of clean energy, given its significant solar and wind potential available throughout the year.

There are other reasons why actions on green investments for Morocco are crucial during the economic crisis:

- Wind energy cost has become competitive without subsidies, nor taxes or ecological taxes on oil.
- Alternative energy produced locally are more resilient to global supply disruptions.
- Green economy and smart sector integration create more new jobs.

Morocco could bounce back better from this pandemic because it has made renewable energy the cornerstone of its ET. By using the ET as an objective, Morocco could turn the crisis of this pandemic into an opportunity to rebuild its economy differently and make it more resilient.

Last but not the least, the world after the Covid-19 crisis will be resolutely more ecological, Morocco commits to it.

7 Conclusion

Controlling energy sources, the main motor of economic and social growth, is a priority objective for all countries in the world. More particularly for Morocco that is strongly dependent on foreign imports for fossil fuels, which is around 93% in 2019.

Linked to its industrial activity, electricity consumption has increased on average by 5.7% since 1996. In 2009, Morocco defined a new energy strategy. It aims to secure its energy supply by diversifying its energy sources from developing renewable energies and accelerating the rhythm of oil and gas exploration.

Morocco is characterized by an unequal distribution of water resources over time. It has built more than 145 large dams with a storage capacity of 18.67 billion m^3. It has also developed great expertise in dam design and construction, recognized worldwide.

As far as renewable energies (RE) are concerned, Morocco benefits from a strategic location in the north of Africa and is located only 14 kilometers from Europe, with varied relief and 3,500 km of coastline. Fortunately, Morocco has significant solar and wind potential. In Ouarzazate, for example, the average annual daily sunshine ranges from 7.9 to 11.2 hours/day. Concerning the wind potential, the average annual wind speed is at 10 meters; for Tétouan in the north, it could reach 11 m/s; but for Dakhla in the south, it could exceed 8 m/s.

Morocco has set itself the objective of bringing the contribution of RE to the installed electrical power to 52% by 2030. In 2020, solar energy is 962 MW. It is one of the first African investors in RE sector and the first user of solar energy. For "Noor-Ouarzazate" (582MW), one of the largest solar power plants in the world, the price per KWh reached $ 0.19, a competitive price compared with the fossil fuels. Regarding wind energy, Morocco (1360 MW) is the second after South Africa (2295 MW).

RE is a promising alternative economic sector. The share of renewable energies in the installed electricity mix is 35% in 2019. Many international organizations regard Morocco as a pioneering African reference, thanks to its experience and achievements in the RE sector.

At the United Nations meeting on climate change (CC) held in Marrakech in 2016, Morocco committed to reducing its greenhouse gas (GHG) emissions by 32% before 2030, by a smart digital integration of energy-efficient equipment and devices. Energy efficiency (EE) is one of the keys for achieving this goal. It is also one of the missions of the Moroccan Agency for Energy Efficiency (AMEE). AMEE has identified an energy saving potential of 25% before 2030, all economic sectors combined, by providing the same service and remaining competitive.

In the globalization context and the constant search for continuously increasing growth, most countries are looking to develop their income-generating economic sectors, notably the industrial sector, by focusing mainly and vitally on energy sources capable of meeting their energy consumption needs with the lowest cost. Currently, this is far from what renewable energies offer as an alternative to fossil energy.

Regarding the geopolitacal dimension of energy in Morocco, if energy constitutes 3% of the Moroccan gross domestic product (GDP), then 97% depends entirely on it. This means that without stable energy security, the problem risk of the Morocco's economy fragility remains. This energy imbalance is one of the reasons for the Moroccan economy weakness. The energy model to be adopted by Morocco must first aim at reducing its energy dependence without giving up the fight against CC.

The energy transition (ET) in Morocco is necessary and profitable, not only for environmental but also for economic reasons because Morocco is a non-oil-producing country. According to the World Economic Forum (WEF), Morocco tops the list of the Arab and African ET index for 2020, thanks to its energy strategy and the implementation of accompanying legislative measures to produce clean electricity.

According to Rifkin, an American economist, a prospective specialist and ardent defender of smart electric grids, "if we set up the energy Internet, it means that all humanity will produce its own energy locally from the solar and wind and shared through all the continents by giving up fossil fuels. In addition, the establishment of the transport internet means that we could eventually share electric vehicles and reduce our ecological imprint. The new society at zero marginal cost is the only economic plan, I know, capable of responding in time to global warming, whose consequences would be terrible for humanity. This ET must be accomplished quickly, with skill and determination if we want to provide a viable future for the future generations."

Once, the pandemic Covid-19 health crisis has affected the majority of the world countries. During the confinement, the health crisis had priority over the economic crisis. In a teleconference, Guterres, the UN Secretary General, addressed the world and warned that the crisis caused by Covid-19 could result in a loss of $ 8.5 trillion for many sectors of the global economy, particularly energy, transport, and tourism.

Because of the Covid-19 and according to the International Energy Agency (IAE), the world energy demand should drop by 6% in 2020, the biggest oil shock since the 1973 one. On the other hand, the IAE estimates that renewable electricity production will increase by almost 5% and the global CO_2 emissions are expected to drop by almost 8% in 2020, the largest decline in history.

The fight against CC will take more time because Covid-19 will slow down some green activities; oil with a low price makes green energy less competitive. Some investors will defer their investments from green technology. Much of the green funding will be for scaling up the technologies that are already operating and profitable such as solar or wind.

Guterres suggested that the huge sums of money which would be spent to overcome the Covid-19 pandemic must create new businesses as part of a clean and ecological transition to a green and sustainable economy. GHGs, like the virus, do not respect national borders. He stressed that the solution resides in building a healthy and resilient future for humanity as well as for the planet.

Oil is a part for both local and international stakes, especially for non-oil-producing countries, such as Morocco. At the same time, energy is a key element of a country's competitiveness and sovereignty. Most of the developed countries in transition to a postepidemic world will try to position themselves advantageously in Africa. In addition, being a large consumer market, Africa is rich in primary resources. Energy is at the top of its priority pyramid. It is an essential lever for its development. For this reason and in the light of the political and economic security challenges facing Africa, it is appropriate and in the interest of Morocco to bet on scientific research in renewable energies and investment in the green economy.

The wind energy cost has become competitive with fossil fuel, without subsidies, taxes, or environmental taxes on oil. The alternative energies locally produced are more resilient to the global supply disruptions. The green economy and the integration of smart digitalization create more jobs.

Once Covid-19 has started to spread rapidly in Europe and the World Health Organization (WHO) has declared the alert on the virus, Morocco also announced a state of health emergency with the closing of its borders for 3 months which caused major damage for certain economic sectors, particularly tourism and transport, and led to an economic recession. Morocco could bounce back better from this pandemic because it has made renewable energies the cornerstone of its TE.

Faced with these challenges, there are key factors to boost research and development, such as revising national priorities (R&D roadmap for RE with an orientation law) and setting up the infrastructure for a national smart grid.

Today, Morocco is faced with a decisive historical challenge which is the necessity of winning the bet of modernization as well as scientific and intellectual progress. It is required to double its efforts to reach it, so that it fills the gap that separates it from the West in this domain. It has no choice but to be up to date to keep rhythm with the scientific revolution of the West and the profound technological transformations that the modern world is experiencing in terms of digital culture and computer sciences.

The world after the Covid-19 crisis will be resolutely more ecological, Morocco is committed to it.

References

1. https://iea.li/3d5KVHf
2. Rifkin J (2019) The Green New Deal: Why the Fossil Fuel Civilization Will Collapse by 2028, and the Bold Economic Plan to Save Life on Earth, Griffin; Édition.
3. Rifkin J (2015) The Zero Marginal Cost Society: The Internet of Things, the Collaborative Commons, and the Eclipse of Capitalism, Griffin.; Édition.
4. (2016) Transition énergétique: Opportunités économiques et environnementales pour le monde arabe, www.imarabe.org
5. Stéta A (2014) Le mystère de la valeur, Revue des deux mondes, asteta@hotmail.fr
6. Annuaires statistiques du Maroc (1988-2017). https://www.hcp.ma/Direction-de-la-statistique_a716.html.
7. www.masen.gov.ma; www.one.gov.ma; www.mem.gov.ma; www.amee.ma; Direction de la Météorologie Nationale (DMN). http://www.marocmeteo.ma/
8. Morocco (2014) Energy policies. Beyond IEA countries, https://www.iea.org/publications/freepublications/publication/Morocco2014.pdf
9. Nfaoui H. (2019) Cite as: New horizon of renewable energy in Morocco and Africa - Renewable Energy and Sustainable Buildings, pp 551-565, Editor: Sayigh Ali, Springer International Publishing AG
10. Bahraoui-Buret J., Bargach M. N. and Ben Kaddour M. L. (1983) Le gisement solaire marocain, Editions SMER
11. Nfaoui H (2017) What will remain of the climate challenge conference: the fate of the planet between Paris and Marrakech, http://www.renewableenergyworld.com/articles
12. Nfaoui H (1986) Trois années de mesures d'ensoleillement à Rabat: étude statistique et présentation des données Thèse de 3ème cycle, Faculté des Science de Rabat
13. Buret-Bahraoui J (1986) Caractéristiques du gisement solaire marocain, Etude spécifique du site de Rabat, Thèse de Doctorat d'Etat, Faculté des Sciences de Rabat
14. Knidri F, Laaouina A (1986) L'Energie Eolienne au Maroc. CDER, Marrakech
15. Nfaoui H(2020) Wind Energy in Morocco, Resources, Potential and Progress, Chapter of a book: "Wind Energy- Progress and Future Directions from a Global Perspective," Ali Sayigh and David Milborrow Editors, Springer International Publishing AG
16. Nfaoui H (2004) Caractéristiques du gisement éolien marocain et optimisation d'un système aérogénérateur/groupe électrogène pour l'électrification des villages isolés, Thèse de Doctorat d'Etat, Faculté des Sciences de Rabat
17. Enzili M, Nayssa A, Affani F (1995) Le Gisement Eolien Marocain, le Centre Développement des Energies Renouvelables (CDER), Marrakech.
18. www.weforum.org
19. www.cg.gov.ma
20. António Guterres (2020) Impacts of the Coronavirus pandemic: COVID-19. https://www.un.org
21. https://docplayer.fr/51193702-La-politique-et-les-bienfaits-des-barrages-au-maroc-presentee-par-mr-khalid-el-ghomari-directeur-des-amenagements-hydrauliques.html.
22. https://www.theguardian.com/global-development/2016/nov/17/cop22-host-morocco-lights-way-africa-renewable-energy-2020
23. https://www.theguardian.com/environment/2018/apr/11/day-zero-water-crises-spain-morocco-india-and-irak-at-risk-as-dams-shrink.
24. Enzili M(2008) Renewable energy, energy efficiency in Morocco, situation and prospects. Cairo (Egypt). http://www.needs-project.org.
25. https://cmconjoncture.com/
26. www.who.int
27. Taherian Z (2020) Green Deal is the 'Motor For The Recovery' for Europe, https://www.forbes.com/sites/suzytaherian/2020/05/08/green-deal-is-the-motor-for-the-recovery-for-europe/#19f597c82b83
28. https://ec.europa.eu/

Climate Change Adaptation Amid COVID-19: Sustainable Transport Guidelines to Mitigate CO_2 Emissions, Enhance Livability in Cities, and Attain SDG 7, 11, 12, and 13

Mai M. Barakat and Mohsen M. Aboulnaga

1 Introduction

The world is witnessing a huge breakdown in the national economy and environmental balance. These perspectives have been announced according to the disasters caused by climate change and the spread of dangerous infections related to it as well as the spread of Coronavirus (COVID-19) pandemic and crisis in 2020 [1]. Some precautions to these demonstrations led to a conflict in the quality of human life [2]. Transport is one of the key sectors that contribute to climate change (CC) impacts that is increasing and rapidly manifested in cities worldwide [3]. Global CO_2 emissions from transport are estimated to be about 30% of the total generated emissions in cities [4]. Appropriate measures are not implemented at the same pace and are not effective enough to mitigate such impacts to meet Paris 2015 Climate Agreement targets [4]. Currently, there are about 750 million vehicles in the world that are powered by petroleum-based fuel and predicted to be doubled in the next 30 years due to the spiking urban population increase in cities by 2050 [5]. Hence, more actions should be taken to adapt and overcome these challenges.

One of these measures or actions is prompting sustainable transport in cities that are clean and effective in mitigating CO_2 emissions and reducing air pollution [6]. Among the neglected factors that directly affect CC is transport which has a direct link that is related to the international economy. Transport and transportation modes are key factors that affect sustainability and ensure livability (better quality of life) in cities as well as attain sustainable development goals (SDGs). Sustainable development (SD) seeks better social development, environmental protection, and economic efficiency reflecting a better life as well as livable and balanced community [7].

Relative to congestion, people are spending extra time commuting between their residence and workplace though this has been dramatically reduced since February 2020 due to COVID-19 pandemic. An important factor behind this trend is related to residential affordability as housing located further away from central areas (where most of the employment remains) is more affordable. Therefore, commuters are exchanging commuting time for housing affordability. However, long commuting is linked with several social problems, such as isolation (less time spent with family or friends) as well as poorer health (obesity). Time spent during commuting is at the expense of other economic and social activities. Information technologies have, however, allowed commuters to perform a variety of tasks while traveling [8].

Some challenges related to urban development take place when urban transport systems, for a variety of reasons, cannot adequately satisfy the requirements of urban mobility. These challenges are summarized in traffic congestion, insufficient parking spaces, public transport inadequacy, and communication difficulty as well as loss of public spaces, impediment for non-motorized transport, environmental impacts and energy consumption, high infrastructure maintenance costs, and safety [8, 9]

The Coronavirus (COVID-19) pandemic is causing huge impacts on the world's economy including transport and transportation in cities all over the world since February 2020 [9, 10]. Enacted government policies on lockdown and citizens' confinement for movements (land, sea, and air) are having immense impacts on public and private transport that were halted for more than 100 days in most of the global cities worldwide due to the lockdown. According to the Mobility Market

M. M. Barakat
Department of Architecture, Faculty of Engineering, Misr University for Science and Technology – MUST, Giza, Egypt

M. M. Aboulnaga (✉)
Department of Architecture, Faculty of Engineering, Cairo University, Giza, Egypt
e-mail: maboulnaga@eng.cu.edu.eg

© The Author(s), under exclusive license to Springer Nature Switzerland AG 2022

A. Sayigh (ed.), *Sustainable Energy Development and Innovation*, Innovative Renewable Energy,
https://doi.org/10.1007/978-3-030-76221-6_65

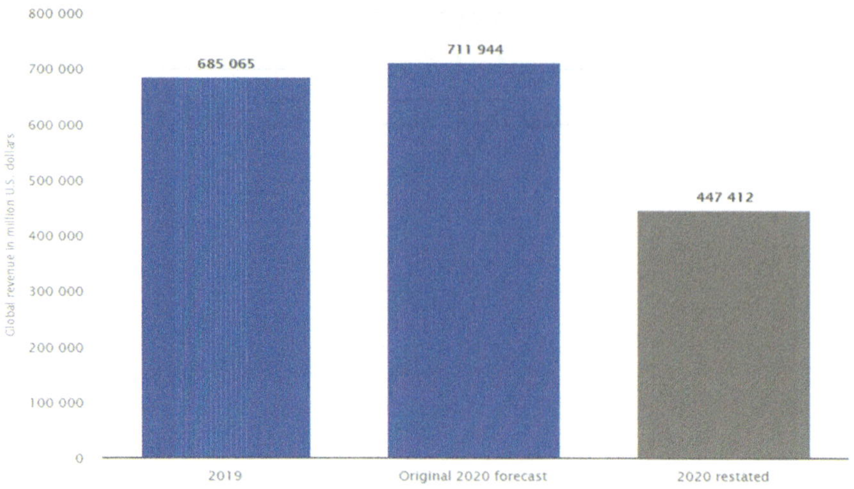

Fig. 1 Forecasted change in revenue from the travel and tourism industry due to the Coronavirus (COVID-19) pandemic worldwide (2019–2020). (Image source: https://www.statista.com/forecasts/1103426/covid-19-revenue-travel-tourism-industry-forecast)

Outlook on COVID-19, the global revenue for the travel and tourism industry is estimated at 447.4 billion U.S. dollars in 2020 – a decrease of around 34.7% from the previous year. Additionally, this is significantly lower than the original 2020 forecast of around 712 billion US$. Figure 1 shows the revenue of transport between 2019 and 2020 due to COVID-19 [11–13].

Amid such global and national challenges, sustainable and clean actions specifically in transport and transportation create healthier, smart, resilient, and sustainable cities. This paper presents a study conducted on the importance of sustainable transport design and how it reflects on the economy regarding its impact on CC and how can applying the categories and dimensions of sustainable development (environmental, social, and economic) strategies affect its development.

2 Objectives

This study aims at proposing a new environment and social safeguards that help in mitigating CC by assessing pollution types and their effects. In addition to how it would reflect on the global economy and CC. Transforming transport to be clean and low-carbon ought to be a priority through direct and easy daily routines integrating action plans with SDG 3, 7, 9, 11, 12, and 13. Figure 2 shows the global objectives of transportation. In this study, the focus is on sub-objectives 1 and 2 shown in Fig. 2.

3 Challenges and Aspects of Sustainable Transport

It has been indicated that *'the goal of sustainable transportation is to ensure that environmental, social, and economic considerations are factored into decisions affecting transportation activity'*. Besides, it stated that *'Sustainable transport goal is all about factoring environment, Social and economic considerations into decisions that will affect transport activity'* [14]. Transportation progress involves improving all existing models, not just the newest model.

Nowadays, in many cities, the most beneficial strategies may involve improving walking and cycling, more support for public transit, and restricting automobile travel in congested urban areas. This does not confirm that sustainable transport should mean faster travel or more mileage, sustainable may seek for more safety and comfort, provide cost savings, or even reduce the total need for travel. Figure 3 summarizes the effect of transport on sustainability.

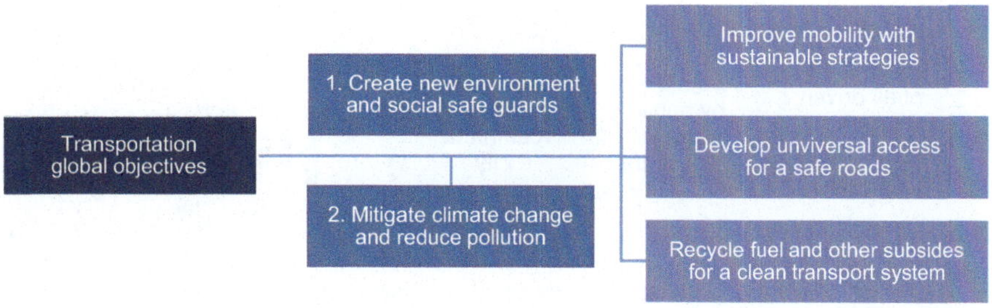

Fig. 2 Transportation global objectives. (Source: Developed by authors)

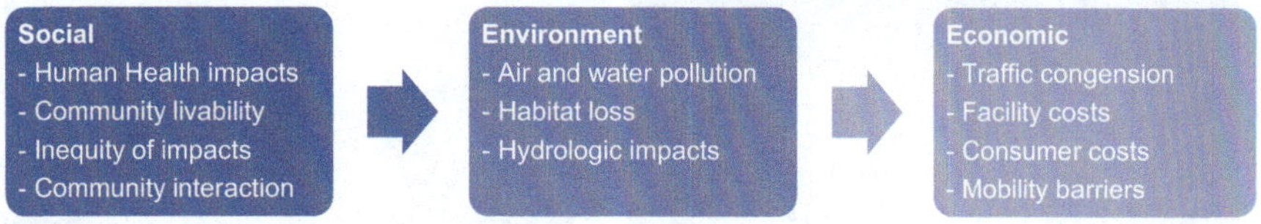

Fig. 3 Transportation influences on sustainability

4 Sustainable Transport

According to the United Nations, clean fuels and clean energy based on advanced technology is of high priority, especially concerning the current scale of the health and climate challenges in cities [15]. The provision of services and infrastructure for the mobility of people and goods – forced and encouraged economic and social development to benefit current and future generations – in a manner that is accessible, safe, affordable, efficient, and resilient as well as mitigate carbon emissions and lessen environmental impacts [16]. The key is meeting the needs of people in their personal and economic lives while respecting the ability of future generations to meet their needs: the essence of sustainable development. Transport is responsible for 23% of global energy-related greenhouse gas emissions, so it would be impossible to address CC without promoting sustainable transport [17].

Transport planners must work in a robust concert with several other sectors to ensure sustainable and clean transport for green and resilient cities. This includes many issues such as energy, technology, and mobility as well as accessibility, affordability, connectivity, and health. For example, Copenhagen's long-term vision is centered on three main plans: a) having at least one-third of all driven traffic in the city should be made by bicycle, b) at least one-third by public transport, and c) not more than one-third by car (Fig. 4) [18]. In Dubai – UAE, the Roads and Transport Authority (RTA) is targeting the share of public transport trips to be 20% of total trips in 2020 and 30% in 2030 – a significant increase from the public transport modal share of 15% of 2015 as shown in Fig. 5 [19].

Box 1 Clean energy and clean transport
Clean fuels and clean energy is a high priority, and when considering the scale of the health and climate challenges, it is imperative.

United Nations, 2016.

In Casablanca, Morocco, the electrical tram has been in operation since 2015 [20] (Fig. 6). Targets can also be set at the national level to guide local mobility planning. For example, Malaysia has set a goal to achieve a public transport target of 40% of all trips in urban areas by 2030 [21, 22], as shown in Fig. 7. Progress in greener transport will be considered meaningful only with concurrent progress in alternative energy production, and as transport technologies advance, the policy frameworks need to adapt and respond to the new innovative technology landscape. Figures 8, 9, 10, and 11 illustrates the sustainable transport modals and aspects [23].

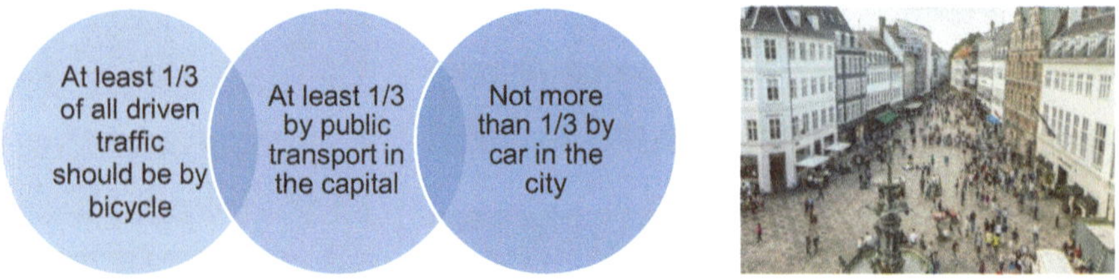

Fig. 4 Copenhagen's long-term vision on sustainable transport. (Image source: https://www.theguardian.com/cities/2015/apr/02/where-is-the-worlds-greenest-city-ecofriendly)

(a) Dubai Metro lines

(b) New electrical tram in Dubai, UAE

Fig. 5 Dubai vision on sustainable transportation. (**a**) Dubai metro lines. (Image credit: Authors (Aboulnaga)). (**b**) New electrical tram in Dubai, UAE. (Image source: http://gulfnews.com/polopoly_fs/1.1408579!/image/3073944895.jpg_gen/derivatives/box_475/3073944895.jpg)

(a) Clean Transport

(b) Electrical tram

(c) passengers' mobility

Fig. 6 Public and clean transportation in Casablanca, Morocco, (**a**) Clean transport. (**b**) Electrical tram. (**c**) Passengers' mobility**.** (Images' Credit: Dr. Mohsen Aboulnaga 2016)

(a) Speedy low-emitting trains for BRT

(b) Enviro 500 buses

(c) Electric buses

Fig. 7 Clean public transportation modal in Malaysia. (**a**) Speedy low-emitting trains. (Images source: https://www.nst.com.my/news/nation/2020/01/555246/good-public-transport-still-distant-dream), (**b**) Enviro 500 buses. ((Images source: https://www.thestar.com.my/opinion/letters/2019/04/24/grouses-of-bus-companies)**,** (**c**) Electric buses for BRT. (Images source: https://www.sciencedirect.com/science/article/pii/S1878029615002686)

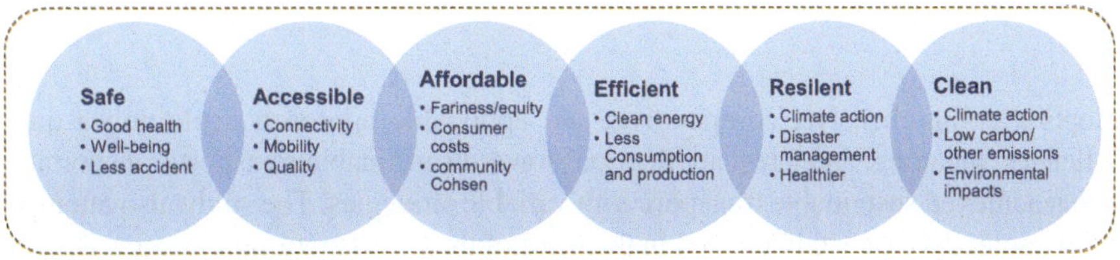

Fig. 8 Aspects of sustainable transport. (Source: Developed after Litman and Burwell)

Fig. 9 Clean transportation in European cities . (**a**) Rental bicycles in Brussels, Belgium. (**b**) Rental bicycles in London, United Kingdom. (**c**) Bicycles in Linköping, Sweden. (Images' credit: Dr. Mohsen Aboulnaga 2010–2011)

Fig. 10 Clean transportation in Barcelona, Spain. (**a**) Rental bicycles. (**b**) Bicycle lanes. (**c**) Motorcycles. (**d**) Electrical cars. (Images credit: Dr. Mohsen Aboulnaga 2017)

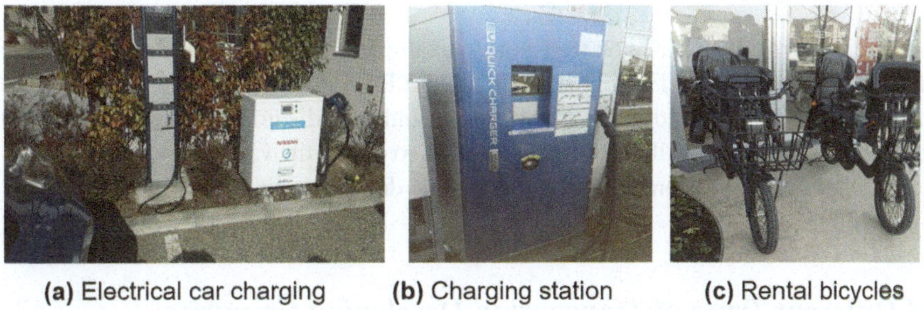

Fig. 11 Clean transportation in Tokyo metropolitan area, Japan. (**a**) Electrical car charging. (**b**) Charging station. (**c**) Rental bicycles. (Images' credit: Dr. Mohsen Aboulnaga 2018)

5 Methodology

The methodology adopts a comprehensive, comparative, and applicable approach exploiting a qualitative assessment to underestimate the challenges, meaning, and latest applications toward sustainable transport/transportation. The study presents different case studies that achieved sustainable transport with variable strategies. The study also attempts to apply the findings and recommendations in a selected case study located in the heart of the Cairo metropolitan area. Also, some guidelines ought to be deduced from the assessment of the selected case study to envisage the economic, social, and economic aspects of sustainable development and related SDGs.

6 International Successful Case Studies of Sustainable Transportation

Communities, cities, schools and universities, and business companies are rethinking and redesigning the way they transport people and goods. The COVID-19 pandemic also forced governments and citizens globally to rethink using less transportation and working from home [24]. These approaches are helping to a significant reduction of greenhouse gas (GHG) emissions and improving air pollution, and less traffic congestion in cities [25]. Such actions are also strengthening communities through interaction (only families) and improving individual health through more active living. Table 1 presents various case studies on how these examples succeeded in achieving sustainability in transport in different selected cities. The three case studies were selected according to their similarities to the applied case study in addition to the availability of the data. Copenhagen is considered one of the greatest European cities that use cycling instead of cars daily as a way to minimize GHG emissions, whereas, Moscow has developed 148 new routes. In 2018 congestion was decreased by 25% and still decreasing till now. Concerning India, the federal government has shortlisted the establishment of '109 smart cities' in India. The key idea of smart cities is the alliance of public services with an integrated public transport system. All the three selected case studies share the same huge number of population, the deflected roads, and the restricted policies [26, 27].

7 Applied Case Study – Period and Data

The applied case study is located in the heart of downtown Cairo, where it connects 26th July Street with Cornice el-Nile, exactly around Maspero triangle [28]. The site is easily reached through the 15th of May or October Bridge, 30 minutes from October city, and 45 minutes from new Cairo as shown in Fig.12 (b). It is situated in the west Cairo Bulaq district overlooking the Nile River. The site went through economic and upgrading transformation as spaces of capital accumulation and central business districts [29, 30]. The study was conducted in 2018 before demolishing of the area due to an upgrading plan. The study was carried on the selected site–Maspero due to the unique location and historical value. Figure 12 shows the exact location of study site [30, 31].

7.1 Transport Sustainable Indicators' Framework

In this assessment, the focus is mainly directed on transport to examine the three dimensions: economic, social, and environmental initiatives of the European Union (EU), The World Bank (TWB), and other international organizations [35]. More than 40 indicators have been used in the study with 11 categories and 41 indicators selected and integrated into the following framework structure as shown in Table 2. All presented indicators to detect the variables and challenges needed to be applied to achieve sustainable transport with the constraints of sustainable development goals [36–38].

7.2 Statistics of Different Types of Transport in Developing Countries

Regardless of the technology and welfare of occupants in developing countries, the GHG emissions are nearly double that in the developed countries [39]. It was discovered that when lifestyle and welfare improved, the number of vehicles were increased, consequently led to less shifting to public transport and higher GHG emissions, specifically in developing countries. Figures 13 and 14 illustrates the results of this study [40].

Table 1 Sustainable transport of selected case studies

Case study	Copenhagen, Denmark	Moscow, Russia	Delhi, India
Description	2025 World's best city for cyclists 2025 Eco-metropolis of the world Green accounts publications for raising awareness 	A comparative and integrated set of policies to reduce the number of cars by 25% New gridlock (pavement) for most of the streets 	Roads' network covers 210 kilometers and is about 325 km long in 2017 Delhi Metro carries 2.8 million passengers daily replacing 400,000 vehicles 
	Images' source: https://sustainabledevelopment.un.org/index.php?page=view&type=400&nr=2375&menu=1515		
Achievements Till 2015	At least 50% of people to cycle to their workplace or educational institution in Copenhagen Noise-reducing asphalt, traffic planning, and speed reduction noise in general in the city The municipality subsidizes noise-insulating windows in urban renewal projects 96% of Copenhagen can walk to a large green or blue recreational area in less than 15 minutes [32]	The introduction of 67,000 paid parking spaces generating EUR 90 million in revenues since 2012 Development of the public transport network and the introduction of an electronic travel card facilitating intermodal trips in the city Cycling stated by the addition of over 200 km of bicycle lanes to the city that had just 2.3 kilometers of such paths in 2011 [33] Moscow received the 2016 Transport Achievement Award of the International Transport Forum	Saving of 300,000 tons of oil import per year, and preventing 70 tons of pollutants being let into the air every day [34, 35]

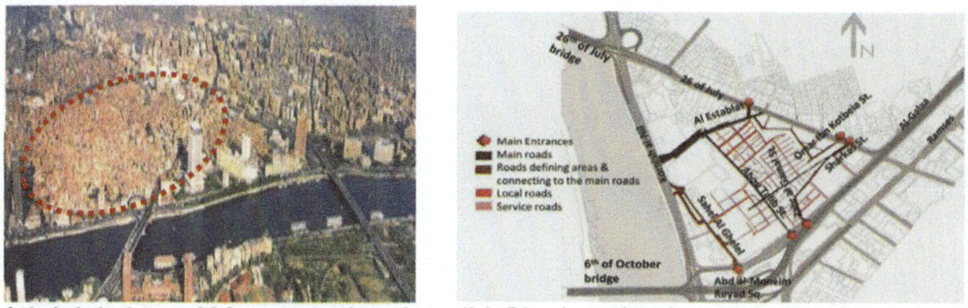

(a) Arial view of Maspero triangle (b) Site location in proximity of 4 main axes

Fig. 12 The case study of Maspero triangle area and location. (**a**) Arial view of Maspero triangle. (**b**) Site location in proximity of 4 main axes. (Image source: https://invest-gate.me/news/govt-extracts-97-of-rubble-in-maspero-triangle/)

7.3 Applying Sustainable Transport Indicators over the Selected Case Study

Selecting some important indicators and categories to determine the condition of the selected case study is imperative to assess the selected area in Maspero triangle. Data collection was conducted for 4 days. By analyzing the collected data through the case study and detecting the numbers of the cars passing in and out, in addition to their types was important to find out the related GHG emissions. For the environmental dimensions, indicator' themes are selected such as air emissions and noise pollution. For economic dimension, these include: land-use impacts, transport planning, and policy and equity. Concerning the social dimension, the themes are: travel activity and overall accessibility. Nevertheless, Table 3 lists the selected indicators for the applied case study of Maspero area in downtown Cairo, Egypt [41, 42].

Table 2 The dimensions, categories, and indicators (1–2)

Dimension	Category	Indicator
Economic	Transport demand and intensity	1.The volume of transport relative to GDP (tonne-km; passenger-km)
		2. Road transport (passenger, freight; tonne-km and passenger-km)
		3. Railway transport (passenger and freight; tonne-km and passenger-km)
		4. Maritime transport for goods and passengers (tonne-km and passenger-km)
		5. Inland waterway transport (passenger and freight; tonne-km km and passenger-km)
		6. Air transport (passenger, freight; tonne-km and passenger-km)
		7. Intermodal transport (tonne-km and passenger-km)
	Transport Costs and Prices	8. Total per-capita transport expenditures (vehicle parking, roads and transit services)
		9. Motor vehicle fuel prices and taxes (gasoline and gas/diesel)
		10. Direct user cost by mode (passenger transport)
		11. External costs of transport activities (congestion, emission costs, safety costs) by transport mode (freight and passenger)
		12. Internalization of costs (implementation of economic policy tools with a direct link with the marginal external

(continued)

Table 2 (continued)

		costs of the use of different transport modes)
		13. Subsidies to transport
		14. Taxation of vehicles/ vehicle use
		15. % of GDP contributed by transport
		16. Investment in transport infrastructure (per capita by mode/ as a share of GDP)
	Infrastructure	17. Road quality - paved roads, fair/ good condition
		18. Total length of roads in km by mode
		19. Density of infrastructure (km-km^2)
Social	Accessibility and Mobility	20. Average passenger journey time
		21. Average passenger journey length per mode
		22. Quality of transport for disadvantaged people (disabled, low incomes, children)
		23. Personal mobility (daily or annual person-miles and trips by income group)
		24. Volume of passengers
Social *(Cont'd)*	Risk and Safety	25. Persons killed in traffic accidents (number of fatalities 1000 vehicle km; per million inhabitants)
		26. Traffic accidents involving personal injury (number of injuries – 1000 vehicle km; per million inhabitants)

(continued)

Table 2 (continued)

	Health Impacts	27. Population exposed to and annoyed by traffic noise, by noise category and by mode associated with health and other effects
		28. Cases of chronic respiratory diseases, cancer, headaches. Respiratory restricted activity days and premature deaths due to motor vehicle pollution
	Affordability	29. Private car ownership
		30. Affordability (a portion of households income devoted to transport)
Environmental	Energy Efficiency	31. Energy consumption by transport mode (tonne-oil equivalent/ vehicle km)
		32. Fuel consumption (vehicles-km by mode)
		33. Habitat and ecosystem disruption Impacts on Environmental Resources
		34. Land take by transport infrastructure mode
		35. Polluting accidents (land, air, water)
	Renewables	36. Use of renewable energy sources (numbers of alternative-fuelled vehicles) - use of biofuels
	Environmental Risks and Damages	37. Hazardous materials transported by mode
	Transport Emissions	38. O_3 concentration (per capita)
		39. CO_2 emissions (per capita)
		40. N_20 emissions (per capita)
		41. CH_4 emissions (per capita)

Source: developed by the authors

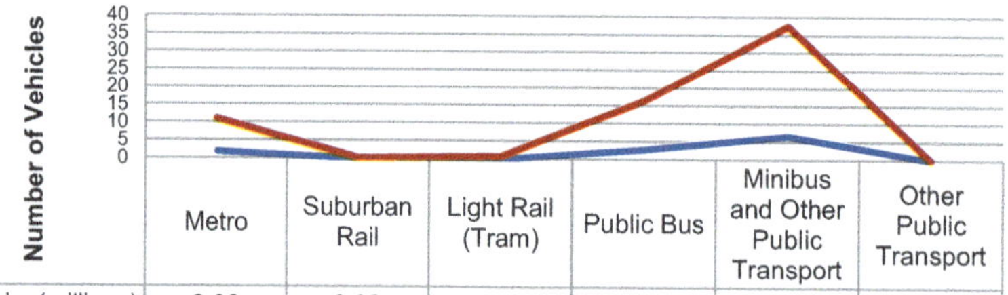

	Metro	Suburban Rail	Light Rail (Tram)	Public Bus	Minibus and Other Public Transport	Other Public Transport
Daily Trips(millions)	2.06	0.08	0.18	3.06	6.93	0.14
% of Total Trips	11.2	0.4	1	16.7	37.8	0.1

Fig. 13 Rate of daily transport in developing countries. (Image Source: www.who.int/mediacentre/news/releases/2014/air-pollution/en)

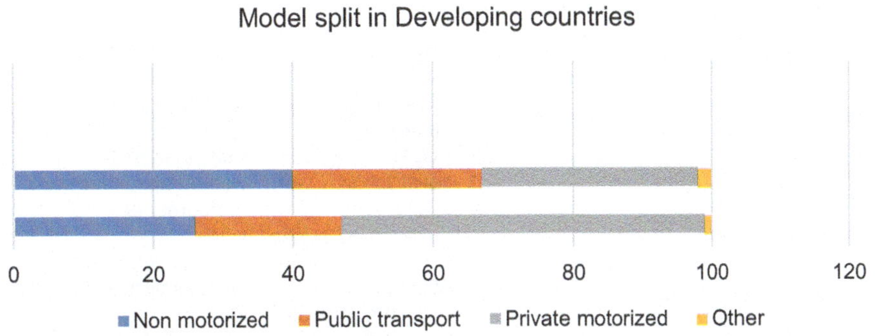

Fig. 14 Average modal split in 42 developed cities and 10 developing cities. (Image source: https://sustainabledevelopment.un.org/topics/transport/decisions)

Table 3 Sustainable transport in the case study – Downtown Cairo, Egypt

Dimension	Category	Sub-category	Indication
ENVIRONMENTAL	Air pollution emissions	- Emissions - Air pollution exposure - Climate change - Embodied energy	- Motor Vehicle Emissions=119889240 g/hr. - Emissions of CO_2 and CH4=720.5kg/year - Days of Exposure/year=270
	Noise pollution	-Traffic Noise	- People Exposed to Traffic above 55LAeq, From Workshop machines noise in addition to vehicles noise=85%
ECONOMIC	Land use impacts	- Land use impacts - Sprawl - Transport and consumption - Ecological and cultural degradation	- Per capita impervious surface area=0.17m² per person - Land devoted to transport Facility =10% of the total land - Habitat and cultural sites degraded by transportation facilities = 40%
	Travel policy and Planning	- Pricing Efficiency	- Cost-based pricing (Public

(continued)

Table 3 (continued)

		- Strategic Planning - Planning efficiency - User satisfaction	transportation only) - The degree of individual planning decisions=30% resulting in informal settlements
	Equity	- Affordability housing - Affordability transport - Basic accessibility - Portions of the household budget	- Affordable housing accessibility = 60% of the units existing
SOCIAL	Overall Accessibility	- Affordability housing - Affordability of transport - Basic accessibility - Portions of the Household budget	- Quality of walking, cycling, and taxi= No walkability, the only car is driven - Accessibility - Internet service=only 40% access net mainly cornice buildings
	Travel activity	- Vehicle - Mobility - Mode split	- Motor Vehicle Ownerships = 43.5 - Motor Vehicle travel = 3.06 - The portion of Trips = 5.26%
	Traffic Risks	- Crash casualties - Crashes - Crash costs	- Crashes Death and injuries=2% of Egypt crashes - Policy reported crashes=1% - Traffic crash economic costs = 4000000

8 Results and Discussions

It is clear after analyzing the case study of Maspero triangle district that efforts should be paid toward enhancing roads network inside the triangle in addition to the importance of creating multitypes of public transport that serve the area. Public spaces and green areas should be at least triple the spaces that exist to enhance walkability. Moreover sub-roads should be designed for bikes and golf cars in order to reduce noise, CO_2 emissions, and air pollution. Also, land devoted to transportation, which is stated by 10% should be increased by 30% to serve public means of clean and eco-transportation. All the above will reflect on minimizing car accidents, enhance air quality, and minimize air pollution as well as mitigate GHG and lessen traffic congestion.

8.1 Lessons Learned from the Case Studies

It is important to state the main recommendations deduced from the above-examined case studies and the applied local one to reinforce the dimensions of the assessment and to present guidelines to achieve sustainable transport in developing countries. These are as follows:

- Providing comfortable and eco-friendly ride is not a car, but it can be a bike or a friendly public transport with low carbon emissions.
- Ensuring better roads and well-organized ones cannot solve traffic congestion without other ways to minimize the pressure on these roads or else that would be only a matter of time to face the same traffic problems.
- Enhancing the fuel supply and replacing them by eco-friendly ones will minimize greenhouse gas emission and decrease pollution rate.
- Encouraging plans and policies for restrict systems and rules along with residents cooperation for a better environment and creating a better response toward sustainability.
- Creating better public spaces with better surroundings to encourage people to walk more or use their bikes instead of cars [43–45].

8.2 Recommendations for Better Transport Conditions in Developing Countries

It is imperative to highlight the summarized recommendations that resulted from the provision of services and infrastructure for the mobility of people and goods as well as integrating economic and social development to benefit today's and future generations; yet conclude a safe, affordable, accessible, efficient, and resilient transport, while reducing wasted energy use and GHG emissions. These recommendations are:

- Make transport planning, policy, and investment decisions based on the three sustainable development dimensions – social development, environmental (including climate) impacts, and economic growth – and a full life cycle analysis.
- Integrate all sustainable transport planning efforts with an appropriately balanced development of transport modals: integration vertically among levels of government and horizontally across modes, territories, and sectors.
- Create supportive institutional, legal, and regulatory government frameworks to promote effective sustainable transport.
- Build technical capacity of transport planners and implementers, especially in developing countries, through partnerships with international organizations, multilateral development banks, and governments at all levels to ensure equitable access to markets, jobs, education, and other necessities.
- Reinforce efforts toward preventing road traffic deaths and injuries.
- Foster an informed, engaged public as a crucial partner in advancing sustainable transport solutions.
- Establish monitoring and evaluation frameworks for sustainable transport, and build capacity for gathering and analyzing sound and reliable statistics.
- Increase international development funding and climate funding for sustainable transport.
- Promote sustainable transport technologies through government investment and guiding policies that encourage private sector investments and actions through various incentives' structures.
- Construct high quality street design standards for sidewalks and bicycles' lanes.
- Set up public transport associations, integrating timetables, and tickets.
- Integrate urban and transport developments for better planning and policy strategies [46].

9 Conclusions

Sustainable transport and transportation are key factors in developing healthier, resilient, and sustainable cities. With the COVID-19 pandemic, healthier and resilient transport and transportation are even becoming more important and of high significance to eliminate the spread of the Coronavirus worldwide. Many examples of cities, where sustainable transportation models are implemented and used were highlighted.

A case study was analyzed to illustrate the main categories, indicators, and dimensions that affect transport and its consequents. Sustainable transportation indicators were useful in assessing environmental, economic, and social aspects of the case study. It was deduced that the integration of urban and transport policies is a must to minimize traffic crash injuries and deaths and consequently reduce the crashes' finance loss which affects the economy directly. The study also indicates that a new direction for pedestrians is essential to reduce the life load over transport facilities, mitigate GHG emissions, and enhance livability in such a vital and vibrant area in downtown Cairo.

There is a need for creating several initiatives within concerned entities to raise awareness about sustainable transport/transportation, to assist the whole communities to take serious shifts and steps toward a green and sustainable society. New gridlocks for the ancient streets with constant yearly finance are essential to guide us for a better lifestyle and reapplying the previous case studies recommendations will foster the transport network and consequently enhance the local and international economy.

References

1. Evans S, Brief C (2020) Analysis: Coronavirus set to cause largest ever annual fall in CO2 emissions. Available at: https://www.carbonbrief.org/analysis-coronavirus-set-to-cause-largest-ever-annual-fall-in-co2-emissions. Accessed 9 July 2020
2. United Nations Habitat, Analysis of the transport relevance of each of the 17 SDGs (2015). Available at: https://sustainabledevelopment.un.org/content/documents/8656Analysis%20of%20transport%20relevance%20of%20SDGs.pdf/. Accessed 14 April 2020
3. Merry corps (2019). Available at https://www.mercycorps.org/blog/climate-change-poverty/. Accessed 14 April 2020
4. Issues in sustainable transport (2006) Available at https://www.researchgate.net/publication/5172232_Issues_in_sustainable_transportation/. Accessed 16 April 2020
5. Sustainable development goals (2017). Available at https://sustainabledevelopment.un.org/topics/transport/publications/. Accessed 16 April 2020
6. Hala Abou Ali and Alban Thomas, Regulating traffic to reduce air pollution, working paper No.664, Economic research forum, 2011. Available at: https://erf.org.eg/wp-content/uploads/2014/08/664.pdf/. Accessed 18 April 2020
7. Oxfam International (2019) 5 natural disasters that beg for climate change. Available at: https://www.oxfam.org/en/5-natural-disasters-beg-climate-action/. Accessed 9 July 2020
8. Jean-Paul (2019) The geography of transport system, urban transport challenges. Available at: https://transportgeography.org/?page_id=4621/. Accessed 9 July 2020
9. Mckinsey and company (2018) Elements of success: Urban transportation systems of 24 global cities. Available at: https://www.mckinsey.com/~/media/McKinsey/Business%20Functions/Sustainability/Our%20Insights/Elements%20of%20success%20Urban%20transportation%20systems%20of%2024%20global%20cities/Urban-transportation-systems_e-versions.ashx/. Accessed 9 July 2020
10. International Monitory Fund - IMF (2020) Policy tracker. Available at: https://www.imf.org/en/Topics/imf-and-covid19/Policy-Responses-to-COVID-19/. Accessed 9 July 2020
11. Lock S (2020) Global change in travel and tourism. Available at: https://www.statista.com/forecasts/1103426/covid-19-revenue-travel-tourism-industry-forecast/. Accessed 9 July 2020
12. World Bank, 2020, The global economic outlook during COVID-19, available at: https://www.worldbank.org/en/news/feature/2020/06/08/the-global-economic-outlook-during-the-covid-19-pandemic-a-changed-world/. Accessed: July 9, 2020
13. Congressional Research Service (2020) Global economic effects of COVID-19, Available at: https://fas.org/sgp/crs/row/R46270.pdf/. Accessed 9 July 2020
14. Batta RN (2008) Economics on the road transport, Available at https://books.google.com.eg/books?uid=116682365015505826346&source=gbs_lp_bookshelf_list/. Accessed 9 July 2020
15. United Nations, Sustainable development goals, Goal no. 7, Available at: https://www.un.org/sustainabledevelopment/energy/. Accessed 9 July 2020
16. United Nations (2016) Mobilizing sustainable transport for development. Available at: https://sustainabledevelopment.un.org/content/documents/12453HLAG-ST%20brochure%20web.pdf/. Accessed 9 July 2020
17. Sevrige S (2020) Safe guarding for the future. Available at: https://sweden.se/climate/. Accessed 9 July 2020
18. Timpothy Papandreou, IBTTA (2020) Collection of articles on transport in a post-COVID 19. Available at:https://www.ibtta.org/sites/default/files/documents/2020/Coronavirus/A%20Collection%20of%20Articles%20on%20Transportation%20in%20a%20Post%20Covid-19%20World%202020.04.28.pdf. Accessed 9 July 2020
19. Intelligent transport (2019) Exploring Dubai's plans for the future of mobility. Available at: https://www.intelligenttransport.com/transport-articles/82445/dubai-mobility-rapid-ncrease-use/. Accessed 9 July 2020
20. IAEA (2018) Country nuclear power profiles, Morocco. Available at: https://www-pub.iaea.org/MTCD/Publications/PDF/cnpp2018/country-profiles/Morocco/Morocco.htm. Accessed 9 July 2020
21. C40, Case study (2011). Available at: https://www.c40.org/case_studies/city-of-cyclists-reduces-approximately-90000-tons-of-co2-emissions-per-year-and-has-over-50-of-the-citys-population-cycling-to-work-everyday. Accessed 9 July 2020
22. Building design and construction, COVID-19, and real estate: How the Coronavirus is impacting the AEC industry (2020) Available at: https://www.bdcnetwork.com/covid-19-and-real-estate-how-coronavirus-impacting-aec-industry. Accessed 8 July 2020
23. Cleaner and safer transport in cities (2020) Smart choice for cities. Available at: https://ec.europa.eu/transport/sites/transport/files/cycling-guidance/smart_choices_for_the_city_cycling_in_the_city_0.pdf/. Accessed 8 July 2020
24. European Commission (2005) Walking and cycling as transport mode. Available at: https://ec.europa.eu/transport/road_safety/specialist/knowledge/pedestrians/pedestrians_and_cyclists_unprotected_road_users/walking_and_cycling_as_transport_modes_nl/ Accessed 9 July 2020

25. Sustainable transportation Case studies (2010). Available at: https://www.tigurl.org/images/tiged/docs/activities/953.pdf/. Accessed 9 July 2020
26. Alberto bull, United Nations publications (2004) Traffic congestion the problem and how to deal with it. Available at: https://repositorio.cepal.org/bitstream/handle/11362/37898/LCG2199P_en.pdf. Accessed 9 July 2020
27. Sustainable development goals (2011) Sustainable transport. Available at: https://sustainabledevelopment.un.org/topics/sustainabletransport/. Accessed 9 July 2020
28. Gamil A (2017) MSA University, Graduation project, Available at: https://issuu.com/abdelrahmaangs/docs/culture_center_-_maspero_triangle. Accessed 9 July 2020
29. Shakran K, The Uneven City: planning insurgencies in Ramlet Bulaq and Maspero Triangle, Master thesis. Available at: https://escholarship.org/content/qt2441k3s4/qt2441k3s4.pdf?t=oeipr4&nosplash=f040f62a96543b415072569853d2c50b. Accessed 9 July 2020
30. el Kadi G (2005) Modern heritage in Cairo, UNESDOC. Available at https://unesdoc.unesco.org/ark:/48223/pf0000140930. Accessed 9 July 2020
31. UNECE (2011) Climate change and sustainable transport. available at: https://www.unece.org/trans/theme_global_warm.html. Accessed 9 July 2020
32. Sustainable Urban Transport Plans (2007) Preparatory Document in relation to the Thematic Strategy on the Urban Environment. Available at: https://ec.europa.eu/transport/sites/transport/files/themes/urban/studies/doc/2007_sutainable_urban_transport_plan_annex.pdf. Accessed 9 July 2020
33. Havns K (2011) City of Cyclists. Available at: https://urbandevelopmentcph.kk.dk/artikel/city-cyclists/. Accessed 9 July 2020
34. Wahba D (2020) Urban rights and local politics in Egypt: The Case of the Maspero Triangle. Available at: https://www.arab-reform.net/publication/urban-rights-and-local-politics-in-egypt-the-case-of-the-maspero-triangle/. Accessed 9 July 2020
35. Competitions Archi (2015) Urban re-planning of Maspero triangle, 2015. Available at: http://competitions.archi/competition/urban-re-planning-of-maspero-triangle/. Accessed 9July 2020
36. Hoff J (2020) Developing and Implementing Green National Accounts and the Green GDP-University of Copenhagen. Susy.ku.dk. Retrieved 17 July 2021, from https://susy.ku.dk/research/developing-and-implementing-green-national-accounts-and-thegreen-gdp/
37. CPI, Cycle infrastructure in Copenhagen (2016). Available at: https://www.centreforpublicimpact.org/case-study/cycle-infrastructure-copenhagen/. Accessed 9 July 2020
38. Sustainable Urban Transport Plans (2007) Preparatory Document in relation to the Thematic Strategy on the Urban Environment. Available at: https://ec.europa.eu/transport/sites/transport/files/themes/urban/studies/doc/2007_sutainable_urban_transport_plan_annex.pdf. Accessed 9 July 2020
39. Archdaily, Living pedestrian bridge over the Nile (2020). Available at: https://www.archdaily.com/932511/living-pedestrian-bridge-over-the-nile-competition-cairo-egypt-tamayouz-excellence-award-2020/. Accessed 9 July 2020
40. International Energy Agency (2015) Available at: CO2 Emissions from Fuel Combustion – Highlights www.iea.org/publications/free publications/publication/CO2EmissionsFromFuelCombustionHighlights2015.pdf. Accessed 14 April 2020
41. United Nations, 2016: Habitat III – zero draft of the new urban agenda. Available at: www.habitat3.org/bitcache/3d1efc3b20a4f563ce67367 1f83ea0f767b26c10?Vid=578792&disposition=inline&op=view. Accessed 14 April 2020
42. World Health Organization (WHO) (25 March 2014) News Release: 7 Million Premature Deaths Annually Linked to Air Pollution. Available at: www.who.int/mediacentre/news/releases/2014/air-pollution/en/. Accessed: 16 April 2020
43. Smith K (2016) International Railway Journal, Available at: https://www.railjournal.com/policy/un-report-offers-10-points-to-achieve-sustainable-transport/. Accessed 16 April 2020
44. The IHS Library (2020) Available at: https://www.ihs.nl/en/master/urban-land-governance-sustainable-development?gclid=CjwKCAjwhOD0 BRAQEiwAK7JHmHXprfjo5Kr1YoACnw-7qKPqH2sQruhe98siD9zNRlJ4NAosrpIALRoCMPQQAvD_BwE/. Accessed 16 April 2020
45. Indicators to Assess Sustainability of Transport Activities (2007). Available at https://core.ac.uk/download/pdf/38617455.pdf/. Accessed 16 April 2020
46. Sustainable transport (2017) Available at https://sustainabledevelopment.un.org/topics/transport/decisions/. Accessed 16 April 2020

Achieving Sustainable Electricity Through Renewable Energy Development in the Gambia: Regulatory and Policy Considerations

Nana Asare Obeng-Darko

1 Introduction

The sustainable development goals (SDGs) has been adopted by international community as a world vision agenda to bring about sustainable global development by the year 2030. This international development agenda (2030 Agenda), adopted in 2015, consists of 17 broad aims. These vary from eradicating poverty and hunger, energy, promotion of health and well-being, gender equality, combating climate change to revitalising global partnership for sustainable development [1]. Among these 17 goals is the call to ensure access to affordable, reliable, sustainable and modern energy for all. This is the goal seven of the SDGs (SDG 7) [1]. Vital to achieving SDG 7 is the appeal to increase significantly the share of renewable energy (RE) in the global energy mix and it relies upon 'enhance international cooperation to facilitate access to clean energy research and technology, including RE, energy efficiency and advanced and cleaner fossil-fuel technology, and promote investment in energy infrastructure and clean energy technology' [1].

The year 2020 marks the fifth year into the 15-year goal plan of the 2030 Agenda and the general question being observed thus far is whether the global community is on course to achieve the SDGs by 2030. More specifically, can SDG 7 be achieved by least developing countries such as The Gambia by 2030? Recent data show that the world is making progress towards achieving SDG 7, but will fall short of meeting the targets by 2030 at the current rate of ambition [2]. It is projected that 92% access rate would be reached by 2030 [2]. Without sustained and improved policies and strategies, it is estimated that 650 million people will be without access to electricity in 2030 with 89% of those living in Sub-Saharan Africa [3, 4]. Africa as a whole has the richest solar resources in the world but has only installed 5 gigawatts (GW) of solar photovoltaics (PV), which is less than 1% of global capacity [5, 6]. Can Gambia through its RE policy and law development provide access to affordable, reliable, sustainable electricity to its citizens? In order to attempt an answer to this question, an examination of the Gambian RE legal, regulatory and policy frameworks have to be conducted and the issues therein examined. The purpose of this paper thus is to examine whether The Gambia can achieve its RE development target from regulatory perspective. Following this introduction, Sect. 2 presents the electricity outlook of Gambia. Section 3 takes an overview of the Gambian RE law and policy while Sect. 4 examines regulatory issues that are impeding the achievement of the country's RE development goals. Conclusion and recommendations are drawn in Sect. 5.

2 The Electricity Outlook in Gambia

The Gambia is the smallest country on mainland Africa. It is situated in West Africa on the Atlantic coast and surrounded by Senegal. The main source of energy supply in The Gambia is biomass (wood-fuel and charcoal). However, the main energy sources for electricity production in the country is heavy fuel oil (HFO) and diesel oil that are mostly imported into the country [7]. Due to the importation of these energy products for the production of electricity in the country, The Gambia has one of the highest electricity tariffs within the sub-Sahara Africa [7, 8]. Current electricity tariff is at $0.26/ kilowatt hour (kWh)

N. A. Obeng-Darko (✉)
Center for Climate Change, Energy and Environmental Law (CCEEL) at the UEF Law School, University of Eastern Finland, Joensuu, Finland
e-mail: nana.asare.obeng-darko@uef.fi

© The Author(s), under exclusive license to Springer Nature Switzerland AG 2022

A. Sayigh (ed.), *Sustainable Energy Development and Innovation*, Innovative Renewable Energy,
https://doi.org/10.1007/978-3-030-76221-6_66

[9]. Foreign exchange is also a major contributing factor to the high electricity tariff; and the country, due to import dependency, is susceptible to price volatility in global oil prices.

The World Bank reports that 'The Gambia's inadequate energy supply is a binding constraint on inclusive growth and competitiveness, and businesses in general often report unreliable and expensive electricity as one of the major constraints of growth, especially hotels, who are forced to depend on expensive backup generators' [10]. There is a wide gap between installed capacity and available electricity supply in the country. Installed capacity countrywide is 99 megawatts (MW) of which 55 MW is available. For the greater Banjul area, installed capacity is 88 MW yet 45 MW remains available [9]. Electrification in The Gambia has an estimated rate of 35% for a population of 2.1 million people [9]. For the Banjul area, 60% of the population has access to electricity whereas only 6% has access in the outlying provinces such as in the North Bank and Central River regions [11]. The perennial inadequacy of electricity supply in The Gambia has led to widespread blackouts, with some parts of the greater Banjul Area and other provinces throughout the country reporting only 2–3 hours of power per day [8]. Insufficient electricity supply and frequent blackouts continue to hinder effective governance and impede private sector growth, and hence, economic and social growth are being stifled [12].

To improve accessibility to electricity and the security of its electricity supply, The Gambia is turning to renewable energy sources (RES) to diversify and bolster its energy supply sources for electricity production. RE represents a tremendous opportunity for The Gambia, but it is currently underexploited. The government, however, intends to utilise the RES potential for electricity production. In this regard, the nation in 2005 adopted a policy instrument—the Energy Policy (2005)—setting the targets for RE integration into the nation's energy mix for electricity production. In 2018, The Gambian Government adopted its National Development Plan (NDP) 2018–2021 [9], which sets the national agenda and goals—including RE targets—to be achieved by the end of 2021. To provide both legal and regulatory environments within which policy goals for RE will be achieved, The Gambian RE law—*Renewable Energy Act 2013* (RE Act 2013) was adopted in 2013 [13].

3 The Gambian RE Law and Policy

3.1 Energy Policy (2005)

The objectives of The Gambian Government concerning the energy sector and the RE sub-sector are contained in the Energy Policy (2005). The primary objective for RE development is to 'ensure the promotion and utilisation of renewable energy in support of sustainable development in the country' [7]. To achieve this overarching aim, the following specific objectives are to be achieved:

- Promoting the utilisation of renewable forms of energy such as solar, wind and biomass.
- Promoting the use and develop, to the extent possible, a domestic production capacity for RE fuels and technologies.
- Ensuring the sustainable supply of RE fuels/device/technologies at competitive prices through private sector participation [7].

3.2 The Gambia National Development Plan (NDP) 2018–2021

The goal set by the Gambian Government regarding RE integration in its NDP is to increase the share of RE in total electricity generation from 2% to 40% between 2018 and 2021 [9]. According to the NDP, the government aims to achieve its RE target by the use of appropriate technology, research and development and other efficient measures. RE, according to the government represents an area of tremendous opportunity for The Gambia and as such, it is committed to promote the development of RE and will actively work to increase the RE share in the national energy mix. The government says it will prioritise RES in order to have a balanced energy mix [9].

3.3 Renewable Energy Act 2013

The Gambian *Renewable Energy Act 2013* (RE Act 2013) was passed by the President and National Assembly in December 2013 to set the 'legal, economic and institutional basis to promote the use of renewable energy resources and for connected matters' [13] in bid to achieving the nation's RE target. Crucial to the achievement of RE target, the RE Act 2013 specifically instructs the Ministry (responsible for energy) to 'recommend middle and long-term national targets for the use of RE resources in electricity generation, which may include targets related to geographic location and diversity' [13]. In addition to the RE Act 2013, the production of electricity for RES is promoted in the *Electricity Act, 2005* of The Gambia [14]. As part of the objectives of The Gambian Electricity Act, Article 3(l) encourages the production of electricity through the use of RE.

3.4 Regulators

The regulatory agencies responsible for the regulation and implementation of the RE Act 2013 are the Ministry of Petroleum and Energy, the Public Utilities Regulatory Authority (PURA) and the National Water and Energy Company (NAWEC). The Ministry, serving as the policy regulator is primarily responsible for overseeing the overall implementation of the RE Act 2013. This is firmly established under section 19 of the Act. The PURA, serving as the economic regulator, is responsible for formulating feed-in tariff (FIT) rules and managing the 'Renewable Energy Fund' which is established under section 5 of the RE Act 2013 [13]. Occupying the role of the technical regulator, NAWEC is predominantly responsible for determining the safety and technical capability of the grid to connect electricity generated from RE resources [13]. It is also to undertake such reporting duties as are required by this Act, in particular section 18(2) of the RE Act 2013.

4 Regulator, Legal and Policy Issues

4.1 Lack of Regulatory Measures

The RE Act 2013 under Section 19 instructs the Minister to make regulations for the implementation of the law and under Section 20, the Minister is to work closely with the NAWEC on such aspects that pertain to electricity generation. Essential to the instructions in both Sections 19 and 20 is the adoption of the proposed Biomass Strategy by the Minister following the conduct of an impact assessment a year after the RE Act 2013 come into effect. At the time of writing, such a strategy is yet to be adopted. It may be assumed at this point that the impact assessment of the use of biomass for electricity generation is yet to take place. While this is only an assumption, it helps to throw light on the zero rate of RE integration in the country's electricity generation mix [9]. Energy mix for electricity production in the Gambia is 100% HFO [9]. Section 15(3) of the RE Act 2013 bars projects generating electricity exceeding 1 MW from biomass energy from benefiting from feed-in tariffs or any other incentives until such a time when the impact assessment on biomass is undertaken by the Energy Ministry. This is a serious deterrent to generating electricity from biomass sources to bolster electricity access in the country since project investors whose electricity production from biomass may exceed 1 MW may feel discriminated against and ultimately decide against investing in the RE sector. It is important to note at this point that the Ministry is yet to make any regulations to give effect to the various provision of the *RE Act 2013*. When the RE Act 2013 is compared to similar RE laws in the same geographical region (sub-Saharan Africa) such as the Ghanaian *Renewable Energy Act 2011* (RE Act 2011) [15], regulatory measures are quite limited and are not addressed to any specific aims or objectives. For instance, unlike the regulatory mechanism in the Ghanaian RE sector where the law defines specific objectives for which legislative instruments are to be adopted in implementing the provisions of the RE Act 2011, it is not the case in The Gambian RE sector. Under section 19 of the RE Act 2013, the Ministry of Petroleum and Energy is only to make regulations for the implementation of the Act without identifying specific aims to which regulations are targeted.

4.2 Weak Licencing Regime

The Gambian RE Act 2013 provides for a permitting process for the facilities or projects producing electricity from RES. The RE Act 2013 contains a title called 'streamlined permitting', which essentially aims to make the application and process of issuing permits to renewable electricity project operators as smooth as possible. Section 16 of the RE Act 2013, for example, instructs the Ministry (of Petroleum and Energy) to issue clear guidelines for application processing and to collaborate with other ministries in simplifying permitting process regarding such things as environmental impact land use and water use. The streamlined permitting title of the RE Act 2013 does not, however, provide qualification standards for obtaining a licence or permit, conditions under which a permit may be awarded to a renewable electricity producer, the duration of a permit, conditions for renewal and more importantly, the agency responsible for the issuance of licences. The law only mandates the Ministry, working with other agencies, to issue clear and simple guidelines on the processing of applications but not the power to issue permits. In fact, the RE Act 2013 does not elect or grant the power to award licences to any of the regulatory agencies in the RE sector of The Gambia. This raises the question of clarity in the responsibilities of the regulatory agencies in the Gambian renewable energy sector.

The Gambian RE Act, in comparison to the Ghanaian RE law, lacks depth when it comes to its licensing regime. The Ghanaian RE Act 2011, for example, has provision for the requirement for licences, specifying in details the qualification, conditions for grating, conditions for the suspension or cancellation, conditions for renewal, the responsible authority and a mechanism for settling disputes. All of these are missing in the Gambian RE law. In effect, there is no licensing regulation in the Gambian RE sector.

4.3 Ineffective Regulation

The issue of ineffective regulation in the Gambian regulatory regime pertains to non-performance by the elected regulatory agencies in the RE sector. As mentioned earlier the Minister for Petroleum and Energy has failed to conduct the biomass strategy impact assessment as instructed in section 15 of RE Act 2013. Besides this, the Minister is yet to adopt any regulations to implement the feed-in tariff provisions of the Act. Section 11(1) of the RE Act 2013 provides that a feed-in tariff (FIT) system is to be established to help accelerate the development of RE.

Prior to the adoption of regulations implementing the feed-in tariff provision, the RE Act 2013 instructs the PURA to formulate, with the approval of the Minister for Petroleum and Energy, feed-in tariff rules that shall conform to the conditions stipulated under section 11(3) of the RE Act 2013 within 6 months of the law coming into force. Such feed-in tariff rules are yet to be formulated and made accessible. It may be argued that the Minister cannot adopt necessary regulations implementing section 11 of the RE Act 2013 because the PURA has not yet formulated the rules needed for the feed-in tariff system to come into effect. At this point, it is not clear whether the PURA has simply failed to formulate the required rules, or it might have done so but has not met the approval standards of the Minister. What is certain in any case is that there is not an existing feed-in tariff system in place to support the development of renewable electricity in The Gambia.

4.4 Lack of Regulatory Assessment

Since the enactment of the Gambian RE Act in 2013, there has not been assessment of the regulatory measures to determine if regulation of the nation's renewable sector is effective or not. A regulatory fitness assessment of the regulatory bodies and the various regulatory mechanisms needed in regulating the RE sector could provide necessary information pertaining to whether regulatory goals can be achieved, a restructuring of the regulatory agencies may be necessary or if regulatory measures may need adjustments so that goals could be achieved [16].

In fact, section 4(2) of the RE Act 2013 provides that the Ministry of Petroleum and Energy is to prepare and submit an annual report on the progress towards the achievement of the RE target to the Cabinet. These reports are also to include a review of the performances of the incentives provided in the Act. This clearly provides a basis for conducting regular regulatory assessment, particularly the support schemes under the regulatory structure. Nevertheless, such an assessment has never been undertaken. This, predominantly, might be as a result of the persistent lack of regulatory instruments that are needed to implement the various aspect of the Act, such as the absence of feed-in tariff system and its rules.

4.5 Investment Risk

RE projects require considerable financial investment and this acutely exposes investors to higher risk if the project fails to be successful [17]. It is also thought that investments in RE projects for generating electricity are based on public financial support [18]. Likewise, investments that are based on support schemes primarily rely on the policies and regulations of a country's RE sector and the ability of regulators to sustainably implement those support systems [19, 20]. In this sense, regulatory risks such as changes to regulatory goals and activities expose investors to higher degree of uncertainty in the market.

The Gambian Government's intended approach to attracting private investment into its RE sector is mainly through the provision of support schemes in the regulatory systems. Except for tax exemptions and breaks, feed-in tariff is the main support mechanism in the Gambian RE market. Since The Gambia does not have a feed-in tariff system currently in place, an investment into a renewable electricity project in the country would certainly be an extremely risky undertaking with no guarantee of return on the investment. This is because the nation's transmission and distribution system operator, NAWEC, would not be able to purchase the renewable electricity at a guaranteed fixed price that is established to ensure that project investors would be able to recoup their investment cost and if possible, with fair profit margins. Potential renewable electricity producers cannot be certain of guaranteed payment without an established and working feed-in tariff, which is presently the case in the Gambia.

4.6 *Infrastructure Issues*

Infrastructure deficiencies could be another impeding factor to integrating renewable electricity into the national generation mix in the Gambia. According to the Gambian Government, a quarter of grid electricity transmission was lost in 2017 [9] through inefficient transmission systems and aged electricity infrastructure and is expected to increase to 29% in 2018 [21]. These losses increase the cost of electricity supply on a per kWh basis. The current Gambian national grid system is aged and underdeveloped. This poses technological and technical capabilities issues for the system and may therefore not be able support uptake of renewable electricity. Infrastructure inadequacies also imply that it is not likely for investors to put money into renewable electricity projects knowing that additional capacity from those projects cannot be integrated into the nation's transmission and distribution lines.

4.7 *Overambitious Policy Goals*

It is the aim of the Gambia government to increase the integration rate of RE into the total electricity generation mix from less than 2% in 2018 to 40% by the end of 2021 [9]. While this is a laudable idea, it appears to be an overambitious aim. Due to present technological and technical impediments to uptake renewable electricity because of outdated grid systems and substantial financial investment needed to upgrade them [9], it is highly improbable that an integration rate of 40% could be achieved in the space of 3 years. In fact, the African Development Bank Group reports that The Gambia had 100% HFO energy mix for electricity generation as of the end of 2018 [22].

It is certainly not impossible to achieve such a feat, but the chances of it being achieved considering all the factors—inefficient regulation, ageing infrastructure, financial and technical constraints—seems unlikely. A modest, yet realistic and achievable target may prove to be a catalyst for attracting investment into the sector. The reason is that, should The Gambian Government fail to deliver a 40% RE integration by 2021 – moving the goalpost as it was, investors may tend to believe that the government is not committed to its RE developmental target for the nation and may deter prospective investors from investing into the RE sector.

5 Conclusion and Recommendations

This paper aimed at examining the regulatory issues in the Gambian RE developmental agenda. The Gambian Government aims to increase and integrate the share of RE into the country's total electricity generation mix from 2% to 40% between 2018 and 2021. It is also the goal of the government to increase the percentage of the population with access to electricity from 35% to 60% by 2021 through renewable electricity [9]. By achieving this RE integration target, increasing the share of the population with access to affordable and sustainable electricity, the Gambian Government hope to fulfil its commitment to SDG 7 under the Agenda 2030 agreement [1].

While policy, regulatory and legal objectives have been established by the Gambian Government regarding the development of RE for electricity generation, a number of key issues that have the potential to impede any efforts by the country towards achieving its integration targets have been identified and examined in this paper. This paper shows that the current lack of regulatory measures and tools by any of the elected regulatory agencies in the sector such as the feed-in tariff system and the biomass strategy means that the various sections of the RE Act 2013 that need to be implemented for the effective functioning of the RE law are not implemented. Connected to the issue of lack of regulatory measures is the issue of non-performance by the regulatory agencies in the Gambian RE sector. The PURA is yet to prepare FIT rules as prescribed by the law, the Minister has also failed to conduct the required biomass strategy impact assessment necessary for the adoption of the biomass strategy regulatory tool as instructed in section 15 of RE Act 2013. The licensing regime of the regulatory framework in the Gambian RE sector also appears to be weak and lacking details such as the requirement for licences, conditions for granting, conditions for the suspension or cancellation, conditions for renewal, the responsible authority and a mechanism for settling disputes.

The examination of the Gambian RE sector development has also shown that there has not been an assessment of the regulatory framework since the country adopted its RE Act since 2013. While the section 10(1) of the RE Act 2013 specifically instructs the regulatory agencies to provide an annual progress report including an assessment of regulatory measures to the Cabinet, no such assessment has been conducted and hence, no progress report has been made nor recorded since 2013. There is also a high investment risk and uncertainty associated with the present regulatory issues in the Gambian renewable energy market. The FIT support system is the only government's designed approach to attract private investment into the sector. However, the feed-in tariff system is yet to be instituted. Hence, any project investment in the sector would undoubtedly be an extremely risky venture with no guarantee of return on such investments. The issue of an overambitious integration target might be a deterrent to attracting investment into the nation's renewable energy market. This is in consideration of the fact that the current infrastructure, technical and technological inadequacies in the country's entire electricity sector and the significant financial investments needed to address them implies that a renewable electricity integration rate of 40% by 2021 from zero in 2018 is highly improbable.

In going forward, the Gambian Government should seek to hold regulatory agencies accountable for failing to perform their mandated duties under the RE Act 2013. The lack of regulations necessary for the effective implementation of the specific provision of the RE law means that the overall objective of the RE Act is in danger of not being achieved and may subsequently render the law ineffectual. Accountability and performance of the regulators may send a signal to prospective investors that the Gambian Government is committed to the policy goals of the RE sector. This also helps to improve credibility and trust in the institutions in the regulatory framework [16].

Although the Gambian Government in an attempt to provide access to affordable, reliable, sustainable electricity for all of its citizens in the shortest possible time, has set an overambitious RE integration goal, it has failed to attract investment into the market. A more realistic and easily achievable goal would help attract private investment in the country's RE sector. It would be recommended that a revised RE integration target should be set following a thorough assessment of current regulatory measures and policy strategies to achieve policy goals. In addition, an updated integration target could perhaps have a timeframe that extends to at least 2030. This will coincide with the target year of SDG 7 of the 2030 Agenda. Essentially, a new and separate RE policy document with updated targets and strategies to achieving those targets with a more realistic time period might be necessary and may help boost private investment into the Gambian RE market.

References

1. UNGA Res 70/1 (21 October 2015) UN Doc A/RES/70/1, 2015
2. IRENA (2019) Tracking SDG7: The energy progress report, international renewable energy agency, Washington, D.C. https://www.irena.org/-/media/Files/IRENA/Agency/Publication/2019/May/2019-Tracking-SDG7-Report.pdf. Accessed 22 Jan 2020
3. Obeng-Darko NA (2020) Legal, policy and regulatory aspects of integrating electricity from renewable energy in Sub-Sahara Africa. The case of Ghana, The Gambia, and Nigeria. In: Hunter T, Herrera I, Crossley P, Alvarez G (eds) Routledge handbook of energy law, 1st edn. Routledge, London, pp 489–506. https://doi.org/10.4324/9780429450891
4. Nhamo G, Nhamo S, Nhemachena C, Nhemachena CR (2020) Africa and the 2030 sustainable energy goal: a focus on access to renewables and clean fuels for cooking. In: Nhamo G, Odularu GOA, Mjimba V (eds) Scaling SDGs implementation Emerging Cases from State, Development and Private Sectors. Springer International Publishing, Cham, pp 39–57. https://doi.org/10.1007/978-3-030-33216-7_3
5. Wahlén CB (2019) IEA report presents scenarios for sustainable energy, SDG 7 Outlook for Africa, SDG Knowledge Hub. https://sdg.iisd.org:443/news/iea-report-presents-scenarios-for-sustainable-energy-sdg-7-outlook-for-africa/. Accessed 22 Jan 2020
6. IEA, Africa Energy Outlook (2019) International Energy Agency, Paris, 2019. https://www.connaissancedesenergies.org/sites/default/files/pdf-actualites/Africa_Energy_Outlook_2019.pdf. Accesesed 22 Jan 2020.
7. UNDP (2012) Sustainable energy for all: rapid assessment and gap analysis – The Gambia. United Nations Development Programme, New York. https://www.seforall.org/sites/default/files/Gambia_RAGA_EN_Released.pdf. Accessed 30 Nov 2018
8. IRENA (2013) Gambia renewables readiness assessment. International Renewable Energy Agency, Abu Dhabi. https://www.irena.org/-/media/Files/IRENA/Agency/Publication/2013/RRA_Gambia.pdf. Accessed 30 Nov 2018
9. The Gambian Ministry of Finance and Economic Affairs, The Gambia National Development Plan (2018-2021) The Gambian ministry of finance and economic affairs, Banjul, 2017. http://www.thegambiatimes.com/wp-content/uploads/2018/02/1.-The-Gambia-National-Development-Plan-2018-2021-Full-Version.pdf Accessed 3 Dec 2018

10. World Bank, International Development Association Project Appraisal Document on a Proposed Grant in the Amount of SDR 28.4 Million (US$41 Million Equivalent) to the Republic of the Gambia for the Gambia Electricity Restoration and Modernization Project, The World Bank, Washington, DC, 2018. http://documents.worldbank.org/curated/en/171661526614264416/pdf/GAMBIA-PAD-05042018.pdf. Accessed 23 Jan 2020
11. World Bank (2015) The Gambia: policies to foster growth: Volume 2. Macroeconomy, Finance, Trade and Energy, World Bank Group, Washington, DC. https://openknowledge.worldbank.org/handle/10986/22382. Accessed 22 Jan 2020
12. Mbai PN (2017) Gambia: what is Government's policy on renewable energy?, freedom Newsp. https://www.freedomnewspaper.com/2017/08/18/gambia-what-is-governments-policy-on-renewable-energy/ (accessed November 30, 2018)
13. Renewable Energy Act 2013 of the Republic of The Gambia. 12 Dec 2013
14. Electricity Act (2005) 2005 of the Republic of the Gambia, 2005
15. Renewable Energy Act, 2011 (Act 832) of the Parliament of the Republic of Ghana, 31 Dec 2011
16. Obeng-Darko NA (2019) Why Ghana will not achieve its renewable energy target for electricity. Policy, legal and regulatory implications. Energy Policy 128:75–83. https://doi.org/10.1016/j.enpol.2018.12.050
17. Schwerhoff G, Sy M (2017) Financing renewable energy in Africa – key challenge of the sustainable development goals. Renew Sust Energ Rev 75:393–401. https://doi.org/10.1016/j.rser.2016.11.004
18. Obeng-Darko D (2016) Renewable energy in Ghana and the lessons from the EU: an examination of the regulatory support schemes. Renew Energ Law Policy Rev 7:46–60
19. Talus K (2016) Introduction to EU energy law. Oxford University Press, Oxford
20. Boute A (2011) A comparative analysis of the European and Russian support schemes for renewable energy: return on European experience for Russia. J World Energ Law Bus 4:157–180. https://doi.org/10.1093/jwelb/jwr002
21. World Bank Group, Projects: Gambia Electricity Support Project (2016). http://projects.worldbank.org/P152659/?lang=en&tab=documents&subTab=projectDocuments Accessed 5 Dec 2018
22. African Development Bank Group (2019) Republic of the Gambia – The Gambia Electricity Access Project (GEAP) – Project Appraisal Reports, African Development Bank – Building Today Better Africa. Tomorrow. https://www.afdb.org/en/documents/republic-gambia-gambia-electricity-access-project-geap-project-appraisal-reports Accessed 4 Mar 2020

The Role of Circular Economy in Mitigating Climate Change Impacts: Thermal Insulation Materials' Assessment in Developing Countries

Sarah A. GadAllah and Mohsen M. Aboulnaga

1 Introduction

Climate change is a major challenge to all nations due to high greenhouse gas (GHG) emissions, especially generated from the building and transport sectors. The building sector is responsible for 40% of total energy consumption and one-third of GHG emissions [1]. A tremendous amount of embodied energy is consumed by conventional thermal insulation materials that contribute to GHG emissions and consequently climate change impacts. A research by CSIRO (Commonwealth Scientific and Industrial Research Organisation) found out that an average house contains about 1000GJ of energy embodied in the materials exploited in the construction [2]. Such amount is equivalent to approximately 15 years of normal operational energy use [2]. Ramesh et al. [3] reported a share of 10–20% for embodied energy, based on an article review effort which engrosses 73 conventional residential and office buildings [4].

Also, Sartorial and Hestnes [5] examined the data on 60 conventional and low-energy buildings studied by the extant literature on energy use in buildings and concluded that embodied energy constitutes 2–38% of total energy use in conventional buildings and 9–46% in low-energy buildings [4]. Another study conducted by Chastas et al. in 2016 on 90 residential buildings, where the share of embodied energy accounted from 6% to 20% in conventional buildings, 11–33% in passive buildings, 26–57% in low-energy buildings, and 74–100% in net-zero energy buildings [4] as shown in Fig. 1.

Insulation materials represent an integral component of the building envelope and lower energy demand in buildings [6, 7], where inorganic fibrous materials account for 60% of the European market [8]. In a review of previous articles, Ding suggests that the embodied energy use in buildings ranges from 3.6 to 8.76 Giga Joule (GJ)/m^2 of gross floor area (at a mean of 5.506 GJ/m^2) in residential buildings and from 3.4 to 19 GJ/m^2 of gross floor area (at a mean of 9.19 GJ/m^2) in commercial buildings, whereas Aktas and Bilec reported an initial embodied energy (i.e., associated with pre-use stage) range of 1.7–7.3 GJ/m^2 (at a mean of 4.0 GJ/m^2) and 4.3–7.7 GJ/m^2 (at a mean of 6.2 GJ/m^2) for conventional residential buildings and low-energy residential buildings respectively [8].

Despite the weakness points regarding thermal insulation materials that are presented in the deficiency of information of the developed insulation materials, in addition to the absence of buildings review to check their compliance with the code, Egypt progressed in the field of energy conservation awareness [9]. In fact, thermal insulation materials' importance is realized in a few buildings in Egypt; these buildings are hotels or commercial (offices and banks) buildings, where there are almost no buildings with natural or unconventional thermal insulation materials. Nevertheless, it has been reported by many consultants (Raafat Miller Consultants, Engineering Consultants Group (ECG) and EHAF; that the number of buildings with external walls' thermal insulation materials is low. Another important outcome of these meetings reveals that many projects by leading developers lack the incorporation of thermal insulation in residential buildings such as Palm Hills, Sodic and Mivida Compounds [10].

The impetus of natural thermal insulation materials would create a path to mitigate carbon emissions. Therefore, the utilization of natural thermal insulation materials as agricultural waste that reflects the concept of circular economy is an optimum solution to reduce energy consumption and embodied energy.

The concept of circular economy, particularly the usage of thermal insulation materials, is important in developing countries. A respectful number of studies have concentrated on attempting to minify the usage of petrochemicals in the

S. A. GadAllah (✉) · M. M. Aboulnaga
Department of Architecture, Faculty of Engineering, Cairo University, Giza, Egypt
e-mail: maboulnaga@eng.cu.edu.eg

© The Author(s), under exclusive license to Springer Nature Switzerland AG 2022

A. Sayigh (ed.), *Sustainable Energy Development and Innovation*, Innovative Renewable Energy,
https://doi.org/10.1007/978-3-030-76221-6_67

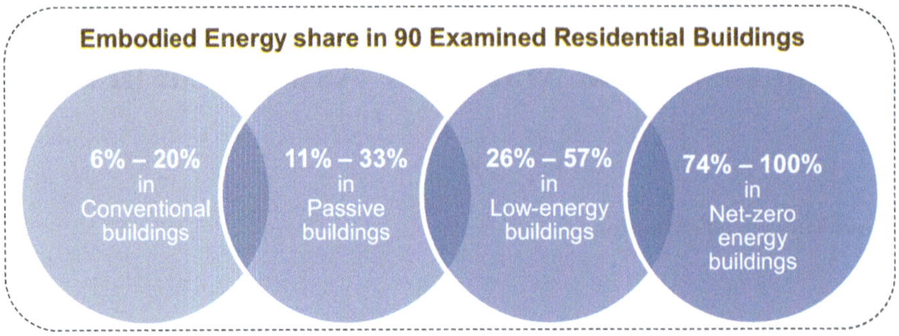

Fig. 1 Share of embodied energy residential buildings [4]. (*Source:* https://www.sciencedirect.com/since/article/pii/B9780128117491000043)

manufacturing of buildings' thermal insulation materials, and concentrated on finding replacements for these nonrenewable sources from natural or recycled content. That leads to a substantial link with circular economy principles that are based on the recirculation of building materials that can aid in minimizing the usage of depleted energy resources [11]. In other words, circular economy "CE" in materials can be defined as a closed-loop system, where products can be retrieved from wastes to be raw materials. Manufacturing recognizable value-added new green or natural materials and products is the ultimate target of circular economy [12].

This paper investigates different types of buildings to review the status of thermal insulation materials in Egypt and to explore the concept of CE in reducing energy demand and mitigating CO_2 emissions.

2 Objectives

The aim of this work is to examine the thermal insulation materials' status in Egypt and to explore different examples incorporating thermal insulation materials to identify the gap in this sector. The study also exhibits the concept of Circular Economy (CE) as new principle to exploiting natural thermal insulation materials in buildings through identifying Egypt's strategic crops' areas that can be used in the manufacturing of natural thermal insulation materials.

3 Methodology

In order to investigate the status of thermal insulation materials, a methodology is developed based on analytical approach that utilizes a qualitative assessment. This was carried out in six phases, including: (a) examining different types of buildings that incorporated thermal insulation materials in Egypt. To pursue this study, the second stage was conducted by considering the circular economy (CE) concept. This was followed by the third stage to review and discuss agricultural waste to be utilized in the production of natural sustainable thermal insulation materials. In order to pursue this assessment, it was vital to conduct stage 4 to determine leading agricultural crops' areas that can be used for this purpose in Egypt. Hence, 28 governorates were examined in terms of their potential crops and significance of residual waste to be used for natural thermal insulation materials (stage 5) and then finally carry out stage 6 to mapping leading crops and select the most potential and significant ones in the production of natural thermal insulation materials in Egypt. The methodology analytical approach for qualitative assessment is illustrated in Fig. 2. The building types that are reviewed include Healthcare, Hospitality, Exhibitions, and Commercial, which has two categories: office buildings and banks as presented in Fig. 3.

When it comes to marking the crops' regions that can aid in the manufacturing of natural thermal insulation crops, mapping of these crops is required to identify and classify the areas of potential crops in each governorate. The assessment methodology of total average area of all crops depends on the areas of Egypt's four major crops in different climatic zones, divided by their number.

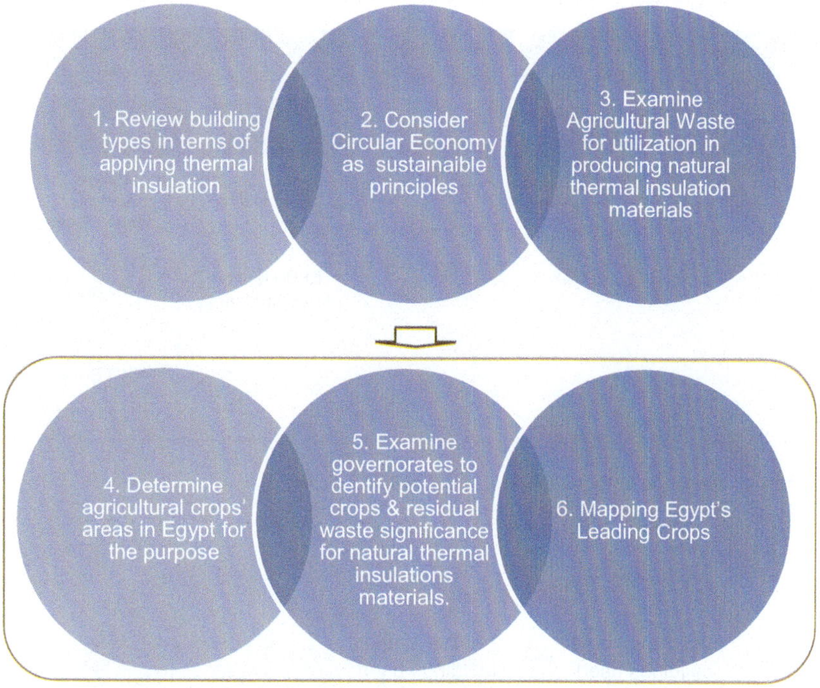

Fig. 2 Methodology analytical approach for qualitative assessment of the study

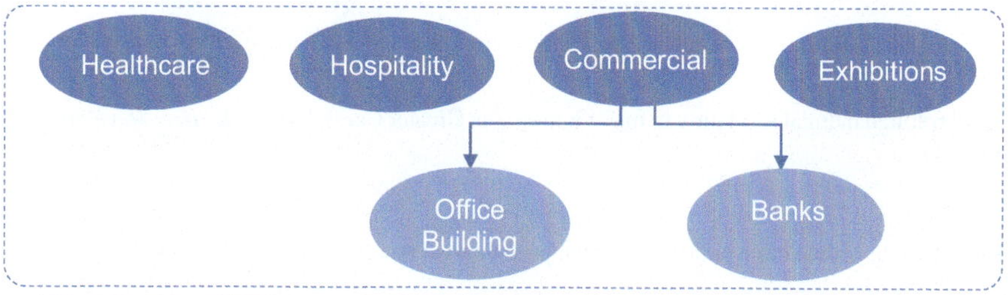

Fig. 3 Building types under examination in the study

4 Egyptian Buildings with Thermal Insulation Materials

In the first phase of the assessment, it was important to review the building types depicting buildings' examples that incorporate thermal insulation materials. These buildings represent many types such as healthcare, hospitality, exhibitions, and commercial buildings. Eight major buildings were selected and reviewed to find out that the specific type of thermal insulation materials stipulates their buildings' envelops; in attempt to map the most common used thermal insulation.

Figure 4 presents these buildings' types that were selected based on set of criteria such as data, access availability, and newly constructed buildings as well as incorporating insulation materials.

Based on the review, it is imperative to state that there is almost no building with natural or unconventional thermal insulation materials, yet in Egypt. Subsequent to mentioning the eight examples of buildings with external thermal insulation in Egypt, it is found out that almost 62.5% of the studied buildings tend to use one of the types that is extruded polystyrene (five buildings) as the best choice for thermal insulation as shown in Fig. 5 and Table 1. It is imperative to state that even Credit Agricole Bank Headquarters in Cairo, which is a LEED certificate Green building applied the same thermal insulation [25], in addition to rock wool, neglecting the opportunity or possibility of using natural insulation materials.

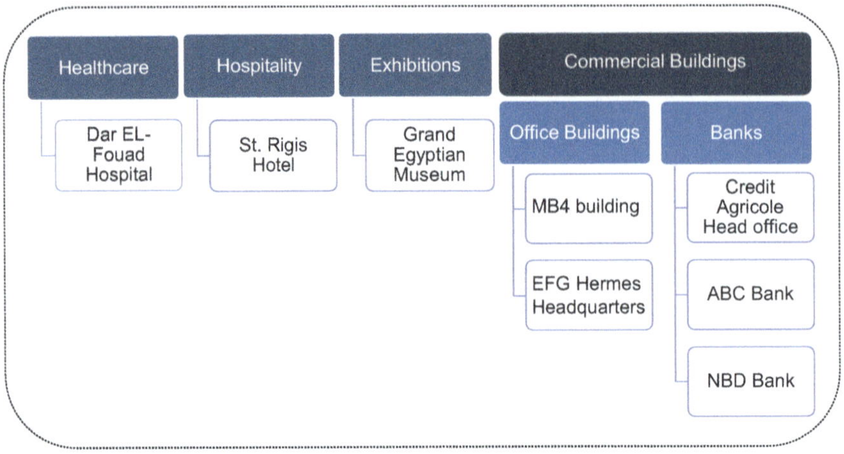

Fig. 4 List of buildings with thermal insulation in Egypt. (*Source: Developed by Authors*)

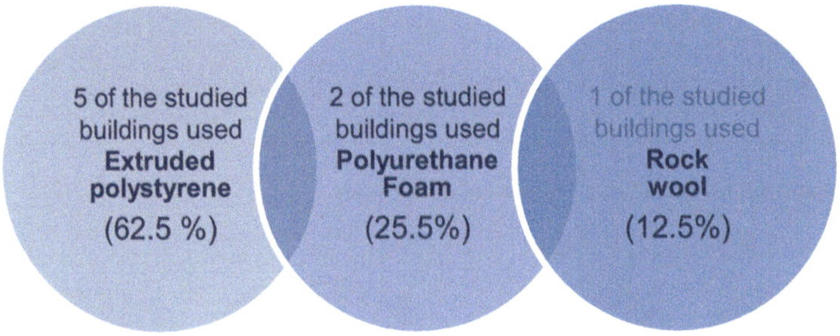

Fig. 5 Exploited thermal insulation in major building in Greater Cairo, Egypt. (*Source: Developed by Authors*)

Table 1 List of studied main buildings for mapping thermal insulation materials in Egypt

Building Name	1. Dar Elfouad Hospital	2. St. Rigis, Nile Corniche Hotel	3. Grand Egyptian Museum	4. MB4 Building	5. EFG Hermes	6. Credit Agricole	7. ABC	8. NBD
Building Image	 [11]	 [14]	 [17]	 [19]	 [21]	 [23]	 Credit: Author 2018	 [28]
Building Type	Healthcare	Hospitality	Exhibitions	Commercial (Office Buildings)		Commercial (Banks)		
Designer	RBSD architects [11]	Michael Graves [14]	Heneghan Peng Architects [17]	ECG [19]	ECG [21]	Atelier Tom Sheehan & Partners [23]	ECG [26]	ECG [28]
Contractor	Orascom Construction [12]	Consolidated Contractors Company (CCC) [15]	Orascom Construction and BESIX [17]	Shuttering Construction [19]	Orascom Construction [22]	Orascom Construction [24]	Redcon Construction [26]	Orascom Construction [28]
Total Built-up Area	61,000 m² [12]	200,000 m² [15]	100,000 m² [17]	24,000 m² [19]	28,000 m² [21]	42,000 m² [25]	12,700 m² [27]	12,500 m² [28]
Used Thermal Insulation types	Rock wool [13]	Extruded Polystyrene [16]	Polyurethane Foam [18]	Polyurethane Foam [20]	Extruded Polystyrene [21]	Extruded Polystyrene/ Rock wool	Extruded Polystyrene [27]	Extruded Polystyrene [28]

Source: Developed by authors

5 Thermal Insulation Is Not Generalized in Egypt

In fact, Egypt faces profound challenges in implementing the developed energy efficiency codes in buildings, where efficient national governmental leadership is essential to develop basic compliance with the code and the thermal insulation standard; due to their limited implementation [31]. Consequently, upon reviewing the Ministry of Housing residential projects, such as "Janna," "Egypt's Housing," and "Social Housing" projects, it is ascertained that there is an absence of external thermal insulation [32]. This is may be due to the cost of applying thermal insulation materials.

As mentioned above and as a result of the non-enforcement of laws regarding environmental efficiency that includes adequate thermal insulation materials, thermal insulation is thus not generalized in Egypt due to six substantial reasons [33]. Figure 6 illustrates these reasons.

6 Circular Economy

Subsequent to introducing the reasons for not generalizing thermal insulation materials in Egypt, this part of the study introduces the concept of Circular Economy (CE) and its relation with the usage of thermal insulation materials. A respectful number of studies have concentrated on attempting to minify the usage of petrochemicals in the manufacturing of buildings' thermal insulation materials. These studies also focus on finding replacements for these non-renewable sources from natural or recycled content [11]. That leads to a substantial link with CE principles that are based on recirculation of building materials that aid in minimizing the usage of depleted energy resources [11]. Alternatively, CE in materials can be defined as a closed-loop system, where products can be retrieved from wastes to be raw materials [12].

Sustainable CE can be achieved by the successful management of the world's waste materials. The idea of materials' production from waste is generated by rethinking of the nature of waste, cooperating with industry partners and creating direct contact between potential end users and waste sources. It can be clearly stated that CE contributes in solving the global environmental problems [12].

Figure 7 illustrates the concept of CE in terms of waste materials' four principles (reduce, reuse, recycle, and remanufacture).

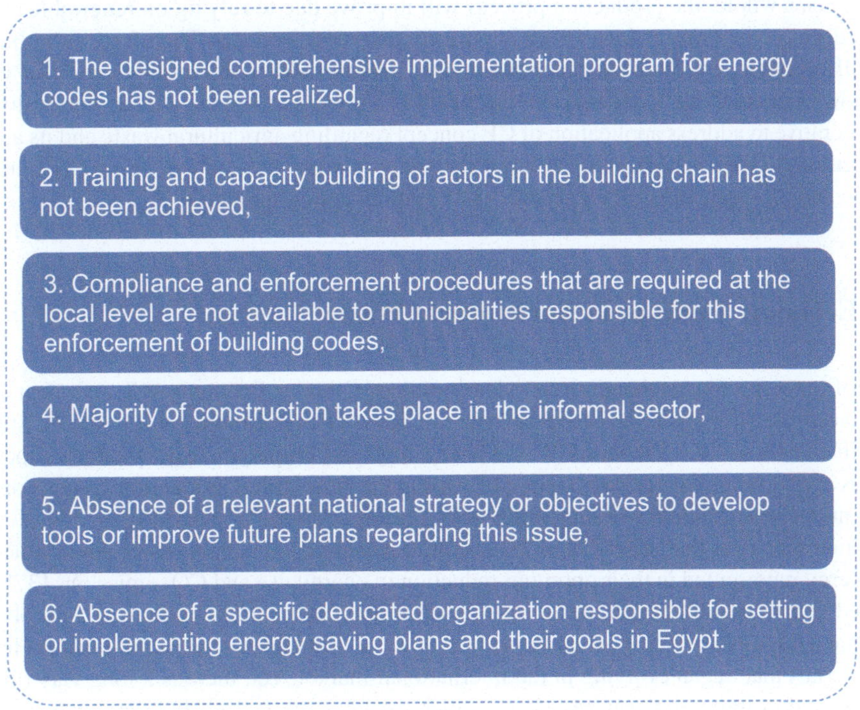

Fig. 6 Reasons why thermal insulation is not generalized in Egypt. (*Source: World Bank 2010*)

Fig. 7 Circular economy—waste materials concept. (*Source: Developed by authors, after: European Materials Research Society 2018*)

In the present time, polystyrene, polyurethane, and stone wool are used as the main conventional thermal insulation materials, where they are non-renewable, with their downside represented in the usage of fossil fuels, increasing environmental pollution and global warming. It is commonly acknowledged that CE is necessary to reverse this trend. Circular economy concept is clearly represented in the use of natural insulation materials as substitutes to conventional ones. In fact, its goal is to dissociate the economic activity from the non-renewable energy resources consumption [34].

7 Benefits of Applying Circular Economy Concept in Egypt

When it comes to tackling the issue of applying CE concept in Egypt; theoretically, anything can be applied. However, realistically, the dynamics that will govern such a concept are quite different than on paper. Instead of the disposal of the product after a certain life span, the concept of CE replaces the disposal concept with the repair and recycle of the material, giving a circular criterion to the life cycle of the product. With the flotation of the Pound already in place, and the purchasing power of the average Egyptian that has remarkably reduced due to rising prices; CE seems like a successful solution. For producers and manufacturers, it's still not that widespread. Given the right business environment, major manufacturers and producers will be on board. It's important that for such a concept to be applied, that manufacturers and producers align themselves with their respective supply chains to create options for recycling and repairing of their specific brands to give their customers the option to reuse them [35].

When adopting CE concept and principles in Egypt, the Egyptian economy will realize some major benefits. Egyptian businesses will witness a major breakthrough regarding cost minimization; where the cost of recycling or reusing components from their previously made products will be less than producing these components or buying them from their respective supply chains. Second, energy efficiency will be maximized, owing to the reduction in the usage of energy sources and encouraging the use of renewable energy, which would be cleaner and less costly. As for the consumers, their purchasing power will be developed; given that they will reuse and recycle their products, instead of buying new ones [35]. As for the Egyptian economy as a whole, the production process will improve due to relying on renewable energy sources in addition to the significant decrease in harmful waste from previously disposed products. Therefore, air pollution will be reduced and quality of air will be improved. With revenues maximized from production, the application of CE concept can increase investments' inflows and create new business opportunities and growth in Egypt, adding more value into the economy [35]. It is therefore imperative to address application of CE concept regarding agricultural waste and its application on the residues of Egyptian agricultural crops. The next sections of the study address the application of CE concept/principles regarding agricultural waste and its application on the residues of Egyptian agricultural crops.

8 Circular Economy Concept in Agricultural Waste

The production of agricultural merchandise can clearly reflect Circular Economy (CE) concept. Employing this concept perspective to investigate the whole agri-food system can uncover opportunities at all stages, starting from fundamental manufacturing till the phase of agricultural wastes' recycling and employment to produce food, materials, and energy. Resources such as agricultural crop residues cannot be used in the agricultural production systems only; it can also be utilized to produce food and materials, while still remaining within the agricultural system [36].

In fact, insulation materials that are made of natural or recycled contents have lower environmental costs, in addition to low embodied energy represented in their optimum reduction in energy use and CO_2 emissions [37]. Hence, building insulation manufactured from agricultural residues is an important approach. After a review of agricultural wastes background, despite the fact that a lot of research papers addressed the issue of using agricultural crop residues in insulation and the natural insulation materials that began evolving in the international market. Yet, there is a long way till these materials become from the commonly used items in the market [38], which explains the lack of building examples that actually use these materials in insulation.

In light of these studies, the next section tries to spread the usage of low impact thermal insulation materials to achieve sustainability of buildings through the usage of agricultural crop residues in Egypt, aiming to apply the concept of CE by exhibiting the map of Egyptian major crops in each governorate, along with choosing the adequate ones in Egypt that possess crops which can be converted into natural thermal insulation materials to lower the embodied energy of materials through the saved energy in the transportation by the localizing of materials, which is looked upon as the fundamental element in embodied energy [39].

9 Mapping Egypt's Leading Crops

Although Egypt cultivates a lot of crops, it is specialized with four main crops, wheat, cotton, sugarcane, and rice, which are beneficial in producing natural thermal insulation materials. Table 2 illustrates these cultivated crops areas in feddans [40], showing the largest and slightest areas of crops in each governorate. In fact, climate is a key factor in crop production. Weather conditions and water availability directly affect and define the agricultural land area [41]. Hence, two governorates in two opposing climatic zones are chosen, due to the variance in their crops areas and availability. The assessment methodology of total average area of all crops mentioned in Table 2 depends on the areas of the four crops in different climatic zones, divided by their number.

10 Results and Discussions

Based on Table 2 average crops' areas and Fig. 8 as well as the area of each crop that will be beneficial, Dakahlia governorate turned to be the pioneer of all other governorates in the agriculture of major crops, followed by ElSharqia, Kafr El Sheikh, and Beheira governorates, respectively. However, these three governorates have humble areas of sugarcane crop, which is important in manufacturing natural thermal insulation material out of it. For that reason, Dakahlia is the perfect location for creating natural thermal insulation materials out of its main crops, represented in Fig. 9. Meanwhile, it is considered in the northern Egypt, which makes Qena the pioneer governorate in its average area of crops in South Egypt, with a respectable area of wheat and sugarcane crops, followed by Sohag governorate. Although Qena does not cultivate cotton, it has the largest area of sugarcane crop, as shown in Fig. 10. As a result, Qena is the chosen governorate in South Egypt for the production of natural thermal insulation materials out of its crops.

11 Conclusions

According to some of the examples of buildings' thermal insulation materials used in Egypt and Table 1, extruded polystyrene is found out to be used in most of the buildings with external walls' thermal insulation, despite its mentioned harmful impacts. On top of that, after reviewing some private and public residential projects, it is found out that neither the private sector nor the Ministry of Housing residential projects apply external walls thermal insulation in their residential units.

Considering the fact that Egypt's development in thermal insulation materials is not put in a practical application around the country, it leads to highlighting six main reasons for not propagating the use of thermal insulation materials in Egypt. Thereby, as an attempt to spread the usage of low impact thermal insulation materials, minimize the usage of nonrenewable energy sources, and accordingly minimize the buildings' embodied energy and carbon foot print, circular economy significance that lies in the reuse of materials, is studied in relevance to the usage of agricultural waste. That leads to studying CE in Egypt and its main crops' map, where climate controls the crop production and defines the agricultural land area. That leads to marking the convenient governorates in North and South Egypt due to the variance in their weather; that directly affects their crops' areas and the possession of agricultural crops which can be converted into natural thermal insulation materials; where Dakahlia and Qena are proved to be the best governorates for this goal, due to their vast average areas of crops that can be utilized for the purpose of thermal insulation.

Table 2 Egyptian governorates' strategic crops' areas in feddans[a]

**Feddan = 1 Acre = 4800 Sq.m*

Governorate Name	Wheat	Cotton	Sugarcane	Rice	Total Average area of all crops
Egypt climatic zones:		0	14	3	24.5
Alexandria	78297	5786	6	2046	21533.75
Port Said	16704	1372	0	19985	9515.25
Suez	4664	0	14	0	1169.5
Damietta	29604	6853	30	66922	25852.25
Dakahlia	303291	40924	570	454697	199870.5
Sharqia	425040	39796	28	262820	181921
Qalyubia	53079	293	758	15439	17392.25
Kafr El Sheikh	236281	90238	85	290128	154183
Gharbia	151032	16841	1439	147042	79088.5
Monufia	126592	3190	22	1308	32778
Beheira	321524	84379	187	199888	151494.5
Ismailia	53798	1879	0	6148	15456.25
Giza	40538	0	1870	0	10602
Beni Suef	126125	6865	676	1076	33685.5
Faiyum	174828	20349	539	746	49115.5
Minya	218657	1438	39133	0	64807
Asyut	190899	4431	1798	0	49282
Sohag	175041	1941	16258	0	48310
Qena	106120	0	117739	0	55964.75
Aswan	52253	0	82282	0	33633.75
Luxor	40360	0	61907	0	25566.75
Red Sea	0	0	0	0	0
New Valley	103329	32	0	8344	27926.25
Matruh	4490	0	0	0	1122.5
North Sinai	251	0	0	0	62.75
South Sinai	211	0	0	0	52.75
Nubaria	127571	6753	450	278	33763

North Coast region
Delta and Cairo Region
North Upper Egypt region
Southern Upper Egypt Region
East Coast region
Altiplano Region
Desert region
Southern Egypt Region

Source: Developed by Authors after: www.arabgeographers.net
[a]*Feddan = 1 Acre = 4800 Sq.m*

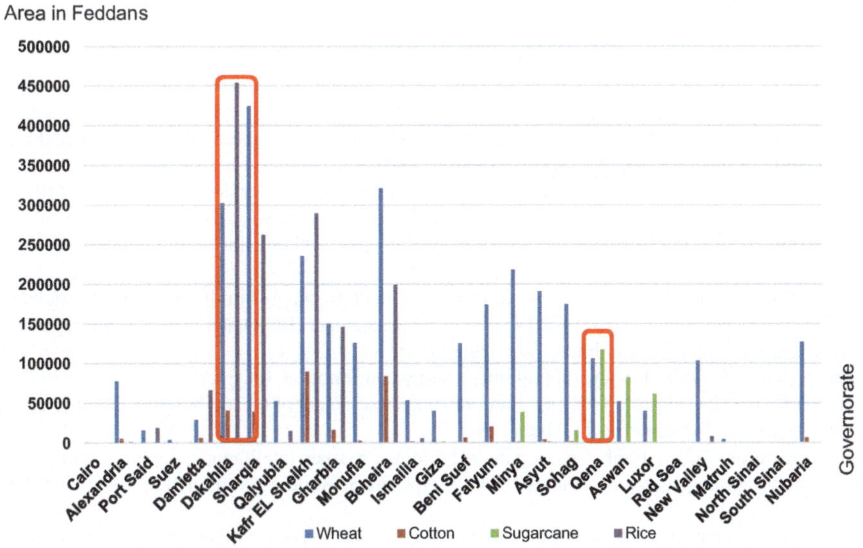

Fig. 8 Governorates' crops' areas in Egypt. (*Source: Developed by authors after:* www.arabgeographers.net)

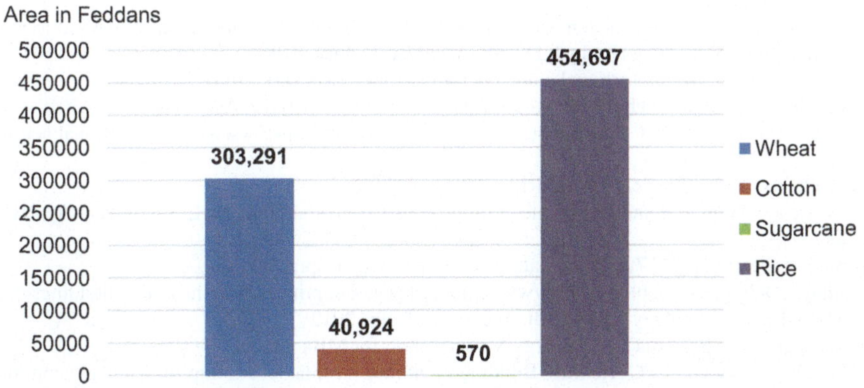

Fig. 9 Dakahlia Governorate's leading crops—Egypt. (*Source: Developed by Authors after:* www.arabgeographers.net)

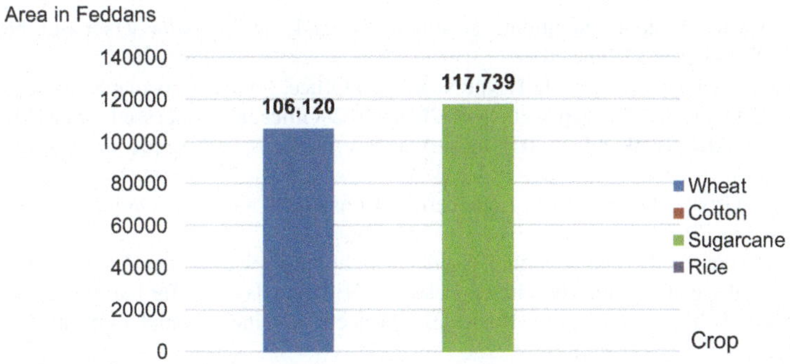

Fig. 10 Qena Governorate's leading crops areas in Egypt. (*Source: developed by authors after:* www.arabgeographers.net)

References

1. Asdrubali F, D'Alessandro F, Schiavoni S (2015) A review of unconventional sustainable building insulation materials. Sustain Mater Technol 4:1–17
2. Australian Government; Your home - Australia's guide to environmental sustainable homes; embodied energy [Online]. Available at: https://www.youhome.gov.au/materials/embodied-energy/. Accessed 4 May 2020
3. Ramesh T, Prakash R, Shukla KK (2010) Life cycle energy analysis of buildings: An overview. Ener Build 42(10):1592–1600
4. Azari R (2019) Life cycle energy consumption of buildings; embodied + operational, Book Chapter 5, Sustainable construction technologies; life-cycle assessment. Elsevier, p 123–144. Available at: https://www.sciencedirect.com/since/article/pii/B9780128117491000043/. Accessed 4 May 2020
5. Sartori I, Hestnes AG (2007) Energy use in the life cycle of conventional and low-energy buildings: a review article. Ener Build 39(3):249–257
6. Cabeza LF, Rincón L, Vilariño V, Pérez G, Castell A (2014) Life cycle assessment (LCA) and life cycle energy analysis (LCEA) of buildings and the building sector: a review. Renew Sustain Energy Rev 29:394–416
7. Perez-Lombard L, Ortiz J, Pout CA (2008) Review on buildings energy consumption. Inf Energy Build 40(3):394
8. Sierra-Pérez J, Boschmonart-Rives J, Gabarrell X (2016) Environmental assessment of façade-building systems and thermal insulation materials for different climatic conditions. J Clean Prod 113:102–113
9. International Finance Corporation, World Bank Group, Green buildings market intelligence, Egypt Country Profile. [Online]. Available at https://www.edgebuildings.com/wpcontent/uploads/2017/09/Egypt-Green-Building-Market-Intelligence-EXPORT.pdf/. Accessed 15 Jan 2018
10. Elagamy H, Badr M, Magdy A, Shahin W. Interviewees, thermal insulation in Egyptian residential buildings. [Interview]. 12 Dec 2019
11. Eberhardt L, Birgisdottir H, Birkved M et al (2019) Mater Sci Eng 471:092051
12. European Materials Research Society (2017) Materials innovation for the global circular economy and sustainable society. In VI World Materials Summit, Strasbourg
13. Sika. Dar ELFouad Hospital and Medical Tower. [Online]. Available at: http://www.sika.com/dms/getdocument.../saw_Dar_EL_Foad_Hospital_egypt.pdf./. Accessed 14 Jan 2018
14. Orascom Construction (2017) Orascom Construction Annual Report 2017. [Online]. Available at: http://www.orascom.com/media/cms_page_media/47/OCL%20Annual%20Report%202017_zLy4gD9.pdf. Accessed 16 June 2019
15. I. c. f. i. (Rockal). Previous projects list. International company for insulation (Rockal), Cairo, 2015/2016
16. 15 7 2018. [Online]. Available: www.michaelgraves.com/portfolio/nile-corniche-st-rigis/. Accessed 28 Dec 2017
17. CNL International, Nile Corniche Project- Cairo, Egypt, [Online], available at: https://www.cmlinternational.net/projects/ncr/. Accessed 13 June 2019
18. ECG Engineering Consultants Group-Cairo (2015) "St Rigis site report," Cairo
19. Design Build-Network. Design build. [Online]. Available at: https://www.designbuild-network.com/projects/grand-egyptian-museum-cairo/. Accessed 14 Aug 2018
20. Sika Egypt for construction chemicals (2017) Product data sheet-Sika PU foam spray
21. Shuttering. MB4 building. [Online]. Available at: http://www.shutteringeg.com/projects/administrative-buildings-contracting-steel-structure-companies-in-egypt/34-mb4-new-technology-area-maadi. Accessed 13 June 2019
22. ECG Engineering Consultants Group-Cairo (2012) Energy Model Report for MB4 project
23. ECG-Engineering Consultants Group. EFG Hermes Building. [Online]. Available at: http://www.ecgsa.com/project/113. Accessed 14 Aug 2018
24. United Import & Export Co. Sample list of supplied projects. [Online]. Available at: http://www.uiecable.com/pdf/image%20(1).pdf/. Accessed 16 June 2019
25. Ongreening. "Ongreening," [Online]. Available at: https://ongreening.com/en/Projects/credit-agricole-egypt-headquarters-1211#info. Accessed 13 June 2019
26. Master Builders Solutions. Master Builders Solutions. [Online]. Available at: https://assets.master-builders-solutions.basf.com/en-eg/credit%20agricole%20bank.pdf. Accessed 13 June 2019
27. ECG-Engineering Consultants Group. Credit Agricole Egypt New Head Office. [Online]. Available at: http://ecgsa.com/backoffice/Content/project/8/pdf/Cr%C3%A9dit%20Agricole%20Egypt%20New%20Head%20Office.pdf/. Accessed 3 Mar 2018
28. Redcon Construction Company. ABC Bank Egypt. [Online]. Available at: https://www.redconcon.com/ar/project/abc-bank-headquarter/. Accessed 13 June 2019
29. ECG-Engineering Consultants Group. Projects-ECG-Engineering Consultants Group. [Online]. Available at: http://www.ecgsa.com/project/4/-1/-1/-1. Accessed 14 Aug 2018
30. ECG-Engineering Consultants Group. NBD Bank. [Online]. Available at: http://www.ecgsa.com/project/4346/-1/-1/-1. Accessed 14 Aug 2018
31. Attia S, Wanas O (2011) The database of Egyptian building envelopes (DEBE): a database for building energy simulations
32. Ministry of Housing, Utilities & Urban Communities. Residential Projects. [Online]. Available at: http://www.mhuc.gov.eg/programs/programs/2. Accessed 16 Oct 2019
33. World Bank Working Paper No. 204 (2010) Mainstreaming building energy efficiency codes in developing countries. The World Bank, Washington
34. Better World Solutions. Circular insulation materials for a better future. [Online]. Available at: https://www.betterworldsolutions.eu/circular-insulation-materials/. Accessed 24 Sept 2019
35. Egypt Business Directory. Applying circular economy in Egypt. 2 2018. [Online]. Available at: https://www.egypt-business.com/news/details/1806-Applying-Circular-Economy-in-Egypt/217479. Accessed 17 Nov 2019
36. Ward SM, Holden NM, White EP, Oldfield TL (2016) The 'circular economy' applied to the agriculture (livestock production) sector. In Workshop on the sustainability of the EU's livestock production systems, Brussels
37. Praktijkonderzoek Plant and Omgeving (Applied Plant Research) (2015) Health, comfort, energy use and sustainability issues related to the use of biobased building materials

38. Balador Z, Gjerde M, Saacs N, Imani M (2018) Thermal and acoustic building insulations from agricultural wastes. In Handbook of eco-materials. Springer International Publishing, Switzerland
39. Sattary S, Thorpe D (2012) Optimizing embodied energy of building construction through bioclimatic principles. In 28th Annual ARCOM Conference, Edinburgh
40. Arab Geographers, "Arab Geographers," [Online]. Available at: https://www.google.com/imgres?imgurl=http%3A%2F%2Fwww.arabgeographers.net%2Fvb%2Fattachments%2Fattachments%2Farab3393%2F&imgrefurl=http%3A%2F%2Fwww.arabgeographers.net%2Fvb%2Fthreads%2Farab19159%2F&docid=W_GdrJB49bKilM&tbnid=ZCO1st3XeKl17M%3A&vet=10ahUKEw. Accessed 21 Oct 2018
41. Khalil H, Ouda S, Osman NA, Khamis AE (2011) Determination of agro-climatic zones in Egypt using a robust statistical procedure. In Fifteenth International Water Technology Conference, Alexandria

Future of Renewable Energy in India

Pradeep Chaturvedi

1 Introduction

India is poised for a big shift in its energy supply over the next decade which also coincides with the targeted date of 2030 for attaining sustainable development goals. The pandemic COVID-19 has impacted all countries in the world with uncertainties of future economic development. This has adversely affected national and international efforts to move toward "sustainable energy supply in an affordable, reliable and modern energy form for all." The impact of accelerated progress over the past decade has lost momentum. Only heightened economic progress with sufficient safeguards can ensure attaining SDG 7 which aims at "energy for all."

The silver lining is that a way to boost a photovoltaic cell's efficiency has been observed by researchers, and efficiency of 47.1% in laboratory conditions has been reported. With further work, they reckon, efficiency of more than 50% should be possible. John Geisz and his colleagues at the National Renewable Energy Laboratory in Colorado, USA, have produced six junction cell containing various III–V mixtures, each with different light-absorbing properties. It is claimed that the efficiency of the cell rises as more light is concentrated on it. Laid out in standard solar platform, it would manage just under 40%. This practical application has still to be verified; however, it opens an opportunity for bringing down the solar photovoltaic electricity cost.

2 India's Commitment for Renewable Energy Use

As of March 31, 2020, 36% of India's installed electricity generation capacity is from renewable sources, generating 21.22% of total utility electricity in the country.

In the Paris Agreement, India has committed to intended nationally determined contribution of achieving 40% of its total electricity generation from non-fossil fuel sources by 2030. India has further raised its target of 50% of the total electricity capacity from renewable sources by 2027 in Central Electricity Authority's Strategy Blue Print. According to 2027 Blue Print, India aims to have 275 GW from renewable energy, 72 GW of hydroelectricity, 15 GW of nuclear energy, and nearly 100 GW from other zero emission sources. India has further raised its desire to have higher capacity of renewable energy-based electricity generation in India.

In 2019 at UN Climate Summit, India announced that it will be more than doubling its renewable energy target from 175 GW by 2022 to 450 GW by 2030. Newer renewable electricity sources are projected to grow massively by near term 2022 target, including a more than doubling of India's large wind power capacity and an almost 15-fold increase in solar power from April 2016 levels. These targets would place India among the world leaders in renewable energy use and place India at the center of its "Sunshine Countries" International Solar Alliance Project promoting the growth and development of solar power internationally to over 120 countries.

On November 30, 2015, the Prime Minister of India and the President of France launched the International Solar Alliance. ISA is an alliance of 121 solar-rich countries lying partially or fully between the Tropic of Cancer and Tropic of Capricorn;

P. Chaturvedi (✉)
Indian Association for the Advancement of Science, New Delhi, India

© The Author(s), under exclusive license to Springer Nature Switzerland AG 2022
A. Sayigh (ed.), *Sustainable Energy Development and Innovation*, Innovative Renewable Energy,
https://doi.org/10.1007/978-3-030-76221-6_68

several counties outside of this area are also involved with the organization. ISA aims to promote and develop solar power amongst its members and has the objective of mobilizing US$ 1 Trillion of investment by 2030.

An expectation of installing 450 GW of renewable power by 2030 means that India has to add 375 GW in the next 10 years, which is equivalent to the present total installed capacity from all sources in the country. It will pose major challenges that will include land acquisition, equipment availability and supply chain, and the grid availability and connectivity. Physical execution challenge is very high and a pragmatic assessment has to be undertaken to understand how the market capacity can be enhanced to meet the target. The power demand will continue to grow and even beyond 2030 India will need to have a long-term policy, financial arrangements, appropriate manufacturing infrastructure and supply chain and execution policy in place both at the central government level as well as the state governments. It is necessary that the industry gets all support from the government.

Renewable energy prices have been declining over time and it is to be seen as to how low these will go. Opportunities do exist as the prices come down and that of the conventional sources are on the increase. The quality of installations is the determining factor for larger acceptability of renewable energy sources. Contribution of renewable energy sources should not be considered as base load supply for which sufficient thermal capacity is available. The storage capacity required for renewable-based electricity supply also depends on notice time required for ramping-up power supply from thermal sources and as such 1–2 hours of storage may be sufficient for balancing power grid supply backed-up by thermal-based electricity supply.

India was the first country in the world to set up a Ministry of Non-Conventional Sources of Energy, in the early 1980s, and its Public Sector Undertaking, Solar Energy Corporation of India is responsible for the development of solar energy projects in India. The Indian Renewable Energy Development Agency is responsible for the development of solar energy industry in India by arranging finance.

The development of wind power in India began in 1990s and has significantly increased in the last few years. India has become the fourth largest wind power capacity in the world. The government announced a new wind-solar hybrid policy in May 2018. This means that the same piece of land will be used to house both the wind farm and the solar panels.

Much of the countries did not have an electrical grid, so one of the first applications of solar power was for water pumping, to begin replacing India's 4–5 million diesel-powered water pumps, each consuming about 3.5 Kw, and off-grid lighting. Some large projects have been proposed, and 35,000 Sq Km area of the Thar Desert has been set aside for solar power projects, sufficient to generate 700–2100 GW. Solar power in India has been growing at a rate of 113% year on year. Cost of solar-based electricity has now dropped to less than US 3 cent, which is around 50% less than the average price for electricity generated by coal-fired plants. India is also richly endowed with biomass resources. The country's vast agricultural potential provides huge agro-residues which can be used to meet energy needs, both in heat and power applications. According to Indian Renewable Energy Development Agency, biomass is capable of supplementing coal to the tune of about 260 mt.

3 Large Provider of Jobs

After China and Brazil, India is the third largest job provider in renewable energy sector. It provides 8.24 lac jobs and finds a study by International Renewable Energy Agency.

Solar photovoltaic segment provides the largest numbers of both direct and indirect jobs—2,04,000 jobs—in India. On-grid solar sector provides an estimated 1,09,000 jobs while another 95,000 people were employed in the off-grid sector.

In case of hydropower, the country accounts for 19% of the 1.93 million people employed worldwide. According to IRENA, India employs around 3,67,000 people in hydropower segment.

India is the fourth largest market in wind energy in terms of capacity. Around 63,000 jobs have been created by wind energy sector. However, the slowing pace of new installations has resulted in extensive competition and oversupply of equipment. The number of wind turbine makers has come down significantly in recent times and this has also led to job losses.

Solar heating and cooling solutions have offered employment to over 23,000 people. Though India is not a big player in liquid biofuels globally, the segment provides 35,000 jobs. Solid biomass and biogas too have contributed to the total employment generated in the renewable space.

According to IRENA, decentralized renewable energy solution companies across different segments directly employed 95,000 workers in India and that number is expected to increase in time to come. Decentralized applications have the potential to create up to five times more jobs in local communities than direct formal employment.

4 Transition of Renewable Energy Helps in Reducing SO_2 Emissions

India witnessed 6% decline in its sulfur dioxide (SO_2) emissions in 2019. It was the steepest decline in SO_2 emissions in 4 years, which came on account of India's shift toward renewable energy initiatives. This reduction was realized mainly due to reduction in coal usage and thereby the air quality and health.

India, however, has continued to occupy the top spot in the world in SO_2 emissions for the fifth consecutive years. SO_2 is a poisonous air pollutant that increases the risk of stroke, heart disease, lung cancer, and other health issues that can lead to premature death.

"In 2019," India emitted 21% of global anthropogenic (human made) SO_2 emissions, nearly double that of second rank global emitter, Russia.

While COVID-19-induced lockdown did give a breather to the overexploitation of the environment, the pressure to resume operations at a faster pace to make up for lost business and time is bringing back the same level of pollution.

5 Coal-Fired Power Plants Being Replaced with Renewables

India is planning to replace retiring coal-fired power plants with renewable generating capacity in an effort to cut the nation's carbon footprint. Some coal-fired plants have already retired and 29 more plants are going to retire, and all that space will be occupied by renewable energy.

6 Government Interventions

India's current annual installed and functional manufacturing capacity of solar modules is around 10 GW while that of solar cells is only around 2.5 GW. The government has the plan to raise the capacity to around 25 GW per annum for solar cells and modules in another 2 years' time. To protect local industry from cheaper imports, India has extended Safeguard Duty on imports of solar cells and modules by another year (i.e., end-2021), and is planning an import tax on solar cells and modules.

The government has also decided to increase the level of import duty on solar invertors, which are connected to solar panels to convert the direct current power into usable electricity. India is also trying to create manufacturing hubs for renewable energy equipment with the help of private sector industry.

The Ministry of New and Renewable Energy has been announcing schemes with domestic content requirement (DCR); for example, the KUSUM scheme for solar pumps that promises a solar capacity over 25 GW by 2022. The Central Public Sector Undertakings (CPSU) are planning second phase with a target of 12 GW by 2022, the residential Rooftop Solar Program with provision of government subsidy, etc.

7 Challenges of Solar Cell Manufacturing Industry

Solar cell is a high technology product and continuous improvement has been taking place all over the world to minimize material requirement, to increase efficiency, and to improve reliability with the sole objective to bring down the cost. Setting up a modern cell manufacturing facility is an expensive proposition with an estimated cost of around Rs. 80 crores for the equipment part only of a 100 MW solar cell manufacturing facility. Over and above one requires investment for land and infrastructure. The final product cost needs to be reasonable in the highly competitive market where the cell prices have come down drastically. To bring economies of scale, it is therefore necessary to set up a very large manufacturing facility with the latest technology and with necessary provision of changing the production line continuously with ever-emerging improvements that will be taking place over the life cycle of the plant. But this means further investments in production facility. The related infrastructure and manufacturing ecosystem also needs to be in place to enable the production to be competitive. Also there must be very efficient outgo of the products in the country and elsewhere so that the unit becomes financially viable.

To make this possible, government intervention is essential, and it must be a concerted intervention with long-term policy for the development of manufacturing ecosystem, bringing low-cost finance as is available in international markets, provision of incentives to keep the product at the best all the time with necessary changes in the production line, ensuring world class testing and standardization facility, etc.

8 Finance an Important Element

India has not been able to provide comfort to lenders and investors on enforceability of long-term contracts. "Off-taker risk" with regard to "must-run status" has also not been fully enforceable. The above two factors increase the risk premium on financing.

India has extreme potential to export renewable energy equipment to Africa. ISA is working very closely to aggregate the demand and facilitating deployment on the ground.

9 Conclusion

With regard to accelerating renewable energy implementation, following steps will be necessary:

(i) Feed in tariff for smaller projects say 10 MW to open up retail market.
(ii) Waiver or rationalization of open access charges for captive/third-party sale shall open up huge market
(iii) For interstate transmission charges long-term visibility should be there.
(iv) For manufacturing support in the form of uninterrupted power supply, water supply, environment clearance, etc., should be provided

Renewable energy is key to achieving SDG 7 and building resilient, equitable, and sustainable economies in a post-COVID-19 world. Now more than ever is the time for bold international cooperation to bridge the energy access gap and place sustainable energy at the heart of economic stimulus and recovery measures. Solar energy alliance established by the Government of India, headquartered in India, has important role to play along with International Energy Agency of the International Renewable Energy Agency.

Accelerating the pace of progress will require stronger political commitment, long-term energy planning, increased public and private financing, and adequate policy and fiscal incentives to spur faster deployment of new technologies. The Government of India has strongly promoted new manufacturing facilities to manufacture solar cells and other renewable energy equipment. This new policy thrust is aimed at producing not only for India but also for other developing countries and providing them equipment at an affordable cost so that sustainable energy supply is provided to all by 2030, and SDG7 is successfully attained.

Estimation of Daily Direct Solar Radiation for Rabat

Mohammed Benchrifa, Hajar Essalhi, Rachid Tadili, and Hassan Nfaoui

1 Introduction

The direct solar radiation designates the energy flow coming directly from the solar disk received per unit area. This flow varies significantly depending on many factors, including the distance Earth-Sun, the sun's height, the state of purity of the atmosphere, and the latitude of the site considered.

During the passage of solar radiation in the atmosphere, it is influenced by several attenuation phenomena such as diffusion by air molecules (Rayleigh effect), diffusion and reflection by aerosols present in the atmosphere, and selective absorption by atmospheric gases. To these influences, we must add that of clouds according to their nature and their optical density. Therefore, knowing the solar radiation on the ground in the presence of clouds requires systematic and regular measurements. However, in the absence of measurement stations, methods of estimating solar radiation are used.

In this work, we will focus more specifically on the direct component H_b of solar radiation and we will develop a method for estimating daily direct irradiation valid for Rabat and neighboring sites.

Indeed, this variable is very important for the study and the functioning of high-temperature solar thermal systems and concentrated photovoltaic systems. Consequently, several models have been made to predict this component such as the Bargach [1] model which is based on the duration of sunshine, the Khtira model [2] which is based on satellite data, and the Nfaoui model [3] which is based on the clarity index.

On the other hand, other models were based on the estimation of the diffuse to draw the direct one such as the model of Ruth and chant [4] based on the data of the Canadian sites and the model of Collares-Pereira and Rabl [5] based on the measures of five stations in the United States.

Based on these works and using the Rabat data, we developed three models based on three types of equations. Then we compared the new models as well as the existing models with measurement through a validity study. As results, the model that represents the smallest error can be used to generate the daily values of direct irradiation.

2 Description of Data

2.1 *Description of Rabat's Measurement Station*

2.1.1 Rabat Climate Description [6]

The city of Rabat is the capital of Morocco. It is located on the edge of the Atlantic and has a pleasant climate characterized by four seasons. Winters are rainy and cold, and summers are hot. As for spring and autumn, it's a compromise between the other two seasons. The lowest temperatures are recorded in December, January, and February. The average temperature in Rabat is 17.9 °C, and the average annual rainfall is 523 mm per year.

M. Benchrifa · H. Essalhi (✉) · R. Tadili · H. Nfaoui
Laboratory of Solar Energy and Environment, Mohammed V University in Rabat, Faculty of Sciences, Rabat, Morocco

© The Author(s), under exclusive license to Springer Nature Switzerland AG 2022

A. Sayigh (ed.), *Sustainable Energy Development and Innovation*, Innovative Renewable Energy,
https://doi.org/10.1007/978-3-030-76221-6_69

2.1.2 Characteristics of the Station [7]

Since 2008, the solar energy and environment laboratory (LESE) has renovated its station for measuring the components of solar radiation. The station is placed at the top of the building of the Faculty of Sciences of Rabat (Latitude: 34°N, Longitude: −6°45'W).The station site is completely clear of any shading effect. It measures all the components of solar radiation and climatic variables. The instruments used for the measurement of global and direct irradiation are:

- The SP-Lite Pyranometer Kipp & Zonen mark for measurements of global solar radiation (2% accuracy) [8].
- The Model NIP 31820 E6 pyrheliometer, Eppley type mounted on a tracker to measure direct solar radiation (accuracy 0.5%) [9].

All the instruments are connected to a data acquisition center (CR10X) fitted with a storage module. The collected measurements are then processed and subjected to validity tests to detect and eliminate the erroneous values and replace the missing values.

3 Existent Model in Rabat

Given the importance of direct irradiation, several models have been made to predict this component such as:

3.1 The Nfaoui Model [3]

On the basis of 3 years of measurements in Rabat (1982–1985) **Nfaoui** and **Buret** produced a model for estimating direct irradiation based on K_t. The model is expressed in the form:

$$\frac{H_b}{H_{oh}} = \begin{cases} 2.02K_t^2 + 0.95K_t + 0.13, & K_t > 0.25 \\ 0 & , K_t < 0.25 \end{cases}$$

$$R = 0.96$$

3.2 Khtira's Model [2]

On the basis of a year of measurements in Rabat (1982), Khtira carried out an estimation model of direct irradiation based on K_t. The model is expressed in the form:

$$\frac{H_b}{H_h} = \begin{cases} 1.384K_t^2 + 0.105K_t + 0.026, & K_t > 0.25 \\ 0 & , K_t < 0.25 \end{cases}$$

$$R = 0.9$$

4 Development of New H_b Estimation Models

The direct component of solar radiation is an important factor in the study and design of solar systems. The estimation of this solar component is mainly done by two categories of model [10]:

- Parametric Models
- Decomposition Models

The parametric models are based on the estimation of solar irradiations using meteorological parameters such as the optical air mass, the fraction of sunshine, the Linke turbidity factor, etc.

The decomposition models are based on the estimation of a solar component as a function of another component. Generally, this type of model uses global irradiation to estimate other components such as direct or diffuse radiation. Depending on the nature and the number of meteorological data available on a site, one of the two types of models will be used.

Given the availability of measurements of solar irradiation, we chose to use in the present work the decomposition models for developing the equations to estimate direct radiation.

Considering that the irradiations outside the atmosphere vary with declination and latitude, while the irradiations on the ground depend not only on declination and on latitude but also on climatic conditions. So, to reduce the effect of astronomical and climatic variables in the estimation of direct radiation, we have introduced the clarity index K_t, which characterizes the transparency of the atmosphere. This variable indicates the percentage of the energy reaching the ground compared to that received at the limit of the atmosphere.

Thus, using the data measured in Rabat from the year 2016, we plotted H_b/H_{oh} as a function of K_t, which gave us Fig. 1.

From Fig. 1, we can see that the point cloud produced looks like a parabola. To obtain the best correlation, three types of equations have been developed. The first type is in the form of a fourth-order polynomial, the second is in the form of an exponential, while the third is in the form of a power function. Figure 2 gives a graphical representation of the three types of equations.

Using the least-squares method, we determined the coefficients of each type of equation, which gave us the following models:

Polynomial Model

$$\frac{H_b}{H_{oh}} = \begin{cases} -41.655K_t^4 + 80.108K_t^3 - 53.534K_t^2 + 15.508K_t - 1.6384, & K_t > 0.3 \\ 0 & , K_t < 0.3 \end{cases}$$

$$R = 0.87$$

Exponential Model

$$\frac{H_b}{H_{oh}} = \begin{cases} 0.0018e^{8.17K_t} & , K_t > 0.3 \\ 0 & , K_t < 0.3 \end{cases}$$

$$R = 0.9$$

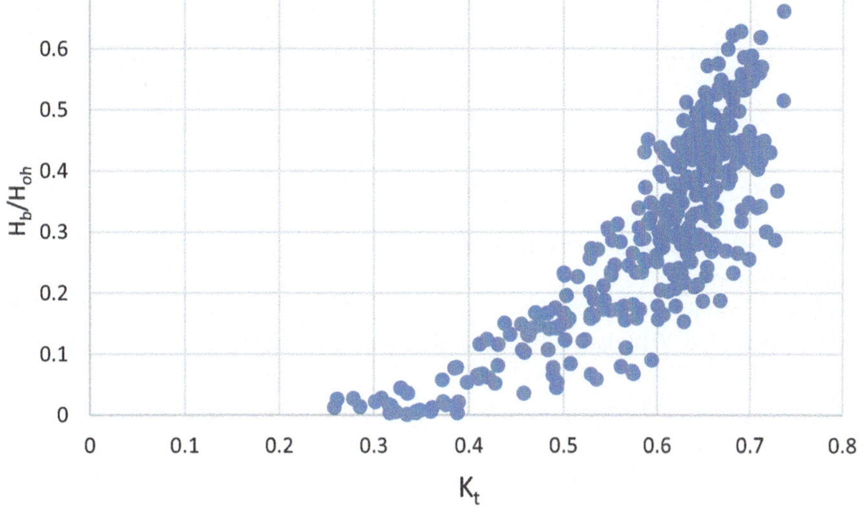

Fig. 1 Graphical representation of H_b/H_{oh} as a function of K_t

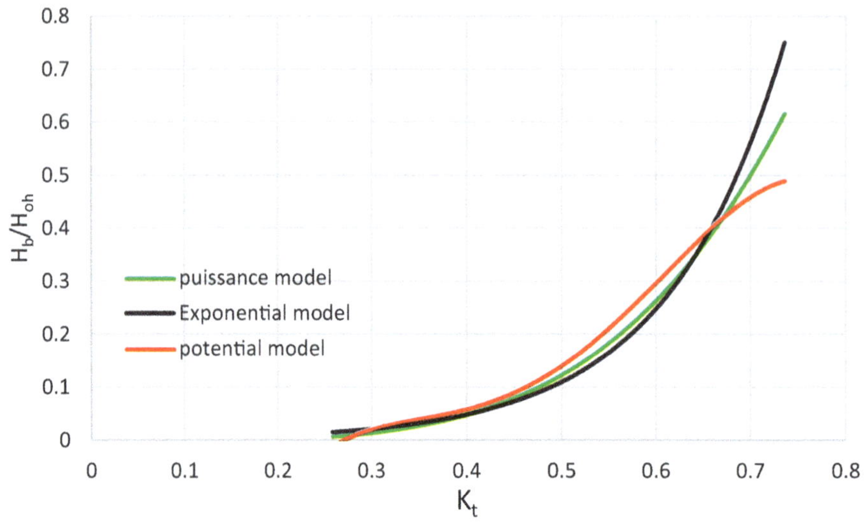

Fig. 2 Variation of daily direct radiation as a function of the clarity index. Representative curves of the new estimation models

Power Model

$$\frac{H_b}{H_{oh}} = \begin{cases} 2.207K_t^{4.1483} & , K_t > 0.3 \\ 0 & , K_t < 0.3 \end{cases}$$

0.91

5 Validation and Evaluation of Model Performance

The validation of a solar radiation estimation model is done by two types of evaluations:

- The qualitative evaluation which is based on the graphical representation of the estimated data as a function of the measured data and to visually draw the bias (systematic error) and the dispersion (random error) of the predicted values compared to the measured values [11].
- Quantitative evaluation which uses statistical indicators such as [12]:

$$\text{RMBE} = 100\frac{\sum_{i=0}^{n}(p_i - m_i)}{\sum_{i=0}^{n} m_i}, \ \text{MAPE} = \frac{1}{N}\sum_{n}^{i=0}\left|\frac{p_i - m_i}{m_i} * 100\right|$$

$$\text{RMSE} = \frac{100}{\overline{m}}\sqrt{\frac{\sum_{i=0}^{n}(p_i - m_i)2}{N}}$$

where p_i and m_i are the ith calculated and measured values.
N is the number of observations.
$\overline{m}$ is the mean of measured values.
For this validation study, we used 2018 data collected at Rabat measurement station.

6 Results and Discussions

6.1 Validations and Choice of Models

To choose the best model for the estimation of direct irradiation, we compared the daily data generated by the different methods with the measurements collected during the whole year of 2018. This comparison is made using the statistical indicators presented in Table 1 and the graphical representation of the estimated data as a function of the measured data presented in Fig. 3.

From Table 1, we note that the exponential and most precise model in the estimation of direct radiation, because it represents the smallest errors. Follow-up of the Nfaoui model which has a small relative error but a fairly high quadratic error, which explains the dispersion of the points generated by this model.

Table 1 Performance of estimation models for daily direct irradiation in Rabat

Models	RMBE (%)	MAPE (%)	RMSE (%)
Exponential model	−6.54	13.39	14.69
Polynomial model	−9.21	15.13	17.61
Power model	−10.07	14.68	17.47
Nfaoui model	−7.27	16.35	19.25
Khtira model	−10.43	16.49	23.27

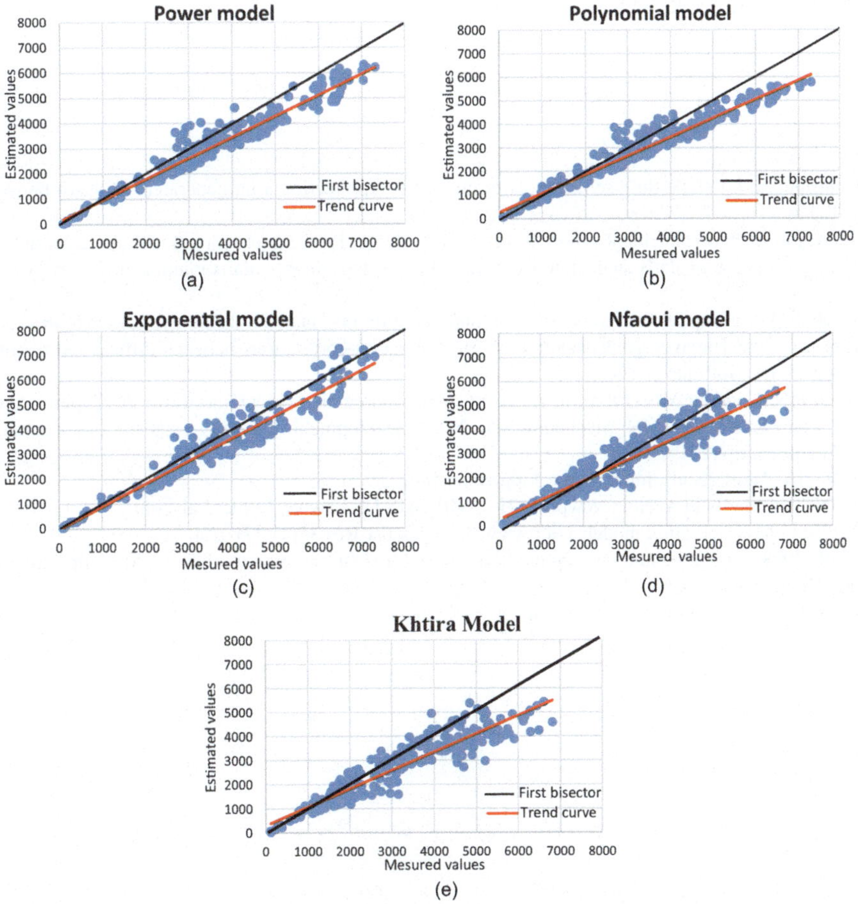

Fig. 3 Graphical representation of the estimated values according to the measured values of direct radiation estimation models: (**a**) power model, (**b**) polynomial model, (**c**) exponential model, (**d**) Nfaoui model, (**e**) Khtira model

Besides, from the comparison of the graphs in Fig. 3, we note that Figure (c) presents a good approximation between the first bisector and the trend line of the estimated data, which approves the results obtained in Table 1.

These findings lead us to conclude that the exponential and most adapted model for the estimation of direct radiation for the Rabat site.

7 Conclusion

The values of direct solar irradiation constitute an important data in developing and studying the high temperature solar thermal systems such as CSPs and the concentrated photovoltaic systems such as heliotropes.

When measurement is unavailable, predicting radiation by estimation models is the most suitable solution. To make these models easy to use, they must be simple and employ commonly measured variables.

The aim of this work is to develop a method for estimating daily direct irradiation valid for Rabat and neighboring sites.

To meet these obligations, we have developed several models that estimate direct irradiation on a daily scale. These models were made in the form of three types of equations (exponential, power, polynomial) according to the clarity index K_t.

The new models, as well as the existing models, were validated by qualitative and quantitative evaluation using measurements collected in Rabat.

The validation study results led us to choose the exponential model to estimate the daily direct irradiation because it presents the lowest errors, which do not exceed −6.45% as a relative error, 13.39% as an absolute error, and 14.69% as the mean square error, and the highest coordination between the first bisector and the trend line of the estimated data as a function of the measured data.

Finally, using the chosen model and the data from the daily global irradiation of Rabat, we were able to generate the daily direct irradiation data for Rabat and neighboring sites.

References

1. Bargach MN (1983) Contribution à l'étude du gisement solaire marocain. Application à un générateur photovoltaïque. Thèse de 3ème cycle, Faculté des sciences de Rabat
2. Bahraoui-buret J, Khtira A. Étude du rayonnement solaire direct à rabat Maroc. Revue internationale d'heliotechnique, 7–11, 1984
3. Nfaoui H, Buret J (1993) Estimation of daily and monthly direct, diffuse and global solar radiation in Rabat (Morocco). Renew Energy 3(8):923–930
4. Ruth DW, Chant RE (1976) Relationship of diffuse radiation to total radiation in Canada. Sol Energy;(United States) 18(2):153–154
5. Collares-Pereira M, Rabl A (1979) The average distribution of solar radiation-correlations between diffuse and hemispherical and between daily and hourly insolation values. Sol Energy 22(2):155–164
6. Bahraoui-Buret J, Bargach MN, Kaddour MB (1983) Le gisement solaire marocain. Société marocaine des éditeurs réunis
7. Benchrifa M, Essalhi H, Tadili R, Bargach MN, Mechaqrane A (2019) Development of a daily databank of solar radiation components for Moroccan territory. Int J Photoenergy(2019):1–20.
8. Kipp & Zonen B.V. Delft - The Netherlands. https://www.kippzonen.com
9. The Eppley Laboratory INC, 12 Sheffield Avenue Newport, RI 02840-1618 USA. www.EppleyLab.com
10. Ahmad MJ, Tiwari GN (2011) Solar radiation models—a review. Int J Energy Res 35(4):271–290
11. Gueymard CA, Myers DR (2008) Validation and ranking methodologies for solar radiation models. In: Modeling solar radiation at the earth's surface. Springer, Berlin, Heidelberg, pp 479–510
12. Iqbal M (1983) An introduction to solar radiation. Academic press, Canada

Part VI
Energy Meteorogy

Darrieus Turbine for Applications in Urban Environments

Carlos V. M. Labriola and Jorge Lassig

1 Introduction

Since approximately 2005, commercial applications of wind turbines have been carried out in urban environments and/or residential or office buildings. The applied technology is varied, ranging from small devices (1–10 kW), through medium-sized, to large turbines (hundreds of kW). Several have been of vertical axis (VAWT) (Savonius, Darrieus) and others of horizontal axis (HAWT). Not all installations have been successful projects [1]. The main causes were due to lack of previous noise and vibration impact studies, material fatigue, and fundamentally lack of the wind resource in the building studies.

Regarding the wind resource in urban environments, we can say that the structure and height of the buildings allow increasing the average wind speed in the terraces in general, providing an additional 1–4 m/s to the average meteorological speed of the place [2]. In addition, depending on the design of the building or house, the solar sources (photovoltaic panels) and microhydraulics can be integrated. Last one uses accumulated gray water in intermediate tanks and subsequent turbination.

Horizontal axis turbines of possible application in urban buildings (small wind turbines 1–65 kW, Standard 61400.2) can have two, three, and multi-blade (Fig. 1), with Pitch control (change of angle of attack by centrifugal action) or Stall control (loss of variable load along the blade).

Pitch control requires periodic maintenance, which is sometimes difficult in tall buildings. Stall control with plastic blades produces a lot of noise (buzz) when gusts push and pull the rotor. This noise can affect up around 100 m.

If we analyze several commercially available wind turbines from 2014, we can appreciate the following situations [3]:

- Specific weight dispersion (ratio between the weight of the wind turbine and the surface swept by the rotor) as a function of the diameter of the rotor. There is no trend of diameter growth with the specific weight referred to the swept area.
- Dispersion of the variation of cost/m^2 of swept area as a function of the nominal power of the turbine. This dispersion may be due to the variety of different materials to make the blades (different extrusion plastics, molded FVPR, aluminum, steel, etc.), the number of blades for equal power, whether or not the turbine includes the tower (10–20 m) and if it is provided with a DC/AC or AC/DC/AC converter for energy accumulation. In general, accumulation batteries are not included.
- Diameter dispersion for the same power. The cause of this dispersion is the existence of different nominal speeds from different manufacturers for different turbines of equal power.

Argentina adheres to IEC standards, and today, this institute has more than 20 international standards referred to wind turbines. Besides, there are different regulations and guidelines from different countries and companies. It is particularly interesting to compare the IEC with regulations of wind energy application of other countries.

Table 1 shows that the Japanese Standard is the most restrictive of all and is the only one that includes considerations for vertical axis wind turbines (VAWT).

In the case of VAWT, it is necessary to distinguish those of solidity close to 1 (Savonius) and those of very low solidity (Darrieus, $\sigma \leq 0,1$)). Both, and even those with a horizontal axis, have a power extraction coefficient (Cp) between 0.15 and

C. V. M. Labriola (✉)
Department of Electrotecnia, Faculty of Engineering, National University of Comahue, Neuquen, Argentina

J. Lassig
Department of Mechanics, Engineering Faculty, National, University of Comahue, Neuquen, Argentina

© The Author(s), under exclusive license to Springer Nature Switzerland AG 2022

A. Sayigh (ed.), *Sustainable Energy Development and Innovation*, Innovative Renewable Energy,
https://doi.org/10.1007/978-3-030-76221-6_70

Fig. 1 Horizontal axis turbines for urban environments

Table 1 Comparison of IEC 61400-2 and complementary with other standards from other countries [4]

	Standard/Regulation/Normative		
Parameter value to consider	USA regulation AWEA 9.1	Renewable UK small wind turbine	Japan regulation JSWTA-0001
Power and energy; voltage and current	Rated power of small wind turbine at 11 m/s	Maximum power: maximum output power (being the one minute definition, P60) as defined in IEC 61400-2 ed 3.	It follows the IEC but includes somewhat different electrical tests than the IEC 61400-2 standard.
	Rated annual energy at 5 m/s	Maximum voltage: maximum output voltage (being the one minute definition, U60) as defined in IEC 61400-2 ed 3.	
	P vs. v is extended more than 5 m/s, taking into account speed reached at 85% of rated power	Maximum current: maximum output current (being the one minute definition, I60) as defined in IEC 61400-2 ed 3.	
Other parameter and tests	In the duration test, it is necessary to have 10 minutes above 15 m/s regardless of the class.	It incorporates as a novelty a technical note about the inverters and how they can be used on certified systems.	It has some more restrictive points in structural part. Blade test is an essential condition.
	Sound level at 5 m/s		They are proposing things for VAWT.

0.25. This decrease in Cp in fast wind turbines is due to the losses induced by its size. Darrieus turbines also do not generally have an auto-start.

Particularly in applications in urban sites, the following aspects must be taken into account in addition to the parameters and characteristics (nominal power, annual energy, cost/Kw, etc.) of turbine offered by the manufacturer to convince users:

- Wind resource studies in buildings or houses to install the wind turbine.
- Maximum dimensions of the wind turbine for a given installation site.
- A single turbine covering all power demand or several summing the total to be installed.
- It has an auto-start, or starts from the network as a motor and then change generation mode.
- The equipment is self-supporting by the main axis or not (braced or with additional turnbuckles).
- Noise level produced during operation and level of vibrations transferred to the supports
- Schedule of maintenance and availability of parts.

2 Comparison of Horizontal and Vertical Axis Wind Turbines

In the introduction the most usual forms and types of turbines available for applications in urban environments have been described. Based on these turbines, Table 2 shows its comparison:

As we can see, VAWT, particularly Darrieus turbines, have more constructive advantages than HAWT for applications in urban environments.

Table 2 Comparison of the application of turbines in urban environments [5]

Horizontal axis (HAWT)	Vertical axis (VAWT)
With two or three blades, the rotor emits noise during gust of wind	It works at lower specific speed at equal power than HAWT, they emit less noise and vibrations to the building
Multi-blade turbines emit less noise and transmit less vibration to building structures	Rotors are made of different materials and usually with one to five blades). Helical rotors (Fig. 2 right) reduce noise and alternation of torque and power considerably
Cp is not 0.4 as in large turbines, but does not exceed 0.25	
Sometimes it is necessary to limit orientation with stops to operate in the prevailing wind direction. It can be a source of noise	Savonius types have Cp: 0.15–0.2 and their bracing must be done carefully particularly in places with turbulences due to their high solidity
Heavy and mobile parts are at top of tower. They need orientation systems	The moving parts are at floor level facilitating maintenance. They are omnidirectional
Before fixing it, if turbulence on the terrace has not been studied, it can reduce the useful life (parts break prematurely) if it's not located in the right place	It can be self-supporting with axes embedded in the base or they are braced with two or three tensioners from the upper end of axis, which must be verified in periodic maintenance
Fixed cornice turbines are horizontal axis limited to the preferred wind direction and have a Cp: 0.15–0.2 which is very interesting to combine with other renewable sources	Cornice turbines can be integrated in the horizontal edge of the terrace that faces the preferential wind of the location. They provide additional power to other renewable energy devices installed

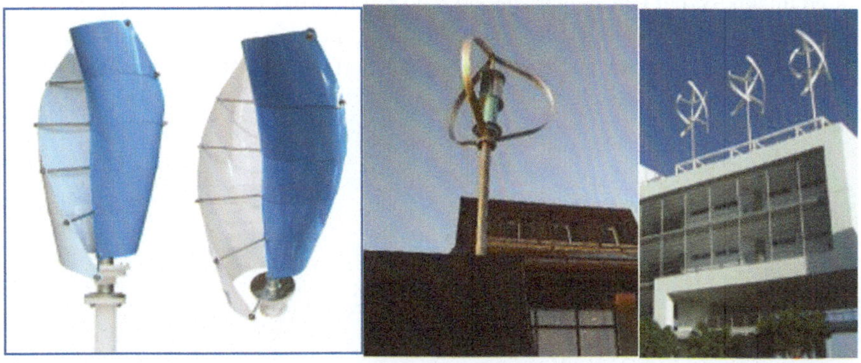

Fig. 2 VAWT: Savonius (left), Darrieus (center and right) used in urban environment

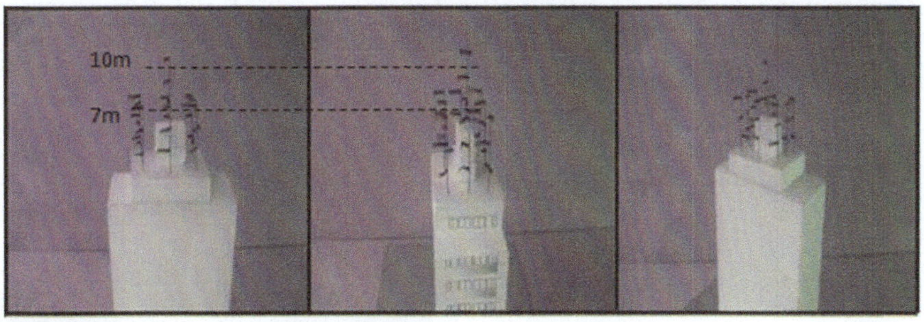

Fig. 3 Building in Rosario City, Province of Santa Fe, prevailing east wind and turbulence-free zone according to winds between 7 and 10 m, with an average wind speed of 7 m/s (2 m/s more than the average of the place) [2]

3 Wind Resource Study in Urban Environment

In wind resource study in urban environment, a distinction must be made between an application for an existing building or for a building to be constructed.

The design of a new building should include installation, connectivity, and even aesthetic considerations for a better integration of the renewable energy sources available on the site, in particular wind power. Once the design is defined, it is necessary to measure or estimate the wind speed in the terrace by correlation with other points near the place for which data are available. It is also necessary to be able to model the building and study the turbulence on the terrace for different design options, seeking the convergence between optimal energy production, aesthetics, and the safety of people, equipment, and surroundings buildings.

When the building is raised, the velocity profile can be measured on the terrace, but it is not generally designed to have wind turbines on it. For this reason, it is necessary to carry out a detailed study of wind resource on terrace, especially a test in a wind tunnel (Fig. 3). The aim is to define turbulence-free places, value of the available average speed and height which this value is located. This allows to define size, power, and number of turbines to be installed.

4 Constructive Considerations for Darrieus Wind Turbines in Urban Environments

Usually, Darrieus turbines use biconvex profiles, which are easier to builld than curved ones. But they do not allow the rotor to auto-start because they are symmetrical; it only occasionally starts when the wind speed exceeds 2 or 3 tenths of m/s [6].

It is relevant to study low Reynolds number (Re) profiles, to give Darrieus auto-start, improving the Cl/Cd ratio. In this work, the profiles that were analyzed are type S designed by the National Renewable Energy Laboratory of the United States (NREL). The profiles studied are S805, S809, S813, S819, S820, and S821, which are simulated with free Q-Blade software and in the range between 10000 to 50000 of Reynolds Number, according to the range of turbine rotor speeds up to 10KW as Table 3 shows:

Only the points of the linear section closest to the Cl axis were selected. The best results are obtained for profiles S805, S813, and S820, which have a reduced maximum thickness and a low asymmetry compared to the rest. The S805 profile was simulated with increased thickness, but Cl/Cd characteristic did not improve (Fig. 4).

H-Darrieus prototype was built with S813 profile and it was tested in wind tunnel [7]. Self-starting was achieved at 4 m/s. In the physical test different angles of attack and rpm were tested, and number of blades as well.

Table 3 Cl/Cd for profiles series S considered

Perfil	C_d min.	C_l max.	C_l (α=0)	α (C_l=max)	ΔC_d	ΔC_l	$\Delta C_l/\Delta C_d$
S805	0,0183	1,115	0,252	8	0,01442	0,863	59,84
S809	0,02706	1,054	0,173	8	0,00843	0,881	104,5
S813	0,02178	1,208	0,331	8	0,01289	0,876	67,95
S819	0,0218	1,302	0,245	10	0,02217	0,773	34,86
S820	0,02108	1,222	0,345	8	0,01392	0,873	62,71
S821	0,0248	1,514	0,38	10	0,02061	0,566	27,46

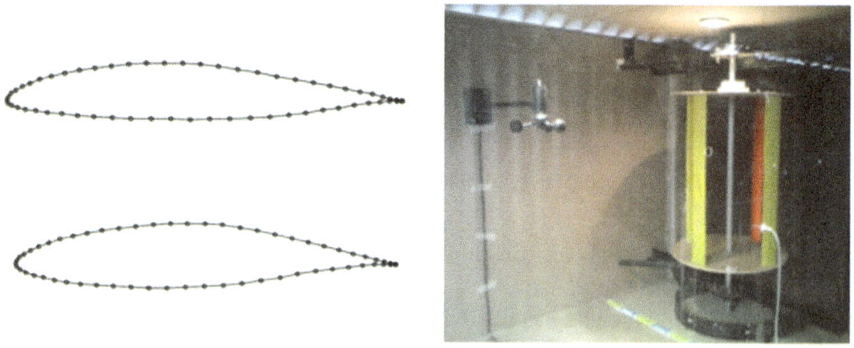

Fig. 4 S805 profile (left), S813 profile (below), view of H-Darrieus with S813 profile, Vstart at 4 m/s (right)

Fig. 5 Photovoltaic parasols (left). Turgo turbines (center and right)

5 Combination with Other Energy Sources

Solar energy: Using photovoltaic panels (PF), it is relevant that one of the building fronts faces north (southern hemisphere) and without any shadows from other buildings of surroundings. The market offers PF from 280 to 405 W/panel and sizes from 1.6 m^2 to 2 m^2 per 1 m width. These can be placed as parasols (Fig. 5) on the fronts at a convenient angle. Installed power will be: (No. of panels* front width * number of floors).The angle can be optimized using software (e.g., PVSyst™).

Microturbines that use gray water (GW): storage tanks can be placed every three or four floors to accumulate gray water. Once the sizes and locations of the intermediate AG tanks have been defined, Turgo type turbines can be used in confined flow mode, in the range of 1–5 kW. If there are three tanks, the uppermost turbine only turbines the water of first tank, the middle one turbines water of two tanks, and the last one turbines the water of three tanks. The turbines should work at peak consumption hours, which in Argentina are from 6 to 10 am and from 6 to 10 pm.

6 Conclusions and Recommendations

Standards and Normative today is vast (IEC, AWEA, UKSWT, JSWTA-001, etc.), but budget of an entrepreneur or small and medium company that have to do the necessary tests [4] are not considered in them.

Vertical axis wind turbines are better than horizontal ones for applications in urban environments because of its qualities: omnidirectionality, less noise and vibration impact, ease of assembly and maintenance.

Darrieus turbines with three or more blades are recommended to reduce the possibility of noise and/or vibrations. The profile should be of low Re (type S: e.g. S803, S813). Since they will rotate slower, due to the greater number of blades, a synchronous multipolar generator with AC/DC/AC conversion [8] without any gearbox multiplier is recommended.

Small wind Darrieus turbines, in combination with other energy sources (PV and microhydraulics with GW), for an application on a building can supply between 40% and 80% of the energy consumed in its common spaces (lights) and services (lift, pumps for drinking water, etc.) [9]. Optimized application is achieved when RES are included in the building design stage.

References

1. Bahrain World Trade Center - veredict designbuild, 2008. www.designbuld-network.com/projects/bahrain-world-trade-center/
2. Lassig J, Jara U, Sosa JJV, Palese C (2019) Urban environment: characterization of the wind in flat roofs. In: Sayigh A (ed) Wind energy for all. Springer, Cham
3. Cruz I, CEDER, 2014: study on 28 small wind turbines from Europe
4. Standards of small wind turbines (2019) Author: Dr. Luis Cano, Wind Energy Unit, CIEMAT (Spain). Course on Distributed and Isolated Wind Systems Montevideo. Uruguay. October 10, 2019
5. Labriola C (2000) Aerodynamics brakes for vertical axis wind turbine Darrieus-Troposkien for Patagonia, Argentina. Thesis of Master in Renewable energy and the environment 1999–2000, University of Reading, Berkshire, United Kingdom
6. Ibañez D (2019) Final report for lecture "Renewable energy and the environment", course 1st. semester 2019, Professor: MSc. Carlos Labriola. National University of Comahue
7. Lassig J, Sosa JV, Jara U (2016) Wind tunnel testing of a small wind turbine Darrieus-H, with asymmetric airfoil, for auto start. IOSR J Mech Civil Eng 13(5 Ver. III):63–69
8. Labriola C (2019) Seminar included in training plan of red CYTED "Microeolo", Lecturer: Prof. Carlos Labriola, Electrical Engineering Institute – CONICET, National University of San Juan, Argentina
9. Labriola C (2014) Use of renewable energy sources intrinsic and external to a multi-storey building. Authors: Labriola C, Lassig J, Padilla P, Solar Cities Congress, Buenos Aires, September, 2014

Part VII
Wind Energy Technology

Air Pollution By Fuel and Energy Complex and Prospects for Wind Power Production Development in Belarus

Alexandra Kurachkina

1 Introduction

According to the obligations under the Paris Agreement, Belarus shall reduce greenhouse gas (GHG) emissions by 28% by 2030 compared with 1990 (it is required not to exceed the level of 96.1 million tons of CO_2 eq) [15].

The energy sector of Belarus accounts for the largest amount of greenhouse gas emissions – 61.4% (2017) [6]. When fuel is burned at thermal power plants, carbon dioxide, nitrogen oxides, water vapor, and other substances causing the greenhouse effect are formed. Typical pollutants are sulfuric and sulfur dioxide, nitric oxide, and solid particles. In Belarus natural gas and fuel oil are mainly used for electricity production, but peat and firewood are also burned in small boiler houses. Since 2017 there has been a decrease in the share of use of natural gas in boiler and furnace fuel, an increase in the use of fuel oil, peat and wood, which entails an increase in air pollution, especially by particulate matter. The problem of reducing emissions of pollutants and greenhouse gases is of great importance for Belarus. The solution to this problem is based on the diversification of fuel and energy companies, in which the transition to the use of renewable energy sources has great advantages for solving these problems. A spatial analysis of the environmental problems of the Belarusian fuel and energy complex in the context of identifying the prospects for the development of wind energy has been the goal of this study.

2 Results and Discussion

Electricity, gas, and hot water supply account for 13.7% of all pollutant emissions from stationary sources (e.g., solid particles, nitrogen oxides, sulfur dioxide, hydrocarbons, and non-methane volatile organic compounds) compared with other types of economic activity, 18.5% of emissions accounted for the oil refining industry. The volume of pollutant emissions in the areas where the largest thermal power plants (TPPs) and oil refineries are located is higher than in other regions of Belarus [19].

Particulate matter (PM) produced by combustion of fuel oil has a negative impact on the health of people. Their main components are sulfates, nitrates, soot, and mineral dust. Particles with a diameter less than 10 microns (≤ PM10) can penetrate deep into the lungs, however, particles with a diameter less than 2.5 microns (≤ PM2. 5) have more negative impact on the human body, they can get into the blood stream. Chronic exposure to particulates increases the risk of developing cardiovascular and respiratory diseases [1].

In addition, sulfur compounds emitted into the atmosphere from the combustion of fuel in thermal power plants have negative impact on human health and the environment. Also they affect the fauna negatively and cause corrosion of plant equipment [13].

The main greenhouse gas associated with the operation of energy in the Republic of Belarus is carbon dioxide (CO_2), whose share in emissions of greenhouse gases amounted to 97.7% in 2017, followed by methane (CH_4) – 1.9% and nitrous oxide (N_2O) is 0.4%. In addition, suppliers of greenhouse gases are the enterprises of oil refining and chemical industries and agriculture. Total GHG emissions are 94 million tCO_2-equivalent per year, and a decline in 2017 compared to 1990, when

A. Kurachkina (✉)
Lomonosov Moscow State University, Faculty of Geography, Department of World Physical Geography and Geoecology, Moscow, Russia

© The Author(s), under exclusive license to Springer Nature Switzerland AG 2022

A. Sayigh (ed.), *Sustainable Energy Development and Innovation*, Innovative Renewable Energy,
https://doi.org/10.1007/978-3-030-76221-6_71

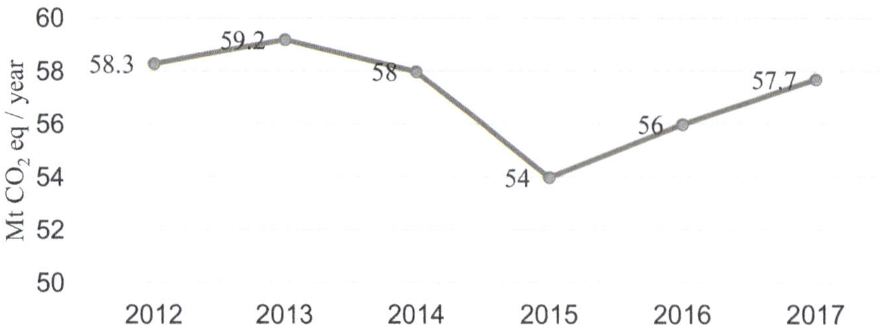

Fig. 1 Greenhouse gas emissions in the energy sector of Belarus [6]

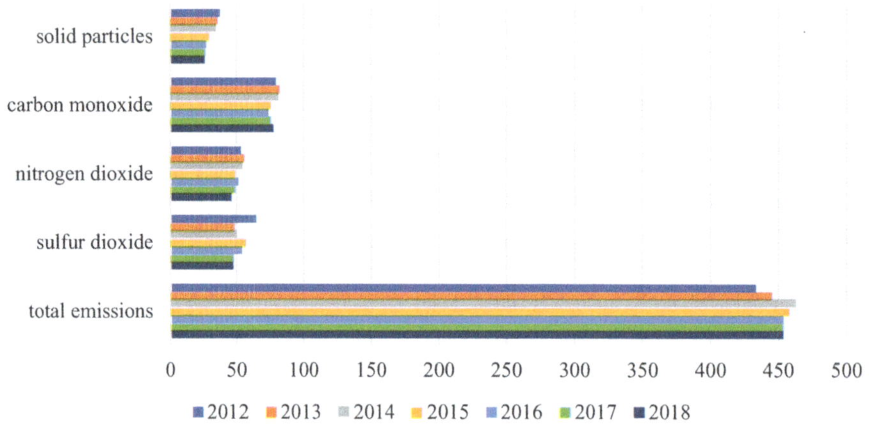

Fig. 2 Emissions of pollutants from stationary sources in Belarus (thousand tons) [6]

they amounted to 140.7 million tons of CO_2-equivalent per year by 33.2%. However, it should be noted that the greenhouse gas emissions generated in the energy sector has been growing since 2015 (Fig. 1).

The decrease in emissions in 2015 is associated with the reduction of electricity consumption; this figure was 36,854 million kWh (2014 – 38,055 million kWh). However, in 2017, electricity consumption rose again to a 37,107 million kW/h, which led to the growth of greenhouse gas emissions in the energy sector [6].

The total emissions of pollutants from stationary sources increased compared with 2012, the maximum emissions occurred in 2014–2015 (Fig. 2).

The increase in total fuel consumption and fuel oil consumption at power facilities led to an increase in total emissions of pollutants into the air from stationary sources in 2018 compared to 2017. In the energy sector emissions in 2018 increased by 28,000 tons compared with the previous year. An increase in emissions is observed for all major pollutants.

The state atmospheric air reported gives a general picture of emissions, while the most common pollutants from stationary sources are carbon monoxide and sulfur dioxide.

In order to reduce the amount of pollutant emissions, automated continuous monitoring systems for emissions of major pollutants into the air (ASK) were put into operation by:

- Pinsk CHPP of the Pinsk Heating Networks branch of RUE "Brestenergo"
- Gomel TPP-1 of the Gomel Heating Networks branch of RUE "Gomelenergo"
- Lida CHPP of the Lida Heating Networks RUE "Grodnoenergo branch"
- Borisovskaya CHPP of the branch "Zhodinskaya CHPP" RUE "Minskenergo"

They also carried out works related to the introduction of ASC in the branch "Grodno CHP-2" RUP, "Grodnoenergo," and RK "Kedyshko" branch of "Minsk heating networks" of RUE "Minskenergo" [5].

High pollutant emissions (e.g., sulfur dioxide, particulates, and nitrogen oxides) in Belarus are in the Minsk and Vitebsk regions. In Minsk region the main share of emissions accounted for a number of large industrial enterprises and energy facilities: Minsk CHP-4, the branch of "Minsk heating networks" of RUE "Minskenergo," Minsk CHP-3. In Vitebsk region there was a large part of emissions from CHP 14, the largest in Belarus Lukoml and Novopolotsk oil refinery. Thus, the burning and processing of fossil fuels make a significant contribution to air pollution [5].

In addition to atmospheric pollution, the power plants are associated with water pollution and noise pollution. So the electricity, gas, and hot water accounts for 11.5% of wastewater discharges into surface water bodies.

In addition, the noise level of mini TPP with a capacity of 1000 KW is 84 to 95 dB. The noise from the wind farm (10 turbines) is at a distance of 350 m 35–45 dB, which corresponds to the background noise in rural areas. The noise produced by 10 wind turbines at a distance of 350 m from them can be assessed as negligible, that is, indistinguishable from the other noises in everyday life [17].

3 Prospects for the Development of Wind Energy to Solve Environmental Problems

A global trend in the energy sector in the context of transition to "green economy" is the increased use of renewable energy sources (RES). Its main advantages are the almost inexhaustible resources, environmental friendliness, ergonomics (ease of use in many areas, quick and easy dismantling), and creating conditions for the development of advanced technologies [12]. The experience of many countries of Europe, with similar natural conditions of Belarus (Poland, Germany) shows the potential of wind energy development.

The development of wind energy contributes relatively favorable natural conditions: the average wind speed at height 100 m in most parts of the territory of Belarus is 5–7 m/s [2]. However, when choosing specific places for wind turbines, it should take into account not only wind speed but also a number of other factors associated with the local energy potential of wind and with many ecological and geographical characteristics of the territory [3, 10, 11].

According to the data for 2019 in Belarus there are 125 wind turbines with a total capacity of 218 MW. Most of wind generator in Belarus are located at higher elevations, where the wind speed at the height 100 m is greater than 6 m/s (e.g., Novogrudok, Minsk, Orsha, and Oshmiany Uplands). Taking into account the topography and average wind speeds the most promising for wind power development are agricultural areas with an altitude of 200 m and more above the sea level. In these areas at a height of 80–100 m, it is expedient to arrange the axis of the rotors of wind turbines with installed capacity of 1.5–2.5 MW. At altitudes less than 200 m profitable to place a wind turbine less power [22].

The active development of wind energy is due in large part to environmental benefits. Two-thirds of global greenhouse gas emissions accounted for traditional energy. Wind turbines in the process of operation does not consume fossil fuels, reducing atmospheric pollution, and energy production, which is important for Belarus [10]. However, wind farms also have an impact on the environment. The design life of a modern wind farm is 25 years. In operation the wind farm has an impact on humans, flora and fauna, atmospheric air, water objects, land use in the form of noise, vibration, electromagnetic radiation, optical effects, mechanical effects, and waste of operation. However, the increase in the share of renewable energy in the energy mix will reduce emissions of pollutants into the atmosphere. Greenhouse gas emissions in the production of energy for wind power is four times lower than solar power and is ten times lower than when using gas, coal, and oil [14]. The technical capacity of wind energy in Belarus is estimated at the level of 1.9–2.0 million t.c.t. Electricity generation for small hydro power plants and the savings of fossil fuels is around 0.11–0.15 million t.c.t. The potential energy in solid waste treatment is estimated at 470 thousand t.c.t. The natural annual growth of wood is 6.6 million t.c.t [10].

Wind energy reduces the use of traditional energy sources in Belarus to 82.75 tons of conventional fuel per year. Wind energy can reduce emissions of pollutants into 485.9 thousand t/year.

4 The Economic Benefits of Wind Energy Development: Diversify Energy Resources and Reduce the Cost of Transporting Electricity

Wind power in Belarus allows to diversify energy resources, which Belarus exports by 88%. To solve this problem it is necessary to increase the share of renewable and local energy sources. Wind energy is a promising area for diversification of energy resources in Belarus in general because of favorable natural conditions: sufficient wind speed, the terrain, the low proportion of days with calm weather (in 2018 their share amounted to 11.3%).

The construction of wind farms in areas where there are large CHP plants but there is a high wind power potential will reduce the cost of transportation of electricity and make it more accessible to the consumer [4].

5 Institutional Factors of Wind Energy Development

The need for the development of renewable energy in the country is enshrined in the regulatory documents: the Law of the Republic of Belarus "On renewable energy," the Energy Security Concept of the Republic of Belarus, the national program for the development of local and renewable energy sources, etc., which recognize wind energy as a promising renewable energy sector in Belarus.

Currently, more than 96% of electricity generated from renewable sources (RES) comes to the state electric grid of the Republic of Belarus. The cost of energy due to increasing coefficients on the sale of excess energy is many times larger than the cost of energy from traditional sources. The issuance of the Decree of the President of the Republic of Belarus dated May 18, 2015 № 209 "On the use of renewable sources of energy" is a step toward this problem.

On the territory of Belarus 1840 areas for wind turbines were identified. They have theoretically possible energy potential of 1600 MW and annual power generation of 2.4 billion kWh. The Decree № 209 provides that the creation of new, modernization, reconstruction of existing plants for the use of RES must be within the established quotas: using solar energy – 20 MW, wind – 11 MW, sun – a 1.55 MW, the energy of motion of water flows of 73.59 MW, biomass (firewood, chips) – 11.28 MW [25]. These quotas were set for 2017–2019. In 2020 the quota of wind energy is 19.8 MW; however, this quota is set in the framework of the international technical assistance project "Removing barriers to wind power development in the Republic of Belarus" [20].

Possible modernization of the energy system as a whole, the admission of new participants to the markets for the sale of electricity to markets, and operating networks can lead to competitive relations. This will allow to abolish the quota system, which hinders the development of renewable energy in the country.

6 Conclusion

Traditional energy is the main supplier of greenhouse gases and pollutants into the atmosphere in Belarus from stationary sources, fuel, and energy complex accounts for more than 60% greenhouse gas emissions from stationary sources, due to which many of the environmental problems of the country require decisions.

A significant contribution to reducing air pollution and emissions of carbon dioxide may develop wind power in Belarus.

Technical wind energy resources of Belarus can provide the production of 280 billion kWh.h of electricity. In addition, great potential is the development of small wind power (wind turbines with a capacity of 100–150 kW) [3, 4].

Wind energy development will contribute to Belarus international obligations under the Paris Agreement, to improve the environmental situation: reduction of greenhouse gas emissions by 28% by 2030.

In addition, the development of wind energy will increase the share of renewable and local energy sources, which will contribute to maintain energy security of the country. Today Belarus exports 88% of energy resources.

Despite the existing selection of places for construction of wind turbines, it should be noted a weak elaboration of ecological-geographical support, which leads to an incomplete account, primarily, of environmental factors.

The positive experience of the exploitation of the continental turbine in similar climatic conditions of Belarus, Germany, Denmark, Poland, and the Baltic States can count on the fact that in Belarus, electricity production can successfully occur at the wind farms.

References

1. Air quality and health [electronic resource] // World Health Organization: [site]. URL: https://www.who.int/en/news-room/fact-sheets/detail/ambient-(outdoor)-air-quality-and-health. Accessed 08.02.2020
2. Bezrukih PP (2015) Efficiency of renewable energy: myths and facts // Bulletin of agrarian science of the don 1(29):5–17
3. Yu BM, Sinyugin OA (2017) Intra-industry competition and renewable energy sources. Innov Agric 3:159–165
4. Degtyarev KS, Koshkin SP, Sangadzhiev MM (2016) Economic and socio-geographical aspects of renewable energy development in the Republic of Kalmykia. Powerman (8):32–36
5. Ecology [electronic resource] // Belenergo: [site]. URL: http://www.energo.by/content/deyatelnost-obedineniya/ekologiya/. Accessed: 02/08/2020
6. (2019) Environmental protection of the Republic of Belarus: statistical compilation. Medvedev IV (ed). Minsk: National Statistical Committee of the Republic of Belarus, p 200

7. Kovalev MM (2018) In: Kovalev MM, Kuznetsov AS (eds) The future of Belarusian energy on the background of global trends: monograph. BSU Publishing Center, Minsk, p 223, [4] p
8. Koshkin SP (2019) Assessment of the potential of wind energy development at the regional level (on the example of Rio Grande do Norte, Brazil) / / Bulletin of the Moscow University. Series 5: Geography 3:93–98
9. Krasovskaya T, Koshkin S Windfarms – combining energy and ecological performance in Crimea. Geogr Environ Sustain 12(1):88–103
10. Lavrentiev NA, Zhukov DD (2005) On the use of wind energy resources in Belarus/ Lavrentiev NA, Zhukov DD / / Materials of the international conference "Energy of Belarus: ways of development." p 61–71
11. Mikhalycheva EA, Trifonov AG (2013) Environmental aspects of construction and operation of wind power stations / / Bulletin of the Brest state technical University. 2:121–122
12. PARIS Climate agreement [electronic resource] // Ministry of Natural Resources and environmental protection of the Republic of Belarus: [site]. URL: http://www.minpriroda.gov.by/ru/paris-ru/. Accessed: 02/08/2020
13. Ryzhenkov MA, Ermolenko BV, Ermolenko GV Environmental aspects of wind energy. URL: http://windpower.ru/wp-content/uploads/Ermolenko.pdf
14. Medvedev IV (ed) (2019) The energy balance of the Republic of Belarus: statistical compilation. National Statistical Committee of the Republic of Belarus, Minsk, p 154
15. The project of international technical assistance "Removing barriers to the development of wind energy in the Republic of Belarus" [Electronic resource] // Wind energy in the Republic of Belarus. URL: http://www.windpower.by/ru/project/. Date of treatment 02.15.2020
16. The State Cadastre of Renewable Energy [electronic resource] // Ministry of Natural Resources and Environmental Protection of the Republic of Belarus: [site]. URL: http://minpriroda.of.by/Cadastre/Map. Accessed 02.02.2020
17. Zakrevsky VA (2017) Renewable energy "for" and "against" / / Energy Strategy 1(55):11–13

Floating Offshore Wind Technology Development: Current State and Future Vision

Abdul Salam Darwish

1 Introduction

Climate change is severely affected by the excessive use of fossil fuels threatening human security and development. This effect has continuously put forward a significant demand for alternative energy. An abundant, clean, alternative and renewable energy sources have been utilised with considerable amount worldwide. Wind power has been developing with reduced costs over recent years. Floating wind turbine technology has been developed in a way that resulted in high capacity and low in cost. Figure 1 shows global cumulative installed wind power capacity (MW), which reached about 650.8 GW in 2019 [1], and from this figure's trend, it is expected to reach around 700 in 2020. Figure 2 shows the cost in comparison with other power sources [2]. Wind power in the UK has recently exceeded the nuclear power over 3 months in 2018, and it is repeated on many different occasions, as shown in Fig. 3 [3].

There are some restrictions to the use of the onshore wind farm, such as the limits for land availability. Wind turbine noise and their visual impacts are the problems and the main reasons for people to refuse to accept the building of onshore wind turbines close to residential areas. In contrast, offshore wind farms have the same operation process as onshore wind farms. Still, they have several advantages: There is plenty of land and space, the wind is steadier, and no visual and noise problems resulting in fewer complaints. Offshore wind farms are, therefore, the future of renewable energy, and the floating wind farms are believed to be the most applicable technology as the experts foresee offshore wind growing by more than 20% each year over the next several years. Floating wind farms will open up entirely new growth opportunities [4]. An overview of present and future trends in floating offshore wind technology has been presented to structure the different concepts and clarify the advantages and disadvantages of each [5].

The potential of offshore wind suitable for implementing floating technology worldwide can be seen in Fig. 4 [6].

A steep and continuous fall in the cost of offshore wind, higher and steadier wind speeds are available in deeper waters. These turbines are experiencing lower offshore wind turbulence enjoying longer farm life ~25–30 years. Moreover, connecting the floating wind farm to the electric grid using subsea alternating current (AC) or high voltage direct current (HVDC) cables has recently proved cheaper and reliable for the installation process [6]. In this paper, the author is to explore the arguments and benefits that are making the floating wind technology more attractive for investing with the vision to replace onshore wind farms and even the coastal ones.

2 The Technical Potential for Floating Wind in Major Economies

As mentioned in an earlier section, there exist plenty of offshore areas worldwide with high potential for floating wind farms. Many of the major economies' governments have started to consider utilisation. Europe, USA and Japan specified shares of offshore wind resources requiring a floating facility (>60 m depth) and the technical potential for floating wind farms, as shown in Table 1 [7]. In addition, many other areas over the world have the potential such as south-east Asia, Oceania and Northern Europe [8].

A. S. Darwish (✉)
School of Civil Engineering, University of Bolton, Bolton, UK
e-mail: a.darwish@bolton.ac.uk

© The Author(s), under exclusive license to Springer Nature Switzerland AG 2022

A. Sayigh (ed.), *Sustainable Energy Development and Innovation*, Innovative Renewable Energy,
https://doi.org/10.1007/978-3-030-76221-6_72

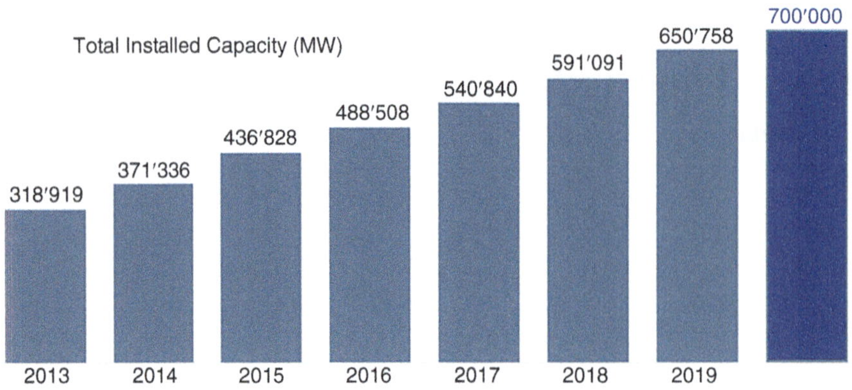

Fig. 1 Global cumulative installed wind power capacity (MW)

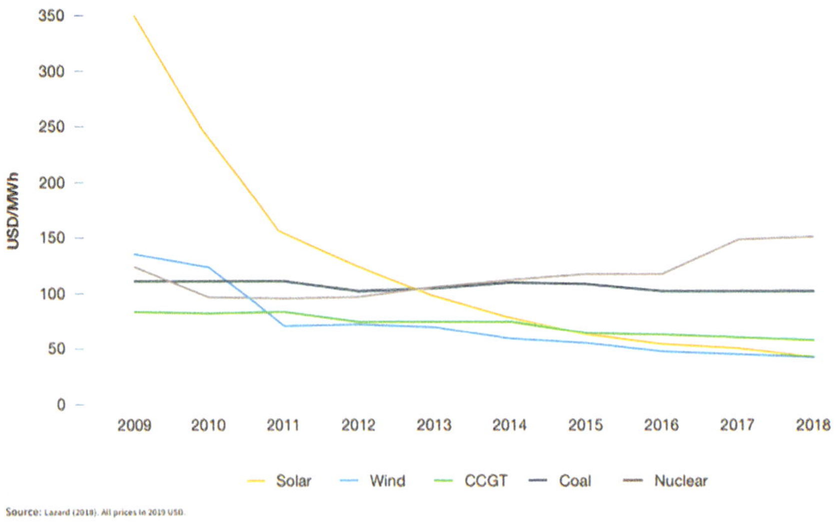

Fig. 2 Cost in comparison with other power sources [2]

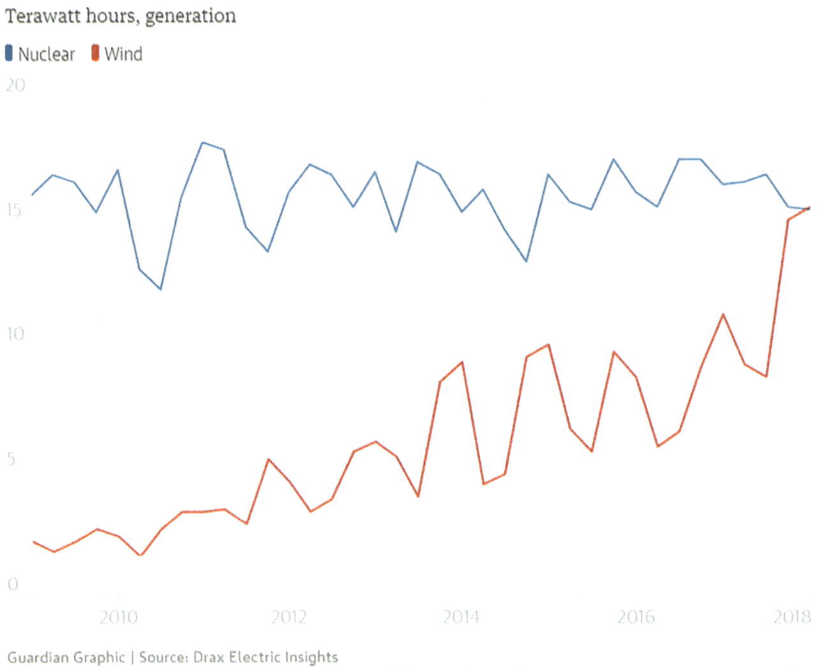

Fig. 3 Wind power (UK) beats nuclear power over 3 months in 2018 [3]

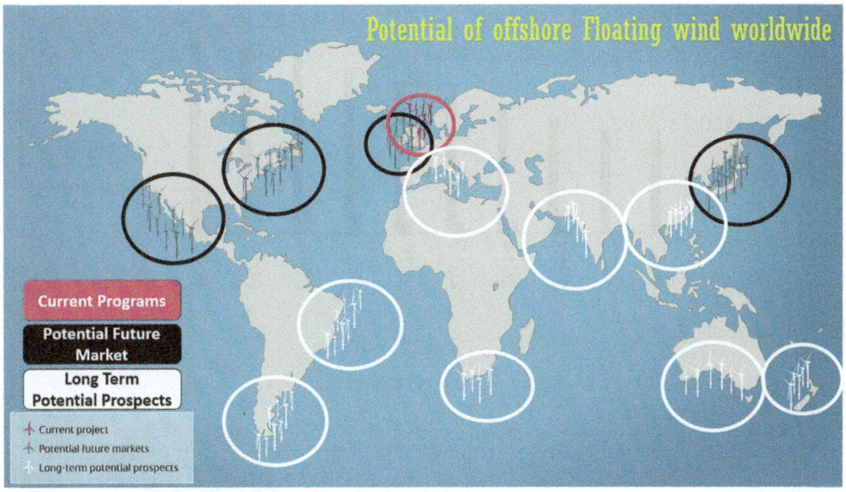

Fig. 4 Vast potential of possible floating wind technology sites worldwide [6]

Table 1 Technical potential for floating wind in major economies [7]

Country/ region	Shares of offshore wind resource requiring floating facility (>60 m depth)	Technical potential for floating wind [GW]
Europe	80%	4000
US	60%	2450
Japan	80%	500

3 Offshore Wind Turbine Evolution

Offshore wind turbine technology has been recently developed in a way that encouraged significant investments, and the installation capacity has sharply increased, see Fig. 5 [9]. It is now the front line to fight against climate change and decarbonising the energy system, with installed capacity growing from 1.4 GW in 2008 to 24 GW in 2018 [10]. It reached around 30 GW globally in 2019 [11].

The exponential growth of the offshore wind farm started in 1991, a 4.9 MW wind farm was installed in Denmark that composed of 11 wind turbines with a unitary capacity of 450 kW along with an increased understanding of wind turbine design and operation [12]. Since 2007, offshore was developed and highly utilised, see Fig. 6, presenting the cumulative installed capacity to 2022 by nationality [13]. The machine size and performance have been dramatically progressed in technology, attracting a large scale of investment.

Record-capacity installations shown by circle area and are sized proportionality.

Offshore wind farms are more complicated than onshore in design point of view and construction. They are located about 10 km off the coast in water depths of about 10 m. Main condition is that the turbine should be above the crest level of the highest waves and connected to the seabed by structure [9].

Offshore wind turbines have comprised turbines with fixed foundation up to a depth less than 50 m and for higher than 50 m floating foundation is technically advanced and proved effective. For less than 50 m gravity base, monopile, tripod and jacket foundations are commonly used. Figure 7 shows some examples of turbine support for different depths [14].

4 Floating Wind Turbine Technological Advancement

The motion of the floating platform includes three translational components (heave in the vertical, sway in the lateral and surge in the axial) and three rotational components (yaw about the vertical axis, pitch about the lateral and roll about the axial) motion as shown in Fig. 8. Therefore, the additional effect of the wind contribution, which is transmitted to the rotor due to the platform motion needs to be considered [15].

X′, Y′ and Z′ are the blade coordinate system and x, y and z are the inertial coordinate system.

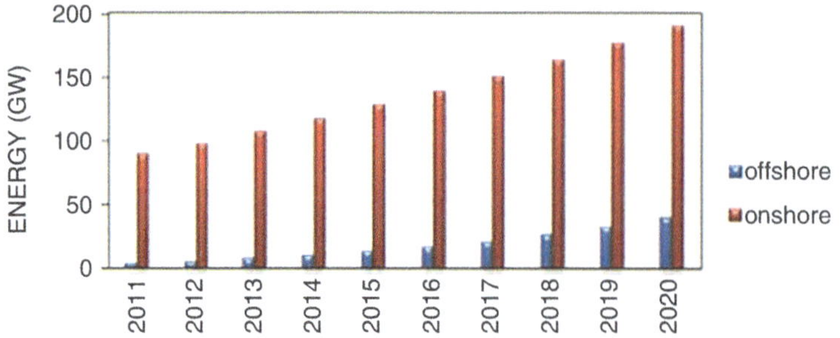

Fig. 5 Wind energy installation in years [9]

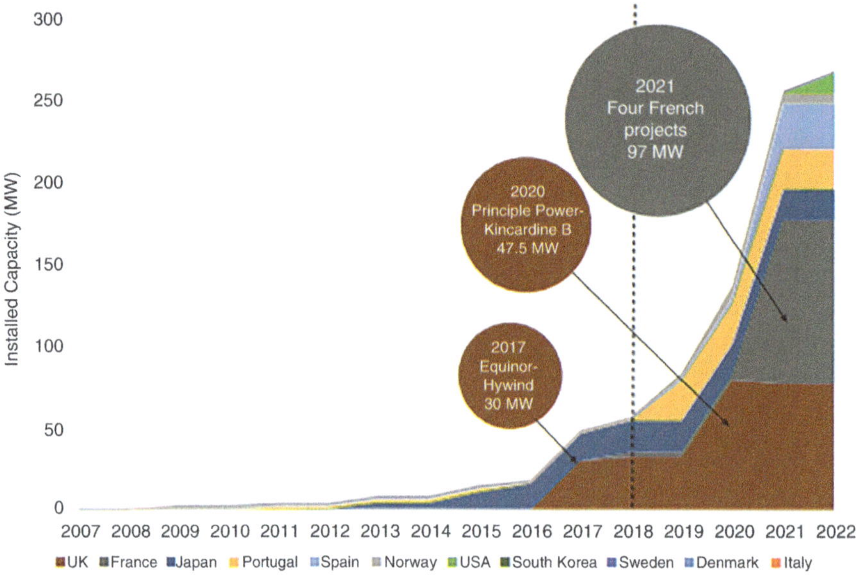

Fig. 6 Cumulative installed capacity to 2022 by nationality [13]

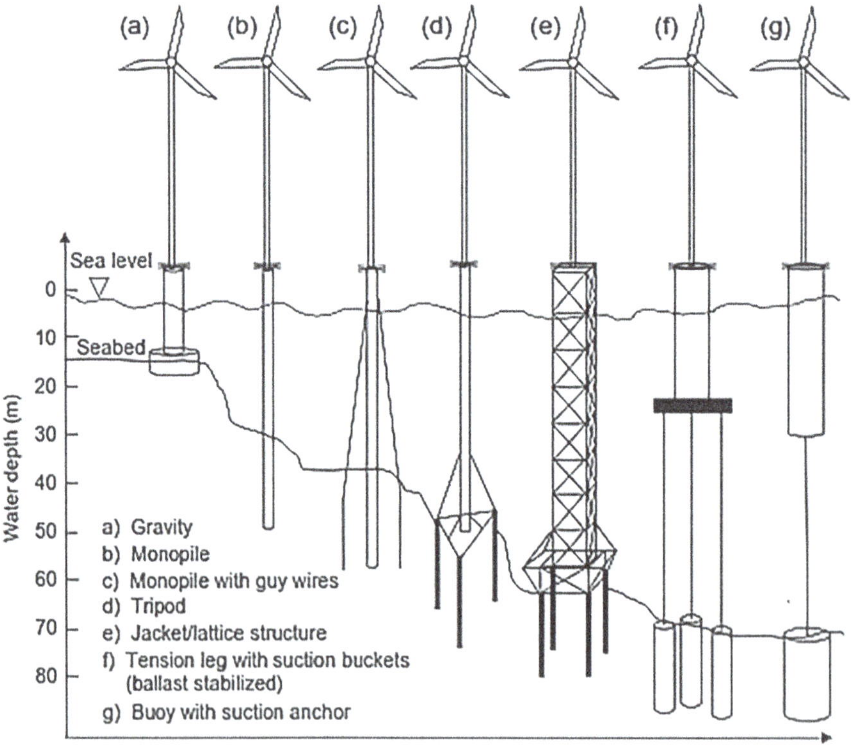

Fig. 7 Support structure options showing typical ranges of water-depth application [14]

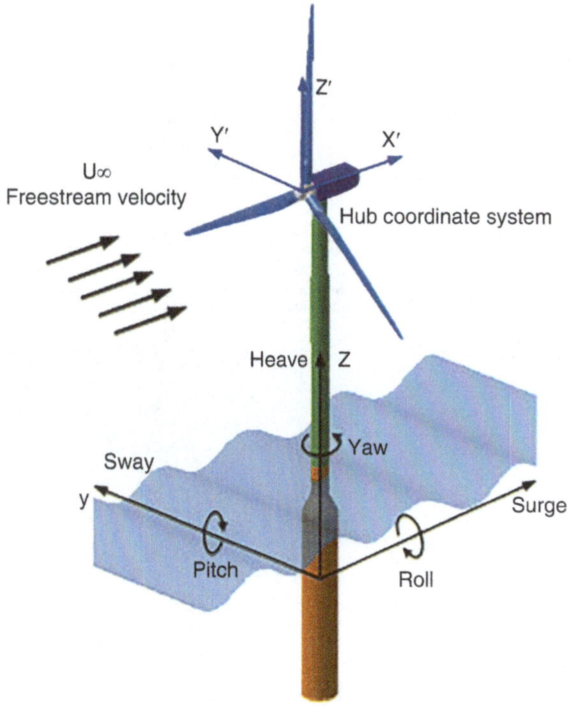

Fig. 8 The degrees of freedom of an offshore floating wind turbine platform [15]

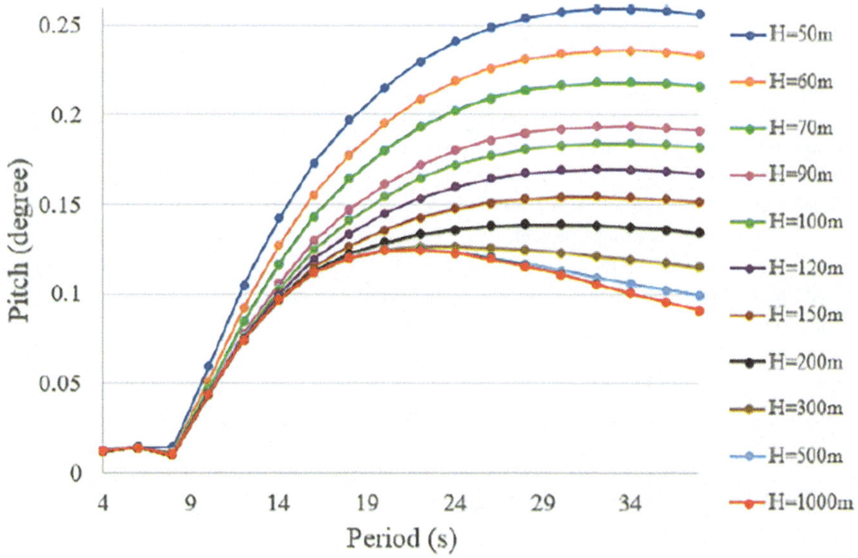

Fig. 9 Influence of water depth on pitch motion of a floating wind turbine [16]

Floating wind turbine foundation is a game-changer as it decides on whether the stability is totally achieved with regards to wave and tide. Pitch motion, which is a key stability design factor may have a big influence on the power generation efficiency of a floating wind turbine which increases with decreasing water depth H (Fig. 9) [16]. This is why floating wind turbines were erected in deep water rather than in shallow water [16]. Consequently, three types of floating wind turbines are therefore classified: spar-supported turbines, semi-submersible floater (SSF) and tension leg platform (TLP), as shown in Fig. 10 [15]. Table 2 gives existing floating wind turbines classified according to their foundations [16]. As a result, the critical factor in this game is the floating wind turbine foundation which will be leading the way for rapid future growth in the offshore wind power market.

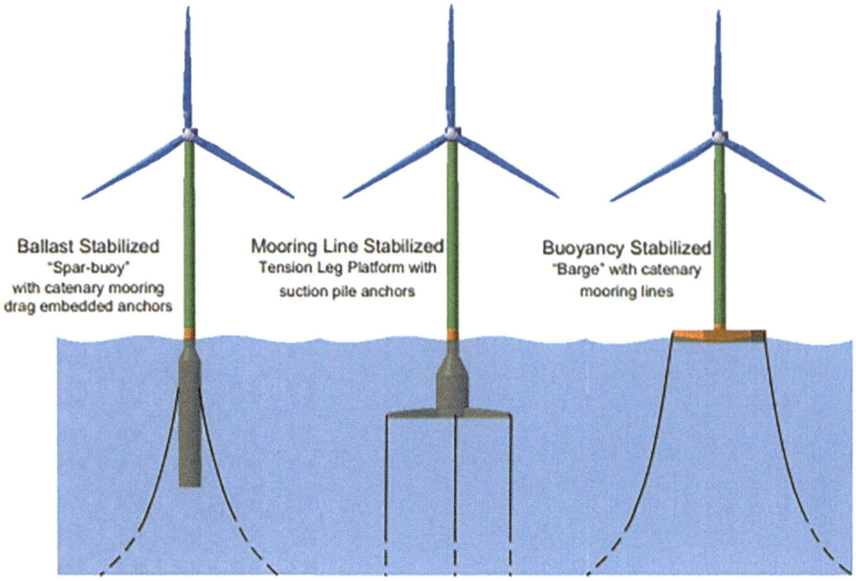

Fig. 10 Principle floating wind turbine foundation concepts [15]

Table 2 An existing floating wind turbines classified according to their foundations [16]

Concept	Target Water Depth	Current Status	Type of Supporting Platform
Hywind	Prototype, 220 m	In operation	Spar
	Scotland, 95–120 m		
	Target, 60–300 m		
Sway	1:6 scale prototype, 25 m	Prototype sank in 2011	
Fukushima Hamakaze	120 m	In operation	
WindFloat	Prototype, 40–50 m		Semi-submersible floater (SSF)
	Portugal project, 100 m	Prototype is in operation	
Fukushima Mitai	>150 m	In operation	
Fukushima Shimpuu	120 m	In operation	
Blae H	113 m	In operation	Tension leg platform (TLP)
PelaStar	>50 m	in plan	

5 Floating Wind Turbine Technical Challenges

The floating wind turbines floating structures have to withstand harsh conditions. Due to the fact that the floating wind turbine is to be installed in very deep areas, and in order to ensure that most benefits are gained, a massive size weight of structure has consequently resulted. This has added many engineering design challenges, some of which are as follows [17, 18]:

- Massive steel structure mounting techniques, strong equipment and cost effectiveness, to keep stability at different weather conditions of high wind, wave and tidal, and aerodynamic forces might minimise the turbine efficiency. As the wind is stable in an open sea, than near the coast, it could minimise some of these challenges.
- Anchorage – Anchoring in shallow water could be difficult and expensive. The commonly used catenary moorings could perform poorly.
- Wave loads – The large areas projected by multiple-turbine-floaters to the water surface could result in very high wave loads, which might be beyond the structural bearing and require increased maintenance.
- Composite materials are to be manufactured with a high-tech manufacturing process as well as research and development.
- Stability – The turbine thrust multiplied by the tower moment arm could result in significant overturning moments.

- Motion response – To minimise the fatigue damage due to gyroscopic loads, floater motion.
- Responses must be minimised. It was noted that this requirement was contradictory to the requirement of minimising structural loads.
- The other major challenge is environmental. A turbine's blades sweep through a large area, creating a potential hazard for birds. This means that farms cannot be placed in areas where there are large bird migrations.

In order to lower the costs, it is recommended to keep the turbine operating close to its optimum point of operation under partial load, guaranteeing reliability by reducing structural loads and regulating the power generated in strong wind regimes [19]. Floating wind turbines' foundations are generally offering environmental benefits in contrast with the fixed concrete foundations as there are less hard activities on the seabed during installation.

6 Existing Floating Wind Farms

The world-first floating wind farm is the Hywind in Scotland with 30 MW capacity which consists of five 6 MW turbines. The floating foundation is a spar buoy substructure commissioned in 2017 [20]. At the end of the year 2018, nine floating wind farms were installed over the world five in the UK (Europe), four in Japan with a total capacity of 50 MW. In addition, another 13 projects are supposed to be started (3 in the Republic of Korea and Japan, 9 in France, Portugal and the UK and 1 in the USA) [21].

7 Future Perspectives of Floating Wind Technology

It is expected to have a massive deployment for the floating offshore wind technology by and after 2030 in Europe to meet their target of 450 GW [8]. 18 GW is to be installed in Japan by 2050 [22]. Table 3 listed the 5 years planned pre-commercial floating offshore wind farms projects in Europe [23]. The energy future as 2050 is totally affected by how much the CO_2 emissions are reduced with the aid of the alternative energy technology systems, which is thought to be as solar and wind. The wind has been deployed in an increasing trend. Floating wind technology is the key perspective. Many drivers are behind this argument shown in Fig. 11 [8].

Floating wind turbine technology has been tested for a couple of years now and the costs have started to come down, so it is expected soon that more locations will be added, including the northern part of the UK, USA and Japan [24].

Table 3 Europe 5 years plan, floating wind farms projects [23]

Wind farm name	Capacity	Country (MW)	Commissioning date
Hywind Scotland	United Kingdom	30	2017 (in operation)
Windfloat Atlantic	Portugal	2	2019
Flocan 5 Canary	Spain	2	2020
Nautilus	Spain	5	2020
SeaTwirl S2	Sweden	1	2020
Kincardine	United Kingdom	49	2020
Forthwind Project	United Kingdom	12	2020
EFGL	France	24	2021
Groix-Belle-Ile	France	24	2021
PGL Wind Farm	France	24	2021
EolMed	France	25	2021
Katanes Floating Energy Park –Array	United Kingdom	32	2022
Hywind Tampen	Norway	88	2022

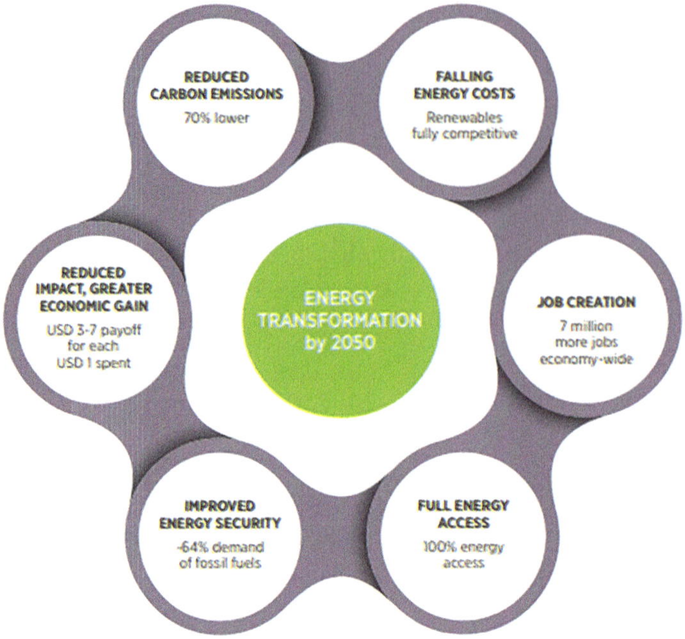

Fig. 11 Pressing needs and attractive opportunities are driving the transformation of the world's energy system [8]

8 Strategies and Recommendations for Advanced Future Expansion

For the reasons argued in this paper, the floating wind technology and projects should have excellent attention from the Governments in many aspects. These aspects are to encourage the development of a successful floating wind industrial sector and make a significant contribution to global energy. The following are some recommendations that are as follows:

- Specify clear targets to compromise the use of conventional fuel.
- Educating the public to appreciate the low carbon electricity generation.
- "Domestic energy providers are required with each bill, and on their home page, to clearly advertise the mix of electricity production between zero-emission and carbon-emitting sources" [25].
- More investment should be encouraged with Governments' grants for maintaining industrial facilitation and research and development to advancing this type of technology.
- Market development should well be expanded and performed in a wide range.
- "The Offshore Wind Innovation Hub should be further developed to improve accessibility to the Technology Roadmaps and build on the success of the Innovation Challenge program" [25].
- A global floating wind technology organisation is to be formed in order to lead the advances in technology, cost reduction, market development and research & development to organise this type of technology.

9 Conclusions

In response to the severe change in the climate, an increasing effort is put to use alternative energy sources. Wind energy has had an encouraging trend in technology development, cost and rate of investment. Floating wind technology has had successful resulted proposals indicating that it is the right technology for utilising wind energy and could replace the onshore and coastal offshore farms for many advantages. These advantages are the reduced, visual, noise, plenty of available lands, steadier wind and possible for large wind machines with higher energy production. Potential floating wind sites are located around many countries supporting the concept of possible integration and backs up any future investment decisions. Cumulative installed capacity of the floating wind turbines and the capacity to be installed were discussed showing possible areas over the world. This paper reviewed the recent developments technically, current and near-future installations worldwide. Floating wind turbine technological advancements were reviewed, highlighting the main significant challenges and what would be the suggested solutions to some of them. Foundation structure and stability of the floating wind turbine is the game-changer of this technology as it decides on whether the stability is totally achieved with regards to wave and tidal. One of concluding remarks is that in order to lower the costs, it is recommended to keep the turbine operating close to its optimum point of operation under partial load, guaranteeing reliability by reducing structural loads and regulating the power generated in strong wind regimes. Future perspectives of wind technology with attractive opportunities were discussed and the existing drivers for the transformation of the world's energy system are presented. Strategies and recommendations were put forward for achieving the expected future expansion.

References

1. Wwindea.org (2020) World wind capacity at 650.8 GW, corona crisis will slow down markets in 2020, Renewables to be core of economic stimulus programmes – World Wind Energy Association. [Online] Available at: https://wwindea.org/blog/2020/04/16/world-wind-capacity-at-650-gw/. Accessed 16 May 2020
2. Slideshare.net (2020) Global market outlook for solar power 2019–2023. [Online]. Available at: https://www.slideshare.net/SolarPowerEurope/solar-power-europeglobalmarketoutlook20192023?from_action=save. Accessed 16 May 2020
3. Vaughan A (2020) Wind power overtakes nuclear for first time in UK across a quarter. [Online] the Guardian. Available at: https://www.theguardian.com/environment/2018/may/16/wind-power-overtakes-nuclear-for-first-time-in-uk-across-a-quarter. Accessed 17 May 2020
4. StormGeo (2020) The future is bright for floating wind turbines. [Online] Available at: https://www.stormgeo.com/solutions/renewables/articles/the-future-is-bright-for-floating-wind-turbines/. Accessed 17 May 2020
5. Gonzalez S, Diaz-Casas V (2016) Present and future of floating offshore wind. In: Floating offshore wind farms. Green Energy and Technology. Springer, Cham, pp 1–22
6. Darwish A, Al-Dabbagh R (2020) Wind energy state of the art: present and future technology advancements. Renew Energy Environ Sustain 5:7
7. Carbon Trust (2015) Floating offshore wind: market and technology review, London. Available at: https://prod-drupalfiles.storage.googleapis.com/documents/resource/public/Floating%20Offshore%20Wind%20Market%20Technology%20Review%20-%20REPORT.pdf
8. IRENA (2019) Future of wind: Deployment, investment, technology, grid integration and socio-economic aspects (A Global Energy Transformation paper). International Renewable Energy Agency, Abu Dhabi
9. Wu X, Hu Y, Li Y, Yang J, Duan L, Wang T, Adcock T, Jiang Z, Gao Z, Lin Z, Borthwick A, Liao S (2019) Foundations of offshore wind turbines: a review. Renew Sustain Energy Rev 104:379–393
10. IRENA (2019a) Dashboard - capacity and generation. Available at: https://www.irena.org/publications/2018/May/Leveraging-Local-Capacity-for-Offshore-Wind. Accessed 20 May 2020
11. Offshore Wind (2020) GWEC: global offshore Wind capacity reaches 29GW | offshore Wind. [Online]. Available at: https://www.offshorewind.biz/2020/03/19/gwec-global-offshore-wind-capacity-reaches-29gw/. Accessed 19 May 2020
12. Danborg F, Clausen NE (2013) Offshore wind energy in Denmark. In Tonderski A, Jdrzejewska A (eds) Offshore wind energy in the South Baltic Region – challenges and opportunities. Drukarnia WL, p 57–60. Available at: https://backend.orbit.dtu.dk/ws/files/57802137/South_Baltic_OFFER.pdf
13. Hannon M, Topham E, MacMillan D, Dixon J, Collu M (2019) Offshore wind, ready to float? Global and UK trends in the floating offshore wind market. University of Strathclyde, Glasgow. Available at: https://strathprints.strath.ac.uk/69501/13/Hannon_etal_2019_Offshore_wind_ready_to_float_global_and_uk_trends_in_the_floating_offshore_wind_market.pdf
14. O'Kelly BC, Arshad M, O'Kelly BC, Arshad M (2016) Offshore wind turbine foundations - analysis and design. In: Ran CNAL (ed) Offshore wind farms: technologies, design and operation. Woodhead Publishing, London, pp 589–610
15. Tran T, Kim D (2015) The platform pitching motion of floating offshore wind turbine: a preliminary unsteady aerodynamic analysis. J Wind Eng Ind Aerodyn 142:65–81
16. Yang W, Tian W, Hvalbye O, Peng Z, Wei K, Tian X (2019) Experimental research for stabilising offshore floating wind turbines. Energies 12(10):1947
17. StormGeo (2020) The future is bright for floating wind turbines. [Online] Available at: https://www.stormgeo.com/solutions/renewables/articles/the-future-is-bright-for-floating-wind-turbines/. Accessed 26 May 2020
18. Shipping AB (2012) Floating wind turbines. US Department of the Interior, Herndon. Available at: https://www.bsee.gov/sites/bsee.gov/files/tap-technical-assessment-program//669ab.pdf
19. Salic T, Charpentier J, Benbouzid M, Le Boulluec M (2019) Control strategies for floating offshore wind turbine: challenges and trends. Electronics 8(10):1185. Available at: https://www.mdpi.com/2079-9292/8/10/1185/pdf
20. Equinor (2018) Energy perspectives 2018: long term macro and market outlook, Stavanger, Norway. Available at: https://www.equinor.com/content/dam/statoil/documents/energy-perspectives/energy-perspectives-2018.pdf
21. GlobalData (2019) Floating foundations: the future of deeper offshore wind. GlobalData, London
22. POWER Magazine (2020) Floating offshore Wind buoyant on new developments, Projects [online] Available at: https://www.powermag.com/floating-offshore-wind-buoyant-on-new-developments-projects/. Accessed 28 May 2020
23. WindEurope (2020) Floating offshore wind energy: a policy blueprint For Europe | Windeurope. [Online]. Available at: https://windeurope.org/policy/position-papers/floating-offshore-wind-energy-a-policy-blueprint-for-europe/. Accessed 28 May 2020
24. Offshore Wind (2020) Floating wind turbines, promising future | Offshore wind. [Online]. Available at: https://www.offshorewind.biz/2019/05/01/floating-wind-turbines-promising-future/. Accessed 26 May 2020
25. Whitmarsh M, Canning C, Ellson T, Sinclair V, Thorogood M (2019) The UK offshore wind industry: supply chain review. The UK Offshore Wind Industry: Supply Chain Review. Offshore Wind Industry Council. Available at: https://cdn.ymaws.com/www.renewableuk.com/resource/resmgr/publications/supply_chain_review_31.01.20.pdf

Energy Status and the Importance of Wind Energy Resources in Sri Lanka

Vidya Amarapala, Abdul Salam Darwish, and Peter Farrell

1 Introduction

For centuries, people have harnessed wind energy for different purposes such as seafaring, irrigation/cultivation, grinding various types of grains, etc., [1, 2]. According to the records, in July 1887, James Blyth (a Scottish electrical engineer) introduced the potential wind turbine to generate electricity. However, the working model was put into operation only after 25 years [3]. At about the same time, Charles F Brush of Ohio USA built a 12 KW wind turbine to meet some of the energy needs at home which had a rotor diameter of 50m with 144 rotor blades. Even though the wind-generated electricity was introduced in the early 1900s, it was not commercialised till the 1980s. The first offshore commercial wind farm was introduced in 1991 in Denmark with 11 turbines with a capacity of 450KW each [4] (see Fig. 1).

James Blyth built the first wind power plant, and it had cloth sails instead of the wind blades used in modern wind plants.

Sri Lanka had envisaged plans to generate 10% of its total energy need with non-conventional renewable energy (NCRE) by 2016, and according to the Sustainable Energy Authority, it had achieved the 10% target by 2015. Progressively, the next goal is to reach its 20% by 2020 [5].

The next most ambitious target is to be Carbon neutral by 2050 [6]. Presently over 50% of the electricity is produced by an imported energy source, which is costing a colossal amount of foreign exchange. Therefore, it is a substantial burden on the Government's budget, and in turn, the consumers have to meet such escalating costs directly or indirectly by way of numerous taxes. On the other hand, this component of energy generation with hydrocarbons naturally increases the carbon footprint drastically, even though the present per capita CO_2 emission is less than 1 ton in Sri Lanka [7].

In 1895, M/S Boustead Brothers made Sri Lanka's first public electricity supply network in Colombo under the name of "Colombo Electric Tram Ways and Lighting Company" and then in Kandy by the "Kandy Lighting Company" which met the electrical needs in their spheres in the first decade of the twentieth century [8]. In 1950, the first Hydro Power station was commissioned by D. J. Wimalasurendra, who is known as the father of Hydropower in Sri Lanka [9].

Historically in the plantation sector, managed by the British during the period of colonisation, the electricity was generated with Hydropower, mainly by run-off streams. The other significant source of energy was biomass. In ancient days, the kitchens were powered by biomass for the daily cooking needs, and it was only after the industrialisation that fossil fuel was used as a source of energy in domestic kitchens. However, it is recorded that Sri Lanka has used wind power for industrial applications, in iron-ore furnaces dating back to third century BC. There are several compelling papers and articles about the archaeological research and findings of this unique furnace in Sigiriya and Balangoda. Figure 2 is admired and praised by many archaeologists around the world of this great invention, as to how the ancient Sri Lankans utilised the wind energy in the industrial application of iron smelting as early as 3000 BC for the first time in the world. However, the archeologists had rejected the possibility of the wind-powered iron-smelting furnace, as the wind blowing into the charcoal fire chambers will not sustain the consistency of the furnace temperature. Figure 2 the replica built based on the remains of the ancient furnaces reviled the monsoon winds were able to continuously maintain the required temperature in the smelting furnace and about half the metal produced in the trails has been high quality [10–12].

V. Amarapala · A. S. Darwish (✉) · P. Farrell
School of Engineering, The University of Bolton, Bolton, UK
e-mail: vda1res@bolton.ac.uk; a.darwish@bolton.ac.uk; P.Farrell@bolton.ac.uk

© The Author(s), under exclusive license to Springer Nature Switzerland AG 2022

A. Sayigh (ed.), *Sustainable Energy Development and Innovation*, Innovative Renewable Energy,
https://doi.org/10.1007/978-3-030-76221-6_73

Fig. 1 Blyth's "windmill" at his cottage in Mary Kirk in 1891 [3]

Fig. 2 is an image of the model of iron smelting furnaces used in Sri Lanka dating back to the third century BC [14]

2 Present Energy Status in Sri Lanka

Sri Lanka is perhaps the only country in the South Asian Region (SARC) which has the near 100% electricity penetration with quality uninterrupted service [15]. It is an isolated island grid with an installed capacity of just over 4000MW, according to the current data published by the Ceylon Electricity Board (CEB) [16]. The 220kV, 132kV transmission network and grid substations together with main power plants of hydro, wind and thermal is marked in Table 1.

From Table 1, the figures of generation, it is noted that of the total evacuated energy is approximately 49% of the renewable energy (RE) sources. However, the total energy evacuated from the NCRE sources is only about 3% of the total power generated. Table 1 shows that the CEB generation statistics are extracted from the CEB Annual report as a reference, and to elaborate, the quantum of energy evacuated from hydro, thermal and NCRE is recorded for this purpose [16].

The government has also coined the idea of converting some of the heavy fuel oil (HFO) powered plants into liquefied natural gas (LNG), thereby reducing the carbon footprint to a remarkably low figure. In the year 2015, the said idea was cast into a policy document by a publication of the Ministry of Power & Energy "Sri Lanka Energy Sector Development Plan for a Knowledge-based Economy" [15]. Aforesaid the plan was to be implemented on HFO-powered plants such as 360MW, AES Kelanitissa Power Plant and 300MW Yugadanavi Power plant at Kerawalapitiya. This move would have brought environmental benefits, energy security and economic benefits by subsequently switching to indigenous LNG resources. Even though this was planned, the LNG upstream development in Manna and Cauvery basins has encountered unforeseen delays and as such, the overall LNG component was not accomplished as per the energy policy. Hence this has caused a substantial negative impact in the balance sheet of CEB and the National annual Environment statistics. Figure 3 illustrates the transmission network of Sri Lanka with the locations of power plants [17].

Table 1 CEB generation statistics

	Energy source	Installed capacity MW	Total energy generated GWH
01	Thermal power	2,115	9,507
02	Hydro power	1,726	4,220
03	NCRE	177	421

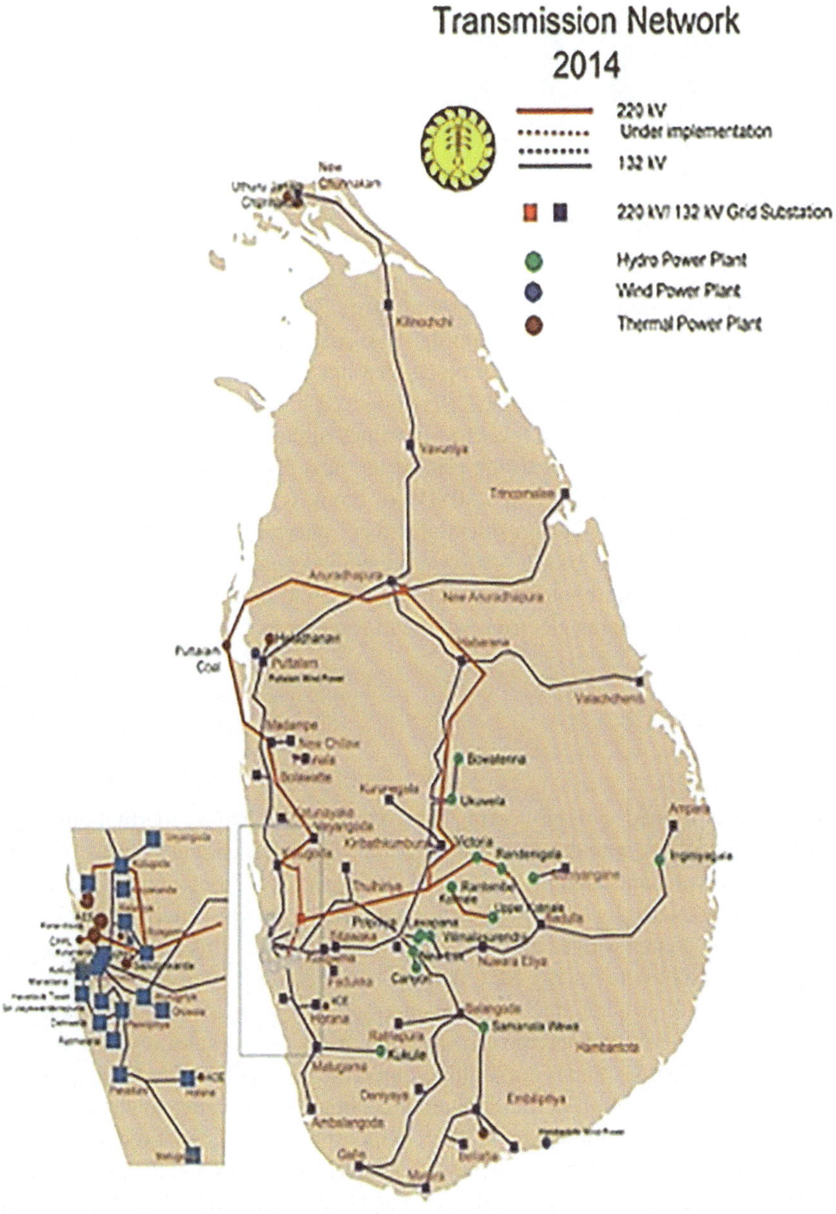

Fig. 3 The transmission network of Sri Lanka with power plants [17]

It is essential to understand the Generation Plan of Sri Lanka, together with the National Energy Policy to get a clear picture of the present energy situation and the way forward. The Ceylon Electricity Board (CEB) is responsible for providing adequate electricity to meet the electricity demand in the country. At the same time, Sustainable Energy Authority (SEA) is tasked to manage the operation demand, and the Public Utility Commission of Sri Lanka (PUCSL) is regulating the licensees for Generation, Transmission and Distribution of electricity. In order to achieve the equilibrium among these institutions, the Government has developed national policies, strategies and the actions to address to mitigate climate change impacts etc.

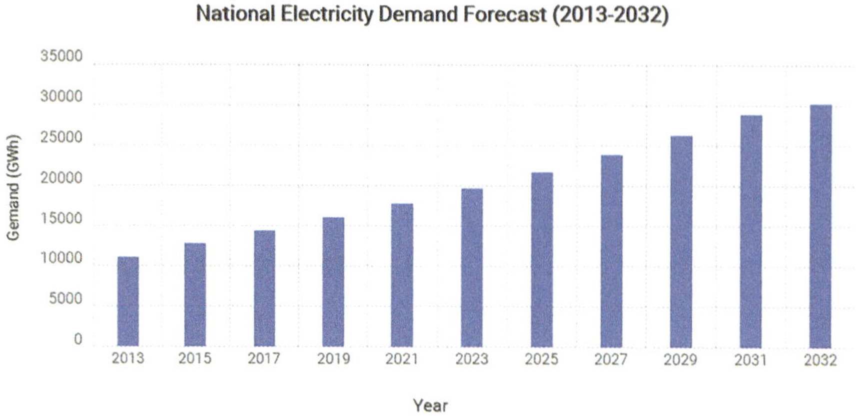

Fig. 4 National electricity demand forecast (2013–2032) [8]

Thereby meeting the increasing electricity demand integrating more NCRE projects with thermal energy sources. As part of its action plan, the Government is on the path to developing large scale wind power projects, solar power projects, biomass and waste to energy in addition to mini-hydropower projects that have been connecting to the national grid throughout the years. It is expected to increase the grid-connected NCRE above 20% of the installed capacity from 2020 and further to increase the annual renewable energy absorption level to four times higher compared to the previous 12 years [5]. As such, the expected new addition of renewable energy to the grid in the next 20 years is estimated at 1205MW of wind energy, 1232MW of solar power, 200MW of mini-hydro and 80MW of biomass. It is expected to commission several small-scale solar energy projects and a substantial amount of residential and commercial rooftop solar energy injection, further enhancing the renewable energy component in the grid. It is also interesting to note a critical comment in the long-term generation plan published in 2017 "LNG infrastructure must be established by 2020 in order to gain the maximum benefit of environmental impact mitigation" [18]. The growing electricity demand forecast in GWh is illustrated in Fig. 4 from 2030 to 2032.

3 Energy Mix

The Energy Mix of Sri Lanka is formed with pioneering Hydropower followed by thermal power generation, initially with furnace oil (FO), heavy fuel oil (HFO), auto diesel and finally coal. The Independent Power Producers (IPPs) mostly consist of diesel and combined cycle power plants along with a significant capacity of NonConventional Renewable Energy (NCRE). The NCRE component is mainly owned by the private sector comprising of wind energy, solar energy, biomass projects and the waste to energy projects to be commissioned and connected to the grid later this year. Even though I have discussed about the importance of introducing LNG to the Energy Mix, it is yet to be realised.

A well-balanced energy mix will provide much-needed energy security for a nation. Sri Lanka is currently generating a greater proportion of energy from fossil fuel and the rest from renewable sources; however, fossil fuel is completely import-dependent. As we know, thermal energy sources are highly vulnerable to world market pricing and availability. Therefore, it is undoubtedly vital to have the LNG option in the Sri Lankan energy mix in order to strengthen the energy security. This idea was also recorded in long-term generation plan of 2017, to incorporate the LNG-fired power plants to Sri Lankan power system. However, they have also considered the present trend of LNG fuel prices and the possibility of recovering the capital cost of LNG infrastructure with the options of adopting a land base LNG terminal or floating storage regasification unit (FSRU). It is also noted that according to the CEB Generation plan, "Breakeven price of local Natural Gas was determined by comparing 4x300MW coal power plants with a coal handling jetty with 4x300MW LNG power plants. The breakeven price is determined as 6.9 US$/MMBTU. It is slightly higher than the imported LNG breakeven price due to the reason that a terminal or FSRU infrastructure cost need not be incurred in the local LNG Scenario" [18]. Figure 5 interprets the 2015 energy mix and provide you with a fair idea as to how energy over 50% is import-dependent.

Even though the generation plan quoted the present trend of LNG fuel prices, it would make sense to take a deferent view as the LNG market statistics are not critical components of this matter as the country is expected to use indigenes LNG resources in the foreseeable further. The energy policies and strategies of Sri Lanka states, "the main objective of the national energy policy strategies declared here is to ensure convenient and affordable energy services are available for the equitable

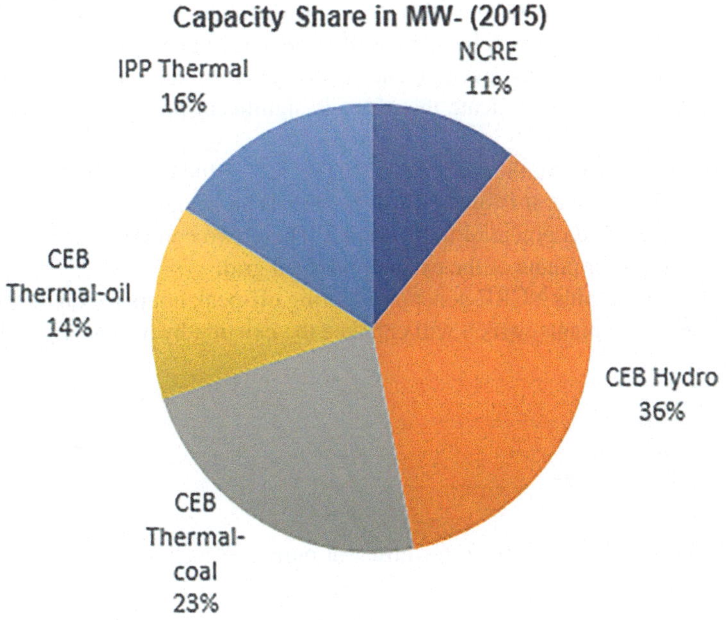

Fig. 5 Developed as per the data of CEB [19]

development of Sri Lanka using clean, safe, sustainable, reliable and economically feasible energy supply. This policy is formulated in alignments with the future goals of Sri Lanka, current global trends in energy and goal 7 of the sustainable development goals of the United Nations. This policy will impact the vast realm of social economic and environmental spheres and pave the way to realise the vision of Sri Lanka in achieving carbon neutrality and complete transition of all the energy value chains by 2050. This document also contains results delivery framework assigning responsibilities of meeting the stated policy goals to relevant institutions. A draft of this national energy policy and strategies was amended after being subjected to public scrutiny and a review by a committee of experts related to the energy sector before publication" [5].

Hence, the energy evacuated to the national electricity grid should be clean, safe, sustainable, reliable and economically feasible. It further indicates a timeline target of 2050 in achieving carbon neutrality, at the same time, setting a mid-course target to achieve 50% of primary energy supply and to reduce the specific energy use across end-users by 2030. It is also noted that the significant hydropower potential has already been tapped to a great extent and the growing energy demand is met with an increasing amount of thermal energy, as such, it is inevitable that renewable energy percentage in the energy mix is due to a decline in the future. Therefore, the attention must be directed towards the development of NCRE's without any delays.

4 Energy Self-sufficiency by 2030

Self-sufficiency in energy is of utmost importance to an independent nation, as much as the financial and political stability [20, 21]. It is interesting at this juncture to refer to an Official publication of The Ministry of Power & Energy in 2015, "Sri Lanka Energy Sector Development Plan for a knowledge base economy" states "to make Sri Lanka an energy self-sufficient nation by 2030" [15]. It looks like a very optimistic target just by looking at the physical picture of Sri Lanka today. However, the Sri Lanka national energy policy published in 2019 has further committed to the 2030 timeline. It has set significant milestones for some of the essential activities for the nation to be self-sufficient in power and energy. Section 4.5 of the said document illustrates this under the heading "enhancing self-reliance". "Indigenous energy resources will be developed to the optimal levels to minimise dependence on imported resources and to attain a higher degree of reliance in the energy sector." The following targets and milestones will have to be met by the institutions to which responsibilities are assigned [5]. The subsections from 5a to 5f of the Energy Policy of Sri Lanka outlines the responsibilities above of which 5b is the most important for the above subject matter.

Wind energy has been identified as the second most promising renewable energy resource after hydropower, and the highest priority is given to wind energy projects to achieve a minimum of 20% share of electricity demand from renewable energy sources other than from major hydro by 2020. The other segment which can be seriously considered is to utilise the NCRE resources to its maximum capacity in order to achieve the self-sufficiency in energy.

Sri Lanka is yet to embark on the upstream development of LNG gas resources "Barracuda and Dorado". The preliminary studies in the Cauvery and Manna Basins indicate that there is significant potential in discovering hydrocarbon deposits [22, 23].

Sri Lanka has only 9 years left to achieve the set target of energy self-sufficiency. Therefore it is necessary to revise action timelines to achieve this critical objective. It is of the view that this is an achievable target, realising it can only be done by investing mainly in NCRE and the LNG potential in Sri Lanka. These two components will be able to replace most of the import-dependent energy-based power plants of the existing national grid.

It is also essential to utilise the surplus NCRE generation during off-peak hours to run the pump storage in all of the reservoir-based hydropower generation plants, which will enhance the existing hydropower generation of the country to more than its current capacity [18].

5 Wind Potential in Sri Lanka

For centuries, people have harnessed wind energy for different purposes such as seafaring, irrigation/cultivation, grinding various types of grains, etc. [1, 2].

The National Renewable Energy Laboratory estimates that there are nearly 5000km^2 of windy areas with good to excellent wind resource potential in Sri Lanka. Above 4,100km^2 of the total windy area is land and about 700km^2 is the lagoon. The windy land represents about 6% of the total land area of 65600km^2 being the landmass of Sri Lanka. Using a conservative assumption of 5MW per square kilometre, this raw land could support approximately 20,000MW of potential installed capacity. If the windy lagoon areas are taken into account, this potential could increase to 24,000MW. If the areas with moderate wind resource potential are also considered to estimate the total windy land area, this will increase to approximately to10,000km^2 or almost 15% of the total land area of Sri Lanka. This would interpret to more than 50,000MW of installed capacity. The National Renewable Energy Laboratory (NREL) has done a fair amount of research into Wind Energy potential in Sri Lanka [24]. NREL of the USA used the services of a company called True Wind Solutions (TWS) of New York, USA. TWS used its propriety MesoMap system to calculate the wind power density values in Sri Lanka. The MASS model simulated weather conditions over Sri Lanka for over 366 days randomly selected from 15 years data. The random sampling was satisfied so that each month and season was represented equally. Each simulation generates wind and other meteorological variables throughout the model domain for a particular day and stores the information at hourly intervals. The simulations used a variety of meteorological and geophysical data. The main geophysical inputs in MASS are elevation, landcover, the greenness of vegetation and soil moisture. The MASS translates both land cover and vegetation greenness into important surface parameters such as surface roughness [24].

The pilot project by CEB was a good approach to drive the fear away from Sri Lanka on wind energy projects even though the installed capacity was mere 3MW with five turbines of 600KW each and did not enjoy an excellent plant factor [25]. It is estimated that inland of Sri Lanka has a potential wind energy capacity of 24,000MW and of which, Sri Lanka has only commissioned around 100MW capacity of Wind projects mainly in the northwestern region of Sri Lanka. The good news is that there are several projects in the pipeline, including the 100MW wind power project in Manna Island. The most promising areas in the Island for wind projects are marked in Figs. 6 and 7 is the detail survey map of Mannar Island and demarcated area for the proposed 100MW of the total of 300MW wind project by the Government of Sri Lanka.

Figure 7 is the Map of Manna island Reserved for the proposed 300MW wind power project by the Government of Sri Lanka, and the project development ownership is given to CEB based on competitive bidding for public-private partnership (PPP) basis, and the first project for 100MW is in the pipeline to call for competitive bids [16].

Figure 8 provides wind data across the year with reference to three prominent inland locations with high wind potential in Ambewela, Mannar and Hambantota.

Ambewela wind potential will be high in the months of June to August and good in the months of May, September and October, but a challenging location for logistics and as such, the project capacity is much limited. The Mannar wind potential is much more interesting as the full summer through to autumn is very good, and therefore plant factor would be greater than other locations in general [27].

It is often debated that wind energy is not dispatchable and it will cause instability of the grid. Hence it is proposed to have expensive spinning reserves or a source that could be swiftly synchronised to the grid in order to take over such emergency loads in the event of a sudden loss of wind energy in the grid. The present arrangement with the system control is to use the Coal as the baseload. The peak demand is met with other plants, especially with hydropower plants. In order to evacuate higher capacity of wind energy, it may be required to introduce pump storage where extra wind energy beyond the safe

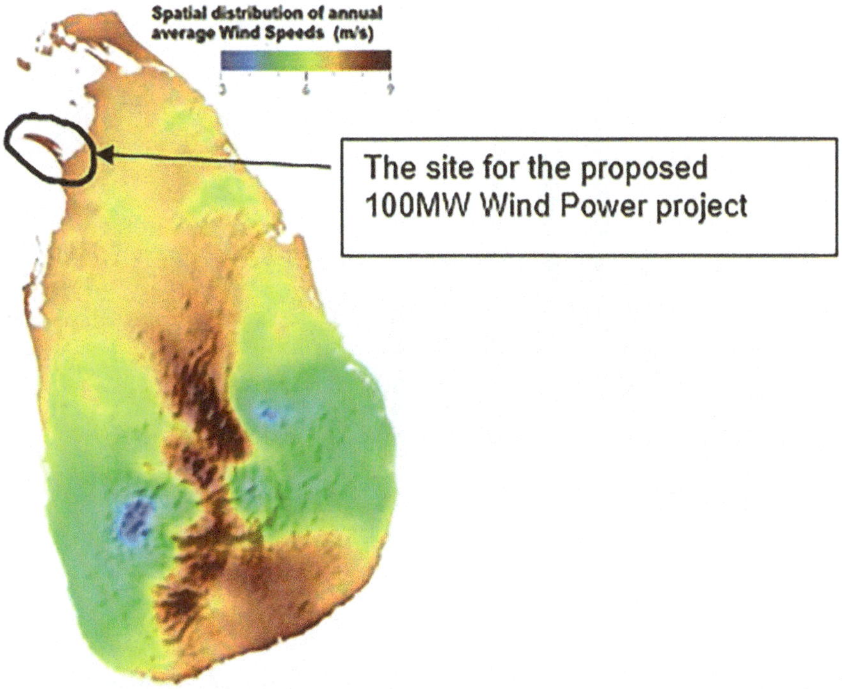

Fig. 6 Overview of the spatial distribution of annual wind speeds (m/s) [26]

Fig. 7 Map of Manna island reserved for the proposed 300MW

margin of the grid capacity is utilised for this and thereby increasing the hydropower capacity or in other alternative energy storage methodologies such as hydrogen, heat storage or any other energy storing methods [28, 29].

Sri Lanka is known to have recorded a very high cost of energy in the whole of South Asia, but Sri Lanka enjoys 24/7 uninterrupted quality electricity, which is only guaranteed in Sri Lanka in the whole of SARC region. Despite quality and reliability, the Sri Lankan population is demanding for more affordable energy while the commercial operations and industries finds it tough to compete internationally due to the high cost of energy. Hence it has become increasingly important to find other sustainable energy sources to be a responsible and a competitive nation in many ways.

As such, in addition to the onshore wind potential, it is interesting to look at the feasibility of offshore wind potential to achieve multiple objectives such as economy, environment and sustainability. Europe has been a pioneer and leader in offshore wind energy since the 1990s. The offshore wind continued to expand throughout the Baltic Sea, Southern North Sea

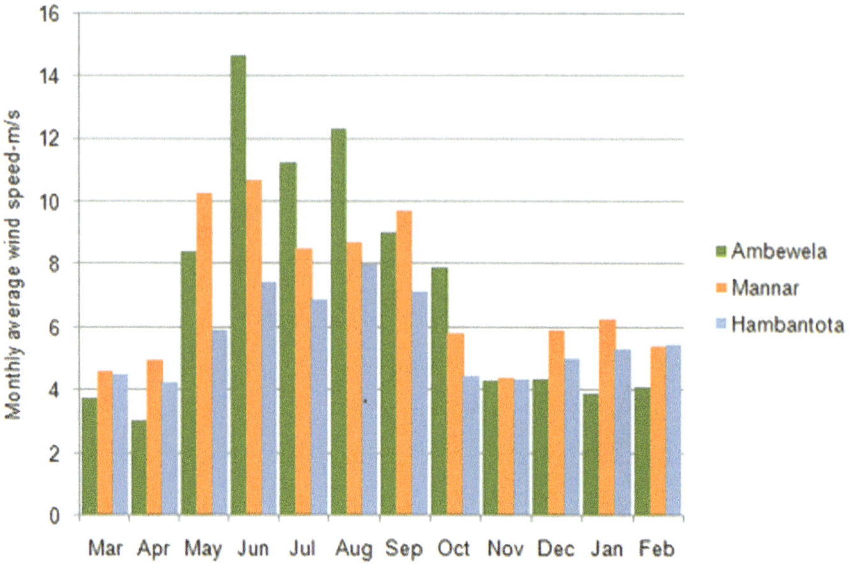

Fig. 8 Wind data across the year for Ambewela, Mannar and Hambantota [26]

of UK and extended to the Irish Sea [4]. It is further noted by the end of 2019, and estimated that the global offshore wind capacity stands at around 25GW [2]. For many years' the offshore wind was quite expensive compared to onshore [30, 31]. This is changing dramatically due to the advancement of technology and the competition among global manufactures, including economies of scale, the maturity of the supply chain, better procurement strategies etc. However, the offshore wind industry has remained limited mainly to Europe and China as prices continue to drop, offshore wind is increasingly gaining access in emerging markets. Projections are for offshore wind to add about 10GW per year from 2019 to 2024 and 15 to 21GW per year from 2025 to 2030 in the world [32].

A World bank publication about expanding offshore wind to emerging markets presents eight case studies on the potential offshore wind in Brazil, India, Morocco, Philippines, South Africa, Sri Lanka, Turkey and Vietnam. Considering the offshore areas within 200km of the cost of these eight countries have a total technical potential of 3.1 Terawatts, including 1,016GW of fixed capacity and 2,066GW of floating capacity. For most countries, offshore estimates represented multiples of their total current installed generation capacity. This means that offshore wind can play a transformational role in meeting national goals ranging from expanding electricity access to increase the mix of renewable resources in the energy mix [33].

However, it is left to those referred countries to transform the said technical potential into commercial potential facing the challenge of managing the technical, financial and environmental complexities that are prevailing in offshore wind projects. It is essential for Sri Lanka to take a very bold step towards expanding to offshore wind projects. It was thereby reaching its full wind energy capacity for the benefit of the future generation, especially in the area of energy, environment and the economy. According to the statistics, Sri Lanka is blessed with 92GW of potential wind energy, and it is essential to convert it to power and evacuate it to the grid as soon as possible [32].

Figure 9 provides a clear impression of offshore wind energy potential in Sri Lanka together with a recommendation for fixed and floating turbine areas marked. However, there has not been adequate research or commitment so far in developing offshore wind energy projects.

Having said that, it is vital to recognise the challenges ahead of every step towards realising this. Even though 92GW against mere National demand of 4,000MW is unjustifiable, energy is a very precious commodity in the world. In order to overcome some of the technical limitations, experts in the industry believe that the extra capacity of offshore wind energy evacuation is only possible by implementing a SARC grid with Sri Lanka having the DC connectivity with India and Fig. 10 has the simulated model of the Indian power grid and the visual of the proposed DC connectivity with Sri Lanka.

India, Sri Lanka grid connectivity feasibility was conducted by the power Grid Cooperation of India, which is outdated due to the present scenario. According to the conclusion of "The cross border electricity trade between India and Sri Lanka", "This study finds that Cross Border Electricity Tariff (CBET) between India and Sri Lanka in 2025 presents an opportunity to reduce the cost of providing electricity, reduce RE curtailment in southern India and improve system reliability through the coordinated use of energy resources" [35].

The energy demand of Sri Lanka is growing like any other developing country in the world, and it is essential to follow the worldwide trend and the growth rate of renewable energy. The energy demand forecasted data are provided in Fig. 10.

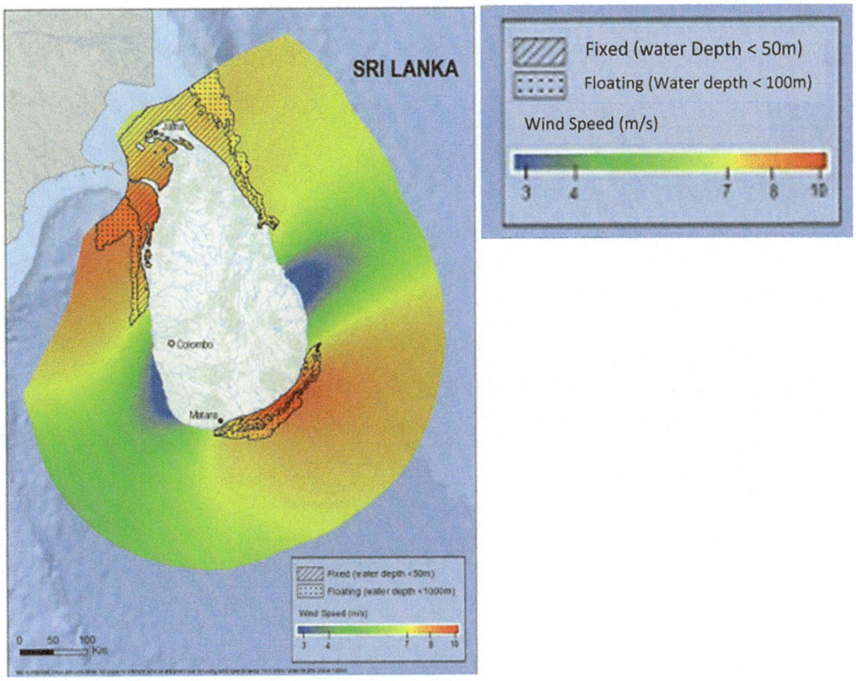

Fig. 9 recorded offshore wind speed data illustrated with water depths marked in the most potential areas [33]

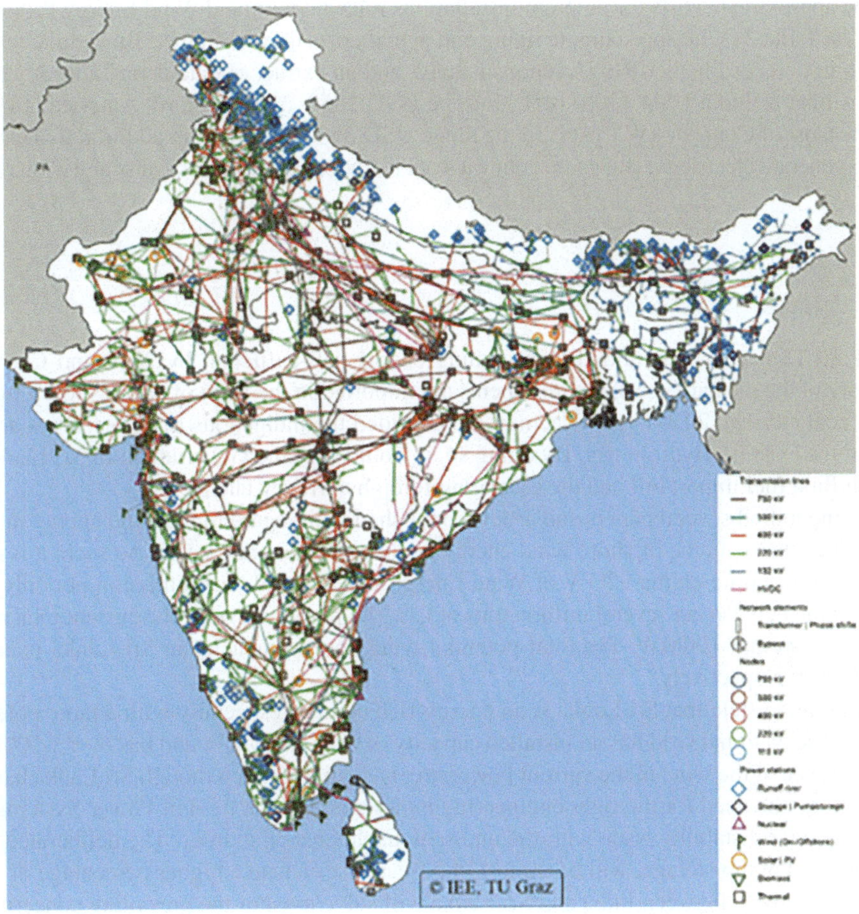

Fig. 10 provides an overview of the physical layer in the simulation model ATLANTIS_India [34]

Table 2 Approximate costs for deferent installations relevant to inland Wind Energy Projects in Sri Lanka [13]

Criteria	Southeast coast	Kalpitiya peninsula	Ambewela cattle farm
Capital cost (Per kW)	$1,195	$1,220	$1,325
Fixed charge Rate	9.5%	9.5%	9.5%
50-m wind speed (m/s)	7.0	7.6	7.9
Capacity factor	26%	34%	32%
O&M per kWh	$0.015	$0.015	0.015
COE estimate ($/kWh)	$0.065	$0.054	$0.60

Table 3 Actual costs in 2010 across the world as recorded iby IRENA report in 2012 [36]

	Installed cost (2010 USD/kW)	Capacity factor %	Operations and maintenance (USD/kWh)
China/India	1300 to 1450	20 to 30	N.A.
Europe	1850 to 2100	25 to 35	0.013 to 0.025
North America	2000 to 2200	30 to 45	0.005 to 0.015

6 Cost of Wind Energy in Sri Lanka

Table 2 illustrates the approximate costs for deferent installations relevant to inland Wind Energy Projects in Sri Lanka [13]. The Cost of per kW on Table 2 is the approximate figure and actual costs may be higher. But actual costs in 2010 across the world has been taken in to record in the IRENA report in 2012, and an extract of data is on Table 3, [36]. It is recorded that the price per kW in China & India USD 1300–1450, Europe USD 1850–2100 & North America is USD 2000–2200. It is interesting to note the Cost of Coal per kW Cost is in the range USD 3500–3800 and in addition to the capital cost, this plant has an energy cost to generate electricity. But some countries opt to choose Coal against zero-emission renewable energy [01]

7 Conclusion

The Government of Sri Lanka has committed to becoming energy self-sufficient by 2030 and Carbon neutral by 2050. Currently, the majority of the primary energy requirements are imported in the form of various fossil fuels such as crude oil, HFO, diesel, petrol, coal etc. The 2030 target is achievable provided the indigenous LNG resources are harnessed together with NCRE and with the existing Hydropower. By doing so, it would ease the Government of Sri Lanka from the immense pressure of over $4 billion fuel import bill and the escalating foreign currency tariffs.

It is very encouraging to find several papers and articles published with regard to the wind energy in Sri Lanka. However, the involvement and the interest in the offshore wind energy is not prominent at present. As such, it is of paramount importance to prioritise and concentrate on the 92GW of Wind Energy which could decommission not only several black power plants in the SARC region but also cut several million tons of CO_2. As per the records, the current total installed wind energy capacity in Sri Lanka is around 100MW. The total potential wind energy capacity in Sri Lanka is around 24,000MW & 92GW inland and offshore respectively.

Hence, it is appropriate to state that Sri Lanka is an energy-rich country especially with a large potential of onshore and offshore wind energy. The National Grid has an installed capacity of over 4,000MW and has over 1,700MW of Hydropower. Unfortunately, the presence of the wind in the current power mix is negligible even though Sri Lanka has a substantially high wind potential. Therefore, the need for the right engineering intervention in the field of Power Systems is essential to overcome the pressure on the Grid stability issue with the undispatchable energy sources. The deliberated points in the article, such as energy storage and pump storage, will be utmost important to evacuate a higher percentage of wind energy into the CEB grid. The grid connectivity between India and Sri Lanka will solve most of the prevailing concerns and stability issues in the Sri Lanka electricity grid. However, the untapped wind energy capacity can only be evacuated into the grid with a strong commitment from the Government of Sri Lanka and further empowering the National Energy Policy enacted as an Act by the Parliament.

References

1. Wikiepedia (2020, June 06) History of Wind Power. Retrieved 07 June 2020, from Wiliepedia: https://en.wikipedia.org/wiki/History_of_wind_power
2. U.S.Energy Information Administration (2020, March 24). US Energy Information Administration. Retrieved 15 May 2020, from Wind Explained-History of Wind Power. https://www.eia.gov/energyexplained/wind/history-of-wind-power.php
3. Wikipedia (2009, October 09) James Blyth (engineer). Retrieved 09 June 2020, from https://en.wikipedia.org/wiki/James_Blyth_(engineer)
4. Nixon, N (2008, October 17) The Guardian. Retrieved 10 May 2020, from The Guardian. https://www.theguardian.com/environment/2008/oct/17/wind-power-renewable-energy
5. Karunanayake R (2019) Gazette of the Government of Democratic Socialist Republic of Sri Lanka-National Energy Policy of Sri Lanka (No.2135/61. ed., Vol. 2135/61). Colombo, Sri Lanka: Ministry of Power, Energy & Business Development
6. World Bank (2019) Sri Lanka Energy InfraSAP. World Bank, Washington DC
7. World Bank and IFC (2019) Sri Lanka Energy Infra SAP. World Bank Group, Washington DC
8. USAID (2018, June) South Asia Regional Initiative for Energy Integration. (IRADe). Retrieved 3 May 2020, from USAID. https://sari-energy.org/oldsite/PageFiles/Countries/Sri_Lanka_Energy_detail.html
9. Wikipedia (2013) D.J.Wimalasurendra. Retrieved 17 Sept 2010. from https://en.wikipedia.org/wiki/D._J._Wimalasurendra
10. Juleff G (1996, January 04) An ancient wind-powered iron smelting technologyin Sri Lanka. Nat Res J *379*(6560):60–63
11. Kumbalathara et al (2015) Simulation of ancient wind-driven iron smelting furnaces of Sri Lanka. 2015 Moratuwa Engineering Research Conference (MERCon). IEEE, Moratuwa, Sri Lanka
12. Freese (1996, September/October) Wind Powered Furnaces. Archaeol Arch. 49(05).
13. Young M (2003) Sri Lanka Wind Farm Analysis and Site Selection Assistance. NREL, Washington DC
14. Sustainable Energy Authority of Sri Lanka (2016) Sri Lanka sustainable energy authority. Retrieved 20 April 2020, from http://www.energy.gov.lk/en/renewable-energy/technologies/wind-power
15. Ministry of Power & Energy (2015) Sri Lanka energy sector development plan for knowledge-based economy, vol 1, 1st edn. Ministry of Power & Energy, Colombo
16. Ceylon Electricity Board (2019) Annual Report 2016. Ceylon Electricity Board. Ceylon Electricity Board, Colombo
17. Office of the AGM Transmission, CEB (2014) CEB Transmission Network. Retrieved 28 April 2020, from Ceylon Electricity Board: https://www.google.com/search?q=ceb+transmission+network&sxsrf=ALeKk029iV_X-JvDPMoyEa6DpD7tIyajfw:1588600352419&tbm=isch&source=iu&ictx=1&fir=xFt3qKbFU-ZH0M%253A%252C3XoLvrKmBVhgOM%252C_&vet=1&usg=AI4_-kTyitY3SWub5W3NLgcmuRd0umg5cg&sa=X&ved=2ahUKEwjqpsi9r
18. Ceylon Electricity Board (2018) Long Term Generation Expansion Plan 2018-2037. Ceylon Electricity Board. CEB, Colombo
19. Ceylon Electricity Board (2016, May) Sri Lanka's Power Generation Past, Present and Future published by Ceylon Electricity Board. Retrieved 06 May 2020, from Ceylon Electricity Board. https://www.google.com/search?q=Sri+Lanka%E2%80%99s+Power+Generation+Past%2C+Present+and+Future+published+by+Ceylon+Electricity+Board.&oq=Sri+Lanka%E2%80%99s+Power+Generation+Past%2C+Present+and+Future+published+by+Ceylon+Electricity+Board.&aqs=chrome..69
20. Wikipedia (2011, March 10) Energy security. Retrieved 06 May 2020, from https://en.wikipedia.org/wiki/Energy_security
21. Oxford Policy Management (2018, November 15) The Importance of Energy Security. Retrieved 06 May 2020, from Oxford Policy Management. https://www.opml.co.uk/blog/importance-energy-security
22. Premarathne DS (2013) A Petroleum System in the Gulf of Mannar Basin, Offshore Sri Lanka. Geological Socity of Sri Lanka. ResearchGate, Colombo, pp 09–10
23. Piyadasa H (2014, September 22) World Maritime University. Retrieved 09 June 2020, from World Maritime University: https://commons.wmu.se/
24. Elliott D, Schwartz M, Scott G, Haymes S, Heimiller D, George R (2003) Wind Energy Resource Atlas of Sri Lanka & Maldives. National Renewable Energy Laboratory. National Renewable Energy Laboratory, Washington DC
25. Atputharajah AT (2007, January 03) Wind power develoment and its status in Sri Lanka. pp 8,9
26. Pathmasiri M (2017, June 08) Asian Development Bank. Retrieved 20 April 2020, from Asia Clean Emergy Forum. https://asiacleanenergyforum-revision.pi.bypronto.com/2/wp-content/uploads/sites/2/2016/03/5aPathmasiri-Wind-Development-Status-Sri-Lanka.pdf
27. Kaluthanthrige RW (2017) Wind resource assesment and turbine selection: case study for Mannar, Sri Lanka. IESL (p. 06). ResearchGate, Moratuwa
28. Abo-Khalil A (2013, March) Intech Open. Retrieved 15 May 2020, from https://www.intechopen.com/books/modeling-and-control-aspects-of-wind-power-systems/impacts-of-wind-farms-on-power-system-stability
29. Sreedevi JM (2017, February 09) IEEE Xplore. 2016 IEEE Region 10 Conference (TENCON). 2159-3450. Singapore: IEEE. Retrieved May 14th, 2020, from IEEE. https://ieeexplore.ieee.org/document/7848065
30. Gielen (2012) Renewable energy technologies: cost analysis series. IRENA. IRENA, Abu Dhabi
31. Lee A (2019, 10 22) *RECHARGE.* Retrieved 07 30, 2020, from https://www.rechargenews.com/transition/offshore-wind-power-price-plunges-by-a-third-in-a-year-bnef/2-1-692944
32. Dutton A (2019) Expnading Offshore Wind to Emerging Markets. The World Bank Group, Washington DC
33. The World Bank's Energy Sector Management Assistance Program (ESMAP), The International Finance Corporation (IFC) (2019, October 31) Going Global: Expanding Offshore Wind to Emerging Markets. The World Bank, Washington DC. Retrieved 07 May 2020, from https://www.worldbank.org/en/topic/energy/publication/expanding-offshore-wind-in-emerging-markets
34. Bhat KB (2019, December 05) A techno-economic model-based analysis of the renewable energy transition in the Indian subcontinent region. Retrieved 20 April 2020, from https://link.springer.com/article/10.1007/s00502-019-00773-wSri+Lanka%E2%80%99s+Power+Generation+Past%2C+Present+and+Future+published+by+Ceylon+Electricity+Board.&aqs=chrome..69
35. Rose A, McBennett B, Palchak D, Cochran J, Wijekoon HM, Samarasekara B, Wijekoon R (2018) Cross-border electricity trade between India and Sri Lanka: impact on power system oprations. NREL, Washington DC
36. Taylor (2012) Renewable Energy Technologies: Cost Analysis Series. International Renewable Energy Agency. Bonn, IRENA

Wind Power Resources Assessment Under Varying Topography

Luai M. Alhems, Shafiqur Rehman, Narayanan Natarajan, and Vasudevan Mangottiri

1 Introduction

The growing concern of environmental pollution and awareness among people from all walks of life has lead the researchers, developers, engineers, and even the politicians to think about new, clean renewable sources of energy to combat the increasing concentrations of greenhouse gases in the atmosphere.

Depletion of conventional energy sources and environmental sustainability issues have created challenges for researchers and policy makers to explore and discover renewable energy sources such as solar, wind, geothermal, tidal, biogas, biomass, biofuels, etc. Among these, wind energy has become a promising source due to its competitive nature abundance, ease of harvesting, inexhaustible and affordable resources [1, 2]. Irrespective of irregularities existing in the global renewable energy market, the overall installed capacity of wind power has increased over 104% during the last decade [3]. It is evident from the increasing global wind power capacities (591.5 GW end 2018 with new addition of 51.3 GW and increase of 10.9% compared to 2017) that it is a well-established technology.

For the development of wind power at any location, statistical analysis of wind data serves as a key tool for assessing wind energy potential [2]. Extensive studies have been conducted on wind characteristics and wind power potential in many countries around the world [4–12].

To accentuate the importance of evaluation of wind energy potential, a few studies have been briefly mentioned here. Alaydi [13] analyzed the wind data at two sites in the Gaza strip in Rafah and recommended for wind energy exploitation. Ajayi et al. [14] assessed the potential of wind resources in Jos, Nigeria for power generation based on the monthly mean wing speeds of 1987–2007 and concluded that Jos was a suitable site for wind farm projects. Allouhi et al. [15] evaluated the potential of wind energy for six coastal locations in Morocco and that Dakhla was the most suitable location for harnessing the wind power while Laayoune the second suitable site.

In the Indian context, wind power development in India has gained momentum over the last few decades. Ramachandra et al. [16] evaluated the wind energy availability and its characteristics in the Uttara Kannada district of Karnataka and concluded that coastal taluks such as Karwar and Kumta have good wind energy potential. Ganesan and Ahmed [17] reported annual energy yields of 3.712 GWh and 4.431 GWh corresponding to hub heights of 50 and 70 m for Bhopal in India. In another interesting study, Singh and Prakash [18] evaluated the availability of wind energy for electricity production at Ranchi, Jamshedpur, Devghar, Lohardaga, and Chaibasa in Jharkhand, India, and reported that the sites were unsuitable for wind power generation on a large scale. Phadke et al. [19] reported Tamil Nadu as one of the potential states for harvesting 95% of the nation's wind energy potential.

In this paper, wind characteristics and energy potential of three cities, namely Chennai, Erode, and Coimbatore, located at different elevations above mean sea level, with different topographical features and distances from the shore, have been

L. M. Alhems · S. Rehman (✉)
Interdisiplinary Research Center for Renewable Energy and Power Systems (IRC-REPS), The Research Institute, King Fahd University of Petroleum and Minerals, Dhahran, Saudi Arabia
e-mail: luaimalh@kfupm.edu.sa; srehman@kfupm.edu.sa

N. Natarajan
Department of Civil Engineering, Dr. Mahalingam College of Engineering and Technology, Pollachi, Tamil Nadu, India

V. Mangottiri
Department of Civil Engineering, Bannari Amman Institute of Technology, Sathyamangalam, Erode, Tamil Nadu, India

© The Author(s), under exclusive license to Springer Nature Switzerland AG 2022

A. Sayigh (ed.), *Sustainable Energy Development and Innovation*, Innovative Renewable Energy,
https://doi.org/10.1007/978-3-030-76221-6_74

studied. Specifically, the study provides the diurnal variation of the mean wind speed, wind directional analysis, wind speed probability distribution, monthly and annual mean variation of the wind speed, wind energy yield, and capacity factors using five different wind turbines of 2.0 MW rated capacities. Although several studies have been reported on the potentiality of the wind speed as a direct indicator of wind power density, studies considering the combined effect of topography, that is, distance from the shore and the elevation are limited. This study is the first attempt of its kind, where the combined effect of the above factors has been considered in estimating the wind energy potential of a location.

2 Study Area and Data Description

The study focuses on analyzing the wind power potential of cities located in the Tamil Nadu State of India. The cities considered for the study are Chennai, Coimbatore, and Erode located at 6m, 183m, and 411m above mean sea level (MSL). The map showing the locations of these cities (in rectangular boxes) is provided in Fig. 1. The summary of the site-specific details (geographical coordinates, data collection period, long-term mean ambient temperature, pressure, and air density) is provided in Table 1. The wind speed and other parameters were used for a period of 38 years (January 1980 to June 2018).

3 Results and Discussion

The wind speed characteristics (site-dependent, annual, monthly, and diurnal scales) over a long period of almost 38 years are analyzed and discussed in the following sub-sections. These characteristics include the wind speed, wind power density, energy yield, and plant capacity factor using 5 wind turbines each of 2.0 MW rated power.

3.1 Long-Term Variation at Three Sites

The long-term average wind speed values at 50 m above ground level over the entire data collection period at Chennai, Erode, and Coimbatore are compared in Table 2. As one moves away from the coastal site (Chennai) toward inland (Erode and Coimbatore), the mean wind speed decreases. Also, the mean maximum wind speed of 21.42 m/s is found at Chennai over the data collection period and a mean minimum of 13.97 m/s at Erode. A maximum wind power density of 129 W/m^2 is obtained for Chennai and a minimum of 76 W/m^2 for Erode. The mean energy content for the three cities is 1127 kWh/m^2/year for Chennai, 666 kWh/m^2/year for Erode, and 849 kWh/m^2/year for Coimbatore. The energy pattern factor varies from 1.418 to 1.705. The higher values of mean wind speed, wind power density, and mean energy content at Chennai may be due to its being the coastal site which is open to sea on the eastern side and is also relatively flat land. The above parameter values are the least at Erode measurement site, may be due to inland location though having gentle topographic features. However, Coimbatore station is relatively better as far as wind characteristics are concerned though locked by mountains in the north and south and to a certain extent in the west as well.

Higher magnitudes of 6.0 m/s and more are found to be more prevalent from northeast and southwest directions at the Chennai data collection station (Fig. 2). The analysis also reflected that more than 50% times the wind blows from southwest direction and around 30% of the times from north-east at Chennai. At Erode station, the higher magnitude winds between 5.0 m/s and 6.0 m/s blow from west-south and west-north directions. At Coimbatore, wind speeds of 5.0 m/s to 6.0 m/s blow from the west and west-south directions. It is evident from the analysis that wind speed remained around 70%, 60%, and 70% between 4.0 m/s and 8.0 m/s at Chennai, Erode, and Coimbatore, respectively. This means that wind power can be generated at these sites above these percentages of times.

3.2 Long-Term Annual Variation

The annual variation of wind speed over a longer period and other parameters are important to have confidence on the annual availability of wind for power generation. At Chennai, the annual mean wind speed varied between a minimum of 5.1 m/s in 2008 and a maximum of 5.9 m/s in 1983 while remained around 5.2 m/s during most of the years (Fig. 3). At Erode station,

Fig. 1 Locations of the three sites used for wind power resources assessment

Table 1 Site-dependent data summary

Variable	Location		
	Chennai	Erode	Coimbatore
Latitude	N 13.000	N 11.5000	N 11.000
Longitude	E 80.000	E 77.5000	E 76.875
Elevation (m)	6	183	411
Start date	1/1/1980	1/1/1980	1/1/1980
Distance from sea shore (km)	10	227	318
End date	6/1/2018	6/1/2018	6/1/2018
Mean temperature (°C)	27.9	26.2	24.4
Mean pressure (kPa)	100.3	96.23	94.93
Mean air density (kg/m^3)	1.152	1.102	1.104
Power density at 50m (W/m^2)	129	76	97
Wind power class	1-(Poor)	1-(Poor)	1-(Poor)

Table 2 Wind speed sensor summary

Variable	Location Chennai	Erode	Coimbatore
Mean wind speed (m/s)	5.40	4.34	4.70
Median wind speed (m/s)	5.49	4.15	4.55
Max wind speed (m/s)	21.42	13.97	15.50
Mean power density (W/m^2)	129	76	97
Mean energy content (kWh/m^2/yr)	1,127	666	849
Energy pattern factor	1.418	1.705	1.694
Data recovery rate (%)	100	100	100

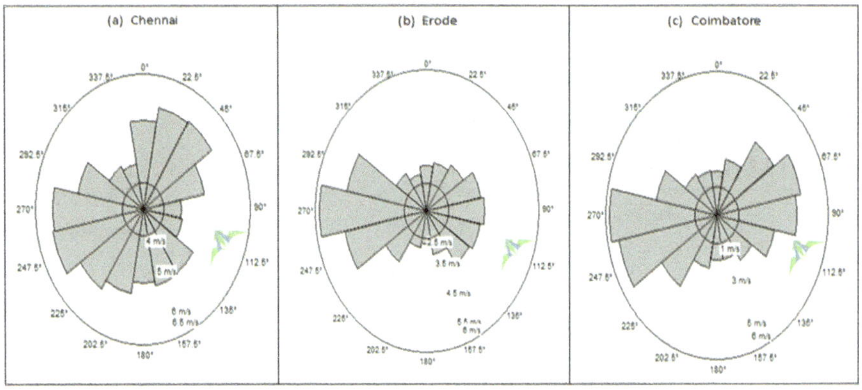

Fig. 2 Example of directional variation of long mean wind speed at Chennai station

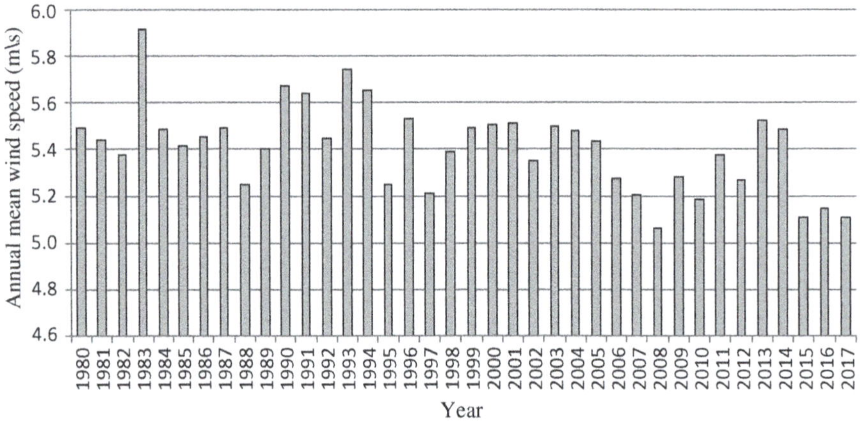

Fig. 3 Long-term mean wind speed variation at Chennai, an example

the annual mean wind speed showed a consistent behavior with minimum and maximum wind speeds of 4.0 m/s and 4.5 m/s. At Coimbatore the wind speed always remained more than 4.4 m/s with a minimum of 4.3 m/s in 2015 and a maximum of 5.1 in 1990.

3.3 Long-Term Monthly Variation

For the quality of power and grid stability management, the knowledge of the monthly variation of mean wind speed over an extended period is important. Monthly mean wind speed is found to increase from January to mid of the year, reaching a maximum of 6.23 m/s in June, and then decreased toward the end of the year at Chennai (Table 3). The overall mean wind

Table 3 Monthly mean values of all the meteorological parameters at Chennai (1980–2018)

Data column	Units	Jan	Feb	Mar	Apr	May	Jun	Jul	Aug	Sep	Oct	Nov	Dec	All
Speed_50m	m/s	4.95	4.57	4.85	5.48	5.85	6.23	6.06	5.72	4.63	4.43	5.92	6.14	5.4
Direction_50m	°	60.4	97.9	140	161	193	242	246	246	228	61.3	41.3	43.9	164
Temperature_2m	^{0}C	23.9	25.6	28.3	30.6	31.6	30.4	29.5	28.9	28.3	27.3	25.6	24.2	27.8
Pressure_0m	kPa	101	101	101	100	100	99.9	99.9	100	100	100	101	101	100
Air Density	kg/m^3	1.17	1.17	1.15	1.14	1.13	1.14	1.14	1.14	1.15	1.15	1.16	1.17	1.15
Speed_50m WPD	W/m^2	95	78	87	117	159	185	172	143	85	82	165	173	129

speed of 5.4 m/s is observed over the entire data collection period. At Erode and Coimbatore, the maximum wind speeds were observed in the month of July (6.35 m/s) and June (6.94 m/s), respectively. The highest values of WPD of 185 W/m^2, 190 W/m^2, and 234 W/m^2 were observed at Chennai (June), Erode (July), and Coimbatore (June), respectively.

3.4 Long-Term Diurnal Variation

Knowledge of diurnal changes in mean wind speed over extended periods is critical to match the peak load on the grid. At Chennai, the diurnal mean wind speed shows two peaks, one in the early morning at 4 o'clock and the other at 14:00 hours in the early afternoon (Fig. 4). So, if the local peaks occur at these timings, then this diurnal wind profile can be best utilized to supplement the peak load through wind power. At Erode, there is a clean-cut peak wind availability at around 15:00 hours which could also best utilized for peak load saving through wind power if the peak load also occurs at this time. However, there is a flat peak between 5:00 and 12:00 hours at Coimbatore which could be best utilized for flat load demands during this specific duration.

3.5 Wind Power Generation and Capacity Factor Analysis

In order to study the power generation and capacity factor, five wind turbines, each of 2.0 MW rated capacity and pitch controlled type, are chosen. The technical specifications of the chosen turbines are summarized in Table 4. For calculating the net energy generation and net capacity factor calculations, various equivalent losses of 11.4527% (availability 2%, wake effect 4%, turbine performance 2%, electrical 0%, environmental 2%, curtailment 0%, and others 2%) are considered.

Annual energy yields and capacity factors are calculated using the site-specific wind speed and wind turbine technical data. Wind turbines WT3 and WT5 produced the highest energy while WT1 and WT4 produced lesser energy at Chennai. Wind turbine WT2 performed the worst compared to the other four turbines. Almost same performance is observed at Erode and Coimbatore (figures are not included). Highest capacity factors are obtained in 1983 and the lowest in 2008, irrespective of the turbine type.

The average annual energy yield of > 3000 MWh over the entire data reporting period is obtained at Chennai based on turbines WT1, WT3, WT4, and WT5 (Fig. 5). Similarly, the higher capacity factors between 17 and 19% are obtained at Chennai corresponding to the same turbine types as above while at Erode the capacity factors remained around 10% (Fig. 6). However, at Coimbatore, the capacity factor values remained between 12 and 14%. In terms of turbine performance, WT3 and WT5 produced the maximum energy at all the locations. In terms of location performance, Chennai is found to be the best, Coimbatore the second best while Erode the least performer.

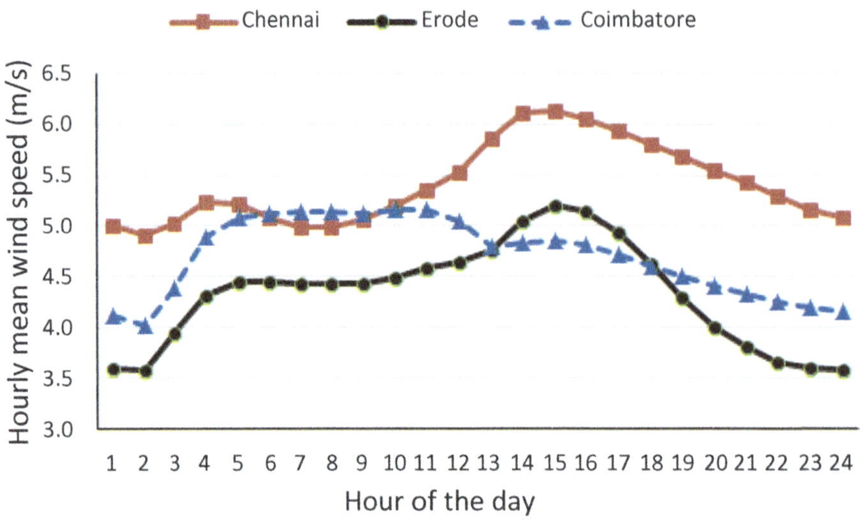

Fig. 4 Diurnal variation long-term hourly mean wind speed at three locations

Table 4 Technical specifications of pitch controlled 2.0 MW rated capacity wind turbines

Wind turbine generic names	Rotor diameter (m)	Cut-in speed (m/s)	Rated speed (m/s)
WT1	82	4	15
WT2	80	3	13
WT3	90	3	14
WT4	87	3	13
WT5	93.3	3	11.2

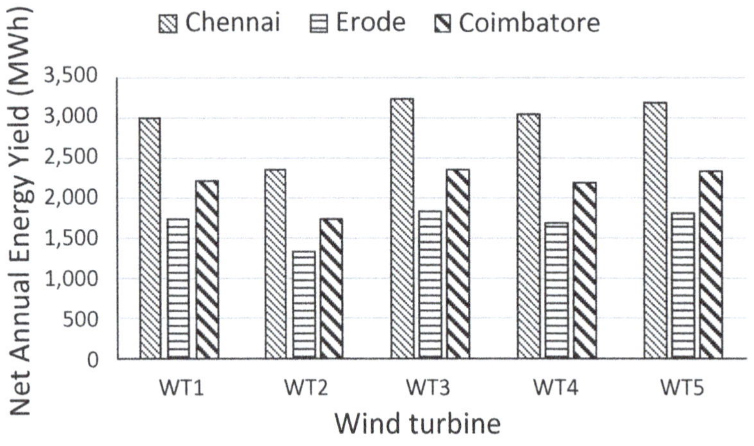

Fig. 5 Annual energy yield

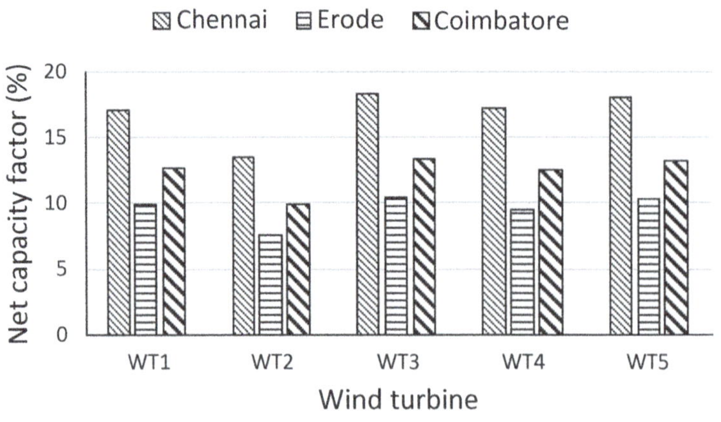

Fig. 6 Annual capacity factor

4 Conclusions

For the wind resource characterization, 38 years of wind data was statistically (diurnal, monthly, and annual wind speed) analyzed. The mean wind speed for Chennai, Erode, and Coimbatore are 5.4 m/s, 4.34 m/s, and 4.7 m/s while the mean maximum are 21.42 m/s, 13.97 m/s, and 15.50 m/s. The wind speed decreases for cities located inland. The mean wind power density for Chennai, Erode, and Coimbatore are 129W/m2, 76 W/m2, and 97 W/m2 and the mean energy content 1127 kWh/m2/year, 666 kWh/m2/year, and 849 kWh/m2/year. The mean wind speed remained around 70%, 60%, and 70% between 4m/s and 8 m/s at Chennai, Erode, and Coimbatore.

The monthly variation of the mean wind speed showed an increasing trend from Jan to mid of the year and a decreasing trend toward the end of the year at all the locations. The peak frequency of wind speed is observed between 5 m/s and 6 m/s at Chennai, between 4m/s and 5 m/s at Erode and Coimbatore. Highest capacity factors are obtained in 1983 and the lowest in 2008 irrespective of the type of turbine. In terms of turbine performance, WT3 and WT5 produced the maximum energy at all the locations. Chennai is found to the best site for wind energy development followed by Coimbatore and Erode has the least prospects.

Acknowledgment This work was supported by the Deanship of Research at King Fahd University of Petroleum & Minerals, Saudi Arabia, under project number SB181005.

References

1. Baker JR (2003) Features to aid or enable self-starting of fixed pitch low solidity vertical axis wind turbines. J Wind Energ Ind Aerodyn 15:369–380
2. Rehman S, Ahmad A (2004) Assessment of wind energy potential for coastal locations of the Kingdom of Saudi Arabia. Energy 29(8):1105–1115
3. Global Wind Report (GWEC). Available online: http://www.gwec.net/wp-content/uploads/vip/GWEC_PRstats2016_EN_WEB.pdf. Accessed on 20 Apr 2019
4. Kose R (2004) An evaluation of wind energy potential as a power generation source in Kutahya, Turkey. Energ Conver Manage 45:1631–1644
5. Rehman S, Halawnai T, Mohandes M (2003) Wind power cost assessment at twenty locations in the Kingdom of Saudi Arabia. Renew Energy 28:573–583
6. Rehman S, El-Amin IM, Ahmad F, Shaahid SM, Al-Shehri AM, Bakhashwain JM (2007) Wind power resource assessment for Rafha, Saudi Arabia. Renew Sustain Energy Rev 11:937–950
7. Al-Abbadi NM (2005) Wind energy resource assessment for five locations in Saudi Arabia. Renew Energy 30:1489–1499
8. Elamouri M, Ben AF (2008) Wind energy potential in Tunisia. Renew Energy 33:758–768
9. Oyedepo SO, Adaramola MS, Paul SS (2012) Analysis of wind speed data and wind energy potential in three selected locations in south-east Nigeria. Int J Energy Environ Eng 3:1–23
10. Baseer MA, Meyer JP, Rehman S, Alam M (2015) Wind power characteristics of seven data collection sites in Jubail, Saudi Arabia using Weibull parameters. Renew Energy 102:35–49
11. Rehman S, Ahmed MA, Mohamed MH, Al-Sulaiman FA (2017) Feasibility study of the grid connected 10 MW installed capacity PV power plants in Saudi Arabia. Renew Sustain Energy Rev 80:319–329
12. Rafique MM, Rehman S, Alam M, Alhems LM (2018a) Feasibility of a 100 MW installed capacity wind farm for different climatic conditions. Energies 11(8):2147
13. Alaydi JY (2011) Assessment of wind energy potential in Gaza strip. Front Energy 5(3):297–304
14. Ajayi OO, Fagnenle RO, Katende J, Okeniyi JO (2011) Availability of wind energy resource potential for power generation at Jos, Nigeria. Front Energy 5(4):376–385
15. Allouhi A, Zamzoum O, Islam MR, Saidur R, Kousksou T, Jamil A, Derouich A (2017) Evaluation of wind energy potential in Morocco's coastal regions. Renew Sustain Energy Rev 72:311–324
16. Ramachandra TV, Subramanian DK, Joshi NV (1997) Wind energy potential assessment in Uttara Kannada district of Karnataka, India. Renew Energy 10(4):585–611
17. Ganesan S, Ahmed S (2008) Assessment of wind energy potential using topographical and meteorological data of a site in Central India (Bhopal). Int J Sustain Energy 27(3):131–142
18. Singh R, Prakash O (2018) Wind energy potential evaluation for power generation in selected districts of Jharkhand. Energy Sources Part A 40(6):673–679
19. Phadke A, Bharvirkar R, Khangura J (2012) Wind potential estimates for India: economic and policy implications. Environ Energy Technol Division, Berkley Lab Report:1–24

Development of a Micro Horizontal Axis Wind Turbine Blade for the Semi-Arid Region of Nigeria

G. M. Ngala and M. Shuwa

Nomenclature

P	The Extractable power from the wind
A	Swept area of the rotor
ρ	Air density
v	Mean wind speed
v_{ulti}	Ultimate wind speed
d	Rotor diameter
r	Rotor radius
C	Chord length
ω	Angular velocity of the rotor 2
θ	Angle of attack
ϕ	Wind relative angle
η	Efficiency of the blade on the wind turbine
L	Blade span
$\dot{r}$	Radial length of the blade element
$d\dot{r}$	Increase in blade span length
T	Rotor torque
σ	Solidity
λ_r	Tip speed ratio
C_P	The power coefficient
C_L	Coefficient of lift
C_D	Coefficient of drag
F_L	Lift force
F_D	Drag force
a	Axial induction factor
$\dot{a}$	Radial induction factor
Re	Reynolds number
F_R	Resultant force
B	Number of blades
P_T	Power developed by the turbine
N	Speed of the rotor

1 Introduction

Nowadays, prompt attentions have continuously been given to renewable sources of energy as demand in areas with potentials and effects of climate change (carbon emissions) are increasing. The semi-arid region of Nigeria located between Longitudes 3° and 15° East and Latitudes 9° and 14° North; it is one region of the country where energy demand exceeds that provided by the national

G. M. Ngala (✉) · M. Shuwa
Department of Mechanical Engineering, University of Maiduguri, Maiduguri, Nigeria
e-mail: garbangala68@unimaid.edu.ng

© The Author(s), under exclusive license to Springer Nature Switzerland AG 2022

A. Sayigh (ed.), *Sustainable Energy Development and Innovation*, Innovative Renewable Energy,
https://doi.org/10.1007/978-3-030-76221-6_75

grid; besides, the region is vast, rural areas are scattered, and most are off-grid or far from the point of connection. However, investigators have shown that the region has exceptionally reliable not fully exploited wind energy resource. Thus, the need for a stand-alone wind turbine specifically developed for the region using the available appropriate technology. This can only be achieved by developing a wind turbine blade that will withstand gust and function optimally within a predetermined safe wind speed range in the region and also provide electrical energy from the renewable energy source (wind) to rural areas of the region [1, 5–7, 9, 13]. However, despite its wind energy potential which is not fully exploited the semi-arid region of Nigeria, it is associated with stochastic wind typical of high wind speed phenomena. This phenomenon is destructive to wind turbine blades not specifically designed for the region, thus causing a potential risk of rotor failure. Besides, the design of a wind turbine blade largely depends on the wind condition of the site and it is the most important component required in harvesting some of the kinetic energies in the wind [8, 17, 20].

Wind is a free-flowing fluid phenomenon due to temperature difference on the earth surface caused by solar radiation. The free-flowing stream possesses kinetic energy, and a certain quantity of this energy can be extracted by the turbine blade and converted to other useful means, especially electricity by the wind turbine [3, 14]. A wind turbine is a rotor dynamic energy conversion system classified either as horizontal with axis of rotation of their blades horizontal to the ground and almost parallel to the wind or vertical axis wind turbine with axis of rotation and blades vertically oriented [15, 21]. Most wind turbines fall under the category of HAWT, because of their distinct advantages that include low cut-in wind speed, easy furling, and relatively high-power coefficient [21]. A horizontal axis wind turbine consists of five main parts: rotor, power transmission system, control system, tower, and generator. The rotor which is the most important and complex part consists of the blade and the hub. The blade is a key element of the HAWT that converts the kinetic energy in the wind into electricity through a power transmission system (rotor shaft for direct or gearbox for indirect transmission) and a generator [3, 18].

In this work, the focus is on the development of a HAWT blade to extract the maximum kinetic energy from the wind and determine the optimal design parameters of the HAWT blade. The HAWT blade airfoil was developed and analyzed for optimization using computational fluid dynamic (CFD) code ANSYS Fluent 6.18. base on 32 years (1985–2016) correlated wind data of the semi-arid region of Nigeria obtained from the Nigerian Meteorological Agency. The National Advisory Committee for Aeronautics (NACA) blade airfoil 4412 is selected and modified. NACA (National Advisory Committee for Aeronautics) is one of the international bodies that set standard for airfoil sections. NACA specifies the features of series airfoils by numbers. In a four-digit specification, the first number denotes the maximum camber of the airfoil at the chord line in percent of chord, the second number gives the location of the point of maximum camber from the leading edge in tenth of the chord, and the third and fourth numbers indicate the maximum thickness in percent of the chord [14]. Thus, a NACA 4412 airfoil has maximum camber of 4%, located at 0.4 times the chord length from the leading edge and the maximum thickness is 12% of the chord. The NACA 4412 blade airfoil is modified using Airfoil Design Workshop code to determine an appropriate airfoil section for the region with a defined optimal lift to drag coefficient ratio. Modified airfoil parameters that determine geometrical and aerodynamic optimization of the blade based on the predefined wind data are obtained using blade element momentum (BEM) theory. These parameters include the chord length, angle of attack, tip speed ratio, blade span, axial induction factor, radial induction factor, lift and drag forces, coefficient of lift, coefficient of drag, and wind relative angle. Blade element momentum (BEM) theory method consists of dividing the blade into annular control volumes, by applying momentum balance and energy conservation in each control volume. The annuli are bounded by stream surfaces that enclose the blade and extend it from upstream to downstream. This is used to obtain the distribution of the angle of relative wind which consists of pitch angle and angle of attack on the blade sections (airfoil) and the aerodynamic forces acting on the airfoil section such as thrust and torque. Both are generated by lift and drag forces, with the lift and drag being functions of the angle of attack and the Reynolds number. Moreover, the performance of the developed blade airfoil is predicted using the CFD commercial code ANSYS Fluent 6.18. The airfoil is created code's workbench within a computational domain, and the domain is meshed and the boundary condition specified. Wind flow within the domain and around the blade is analyzed numerically based on the Navier-Stroke Equations using ANSYS Fluent 6.18 Spalart Allmaras and $k-$ epsilon turbulence models. Moreover, blade parameters that determine geometrical and aerodynamic optimization of the blade based on the predefined wind data are obtained using blade element momentum (BEM) theory. These parameters include the chord length, angle of attack, tip speed ratio, blade span, axial induction factor, radial induction factor, lift and drag forces, coefficient of lift, coefficient of drag, and wind relative angle.

2 Material and Method

2.1 Wind Data Analysis

Thirty-two (1985–2016) years' correlated wind data for eight locations at 10 m height in the semi-arid region of Nigeria as in Table 1 were obtained from the Nigeria Meteorological Agency Lagos and used in the development of the horizontal axis wind turbine blade. These wind data are also used in carrying out wind tunnel test of a developed HAWT blade model based on the prototype design to experimentally determine the performance of the HAWT blade.

Table 1 Annual monthly wind speeds from 1985–2016 in eight locations in the semi-arid region of Nigeria

Location	Jan	Feb	Mar	Apr	May	June	July	Aug	Sept	Octo	Nov	Dece	AMM-WS
Gusau	8.14	5.94	3.83	2.84	5.26	5.17	5.11	7.70	7.91	7.05	7.46	9.23	6.30
Kano	9.30	5.76	3.58	2.88	3.43	4.49	6.68	6.06	5.89	6.03	10.03	10.59	6.23
Katsina	5.78	4.53	3.54	3.06	4.51	5.48	5.39	5.43	5.75	6.27	6.85	6.90	5.29
Yalwa	4.27	3.65	3.07	2.57	2.97	3.67	4.16	3.90	3.80	4.05	4.59	5.49	3.85
Sokoto	1.26	10.14	4.27	2.73	5.13	5.17	5.11	7.81	7.90	7.30	9.23	10.71	7.23
Nguru	5.60	4.16	3.10	2.81	3.41	4.05	4.08	4.08	4.61	5.06	5.81	6.38	4.43
Maiduguri	6.54	4.16	3.10	3.218	4.52	4.28	4.76	6.213	4.61	5.40	6.60	7.17	5.05
Bauchi	6.83	5.37	3.08	3.23	4.79	4.28	4.76	6.21	4.61	5.54	6.57	7.00	5.19

Source: NIMET [16]
AMM-WS Annual Monthly Mean Wind Speed

2.2 *Designing the HAWT Blade*

The blade design parameters include airfoil shape, design angle of attack, and design tip speed ratio, which are determined based on the extractable rated wind speed of the ten selected sites of the semi-arid region of Nigeria. These parameters are considered in the HAWT blade aerodynamic structural design and based on the blade element momentum (BEM) theory. The amount of wind energy that can be extracted by a wind turbine blade is obtained by half the air density, the cross-sectional area of the rotor, and the cube of the free steam velocity as given in Eq. 1 [19, 22]. The rotor diameter (D_r) and the blade span are calculated from the same equation at the rated wind power and wind speed

$$P_w = \frac{1}{2}\rho A_r v^3 \tag{1}$$

Where: P_w is the wind energy extracted, is the air density of the region, A_r is the swept area of the rotor and is given by $\frac{\neq D^2}{4}$, and υ is the air velocity. Therefore, from Eq. 1, the rotor diameter D_r will be

$$D_r = \sqrt{\frac{8P_w}{\pi \rho v^3}} \tag{2}$$

And the blade span (L) is calculated from Eq. 3

$$L = \frac{D_r}{2} - \dot{r} \tag{3}$$

Where $\dot{r}$ is the radius of the hub.

The chord length (C) is the length from the leading edge to the trailing edge of a blade cross-section that is parallel to the vertical axis of symmetry as shown in Fig. 1 and is given by Eq. 4 and the wind relative angle (ϕ) is calculated from Eq. 5 [4].

$$C = \frac{8\pi r}{BC_L}(1 - \cos\phi) \tag{4}$$

$$\phi = \frac{2}{3}\tan^{-1}\left(\frac{1}{\lambda_r \frac{\dot{r}}{r}}\right) \tag{5}$$

Where $\dot{r}$ is the radial length of the blade and r is the radius of the rotor

The tip speed ratio (λ_r), is the ratio of the blade tip speed over wind speed. It is a significant parameter for wind turbine design and is given by Eq. 6 [2, 3]

$$\lambda_r = \frac{\omega r}{v} = \frac{2\pi N r}{60 v} \tag{6}$$

The blade angle of attack (θ) is the angle between the incoming flow stream and the chord line of the airfoil. The angle of attack is given by Eq. 7 [2].

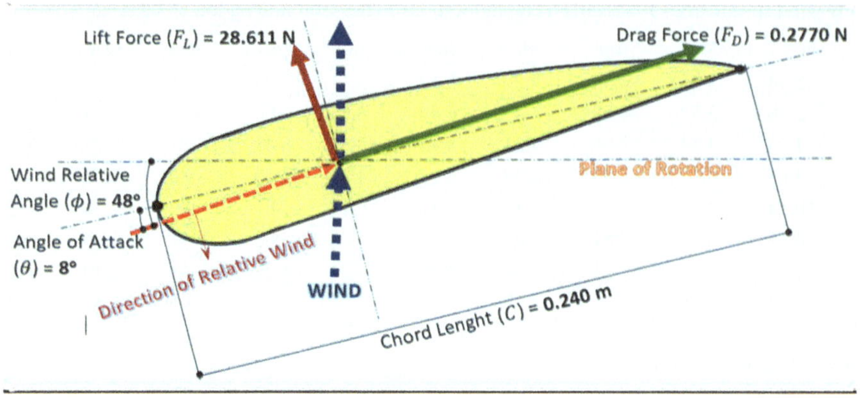

Fig. 1 Modified NACA 4412 HAWT blade geometry design

$$\theta = \tan^{-1}\frac{2}{3\lambda_r} \tag{7}$$

Solidity ratio (δ) is the ratio of the area occupied by the blade to the available free space and is given by Eq. 8 [3].

$$\delta = \frac{B+C}{2\pi r} \tag{8}$$

The ultimate wind speed (v_{ult}) is the maximum wind speed the blade is expected to withstand based on the blade design and is given by Eq. 9. As well, the probability of collapse (Cprop) that is the probability that the blade can with stand such a wind speed without failing considering 1.5 as factor of safety (γ_F) is given by Eq. 10 [22].

$$v_{ulti} = v_{max}\gamma_F C_{prop} \tag{9}$$

$$C_{prop} = \left[\frac{1-kx\text{In}(-\text{In}(1-0.01))}{1-kx\text{In}(-\text{In}(0.98))}\right]^n \tag{10}$$

Where n is the number of events, i.e. the blade may or may not collapse, under this situation is 0.5 and k is a constant 0.2. The lift force (F_L) given by Eq. 11 above is the force acting on the blade perpendicular to the undisturbed wind flow and the drag force (F_D) given by Eq. 12 below is the force acting on the blade in the direction of the undisturbed wind flow [3, 4].

$$F_L = \frac{1}{2}\rho v^2 CLC_L \tag{11}$$

$$F_D = \frac{1}{2}\rho v^2 CLC_D \tag{12}$$

Where C_L and C_D are the lift and drag coefficients respectively. Based on the airfoil geometry C_L and C_D are determined from the Airfoil Design Workshop code at a predetermined Reynolds number (Re) and angle of attack (θ). The shear force ($S.F.$) on the blade is given by Eq. 13, while the bending moment ($B.M.$) at a distance x along the span of the blade is by Eq. 3.14 [22].

$$\text{S.F.} = -F_R x \tag{13}$$

$$\text{B.M.} = -\frac{1}{2}F_R(L-x)^2 \tag{14}$$

Where F_R is the resultant force acting on the blade and x is any distance along the blade span L given by Eq. 15 [22].

$$F_R = \sqrt{F_L^2 + F_D^2} \tag{15}$$

Therefore, the torque on the rotor as a result of the thrust developed by the HAWT blade is given by Eq. 16 [4].

$$T = F_R r\omega \tag{16}$$

Where F_R is the resultant thrust, r is the radius of the rotor, and ω is the angular velocity of the rotor determined from Eq. 17 [4, 22].

$$\omega = \frac{2\pi N}{60} \tag{17}$$

The applied torque is a factor that depends on the indication axial induction factor (a) and is given by Eq. 18 [2, 3]

$$a = \frac{1}{\frac{4F_R \cos\phi}{\delta C_L}} \tag{18}$$

And the rotating annular element angular induction factor ($\dot{a}$) is determined from Eq. 19 [2, 3]

$$\dot{a} = \frac{1}{\frac{4\cos\phi}{\delta C_L}} \tag{19}$$

The amount of energy produced by the rotor depends on the speed of the rotor and the developed toque by the rotor is determined from Eq. 20 [4].

$$P_T = \frac{2\pi NT}{3600} \tag{20}$$

The power coefficient is a measure of the mechanical power delivered by the rotor to the turbine's shaft. It is defined as the ratio of the mechanical power to the power available in the wind and is given by Eq. 21 [2, 3, 10].

$$C_P = \frac{2P_T}{\rho\pi r v^3} \tag{21}$$

The German engineer Betz showed that the maximum power extracted from an air stream is 16/27 or 0.59 of the theoretical available power. Only 59% of the wind power can be captured by a wind turbine even if the power losses are neglected. Therefore, the HAWT theoretical efficiency η is determined from Eq. 22 [2, 3, 11, 12].

$$\eta = C_P \frac{16}{27} \tag{22}$$

2.3 CFD Analysis of the HAWT Blade

To reduce the cost of the HAWT blade optimization and in order to solve and determine the wind flow field characteristic, interaction, and influence of aerodynamic loads on the HAWT blade, a CFD analysis is carried out. This is to properly design, predict the structural integrity, and the behavior of the blade under predefined ultimate wind speed condition. The CFD code selected for the HAWT blade aerodynamic analysis is the commercial ANSYS Fluent 6.18. The blade geometry is prepared based on NACA 4412 blade airfoil shape, which is modified using the Design Foil Workshop CAD software, where the appropriate lift and drag coefficients are determined. The modified blade geometry is then created in ANSYS Fluent 6.18 geometry interface within a defined computational domain and the domain meshed. A pressure-based CFD model is selected, governing equation defined, boundary conditions identified and analysis carried out using Spalart Allmaras and $k-\varepsilon$ turbulence models. The aerodynamic analysis is carried out based on the semi-arid region's ultimate wind speed (υ_{ulti}).

2.4 Blade Airfoil, Computational Domain, and Grid Generation

The HAWT blade airfoil is developed based on modified NACA 4412 airfoil using Design Foil Workshop CAD software. Prediction of the modification is that it should satisfy the Bernoulli Effect and Newton's Third Law based on the expected wind speed of the semi-arid region of Nigeria. The two dimension (2D) x and y coordinates vertices were imported from the Design Foil Workshop to ANSYS workbench interface and the airfoil is created within a defined computational domain at the optimal angle of attack determined in the Design Foil Workshop. The domain boundaries are selected with AB as Farfield 1, CID as Farfield 2, AEC and BGD as upper and lower surfaces of the domain while the blade was considered as a wall within the computational domain as shown in Fig. 2. The scale used in this development is in meters (m).

The CFD analysis accuracy and duration depend on the mesh size and orientations, with the increase in mesh size decrease but the accuracy increases [14]. The grid generated is unstructured consisting of quadrilateral cells because of the complex shape of the blade airfoil. In an unstructured grid, the grid is not uniquely identified by indices i and j, but instead cells are numbered in another way internally in the CFD code. The grid is optimized by means of two-dimensional (2D) simulations sufficient to model the boundary layer with the help of wall functions. Also, the mesh is refined along both span wise and axial domain direction. Boundary and operating conditions are selected as well as the simulation model. The mesh generated as in Fig. 2 is uniformly distributed around airfoil for accurate prediction using Spalart-Allmaras and $k{-}\varepsilon$ turbulence models.

2.5 Boundary Conditions

The domain boundaries are defined with Farfield 1 (AB) as velocity inlet, Farfield 2 (CID) as pressure outflow, and upper (AEC) and lower (BGD) surfaces of the domain as symmetry walls 1 and 2, while the blade was considered a default interior wall within the computational domain. The computational zone is considered air with a density of 1.225 Kgm^{-3} and the computational domain model is the stationary type with a pressure-based CFD model. The scheme of the discretized computational zones is computed for standard pressure, first-order upwind momentum, first-order upwind turbulent viscosity, and first-order upwind energy based on a define Scheme.

2.6 Governing Equation

The 2D geometry of the HAWT blade created within the computational domain is analyzed for aerodynamic performance under applied fundamental laws of fluid mechanics (Newton's Third Law and Bernoulli Effect) and solved CFD Navier-Stokes system of equations (Continuity, Momentum and Energy Equation) shown in Eqs. 23, 24, 25, and 26 respectively in ANSYS Fluent 6.18.

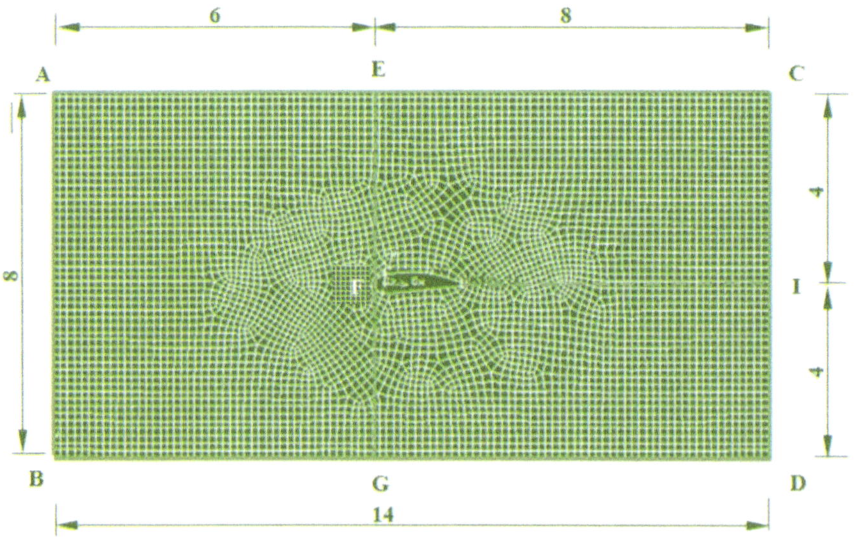

Fig. 2 The HAWT blade within the meshed computational domain

$$\frac{\partial \rho}{\partial \rho} + \nabla \cdot \left(\rho \vec{V} \right) = 0 \tag{23}$$

Momentum Equation

$$x - \text{component}: \frac{\partial (\rho u)}{\partial t} + \nabla \cdot \left(\rho u \vec{V} \right) = -\frac{\partial p}{\partial x} + \rho f_x \tag{24}$$

$$y - \text{component}: \frac{\partial (\rho v)}{\partial t} + \nabla \cdot \left(\rho v \vec{V} \right) = -\frac{\partial p}{\partial y} + \rho f_y \tag{25}$$

Energy Equation

$$\frac{\partial}{\partial t} \left[\rho \left(e + \frac{V^2}{2} \right) \right] + \nabla \cdot \left[\rho \left(e + \frac{V^2}{2} \right) \vec{V} \right] = \rho \dot{q} - \frac{\partial (up)}{\partial x} - \frac{\partial (vp)}{\partial y} + \rho \vec{f} \cdot \vec{V} \tag{26}$$

Where e is the specific internal energy, t is the time advancement and $\dot{q}$ is the heat flux term and is given by $\dot{q} = -k\frac{\partial T}{\partial x_i}$; k is the coefficient of thermal conductivity and $\vec{f}$ and T are the temperature and the external force of the Newtonian fluid respectively.

2.7 *Turbulence Model*

Turbulence is a flow feature that is described by a chaotic state of motion, by which the velocity and pressure of the flow continuously change with time within a bounded region of the flow [21]. The flow is characterized by velocity fluctuations and highly disordered motion. The governing equations are numerically computed based on two turbulence models Spalart Allmaras and $k-\varepsilon$ (Standard or RNG). After setting the computational material and convergence criteria, the simulation is iterated and after convergence of the solution, results of the activity within the computational domain were displayed and studied in terms of static pressure and velocity magnitude. These results were interpreted in a post processor (Microsoft Excel) and the blade energy extraction ability is determined.

3 Result and Discussion

3.1 *The Aerodynamic and Structural Design of the HAWT Blade (Table 2)*

Table 2 Results of the aerodynamic and structural design of the HAWT blade under extreme wind speed condition

Parameter	Value
Extractable power from the wind	1.083 kW
Resultant force	28.612 N
Probability of collapse of the blade	1.04
Factor of safety	1.7
Ultimate wind speed	18.72 m/s
Maximum mean wind speed	
Monthly mean wind speed	5.45 m/s
Tip speed ratio	5
Lift force	28.611 N
Drag force	0.277 N
Centrifugal force	129.89 N
Gravitation load	60.822N

(continued)

Table 2 (continued)

Parameter	Value
Shear force	−22.89 Nm
Bending moment	−10.19 Nm^2
Solidity ratio	0.08
Axial induction factor	0.03458
Radial induction factor	0.00121
Wind relative angle	48°
Generated torque	449.35 Nm
Power produced by the turbine	78.43 W
Power coefficient	0.387
Theoretical turbine efficiency	33%

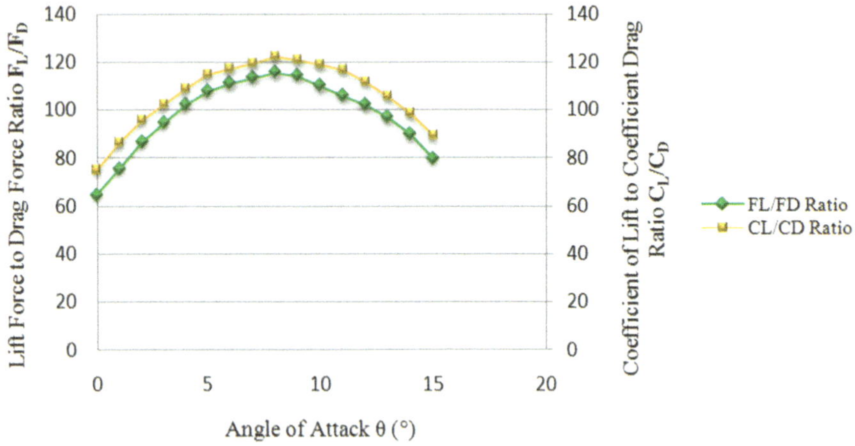

Fig. 3 Lift to drag force ratio and coefficient of lift to coefficient of drag ratio at various angles of attack

The aerodynamic design shows an optimum angle of attack of 8 degrees with an optimum lift to drag coefficient ratio of 124 at Reynolds number (R_e) 3 × 10^6 and lift to drag force ratio of 118 as shown in Fig. 3.

3.2 *Analysis of the HAWT Blade Aerodynamics*

Rajakumar and Ravindran [21] in their work analyze lift and drag forces at different sections of a blade for angles of attack from 0 to 12 degrees. Their analysis showed that the angle of attack of 5 degrees has a high lift to drag force ratio. [14] also predicted the performance of a HAWT blade and showed that the optimal performance of the blade occurs at angle of attack between 7 and 9 degrees, hence the optimal angle of attack of 8 degrees recorded for this design is in agreement with the prediction of [21] and also that of [14]. Figure 4 shows the performance analysis of the blade determined in terms of its ability to extract energy from the wind for wind speeds from 0 to 10 m/s based on a 1.5 m blade span at 8 degrees angle of attack. The result shows that the blade extracts 33.4% of the power available in the wind. Maximum predicted power available in the wind at 6 m/s is 155 Watts and the maximum power extracted by the blade considering a rotor diameter of 3 m. Figure 4, Peter and Richard [19], indicates that physical limit exists to the quantity of energy that can be extracted which is dependent on the blade design. The magnitude of energy harnessed is a function of the reduction in wind speed over the turbine. This principle is also reported by Kale and 11.

Varma, 2014, and is known as the Betz limit, which states that the efficiency of a wind turbine cannot exceed 59.3%.

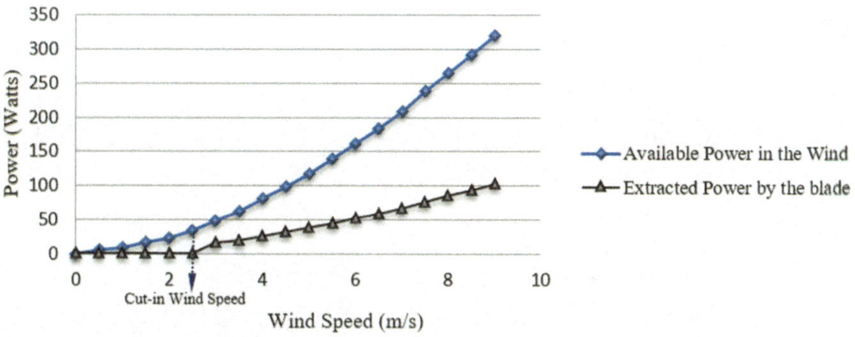

Fig. 4 Available power in the wind and extracted power by the blades

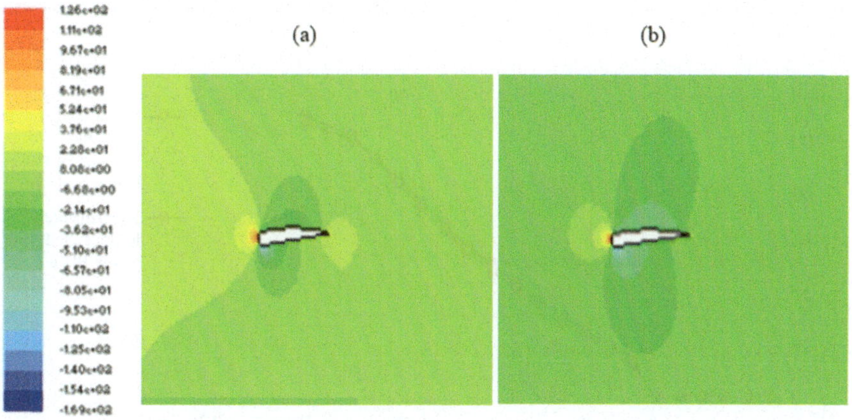

Fig. 5 Contours of static pressure (Pascal) under ultimate wind speed condition. (**a**) Spalart Allmaras Model. (**b**) *k*-epsilon turbulence model

3.3 The Computational Analysis

The computational fluid dynamic (CFD) result revealed contours of pressure and velocity distribution of Spalart-Allmaras and k – epsilon turbulence models. Results of static pressure and velocity magnitude distribution using Spalart-Almaras and *k*-epsilon turbulence models for maximum wind speeds of 10 m/s, as determined for the semi-arid region of Nigeria, are shown in Figs. 5 and 6 respectively. In both Spalart Allmaras and *k*-epsilon turbulence models under the eight different wind speeds, contours of static pressure show that the bottom side of the airfoil experiences a higher pressure than the top, resulting in a higher lift than drag. Also, the head of the airfoil is exposed to low-pressure and high-velocity zone, thus allowing the higher air velocity to drive the blade; however, the airfoil tail does not experience any resistive air pressure. The static pressure around the blade airfoil is lower at the top stream than the bottom, and thus satisfies the Bernoulli's effect which state that lift is generated when the air pressure around an airfoil is lower at the top than the bottom and is accompanied by a speed difference of the air around the airfoil. On the other hand, results of contours of velocity, magnitude show high rate of air movement at the top of the airfoil for both Spalart-Almaras and *k*-epsilon turbulence models.

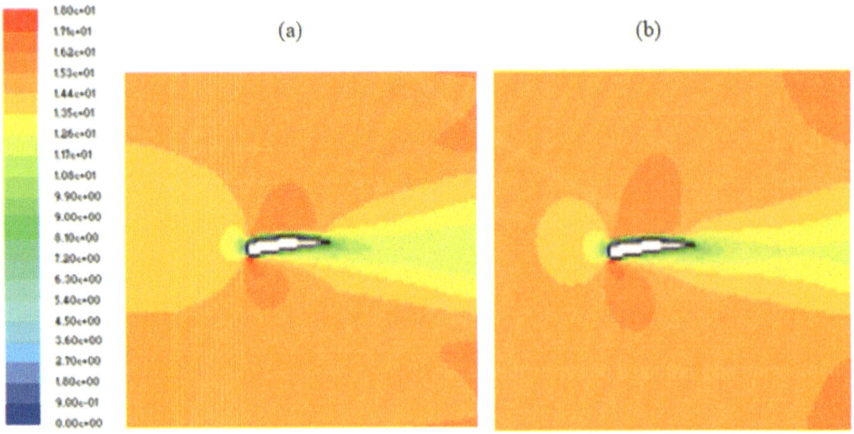

Fig. 6 Contours of velocity magnitude (m/s) under ultimate wind speed condition. (**a**) Spalart Allmaras Model. (**b**) *k*-epsilon turbulence model

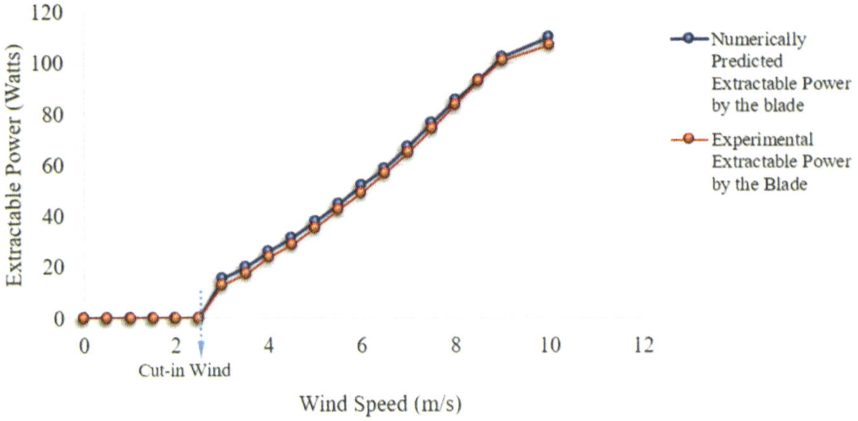

Fig. 7 Experimental and numerically predicted extractable power by the HAWT blade

3.4 *Validation of Results*

A model of the HAWT blade was developed based on the numerically result and was tested for performance in an open circuit wind tunnel at the Department of Mechanical Engineering, University of Maiduguri. The experiment shows that the blade cut-in wind speed is 2.5 m/s at 8 degree angle of attack. The performance test results in terms of extractable power from the wind by blade were compared with the numerical results for validation. The validation result shows that there was an agreement between the results as shown in Fig. 7. Moreover, the F-test statistical tool in Microsoft Excel is also used to test for significant difference between the two results. The test result shows that there is no significant difference between the two results with a test value of 0.98400234.

4 Conclusion

The horizontal axis wind turbine (HAWT) blade has been developed based on the modified National Advisory Committee on Aeronoutics (NACA) 4412 blade airfoil profile. The developed blade has an optimum angle of attack of 8°, maximum chord length of 0.240 m, and a blade span of 1.500 m. The blade performance has been numerically predicted with 51 Watts extracted from the wind at a wind speed of 6 m/s and at an efficiency of 33.4%. A model of the blade is also developed on a scale factor of 1:11 and its energy extraction ability is also determined experimentally in a wind tunnel. The blade extracted

50 Watts at a wind speed of 6 m/s with a blade efficiency of 33.2%. However, the numerical performance prediction of the blade shows that the blade can extract 51 Watts of power from the wind at a wind velocity of 6 m/s. The prediction based on the result of the static pressure and velocity magnitude in both Spalart Allmaras and $k-\varepsilon$ turbulence models also shows that the blade airfoil has satisfied the Bernoulli's effect and Newton's Third Law. Furthermore, the cut-in wind speed of the blade has been established to be 2.5 m/s in the numerical prediction and the experimental test.

References

1. Ahmed A, El-Suleiman A, Nasir A (2013) An assessment of wind energy resource in North Central Nigeria plateau. Sci J Energy Eng 1:13–17
2. Ajao KR, Adegun IK (2009) Development and performance test of a small three-bladed HAWT. J Am Sci 5:71–78
3. Bai C, Hsiao F, Li M, Huua G, Chen Y (2013) Design of 10 kW horizontal axis wind turbine (HAWT) blade and aerodynamics investigation using numerical simulation. Elsevier Publishers Ltd 5(2):34–40
4. Chandrala M, Choubeg A, Gupta B (2012) Aerodynamic analysis of horizontal axis wind turbine blade. Int J Eng Res Appl 2(6):1244–1248
5. Emmanuel OO (2016) Analysis of renewable energy potentials in Nigeria for national development. Int J Eng Res Rev 4:15–19
6. Fadare DA (2010) The application of artificial neural networks to mapping of wind speed profile for energy application in Nigeria. Appl Energy 87:934–942
7. Fadere DA (2008) Statistical analysis of wind energy potential in Ibadan, Nigeria, based on Weibull distribution function. Pacific J Sci Technol 9(1):110–119
8. Fagbenle RO, Karayiannis TG (1994) On wind energy resource of Nigeria. Int J Energy Res 18(5):493–508
9. Ifabiyi IP, Ajao K, R. and Adeniyi, A. (2017) Analysis of the potential of wind energy generation at different rotor diameters in Sudano-Sahelian zone of Nigeria. Int J Soc Sci Humanit Res 5(2):430–443
10. Kale S, Varme R (2014) Aerodynamic design of horizontal axis micro wind turbine blade using NACA 4412 profile. Int J Renew Energy Res 4(1):21–28
11. Maheswari RU, Tamilvendhan J (2012) Analysis of modeling of active stall control and active pitch controlled variable speed wind turbines. Int J Mod Eng Res 2:2662–2667
12. Manwell JF, MaGowan JG, Rogers AI (2002) Wind energy explained - theory, design and application. Wiley, New York
13. Ngala GM, Alkali B, Aji MA (2007) Viability of wind energy as a power generation source in Maiduguri, Borno State, Nigeria. Elsevier Int J Renew Energy 7:2242–2246 14
14. Ngala GM, Shuwa M, Oumaru MB, Muhammad AB (2016) A CFD analysis of a micro horizontal Axis wind turbine blade aerodynamics. Arid Zone J Eng Technol Environ, 11:13–23
15. Ngo C, Natiwitz J (2009) Resources, alternatives and the environment. Wiley, New Jersey
16. NIMET (2017) Meteorological data. Nigeria Meteorological Agency, Lagos
17. Ojosu JO, Salawu RI (1990) An evaluation of wind energy potential as a power generation source in Nigeria. Sol Wind Technol 7:663–673
18. Petal H, Daminia S (2013) Performance prediction of horizontal axis wind turbine blade. Int J Innov Res Sci Eng Technol 2(5):1401–1406
19. Peter S, Richard C (2012) Wind turbine blade design. Energies Publishers 4(1):21–32
20. Raja A, Srivastava A, Dwivedi M (2006) Non-conventional energy resource and utilization power plant engineering. New Age Int. Publishers, New Delhi
21. Rajakumar S, Ravindran D (2010) Computational fluid dynamics of wind turbine blade at various angles of attack and low Reynolds number. Int J Eng Sci Technol 2(11):1243–1248
22. Shuwa M, Ngala GM, Maina M (2016) Development and performance test of a micro horizontal axis wind turbine blade. Int J Eng Res Appl, 6(2):11–17

Popular Mistakes About Renewables

Don Swift-Hook

1 Introduction

Global warming is an emotive subject. Those who question its theoretical basis are accused of being "in denial" and there is of course no experimental basis: we shall not know for sure the global temperatures at the end of the century until the end of the century arrives. Meanwhile, there are many popular mistakes abounding in the subject and this paper will point out a number of them.

The reports produced by the United Nations Intergovernmental Panel on Climate Change are the most authoritative summaries of the scientific work available. The most recent IPCC report appeared in 2014 and the next major IPCC report is due in 2022. Many of the misunderstandings about global warming arise because the IPCC reports are misused or ignored.

Mistake 1 CO_2 levels have increased dramatically since the Industrial Revolution. [Wrong!]

2 Historical Measurements of CO_2

CO_2 levels can be measured going back for many centuries in the air that is trapped in Antarctic snowfall. Snowflakes are light and airy, so a good deal of air is trapped in the ice that is formed as snow on the ground is compressed by the weight of successive snowfalls. Ice cores can be bored which then carry air samples from successive depths corresponding to successive years of their deposition. Figure 1 is a well-known graph which shows CO_2 concentrations measured in Antarctic ice cores. This shows that CO_2 levels have been rising from the beginning of the Industrial Revolution around 1750.

It is claimed that CO_2 levels are rising dramatically. "Rocketing up" is a phrase that is often used. Certainly, the levels are rising and indeed the rate of rise is accelerating. However, plotting the same data on a graph with a zero gives a very different impression, see Fig. 2. The total rise is only 40% over 2½ centuries, which hardly deserves descriptions such as "dramatically" and "rocketing up". A fairer description is that CO_2 levels have stayed fairly constant.

Mistake 2 Global warming was on-going after Kyoto. [Wrong!]

D. Swift-Hook (✉)
Kingston University, London, UK

Secretary of WREN, Woking, Surrey, UK
e-mail: donald@swift-hook.com

© The Author(s), under exclusive license to Springer Nature Switzerland AG 2022

A. Sayigh (ed.), *Sustainable Energy Development and Innovation*, Innovative Renewable Energy,
https://doi.org/10.1007/978-3-030-76221-6_76

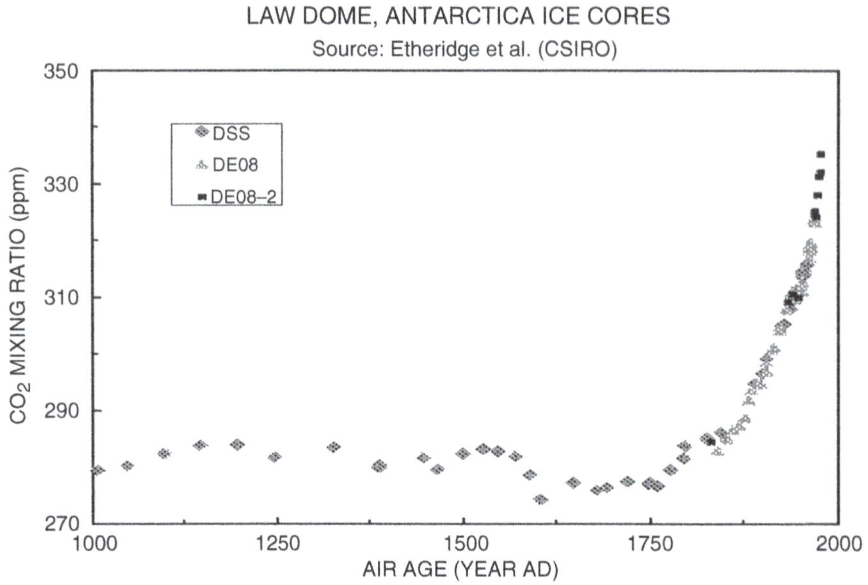

Fig. 1 Historical measurements show CO_2 levels increasing after 1750, the start of the Industrial Revolution

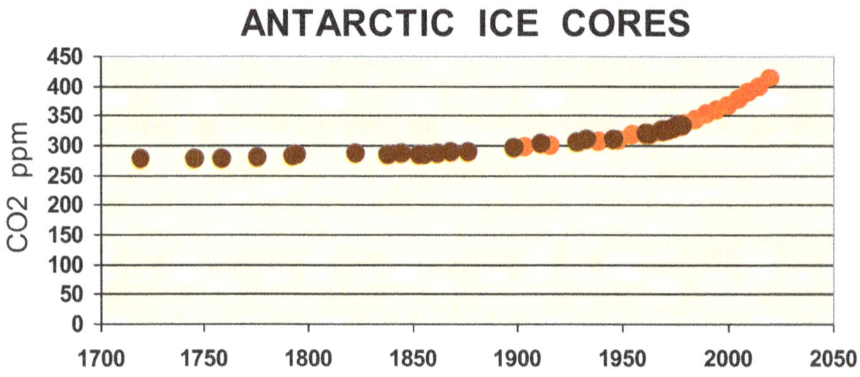

Fig. 2 Re-plotting Fig. 1 ice core data (brown points) shows CO_2 levels have stayed fairly constant. Orange points are recent atmospheric measurements on Hawaii

3 The Hiatus in Global Warming

Perhaps the most glaring mistake concerns on-going global warming itself. Ironically, the most recent IPCC report in 2014 showed that global warming was not on-going but had in fact ceased. The global temperature peaked in 1998, the year following the Kyoto Agreement (1997) and showed no signs of an increase after that, see Fig. 3. It was a mistake to suggest that Global Warming was on-going after Kyoto (Mistake 2).

This levelling off in temperature was referred to as a "hiatus" (which means a temporary stand-still) in the belief that global warming would continue at some time in the future. There are some recent indications that this is now happening but there is still no explanation of this stand-still phenomenon following the pre-Kyoto rise from 1980 to 1998. A hiatus, whether temporary or otherwise, is not predicted or explained by any of the models which are used to predict global warming.

It is interesting to note the response of the global warming community to this good news. The cessation of global warming should surely have been widely welcomed even if it proves to be only temporary. Instead, it has been widely ignored. However, the designation "climate change" is now used far more often than "global warming" and this is sensible when there hasn't been any.

In more recent years since 2014, some higher global temperatures have been recorded. These may well indicate a resumption of global warming but it is far too early to tell. The next major report by the IPCC is not due until 2022 and

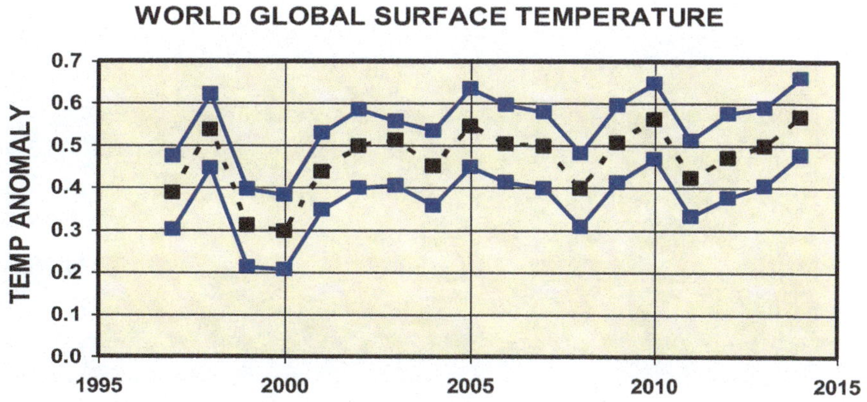

Fig. 3 There was a hiatus in the rise of global surface temperatures after Kyoto 1997

measurements since 2014 will need to be reviewed, confirmed and approved in that new report to achieve the same weight as those shown in the 2014 report in Fig. 3, which are backed by the whole weight of the IPCC.

4 How to Measure a Global Temperature

It should not be necessary (but it is) to say what global warming means. Global warming is an increase in global temperature.

Some enthusiasts, when faced with the recent hiatus, assert that global warming is a process that was continuing steadily despite the fact that the global temperature was not changing. Such double-speak, altering the meanings of words, should have no place in sensible scientific discussions of the subject.

However, the global temperature needs to be clearly identified and there is no global thermometer that we can point to which measures the global temperature. There are many meteorological stations around the world including satellites, whose data is collected by two principal organisations, NASA in the USA and the Meteorological Office in the United Kingdom. The Americans are regarded as the experts when it comes to satellite measurements of surface temperatures while the British have a history of collecting data from meteorological stations all-round the globe, often in remote locations, dating back to the time when they had "an Empire upon which the sun never set".

The meteorological data is sent back to these two organisations who average it spatially (around the globe) and temporarily, i.e., time-wise (annually) to arrive at a single figure which is effectively jointly agreed and published. That is the only meaningful global temperature. Everybody has to use the same one and it is the one used by the IPCC.

Attempts are often made to use temperatures averaged over different regions or periods, for example, Arctic or desert areas. It cannot be emphasised too strongly that such alternative temperatures cannot be firmly relied upon and certainly such work cannot carry the same weight as that published by the IPCC.

Mistake 3 The Arctic ice cap is melting and we shall all be flooded. [Wrong!]

5 Floating Ice

The Arctic ice cap is melting. Every year about half of the floating ice melts and most of it then re-freezes but the minimum ice cover (at mid-summer) has been lessening over the last few years. This means, for example, that the Northwest Passage from the Atlantic Ocean to the Pacific now opens up each year as more ice melts.

The media follow these developments with predictions that we shall all be flooded but such assertions are totally mistaken. It was Archimedes, the Ancient Greek gentleman who leapt from his bath to run naked down the streets of Syracuse shouting "Eureka" when he discovered his principle of buoyancy. This principle means that melting ice will fill the volume of water it displaces, so there will be no change in sea level when floating ice melts (Fig. 4).

Also, recently, when the Arctic ice melt was at a maximum and the Northern sea-ice cover was at a minimum, there was the greatest coverage of Antarctic sea ice on record. This is perhaps not quite as impressive as it sounds. The record needs to

Fig. 4 Floating sea ice melting

be from satellites and observations only started in 1979. Nevertheless, it shows that local conditions cannot be extrapolated for global effects.

Mistake 4 We (UK) need more generation to meet the relentless increase in air demand for electricity. [Wrong!]

6 The Relentless Increase in Electrical Power Consumption Has Come to an End

Ever since power systems were invented, all countries used to experience a steady increase in their power consumption. Power generation is a major source of the greenhouse gases that are thought to cause global warming.

This increase has seemed to be relentless and inexorable as civilisation expands. Indeed, electrical power consumption has often been regarded as a surrogate for economic development and standard of living.

It is with a great deal of surprise for many people to discover that electrical power consumption and, indeed, the consumption of energy as a whole have been falling for many years lately in many countries.

In the UK, peak power consumption of 62 GW was reached in 2005 and it had fallen by 20% to 50.4 GW in January 2018. Primary energy consumption is also falling: it is down 10% since 2000 and is back below the level that it was half a century ago. See Fig. 5.

Clearly, the supposedly relentless increase in power consumption has come to an end and is no longer inexorable, so the need for more generating capacity cannot be taken for granted (Mistake 4).

One of the implications of this is that we can expect the emission of greenhouse gases, which cause global warming, to reduce correspondingly, quite apart from any more positive action that may be taken.

Mistake 5 Subsidies are withdrawn because wind and solar are out of favour. [Wrong!]

7 Subsidies Are Withdrawn Because Wind and Solar Are the Cheapest

In recent years, the prices of wind turbines and photovoltaic panels have fallen dramatically. The third Contracts for Difference auction round held by the UK Government in September 2019 resulted in prices as low as £39.65/MWh for 6GW of offshore wind power.

Solar prices halved in 2 years (see Fig. 6). These two technologies are now the cheapest available in many countries, cheaper than gas turbines and coal-fired plant (see Fig. 7). There is more wind and solar capacity being installed today than any other type of power-plant. This has led to many misunderstandings. For example, when wind and solar are the cheapest type of generation it is no longer necessary or appropriate to subsidise them in order to encourage their use. Governments therefore are widely withdrawing their subsidies. The media frequently misinterpret this and report that Governments no longer favour wind or solar (Mistake 5) when the reverse is true and governments are recognising that wind and solar are the cheapest and most attractive methods of generation available, which are economic in their own right without the need for subsidies or other Government assistance.

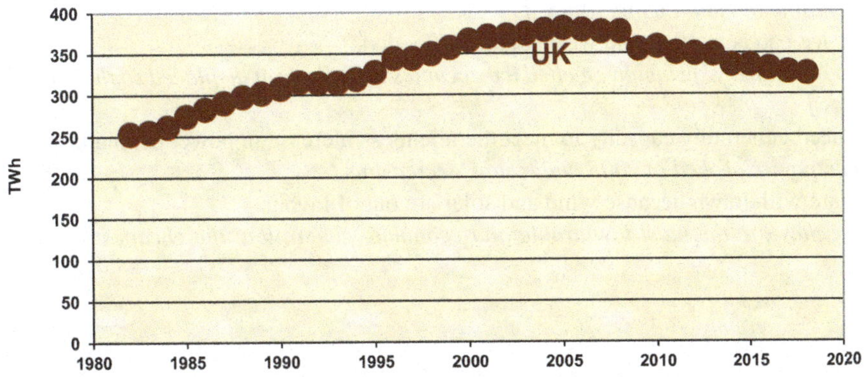

Fig. 5 *After growing relentlessly up to year 2000, UK electricity consumption peaked in 2005 and is now falling*

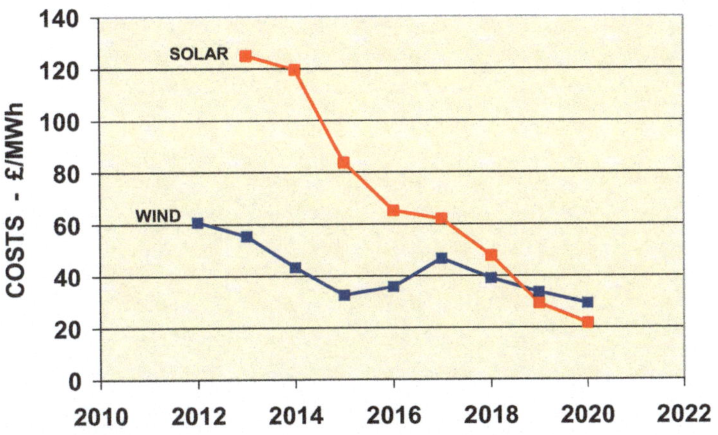

Fig. 6 ***Average auction prices for wind and solar power***

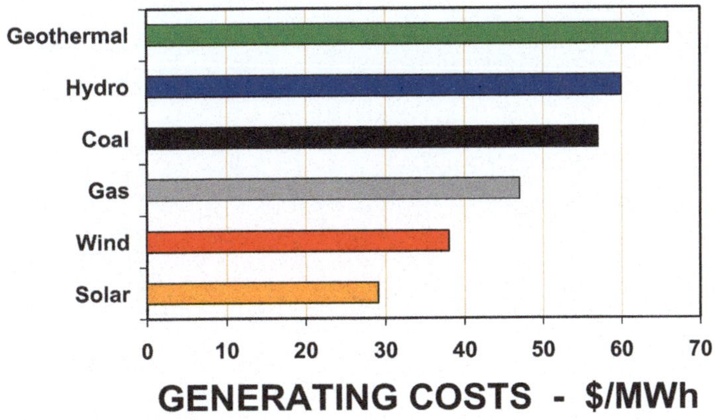

Fig. 7 *Auction prices in Chile, August 2016, showing that wind and solar were already the cheapest form of generation there*

8 Conclusions

These Mistakes are wrong for the reasons given:

- Mistake 1. Man-made CO_2 levels have increased dramatically since the Industrial Revolution *[Wrong! There has only been a 30% increase in 250 years.]*
- Mistake 2. Global warming was on-going after Kyoto.

[Wrong! There was a hiatus in global temperature.]

- Mistake 3. The Arctic ice cap is melting and we shall all be flooded.
  [Wrong! According to Archimedes' principle, melted ice occupies the volume it displaced so there is no change in sea level when floating ice melts.]
- Mistake 4. We [UK] need generating capacity to meet the relentless increase in power demand.
  [Wrong! UK maximum power peaked in 2005 and is now decreasing.]
- Mistake 5. Subsidies are withdrawn because wind and solar are out of favour.
  [Wrong! It is because they are the most favourable and economic generation, the cheapest in their own right without subsidies.]

Environmental Sustainability Assessment of Small Hydropower Plants: A Case Study from Ceyhan River Basin in Turkey

Serhat Kucukali, Ahmet Alp, Adil Akyüz, and Omar Al Bayatı

1 Background

Small hydropower (SHP) projects receive considerable amount of funds and financial support from governments and financial institutes. There has been a boost in SHP development prompted by financial subsidies from national renewable energy legislations. Aside from their doubtless advantages, SHP projects can lead to several environmental impacts such as blockage of fish migration and sediment transport, change in downstream morphology, and biodiversity loss. Implementation of WFD aims to achieve ecological continuity in rivers (i.e., continuity of flow, sediment, and fish passage). One of the main environmental challenges of small hydropower development is related to fish passage [1]. Migrations are ecological imperatives for fish populations creating the need for efficient fish passage systems. Under WFD, it is hydropower plant operator's responsibility to provide a suitable passage for migrating fish. For instance, recognizing that environmental sustainability is a global issue, researchers from the USA, Australia, and Lao PDR have joined forces in a collaborative effort to improve safe fish passage through hydropower systems in Mekong and Murray-Darling Basins focusing on local fish species [2]. Moreover, France Hydro Electric introduced a guide titled "Towards a hydroelectric plant of the 21st century", for the development of small hydroelectric plants with due respect to the natural environment [3]. In the USA, Green-e offers certification program for SHP plants which is based on the following environmental criteria: river flows, water quality, fish passage and protection, watershed protection, and threatened and endangered species protection. The Low Impact Hydropower Institute (LIHI) in the USA dedicated to reducing the impacts of hydropower generation through the certification of hydropower projects that have committed to environmental, cultural, and recreational stewardship pursuant to the Low Impact Hydropower Institute's criteria. Kucukali [4] assessed the environmental risks associated with a small plant on the basis of The European Bank for Reconstruction and Development (EBRD) standards for Sustainable Energy Financing for Small Hydropower. But, Kucukali [4] did not use the environmental monitoring data for the risk assessment. Gronbrekk et al. [5] and Tahseen and Karney [6] reported literature review on international sustainability standards and guidelines on hydropower. Their literature review revealed that most of the hydropower sustainability guidelines (i.e., EBRD and Equator principles) ignored the erosion and sedimentation issue except for IHA Sustainability Assessment Protocol and IEA framework [7]. However, sedimentation is a major concern for the life cycle of a hydropower plant and a serious environmental issue on a global scale [8]. Sustainable hydroelectric generation cannot be measured with the efficiency of the plant and the availability of the resource. Different parameters from a variety of disciplines should be taken into account and integration of those parameters will increase the quality and sustainability of a hydropower project. In this context, the International Hydropower Association

S. Kucukali (✉)
Hacettepe University, Civil Engineering Department, Ankara, Turkey
e-mail: serhatkucukali@hacettepe.edu.tr

A. Alp
K. Maraş Sütçü İmam University, Fisheries Department, K. Maraş, Turkey
e-mail: aalp@ksu.edu.tr

A. Akyüz
K. Maraş Sütçü İmam University, Biosystem Engineering Department, K. Maraş, Turkey
e-mail: adilakyuz@ksu.edu.tr

O. Al Bayatı
Çankaya University, Computer Engineering Department, Ankara, Turkey

© The Author(s), under exclusive license to Springer Nature Switzerland AG 2022

A. Sayigh (ed.), *Sustainable Energy Development and Innovation*, Innovative Renewable Energy,
https://doi.org/10.1007/978-3-030-76221-6_77

Table 1 Sustainability assessment framework of International Hydropower Association

Cross-cutting	Environmental	Social	Technical	Economic/financial
Climate change	Downstream flow regimes	Resettlement	Siting and design	Financial viability
Human rights	Erosion and sedimentation	Indigenous peoples	Hydrological resource	Economic viability
Gender	Water quality	Public health	Infrastructure safety	Project benefits
Livelihoods	Biodiversity and invasive species	Cultural heritage	Asset reliability and efficiency	Procurement

(IHA) launched the Hydropower Sustainability Assessment Protocol in June 2011 to assess the sustainability of hydropower projects in each plant stage of planning, construction, and operation [7]. The framework has been widely used in the hydropower sector around the world by project developers, financial institutions, and hydro-turbine suppliers [9]. The protocol offers a way of assessing the sustainability performance of hydropower in five aspects: cross-cutting, environmental, social, technical, and economic/financial (Table 1). Each aspect has four different criteria. For the environmental aspect, the criteria are: (i) downstream flow regimes, (ii) erosion and sedimentation, (iii) water quality, and (iv) biodiversity and invasive species. It is crucial to assess SHP plants environmental sustainability adequately by using worldwide accepted reliable criteria. Accordingly, this study aims to evaluate the environmental sustainability of SHP plants during operation by using environmental monitoring data. The environmental criteria are taken from the internationally recognized Hydropower Sustainability Assessment Protocol of the International Hydropower Association (IHA). Moreover, several studies focused on the sustainability assessment of hydropower plants in the planning phase instead of the operation phase. Accordingly, this study aims to evaluate the environmental sustainability of a hydropower plant during operation by using environmental monitoring data, site surveys, and documented evidence considering the abovementioned environmental criteria in an integrated approach.

2 Assessment of Ceyhan River Basin and Çataloluk SHP

This study was carried out at Çataloluk SHP in Tekir Stream of Ceyhan River. Ceyhan River Basin located in the eastern Mediterranean region of Turkey, between latitudes of 36°35′ and longitudes of 35°34′. Ceyhan River Basin is bordered by the Seyhan Basin in the west and northwest, Asi in the south, and Euphrates in the east and north (Fig. 1). The Ceyhan River has its source at a location called Pınarbaşı 3 km on the Nurhak Mountains (2000-2500 m) of the Eastern Taurus Mountains range, southeast of the town of Elbistan in the Kahramanmaraş province of Turkey. The Ceyhan River discharges into the Mediterranean Sea at Karataş. The river basin covers an area of 20,670 km^2. The mean annual flow volume is 7.18 billion m^3 and this corresponds to 4% of the total runoff of Turkey. In recent years, Ceyhan River Basin has undergone great changes due to construction of high dams such as Aslantaş, Berke, Sır, Kılavuzlu, Menzelet, and Suçatı. By the year 2016, 57 SHP plants have been in operation in Ceyhan River Basin. From these, only 16 SHP plants include fish passages. These fish passages are pool-weir type (5 fish passages), vertical slot type (10 fish passages), and Denil type (1 fish passage). Fish fauna of Ceyhan River Basin is constituted by 20 species belonging to 10 families [10]. The location of Çataloluk Hydropower Plant in Ceyhan River Basin is shown in Fig. 1, its reservoir volume varies from 1,500,000 to 2,000,000 m^3 and active reservoir storage is 500,000 m^3. The characteristics of Çataloluk SHP are as follows: design discharge = 10 m^3/s, net head = 106.6 m, diversion length = 2.68 km, installed capacity = 2 × 4.92 MW.

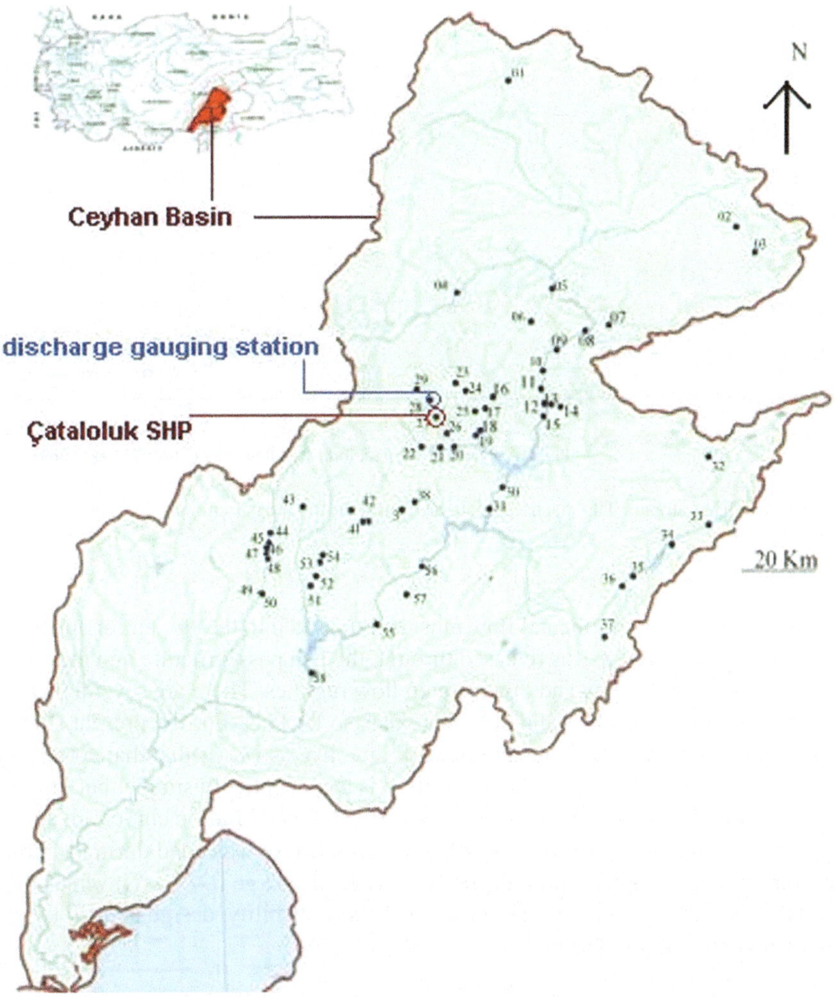

Fig. 1 Ceyhan River Basin and distribution of hydropower plants. Çataloluk SHP is shown at Point 27 and DSI discharge gauging station no. 2054 is located at the upstream of Point 28

3 Environmental Sustainability Assessment of Çataloluk SHP

Environmental sustainability of Çataloluk SHP has been assessed by taking into account: (i) environmental flow, (ii) local fish species and fish passage performance, and (iii) sediment management.

3.1 Environmental Flow

Under scope of this study, the discharge data of the gauging station no. 2054 of DSI were used in order to evaluate the environmental flow. Figure 2 shows the time series of the discharge recordings from October 2001 to September 2011. The daily discharge varied between 0.433 and 50.6 m^3/s and mean daily discharge was calculated as 5.37 m^3/s. Minimum daily discharge values within the years were as follows: 0.743 m^3/s in 2002, 0.566 m^3/s in 2003, 0.743 m^3/s in 2004, 0.433 m^3/s in 2005, 0.875 m^3/s in 2006, 1.140 m^3/s in 2007, 0.743 m^3/s in 2008, 0.743 m^3/s in 2009, and 1.250 m^3/s in 2011. Design discharge of the facility is 1.86 times higher than corresponding annual average river flow and the river flow is diverted into an artificial channel and tunnel through the forest with a total length of 2.67 km. Çataloluk SHP totally blocks the entire cross-section of the river and the dam body of the facility acts as a barrier, reducing flow velocities and creating a reservoir at the upstream of the facility. Thus, original stream flow regime between the weir and hydroelectric power plant was modified. A

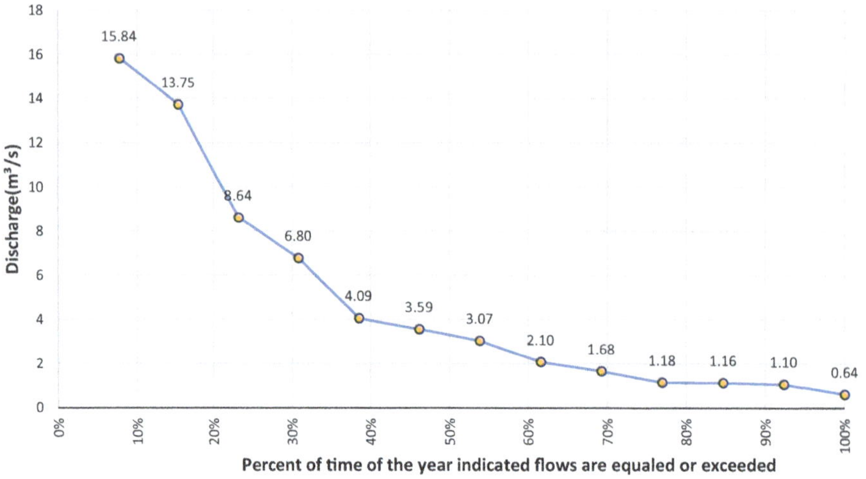

Fig. 2 Flow duration curve for Tekir stream. The discharge data are based on the measurements at Gauging Station 2054 covering the period of 2001–2012

certain amount of water known as environmental flow must be released into the stream for aquatic life and stream ecosystem health, continuously. Environmental flow was released through the fish pass structure and there is a monitoring gauging station in order to control environmental flow and downstream flow regimes. There are several methods and formulas to calculate environmental flow, but no universally valid solution exists so far [11]. For the present study, environmental flow rate, Q_e, was computed according to European Water Framework directive as $Q_{0.95}$ value that corresponds to 95% probability of exceedance on minimum discharge [12]. Flow duration curve is a conceptually simple, but highly informative way to summarize the temporal variability of streamflow at given location [13]. Flow duration curves are cumulative frequency distributions that show the percent of time that a specified discharge is equaled or exceeded during a period of interest. In statistical terms, the flow duration curve is a graph plotting the daily average discharge (Q, y-axis) versus the fraction of time, or probability, that Q exceeds any specified value $Q = Q_{ep}$ (x-axis). This probability, designated EP (Q_{ep}), is called the exceedance probability and is defined in probability terms as:

$$\mathrm{EP}\left(Q_{\mathrm{ep}}\right) \equiv \Pr\left\{Q > Q_{\mathrm{ep}}\right\} \tag{1}$$

where Pr{ } denotes probability of condition with braces. $Q_{0.95}$ flow is most often used as low flow index in academic sources [13]. Daily flows within the 10 years exceeding 95% of the time were accepted as minimum environmental flow. Previous studies used monthly discharge values for flow duration curve establishment and environmental flow calculation [8]. It is proposed that using daily discharge values will be more meaningful because in some periods of year daily discharge value can be less than environmental flow. By computing the percentage of time when various flow rates are equaled or exceeded, flow duration curve was developed from October 2001 to September 2012 at gauging station no. 2054. Accordingly, for Çataloluk SHP, $Q_{0.95}$ was found to be 0.92 m^3/s (Fig. 2). Environmental flow should ideally be adapted to site characteristics, for example, by considering habitats and/or ecological criteria. Also, in previous studies, there is no analysis available regarding relationship between assessed minimum water flow and water needed for maintaining aquatic life (e.g., one of the parameters could include water depth required for fish species of the river). From discharge rating curve of environmental monitoring station, for an environmental flow of 0.92 m^3/s, flow depth and average velocity are calculated as 0.22 m and 0.33 m/s, respectively. There is actually no data about depth and flow velocity requirements of local fish species (*Capoeta damascina and Alburnus adanensis*). But there is some information that depth and flow velocity should be 0.20 m and 0.3–0.4 m/s for some Cyprinid species. If we consider an estimated environmental discharge of 0.92 m^3/s, depth and flow velocity will not create high negative impact on fish species, macrobenthic fauna, and riparian zone. It should be noted that in Çataloluk SHP the instantaneous flows within the day may decrease under determined environmental flow. Hence, plant operator can make an energy generation plan taking into account water storage in the reservoir of Çataloluk SHP. Also, operator can interrupt or reduce energy production, when daily discharge is less than the environmental flow.

3.2 *Local Fish Species and Fish Passage Performance*

Çataloluk SHP contains a fish passage structure and it was designed properly in order to enable effective fish passage. Moreover, information was gathered on local fish populations in the project area. Five fish species were determined in Tekir stream and these were *Capoeta damascina*, *Alburnus adanensis*, *Salaria fluviatilis*, *Capoeta erhani*, and *Garra rufa*. *Capoeta damascina,* and *Alburnus adanensis* inhabit all sections of Tekir stream, while other species inhabits lower sections (near Suçatı reservoir) of the stream and they were not present in the region of Çataloluk SHP. According to the International Union for Conservation of Nature (IUCN), *Capoeta damascina* is in the DD class (data deficient), whereas *Salaria fluviatilis* is in the LC class (least concern), and other fish species are in the NE class (not evaluated). For the project area, *Capoeta damascina*, and *Alburnus adanensis* were determined as migratory species. The fish passage in Çataloluk SHP is a vertical slot-type passage (Fig. 3). The structural and hydraulic characteristics of the fish passage were examined and they were compared with the international standards [14]. Vertical slot fish passage has a total length of 126 m and has a head difference of 14.10 m with a bed slope of 8.5%. Vertical slot fish passage has two water inlets from the upstream (first of them is located at 837.50 m level and second is located at 836.50 m level). The fish passage has 63 pools. The characteristics of fish passage are as follows: pool length = 2 m, pool width = 1.4 m, water level differentiation between the pools = 0.2 m, slot opening = 0.19 m, energy dissipation per unit volume $\Delta P = 200$ W/m^3. The energy dissipation per unit volume is calculated as:

$$\Delta P = \frac{\gamma Q S_0}{Bd} \tag{2}$$

where γ is the specific weight of the water, Q is the discharge, S_0 is the fish pass structure bed slope, B is the channel width, and d is the flow depth. In international standards, the energy dissipation rate per unit volume is taken as the turbulence indicator [14]. The hydraulic characteristics of the Çataloluk fish passage comply with the international standards [14]. However, the downstream entrance of the fish passage was not adequately designed and this technical problem was eliminated by establishing boulder rock ramp (Fig. 3).

Water temperature was recorded continuously in fish passage pool by using Hobo onset sensor which was installed in the fish passage. The time series of hourly water temperature records are shown in Fig. 4. The water temperature was 14 °C during first days of fish migration and it reached to 24 °C at the end of migration period (May–July). The results demonstrated that in addition to flow, water temperature is found to be key parameter for fish migration timing which is in agreement with findings of Pettersen [15]. In order to determine the performance of fish passage in Çatloluk SHP, a fish trap was placed at the exit of fish passage. The size of the trap was prepared according to the dimension of the fish passages and it was made of a metal frame box, including a cone, and 0.5 cm stretched mesh size. The trap was controlled in a few days and the fish in the traps were taken in a tank and anesthetized by 2-phenoxyethanol. After recording fish number, length, weight, and sex, fish were released to the upper pool. Accordingly, a total of 584 fish were recorded in the trap during fish migration period from 29 April 2012 to 3 July 2012. Of total, 357 fish were *Capoeta damascina* and 227 were *Alburnus adanensis*. The length of *Capoeta damascina* in the traps varied between 13 and 43 cm and their mean length was 21.3 ± 3.9 cm. The length of *Alburnus adanensis* in the traps varied between 10 and 23 cm and their mean length was 14.6 ± 1.6 cm [10]. According to

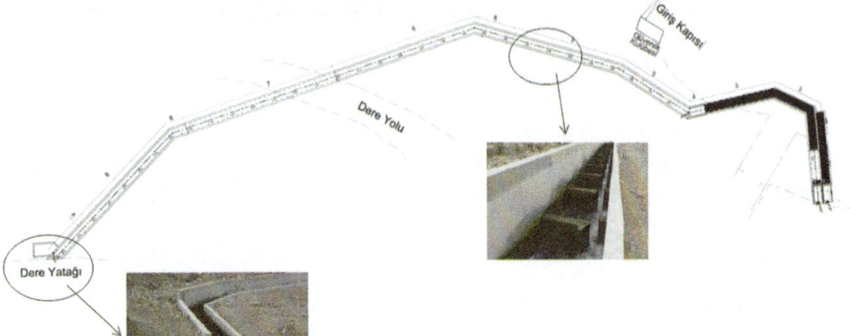

Fig. 3 Plan view of the fish passage in Çataloluk SHP with photos from different sections

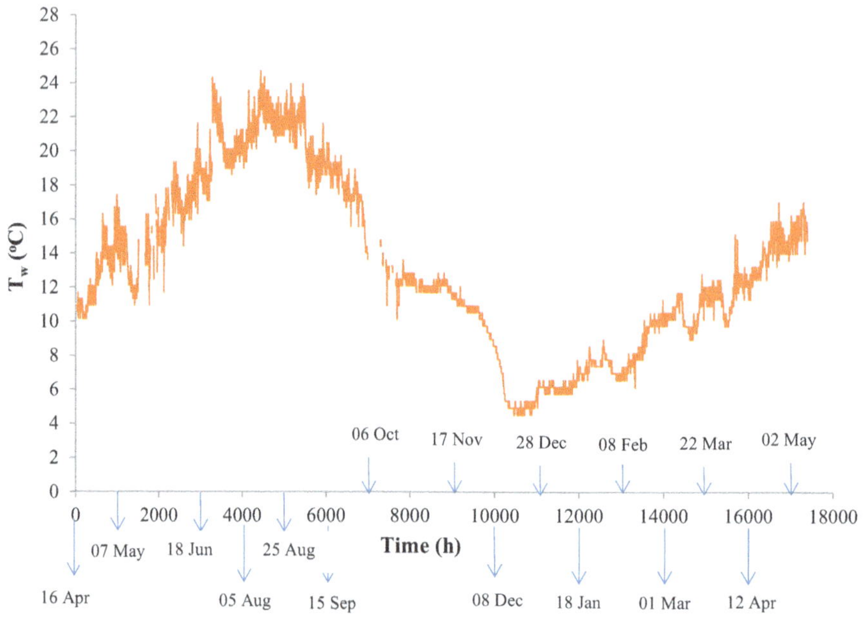

Fig. 4 Time series of water temperature for Tekir stream. The samples were collected at the fish passage pool of Çataloluk SHP for each half an hour

Fig. 5 Settling basin in Çataloluk SHP and different images from fine sediment accumulation

the fish monitoring data, it can be accepted that fish passage in Çataloluk SHP works effectively and provides migration corridors for fish in Tekir stream.

3.3 *Sediment Management*

The Tekir stream basin has a rich forest cover and there is no high siltation load in the region. The bottom substrate of the stream is generally in gravel and stone. There is also another SHP (Kısık HPP) which is located 200 m upstream of the Çataloluk Reservoir. Çataloluk SHP interrupted the sediment transport mechanism of the river. Natural flow velocity is reduced due to the presence of the weir, resulting in the deposition of transported sediments in the reservoir and in the settling basin. Hence, the weir caused upstream sediment surplus and downstream sediment deficit. Therefore, natural sediment dynamic is altered and transport is disrupted, especially of coarse sediments. This may lead to changes in substrate composition and altered morphological processes in the river downstream. In SHP plants, settling basins (sediment settlement pools) are established in order to ensure sediment-free water flowing to the powerhouse. Çataloluk SHP has a settling basin which has a length of 56.25 m, width of 8 m, and depth of 3 m. The settling basin was evacuated and cleaned once a year. During the 1 year period, sediment accumulation in the pool was investigated. In this investigation, forestry wastes and wood chips were determined and fine sediment was observed in the bottom of the pool (Fig. 5). In the Çataloluk SHP, mobilizing flows should be implemented at the bottom settling basins to deliver fine sediments to downstream river sections as a sediment management strategy. Flushing flows from the bottom outlet of settling basins lead high peaks with high loads of fine sediments, which is in many cases a critical alteration for river reaching downstream. Also, some areas in the settling basin are not cleaned from sediments during flushing. Accordingly, environmentally friendly sediment removal operations (i.e., hydrosuction removal system) should be used to mitigate adverse ecological consequences below SHP plant and sediment should be released to the river back for ensuring required sediment for maintaining aquatic habitats. The current sedimentation management of the Çataloluk SHP plant is not environmentally sustainable and does not comply with the Water Framework Directive. Accordingly, a suspended sediment monitoring station should be established at the downstream of the diversion weir and suspended sediment concentration (SSC) should be sampled continuously. For instance, based on the recent French regulation, the SSC must not exceed 15 g/L during flushing in order to make flushing ecologically optimized [16].

4 Conclusions

In order to consider SHP plants as a sustainable and renewable energy source, SHP plants have to meet ecological requirements of river systems. Accordingly, environmental responsibility and accountability are crucial issues for hydropower plant operators to generate hydroelectricity in a sustainable way. The most obvious alteration on river ecosystems caused by SHP plant operation interrupted river continuity. The continuity for aquatic organisms can be interrupted both upstream and downstream of weirs. The impacts are particularly significant on migratory species. Hence, functional fish passage structure, effective sediment management, and adequate environmental flow are mandatory to mitigate the ecological impacts of SHP plants. The recent trends in SHP development and findings of the case study indicated that in order to develop small hydropower in an environmentally sustainable manner, a monitoring system must be established at the SHP plant to control the environmental flow, sediment, and fish passage. By using the proposed environmental sustainability evaluation method, SHP operators and responsible government agencies can quantify the plant ecological impacts during operation and they can develop strategies to mitigate those ecological impacts. The proposed evaluation tool can be applied to other SHP plants around the world by following the procedure and by employing the environmental monitoring data. The main merit of the proposed assessment is based on the measured data and documented evidence. Since Turkey is an European Union (EU) candidate, it has begun to adopt the Water Framework Directive, which aims good ecological status in rivers, in its legislation system by 2012. In the near future, it is highly possible that small hydropower plants may sustain their current operational conditions and be largely developed in Turkey if they generate electricity with respect to the environment and meet the ecological requirements.

References

1. Deng Z (2014) Monitoring and evaluation technologies related to fish and hydropower. In: 4th seminar on strategies for conservation of Minas Gerais fishes, Belo Horizonte
2. Thorncraft G, Baumgartner GL, Boys C, Brown R (2013) Merging hydraulics with biology. Water Power & Dam Construction 65:42
3. Albanel H (2012) Towards a hydroelectric plant of the 21st century. In: Proc., 2012 HidroEnergia Conf., Wroclaw
4. Kucukali S (2014) Environmental risk assessment of small hydropower (SHP) plants: a case study for Tefen SHP plant on Filyos River. Energy Sustain Dev 2014(19):102–110
5. Gronbrekk W, Barton H, Khoury RH (2010) International sustainability tools for hydropower role, relevance and industry reporting trends. In: Proc., Hydro 2010 Conf., Lisbon
6. Tahseen S, Karneey BW (2017) Reviewing and critiquing published approaches to the sustainability assessment of hydropower. Renew Sust Energ Rev 2017(67):225–234
7. IHA – International Hydropower Association (2016) Hydropower sustainability assessment protocol. http://www.hydrosustainability.org/Protocol-Assessments.aspx. Accessed 15 Nov 2019

8. UNESCO-IRTCES (2011) Sediment issues & sediment management in large river Basins interim case study synthesis report
9. Schuol J, Fischer-Aupperle B, Riedel N (2012) Technical solutions to enhance sustainability in hydro power. In: Proc., Hydro 2012 Conf., Bilbao
10. Alp A, Akyüz A, Özcan M (2015) Efficiency and suitability of fish passages in River Ceyhan, Turkey. In: Proc., Fish Passage 2015 Conf., Groningen
11. Hetherington D, Artza IL (2016) A holistic system-focused approach to environmental flow regime investigations and design of mitigation. In: Proc., Hydro 2016 Conf., Montreux, (CD)
12. EC (2015) Ecological flows in the implementation of the Water Framework Directive. 635 Technical Report no. 086, Guidance document no. 31, ISBN 978-92-79-45758-6
13. Dingman SL (2009) Fluvial hydraulics. Oxford University Press, New York, USA
14. DWA - German Association for Water, Wastewater and Waste (2014) Merkblatt DWA-M 509: Fischaufstiegsanlagen und fischpassierbare Bauwerke – Gestaltung, Bemessung, Qualitätssicherung. Hennef (in German)
15. Pettersen V (2014) Defining e-flow through trial regulation: the example of the Suldalslagen river in Norway. In: Proc., Hydro 2014 Conf., Lake Como, (CD)
16. Bollaert E, Diouf S, Barras J (2016) Pressurized flushing of Verbois reservoir: a first experience of environment-friendly sediment flushing. In: Proc., Hydro 2016 Conf., Montreux, (CD)

Part VIII
Water, Hydropower and Ocean Energy

Finding the Most Suitable Irrigation Dams for Hydropower Development in Turkey Using Multi-Criteria Scoring and GIS Spatial Analysis Tool

Omar Al Bayatı, Serhat Kucukali, and Hakan Maraş

1 Introduction

Renewable energy including hydropower, wind as well as solar power have attracted much attention mainly due to the negative effects resulting from non-renewable energy [1]. Hence, there has been a dire need to find ways in which renewable energy can be utilized to its full capacity. One of the most feasible and viable solutions for achieving a long-term, sustainable and cost-effective renewable energy resource is through the implementation of small hydropower projects [2]. While in many countries, large storage and moderate-sized dams have been used for power production, small dams have served only for the purpose irrigation. It is no wonder that many countries are considering to transform existing irrigation reservoirs for energy production, as the main idea is to collect the largest amount of hydropower from many small and medium irrigation dams to achieve the highest level of sustainability as it is an available source with low cost that feeds the network in case of emergency and with continuous investment, gradually fulfils the need for large thermal energy projects.

In most cases, the preliminary step to turning existing reservoirs to hydropower schemes is to carry feasibility tests which determine the total capacity that can be created [3]. In doing so, the ultimate aim is to identify the most suitable existing reservoir sites where the hydropower schemes can be implemented. GIS, known as the geographic information system which includes geodatabase and spatial raster maps like digital elevation raster, has been widely used to extract information to assess the potential hydropower schemes [4]. The use of GIS spatial analysis tools can help in performing analyses on spatial data, and can provide answers to spatial questions such as "How steep is this location?" and "What is the distance between these two locations?" It also provides answers to more complex spatial questions such as "Which is the nearest facility for a particular location?" and "What is the least costly path between two locations?" [5], this can be very important in planning of hydropower projects. Suitability mapping involves using a variety of data sources which are assigned to multi-criteria. Data are often imported into GIS, which combines potentially unrelated data in a meaningful manner [6]. Weights that emphasize the relative importance of one criterion to another are often determined by research specialists or stakeholders to enhance decision making. The choices are often sorted according to its suitability of meeting the criteria using fuzzy logic [7].

2 Background

In Turkey, it is expected that there will be an increase in demand for electricity due to economic growth. According to the European Environment Agency Report 2017 of changes in Turkey land cover for years 2006–2012 [8], the types of land use that increased are residential, industrial and water bodies, which proves the increase in electricity demand, at the same time increasing the possibility of exploiting water bodies' increase represented by irrigation dams in hydropower generation. The laws in recent years succeeded in promoting the utilization of renewable energy for electricity generation [9]. With an annual growth rate of 1%, the population is expected to reach 91 million by 2030 [10]. As seen in Fig. 1, most of the water is

O. Al Bayatı · H. Maraş
Çankaya University, Computer Engineering Department, Ankara, Turkey
e-mail: c1679006@student.cankaya.edu.tr; hhmaras@cankaya.edu.tr

S. Kucukali (✉)
Hacettepe University, Civil Engineering Department, Ankara, Turkey

© The Author(s), under exclusive license to Springer Nature Switzerland AG 2022

A. Sayigh (ed.), *Sustainable Energy Development and Innovation*, Innovative Renewable Energy,
https://doi.org/10.1007/978-3-030-76221-6_78

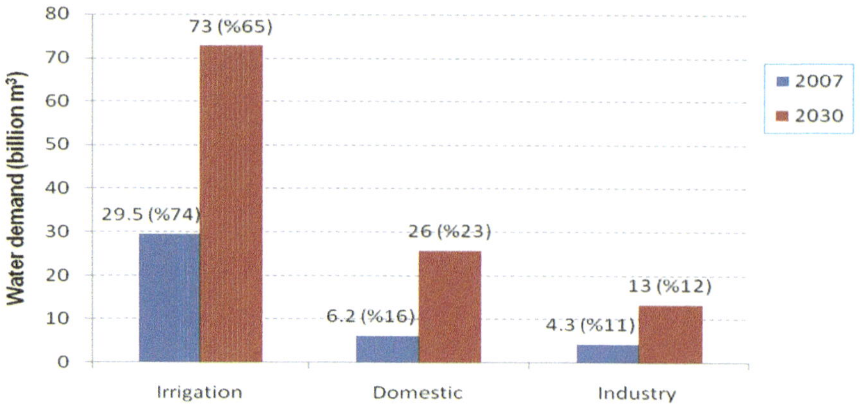

Fig. 1 Water demand in Turkey by sectors, for 2007 and 2030 (forecast), data source: Turkish General Directorate of State Hydraulic Works (DSI)

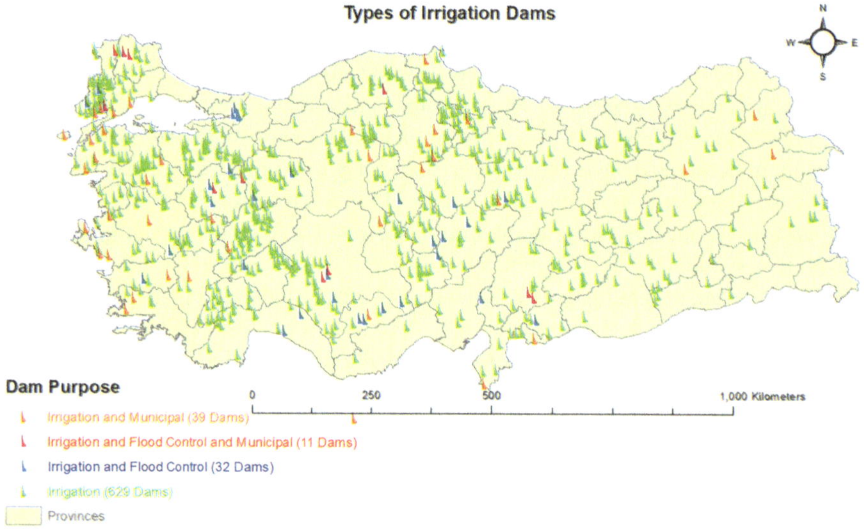

Fig. 2 Projection of 711 irrigation dams including single and multi-purpose (not hydropower dams) over 81 provinces in Turkey by 2018

consumed through irrigation. Turkey has 25.85 Mha (million hectares) of irrigable area (4.3 Mha is currently being irrigated) and 106.6 km^3 per year total available water resources, most of the irrigation systems (96%) depend on dams [11]. According to that and by the fact of 711 existing irrigation and multipurpose dams in Turkey until 2018 which are projected on the map in Fig. 2 could be used to install hydropower plants and collect the highest amount of sustainable energy from these dams to avoid cost of constructing new hydroelectric dams [4], or thermal power plants which have a negative impact on the environment [12].

A prior research has shown that hydropower can be produced from 43 domestic water dams in Turkey, where the expected power is 173 GW/year [13]. Based on all the above, it should be possible to collect around 500 GW/year of power by choosing the most suitable from the existing 711 irrigation dams for hydropower development with the lowest investment cost according to the technical criteria that achieve the highest energy and the spatial criteria of best grid connection to about 692 existing power substations, which are projected on the map in Fig. 3.

A variety of studies planning the resources of the renewable energy have been undertaken using GIS multi-criteria modelling techniques. There have been many studies based on different criteria in Germany, Turkey and Switzerland that have successfully identified a number of potential sites for small hydropower production [12, 14, 15], and several studies have successfully evaluated a number of existing paths based on hypothetical path or evaluated the criteria for power lines routing for grid connection in Turkey and Spain [16, 17].

Unlike previous studies, in the present study modern GIS tools and fuzzy logic approach are used together to achieve the objectives by combining grid connection and dam criteria. A second step of spatial analysis is about electric route line optimization between the dam and the substation to avoid unbalanced results as in some energy projects where the cost of grid

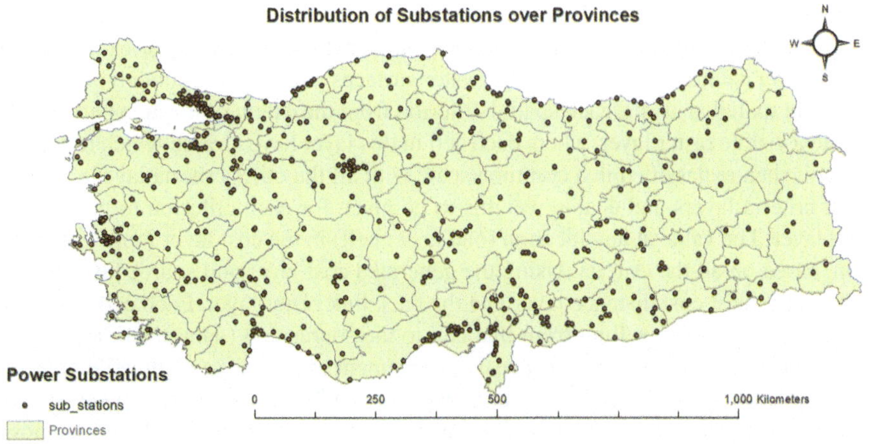

Fig. 3 Projection of 692 substations over 81 provinces in Turkey by 2018

connection exceeded the benefits of generated power [18]. Then by using fuzzy logic approach, the selected dams were ranked according their suitability degree of meeting multi-criteria which include technical, spatial and environmental risks. A second phase of spatial analysis was applied by finding the least cost path to link the energy produced from the selected most suitable dams to existing electricity grid according to criteria that are related to the cost of establishing and maintaining new power lines, the lands deduction costs, EIA requirement and the natural and artificial obstacles.

3 Methodology

A geodatabase containing the following objects was built using (ESRI) ArcGIS software to apply database query and multi-criteria spatial analysis and to save the resulting data. The data table of 751 dams (not hydropower dams) with the following fields: dam name, start operation year, river, province, purpose, dam body type, reservoir capacity, dam height from the base, dam height from the thalweg (river bed), normal water elevation at sea level, dam crest elevation at sea level, full spillway and dam (longitude X, latitude Y and elevation Z) geographic coordinates. In addition to the data table of 692 substations with the following fields: substation name, operational status, voltage level and substation (longitude X, latitude Y and elevation Z) geographic coordinates. And map of Turkey provinces using World Geodetic System (WGS) 1984 Lambert Conformal Conic geographical coordinate system.

To select the suitable irrigation dams for hydropower development, the following technical and spatial criteria were defined for geodatabase query: dam height from the base ≥15 m, reservoir storage capacity ≥3 hm^3, dam years in operation <80, dams with irrigation purposes, in addition to the spatial criterion of maximum 40 km straight line distance (Break-Even distance) from each dam's nearest substations. The criterion of nearest substation distance was used to ensure that the benefits of potential power are consistent with the cost of grid connection. For the grid connection spatial analysis, step-up transmission substations in operation with a voltage of 154 (kV) and higher were selected.

After selecting the suitable dams, it is important to determine which are most suitable, especially in view of the variation in the criteria values for the selected dams. The selected dams are arranged based on their suitability scores into three categories (high, medium and low) using a multi-criteria fuzzy logic that calculates the score separately for each criterion and then aggregates them into an overall suitability score. Six criteria are assessed: normal water level, reservoir storage capacity, dam purpose, years in operation, nearest substation distance and environmental impact assessment requirement. The equation below was derived to calculate normal water level (NWL) field values from the available data, where:

$$\text{NWL} = \left[\text{dam height-(crest elevation-normal water elevation)}\right]$$

Based on the criteria that achieve least cost, lowest power losses and lowest environmental impact for power lines routing, it is possible to find the best grid connection path from any of the most suitable dams to its nearest substation using the multi-criteria GIS spatial analyst, where the following spatial criteria were selected: shortest distance, suitable elevation, least slope, near to roads and near to existing power lines, which are related to the risks, lands confiscation and costs of establishing and maintaining new power lines. An important criterion related to land costs, lands confiscation and natural and artificial obstacles, which is land use, was also taken into account, including avoiding or excluding certain types of public and private property, avoiding natural reserves and large water surfaces and crossing some types of lands.

Using GIS multi-criteria spatial analysis, each spatial criterion from the extent of the area between the dam and its nearest substation translated into a visible raster layer, where each cell in this layer have a value which reflects its criterion suitability degree. This raster layer is then reclassified in a common scale with all the other criteria layers. This provided the ability to combine with the other criteria layers and enable weights to be given for each criterion to produce a new layer called the suitability layer or cost layer. The value at any location (one raster cell) in the cost layer represents the cost or the suitability degree of that location, based on all the criteria. Using the generated cost layer and source layer (dam position), two layers were produced, which are the least cost distance layer and the back link raster layer. Each cell in the cost distance layer represents the lowest cumulative cost of travelling from that cell to the source cell (dam position), while each cell in the back link layer represents the direction code for the lowest cost of movement to one of the eight neighbour cells back to the source cell (dam position). Using least cost distance layer, back link layer and destination layer (nearest substation position), the least cost path raster layer (best path) was calculated. This layer represents the path with the lowest possible cost of directing a new power line from the dam to its nearest substation, based on all criteria. The path avoids steep slopes and certain types of lands that are considered costlier for constructing power lines; the path also seeks to be near to roads and existing power lines and in a preferable land use. The path raster was converted to vector feature class and projected on 3D surface of satellite imagery of the area to enable controlling the design of power line route, placement of pylons, apply topology rules of power lines routing and visualize the path measurements in the geodatabase table of the path.

4 Case Studies

Two from the selected suitable dams were chosen as case studies, Karadere and Karaçomak. For these dams, technical and spatial criteria are evaluated and the potential power and economic benefits are estimated. A least cost path methodology was applied to find the best grid connection path for each of these dams to its nearest substations.

4.1 Karadere Dam

The dam is only for irrigation purpose with 11 years in operation, 90 m height from the foundation, 64 m normal water level, 26.08 hm^3 reservoir capacity and 12.07 km distance to nearest substation. The dam got 78% overall suitability score; this dam was therefore from the high suitability class. By applying the methodology for finding the best path for grid connection, the least cost path to nearest substation was found with a length of 12.789 km, which is more than the straight line distance of 12.07 km (Fig. 4).

4.2 Karaçomak Dam

The dam is for irrigation, flood control and municipal purposes with 42 years in operation, 70 m height from the foundation, 43 m normal water level, 23 hm^3 reservoir capacity and 8.56 km distance to nearest substation. The dam got 58% overall suitability score; this dam was therefore from medium suitability class. By applying the methodology for finding the best path for grid connection, the least cost path to nearest substation was found with a length of 8.607 km, which is more than the straight line distance of 8.56 km (Fig. 5).

5 Conclusion

A total of 265 of 711 irrigation dams were found to be suitable for hydropower development, and 183 suitable substations were found within a distance of 40 km from those dams for the purpose of grid connection (Fig. 6).

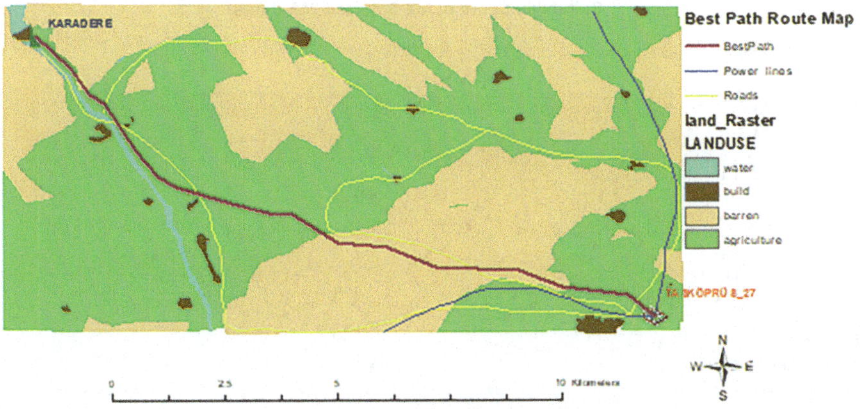

Fig. 4 Least cost path for grid connection between Karadere dam and Taşköprü substation

Fig. 5 Least cost grid connection path between Karaçomak dam and Kastamonu substation

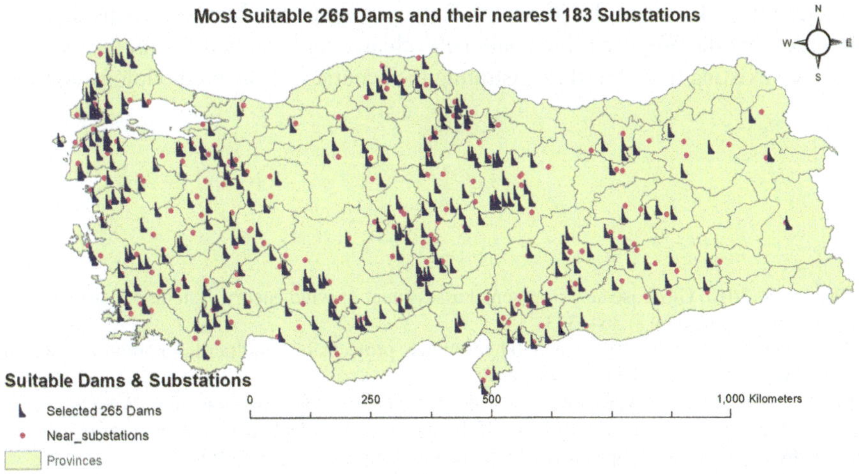

Fig. 6 The selected 265 suitable dams and their nearest 183 substations within 40 km

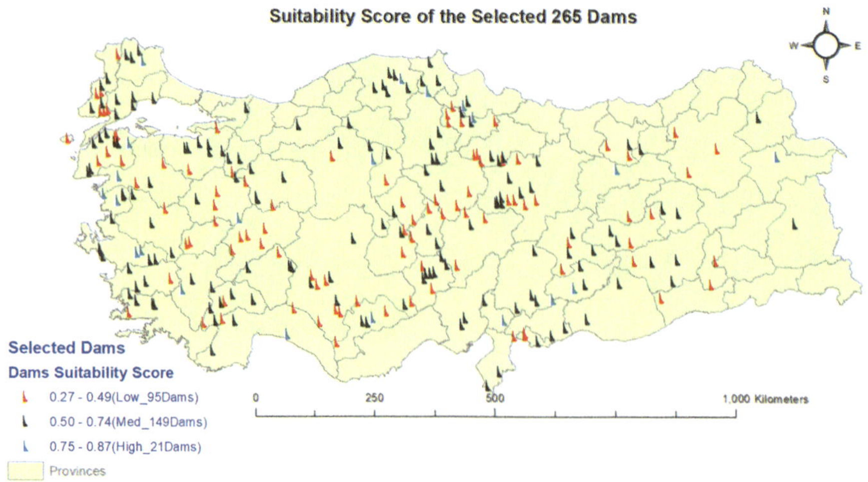

Fig. 7 Distribution of the selected most suitable 265 dams with their suitability classes

A variation was observed in the preferences for any selected dam through the results of the statistical indicators of its criteria. For example, the dam with the highest preference for the criterion of distance to the nearest substation, which is 720 m, had low preferences in relation to its other criteria where the reservoir capacity was 3.7 hm^3. Therefore, the selected dams were arranged according to their overall suitability score (Fig. 7). Three categories were identified: 21 dams with high suitability (suitability score greater than 75%), 149 dams with medium suitability (suitability score between 50% and 74.9%) and 95 dams with low suitability (suitability score of less than 50%).

By comparing the two case studies, it can be observed how the differences in suitability score effect the potential power and the annual benefit for each dam. It can also be noted that the length of the proposed least cost grid connection path is related to the straight line distance from the dam to its nearest substation where the least cost path is close to the straight line distance when the substation is close to the dam, as in the Karaçomak case study (Fig. 5), and it is more different from the straight line distance when the substation is far from the dam, due to the effect of taking the spatial criteria into account and avoiding obstacles in wide areas, as in the Karadere case study (Fig. 4).

The criterion of near to rivers was not one of the criteria examined while finding the best grid connection path in this study. However, in Karadere dam case study, the proposed path was found close to Karadere river. This can be attributed to the other criteria that take into account the important factors that affect the cost of grid connection such as land confiscation and slope, which always have low values near rivers in Turkey. This result can be considered a good indicator of the accuracy of the method (Fig. 4).

In Karaçomak dam case study, the criterion of near to existing power lines was deliberately not taken into account while finding the best grid connection path, due to the presence of an existing power line nearby the dam and goes to its nearest substation. This was in order that the proposed line could be evaluated and compared with the existing line. The proposed best path almost matched the existing line, thus demonstrating the accuracy of the methodology and criteria used to find the best route (Fig. 5).

References

1. Aydin NY, Kentel E, Duzgun S (2010) GIS-based environmental assessment of wind energy systems for spatial planning: a case study from Western Turkey. Renew Sust Energy Rev 14(1):364–373
2. Adhikary P, Roy PK, Mazumdar A (2015) Selection of small hydropower project site: a multicriteria optimization technique approach. ARPN Journal of Engineering and Applied Sciences 10(8):3280–3285
3. Haktanir T, Cobaner M (2007) Feasibility of hydropower plant installation to existing irrigation dams. Ware Int. 32(2): 254–264
4. Fitzgerald N, Arántegui RL, McKeogh E, Leahy P (2012) A GIS-based model to calculate the potential for transforming conventional hydropower schemes and non hydro reservoirs to pumped hydropower schemes. Energy 41: 483–490
5. ESRI - Environmental Systems Research Institute (2005) Using ArcGIS spatial analyst. Esri Press, USA
6. Sánchez-Lozano JM, García-Cascales MS, Lamata MT (2016) GIS-based onshore wind farm site selection using Fuzzy multi criteria decision making methods. Evaluating the case of Southeastern Spain. Appl Energy 171:86–102

7. AK M, Kentel E, Kucukali S (2017) A fuzzy logic tool to evaluate low-head hydropower technologies at the outlet of wastewater treatment plants. Renew Sust Energy Rev 68:727–737
8. European Environmental Agency (2017) Turkey land cover 2012, Country fact sheet. https://www.eea.europa.eu/themes/landuse/land-cover-country-fact-sheets/land-cover-country-fact-sheets-2012/tr-turkey-landcover-2012.pdf/view
9. MENR - Ministry of Energy and Natural Resources of Turkey (2009) The Strategic document of Electricity Energy Market and Security of Supply. Ankara
10. OECD - Organization for Economic Co-operation and Development (2009) OECD Fact book ISBN 92-64-05604-1. http://oberon.spourceoecd.org
11. Cakmak EH (2010) Irrigation management and water pricing in Turkey. Int J Soc Ecol Sust Dev (IJSESD) 1:13–26
12. Kucukali S (2014) Finding the most suitable existing hydropower reservoirs for the development of pumped-storage schemes: an integrated approach. Renew Sust Energy Rev 37:502–508
13. Kucukali S (2010) Hydropower potential of municipal water supply dams in Turkey: a case study in Ulutan Dam. Energy Policy 38(11):6534–6539
14. Schaefer P, Thapa P, Schlüter J, Palt S, Vennemann P (2011) Identification of potential pumped-storage sites in large areas based on a multi-criteria GIS-model, Proc., HYDRO 2011 Conference, Prag
15. Bousquet C, Samora I, Manso P, Rossi L, Heller P, Schleiss AJ (2017) Assessment of hydropower potential in wastewater systems and application to Switzerland. Renewable Energy 113:64–73
16. Dedemen Y (2013) A multi-criteria decision analysis approach to Gis-based route selection for over head power transmission lines. Middle East Technical University, Ankara
17. Monteiro C and others (2005) Gis spatial analysis applied to electric line routing optimization. IEEE Transactions On Power Delivery 20(2): 934–942
18. Kumar P. (2015) Nepal-scaling up electricity access through mini and micro hydropower applications : a strategic stock-taking and developing a future roadmap. Washington, D.C. : World Bank Group.

Large-Scale H_2 Production By Electrolysis Based on Renewable Electricity

Flávia Franco, R. P. Borges, Fátima Nunes Serralha, Paulo Martins, Ricardo Aguiar, Isabel Cabrita, and Paulo Partidário

1 Introduction

Portugal was the first country in the world to assume the compromise of achieving carbon neutrality by 2050. The European Union, in the framework of the Paris Agreement, has set up the Energy Union [1], 'to give EU consumers – households and businesses – secure, sustainable, competitive and affordable energy'. The Energy Union has assumed ambitious targets for the reduction of greenhouse gas emissions, to be achieved, in the period 2021–2030, through the implementation of National Energy and Climate Plans (NECP) by all member states. In the NECP [2], Portugal made compromises on several mandatory targets to be achieved by 2030: 47% renewables in gross final energy consumption (30.3% in 2018), 20% renewables in transport (9.0% in 2018), 80% renewables in electricity generation (52.2% in 2018) and 38% renewables in heating and cooling (41.2% in 2018) [3]. After a preliminary study [4], Portugal has recently adopted a National Hydrogen Strategy (NHS) [5] that serves as a complement to the National Energy and Climate Plan (NECP). The major targets of the NECP are not changed by the National Hydrogen Strategy, however the introduction of the H_2 vector will provide a wider range of technological options for achieving those targets and simultaneously improve the integration and resilience of different (and hitherto separate) parts of the energy system. Hydrogen is a key energy vector in the efforts for decarbonizing modern economies that are strongly dependent on the use of fossil fuels. One of the major attractions of this fuel is the potential to promote sector coupling between the gas, electricity and transport infrastructures.

One of the anchor projects announced in the National Hydrogen Strategy is a large-scale renewable H_2 production plant, to be located in the Sines region. The renewable H_2 plant is expected to function on the basis of water electrolysis, with the power to the electrolysers being supplied by a dedicated renewable energy infrastructure. This chapter presents a study of the technical e-economic operation of such plant [2].

2 Method

The simulation of the functioning of the plant is based on the *energyPLAN* software [6]. *EnergyPLAN* allows the simulation of an energy system during a year with 1-hour interval, although the software provides no spatial resolution so that the system is studied as a single point in space. *EnergyPLAN* is an appropriate tool for modelling energy systems with a high penetration of renewables and it has recently been used in a number of studies on the Portuguese energy system: for analysing the technical and economic feasibility of the substitution of solar photovoltaic generation for coal-fired power plants [7]; to

F. Franco (✉)
Escola Superior de Tecnologia do Barreiro, Instituto Politécnico de Setúbal, Setúbal, Portugal

Divisão de Estudos, Investigação e Renováveis, Direção-Geral de Energia e Geologia, Lisbon, Portugal

R. P. Borges · P. Martins · R. Aguiar · I. Cabrita · P. Partidário
Divisão de Estudos, Investigação e Renováveis, Direção-Geral de Energia e Geologia, Lisbon, Portugal

F. N. Serralha
Escola Superior de Tecnologia do Barreiro, Instituto Politécnico de Setúbal, Setúbal, Portugal

© The Author(s), under exclusive license to Springer Nature Switzerland AG 2022

A. Sayigh (ed.), *Sustainable Energy Development and Innovation*, Innovative Renewable Energy,
https://doi.org/10.1007/978-3-030-76221-6_79

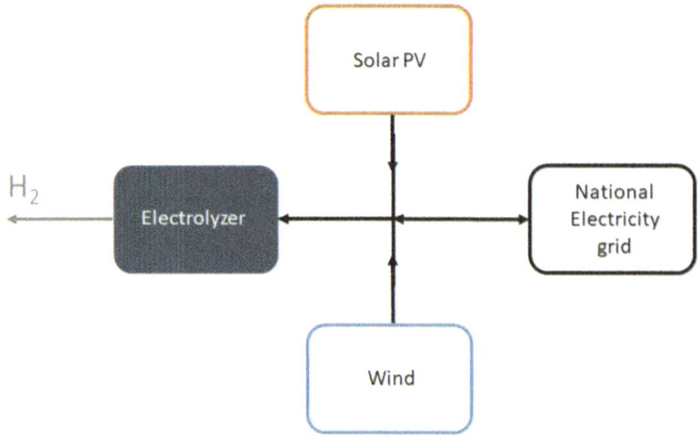

Fig. 1 Schematic diagram of the H_2 production plant and associated power sources

Table 1 Description of the conditions considered in each scenario

Scenario	Description	Electrolyser capacity (GW)	Solar PV capacity (GW)	Wind capacity (GW)	Weather conditions	Exchanges with electricity grid
S	Self-sufficient				Sunny/windy	Exports
L	Minimum electrolyser load of 20%	1–3	1–3	1–3	Sunny/windy	Exports and imports

forecast the utilization of storage capacity, in the context of an increase in the penetration of renewable [8] or to evaluate the extent to which electric cars can contribute to stabilize the electricity grid and obviate the need for gas power plants [9].

For this study, the H_2 plant is treated as a simplified electricity grid whose configuration is shown in Fig. 1. The electrolyser is powered either by dedicated solar PV or wind farms. Exchanges with the national electricity grid are possible in both ways. No assumptions are made about the national electricity grid that in the present model (when exchanges with the national grid are considered) is treated as a 'reservoir' of electricity to supply the needs of the H_2 plant or receive excess electricity production. In the text below, exchanges with the national electricity grid are referred to as 'imports' when the electrolyser uses electricity from the national grid, and 'exports' when production by solar PV or wind exceeds electrolyser load and this excess production is injected into the national grid. Electrolyser capacity is always considered as electric input capacity.

The operation of the plant is tested under two different weather conditions: a sunny year and a windy year. For that purpose, we simulate the production of each renewable technology by using the respective national production curves, obtained from REN (*Redes Energéticas Nacionais*), the national operator of the transport infrastructure and overall manager of the national electric system. Production curves for characteristic years with those weather conditions were used: the year 2012 for a windy year and 2016 for a sunny year.

The combination of electrolyser capacity, RES capacity, operation regimes and weather conditions provide 72 different configurations. Capacity values for the electrolyser and the RES sources are taken as integer multiples of 1 GW. The overall characteristics of each scenario are summarized in Table 1. Each one of them was analysed from the point of view of total annual H_2 production, exchanges with the electricity grid, equivalent full load hours and levelized cost of hydrogen (LCOH).

3 Technical Analysis

In scenario S, the operation of the plant is analysed considering a self-sufficiency mode of operation. The electrolysers use only the power provided by the associated renewable energy sources. If there is no renewable power, the electrolysers are switched off; if renewable power exceeds the nominal electrolyser capacity by more than 50%, the excess electricity is exported to the electricity grid.

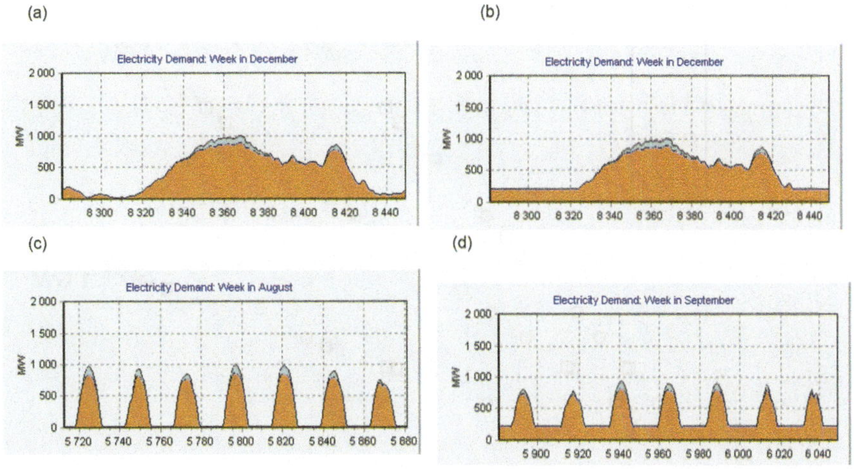

Fig. 2 Renewable energy production (light blue) and electrolyser load (orange). For wind production: (**a**) in self-sufficiency mode (scenario S) and (**b**) with minimum electrolyser load of 20% of nominal capacity (scenario L). For solar PV production: (**c**) in self-sufficiency mode (scenario S) and (**d**) with minimum electrolyser load of 20% of nominal capacity (scenario L)

In scenario L, the operation of the plant is analysed considering that the electrolysers must operate at a minimum load of 20% of nominal capacity. The same operating conditions of scenario S are used except that now the electrolysers are not allowed to switch off. If power from RES falls below 20% of electrolyser nominal capacity, the system is allowed to import from the grid the necessary power to maintain a minimum of 20% load at the electrolyser.

Figure 2 shows a set of characteristic production and load curves of the system in each scenario and for each electricity production technology. The use of solar PV as the RES for feeding the electrolysers requires a strong dependence on the national electricity grid if the electrolysers are not to be turned off during night-time (see Fig. 2c, d). This problem is relevant from the point of view of choosing the appropriate electrolyser technology. While alkaline electrolysers are capable of switching on and off regularly, they have a slow response time (ca. 1 hour to restart). PEM electrolysers, on the other hand, have fast response time (fluctuations in capacity in seconds and less than 5 minutes to cold start) and are best adapted to variable power capacity systems. Nevertheless, to avoid degradation of efficiency and lifetime, electrolysers are best operated in a continuous mode (avoiding full shutdowns) and with variable capacity kept above 20% of the installed capacity. From the point of view of the simulation, the choice of the electrolyser technology is not a relevant problem, the significant parameter is the conversion efficiency, which in this study was set at a value of 73%.

The total annual production of the H_2 plant depends on the combination of all the different factors. The plot of the total annual H_2 production as a function of electrolyser full load hours (Fig. 3) shows that the results split very neatly along three lines, one for each value of electrolyser capacity. It should be noted that the points on the graph include all values of renewable capacity considered in this study. In general, feeding the electrolysers with wind power provides higher production values. Also, the operation in self-sufficiency mode (scenario S) is generally less productive, with the exception of a few combinations that use wind power as the source of electricity.

Total H_2 annual production is not the exclusive figure of merit for comparing the different combinations of capacities, technologies and operating conditions. A decision on implementation must be based also on an economic analysis of the plant operation, and the ability to produce renewable hydrogen at a cost that makes it competitive with current production processes.

4 Economic Analysis

Economic analysis of the H_2 production plant was based on an estimation of annualized investment costs and annual operation and maintenance costs (O&M) associated with the equipment (electrolysers and renewable electricity generators) and variable costs of water and grid electricity. A discount rate of 3% and an average annual inflation of 1.5% were considered. Investment costs and O&M costs for 2030 were adapted from the EU Reference Scenario cost projections. The investment and O&M costs considered in the study are represented in Table 2.

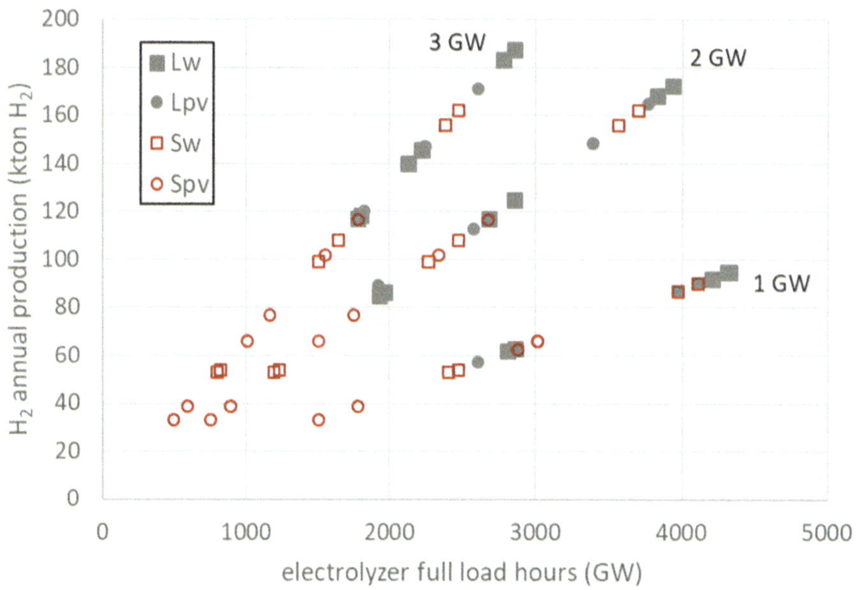

Fig. 3 Total annual H_2 production for different scenarios and RES technology. In the legend, the capital letter identifies the scenario of self-sufficiency operation (S) or minimum 20% load (L). The small letters identify the RES source in each case: wind power (w) or solar photovoltaic power (pv)

Table 2 Investment and O&M costs for the technologies considered in this study

Technology	2030 Investment costs (k€/MW)	2030 O&M (k€/MW)	Duration of the equipment
PEM electrolyser	466	15.0	50,000 hours
Wind onshore	1159	14.0	25 years
Solar PV	550	12.6	25 years

Table 3 Variable costs considered in this study

	2030 Variable costs
Water	2.37 €/m^3
Access to water supply	25,000 €/year
National grid electricity	38 €/MWh
Access to electricity network for export of excess production	0.5638 €/MWh

Table 3 provides information on the different variable costs considered in the analysis. It is assumed that the water used by the electrolysers is of drinking water quality. Water costs were obtained from the 2020 consumer price of drinking water supplied by Sines water utility company (*Águas de Santo André*) and by applying the average annual inflation rate of 1.5% to 2030. The same procedure was used to estimate the annual cost of connection to the water grid [10]. The cost of electricity from the grid was based on contracts in the futures market of MIBEL for 2030 [11].

The levelized cost of hydrogen (€/kgH_2) is obtained by dividing the total costs (annualized investment costs, O&M and variable costs) by the annual H_2 output for each configuration. The results are presented as a function of annual production in Fig. 4, where all different combinations of capacities and technologies are displayed. It is possible to observe that electrolyser investment and operation costs are the determining costs of the project, as the LCOH values spread over three clearly separated regions.

According to IRENA, the production costs of renewable hydrogen should be lower than 2.5 USD/kgH_2 for it to be competitive with blue hydrogen [12]. Figure 4b shows that there is a number of technological combinations that result in production costs between 1.5 and 3 €/kgH_2. These are achieved only for configurations that use 1 GW of electrolyser capacity, and as can be seen from Fig. 4a, the electrolyser cost is the determining factor in achieving competitive costs of production. An

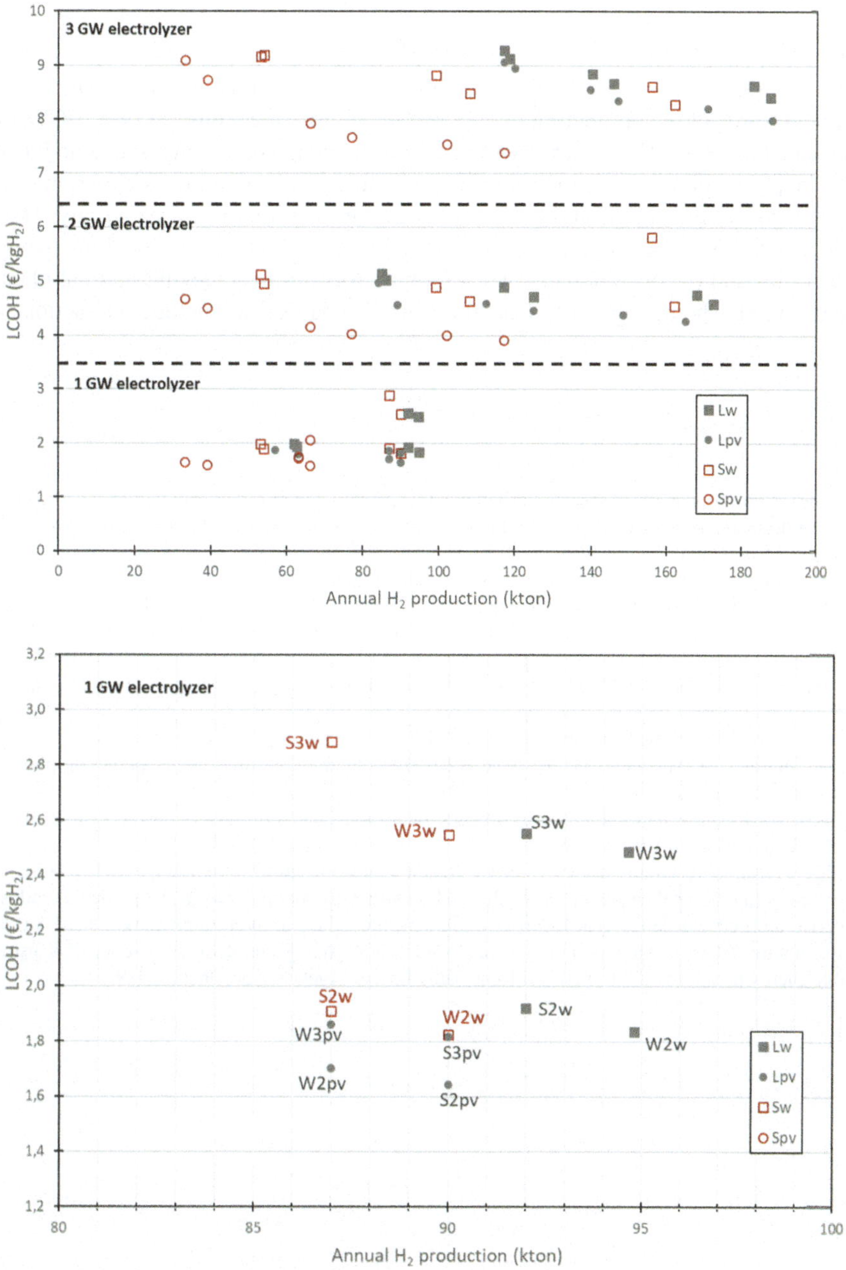

Fig. 4 Levelized cost of hydrogen (LCOH) as a function of total annual H_2 production for the different scenarios and RES technology. In the legend, the capital letter identifies the scenario of self-sufficiency operation (S) or minimum 20% load (L). The small letters identify the RES source in each case: wind power (w) or solar photovoltaic power (pv). (**a**) LCOH of all configurations of the H_2 plant considered in this study; (**b**) enlarged representation of the area inside the dotted box of (**a**). The capital letter in the data labels represent the weather condition (S for sunny year and W for windy year), the number and small letters represent the capacity and technology that provide electricity. So W3pv stands for LCOH of the configuration with 3 GW of solar PV power in a windy year

electrolyser with 1 GW, combined with 2–3 GW of renewable solar PV power, assisted by imports from the grid, allows production at the lowest cost (1.6–1.9 €/kgH_2). A combination of 1 GW electrolyser with 2 GW wind power and electricity imports from the national grid achieves the largest production (92–95 kton H_2 per year) at a low cost (1.8–2.0 €/kgH_2). Figure 4b is a magnification of the area inside the rectangular dashed line of Fig. 4a and provides a map of the area where equipment characteristics and operation can be fine-tuned to further optimize H_2 production and costs.

5 Conclusion

We show the possibility of industrial scale production of renewable H_2 at a competitive cost. The best options from the point of view of the economic rewards of the production of H_2 seem to be those that consider only 1 GW of electrolyser capacity and 2 GW of RES dedicated capacity. The electrolyser represents the main cost of investment and its capacity must be carefully adjusted to make the production cost of renewable H_2 competitive with other production routes, namely those based on fossil fuels. Production costs are not significantly affected by the limited use of grid electricity. The main conclusions are twofold: (i) methodologically, it is shown that *energyPLAN* provides a useful tool for modelling the operation of such large-scale H_2 production facilities and (ii) operationally, when looking forward to achieving optimal plant design and operation conditions, its electricity supply should rely on dedicated renewable sources and include interactions with the grid.

References

1. Communication from the Commission to the European Parliament, the Council, the European Economic and Social Committee, the Committee of the Regions and the European Investment Bank a Framework Strategy for a Resilient Energy Union with a Forward-Looking Climate Change Policy /* COM/2015/080 final */
2. https://ec.europa.eu/energy/topics/energy-strategy/national-energy-climate-plans_en
3. Directorate General for Energy and Geology, https://www.dgeg.gov.pt/pt/estatistica/energia/indicadores-energeticos
4. Partidário P, Aguiar R, Martins P, Rangel CM, Cabrita I. Int J Hydrogen Energ. Available online 13 Nov 2019. https://doi.org/10.1016/j.ijhydene.2019.10.132
5. Resolução do Conselho de Ministros n° 63/2020 de 14 de Agosto, Estratégia Nacional para o Hidrogénio, https://dre.pt/application/conteudo/140346286
6. EnergyPLAN, Advanced energy system analysis computer model, https://www.energyplan.eu/
7. Figueiredo R, Nunes P, Meireles M, Madaleno M, Brito MC (2019) Replacing coal-fired power plants by photovoltaics in the Portuguese electricity system. J Clean Prod 222:129–142
8. Ramos R, Castro R (2020) Modeling renewable energy integration in the 2030 Portuguese power system: the role of energy storage. J Electrochem Energy Conv Stor 17(1):011013. (9 pages)
9. Nunes P, Brito MC (2017) Displacing natural gas with electric vehicles for grid stabilization. Energy 141:87–96
10. de Santo André Á, the prices to industrial customers were checked on 9/9/2020 in http://www.adsa.pt/clientes-industriais/
11. MIBEL, Mercado Ibérico de Eletricidade (Iberian Electricity Market), the values in the futures market were checked on the 9/9/2020 in the market operator OMIP in https://www.omip.pt/pt/dados-mercado?date=2020-07-08&product=EL&zone=ES&instrument=FTB
12. Hydrogen: a renewable energy perspective, International Renewable Energy Agency, Abu Dhabi, 2019

Hydrogen: From a Breakthrough Innovation Towards a Wide Adoption

Paulo Partidário

1 An Energy Transition Is Under Way

The World is progressing towards a global deep decarbonised society where without major advances on the reduction of greenhouse gas (GHG) emissions, net-zero emissions targets will hardly be achievable. The global energy value chain is thus being challenged to embracing a new energy transition that regards a system wide innovation-based transformational switch away from fossil fuels to renewable energy sources (e.g. solar, wind, water, geothermal, sea waves). That innovative transformation has already shown evidence of how it is disrupting the *status quo* along a path where not one but a mix of renewable energy sources and green energy vectors are emerging according to desirably sustainable contexts and leading to a structural and permanent change in supply, demand, energy mix and prices. Two primary enablers – renewables and electrification, are supporting this push towards a more sustainable future.

Innovation processes are key to fostering advances in technology for decarbonisation, from new ideas to prototypes, and from making move them – within demonstration pilots, to the context of early adopters. The early adoption stage is critical for the success or failure of a given technology. At this stage, there are still cost and performance gaps – when compared to mature well-established competing technology, which call for policy attention while the deployment process evolves. That is the case how electricity from solar energy is currently becoming the cheapest ever, thanks largely to technological improvements and policies that reduce the risk of investing in renewable energy [1].

More than a decade has past following to Marchetti's keynote on 'Half a century of pushing the Hydrogen Economy idea, some personal memoirs' [2]. By then, based on his studies of the history of basic inventions and innovations in the context of the social and economic system, we were called to remember that: '*…ideas, experiments, diffusion, and success are not only quantifiable in time with neat logistic equations, but also occur under a periodicity of about 55 years called a Kondratiev cycle*' [3], which might be represented in Fig. 1.

2 The Innovation Challenge and Its Scale

This energy transition, and the driving forces therein of cheaper renewable technologies and more aggressive government policies, are disrupting the entire fossil fuel system, with profound consequences for financial markets and geopolitics [4]. That is the particular case of a strong and unprecedented momentum being witnessed currently, as both governments and private players speed up their plans to adopt H_2 as a green or low-carbon energy vector despite COVID-19 creating challenges along the path.

At this stage, it is useful to revisit the disruptive innovation definition, as H_2 is currently providing all the elements related to this particular innovation type [5]. Not being radically new, it enables a significant change at a system level through the combination of both existing and new system elements, and challenges as well users/adopters, in order to change significantly their behaviour to use that innovation. By doing so, H_2 is leading to the creation of new markets and value networks,

P. Partidário (✉)
General of Energy and Geology – Research & Renewables Division, Lisbon, Portugal
e-mail: Paulo.Partidario@dgeg.gov.pt

© The Author(s), under exclusive license to Springer Nature Switzerland AG 2022

A. Sayigh (ed.), *Sustainable Energy Development and Innovation*, Innovative Renewable Energy,
https://doi.org/10.1007/978-3-030-76221-6_80

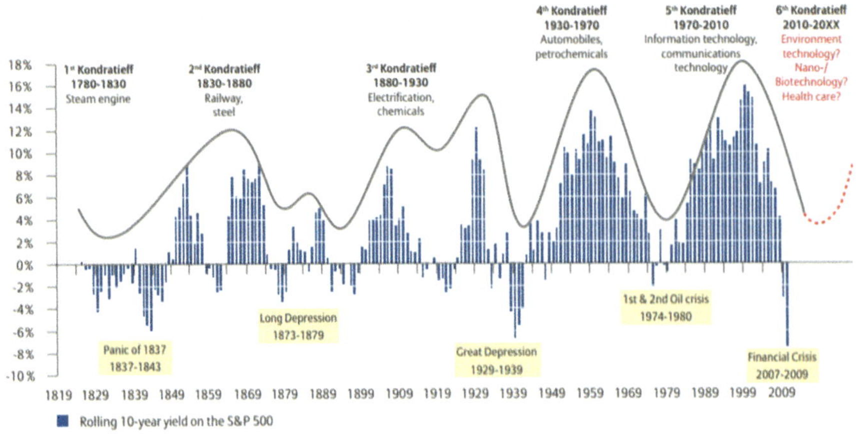

Fig. 1 Kondratiev long waves and financial returns – prosperity cycles and recession valleys since 1814 till March 2009, in (%) per annum

as well as to disrupting an existing market and value networks, putting in danger or displacing established market leading firms, products and alliances.

That being said then both competitiveness and technology changes need to be managed within a strategic framework and well aware of a frequent problem with conflating a disruptive innovation with any type of breakthrough, which might change competitive patterns in a given sector. Adding to that, it is key to acknowledge not just that the technology-strategy interaction affects the firm and its resource-base, but its business model as well [6, 7]. Therefore, the H_2 approach has not just significant systemic opportunities but holistic requirements as well.

Having into consideration that a significant cost-effective decarbonisation can only be achieved through an integrated sectoral approach using both the electricity, gas and heat infrastructures, H_2 is effectively being recognised worldwide as a potential disruptive energy carrier, fuel and feedstock that will shape the global markets of tomorrow.

3 The Green Hydrogen Role in Energy System Integration

An integrated energy system means:

(a) A key prerequisite for cost-efficient decarbonisation and ensuring the security of supply, stability and flexibility of the energy system.
(b) Designing the energy system as a whole, not as a sum of individual sectors or energy carriers. It is about interlinking the various energy carriers (electricity, gas, liquids, heat and cold), as well as with the end-use sectors (buildings, mobility, industry, agriculture) for a more sustainable and cost-efficient use of energy.
(c) Building bridges between all energy carriers, using renewable electricity to produce decarbonised gases like H_2, and fuels which can play a key role in the hard-to-decarbonise sectors.
(d) A circular approach to the energy system, such as mobilising waste streams for heating, or production of renewable gases from waste.

H_2 is very often been referred as a key missing link in sector coupling and in energy system integration. Playing an important role at a system level, provided the scaling up of production will be achieved in order to bring down costs, it will be enabling: competitive transportation methods, the decarbonisation of natural gas, the grid balancing and seasonal storage, and the implementation of regulations (e.g. safety) and best practices.

It is considered a promising system facilitator to support a carbon neutral economy system. Such role has a priority request for renewable H_2 or, at least in a transition phase, for decarbonised H_2 made by combining production from fossil sources with carbon capture and storage (CCS).

4 Green Hydrogen from Renewable Energy Sources – How Far Are We Progressing?

There is a growing global consensus that green H2 will be key for decarbonisation, which might be expressed by an increasing number of hydrogen strategy roadmaps and concrete actions being announced by the major global economies.

The current and expected H_2 supply costs from solar and wind-based renewable energy production costs can be compared to fossil fuel options with CCS (Fig. 2). In the fossil fuel H_2 production, a small part of CO_2 is not captured in the CCS facility for which carbon prices were considered. On those calculations:

(a) Electrolyser costs considered, were the following: 770 US$/kW (2020); 540 US$/kW (2030); 435 US$/kW (2040); and 370 US$/kW (2050).
(b) Carbon dioxide costs considered, were the following: US$ 50 per tonne (2030); US$ 100 per tonne (2040); US$ 200 per tonne (2050).

Figure 2 shows the forecast for green H_2 production costs from low-cost solar and wind, which are expected in the next 5 years and for the best regions to keep consistently below those for low-carbon fossil fuel-based H_2. By 2035, average cost RES will also become competitive. Pricing of CO_2 emissions from fossil fuels further improves the competitiveness of green H_2. In the best locations, renewable H_2 is competitive in the next 3–5 years compared to fossil fuels.

Developing low-carbon H_2 production is often addressed in the international agenda. However – and above all, taking a lifecycle perspective in order to avoid the risk of carbon leakage and transfer of negative effects between different parts of the World, a fully clean H_2 production technological path is key to support a clean energy transition.

The two main H_2 production paths are considering: on the one hand, the conventional technology coupled to an end-of-pipe CCS approach, and on the other hand, the water electrolysis.

At present, 95% of H_2 production is based on emissions-intensive natural gas reforming and coal gasification, resulting in CO_2 emissions equivalent to the UK and Indonesia combined. In that respect, however, despite some countries are still adopting a low-carbon approach with a 'clean' H_2 approach via natural gas and CCS, for many others, the path is fully green, which means that H_2 is entirely produced from renewable power.

When produced from renewable electricity through electrolysis, H_2 has the potential to contribute to large-scale energy storage, especially longer term storage, and to the cost-effective integration of renewables from the electricity sector into other economic sectors, while supporting a more effective planning and operation of electricity, gas and heat infrastructures.

Electrolysers technology enables the production of H_2 either from low-carbon or from renewable electricity and water. In the latter, conditions are settled to obtain a fully clean H_2. During 2019, there was a record year for electrolysis capacity (Fig. 3) becoming operational, and expressive announcements were made for the upcoming years targeting different application sectors. Electrolysers are a well-known and long-used technology in a variety of industrial sectors. Among them, the fastest-growing market is for uses that serve energy and climate objectives: vehicle fuelling; H_2 injection into the gas grid; using H_2 as a cleaner input for industrial processes; electricity storage; and synthetic fuel manufacturing.

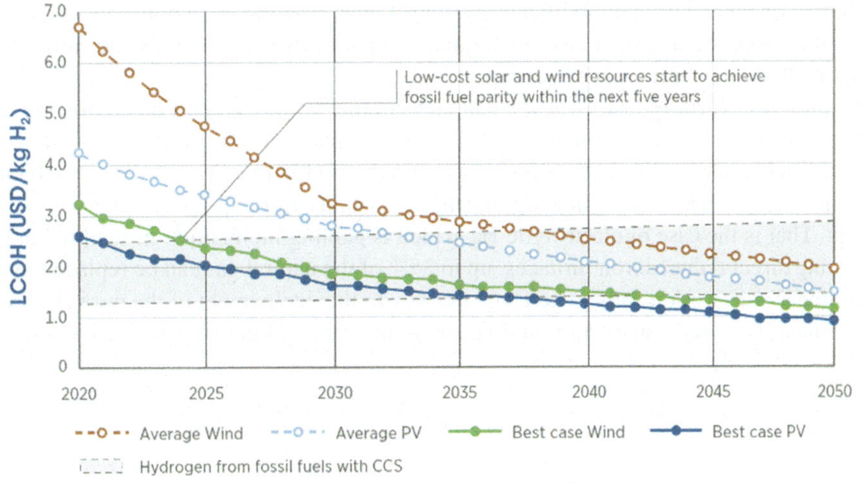

Fig. 2 The H_2 production costs from solar and wind versus fossil fuels (In: IRENA [8])

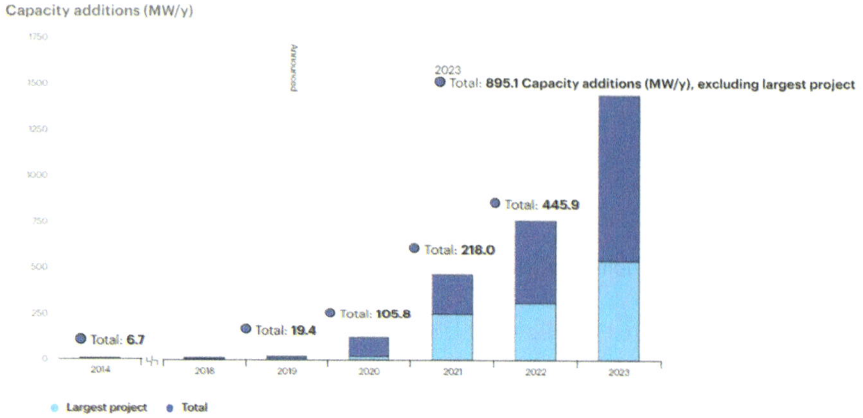

Fig. 3 The evolution of electrolysis capacity worldwide – A IEA analysis (2014–2023) (In: Adapted from IEA [1])

With its associated technologies such as fuel cells, H_2 constitutes a potential decarbonisation pathway for a wide range of end-use sectors. This is relevant in particular for the hard-to-decarbonise industrial sectors (e.g. steel, chemical, refining, etc.), parts of the heating sector, and heavy-duty or long-distance transport (buses, trucks, ships and rail). In these areas, direct use of electricity is less evident and a greater role for renewable gases or synthetic fuels is expected.

Regarding the fuel cell electric vehicle (FCEV) market, its deployment almost doubled in the period 2017–2019, catalysed by an outstanding expansion in China, Japan and Korea.

Regarding vehicles refuelling stations, 470 H_2 refuelling stations were in operation worldwide in 2019, which represented an increase higher than 20% when compared to the previous year. Japan is leading in number of refuelling stations in operation (113), followed by Germany (81) and the United States (64). Similarly, to FCEVs, in 2019, the number of refuelling stations increased threefold in China (from 20 to 61), giving China the fourth largest number of stations, followed by Korea and France. In non-road vehicles, new applications are gaining recognition. At the end of 2018, two fuel cell trains produced by Alstom became operational in Germany, and successful trials led to the announcement that another 14 will be put into service in 2021. The United Kingdom and the Netherlands have also shown interest in Alstom H_2 trains, and a fuel cell tram began operating in Foshan (China) in 2019, with China exploring further possibilities for H_2-fuelled rail.

H_2 injection into the gas distribution grid, like the injection of other renewable gases with the required input specifications, is a strong option for those economies using already existing infrastructures. Taking profit of the existing but adapted assets, H_2 and gas integration can ease the transition towards a deep decarbonisation, thanks to the ability of the gas grid to cover varying geographies and scales (e.g. clusters, regions, countries) as well as to the H_2 blending possibilities in the existing gas grid via centralised and decentralised supplying models. In addition, the gas infrastructure is able to transport larger volumes over longer distances at lower costs when compared to the electricity grid. In such a way, it is about an opportunity for the decarbonisation of hardly-to-decarbonise sectors (buildings, industry). Blending H_2 up to 20% on a volumetric basis into the gas grid requires minimal or potentially no modifications to grid infrastructure or to domestic end-user appliances. Beyond those blending levels, it is more challenging due to material incompatibilities at high pressures and a lower H_2 concentration tolerance in the blending that industrial users can accept.

H_2 is a way of helping transform industry for a clean and circular economy, and of mobilising research and fostering innovation. Industry is the sector having the highest H_2 demand, in particular for the refining, chemical and high-energy-intensive subsectors like steel making. Currently, those industries use high-carbon content H_2, which means that by replacing it with low-carbon content H_2 would consist in a significant opportunity to ramp up demand while decreasing GHG emissions in the short term. That is the case of electrolytic H_2, which is gaining momentum in steelmaking. Without making any major changes to existing direct reduced iron furnaces, up to 35% of the natural gas can be replaced by H_2. Several steelmakers are pursuing blending as a transition strategy to forge the way to the deployment of the pure H_2 direct reduced iron route, for which a large pilot plant is under construction in Sweden, in the frame of the HYBRIT project (Fig. 4).

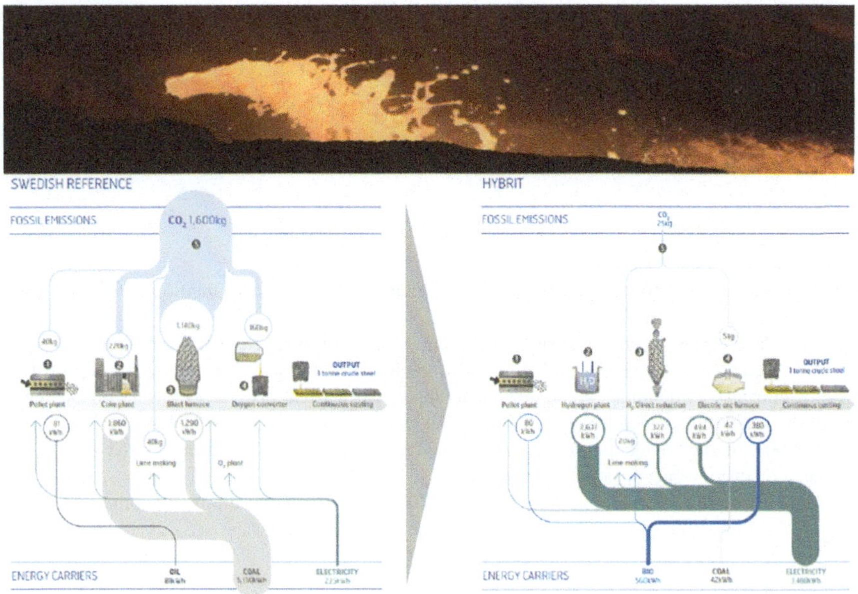

Fig. 4 The HYBRIT project: The fossil-free steel goal by the SSAB/LKAB/VATTENFALL partnership (In: Adapted from Vattenfall [9])

5 Concluding Remarks

Green H_2 produced from wind and solar electricity could become a game changer in global efforts to decarbonise the economy, and adding to that, falling renewable costs, a clear policy direction, as well as a significant private participation could pave the way for green H_2 to change the current GHG emissions landscape.

The H_2 value chain is able then to emerge, with different configurations and variants according to the application markets, and where multiples risks, especially for new network infrastructure and first movers, need to be addressed adequately. There are currently few applications for which low-carbon H_2 are a cost-effective fuel or feedstock. Therefore, the incentives for investment are low and the path to cost reductions and competitiveness is unclear. In addition, there is a strong link between the widespread production and use of low-carbon H_2 and climate change ambition, which will continue to be its main driver. Along each particular chain, and depending on the maturity of the technology, it is possible to identify contrasting TRLs that determine either more R&D efforts, strategic niche approaches, or mainstream conditions.

To have real progress on H_2 approaches, it is key to set clear goals in the long-term policies and strategies, in order to provide investors with the required conditions (stability, predictability) to analyse opportunities and strategies, and finally plan future investments on H_2. Governments have a central role in building an effective environment for H_2 value chains to develop and get into real-life conditions with adequate business models so that they will contribute to climate targets and other policy objectives (e.g. air quality, RES integration, circular economy and energy security). The setting up of a robust legal and regulatory framework is a key enabler of innovation, investment initiatives along the value chains, across the energy system, as well as on international cooperation. It is our strong belief that accelerating the innovation cycle is critical to reaching climate neutrality, therefore, the R&D funding needs to be applied along the TRL scale and assisted by a balanced technology mix and adequate cost-benefit analysis.

Regarding Portugal, within both the national and international framework different initiatives have already been taken, namely by having developed and launched the National Strategy (EN-H_2) in line with its NECP2030 and the CNR2050, by the approval of the legal and regulatory framework that allows H_2 production, distribution, transport and storage, by having planned up to 2.5 GW of electrolyser capacity planned until 2030, and by giving preparation steps for PT participation in different value chains, according to TRL, market availability and cost-benefit analysis. Public calls for project proposals will be launched on an annual basis, until 2030, and the first public call for expressions of interest was already launched: 74 proposals were submitted by partnerships incl. Industry, research centres and other players. After screening, 37 were selected to the next stage, involving investment of 79000 Milion Euro invested from companies until 2030, to which 40 M€ of public investment should be added. Auctions are being designed to support temporarily CAPEX and OPEX, to promote H_2

production in PT. Public investment will represent 900 M€ until 2030. International partnerships are being prepared at an EU level and beyond, as PT looks forward to teaming up with countries worldwide that are ready to establish robust and integrated partnerships.

References

1. IEA (2020) World energy outlook. www.iea.org/weo
2. Marchetti C (2007) Half a century of pushing the Hydrogen Economy idea, some personal memoirs, Keynote speech at the World Hydrogen Technology Convention, Montecatini Terme, November 6th
3. Marchetti C (1980) Society as a learning system: discovery, invention, and innovation cycles revisited. Technol Forecast Soc Chang 18:267–282
4. Carbon Tracker (2020) Decline and fall: the size & vulnerability of the fossil fuel system, carbon tracker report, June 4th
5. Christensen C (1997) The innovator's dilemma – when new technologies cause great firms to fail. Harvard Business School Press, Boston
6. Bower J, Christensen C (1995) Disruptive technologies: catching the wave. Harv Bus Rev 73(1):43–53
7. Walsh S (2004) Roadmapping a disruptive technology: a case study. Technol Forecast Soc Chang 71:161–185
8. IRENA (2019) Hydrogen: a renewable energy perspective, Sept 2019
9. Vattenfall (2019) HYBRIT, IEA Hydrogen meeting, Paris Feb 11st, 2019. www.hybritdevelopment.com/

Part IX
Hydrogen, Fuel Cells and Transport

An Analysis for Benefits of Shared Community Energy Storage for three Real Settlements in the UK

Sanjana Ahmed, Mark Sumner, and Arthur Williams

1 Introduction

As the world slowly transforms into a low-carbon economy, the advent of newer energy technologies, depletion of fossil fuel and regulations on climate change are causing the energy systems across the globe to go through a radical transformation. A large number of distributed energy resources (DERs) such as solar, wind and storage devices are making their way into the energy system [1, 2]. Community Energy is a development that allows local communities to optimize their energy system based on community owned and managed renewable energy technologies, onsite conditions and energy prices. A better understanding of the possible benefits of deployed DERs such as community energy storage systems is therefore required, along with an active energy management system within the control framework of community smart-grids, in order to integrate and optimize the operation of flexible resources within the energy communities. With new small-scale residential battery energy storage system (BESS) such as Tesla Powerwall providing higher capacity energy storage systems (e.g. 13.5 kWh), the way to maximize the benefits through community operation seems even more promising.

In [3], the authors present a community power flow control (PFC) strategy which reduces peak grid demand, and increases self-consumption of renewable energy, by controlling demand side management (DSM)-based direct load control and the battery storage, resulting in energy cost savings in smart communities with grid-connected photovoltaic (PV) systems. However, for the residential community model, the study relied on simulation of domestic demand created using the Centre for Renewable Energy Systems Technology (CREST) tool [4] to generate 100 residential power profiles. The energy prices are based on real-time pricing (RTP) which is not offered to electricity consumers in the UK as yet and therefore a realistic scenario cannot be derived from this tariff scheme.

Similarly, [5] also discusses the community power flow concept in designing a hierarchical control structure for optimal dispatch of distributed energy resources within different levels of the distribution network and optimizing the operation of flexible resources at the community cell level. The electricity demand model is again obtained through a bottom-up model created in MATLAB which generates demand profiles for every individual domestic load and the overall demand for a dwelling with a resolution of 1 minute. The active occupancy of residents is derived through a Markov Chain Technique which is then used to predict their appliance usage. The Heat Demand data and the Micro Combined Heat and Power (μCHP) data within the study are also derived through models based on other literature.

A similar study to this paper is undertaken by Panagiotou et al. [7], who quantify the financial benefits for end-users and energy management authority by simulating different operating scenarios for power flow to and from the battery storage and the elements of a community smart grid. The study uses real load data as CREST input for the consumption and generation profiles of the houses but bases the assessment on only a single community consisting of 8 houses. The battery modelling used is specific to lead-acid technology, and therefore cannot be applied to the general battery storage profiles or the increasingly popular and market-available larger capacity lithium-ion battery technologies such as the Tesla Powerwall. The tariff scheme used in this work is the Economy-7 [8–10], which has been previously used in other studies, however, this tariff

S. Ahmed
Watershed Technologies, North York, Toronto, Canada
e-mail: sanjana.ahmed@watershed.ca

M. Sumner · A. Williams (✉)
Department of Electrical and Electronic Engineering, University of Nottingham, Nottingham, UK
e-mail: mark.sumner@nottingham.ac.uk

© The Author(s), under exclusive license to Springer Nature Switzerland AG 2022

A. Sayigh (ed.), *Sustainable Energy Development and Innovation*, Innovative Renewable Energy,
https://doi.org/10.1007/978-3-030-76221-6_81

structure does not emphasize the difference in ToU rates of peak and off-peak hours and therefore neglects the potential of cost benefits through energy-shifting under such a scheme.

As such, a gap exists in current literature for using data from real communities rather than using data acquired from simulated models. Also, results have not been tested across different communities to validate the methods. This paper presents a management scheme for community battery storage systems that was developed based on analysis of real community load and PV data collected from a community in Corsham, UK and a currently available ToU tariff scheme known as the TIDE tariff provided by Green Energy UK [6]. The paper assesses the performance of the developed control framework in efficient energy shifting using data from two additional communities in the UK. Technical and economic evaluation is carried out over a period of 1 year, on these real communities, to demonstrate the impact of the proposed energy management.

This paper is structured as follows: Sect. 2 gives details of the communities whose data are used in this paper, Sect. 3 describes the examined energy management system, Sect. 4 gives details of the achieved benefits through energy-shifting and import cost savings due to the community BESS. In Sect. 5, comparisons are made with baseline scenarios where the community is considered to be operating on an energy plan with flat-rate tariff scheme and no battery and/or no PV generation, as opposed to the same community operating with the TIDE tariff and community power flow (CPF) battery management system (BMS) [12]. Section 6 assesses the operation of loads clustered as a community as opposed to individual users. Finally, Sect. 7 summarizes the conclusions.

2 Summary of the Communities

Consumption and PV Generation data for several properties in different communities were collected in order to form the inputs for the design and evaluation of the CPF battery management algorithm. Since the model was originally developed based on the real-life data from the Corsham community in the UK, further real-life households and communities are studied for the development and testing of the CPF controller algorithm taking into account different operating characteristic and limitations of each of the participatory DER.

The Corsham trial data were collected over a 1-year period from May 2015 to June 2016, which investigated the energy management of a community of 11 properties, comprising of 9 residential houses with a varying number of occupants, a school and an office building. Based on the findings, an algorithm was developed in the form of a community power flow (CPF) Controller which allows the capture of available energy from dispatchable sources for peak time usage as well as charging the battery at off-peak times to meet demand during peak periods. The aim is to generate economic benefits for the electricity end-users working together as a community.

A project by the University of Loughborough which analysed electricity use at 1-minute resolutions in 22 dwellings in Milton Keynes, UK, over the complete year of 2008 has contributed to the battery algorithm inputs of consumption and generation data [4]. For the measured data, each dwelling was fitted with a single high-resolution meter to monitor electricity use of the whole dwelling, whereby, the 1-minute resolution allowed the switching of individual domestic appliances to be observed. The meters measured true active power and a survey form for each dwelling detailed its characteristics and the types of appliances and lighting installed.

A second source of data came from collaboration with an energy consultant company (Carbon Legacy). This resulted in the acquiring of client data in the East Midlands, UK, for hourly consumption and PV generation. The annual data of the four residential properties investigated spanned over October 2016 to September 2017 monitored by PowerRouter from Nedap [11]. This set of data was very reliable in terms of consistency of data and unbroken communication to a storage server at mypowerrouter.com.

Due to the consistency and unbroken length of data, for each of these two latter data sets, analysis has been carried out for individual properties as well as the properties grouped together as a community to test and evaluate the validity of the developed CPF controller scheme on wider communities. To be considered as a community, the individual consumption and generation data were summed to give the combined kWh for the community, and the battery was sized and managed accordingly. The communities can be considered as 'virtual', since not all the properties within the communities are located in the same area geographically.

3 Design Methodology of CPF-BMS

The battery energy management scheme developed in this work is based on the concept of CPF [12] and relies primarily on the different Time-of-Day periods of fixed tariff levels offered by Green Energy UK TIDE scheme. This multi-level tariff scheme encourages consumers to buy electricity during off-peak periods. Those with energy storage can buy cheaper energy in advance to consume later during high tariff (peak) periods. Figure 1 shows the prices in p/kWh offered by TIDE scheme for the three different tariff bands. The management scheme also requires historical load and generation data of properties in the community and parameters defining the efficiency of the battery and power converter, along with the current state of charge (SOC) of the battery and its limits.

The battery in this work is modelled as a generic energy source and sink, with a finite capacity in Wh and a fixed one-way efficiency of 90%. The battery operation is limited by its state-of-charge (SOC) having to remain between a maximum of 90% and a minimum of 10% throughout the day after each charging or discharging instance. A bidirectional power converter is employed to bridge the gap between the AC grid and the DC battery. A converter model is therefore incorporated into the battery management algorithm using equations which represent converter loss in terms of converter loading, since losses can significantly vary with converter loading.

In this work, the historical data are considered to be available in advance for the whole year, which achieves 'perfect prediction' of excess available PV of the day and accordingly sets the SOC of the battery at different periods. Inputs to the BMS are also instantaneous load demand from individual properties or community, along with instantaneous PV generation. The net demand of the consumers is found by subtracting PV generation from load demand, as given by (1). When net demand is negative, there is excess PV generation which should either be stored in the battery for use during peak demand periods or exported to the grid if the battery is full. If the net demand is positive, the unmet load is supported either through battery discharge (during peak tariff period) or by the grid if the battery has reached its minimum SOC.

$$P_{\text{Net},t} = P_{\text{Load},t} - P_{\text{PV},t} \tag{1}$$

A weekday can be divided into four distinct segments based on the variable tariff levels, and the energy management algorithm for each period is described below, with Table 1 summarizing the role of battery and the grid at each interval. The supporting power flow equations denote the battery charging ($P_{\text{Batt,ch}}$), battery discharging ($P_{\text{Batt,dch}}$) and grid power (P_{grid}) during each interval where 't' is instantaneous time, d is the day in the year, P_{Net} is the net load, P_{Load} is the power demand and P_{PV} is the generated PV power.

- *00:00–07:00 hrs (Low Tariff)*

  (a) The grid must meet net load demand at this interval since import tariff is the lowest at this time. The equation for grid power is given by (2).

WEEKDAY			
13%	Low Tide	00:00 – 07:00	6.41p
24%	Tide Weekday	07:00 – 16:00	14.02p
19%	High Tide	16:00 – 20:00	29.99p
16%	Tide Weeknight	20:00 – 00:00	14.02p
WEEKEND			
5%	Low Tide	00:00 – 07:00	6.41p
24%	Tide Weekend	07:00 – 00:00	14.02p

Fig. 1 Multi-level TIDE tariff rates in p/kWh offered by Green Energy UK [15]

Table 1 Summary of the role of battery and grid in different time periods defined by TIDE tariff

Time-of-Day	Battery	Grid
00:00–07:00 hrs (low tariff)	Charge from grid to maximum SOC less excess PV energy available for the day. 	Charge battery Meet net load 
07:00–16:00 hrs (mid Tariff-1)	Charge from excess PV. 	Meet net load 
16:00–20:00 hrs (peak tariff)	Discharge to min. SOC Charge from excess PV. 	Meet net load only if battery SOC reaches minimum limit. 
20:00–00:00 hrs (mid Tariff-2)	Discharge any remaining capacity. Charge from excess PV. 	Meet net load only if battery SOC reaches minimum limit. 

$$P_{grid,t} = P_{Net,t} + P_{batt,ch,t} \tag{2}$$

(b) The grid must charge battery to maximum capacity less the predicted PV energy export for the day after 7 am, as given by (3). This will ensure that the battery has enough capacity remaining to accommodate the free 'export' PV energy available during the day. PV forecasting will play a major role here to determine how much of the battery capacity is to be left empty for PV charging throughout the day.

$$P_{\mathrm{batt}_{\mathrm{ch}},\mathrm{t}} = \frac{\left(\left(\mathrm{SOC}_{\max} - \mathrm{SOC}_{\mathrm{t}}\right) \times \mathrm{BatteryCapacity}\right) - \left(E_{\mathrm{export,d}}\right)}{7\,\mathrm{hours} - t} \tag{3}$$

(c) No battery discharge should take place during this interval since grid import prices are low, and because the battery should primarily hold charge to meet peak tariff load.

$$P_{\mathrm{batt}_{\mathrm{dch}},\mathrm{t}} = 0 \tag{4}$$

- *07:00–16:00 hrs (Mid-Tariff)*

(a) The grid should supply the net positive energy consumption of the community defined by (5).

$$P_{grid,t} = P_{Net,t} \text{ iff } P_{Net} > 0 \quad (5)$$

(b) Battery should charge up using any unused PV energy, that is, the PV exports defined by (6).

$$P_{batt_{ch},t} = P_{Net} \text{ iff } P_{Net} < 0 \ \& \ \text{Battery SOC} \leq 90\% \quad (6)$$

(c) No battery discharge should take place during this interval, as shown by (7), since the battery should primarily hold charge to meet peak hour load.

$$P_{batt_{dch},t} = 0 \quad (7)$$

- *16:00–20:00 hrs (Peak Tariff)*

(a) Battery should be discharged to the minimum allowed SOC to meet any load demand, defined by (8).

$$P_{batt_{dch},t} = P_{Net} \text{ iff } P_{Net} > 0 \ \& \ \text{Battery SOC} > 10\% \quad (8)$$

(b) The grid supplies load demand if and only if the battery has reached minimum SOC, or power exceeds maximum battery system power, given by (9).

$$P_{grid,t} = P_{Net,t} \text{ iff } P_{Net} > 0 \ \& \ \text{Battery SOC} < 10\% \quad (9)$$

(c) Battery should charge up using any PV exports, given by (10).

$$P_{batt_{ch},t} = P_{Net} \ iff \ P_{Net} < 0 \ \& \ \text{Battery SOC} \leq 90\% \quad (10)$$

- *20:00–00:00 hrs (Mid-Tariff-2)*

(a) Any remaining battery energy should be discharged to meet load.

$$P_{batt_{dch},t} = P_{Net} \text{ iff } P_{Net} > 0 \ \& \ \text{Battery SOC} > 10\% \quad (11)$$

(b) Charge battery if there is surplus PV

$$P_{batt_{ch},t} = P_{Net} \text{ iff } P_{Net} < 0 \ \& \ \text{Battery SOC} \leq 90\% \quad (12)$$

(c) Grid supplies load consumption if and only if battery SOC falls lower than 10%.

$$P_{grid,t} = P_{Net,t} \text{ iff } P_{Net} > 0 \ \& \ \text{Battery SOC} < 10\% \quad (13)$$

The simulation model is developed in MATLAB and works by considering a time-step of 1-minute resolution. In each time-step '*t*', it attains the $P_{Load,t}$, $P_{PV,t}$ from input datasets and the current battery SOC_t is passed from the previous time-step which was found using (14). The algorithm then calculates $P_{Net,t}$ using (1) and decides whether to charge or discharge the battery based on the calculated $P_{Net,t}$, the time of the day and TIDE tariff band corresponding to 't', the current battery SOC_t and resulting battery SOC_{t+1}, and the SOC limits. The battery does not charge if the current or resulting SOC is or goes above 90% and it does not discharge if the resulting SOC from this time-step is smaller than or equal to 10%. At the end of the charge/discharge process, the new battery SOC_{t+1} is calculated according to (14) again. A 30% initial SOC is considered for the first minute of each simulation. In this way, the battery control algorithm is used to determine system behaviour.

$$SOC_{t+1} = SOC_t + \left(\frac{\text{Charging or Discharging Energy}}{\text{Battery Capacity}} \times 100 \right) \quad (14)$$

A battery size of 26 kWh for the Milton Keynes community and 27 kWh for the Carbon Legacy community is applied, along with a power converter size of 6 kW and 5 kW respectively, which have been derived through an optimization method that balances the trade-off between increasing battery capacity and reducing system cost for each community.

4 Energy-Shifting Patterns Due To CPF BMS

Load consumption and PV generation data from the properties in the communities of Milton Keynes and Carbon Legacy, along with the battery and converter ratings stated in the previous section were fed into the CPF algorithm in order to test the effectiveness of the CPF BMS in energy shifting (load and PV shift) and achieving economic benefits through battery deployment. Figure 2 shows the summary of energy shifting achieved for the two communities of Milton Keynes and Carbon Legacy after running the CPF BMS for 1 complete year for each of the communities, analysed according to the tariff periods of the TIDE scheme. The losses shown with the 'yellow' bars are due to the non-unity roundtrip efficiencies of the battery and converter system.

It can be seen from Fig. 2 that for both communities, during the low tariff band, the grid import increases due to CPF BMS (shown by purple bars) compared to the original grid import (shown by green bars), as a result of heavy battery charging (represented by the indigo bars) during this period. A total import increment of 2311 kWh for Milton Keynes and 4181 kWh for Carbon Legacy community has taken place during the low tariff band. This increment is much lower than the import

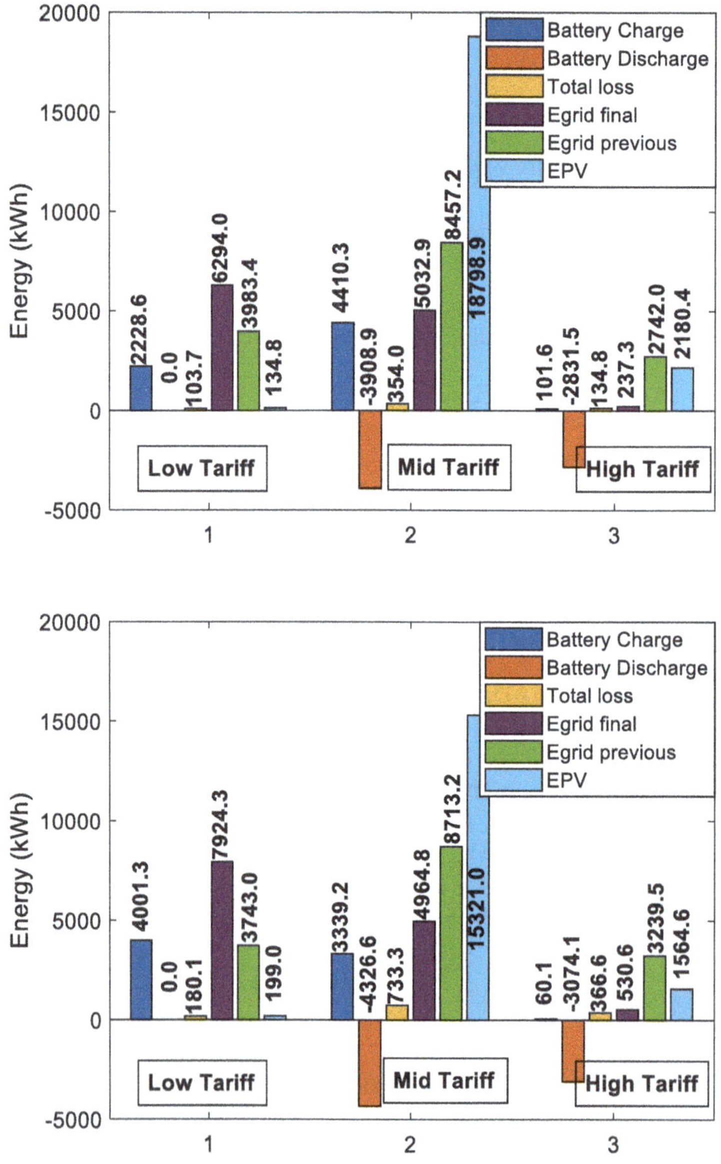

Fig. 2 Annual energy imported from the grid in each tariff period of TIDE using CPF BMS (Egrid final) alongside the battery charging and discharging energy in each of these period, contrasted with the import energy in the scenario without a battery (Egrid previous), for communities of Milton Keynes (top) and Carbon Legacy (bottom)

reductions which have occurred due to load shifting and PV self-consumption in the two other tariff bands, as will be discussed in the following paragraphs. There is no battery discharge (represented by orange bars) for either community during the low tariff period. This shows that the CPF BMS is fulfilling its aim for the low tariff band operation throughout the whole year for both the communities.

For the high-tariff band, the import energy can be seen to be drastically reduced by CPF BMS for both the communities. A total import of **2505** kWh for the Milton Keynes community and **2709** kWh for the Carbon Legacy community are shifted away from the peak period. Ideally, the import during high-tariff region should be zero, however, due to limiting battery size and converter power rating, there will be some remaining load in high-tariff band, which would have to be supplied by the grid.

Import from the grid is also reduced during mid-tariff band as seen from Fig. 2. A total import of **3424** kWh for the Milton Keynes community and **3748** kWh for the Carbon Legacy community are reduced or shifted away from the mid-tariff band. Both charging and discharging is taking place to/from the battery during this period. Since there is a large amount of PV energy (sky blue bar) available during this period, most of the charging energy is attributed to PV charging. The discharge mainly occurs during the second half of the mid-tariff band (8 pm to 12 midnight) when any remaining charge in the battery is drained out to take the battery to minimum SOC, ready to begin charging the next day with lower import prices.

Table 2 tabulates the total import costs for each community, with the new CPF BMS, compared with no battery deployment as well as against the original battery management system of the Carbon Legacy clients. For the Milton Keynes community, a total of **3619 kWh** (15183–11,564 kWh) of grid import can be reduced over a year by CPF BMS, along with a saving of **£1084** (£2263 – £1180) for the year by means of improved PV self-consumption and load shifting throughout the days. PV self-consumption is increased by **4706 kWh** over the year as calculated from the table.

For the Carbon Legacy community, a total of **2276 kWh** (15696–13,420 kWh) of grid import can be reduced over a year by the CPF BMS, compared to no battery employment in the community, along with a saving of **£1070** (£2433 – £1363) for the year by means of improved PV self-consumption and load shifting throughout the days. PV self-consumption is enhanced by **3556 kWh** over the year as calculated from the table. The differences in the grid import reduction and self-consumption increment are a result of the combined battery and power converter losses. Hence, without the battery and converter losses, the self-consumption increment for the communities would have been the same as the grid import reduction.

It should be noted that the exports under CPF BMS are not reduced to zero due to the limitations of battery and converter size. To absorb all of the PV surplus generation and achieve zero export, would require an impractically large and uneconomic size of battery capacity and power converter rating.

Another comparison can be made for the results of CPF BMS for Carbon Legacy community by comparing the CPF-based BMS to the behaviour of the commercial BMS embedded in each of the batteries themselves. The CPF BMS achieved a PV self-consumption increment of **2534 kWh**, and by load shifting away from peak period, an import price reduction of **£939** over a year is accomplished. Grid import is also reduced by **1789 kWh** by the CPF BMS compared to the original BMS deployed for the clients of Carbon Legacy.

A closer look is taken into the energy traces for the communities over a small duration, that is, 4 days, in order to understand the flow of energy within the CPF BMS. Figure 3 shows the battery and import energies for the Milton Keynes community over 4 days in the month of January. A battery size of 26 kWh and converter rating of 6 kW is used for the CPF BMS.

The patterns in the graph show how the battery is charging and discharging according to the net consumption ($P_{netgrid}$) of the community, which is given by load consumption minus total PV generation at each sample time. For the first day, it is

Table 2 Summary of import and export kWh and cost (£) over 1 year for different battery management scenarios for Carbon Legacy and Milton Keynes communities

Community			Carbon Legacy	Milton Keynes
Import	Original Battery	kWh	15203	-
		£	2302	-
	CPF BMS	kWh	13420	11564
		£	1363	1180
	w/o Battery	kWh	15696	15183
		£	2433	2263
Export	Original Battery	kWh	3892	-
	CPF BMS	kWh	1358	9705
	w/o Battery	kWh	4914	14411

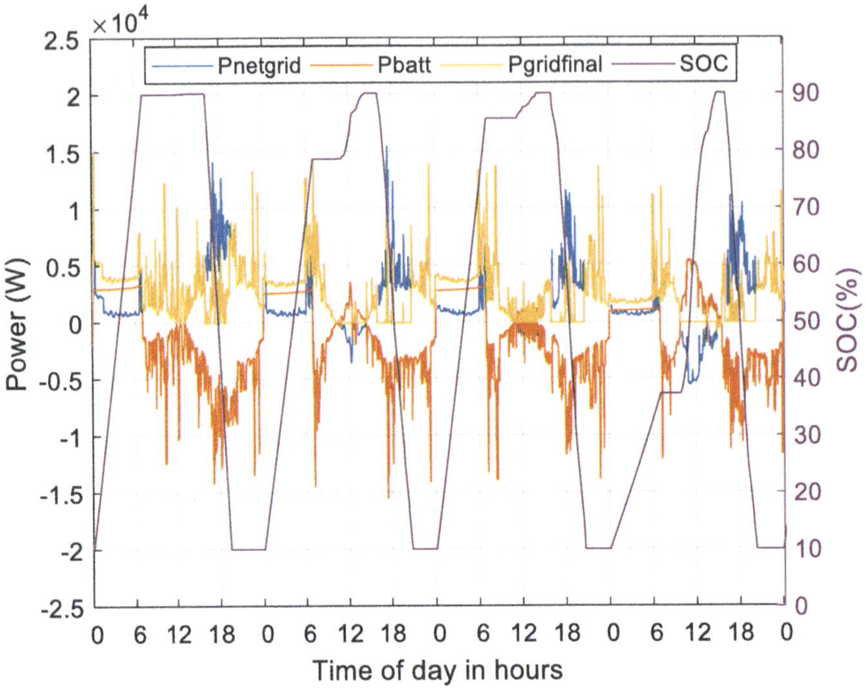

Fig. 3 Energy flow from grid and to and from battery for Milton Keynes community over 4 days in the month of January

seen from Fig. 3 through perfect prediction that there are no projected PV exports during the day (i.e. $P_{netgrid}$ is always positive). As a result, the battery charges throughout the low tariff period from 00:00 midnight to 7 am in the morning, resulting in the battery nearly reaching its maximum SOC at 90%. This energy is held throughout the mid-tariff band and is only discharged during the peak period starting from 4 pm until the battery reaches its minimum SOC at 10% close to 8 pm. There is no more discharging throughout the remainder of the day since the battery has reached its minimum SOC, and there is no charging until the beginning of the next low tariff period after midnight.

Over the next 2 days as shown in Fig. 3, there is some excess PV projected during the day. During the low tariff period, the battery therefore charges to a SOC which leaves enough capacity to be filled by the excess PV during the middle part of the day. Between 10 am to 4 pm, we can see the battery charges up through the excess PV and reaches its maximum 90% SOC, which is then discharged over the peak period until the battery is drained to minimum SOC.

For the last day shown in Fig. 3, there is abundant excess PV during the day which can cover half of the battery charging in order to meet the peak-period consumption of the day. As a result, due to considering perfect prediction, the battery charges only up to approximately 40% SOC during the midnight to 7 am off-peak hour, and keeps its remaining capacity available to be charged by PV energy. Between 10 am to 4 pm, it can be seen that the battery charges up to 90% SOC through all the available PV charging, which is then released to support all the load consumption during peak tariff period.

It can therefore be concluded, looking at the charging/discharging patterns and the SOC of the battery in relation to TIDE tariff bands and net consumption that the CPF BMS is successfully performing the energy shift for the communities according to the established set of rules and aims.

5 Baseline Comparison of CPF BMS

This section compares the grid import and electricity cost incurred by a community installed with PV generation and battery having CPF BMS versus the same community over the same days running with the same PV generation but without a battery. The former utilizes the TIDE tariff scheme while the latter is assumed to operate on a standard flat-rate import tariff offered throughout the whole day, which is taken from an average of the rates offered by major UK electricity companies. A tariff rate for exports is also considered which allows the households/community to receive a flat-rate payment for every excess unit of generated PV that goes back into the grid.

The FIT scheme was a UK government programme designed to promote the uptake of small-scale renewable and low-carbon electricity generation technologies, which is no longer available for new PV connections [13]. Therefore, we assume no FIT is applied but a fixed rate is paid for exports to the grid. Table 3 lists the different electricity providers and the flat-rate import tariff rates they offer, along with the export tariff from Ofgem [13, 14]. This assumes the use of net metering, which is not usual for residential PV in the UK. Export tariff is applied to a nominal 50% of PV generation, which would further improve the cost-effectiveness of the battery.

Using the average tariff rate from Table 3 and having calculated the total net consumption of communities from their available generation and consumption data, the total import cost, export revenue and using these, the net electricity cost is calculated according to Eqs. (15)–(17).

$$\text{Total Import Cost} = \text{Avg.std.unit rate} \times \text{Import kWh Energy} \tag{15}$$

$$\text{Total Export Cost} = \text{Standard export tariff} \times \text{Export kWh energy} \tag{16}$$

$$\text{Net Electricity Cost} = \text{Total Import Cost} - \text{Export Revenue} \tag{17}$$

Table 4 compares import energy and net electricity cost for both the scenarios of a community without a battery and the same community with CPF BMS-driven battery. It can be seen that using a battery with CPF BMS and TIDE tariff saves a total of **£1001** and **£959** over a year on electricity costs for Milton Keynes and Carbon Legacy communities respectively. It can also be seen from the table that the net usage actually increases for battery deployment in the community (due to battery losses), however, due to successful energy shifting from higher to lower tariff bands and by increasing self-consumption, the total import cost is reduced in reality.

The reduction in electricity costs achieved by employing the CPF BMS in conjunction with the Green Energy UK TIDE tariff is a significant value and therefore can justify the incorporation of this energy management system in communities for notable economic benefits to users.

Another analysis has been made to see the effect of CPF BMS employment in the two communities considered when they do not have any PV generation. The results are displayed in Table 5. It is seen that although CPF BMS increases kWh imported to the community from the grid (due to charging and discharging losses of battery and power converter) for both

Table 3 Standard electricity import flat-rate tariff offered by different energy suppliers along with yearly standing charges and tariff for export

Provider	Standing Charge £/year	Standard Unit Rate p/kWh
British gas	90.41	14.350
E.on	57.09	16.520
EDF Energy	80.92	16.670
nPower	55.01	17.620
Scottish power	95.23	15.405
Average import Price	75.73	16.113
Export (Ofgem)	–	5.24

Table 4 Comparison of annual import, export and net electricity cost and energy for communities without battery (standard flat-rate tariff) and with CPF BMS (TIDE tariff)

Community	Battery System	IMPORT		EXPORT		Net	
		kWh	£*	kWh	£	kWh	£
Milton Keynes	CPF BMS	11564	1275	9705	508.5	1859	766
	w/o Batt	15183	2522	14411	755	772	1767
	Difference	3619	1247	4706	246	-1087	1001
Carbon Legacy	CPF BMS	13420	1458	1358	71	12062	1387
	w/o Batt	15696	2604	4914	258	10782	2346
	Difference	2276	1146	3556	187	-1280	959

* Includes standing charge in £/year. Standing charge for Baseline scenario is £75.73 per year and for TIDE tariff is 26.02 p/day [15]

*Includes standing charge in £/year. Standing charge for Baseline scenario is £75.73 per year and for TIDE tariff is 26.02 p/day [15]

Table 5 Effect of CPF BMS on community annual import in kWh and in £ for a scenario without PV generation. A standard flat-rate import price is used for the case 'without battery'

Community		Import		
		w/o Battery	CPF BMS	Difference
Milton Keynes	£	3602	2305	-1297
	kWh	21885	22550	665
Carbon Legacy	£	4566	3209	-1357
	kWh	27866	28568	702

the Milton Keynes and the Carbon Legacy communities by **665 kWh** and **702 kWh** respectively, it significantly reduces the import price by **£1297** and **£1357** for the two communities by shifting energy to cheaper periods. It can therefore be inferred that the employment of CPF BMS has a notable effect on the communities in reducing electricity import prices regardless of the community being equipped with distributed generation.

6 Community Versus Individual Results

The energy saving established by the battery algorithm when paired with the optimum battery size and converter rating for individual properties as well as relative energy savings as a community is evaluated in this section to show merits of the battery algorithm on its own and further benefits for community energy. A comparison is made between the savings achieved by individual households of the two different communities of Milton Keynes and Carbon Legacy as opposed to having the CPF BMS run on the properties summed together as a community, allowing a reduced number of higher capacity batteries and converters. This will also demonstrate the case for community energy being beneficial in terms of overall energy and cost savings and increased utilization of renewable generation within the community.

Table 6 lists the optimum battery and converter size of each household in the community found using optimal sizing techniques. The sum of these individual battery sizes of the properties of Milton Keynes community amounts to 35 kWh battery capacity and 13 kW total converter kWs. For the Carbon Legacy community, this number amounts to 33 kWh total battery capacity for all the properties and a total of 7.5 kW converters. For the Milton Keynes community, this is an increase of 35% for the battery size in kWh and a 117% increase for the converters kWs. For Carbon Legacy community, the increase in battery size for individual operation as opposed to a community operation is 22% and the converter kWs is 50% higher.

In terms of energy and cost savings, the community working together as a single unit in the BMS can make better utilization of the battery and PV generation by self-consumption. It is likely that during the PV generation hours, even if there is no active occupancy at one given house, there could be load demand in another house within the community which can make use of the direct PV generation without having to export the PV back to the grid due to no demand, or without having to store the PV in the battery resulting in roundtrip losses.

Table 6 Energy and cost saving comparison for individual properties in a community versus saving achieved by the community as a unified system for Milton Keynes community (top) and Carbon Legacy community (bottom)

Property	Batt size, Conv size	Import £	Import kWh	Export kWh
M3	5 kWh, 3 kW	86	960	2840
M7	7 kWh, 2 kW	176	2130	2155
M13	6 kWh, 2 kW	216	2796	2121
M14	8 kWh, 3 kW	259	2884	1607
M19	9 kWh, 3 kW	296	3660	1430
Sum	35 kWh, 13 kW	1033	12430	10153
MK Community	26 kWh, 6 kW	1180	11559	9705

Property	Batt size, Conv size	Import £	Import kWh	Export kWh
AH	11 kWh, 2 kW	600	5183	255
ES	5 kWh, 1.5 kW	157	1567	1008
JF	11 kWh, 2 kW	527	5638	318
W123	6 kWh, 2 kW	281	2725	1338
Sum	33 kWh, 7.5 kW	1565	15113	2919
CL Community	27 kWh, 5 kW	1363	13414	1358

This can be seen in the values tabulated in Table 6 where the community operating as a whole over 1 year can reduce exports by 448 kWh (Milton Keynes) and 1561 kWh (Carbon Legacy) compared to individual BMS operation. Having a large overall battery for the community also allows the battery to meet more of the peak-period demand. Grid import energy is also reduced due to enhanced self-consumption, and is 871 kWh and 1699 kWh lower for community operation of the Milton Keynes and Carbon Legacy properties respectively. These energy savings are reflected in the grid import costs which are £147 (Milton Keynes) and £202 (Carbon Legacy) lower over the year compared to running the CPF BMS on a TIDE tariff scheme for individual properties.

Since battery and converter purchase cost per kW are significant in any given BMS, and the community operation causes a notable percentage reduction in both of these costs, it can be said that differences in import energy cost mentioned above are increased further if the savings in battery and converter purchase are considered in terms of monetary saving. However, since the price of batteries and converters in the market are subject to constant change, the gains are left in percentages for this discussion.

7 Conclusion

A community battery management system has been assessed in this paper based on a community power flow controller algorithm and its benefits verified through real communities in the UK equipped with grid-connected PV and BESS. The management algorithm controls the power flow between the electricity end-users, the communal BESS and the power grid, based on the ToU TIDE tariff, perfectly predicted PV generation data, alongside instantaneous PV and load consumption data. The battery management scheme is able to deliver the following outcomes for a community.

- Increment of energy self-consumption.
- Reduction of energy consumption at peak tariff rates.
- Reduction of electricity import cost for end-users by shifting energy import away from peak period.

The annual energy consumption for the Milton Keynes and Carbon Legacy communities have been evaluated based on a battery capacity and converter rating sized according to optimum cost saving of the respective communities. The superiority of the CPF BMS results has been demonstrated in comparison with the originally deployed batteries or with scenarios without any batteries. The results show effective energy shifting from peak period to the off-peak period as well as enhanced self-consumption, resulting in both a decrease in grid import as well as a decrease in import cost achieved by energy shifting.

Two control scenarios have been considered where the baseline community is considered to be operating on an energy plan with flat-rate tariff scheme and no battery, as opposed to the same community operating with the TIDE tariff and CPF BMS. The results have confirmed that £1001 and £959 are saved for the Milton Keynes and Carbon Legacy community respectively by employing the CPF BMS in conjunction with the TIDE ToU tariff scheme.

Another scenario is considered where a comparison is made between a community without PV generation, operating on a flat-rate import tariff and no storage system, as opposed to the same community operating under the CPF BMS and variable TIDE tariff. The results show that despite an increase in import energy in kWh, load shifting alone by successful operation of the CPF BMS reduces import costs significantly over a year. This amount is approximately £1300 for the two studied community and can be deemed as a notable saving.

Lastly, communities working together can save energy and reduce bills by employing community energy storage which in turn can better accommodate collective or individual renewable energy generation sources, and utilize ToU tariff differences between peak and off-peak electricity prices by load shifting to beneficial tariff bands. It also allows for smaller optimum battery and converter sizes per household, which in turn allows higher utilization of the battery. This has been demonstrated in the final section of the paper where a comparison is made between the savings achieved by individual households of the two different communities as opposed to having the CPF BMS run on the properties summed together as a community, resulting in lower grid imports, lower overall import cost (£147 for Milton Keynes and £202 for Carbon Legacy) and higher PV self-consumption. This establishes the case for community energy being beneficial in terms of energy and cost savings.

References

1. Klaimi J, Rahim-Amoud R, Merghem-Boulahia L, Jrad A (2016) Energy management algorithms in smart grids: state of the art and emerging trends. Int J Artif Intell Appl 7:25–45
2. Koirala BP, Koliou E, Friege J, Hakvoort RA, Herder PM (2016) Energetic communities for community energy: a review of key issues and trends shaping integrated community energy systems. Renew Sustain Energy Rev 56:722–744
3. Pholboon S, Sumner M, Kounnos P (2016) Community power flow control for peak demand reduction and energy cost savings. In IEEE PES Innovative Smart Grid Technologies Conference Europe (ISGT-Europe), Ljubljana
4. Richardson I, Thomson M (2010) One-minute resolution domestic electricity use data, 2008–2009 [computer file]. UK Data Archive [distributor] Study Number 6583, Colchester, Essex
5. Fazeli A, Sumner v, Johnson CM, Christopher E (2012) Coordinated optimal dispatch of distributed energy resources within a smart energy community cell. In 3rd IEEE PES Innovative Smart Grid Technologies Europe (ISGT Europe), Berlin,
6. Green Energy UK. Take control with TIDE, our new smart time of day tariff. Prices frozen until March 2018. 10 August 2017. [Online]. Available: https://www.greenenergyuk.com/Tide. Accessed 10 Aug 2017
7. Panagiotou K, Klumpner C, Sumner M (2017) Being a member of an energy community: assessing the financial benefits for end-users and management authority. In IEEE 26th International Symposium on Industrial Electronics (ISIE), Edinburgh
8. Parra D, Gillott M, Norman SA, Walker GS (2015) Optimum community energy storage system for PV energy time-shift. Appl Energy 137:576–587
9. Li J, Wu Z, Zhou S, Fu H, Zhang X-P (2015) Aggregator service for PV and battery energy storage systems of residential building. CSEE J Power Energy Syst 1(4):3–11
10. Parra D, Walker GS, Gillott M (2016) Are batteries the optimum PV-coupled energy storage for dwellings? Techno-economic comparison with hot water tanks in the UK. Energ Buildings 116:614–621
11. Nedap. PowerRouter. 2 May 2018. [Online]. Available: https://www.powerrouter.com/en/downloads-uk.html
12. Fazeli A (2014) The application of a novel deterministic algorithm for controlling the power flow through real time dispatch of the available distributed energy resources within a microgrid, PhD Thesis, University of Nottingham
13. Ofgem. Feed-In Tariff (FIT) rates, 6 Jul 2018. [Online]. Available: https://www.ofgem.gov.uk/environmental-programmes/fit/fit-tariff-rates. Accessed 20 Jul 2018
14. UK. Power, "Everything you need to know about fixed rate energy tariffs," 19 7 2018. [Online]. Available: https://www.ukpower.co.uk/home_energy/fixed-rate-tariffs. Accessed 20 7 2018
15. Green Energy UK. Take control with TIDE, our new smart time of day tariff. Prices frozen until March 2018. 10 August 2017. [Online]. Available: https://www.greenenergyuk.com/Tide

Benchmarking Sustainability of Megaprojects: A Review

Shafak Ahmed, Abdul Salam Darwish, and Peter Farrell

Abbreviations

GDP Gross domestic product
GHG Greenhouse gas
IDP Integrated design of the project
IT Information technology
KPI Key performance indicators
LCC Lifecycle cost
LCOE Levelised cost of energy
MSR Megaproject social responsibility
MW Megawatt
PV Photovoltaic
UAE United Arab Emirates
UK United Kingdom

Nomenclature

Ct the capital cost in period t (including decommissioning)
D the discount rate
Et the energy generated in period t
n the final year of operation including decommissioning
Ot the fixed-operating cost in period t
T the period ranging from year 1 to year *n*,
Vt the variable operating cost in period t (including fuel cost, and sometimes taxes, carbon costs, etc.)

1 Introduction

"Megaprojects face risks and uncertainties from numerous sources, such as the increasing developments in technology, climate change and social-economics, which causes results in them to overrun in cost and time" [1, P.1].

Megaprojects are unique. There are no clear guidelines or practices that are used in megaprojects for their sustainability and structure. Therefore, there is a need for a detailed initiatives framework with all aspects of guidelines to comprehend the megaproject phenomenon better. In this costly and consequential areas of business and government, these guidelines will support the advice policy, practice, and public debate [2].

Megaprojects are defined as significant and complex projects [1]. They have developed into a global multi-trillion-dollar business; sometimes, these project's capital costs are equivalent to the GDP of nations [1, 2]. The complexity and scale of megaprojects make it difficult to fully comprehend their transformational effects because they affect all aspects of people's

S. Ahmed (✉) · A. S. Darwish · P. Farrell
School of Civil Engineering, The University of Bolton, Bolton, UK
e-mail: a.darwish@bolton.ac.uk; P.Farrell@bolton.ac.uk

© The Author(s), under exclusive license to Springer Nature Switzerland AG 2022

A. Sayigh (ed.), *Sustainable Energy Development and Innovation*, Innovative Renewable Energy,
https://doi.org/10.1007/978-3-030-76221-6_82

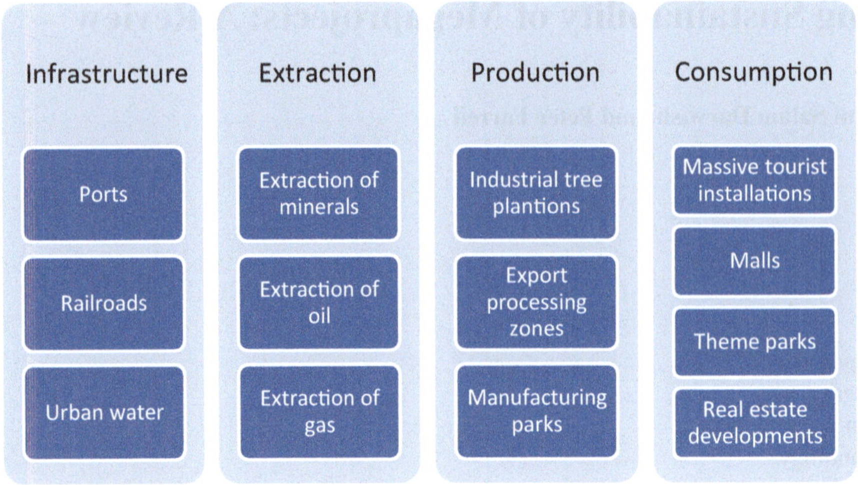

Fig. 1 The four types of megaprojects with examples. (Adapted from: [3])

lives, from their electricity bill to how they shop, do on the Internet and how they commute [1, 2]. Megaprojects are mainly reserved for the rich and developed nations [2]. According to Gellert and Lynch (2003), megaprojects can be divided into four types (see Fig. 1). These project types regularly appear in combination with each other.

Flyvbjerg and Turner (2018) argue that megaprojects have proved to be remarkably recession-proof [2]. In 2008, during the downturn, the megaproject business grew further by pouring stimulus spending on megaprojects from transportation infrastructure to IT [2]. On the contrary, Rickaby, Krystallis and Daviews (2018) claim that megaprojects fundamentally have high levels of uncertainty. They have the nature of being highly challenging, fragile, and vulnerable to developing threats. This is because of the megaproject's scale, complexity, and heterogeneous one-off delivery nature. If something goes wrong in megaprojects, the effects can lead to diminishing their return on investment [1]. Other major uncertainties that put megaprojects at risk are increasing developments in technology, climate change, and social-economics, which then results in the megaproject to overrun in cost and time [1].

Several planned megaprojects around the world have the trait to increase economic growth. With economic growth comes a substantial toll on local communities and their natural environment [4]. Governments have a critical role in the sustainability of planned projects because they can implement incentive policies that positively impact the sustainability of megaprojects by encouraging investors and financial institutions [5].

For both scholars and practitioners globally, sustainability has become a fundamental concern. Megaproject social responsibility (MSR) goes beyond the traditional project success criteria, such as cost, time, and quality [4]. MSR has developed indicators that accomplish the sustainability of megaprojects, addressing important demand for social and ethical criteria in megaprojects. The four categorised indicators of MSR are economic, legal, ethical, and political responsibilities. These indicators do not only include the main international sustainability metrics such as economy, environment, and society but also include performance-based indicators like quality, health and safety, resource utilisation, and characteristics related to megaproject management [4]. Table 1 shows these indicators on the project level.

Megaprojects have the nature to have many essential stakeholders with diverse and mixed motives for their decisions and actions. Therefore, making it clearer for each stakeholder their role in the MSR will nurture megaprojects' sustainable development by offering a way to assess stakeholders' commitments and accomplishments as a total [4].

Due to the complexity of megaprojects, it is hard to grasp and focus on what the objective is. When using KPIs to benchmark, the chance increases that all stakeholders will eventually meet their objectives. Then the individual objectives are aligned to reach the goal of having a sustainable megaproject.

Several papers that discuss either sustainability in construction or the social responsibility of megaprojects have been published. A few examples are as follows:

- Indicator system for evaluating megaproject social responsibility by H. Lin et al.
- Sustainable building assessment tool development approach by Saleh H. Alyamia and Yacine Rezguib.
- Sustainable construction key indicators by Catarina Araújo et al.

These papers discuss different parts of the objective of this paper. Therefore, it is important to include these papers in this review but combine and add to them to reach the intended outcome.

Table 1 Key indicators of MSR at the project level

Lifecycle of megaprojects	Social responsible dimension (level 1)	Indicator (level 2)
Initialisation	Economic responsibility	Decision making on the project's economic/technology feasibility
		Concern over the economic impact of the stakeholders
	Legal responsibility	Transparent information disclosure
		Promote public engagement
		Obey laws and observe disciplines
		Independent and impartial reporting
	Ethical and environmental responsibility	Concern over environment, ethical and environmental problems, and ecology
	Political responsibility	Concern over influence on (local) communities, public requirements, and relationship improvement
Design	Economic responsibility	Design scheme's quality and economic feasibility
		Innovation and technology progress
		Control design cost
	Legal responsibility	Compliant with industrial standards
		Design scheme's information disclosure
	Ethical and environmental responsibility	Green designing
	Political responsibility	Concern over community requirements
		Design scheme's public engagement
Construction	Economic responsibility	Normative project governance mechanism
		Ensure reasonable return on investments
		Ensure construction quality (of construction materials) and safety
		Control construction cost and schedule Construction technology innovation and progress
		Supervising quality, safety, and the economic rights of construction workers
	Legal responsibility	Constructing following the law and industry standards
	Ethical and environmental responsibility	Adoption of green construction
		Efficiently utilise resources
		Protect local ecological and community environment
		Supervising environmental protection
		Use and promote green materials
	Political responsibility	Concern over the community and public requirements
		Maintain social stability and favourable relationship with local communities
		Properly handle emergent public events
Operation	Economic responsibility	Routine maintenance of the project
		Control operation cost and ensure safety
	Legal responsibility	Operating in accordance with the law and industry standards
	Ethical and environmental responsibility	Protect local ecological and community environment
		Efficiently utilise resources
	Political responsibility	Maintain favourable relationship with local communities

Adapted from: [4]

This paper intends to understand, benchmark, and review different existing megaprojects. This review concluded that the sustainability assessment of Megaprojects involves deeply interrelated analyses of the main key performance indicators (KPIs). The benchmarking process is done by first identifying which KPIs are relevant, then gathering data regarding existing megaprojects to compare them then and discuss the outcome.

2 Examples of Analysed Megaprojects

The chosen megaprojects to be analysed are Walney Offshore Windfarms, UK and Masdar City, UAE. These projects promote sustainability differently. An analysis is needed to demonstrate better the importance of benchmarking. Moreover, having existing megaprojects as examples helps show the significance of the use of KPIs. Due to the nature of this paper, the chosen megaprojects have a major part in it.

2.1 *Walney Offshore Windfarms*

Due to the increasing population and urbanisation, energy demand has been rising. The global economy and society are radically dependent on energy availability because energy touches every aspect of human life and its activities [6]. Wind energy has become an essential energy resource for numerous developed and developing countries [7]. Offshore wind is most likely to become an essential element in the global future energy mix in general [8].

Walney Offshore Windfarms is a group of three wind farms that were built in several stages off the Cumbrian coast. Figure 2 shows the geographical location of the windfarms. First, Walney 1 and 2 were fully operational since 2012, with a cost of £1bn. Each of them consists of 51 turbines and the total combined capacity of the wind farms is 370 MW [9]. In 2018, the Walney extension was completed with 87 turbines and a capacity of 660 MW. It is the world's largest windfarm and has the power of the equivalent of almost 600,000 homes [10, 11]. According to Matthew Wright, the UK managing director of Danish energy firm Ørsted, which is the project's developer of the Walney Extension, the project is another benchmark in terms of scale. This is because the turbines are more significant and further out from each other (Vaughan 2018).

2.2 *Masdar City*

In the UAE, Abu Dhabi and Dubai have set targets of 7% and 5% of their total energy needs to be from renewable energy by 2020. Abu Dhabi has made a big move on this by building the carbon-neutral town, Masdar City [12]. Masdar City (masdar is Arabic for "source") has recognised the unsustainable impacts that people have on the environment and are acting to effect

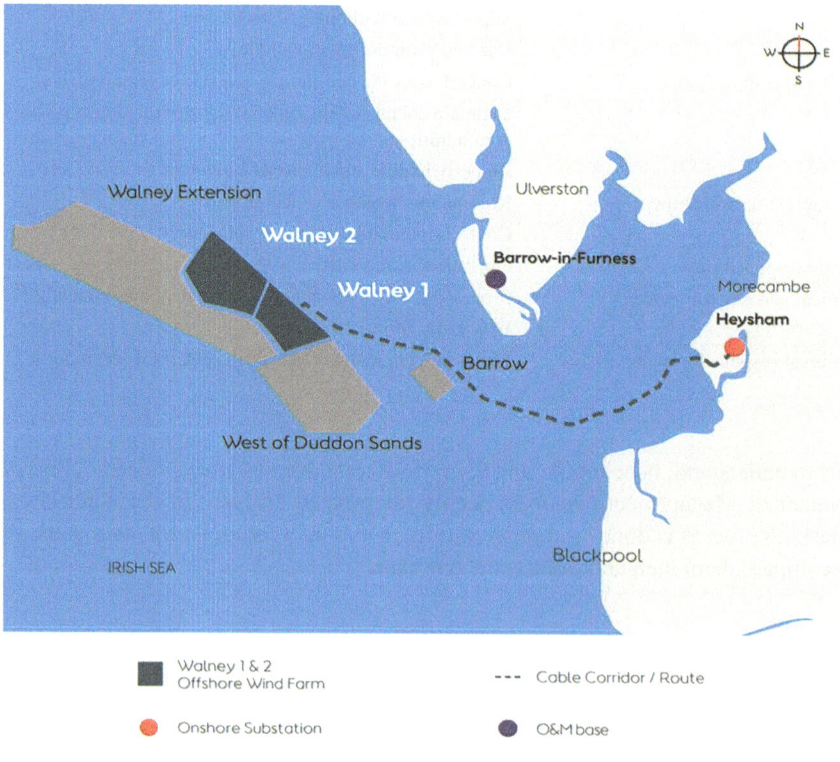

Fig. 2 The Walney Offshore Windfarms are located off the Cambrian coat [9]

change [13]. This megaproject had an initial investment of $22 billion, but after changing their plans to save cost, the megaproject investment went down to $16 billion [14, 15].

Masdar City is one of the most sustainable urban communities in the world that is based on the three pillars of economic, social and environmental sustainability. It is a low-carbon expansion that is made up of a fast-growing clean-tech cluster, business-free zone and residential neighbourhood that has restaurants, shops and public green spaces [16]. Figure 3 shows the geographical location of Masdar City.

Masdar City is considered a 'greenprint' for sustainable development of cities by applying real-world solutions in energy and water efficiency, mobility and waste reduction [16]. This is a good example of what a country that produces oil can do to create a balance between hydrocarbons and renewable energy while simultaneously concentrating on climate change and energy security [12].

3 Key Performance Indicators

Key performance indicators (KPIs) are targets that measure the progress in the achievement of critical success factors [18]. KPIs can be used in to measure different objectives and goals. Also, different stakeholders can see if the megaproject is on track by comparing the measure KPIs to the goals.

The commitment of accomplishing a high level of sustainability needs to be shared and communicated to all stakeholders, otherwise, the chance of reaching the objective of sustainability derails [19]. As explained earlier, sustainability has three core matrices (environment, society, and economy). Based on these matrices, 5 KPI's are formed to measure the sustainability of each megaproject (Table 2).

3.1 Environment

Two indicators measure how megaprojects are being sustainable in the environmental element. First, greenhouse gases (GHGs) have this name since they contribute towards the greenhouse effect. In the greenhouse effect, certain gases in the atmosphere increase the earth's temperature because they can trap heat just like glass traps heat in a greenhouse. According to the Kyoto Protocol, there are six main GHGs; carbon dioxide – CO_2, methane – CH_4, nitrous oxide – N_2O, hydrofluorocarbons – HFCs, perfluorocarbons – PFCs, and sulphur hexafluoride – SF6 [20].

In order to measure the GHG emissions in a construction setting, all operations associated with extraction, manufacturing, transporting, installing, maintaining, and disposing of construction materials and products should be considered and measured [21].

Scientific agreements show that the increase in these emissions has an evident effect on the climate. Elevated global temperature is likely to cause a climatic disturbance, desertification, rising sea levels, and spread of disease [21].

The second indicator for the environmental dimension is responsible for sourcing materials. When a sustainable building is designed, a crucial aspect is the consideration of the materials that are used for construction because it determines the building's sustainable performance. Also, something to consider is the material wastage because most raw materials that are used in construction come from non-renewable resources [22].

Table 2 Overview of the five indicators

Dimension	Indicator
Environment	Greenhouse gases (GHGs) emissions
	Responsible sourcing materials
Society	Integrated design of the project
	Users health and comfort
Economy	Lifecycle cost (LCC)

3.2 Society

To have an integrated design of the project (IDP), excellent communication must remain between all stakeholders. An IDP starts with the design phase but remains throughout the entire project life [19]. After being designed, the project needs to be constructed, commissioned and operated in a manner that supports the sustainable objective. Having an IDP multiplies the design possibilities and creates projects that are better performing and more cost-effective [19]. Figure 3 gives a good overview of how IDP benefits projects. The graph shows that it is easier to achieve major accomplishments if integrating the design starts in the earliest stages. When proceeding further into the more advanced stages, the impact/benefit starts to decline. The main idea behind this graph is to do as much as possible upfront to have the most impact and least cost to realise [19] (Fig. 4).

The second KPI to measure the social dimension is users' health and comfort. Users' health and comfort is a way to look at what the megaproject does for society. This KPI can be seen in many ways. Generally, some megaprojects have a limited number of users. Therefore, it is important to look at the individual's health and comfort. This includes the health and comfort of both the staff and users. When looking into people's health, considering safety automatically becomes a key priority to maintain it. This is especially important for staff and workers of the megaproject since, during construction, the probability of accidents is high. As for the individual's comfort, it looks at how easy it is to use the facilities within the megaproject.

3.3 Economy

The KPI to measure how sustainable to megaproject is based on the economical dimension is to calculate its Lifecycle Cost (LCC). The LCC shows how much the entire megaproject costs. It is the sum of all recurring and non-recurring costs over the full life span or a specified period of the megaproject (Fig. 5).

3.3.1 Levelised Cost of Energy (LCOE)

The LCC analysis is typically in the form of the levelised cost of energy (LCOE) for a project that is related to energy production. The LCOE measures all the costs over the lifespan of a plant by determining how much it costs to produce an amount of energy (usually per MWh) [24].

Fig. 3 Location of Masdar City [17]

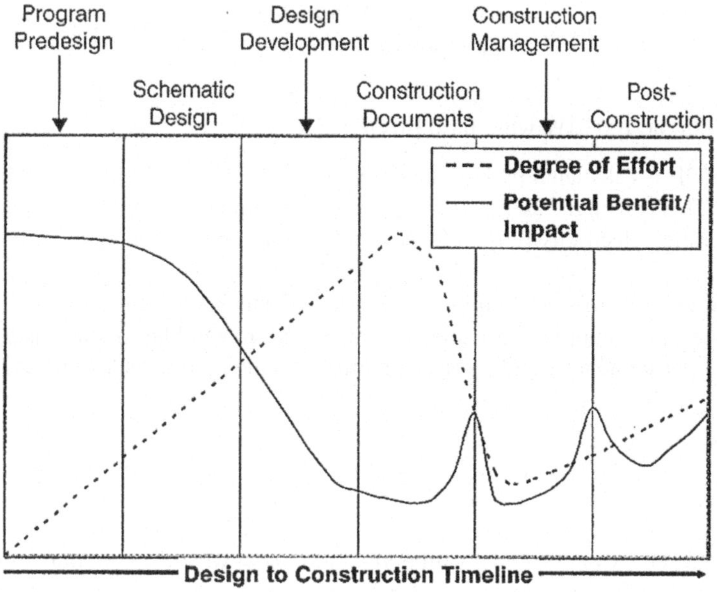

Fig. 4 Benefits of early design collaboration [19]

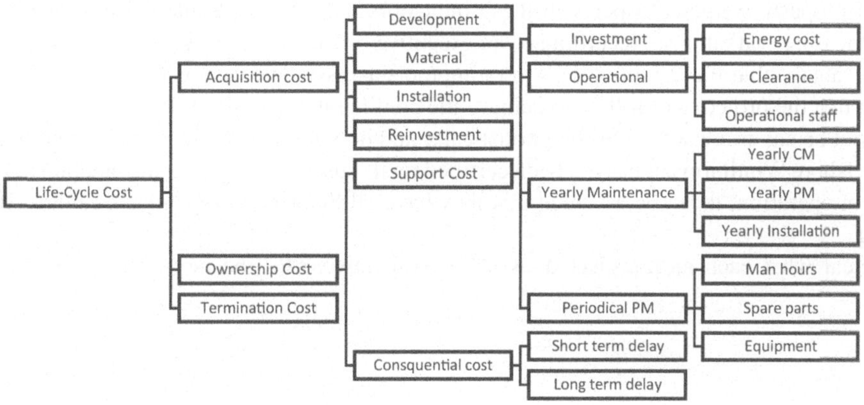

Fig. 5 A visual explanation of the LCC calculation. (Adapted from [23])

For example, to make a wind farm project more attractive for investors, it is advised to keep the LCOE as low as possible. Moreover, the LCOE is an excellent way to inform better policy choices [8].

Equation 1 shows how LCOE can be calculated. The meaning of the symbols can be found in the terminology.

$$\text{LCOE} = \frac{\text{NPV}_{\text{cost}}}{\text{NPE}} = \sum_{t=1}^{n} \frac{C_t + O_t + V_t}{(1+d)^t} / \sum_{t=1}^{n} \frac{E_t}{(1+d)^t}$$

Equation 1 Levelised cost of energy (LCOE) [8]

The following simple example is about a 100 MW wind farm:

Total Capital Cost = \$1,400/KW
Fixed O&M Cost = \$45/KW per year
Capacity Factor = 40%
Useful Life = 30 years
Discount Rate = 6%

$$\text{LCOE} = \frac{\$140\,\text{MM} + 13.76 \times \$4.5\,\text{MM}}{350,400\,\text{MWh} \times 13.76} = \$41.87\,/\,\text{MWh}$$

Equation 2 Example of a simple 100 MW wind farm LCOE [25]

As shown above, the LCOE is calculated by looking at the present value (13.76) of the lifetime cost with the PV of the lifetime energy generation. This is with the assumption that in the beginning, all capital expenditure is incurred and the projects immediately start generating electricity [25]. This calculation shows that every megawatt per hour generated has a cost of $41.87.

To have a lower LCOE for wind farms, it is required to continue the technological evolution, efficiency in operating practices and lower discount rates. An example of a technological evolution could be the increasing capacity of wind turbines, which allows more capacity per foundation and so be economies of scale in manufacturing and installation [8].

4 Benchmarking

In order to benchmark the chosen megaprojects, it is important to look at what they are doing in each KPI.

Masdar City's buildings are meant to have at least a 40% reduction in water and energy consumption following LEED and the Estidama Pearl Building Rating System baselines. They have achieved a '3 pearl' rating under the Estidama guidelines [32]. While the developers at Masdar City are using many KPIs for sustainability to achieve their targets, it is not their aim to achieve maximum results at the Estidama and LEED indicators. This is due to the fact that they put more emphasis on matters that are not properly weighted aspects in these indicators [31]. For example, LEED does not look at water and Estidama does not measure walkability. According to an executive of Masdar City, the rating tools have a lack of a holistic approach. The disadvantage that the executive sees in rating tools/assessment systems is that it predefines the sustainability criteria and the path that the project will follow to achieve the credits/ratings desired. If the focus is only put on the criteria, then the focus will not be any more at the building holistically and how it reacts to the environment holistically [31].

As for Walney Offshore Windfarms, it has not been certified on its sustainability by a sustainability certification. However, since the project is in operation, it can be assumed that they have all the needed sustainable ratings. If not, they would not have been operating.

Now that it is clear what each megaproject does in the respective KPI, the next step is to benchmark them against each other.

5 Discussion

There are many KPIs in megaprojects that can be measured, but it depends on what the objective is of the megaproject. In this case, the focus was on sustainability, which can be classified into three categories. As shown in Table 1, many indicators can be chosen based on the objectives of the megaproject. Due to the great number of indicators, the chosen to focus on five indicators that can be seen as general KPI's for megaprojects.

To test those five indicators, there were two megaprojects selected to review.

The activities that are described in Table 3 are scored in Table 4 on how sustainable their activities are. The scaling goes as follows: 5 = excellent, 4 = very good, 3 = good, 2 = satisfactory, 1 = poor, 0 = very poor.

What can be seen in this review is that although both projects are seen as sustainable megaprojects, they act upon this in very diverse manners. For example, when looking at the KPI Responsible Sourcing Materials, Walney Offshore Windfarms does rely on natural resources during construction, but it is not one of their main goals. On the contrary, Masdar City is very keen on Responsible Sourcing Materials in all aspects of the megaproject. Everything has been taken into consideration, from construction design to construction materials. Moreover, later on, when the project is all up and running, their designs are made to reduce water and energy. This makes Masdar City score better compared to Walney Offshore Windfarms.

Something that should be taken into consideration is that occasionally it is hard to find information on certain details from megaprojects. Thus, it will not be a just or objective benchmark since the lack of information cannot be evaluated.

Table 3 Activities of the megaproject in each KPI

	KPIs				
Megaprojects	GHGs emissions	Responsible sourcing materials	Integrated design of the project	Users health and comfort	Lifecycle cost
Walney Offshore Windfarms	Walney Offshore Windfarms support Great Britain to fulfil its international commitment to reduce GHGs emissions. The construction of the windfarms could have some degree of negative environmental impacts, but a balanced evaluation based on the Environmental Statement approves the construction of the Walney Offshore Windfarm (Dong Energy Ltd., 2006)	Walney Offshore Windfarms rely on natural resources, such as their construction materials [26]	The Walney Offshore Windfarms are designed up to a standard where all specialists' objectives are considered	Walney Offshore Windfarms give great positive benefits to the British society [27]. Also, the turbines are not placed close to the shore where it might bother people	According to Aldersey-Williams, Broadbent and Strachan, 2019, Walney has an LCOE of 120.37 (£/MW h) and the Walney Extension has an LCOE of 100.24 (£/MW h)
Masdar City	The main focus of Masdar City's developers lays in the reduction of CO_2 emissions. This can be capitalised through the development and commercialisation of clean technologies that are invented to reduce the carbon emission of urban environments [28]	The cement used to build the buildings in Masdar City is considered low-carbon cement because it utilises 90% recycled aluminium and is designed to reduce energy and water consumption by at least 40% compared with the American Society of Heating, Refrigerating and Air-Conditioning Engineers (ASHRAE) standards. Another way where Masdar City is strongly focusing on sustainable development is through the reduction of demand for energy and water and recycling and reuse of waste material, particularly during the construction activities	The Estidama Pearl Rating System is mandatory for all buildings that are designed and built in Abu Dhabi. One of the categories that is included is: "Integrated Development Process: Encouraging cross-disciplinary teamwork to deliver environmental and quality management throughout the life of the project" [29] Masdar City scores a four-pearl rating on their eco-villas [30]	Masdar City provides a high-quality living and working environment while having the lowest possible ecological footprint. Also, Masdar City strives to be a model of sustainable development where residents desire to live, work, and play [29]	One of the important indicators that Masdar City has emphasised is to minimise the cost of construction [31] Not all components are available to calculate the LCC for Masdar City. This is what is available: The initial investment was \$22 billion but after changing their plans to save cost, the megaproject investment went down to \$16 billion [14, 15]. The per capita cost will be \$320,000 when the investment is divided by 50,000 residents [15]

Table 4 Megaproject activity scores

	KPIs				
Megaprojects	GHGs emissions	Responsible sourcing materials	Integrated design of the project	Users health and comfort	Lifecycle cost
Walney Offshore Windfarms	2	2	N/A	3	4
Masdar City	5	5	5	4	2

6 Conclusion

Sustainable development of megaprojects is essential as the built infrastructure has a crucial role to play in both the continued and future development of communities, societies, the world economy, and the ecosystem.

KPIs in megaprojects, in general, are important but especially in sustainable megaprojects. The five chosen KPIs are an excellent example of a basic analysis for a sustainable megaproject. After analysing the two megaprojects and benchmarking them, it becomes clear that some megaprojects score better on certain KPIs than others. This is due to their commitment and focuses on certain areas of sustainability. Also, another factor to keep in mind is that some megaprojects are more commercial than others. Hence, some megaprojects focus on certain sustainability indicators that are easily marketable. For example, Masdar City has a great focus on the reduction of CO_2 emissions. This can be capitalised through the development and commercialisation of clean technologies that are invented to reduce the carbon emission of urban environments [28].

It is best to consider KPIs that cover all three pillars of sustainability during the first stages of the megaproject. It is easier to achieve major accomplishments if integrating the design and KPI objectives starts in the earliest stages.

There are no clear guidelines or practices that are used in megaprojects. It is important to understand, benchmark, and review different existing megaprojects.

To conclude from this review, when clear KPIs are set at the beginning of a megaproject for each stakeholder involved, an effective management framework for the sustainability of the megaproject will endure.

References

1. Rickaby B, Krystallis I, Daviews A (2018) Managing uncertainty (opportunities and threats) in megaprojects' decisions. In: 14th international research network on organizing by projects (IRNOP) conference
2. Flyvbjerg B, Turner JR (2018) Do classics exist in megaproject management? Int J Proj Manag. 36(2):334–341
3. Gellert PK, Lynch BD (2003) Megaprojects as displacements. Int Soc Sci J. 55(175):1–1
4. Lin H, Zeng S, Ma H, Zeng R, Tam VWY (2017) An indicator system for evaluating megaproject social responsibility. Int J Proj Manag [Internet] 35(7):1415–1426. Available from: https://doi.org/10.1016/j.ijproman.2017.04.009
5. Wu G, Qiang G, Zuo J, Zhao X, Chang R (2018) What are the key indicators of mega sustainable construction projects? -A stakeholder-network perspective. Sustainability. 10(8):2939
6. Motoasca E (2019) Energy sustainability in built and urban environments [Internet]. Available from: http://link.springer.com/10.1007/978-981-13-3284-5
7. Darwish AS (2015) Advancements and challenges affecting wind turbine implementation in the member states of the Cooperation Council for the Arab States of the Gulf (GCC Countries). In: The age of wind, Springer, UK, pp 175–192
8. Aldersey-Williams J, Broadbent ID, Strachan PA (2019) Better estimates of LCOE from audited accounts – A new methodology with examples from United Kingdom offshore wind and CCGT. Energy Policy. 128(April):25–35
9. Ørsted (2019) Walney Offshore Wind Farm
10. Vaughan A (2018) World's largest offshore windfarm opens off Cumbrian coast. The Guardian [Internet]. pp 1–2. Available from: https://www.theguardian.com/environment/2018/sep/06/worlds-largest-offshore-windfarm-opens-cumbrian-coast-walney-extension-brexit
11. Ørsted (2019) Walney extension offshore wind farm [Internet]. Available from: https://orstedcdn.azureedge.net/-/media/WWW/Docs/Corp/UK/Walney-extension/180822_Walney-Extension-Project-Summary-V4.ashx?la=en&hash=4F8D0C0F6EE86D9300B36D2DEFB3462DA4A9C419&rev=1eb41ae526d7498b97f0311663d4e530
12. Haraksingh I (2014) Renewable energy against an oil and gas background: challenges, drivers, and case studies. In: Renewable energy in the service of mankind Vol II, Springer, UK, pp 311–323
13. Gharib R, Tok ME, Zebian M (2016) Neoliberal urbanization and smart cities in the Gulf region: the case of Abu Dhabi's Masdar City. In: Gateways to the world: port cities in the Persian Gulf
14. Carlisle T (2010) Masdar City clips another $ 2.5bn from price tag. The National. pp 1–4
15. Jeppesen H (2010) Built on sand: Masdar City to become eco-city of the future. DW. pp 1–2
16. Masdar (2020) The City [Internet]. Available from: https://masdar.ae/en/masdar-city/the-city
17. French K (2018) Developing a smart and sustainable city: Is Masdar City the answer?
18. McCabe S (2008) Benchmarking in construction. Wiley, New York, 306 p
19. Lewis M (2004) Integrated design for sustainable buildings. ASHRAE J 46(9):s22–s30
20. DEFRA (2006) Environmental key performance indicators
21. Danso H (2018) Identification of key indicators for sustainable construction materials. Adv Mater Sci Eng 2018
22. Atombo C, Cudjoe J, Dzantor K, Agbo AA (2015) Integration of sustainable construction in project management: a case study in Ghana. Int J Constr Eng Manag [Internet] 4(1):13–25. Available from: http://journal.sapub.org/ijcem
23. INFRALERT (2017) Report on RAMS & LCC integrated models . Available at: http://infralert.eu/wp-content/multiverso-files/21_572b4a01a6b67/INFRALERT-D5.2-RAMS_LCC_Integrated_Models_V4.0-FINAL_corrected-Cover.pdf
24. Lyndon PC, Donev J (2018) Life-cycle cost. Energy Education. pp 1–2
25. Raikar S, Adamson S (2019) Renewable energy finance: theory and practice. Academic, Amsterdam
26. Ørsted (2018) Ørsted ESG performance report 2018. pp 1–45

27. Dong Energy Ltd. (2006) Walney Offshore Windfarm environmental statement non-technical summary [Internet]. Available from: http://www.bei.org/attachments/pipeline/20090322_nts3_en.pdf
28. Cugurullo F (2016) Urban eco-modernisation and the policy context of new eco-city projects: where Masdar City fails and why. Urban Stud 53:2417–2433
29. Alzaabi M, Rizk Z, Mezher T (2019) Linking smart cities concept to energy- water- food nexus: the case of Masdar City in Abu Dhabi, UAE. In: Smart cities in the Gulf - current state, opportunities, and challenges
30. Masdar (2020) Sustainability - Masdar City is one of the world's most sustainable urban communities [Internet]. Available from: https://masdar.ae/en/masdar-city/the-city/sustainability
31. Azar E, Raouf M (2017) Sustainability in the Gulf: challenges and opportunities. Milton Taylor and Francis
32. Masdar. Masdar City at a Glance. Madar Media

Part X
Renewable System Integration

The Energy Performance of Hospital Buildings in the South Balkan Region: The Prospects for Zero-Energy Hospitals

A. Dimoudi, A. Kantzioura, P. Toumpoulides, St Zoras, D. Serghides, St Dimitriou, St Thravalou, M. Metaj, E. Mara, and A. Dorri

1 Introduction

One of the basic energy objectives of EU is to decouple energy consumption from economic development. To meet the challenge, the EU is setting specific goals for the years 2020, 2030 and 2050. The energy efficiency (EE) goal for the year 2020 is to achieve 20% reduction in energy consumption, 20% increase of the share of RES in primary energy and 20% reduction in CO_2 emissions with respect to the 1990 levels. The corresponding goals for the year 2030 are 27–27- 40%. For the year 2050, the set goal is to achieve a reduction of the CO_2 emissions to 85% with respect the 1990 levels [1].

The building sector accounts for 40% of the total energy consumption and 36% of total CO_2 emissions [2]. Thus, the sector has a very significant role to perform in implementing the EU energy efficiency objectives. The EU set up a policy framework focused on reducing energy consumption and obtaining important savings from buildings. One important policy action is the Energy Performance of Buildings Directive recast (EPBD recast, Directive 2010/31/EC). The EPB Directive introduces the notion of zero-energy buildings (ZEB) which are buildings of high-energy performance and low-specific energy requirements. According to the EPBD recast, new buildings occupied by public authorities and properties had to be NZEBs by December 31, 2018 and all new buildings by December 31, 2020 [3, 4]. Non-residential buildings constitute a heterogeneity sector characterized by high-energy consumption and various building types, sizes and energy characteristics over Europe. Non-residential buildings account on average for 25% energy consumption of the total European building stock, representing a heterogeneous sector compared with the residential [5]. Large buildings such as hospitals, consume a lot of energy since the different services (for example, space heating and cooling, ventilation system, lighting, electrical energy for refrigerators, etc.). In particular, hospitals have a continuous electricity, heating and cooling energy demand because thermal comfort, air quality levels and specialized services for the patients have to be guaranteed without interruptions to avoid discomfort during the entire year. Worldwide, about 6% of the total energy consumption in the buildings sector is represented by energy usage in medical centres. Studies from REHVA show that only 25–40% of the energy use by academic hospitals is for appliances and medical equipment. The rest of the energy is used by building and building services related to achieve the required indoor conditions [6].

As a result, hospitals present a great potential for energy and cost savings because the simultaneous demand of energy uses for the entire year allows several actions for energy efficiency improvement to be studied and planned.

A. Dimoudi (✉) · A. Kantzioura · P. Toumpoulides · S. Zoras
Laboratory of Environmental and Energy Design of Buildings and Settlements, Department of Environmental Engineering, Democritus University of Thrace, Xanthi, Greece
e-mail: adimoudi@env.duth.gr

D. Serghides · S. Dimitriou · S. Thravalou
Energy, Environment and Water Research Center (EEWRC), The Cyprus Institute, Aglantzia, Cyprus

M. Metaj · E. Mara
ALBAFOREST, Tirana, Albania

A. Dorri
Polytechnic University of Tirana, Tiranë, Albania

© The Author(s), under exclusive license to Springer Nature Switzerland AG 2022
A. Sayigh (ed.), *Sustainable Energy Development and Innovation*, Innovative Renewable Energy,
https://doi.org/10.1007/978-3-030-76221-6_83

2 Methodology of the Project

The technical analysis required is based on energy performance assessment of hospital buildings through extensive energy auditing of such buildings. Buildings will be examined under different climate conditions and indoor climate conditions. Technologies applicable to the regional characteristics of the southern Balkan area will be identified and described. The Methodology of the Project includes the following steps:

- Collection of energy data of hospitals.
- Energy audits to obtain actual data and determine the operational characteristics of the selected hospital.
- Identification of the most prominent energy efficient (EE) and renewable energy sources (RES) technical solutions applicable to the energy consumption profiles of hospitals.

3 Analysis of Current Hospitals Energy Performance

3.1 Energy Performance of Hospitals in Greece

The stock of Hellenic hospitals and clinics exceeds 360 facilities (not including military hospitals). Over 30% of the total number of medical facilities are located in the greater metropolitan area of Athens. A good number of the Hellenic hospitals are in over-aged buildings. In terms of ownership, most of the Hellenic hospitals are public and they serve the great majority of the population. The total number of available operative rooms in Hellenic hospitals is rather small and space availability is problematic in most cases [7].

Health care buildings, including hospitals, clinics and health centres, represent the lowest percentage, 0.05%, of the total Hellenic non-residential building stock, while they have the highest energy consumption per unit floor area compared to the other main categories of these buildings. Over the last three decades, practically all health care buildings have been equipped with some kind of AC, thus, further increasing their energy consumption [8]. The main energy types of fuels which are used in Greece are electricity, natural gas and oil and very limited district heating. The inspections for energy certification according to the EPBD legislative framework have shown that hospitals have low energy performance levels. The average energy consumption of buildings inspected ranged from 736,36 kWh/m^2/y (Climatic Zone C) to 1.555,90 kWh/m^2/y (Climatic Zone A) [9].

The Greek territory is divided into seven Health Regions (HR) (Law 4052/2012) and four Climatic Zones according to the Greek Regulation of Energy Efficiency Buildings (KENAK) (Fig. 1).

In all Greek hospital buildings, the primary energy consumption per square metre declines from the target proposed for existing nZEB in tertiary sector, which was defined at 90 kWh/m^2 [10]. The deviation of the primary energy consumption per m^2 of each hospital from the desired target value for nZEB tertiary buildings is about 81.8%. This marks the great challenge for interventions to improve the energy efficiency behaviour of buildings and their upgrading into nZEB Hospitals (Fig. 2).

In Fig. 3, the energy consumption for each Health Region is observed. The consumption in the fourth HR is the higher than in the other HRs due to the heavily winter weather conditions. This region corresponds to C' and D' Climatic Zones of Greece. Also, the fourth HR covers the higher surface (10.8%) of the total surface of Greek hospitals. Follows the first HR (A' Climatic Zone) which is located in the greater metropolitan area of Athens and the total covered surface is 8.3% of the total surface of Greek hospitals.

The energy demand in Greek hospitals for electricity is 44%, natural gas 32%, oil 22% and district heating only 1%. Regarding the energy cost, more than 50% is spent for electricity use (Fig. 4). Figure 5 presents the fuel percentage at the total energy consumption and Fig. 6 presents the cost percentage at the total annual energy cost, for each HR. The district heating is only used in the third HR, the oil is mainly used in the sixth and seventh HR, the natural gas is mainly used in the first and second HR and the electricity in the second, third and sixth HR.

So, important factors that affect the energy consumption and cost of the hospitals are, among others, the total covered surface of the hospital, the available energy sources, the prevailed extremely weather conditions during winter in some areas or the increased cooling needs during summer in others.

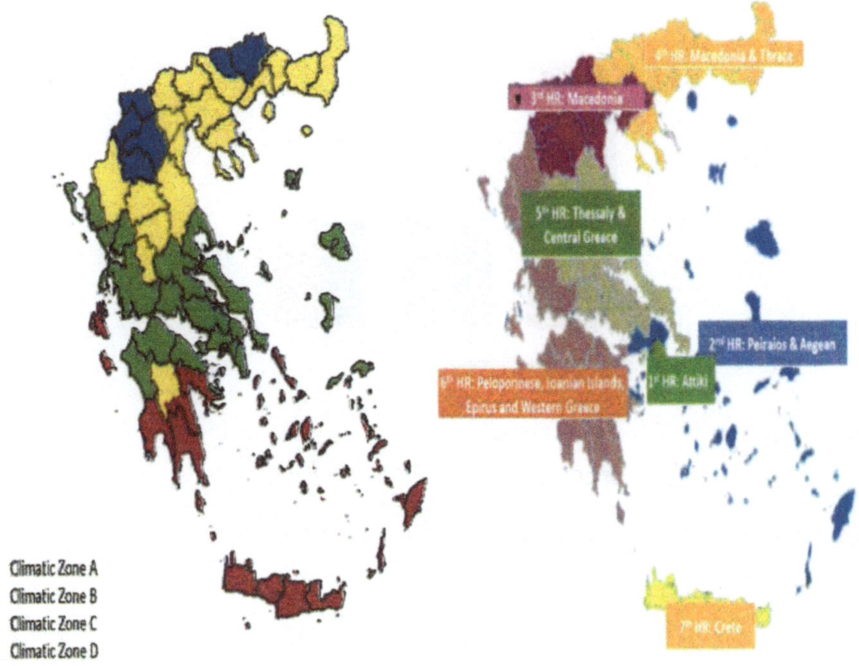

Fig. 1 The climatic zones (KENAK) and health regions of Greece

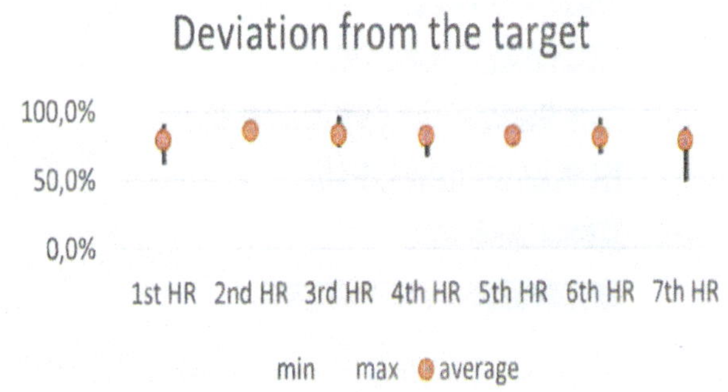

Fig. 2 The deviation of the primary energy consumption per m^2 (%) for each HR

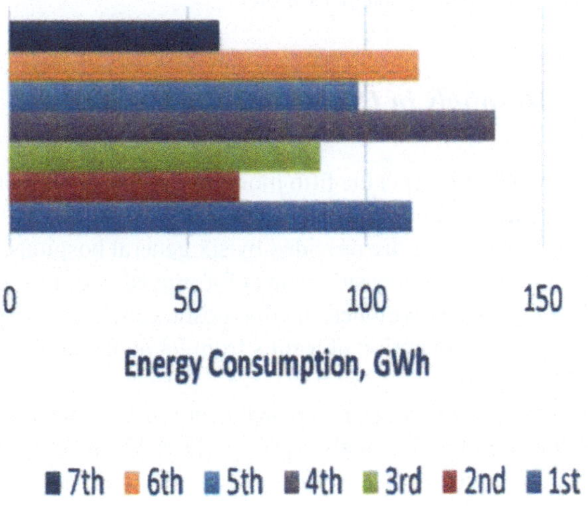

Fig. 3 Total consumption of each HR

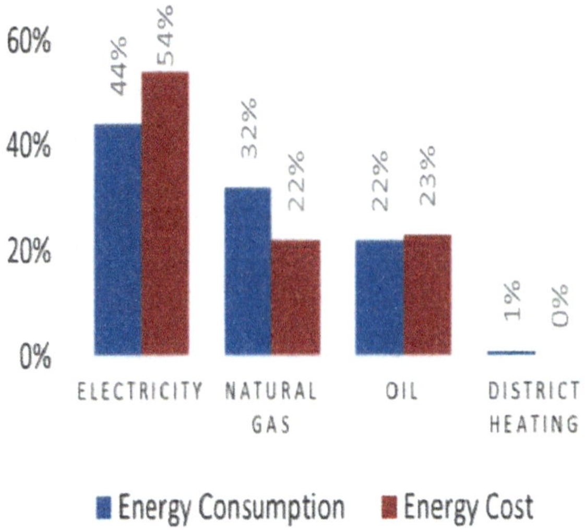

Fig. 4 Energy consumption and energy cost of each fuel type in Greek hospital buildings

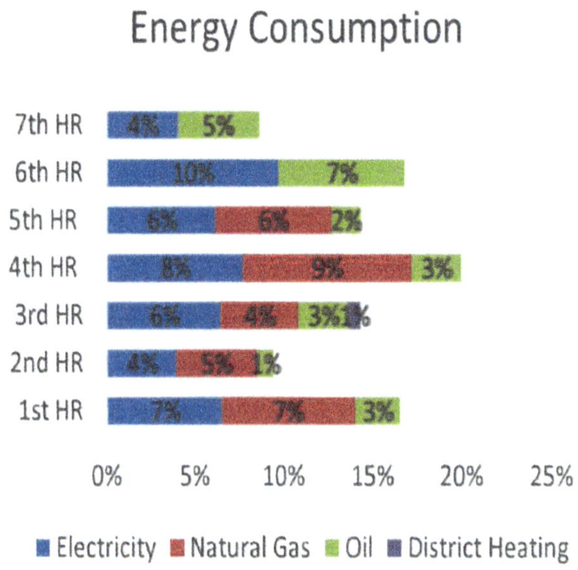

Fig. 5 Fuel percentage at the total annual energy consumption for each HR

3.2 *Energy Performance of Hospitals in Cyprus*

The health provision sector (hospitals and clinics) is the fifth most energy consuming one in Cyprus (excluding housing) as a whole, while being the fourth most energy demanding one per m^2 [11].

In Cyprus, Health Services in the public sector are provided by six general hospitals, three specialized centres, two small rural hospitals and 39 health centres, as well as many sub-centres for provision of primary health care services. In 2016, the private sector included a total of 74 hospitals, polyclinics and diagnostic centres, all profit organizations. The energy source breakdown per energy need refers to an agglomeration of results from 83 public and private hospitals and clinics of an average area of 17.354 m^2 [11] (Figs. 7, 8, and 9).

The average national final energy consumption of private healthcare institutions in Cyprus is 386.30 KWh/m^2 [11]. The average primary energy consumption of public hospitals rises to 509 KWh/m^2 [12]. This indicates the great potential of energy savings that lays within the public and private healthcare sector. Both in the case of public and private hospitals, primary energy consumption can be reduced by up to 86% if a nearly zero-energy refurbishment is implemented [12, 13].

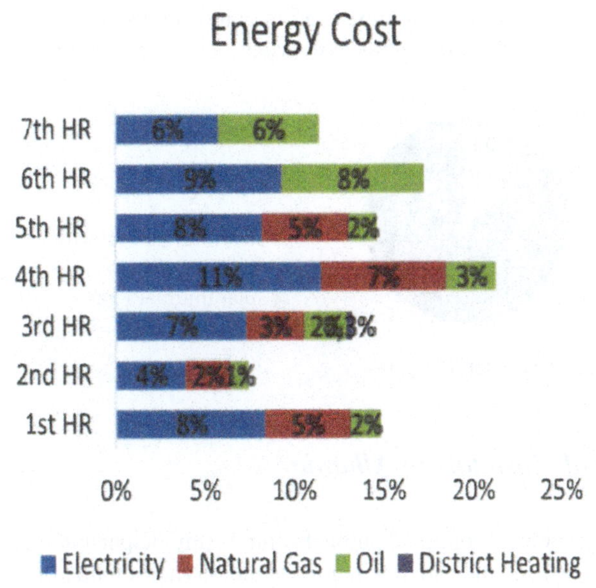

Fig. 6 Cost percentage at the total annual energy cost for each HR

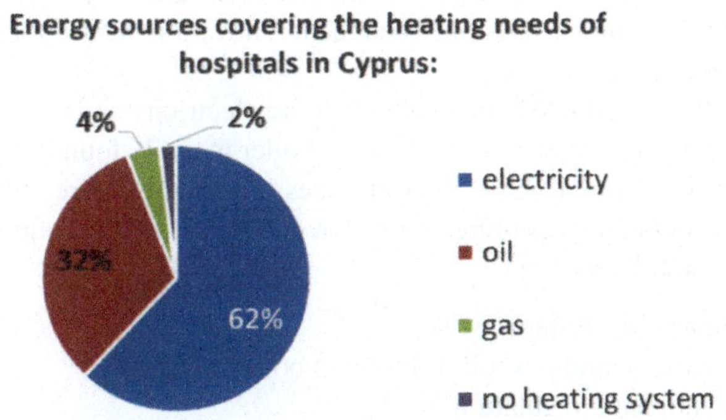

Fig. 7 Energy sources covering the heating needs of hospitals in Cyprus

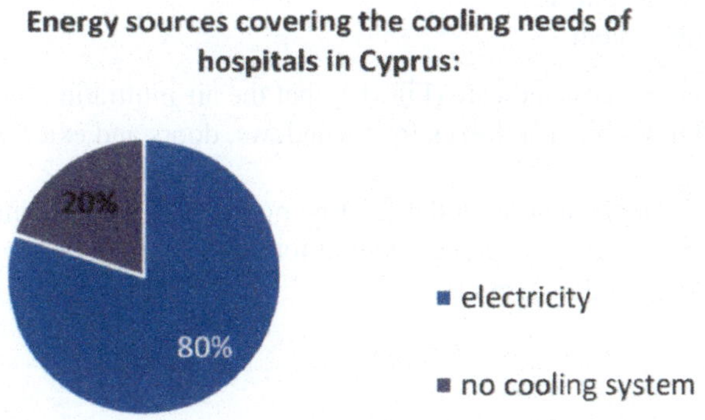

Fig. 8 Energy sources covering the cooling needs of hospitals in Cyprus

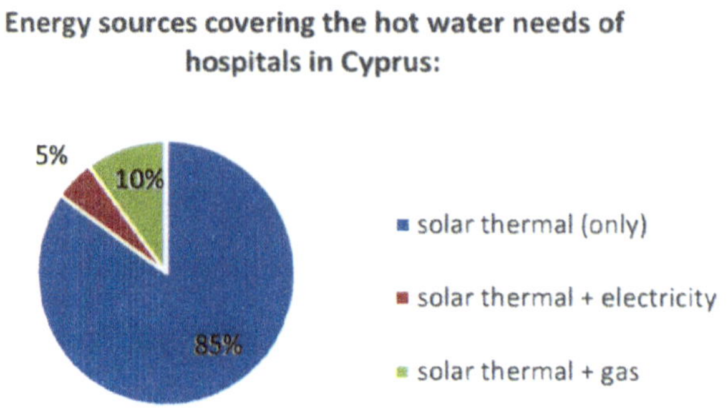

Fig. 9 Energy sources used for hot water in hospitals in Cyprus

3.3 Energy Performance of Hospitals in Albania

In Albania, hospital services are provided through 42 public hospitals (municipal and regional), the most of which were built before 1990 (between 1960 and 1990) and they are not in good situation (non-quality of construction materials, lack of maintenance and investment). Some of them have been renovated in the last decade. The renovations could be categorized into two phases. In the first phase, the old frames and windows were replaced by single-glazed aluminium framed windows and recently in the second phase by double-glazing windows.

The typical material for wall construction is plastered solid brick (U-value = 1.3–1.6 W/m^{2}K). Most buildings have flat roofs or unused loft space with no or limited insulation. The recent renovation intervention includes insulation reinforcement of some flat roofs in Albanian hospitals.

The main energy sources which are used in Albanian hospitals are electricity and oil, especially for heating (boiler) during the winter or domestic hot water (DHW) production while a gas boiler is rarely found. In the last years, pellet is another fuel that is used for heating or DHW production. Also, in cold zones, wood stoves are still in use in medical care buildings. Energy is used to cover the needs for heating, cooling, hot water, cooking, electrical equipment and lighting. The main characteristics of HVAC systems are the follows:

- Both decentralized and central heating systems exist.
- Recently renovated buildings have central gas, oil or biomass boilers.
- Mostly central system has no automatic regulation.
- DHW mostly is made through heating system boilers or decentralized electric boilers.
- No mechanical ventilation exists (ventilation is mainly done through open doors and windows).
- Many public hospitals have a central heating system but no cooling systems.
- Cooling is mostly realized through split AC.
- Only surgery rooms have HVAC system.

The data from hospitals auditing process indicate (Fig. 10) that the air infiltration is mainly responsible for the majority of heat losses (about 35%) and follows the heat losses from windows, doors and external walls (about 25%) due to insufficient insulation.

Concluding, the physical and thermal situation of the existing public Albanian hospitals (new and old stock) is generally not good. For this reason, this stock represents a great potential for energy saving by taking energy efficient measures (thermal insulation, efficient window/door, lighting, etc.).

4 Conclusions

Worldwide, about 6% of the total energy consumption in the buildings sector is represented by energy usage in medical centres. Hospitals present a great potential for energy and cost savings because the simultaneous demand of energy uses for the entire year allows several actions for energy efficiency improvement. The EPBD introduces the notion of zero-energy buildings (ZEB) and it can apply to non-residential buildings and is specific to hospitals, in order to achieve high-energy performance and low-specific energy requirement.

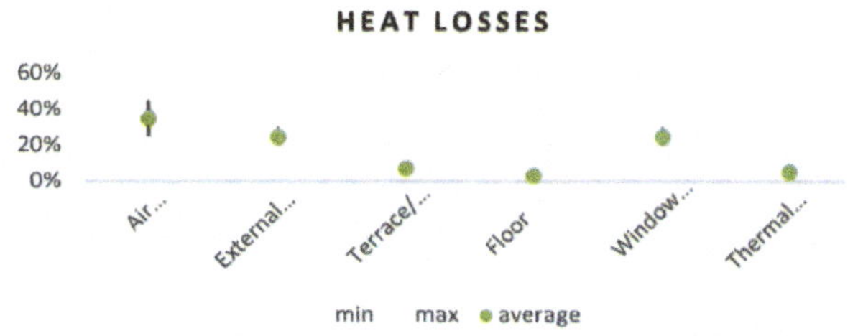

Fig. 10 Heat losses of hospital buildings in Albania

In Greece, the deviation of the primary energy consumption per m^2 of each hospital from the desired target value for nZEB tertiary buildings is about 81.8%. The energy demands are particularly high for electricity use (44%), and follows for natural gas (32%), for oil (22%) and for district heating (only 1%). In Cyprus, electricity covers the majority of energy needs for heating (62%) and cooling (80%). For heating, oil (32%) and gas (4%) are used. For domestic hot water, 85% of hospitals use solar collectors and 15% use electricity.

In Albania, the main energy sources are electricity and oil. Fuel oil is used to cover space heating (boiler) and domestic hot water (DHW) needs. Hospital buildings in cold zones still use wood stove . Also, recently renovated buildings have central gas, oil or biomass boilers. It should be noticed that in most cases, no mechanical ventilation exists; ventilation is mainly through open doors and windows and only surgery rooms have HVAC system. So, this research indicates the great challenge for interventions to improve the energy efficiency behaviour of buildings and their upgrading into nZEB Hospitals.

Acknowledgements This work was performed in the frame of the "ZenH Balkan" project and was co-funded by the Interreg Balkan-Mediterranean Programme and National Funds.

References

1. Commission Staff Working Paper (2012) Analysis of options beyond 20% GHG emission reductions: member state results
2. EU Publications (2018) High energy performing buildings, Support for innovation and market uptake under Horizon 2020 energy efficiency
3. European Commission (2002) Energy performance of buildings. Directive, 2002/91/EC of the European Parliament and of the Council. Official Journal of the European Community, Brussels
4. D'Agostino D, Mazzarella L (2019) What is a nearly zero energy building? Overview, implementation and comparison of definitions. J Build Eng 21:200–212
5. D'Agostino D, Cuniberti B, Bertoldi P (2017) Energy consumption and efficiency technology measures in European non-residential buildings. Energ Buildings 153:72–86
6. Maassen W (2017) (nearly) Zero Energy Hospital Buildings, REHVA, TVVL, Royal HaskoningDHV
7. Balaras C, Dascalaki E, Gaglia A (2007) HVAC and indoor thermal conditions in hospital operating rooms. Energ Buildings 39:454–470
8. Gaglia AG, Balaras C, Mirasgedis S, Georgopoulou E, Sarafidis Y, Lalas D (2007) Empirical assessment of the Hellenic non-residential building stock, energy consumption, emissions and potential energy savings. Energ Conver Manage 48:1160–1175
9. Papamanolis N (2015) The first indications of the effects of the new legislation concerning the energy performance of buildings on renewable energy applications in buildings in Greece. Int J Sustain Built Environ 4(2):391–399
10. MEE (2017) National Plan for increase of buildings' number with nearly zero energy consumption, Ministry of Environment and Energy (Article 9, Directive 2010/31/EE for the 'Energy Performance of Buildings').
11. Economidou M, Zangheri P, Paci D (2017) Long-term strategy for mobilizing investments for renovating Cyprus national building stock. (Technical report JCP No. D1.8)
12. Ministry of Energy, Trade and Commerce (MECIT) (2015)
13. Serghides D, Dimitriou S, Kyprianou I (2019) Paving the way towards zero energy hospitals in the Mediterranean region. Mediterranean Green Buildings and Renewable Energy Forum, Florence

Proposed Solar Kiosk for COVID-19 and Other Infectious Disease Screening in Nigeria

Amina Batagarawa and Yakubu Aminu Dodo

1 Background

Coronavirus disease 2019 (COVID-19) is an exceptionally pathogenic human coronavirus and is primarily transmitted by person-to-person contact in community and healthcare settings [24]. Person-to-person transmission has been recognised among individuals who are in close interaction with one another (a range of 2 m/6 ft), predominantly through respiratory droplets. Droplet communication occurs when respiratory droplets produced via coughing, sneezing or talking contact susceptible mucosal surfaces, such as the eyes, nose or mouth. Transmission may also occur indirectly via contact with contaminated fomites with hands and then mucosal surfaces. Respiratory droplets are huge and are not able to remain suspended in the air; in this way they are more often than not scattered over brief separations [19].

As of September 15, 2020, Nigeria has 69 testing laboratories covering its 36 states and the Federal Capital Territory (FCT). There is an average of one laboratory in most states, with Enugu having the most laboratories at only six [18]. The Nigerian government has ranked testing as one of the key processes towards finding a sustainable solution for the COVID-19 pandemic with plans to rapidly scale diagnostic testing to cater for the increasing number of infections across the country.

On a global scale, worldwide deficiencies of diagnostic kits and test centre consumables progressively affect the ideal usefulness of the research facility framework in Nigeria; therefore, innovative testing approaches are required.

2 Literature Review

COVID-19 survives on contaminated dry surfaces and fomites, which facilitate hand to mucous membranes (of the mouth, nose and eyes) spread. The survival of the virus's ability to survive on inanimate surfaces poses challenges in the site of screening for the infection [26], which require mechanisms to collect specimens safely and efficiently while minimize needs for personal protection equipment. Self-collection of samples is widely used for monitoring of other infectious diseases and health conditions; however, more research is required to test the viability and feasibility of such an approach, especially in a developing country such as Nigeria.

Based on portable COVID-19 screening models such as COVSACK [13], patient screening centres sometimes resemble phone booths, and allow people to get professional medical counsel, along with a COVID-19 swab test, contact-free. They may be housed under medical tents, usually found just outside of emergency rooms, which can be manufactured quickly, then deployed to hospitals in need. A Nigerian version, the Elite Solar Kiosk [30], provides 100% off-grid, one-stop, multi-user quick-build that is easy to assemble when required. While Nigerian protocol presently limits screening and treatment in the domain of the Nigerian Centre for Disease Control, neighbouring countries like Senegal are expanding into the use of rapid result test kits [28].

A. Batagarawa (✉) ·
Department of Architecture, Faculty of Environmental Design, Ahmadu Bello University, Zaria, Nigeria
e-mail: amina.batagarawa@bazeuniversity.edu.ng

Y. A. Dodo
Department of Architecture, Faculty of Engineering, and Architecture, Istanbul Gelisim University, Istanbul, Turkey
e-mail: yadodo@gelisim.edu.tr

© The Author(s), under exclusive license to Springer Nature Switzerland AG 2022

A. Sayigh (ed.), *Sustainable Energy Development and Innovation*, Innovative Renewable Energy,
https://doi.org/10.1007/978-3-030-76221-6_84

The logistics of testing is identified as a major risk in the testing up-scaling plan for Nigeria. With a highly contagious virus, robust plans for safe access to testing facilities, storage and transportation of clinical samples are vital. There is a plan to harness traditional logistical capabilities alongside new digital solutions while providing people with clear information about what is happening throughout the process [18]. Other risks related to sample collection include simple technical errors, inadequate collection, improper handling and shipping [24].

The Nigeria Centre for Disease Control (NCDC) implements COVID-19 diagnosis by real-time reverse transcriptase-polymerase chain reaction (rRT-PCR) on respiratory specimens such as nasopharyngeal swabs [18, 24]. For specimen collection, healthcare workers MUST wear gloves, N95, face shield/goggles and apron [19]. None of these actions requires electricity unless night-time collection is specified.

Within the sample collection process itself, there is minimal need for electricity, except for refrigerating the samples [31]; an alternative to refrigeration is the use of ice packs in insulated boxes. The actual testing process is one nasal and one oro-pharyngeal swab, which involves inserting a swab into the mouth and rubbing the swab tip against the back of the nose and throat for 20 seconds on the left side, then 20 seconds on the right side while avoiding touching the tongue, teeth and gums. The swab is inserted back into the tube of viral transport media, after breaking the swab at the score and capping the tube [19, 26]. Sputum samples can be collected if a patient has a productive cough.

The WHO [25] has recently approved the adoption of the rapid testing that detects IgG and IgM anti-SARS-CoV-2 in blood samples that may be combined with any cloud-based platform that allows remote monitoring of tests, accessing results and collaboration. The E-HUBB screening model is operated in a three-step procedure comprising screening, treatment and tracking [25].

The NCDC policy is to report any positively tested patient to the isolation centres in the state, thereby reducing the need for treatment in any portable screening option. The sample, however, needs to be registered [9, 16, 17, 29], which may require ICT equipment for electronic or remote approaches, or may be done manually.

Packaging of the sample, donning and doffing of personal protective gear (PPE) using CDC sample guidelines uses no electricity directly [15]. In the event that night-time sample collection is specified, lighting is required.

Similarly, there is no electricity requirement for performing hand hygiene according to the CDC [3, 4]. The CDC recommends hand sanitisation using solutions other than soap and water in most clinical circumstances due to the proven superior compliance compared to cleanser and water. A tank may be incorporated using gravity to supply water mounted above the kiosk to supply the kiosk with water in the event of soiled hands for the collectors and patients.

A design of an adequate ventilation diffuser system to ensure proper mixture of air without turbulence and proper differential pressures are maintained between spaces and within the building envelope is required [14]. Poorly designed air supply and exhaust may lead to increase in personnel infection [7]. Design considerations for ventilation include:

- To place the laboratory exhaust fans on emergency power
- To split the exhaust capacity into more fans with lower horsepower to lower in-rush current as each fan is started
- To design a fan system consisting of a minimum of two or more fans for better reliability if one fan fails

Due to the portable and mobile nature of kiosks suitable for the screening of patients for COVID-19 and other epidemic infectious diseases, energy demand and supply is vital for the following processes as specified by the NCDC [16, 17]:

- Sample collection, packaging and transportation
- Donning and doffing of PPE
- Infection prevention and control (IPC)
- Waste management
- Filling of case investigation forms (CIF), among others

This research aims to explore the potential of solar PV technology for powering a screening kiosk for infectious diseases like COVID-19 in Nigeria.

3 Methodology

The study employs a quantitative approach. Electricity consumption data are collected from primary and secondary sources. The data from primary sources are based on interviews with professionals in the health sector who are directly linked to the COVID-19 response in Nigeria, while the data collected from secondary sources is collected through a review of published literature from world renowned establishments like the World Health Organisation (WHO), Centre for Disease Control

(CDC) and Nigeria Centre for Disease Control (NCDC). The data collected from both primary and secondary sources include:

1. Type of equipment
2. Energy consumption of equipment
3. Hours of usage of equipment

Electronic questionnaires, employing a snowballing approach, are answered by 20 respondents. This activity is conducted to establish the sample collection procedures employed or recommended in Nigeria and the type of equipment and their usage. These professionals are virologists and other medical personnel. From the data, a list of potential electricity-consuming appliances is developed and average consumption of the appliances is estimated through quantitative analysis of the demand-side data.

The spaces required in the sample collection kiosk include:

1. Entrance
2. Clock-in and out
3. Changing room
4. Toilet
5. Main sample collection space (collector/registration)
6. Main sample collection space (patient)

Considering the ready availability of storage containers in Nigeria, they are recommended for retrofitting to mobile testing kiosks [12, 22]. The kiosk roof space is considered as the basis to estimate electricity-generating capacity of the 20 and 40 ft kiosk called $Solarkiosk_{20}$ and $Solarkiosk_{40}$ respectively. The design of PV system for the $Solarkiosk_{20}$ is conducted using a system configuration of 5 and 11 units of 300 watts mono solar panels. This wattage is most prevalent in the Nigerian market (NASENI 2020). The system is tested at an optimal 15° tilt angle facing at an azimuth of 10° [2], whereas losses from cables and inverters are assumed to be 14% which is given by the simulation tool.

The estimated values from average consumption and electricity generated are compared and the percentage of electricity that can be supplied through solar PV panels is calculated using The Joint Research Centre (JRC) PVGIS-5 [27] with the configurations below:

Provided Inputs
Latitude/Longitude: 9.063, 7.462
Horizon: calculated
Database used: PVGIS-SARAH
PV technology: crystalline silicon
PV installed: 1.5 kWp
System loss: 14%
Slope angle: 15 (opt)°
Azimuth angle: 10 (opt)°

4 Results

This section covers the estimated demand of the kiosks and the estimated supply from the solar PV design. The estimated demand and supply are compared to measure what component of the demand can be satisfied by the estimated supply.

4.1 Estimated Energy Demand in Existing Sample Collection Centres

Based on the identified spaces of the solar kiosk, the appliances that consume electricity, their usage and average electricity consumption is estimated in Table 1. The estimated daily demand is 9 kWh, while the annual demand is 3131 kWh.

The kiosks are designed to employ LED lights, supply and exhaust ventilation fans, mechanical cooling and refrigeration of samples, remote registration and automated temperature testing. Following the recommendations of the WHO, the proposed solar kiosks would provide room for rapid testing that detects IgG and IgM anti-SARS-CoV-2 in blood samples in a

Table 1 Power consumption of laboratory appliances [10, 23, 25]

Spaces	Appliances	Energy rating (kW)	Quantity	Usage (hr/day)	Average total consumption (kWh/day)	Average annual consumption (kWh/yr)
Entrance/	Lighting	0.011	2	9	0.198	72.27
Clock-in and -out					0	0
Changing room					0	0
Toilet	Lighting	0.011	1	9	0.099	36.135
Main sample collection space (collector)	Ventilation					
	Supply fans	0.0273	2	12	0.6552	239.148
	Exhaust fans	0.0275	2	12	0.66	240.9
	Printer					
	Print mode	0.0365	1	3	0.1095	39.9675
	Idle mode	0.0168	1	15	0.252	91.98
	Laptop	0.4	1	9	3.6	1314
	Fridge	0.5	1		0.6356	232
	Lighting	0.025	2	9	0.45	164.25
	Power outlet (phone charging)	0.006	1	3	0.018	6.57
Main sample collection space (patient)	Cooling	1.103	1		0.90	330
	Internet connection	0.01	1	18	0.18	65.7
	Security lighting	0.028	4	6	0.672	245.28
	Thermal temperature test	0.008	1	18	0.144	52.56
Total					*9*	*3131*

cassette format [5] in combination with any cloud-based platform that would allow remote monitoring of runs, access results and collaboration.

4.2 Estimated Supply from the Solar PV System Configuration

The configuration of SolarKiosk_{20} and SolarKiosk_{40} is shown in Figs. 1 and 2.

Solarkiosk_{20} and Solarkiosk_{40} are designed with 5 and 11 mono-crystalline solar panels running on 24 V and 300 W generating 1.5 kWp and 3.3 kWp installed nominal power respectively. The yearly PV energy production predicted after losses for Solarkiosk_{20} and Solarkiosk_{40} is 2322.1 kWh and 5108.61 kWh as shown in Tables 2 and 3. The yearly irradiation on the 15° incidence plane specified is 2109.98 kWh/m^2.

The predicted monthly performance from the simulations is shown in Table 3. Figures 3 and 4 show the monthly pattern of supply with lowest output corresponding to the peak of the rainy season experienced in Abuja.

Where:

- *E_m_{20}: Average monthly electricity production from the 20 foot container [kWh]*
- *E_m_{40}: Average monthly electricity production from the 40 foot container [kWh]*
- *H(i)_m: Average monthly sum of global irradiation per square meter received by the modules of the given system [kWh/m^2]*
- *SD_m_{20}: Standard deviation of the monthly electricity production from the 20 foot container due to year-to-year variation [kWh]*
- *SD_m_{40}: Standard deviation of the monthly electricity production from the 40 foot container due to year-to-year variation [kWh]*

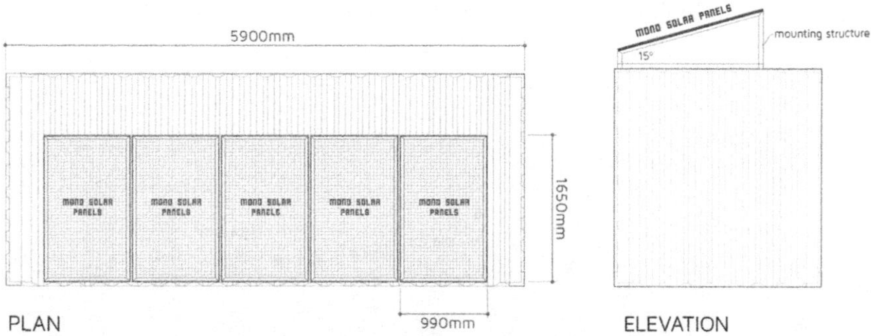

Fig. 1 Solar panels configuration on 20 ft container kiosk. (Source: Author)

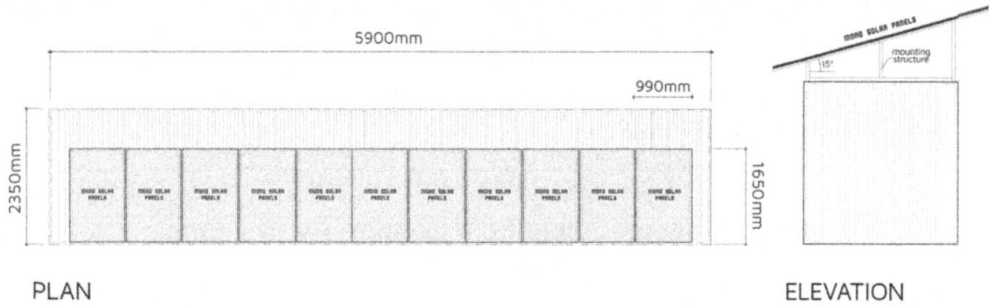

Fig. 2 Solar panels configuration on 40 ft container kiosk. (Source: Author)

Table 2 Predicted simulation output for $SolarKiosk_{20}$ and $SolarKiosk_{40}$

Simulation outputs	$SolarKiosk_{20}$	$SolarKiosk_{40}$
Yearly PV energy production	2322.1 kWh	5108.61 kWh
Yearly in-plane irradiation	2109.98 kWh/m^2	
Year-to-year variability	47.01 kWh	103.43 kWh
Angle of incidence	−2.8%	
Spectral effects:	0.66%	
Temperature and low irradiance	−12.8%	
Total loss	−26.63%	

Table 3 Predicted monthly performance for $SolarKiosk_{20}$ and $SolarKiosk_{40}$

Month	E_m_{20}	E_m_{40}	SD_m_{20}	SD_m_{40}	H(i)_m
January	239.4	526.7	8.5	18.7	219.7
February	211.5	465.3	9.6	21.2	196.3
March	218.2	479.9	9.9	21.7	203
April	195.7	430.4	9.5	20.8	180.3
May	179.3	394.6	7.5	16.4	162.5
June	156.7	344.7	7.9	17.4	139.4
July	149.6	329.1	9.7	21.4	132.3
August	141.6	311.5	8.3	18.2	124.4
September	164.9	362.9	9.2	20.3	146.5
October	198.9	437.6	12	26.5	179.1
November	227.1	499.7	6.2	13.7	207.7
December	239.2	526.2	5.5	12.1	218.5
Total	*2322*	*5109*	*104*	*228*	*2110*
Average	**194**	**426**	**9**	**19**	**176**

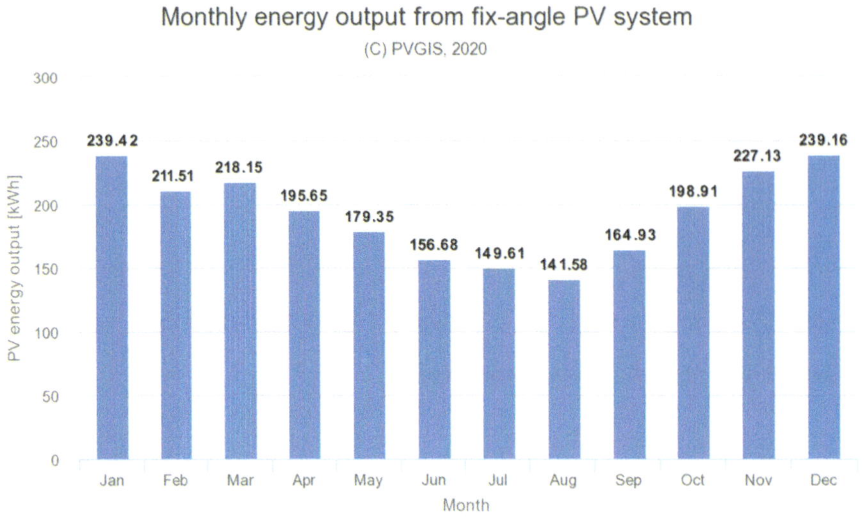

Fig. 3 SolarKiosk$_{20}$ monthly energy output. (Source: PVGIS)

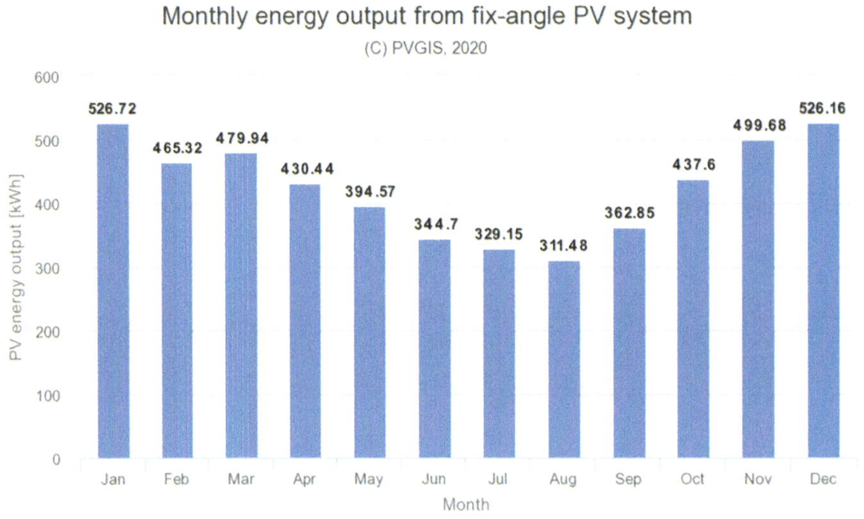

Fig. 4 SolarKiosk$_{40}$ monthly energy output. (Source: PVGIS)

5 Conclusion

Due to the portable and mobile nature of kiosks suitable for the screening of patients for COVID-19 and other epidemic infectious diseases in Nigeria, energy demand and supply is vital. This research aims to explore the potential of solar PV technology for powering a screening kiosk for infectious diseases like COVID-19 in Nigeria.

The total electricity demand estimated for the activities in the solar kiosks is 3131 kWh/yr for a combined sample collection, patient administration, medical personnel administration, HVAC and lighting.

Considering the ready availability of storage containers in Nigeria, they are recommended for retrofitting to mobile testing kiosks. The kiosk roof space is considered as the basis to estimate electricity-generating capacity of the 20 ft and 40 ft kiosk called Solarkiosk$_{20}$ and Solarkiosk$_{40}$ respectively. The design of PV system for the Solarkiosk$_{20}$ is conducted using a system configuration of 5 and 11 units of 300 watts mono solar panels. This wattage is most prevalent in the Nigerian market. The system is tested at an optimal 15° tilt angle facing at an azimuth of 10°. While losses from cables and inverters are assumed to be 14%, which is given by the simulation tool, electricity generated are compared and the percentage of electricity that can be supplied through solar PV panels is calculated using The Joint Research Centre (JRC) PVGIS-5: The predicted supply is 2322 kWh and 5109 kWh for Solarkiosk$_{20\ \text{and}}$ Solarkiosk$_{40}$ respectively. In comparison to the demand, utilising the configuration of Kiosk$_{40}$ is better suited to supply the required power.

References

1. ASHRAE Technical Committee 9.10 (2018) Classification of laboratory ventilation design levels. ASHRAE, Atlanta. Retrieved from https://www.ashrae.org/File%20Library/Technical%20Resources/Free%20Resources/Publications/ClassificationOfLabVentDesLevels.pdf
2. Batagarawa A, Abdulkarim YA, Sagada ML (2020) Optimizing building form for integration of solar photovoltaic in the design of a textile industry in Katsina, Nigeria. In: Sayigh A (ed) Innovative renewable energy, vol 7. Springer, pp 257–279. https://doi.org/10.1007/978-3-030-18488-9_20
3. Centers for Disease Control and Prevention (2020) (U. D. Services, Editor). Retrieved from https://www.cdc.gov/about/default.htm
4. Centers for Disease Control and Prevention (2020) Guidelines for submitting specimens to CDC. Retrieved from COVID-19-Specimen-Submission-Guidance: https://www.cdc.gov/coronavirus/2019-ncov/downloads/COVID-19-Specimen-Submission-Guidance.pdf
5. Covid Tests Shop (2020) COVID-19 rapid antibody IgG/IgM test (Whole. Blood/Serum/Plasma). Pack of 2. FDA and CE. Retrieved from COVID Tests Shop: https://www.covidtests.shop/product/covid-single-test-cassette/
6. DaklaPack Medical (2020) UN 3373 Category B schematic for packaging. (DaklaPack Medical). Retrieved from UN3373.com: https://www.cdc.gov/coronavirus/mers/downloads/lab/UN3373-packaging-schema.pdf
7. D'Cruz, K (1999) Design guidelines for laboratory exhaust fans and stacks. 1999 ASHRAE Winter Meeting. Chicago: American Society of Heating, Refrigerating and Air-Conditioning Engineers, Inc (ASHRAE). Retrieved 7 May 2020, from https://www.aivc.org/sites/default/files/airbase_11909.pdf
8. Department of Facilities Services, University of North Carolina (2013) Laboratory design guidelines. University of North Carolina, North Carolina. Retrieved from https://facilities.unc.edu/files/2016/03/Laboratory-Design-Guidelines.pdf
9. Department of Health, Rhode Island (2020) 2019-novel coronavirus (COVID-19) specimen collection kit. Department of Health, Rhode Island, Rhode Island. Retrieved from https://health.ri.gov/publications/instructions/COVID-19-Specimen-Collection-Kit.pdf
10. EnergyStar (2020) Refrigerators. Retrieved from https://www.energystar.gov/most-efficient/me-certified-refrigerators/results
11. Environmental Protection Agency (2020) List N: disinfectants for use against SARS-CoV-2. United States. Retrieved from https://www.epa.gov/pesticide-registration/list-n-disinfectants-use-against-sars-cov-2
12. Hellenic Shipping News Worldwide (2020) Nigerian ports: where there's no ease of doing business. Retrieved from Hellenic Shipping News Worldwide: https://www.hellenicshippingnews.com/nigerian-ports-where-theres-no-ease-of-doing-business/
13. Joshi JR (2020) COVSACK: An innovative portable isolated and safe COVID-19 sample collection kiosk with automatic disinfection. Trans. Indian Natl. Acad. Eng. 5:269–273
14. Laboratories for the 21st Century (2011) Laboratories for the 21st century: best practices. National Renewable Energy Laboratory. US Department of Energy
15. National Center for Immunization and Respiratory Diseases (NCIRD), Division of Viral Diseases (2020) Using personal protective equipment (PPE). Retrieved from Centre for Disease Control and Prevention: https://youtu.be/of73FN086E8
16. NCDC (2020) A step by step guide on setting up a sample collection centre. Nigeria Centre for Disease Control, Abuja
17. NCDC (2020) National interim guidelines for clinical management of COVID-19. Nigeria Centre of Disease Control, Abuja
18. Nigeria Centre for Disease Control (2020) National strategy to scale up access to coronavirus disease testing in Nigeria. Nigeria Centre for Disease Control, Abuja
19. Nigeria Centre for Disease Control (2020) National interim guidelines for clinical management of COVID-19. Nigeria Centre for Disease Control, Abuja
20. Panasonic (2020) WhisperFresh Select™ Fresh Air Supply Fan
21. Panasonic (2020). WhisperLine™ Remote Mount In-Line Fan 120 CFM
22. Quality Storage Container (2020) Shipping containers available in different dimensions. Retrieved from Quality Storage Container: https://www.qualitystoragecontainer.com/dimensions.html
23. Roth K, Goldstein F, Kleinman J (2002) Energy consumption by office and telecommunications equipment in commercial buildings, U.S. Department of Commerce, National Technical Information Service (NTIS). Arthur D. Little, Inc, Cambridge, MA. Retrieved from http://biblioite.ethz.ch/downloads/Roth_ADL_1.pdf
24. Shidham VB, Frisch NK, Layfield LJ (2020) Severe acute respiratory syndrome coronavirus 2 (the cause of COVID 19) in different types of clinical specimens and implications for cytopathology specimen: An editorial review with recommendations. CytoJournal 17(7):1–9
25. Solar Kiosk (2020) Solarkiosk Solutions GmbH. Retrieved from Solarkiok: solarkiosk.eu
26. Sullivan PS, Sailey C, Guest JL, Guarner J, Siegler AJ, Valentine-Graves M et al (2020) Detection of SARS-CoV-2 RNA and antibodies in diverse samples: protocol to validate the sufficiency of provider-observed home-collected blood, saliva and oropharyngeal samples. JMIR Public Health and Surveillance
27. The Joint Research Center (JRC) (2019) PVGIS-5. Retrieved 24 Aug 2020, from The European Commission's science and knowledge service: https://ec.europa.eu/jrc/en/PVGIS/releases/pvgis5
28. The World News (2020) Researchers in Senegal are developing a coronavirus test kit to be used across Africa. Retrieved from TheWorld: www.pri.org/stories/2020-05-11/Researchers-in-Senegal-are-developing-a-coronavirus-test-kit-to-be-used-across-Africa
29. US Department of Health and Human Services (2020) Influenza specimen collection. Centres for Disease Control and Prevention. Retrieved from https://www.cdc.gov/flu/pdf/professionals/flu-specimen-collection-poster.pdf
30. Vanguard News (2020) Agency develops Solar powered kiosk for Covid-19 quick test centre. Retrieved from Vanguard: https://www.vanguardngr.com/2020/07/agency-develops-solar-powered-kiosk-for-covid-19-quick-test-center/
31. World Health Organisation (2006) Collecting, preserving and shipping specimens for the diagnosing of avian influenza A(H5N1) virus infection: Guide for field operations. WHO

Optimising Building Energy Consumption Using Energy Monitoring and Management: Case Study in University of Reading, UK

Shan Zhou, Baizhan Li, Hong Liu, Yuxin Wu, and Runming Yao

1 Introduction

The UK government has committed the UK to legally binding targets to reduce carbon emissions by 80% by 2050 [1]. Considering targets of saving energy and reducing carbon emission, building energy consumption has become topical, and the significant amount of building energy consumption needs serious attention. One of the drivers for reducing carbon emission and improving energy efficiency is 'reduced operating costs through lower energy use' [2]. In this regard, energy monitoring and management is a powerful strategy to record energy usage data and then find potential ways for energy saving. Energy monitoring and management has benefited significantly from the recent availability of low-cost and easy-to-install sensors, and advances in data storage, analysis and visualisation.

Sensor network technologies enable providing basic data for statistical analysis [3], and savings of 5–10% of energy use are typically attributed to effective metering supported by monitoring and corrective action [2]. Monitoring sensors can collect data illustrating where and how energy is used in order to provide a defined baseline of energy consumption. Further, opportunities can be identified for potential savings, and early warnings of unexpected excess consumption can be given [4]. The parameters that can be monitored mainly include electricity consumption, environmental parameters and occupant activities. Metering electricity consumption is widely used, and it is directly metered using an ammeter and a voltmeter to measure electric current and voltage respectively. In addition to traditional monitors above, nowadays wireless sensor networks (WSNs) can also be used as communication technologies because of its multiple advantages. However, cyber security problem arises because of the risk of information leakage [5], and the strategic functions, including energy metering, monitoring and targeting, are underutilised [2].

Many researches [6] studied data analysis for building energy management systems (BEMS). The system usually consists of a set of high-quality sensors, data acquisition software and hardware and data visualisation software, including a web-based remote access system, that can be used to identify control problems and equipment faults. The CIBSE Guide F [2] gives directions for energy data analysis with three main data sources: energy invoice data (bills), half-hourly energy data and sub-meter readings. The guidance [2] are as follows: (1) Energy invoice data can be used digitally with half-hourly or smaller intervals to indicate energy use with electricity, gas or oil. The data can be analysed into histograms, averages, day/night differences, seasonal differences and baseloads, etc. with occupancy, day/night ratios and seasonal factors. (2) Half-hourly energy data are another data type generated from energy suppliers that is available for electricity and gas

S. Zhou (✉) · B. Li · H. Liu · Y. Wu
Joint International Research Laboratory of Green Buildings and Built Environments (Ministry of Education), Chongqing University, Chongqing, China

National Centre for International Research of Low-carbon and Green Buildings (Ministry of Science and Technology), Chongqing University, Chongqing, China
e-mail: shanzhou@cqu.edu.cn

R. Yao
Joint International Research Laboratory of Green Buildings and Built Environments (Ministry of Education), Chongqing University, Chongqing, China

National Centre for International Research of Low-carbon and Green Buildings (Ministry of Science and Technology), Chongqing University, Chongqing, China

School of the Built Environment, University of Reading, Reading, UK

© The Author(s), under exclusive license to Springer Nature Switzerland AG 2022

A. Sayigh (ed.), *Sustainable Energy Development and Innovation*, Innovative Renewable Energy,
https://doi.org/10.1007/978-3-030-76221-6_85

consumptions for buildings and sites. The time intervals are also typically half an hour or 15 minutes. The data can be used on a real-time basis with monitoring sensors and energy systems to alarm unexpected high load immediately and take measures promptly. However, energy data are meeting the challenge of occupant security because of the availability of real-time energy use. It can also be translated into spreadsheets with historical data to compare seasonal/weekly changes, daily trends and differences regarding different energy factors. (3) The third way to use energy data is sub-meter readings to obtain more detailed energy use report. Compared with the first two methods, in this way, the building/site energy consumption is given in a more accurate and specific way by breaking down the total energy consumption into different using categories and building sections. Considering information security issue and existing monitoring system installed in University of Reading, the second way is chosen as the analysing bases.

Accordingly, the aim of this study is to demonstrate the capability of sensor metering and data analysis to achieve reduced energy consumption and lower carbon emission. Energy management methods are introduced with a case study in Chancellor's building of University of Reading with both constant and variable electricity consumption. After statistical analysis, the potential ways to achieve lower energy use and lower operating costs can be found. Finally, recommendations are proposed to reduce the electricity consumption according to the data analysis.

2 Case Study

The case study selected Chancellor's building of University of Reading, UK, as the research building. The Chancellor's building was rated "B" in display energy certificates (DEC) [2], which was far below the average energy consumption level. Also, it consumed lower electricity compared which other buildings in the campus.

Chancellor's building has two storeys, and consists of classrooms, computer labs, staff offices and other rooms/spaces for meeting, resting and talking. On weekdays, the building is used for courses and lectures. As the same time, it has many individual offices for staff to work. The building is also open on weekends, when the classrooms and computer labs are available for students to study. As a result, the main function of this building is providing space for study and work to students and staff.

The energy sources of this building are electricity and central heating. In winter, central heating is provided through radiators to heat the spaces. In comparison, electric plugs supply energy for other energy demands, including using media devices, using office facilities and lighting.

3 Results and Discussion

The half-hourly energy data are displayed on the University of Reading sustainability services website [7]. The daily electricity consumption data are shown in Fig. 1 which are divided into constant and variable load with grey and green colour, respectively, due to hourly fluctuation during the day. Constant electricity consumption (baseload) means the daily lowest half-hourly electricity consumption to maintain the basic operation, for example, basic lighting, device standbys and working of sensors, while variable electricity consumption is the area of electricity consumption above the baseload line (see Fig. 2). The duration from 1 January 2018 to 30 April 2018 was selected due to good data continuity. Constant and variable electricity consumption data are analysed separately in this study to provide targeted energy saving solutions. After excluding data outliers, the average daily constant loads of January, February, March and April were 160, 187, 167, and 137 kWh, respectively. Similarly, the average daily variable load of the four months were 354, 374, 322, and 191 kWh.

3.1 Constant Load (Baseload)

Baseloads are shown in Fig. 2 as the horizontal lines. The lowest electricity consumption usually appeared between 9 p.m. and 3 a.m. according to the monitoring data when occupancy rate was the lowest during a day, especially between 10 p.m. and 12 p.m. One-way analysis of variance (ANOVA) showed that there was no statistically significant difference among January, February and March (P-value >0.05). By contrast, a significant difference existed among the 4 months (P-value <0.01), which means the baseload witnessed a distinct decline in April. Tracing back to April 2018, most of the days were school holidays of spring term. Supposing less staff and students were using the building, more electronic devices (printers,

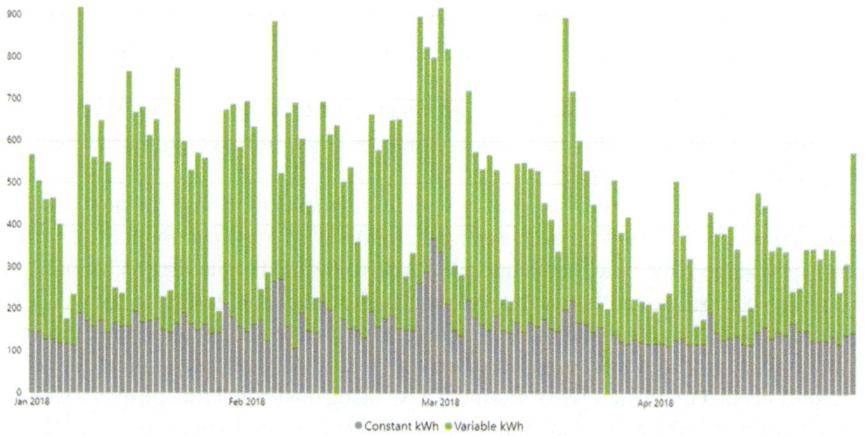

Fig. 1 Constant and variable electricity consumption in Chancellor's building from 1 January 2018 to 30 April 2018

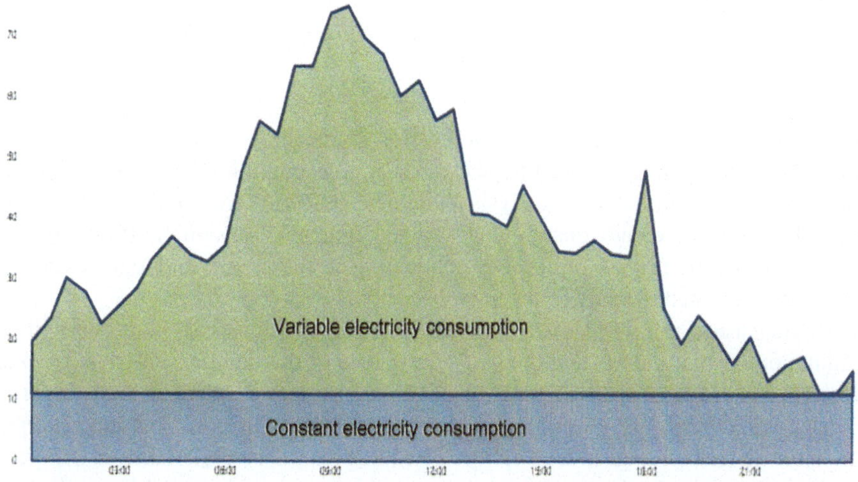

Fig. 2 An example of variable and constant electricity consumption on 26 February 2018 (Monday) (Unit: kW)

copying machines and desktop computers) were shut down instead of standby in school days. The power consumption of printers on standby mode could be 25–60 Watts, desktop computers being about 70 Watt and copiers being 40–300 Watts with the rollers are usually kept hot. Assuming 10 printers, 10 desktops and 5 copiers were completely shut down in April in this building, then at least the daily baseload can decrease by 27.6 kWh, which matches the electricity consumption data difference between March (167 kWh) and April (137 kWh). Thus, it can be identified that there is an energy waste as a result of daily standby of office machines. Therefore, facilities on standby are notable for power consumption, considering they are always forgotten by operators. The cheapest and most effective way to reduce the energy consumption of computers, printers and copying machines is to switch them off when they are not being used [8]. In this way, the baseload can be even lower than the present situation.

3.2 *Variable Load*

Consistent with baseload data is the condition of variable load. As a whole, it was obvious in Fig. 1 that variable power consumptions in weekdays were much higher than those on weekends, and Monday always had the highest electricity usage among weekdays. One of the reasons for more consumption on Monday especially early hours of Monday was that students needed more time to prepare for their assignments or group work with due date being the next morning. Similar ANOVA results were found in the variable electricity consumption among the 4 months as the baseload. In April, significantly less variable electricity was used compared with the previous 3 months. There was no significant difference in variable electricity consumptions among January, February, and March. One reason is the changing weather condition and sunshine duration of these months. The average daytime hours from January to April were 8.5, 10, 12 and 14 h respectively [9]. The light intensity also increased with time in this period [10]. Due to extended daytime hours and increased solar intensity in April, the need for artificial lighting was reduced. Therefore, the lighting system had a lower proportion, consuming less electric energy. For the same reason of the spring term holiday, with less courses conducted, less room occupied and reduced hours of use, it is evident that the variable power demand was lower.

As a result, students who need to study at night especially with assignments to submit the next day are suggested to study in the university library which is open all night on Sunday nights. In this way, a large amount of electricity can be saved due to decreased lighting hours, and at the same time, both chancellor's building and the library can become more energy efficient.

What's more, a great difference was found between weekdays and weekends. During every weekend, hourly electricity consumption always kept near to the baseload and did not change a lot until 9 a.m. After 9 a.m., electricity usage increased dramatically. By contrast, similar increased profile appeared around 6 a.m. on weekdays, sometimes even ahead of 3 a.m. It may be because cleaning staff needed to do their work around 6 a.m. using dust collectors on every weekday with the lighting on. Some students may go to class much earlier than 9 a.m. so that the classroom lights were on for longer hours. With considerable electric power used between 3 a.m. and 6 a.m., it is recommended that cleaning work and logistics are conducted after 6 a.m. and to be finished within 3 hours so that it has no time confliction with courses.

4 Conclusion

This research analysed the half-hourly energy consumption data for Chancellor's building in University of Reading from 1 Jan 2018 to 30 April 2018 in terms of constant and variable electricity consumption. The result exhibits a significant relation between energy usage and occupancy. Further, data analysis method was used to give solutions for reducing the baseload and variable load respectively. To reduce baseload consumption, it is recommended to switch off the office facilities when they are not being used. Variable consumption can be reduced by conducting cleaning work and logistics after 6 a.m., and students who need to study at night are suggested to study in the university library.

The results of this study are a useful reference when managing or designing university buildings with both classrooms and offices. However, the limitation of this study is that the half-hourly data of the building only have two categories: baseload and variable load. It is hard to analyse contribution from each energy-consuming part, for example, lighting, electricity plugs, equipment in computer labs, staff offices, etc. Therefore, to clarify energy use can assess building energy performance more precisely. The energy monitoring for buildings in University of Reading should be undertaken in a more detailed and precise way. If the data could be logged into the database using the sub-meter monitoring, they could provide more detailed information about what happened in different spaces, systems and time. In this way, managers could have better inspirations reducing energy consumption.

References

1. The National Archives (2008) Climate Change Act 2008. URL: http://www.legislation.gov.uk/ukpga/2008/27/contents [23 November 2019]
2. CIBSE (2012) Energy efficiency in buildings: CIBSE Guide F. URL: https://www.ihsti.com/tempimg/538d075-CIS888614800315138.pdf [23 November 2019]
3. Kim WH, Lee S, Hwang J (2011) Real-time energy monitoring and controlling system based on ZigBee sensor networks. Procedia Comput Sci 5:794–797
4. Kleissl S, McKinney P, Lounsbury E, Macleod A (2019) Digital technologies for energy management. URL: https://www.carbontrust.com/resources/reports/technology/digital-technologies-for-energy-management-report/ [23 November 2019]
5. Jaradat M, Jarrah M, Bousselham A, Jararweh Y, Al-Ayyoub M (2015) The internet of energy: smart sensor networks and big data management for smart grid. Procedia Computer Science 56:592–597
6. Piette MA, Kinney SK, Haves P (2001) Analysis of an information monitoring and diagnostic system to improve building operations. Energ Buildings 33(8):783–791
7. University of Reading sustainability services (2019) Chancellor's building energy data. URL: https://sites.reading.ac.uk/sustainability/energy-map/whiteknights/chancellors-building/ [23 November 2019]
8. Carbon Trust (2006) Office equipment: introducing energy saving opportunities for business. URL: https://www.carbontrust.com/media/13113/ctv005_office_equipment.pdf [24 November 2019]
9. Sunrise-and-sunset.com (2019) Sunrise and sunset Reading, United Kingdom. URL: https://www.sunrise-and-sunset.com/en/sun/united-kingdom/reading [24 November 2019]
10. Garnham C (1999) An investigation into light intensity in the UK. Br Cactus Succulent J 17(3):154–156

Part XI
Energy and Gender

Key Challenges for Solar Electrification of Rural Schools and Health Facilities in Sub-Saharan Africa

Terence Cook and David Elliott

1 Introduction

A country's infrastructure of roads, railways, water supply, sewers, power supplies, electricity grid and telecommunications are what makes an economy function. For Africa there is still a need for adequate infrastructure for the economy to grow. Although investment in infrastructure accounts for over half of the recent improvement in economic growth in Africa, there is also a marked urban/rural divide with electrification rate at 65% and 28% respectively [1]. The absence of the electricity grid in rural areas is an opportunity for new renewable energy technologies to provide decentralized solutions as the least-cost way to provide power to rural populations and to social institutions – schools and health clinics [2].

However, the willingness of the private sector to get involved with electrification projects related to schools and health clinics in sub-Saharan Africa will largely depend on the capacity of each country to attract private investments into the off-grid sector. This requires a de-risking mechanism to address the high level of perceived investment risk. For example, investor confidence may require strong guarantees that the projects have operational sustainability and that electricity bills will be paid to ensure continued revenues. Certainly, so far, there has been more reliance on public sector finance and also on international grant aid, both of which can be problematic.

To try to improve this situation, new business models for funding electricity and internet access for rural schools and health clinics in sub-Saharan Africa have been proposed that generate their investment capital from various sources like commercial or concessional loans and equity, in addition to grants. They attempt to move away from the 100% grant funding from international donors which, it is argued, is not sustainable.

Most social infrastructure rural electrification projects need support to reduce initial capital costs, in the form of grants, subsidies, loans or public private partnerships. A number of scenarios for business models which combine public and private resources are investigated and presented in this chapter. The business model scenarios discussed present a model with a shared public sector CAPEX investment combined with private sector CAPEX investment and incentive support via results-based financing (RBF). Other options presented include 100% CAPEX payment from the public sector supported by international donor funds in ways to increase the commitment of the private sector as energy provider but at the same time to set tariffs that can be realistically paid by social institutions such as schools and clinics. They also face operation and maintenance costs, a key issue for long-term project sustainability.

This chapter examines emerging business models with reference to two case studies – an EU-funded feasibility study for solar electrification of rural primary and secondary schools in Rwanda and an Open University-led study of low-cost energy-efficient, diagnostic health devices for remote rural health clinics in Kenya funded by the Engineering and Physical Sciences Research Council (EPSRC) and the Department for International Development (DFID).

T. Cook (✉) · D. Elliott
Faculty of Science, Technology, Engineering & Mathematics, School of Engineering & Innovation, The Open University, Milton Keynes, UK
e-mail: Terry.Cook@open.ac.uk; D.A.Elliott@open.ac.uk

© The Author(s), under exclusive license to Springer Nature Switzerland AG 2022

A. Sayigh (ed.), *Sustainable Energy Development and Innovation*, Innovative Renewable Energy,
https://doi.org/10.1007/978-3-030-76221-6_86

2 Electrification Issues

There are limited national off-grid electrification strategies for schools and health clinics in most sub-Saharan Africa countries. This means that individual initiatives exist instead of a strategic approach. A fully grant-based approach is understandably justified in the poorest countries for last-mile rural health facilities based on a basic kit for critical obstetric and surgical lighting. In rural Zimbabwe, for example, women are required to provide six candles for their own maternity birthing kit. But in the case of larger health clinics, a strategic approach has to be investigated if multiple schools / clinics are to be electrified by a single activity.

A business model for electrification of social institutions with minimal private sector participation means the private company is responsible only for the installation of the solar systems. The operation of the systems will be the responsibility of the public electricity utility which is also responsible for operation and maintenance (O&M). The schools or clinics pay the government electricity tariff. If this money is not enough to ensure long-term maintenance of the systems and especially the solar system battery change, which is the most expensive component, the government will ultimately be responsible for continued support. So, there may be problems, especially if the project is heavily reliant on grants.

By contrast, financially sustainable business models for incentivizing the private sector to expand their activities beyond equipment provision would obviously have to provide the companies involved with quid pro quo. The private company may only make provision for O&M over the project lifetime if the project framework is changed. For example, the installed solar systems may be configured as part of a local micro-grid under the companies' control. In which case, the private sector might have the right to use the school systems as part of a larger energy system in order to provide electricity or electricity services to the surrounding communities. That might initially involve rechargeable lights and mobile phone charging, but then move up to mini-grids for supplying productive loads. There could be some conflicts between these end uses, but the funding model could seek to limit them. For example, at all times, the company could be made responsible for providing the contracted service to the schools, tied in via a results-based financing Scheme [3].

Clearly, since systems based on renewables are loaded to the initial capital cost of the equipment, an upfront grant de-risks the investment considerably. However, there are shortcomings with grants, not least the fact that they do not provide funding into the future for operational sustainability. Self-funding from revenue is obviously the ideal, but whether public or private, an operator will need to have enough income to be able to maintain the system. Below we look at some examples of projects that have tried to respond to these challenges.

3 Project Case Studies

3.1 *Rural Healthcare in Kenya*

Long distances to health facilities, lack of public transport and poor roads limit access to health facilities in rural areas in Kenya. The Kenyan government has committed itself to fully subsidize the cost of healthcare for the poorest rural households through donor support. Pro-poor health reforms have improved access to healthcare for the poor to some extent, but the rural poor in Kenya still have reduced access to healthcare services because of geographical access. For rural healthcare, particularly maternal healthcare, to be maintained and sustained, there has to be community access to low-cost, solar-powered, off-grid medical systems.

The fieldwork for the investigation of off-grid health facilities in Kenya was conducted for an Open University-led research project funded by EPSRC-DFID in a 3-year programme, which started in January 2015. The research investigated renewable energy as the central theme to understand ways to benefit "bottom of the pyramid" segments in poor rural communities. The focus was on low-cost energy-efficient technologies, in particular health diagnostic products that could be deployed to have a large-scale impact. A survey was conducted among 400 nurses who had medical triage training but limited medical equipment working in predominately rural clinics across four locations in Kenya (Kirinyaga, Marsabit, Malindi and Kisumu) [4]. The health surveys proved difficult to arrange due to the remoteness of the rural health clinics.

Surveys were also undertaken on the choice of low-cost energy-efficient technologies in the health sector, in order to develop sustainable business plans for providing low-cost energy-efficient technologies and measure the impact of the products over a 3-year period.

The target end users were nurses and mobile heath worker at rural clinics – rather than doctors. The priority was point-of-care diagnostic devices, telemedicine and ultrasound. The supporting technologies were standalone power, digital connectivity, vaccine refrigeration and water sterilization. The survey respondents included key users (the nurse practitioner) and other stakeholders (patients, doctors, etc.) depending on the devices/technology solution. Point-of-care (POC) diagnostics devices are emerging which can be used by people with much less training to conduct tests at the point of care, that is, during patient consultation.

It is virtually impossible in rural off-grid areas to equip clinics with conventional expensive medical devices because of cost and the need for reliable grid-connected electricity. Only 19% of Kenyan medical facilities have reliable energy access, with the greater proportion of population and facilities without reliable supply being in rural areas. The Open University research programme portfolio of new low-cost, low-energy medical devices such as solar-powered infant incubators and solar-powered vaccine refrigerators could make it possible to build healthcare capacity at a fraction of cost. This would allow women to give birth at clinics that are safer and much better equipped than homes where the majority of rural births take place.

The new neonatal medical devices such as an infant incubator in these clinics will lead to faster and more effective intervention during birth emergencies. Further, a device such as the solar-powered vaccine fridge would also enable the care of the newborn through an ongoing, stable vaccination regimen throughout their childhood years.

To support services like this, the government has made reforms intended to expand health services with a specific focus on the poor to reduce the financial burden. For example, the Kenya healthcare reforms include the provision of free maternity services. Because primary healthcare services are free, rural and public clinics depend on government funding to sustain them. But delays in receiving funds for up to 6 months make it difficult to operate. So, there can be problems. In the next sections, we look at other approaches that demonstrate how health centre electrification can be delivered in a commercially sustainable manner with public-private sector participation. Indeed, the healthcare facilities can act as 'anchors' for distributing electricity via mini-grid generation systems.

3.2 Solar for Off-Grid Schools in Rwanda and Elsewhere

An EU feasibility study for solar electrification of 400 rural schools in Rwanda was undertaken between 2019 and 2020 to deliver a €23 million programme for off-grid solar electrification for 'smart classrooms'. Limited access to electricity has made it impossible for some schools to use computers, have access to digital content and networking and to access other information beyond what the teachers offer in classrooms.

Certainly, not all schools in Rwanda have electricity, especially rural schools. For example, over half of the schools in Kamonyi District in Rwanda's Southern Province do not have access to electricity. Moreover, less than 5% of Kamonyi District's children have electricity at home, so children are not able to study and families do not have basic access to lighting in the evening. Of Rwanda's 12.6 million people, 82% live in rural areas.

Rather than extending the grid to all areas, the Government of Rwanda has developed the National Electrification Master Plan [5] that identifies future areas that will have grid extension and those areas that will remain off-grid. These off-grid areas are designated for distributed renewables generation.

Under the plan, secondary schools would be provided with AC standalone solar systems to power two smart classrooms with up to 50 computers, and primary schools with solar systems to power laptops. The estimated CAPEX (capital cost) of the systems of €18 million would be an EU grant and the estimated OPEX (operating cost) of €five million would be paid for from an electricity tariff comparable with the electricity tariff of grid-connected schools. The estimated level of EU subsidy required is based on the hypothesis that the CAPEX is a grant and the OPEX is paid for by sale of electricity.

The fieldwork survey of schools in Rwanda estimated the average annual consumption per (primarily/secondary) pupil at 7.5 kWh, resulting in an average sizing of the PV/battery systems at 2 kWp for off-grid school sites. However, Rwanda's proposal for smart classrooms increases the load considerably. In the case that the school has two smart classrooms, which is the national target, the daily consumption will rise to 50 kWh/day [6]. The size of the required system to meet this load falls under the category of a mini-grid, even though an actual grid might not be installed eventually (the system will supply only the school premises). Smaller (1kWp) systems that provide electricity also exclusively for lighting have a CAPEX circa € 18,000 with an OPEX of € 400 per year. The most sophisticated system with grid-like performance allowing for two smart classrooms equipped with 55 W laptop computers would require a 14 kWp installation, with system cost per school of circa €80,000 and operating cost of €2000 per year.

The economics is being improved by the advent of high-quality lithium batteries with warranty of 10 years which is a major change because the lifespan of previous batteries used for off-grid PV systems has been recurrently identified as the main barrier to the system's long-term functioning.

Software also plays an important role in determining the actual energy consumption of a computer. Modern operating systems, such as Windows 10, offer various profiles including power efficiency. If an energy efficiency profile is chosen, then the actual energy consumption of the computer can be decreased quite substantially. The USAID Powering Health: Electrification options for rural health centres report cite a 15–20 W real energy consumption from an energy-efficient laptop [7]. So, for the purposes of the Rwanda study, the assumption was made that the average power draw of each energy-efficient laptop configured with a high energy efficiency profile in software was assumed to be 17.5 W.

Accountability for what is going on over the lifetime of the solar system is clearly essential to prevent stranded assets. For the Rwanda schools project, to keep it going, the government has to balance, and account for, the heavy CAPEX upfront expenditure on solar equipment (€18 million), funded by a 100% grant from the EU, and the operation and maintenance costs of the solar systems (€five million over 15 years).

Rwanda is not alone in having projects for local solar in remote areas. For example, there is an ongoing rural solar electrification programme in Nigeria – the Solar Nigeria Programme – funded initially by DFID and now funded by the EU for €30 million and implemented by DFID in rural areas. It reflects the limited reach of the grid and limited prospect of connection to the grid in the near future. The Nigeria solar project is aimed at increasing sustainable electricity supply using renewable energy for health and education facilities serving poor communities in off-grid areas of Kano, Kaduna and Adamawa, three of the largest states in Nigeria. There is poor power supply in the entire northern states which includes the conflict affected areas in Northern Nigeria.

There are innovative business models for funding electrification being adopted in countries that focus on wider developmental impacts for local communities linked to schools and health facilities. For example, there is an EU project in the Gambia with a scheme proposed to provide electricity to rural communities, not only in terms of electrification of schools and health centres but also to cover some small community activities using excess electricity, leading to improving living conditions, job creation and income generation for the surrounding community.

There are examples of the Gambia model being adopted in other countries, Sierra Leone, for example, with school's solar generation systems linked to mini-grids, to distribute electricity to the surrounding community. The Democratic Republic of Congo (DRC) has a concession approach which has stimulated significant private sector and investment interest in this kind of at-scale public-private mini-grid model.

4 Discussion of Issues

4.1 Social Infrastructure and Sustainable Development Funding

There have been increasing provisions made for funding solar energy uses in Africa, including for schools and healthcare. Infrastructure investment funds have been set up in some countries, Ghana, for example, to attract private investors to support social infrastructure with the recognition that there will be lower profitability but significant social impacts. The Ghana Infrastructure Investment Fund (GIIF) is a sovereign wealth fund established to help Ghana develop commercially viable infrastructure projects through private sector participation in West Africa [8]. A new initiative for the fund is a partnership with the Agence Française de Développement (AFD) for a scheme where 50% of the credit facility is for lending to a pipeline of social projects in health, education, clean water, sanitation and social housing.

The approach to social infrastructure is normally set within a wider macro-economic context – that public spending on social infrastructure like education can be macro-critical. The IMF this year announced a change in approach urging countries to protect spending on education and highlighting the impact of social spending on growth and stability. A model sustainability framework for social infrastructure was launched in 2019, with a new phase of the Solar for Health Initiative. It is supported by 20 governments from developing countries worldwide and the United Nations Development Programme (UNDP) [9] and aims to install solar systems in health centres and clinics in underserved communities in rural areas. A model sustainability framework was developed under this initiative.

4.2 Operation and Maintenance Costs

Within that general funding context, this chapter has looked at some specific solar project development issues. The solar industry has had large-scale growth over the past few years as the costs of photovoltaic (PV) cells have gone down. But this means an ever-growing demand for operations and maintenance (O&M) for solar systems.

In the study of the Rwanda schools off-grid electrification, the level of EU subsidy was based on the upfront capital cost of the solar hardware being a grant, but with the OPEX (the ongoing operating costs) being paid by sale of electricity fixed at the same tariff as grid electricity. In Kenya, the Green Mini-Grid (GMG) Facility [10] provides a flexible model for results-based incentives. This Facility is a €30 million venture grant facility supported by DFID and the European Union with the Agence Française de Développement (AFD) as the implementing partner, providing investment grants and output-based grants to encourage private investment in renewable energy-based mini-grids.

Clearly, business model innovation to fund social infrastructure in emerging economies requires new ways for structuring relationships between suppliers and consumers of energy. One new business model option looked at in the EU Rwanda Schools study was a mini-grid component linked to the local community with the school as the anchor client, but where the whole community can benefit. The advantage of a school as an anchor client is a predictable load profile. It is preferable to have an anchor that will be part of the community for many years. With this model the anchor load is predictable and offers a guaranteed source of revenue for the project developer, as well as small business and community members becoming the customers for the project. This would be private sector led and provide viability gap subsidies to mini-grid developers with performance-based grants in the form of results-based financing (RBF), a key element of many of the proposed business model options by donors.

The RBF approach is used within Energising Development (EnDev) [11], an energy access partnership financed by six donor countries: the Netherlands, Germany, Norway, United Kingdom, Switzerland and Sweden. EnDev promotes sustainable access to modern energy services that meet the needs of the poor – long lasting and affordable – that uses financing as a performance incentive rather than a grant payment. The key difference from other schemes is that payment is upon delivery of results. Private investors are expected to take the full risk until delivery of the contracted results. For that to be acceptable, it is vital to give the private sector confidence that they will be paid when they deliver their service to schools.

Certainly, under whatever payment scheme, moving away from grants funding for both CAPEX and OPEX, there is a need for guaranteed revenues from electricity tariffs for sustainable cash flows to pay for the operation and maintenance. The sustainability of the system over the asset's lifetime requires monitoring and replacement of inverters and system batteries (depending on use patterns, care batteries may need to be replaced) to avoid stranded solar assets no longer functioning at capacity. It is one of the short comings of capital costs paid upfront by a grant that Africa has become a cemetery of solar projects. The challenge is to encourage governments to create value throughout the lifetime of the project and make procurement structured in a way to be O&M focused so there is commitment to system sustainability.

5 Conclusions – The Need for Action

Many schools and health centres in sub-Saharan Africa have been waiting for electricity for years with no certainty when if ever the grid will be extended. Addressing needs now means solutions other than grid extension are needed. For women to give birth by candlelight with no life-saving machines in case of an emergency is contrary to the UN third sustainable development goal SGD3 (good health and well-being). To deliver education to children in this era of digital learning and release them from sharing very old textbooks will address the UN fourth sustainable goal SGD4 to 'ensure inclusive and equitable quality education and promote lifelong learning opportunities for all'. Finally, the provision of access to energy to enable the productivity of public institutions fulfils the UN seventh sustainable development goal SD7 "to ensure access to affordable, reliable, sustainable and modern energy for all".

So, resolution of the issues discussed in this chapter is urgent. To deliver a sustainable system of solar electrification, there has to be an 'operational sustainability policy', that is, more accountability from government with different models for operations and maintenance, training for district technicians on O&M, more resources allocated to remote monitoring technologies and a focus on providing a service rather than just the solar PV assets. This also suggests an expanded role for the private sector, not only limited to equipment provider but also as a provider of energy 'services' to power vital social infrastructure – health services and rural schools [12].

It is problematic when business models for funding the solar electrification of schools and clinics is grant based only. The concern is that the project should get fully funded in all aspects because, although the CAPEX is the largest part of the cost, there is still the OPEX, the operating costs and the maintenance costs over the life cycle of the project. If the private sector is to step in and invest capital to deliver energy services to hundreds of schools and clinics and to work with public institutions, it needs to have confidence that there is a risk mitigation cushion for non-payment of electricity tariffs. A risk mitigation model is needed to help the private sector and government to work together to address the solution at scale.

References

1. African Development Bank: Tracking Africa's Progress in Figures. Report 2014
2. Elahi R; Srinivasan R, Mukurazhizha T (2020) Increasing human capital by electrifying health centers and schools through off-grid solar solutions. Live Wire;2020/104. Washington, DC, World Bank. © World Bank
3. AfDB Financing Africa's Infrastructure: New Strategies, Mechanisms, and Instruments. https://www.afdb.org/fileadmin/uploads/afdb/Documents/Publications/2018AEO/African_Economic_Outlook_2018_-_EN_Chapter4.pdf
4. Cheruiyot M (2016) Low cost technologies survey report, African Centre for Technology Studies
5. Rwanda Energy Group (REG) (2019) The National Electrification Plan: Report on definition of technologies (On-grid and off-grid) at village level
6. Kyriakarakos George Dr, Solar PV expert for EU TAF Rwanda Pre-Feasibility Study, 2020
7. Laender D (2018) Powering health: electrification options for rural health centers. USAID, Washington, DC
8. The World Economic Forum: The world is facing a $15 trillion infrastructure gap by 2040 https://www.weforum.org/agenda/2019/04/infrastructure-gap-heres-how-to-solve-it/
9. UNDP Solar for Health initiative. https://stories.undp.org/solar-for-health#!
10. Green Mini Grid Facility Kenya (GMG Facility). https://www.gmgfacilitykenya.org/
11. EnDev Rwanda Energizing Development. https://endev.info/content/Rwanda
12. Thomas Day, Marie-Jeanne Kurdziel. NewClimate Institute. The role of renewable energy mini grids in Kenya's electricity sector. Evidence of a cost-competitive option for rural electrification and sustainable development (November 2019)

Part XII
Late Arrival Papers

Contribution to the Improvement of Oscillating Water Column Efficiency Using L-Shape Front Wall

Mahmoud M. Samak, Hassan Elgamal, and Ahmed M. Nagib Elmekawy

1 Introduction

The simple OWC consists of a partially submerged chamber with an open in the front wall to let incident waves excite the trapped water in the chamber and force it to oscillate. The oscillatory motion of the water inside the chamber forces the air to drive a turbine which is connected to an electrical generator. The early researches were based on simplified theoretical models such as McCormick [1], Evans [2, 3], Sarmento and Falcao [4], and Falnes and McIver [5]. Experimental and numerical researches were carried out to study the effect of various parameters on the device efficiency such as front wall geometry [6], front wall draught [7–10], front wall thickness [8–10], front wall orientation [11], chamber width [7], air column height [8], bottom slope [7, 8], duct diameter [7, 12, 13], power take-off damping [8, 14], wave conditions [8], water depth [9], wave steepness [15], and incident wave amplitude [7].

Many modifications were made on the simple OWC to improve its performance. U-shaped OWC improved the efficiency as it increases the Eigen period and the amplitude of the pressure fluctuations [16]. An inclined bend-free OWC device was tested experimentally to investigate the effect of the inclination angle on the performance and the results show that the air velocity increases as the inclination angle decreases [17]. A fully submerged barrier with a gap is used instead of the simple front wall and the results show that the efficiency can be increased by the appropriate choice of the geometrical parameters [18]. L-shaped duct oscillating water column was tested experimentally to investigate the effect of inlet duct cross-sectional area [19] and the immersion depth [20] on the performance. The results show that the variation in inlet geometry provided low deviation in results and the performance of the device is enhanced by reducing immersion depth.

The present work is an attempt to improve the efficiency of OWC. The simple front wall is replaced with L-shape front wall with the horizontal edge extending inside the chamber to make a contraction in the water column. Numerical simulations were carried out by ANSYS FLUENT 19.1 on a 2D numerical wave tank. First, the experiments which were carried out by Morris-Thomas et al. [10] were used to validate the numerical model. Second, the effect of horizontal edge length on the performance was investigated.

2 Problem Statement and Governing Equations

The flow is a multiphase flow: primary phase (air) and secondary phase (water) [21]. Volume of fluid (VOF) method is used to track the free surface [22]. The volume fraction of each phase in each control volume is determined using geo-reconstruction method [6]. The flow is turbulent because of the turbulence around the front wall [23]. K-ε model is used to describe the turbulence phenomenon in the water and air dynamic motions as suggested by other researchers [12]. The chamber length, in the present research, is small with respect to incident wavelengths so the air can be assumed incompressible [24]. Equation (1) represents the second-order wave theory which is an appropriate theory to describe the profile of the incident waves [25].

M. M. Samak (✉) · H. Elgamal · A. M. Nagib Elmekawy
Department of Mechanical Engineering, Faculty of Engineering, Alexandria University, Alexandria, Egypt
e-mail: mahmoudsamk@alexu.edu.eg; a.nagib@alexu.edu.eg

© The Author(s), under exclusive license to Springer Nature Switzerland AG 2022

A. Sayigh (ed.), *Sustainable Energy Development and Innovation*, Innovative Renewable Energy,
https://doi.org/10.1007/978-3-030-76221-6_87

$$\eta = \frac{H}{2}\cos(kx-\omega t) + \frac{H^2k}{16}\frac{\cosh(kh)}{\sin h^3(kh)}(2+\cosh 2kh)\cos 2(kx-\omega t) \tag{1}$$

where η is free surface elevation from the still water level, H is incident wave height, h is the water depth, k is the wave number, and ω is the wave frequency. The dispersion relation which relates wavelength ($\lambda = \frac{2\pi}{k}$), wave period ($T = \frac{2\pi}{\omega}$), and water depth can be formulated as:

$$Kh = k\,h \tanh(k\,h) = \frac{h\,\omega^2}{g} \tag{2}$$

where Kh is the infinite water depth parameter and g is the gravitational acceleration. The incident wave power per unit width of the wave tank (P_{in}) in the case of second-order wave theory can be calculated from Eq. (3) [6]:

$$P_{\text{in}} = \frac{1}{16}\rho g H^2 \frac{\lambda}{T}\left(1 + \frac{2kh}{\sin h(2kh)}\right)\left(1 + \frac{9H^2}{64k^4h^6}\right) \tag{3}$$

where ρ is the water density. The average output power per unit width of the chamber (P_{out}) can be calculated from Eq. (4) [13]:

$$P_{\text{out}} = \frac{1}{T}\int_0^T P_c(t)Q(t)dt = \frac{1}{T}\int_0^T P_c(t)\,.V(t).b\;dt \tag{4}$$

where P_c is the air pressure in the chamber, Q is the air volume flow rate through the vent, V is the free surface vertical velocity, and b is the chamber length in the wave direction. The average efficiency (e) can be calculated from Eq. (5):

$$e = \frac{P_{\text{out}}}{P_{\text{in}}} \tag{5}$$

3 Numerical Model

Figure 1 shows the computational domain and the principal dimensions. L_{tank} denotes to the wave tank length from the inlet to the front wall. The wave tank length should be not only as short as possible to reduce the simulation time but also should be long enough to prevent the interaction between incident waves at inlet and the reflected waves from OWC walls [8]. It is found that the optimum value of wave tank length is four to five times of the wavelength. Figure 2 shows the domain discretization to small quad control volume with $\Delta x = 0.025$ mm [14, 15, 26]. The computational domain is divided into three regions: region (i) represents the region of the free surface which is divided into 20 elements in y-directions [27], region (ii) represents the region of the air column inside the chamber which is divided into 30 elements in y-direction, and region (iii) represents the rest of the computational domain and has cells with $\Delta x = \Delta y = 0.025$ mm.

Open-channel wave boundary condition is used to generate waves at the inlet [21], atmospheric pressure is considered at the top of the numerical tank and vent outlet, and no slip boundary condition is used at the sea bed and all the walls.

PISO algorithm is adopted for pressure-velocity coupling. The spatial discretization for pressure and momentum is determined by PRESTO and second-order upwind, respectively. The time discretization depends on the first-order implicit approach. It is found in pervious numerical works that the time step should be less than $\frac{T}{50}$ to ensure that the numerical solution is independent on the step time [9]. In the present work, step time equals: $\Delta t = 0.01$ s is used in all the numerical simulations.

4 Validation

The experiments which were carried out by Morris-Thomas et al. [10] and the detailed results which were shown by Kamath [14, 15] are used to validate the numerical model in the present work. The experimental model was tested with 0.92 m water depth (h), 0.12 wave height (H), and has the following dimensions: $a = 0.15$ m, $b = 0.64$ m, $c = 0.04$ m, bv = 0.005 m,

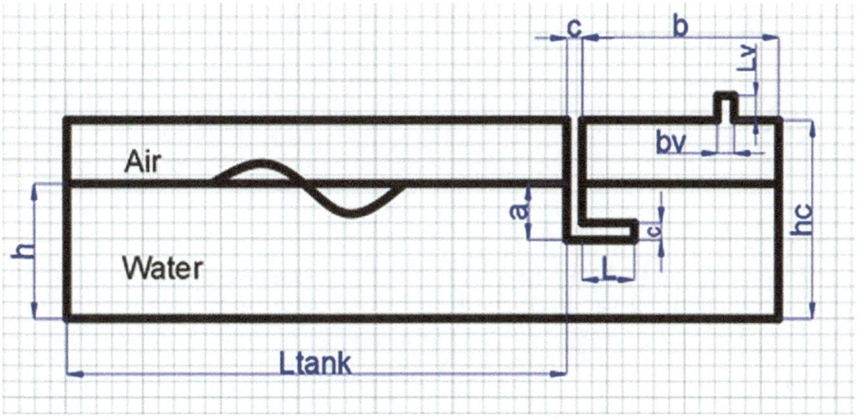

Fig. 1 Computational domain

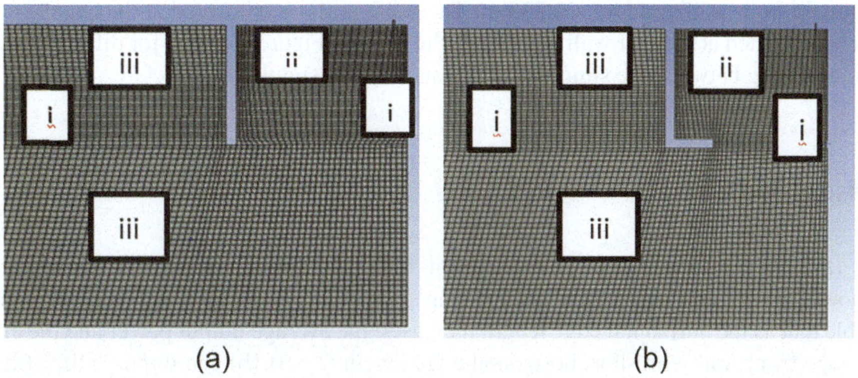

Fig. 2 Mesh (**a**) simple front wall OWC and (**b**) L-shape front wall OWC

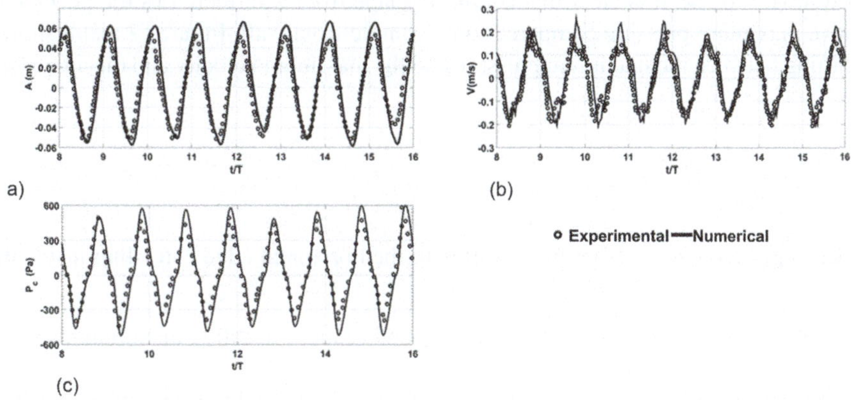

Fig. 3 Comparison between numerical and experimental results for Kh = 0.93. (**a**) Free surface elevation inside the chamber A(m). (**b**) Free surface vertical velocity V(m/s). (**c**) Air pressure in the chamber P_c(Pa)

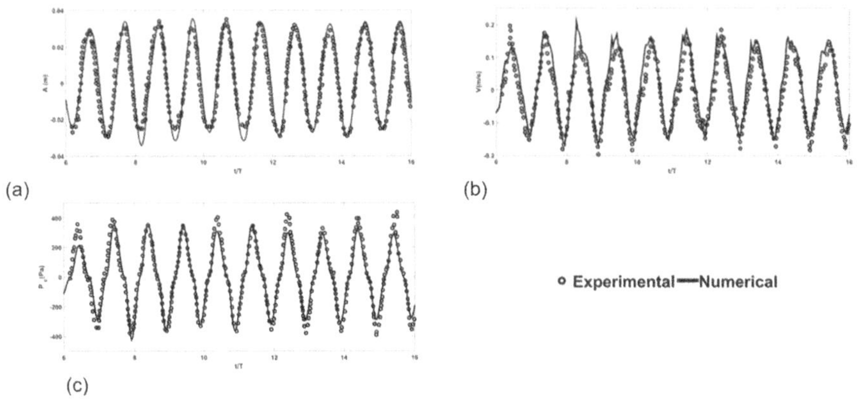

Fig. 4 Comparison between numerical and experimental results for Kh = 1.92. (**a**) Free surface elevation inside the chamber A(m). (**b**) Free surface vertical velocity V(m/s). (**c**) Air pressure in the chamber P_c(Pa)

Lv = 0.025 m, $L = 0$ (simple front wall), and hc = 1.275 m. The wave probe was located at the center of the chamber while the pressure transducer was located at the rear wall 5 cm from the ceiling. Figures 3 and 4 for different values of water depth ratio (Kh) show a good matching between the simulation data and current simulations.

5 Results and Discussion

The simple front wall ($L = 0$) is replaced with L-shape front wall with various horizontal edge lengths (L = 0.1b and L = 0.25b). Figure 5 shows the average efficiency and the average output power of OWC with simple front wall and L-shape front wall. It is noticeable that as the horizontal edge length increases, the average output power and the efficiency decreases. It is also clear that L-shape front wall with short horizontal edge length (L = 0.1b) can improve the efficiency of the OWC and increase the output power compared to the simple front wall. Figure 5 also reveals that the best front wall configuration for short waves (Kh = 2.5) is the simple front wall. Figure 6 shows the time series of air pressure in the chamber for Kh = 0.52, Kh = 0.93, Kh = 1.26, and Kh = 2.5 and Δp is the pressure difference between the peak points of simple front wall and L-shape front wall (L = 0.1b). It is clear that adding L-shape front wall increases the positive maximum pressure and decreases the negative maximum pressure compared to the simple front wall. Figures 7 and 8 show that L-shape front wall reduces the time shifting between the maximum pressure and the maximum velocity and this increases the output power.

6 Conclusions

The simple front wall was replaced with L-shape front wall with the horizontal edge extending inside the chamber and the results show that:

1. L-shape front wall increases the positive maximum pressure and reduces negative maximum pressure compared to the simple front wall.
2. L-shape front wall decreases the shift angle between the air pressure curve and the free surface vertical velocity curve.
3. The increase in the horizontal edge length leads to a reduction in the air pressure in the chamber, output power, and the efficiency.
4. L-shape front wall with short horizontal edge length (L = 0.1b) can increase the average output power and the efficiency in all the study cases compared to the simple configuration except in case of short waves.
5. The simple front wall is the best configuration when the incident waves have short wavelength.

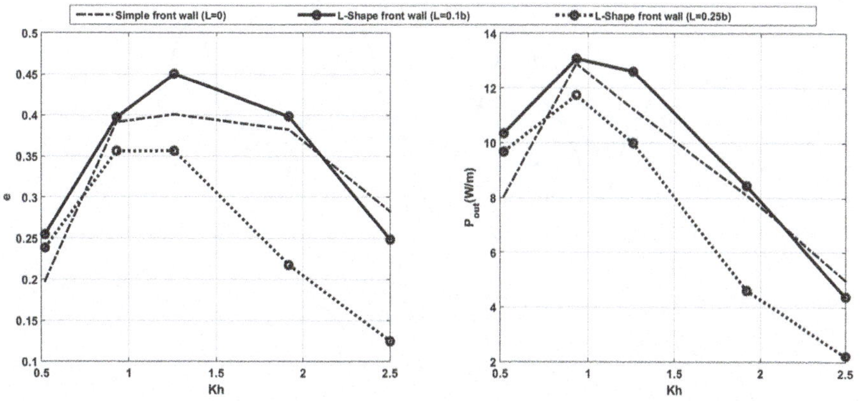

Fig. 5 Average efficiency and average output power of the OWC for various wavelengths

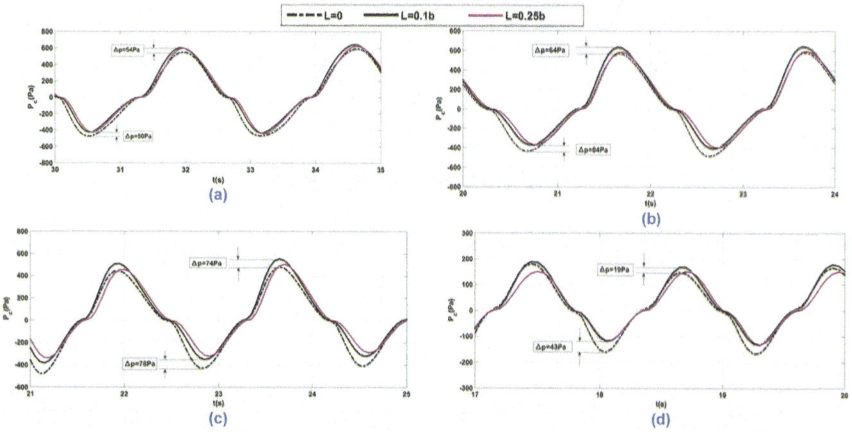

Fig. 6 Time series of the air pressure in the chamber. (**a**) Kh = 0.52. (**b**) Kh = 0.93. (**c**) Kh = 1.26. (**d**) Kh = 2.5

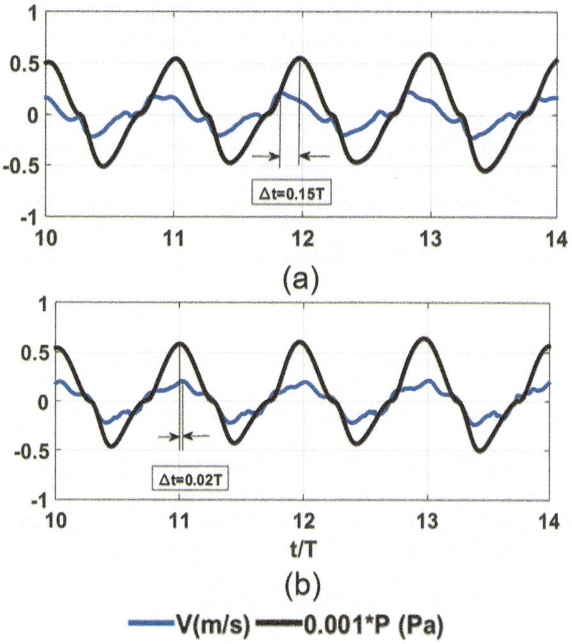

Fig. 7 Shift angle between pressure curve and velocity curve for Kh = 0.52. (**a**) Kh = 0.52, L = 0. (**b**) Kh = 0.52, L = 0.1b

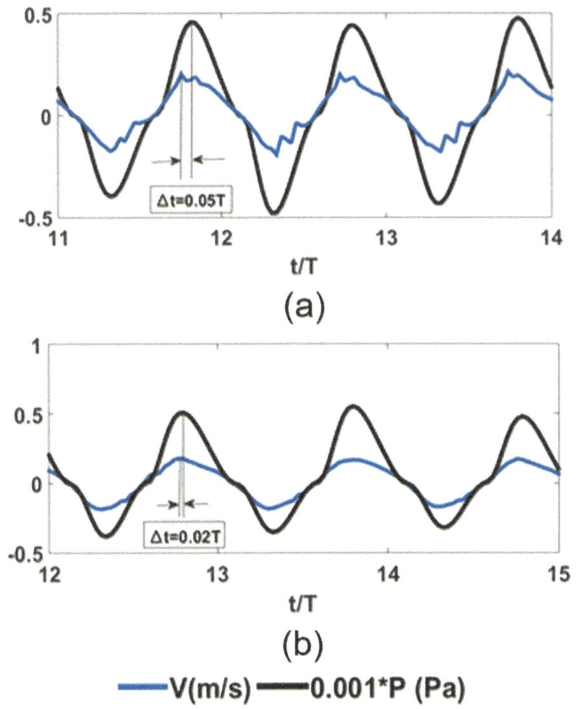

Fig. 8 Shift angle between pressure curve and velocity curve for Kh = 1.26. (**a**) Kh = 1.26, L = 0. (**b**) Kh = 1.26, L = 0.1b

References

1. McCormick ME (1976) A modified linear analysis of a wave-energy conversion buoy. Ocean Eng 3(3):133–144
2. Evans D (1978) The oscillating water column wave-energy device. IMA J Appl Math 22(4):423–433
3. Evans D (1982) Wave-power absorption by systems of oscillating surface pressure distributions. J Fluid Mech 114:481–499
4. Sarmento AJ, Falcão ADO (1985) Wave generation by an oscillating surface-pressure and its application in wave-energy extraction. J Fluid Mech 150:467–485
5. Falnes J, McIver P (1985) Surface wave interactions with systems of oscillating bodies and pressure distributions. Appl Ocean Res 7(4):225–234
6. Horko M (2007) CFD optimisation of an oscillating water column wave energy converter. University of Western Australia
7. Ning D-Z et al (2016) An experimental investigation of hydrodynamics of a fixed OWC wave energy converter. Appl Energy 168:636–648
8. Bouali B, Larbi S (2017) Sequential optimization and performance prediction of an oscillating water column wave energy converter. Ocean Eng 131:162–173
9. Liu Z et al (2010) Practical simulation of oscillating water column chamber for wave energy conversion. Int J Green Energ 7(3):337–346
10. Morris-Thomas MT, Irvin RJ, Thiagarajan KP (2007) An investigation into the hydrodynamic efficiency of an oscillating water column. J Offshore Mech Arct Eng 129(4):273–278
11. Bouali B, Larbi S (2013) Contribution to the geometry optimization of an oscillating water column wave energy converter. Energy Procedia 36:565–573
12. Liu Z, Hyun B-S, Hong K (2011) Numerical study of air chamber for oscillating water column wave energy convertor. China Ocean Eng 25(1):169–178
13. Zhang Y, Zou Q-P, Greaves D (2012) Air–water two-phase flow modelling of hydrodynamic performance of an oscillating water column device. Renew Energ 41:159–170
14. Kamath A, Bihs H, Arntsen ØA (2015) Numerical modeling of power take-off damping in an oscillating water column device. Int J Marine Energ 10:1–16
15. Kamath A, Bihs H, Arntsen ØA (2015) Numerical investigations of the hydrodynamics of an oscillating water column device. Ocean Eng 102:40–50
16. Boccotti P (2007) Comparison between a U-OWC and a conventional OWC. Ocean Eng 34(5–6):799–805
17. Ram KR et al (2010) Experimental studies on the flow characteristics in an inclined bend-free OWC device with parallel walls. In Proceedings of Renewable Energy 2010 Conference and Exhibition (Wind Energy-Advanced Technology Paths to Global Sustainability
18. Şentürk U, Özdamar A (2012) Wave energy extraction by an oscillating water column with a gap on the fully submerged front wall. Appl Ocean Res 37:174–182
19. Howe D, Nader J-R (2017) OWC WEC integrated within a breakwater versus isolated: experimental and numerical theoretical study. Int J Marine Energ 20:165–182
20. Rezanejad K, Souto-Iglesias A, Soares CG (2019) Experimental investigation on the hydrodynamic performance of an L-shaped duct oscillating water column wave energy converter. Ocean Eng 173:388–398

21. ANSYS, I., ANSYS fluent user's guide, release 19.0. Equation, 2018(6.68)
22. Hirt CW, Nichols BD (1981) Volume of fluid (VOF) method for the dynamics of free boundaries. J Comput Phys 39(1):201–225
23. Iturrioz A et al (2015) Validation of OpenFOAM® for oscillating water column three-dimensional modeling. Ocean Eng 107:222–236
24. Sheng W, Lewis A (2016) Wave energy conversion of oscillating water column devices including air compressibility. J Renew Sustain Energ 8(5):054501
25. Huntley D, Le Méhauté B (1976) An introduction to hydrodynamics and water waves. Springer-Verlag, New York, viii+ 323 p. $24.80. Limnol Oceanogr 22(5):974–975
26. Kharati-Koopaee M, Fathi-Kelestani A (2020) Assessment of oscillating water column performance: influence of wave steepness at various chamber lengths and bottom slopes. Renew Energy 147:1595–1608
27. Teixeira PR et al (2013) Numerical simulation of an oscillating water column device using a code based on Navier–stokes equations. Energy 61:513–530

Development of Hydrogen Production from Solar Thermal Energy Application Model for the GCC Region

Abdulrahman Joubi, Yutaro Akimoto, and Keiichi Okajima

1 Introduction

The Gulf Cooperation Countries (GCC) and the Middle East are facing a surge in energy demand and growing CO_2 emissions [1], directing the region into seeking new sources of energy with low carbon footprint.

Solar energy presents a suitable fit to the region's climatic advantage [1], from sunlight abundance to the vast open areas of desserts. Few projects and plans are already on foot, such as concentrated solar plants (CSP), to exploit the GCC's solar potential, but addressing its full capabilities is still far to achieve, due to the large economic investment that does not yet prove feasible compared to the current traditional applied technologies.

In this paper, we consider using the excessive energy from two different scenarios of large-scale CSP application in the GCC for producing liquid hydrogen and using it for export, using the profits as an incentive for CSP large application, and keeping the Gulf Cooperation Countries as a major energy export. Although few studies addressed the potential of renewable energy and hydrogen production of the GCC, along with their possible future cost, none fully compares different scenarios of CSP application or investigates the ability of the current infrastructure to produce and deliver the produced H_2 to possible export areas.

2 Method and Data

In this study we investigate the technical potential of liquid hydrogen production from surplus electricity of three different scenarios of CSP application in the four gulf countries, which are United Arab Emirates, Kuwait, Oman, and Saudi Arabia, based on our previous study [2], where the potential of solar thermal energy in these countries is calculated in three different cases, depending on the distance from roads, electrical grid, and water areas. The cases include 10, 25, and 40 km distances. The results show that only two cases with the distances of 25 km and 40 km from mentioned parameters can leave extra electricity for hydrogen production after satisfying the future demand of the GCC countries for the year 2023, while other countries have excessive energy for all cases for the same year.

Two hydrogen delivery scenarios are considered in this paper depending on the location of H_2 plants, the first one assumes they are at the same location of CSP, where 6 main spots are assumed for the location of the plants for the UAE, 5 for Kuwait, 8 for Oman, and 14 for Saudi Arabia, covering all areas suitable for CSP application in all application cases, also with access to main roads and high voltage electricity grid of the GCC. In the first scenario, the H_2 is to be delivered using trucks toward the main seaports for export in the GCC [3]; shortest distance from plants to ports is analyzed using geographical information system (GIS) to investigate the applicability of this scenario, where road network data from OpenStreetMap [4] are used and the truck speed is set to 80 km/h.

The second location scenario assumes hydrogen production plants be placed at the ports, where excessive electricity from CSP is transmitted through the electrical grid toward export ports, presuming that 3% of electricity is lost in the grid for every

A. Joubi (✉) · Y. Akimoto · K. Okajima
Graduate School of Science and Technology, University of Tsukuba, Tsukuba, Japan
e-mail: s2030126@s.tsukuba.ac.jp

© The Author(s), under exclusive license to Springer Nature Switzerland AG 2022

A. Sayigh (ed.), *Sustainable Energy Development and Innovation*, Innovative Renewable Energy,
https://doi.org/10.1007/978-3-030-76221-6_88

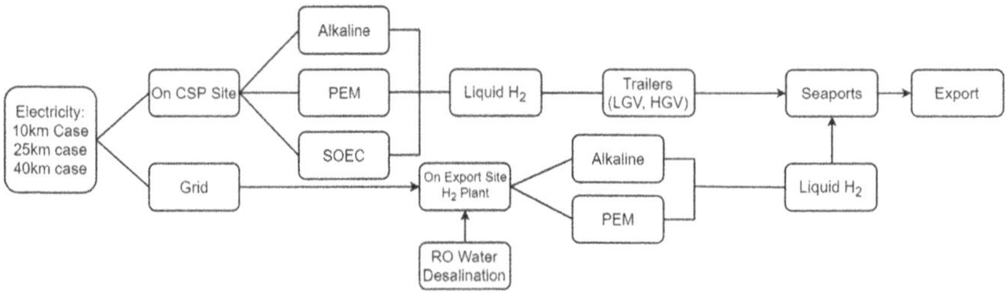

Fig. 1 Scenarios layout

1000 km traveled [5]. The shortest routes of electrical grid are analyzed with GIS to ensure the least loss of electricity possible in the network in this scenario. The layout of study methodology is shown in Fig. 1.

We compare three different water electrolysis technologies for hydrogen production. The first is alkaline electrolysis, which is the most common technology [6] and has been used by industry for nearly a century; therefore, it is the most mature way to produce hydrogen. It uses potassium hydroxide (KOH) [6], which adds extra expenses for the production process. The second technology is proton exchange membrane electrolyzers (PEM), where the electrolyte is a polymer membrane which is only permeable for hydrogen ions. Today they are commercially available. The last technology is solid oxide electrolyzers (SOEC), and the electrolyte is a solid ceramic membrane. All electrolyzers assume higher heating value (HHV), where alkaline and PEM work at temperatures between 60 °C and 90 °C, while the SOEC that use high-temperature steam electrolysis (HTE) work at temperatures in the range of 700 °C to 800 °C; this technology could be coupled with CSP, but it is still at development phase. Chosen efficiencies for each technology are explained in Table 1.

For the second scenario where the hydrogen plants are located at the ports, only alkaline and PEM are considered, since SOEC would require extra energy consumption at the ports. Also, the water source for electrolysis in this scenario is assumed to come from desalinated seawater using reverse osmosis (RO) desalination, and electricity used for water desalination is added to the energy demand of the electrolysis process. The energy needed can be determined by multiplying the energy need per cubic meter with the amount of water that is necessary, which is as follows [6]:

$$m_{\dot{H}_2O} = \frac{\dot{m}}{M \times \eta} \times M_{H_2O} \tag{1}$$

where $\dot{m}$ (kg/s) is the production rate of hydrogen, M is the molar mass of hydrogen, η is the efficiency of water use, and M_{H_2O} is the molar mass of water.

3 Results and Discussion

3.1 Hydrogen Plants at the Same Site as CSP Scenario

Based on the assumptions mentioned in the previous section, in this scenario we first calculate the hydrogen for all different technologies for the two energy supply scenarios we have. SOEC has the biggest amount of hydrogen in both energy cases, as shown in Table 2.

For a better understanding, Fig. 2 shows a comparison between the amounts of hydrogen produced in all countries.

SOEC results were shown in pattern style to express that this technology is still at lab level, whereas the others are at industrial level.

The next step is the geospatial network analysis to find the shortest routes for the hydrogen delivery from production sites to the closest ports. Figure 3 shows that the shortest routes analyzed deliver the hydrogen to the ports in all countries; not all ports were chosen as they were far from the plants compared to other ports.

The distances calculated can be used in future cost analysis to compare scenarios feasibility.

To have a better undertanding of the size of hydrogen production in the GCC, we compare the amounts of hydrogen calculated in this scenario for the 40 km case in Fig. 4, which shows the high potential for H_2 in Saudi Arabia compared to the rest of GCC countries.

Table 1 Electrolyzers' efficiencies [7–9]

Electrolyzer	Electricity consumption	Efficiency
Alkaline	53.5 kWh/kg-H_2	73%
PEM	49.8 kWh/kg-H_2	78.9%
SOEC	39.4 kWh/kg-H_2	82%

Table 2 Produced hydrogen from different technologies in the first scenario in all energy cases (kg x 10^6)

Country	Energy case	Alkaline	PEM	SOEC
UAE	25 km case	4784.08	5139.524	6496.15
	40 km case	8918.5	9581.11	12,110.134
Kuwait	10 km case	434.44	466.72	590
	25 km case	2801.078	3009.2	3803.5
	40 km case	3238.72	3479.35	4397.75
Oman	10 km case	5726.11	6151.54	7775.3
	25 km case	18,621.86	20,005.42	25,286.04
	40 km case	28,413	30,524	38,581
Saudi Arabia	10 km case	2361.56	2537.2	3206.7
	25 km case	193,863.24	208,266.74	263,240.7
	40 km case	308,675.13	331,608.82	419,140.1

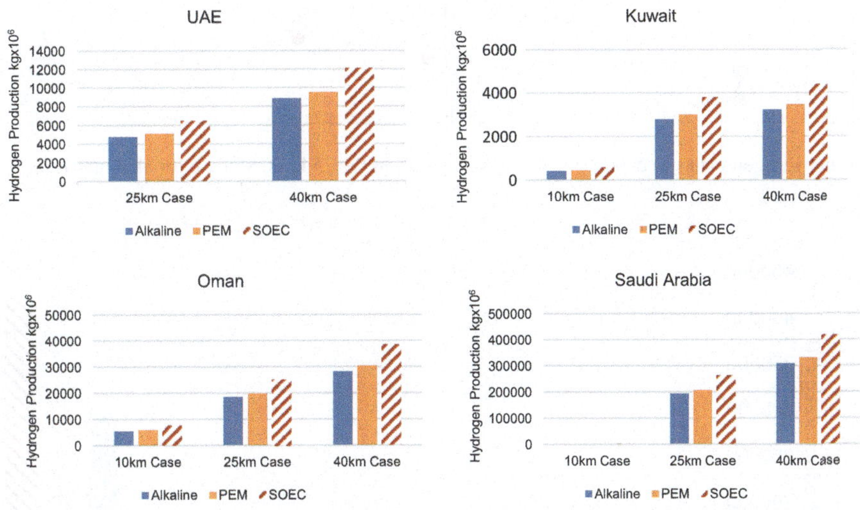

Fig. 2 Comparison of hydrogen production in all technologies for 10 km, 25 km, and 40 km energy supply cases in plants at the same location as CSP sites scenario for all countries

3.2 Hydrogen Plants at Ports Sites Scenario

For this scenario, we first start with the network analysis to define the shortest distances of electrical grid to deliver the surplus electricity in all cases from CSP to ports where we want to produce the hydrogen. Figure 4 shows the GIS analysis results of the shortest routes of hydrogen delivery in all four countries, like the first scenario, some ports were not selected due to being far from the assumed locations of concentrated solar plants.

Analysis shows that the sum of distance traveled in electrical grid is 1031.24 km for UAE, 270.4 km for Kuwait, 1266.5 km for Oman, and 5083.37 km for Saudi Arabia; therefore, the losses in the grid would be up to 3.1%, 0.81%, 3.8%, and 15.25% for each country consequently of the total energy transmitted. The remaining energy is used to desalinate water and produce hydrogen at the two chosen ports. Electricity consumption for desalinating the water needed for 1 kg of H_2 from Eq. (1) is 1.8 kWh, which is added to elctrolyzer's consumption. In this scenario, only alkaline and PEM electrolyzers are used as

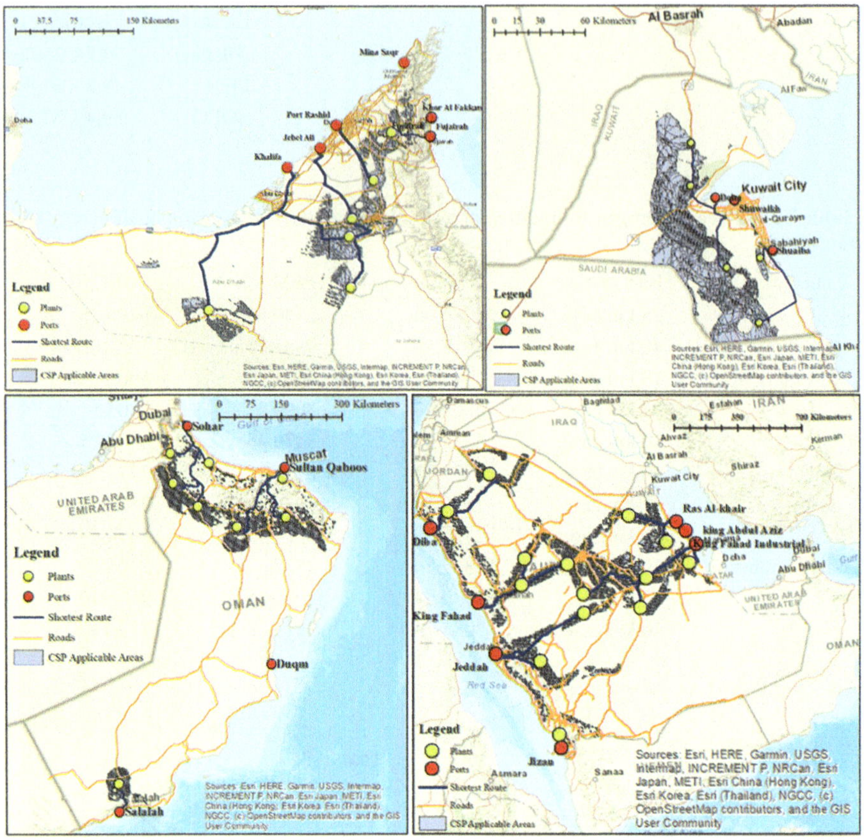

Fig. 3 Shortest routes from hydrogen plants to export ports analysis by GIS in the 25 km CSP case for all countries

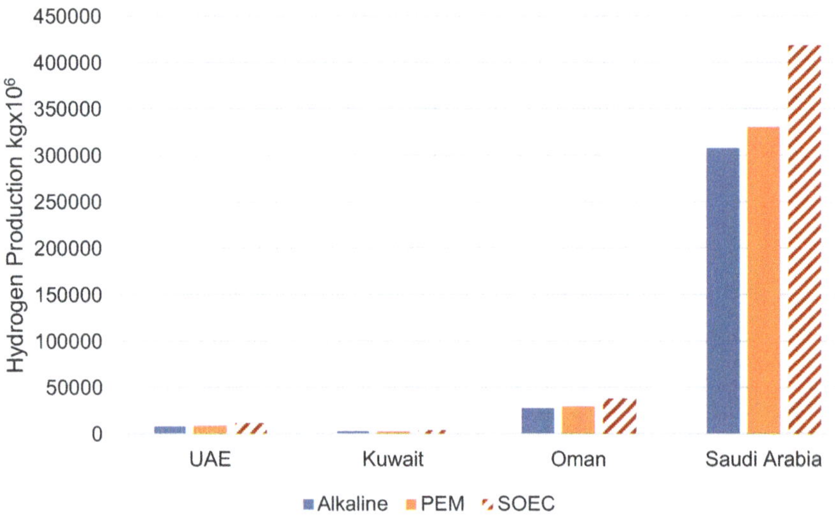

Fig. 4 Comparison of hydrogen production in 40 km energy supply cases in plants at the same location as CSP sites scenario for all countries

Table 3 Hydrogen produced from different technologies in the second scenario in both energy cases for all countries (kg × 10^6)

Country	Energy case	Alkaline	PEM
UAE	25 km case	4635.77	4806.471
	40 km case	8642.018	8960.23
Kuwait	10 km case	417	446.78
	25 km case	2687.92	2880.66
	40 km case	3107.88	3330.73
Oman	10 km case	5329.218	5711.35
	25 km case	17,331.13	18,573.87
	40 km case	26,443.51	28,339.66
Saudi Arabia	10 km case	1936.27	2075.12
	25 km case	158,951.2	170,348.87
	40 km case	253,087.1	271,234.81

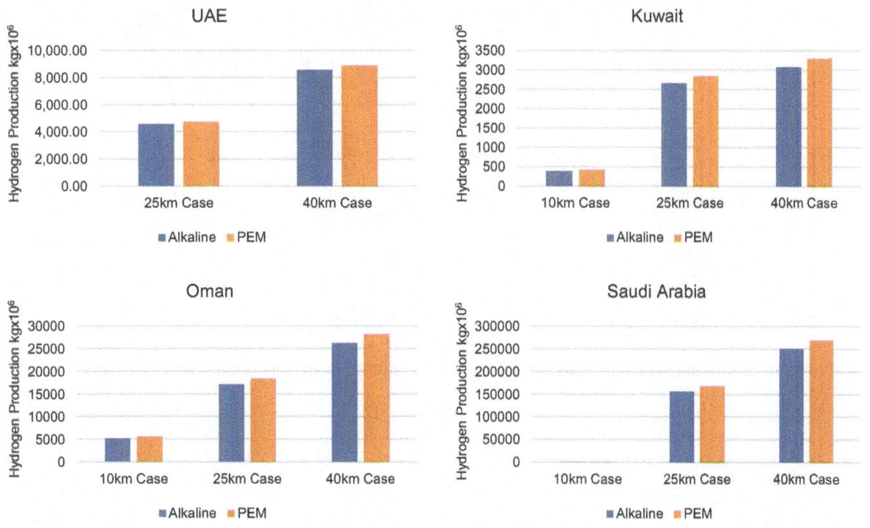

Fig. 5 Comparison of hydrogen production in all technologies for 10 km, 25 km, and 40 km energy supply cases in H_2 plants at the seaports sites scenario for all countries

SOEC works better coupled to steam source such CSP. Table 3 shows a comparison between the amounts of hydrogen produced in this scenario for all energy cases in all countries, while Fig. 5 compares the results (Fig. 6).

Again we compare these results for all countries in the 40 km energy supply case, as shown in Fig. 7.

3.3 *Location Scenarios Comparison*

Finally, we compare produced hydrogen in all scenarios for the 40 km energy case to have a better understanding of the results, as shown in Fig. 8. The results show higher amounts of H_2 in the first scenario due to the energy losses in the electrical grid, and also SOEC has the highest potential between all eletrolysis technologies.

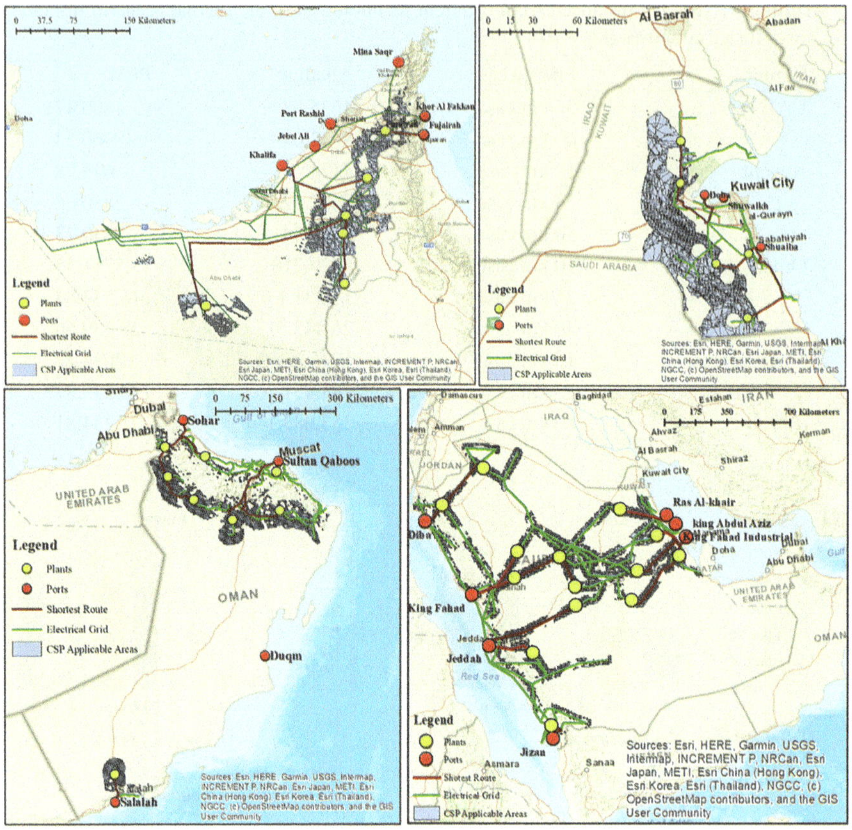

Fig. 6 GIS analysis results for the shortest electrical grid routes from CSP plants to export ports in the 25 km CSP case for all countries

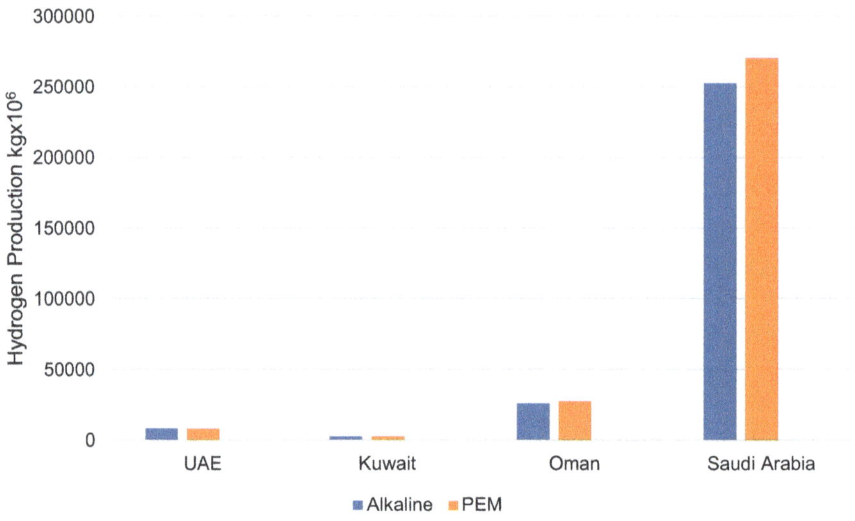

Fig. 7 Comparison of hydrogen production in 40 km cases in plants at seaports for all countries

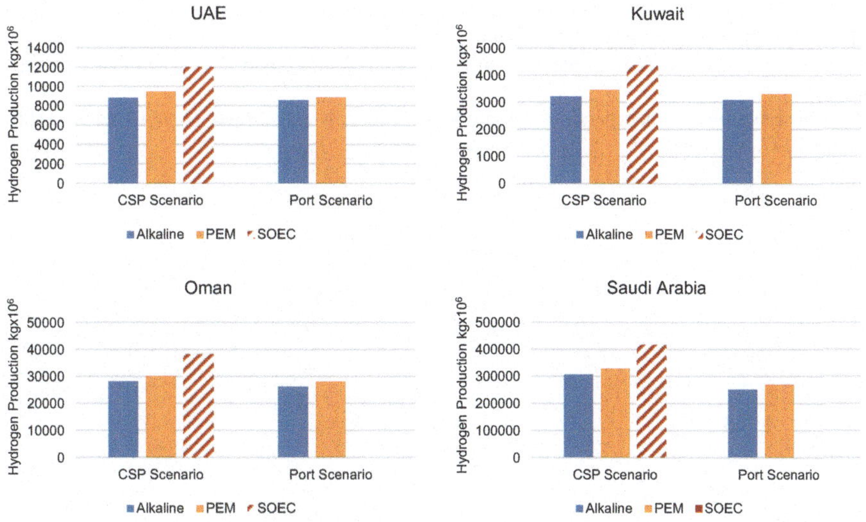

Fig. 8 Comparison of hydrogen production in all delivery scenarios for the 40 km energy supply cases, in all countries

4 Conclusion

In this study, we evaluate hydrogen production from different technologies in two location scenarios based on excessive electricity from three different energy cases of CSP application in the GCC. The hydrogen produced is planned to be exported as a means of profit to make the investment in large-scale CSP application in United Arab Emirates, Kuwait, Oman, and Kingdom of Saudi Arabia feasible.

The resulted from the two delivery pathways show that most ports can be used for export from various CSP, providing more flexibility to delivery options. While in the second scenario, fewer ports are chosen as the closer option for electricity transmission and hydrogen production sites.

Production results show SOEC elctrolyzer as the best technology for large-scale H_2 production in the first scenario when coupled to CSP, while PEM seems like the better fit for the H_2 production at seaports scenario, when using desalinated water from RO technology.

Saudi Arabia shows the largest amounts of hydrogen due its high potential of CSP application and its vast areas compared to the other GCC countries, followed by Oman, UAE, and then Kuwait. Saudi Arabia has the highest losses of electricity in the grid as the energy travels longer distances making the first scenario of delivery more attractive.

A future cost analysis is needed to provide a wider perspective and better understanding for what is a better scenario and delivery path.

Acknowledgement The first author acknowledges funding support from the Japanese government, Ministry of Education, Culture, Sports, Science and Technology, under their international program (MEXT) scholarship.

References

1. El-Katiri L (2014) A roadmap for renewable energy in the Middle East and North Africa. Oxford Institute for Energy Studies, Oxford, The United Kingdom.
2. Joubi A, Okajima K (2020) Development of an energy potential estimation model for concentrated solar plants penetration in the GCC region. Int J Smart Grid Clean Energ 9(6):1000–1010
3. The United Arab Emirates' Government Portal: https://u.ae/en/information-and-services/infrastructure/civic-facilities/seaports. Accessed on 20 August 2020
4. Vaillancourt K, Simbolotti G, Tosato G (2014) Electricity transmission and distribution. Technology Brief, pp.1–16. https://www.geofabrik.de/geofabrik/openstreetmap.html. Accessed 25 August 2020
5. Vaillancourt K, Simbolotti G, Tosato G (2014) Electricity transmission and distribution. IEA ETSAP-Technol Brief E 12
6. Meier K (2014) Hydrogen production with sea water electrolysis using Norwegian offshore wind energy potentials. Int J Energy Environ Eng 5(2–3):104

7. Monnerie N, von Storch H, Houaijia A, Roeb M, Sattler C (2017) Hydrogen production by coupling pressurized high temperature electrolyser with solar tower technology. Int J Hydrog Energy 42(19):13498–13509
8. Ivy J (2004) Summary of electrolytic hydrogen production: milestone completion report (No. NREL/MP-560-36734). National Renewable Energy Lab, Golden
9. Hamdan M (2013) PEM Electrolyzer incorporating an advanced low-cost membrane (No. DOE/GO/18065-22). Giner, Inc./Giner Electrochemical Systems, LLC, Newton

Sustainable Development of Higher Education Institutions – An Analytical Study for SD Projects' Retrofitting of Selected Global Universities' Buildings

Tahany Abdelmawgood and Mohsen Aboulnaga

1 Introduction

Higher education institutions play a major role in attaining sustainable development and SDGs in cities. Higher education buildings (universities/research centers) worldwide are considered the highest in terms of energy use and GHG emissions, yet the third of the total utilized areas. Hence, sustainable transforming of universities' buildings is imperative for governments and policy makers due to high energy use intensity/m^2 according to American National Energy Audit by CBECS [1]. The building sector is one of the biggest energy-consuming sectors and carbon producers worldwide, which is estimated to be nearly 40% of the global energy consumption and around 39% of carbon emissions [2]. The International Energy Agency (IEA) 2018 report stated that the final energy demands in buildings will increase to 50% by 2050 compared to 2015 level [3, 4]. Energy-related CO_2 emissions from buildings have increased in recent years. In 2019, direct and indirect emissions from electricity and commercial heat used in buildings rose to 10 $GtCO_2$, the highest level ever recorded [5]. It is estimated that 50 million tons of oil equivalent (Mtoe) are used in the building sector in the MENA region, roughly one-third of the region total annual energy consumption; and without appropriate countermeasures the sector is expected to grow to 70 Mtoe (+40%) by 2030 [6]. Combining comprehensive energy efficiency requirements for new building and extensive retrofit of existing buildings would reduce the total final energy consumption of the building sector in the MENA region by 600 TWh by 2030 and 900 TWh by 2050 if all the energy programs start to be implemented by 2020 [7]. Thus, sustainability of existing buildings is vital to lower energy consumption and mitigate CO_2 emissions. The Global Alliance of Buildings and Construction Sector (GABCS) at COP 22 in Marrakech reported that upgraded buildings (commercial, educational, and hospitals) have economic and social values and will have the greatest impact on lowering energy use and mitigating CO_2 emissions [8].

The UNEP global status report 2019 for buildings and construction indicates that of the total global spending on buildings' construction and renovations (US$ 4.5 trillion), about 3.1% of this amount (US$ 139 billion) is spent globally on energy efficiency of buildings as of 2018 [9]. In fact, Universities buildings are among the highest in energy use intensity buildings due to the large floor areas and the high occupancy rates and long hours of operations as well as the activities all year around [10]. Since natural resources are scarce, enhancing higher education buildings to attain sustainable development is primarily important economically, socially, and environmentally.

In the time of Coronavirus (COVID-19) pandemic, higher education buildings (universities and research centers) has become the focal point of governments' policies due to the confinement measures and lockdown [11]. Thus, transforming universities' buildings to be energy efficient, low-carbon, and excellent indoor environmental quality (indoor air quality – IAQ; very good ventilation rate and thermal comfort) is essentially vital to meet the Coronavirus (COVID-19) precautionary measures, adapt to climate change, and attain sustainable development goals (SDGs), mainly SDG 4, 7, 11, 12 13, and 17.

There are many universities' buildings worldwide that adopted sustainable development (SD) and integrated SDGs in their projects' upgrading [12, 13]. A study by Ma (2012) established six key criteria that influence buildings' retrofitting as shown in Fig. 1 [14].

T. Abdelmawgood (✉)
Environmental Design & Energy Efficiency Program, Department of Architecture, Faculty of Engineering, Cairo University, Giza, Egypt

M. Aboulnaga
Department of Architecture, Faculty of Engineering, Cairo University, Giza, Egypt
e-mail: maboulnaga@eng.cu.edu.eg

© The Author(s), under exclusive license to Springer Nature Switzerland AG 2022

A. Sayigh (ed.), *Sustainable Energy Development and Innovation*, Innovative Renewable Energy,
https://doi.org/10.1007/978-3-030-76221-6_89

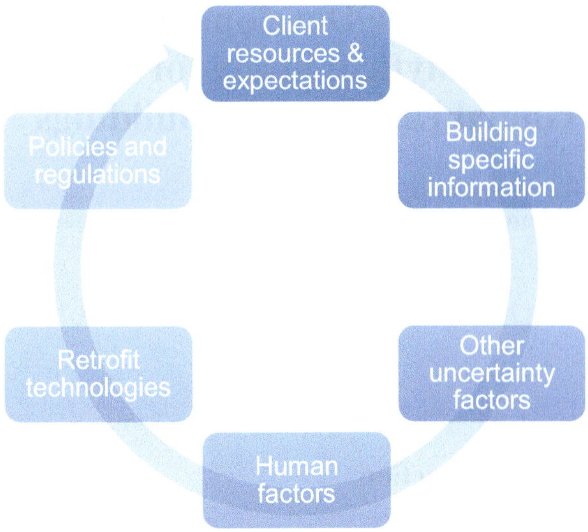

Fig. 1 Key criteria influencing buildings' retrofitting. (*Source*: Developed by authors based on Ma [14])

2 Objectives

There are many sustainable development (SD) projects for universities' buildings worldwide. The work discussed in this chapter has reviewed and developed the categories, key elements, and indicators for sustainable development projects (SDP) by analyzing SDP in the United States, the European Union, Australia, and Singapore. Monitoring the SD main categories, key elements, and indicators that affect energy efficiency and indoor environmental quality that were achieved in these universities' projects is vital to understand and inform the assessment.

This study focuses on examining the categories affecting the SDP of international universities' buildings in various parts of the world in an attempt to assess these universities' buildings to determine the most influential retrofit categories, key elements, and indicators that affect sustainability and SD of different universities. Hence, the main objectives of the study are as follows:

- To determine the common key elements and indicators that affected these projects through the different applications of technologies in university existing buildings in order to improve energy efficiency and counteract the impacts of climate change, and those factors are (building's energy consumption, cost, regulations and laws, the Indoor environment and comfort rates, global environment and emissions ratios, and other factors) based on the study by Ma 2012 and Kolokotsa, 2009 [14, 15].
- To establish a list of the most influential retrofit categories and indicators that are used to apply the sustainable development actions in each project, in terms of the goals of saving energy, carbon, water, total cost of building development, measures, comfort rates, legislation, and other factors, with the aim of arriving at a list of factors, applications, and standards (Fig. 2).

3 Methodology

The methodology depends on two approaches inductive and analytical approaches. The first approach includes a review of the categories affecting sustainable development (SD), whereas the second approach is an analytical study conducted to investigate the most influencing retrofitting categories and key elements of (SD) projects of universities' buildings. Such analytical study consists of three phases, namely:

1. Selecting the projects under study and collecting the data.
2. Checking the categories and key elements affecting sustainable development projects in order to determine the most influencing key elements of all categories that affect sustainable development projects of university buildings.
3. Analyzing the results and finding indicators that determine the effectiveness and the percentage of each category affecting sustainable development in the projects' upgrading and the level of achievement.

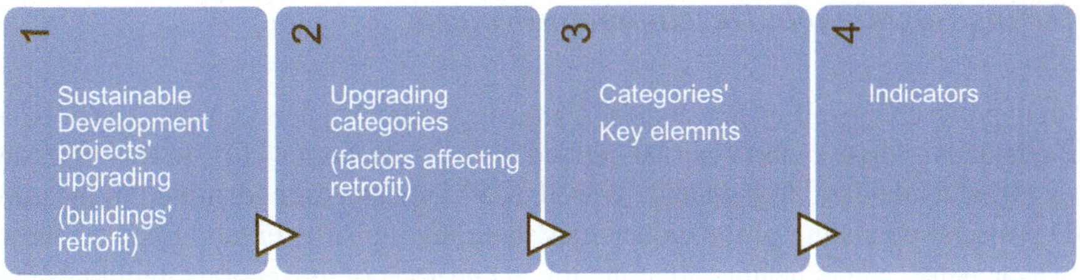

Fig. 2 Simplified workflow of the research. (*Source*: Developed by authors based on Ma and Kolokotsa [14, 15])

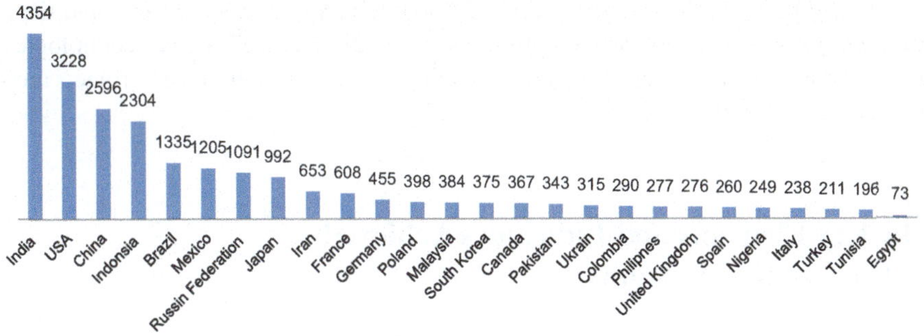

Fig. 3 Number of universities worldwide by country, 2020. (*Source*: https://www.statisa.com/statistics/918403/number-of-universities-worldwide-by-counrty/)

Table 1 Selected universities for the analytical study

Country	USA		Denmark	Germany		Singapore	Australia	Greece	Poland
University	Harvard University Campus	Harvard (LHRRB)	Copenhagen university	Library Building University of Bremen	Research Building University of Stuttgart	Nanyang Technological University	Woolley Building University of Sydney	N.T.U.A. University Athens	Faculty of Electrical Engineering Poznan
Level of retrofitting	Deep retrofit	Deep retrofit	Deep retrofit	Whole retrofit	Deep retrofit	Whole retrofit	Deep retrofit	Whole retrofit	Deep retrofit

Source: Developed by authors

3.1 Selecting the Projects Under Study and Collecting the Data

According to the Center for World University Rankings (CWUR), there are less than 20,000 universities in the world as of 2019 [16]. Nevertheless, this number could be higher in 2020 (23,045 universities) as shown in Fig. 3 [17, 18]. Based on the objectives of the study, the selected cases of the universities' buildings were chosen to be from different universities all over the world that have conducted sustainable development (retrofitting) projects. Also, the level of retrofit should include three types of retrofit (standard, deep, and whole building retrofit), either for one building or for the full campus (Table 1). Besides, the selection of the universities was carried out based on the availability of the data and information about the factors that affected the sustainable development, including six main factors (energy efficiency, legislations and laws, cost, indoor environmental quality, global environment, and variable factors). This means that the projects were selected based on the availability of information and data on the indicators used in the developed projects, the data are; standards for measuring energy savings, information about funding sources used in the project, targets for energy reduction and carbon mitigation, or any one of them, also the cost of technologies exploited and the percentage of savings achieved as well as considering the key elements related to every category of the indoor environment and the global environment of the project. Thus, nine sustainable development projects of universities' buildings were chosen and analyzed as listed in Table 1.

3.2 Criteria Affecting Sustainable Development Projects

The criteria pursued to improve energy efficiency and energy management in a new construction or buildings upgrade can be either quantitative or qualitative, these criteria can be divided into six categories [14]. In this assessment, 6 categories are represented by 13 key elements that are investigated, as well as 50 indicators are identified to measure these key elements verification in the SD projects' upgrading of the universities' building globally. Figure 4 depicts these criteria and Table 2 lists the categories, key elements, and indicators affecting the SD projects of universities' buildings in details [14, 15], [19]. The selected universities (Fig. 5) are analyzed by documenting the buildings' construction date, the development date, and the type of building under upgrading including the area and the number of occupants in such building as well as identifying the factors that affect the sustainable SD by inducing the categories of SD and key elements that influence the project measures and metrics used to elaborate on the progress (Table 2), a summary of the SD project and the results of post-retrofit survey, the procedures, energy savings, and carbon emissions' mitigation as well as the technologies used to improve the building's performance. The influencing categories and indicators monitored in these SD projects are shown in Appendices (Tables 4, 5, and 6 in Appendices A, B, and C, respectively).

4 Analysis of the Key Elements and Indicators Achieved in the Global Universities' Retrofits

In university buildings, the air conditioning and lighting systems are deemed the most energy-consuming elements for which development occurs, followed by control systems and building management systems, and then building envelope upgrades and renewable energy technologies [21]. Most of the upgrades are considered to be adaptation strategies rather than mitigation ones. The heating and ventilation systems in these projects included improvements to refrigeration and heating systems, change of air handling units (AHU) and filters, separation of heating and cooling, improvements to lighting systems and replacement of lamps with more efficient lamps, enhancements related to control systems in building including conducting building control systems such as DDC and BMS, and installation of lighting and heating occupancy sensors in addition to the dependence on renewable energy technologies in buildings, especially photovoltaics and solar heating [22–36]. Tables 4, 5, and 6 are listed in the Appendices (Appendices A, B, and C) highlighting the assessment of the 50 indicators. Also, Fig. 6 presents the achieved indicators and percentage of implementation in the selected nine universities' buildings.

Based on Tables 4, 5, and 6 in the Appendices A, B, and C and Fig. 7 as well as the indicators' analysis, the following can be deduced:

(a) Category 1 – Energy consumption (22 indicators, 1–22).

- It is possible to save 20% of energy consumption only by improving the behavior of users in buildings as indicated in the University of Copenhagen and University of Stuttgart [22, 28, 30, 31, 34]. There is a potential of saving 15% of energy consumption by enhancing the building's envelope, whereas energy-efficient building and efficient equipment can cause savings up to 45% [33].
- The use of passive and low-cost retrofit measures has an impact on improving the university's building energy use if these measures are coupled together. Passive measures are considered among the best upgrading strategy options for universities' buildings due to positive effect on the quality of the internal environment and comfort [33].
- Laboratories and research buildings consume higher amount of energy than that of libraries and teaching rooms or lecture halls of university's campus.
- The building size and area is not a barrier for upgrading development and retrofit, but the most important factors are the type of the building, activities, and occupancy rates as well as the number of operating hours since energy intensity increases based on the number of operating hours of the university buildings as shown in Fig. 7a.

(b) Category 2 – Cost (7 indicators, 23–29).

- Universities' sustainable development and upgrading projects are financed by internal investment funds, green expenses for students, or green revolving fund established by the university. The total direct and indirect cost and the returned investments (RoI) are the most important indicators that signify the cost factor, and annual maintenance cost is vital (Fig. 7b).

(c) Category 3 – Legislations and laws (4 indicators, 30–33).

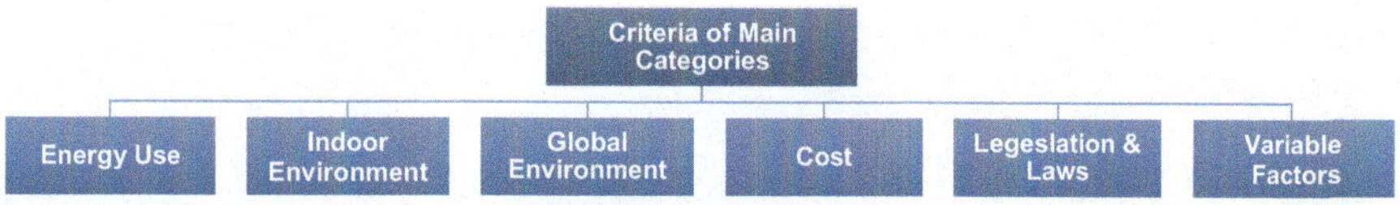

Fig. 4 Main categories used for SD assessment of the selected global universities' buildings [14, 15]

Table 2 Categories, key elements, and indicators affecting the SD projects of global universities' buildings

Category	Key elements	Indicators
Energy consumption	Energy improvement technologies	Lighting efficiency HVAC efficiency improvements Separating heating and cooling systems Ventilation efficiency Equipment efficiency User behavior Improving building operation Using building control (BMS and DDC) Using (motion, lighting, and temperature) sensors and scheduling occupation Energy consumption monitoring
	Building envelope technologies	Wall/roof insulation Replacing/insulating windows/WWR External shading
	Renewable energy technologies	Photovoltaics Solar heating Wind energy Purchasing renewable energy
	Energy measures	Annual energy savings Percentage of energy savings from technologies Heating cooling energy consumption Embodied energy
Cost	Internal financing	Owner financing Green revolving fund Students green expenses
	Cost measures	Direct and indirect cost Payback periods Annual total savings Maintenance cost Interest rate
Legislations and laws	Legislation measures	Legislations and laws of buildings upgrade in a countr State subsidies for buildings development projects Supporting renewable energy projects Carbon reduction plans
Indoor environmental quality	Building considerations	Building shape, direction, and number of floors Activity and building type Building age Occupancy hours
	Indoor air quality and comfort rates	Visual and acoustical comfort Indoor air quality (CO_2 factor and ventilation rate) Thermal comfort (air temperature, humidity, and thermal gain) Pre/post-development survey occupant satisfaction
Global environment	Carbon reduction	Measure the amount of emissions Emission reduction measurement Water consumption rate Efficient water fixtures
	Water reduction	Grey water
	Waste management	Recycling waste
Variable factors	Building control, functionality, and time	Control the building elements and parts using the building's smart energy management systems Functionality of retrofit technologies (user satisfaction) Duration of the development procedure

Source: Developed by authors based on Ma, Kolokotsa, and Liu [14, 15, 19]

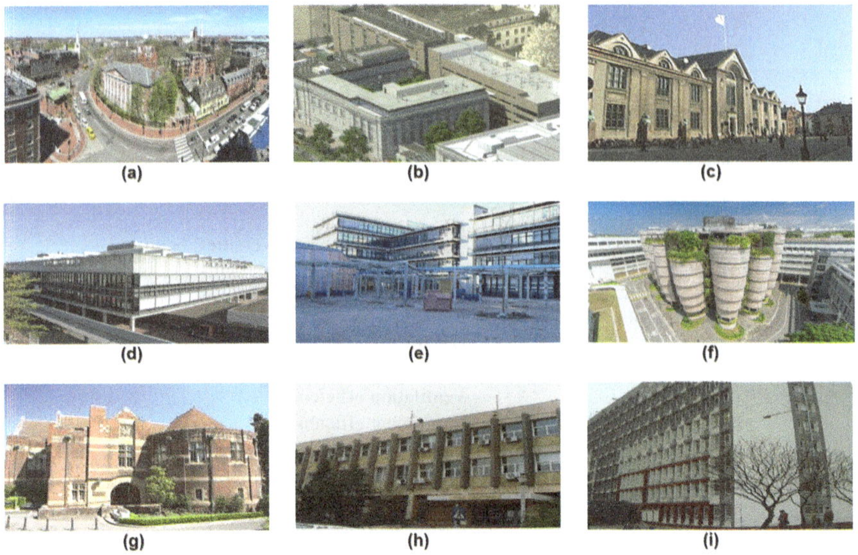

Fig. 5 Selected Universities' buildings for the sustainable development projects' assessment [20]. (**a**) Harvard University Campus, USA. (**b**) Harvard (LHRRB), USA. (**c**) Copenhagen University, Denmark. (**d**) Library Building, University of Bremen, Germany. (**e**) Research building University of Stuttgart, Germany. (**f**) Nanyang Technological University, Singapore. (**g**) Woolley Building, University of Sydney, Australia. (**h**) N.T.U.A. University of Athens, Greece. (**i**) Faculty of Electrical Engineering Poznan, Poland

Table 3 The most influencing indicators in each category

Category	Indicator	Iterations
Energy consumption	Annual energy consumption/savings	6
	Savings from the technology	9
	Reduction in cooling/heating energy consumption	2
Cost	Total direct/indirect cost	7
	Payback periods	6
	Total annual savings	4
	Maintenance cost	6
Legislations and laws	Legislations and laws of development buildings	7
	State subsidies for buildings development projects	1
	Supporting renewable energy projects	6
	Carbon reduction plans	6
Indoor environmental quality	Building shape, direction, and number of floors	9
	Building age	9
	Occupancy hours	3
	Indoor air quality and thermal comfort (air temperature, humidity, and thermal gain)	6
Global environment	Measure the amount of emissions	3
	Emissions reduction measurement	2
	Water consumption rate	1
Variable factors	Control the building elements and parts using the BEM smart systems	9
	Functionality of retrofit technologies (user satisfaction)	4
	Duration of the development procedure	1

- The sustainable development of university buildings depends on the laws that are enacted by the state or endorsed by the university senate and higher administration. These laws are subordinate to the university's goals to mitigate carbon emission and enhance energy efficiency. In most countries, governments often encourage universities to limit high-carbon fuel used and promote renewable energy projects in the upgrading process as illustrated in Fig. 7c.

(d) Category 4 – Indoor air quality (8 indicators, 34–41).

- The user satisfaction questionnaire must be carried out before and after the development/ upgrading retrofit. User satisfaction can be pursued by considering the visual, thermal, and acoustical comfort rates and internal environment quality rates with a strong reliance on occupancy. The best results, nevertheless, are reached when contracts with companies are signed in order to achieve comfort rates and the quality of the internal environment as shown in Fig. 7d.

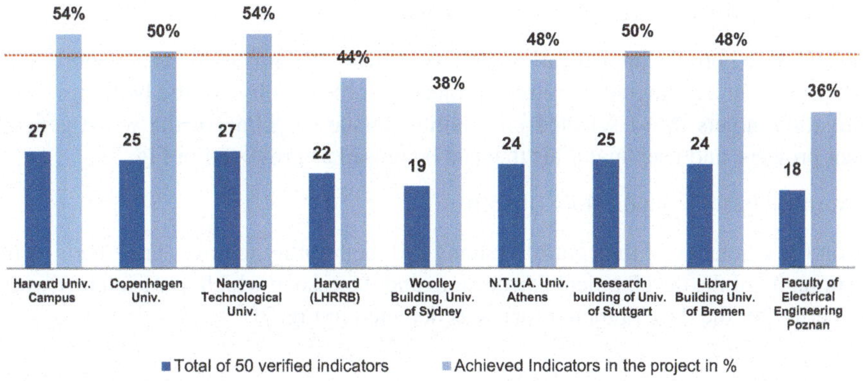

Fig. 6 Achieved indicators and percentage of implementation in selected universities' buildings

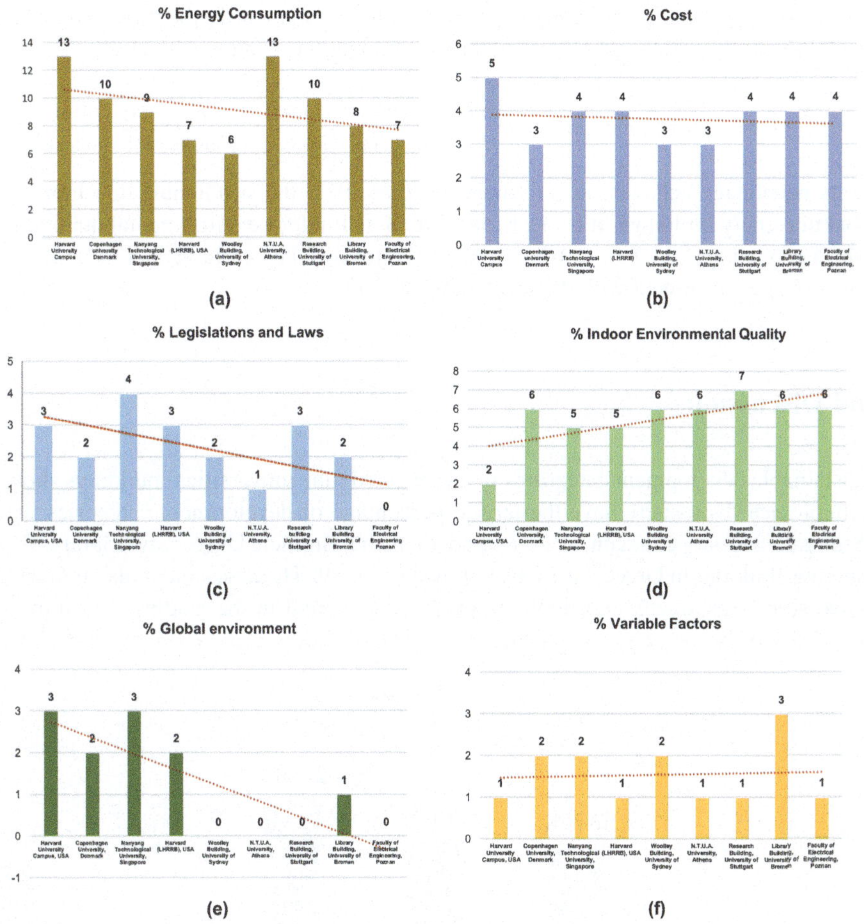

Fig. 7 Verification percentage of the selected nine global *university buildings* according to categories. (**a**) Energy consumption verification. (**b**) Cost verification percentage. (**c**) Legislations and laws verification percentage. (**d**) Indoor environmental quality verification percentage. (**e**) Global environment verification percentage. (**f**) Variable factors verification percentage

(e) Category 5 – Global environment (6 indicators, 42–47).

- Universities measure both the annual carbon emissions from the university as a whole and carbon emissions mitigation after development and upgrading retrofit projects in order to observe the buildings' footprint and verify desired carbon mitigation targets. Most selected universities pursue recycling of the waste generated from the university, development projects, and keep materials that can be reused as presented in Fig. 7e.

(f) Category 6 – Variable factors (3 indicators, 48–50).

- It depends on the efficiency of the control systems and their ability to cover large parts of the building. It also relies on the functionality of the developmental procedure and the ease in which users interact with such procedures as well as the time period for the development process as depicted in Fig. 7f.

4.1 *Category Verification Percentage*

To calculate the verification percentage through the previous equation to identify the most achieved factors using Eq. 1, a list of sustainable development categories and criteria and measurement has been reached [37].

$$\text{Category verification percentage} = \frac{\text{Number of practicies (indicators) varified}}{\text{Number of universities x number of the indicators}} \quad (1)$$

Figure 8 presents percentage achieved in six categories that affect the sustainable development projects upgrading the selected nine global university buildings. It is clear from Fig. 8 that indoor environmental quality, legislations and laws, and cost are 23% (highest percentage), 19.1%, 18.6%, respectively, followed by variable factors and energy consumption (17.9% and 1.4%), and the global environment with the lowest effect (7%).

5 Results and Discussions

The analyses of sustainable development categories in the selected nine universities' buildings show that Harvard University and Nanyang Technological University have the largest percentage of development categories with 27%, followed by the Universities of Stuttgart and Copenhagen at 25%, and then the Bremen University Library Building and the National University Engineering Building in Greece at 24% as shown in Fig. 9. The assessment also indicated that the most influencing factors for universities' sustainable development projects are found in the leading four universities (Harvard in USA, Technological University of Nanyang in Singapore, University of Stuttgart in Germany, and University of Copenhagen in

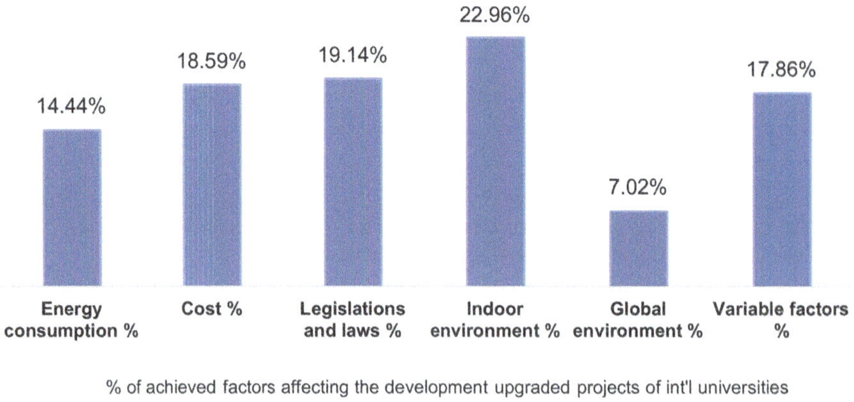

Fig. 8 Categories' assessment percentage of the main categories

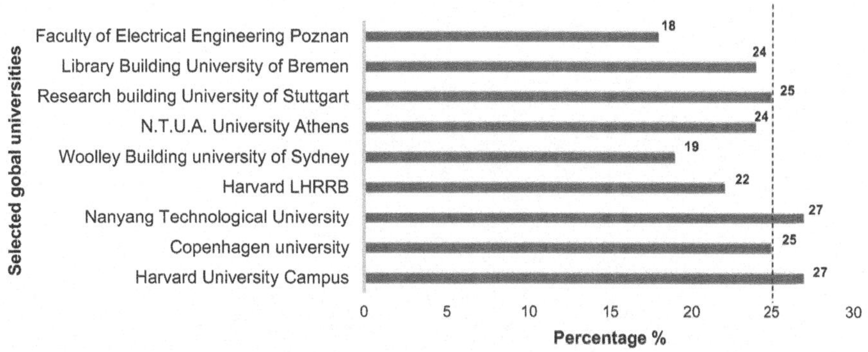

Fig. 9 Verification of development categories in the international universities. (*Source*: Authors)

Denmark) as shown in Figs. 6 and 9. The reason for the low rates is due to the large number of applications per category and the multiple measurement standards associated with them. This is further explained as follows:

- The results are attributed to the dependence of these universities on improving energy efficiency through different systems and technologies, particularly building control systems (DDC & BMS).
- Relying on university green fund financing and calculating all cost elements such as the redemption period, total annual savings costs, and maintenance cost.
- There are country-specific legislations to reduce emissions by a certain percentage, and the universities' own reduction goals in order to enhance its environmental performance. In all cases, high-carbon fuel reduction plans are supported by either substitution/purchase of renewable energy.
- The quality of the indoor environment has been taken into consideration; the CO_2 coefficient has been measured in the voids and the ventilation rate for air-conditioned places. Each university also measures and limits the emissions from it to preserve the surrounding environment or plans to recycle waste and reduce water use or recycling.
- The building monitoring with smart control systems is a key factor, then the functionality of retrofit procedures and a post-development questionnaire to measure user satisfaction (Table 3).

6 Conclusions

This research reviewed a group of selected international universities' buildings/campus that implemented sustainable development projects and retrofitting to improve energy efficiency and mitigate carbon emissions, and these universities were chosen among the leading universities in the world (North America, Europe, Asia, and Australia) that implemented upgrading projects to attain sustainable development (SD). The analytical study of the university campus or buildings included nine universities: Harvard University in USA; University of Stuttgart in Germany; Copenhagen University in Denmark; Nanyang Technological University in Singapore; Harvard Medical College Building (LHRRB), USA; College of Literature and Science, University of Australia; Engineering Building (NTUA), National University in Athens, Greece; Research Building University of Stuttgart in Germany; Library Building, Bremen University in Germany; and Electrical Engineering Building, University of Poznan in Poland.

Through the study, the categories of sustainable development (SD) projects in the nine selected universities' buildings as well as the indicators and their associated technologies and practices were monitored and analyzed to identify and induce the most influencing practices and technologies in these campuses such as air conditioning improvements, lighting systems, and building control systems enhancement. It turns out that the indoor environment (IAQ and comfort) is the most important category (23%), followed by the cost (18.6%), legislations and laws (19%), the variable factors (18%), and energy consumption (14.4%), and external environment is low (7%).

The verification of development categories in the international universities indicated that only four out of the nine selected university buildings achieved 25% or above (Research Building, University of Stuttgart in Germany and Copenhagen University in Denmark [25%] and Nanyang Technological University in Singapore and Harvard University in USA [27%]). The deduced list of indicators and criteria for measuring the development categories of the global universities' buildings could be exploited to measure their availability in higher education buildings when conducting SD projects in Egyptian universities by selecting the most achieved and appropriate indicators for Egypt from the applicability view point.

Appendix A

Table 4 The percentage of verification of energy consumption in the selected universities' projects

Indicators		University	Harvard University Campus	Copenhagen university	Nanyang Tech-nological University	Harvard (LHRRB)	Woolley building, University of Sydney	N.T.U.A. University, Athens	Research building, University of Stuttgart	Library Building, University of Bremen	Faculty of Electrical Engineering Poznan	Sum of indicators	Total
Category		Energy consumption											
Energy efficiency measures	1	Lighting efficiency	✓✓	✓	✓	✓	✓	✓	✓	✓		8	14.4
	2	HVAC efficiency	✓		✓	✓	✓	✓	✓	✓		7	
	3	Separate cooling/ heating	✓					✓	✓	✓	✓	5	
	4	Ventilation efficiency		✓	✓			✓		✓	✓	5	
	5	Equipment efficiency			✓							1	
	6	User behavior		✓					✓			2	
	7	Improve building operation				✓		✓	✓			3	
	8	Use building control modules DDC/ BMS	✓	✓	✓	✓		✓	✓			6	
	9	Sensors (movement, lighting, and temperature) occupancy schedules	✓	✓	✓	✓	✓	✓	✓		✓	8	
	10	Energy consumption monitoring		✓					✓			2	
Building envelope	11	Wall/roof insulation		✓			✓			✓	✓	4	
	12	Replace and insulate windows		✓			✓			✓	✓	4	
	13	External shading		✓				✓				2	
Renewable energy	14	Photovoltaic cells PVC	✓		✓			✓				3	
	15	Solar heating	✓					✓				2	
	16	Wind energy	✓									1	
	17	Biogas energy	✓									1	
	18	Geothermal energy	✓									1	
	19	Purchase clean energy	✓									1	
Energy metrics	20	Annual energy* consumption/savings*	✓		✓	✓		✓	✓		✓	6	
	21	Technology savings*	✓	✓	✓	✓	✓	✓	✓	✓	✓	9	
	22	Reduction in cooling/heating energy consumption*						✓		✓		2	
		Total	13	10	9	7	6	13	10	8	7		

**Most influencing indicators/metrics*
✓Existence in the project

Appendix B

Table 5 The percentage of verification of the cost, legislations, and indoor environmental quality

Indicators		University	Harvard University Campus	Copenhagen University	Nanyang Technological University	Harvard (LHRRB)	Woolley Building University of Sydney	N.T.U.A. University, Athens	Research Building University of Stuttgart	Library Building University of Bremen	Faculty of Electrical Engineering Poznan	Sum of indicators	Total %
Category		**Cost**											
Funding	23	External financing Loans/investments										0	18.6
	24	*Internal financing* Green financing fund	✓	✓	✓	✓	✓	✓	✓	✓	✓	9	
	25	Green expenses for students	✓			✓						2	
Cost and savings	26	Total direct/indirect cost	✓		✓		✓	✓	✓	✓	✓	7	
	27	Payback periods	✓	✓	✓	✓	✓			✓		6	
	28	Total annual savings	✓			✓			✓		✓	4	
	29	Maintenance cost		✓	✓			✓	✓	✓	✓	6	
		Total	5	3	4	4	3	3	4	4	4		
Category		**Legislations and laws**											
Binding legislations and laws	30	Legislations and laws of development buildings	✓	✓	✓	✓	✓		✓	✓		7	19.1
	31	State subsidies for buildings development projects			✓							1	
	32	Supporting renewable energy projects	✓		✓	✓	✓	✓	✓			6	
	33	Carbon reduction plans	✓	✓	✓	✓			✓	✓		6	
		Total	3	2	4	3	2	1	3	2	0		
Category		**Indoor environmental quality**											
Building considerations	34	Building shape, direction, and number of floors		✓	✓	✓	✓	✓	✓	✓	✓	8	22.9
	35	Activity and building type	✓	✓	✓	✓	✓	✓	✓	✓	✓	9	
	36	Building age	✓	✓	✓	✓	✓	✓	✓	✓	✓	9	
	37	Occupancy hours		✓					✓	✓		3	
Comfort rates	38	Visual and acoustical comfort		✓		✓	✓	✓		✓		5	
	39	Indoor air quality (CO_2 factor and ventilation rate)		✓	✓		✓	✓	✓		✓	6	
	40	Thermal comfort (air temperature, humidity, and thermal gain)			✓	✓			✓	✓	✓	4	
	41	Pre–/post-development survey (occupant satisfaction)					✓	✓	✓		✓	4	
		Total	2	6	5	5	6	6	7	6	6		

**Most influencing indicators/metrics*

✓ *Existence in the project*

Appendix C

Table 6 The percentage of verification of the global environment and variable factors

Indicators		University	Harvard University Campus	Copenhagen university	Nanyang Technological University	Harvard (LHRRB)	Woolley Building university of Sydney	N.T.U.A. University, Athens	Research Building, University of Stuttgart	Library Building, University of Bremen	Faculty of Electrical Engineering Poznan	Sum of indicators	Total %
Category		**Global environment**											
Carbon reduction	42	Measure the amount of emissions*	✓	✓		✓						3	7.0
	43	Emission reduction measurement*	✓			✓						2	
Water reduction	44	Water consumption rate*			✓							1	
	45	Efficient water fixtures			✓							1	
	46	Grey water			✓							1	
Waste management	47	Recycling waste	✓	✓						✓		3	
		Total	3	2	3	2	0	0	0	1	0		
Category		**Variable factors**											
Building control, functionality, and time	48	Control the building elements and parts using the building's smart energy management systems (BSES)*	✓	✓	✓	✓	✓	✓	✓	✓	✓	9	17.8
	49	Functionality of retrofit technologies (user satisfaction)*		✓	✓		✓			✓		4	
	50	Duration of the development procedure*								✓		1	
		Total	1	2	2	1	2	1	1	3	1		100
		Total 50 verified indicators	27	25	27	22	19	24	25	24	18		
		Achieved indicators (%)	54	50	54	44	38	48	50	48	36		

**Most influencing indicators/metrics*
✓Existence in the project

References

1. U.S Energy Information Administration. A Look at the U.S. Commercial Building Stock: Results from EIA's 2012 Commercial Buildings Energy Consumption Survey (CBECS). https://www.eia.gov/consumption/commercial/reports/2012/buildstock
2. Global Alliance for Buildings and Construction Sector (2016) Global Roadmap to Climate Action in The Buildings and Construction Sector, available at: https://www.globalabc.org/uploads/media/default/0001/02/9ab4984b5d2e006ad533bca08257a43bdf974fa9/. Accessed 28 August 2020
3. Abergel T, Dean B, Dulac J (2016) Towards a zero-emission, efficient, and resilient buildings and construction sector: global status report 2016. UNEP and International Energy Agency, Paris, France, [online], available at: https://orbit.dtu.dk/en/publications/towards-zero-emission-efficient-and-resilient-buildings-global-st. Accessed 28 August 2020
4. International Energy Agency and United Nations Environment Programme (2018) Global Status Report: Towards a zero-emission, efficient and resilient buildings and construction sector. Available at: https://www.worldgbc.org/sites/default/files/2018%20GlobalABC%20Global%20Status%20Report.pdf. Accessed 28 August 2020
5. International Energy Agency (2020) Energy-related CO_2 emissions from buildings have resin in recent years, [online], available at: https://www.iea.org/topics/buildings/. Accessed 16 November 2020
6. Context – BUILD_ME- buildings in the MENA, [online], available at: www.buildings-mena.com/. Accessed 16 November 2020
7. Krarti M (2019) Evaluation of Energy Efficiency Potential for the Building Sector in the Arab Region, Energies, MDPI, [online], available at: https://www.mdpi.com/1996-1073/12/22/4279
8. Abergel T, Dean B, Dulac J (2017) Global Status Report 2017: towards a zero-emission, efficient and resilient buildings and construction sector, International Energy Agency for the Global Alliance for Buildings and Construction (GABC), United Nations Environment, [online], available at: https://www.forskningsdatabasen.dk/en/catalog/2348897989. Accessed 16 November 2020
9. United Nations Environment Programme – UNEP and International Energy Agency – IEA (2020) Globale Status Report for Buildings and Construction – towards a zero-emissions, efficient and resilient buildings and construction sector, Global Alliance for buildings and Construction, [online], available at: https://wedocs.unep.org/bitstream/handle/20.500.11822/30950/2019GSR.pdf/. Accessed 15 November 2020
10. Khoshbakht M, Gou Z, Dupre K (2018) Energy use characteristics and benchmarking for higher education buildings. Energ Buildi 146. https://doi.org/10.1016/j.enbuild.2018.01.001., available at: https://www.researchgate.net/publiucation/32246897_Energy_use_characteristics_and_benchmarking_for_higher_education_buildings/. Accessed 30 August 2020
11. UNESCO – IEALC (2020) COVID-19 and Higher education: today and tomorrow – impact analysis, policy responses and recommendations, April 9, 2020, [online], available at: https://COVID-19-EN-090420.2.pdf. Accessed 16 November 2020
12. Education for Sustainable Development, Sustainable UCL, University College London, 2020 [Online], available at: https://www.ucl.ac.uk/sustainable/education-sustainable-development/. Accessed 16 January 2020
13. Social responsibility at Oxford Brookes, Oxford Brookes University, 2020 [Online], available at: https://www.brookes.ac.uk/social-responsibility/students/being-sustainable-at-brookes/. Accessed 20 May 2020
14. Ma Z, Cooper P, Daly D, Ledo L (2012) Existing building retrofits: methodology and state-of-the-art. Energ Buildings 55:889–902. available at: https://www.sciencedirect.com/science/article/abs/pii/S0378778812004227
15. Kolokotsa D, Diakaki C, Grigoroudis E, Stavrakakis G, Kalaitzakis K (2009) Decision support methodologies on the energy efficiency and energy management in buildings. Adv Build Energ Res 3(1):121–146. available at: https://www.tandfonline.com/doi/abs/10.3763/aber.2009.0305
16. Center for World University Rankings (2014), [online], available at: https://cwur.org/about.php/. Accessed 14 November 2020
17. Duffin E (2020) Statista, number of universities worldwide in 2020, by country, [online], available at: https://www.statisa.com/statistics/918403/number-of-universities-worldwide-by-counrty/. Accessed 15 November 2020
18. Ranking Web of Universities, 2020, [online.], available at: https://www.webmetrics.info/en/aw/egypt/. Accessed 15 November 2020
19. Liu MZY (2016) Energy efficiency retrofits of tertiary education buildings: an investigation of case studies in New South Wales. PhD Thesis, University of Sydney, [online], available at: https://ses.library.usyd.edu.au/handle/2123/16336/. Accessed 30 August 2020
20. World Top 20 Project, Global Universities Rankings (2019). https://worldtop20.org/global-universities?gclid=Cj0KCQjwh6XmBRDRARIsAKNInDHZP-QYtR-MVsby-P3dcnMQ4dBWTcRIeRF1YiO9MlQmR1r1HnyrgQEaAqrmEALw_wcB/. Accessed 30 August 2019
21. Stankeviciute L, Riqui P (2008) Energy and climate policies to 2020: the impacts of the European "20/20/20" approach. Int J Energ Sector Manag 2(2):252–273
22. Harvard Office for Sustainability (2015) Harvard University Sustainability Plan Fiscal Year 2015-2020. Cambridge, MA, [online], available at: http://www.report.green.harvard.edu/. Accessed 30 August 2020
23. Harvard University (2016) Climate goal report 2006–2016, how Harvard University met its greenhouse gas emissions reduction target,. available at: https://green.harvard.edu/campaign/harvards-climate-action-plan/. Accessed 20 August 2020
24. BCA Building and Construction Authority (2016), Building energy and benchmarking report, Singapore, https://www.bca.gov.sg/GreenMark/others/BCA_BEBR_Abridged_FA_2016.pdf/. Accessed 30 August 2020
25. Emissions Inventory (2016) Harvard has tracked and publicly reported University-wide greenhouse gas emissions since 2000. https://green.harvard.edu/topics/climate-energy/emissions-inventory/. Accessed 30 August 2020
26. Harvard University, Energy efficiency (2016) https://green.harvard.edu/topics/climate-change-energy/energy-efficiency/. Accessed 30 August 2020
27. Harvard University, Renewable energy (2016) https://green.harvard.edu/topics/climate-energy/renewable-energy/. Accessed 30 August 2020
28. University of Copenhagen (2016). Making the University of Copenhagen one of the greenest campuses in Europe, [online],. available at: https://greencampus.ku.dk/. Accessed 30 August 2020
29. Harvard Office for Sustainability (2015).Harvard University sustainability plan fiscal year 2015–2020. Cambridge, MA. http://report.green.harvard.edu/. Accessed 30 August 2020
30. The Danish Government (2013) The Danish climate policy plan towards a low carbon society. Copenhagen, Denmark. https://ens.dk/sites/ens.dk/files/Analyser/danishclimatepolicyplan_uk.pdf/. Accessed 31 August 2020

31. Sophie Y. (2013) Denmark publishes 78 ideas to curb emissions, https://www.climatechangenews.com/2013/08/15/denmark-publishes-78-ideas-to-curb-emissions/. Accessed 31 August 2020
32. Harvard University (2012).LHRRB Upgrade, [online],. available at: https://green.harvard.edu/tools-resources/case-study/lhrrb-upgrade-and-re-commissioning/
33. IEA-ECBCS.Annex36-Retrofitting of Educational Building-REDUCE, (2004) Rural and Surveying Engineering Building, N.T.U.A., Athens, [online], available at: http://www.annex36.com/unia.html/. Accessed 31 August 2020
34. IEA-ECBCS Annex36-Retrofitting of Educational Building-REDUCE, University of Stuttgart, [online], available at: http://www.annex36.com/unis.html/. Accessed 31 August 2020
35. IEA-ECBCS Annex36-Retrofitting of Educational Building-REDUCE, Library Building of the University of Bremen, [online], available at: http://www.annex36.com/unib.html/. Accessed 31 August 2020
36. IEA-ECBCS Annex36-Retrofitting of Educational Building -REDUCE, Faculty of Electrical Engineering Building, Poznan University of Technology, [online], available at: http://www.annex36.com/poznan.html/. Accessed 31 August 2020
37. Paule RC, Mandel J (1982) Consensus values and weighting factors. J Res Natl Bureau Stand 87(5), 377–385, [online], available at: https://nvlpubs.nist.gov/nistpubs/jres/087/5/V87-5.pdf

SO3H Functionalized Mesoporous Carbon as a Potential Catalyst to Obtain Glycerol Oleate Esters from Crude Glycerol

Maria Rosiene A. Arcanjo, Isabela A. dos Santos, Solange A. Quintella, Eduardo Rigoti, Sibele B. Castella Pergher, Enrique Rodríguez-Castellón, Célio L. Cavalcante Jr, and Rodrigo Silveira Vieira

1 Introduction

As a result of increased biodiesel production, there is a large number of resulting coproducts. Among the residues, the production of crude glycerol and its use to add value to this coproduct have been studied in the last years. The current glycerol consumption market is specific and restricted, being essential to intensify the glycerol consumption to make biodiesel production competitive. Therefore, it is necessary to seek technological and product routes for glycerol that make biodiesel production more sustainable [1].

The high level of impurities in crude glycerol limits its application, especially for the food and pharmaceutical industries. However, the number of studies on the application of glycerol as a raw material for other products of greater added value is growing, such as the production of mono-, di- and tri-oleate of glycerol [2], 1,2-propanediol [3], 1,3-propanediol [4], acrolein [5], di- and triglycerides [6], organic acids [7–9], with preferential use of heterogeneous catalysts [10].

Among these routes, the esterification of glycerol with oleic acid can produce numerous products of high commercial value such as glycerol mono-, di- and tri-oleate that have a range of food, pharmaceutical, cosmetic, and industrial applications [11]. Glycerol mono, di-oleates (GMO, GDO) are lipids with amphiphilic, nonionic, and excellent emulsifying properties, widely applied in the food, cosmetic, and pharmaceutical industries [12]. Meanwhile, GDO is mainly used in drug delivery applications and as a safe plasticizer for the polymer industry [13]. Glycerol tri-oleates (GTO) is one of the common biolubricants, with a symmetrical structure to that of the triglyceride. Monounsaturated oleic acid was selected in the synthesis of biolubricant, since the conjugated linkage of oleic acid exhibits lower pour point, cloud point and low temperature stability lubricant. GTOs are widely applied to two-stroke engines, laminated metal, cast aluminum, tire tread, and as a stabilizing oil component. The combination of lubricity, biodegradability, renewability, and nontoxicity of GTO lubricants has the potential to lead to the growth of this market [14].

Currently, industrial esterification processes are carried out in the presence of homogeneous Brønsted acids, such as sulfuric acid or p-toluenesulfonic acid. However, these homogeneous acid catalysts are difficult to recycle and generate serious environmental problems, as corrosion problems. As a result, it is extremely interesting to synthesize esters using heterogeneous catalysts in place of homogeneous liquid catalysts [15].

The use of ordered mesoporous carbona (CMO) or carbon mesoestructured from Kaist (CMK-n) as a catalyst with uniform and ordered pores that favor the diffusion of molecules and contribute to an excellent catalytic performance [16]. In the group of acidic solids, sulfonated carbons have been pointed as a catalyst, with promising results for esterification reactions. The main objective of carbon functionalization is to produce a highly stable catalyst with a high density of –SO3H active sites, presenting good performance in fatty acid esterification reactions, in addition to being able to be reused in other reactions [17].

M. R. A. Arcanjo · I. A. dos Santos · S. A. Quintella · C. L. Cavalcante Jr · R. S. Vieira (✉)
Department of Chemical Engineering, Federal University of Ceara, Fortaleza, Brazil
e-mail: rodrigo@gpsa.ufc.br

E. Rigoti · S. B. Castella Pergher
Institute of Chemistry, Federal University of Rio Grande do Norte, Natal, Brazil

E. Rodríguez-Castellón
Department of Inorganic Chemistry, Faculty of Sciences-University of Malaga, Malaga, Spain

© The Author(s), under exclusive license to Springer Nature Switzerland AG 2022

A. Sayigh (ed.), *Sustainable Energy Development and Innovation*, Innovative Renewable Energy,
https://doi.org/10.1007/978-3-030-76221-6_90

In this context, the objective of this study was to investigate the application of treated glycerol from a biodiesel industry in esterification with oleic acid for the production of glycerol oleate esters, using carbon mesoestructured from Kaist (CMK-n) functionalized with –SO3H as a catalyst. This could be more economically attractive due to two factors: an alternative commercial application for glycerol, and the use a heterogeneous catalyst (CMO), which is advantageous in relation to homogeneous catalysts. All materials were characterized by SEM-EDX. Additionally, the results indicated that the best catalytic performance were obtained in the reuse tests.

2 Experimental

2.1 Catalyst Preparation

The carbon mesoestructured from Kaist (CMK-3 and CMK-8) was synthesized and kindly provided by the Molecular Sieves Laboratory (LABPEMOL), from the Chemistry Institute, located in Natal/RN (Brazil). The carbon synthesis was described by Rigoti [18].

For the sulfonation of the ordered mesoporous carbons, 96% sulfuric acid was used. The methodology was adapted from Ferreira et al. [19]. Nitrogen was used to form an inert atmosphere in the sulfonation reactor. The sulfonation was carried out at 120 °C, for 6 h, in the mass ratio of acid: carbon of 10. After sulfonation, the solids were filtered and washed with distilled water until the washing water had a neutral pH, after drying at 80 °C for 24 h in an oven. The catalysts were named CMK3-SO3H and CMK8-SO3H. The non-sulfonated mesoporous carbons (CMK3 and CMK8) and sulfonated mesoporous carbons (CMK3-SO3H and CMK8- SO3H) were used as catalysts in the reactions.

2.2 Catalyst Characterization

The study of the morphology was carried out by scanning electron microscopy using Inspect S50 FEI, operating at 20 kV acceleration voltage. The preparation of the samples was carried out by dispersing it on a carbon tape and then metalizing it with Au. Energy dispersive X-ray spectroscopy (EDS), measured with the SEM, was used to evaluate the elemental content of the original and modified samples.

2.3 Treatment of Crude Glycerol

The crude glycerol samples used in this study were kindly provided by the FIAGRIL, a company located at Lucas do Rio Verde/MT (Brazil).

The first step in the crude glycerol treatment was the acidification with phosphoric acid solution (4.0 mol/L) according to the methodology optimized by Silva [20]. The acidification aims to remove impurities present in the biodiesel production reaction. The second step treatment was the purification using powdered activated carbon and resins Amberlite IRA 410, and Amberlite IRA 120. For the activation of the resins, solutions of NaOH 4% were used, for the anionic resin and HCl 5%, in the cationic, for 1 hour. The flowchart of the purification process for treating crude glycerol is shown in Fig. 1.

To determine the physicochemical characteristics of crude glycerol and treated glycerol, the following analyses were performed:

1. The glycerol content was analyzed using the sodium periodate method [21].
2. The density of crude glycerol and treated glycerol was determined according to the standard ASTM D 891–95 (2004).
3. The alkalinity of the crude glycerol and the treated glycerol was calculated according to the IUPAC – ACD 1980 (6ª ed.).
4. According to the standard ASTM D – 1293-99 (2005), pH measurement for crude glycerol and treated glycerol was performed at a pH meter of DIGIMED - pH DM22.
5. Similar to the pH measurement, the measurement of conductivity for crude glycerol and for treated glycerol was carried out following the standard ASTM (i.e., ASTM D1125–95 (2005)) using a conductivity meter DIGIMED - pH DM32.
6. The ash content analysis was carried out according to the standard method ISO 2098-1972.

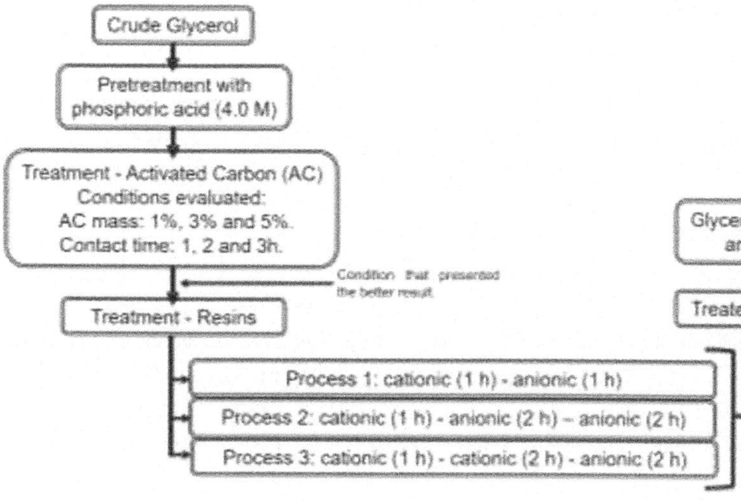

Fig. 1 Flowchart of the purification process for crude glycerol

7. The present humidity was also obtained based on the gravimetric method, which consists of evaluating the difference in mass of the sample after drying (adapted Ohlweiler [22]).

2.4 Catalytic Reaction and Analytical Methods

The esterification of glycerol (GLY) with oleic acid (OA) was carried out in a glass reactor with three connections: the first to measure the temperature, the second connected to N2 to maintain an inert atmosphere during the reaction time, and the third connected a ball condenser to prevent evaporation losses.

The reaction of glycerol with oleic acid was evaluated in different molar proportions GLY:OA (1:1, 1:2 and 1:3), at different temperatures (80, 100, 120, and 140 °C) with variation in the loading of catalyst 2.5 and 5 wt % with stirring for 5 h (Table 1).

The reaction conversion was measured by titration with potassium hydroxide (0.1 mol/L) using the acidity Index method. The evolution of the reaction was monitored by collecting aliquots of approximately 1.5 mL, which were taken out at different time intervals (0.5–5 h). The products were analyzed by gas chromatography (Shimadzu GC) equipped with a flame ionization detector (FID). A DB5 30 m × 0.25 mm × 0.25 µm GC capillary column was used to separate the products.

2.4.1 Recyclability and Stability

The recyclability of the catalyst was evaluated, considering the previously described reaction conditions for the catalysts that presented the best catalytic performance. Between each cycle, the catalyst was filtered, thoroughly washed with hexane and dried at 110 °C for 24 h prior to the next reaction.

3 Results and Discussion

3.1 Physicochemical Analysis of Crude and Treated Glycerol

Analyzing the results of the processes (P1, P2, and P3) depicted in Fig. 1, a maximum glycerol content of 89.9% ± 0.06 was obtained for crude glycerol purified only with 3% AC in a contact time of 2 h. After processes 2 and 3, using the resins, the higher glycerol content was 98.2% ± 0.05 (Process 2), and this purified glycerol was esterified with oleic acid and CMKs as catalyst (Table 2).

Table 1 Reactions parameters and catalysts

Parameters				
Catalyst	Molar ratio (GLY: OA)	Temperature (°C)	Catalyst loading (%)	Time reaction (h)
CMK3; CMK8 (non-sulfonated)	1: 1, 1: 2,	80, 100, 120,	2.5; 5.0	0.5, 1,
	1: 3	140.		2, 3, 4, 5,
CMK3-SO3H; CMK8-SO3H (sulfonated)				6.

Table 2 Physicochemical analyses of the crude and treated glycerol after purification processes

Parameter	Glycerol PA	Crude glycerol	P1	P2	P3
Alkalinity	0.0	86.0	0.001	0.001	0.001
Conductivity (μS/cm)	4.21	4126.0	0.31	0.16	0.37
Density (g/cm^3)	1.26	1.01	1.25	1.26	1.25
Color	Colorless	Brown	Colorless	Colorless	Colorless
Refractive index	1.47	1.39	1.47	1.47	1.47
pH	5.5–8.0	10.71	7.01	7.11	7.06
Water content (g/g of sample)	0.005	4.77	0.042	0.010	0.034
Ash content (g/g of sample)	0.0	0.996	0.16	0.09	0.16
Glycerol content (%w)	99.5	47.84	91.6	98.2	92.1

3.2 *Characterization of the Support*

Micrographs obtained by scanning electron microscopy for CMKs and sulfonated CMKs are shown in Fig. 2. The materials are formed by particles of different sizes and irregular surfaces. Apparently, analyzing the images of the same magnification, the high sulfonation temperature and the time did not cause changes in the carbon structure (Fig. 2). It should be noted that with the use of catalysts (Fig. 2), despite the high reaction temperature (120 ° C), the carbon structure remained stable after the reaction, probably due to the chemical and physical resistance of the activated carbon.

3.3 *Esterification Reactions Using CMKs-SO3H for Production of Glycerol Oleate Esters*

Typically, the esterification reaction of glycerol with oleic acid (OA; a common unsaturated fatty acid with a C18: 1 carbon chain) is a viable and economical method for changing the fatty acid profile of a triglyceride. A mixture of glycerol monooleate (GMO), glycerol dioleate (GDO), and glycerol trioleate (GTO) can be produced on the acid-catalyzed esterification reaction, according to the reaction scheme in Fig. 3. The esterification route catalyzed by heterogeneous acid is known to be more selective and economical. In addition, the heterogeneous acid catalytic system is environmentally sustainable compared to the homogeneous catalyst due to less waste production, easier operation, and possible catalyst recycling.

3.3.1 Effects of Catalyst Loading and Glycerol-to-Oleic Acid Molar Ratio

The catalyst loading and molar ratio effects into glycerol conversion are shown in Fig. 4a, b, respectively. Figure 4a shows the effect of catalyst loading on acid conversion and the highest conversion of 85.3% ± 1.78 was obtained with catalyst loading of 5% for CMK3-SO3H sulfonated carbon. The acid conversion increased with catalyst loading due to the enhanced accessibility of acidic sites. Therefore, 5% was the amount of CMK catalyst for the study of the next parameter (molar ratio).

All experiments were carried out at the temperature of 120 °C with 5% catalyst for 5 h. The effect of the molar ratio for the esterification process was carried out at 1: 1, 1: 2 and 1: 3 glycerol in oleic acid. The effect of the proportion of the initial reagent concentration on the reaction rate is significant. As shown in Fig. 4b, the increase in the glycerol/oleic acid molar ratio from 1: 1 to 1: 3 led to an increase in the conversion of the sulfonated catalysts by 86% ± 2.24 and 79.9% ± 0.09 for the catalysts CMK3-SO3H and CMK8-SO3H, respectively, in a molar ratio of 1: 3. Therefore, 5% was the amount of CMK catalyst, and 1: 3 was the molar ratio selected for the study of the next parameter (temperature).

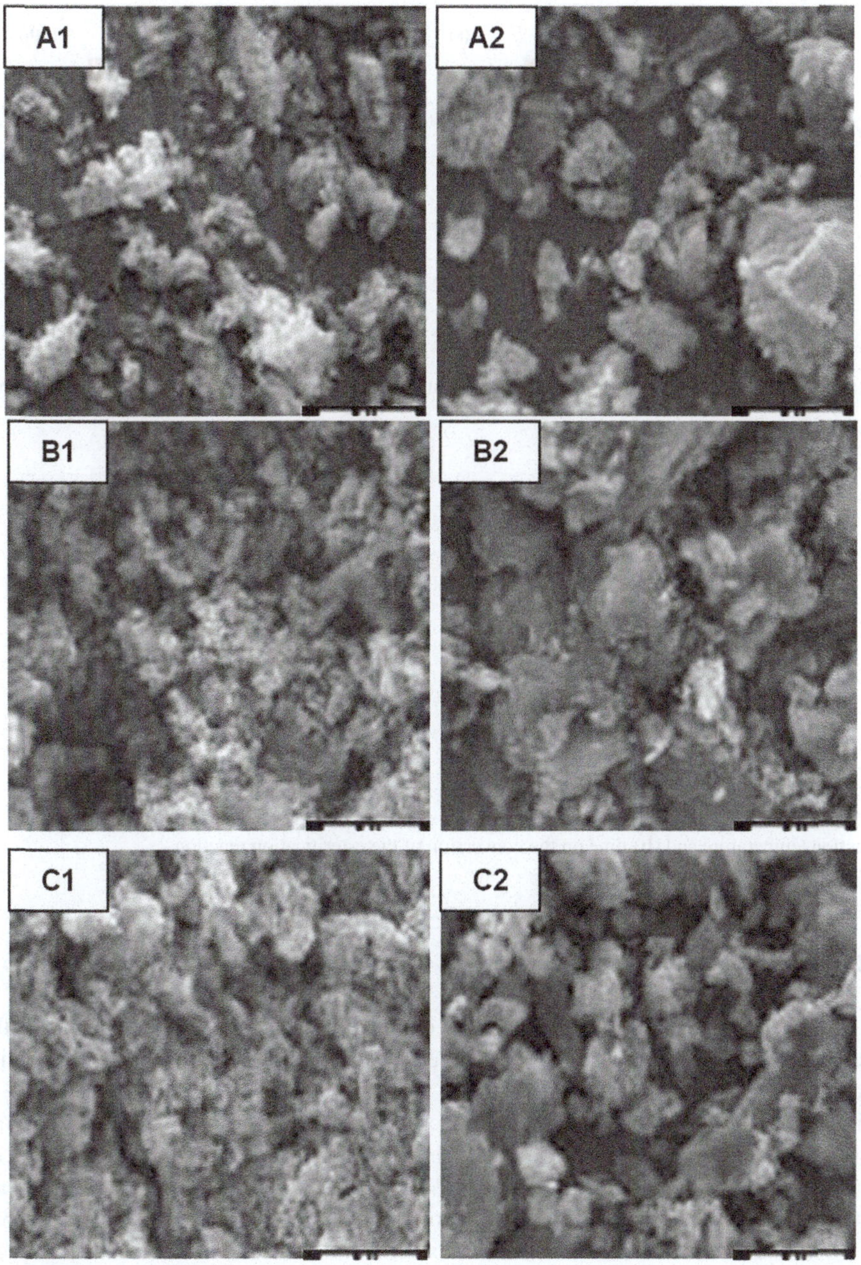

Fig. 2 SEM images: sulfonation-free catalysts (A1) CMK-3, (A2) CMK-8; sulfonation catalysts (B1) CMK3- SO3H, (B2) CMK8-SO3H; spent catalysts (C1) CMK3-SO3H, (C2) CMK8-SO3H, at magnification 4000×

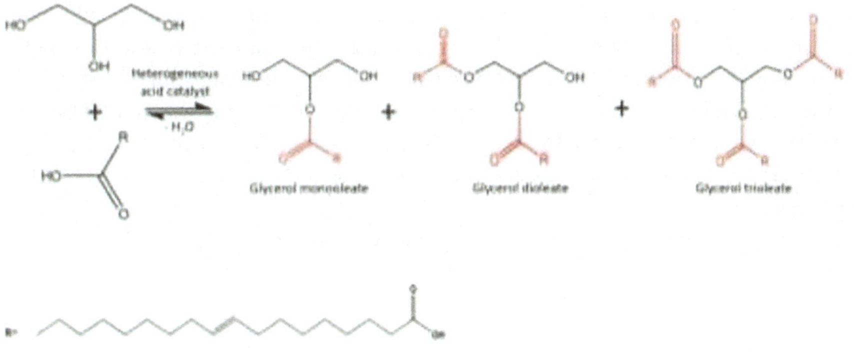

Fig. 3 Reaction scheme for esterification of glycerol with OA. (Modified from Kong et al. [2])

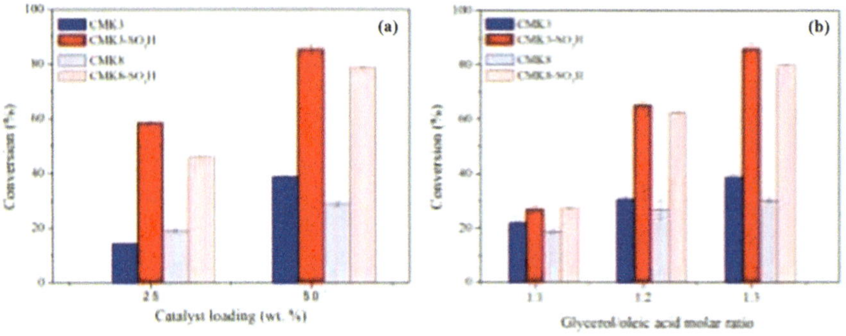

Fig. 4 (**a**) Effect of catalyst loading. (**b**) Glycerol-to-oleic acid molar ratio on the conversion of the esterification reaction

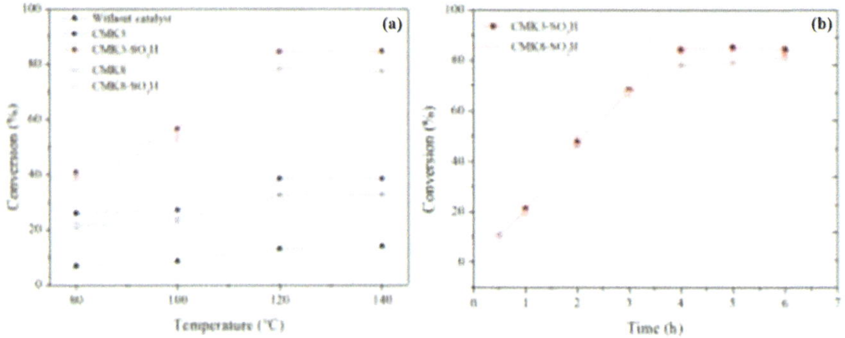

Fig. 5 (**a**) Effect of reaction temperature and (**b**) time on the conversion of the esterification reaction

3.3.2 Effects of Reaction Temperature and Time

It is also possible to observe in Fig. 5a, the conversion results obtained for catalysts without the sulfonation process, in which we obtained lower conversion values of 38.43% ± 0.35 and 32.39% for CMK-3 and CMK- 8, respectively, under the same conditions mentioned above at a temperature of 120 °C.

The effect of temperature was studied by varying the reaction temperature from 80 to 140 °C, applying a 1: 3 glycerol: oleic acid molar ratio and a catalyst quantity of 5% by weight of catalyst for 5 h. Figure 5a shows the effect of temperature on acid conversion. The acid conversion increased from 56.5% ± 0.55 to 84.4% ± 0.38 when the temperature increased from 100 to 120 °C for the CMK3-SO3H catalyst. There was no significant increase in the conversion when the temperature increased to 140 °C (84.5% ± 0.82). The increase in acid conversion occurs due to the increase in the reaction rate by the temperature. The kinetic energy of all molecules in the reagent increases with the temperature reaction, accelerating the mass rate exchange between the oleic components of the acid-glycerol-catalyst phases. It is suggested that the reaction temperature effect for all catalysts is not linear since increasing the temperature even further from 120 to 140 °C showed relatively small changes in the conversion.

From these results with mesoporous carbons, it was possible to perceive the need to apply a sulfonation method as a treatment to provide a more acidic character since these showed potential for the esterification of oleic acid with glycerol. Monitoring the performance of the reaction over time was carried out by collecting samples at 0.5, 1, 2, 3, 4, 5, and 6 h. Figure 5b shows the results obtained in this study for the sulfonated catalysts. During the first 60 minutes of reaction, a conversion of approximately 20% was achieved for both catalysts. The reaction performance gradually increased up to a reaction time of 4 h, with stabilization of the conversion to esters of approximately 84.3% ± 0.43 and 81.2 ± 0.46 for CMK3-SO3H and CMK8-SO3H, respectively.

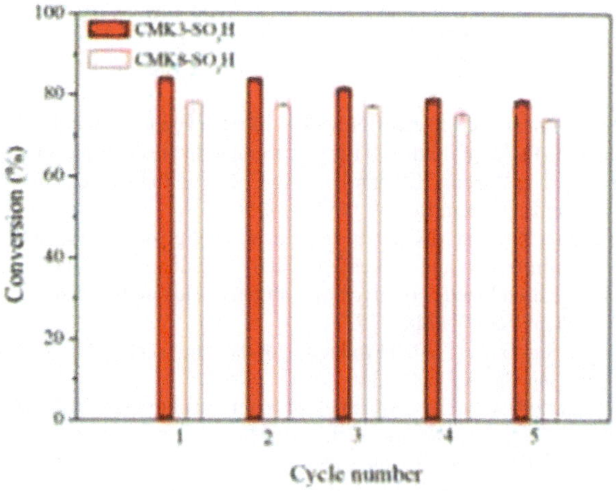

Fig. 6 Effect of repeated use of catalyst on reaction performance

3.3.3 Catalyst Reusability

To make commercialization an economic possibility, the reuse of the catalyst was tested. The reaction conditions of esterification using the sulfonated catalysts were 5% by weight of catalyst, 1: 3 M ratio of glycerol to oleic acid, the temperature of 120 °C and reaction time of 5 h. As can be seen, after the five repeated runs, the conversion rate has not decreased significantly, which means that the catalysts have maintained their high activity even after several uses, suggesting efficient stability of the catalytic structure with glycerol conversion values without significant differences in five consecutive cycles. For these reactions, the highest conversion achieved was 84.39% ± 0.38 with a selectivity of 78% ± 0.22 (mono-, di- and trioleate) (Fig. 6).

4 Conclusion

The glycerol treated with the resins showed a purity level above 90%, proving a good efficiency of the process that was used. The esterification of oleic acid and purified glycerol achieved a conversion of about 84.39%, CMK3-SO3H catalyst, showing that acidic mesoporous carbons may be an ideal catalyst for the esterification of fatty acids. These results showed a good potential to use crude glycerol to generate other products of interest and with other applications. The advantages of using the heterogeneous catalysts studied here are also highlighted, since these catalysts can be reused.

Acknowledgments The authors are grateful for the financial support and scholarship granted by PROCAD/CAPES (88887.284985/2018-00).

References

1. Martins SA, Tapanes NC, Orlandin GR (2020) Study of the properties of an epoxy adhesive with additions of a residue from the biodiesel production process. Int J Adhes Adhes 103:102701
2. Kong PS, Cognet P, Pérès Y, Esvan J, Daud WMAW, ArouA MK (2018) Development of a novel hydrophobic ZrO2–SiO2 based acid catalyst for catalytic esterification of glycerol with oleic acid. Ind Eng Chem Res 57:9386–9399
3. Lee M, Hwang YK, Chang J-S, Chae H-J, Hwang DW (2016) Vapor-phase Hydrogenolysis of glycerol to 1,2-Propanediol using a chromium-free Ni-cu-SiO_2 nanocomposite catalyst. Catal Commun 84:5–10
4. García-fernández S, Gandarias I, RequieS J, Güemez MB, Bennici S, Auroux A, Arias PL (2015) New approaches to the Pt/WOx/Al2O3 catalytic system behavior for the selective glycerol hydrogenolysis to 1,3-propanediol. J Catal 323:65–75
5. Dalil M, Edake M, Sudeau C, Dubois J-L, Patience GS (2016) Coke promoters improve Acrolein selectivity in the gas-phase dehydration of glycerol to Acrolein. Appl Catal A 522:80–89
6. Morales G, Paniagua M, Melero JA, Vicente G, Ochoa C (2011) Sulfonic acid-functionalized catalysts for the valorization of glycerol via transesterification with methyl acetate. Ind Eng Chem Res 50:5898–5906

7. Kondamudi N, Misra M, Banerjee S, Mohapatra S, MohapatrA S (2012) Simultaneous production of Glyceric acid and hydrogen from the Photooxidation of crude glycerol using TiSi2. Appl Catal B 126:180–185
8. Omata K, Matsumoto K, Murayama T, Ueda W (2016) Direct oxidative transformation of glycerol to acrylic acid over Nb-based complex metal oxide catalysts. Catal Today 259:205–212
9. Arcanjo MRA, Paniagua M, Morales G, Iglesias J, Melero J, Silva IJ Jr, Rodríguez-Castellón E, Vieira RV (2020) Temperature effect on pre-treatment of the activated CarbonSupport (Pt/AC and Pd/AC) for glycerin into lactic acid. Ind Eng Chem Res 59:14643–14657
10. Zhang G, Jin X, Guan Y, Yin B, Chen X, Liu Y, Feng X, Shan H, Yang C (2019) Toward selective dehydrogenation of glycerol to lactic acid over bimetallic Pt–Co/CeOx catalysts. Ind Eng Chem Res 58:14548–14558
11. Kong PS, Pérès Y, Daud WMAWD, Cognet P, Aroua MK (2019) Esterification of glycerol with oleic acid over hydrophobic zirconia-silica acid catalyst and commercial acid catalyst: optimization and influence of catalyst acidity. Front Chem 7:1–11
12. Macierzanka A, Szelag H (2004) Esterification kinetics of glycerol with fatty acids in the presence of zinc carboxylates: preparation of modified acylglycerol emulsifiers. Ind Eng Chem Res 43:7744–7753
13. Zhang Z, Huang H, Ma X, Li G, Wang Y, Sun G et al (2017) Production of diacylglycerols by esterification of oleic acid with glycerol catalyzed by diatomite loaded SO42–/TiO2. J Ind Eng Chem 53:307–316
14. Kong PS, Aroua MK, Daud WMAW (2015) Catalytic esterification of bioglycerol to value-added products. Rev Chem Eng 31:437–451
15. Chaari, A.; Neji, S. B.; Frikha, H. H., (2017). *j. Oleo Sci.* 66: p. 455–461
16. Gao Z, Tang S, Cui X, Tian S, Zhang M (2015) Fuel 140:669–676
17. Konwar LJ, Mäki-arvela P, Kumar N, Mikkola JP, Sarma AK, Deka D (2016) Selective esterification of fatty acids with glycerol to monoglycerides over –SO3H functionalized carbono catalyst. Reac Kinet Mech Cat 119(p):121–138
18. Rigoti E (2017) Desenvolvimento de novos materiais catalíticos à base de Mo suportado em sílica e carvão mesoporosos ordenados para reações de hidrodeoxigenação. 101f. Tese (Doutorado). Programa de Pós-Graduação em Ciência e Engenharia de Materiais da Universidade Federal do Rio Grande do Norte, Natal, Rio Grande do Norte
19. Ferreira ARO, Silvestre-Albero J, Maier ME, Ricardo NMPS, Cavalcante CL Jr, Luna FMT (2020) Sulfonated activated carbons as potential catalysts for biolubricant synthesis. Mol Catalysis 488:110888
20. Silva SSO (2019) Estudo do processo de purificação e eterificação do glicerol residual proviniente d a indústria do biodiesel. 100f. Dissertação (Mestrado). Programa de Pós-Graduação em Engenharia Química da Universidade Federal do Ceará, Fortaleza, Ceará
21. Coocks LV, Van Rede C (1966) Laboratory handbook for oil and fat analysis. Academic Press, Londres, p 404
22. Ohlweiler OA (1981) Fundamentos de Análise Instrumental, 1st edn. Livros Técnicos e Científicos, Brasil

Modelling of a Solar Thermal Energy System for Energy Efficiency Improvement in a Ceramic Plant

Miguel Castro Oliveira, Muriel Iten, and Ulisses Fernandes

1 Introduction

The industrial sector is responsible for about 25% of the final energy consumption in Europe [1]. Within European industry, such high energy consumption level is mainly due to thermal energy use, accounting for 70% of the total energy use [2]. Thermal energy is primarily generated by the burning of a fuel (such as natural gas) within the operation of a combustion system [3]. The implementation of energy efficiency improvement relies on the application of several measures [4]. One specific set of measures are the one related to heat recovery, this is, the supply of a determinate quantity of heat to enable savings in the primary heat source (fuel) [5]. Another solution is the use of renewable energy resources (such as solar energy) and the implementation of renewable energy systems, which simultaneously reduces the consumption of fossil fuels and indirectly improves the plant's overall energy efficiency [6].

The improvement of energy efficiency and the promotion of renewable energy resource use in the European Union has been secured through several policies. In specific, the 2020 climate & energy package objectives were updated on the most recent 2030 climate & energy package, in which the minimum targets of 32.5% improvement in energy efficiency, 32% share for renewable energy and 40% cuts in greenhouse gas emissions are established [7].

The research on energy efficiency improvement in the industrial sector subsist on several rigorous methods [8]. The application of these methods results on the development of process and system-level models [9]. Numerical modelling may be applied in a multiscale perspective in order to either develop models for single unit operations in a plant or a building or the whole system which encompasses a set of processes. The Modelica language has been used to develop models and study the implementation of several types of measures [9–11]. Several open source libraries have been developed in Modelica for the simulation of several systems. For instance, the modelling of the overall industrial systems encompassing thermal processes and water networks has been secured by the use of the ThermoPower library [12], with its derivative WaterWatt library being specifically used for the modelling of industrial water circuits [9, 13]. On the other hand, the Soltermica [14] and SolarTherm [15] have been used to model and simulate several solar thermal systems [16, 17].

This chapter presents an innovative approach to improve the energy efficiency of a ceramic plant based on the implementation of a system which combines heat recovery and the use of solar thermal energy. Such implementation is studied through the development of a model of the system assembled using the Modelica language. In a broader perspective, this study is inserted in an overall research which encompasses the promotion of energy efficiency in industry (focusing on the optimization of thermal processes) and which in practice is performed by the development of a modelling and simulation tool. This chapter aims to present the modelling approach (equipment and plant-level), validate the tool and demonstrate the associated potential.

2 Materials and Methods

In this section, the description of a case study undertaken in a ceramic industry plant is presented, the aspects regarding the modelling of the plant's components and the conceptualization of the solar thermal energy system (STES).

M. C. Oliveira (✉) · M. Iten · U. Fernandes
Low Carbon & Resource Efficiency, R&D, Instituto de Soldadura e Qualidade, Porto Salvo, Portugal
e-mail: dmoliveira@isq.pt; mciten@isq.pt; umfernandes@isq.pt

© The Author(s), under exclusive license to Springer Nature Switzerland AG 2022

A. Sayigh (ed.), *Sustainable Energy Development and Innovation*, Innovative Renewable Energy,
https://doi.org/10.1007/978-3-030-76221-6_91

Table 1 Technical specifications of the plant and the two boilers

Plant			
Natural gas consumption (GWh/year)	84.47		
Operational time (h/year)	7784		
Boiler 1		**Boiler 2**	
Natural gas consumption (MWh/year)	137.50	Natural gas consumption (MWh/year)	137.50
Water inlet temperature (°C)	25	Water inlet temperature (°C)	25
Water outlet temperature (°C)	95	Water outlet temperature (°C)	95
Water flow rate (kg/h)	1277.38	Water flow rate (kg/h)	1180.32

2.1 Description of the Case Study

This case-study is introduced in a ceramic brick manufacturing plant located in Portugal. The production line is constituted by two tunnel kilns, one intermittent kilns, two continuous dryers and four hot water boilers. The plant is located in a zone with high levels of solar irradiation over the year.

Currently, there are already implemented a waste heat recovery strategy in the plant which encompasses the use of the waste heat of the exhaust gas streams of the two kilns to be supplied to brick yards. Nonetheless, currently there is a requirement to optimize the performance of the thermal processes within the plant by reducing the natural gas consumption. Another project (which is also being conceptualized at the same time as the project which is the subject of this chapter) encompasses the use of the waste heat potential of these exhaust gases to be used to heat up the water streams at the inlet of two of the boilers, as well as the recycling of hot air from the cooling zone of one tunnel kiln to be used as preheated combustion air in the firing zone of the same kiln. The present study approaches the optimization of the remaining two boilers. The technical specifications of these boilers are presented in Table 1 (for convenience reasons, these boilers are numbered as boiler 1 and boiler 2).

In this prospect, a new strategy must be defined in order to attend for the described requirement. This study approaches a heat recovery strategy which makes use of solar energy, namely through the installation of solar thermal collectors. The solar thermal energy serves as such as the additional heat source for the operation of the boilers.

2.2 Modelling of System Components

The proposed WHR strategy will be studied and assessed through the assembling of a model of the interconnected operations of the plant. The Modelica language is used to model the equipment and simulate the overall operation of the plant. Although a part of the component models necessary to assemble the model of the conceptualized installation are present in existing Modelica libraries (which highly reduces the requirements for model development), some models do not exist in the available libraries (such as the hot water boiler model). As such, some component models had to be assembled to achieve the aims of this research. Two open source Modelica libraries for this prospect: ThermoPower [12] and Soltermica [14]. In Table 2, the libraries used to develop the models presneted in this work are described, with the component models provided by each one of these libraries being identified.

Table 3 presents a brief description of the component models developed from scratch for the aim of this research: the solar collector (modified version of the one from Soltermica) and the boiler.

$$q_{\text{supply}} = M_{\text{fuel,initial}}\text{LHV} \tag{1}$$

$$q_{\text{supply}} = M_{\text{fuel}}\text{LHV} + q_{\text{additional}} \tag{2}$$

$$q_{\text{additional}} = M_{\text{Water}}C_{P,\text{Water}}\left(T_{\text{Water}} - T_{\text{Water,initial}}\right) \tag{3}$$

The total supplied heat in the boiler (q_{supply} (kJ/h)) may be calculated considering the fuel's lower heating value (*LHV*), attending to Eq. (1) (only valid for the case in which there is only one heat source, which is fuel). For the case in which more than one heat source exists to supply thermal energy to the kiln, it is necessary to consider an additional heat parcel as described by Eq. (2), which corresponds to additional thermal energy contained in the inlet water stream ($q_{\text{additional}}$ (kJ/h)), as described by Eq. (3). It is to note the remaining nomenclature used in the set of Eqs. (1), (2), and (3): $M_{\text{fuel, initial}}$ – baseline case fuel mass flow rate (kg/h), M_{fuel} – fuel mass flow rate (kg/h), $M_{\text{Comb. Air}}$ – combustion air flow rate (kg/h), $C_{P,\text{Comb. Air}}$ – air

Table 2 Description of Modelica libraries used to assemble the STES model

Library	Description	Components
ThermoPower [12]	A Modelica library developed for the dynamic simulation of thermal power plants and energy conversion systems. Although it does not contain models for high-level combustion processes, it contains several basic models which may be used for the assembling of these models (such as combustion chambers, gas flow models and water flow models).	Pipes Tank Pump (modified version) Basic components for the development of the tunnel kiln and boiler models
Soltermica [14]	A Modelica library developed for the simulation of low temperature solar thermal systems, focusing on the design of solar water heating systems. Its focus is on residential buildings.	Solar collector (modified version).

Table 3 Description of newly developed components

Name	Description
Solar thermal collector	It consists of a simplified model for a solar thermal collector. The component model was adapted from the solar collector model of Soltermica. Two connectors from ThermoPower library for, respectively, the inlet and outlet of the water stream were added to this model so to make possible the model to be connected to the components assembled using this library.
Boiler	It consists of a hot water boiler. It is assembled according to the mass and enthalpy balances in an industrial boiler (water stream and combustion chamber), as described in Eqs. (1), (2), and (3). The thermal efficiency (ratio between supplied energy and the variation of the enthalpy of the water stream) is considered in the overall mass-enthalpy balance (thus accounting for the overall heat losses).

heat capacity (kJ/(°C.kg)), M_{Water} – water mass flow rate (kg/h), $C_{P,\mathrm{Water}}$ – water heat capacity (kJ/(°C.kg)), T_{Water} – water inlet temperature at the boiler (°C), $T_{\mathrm{Water,\,initial}}$ – baseline case water inlet temperature (°C).

2.3 Modelling of the Solar Thermal Energy System

The proposed solar thermal energy system (STES) was conceptualized considering the layout presented in the research by Kumar et al. [6]: it overall consists of an industrial water circuit (IWC) constituted by two water flows, one which passes by a solar collector and the other that corresponds to the water that is supplied to the boilers. The STES (this is, the whole installation excluding the boilers) is constituted by the solar collector, a centrifugal pump, a water storage tank and several pipelines. The system is presented in Fig. 1.

The model for the system encompassing the water circuit and the two boilers is presented in Fig. 2. This model was created by using the components identified and described in the previous section.

A set of assumptions were considered in the performance of the whole modelling:

- The air-to-fuel ratio in the boiler is constant, not varying according to different operational conditions.
- The ambient temperature is 25 °C and the ambient pressure is 1 bar.

3 Results and Discussion

In this section, the main results obtained by the simulation of the assembled model are presented and analysed. The section is divided in two subsections, the first one being a presentation of the simulations results and model validation and the second one being a primary technoeconomic assessment performed to evaluate the viability of the implementation of the presented project.

3.1 Simulation Results and Model Validation

The assembled plant model was simulated using OpenModelica 1.14.1, an open source distribution of Modelica. The model conceptualized for both components and plants are developed in a basis of dynamic simulation, although the simulation performed for the purpose of model validation was realized in a steady state perspective – the simulation values tend to stabilize at a certain point of the simulation attaining constant values.

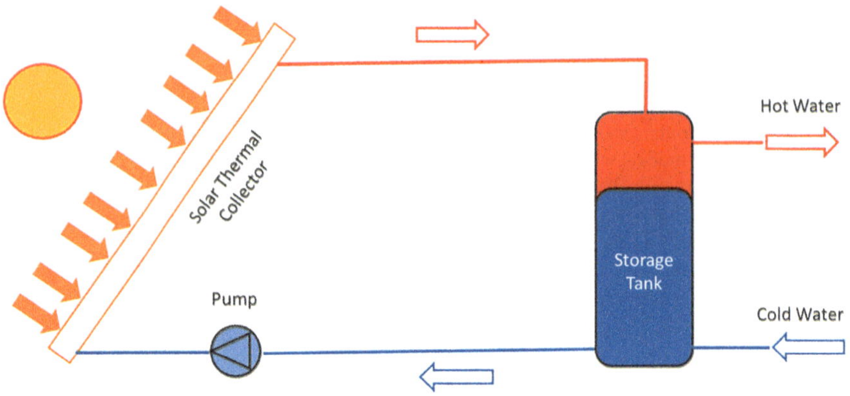

Fig. 1 Flow sheet of a solar thermal energy system including a solar collector and a water storage tank. (Adapted from Kumar et al. [6])

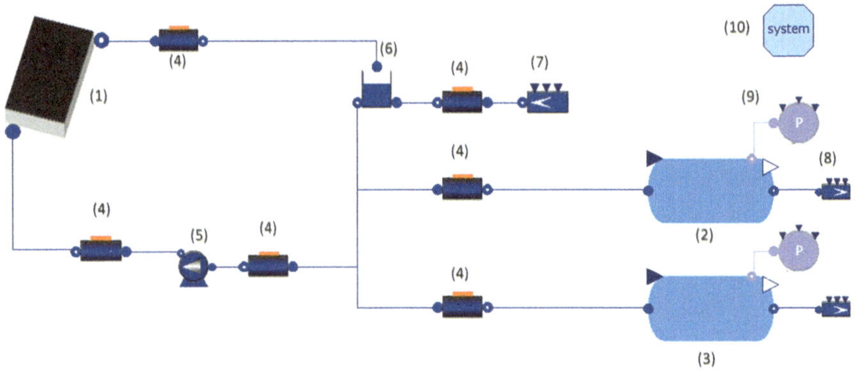

Fig. 2 Assembling of the solar thermal energy system in OpenModelica: (1) Solar collector, (2) Boiler 1, (3) Boiler 2, (4) Pipes, (5) Pump, (6) Water storage tank, (7) Water Source, (8) Water Discharge, (9) Exhaust Gas Discharge, (10) System model

By observing Table 4, the results for the validation of the two boiler models are consistent with real plant measured data. While for the circuit components it is not possible to perform such validation (as it consists of the conceptualized part of the model, not the already existing one), Table 5 presents the outputs of the simulation which serve as the baseline of equipment sizing.

3.2 *Techno-Economic Assessment*

With base on the simulation of the assembled model, it was performed a primary technoeconomic assessment in order to evaluate the viability of the project. This assessment is performed by determining the overall economic savings and calculate the payback time, taking into account an estimative for the investment cost. The economic savings were calculated considering the 2020 price for natural gas and electricity for industrial users (11.64 €/kWh [18] and 0.1654 €/GJ [19], respectively). The investment cost was roughly estimated considering the data present in the technical brief elaborated by the International Renewable Energy Agency (IRENA) and International Energy Agency (IEA)–the Energy Technology Systems Analysis Programme (ETSAP) [20]. Furthermore, a set of assumptions were taken to perform this assessment:

- The economic evaluation overall considers a simple payback, not considering an inflation rate of the investment costs.
- Several investments are considered to be intrinsic to the company, such as the material for repairs and substitution of equipment and the overall maintenance work.

The results for energy and economic savings are presented in Table 6 and the final determination of the payback time is presented in Table 7.

Table 4 Simulation Results and Validation (Boilers)

Component		Variable	Values		Deviation (%)
			Simulation	Real	
Boilers	Boiler 1	Water flow rate (kg/h)	1277.38	1277.38	0.00
		Water outlet temperature (°C)	94.99	95.00	0.01
	Boiler 2	Water flow rate (kg/h)	1180.32	1277.38	0.00
		Water outlet temperature (°C)	94.99	95.00	0.01

Table 5 Simulation results (circuit components)

Component		Variable		Simulation result
Circuit		Water flow rate (kg/h)		5000
		Temperatures (°C)	Before the solar collector	91.68
			After the solar collector	99.68
Pipes	Tank to pump	Inlet pressure (bar)		2.007
		Outlet pressure (bar)		1.989
	Pump to collector	Inlet pressure (bar)		2.057
		Outlet pressure (bar)		2.054
	Collector to tank	Inlet pressure (bar)		2.054
		Outlet pressure (bar)		1.013
	Tank to boiler 1	Inlet pressure (bar)		2.007
		Outlet pressure (bar)		2.006
	Tank to boiler 2	Inlet pressure (bar)		2.007
		Outlet pressure (bar)		2.006
Water storage tank		Inlet pressure (top) (bar)		1.013
		Inlet pressure (base) (bar)		2.007
		Outlet pressure (bar)		2.007
		Water level (m)		10.51
Pump		Inlet pressure (bar)		1.989
		Outlet pressure (bar)		2.057
		Power (kW)		11.65

Table 6 Determination of energy and economic savings

Natural gas consumption				
	Energy consumption		**Energy savings (MWh/year)**	**Boiler economic Savings (1000 €/year)**
	Initial	**Improved**		
Boiler 1	1070.32	390.60	681.72	7.92
Boiler 2	1070.32	218.18	854.14	9.93
Electricity consumption				
	Energy consumption (MWh/year)		**Additional investment (1000 €/year)**	
Pump	90.68		12.43	
Balance				
Boiler economic savings (1000 €/year)	**Additional investment (1000 €/year)**		**Economic savings (1000 €/year)**	
17.85	12.43		5.42	

Table 7 Determination of payback time

Economic savings (1000 €/year)	Total investment cost (€)	Payback time (years)
5.42	57,000	10.52

Regarding the energy balance, it is to note that appreciable economic savings are possibly to be obtained in the case that the economic savings produced by the fuel savings are more significant than the additional investment caused by electricity consumption namely by the circuit pump, which effectively is the case. The economic viability evaluation may be performed by comparing the estimated payback time (10.52 years) with reference values: while for solar water heating projects it may be acceptable (conisdering an acceptable payback time of less than 20 years [21]), comparing with typical energy efficiency improvement projects in the European industry (typical payback time of 2 – 3 years) the conceptualized project may not be economically viable.

4 Conclusion

This chapter presents an innovative framework for the improvement of energy efficiency in the industrial sector, namely through the use of renewable energy resources. This study is included in an overall research which encompasses the promotion of energy efficiency in industry, focusing on the optimization of thermal processes, which uses an innovative modelling and simulation tool developed in Modelica to perform the necessary studies. In this chapter, a case study undertaken in a ceramics plant is analysed and a project to improve the operation of two boilers existing in this plant is assessed.

For the aim of the study, an industrial implementation was projected, consisting of a solar thermal energy systems (STES) for the use of solar energy as an additional heat source in the boilers. The conceptualized system was modelled and simulated, with the main set of results being obtained: the results for the installed boilers (the models were validated and the results were consistent with real plant measured data) and data for the sizing of the water circuit encompassing the solar collector. A primary technoeconomic assessment based on a rough estimate of the investment cost of the installation was performed. An estimated payback time of 10.52 years was determined, considering balanced annual economic savings of 5420 €/year. While the return on investment is acceptable for renewable energy integration project, it far surpasses the acceptable payback time for a heat recovery project. Nonetheless, such project may become economically viable by integrating it within other heat recovery projects which by themselves have low payback times.

As thischapter is based on the validation and potential assessment of the presented modelling tool, it serves as a conceptual basis for the ongoing overall research on thermal process optimization. Future work would then include the dynamic simulation of the conceptualized system and the performance of a life cycle assessment (LCA) of the project.

References

1. Eurostat. Consumption of energy. Available online: https://ec.europa.eu/eurostat/statistics-explained/index.php?title=Archive:Consumption_of_energy. Accessed on 30 March 2020
2. Agathokleous R, Bianchi G, Panayiotou G, Arestia L, Argyrou M, Georgiou GS, Tassou SA, Jouhara H, Kalogirou SA, Florides GA, Christodoulides P (2019) Waste heat recovery in the EU industry and proposed new technologies. Energy Procedia 161:489–496
3. Rajendran K, Ling-Chin J, Roskilly AP (2015) Thermal energy efficiency in industrial processes. Handbook of clean energy systems (John Wiley & Sons, New York, USA), 4:1–30
4. Reddy C, Maidu S, Rangaish GP (2013) Waste heat recovery methods and technologies. Chem Eng (New York, McGraw-Hill, Chemical Week Publishing) 120(1):28–38
5. Fernandes MC, Matos HA, Nunes CP, Cabrita JC, Cabrita I, Martins P, Cardoso C, Partidário P, Gomes P (2016) Medidas Transversais de Eficiência Energética para a Indústria, 1st edn. Direção-Geral de Energia e Geologia, Lisbon, Portugal. ISBN 978-972-8268-41-1
6. Kumar L, Hasanuzzaman M, Rahim NA (2019) Global advancement of solar thermal energy technologies for industrial process heat and its future prospects: a review. Energ Convers Manage 195:885–908
7. EUROPA. Available online: https://ec.europa.eu/clima/policies/strategies/2030_en. Accessed on 20 November 2020
8. Oliveira MC, Iten M, Cruz PL, Monteiro H (2020) Review on energy efficiency progresses, technologies and strategies in the ceramic sector focusing on waste heat recovery. Energies 13(22):6096
9. Oliveira MC, Iten M (2020) Modelling of industrial water circuits with a customised Modelica library. Appl Therm Eng 169:114840
10. Iten M, Oliveira M, Costa D, Michels J (2019) Water and energy efficiency improvement of steel wire manufacturing by circuit modelling and optimisation. Energies 12:223
11. Oliveira MC, Iten M, Matos HA, Michels J (2019) Water–energy Nexus in typical industrial water circuits. Water 11(4):699
12. Politecnico di Milano. ThermoPower Home Page. Available online: https://build.openmodelica.org/Documentation/ThermoPower.html. Accessed on 20 November 2020
13. GitHub. WaterWatt Library. Available online: https://github.com/miguelmcoliveira/WaterWatt-Library. Accessed on 20 November 2020
14. Soltermica Home Page. https://build.openmodelica.org/Documentation/Soltermica.html. Accessed on 20 November 2020
15. SolarTherm Home Page. https://build.openmodelica.org/Documentation/SolarTherm.html. Accessed on 20 November 2020

16. Hernandez-Albadejo G, Urquia A (2018) Modelling of low-temperature solar thermal systems with Modelica. IFAC – PapersOnLine 51(2):783–788
17. de la Calle A, Hinkley J, Scott P, Pye J (2018) SolarTherm: a new modelica library and simulation platform for concentrating solar thermal power systems. SNE 3:101–103
18. PORDATA. Available online: https://www.pordata.pt/en/Europe/Electricity+prices+for+households+and+industrial+users+(PPS)-1479-313033. Accessed on 20 November 2020
19. PORDATA. Available online: https://www.pordata.pt/en/Europe/Natural+gas+prices+for+households+and+industrial+users+(PPS)-1480-313035. Accessed on 20 November 2020
20. Solar Heat for Industrial Processes. Available online: https://www.irena.org/publications/2015/Jan/Solar-Heat-for-Industrial-Processes. Accessed on 20 November 2020
21. Singh S, Anand A, Shukla A, Sharma A (2020) Technical, financial, and environmental feasibility of solar water heater for residential, commercial, and industrial application: a theoretical approach. Mater Sci Energ Technol 3:648–671

Reporting a Survey of the European Ocean Energy Sector in 2018

Ana A. Andrade, Isabel Cabrita, and Ruairi Maciver

1 Introduction

This chapter intends to provide a backstage view of a survey issued in 2019 with the purpose of characterizing some aspects of the ocean energy sector in 2018. It was the first of a series of three surveys to be delivered as part of the OceanSET project.

The OceanSET project was developed to support activities of the SET Plan Implementation Working Group in ocean energy. In this regard, the surveys are organized to assess in what direction the sector evolution is going. In addition to a high-level characterization of the ocean energy policy context in each Member State, the surveys aim to report to the European Commission progression on consenting processes, the supply chain, wave and tidal stream technologies under development, their costs and funding needs.

Detailed descriptions of the survey and its results have been made public through workshops and project deliverables. It is the purpose of this paper to address the design of the first survey and insights gained from respondents' feedback, Focusing on relevant aspects, a first section is devoted to presenting what the SET Plan is and where the OceanSET project fits. Then, survey content, delivery and responses are described in short. The final sections are devoted to analysis and conclusions.

2 Strategic Action from the European Community

2.1 SET Plan Background

The European Strategic Energy Technology Plan (SET Plan) concept was generated in 2006 [1, 2] with the aim of emphasizing the role of technology in the European Community (EC) energy policy. Since then, one specific objective addresses aligning efforts at the Member State (MS) level in order to obtain prioritized and focused technological research and innovation (R&I) in the industrial system. Such alignment was considered as needed not only to address the challenges posed by trying to obtain climate-neutrality by 2050, but also in establishing and maintaining EC leadership in low-carbon energy technologies and energy efficiency.

With this purpose, the SET Plan, coordinated by a Steering Group composed of MS representatives, introduced technology-specific targets, an information system (SETIS, Strategic Energy Technologies Information System), European Industrial Initiatives (EIIs), and the European Energy Research Alliance (EERA) bringing together national research institutes in support of the SET Plan implementation.

In 2015, the SET Plan was acknowledged in relation to the Energy Union pillar on "Research, Innovation and Competitiveness". Based on a commonly agreed roadmap, the SET Plan established priorities and respective key competitiveness actions [3]. For example, "*Number one in renewable energy*" was then defined as core priority 1 and "*Sustain*

A. A. Andrade (✉) · I. Cabrita
Research, Studies and Renewables Division, Directorate-General Energy and Geology, Lisbon, Portugal
e-mail: ana.andrade@dgeg.gov.pt

R. Maciver
Wave Energy Scotland, Inverness, Scotland, UK

© The Author(s), under exclusive license to Springer Nature Switzerland AG 2022

A. Sayigh (ed.), *Sustainable Energy Development and Innovation*, Innovative Renewable Energy,
https://doi.org/10.1007/978-3-030-76221-6_92

technological leadership by developing highly performant renewable technologies and their integration in the EU's energy system" and "*Reduce the cost of key technologies*" became actions 1 and 2, respectively.

In order to identify specific targets, the EC organized a discussion with relevant research and industrial stakeholders and SET Plan Member States, which resulted in an agreement on common targets described in a Declaration of Intent (DoI). The DoI for ocean energy was based on the Ocean Energy Strategic Roadmap and endorsed by the Steering Group in September 2016. The agreed strategic target was to develop cost-competitive ocean energy technologies with high market potential for Europe. The target was further quantified into specific Levelized Cost of Energy (LCOE) reductions for tidal stream and wave technologies. Tidal stream energy is challenged to reduce its cost to at least 15 ct€/kWh by 2025 and to 10 ct€/kWh by 2030. Wave energy should follow the same pathway reaching at least 20 ct€/kWh by 2025, 15 ct€/kWh by 2030 and 10 ct€/kWh by 2035.

To facilitate the delivery of these targets, Temporary Working Groups in each technology area were formed to develop implementation Plans proposing concrete R&I activities and funding opportunities considered essential for the achievement of the agreed SET Plan targets. These plans, 14 in total, cover all the Energy Union Research & Innovation priority areas and the 10 SET Plan actions. The implementation plan for ocean energy was endorsed by the SET Plan Steering Group and the EC in March 2018.

Each Temporary Working Group (TWG) evolved into Implementation Working Group (IWG) – basically the same group, but with the purpose of advancing the respective IP until reaching the agreed technology targets.

2.2 The OceanSET Project

The implementation plan (IP) for ocean energy [4] foresees sector development through three phases (2018–2020): discovery – to devise a collective monitoring approach by the EU Member States and Regions, agree an oversight management process for projects and funding, which outlines achievable interim commercialization targets, determining the likely levels of funding required for phases 2 and 3 (2020–2025); development – with operational arrays demonstrating the ability to meet the technical and financial metrics (2026–2030); and deployment – build the sector's support with large-scale deployments that will drive costs to a commercial level. After 2030, the ambition is to have a sector Delivery at scale, via a commercial market with a developed supply chain.

It further identifies 11 key actions aiming to coordinate MS and Regions to share and track critical information annually; demonstrate the clear development of the ocean energy technologies, stimulate collaboration between the MS, regions and the European Commission to ensure the effective use and appropriate blending of funds to drive large-scale deployment; and annually monitor progress with a review carried out at the end of each phase, to reach a Go/No-Go decision for the next phase. Six of these actions are technical, three financial and two environmental (Table 1).

The OceanSET project was designed to give support to the ocean energy IWG in this endeavour by promoting knowledge sharing, investigating collaborative funding mechanisms and facilitating the implementation of the technology development actions of the implementation plan. Funded by the H2020 programme, the project runs for 3 years from March 2019.

3 The 2018 OceanSET Survey

3.1 Survey Content

One OceanSET activity consists of performing annual mapping and discovery exercises, to chart the progress of the ocean energy sector and the advancement of the 11 IP actions. Part of the information required to monitor this activity is obtained through reporting from relevant groups and IWG members (actions 1.6 2.1, 2.2, 2.3 and 3.1 in Table 1), with the remaining (Technical and Environmental actions) intended to be obtained mainly by survey.

It was decided in OceanSET to follow the two-step model of the mapping exercise performed by the TWG for the IP, which concerned the year 2017 [4]. High level information from MS on their ocean energy sector would be retrieved in a first questionnaire conducted by project partners through the IWG members. Specific information on projects more technologically advanced would be gathered from developers in a second questionnaire.

The exact focus and level of detail deserved special consideration in designing the first year OceanSET survey.

The first questionnaire, called the MS survey, consisted of 27 questions and focused on ocean energy policies, budget and spend on ocean energy, support mechanisms, test facilities, supply chain, permitting processes and funded projects. Answers

Table 1 Operational actions of the ocean energy implementation plan and proposed monitoring mechanisms

Technical		
1.1	Tidal energy – Assist technology development and knowledge building up to TRL6	Register of funded projects
1.2	Tidal energy – Support system demonstration in operational environment and knowledge building in the TRL 7–9 categories	Register of funded projects
1.3	Wave energy – Support technology development, system demonstration and knowledge building up to TRL6	Register of funded projects
1.4	Wave energy – Encourage and support device and system demonstration at early demonstration array scale up to TRL 7–9	Register of funded projects
1.5	Collaborate in the areas of installation, logistics and infrastructure	Register of funded projects, supply chain gap analysis
1.6	Coordinate the development of standards and guidelines for wave technology evaluation and LCOE analysis	Stage gate metrics implemented by MS as standards
Financial		
2.1	Investigate the potential for creation of an investment support fund for ocean energy farms: EU and National Authorities should collaborate in order to create a fund providing flexible capital, and enabling further private capital to be leveraged	Reporting to monitoring group
2.2	Progress the creation of an EU insurance and guarantee fund to underwrite various project risks: This would be targeted at the first ocean energy projects to cover risks such as availability, performance, unforeseen events, failures, etc. consider the provision of a common reserve fund available to multiple projects in the initial farm or plant roll- out, to spread the risk and reduce the cost of providing guarantees support the development of a collaborative procurement model adaptation of the "wave energy	Reporting to monitoring group
2.3	Scotland" approach for wave energy development at EU level using pre-commercial procurement or similar	Annual progress report
Environmental		
3.1	Collaboration on the development of certification and safety standards for the development, testing, deployment of ocean energy devices	Monitoring coordinated through the central standards authorities and committees
3.2	Continue the de-risking of environmental consenting through an integrated programme of measures and in particular through promoting open data sharing	Monitored through the MS Implementation of the marine Strategy directive and local planning permitting procedures in MS

on budget/spend and licensing/consenting time were formulated in terms of ranges. This survey was delivered to IWG members to collect information from the respective MS concerning the year 2018.

The second questionnaire, called the Developer survey, consisted of 22 questions and focused on project funding, jobs created, development areas and technical aspects such as type of technology, installed capacity, energy produced, capacity factor, availability, life time, capital expenditure (CAPEX), operation and maintenance expenditure (OPEX) and LCOE.

The development areas offered as dropdown options in one of the questions were based on ETIP (European Technology and Innovation Platform for Ocean Energy) priority challenges [5]: (1) Developing novel concepts for improved power take-offs; (2) increasing power take-off reliability; (3) investigating novel devices before moving towards convergence of design; (4) increasing device reliability and survivability; (5) developing subsystems related to the balance of plant, for example, electrical connection technology; (6) development of electrical power conditioning and storage solutions; (7) development of moorings, foundations and associated connection subsystems; (8) investigating alternative materials and manufacturing processes for device structures; (9) defining and enforcing standards for stage progression through scale testing; (10) developing and implementing optimization tools; (11) investigating resource availability; and (12) developing environmental impact assessment methodologies and tools. Of these 12 options, only one was required to be selected.

The Developer survey was delivered to Technology Readiness Level (TRL) 7, 8, or 9 project owners identified in the MS survey as receiving funding during 2018. Lower TRL projects were not surveyed because it was considered that the required metrics would not be meaningful.

In both surveys, several questions were made mandatory. A final question was included to allow respondents to explain why they had not answered one or more of the previous questions.

The level of detail in both the MS and Developer surveys was selected from a balance between the wish to collect as much information as possible, going beyond the monitoring mechanisms proposed at the IP (Table 1), and the fear of alienating potential respondents by requesting excessive effort from them.

OceanSET questionnaires are to be sent yearly to MS, Regions, industry and key researchers through the IWG contact points. They are to be updated each year based on the previous year's responses, to determine to what extent any deeper levels of detail in responses are required.

Questions and responses from the 2018 MS and Developer surveys are presented and publicly available in OceanSET's first annual report [6].

3.2 Responses to the MS Survey

This first questionnaire was distributed to the fourteen MS representatives participating in the SET Plan IWG on ocean energy. Eleven representatives sent back their responses (BE, FR, IR, IT, NO, ES, PT, SE, UK, NL, FI).

Table 2 briefly presents some of the information collected in the MS survey, which is fully described in OceanSET's first annual report [6].

3.3 Responses to the Developer Survey

The Developer survey was delivered to 26 project developers. Twenty-two sent back their responses. Seven additional developers who were not identified in the MS survey responded to the survey. Of these, only 12 were eligible to be considered in the final analysis. Those omitted were for the following reasons: projects not active in 2018 or projects of a TRL lower than 7. Responses were aggregated and anonymized before being made public. These 12 respondents comprised both wave (5 out of 12, i.e., 5/12) and tidal technologies (7/12).

Table 2 Main data collected from MS survey

Topic	Data
MS with a policy dedicated to ocean Energy	Yes: BE, IR, IT, ES, PT, UK no: FI, NL, SE, FR, NO
MS with an assigned Ministry/department owner	Yes: BE, FR, IR, IT, NO, ES, PT, SE, UK No: FI, NL
Budget	Estimated total budget for ocean energy (wave, tidal) in 2018: 23 million €[a] estimated total budget for ocean energy (wave, tidal) in 2019: 26 million €[a]
Spend	Total amount spent in ocean energy in 2018: 26.3 million €
Revenue support mechanisms (€/MWh)	Yes – Wave: UK (45), FR (173), BE (50), SE (14), ES (136,93), NL (100) Yes – Tidal: UK (45), FR (173), BE (50), SE (14), IT (100)
	No: FI, IT, IR, PT, NO
Test facilities (number of test facilities)	Yes: BE (1), FR (3), IR (3), IT (3), NO (2), ES (8), PT (2), SE (2), UK (5), NL (2)
	No: FI
Supply chain	Dedicated/self-sufficient: UK
	Part of the chain well complemented by suppliers from other sectors: FI, IT, FR, BE, PT, ES
	Part of the chain hardly complemented by suppliers from other sectors: IR, NL
	Partial lack of supply: SE, NO
Permitting process	Estimated average licencing time in 2018: 3.7 year[a] estimated average consenting time in 2018: 4.25 year[a]
Permitting process requiring Environmental impact assessment	Yes: BE, FR, IR, IT, ES, PT, SE, UK, FI, NL no: NO
Projects funded in each MS in 2018	Tidal: 22 – TRL 1–6: 17[b]; TRL 7–9: 5[b] wave: 57 – TRL 1–6: 50[b]; TRL 7–9: 7[b]
	Other: 11
	7 MS funded ocean energy projects; all 7 were funding TRL 7–9 projects

[a]Metrics estimated because data were collected in terms of ranges

[b]Estimated quantity, should be treated with caution (difference between number of tidal and wave projects identified in MS survey and number of tidal and wave projects identified as being TRL 7–9)

Table 3 Main data from the Developer survey, based on 12 projects with TRL 7–9 active in 2018

Topic	Data		
Project cost and funding (million€)	Total: 95.3 (9/12, i.e. 9 out of 12); MS total: 23.2 (7/12); EC total: 35.9 (9/12) Total tidal: 75.8 (5/5); Total wave: 19.5 (4/7) MS tidal: 16.7 (4/7); MS wave: 6.5 (6.5); EC tidal: 35.9 (4/5); EC wave: 0 (5/7)		
Technology wave/tidal (number of projects)	Tidal: 5/12; wave: 7/12		
Technology development areas (number of projects)	Tidal: Increasing device reliability and survivability (5/5) Wave: Increasing device reliability and survivability (5/7); developing novel concepts for improved power take-offs (2/7)		
Type of technology (number of projects)	Tidal: 7 PTO horizontal axis turbines (4/7) Installation – Fixed gravity base (3/7); floating semi-taut mooring (2/7) wave: 5 PTO – Point absorbers; oscillating water column; other (n.a.[a]) Installation – Fixed gravity base (2/5); floating slack moored (3/5); floating semi-taut moored (1/5); pre-tension with technical anchors (1/5)		
Total installed capacity (MW)	Tidal: 4; wave: 0.6		
		Response rate	**Range of answers**
Jobs created	200, by the 12 projects	75%	0–55
Avg. energy produced (MWh/year)	Tidal: 11.5; wave: n.a.[a]	42%	1.2–6000
Average annual availability (%)	Tidal: 74; wave: 88	50%	4200–8332 h/year
Average capacity factor (%)	Tidal: 32; wave: 29	42%	15–45
Min.-max. Technical lifetime (year)	Tidal: 5–25; wave: 3–25	67%	3–25
Average CAPEX (€/W)	Tidal: 7.9; wave: 12.7	50%	4–6100
Investment – Equipment (€/W)	n.a.[a]	33%	3–10.5
Investment – Installation (€/W)	n.a.[a]	33%	1–3
Average annual OPEX (€/W)	Tidal: 0.1; wave: 0.7	33%	0.088–135
OPEX – Equipment	n.a.[a]	0%	No answers
Average LCOE (€/MWh)	Tidal: 217; wave: n.a.[a]	25%	210–240

[a]n.a. – Metrics could not be presented because the reduced number of responses prevented anonymization

Table 3 briefly presents some of the information collected in the Developer survey which is fully described in OceanSET's first annual report [6].

4 Critical Analysis of Responses

The analysis performed here is based on the response rate, the type of answers and on comments provided by MS representatives before, during, or after the survey.

4.1 Analysis of MS Survey Responses

The questionnaire was answered by MS members of the IWG which were, in general, motivated to provide information. As a result, it was possible to collect good-quality data from almost all countries and from most of the questions.

Questions on public policy for ocean energy, revenue support, funded projects, test site facilities and the supply chain received the highest adhesion. Countries were prone to provide details like the name of the policy, URL, assigned Ministry/ Department owner, specificity for ocean energy, revenue support value, its target technologies, and the number of projects the MS has funded, further indicating if they were tidal or wave and if they were of TRL below, equal to or higher than seven.

On the question about test facilities, which was formulated in terms of "availability of test site facilities for ocean energy prototypes", some countries indicated not only their open sea sites but also their tank and laboratory facilities. The question on the supply chain required a choice between categories, and all MS selected the one considered the most appropriate in describing the respective country case.

Budget/spend and licensing/consenting questions were well answered in general but merit some additional comments. For clarity, an explanation was included at the beginning of the questionnaire on which funding sources could be considered when estimating budget. According to the explanation, budget could include funding for technology, economic and environmental research projects, for ocean energy development projects, and funding for organizations and schemes tasked with delivering, supporting and monitoring the associated funding provision. The explanation section also detailed what constituted a "project", namely projects on technology research and development (sub-systems, devices and technological aspects of arrays, e.g. sub- sea electrical hubs), array development and deployment, environmental and economic characterization, and impact monitoring research. It excluded projects on funding management, research support and monitoring. Following these instructions, all MS provided an answer. However, three included a note on the difficulty in getting a value for budget. These stated that the information provided was determined from funding allocated to programmes open to all renewable technologies, not specifically ocean energy, or the national budget does not contain a specific figure for ocean energy. In the spend question (amount actually spent on ocean energy in 2018), the survey required an exact figure, entered as open text format. When answering, some MS commented on the difficulty in obtaining this kind of data.

Unlike the clarification provided for budget terms, the survey did not include any explanation regarding the meaning of the words "licensing" and "consenting". Answering separately on licensing and consenting time, some MS entered equal values for both, while others provided quite dissimilar responses. Further clarification from MS was required in order to understand how these words mean different procedures in different countries. Furthermore, separate procedures would not necessarily be sequential, thereby preventing obtaining a total time for permitting by only using data collected in this survey.

Questions on installed capacity and electricity injected in the grid received very few answers. Further clarification from MS indicated that data of this kind are not necessarily public or easily accessible, and may even be absent.

4.2 Analysis of the Developer Survey Responses

All 12 respondents provided project data such as the identity of the project lead, whether the project involved a consortium, the project timeframe, and progress towards completion.

The number of completed responses varied for questions on costs and funding. Nine developers indicated total cost and the amount of funding obtained from the EC, while seven indicated how much national funding had been received. Further information on responses per technology is provided in Table 3. Almost all developers answered questions regarding the proportion of project costs sourced from in-kind funding and grant funding. Almost none entered a figure for overall cost of the project.

All responses provided technical information such as type of technology concerning either the device or its installation, the development stage (i.e. TRL) and the development area covered. Concerning the latter, the answers show that of the 12 dropdown options available (Sect. 3.1), only two were selected (Table 3) by all respondents. The options chosen described the case but only in a very general manner. Greater clarity about specific objectives of the projects was obtained in a free-text box, where answers converged on cost reduction, reliability demonstration, validation, and obtaining bankability by certification. Environmental monitoring was mentioned once.

Considering the set of questions concerning jobs created and device technical specifications, the manner of responding resulted in three distinguishable groups: (1) a single respondent answered all these questions; (2) three respondents skipped just one question (jobs or one of the cost topics); (3) three respondents only answered one question (jobs and/or lifetime). The respondents that answered only one question are from the wave energy sector, so data ended up skewed to the tidal energy side. As a result, some of the questions have insufficient responses to make a statistical analysis while preserving anonymity. These questions had to be declared as having non-available answers ("n.a.", as seen in Table 3).

The technical lifetime responses were in the range of 3–5 years, while others were in the 20–25 year range. It was considered that shorter lifetimes referred to devices remaining in the water just for the project duration while the longer lifetimes referred to the design life of first-of-a-kind prototypes or commercial devices.

The questions resulting in the fewest number of responses concerned LCOE and CAPEX or OPEX details: installation cost, equipment cost and equipment O&M cost. We can understand developers' reticence in divulging potentially sensitive commercial information for devices still being tested.

5 Conclusion

The creation of the survey and its data analysis provided some interesting insights above and beyond the responses themselves. These insights have been described and lessons can be learnt for the future.

Both MS representatives and developers responded positively in answering the respective questionnaires. MS representatives seemed to be motivated as a result of their role as members of the SET Plan Implementation Working Group. As the results of the MS survey are public, the MS survey can also be seen as a way to promote MS achievements and technology development context. The Developer survey results are confidential with results published only after being aggregated and anonymized. So, for developers, answering the survey may be motivated by the desire to contribute to the shaping of future policy designs, or by a wish to share experience with others in the sector.

Such good will is best exploited when questions are well structured and clear on the expected feedback. For this purpose, previous knowledge on the matter is required. For example, consenting and licensing are acknowledged as having different meanings in different countries. Launching a survey on this subject would require a previous study on the variety of permitting processes across Europe so that common metrics can be identified. Subsequent iterations of the OceanSET survey are already planned to be different on a number of aspects such as the permitting process, test facilities in a MS, capacity deployment or device lifetime.

Questions on the MS allocated budget or expenditure for ocean energy revealed that this information can be difficult for respondents to gather, in particular where budgets are delegated to different regions. Reasons like the absence of a specific figure for ocean energy in national budgets, or funding programmes open to all renewable technologies (not just for wave/tidal), were advanced by some respondents. If budget and ocean energy spend continue to be figures to monitor, then action would be needed on a country by country basis to make this type of data available.

A similar situation occurs at the MS level with indicators such as total annual electrical energy produced and injected in the grid from wave/tidal devices, or total annual wave/tidal installed capacity. It would appear that accurate centralized data in this domain are unavailable or restricted in some MS. However, these are figures of interest when monitoring the progress of the ocean energy sector. An effort could be initiated in order to have this information be collected and published systematically in MS ocean energy reviews.

In spite of the developers' willingness to provide general technical and financial information on their projects, difficulties arose on potentially sensitive data such as detailed costs and R&I or development area subjects. This kind of information is however particularly important.

Mapping of sector development enables the identification of R&I gaps, and the designing of appropriate funding programmes. In this regard, it should be remembered that the OceanSET survey feeds results into the Implementation Working Group whose assessments have a direct influence on European ocean energy policy.

Cost information is needed to trace the trajectory of the sector. Knowledge of the trajectory will enable the EC and MS to provide the support required to achieve announced renewable energy targets. Ultimately, falling costs would be the signal that investors are waiting for to make more favourable decisions within the sector. Ocean energy stakeholders need to participate in this process by providing relevant information that will help in achieving this goal. OceanSET surveys are a good tool to help accomplish this purpose.

References

1. Towards a European Strategic Energy Technology Plan, C(2006)847
2. A European Strategic Energy Technology Plan, C(2007)723
3. Towards an Integrated Strategic Energy Technology (SET) Plan: Accelerating the European Energy System Transformation, C(2015)6317
4. SET Plan Temporary Working Group Ocean Energy (2018) Ocean Energy – Implementation Plan. https://setis.ec.europa.eu/system/files/set_plan_ocean_implementation_plan.pdf
5. ETIPOcean (2017) Deliverable 2.1: An Integrated Framework of Ocean Energy Sector Challenges. https://www.etipocean.eu/assets/Uploads/170221-Deliverable-2-1-Integrated-challenges-for-distribution4.pdf
6. OceanSET (2020) OceanSET First Annual Report. https://www.oceanset.eu/wp-content/uploads/2020/05/OceanSET_FirstAnnualReport_April2020_with-infographic.pdf

Promoting Energy-Efficient Neighbourhoods: Learning from BedZED

Derya Oktay

1 Introduction

In the era of serious environmental problems threatening the world, cities are often considered 'environmental disasters' and believed to have no place in a sustainable future. However, it is terribly unrealistic to postulate a city-free sustainable future as more than half of the world's population now lives in cities despite the escape from central areas to peri-urban and/or suburban areas. What is needed is the reform of urban areas through research and efforts towards sustainable urbanism [1–4] and remaking cities attractive for living so that the massive outward expansion in cities will be prevented. In this context, city neighbourhoods are the most important components of a city and when they are designed as green- or eco-neighbourhoods, the whole city benefits.

Research in sustainable urban design recommends increased utilising of the energies manifested in the existing urban fabrics, through the adaptive re-use of former industrial (brownfield) sites and upgrading the existing building stock [5]. On the other hand, the so-called contemporary 'green or sustainable buildings', the most referred concept when sustainability at the building scale is discussed, embarks on the significance of the relationship between nature and the architecture in order to embrace the philosophy without resorting to superficial mimicry.

It is important to understand that the buildings cannot be dealt as isolated objects without considering their relationships with their urban or regional contexts. Ensuring energy-efficiency of single buildings alone is not sufficient to solve the complex issues caused by climate change, and the concept of green space between the buildings being the 'lungs of the city' in vast amounts do not help create sustainable environments. Green spaces are important to absorb CO_2, they contribute to biodiversity encouraging flora and fauna, and enhance people's psychology. Moreover, the need for green spaces in the near surroundings has been in the forefront of the debates during the coronavirus COVID-19 pandemic period. However, as advocated by Jane Jacobs [6], too much green space reduces the density of urban areas lessening the viability of public transport and increasing walking distances. Green space, therefore, has to be carefully designed and planned to maximise its contribution to the environment while minimising its effects.

Certain environmental issues have been a concern of housing reformers for much of the twentieth century, and the growing awareness of environmental issues has resulted in international agreements and charters at different intervals. Of all the issues affecting the environment, by far the most important are global warming caused by the greenhouse effect, the loss of the ozone layer, pollution, shortage of water and domestic waste. In this vein, there is a range of environmental issues for the would-be environmentally conscious designers and developers to consider, and certain developments have been exemplary in overcoming these problems.

Some of the precious examples of urban green- or eco-neighbourhood ideas being applied in the urban context are Kronsberg Green District in Hannover, Hammer by Sjöstad in Stockholm, Solar City in Linz-Pichling, GWL District in Amsterdam, Dockside Green in Victoria, British Columbia, Vikki District in Helsinki and Kreuzberg in Berlin (before the fall of the Berlin Wall) [7]. The only full-scale attempt to create a zero-carbon neighbourhood in the UK is BedZED in London. The aim of BedZED (Beddington Zero Energy Development) development was unique in the UK in that the design aimed to use only energy from renewable sources generated on site. It has attracted huge amounts of attention both in professional milieu as a large-scale, mixed-use, sustainable community, and has been an inspiration for low-carbon,

D. Oktay (✉)
Department of Architecture, Faculty of Architecture and Design, Maltepe University, Istanbul, Turkey
e-mail: deryaoktay@maltepe.edu.tr

© The Author(s), under exclusive license to Springer Nature Switzerland AG 2022

A. Sayigh (ed.), *Sustainable Energy Development and Innovation*, Innovative Renewable Energy,
https://doi.org/10.1007/978-3-030-76221-6_93

environment-friendly housing developments around the world. In this chapter, based on the findings of the author's analysis and observation in BedZED as part of the preliminary step of an EC-funded research project carried out at Oxford Institute for Sustainable Development at Oxford Brookes University and the review of the relevant literature, the scheme will be explored through various dimensions.

2 The Case of BedZED

BedZED (Beddington Zero Energy Development), was developed as partnership between the architect Bill Dunster, the BioRegional Development Group and the Peabody Trust and Group, and the Peabody Trust on land provided by the London Borough of Sutton at reduced price. The development, comprising 82 houses, 17 apartments, and just under 1500 m^2 of workspace, was completed in 2002 [8–10].

BedZED has won many awards for green architecture and design. It was shortlisted for the Stirling Prize, the UK's premiere architecture award, in 2003, and been an inspiration for low-carbon, environment-friendly housing developments around the world with its many innovative features.

2.1 Design Principles

BedZED was created as a compact urban development on a brownfield site, a former sewage works. Its compact design creates a walkable environment, and the walkability is enhanced by the design of the street spaces prioritising pedestrians and cyclists (Figs. 1, 2 and 3). BedZED has many innovative features, including its own combined heat and power system burning wood chips, a 'living machine' sewage system, a pool of electric cars and photovolcaic panels mounted on the conservatories of every flat.

In BedZED, all units have their own small garden; the three-storey walk-up type townhouses are highly insulated (Fig. 2). Very high levels of energy and resource efficiency have resulted in greatly reduced running costs. Much of it has stood the test of time but some of its green technologies have proved challenging to adopt.

Fig. 1 Location of BedZED complex. (*Source*: Peabody)

Fig. 2 The layout view of BedZED. (*Source*: http://sustainability-info.blogspot.com/2005/11/bedzed-zero-energy-city.html

Fig. 3 View from a street. (*Source*: Bioregional)

2.2 Energy Saving Ideas

In BedZED, major energy savings are provided by high levels of insulation, airtightness and passive solar heating design. Electricity savings provided by solar photovoltaic arrays integrated into the buildings (Fig. 4). Some other ideas include an on-site car club, ample secure cycle parking, excellent public transport links and 0.6 car parking spaces per home [11, 12].

In the original plan, using energy from renewable sources generated on site was to be achieved through energy efficiency with the south-facing blocks profiled to maximise solar gain while workspace is located on the northern side of the blocks. The two-storey workspaces also included 'live-work' units with people working from home in them throughout the day, and the heat from the occupants' equipment such as office computers was expected to provide much of the space-heating required in BedZED's work areas, which also have generous daylighting from overhead roof lights. Most of BedZED's homes,

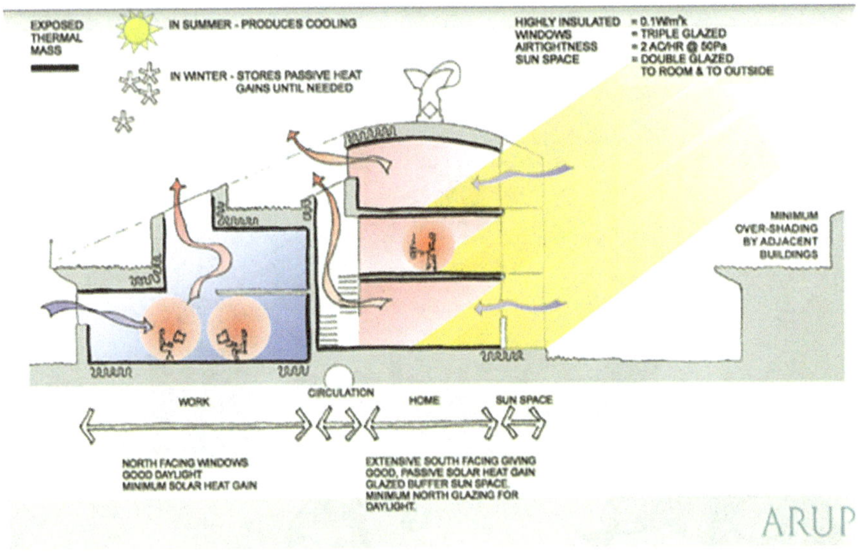

Fig. 4 The typical section of BedZED buildings. (*Source*: Bioregional)

however, have conservatory-like 'sunspaces' behind large, south-facing glazed facades; these are the major solar gain design features.

Natural ventilation is another energy saving aspect of BedZED's homes and workplaces as they do not need any electrically powered fans to do that. When needed, they get ventilated through the windows opening on different sides allowing for cross-ventilation.

Monitoring in 2003 found that BedZED had achieved reductions in energy use of 88% for space heating, 57% for hot water and 25% for electricity (11% of which was produced by solar panels). Water consumption has been reduced by 50% and residents' car mileage by 65%.

2.3 *Sculpture-Like Wind Cowls*

In BedZED, a highly effective ventilation system was targeted to bring in fresh, exterior air into the homes and workspaces and expel stale air. The brightly coloured, eye catching rooftop wind cowls have been the outcome of this intention (Fig. 5). Each of these cowls, which could also be considered a new interpretation of the chimney of the traditional terrace houses in the UK, has a vane and can rotate, so that it always faces into the wind. At the front, a small opening collects fresh, exterior air which moves down into the building and emerges into the homes and workplaces through closeable vents. At the back of the cowl is a larger hole through which the stale, outgoing air (collected through vents in the homes' ceilings) exits. As it leaves, this stale air warms the incoming cool air through a heat exchanger located in the base of the cowl.

2.4 *Hot Water and Additional Space Heating*

While much of BedZED's heat demand is met by the warmth from its occupants and by solar gain, its homes and workplaces still require some additional heat. The homes and workplaces on the blocks' northern sides, which lack their own sunspaces, need some space heating too. This additional heat is provided via BedZED's own small-scale district heating system which circulates hot water around the eco-village through an underground network of insulated pipes. Every home has its own large hot water tank, placed in a vented cupboard, warmed by the heat network. The tank provides all of the household's hot water, year-round, plus background warmth in winter to supplement the lower levels of passive solar heating. The home and work spaces on the northern sides of the blocks have conventional radiators, also heated by this district heating system.

Fig. 5 Sculpture-like wind cowls. (*Source*: https://alchimag.net/en/architecture/green-building-architecture/bedzed-prototype-of-eco-neighbourhoods/)

2.5 *Sustainable Travel and Transport*

BedZED was designed to encourage its residents to use sustainable transportation and to radically reduce greenhouse gas emissions from travel compared to conventional suburban housing development. It includes public transport within less than 10 minutes-walk, incentives for bike use, as well as reduced car parking and road space. Monitored data of the first 8 years occupancy indicate that the use of private car has reduced by 60 percent as compared to the local average. The scheme also includes partnership with City Car Club to provide a car-sharing scheme which now includes an electric vehicle.

2.6 *Sustainable Water Management*

Since BedZED is located in a region which suffers from serious water stress, the complex was designed to be water-efficient and used recycled water when possible. The scheme included a 'Living Machine' system of recycling waste water but this is not currently operating.

2.6.1 Materials

In order to reduce the environmental impacts and carbon footprint of BedZED's construction, the houses are built from sustainable materials sourced from within 35 miles of the site.

2.6.2 Gardens

The great majority of BedZED homes have their own outdoor space in the form of balconies and small gardens at ground, first or second floor level. Several of the homes connect to their gardens by bridges spanning the mews streets.

2.7 Variety of Use and Uses

Owing to a variety of housing both in sizes and market values, shared home ownership and subsidised renting, BedZED is home to a broad social mix. According to the findings of a survey carried out in 2007, there is a strong sense of community in the complex and this is what people liked most about living in BedZED. On average, they knew 20 of their neighbours by name, while two-thirds felt they knew more people living in the eco-village than they did when living in their previous neighbourhoods while in the surrounding neighbourhood of Hackbridge (excluding BedZED), residents knew an average of eight neighbours by name.

2.8 Sustainable Lifestyle

The way our settlements are designed has an important role on the sustainability of human activities and it is the responsibility of those who shape towns and cities to minimise their negative impact on the natural environment. Everything we do as professionals and as human beings in the name of sustainability means very little if we do not actually change environmental behaviour of consumers, companies, communities and governments. Indeed, as highlighted by Asilsoy and Oktay [13], when we evaluate cities that can be considered sustainable, it can be seen that the ecologically concerned inhabitants as citizens, or ecological citizens, are one of the primary dynamics of their sustainability efforts.

It can be stated that BedZED represents a challenge to many urban design orthodoxies as it has become known as the UK's first large-scale, mixed-use ecological development with a 'sense of place' and an increasing 'resale value' [14]. Further, it has become a neighbourhood in which residents can enjoy sustainable lifestyles as ecological citizens as proved through various interviews with its residents. Its management too ascertains its value; in order to support the idea that a sustainable society can be created, the BedZED authorities continue to organise certain activities, that is, regular guided trips and exhibitions in the complex, seminars in the showroom apartments, and educational programmes about sustainable development in the offices belonging to BioRegional Development Group.

3 Conclusion

City neighbourhoods are the most important components of a city and when they are designed as energy-efficient 'green- or eco-neighbourhoods', the whole city benefits. The so-called contemporary 'green or sustainable buildings' dwells on the significance of the relationship between nature and the architecture in order to embrace the philosophy without resorting to superficial mimicry. However, it should not be forgotten that the buildings cannot be dealt as isolated objects without considering their relationships with their urban or regional contexts; the building should conform to its local environment and urban context. Green architecture must encompass a sustainable form of urban development.

Beddington Zero Energy Development (BedZED) designed as a zero-energy development in south London has been an inspiration for energy-efficient neighbourhoods around the world. Further, it has become a neighbourhood in which residents can enjoy sustainable lifestyles as ecological citizens and with a heightened sense of place. BedZED provides an answer to many problems facing planners, designers and developers in the twenty-first century towards more sustainable housing environments including the following: Minimising the consumption of land through more compact development and densification by containing the footprint; creatively using brownfield sites; designing urban quarters where walking and cycling is highly attractive with mixed-uses; promoting transport-oriented development to reduce the use of car; creating a sense of urbanity through increased density and the design of real places, where an urban culture can flourish; developing sites in a way that the consumption of energy, non-renewable materials and pollution is generally reduced; better considering the building's placement on the site, orientation and solar issues, with a focus on high quality public space and proper landscaping; designing developments in ways that increase access to transportation choices; designing spaces with direct access to natural light and air, and an improved orientation to take advantage of passive solar principles, and optimising the orientation and solar exposure to maximise the use of renewable resources in the operation of buildings and districts; and providing a good variety of homes – for sale and rent – on affordable and market terms to attract high and low incomes in order to help create socially inclusive communities.

References

1. Oktay D (2004) Urban Design for sustainability: a study on the Turkish City. Int J Sust Dev World 16, pp 24–35
2. Oktay D (2016) In pursuit of human sustainable urbanism: ecological and cultural integrity (Keynote Paper). Proceedings of the 1st International Conference on Urban Planning / ICUP 2016, Nis, pp 79–86
3. Oktay D (2017) A critical approach to sustainable urbanism: lessons from traditional and contemporary paradigms. In: Hepperle E, Dixon-Gough R, Mansberger R, Paulsson J, Hernik J, Kalbro T (eds) Land ownership and land use development: the integration of past, present, and future in spatial planning and land management policies. vdf Hochschulverlag AG an der ETH, Zurich, pp 295–306
4. Oktay D (2017) Towards smart and human-centred sustainable urbanism: lessons for future cities (Invited Paper). BIT's 7th World Congress of Smart Energy – 2017, Wuxi, 02–04 November 2017
5. Oktay D (2012) Urban regeneration and place-making for sustainability. In: Lamanna C (ed) Recycling common ground. Alinea Editrice, Florence, pp 20–27
6. Jacobs J (1969) The economy of cities. Vintage Books, New York
7. Lehmann S (2010) The principles of green urbanism: transforming the city for sustainability. Earthscan, London, p 6
8. Rudlin D, Falk N (1999) Sustainable urban neighbourhood: building the 21st century home. Architectural Press, Oxford, p 96
9. Dunster B, Simmons C, Gilbert B (2008) The Zed book. Taylor & Francis, London
10. Girardet H (2004) Cities people planet. Wiley-Academy, Chichestee, p 168
11. Schoon N (2016) BedZED. Bioregional Development Group, Retrieved August 2020. https://www.bioregional.com/projects-and-services/case-studies/bedzed-the-uks-first-large-scale-eco-village
12. Rudlin D, Falk N (1999) Sustainable urban neighbourhood: building the 21st century home. Architectural Press, Oxford
13. Asilsoy B, Oktay D (2016) Measuring the potential for ecological citizenship among residents in Famagusta, North Cyprus. In: Salama A, Grierson D (eds)., vol 41, Issue 2 Open house international (Special issue on transformations of architecture and urbanism of cities in the global south), pp 47–55
14. Rowland J (2006) Urban design futures: conclusion. In: Moor M, Rowland J (eds) Urban design futures. Routledge, New York, p 177

Effect of Building Fabric on the Electricity Demand for Space Cooling for Attaining Thermal Comfort in a South Indian Community Building

Kumar Biswajit Debnath and David P. Jenkins

1 Introduction

India has one of the highest amounts of cooling degree days (CDDs) globally [1] predominantly due to geographical location and the adverse effects of climate change. Therefore, with world's second-largest population [2], India has the highest amount of population under the heat stress caused by climate change [3]. However, only 8% of Indian households had air conditioning (A/C) units in 2017 [4], significantly lower than that of China (60%) and USA (90%) [5]. In addition to the climate change and lower penetration of A/Cs, India's affluent middle-class population has been rising rapidly due to economic progress in recent years. Also, under the elevated global temperature scenarios, the number of A/C units in Indian households may increase 42 times (1144 million units) by 2050 as compared to 2016 (27 million units) [6]. Therefore, the elevated demand for cooling for attaining thermal comfort may increase electricity consumption significantly.

There have been several studies on the effect of different building fabrics on the residential building indoor thermal performance and energy use in India [7–9]. However, very few studies took India's household-specific data-driven approach to analyse the effect of building fabric on the residential electricity consumption for cooling to attain thermal comfort, particularly on a community scale. The objective of the study was to assess the performance of residential building fabric on the reduction of daily indoor temperature and cooling electricity demand of a community (Auroville) in south India. Our present work and literature already suggest that cooling might become the most significant energy-intensive appliance in the residential sector in the studied area [10]. The demand reduction assessment in this study demonstrated the effectiveness of building fabric on the community-scale aggregated electricity consumption for cooling. Quantifying the potential for change within a modelling framework is vital for communities which, for reasons of climate, demographics, and technology, are expected to see steep rises in space cooling in the coming decades. The research will further be used to inform the work of the Community-scale Energy Demand Reduction in India (CEDRI) project (funded by DST in India and EPSRC in the UK), to combine energy network and building modelling with behavioural studies to tailor demand reduction strategies for Indian communities.

2 Methodology

As a case study, we focused on a community of Auroville in the warm-humid climate in southern India. The study had three major parts: climate analysis, building thermal model development based on monitored data and different building fabric scenario testing and analysis. Firstly, the climate data for 1989–2019 of the Auroville were analysed to investigate the prevalence of heat stress using Eqs. (1) and (2). Secondly, we developed the building physics model of a community (multi-storey apartment with 15 households with Designbuilder) on the basis of monitored electricity consumption data from 18 selected households for about a year. Then the building model was calibrated through a manually iterative process which was described in [11]. In the third stage, two building fabric scenarios were assessed within the simulation environment (with EnergyPlus)

K. B. Debnath (✉) · D. P. Jenkins
School of Energy, Geoscience, Infrastructure and Society (EGIS), Heriot-Watt University, Edinburgh, UK
e-mail: k.debnath@hw.ac.uk

© The Author(s), under exclusive license to Springer Nature Switzerland AG 2022

A. Sayigh (ed.), *Sustainable Energy Development and Innovation*, Innovative Renewable Energy,
https://doi.org/10.1007/978-3-030-76221-6_94

to find out the effect of different building fabric construction on the indoor temperature and aggregated electricity demand of a residential community building. EnergyPlus was previously used in a significant number of studies regarding building physics and thermal comfort, such as [12–15], which motivated us to use the simulation environment.

2.1 Climate Analysis: Heat Stress

We obtained 30 years' weather data (1989–2019) from the Meteoblue database (www.meteoblue.com) for the heat stress analysis of Auroville's climate. The ambient temperature and relative humidity data were used to calculate the hourly heat index (HI) with the following Eq. (1) adopted from the study cited herein [16]:

$$HI = -42.379 + 2.04901523T + 10.14333127R - 0.22475541TR - 6.83783\times10^{-3}T^2 - 5.481717\times10^{-2}R^2 + 1.22874\times10^{-3}T^2R + 8.5282\times10^{-4}TR^2 - 1.99\times10^{-6}T^2R^2 \tag{1}$$

where, T = ambient dry bulb temperature (°F), R = Relative humidity (integer percentage).

HI was selected for the study, as it has been widely used in several studies to analyse the adverse effect of heat stress, such as [17–20]. For the analysis in this study, we converted the ambient dry bulb temperature into °C with the following Eq. (2) adopted from the study referred here [21]:

$$C = 5/9(F-32) \tag{2}$$

where, C = ambient dry bulb temperature (°C), F = ambient dry bulb temperature (°F)

2.2 Monitoring Household Electricity Consumption and Data Analysis

For the monitoring, three buildings in Auroville – an experimental township established in 1968 in southern India [22] – were selected where 21 and 9 households were chosen to be observed in Citadines (I and II) and Inspiration, respectively. A detailed description of the monitored buildings and the appliances in the households is given in [23]. The household electricity consumption data were collected for November 2018–June 2019. The household electricity consumption data were collected with blink meters. Due to the constraints such as load shedding, technical issues, blink meter specification and human factors, the collected data had gaps. The data collection procedure was described in detail, as well as the data constraints in [23]. To circumvent the missing data issues, we used the 'Missing Data Filling' methodology described in [23]. We used the synthesised 1-minute electricity consumption data of the households to extract the representative cooling appliance operation schedule, which was elaborated in the subsection 'Cooling appliance operation schedule' in the cited study [11].

2.3 Building Physics Modelling of the Community

The 4-storied Citadine-1 building with 15 households (with ~50–100 m^2 area) was modelled (as shown in Fig. 1) in Designbuilder with constructions and materials described in Table 1. Also, from the energy audit of the households, the electricity use for other appliance use apart from heating, ventilation, air conditioning (HVAC) and lighting appliances were

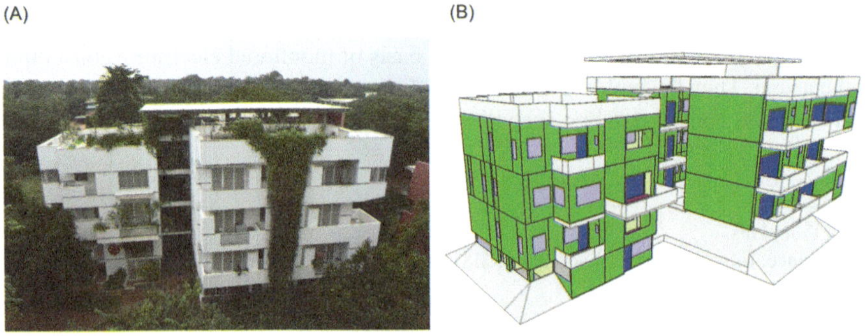

Fig. 1 (**A**) The picture of the modelled and monitored Citadine-1 building; (**B**) Building physics model developed in Designbuilder (Version 6.1.5.004)

Table 1 Construction name, thickness and materials (for the material properties, we used software database)

Name	Thickness (m)	Materials	U-Value (W/m^2-K)
Exterior walls	0.284	Burnt bricks with 0.015 m cement plaster on both sides	1.959
Internal walls	0.155	Brick masonry with 0.015 m cement plaster on both sides	2.301
Roof	0.178	RCC Concrete with 0.013 m cement plaster on both sides	4.019
Windows	0.005	Glass windows with aluminium frames.	1.978
External doors	0.040	Plywood (Heavyweight)	2.479
Internal doors	0.040	Plywood (lightwood)	2.479

7.5–27.1 kWh/m^2 and average 13.66 kWh/m^2, after omitting the outliers average 10.34 W/m^2 (assuming on average 1.5 hours use daily), was used in the simulation. For each fan and A/Cs, 75 W and 32.85 kWh/m^2 were used, respectively. In the case of lighting, the energy use was 0.005 W/m^2 for the simulation. The details of appliance schedule extraction from monitored household electricity demand profile of the building physics model can be found in the studies referred herein [11, 23].

2.4 *Scenario Development*

Burnt clay bricks and reinforced cement concrete (RCC) were commonly used materials for wall and roof construction in Auroville, respectively. The analysis of the electricity consumption showed a mixed-mode use of A/C and electric fans in households with A/C [23]. In the base case and scenarios, we assumed the mixed-use of A/C and electric fans for attaining thermal comfort in all the 15 households in the building. As the bricks have an unsatisfactory performance as an insulating material and the penetration of A/Cs are on the rise [10], building fabrics with lower U-value may offer an opportunity for better insulation to reduce the operative indoor temperature and cooling electricity consumption. For the study, the following scenarios described in Table 2 and illustrated in Fig. 2 were selected to test the effect of building fabric insulation on indoor temperature and cooling demand.

3 Results and Discussion

3.1 *Climate Analysis: Heat Stress in Auroville*

The heat index (HI) during summer 2010 was in extreme caution level, whereas since 2014, the HI during summer regularly crossed into danger level [11]. In 2014 and 2019, the HI was in danger level, which may cause likely heat cramps or heat exhaustion, even possible heatstroke with prolonged exposure and physical activity [24, 11]. The HI analysis of 30 years (1989–2019) is demonstrated in Fig. 3. High prevalence of HI not only during summer but also in rainy season, as compared to 2010, showed that the climate of Auroville has been getting warmer and elevating the risks of heat-related health risks, which may be contributing to the increase in A/C ownership in the community [10], as most the residents of Auroville get the electricity for free [25].

3.2 *Model Validation*

The studied households were modelled and simulated with the energy audit and the electricity consumption schedules and validated with the monitored data. The monitored and simulated 1-minute electricity demand profile for a household with only electric fans (no A/Cs) for 3–9 November is demonstrated in Fig. 4. The average monitored and simulated electricity consumptions were 0.0014 kWh and 0.0021 kWh, respectively. The simulated output was higher (maybe) due to the data collection constraints and limitation of extracting electricity consumption profile of appliances as well as the constraints of simulating the real-world context as described in the study referred in here [26].

Table 2 Construction and U-value of building materials for the base case and other scenarios (Roof and wall composition are illustrated in Fig. 2)

Scenario	Roof and wall composition	Roof materials	U-Value (W/m^2-K)	Wall materials	U-Value (W/m^2-K)
Base	R1W1	RCC, cement plaster	4.019	Burnt bricks, cement plaster	1.959
1	R1W2	RCC, cement plaster	4.019	Concrete blocks, cement plaster	0.646
2	R2W3	RCC, cement plaster, 0.030 m Expanded Polystyrene insulation (EPS) at the outer layer	0.893	Concrete blocks, cement plaster, 0.030 m EPS at the outer layer	0.428

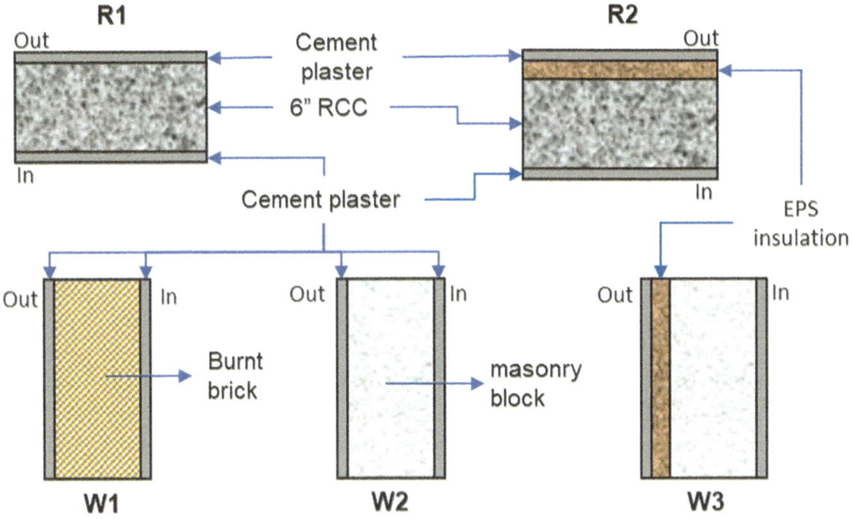

Fig. 2 Building fabric construction for base case (RCC roof with cement plaster and burnt bricks with cement plaster) and other scenarios. Here, Out = Outdoor, In = Indoor

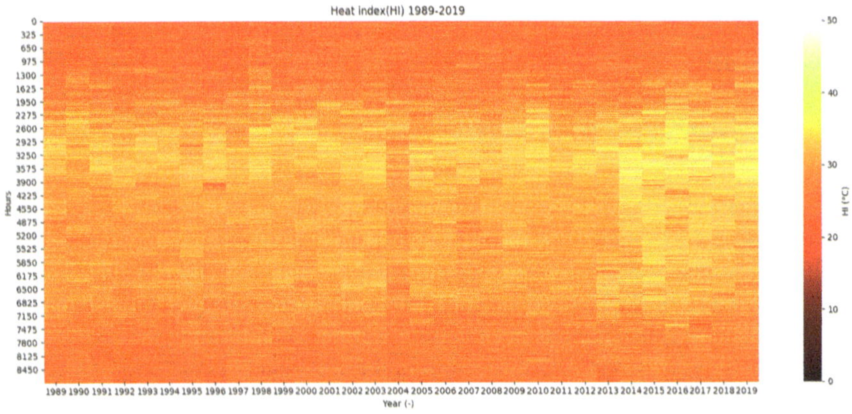

Fig. 3 Hourly HI analysis for 1989–2019

Furthermore, as the studied residential building was situated in a rural context and surrounded by vegetation which could affect the thermal comfort, this may make the modelling and simulation complicated. Also, there were differences in the weather files which work with a representative climate file of the area, whereas the households were monitored during 2018–2019, which demonstrated higher temperature and humidity than prior years (Fig. 3). In addition to the contextual and data constraint, the stochastic user behaviour may cause a difference in monitored data and simulated electricity consumption outputs.

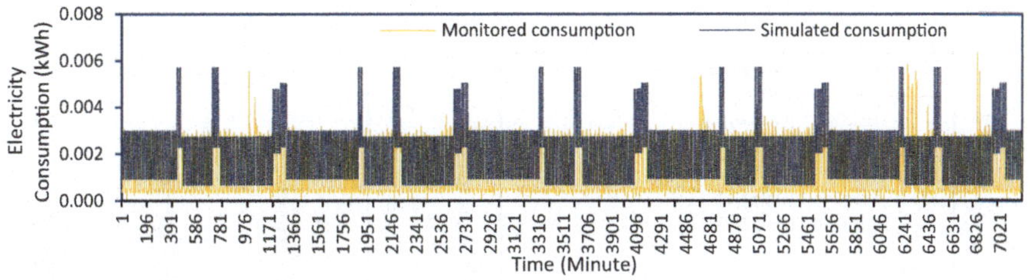

Fig. 4 Minutely electricity consumption profile (monitored and simulated) for a household with fans to retain thermal comfort on 3–9 November 2018

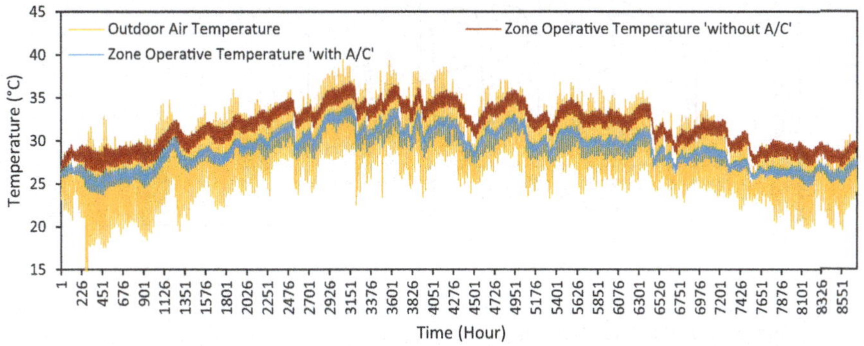

Fig. 5 Comparison of indoor operative temperature with and without A/C (with only fans) in Flat 1: Bedroom and living space

3.3 Indoor Temperature and Aggregated Electricity Demand Due to A/C Usage

The climate analysis showed that residents of Auroville experienced heat stress annually. But the occupants of the households mentioned that they preferred varied indoor temperatures even up to 28 °C [10], which was significantly lower than the outdoor air temperature. Figure 5 showed that during January, February and December, the outdoor temperature stayed close to an acceptable level that mechanical ventilation (predominantly with fans) may provide adequate thermal comfort. However, the fans would not be sufficient for the rest of the 9 months, especially in summer, which may lead to the use of A/Cs for attaining thermal comfort under the high outdoor air temperature and which also resonate with the increase in A/C ownership in Auroville.

In a rough calculation, if the households had high ownership of A/Cs that may significantly increase the aggregated demand for electricity. The aggregated total demand of the whole building community with 15 households with only electric fans and no A/Cs would be 25.69 kWh on 29 April. However, if they all start to use A/Cs, the aggregated total electricity demand increased – approximately 13.7 times – to 352.33 kWh on 29 April.

3.4 Effect of Building Fabric Retrofit on the Indoor Temperature and Cooling Energy Demand

Although all 15 households were simulated for an in-depth analysis on the effect of different building fabrics on its indoor temperature performance in this study, we have selected one flat (Flat 1: Bedroom and living). Also, for analysing the impact of building fabrics on the community's cooling electricity demand, the whole building was considered. Both indoor temperature and aggregated cooling demand were simulated and analysed in two selected days (29 April and 12 May) during the summer period – as 29 April representing approximately the starting of the summer and 12 May for peak summer – under different scenarios. Under the base case (R1W1 wall and roof construction), the average daily indoor temperature was 34.497 °C with only natural ventilation (free running) which may reduce to 31.811 °C (Fig. 6A and B) on 29 April with A/C usage. The indoor air overheated in the room – average daily temperature increased to 1.04% and 1.34% under scenarios 1 and 2, respectively – with only natural ventilation as the R1W2 and R2W3 construction reduced the U-value of the wall and

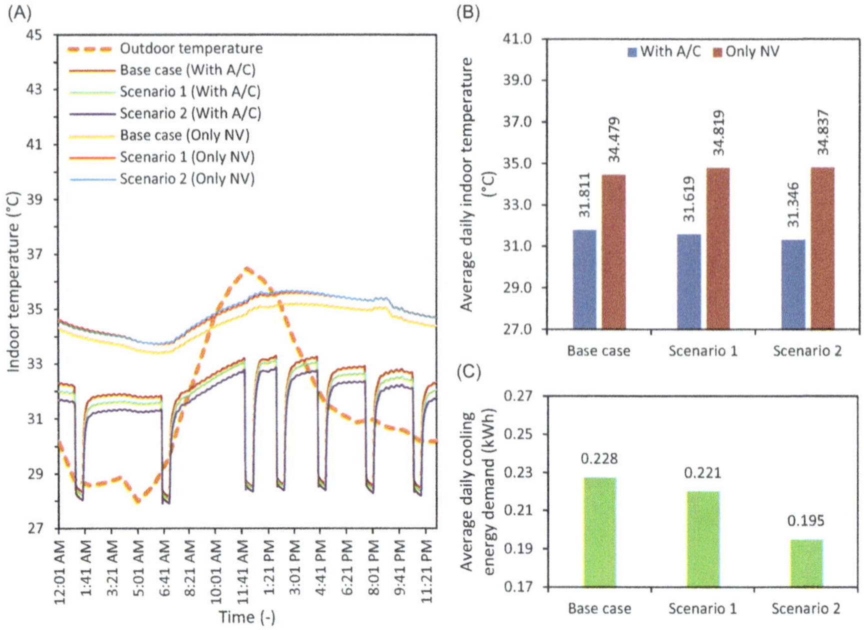

Fig. 6 (**A**) 1-minute indoor temperature: Flat 1 [Bedroom and living space]; (**B**) Average daily Indoor temperature in Flat 1; (**C**) Average daily aggregated cooling energy demand in the building on 29 April

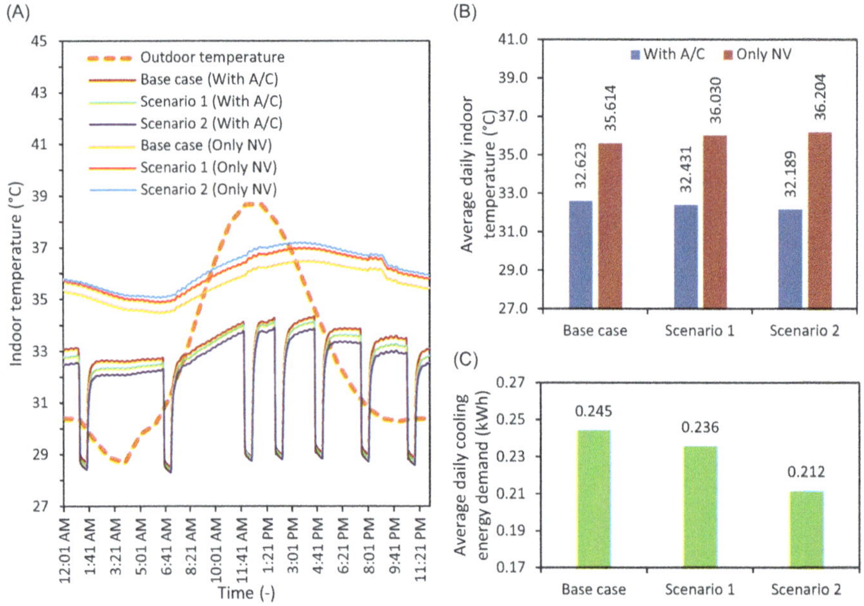

Fig. 7 (**A**) 1-minute indoor temperature: Flat 1 [Bedroom and living space]; (**B**) Average daily Indoor temperature in Flat 1; (**C**) Average daily aggregated cooling energy demand in the building on 12 May

roof. However, the average temperature reduced 0.6% and 1.5% than that of the base case under scenarios 1 and 2, respectively, for 29 April – average outdoor temperature was 31.550 °C with a maximum of 36.500 °C – when A/C was used.

In the case of 12 May, it was a warmer day than that of 29 April with 32.879 °C average daily outdoor temperature and maximum 38.700 °C. In the case of free running building, average daily indoor temperature increased 1.66% and 2.15% than that of base case (35.624 °C) as shown in Fig. 7A and B. The overheating increased significantly due to the increase in outdoor temperature. With the use of A/C, the average daily indoor temperature reduced 1.3% and 1.6% under scenarios 1 and 2, respectively, as compared to base case (32.623 °C). The results from 29 April and 12 May showed a significant overheating risk for buildings with lower insulation in wall and roof construction (scenarios 1 and 2) when operating with only natural ventilation. The building fabric retrofit with better thermal performance (scenarios 1 and 2) only reduces average daily indoor temperature while A/Cs were operational. Therefore, different building fabrics should be used when constructing residential buildings for only natural ventilation or A/C use, rather than using A/Cs in buildings with higher U-value and lower thermal performance which was found to be a common practice.

Use of A/Cs in a residential building – with higher U-value – massively increased the cooling energy demand. The results showed that the average daily cooling energy demand for the whole building was 0.228 kWh and 0.245 kWh on 29 April and 12 May, respectively, under the base case (Figs. 6C and 7C). The cooling energy demand reduced in all the scenarios as compared to the base case. Under scenario 1, average daily cooling energy demand was 3.23% (29 April) and 3.52% (12 May) lower than that of the base case, due to the change in exterior wall material from burnt bricks to masonry blocks. Furthermore, the use of EPS insulation outer direction (Fig. 2), reduced the U-value of the wall and roof construction (Table 2) significantly, which contributed to lowering the cooling demand 14.34% (29 April) and 13.37% (12 May) under scenario 2, as compared to the base case.

As the Indian households have been adopting a high number of A/Cs, the conventional building fabrics with burnt clay brick walls and RCC roof demonstrated a severely higher U-Value, resulting in lower insulation and increased cooling energy demand. Depending on the outputs from our study, it was evident that residential buildings, particularly in the studied area, may need to consider insulating materials in the wall and roof construction to reduce the overheating and cooling energy demand, especially for building with the prospect of users operating A/Cs. The incorporation of high insulation material may negatively affect the indoor temperature if the users use natural ventilation (may include electric fans) for thermal comfort, as it may overheat the indoor air.

The analysis of building fabric retrofit on the cooling demand and indoor temperature offers more scope in testing more material scenarios to examine the effect on the indoor operative temperature as well as the aggregated electricity demand for cooling. This study is in an initial stage of developing calibrated community models to test demand reduction scenarios, which is a part of the Community-scale Energy Demand Reduction in India (CEDRI) project. The next stage of the study would be to create a more robust calibrated model and test more complex demand reduction scenarios with different building fabrics constructions and technologies. Another target would be to extend the community size and validate the simulated results with greater accuracy.

4 Conclusion

The study aimed to develop the simulation model for a community in southern India to investigate the effect of different building fabrics on the indoor temperature and cooling energy demand due to the increased penetration of A/Cs in households for attaining thermal comfort in a climate with high temperature. Initially, climate analysis showed that an unprecedented prevalence of heat stress periods regularly since 2014 may be because of climate change. The high heat stress also may have affected the ownership of A/Cs among the occupants in Auroville as the number of A/C has been increasing.

The results of the simulation study showed that building fabric with insulating materials such as EPS lowered the average daily indoor temperature approximately 1.3–1.5% than that of base case, which consequently would reduce the total cooling energy demand. Also, if all the households start using A/Cs, the daily average aggregated electricity consumption may rise approximately 13.7 times as compared to the use of only electric fans for attaining thermal comfort. Our study showed that the approximately 15% daily reduction in aggregated electricity consumption in the community building in a warm-humid climate of southern India might offer the opportunity of demand reduction by the use of insulating materials in the roof and wall construction. Also, the results suggested that insulation materials should be implemented cautiously, as the use of natural ventilation (may be with electric fans) instead of A/Cs in highly insulated buildings may increase the indoor temperature significantly. The size and nature of the simulation study were not intended for generalizing community-scale conclusions for cooling demands across India or regions from individual simulations. Instead, the study was used to demonstrate a process where measured electricity consumption data combined with the contextual information of community households can be used to enhance understanding on the implications of building fabric with lower U-value on the indoor temperature and aggregated cooling energy demand under increased A/C ownership in a small (residential) building community.

In the future direction of the research, we would like to run aggregated simulation studies on more varied case studies for the community-scale demand reduction with more elaborate and calibrated models. The presented research outputs will also further be used to inform the work of the CEDRI project (funded by DST in India and EPSRC in the UK), where the objective is to combine energy network and occupancy behavioural studies with building modelling to tailor demand reduction strategies for Indian communities.

Acknowledgements This study was funded by the CEDRI project (Grant reference EP/R008655/1) which is funded by the Engineering and Physical Sciences Research Council (EPSRC) and Indian Department of Science and Technology as part of the Newton-Bhabha programme. The authors wish to acknowledge the technical support from Martin Scherfler, Jaswanth Yaddala and others of Auroville Consulting (AVC) for the data collection process.

References

1. Sivak M (2009) Potential energy demand for cooling in the 50 largest metropolitan areas of the world: implications for developing countries. Energy Policy 37(4):1382–1384
2. WB (2019) world development indicators (WDI). [Çevrimiçi]. Available: http://databank.worldbank.org. [Erişildi: 2019]
3. Biardeau LT, Davis LW et al (2020) Heat exposure and global air conditioning. Nat Sustainability 3(1):25–28
4. GoI (2015) India energy security scenarios 2047, Government of India, [Çevrimiçi]. Available: http://iess2047.gov.in. [Erişildi: 2019]
5. IEA (2019) Energy efficiency: cooling, international energy agency. [Çevrimiçi]. Available: https://www.iea.org/topics/energyefficiency/buildings/cooling/. [Erişildi: 2019]
6. IEA (2019) Global air conditioner stock, 1990-2050, International Energy Agency, Paris
7. Sharma V, Rai AC (2020) Performance assessment of residential building envelopes enhanced with phase change materials. Energ Buildings 208:109664
8. Saikia P, Pancholi M, Sood D, Rakshit D (2020) Dynamic optimization of multi-retrofit building envelope for enhanced energy performance with a case study in hot Indian climate. Energy 197:117263
9. Kumar A, Suman B (2013) Experimental evaluation of insulation materials for walls and roofs and their impact on indoor thermal comfort under composite climate. Build Environ 59:635–643
10. Osunmuyiwa OO, Payne SR, Ilavarasan PV, Peacock AD, Jenkins DP (2020) I cannot live without air conditioning! The role of identity, values and situational factors on cooling consumption patterns in India. Energy Res Soc Sci 69:101634
11. Debnath KB, Jenkins DP (2020) Simulation-based assessment of aggregated residential energy demand reduction strategies for cooling in a South Indian community. 5th IBPSA-England Conference on Building Simulation and Optimization 2020, Loughborough.
12. Xu X, Taylor JE, Pisello AL (2014) Network synergy effect: establishing a synergy between building network and peer network energy conservation effects. Energ Buildings 68:312–320
13. Yoon JH, Baldick R, Novoselac A (2014) Dynamic demand response controller based on real-time retail price for residential buildings. IEEE Trans Smart Grid:121–129
14. Al-janabi A, Kavgic M, Mohammadzadeh A, Azzouz A (2019) Comparison of EnergyPlus and IES to model a complex university building using three scenarios: free-floating, ideal air load system, and detailed. J Build Eng 22:262–280
15. Huang S, Wu D (2019) Validation on aggregate flexibility from residential air conditioning systems for building-to-grid integration. Energ Buildings 200:58–67
16. Rothfusz LP, Headquarters NSR (1990) The heat index equation (or, more than you ever wanted to know about heat index), National Oceanic and Atmospheric Administration, National Weather Service, Office of Meteorology, Fort Worth, Texas
17. Luo M, Lau N-C (2019) Characteristics of summer heat stress in China during 1979–2014: climatology and long-term trends. Clim Dyn 53(9-10):5375–5388
18. Modarres R, Ghadami M, Naderi S, Naderi M (2018) Future heat stress arising from climate change on Iran's population health. Int J Biometeorol 62(7):1275–1281
19. Choi G, Lee DE (2020) Changing human-sensible temperature in Korea under a warmer monsoon climate over the last 100 years. Int J Biometeorol 64(5):1–10
20. Opitz-Stapleton S, Sabbag L, Hawley K, Tran P, Hoang L, Nguyen PH (2016) Heat index trends and climate change implications for occupational heat exposure in Da Nang, Vietnam. Clim Serv 2:41–51
21. Fay TH, Hardie KA (2003) Fahrenheit to Celsius: an exploration in college algebra. Math Comput Educ 37(2):207–209
22. Auroville (2017) Auroville in brief. [Çevrimiçi]. Available: https://www.auroville.org/contents/95. [Erişildi: 2019]
23. Debnath KB, Jenkins DP, Patidar S, Peacock AD (2020) Understanding residential occupant cooling behaviour through electricity consumption in warm-humid climate. Buildings 10(4):78
24. NWS (2020) What is Heat Index?, National Weather Service. [Çevrimiçi]. Available: https://www.weather.gov/ama/heatindex. [Erişildi: 2020]
25. Nagy B (2018) Experimented methods to moderate the impact of climate change in Auroville. Ecocycles 4(1):20–31
26. Royapoor M, Roskilly T (2015) Building model calibration using energy and environmental data. Energ Buildings 94:109–120

Analysing Technology Implementation Alternatives in the Portuguese Bioenergy Sector

Ana Andrade, Carlota Duarte, Ricardo Aguiar, and Isabel Cabrita

1 Introduction

The emergence of a low carbon energy system is slow, at the pace of technological discoveries and solid efforts from governments and private investors in capacity deployment. The direction it takes is conditioned by technological innovation and market forces, both driven by push and pull financial instruments administered at a centralized level. Studies are made trying to identify the route to follow, being energy systems modelling one major tool available for this purpose. Other types of studies are those which converge to energy planning. Such studies analyse resource, technical, economic and market potential assessment for each renewable energy technology on the deployment side, or gross domestic product, population and climate trends on the demand side. As a result, energy modelling provides quantitative projections for future energy systems, offering, for example, years-ahead estimations of demand per energy carrier.

The versatility of biomass to be converted into other forms of energy like heat, electricity, biofuels or synthetic fuels makes it a valuable means to contribute to energy sustainability. Several chemical, thermochemical or physical processes allow converting different materials of biological nature into a variety of energetic intermediate or final products. In this regard, in what concerns bioenergy, energy systems planning needs to consider a variety of factors to decide which exact resource to use with which technology and which supply chains to pursue.

This chapter describes an attempt to allocate in an optimized way, the biomass resources available in Portugal (PT) such as forestry residues and oils into a number of bioenergy supply chains, whether they be from technologies already deployed or foreseen to reach maturity. Being biomass a limited resource, its optimal use is of major importance. The outcome quantifies technological paths and resources, according to conversion efficiency or production share criteria and objectives of minimizing imports or costs.

This study focuses on the year 2030. It is considered that identifying best supply chains in the long term can help to improve public policies such as the National Energy and Climate Plan 2021-2030 (NECP) [1, 2].

Besides optimization results, a number of other factors concerning the 2030 context are invoked to further assess the choice of technologies. Diversified stakeholder's views, ranging from socioeconomic to strategic, are articulated with the technological viewpoint through a value-based model obtained from the decision analysis field.

The following section presents the method followed in both approaches. Then, it is described how they were applied to the case under study and the results obtained are presented. Finally, the paper includes a section of conclusions where findings are explained as well as the research lines that the study indicates should be widened.

A. Andrade (✉) · C. Duarte · R. Aguiar · I. Cabrita
Research, Studies and Renewables Division, Directorate-General for Energy and Geology, Lisboa, Portugal
e-mail: ana.andrade@dgeg.gov.pt

© The Author(s), under exclusive license to Springer Nature Switzerland AG 2022

A. Sayigh (ed.), *Sustainable Energy Development and Innovation*, Innovative Renewable Energy,
https://doi.org/10.1007/978-3-030-76221-6_95

2 Methodological Background

This work concerns an optimized choice process between several resources available and several supply chains. It is a complex problem since it involves multiple variables and conflicting selection factors (i.e. criteria). The methodology is drawn from the operations research field, which deals with complex problems through several methods, among them optimization and the weighted sum model, based on the value theory, from multiple criteria decision analysis (MCDA).

2.1 Mathematical Programming

The resource allocation problem seeks to find an optimal allocation of a fixed amount of resources to activities so as to minimize the cost incurred by the allocation, or maximize production, while respecting some constraints. The solution approach to such problem is called mathematical programming or, in this particular case, linear programming [3], since a linear function can describe the objective of the decision-maker: $\sum_{j=1}^{n} cjxj$, where $x_1, \ldots, x_n$ quantities are the n decisions or outputs of each good to produce and c_j a constant related to the objective chosen (e.g. the profit). This problem is further characterized by m linear functions describing the amounts of resources required: $\sum_{j=1}^{n} aijxj$, where $i = 1, \ldots, m$. These amounts may have constraints such as b_i ($\sum^{n} aij\, xj \leq bi$) and positiveness ($x_j \geq 0$).

Such optimization approach has been applied to bioenergy mainly in supply chain design [4, 5].

2.2 Multi-Criteria Decision Analysis

Unlike linear programming where multiple decisions are taken at the same time, in MCDA [6], only one decision is under appreciation. The purpose is to have a set of r alternatives, G_k, being evaluated against u criteria, Cl. For example, to choose one out of the set of alternatives or just to assess the overall performance of each. In the simplest of those methods, the weighted sum model [7], the overall performance of an alternative is $G^T_k = \sum_{l=1}^{u} wlGkl$, for $k = 1, \ldots, r$. That is applicable only when all the data are expressed in exactly the same unit. The weights, w_l, are scale constants and $\sum_{l=1}^{u} wl = 1$, $0 \leq w_l \leq 1$.

Following the value theory [7, 8], in the weighted sum model, the performance of each option is assigned a numerical score on a strength of preference scale and for each criterion. In this way, common preference units are used across all scales. Numerical weights are assigned to define, for each criterion, the relative valuations of a shift between the top and the bottom of the chosen scale. Low scores on one criterion may be compensated by high scores on another. Due to the linear nature of the function, this technique can be used only if the judgements of strength of preference for an option for one criterion are independent from the judgements of strength of preference on another.

MCDA has been applied to the bioenergy field mainly in the choice of technologies or project prioritization and selection [5, 9, 10].

3 Biomass Resource Allocation

3.1 Modelling Assumptions

In spite of decarbonization progresses due to electrification efforts in transports, households, services and industries, the 2030 energy system is expected to still have a large demand for energy carriers in use today such as gasoline, gasoil, natural gas and jet fuels. Those carriers, however, will incorporate products made from biomass through a number of different processes. Biochemical (fermentation, digestion), thermochemical (pyrolysis, gasification, combustion) and chemical (esterification, hydrolysis) methods allow the conversion of biomass resources, with a certain efficiency and if enough stages are added to the process, into either heat, power, fuel or chemicals [11]. This has the format of a resource allocation problem: which types of biomass and in what quantities should be used to make each final product in order to satisfy the respective demand (at least in part) or attain global objectives.

Portugal does not have sustainable bioenergy crops, but does have residual biomass from agriculture, forestry, agro-industry and urbanbiowastes. First-generation biofuels are to be phased out until 2030, so that bioenergy crops do not need to be analysed in this work. Biomass sources that already have competing uses are also not considered for energy conversion. It is, for example, the case of cereals straw, bran or dreche going into animal feed; of rice hush going into aviaries facilities. Fuel wood, pellets and wood chips are assumed to be all used in residential heating. Residues from paper and cork industries yield black liquors are all directed to cogeneration. As a result, only the following types of biomass sources enter in this study: primary forest residues, bush cuttings, pruning residues, coffee grounds, grass cuttings, fruit and vegetables pulp, grape bagasse, wastewaters, organic municipal waste, manure (accounted as municipal waste), animal fat residues and used cooking oils (Table 1).

Biomass potential, that is, the amount of resource available, is a matter of controversy with different studies conveying different estimations [12–14]. This work uses data on total resource mass available as indicated in [14] (Table 1). This work further assumes that biomass potential, whatever the type, will remain approximately constant along the time.

Processes (production technologies) and final products (energy forms or carriers) that may involve biomass in their production process, which were considered for 2030 in the NECP national energy-emissions modelling, are summarized in Table 2.

For each energy product, several processes contributing to its annual production are grouped according to the nature of resource they convert: biological (biomass or biogas-based production technologies); fossil (fossil fuel-based production technologies); and other renewables (renewable energy production technologies apart from those on biomass or biogas). Heat is excluded from the final energy products list because it is related to cogeneration and residential heating (not considered in Table1).

Table 2 presents total annual 2030 production of each final product generated from these types of processes, thus corresponding to the respective total annual demand, specified by the NCEP.

Increasing biomass use only makes sense if it will substitute energy production from fossil-based technologies. So, Table 2 further indicates total annual 2030 production of each final product generated only by both bio- based and fossil production technologies – that is, excluding production from other renewables. That is the target production (or demand to satisfy) in this study.

Total energy available in resources (Table 1) is lower than total energy required at final products (Table 2). For example, the following share of the latter would balance both: gasoil 21,3 TWh (1/2); gasoline 4,3 TWh (1/3); jet fuels 8,5 TWh (1/2); hydrogen 0,3 TWh (1/3); electricity 1,6 TWh (1/6), totalling 36 TWh.

Bio-based technological conversion processes are further detailed, in that transesterification process leads to diesel, fermentation to diesel and gasoline components of advanced biofuels, or hydrogen (dark fermentation), gasification to obtain jet fuels or hydrogen, and anaerobic digestion (AD) to biogas (and then to electricity).

Table 3 presents technology cost and conversion efficiency, estimated for 2030, in relation to each biomass resource considered [11]. Note that the efficiency of a technology varies with the type of biomass resource [11].

Table 1 Biomass resources and respective potential available yearly in Portugal

	Annual mass (ton)	MWh/ton	Energy content (TWh)
Primary forest residues	1,508,941	7.0	10.6
Bushes cuttings	1,025,370	7.0	7.2
Pruning residues	1,778,914	6.0	10.7
Coffee grounds	15,000	5.0	0.1
Grass cuttings	441,000	4.0	1.8
Fruit and vegetables pulp	57,317	3.5	0.2
Grape bagasse	69,840	4.0	0.3
Wastewaters	122,648	2.0	0.2
Organic municipal waste	1,833,000	2.7	4.9
Used cooking oils	20,000	5.0	0.1
Animal fat residues	60,000	5.0	0.3
Total			36.3

Table 2 Energy carriers and respective estimated annual production from several energy conversion technologies active in 2030 in PT

Energy products	Total production in 2030 (TWh)	Other renewable technologies	Bio-based production technologies	Fossil-based production technologies	Bio plus fossil-based production in 2030 (TWh)
Gasoil	42.5	None	1st generation biofuels Advanced biofuels	Fossil fuels	42.5
Gasoline	13	None	1st generation biofuels Advanced biofuels	Fossil fuels	13
Jet fuels	17.4	Synthetic non 'bio'	Advanced biofuels	Oil refinery	17.1
Hydrogen	10	Electrolysis Thermal dissociation	Biomass gasification	None	1
Electricity (power plants)	57	Non-renewable residues (50%) Wind (onshore, offshore) Solar photovoltaic Concentrated solar thermal Hydro (large, small) Wave Geothermal	Woody biomass Biogas Renewable residues (50%) Renewable network gases (10%)	Fossil network gases (90%)	9.4
Total					83

3.2 *Mathematical Programming Runs and Results*

In order to find optimal biomass allocation quantities (decision variables) from the set of resources available (Table 1) to produce gasoil, gasoline, jet fuels, hydrogen and/or electricity (possible constraints), both were related through conversion efficiencies (Table 3). Total quantities converted from each resource were limited to national availability (Table 1). Using Excel Solver software to solve the linear programming problem through the Simplex algorithm [3], various runs were performed in response to several objective functions.

Results of three runs, R1, R2 and R3, are presented in Table 4. The respective objective functions are as follows:

R1. To minimize 'imports', that is, the difference between total energy content of final products and respective total demand (Table 2), without additional restrictions;
R2. To minimize 'imports' while requiring one-tenth of demand for jets, gasoline and diesel to be satisfied;
R3. To minimize both costs and imports (adding objective functions) while requiring approximately one-tenth of demand for jets, gasoline and diesel, one-third of demand for hydrogen and one-sixth of demand for electricity be satisfied.

More runs other than the three above referred were performed if systematic sensitivity analysis was not yet performed. From these, it was verified that maximizing freely the total production conveyed approximately the same results as minimizing 'imports' in R1, but with less resource diversity. Several runs similar to R2 show, for example, that it is possible to satisfy from biomass: one-third of all demand (total production in Table 2) for gasoline through fermentation, but not all of it; one-sixth of electricity demand through biogas combustion, but not all of it; all demand for H2. It was also verified that results are sensitive to changes in the conversion efficiency; however, there is stability in model outcomes to a reasonable extent.

Optimization results (Table 4) inform on detailed resource allocation according to defined objectives, constraints, and parameters. In all cases, all biomass resource available in PT was used. Results show that final products with a demand more difficult to satisfy are gasoil, gasoline and jet fuels due to volume requirements and less alternative production routes. Transesterification for diesel and fermentation for ethanol are routes the model chooses because of technological efficiency and cost.

Electricity was never produced from direct combustion of biomass. Its origin in the results comes from anaerobic digestion and then biogas combustion. Further, it appears only if required by model constraints and because its assumed cost is lower than for direct combustion of biomass.

Due to assumed conversion efficiency, results point to hydrogen being produced by gasification instead of fermentation.

Also due to the assumed cost and efficiency, gasification to generate advanced jet fuels is a preferred route, allocating itself biomass resource in all runs.

Forest-based residues are used to produce jet fuels unless restrictions oblige further diesel or gasoline production. Hydrogen from gasification would use the same type of resource. Pulps, urban biowastes, wastewaters, residues are used in road fuel production, unless electricity production is required.

Table 3 Efficiency (0–1 scale) and cost of conversion technologies

	Transesterification Diesel	AD Electricity	Fermentation Gasoline	Fermentation H2	Combustion Electricity	Gasification H2	Gasification Jet fuels
Primary forest residues	0,30	0,15	0,30	0,05	**0,20**	**0,65**	**0,70**
Bushes cuttings	0,30	0,15	0,30	0,05	**0,20**	**0,65**	**0,70**
Pruning residues	0,30	0,15	0,30	0,05	**0,20**	**0,65**	**0,70**
Coffee grounds	**0,85**	0,15	0,30	0,05	0,12	**0,65**	0,50
Grass cuttings	0,60	**0,20**	**0,50**	**0,10**	0,12	**0,65**	0,40
Fruit and veget. Pulp	0,40	**0,20**	**0,50**	**0,10**	0,08	0,40	0,30
Grape bagasse	0,50	**0,20**	**0,50**	**0,10**	0,08	0,40	0,40
Wastewaters	0,40	**0,30**	**0,50**	**0,10**	0,02	0,30	0,30
Org. municipal waste	0,50	**0,30**	**0,50**	**0,10**	0,02	0,40	0,40
Used cooking oils	**0,85**	0,10	0,10	0,05	0,02	**0,65**	0,70
Animal fat residues	**0,85**	0,10	0,10	0,05	0,02	**0,65**	0,70
Cost (M€/TWh)	30	74	70	78	119	23	45

Table 4 Solutions (TWh) for three objective functions and respective constraints in runs R1, R2 and R3

	Transesterification Diesel			AD Electricity			Fermentation Gasoline			Fermentation H2			Combustion Electricity			Gasification H2			Gasification Jet fuels		
	R1	R2	R3	R1	R2	R3	R1	R2	R3	R1	R2	R3	R1	R2	R3	R1	R2	R3	R1	R2	R3
P. forest res.	0	**7,0**	**7,5**	0	0	0	**2,4**	0	0	0	0	0	0	0	0	**1,5**	0	0	**6,6**	**3,0**	**3,0**
Bushes cutt.	0	0	**6,7**	0	0	0	0	0	0	0	0	0	0	0	0	0	0	**0,5**	**7,2**	0	0
Pruning res.	0	0	0	0	0	0	0	0	**1,8**	0	0	0	0	0	0	0	0	0	**11**	0	0
Coff. grounds	**0,1**	**0,1**	**0,1**	0	0	0	0	0	0	0	0	0	0	0	0	0	0	0	0	0	0
Grass cuttings	**1,8**	**1,8**	**1,1**	0	0	0	0	0	**0,7**	0	0	0	0	0	0	0	0	0	0	0	0
Pulps	0	0	0	0	0	**0,1**	**0,2**	**0,2**	**0,2**	0	0	0	0	0	0	0	0	0	0	0	0
Grape bagas.	**0,3**	0	0	0	0	0	0	**0,3**	**0,3**	0	0	0	0	0	0	0	0	0	0	0	0
Wastewaters	0	0	0	0	0	**0,2**	**0,2**	**0,2**	0	0	0	0	0	0	0	0	0	0	0	0	0
Munic. waste	0	**3,5**	0	0	0	**4,9**	**4,9**	**1,4**	0	0	0	0	0	0	0	0	0	0	0	0	0
Cooking oils	**0,1**	**0,1**	**0,1**	0	0	0	0	0	0	0	0	0	0	0	0	0	0	0	0	0	0
Animal fat res.	**0,3**	**0,3**	**0,3**	0	0	0	0	0	0	0	0	0	0	0	0	0	0	0	0	0	0
Cost (M€/TWh)	48	159	159	0	0	116	240	76	76	0	0	0	0	0	0	23	0	8	769	96	96

4 Technologies Contextual Evaluation

In order to improve our views about the biomass conversion technologies' role in a future energy system, a comparative analysis was additionally performed, taking into account a broader set of criteria, that is, going behind the quantitative assessment in terms of efficiency and cost made in the optimization study. Stakeholders' opinions on that matter were collected and their views were incorporated in this evaluation model.

Alternatives under evaluation are the previous ones (Sect. 2): transesterification to biodiesel (G1); anaerobic digestion to biogas and electricity (G2); fermentation to gasoline substitutes such as bioethanol (G3); dark fermentation to hydrogen (G4); centralized combustion of solid biofuels to electricity (G5); gasification to syngas and hydrogen (G6); and gasification to syngas and advanced biofuels (G7).

The following criteria are considered: expected technology maturity (C1); resource availability and accessibility (C2); contribution to national targets or strategic objectives (C3); contribution of the final product to the security of supply (C4); final product cost (C5); jobs creation and regional development (C6); and conversion efficiency (C7).

Scales were formed in each criterion to compare alternatives in terms of the weighted sum model. Scale levels in C1, C4, C5 and C7 follow usual descriptions in MCDA sustainability assessments. C2, C3 and C6 scales were based on stakeholders' views concerning the national context.

The criterion 'resource availability and accessibility' (C2) was constructed from articulating resource quantity with its geographical distribution. The reason is that both technology deployment and resource availability can be centralized or distributed and only some combinations of these two factors are favourable.

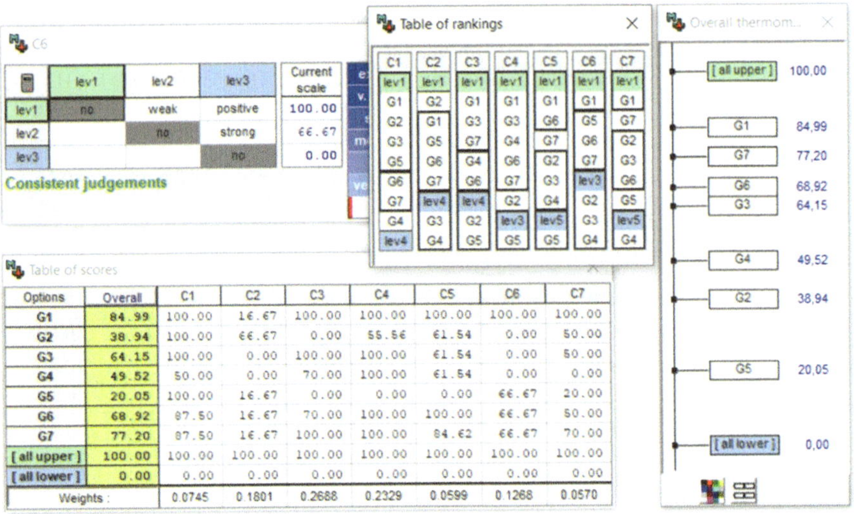

Fig. 1 Software interface showing the performance of alternatives against criteria and their final score

In the criterion 'contribution to national targets or strategic objectives' (C3), the scale was defined in terms of how much the final product helps in reducing the respective imports, or whether it can be obtained from other renewable technologies. The underlying view is that in spite of the 2030 energy system being capable of technically accommodating lower performances in either of these aspects, both are national strategic objectives and their accomplishment is worthy.

In 'jobs creation and regional development' (C6), the scale was constructed on the view that both decentralized and centralized production enables positive socio-economic impacts, in different ways: decentralized is preferred if above a certain "critical mass"; both centralized and decentralized are preferred to a semi- centralized solution with sub-optimal facilities location.

Since these criteria mix quantitative and qualitative performance assessments, value theory is applied so that this performance can be converted into common units of value.

For easiness, M-MACBETH software [6] is used in identifying value scores for criteria levels (e.g. top left in Fig. 1 concerns C6 levels) and weights (bottom left in Fig. 1). Figure 1 exemplifies software interface at different stages of the modelling, including alternatives' performance (top middle and central part of the bottom left board) and its final classification (at right).

A sensitivity analysis was not performed systematically; however, further trials in this regard indicate that results are reasonably stable.

Results of this model point to higher deployment of transesterification to diesel (G1), gasification to jet fuels (G7) and gasification to hydrogen (G6) technologies. These are strong in the more valued criteria, contribution to national targets or strategic objectives (C3) and to supply security (C4). They are followed by fermentation to gasoline technology (G3), similar in C3 and C4, but weaker in resource availability (C2), and regional development (C6).

The least preferred alternatives revealed to be fermentation to hydrogen (G4) due to its low conversion efficiency; anaerobic digestion to electricity (G2); and combustion to electricity (G5). The last two, G2 and G5, involve technologies with low cost-efficiency and reduced strategic interest, because they generate a final product – electricity – which, in 2030, can be plentily obtained from other renewable sources instead of biomass.

5 Conclusions

The main contribution from this work is evidencing how the knowledge of conversion efficiency of biomass technologies allows building an optimization model that specifies detailed resource allocation. This conversion efficiency for a certain technology has to be assessed on a case-by-case basis, and not as a typical or average value independent of the type of resource. This kind of data is poorly known and deserves an enlarged research effort. Knowing better which biomass resources are available at a national level is a further requirement for the quality of the output.

The optimization tool created revealed to be responsive and robust, providing a level of detail useful to plan how different types of biomass can be best used. It can be run over current statistics, or over long-term estimations. Here, estimations were collected from the NECP energy-emissions support modelling, for the year 2030.

Optimization results point as a priority to the use of forest residues in gasification and to produce advanced biofuels or hydrogen. The remaining biomass resources were channelled to advanced biofuels, through fermentation and transesterification. Electricity routes were the least interesting due to unfavourable costs and conversion efficiency.

This allocation priorities weren't substantially changed when using other criteria, such as resource location or regional development. Instead, they were reinforced as a source of security of supply and as a significant contributor towards fulfilment of strategic objectives.

References

1. Resolution of the Council of Ministers ("Resolução do Conselho de Ministros") number 53/2020
2. EC (2019) Portugal – National Energy and Climate Plan 2021 – 2030 (NECP 2030)
3. Hillier FS, Lieberman GJ (2010) Introduction to operations research, 9th edn. McGraw-Hill International
4. Yue D, You F, Snyder SW (2014) Biomass-to-bioenergy and biofuel supply chain optimization: overview, key issues and challenges. Comput Chem Eng 66:36–56
5. Atashbar NZ, Labadie N, Prins C (2018) Modelling and optimisation of biomass supply chains: a review. Int J Prod Res 56(10):3482–3506
6. Greco S, Ehrgott M, Figueira JR (eds) (2005) Multiple criteria decision analysis: state of the art surveys. Springer
7. Fishburn PC (1967) Additive utilities with incomplete product set: applications to priorities and assignments. Operations Research Society of America (ORSA) Publication, Baltimore
8. Gal T, Stewart T, Hanne T (eds) (1999) Multi-criteria decision making. Springer
9. Scoll J, Ho W, Dey P (2012) A review of multi-criteria decision-making methods for bioenergy systems. Energy 42:146–156
10. Özdemir ED, Härdtlein M, Jessen T, Zech D, Eltrop L (2011) A confusion of tongues or the art of aggregating indicators - reflections on four projective methodologies on sustainability measurement. Renew Sust Energ Rev 15:2385–2396
11. Ptasinski KJ (2016) Efficiency of biomass energy: an exergy approach to biofuels, power, and biorefineries. Wiley
12. Bio-based Industries Consortium (2018) Mapping the potential of Portugal for the bio-based industry
13. Ruiz P, Sgobbi A, Nijs W, Thiel C, Longa FD, Kober T, Elbersen B, Hengeveld G (2015) The JRC-EU-TIMES model: bioenergy potentials for EU and neighbouring countries. JRC Science for Policy Report, European Commission
14. Resolution of the Council of Ministers ("Resolução do Conselho de Ministros") number 163/2017

Defining Critical Parameters in a Luminescent Downshifting Layer for PV Enhancement

Maria F. Ayala Barragan, Subhash Chandra, Bill Cass, and Sarah J. McCormack

1 Introduction

While there are several loss mechanisms within a solar cell, an important aspect to consider is the low efficiency of solar cells when absorbing high energy photons. Luminescent downshifting layers (LDS) is a novel way of improving the efficiency of solar cells under high energy ultraviolet radiation [1–3]. LDS consists of a layer that absorbs high energy photons and converts them to a lower energy where the solar cell is able to absorb and convert them to electricity more efficiently. The LDS process is enabled by installing a thin polymer layer on top of the solar cell, thus avoiding the need for performing any internal structural changes on solar cells which could hinder their performance or damage them. Figure 1 is a representation of the optical processes involved in the LDS layer as well as the losses that occur due to the layer on top of the solar cell.

2 Methodology

The luminescent and host materials used in LDS layers have several characteristics that must be carefully selected in order to achieve the maximum potential of the layer. Commercial dyes in the Lumogen -F series. Dyes Violet 570, Orange 240, Red 300, Yellow 083 and Yellow 170 were considered for this analysis. Lumogen-F dyes are produced by (BASF). Dyes O240, R300, Y083 and Y170 are Perylene based, while V570 is Napthalimide based. While it has been shown that Perylene-based dyes have great stability, both types of dyes have high heat stability, which is necessary in dyes used in any type of solar cell or solar concentrator [5]. Lumogen-F dyes are easily incorporated into polymer sheets and have strong luminosity. These dyes are easily soluble in toluene, acetone, isopropyl alcohol, ethyl acetate and xylene, among others [6].

Using data from [7], the normalised absorption and emission spectra of the Lumogen-F series organic dyes were obtained. The data were then used to calculate the Stokes shift and the overlap between the absorption and emission spectra in order to analyse which dye performs best according to these characteristics. Figure 2 presents the normalised absorption and emission spectra for all Lumogen dyes (R300, Y170, V570, O240 and Y083) along with the overlap and the Stokes shift (SS) (the difference between positions of the maxima of the absorption and emission spectra).

Figure 3 depicts the values of the radiative overlap, SS and PLQY of every dye. R300 has the highest radiative overlap (RO) and one of the highest SS. V570 shows excellent characteristics with the smallest RO and the largest SS of all the dyes. Organic dyes do not have a large SS leading to overlap between the absorption and emission bands. This characteristic has allowed for rare earth phosphors (REPs) to be used by taking advantage of their sharp spectra bands. When comparing QDs to organic dyes, QDs absorb in narrow bands generally localised where the SR of the solar cell is low and emit where the EQE of cells is high. Nevertheless, organic dyes have higher PLQY than QDs ultimately being able to convert the energy absorbed more efficiently than QDs. Low PLQY in commercial QDs renders them a less attractive luminescent material for LDS layers.

M. F. Ayala Barragan · S. Chandra · B. Cass · S. J. McCormack (✉)
Department of Civil, Structural and Environmental Engineering, School of Engineering, Trinity College, Dublin, Ireland
e-mail: mccorms1@tcd.ie

© The Author(s), under exclusive license to Springer Nature Switzerland AG 2022

A. Sayigh (ed.), *Sustainable Energy Development and Innovation*, Innovative Renewable Energy,
https://doi.org/10.1007/978-3-030-76221-6_96

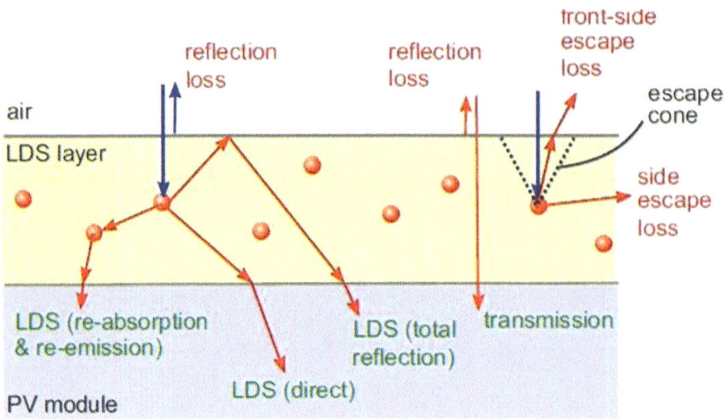

Fig. 1 LDS optical process and losses [4]

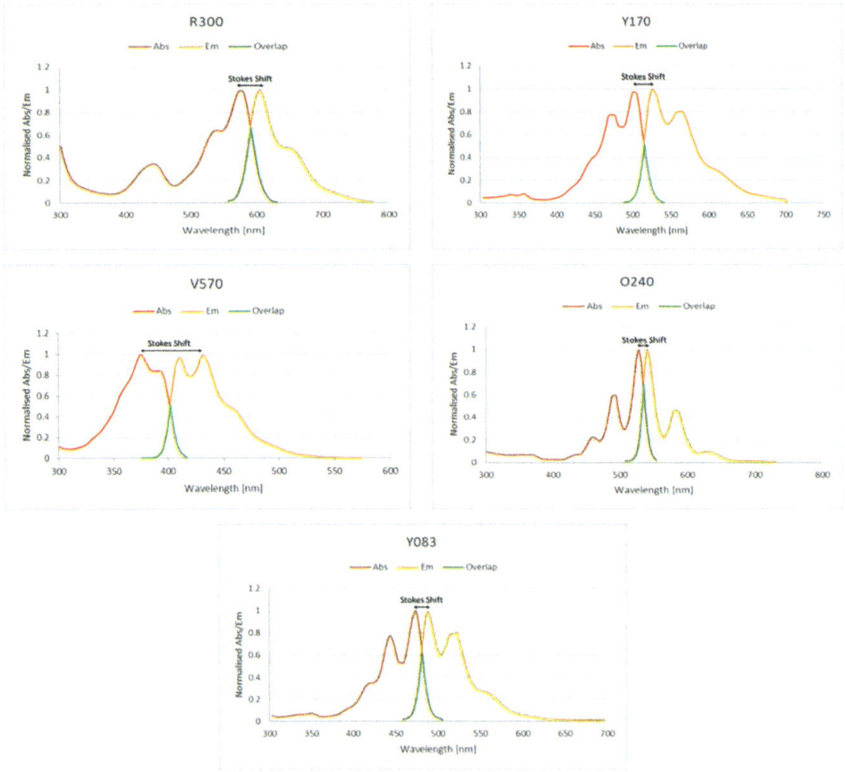

Fig. 2 Absorption, emission and overlap of dye R300, Y170, V570, O240 and Y083

In Table 1, the characteristics of organic dyes as a luminescent material are explored and organised in a colour- coded matrix based on their importance and effect on the performance: Green = Excellent characteristic, Yellow = Satisfactory characteristic; Red = Undesirable characteristic.

The above decision matrix was created by analysing solely the characteristics of the Lumogen-F series of organic dyes as luminescent materials. Each characteristic was ranked independently from each other and based on the literature, it was established whether it was an acceptable characteristic or not. It is clear, based on Table 1, that dyes V570 and Y083 are the dyes which show the best characteristics. Unfortunately, dye V570 has been removed from the market and is not available for purchase anymore. As previously discussed, dye Y083 had an excellent performance due to its properties showing satisfactory absorption and emission spectra, high PLQY, and low RO. Nevertheless, even when luminescent materials show excellent properties previously discussed, the SR of a solar cell plays an important role in the performance of the LDS layer. If the emission of the dye falls within the poor SR range of the cell, the layer will not have a notorious improvement in its EQE and could even provoke further losses due to reabsorption.

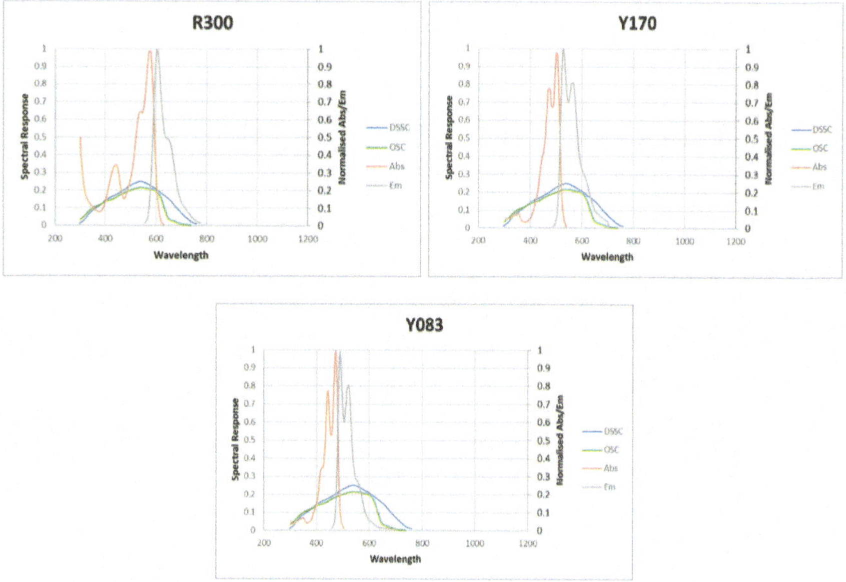

Fig. 3 PLQY, radiative overlap and Stokes shift of Lumogen dyes

Table 1 Lumogen dye performance matrix

	Max Abs (nm)	Max Em (nm)	Stokes Shift (nm)	Radiative Overlap (%)	PLQY (%)
V570	375.92	432.98	57.06	4.62	100
O240	527.77	540.66	12.89	7.86	100
R300	579.81	605.47	25.66	14.16	100
Y083	473.19	488.76	15.57	7.67	96
Y170	500.82	526.99	26.17	7.20	98

In Fig. 4, the SR of different solar cells is plotted against the absorption and emission of each of the Lumogen-F dyes. The solar cells analysed and plotted are c-Si, mc-Si, CdTe, CIGS, DSSC, OSC and gallium arsenide (GaAs). Some dyes may have great ASM and/or ESM for one type of cell, but poorly match another. Such is the case of dye V570 with Si solar cells and CdTe cells. While dye V570 has good ASM with Si cells, it is better with CdTe cells than with either mono or poly-crystalline Si. However, V570 has a poor ESM, as it emits at a range that falls where the SR of CdTe cells is still poor.

Similarly, there are cells that are not compatible with the Lumogen-F dye series, as all dyes absorb in a range where the cell has a high spectral. Response, thus adding a layer would decrease the EQE of the cell. From the graphs in Fig. 4, it can be seen that the SR of DSSC and OSC illustrates how dyes may not be the best luminescent material for this type of cells, as they absorb where the SR of the cell is close to its highest peak.

The maximum values of absorption and emission may give a sense of how well a dye matches a cell, but it is not enough to analyse how compatible the dye is with a type of cell. For this, the complete absorption and emission profiles of each dye were plotted against five types of cells (c-Si, mc-Si, CdTe, CIGS, DSSC, OSC and GaAs) in order to calculate the overlap each spectrum has with the spectral response. The percentage overlap between absorption or emission and the SR of a solar cell was calculated to better visualise the potential behaviour of the dye with the cell.

When analysing absorption and emission spectra, it is expected that the absorption does not interfere with the solar cell absorbing photons, that is, the dye absorbs in a range where the cell has poor to no SR. At the same time, the same dye has to emit the downshifted photons back to the cell at a range where the SR is at its maximum. The overlap between absorption and SR should be low in order to avoid parasitic absorption. Contrarily, the overlap between emission and SR should be high, meaning that the downshifted photons have a higher chance to be absorbed and be efficiently used by the solar cell.

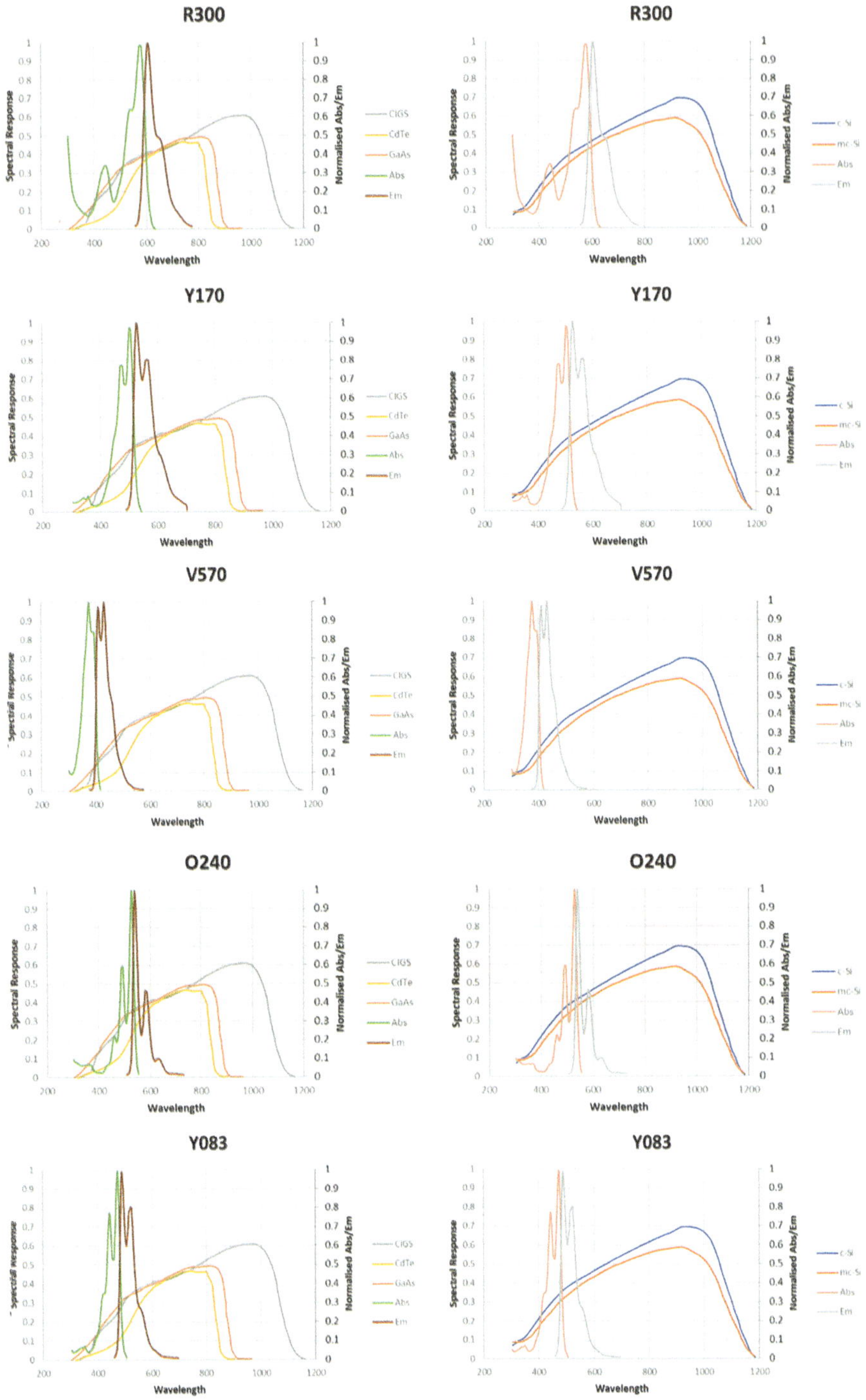

Fig. 4 Absorption and emission profiles with spectral response

Dye V570 was an excellent option to use with Si solar cells due to its absorption in the UV region as seen in Fig. 4 above. However, the emission is still far away from an optimal SR of other solar cells, making it less advantageous for solar cells such as CdTe. Dye R300 does not have an optimal absorption overlap, as its highest peak starts far in the visible range, where most solar cells already have a good absorption. In theory, it could work well for solar cells with high SR in the late visible spectrum such as CIGS, but its absorption interferes with the SR potentially causing PA. The graphs in Fig. 4 depict how dye R300 absorbs at ranges where the solar cell SR is already high and emits either at a good wavelength where the SR is high as seen for CIGS or emits where the SR is already declining as seen for OSC. Similarly, dyes Y083 and Y170 have been

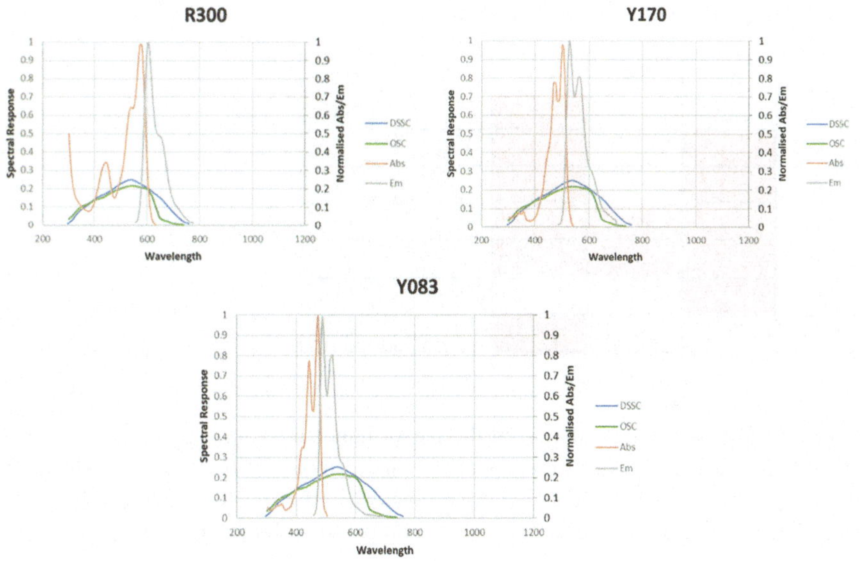

Fig. 4 (continued)

Table 2 Lumogen dye absorption and emission overlap

Absorption Overlap (%)	V570	O240	R300	Y083	Y170
c-Si	15.13	37.94	75.63	32.18	40.95
CIGS	5.2	32.03	61.68	22.34	30.56
GaAs	8.25	32.71	65.01	26.51	34.73
OSC	10.92	25.85	48.06	21.7	27.07
DSSC	9.87	27.71	50.63	22.43	28.38
CdTe	2.72	19.96	42.11	10.02	14.77
mc-Si	12.68	37.11	70.89	28.75	37.35

Table 3 Lumogen dye absorption and emission overlap

Emission Overlap (%)	V570	O240	R300	Y083	Y170
c-Si	28.92	33.64	54.90	39.34	52.65
CIGS	22.22	31.83	50.02	35.54	48.83
GaAs	24.47	31.49	50.09	36.14	48.32
OSC	19.53	22.47	13.86	26.35	28.33
DSSC	20.51	24.50	23.06	28.72	34.51
CdTe	11.43	28.56	48.92	25.80	41.80
mc-Si	25.71	32.44	52.93	36.92	50.08

mentioned in the literature as dyes with good EQE responses when used in LDS layers. Dye O240 has a small SS and an absorption far in the visible range which could cause PA in the solar cell if used as luminescent material. It is especially not the best choice for OSC and DSSC, as its absorption spectrum hinders the SR of these cells as seen in the graphs shown.

The tables show the percentage overlap between the dye spectrum and the SR of the cell. Table 2 only considers absorption values while Table 3 only considers emission values. Table 4 shows a merged version of Tables 2 and 3 with its own colour code. It does not contain values since they are represented in the tables shown. This table is a visual summary of Tables 2 and 3. No dye has an optimal overlap for both absorption and emission. Thus, if a dye had an optimal overlap for either absorption or emission, it was considered as optimal (green) in the final table. If both absorption and emission are sub-optimal, then in Table 4, it was considered as sub-optimal (yellow). Finally, if one of either absorption or emission was non-optimal, regardless of the classification of the other value, then in Table 4, it was considered non-optimal (red) as well.

For absorption: Green = 0–25% (optimal percentage overlap); Yellow = 25–50% (sub-optimal percentage overlap); Red = 50% < (non-optimal percentage overlap). For emission: Green = 50% < (optimal percentage overlap); Yellow = 50–25% (sub-optimal percentage overlap); Red = 25–0% (non-optimal percentage overlap).

Table 4 Lumogen dye performance decision matrix

These matrices provide an estimated representation of the performance dyes have on different solar cells and are not an absolute evaluation of their behaviour. They confirm the analysis made on dye V570 being a good match for c-Si but incompatible with CdTe. With these matrices, it is clearer to compare the SR to the spectrum of each dye. The analysis was made based solely on the absorption and emission spectra of the dyes compared with the SR of different solar cells.

3 Conclusion

After thoroughly analysing the characteristics of organic dyes as luminescent materials, it is concluded that organic dyes fulfil important characteristics for luminescent materials for LDS such as high PLQY, small RO, absorption in or close to the UV range and emission in the visible spectrum. Nevertheless, due to their low SS, organic dyes do not entirely align to the SR of solar cells as ESM and ASM are characteristics that depend on the EQE of the solar cell. The concentration of the dye in the layer also plays an important role in the EQE of the cell with an LDS layer. A highly concentrated layer may cause PA and lower the EQE of the cell, while a layer with a small concentration of dye may not be able to absorb and downshift enough photons to change the EQE of the cell. No dye will match all necessary characteristics nor will it be a perfect match for all types of solar cells. It is the combination of all their characteristics that make them great luminescent materials and an ideal choice for LDS layers.

References

1. Klampaftis E, Ross D, McIntosh KR, Richards BS (2009) Enhancing the performance of solar cells via luminescent down-shifting of the incident spectrum: a review. Sol Energy Mater Sol Cells 93:1182–1194
2. Ahmed H, Doran J, McCormack SJ (2016) Increased short-circuit current density and EQE of silicon and dye sensitised solar cells through plasmonic LDS layers. Sol Energy 126:146–155. https://doi.org/10.1016/j.solener.2016.01.003
3. Ahmed H, Doran J, McCormack SJ (2017) Plasmonic luminescent down shifting layers for the enhancement of CdTe mini-modules performance. Sol Energy 141:242–248
4. Rothemund R (2014) Optical modelling of the external quantum efficiency of solar cells with luminescent down-shifting layers. Sol Energy Mater Sol Cells 120:6
5. Ross D, Alonso-Alverez D, Klampaftis E, Fritsche J, Bauer M, Debije MG, Fifield RM, Richards BS (2014) The impact of luminescent down shifting on the performance of CdTe photovoltaics: impact of the module vintage. IEEE J Photovoltaics 4:8
6. BASF, (1997). Lumogen F Technical Information *Colorants and Graphic Systems,* Alonso-Alvarez, D., Ross, D. & Richards, B. S., (2012). Luminescent down-shifting for CdTe solar cells: A review of dyes and simulation of performance. *8th IEEE Photovoltaic Specialists Conference Photovoltaic Specialists Conference (PVSC).* Austin, TX, USA
7. Crevant C, Lucchesi C, Paire M, Guilmmemoles J-FO (2017) LDS layers using organic dyes and quantum dots. Phys Status Solidi C 14:13–20

Characterisation of Luminescent Downshifting Layer Effects on Silicon Solar Cells

Bill Cass, Subhash Chandra, Arunima Sethi, and Sarah J. McCormack

1 Introduction

While a base Si PV cell can be limited in its ability to absorb photons in the spectral wavelength region between 300 and 560 nm [1–3], the addition of a luminescent downshifting layer allows the higher energy photons found outside of the spectral range in which the Si PV not absorbs efficiently, to be converted to lower energy photons that can successfully interact with the cell underneath the LDS layer (Fig. 1).

The LDS is a thin polymer film doped with a luminescent dye that is applied to the PV cell. The film can increase the capacity for absorption of photons within certain wavelengths, depending on the wavelength of the dye used. While an LDS film can affect the overall efficiency due to parasitic absorption from the host materials, emission at less than unity, additional interface will increase reflection losses [4]. The thickness of the film can be affected by many different attributes of the polymer solution-making process and fabrication methodology. The areas that will be focused on are the dye used, the polymer solution preparation process, speed in which the polymer can mix, the ratio of solvent to polymer and volume of polymer mixture used in the application. The dyes used were Lumogen F Violet570 (V570) produced by BSAF and Methyl Orange produced by Sigma Aldrich.

2 Methodology

The LDS must be prepared using a base PMMA polymer stock solution, which is the solution used throughout the process to avoid any possible variation or undesired variable to occur, mixed with a polymer of choice, again with the same prefix. Chloroform was the solvent chosen to prepare PMMA stock solution to fabricate the LDS film. The solution was prepared by mixing, the Poly(methyl methacrylate) (PMMA) from Sigma Aldrich in the chloroform solvent. The ratio of the solution's concentration (PMMA: chloroform solution) mix is the first process to affect the thickness of the layer. A ratio of 3.83 g PMMA: 10ml of chloroform was first used. This was determined through smaller scale testing on 2 × 2 cm cells through spin coating. Batches of 11.49 g of PMMA and 30 ml of chloroform were made to allow multiple tests of the same polymer solution and produce a more consistent layer; if a single amount of polymer solution was made, the areas that affect the thickness are more exaggerated. These batches were mixed using a magnetic stirrer at 300 rpm for 45–60 mins. When the PMMA was fully dissolved, with no particles floating in solution, a clear and viscous fluid remained. The polymer solution was then mixed with a dye for downshifting depending on dye wavelength. For the dye to be added to the polymer mixture, it was made in a stock solution. This was comprised of chloroform of 15ml and 1 %wt of V570. 0.2 %wt of the dye was used in polymer mixture. The LDS layer with V570 dye is shown in Fig. 2.

A syringe of 10 ml was used to apply the polymer solution to a 12 × 12 cm Si solar cell. A spin coater (G3P-12 Systems) technique was used to coat the polymer solution onto the cell at a max speed of 2000 rpm, a ramping time of 25.5 seconds and a dwell time of 180 seconds. The spin coater allows for a more reliable and repeatable results than drop-casting. The

B. Cass (✉) · S. Chandra · A. Sethi · S. J. McCormack
Solar Energy Applications Group, Dept of Civil, Structural and Environmental Engineering, Trinity College Dublin, Dublin, Ireland
e-mail: bicass@tcd.ie

© The Author(s), under exclusive license to Springer Nature Switzerland AG 2022

A. Sayigh (ed.), *Sustainable Energy Development and Innovation*, Innovative Renewable Energy,
https://doi.org/10.1007/978-3-030-76221-6_97

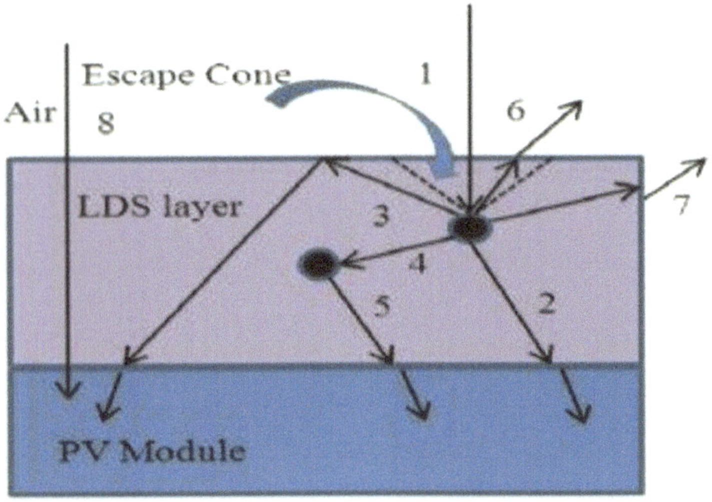

Fig. 1 Diagram of PV modules with LDS layer on top [1]

Fig. 2 LDS layers with V570 dye

purpose of this process ensures that the polymer solution does not begin to polymerise before it can be applied for to the PV cell. The LDS layers produced as shown in Fig. 2 was 20μm. Figure 3 below shows the resulting external quantum efficiency (EQE) of the cells with an LDS layer using V570 applied. It shows that the concentration of 0.2%wt was slightly more efficient compared to the other concentrations. Not all dyes were more effective than the bare cell, however, as shown herein, due to the dye absorbing light and not allowing the higher energy photons to be absorbed into the cell under the cell (in the case of 0.3 and 0.4%wt) or not being able to aid in absorbing higher energies (as seen with 0.1%wt) causing more reflection to occur on the surface of the cell.

The speed in which the spin coater would operate had a varying effect on the LDS layer being coated. The final speed was 2000 rpm, if the polymer mixture is applied at a lower speed, the polymer solution would start to polymerise quickly. This led to the LDS not fully covering the cell underneath causing a decrease in efficiency. This was due to the layer being too thick (>20μm) on the section covered, not allowing light to reach the cell underneath, while the uncovered section would only operate as the 'bare' cell in Fig. 3. Once the layer has been made, it must be left to dry for 3–4 hours (more if possible) to make sure that the layer has fully dried evenly across the PV cell. Figure 4 below shows the absorption and emission ranges of the V570 dye.

Using the same recipe as mentioned previously with the same dye concentration, speed used and volume of solution, a lower ratio of 2.3g of PMMA to 10ml of chloroform would reduce the layer to 5 μm, a reduction of 75%. As mentioned before, a thinner layer would have no absorption benefit but cause extra reflection from the cell (Fig. 5).

Figure 5 shows the effect of change in volume of polymer solution used. The left sample used 15 ml of polymer solution while the right sample used 10 ml of solution.

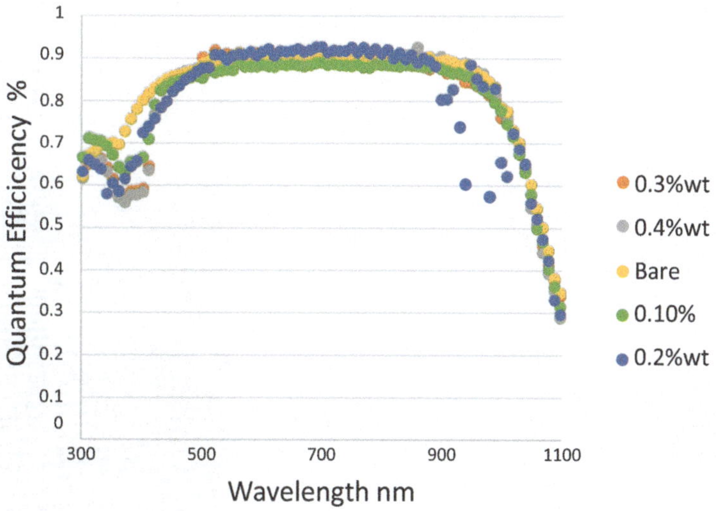

Fig. 3 External quantum efficiency of V570 dye on solar cell

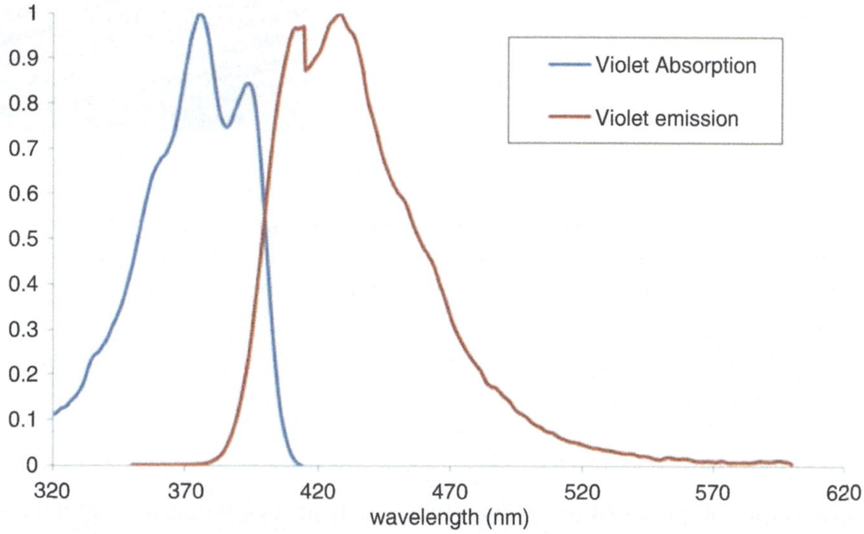

Fig. 4 Absorption and emission of V570 Lumogen dye

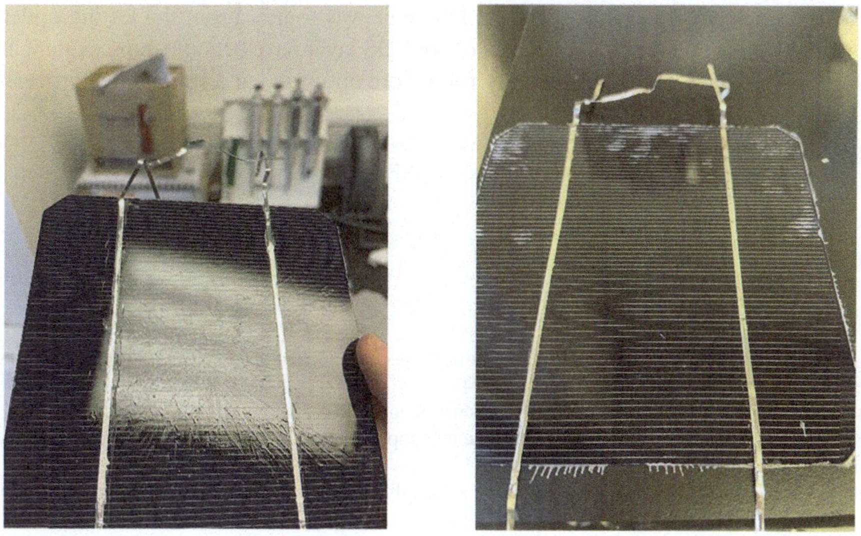

Fig. 5 Effect of volume difference of layer on solar cell

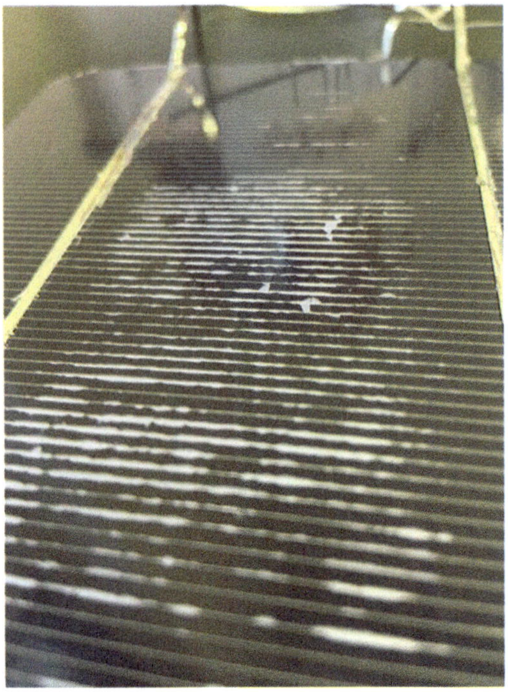

Fig. 6 LDS layer starting to degrade overtime, peeling from base cell

The polymer solution is applied from the syringe onto the cell before being spun. If the solution is not applied evenly and efficiently, the layer begins to degrade overtime or due to adverse weather effects. The layer can begin to peel and lift from the centre of the cell. This affect can take a long time to present itself and can be missed when applying solution if enough care is not taken, as seen in Fig. 6.

3 Conclusions

From repeated testing across different concentrations of dye, ratios of polymer to solution, process of mixing and application process, the characteristics that a LDS layer can demonstrate such as thickness and adhesion can have a varying effect on the efficiency of the base solar cell underneath the layer. During the testing of multiple layers, a key component was be found in the repeatability of the PMMA stock solution. Any variation in recipe or methodology resulted in a layer with different properties, either in thickness, PV cell coverage or quality of the LDS layer. However, through those different variations, a recipe that produced a consistent thickness, coverage and clean LDS layer was found. From the testing undertaken, the areas in production that have a notable impact on the characteristics of the LDS layer have become more evident and allow for further testing with different luminescent dyes and an opportunity for a more optimised and streamlined process.

Acknowledgements The authors would like to acknowledge the European Research Council grant PEDAL: (639760), H2020 IDEAS (815271) and support funding from Science Foundation Ireland (SFI) and insights from the PEARL PV COST Action Network.

References

1. Klampaftis E, Ross D, McIntosh KR, Richards BS (2009) Enhancing the performance of solar cells via luminescent down-shifting of the incident spectrum: a review. Sol Energy Mater Sol Cells 93:1182–1194
2. Gilligan S (2019) Material characterization of luminescent down-shifting layers for solar cells application (November)
3. Ahmed H, McCormack SJ, Doran J (2016) External quantum efficiency improvement with luminescent downshifting layers: experimental and modelling. Int J Spectrosc 2016:1–7
4. Rahman W, McCormack S (2016) Modelling and characterization of metal nanoparticles for plasmonically enhanced luminescent devices for solar cells application

Design and Characterization of a Roof-Mounted Compound Parabolic Concentrator

Anita Ortega, Subhash Chandra, and Sarah J. McCormack

1 Introduction

Photovoltaics convert the solar radiation to electrical energy. The efficiency and power output depend on the solar radiation intensity which reaches the solar cell surface. The amount of solar radiation over the solar cell surface can be increased using concentrators [1]. PV concentrators can be classified into four types: lens concentrator, mirror concentrator, reflector concentrator and static concentrator [2]. An example of reflector concentrator is the compound parabolic concentrator. In this type of concentrator, a large portion of the reflector area can be eliminated without reducing the concentration ratio. Also, they have the potential to reduce the electricity production cost by concentrating the incoming light into a smaller area of PV [3]. However, the temperature in the compound parabolic concentrator becomes an important factor which needs consideration. In this study, an asymmetric compound parabolic concentrator has been designed, constructed and experimentally characterized for roof integration in Dublin, Ireland. In order to reduce the quantity of reflector material, the CPC was truncated by 70% leading to a geometrical concentration ratio of 1.83 as shown in Fig. 1. The system was constructed based on a previous optical performance analysis using race trace technique [4]. The physical and geometrical characteristics are shown in Table 1.

2 Design, Manufacture and Assembly of the CPC

The CPC was composed of a PV absorber, reflector, reflector supports, thermocouples, aperture cover and frame. The materials selection, design and fabrication for each component of the CPC system are described in the following section.

2.1 Reflective Material Employed for CPC System

Aluminium foil with a reflectivity average of 0.76 and 1 mm thickness provided by Easygrow company was chosen as the reflector of the CPC system due to its low cost and acceptable reflectivity properties.

A. Ortega (✉) · S. Chandra · S. J. McCormack
Solar Energy Applications Group, Department of Civil, Structural and Environmental Engineering, School of Engineering, Trinity College, Dublin, Ireland
e-mail: ortegaa@tcd.ie

© The Author(s), under exclusive license to Springer Nature Switzerland AG 2022

A. Sayigh (ed.), *Sustainable Energy Development and Innovation*, Innovative Renewable Energy,
https://doi.org/10.1007/978-3-030-76221-6_98

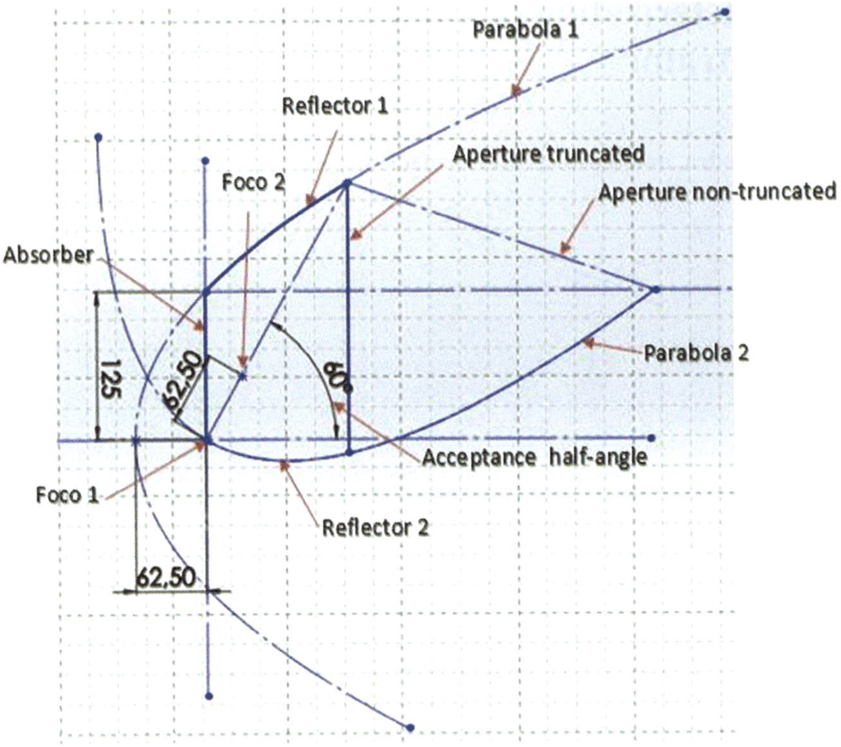

Fig. 1 Parabolas which form the CPC system with an acceptance half angle of 60° generated in SolidWorks

Table 1 Physical and geometrical properties of the CPC

Acceptance-half angle	60°
Absorber (mm)	125
Refl. 1 (mm)	155.34
Refl. 2 (mm)	128.54
Aperture (mm)	228.43
Geometrical Concentration ratio	1.83
Truncation Refl. 2 (%)	70.02

2.2 Design and Material Selection for Reflector Support and Back Plate

The reflector supports and back plate were designed using aluminium. Two designs of reflector support called reflector support south and reflector support north were designed and manufactured from aluminium material with 25 mm thickness in order to provide strength to the bond and bending surface. For each reflector, six pieces were designed using Autocad and SolidWorks, and those designs were transferred to the CNC machine in the workshop at the Mechanical and Manufacturing Department at Trinity College Dublin. For easy assembly, each reflector was fixed to the back plate using two screws M6XØ11 mm. The details of both reflector designs are shown in Fig. 2a. A back plate was designed from aluminium with 6 mm thickness. The dimensions of the back plate were 706 × 525 mm and forty-six holes of Ø11 mm were machined to fix the support reflectors, frame and foam. The detailed design of the aluminium back plate is shown in Fig. 2b.

2.3 Solar Cell Selection and Interconnection

Monocrystalline bare solettes solar cells, 125 mm^2, manufactured by SunPower were selected for the CPC system and PV system. According to technical specification, the solar cell dimensions are 125 × 125 mm, with an efficiency of 17.5%, grade A, short circuit current of 5.15 A, open circuit voltage of 0.515 V and maximum power output of 2.941 W. Four solar cells

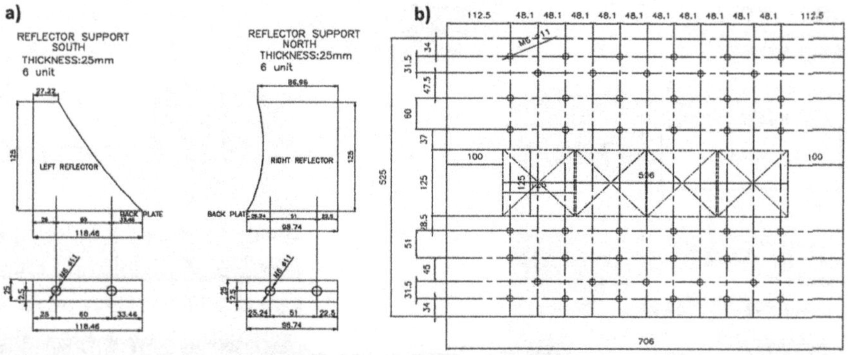

Fig. 2 Design details in Autocad. (**a**) Reflector support (b) Aluminium back plate

in series connection were selected for the CPC system. For the solar cell interconnection, two copper PV ribbons with 2 mm and 5 mm width and 0.2 mm thickness were used. For the soldering process, a leaded solder wire with alloy metal Sn-43Pb-14Bi (Almil Serie: KR-15) of 0.65 mm diameter was used with the soldering iron heated to 260 °C. A flux pen ECO-WORTHY was used to mark the flux of the copper ribbon in order to improve the soldering performance.

2.4 Thermocouple Selection and Characterization

Twenty "K" type thermocouples (chromel-alumel) were used for thermal analysis of the CPC and PV systems. All thermocouples were previously tested all together at 21 °C, 40 °C, 50 °C and 60 °C using a 235 Column Heater Beckman. The results showed a maximum measured deviation of ±0.4 °C compared with a Thermometer HD 2307.0 RTD.

2.5 Electric Circuit, Solar Sensor and Data Collection

Two electric circuits with the same characteristic were used to collect the values of current and voltage for the CPC system and PV system independently. A resistor of 0.2 ohms with maximum power of 15 W was used for each circuit. In order to avoid high current, a RS 257-408 shunt with a conversion factor of 20A-200 mV was connected in series and then connected to the data logger. The horizontal solar radiation was recorded using a Kipp and Zonen Sp Lite 2 pyranometer with a sensitivity of 10.2 V/Wm. A data logger (Agilent 3472A LXI) was used for the electrical and thermal characterization of the CPC and PV systems. Twenty-five channels were programmed and used to measure voltage, current, temperature and solar radiation. The data were recorded every minute.

3 Outdoor Characterization of CPC and PV System

3.1 Roof Assembly

The CPC and PV systems were installed on the roof, mounted to sub-frame rails and then secured with concrete slabs to eliminate the risk of movement in windy conditions. A south orientation was determined using an electronic compass. An insulated box from Campbell Scientific Ltd. was installed in order to host the electric circuit, the data logger and two 220 V plugs. The final location of CPC system and PV system is shown in Fig. 3.

3.2 Power, Efficiency and Temperature of CPC and PV System

The data were recorded on 17 July 2017 from 9:00 to 18:00. The weather during testing was generally sunny. The variation of solar radiation and power with the time are shown in Fig. 4. A combination graph shows the efficiency and temperature behaviour with solar radiation in Fig. 5. As expected, the maximum power and efficiencies values were reached between

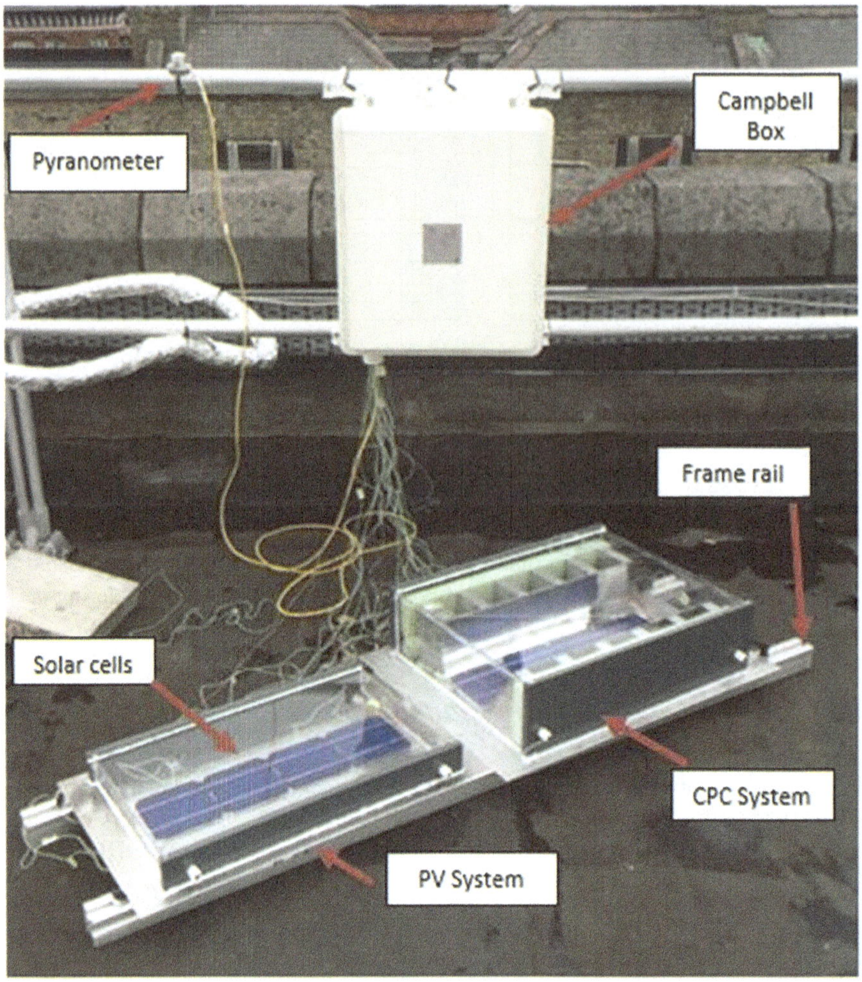

Fig. 3 Final location CPC and PV system on the roof at Trinity College Dublin, Ireland

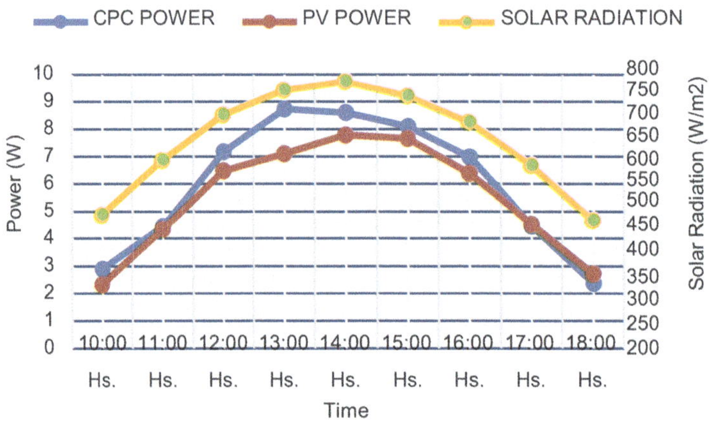

Fig. 4 Variation of solar radiation and power with time on 17 July 2017

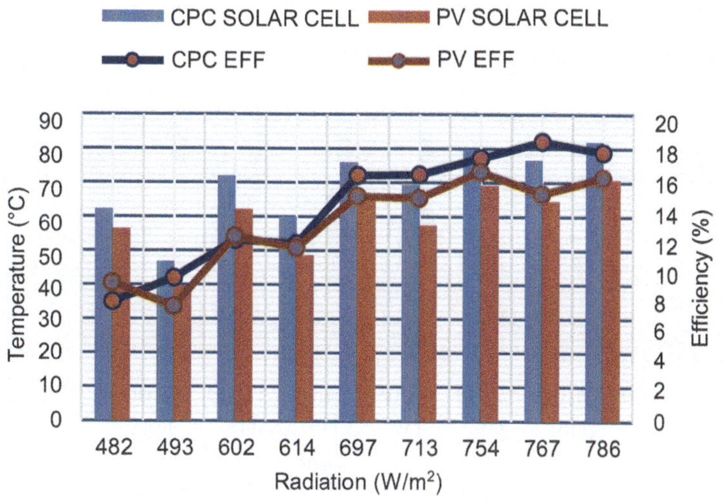

Fig. 5 Average temperature, efficiency with radiation for CPC and PV system

12:00 and 14:00. However, by increasing average solar cell temperature, the efficiency of the system was decreased at high solar radiation. From Fig. 10, it can be seen that at the maximum radiation of 794 W/m^2 at 13:32 for this day, the power achieved for the CPC system and PV system were 8.6 W and 8.3 W, respectively, and the temperatures for the solar cells were 82.6 °C for the CPC system and 71.4 °C for the PV system. A maximum CPC system efficiency of 18.6% was achieved at 744 W/m^2 of solar incident radiation when the power generated by the system was 8.7 W and the maximum power of 8.9 W was achieved at 778 W/m^2 when the efficiency was 18.3%. The test shows that the CPC system was 10 °C higher than the PV panel when both systems were characterized outside.

3.3 Effects of the Temperature in the CPC System

A distribution of the temperature in the CPC system is shown in Fig. 6. From the figure, it can be seen that the solar cells have the highest temperature values in the system.

3.4 Analysis with Convection Cooling System

For this experiment, the aperture cover glass and the two Perspex sheet were removed for the period of 9 h from 9:00 to 18:00 on 18 July 2017. The weather during testing was generally sunny with intermittent cloud cover. In terms of average, the wind speed on the roof was 1.88 m/s and solar cell temperature was 48.5 °C in the cooler CPC system. From Figs. 7 and 8, it can be observed that the efficiency and power improved when the CPC system is cooled by convection. Efficiencies for the cooled CPC system and CPC system were 23% and 17%, respectively, at the solar radiation intensity of 800 W/m^2. At the same radiation, the temperatures were 55 °C for Cooling CPC and 78 °C for CPC system. In addition, the power produced by the cooled CPC was 11 W and 8 W for CPC. These results revealed an improvement in the cooled CPC system over the simple CPC system. At the solar incident radiation of 800 W/m^2 the improvements were of 6% in terms of efficiency and 3 W in terms of power output when the temperatures in the solar cells were reduced by 23 °C with an average wind speed of 1.88 m/s.

3.5 Analysis of CPC, Cooled CPC and PV Systems

Figures 9, 10 and 11 show the average values of temperature, power and efficiency with radiation for each system. It can be clearly seen that the efficiency and power for the cooled CPC system is better than the other systems. At 800 W/m^2 solar incident radiation, the average power achieved for the cooled CPC system was 5 W more than the PV system, while the

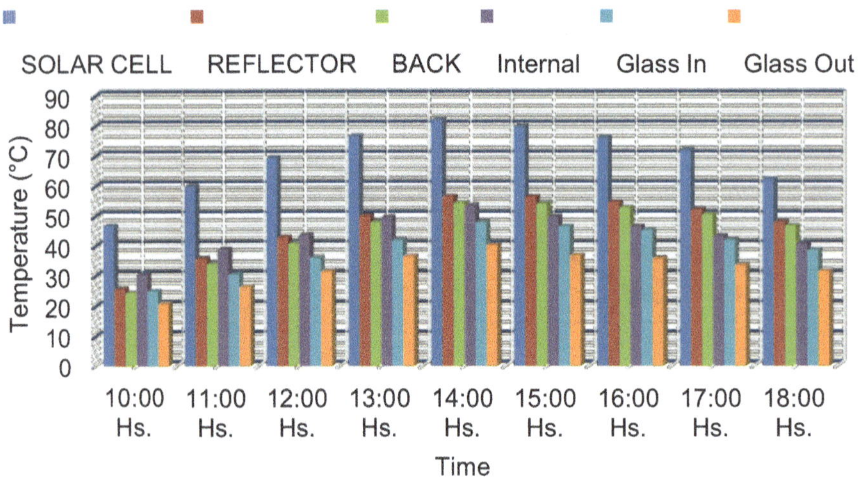

Fig. 6 Average temperature distribution in the CPC system on 17 July 2017

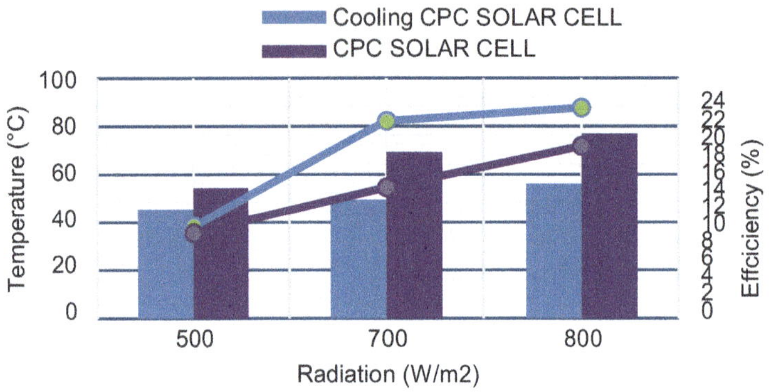

Fig. 7 Average temperature and efficiency with radiation for the cooling CPC system and CPC system

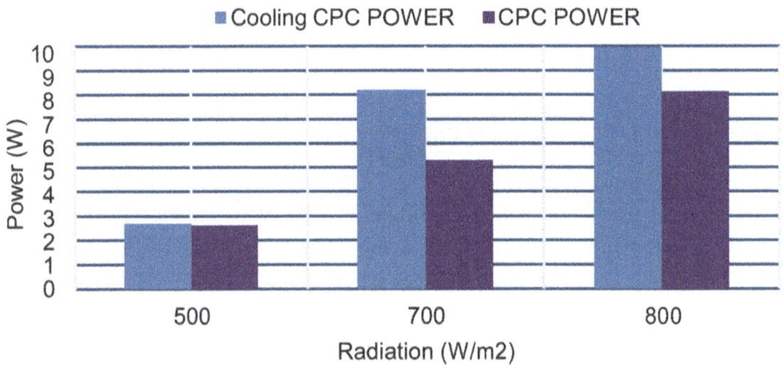

Fig. 8 Average power output with radiation for the cooling

normal CPC was 3 W more than the PV system. In terms of efficiency, the cooled CPC system was 9% more than the PV system, while the normal CPC was 3% more than the PV system.

4 Conclusion

Three different CPC system configurations were manufactured and tested in outdoor conditions and the results were compared with a reference PV system with the same solar cell area and characteristics. The experimental results showed that the efficiency, power and temperature from CPC system were higher than the PV system. The solar cells temperature for the

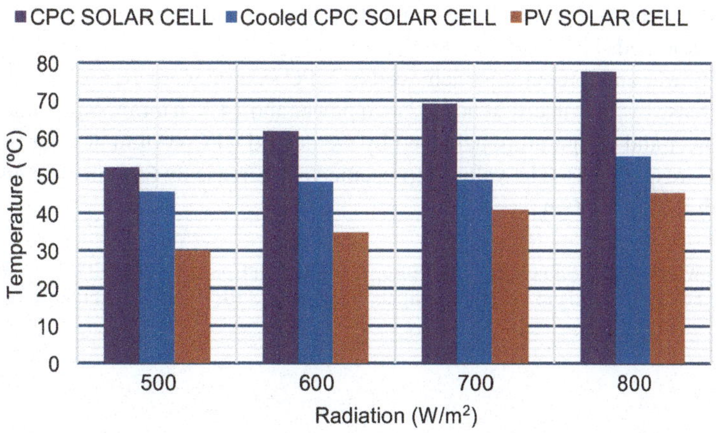

Fig. 9 Average temperature between CPC, cooled CPC and PV systems

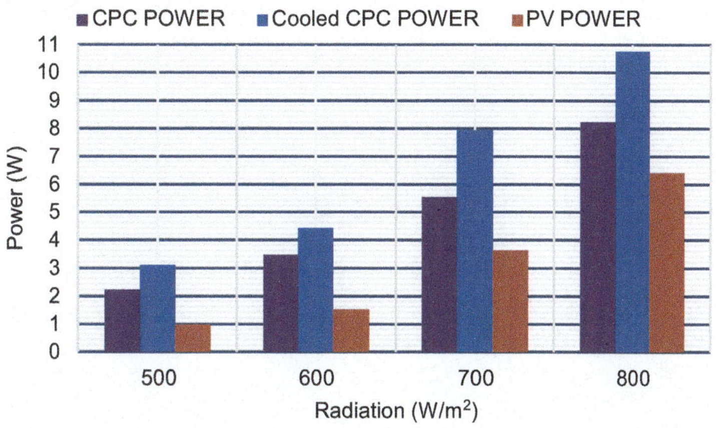

Fig. 10 Average power between CPC, cooled CPC and PV systems

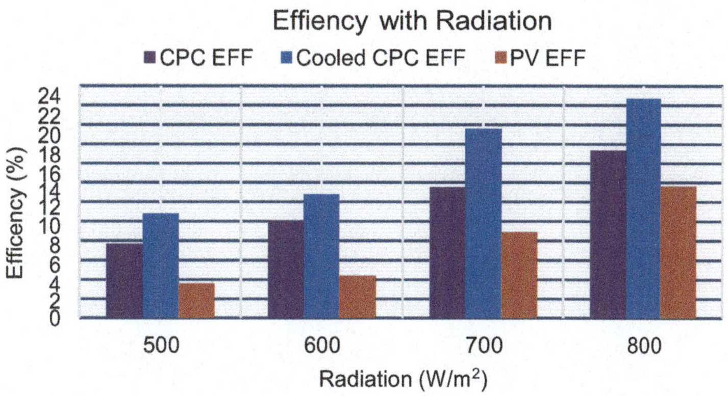

Fig. 11 Average efficiency between CPC, cooled CPC and PV systems

CPC system was decreased when the ambient cooling convection system was added and as a consequence of this, the efficiency and power generated improved.

The maximum efficiency, power and temperature achieved by CPC system were 18.6%, 8.9 W and 83.6 °C, respectively. The values represented an increase by 35% in efficiency, 12% in power and 14% in temperature compared with the PV system.

The average efficiency, power and temperature achieved by the cooled CPC system at solar incident radiation of 800 W/m^2 were 23%, 11 W and 55 °C, respectively. The values represented an increase by 65% in efficiency, 84% in power and 22% in temperature compared with the PV system.

Acknowledgements The authors would like to acknowledge the European Research Council grant PEDAL: (639760), H2020 IDEAS (815271) and support funding from Science Foundation Ireland (SFI) and insights from the PEARL PV COST Action Network.

References

1. Mallick T (2003) Optics and heat transfer for asymmetric compound parabolic photovoltaic concentrator for building integrated photovoltaics. PhD thesis. University of Ulster, Newtownabbey, N.I.U.K
2. McCormack S (2016) PV systems & solar thermal systems. In: J2: solar energy conversion and applications lecture notes. Trinity College Dublin, Dublin
3. Rabl A (1976b) Optical and thermal proprieties of compound parabolic concentrators. Sol Energy 18:497–511
4. Ortega A (2017) Design and characterization of a roof-mounted compound parabolic concentrator with luminescent down sifting layers. Msc Thesis. Trinity College Dublin, Dublin, Ireland

Compound Parabolic Concentrator Design: Assessment of Performance Enhancement for Building-Integrated Photovoltaic Systems

Sander Saers, Anita Ortega, Subhash Chandra, and Sarah J. McCormack

1 Introduction

Photovoltaic (PV) technology allows solar radiation to be converted to electricity, providing means to utilise this near-boundless energy source. Due to continuously improving production cost and financial incentives, PV installations have become increasingly common in domestic and public buildings. Building-integrated PV (BIPV) is considered an effective strategy for increasing the PV system efficiency, as the power generated is in close proximity to the user and no additional land area is required. However, most silicon PV panel systems still only achieve an energy conversion efficiency between 17% and 22%.

Compound parabolic concentrator (CPC) technology can further increase the energy output of solar panels by concentrating solar radiation onto the absorber surface. CPCs contain curved reflectors, which act like a solar funnel, focusing sunlight from a larger inlet surface onto the PV panel. Previous studies have shown the benefits of the application of CPC technology; however, it has not yet been adopted on a large scale.

The aim of this chapter is to assess the technical performance enhancement of the application of BIPV-CPC systems and to create and apply simulation models to estimate the technical performance of the selected CPC types and locations and to determine the economic benefit of the chosen CPCs in electricity savings and feed in tariffs (FIT) in each location, and to present and discuss the simulation model outcomes regarding technical and economic performance.

2 Methodology

Table 1 shows an overview of the reviewed building BIPV-CPC studies (adapted from [1]), and was expanded to include recent studies. It includes the location and publication date of the study, design characteristics regarding symmetry, angle (θ) and the concentration ratio. Most important design parameters are the following:

(i) *Acceptance angle, concentration ratio and truncation:* The acceptance angles (θ) of the CPC dictate the technical performance of the design. A wide θ will result is a low geometrical concentration ratio. Conversely, a narrow θ will result in a high geometrical concentration ratio. Additionally, the CPC can be truncated to reduce the installation size of the system. The truncation length will, however, also influence the performance. The location and orientation should be taken into consideration during the determination of truncation length, and should be based on location and orientation [2].

(ii) *Location and orientation:* Location and orientation of the system globally as well as locally must be included in the process of design decisions. As the research will be performed in Dublin, Ireland, the Irish solar climate conditions such as the ratio of direct and diffuse solar radiation, and the local solar trajectories throughout the year, will dictate these design choices. The desired and available azimuth and the tilt angles for the PV-CPC system are also influencing factors.

S. Saers · A. Ortega (✉) · S. Chandra · S. J. McCormack
Solar Energy Applications Group, Department of Civil, Structural and Environmental Engineering, School of Engineering, Trinity College, Dublin, Ireland
e-mail: ortegaa@tcd.ie

© The Author(s), under exclusive license to Springer Nature Switzerland AG 2022
A. Sayigh (ed.), *Sustainable Energy Development and Innovation*, Innovative Renewable Energy,
https://doi.org/10.1007/978-3-030-76221-6_99

Table 1 Overview of BICPC literature

Year	CPC design	Study type	Conc. Ratio	Half- accept. Angle (θ)	Refs
2000	Dielectric filled symmetrical	Conceptual design	2.46	37	[3]
2000	Dielectric filled asymmetrical	Conceptual design	2.96	36.4	[3]
2004	Asymmetrical	Prototype test outdoor	2.01	50-0	[4]
2007	Asymmetrical	Prototype test outdoor	2.45	37-0	[5]
2014	Dielectric filled symmetrical	Prototype test	2.8	55	[6]
2014	Dielectric filled asymmetrical	Prototype test	2.8	55-0	[7]
2015	Dielectric filled asymmetrical	Prototype test outdoor	2.8	55-0	[7]
2017	Asymmetrical	Prototype test outdoor	1.83	60-0	[8]

Adapted from [1]

(iii) *Temperature and uniformity of radiative flux distribution:* To optimise the efficiency of the CPC, several physical and electrical influences should be considered as well: An increase in temperature will reduce the efficiency of a silicon PV cell. Therefore, the temperature gain should be mitigated. This can be controlled passively through designed cooling ribs or actively with actuated fans. On top of the general negative effect of heating on PV cell efficiency, the uniformity of the distribution of the radiative flux onto the cells also has a significant impact on the efficiency. Because the cells are arranged in series, the cell producing the lowest amount of power will determine the power produced for the entire panel/array. Therefore, an even distribution of radiation is required for optimal performance.

(iv) *Materials and optical gaps:* The optical efficiency of the reflective/refractive materials used for the concentrator should be as high as possible for optimal performance. This increases the amount of radiation focused onto the absorber and reduces heat gain. Furthermore, to channel the maximum amount of solar radiation towards the PV surface, the solar radiation 'leakage' through optical gaps in the design should be minimised, for instance, through the use of dielectric film.

3 Model Development

A simulation approach has been chosen to find the optimal CPC shape and orientation for enhancement of solar radiation onto a photovoltaic surface. To achieve this, separate simulation models were created. In these models, multiple CPC variations were analysed in varying virtual circumstances. The CPC design variations were assessed in simulation conditions representing two cities with different climates. Both direct solar radiation and diffuse (indirect) solar radiation were analysed. Table 2 presents a complete overview of all 22 CPC variations included in the simulations.

Two locations were selected for assessment: Dublin (Ireland) and Ferrara (Italy). 3D models for each CPC type were created for the ray-tracing simulation. *Solidworks* was used as a CAD modelling software to generate these models. Using the software, a 2D profile sketch of each CPC was drawn and subsequently extruded 500 mm into a 3D shape. This 3D model was then imported into a ray-tracing software to perform the ray-tracing analyses. *TracePro*, a software by Lambda Research Corporation, was utilised to perform the ray-tracing analyses for diffuse and direct solar irradiation onto the 3D model.

The parabolic reflector surfaces bordering each side of the PV panel are attributed a 95% reflection and 5% absorption, and PV panel surface was treated as a perfect absorber exit surface. Although this is not realistic for the purpose of simulation of PV panel output, it will allow adequate assessment of the effectiveness of the CPC performance regardless of PV panel quality and efficiency. The inner side panels is perfect mirror material property (100% reflection) to prevent loss of irradiation through the sides: Optimal panel tilt angle (α) per location was calculated using the NASA database (NASA, 2020), which includes the monthly optimal panel tilt angle. The optimal fixed α for the entire year ($\alpha_{opt.year}$) was found through the sum of the product of the monthly optimal angle ($\alpha_{opt.month}$) and the corresponding irradiation ($I_{opt.month}$), consecutively divided by the sum of $I_{opt.month}$. CPC performance was simulated in conditions resembling three times of the year: (i) Summer solstice; (ii) Spring/Autumn equinox; (iii) Winter solstice. From these three points in the year, an extrapolation for an estimated average annual performance was made.

Results from the estimated diffuse and direct irradiance derived from the simulation model were combined and the cumulative irradiance (I_{total}) is estimated per location. The results are shown in Tables 3 and 4 where seasonal and annual average I_{direct}, I_{direct} and I_{total} are presented along with the concentration ratio.

Table 2 CPC variations for ray-tracing simulation

CPC Variations				
Reference code	Acceptance angle θ	Panel orientation	Visual concept representation	CPC axes symmetry
S-H-80	80°	Horizontal		Symmetrical
S-H-100	100°			
S-H-120	120°			
S-H-140	140°			
S-V-80	30°	Vertical		
S-V-100	100°			
S-V-120	120°			
S-V-140	140°			
S-T-80	80°	Tilted (fixed optimal sun facing angle)		
S-T-100	100°			
S-T-120	120°			
S-T-140	140°			
AS-H-40	40° + 19°	Horizontal		Asymmetrical
AS-H-50	50° + 32°			
AS-H-60	60° + 48°			
AS-H-70	70° + 64°			
AS-V-40	40° + 19°	Vertical	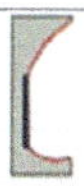	
AS-V-50	50° + 32°			
AS-V-60	60° + 48°			
AS-V-70	70° + 64°			

Table 3 Ray-tracing simulation – Combined daily average irradiation (kWh m^{-2} day^{-1}) in Dublin

Irradiation daily avg (kWh m^{-2} day^{-1})	*Symmetrical Tilted 40°*		REF-T40			S-T40-80			S-T40-100			S-T40-120			S-T40-140		
			WNTR	EQNX	SMMR	WNTR	EQNX	SMMR	WNTR	EQNX	SMMR	WNTR	EQNX	SMMR	WNTR	EQNX	SMMR
	Direct	Season	0.60	1.65	2.08	0.57	4.64	2.00	0.68	3.90	2.12	0.73	3.26	2.15	0.73	2.68	2.13
		Year		1.49			2.96	(198%)		2.65	(177%)		2.35	(157%)		2.05	(137%)
	Diffuse	Season	0.28	1.11	2.09	0.28	1.11	2.09	0.28	1.11	2.09	0.28	1.11	2.09	0.26	1.11	2.09
		Year		1.15			1.15			1.15			1.15			1.15	
	Total	Season	0.88	2.76	4.17	0.85	5.75	4.10	0.96	5.02	4.21	1.01	4.38	4.25	1.01	3.79	4.22
		Year		2.64			4.11	(156%)		3.80	(144%)		3.50	(132%)		3.20	(121%)

Irradiation daily avg (kWh m^{-2} day^{-1})	*Symmetrical Vertical*		REF-V			S-V-80			S-V-100			S-V-120			S-V-140		
			WNTR	EQNX	SMMR	WNTR	EQNX	SMMR	WNTR	EQNX	SMMR	WNTR	EQNX	SMMR	WNTR	EQNX	SMMR
	Direct	Season	0.70	1.32	1.15	1.14	0.51	0.11	1.07	0.93	0.11	0.97	1.16	0.11	0.88	1.29	0.48
		Year		1.12			0.57	(51%)		0.76	(68%)		0.85	(76%)		0.98	(88%)
	Diffuse	Season	0.18	0.72	1.35	0.18	0.72	1.35	0.18	0.72	1.35	0.18	0.72	1.35	0.18	0.72	1.35
		Year		0.74			0.74			0.74			0.74			0.74	
	Total	Season	0.88	2.03	2.49	1.32	1.23	1.46	1.25	1.65	1.46	1.15	1.88	1.46	1.06	2.01	1.82
		Year		1.86			1.31	(70%)		1.50	(81%)		1.59	(85%)		1.72	(93%)

Irradiation daily avg (kWh m^{-2} day^{-1})	*Symmetrical Horizontal*		REF-H			S-H-80			S-H-100			S-H-120			S-H-140		
			WNTR	EQNX	SMMR	WNTR	EQNX	SMMR	WNTR	EQNX	SMMR	WNTR	EQNX	SMMR	WNTR	EQNX	SMMR
	Direct	Season	0.16	0.88	1.77	0.01	0.18	1.81	0.01	0.18	2.02	0.01	0.43	2.06	0.01	0.66	2.01
		Year		0.92			0.54	(59%)		0.59	(64%)		0.73	(79%)		0.83	(90%)
	Diffuse	Season	0.36	1.43	2.69	0.36	1.43	2.69	0.36	1.43	2.69	0.36	1.43	2.69	0.36	1.43	2.69
		Year		1.48			1.48			1.48			1.48			1.48	
	Total	Season	0.52	2.31	4.46	0.37	1.61	4.50	0.37	1.61	4.71	0.37	1.86	4.75	0.37	2.09	4.70
		Year		2.40			2.02	(84%)		2.07	(86%)		2.21	(92%)		2.31	(96%)

Irradiation daily avg (kWh m^{-2} day^{-1})	*Asymmetrical Vertical*		REF-V			AS-V-40			AS-V-50			AS-V-60			AS-V-70		
			WNTR	EQNX	SMMR	WNTR	EQNX	SMMR	WNTR	EQNX	SMMR	WNTR	EQNX	SMMR	WNTR	EQNX	SMMR
	Direct	Season	0.70	1.32	1.15	1.67	2.16	0.11	1.48	2.15	0.63	1.23	2.11	0.76	1.01	1.91	1.08
		Year		1.12			1.53	(136%)		1.60	(143%)		1.55	(138%)		1.48	(132%)
	Diffuse	Season	0.18	0.72	1.35	0.18	0.72	1.35	0.18	0.72	1.35	0.18	0.72	1.35	0.18	0.72	1.35
		Year		0.74			0.74			0.74			0.74			0.74	
	Total	Season	0.88	2.03	2.49	1.85	2.87	1.46	1.66	2.87	1.97	1.41	2.83	2.10	1.19	2.62	2.43
		Year		1.86			2.26	(122%)		2.34	(126%)		2.29	(123%)		2.22	(119%)

Irradiation daily avg (kWh m^{-2} day^{-1})	*Asymmetrical Horizontal*		REF-H			AS-H-40			AS-H-50			AS-H-60			AS-H-70		
			WNTR	EQNX	SMMR	WNTR	EQNX	SMMR	WNTR	EQNX	SMMR	WNTR	EQNX	SMMR	WNTR	EQNX	SMMR
	Direct	Season	0.16	0.88	1.77	0.01	0.20	3.35	0.01	1.00	3.17	0.01	1.05	2.84	0.01	1.22	2.46
		Year		0.92			0.94	(102%)		1.29	(140%)		1.26	(137%)		1.23	(133%)
	Diffuse	Season	0.36	1.43	2.69	0.36	1.43	2.69	0.36	1.43	2.69	0.36	1.43	2.69	0.36	1.43	2.69
		Year		1.48			1.48			1.48			1.48			1.48	
	Total	Season	0.52	2.31	4.46	0.37	1.63	6.04	0.37	2.43	5.86	0.37	2.48	5.63	0.37	2.65	5.15
		Year		2.40			2.41	(101%)		2.77	(115%)		2.74	(114%)		2.70	(113%)

Table 4 Ray-tracing simulation – Combined daily average irradiation (kWh m^{-2} day^{-1}) in Ferrara

	Symmetrical Tilted 60°		REF-T60			S-T60-80			S-T60-100			S-T60-120			S-T60-140		
			WNTR	EQNX	SMMR	WNTR	EQNX	SMMR	WNTR	EQNX	SMMR	WNTR	EQNX	SMMR	WNTR	EQNX	SMMR
Irradiation daily avg (kWh m^{-2} day^{-1})	Direct	Season	1.72	3.70	4.53	0.37	5.38	7.71	1.00	5.16	7.32	1.40	4.84	6.44	1.62	4.51	5.58
		Year		3.43			4.71	(137%)		4.66	(136%)		4.38	(128%)		4.06	(118%)
	Diffuse	Season	0.38	0.95	1.53	0.38	0.95	1.53	0.38	0.95	1.53	0.38	0.95	1.53	0.38	0.95	1.53
		Year		0.96			0.96			0.96			0.96			0.96	
	Total	Season	2.10	4.68	6.06	0.75	6.33	9.24	1.38	6.11	8.85	1.78	5.80	7.97	2.00	5.47	7.11
		Year		4.38			5.67	(129%)		5.61	(128%)		5.34	(122%)		5.01	(114%)

	Symmetrical Vertical		REF-V			S-V-80			S-V-100			S-V-120			S-V-140		
			WNTR	EQNX	SMMR	WNTR	EQNX	SMMR	WNTR	EQNX	SMMR	WNTR	EQNX	SMMR	WNTR	EQNX	SMMR
Irradiation daily avg (kWh m^{-2} day^{-1})	Direct	Season	2.06	2.72	1.64	2.53	0.60	0.12	2.58	0.99	0.12	2.53	1.83	0.12	2.41	2.32	0.18
		Year		2.29			0.97	(42%)		1.17	(51%)		1.58	(69%)		1.81	(79%)
	Diffuse	Season	0.29	0.72	1.15	0.29	0.72	1.15	0.29	0.72	1.15	0.29	0.72	1.15	0.29	0.72	1.15
		Year		0.72			0.72			0.72			0.72			0.72	
	Total	Season	2.34	3.44	2.79	2.82	1.32	1.27	2.87	1.71	1.27	2.82	2.54	1.27	2.69	3.03	1.33
		Year		3.00			1.68	(56%)		1.89	(63%)		2.29	(76%)		2.52	(84%)

	Symmetrical Horizontal		REF-H			S-H-80			S-H-100			S-H-120			S-H-140		
			WNTR	EQNX	SMMR	WNTR	EQNX	SMMR	WNTR	EQNX	SMMR	WNTR	EQNX	SMMR	WNTR	EQNX	SMMR
Irradiation daily avg (kWh m^{-2} day^{-1})	Direct	Season	0.67	2.42	4.16	0.05	0.66	5.67	0.05	1.20	5.64	0.13	2.76	6.60	0.50	2.91	5.69
		Year		2.42			1.76	(73%)		2.02	(84%)		3.06	(127%)		3.00	(124%)
	Diffuse	Season	0.57	1.43	2.30	0.57	1.43	2.30	0.57	1.43	2.30	0.57	1.43	2.30	0.57	1.43	2.30
		Year		1.43			1.43			1.43			1.43			1.43	
	Total	Season	1.24	3.85	6.46	0.62	2.08	7.97	0.62	2.63	7.94	0.70	4.19	8.90	1.07	4.34	7.99
		Year		3.85			3.19	(83%)		3.45	(90%)		4.49	(117%)		4.43	(115%)

	Asymmetrical Vertical		REF-V			AS-V-40			AS-V-50			AS-V-60			AS-V-70		
			WNTR	EQNX	SMMR	WNTR	EQNX	SMMR	WNTR	EQNX	SMMR	WNTR	EQNX	SMMR	WNTR	EQNX	SMMR
Irradiation daily avg (kWh m^{-2} day^{-1})	Direct	Season	2.06	2.72	1.64	4.59	3.83	0.12	4.10	3.59	0.12	3.58	3.82	0.48	2.97	3.87	0.77
		Year		2.29			3.09	(135%)		2.85	(125%)		2.92	(128%)		2.87	(125%)
	Diffuse	Season	0.29	0.72	1.15	0.29	0.72	1.15	0.29	0.72	1.15	0.29	0.72	1.15	0.29	0.72	1.15
		Year		0.72			0.72			0.72			0.72			0.72	
	Total	Season	2.34	3.44	2.79	4.88	4.54	1.27	4.38	4.31	1.27	3.86	4.53	1.63	3.25	4.58	1.92
		Year		3.00			3.81	(127%)		3.57	(119%)		3.64	(121%)		3.58	(119%)

	Asymmetrical Horizontal		REF-H			AS-H-40			AS-H-50			AS-H-60			AS-H-70		
			WNTR	EQNX	SMMR	WNTR	EQNX	SMMR	WNTR	EQNX	SMMR	WNTR	EQNX	SMMR	WNTR	EQNX	SMMR
Irradiation daily avg (kWh m^{-2} day^{-1})	Direct	Season	0.67	2.42	4.16	0.05	3.66	8.41	0.05	3.41	7.70	0.26	3.55	6.79	0.35	3.48	5.69
		Year		2.42			3.94	(163%)		3.64	(151%)		3.54	(146%)		3.25	(134%)
	Diffuse	Season	0.57	1.43	2.30	0.57	1.43	2.30	0.57	1.43	2.30	0.57	1.43	2.30	0.57	1.43	2.30
		Year		1.43			1.43			1.43			1.43			1.43	
	Total	Season	1.24	3.85	6.46	0.62	5.09	10.71	0.62	4.84	10.00	0.83	4.98	9.09	0.92	4.91	7.99
		Year		3.85			5.38	(140%)		5.08	(132%)		4.97	(129%)		4.68	(121%)

For all of the CPC designs assessed in this study, the absorber surface is 125 × 500 mm and the inlet surfaces are larger than this. Therefore, the geometrical concentration ratio is always positive (>100%). The inlet surface, and herewith the geometrical concentration ratio, is increased with decreasing acceptance angles (θ) in the CPC design. Amongst the CPC designs that were assessed, this ratio varies between 133% and 338%.

4 Results and Discussion

The symmetrical CPCs mounted with an optimal fixed tilt show an increase in performance with each design in all locations. Overall, all CPCs appear to have beneficial impact; however, the smaller θ generates more concentration. The highest concentration ratios are achieved in Dublin, where they reach an overall increase of I_{direct} of nearly 200% annually. The simulation shows overall performance losses in every location when the symmetrical design is mounted vertically, for instance, directly onto a façade. The estimated concentration ratios range between 26% and 88%. The smaller θ are paired with the lowest concentration ratios. This poor performance can be attributed to CPC shading. The horizontal orientation of the symmetrical CPC is most beneficial in the southern location, closer to equator. This was to be expected as the maximum solar angle is highest in these locations. An overall increase between 119% for the wide θ and 137% for the narrow θ is estimated. The performance changes in Ferrara depending on θ, and in Dublin, the simulated performance is poor (56–90%). Similarly, this decrease in performance was to be expected due to lower maximum solar angles in locations at higher latitudes. The vertically mounted asymmetrical CPC achieved positive simulation results in most locations. This type appears to be more effective at higher altitudes, as the performance in Dublin was high (131%). Interestingly, the highest concentration in these locations was achieved with intermediate θ, indicating that here the optimal θ is near 50° and 60°. In Ferrara, a moderate concentration was achieved. Unsurprisingly, the vertically mounted asymmetrical CPCs perform best at high latitudes due to lower maximum sun angles. However, the performance is much better when compared to the vertically mounted symmetrical CPCs because the upper reflector is more open, allowing more exposure from above and creating far less solar shading. The horizontally mounted asymmetrical CPC has produced positive simulation results for most acceptance angles and locations. The optimal θ appears to be near 50° in Dublin, with a concentration ratio of 137%. In Ferrara, the concentration ratio reaches up to 163% with θ = 40°. Overall, this type of CPC appears to generate good results everywhere, especially in Southern regions.

5 Conclusion

In this study, technical performance enhancement of BI CPC design types were assessed using ray-tracing simulations for two locations Dublin, Ireland, and Ferrara, Italy, to compare performance. A total of twenty-two 3-D models, each with unique symmetry, acceptance angles and orientations, were subjected to diffuse and direct solar ray-tracing simulations alongside reference models (without the application of CPC). The performance was assessed in each location during three pivotal days in the year (summer solstice; spring/autumn equinox; winter solstice). As a result, average hourly flux (Q) and daily irradiance (I) data were compiled for each CPC type. This information was used to determine annual performance estimations in energy production and economic benefits in terms of electricity savings and feed-in tariffs per location. Combining the results of the diffuse and direct solar ray-tracing simulations shows that the overall performance per m^2 PV surface can be enhanced using several CPC types. All tilted symmetrical CPCs increase annual absorber irradiance in each location (114–156%) and horizontal asymmetrical CPCs are more effective at lower latitudes, whilst tilted symmetrical CPCs are more at higher latitudes. Asymmetrical CPCs have a higher performance than symmetrical CPCs when mounted strictly horizontally or vertically. In Dublin (IRL), CPC type S-T40-80 proved most effective with an estimated increase of annual irradiance from 2.64 to 4.11 kWh m^{-2} $year^{-1}$ (156%). In Ferrara (ITA), CPC type S-T40-80 proved most effective once more with an estimated increase of annual irradiance from 4.38 to 5.67 kWh m^{-2} $year^{-1}$ (129%). The simulation results clearly illustrated the potential effectiveness of application of some CPC types in various environments, providing increased solar irradiance of PV surfaces. The study presents detailed estimations of the technical performance of various CPC types in multiple locations. It provides a pragmatic overview of the estimated performances which can be used to inform decision-making regarding BIPV-CPC design and application.

Acknowledgements The authors would like to acknowledge the European Research Council grant PEDAL: (639760), H2020 IDEAS (815271) and support funding from Science Foundation Ireland (SFI) and insights from the PEARL PV COST Action Network.

References

1. Tian M, Su Y, Zheng H, Pei G, Li G, Riffat S (2018) A review on the recent research progress in (CPC) for solar energy applications. Renew Sust Energ Rev 82:1272–1296
2. Jaaz AH, Hasan HA, Sopian K, Ruslan MHBH, Zaidi SH (2017) Design and development of compound parabolic concentrating for photovoltaic solar collector: review. Renew Sustain Energy Rev 76:1108–1121
3. Zacharopoulos A, Eames PC, McLarnon D, Norton B (2000) Linear dielectric non-imaging concentrating covers for PV integrated building facades. Sol Energy 68:439–452
4. Mallick TK, Eames PC, Hyde TJ, Norton B (2004) Design and experimental characterisation of an asymmetric CPC PV for building façade integration in the UK. Sol Energy 77:319–327
5. Mallick TK, Eames PC (2007) Design and fabrication of low concentrating second generation PRIDE concentrator. Sol Energy Mater Sol Cells 91:597–608
6. Baig H, Sarmah N, Chemisana D, Rosell J, Mallick TK (2014) Enhancing performance of a linear dielectric based concentrating PV system. Energy 73:177–191
7. Sarmah N, Richards BS, Mallick TK (2014) Design, development and indoor performance analysis of a low concentrating dielectric PV module. Sol Energy 103:390–401
8. Ortega A (2017) Design and characterization of a roof-mounted compound parabolic concentrator with luminescent down sifting layers. MSc Thesis. Trinity College Dublin, Dublin, Ireland

Implementation and Optimization of a Luminescent Down-Shifting Photovoltaic System for Use in a Compound Parabolic Concentrator

Aaron Glenn, Subhash Chandra, Bill Cass, and Sarah J. McCormack

1 Introduction

Luminescent downshifting (LDS) is a method of improving the PV cells' solar responsivity through the inclusion of a layer on the cells' surface that re-emits the short wavelength light at longer wavelengths where the solar cells have higher responsivity [1]. It converts photons of high energy (short wavelength) to photons of lower energy (hence downshifting energy) that the photovoltaic materials can more easily absorb and process. Ideally, the dye used within the layer should have a wide absorption band in the range of wavelengths that need to be enhanced and a narrow emission peak in the region where the PV cell already performs very efficiently (500–900 nm). Extensive research has been completed in this area within Trinity College Dublin [2, 3], and it shows that the inclusion of the LDS layer can greatly enhance PV cells' performance, provided the correct dye has been selected.

Figure 1 shows the processes that occur within the LDS layer. Ideally, a dye must be selected with a high transmittance in the 500–1100 nm wavelength range (for silicon PV) in order to avoid losses due to reflection. There are inherent energy loss mechanisms which still exist and are very difficult to avoid. Some examples of these extra loss mechanisms include parasitic absorption, photon emission through the different surfaces of the layer and even reabsorption of the downshifted photon by the LDS itself [4]. Modelling and simulation help substantially, as they can provide an optimum dye concentration and LDS layer thickness, which can reduce the amount of experimentation required within a laboratory tenfold [5]. The modelling process can also greatly reduce costs, as less materials are used for testing extensive ranges of concentrations, for example.

The focus of this study is to build upon previous work [6] and optimize the production of LDS (Fig. 1) cells for application within large-scale solar units such as a compound parabolic concentrator (CPC (Fig. 2) vs flat base (Fig. 3)) which could later be commercialized and utilized around the world to harness the solar energy. The project aims to produce uniform and reproducible layers so that the effects of the dye can be quantified. This optimization task consisted of not only improving the layer production process but also developing the method of connecting the photovoltaic (PV) cells in series. The dye used for this large-scale project was called '4-(Dicyanomethylene)-2-methyl-6-(4-dimethylaminostyryl)-4*H*-pyran)' which was a red dye shortened to 'DCM'. The processes, which will be discussed in the methodology, included finding optimal dye concentrations, finding suitable layer-coating methods, testing that the layer thicknesses are uniform, performing optical characteristic testing (EQE), and then deciding which connection method is best (Fig. 10).

Comparisons are drawn between the different dye concentrations (0.1%, 0.2% and 0.3%) and their performance against the bare and blank-layered PV cell will be checked. The effects of including tabbing with the two different soldering approaches will also be analysed. An optimized final product will be decided and used within the different solar units.

2 Methodology

The methodology approach for this study is as shown in the flowchart in Fig. 4.

A. Glenn (✉) · S. Chandra · B. Cass · S. J. McCormack
Solar Energy Applications Group, Department of Civil, Structural and Environmental Engineering, School of Engineering, Trinity College, Dublin, Ireland
e-mail: aglenn@tcd.ie

© The Author(s), under exclusive license to Springer Nature Switzerland AG 2022

A. Sayigh (ed.), *Sustainable Energy Development and Innovation*, Innovative Renewable Energy,
https://doi.org/10.1007/978-3-030-76221-6_100

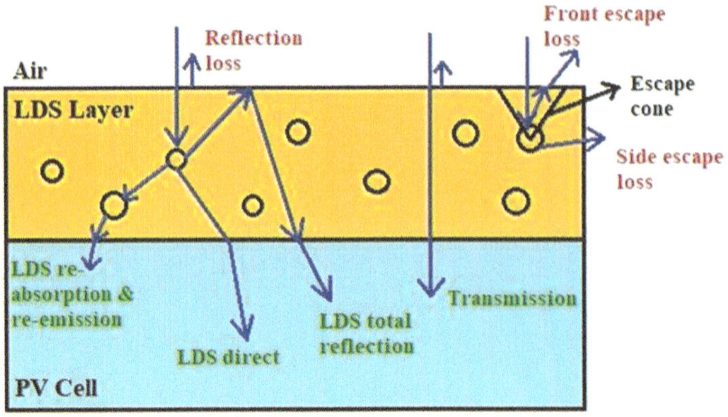

Fig. 1 LDS optical processes and loss mechanisms

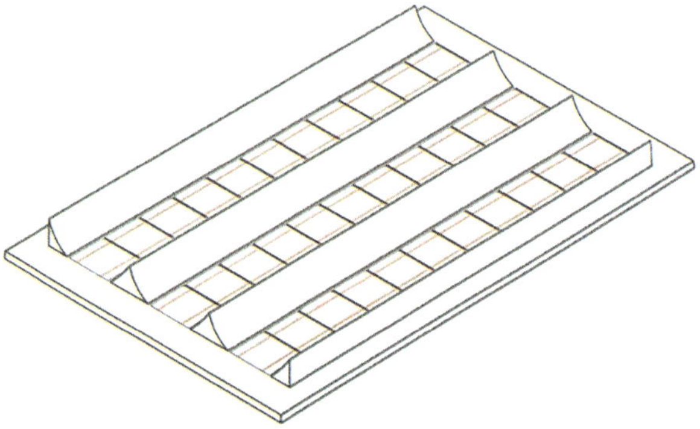

Fig. 2 CPC-based solar panel frame

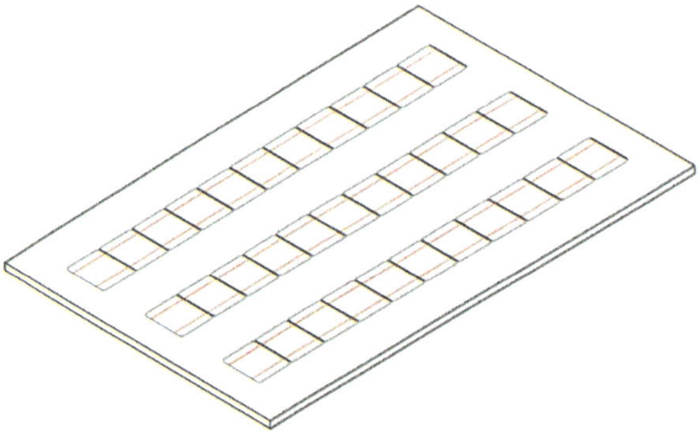

Fig. 3 Flat-based solar panel frame

Blank polymer solution – A chloroform and PMMA mixture is produced in a 0.1% weight solution for comparison with the DCM dye layers to quantify the effect of the dye on the solar power output. 11.5 g of PMMA was added to 30 ml of chloroform and mixed using a magnetic stirrer at room temperature for 30–45 min. Agglomeration of PMMA on the surface was found at higher percentage weights.

Dye solution – A blank solution was prepared using the same method as above and then DCM was included at weight percentage ranges of 0.1%, 0.2% and 0.3%. This range of concentration assumption was based on previous experiments with other lumogen dyes [7] and the results, covered later, show their effectiveness.

Spin coating – A drop casting approach was immediately discarded, as it was clear that the resulting layers would be inconsistent and therefore unsuitable for a comparison study. Numerous tests were performed to determine the amount of solution to dispense on each 12.5 cm^2 PV cell for spin coating. 10 ml of solution was deemed to be optimal, as it had minimal

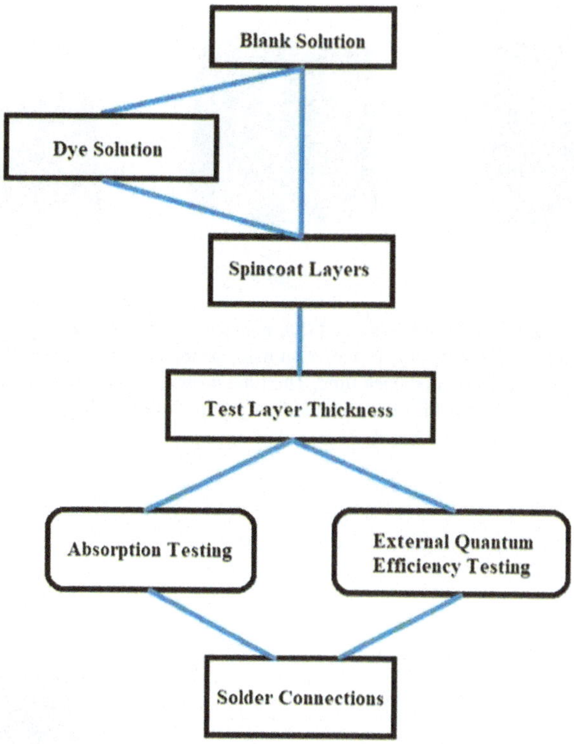

Fig. 4 Methodology flowchart

wastage from solution being spun off the surface whilst also fully covering the face of the cell. This process was finalized after further trials were undertaken relating to the spin coater ramp times, chuck sizes and minimum and maximum revolutions per minute. The final recipe settled upon is an immediate 0–170 rpm for 170 s, followed by a reduction to 500 rpm for 5 s and then finally 300 rpm for another 5 s before coming to a stop.

LDS layer thickness – For a comparison study, there needs to be as little difference between test subjects as possible, and since the solutions poured are exactly the same, the final test to ensure that it is a fair comparison is to check that each layer thickness is the same. This was performed using a digital micrometre at multiple locations across the layer.

Absorption testing – Absorption testing was performed on the solution using a UV/Vis spectrometer. The solution was poured into a cuvette and placed within the device for testing in the 250–650 nm wavelength range (the area of interest). A blank solution of just chloroform was used as a reference so that the resulting graphs produced are a reflection of only the dyes' absorbance and are not influenced by the solvent.

External quantum efficiency – Spectral response testing was performed to measure the PV cells' carriers to photon ratio. The tests were implemented on smaller 2 cm^2 PV cells which, if the results were promising, would lead to upscaling to larger sizes for real-world application. The cells were placed within the device directly below a monochromatic light source and connected at both positive and negative terminals. The system was calibrated before testing using a silicon reference cell supplied by the device manufacturers and the wavelength spectrum of the test was input by the user (300–1100 nm in this case).

Soldering – Two different connection approaches were trialled in this study; the first being to cover the metallic strips with tape before the layer was applied so that the small area could be peeled off to ensure an adequate connection. The second approach was to produce the layers on the cell as normal and then afterwards solder the connection through the polymer surface. The results of these methods will be discussed below.

3 Results and Discussion

Figure 5 shows the progression and improvements made in layer quality after trialling different solution volumes, spin coater recipes and pouring methods.

(a) (b) (c)

Fig. 5 (**a**) Demonstrates the layer quality with no variables decided (i.e. no set solution amount or spin coater recipe selected) and it is clear that it was unsuitable for use within a solar unit. (**b**) Shows some progression made as the decision to use 10 ml of solution was finalized, and this was poured using a syringe so that a fair 10 ml was distributed each time. The final product is displayed in Fig. 5(c), which is the end result of finding a suitable solution volume and spincoater recipe, and it is clear that the final layer is of much higher quality than initial attempts

Fig. 6 Blank layer quality to be compared with dye layer in Fig. 6

Figure 6 shows the quality of layer using blank solution (i.e. without DCM) using the same recipe as before. It is clear that there are minimal defects observable. Each layer produced was checked at multiple locations on their surface for their thicknesses using a digital micrometre, and it was found that all of the layers produced were uniformly 20 microns thick.

Figure 7 demonstrates why placing tape over the connection strips and peeling them off after a layer had been poured on top did not work. No matter how carefully the layer was sliced into with a scalpel and tape peeled back, the cell would either crack or there would be extensive damage to the face of the layer.

After discarding the tape-peeling attempts, an approach of soldering directly through the LDS layer was tested instead. Figures 8 and 9 clearly indicate how this process produced much higher quality layers and connections.

Finally, in relation to cell connectivity, Fig. 10 shows how the PV cells were connected in series.

The metallic strips observed in the figures are tabbing that was included to increase the amount of current that the PV cells would be able to transfer along in series. The effects of the tabbing on the external quantum efficiency measurements will be shown in graphic form. Each PV cell was connected from top surface (positive) to bottom surface (negative) in an alternating fashion.

Included on the following page in Figs. 11 and 12 are the resulting graphs from EQE tests performed on the multiple different cells.

Figure 11 demonstrates why the 0.1% weight DCM solution was selected for the large-scale project. Whilst all 3 concentrations tested performed similarly from approximately 450–1100 nm, the 0.1% solution performed substantially better in the 300–420 nm range, which is where silicon displays its lowest efficiency levels. A bare cell (black line) is included for comparison. The increase in external quantum efficiency for the spectrum of 300–1100 nm tested was 4.5%.

Figure 11 also shows the effect that the inclusion of soldered tabbing had on the cells' efficiency. DCM at 0.1% concentration with tabbing is the best-performing PV cell set-up tested in the group so far. Figure 12 shows more clearly why tabbing

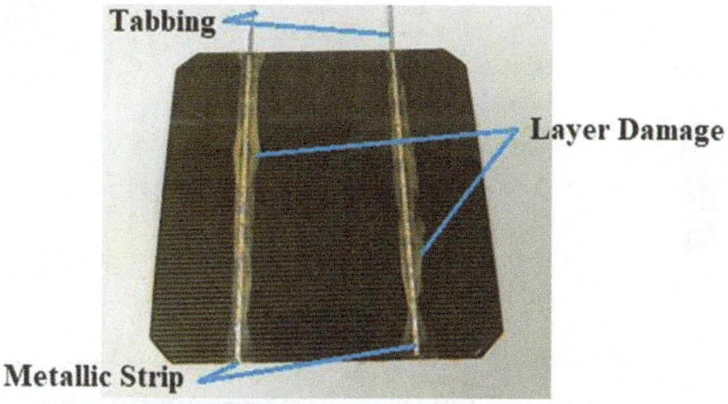

Fig. 7 PV cell after attempting to solder tabbing after peeling off protective tape

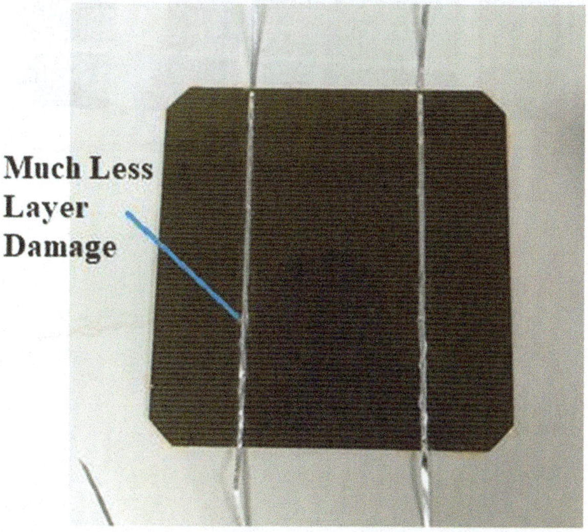

Fig. 8 PV connections made after soldering through the LDS layer

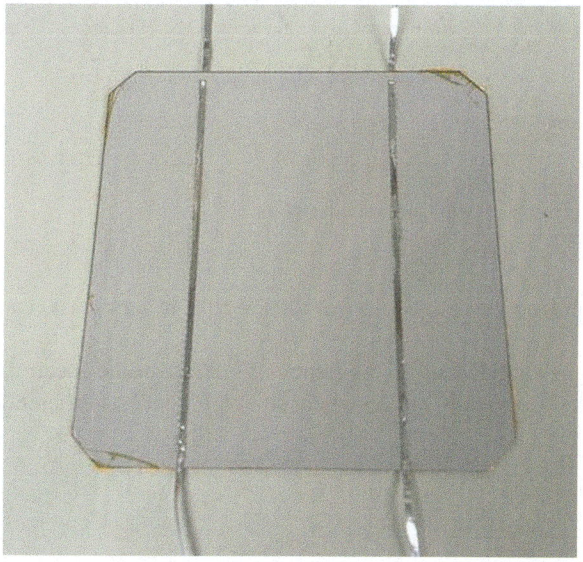

Fig. 9 PV connections made after soldering through the LDS layer

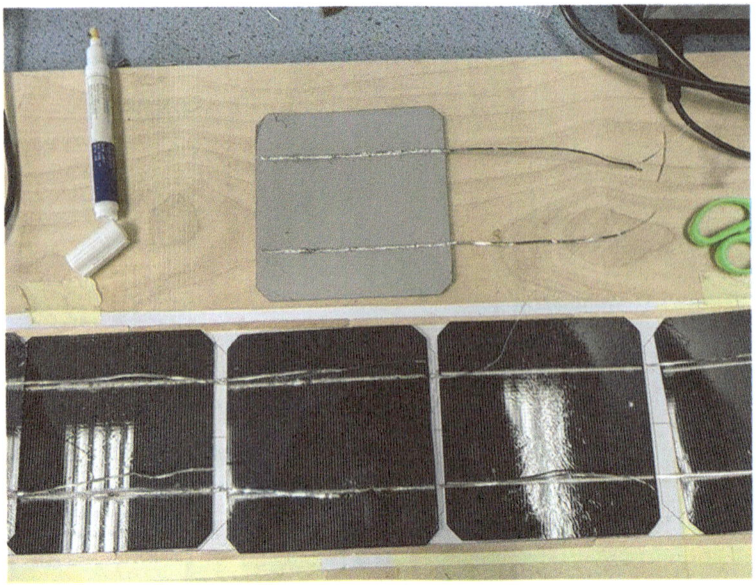

Fig. 10 Series connection through tabbing soldering

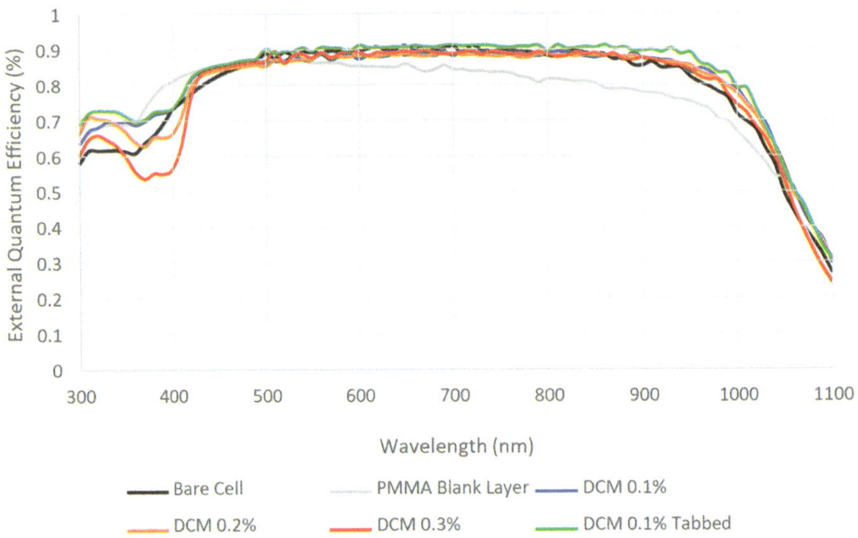

Fig. 11 EQE of DCM in varying weight percentages with and without tabbing

was introduced to the cells, as it improved the connectivity of the PV cells in series. You can see that the 300–380 nm region in particular was enhanced.

The 0.1% DCM tabbing results give an increase in efficiency of 6.8% over the spectral region of interest (300–500 nm) with an enhancement of 10.9% in the wavelength region where silicon PV is least efficient (300–420 nm).

4 Conclusion

To conclude, it is evident that the inclusion of the DCM dye within the LDS layers was a success, as it has enhanced the PV cells' efficiency substantially. 0.1% DCM with tabbing is the best-performing LDS application thus far within the group. Efficiency improvements of 4.5% overall have been calculated across the entire 300–100 nm spectrum, with local enhancements of 6.8% (300–500 nm) and 10.9% (300–420 nm) in the regions where the silicon PV cells are known to perform poorly

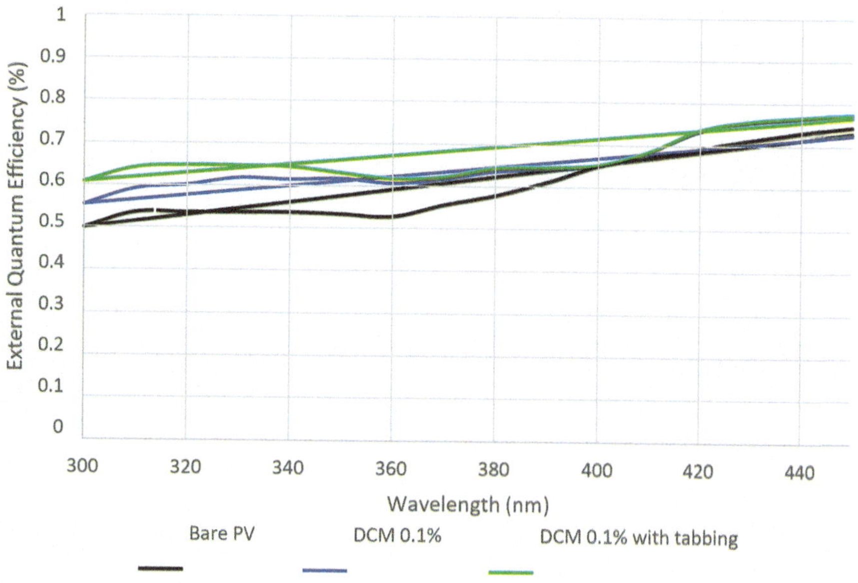

Fig. 12 EQE of 0.1% DCM with tabbing vs bare PV cell

relatively. The DCM results are very promising for application within solar panel commercialization for households. The layer production methods have been optimized, as uniformly reproducible layers have been created, and the benefits of the inclusion of tabbing to the connections have been shown.

Acknowledgements The authors would like to acknowledge the European Research Council grant PEDAL: (639760), H2020 IDEAS (815271) and support funding from Science Foundation Ireland (SFI) and insights from the PEARL PV COST Action Network.

References

1. Polo J, Alonso-Abella M, Ruiz-Arias JA, Balenzategui JL (2017) Worldwide analysis of spectral factors for seven photovoltaic technologies. Sol Energy 142:194–203
2. Ahmed H, Doran J, McCormack SJ (2016) External quantum efficiency improvement with luminescent downshifting layers: experimental and modelling. Int J Spectrosc 2016
3. Ahmed H, Doran J, McCormack SJ (2020) external quantum efficiency measurements and outdoor characterisation for PV luminescent down-shifting devices, from book Renewable Energy and Sustainable Buildings
4. Alonso-Álvarez D, Ross D, Richards B (2012) Luminescent down-shifting for CdTe solar cells: A review of dyes and simulation of performance. In: 2012 38th IEEE photovoltaic specialists conference, pp 000009–000014
5. Rafiee M, Ahmed H, Chandra S, McCormack SJ (2017) Improving the mathematical model for luminescent down-shifting layers by investigating their loss mechanisms. In: 33rd European PV Solar Energy Conference and Exhibitions (EU PVSEC 2017), Amsterdam, 24-29 September 2017
6. Glenn A, McLoughlin C The viability of organic dyes in luminescent down-shifting layers for the enhancement of SI solar cell efficiency'. Front Compos Mater 995:71–76. 4th International Conference on Energy Engineering and Smart Materials (ICEESM 2019)
7. Gilligan S, Ahmed H, Chandra S, McCormack SJ (2018) Optimization and photostability studies of luminescent organic dyes for solar devices

Index

© The Author(s), under exclusive license to Springer Nature Switzerland AG 2022

A. Sayigh (ed.), *Sustainable Energy Development and Innovation*, Innovative Renewable Energy,
https://doi.org/10.1007/978-3-030-76221-6

D

E

H

J

K

L

O

S

T

U

V

X

Z

SPRINGER NATURE

GPSR Compliance

The European Union's (EU) General Product Safety Regulation (GPSR) is a set of rules that requires consumer products to be safe and our obligations to ensure this.

If you have any concerns about our products, you can contact us on ProductSafety@springernature.com

In case Publisher is established outside the EU, the EU authorized representative is:

Springer Nature Customer Service Center GmbH
Europaplatz 3
69115 Heidelberg, Germany

The manufacturer's authorised representative in the EU is Springer Nature Customer Service Centre GmbH, Europaplatz 3, 69115 Heidelberg, Germany. If you have any concerns regarding our products, please contact ProductSafety@springernature.com

Printed and bound by CPI Group (UK) Ltd, Croydon, CR0 4YY
30/09/2026
02225516-0013